International Congress of Mathematicians (ICM 2022)

Sections 12-14
Volume 6 of 7

Online
6-14 July 2022

Editors:

Dmitry Beliaev
Stanislav Smirnov

ISBN: 978-1-7138-9873-3

ICM INTERNATIONAL CONGRESS OF MATHEMATICIANS 2022 JULY 6–14

SECTIONS 12–14

EDITED BY D. BELIAEV AND S. SMIRNOV

EMS PRESS

Editors

Dmitry Beliaev
Mathematical Institute
University of Oxford
Andrew Wiles Building
Radcliffe Observatory Quarter
Woodstock Road
Oxford OX2 6GG, UK

Email: belyaev@maths.ox.ac.uk

Stanislav Smirnov
Section de mathématiques
Université de Genève
rue du Conseil-Général 7–9
1205 Genève, Switzerland

Email: stanislav.smirnov@unige.ch

2020 Mathematics Subject Classification: 00B25

ISBN 978-3-98547-058-7, eISBN 978-3-98547-558-2, DOI 10.4171/ICM2022

Volume 1. Prize Lectures
ISBN 978-3-98547-059-4, eISBN 978-3-98547-559-9, DOI 10.4171/ICM2022-1
Volume 2. Plenary Lectures
ISBN 978-3-98547-060-0, eISBN 978-3-98547-560-5, DOI 10.4171/ICM2022-2
Volume 3. Sections 1–4
ISBN 978-3-98547-061-7, eISBN 978-3-98547-561-2, DOI 10.4171/ICM2022-3
Volume 4. Sections 5–8
ISBN 978-3-98547-062-4, eISBN 978-3-98547-562-9, DOI 10.4171/ICM2022-4
Volume 5. Sections 9–11
ISBN 978-3-98547-063-1, eISBN 978-3-98547-563-6, DOI 10.4171/ICM2022-5
⟶ **Volume 6. Sections 12–14**
ISBN 978-3-98547-064-8, eISBN 978-3-98547-564-3, DOI 10.4171/ICM2022-6
Volume 7. Sections 15–20
ISBN 978-3-98547-065-5, eISBN 978-3-98547-565-0, DOI 10.4171/ICM2022-7

This edition printed by Curran Associates, Inc.
ISBN 978-1-7138-9873-3

e-Book available: https://ems.press/books/standalone/278

No changes have been made to the content of these proceedings. There may be changes to pagination and minor adjustments for aesthetics.

Bibliographic information published by the Deutsche Nationalbibliothek
The Deutsche Nationalbibliothek lists this publication in the Deutsche Nationalbibliografie; detailed bibliographic data are available on the Internet at http://dnb.dnb.de.

Published by EMS Press, an imprint of the

European Mathematical Society – EMS – Publishing House GmbH
Institut für Mathematik
Technische Universität Berlin
Straße des 17. Juni 136
10623 Berlin, Germany

https://ems.press

CONTENTS

VOLUME 1

PRIZE LECTURES

POPULAR SCIENTIFIC EXPOSITIONS

SUMMARIES OF PRIZE WINNERS' WORK

VOLUME 2

SPECIAL PLENARY LECTURES

PLENARY LECTURES

VOLUME 3

1. LOGIC

2. ALGEBRA

3. NUMBER THEORY – SPECIAL LECTURE

3. NUMBER THEORY

4. ALGEBRAIC AND COMPLEX GEOMETRY – SPECIAL LECTURE

4. ALGEBRAIC AND COMPLEX GEOMETRY

VOLUME 4

5. GEOMETRY – SPECIAL LECTURES

5. GEOMETRY

6. TOPOLOGY

7. LIE THEORY AND GENERALIZATIONS

8. ANALYSIS – SPECIAL LECTURE

8. ANALYSIS

VOLUME 5

9. DYNAMICS

10. PARTIAL DIFFERENTIAL EQUATIONS

11. MATHEMATICAL PHYSICS – SPECIAL LECTURE

11. MATHEMATICAL PHYSICS

VOLUME 6

12. PROBABILITY - SPECIAL LECTURE

12. PROBABILITY

13. COMBINATORICS - SPECIAL LECTURE

13. COMBINATORICS

14. MATHEMATICS OF COMPUTER SCIENCE – SPECIAL LECTURES

14. MATHEMATICS OF COMPUTER SCIENCE

VOLUME 7

15. NUMERICAL ANALYSIS AND SCIENTIFIC COMPUTING

16. CONTROL THEORY AND OPTIMIZATION – SPECIAL LECTURE

16. CONTROL THEORY AND OPTIMIZATION

17. STATISTICS AND DATA ANALYSIS

18. STOCHASTIC AND DIFFERENTIAL MODELLING

19. MATHEMATICAL EDUCATION AND POPULARIZATION OF MATHEMATICS

20. HISTORY OF MATHEMATICS

12. PROBABILITY

SPECIAL LECTURE

COMBINATORIAL STATISTICS AND THE SCIENCES

ELCHANAN MOSSEL

ABSTRACT

Combinatorial statistics studies inference in discrete stochastic models. Inference of such models plays an important role in the sciences. We survey research in combinatorial statistics involving the tree broadcast process. We review the mathematical questions that arise in the analysis of this process and its inference via "belief propagation." We discuss the mathematical connections to statistical physics, the social sciences, biological sciences, and theoretical computer science.

MATHEMATICS SUBJECT CLASSIFICATION 2020

Primary 60-02; Secondary 6365, 62-01, 68-02

KEYWORDS

Combinatorial statistics, broadcasting on trees, the reconstruction problem on trees

ICM INTERNATIONAL CONGRESS OF MATHEMATICIANS 2022 JULY 6–14

Proc. Int. Cong. Math. 2022, Vol. 6, pp. 4170–4188
DOI 10.4171/ICM2022/179

Published by EMS Press

1. INTRODUCTION

Discrete probability models are used in many of the hard and soft sciences. Often the scientific challenges lead to novel mathematical questions in combinatorial statistics. The mathematical questions involve inference in such models. The goal of this survey paper is to discuss some of these processes, their inference, and their connections to the sciences. We focus on the tree broadcast process, the mathematical questions that arise in the analysis of this process and its inference via "belief propagation." We review some of the mathematical connections to statistical physics, the social sciences, biological sciences, and theoretical computer science.

1.1. A simple model on trees

In its simplest form, the model in question will be parametrized by four parameters: d, h, q, and θ. The model is defined on the d-ary tree of $h+1$ levels. Level 0 consists of the root, which has d children. Level 1 of the tree consists of the d children of the root. Each of the nodes at level 1 has d children. The collection of d^2 children of the nodes at level 1 makes level 2 of the tree, etc. We will denote the tree by $T=(V,E)$ and the $h+1$ levels by $L_0,\dots,L_h$. We will denote the level of node v by $|v|$. We will denote the root by 0.

We now define a discrete stochastic process indexed by the vertices V of T. We will give two equivalent definitions of this process. First, a recursive definition: the random variable X_0 is chosen uniformly at random from the set $[q]:=\{1,\dots,q\}$. Now, for each child v of the root 0, independently, we toss a coin that lands Heads with probability θ. If it lands Heads, we let $X_v=X_0$. If it lands Tails, we sample X_v independently and uniformly at random from $[q]$. We then apply the same procedure recursively to each node at levels 2, 3, etc.

A moment's thought reveals that the vector $X=(X_v:v\in V)$ has the following probability distribution:

$$\mathbb{P}\big[X=(x_v:v\in V)\big]=\frac{1}{q}\prod_{(u,v)\in E}\left(1(x_u=x_v)\theta+\frac{1-\theta}{q}\right), \tag{1.1}$$

where here and below all edges (u,v) are directed away from the root.

Note that the measure above is well defined for $1\ge\theta\ge-1/(q-1)$, which will always be assumed. The extreme case $\theta=-1/(q-1)$ corresponds to the uniform measure on q-colorings of the tree. Below we will always exclude the frozen measures, where the root color determines all colors, by assuming $\theta<1$, and $\theta>-1$, if $q=2$.

There are many ways in which this process was generalized. Of particular interest are the following two. First, we may consider the process on general rooted trees, random, or deterministic. Second, we may consider more general broadcast processes from parent to child. In particular, assuming the state space is $[q]$, there is no reason that different edges (u,v) will have the same conditional law of X_u given X_v. Moreover, we can consider more general conditional laws of X_u given X_v. Thus for a general, possibly random, finite tree

$V = (T, E)$ and a collection M^e of Markov chains on the state space $[q]$, we may define

$$\mathbb{P}\big[X = (x_v : v \in V)\big] = \pi(x_0) \prod_{(u,v)\in E} M^{(u,v)}(x_u, x_v), \tag{1.2}$$

where π is a given probability distribution on $[q]$. We will always assume all chains $M^{(u,v)}$ are ergodic and that π is not a delta measure.

A lot of what we know about model (1.1) carries over to the more general setting of (1.2). For simplicity, we will mostly discuss (1.1), and sometimes comment on how things generalize.

1.2. Belief propagation and the reconstruction problem

Note that we may define the process $X = (X_v : v \in V)$ recursively also for a d-ary tree $T = (V, E)$ of infinitely many levels. Moreover, if we restrict to $(X_v : |v| \le h)$, it will be distributed according to (1.1).

We are interested in studying if the nodes at level h of the tree are asymptotically independent of X_0 as $h \to \infty$. We first note that by ergodicity of Markov chains we know that X_0 and X_v are asymptotically independent as $|v| \to \infty$, where $|v|$ denotes the level of v (for model (1.2), we need to require a bit more; we will not get into the details). In the language of statistical physics, this means that two point correlations decay exponentially as they do for finite ergodic Markov chains. Instead, we look at point-to-set correlations. More formally, let us denote by X_h the vectors of X_v for v at level h of the tree,

$$X_h = (X_v : |v| = h).$$

Then we are interested in the asymptotic independence of X_0 and X_h as $h \to \infty$. To formalize this question, let

$$Y_0 = \sum_{i=1}^{q} e_i 1(X_0 = i) \in \mathbb{R}^q,$$

where e_i is the ith unit vector.

Definition 1.1. We say that the reconstruction problem is solvable if

$$\lim_{h\to\infty} \mathbb{E}\|\mathbb{E}[Y_0|X_h] - \mathbb{E}[Y_0]\|_2^2 \neq 0. \tag{1.3}$$

In other words, the reconstruction problem is solvable if X_h provides some nonvanishing information on the value of X_0. There are many other equivalent definitions of reconstruction including some involving the limiting mutual information $\lim_{h\to\infty} I(X_0, X_h)$ or softer ones in terms of the tail-triviality of the sequence $X_0, X_1, \ldots$, see, e.g., [34, 43, 92] and the survey [68].

Interestingly, the quantity $f(X_h) = \mathbb{E}[Y_0|X_h]$ can be computed recursively and efficiently as a function of X_h via the belief propagation algorithm. This algorithm is used also for nontree graphical models [88] where it provides an approximation. The accuracy of belief propagation on trees was observed earlier, in specific contexts such as ancestral inference in phylogenetic trees [37, 44] and the study of the Ising model on trees [83]. Note that if

$\mathbb{E}[Y_0|X_h] \to \mathbb{E}[Y_0]$ (say in probability or a.s.) then for large values of h, there is little point in computing $\mathbb{E}[Y_0|X_h]$, as it is most likely trivial. Below we will often write BP instead of belief propagation.

2. LINEAR THEORY AND THE KESTEN–STIGUM BOUND

While there is an easy recursive computation of the function $f(x) = \mathbb{E}[Y_0|X_h = x]$, computing the limiting distribution or the limiting variance of $f(X_h)$ in (1.3) is in general difficult, as f is highly nonlinear and the coordinates of X_h are dependent. To remedy the first difficulty, it is natural to ask if there is a way to linearize the problem so that it is more amenable to analysis.

Interestingly, there are two approaches that lead to studying the same question:

(1) We can introduce an additional noise parameter $\eta > 0$ that will be applied only for the nodes at level h. For a deterministic value x_h of the nodes at level h, define the random vector

$$\tilde{x}_h := (\tilde{x}_v : |v| = h), \tag{2.1}$$

where for nodes v with $|v| = h$, we let $\tilde{x}_v = x_v$ with probability η, and it is independently and uniformly sampled from $[q]$ otherwise. We can then define a new function $\tilde{f}$ of the colors at level h by letting

$$\tilde{f}(x_h) = \frac{d}{d\eta}|_{\eta=0}\mathbb{E}\big[f(\tilde{x}_h)\big].$$

Using the chain rule, it is easy to see that this is a linear function of the variables in x_h. More formally, it is a linear function of the $d^h q$ indicator variables $(1(x_v = i) : |v| = h, i \in [q])$. We can now study the correlation between $\tilde{f}(X_h)$ and X_0 instead of the variance in (1.3).

(2) Perhaps the most natural function of the X_h that one may study is $\sum Y_h := \sum_{v \in L_h} Y_v$, where

$$Y_v := \sum_{i=1}^{q} e_i 1(X_v = i) \in \mathbb{R}^q.$$

Of course, $\sum Y_h$ is just the count of how many of each of the q symbols appear at level h.

It is not hard to see that both approaches lead to studying the correlation between $\sum Y_h$ and X_0, see, e.g., **[68]**. In the work of Kesten and Stigum on multitype branching processes in the 1960s, they proved a law of large numbers for $\sum Y_h$ in **[52]** and then more refined limit theorems **[51]** which in particular imply:

Theorem 2.1. *For model* (1.1),

I. $d\theta^2 \le 1 \implies$ *normalized* $\sum Y_h \xrightarrow[h\to\infty]{(d)}$ a normal law *independent of* X_0.

II. $d\theta^2 > 1 \implies$ *normalized* $\sum Y_h \xrightarrow[n\to\infty]{(d)}$ *a* nonnormal law dependent *on* X_0.

III. *In particular, if* $d\theta^2 > 1$ *then the reconstruction problem is solvable.*

The laws in parts I and II are nondegenerate. The results of [51] are in fact general enough to cover the more general model (1.2) on the d-ary tree if all the M matrices are identical and ergodic. In this case we let $\theta := \max(|\lambda_i| : \lambda_i \neq 1)$, where the λ_is denote the eigenvalues of M. The results further carry to random branching process trees with well-behaved degree distributions, where now d denotes the average number of offsprings.

The original proof of Theorem 2.1 uses the Fourier transform approach though martingale approaches can also be used to prove it as is hinted in [73].

Given Theorem 2.1, it is natural to ask if part I of the theorem implies nonreconstruction when $d\theta^2 \leq 1$. One way for this to work out would be for the higher-order terms in the expansion of BP, $\mathbb{E}[Y_0|X_h]$ to have a bounded contribution in probability. The recursive nature of BP allows proving it in the case of $q = 2$:

Theorem 2.2. *If* $q = 2$ *and* $d\theta^2 \leq 1$, *then the reconstruction problem is not solvable.*

This theorem was first proved by Bleher, Ruiz, and Zagrebnov [10]. Since then many other alternative proofs were presented. In particular, Theorem 2.2 was extended to general infinite trees in [34], where the general definition of d is now in terms of the *branching number* of the tree [56]. See also [81] for the analysis of the critical case for general trees. Proofs by Ioffe [46, 47] are formulated in terms of the FK representation from percolation theory, see, e.g., [41]. There are some recent short proofs based on information inequalities, see, e.g., [1, 82].

Beyond the case $q = 2$, Sly [88] proved nonreconstruction if $d\theta^2 \leq 1$ for $q = 3$ if $d \geq d_{\min}$, where $d_{\min}$ is some constant, and [13] proved it in the case where all the M are identical and given by 2×2 matrices that are almost symmetric.

In terms of the correlation between $\sum Y_h$ and X_0, in the paper [73] it is proven that for all q the distribution of $\sum Y_h$ is asymptotically independent of the root when $d\theta^2 \leq 1$. The paper [49] showed that in the noisy model (2.1), for all q, if $d\theta^2 < 1$, then there exists a constant amount of noise $\eta > 0$ such that $\tilde{X}_h$ is asymptotically independent of X_0 so

$$\lim_{h\to\infty} \mathbb{E}\|\mathbb{E}[Y_0|\tilde{X}_h] - \mathbb{E}[Y_0]\|_2^2 = 0.$$

The last two results say that if reconstruction is possible when $d\theta^2 < 1$ then (1) the information retained about X_0 is not in the count Y_h and (2) the information retained about X_0 is not robust against a fixed amount of noise.

3. NONLINEAR THEORY

Interestingly, for large values of q, reconstruction is possible even for some values of θ such that $d\theta^2 < 1$. First, as $q \to \infty$, the reconstruction threshold θ_q converges to $1/d$ as proven in [67].

Theorem 3.1. *Fix d and let*

$$\theta_q := \inf(\theta' > 0 : \textit{such that reconstruction is possible for parameters } (d,q,\theta), \forall \theta > \theta').$$

Then $\lim_{q\to\infty} \theta_q = 1/d$.

This theorem is proven using branching process techniques. An easy and well-known argument states that reconstruction is impossible when $\theta \leq 1/d$. Indeed, if we consider the branching process, where each node has $\text{Bin}(d,\theta)$ children, then Z_h, the population at level h, counts the number of nodes whose colors have been copied from the root. Moreover, conditioned on Z_h, all other colors are independent of the root. Therefore if $Z_h \to 0$, X_0 and X_h are asymptotically independent. Since the branching process is subcritical ($Z_h \to 0$) if and only if $d\theta \leq 1$ (see, e.g., [7]), it follows that $\theta_q \geq 1/d$ for all q.

When $d\theta > 1$, $Z_h \to \infty$ with positive probability. However, since we do not know from X_h the location of the colors that were copied from the root, it is still possible that X_h and X_0 are asymptotically independent, as is the case when $q = 2$ and $\theta \in (d^{-1}, d^{-1/2})$. The proof of the harder direction of Theorem 3.1 uses the fact that for large q if two recent descendants of the same node have the same color, it is very likely that node has the same color. Thus the proof uses a function that estimates the root to have a specific value i if a certain fractal-like subtree containing the value i at all of its leaves appears in X_h.

On the other hand, taking the asymptotics as $d \to \infty$, Sly [88] proves:

Theorem 3.2. *Fix $q \geq 5$ and let*

$$\overline{\theta}_d := \inf(\theta' > 0 : \textit{such that reconstruction is possible for parameters } (d,q,\theta), \forall \theta > \theta').$$

Then $\lim_{d\to\infty} d\overline{\theta}_d^2 = C_q < 1$.

Similar results are obtained for $\theta < 0$.

Theorem 3.2 is proven by using a central limit theorem to analyze the basic belief propagation recursion and noting that the nonlinear terms shift the threshold.

A special case that attracted a lot of attention is the case of random coloring where $\theta = -1/(q-1)$. Interestingly, again, for large q the relationship between the critical d and θ is almost linear, see [87,89].

4. CONNECTIONS TO STATISTICAL PHYSICS

The reconstruction problem on trees was first studied in statistical physics. The case $q = 2$ corresponds to studying the extremality or tail triviality of the Ising model on the tree [92]. Reconstruction solvability for $q = 2$ when $d\theta^2 > 1$ was proven in [43], the author of which was unaware that a more general result is implied by the results of Kesten and Stigum [51].

Interestingly, nonreconstruction for $q = 2$ when $d\theta^2 \leq 1$ was first proven in a *spin glass* variant [14,15,18]. In this context reconstruction means

$$\lim_{h\to\infty} \mathbb{E}\|\mathbb{E}[Y_0|X_h = B_h] - \mathbb{E}[Y_0]\|_2^2 \neq 0,$$

where B_h are i.i.d. Bernoulli taking each of the two colors with probability $1/2$. The proof in this case is a little easier since in the analysis of the recursion for spin glasses the contributions coming from different subtrees are independent and identically distributed.

The interest in the reconstruction problem in statistical physics saw an explosion as the cavity and replica method played a crucial role in analyzing problems on sparse random graphs, see, e.g., [61–63,78,79]. At a very high level, for many combinatorial problems on sparse random graphs, statistical physics predictions are based on analysis (often nonrigorous) of the reconstruction problem or its variants on a corresponding tree. The connection to the reconstruction problem as defined here was formally made in [61]. In particular, it was conjectured in [61] that the Kesten–Stigum bound predicts the reconstruction threshold for $q = 2, 3$, and does not predict it for $q \geq 5$. As mentioned earlier, this was partially proven [88]. We will not try to summarize the connections between belief propagation and its variants, variants of the reconstruction problem and random constraint satisfaction problems. Some key papers in this area are [20,40,54,65]. This connection is also important in the work leading to the proof of the SAT threshold [21,22,27–29,58].

We will now give more details of one example, the example of *detection* in the *block model*. Here again we will see differences between the linear theory as reflected in the case $q = 2$ vs. nonlinear theory when q is large.

4.1. Detection in the block model

The block model is a random graph model generalizing the famous Erdős–Rényi random graph [33]. The block model is a special case of inhomogeneous random graphs, see, e.g., [11]. The sparse block model may be defined as follows:

Definition 4.1 (The sparse block model). Let $G(n, d, \theta, q)$ denote the model of random, $[q]$-labeled graphs in which each vertex u is assigned (independently and uniformly at random) a label $\sigma_u \in [q]$, and then each possible edge (u, v) is included with probability $(d/n)(1 - \theta)$ if $\sigma_u \neq \sigma_v$ and with probability $(d/n)((1 - \theta) + q\theta)$ if $\sigma_u = \sigma_v$.

We chose this parametrization so that for a fixed node, the distribution of the number of neighbors of each type will asymptotically agree with the distribution of the number of children of each type in model (1.2) with parameters q and θ on a random tree where each node has a Poisson with parameter d number of children.

The block model was studied extensively in statistics as a model of communities [45], see, e.g., [9,85,90], and in computer science as a model to study the average case behavior of clustering algorithms, see, e.g., [19,23,30,50,59] (interestingly there are very few citations between the two communities of papers even in cases where very similar results are proven). The papers above mostly concentrate on cases where the average degree is at least of order $\log n$, where n is the number of nodes in the graph.

The sparse case in Definition 4.1 became a major object of research due to a landmark paper in statistical physics [26] where the authors predicted that

Conjecture 4.2. *For the block model,*

I. *For all q, belief propagation on the graph G predicts the communities better than random if $d\theta^2 > 1$.*

II. *For $q = 2, 3$, it is information-theoretically impossible to predict better than at random if $d\theta^2 < 1$.*

III. *For $q \geq 5$, it is information-theoretically possible to predict better than at random for some θ with $d\theta^2 < 1$, but not in a computationally efficient way.*

These predictions were based on a linearization of belief propagation for the tree model.

4.2. $q = 2$—linear theory

A major challenge in establishing the algorithmic efficiency of belief propagation for block models stems from a fundamental difference between the application of belief propagation to trees and block models. When applied to trees, the input to belief propagation is the actual colors of the leaves. However, in the block model application, the colors are unknown. So here belief propagation is applied to random colors at all nodes that are independent of the actual colors.

In [55] it was conjectured that the global nonlinear operator that described one iteration of belief propagation on the graph should be linearized around its trivial fixed point to lead to a linear algebra based method to detect the communities. The resulting operator is not normal and its spectrum is complex. It is closely related to the operators used to analyze nonbacktracking walks [6, 39, 42, 91].

This suggestion was followed up by an extensive body of work, including [2, 5, 12, 72], that led proofs that linearized versions of BP detect communities better than at random when $d\theta^2 > 1$, which is in the spirit of part I of Conjecture 4.2.

The original statement of part I of Conjecture 4.2 states, furthermore, that belief propagation is optimal for the problem in the stronger sense that it minimizes the fraction of misclassified nodes. A combination of linearized belief propagation and belief propagation is used in [71] to obtain an efficient algorithm that minimizes the misclassification error when $q = 2$ and $d\theta^2 > C$ for some big constant C [71]. The main ingredient is proving that the estimator in the noisy model (2.1) asymptotically agrees with the original model (1.1):

$$\lim_{h\to\infty} |\mathbb{E}[Y_0|\tilde{X}_h] - \mathbb{E}[Y_0|X_h]| = 0. \tag{4.1}$$

Even earlier, part II of Conjecture 4.2 was partially established as it was shown in [70] that for $q = 2$ it is information-theoretically impossible to detect better than at random if $d\theta^2 < 1$ based on coupling of the graph and tree processes. The case $q = 3$ is still open.

4.3. Nonlinear theory

Parts of the nonlinear predictions in part III of Conjecture 4.2 were confirmed in [8] and [4] which provided exponential-time algorithms to detect for some parameters when

$d\theta^2 < 1$ when $q > 5$ (also when $q = 4$ and $\theta < 0, d\theta^2 < 1$). Of course, we rarely know how to prove that computational problems cannot be solved efficiently, so the support we have for the predicted computational-statistical gap is quite limited, see [3] for a more detailed discussion.

5. CONNECTIONS TO MOLECULAR BIOLOGY

The broadcast process on the tree was independently introduced in mathematical biology as a model of evolution of genetic information such as DNA sequences [16, 36, 77].

Naturally, the reconstruction problem is interesting in this context. Given the detailed evolutionary tree of some species, we want to infer as much as possible about the genetics of extinct species from the genetics of extant species.

An even more interesting question from a biological perspective is *recovering the species tree from genetic data*. Note that the details of this tree are required to study the reconstruction problem and infer ancestral genetic data.

Since Darwin's Origin of Species [24], a major goal of evolutionary biology is recovering the relationship between different species. Since the 1970s, this is most often done using genetic information collected from extant species. The models introduced in [16, 36, 77] assume that the genetic distribution of traits $(X_v : v \in V)$ is determined by a binary (rooted) tree $T = (V, E)$ and a collection $\theta_E = (\theta_e : e \in E)$ via the following variant of (1.1):

$$\mathbb{P}\big[X = (x_v : v \in V)\big] = \frac{1}{q} \prod_{e=(u,v)\in E} \left(1(x_u = x_v)\theta_e + \frac{1-\theta_e}{q}\right). \tag{5.1}$$

Note that T and θ_E determine a distribution of traits X and therefore the distribution $D(T, \theta_E)$ of $X_L = (X_v : v$ is a leaf). A major goal of the *phylogenetic reconstruction problem* is to estimate T and θ_E from independent samples from the distribution $D(T, \theta_E)$. In particular, we are interested in knowing how many samples are needed to recover T with good probability, as this translates to the data requirements needed for accurate estimation.

This ideal model that was introduced in the 1970s has since been generalized to account for many additional biological factors and mechanisms. Key theoretical results in this area include the identifiability of phylogenetic models [17] and efficient polynomial time algorithms to reconstruct phylogenetic trees [31, 32, 74]. See, e.g., [38, 86, 93] for general references on the phylogenetic problem.

The connection to the reconstruction problem was predicted by Steel [94] who conjectured that the amount of data needed to reconstruct phylogenetic trees crucially depends on the reconstruction problem.

The easiest setting to understand the connection between phylogenetic inference and the reconstruction problem is when $q = 2$ and the trees are very symmetric. We call a tree an h-level full binary tree if all the leaves are at level h. In the symmetric phylogenetic problem, we assume that the tree is an h-level full binary tree and that $\theta_e = \theta$ for all $e \in E$.

To understand what is inferred in this setup, let us fix $h = 2$. In this case the data given is

$$\left(X_v^i : v \in \{a, b, c, d\}, 1 \le i \le n\right). \tag{5.2}$$

This data can be thought of as four genetic sequences of length n, i.e., the genetic content of species a, b, c, d, where a, b, c, d are the leaves of the tree. Alternatively, the two-dimensional array (5.2) can be viewed as n i.i.d. samples from the process at the four leaves a, b, c, d. The main goal of inference in this simple case is to determine which species are siblings and which are cousins. The three possible sibling relationships are

$$\{\{a, b\}, \{c, d\}, \quad \{\{a, c\}, \{b, d\}, \quad \{\{a, d\}, \{b, c\}.$$

Of course, when h is bigger, we want to determine not just the sibling relation but also higher-order cousin relations.

Theorem 5.1. *Consider the symmetric phylogenetic problem with $q = 2$ and n independent samples from the distribution $D(T, \theta_E)$, where T is an h-level full binary tree and $\theta_e = \theta$ for all $e \in E$.*

I. *If the reconstruction problem is solvable for binary trees at the parameter θ (i.e., when $2\theta^2 > 1$) then there is an efficient algorithm that, given $n = O(h)$ samples, returns the correct tree with probability $1 - \exp(-\Omega(h))$.*

II. *If θ is strictly below the reconstruction threshold, $2\theta^2 < 1$, then it is information-theoretically impossible to infer the correct tree with probability $\ge 1/2$ unless $n \ge \exp(\Omega(h))$.*

The theorem above was first proven in **[68]** in a more general (and biologically relevant) setting.

For the proof of part I, the basic idea is that we may use correlation between different coordinates in samples from $D(T, \theta_E)$ to identify siblings, cousins, etc., in the tree. We may then estimate the state of their ancestor somewhat accurately since the reconstruction problem is solvable. We then use these estimates to find close relationships between the newly identified nodes and continue recursively.

For part II, one proves that for a node v at distance εh from the root, X_v has an exponentially small in h correlation with X_h. By taking ε sufficiently small, this implies that the same is true for the correlation between $X_{\varepsilon h}$ and X_h. Finally, this allows showing that, unless $n \ge \exp(\Omega(h))$, it is impossible to distinguish between the true tree and modifications of it permuting the nodes at level εh and the trees below them.

More realistic phylogenetic problems are not symmetric and much of the work in **[25, 68]** and follow up work was devoted to extending part I of the theorem to asymmetric cases.

Note, moreover, that the proof sketch of part I extends to all (q, θ) such that the reconstruction problem is solvable. However, the proofs in **[25, 68]** that do not have such a strong symmetry assumption do not extend to all such (q, θ) as they require robustness in various steps (the results trivially extend to even q when $2\theta^2 > 1$). Interestingly, the results of part I were extended in **[75]** to large q for some values of θ where $2\theta^2 < 1$ based on the root estimator in **[67]**. The results of **[75]** require the tree to be symmetric but not that $\theta_e = \theta$ for all e. The paper **[75]** also provided an extension of part II of Theorem 5.1 for (q, θ) when $\theta < \theta'$ and there is no reconstruction for parameter (q, θ').

It is natural to ask if there is a computational or information-theoretical barrier to extending the more realistic phylogenetic results of **[25, 68, 84]** to all θ's above the reconstruction threshold when $q \geq 5$. An analog of Theorem 5.1 for the limiting case $q = \infty$ is established in **[76]** for general (asymmetric) trees where the critical value of θ is $1/2$.

6. CONNECTIONS TO THEORETICAL COMPUTER SCIENCE

We have already seen many connections of the reconstruction problem to theoretical computer science. The connections included the role it played in algorithms and determining the satisfiability thresholds of random clustering, random graph, and random constraint satisfaction algorithms in Section 4, and the role it played in the information theoretic and algorithmic analysis of phylogenetic reconstruction in Section 5. Moreover, as belief propagation is a widely used algorithm, the analysis of the reconstruction problem and the robustness of this algorithm provide average case understanding of this important algorithm.

In this section we briefly discuss the computational complexity of the problem of estimating X_0 from X_h or approximately computing $\mathbb{E}[Y_0|X_h]$. Furthermore, we review the connections between this problem and the classical theory of noisy computation and its connection to deep inference.

This question might seem strange as the belief propagation algorithm computes $\mathbb{E}[Y_0|X_h]$ exactly in linear time. Note, however, that despite the linear running time it has two complex features:

(1) It uses real numbers. Indeed, the complexity is measured in terms of real arithmetic, but the model we are interested in is discrete.

(2) It is recursive. In other words, the circuit that computed BP has some depth. Is the depth necessary?

6.1. Recursive bounded memory algorithms

Here we only consider the simple model (1.1) with $q = 2$. In this case we know that the reconstruction threshold is given by $d\theta^2 = 1$ and that $\sum Y_h$ provides a good estimator of X_0 when $d\theta^2 > 1$.

Since the definition of the distribution of X_h is recursive, it is natural to ask if there is a simple recursive algorithm that estimates X_0 in a bottom up fashion, i.e., by a recursion of the form $\hat{X}_v = f(\hat{X}_w : w \in L_t(v))$, where $L_t(v)$ is the set of d^t descendants of v exactly t levels below v. The algorithm begins by initializing $\hat{X}_v = X_v$ for nodes v at the bottom level L_h and terminates by estimating X_0 by rounding $\hat{X}_0$ in some fashion.

As mentioned earlier, belief propagation can be written in this way for some real valued function f. The majority estimator $\operatorname{sgn}(\sum Y_h)$ can also be written in this way by computing the sum recursively. However, both of these require the domain of f to be unbounded. Is it possible to estimate X_0 in such recursive fashion using a function f that takes at most a constant B values and a bounded t?

The case $B = 2$ was studied in [66] assuming f is antisymmetric. In this case we can find the optimal function: f is the majority function. And the overall estimator of X_0 is a recursive majority function applied to X_h. This in turn allows computing for each t a critical threshold θ_t such that $\hat{X}_0$ is correlated with X_0 asymptotically if $\theta > \theta_t$ and is uncorrelated if $\theta < \theta_t$. The computation in [66] shows that for all t, $\theta_t < d^{-1/2}$ and $\lim_{t\to\infty} \theta_t = d^{-1/2}$.

There is an interesting connection between the derivation of the thresholds $\theta_1(d)$ and a derivation of von Neumann in the context of noisy computation [35, 95]. In his work on noisy computation, von Neumann considered circuits with noisy gates with the goal of designing circuits that, by duplicating inputs and applying majority gates to correct intermediate computations, are robust to some amount of noise. The derivation of the amount of noise that can be tolerated reduces to the question if the noisy recursive majority function with the all 1 inputs has limiting expectation bounded away from $1/2$. Interestingly, the broadcast model and the noisy computation model yields the same recursion and therefore we derive the same threshold for $\theta_1(d)$. In the noisy computation setting, the case $d = 3$ was derived by von Neumann [95], and was generalized to all d in [35]. The same recursion also appears in other models of noisy broadcast, see, e.g., [57].

In the context of the reconstruction problem, it was conjectured in [34] that any algorithm with bounded B cannot achieve the reconstruction threshold. This was recently established in [48] where it is shown that with B bits of memory, the critical θ_B satisfies $B^{-C} \le \theta_B - \theta \le B^{-c}$, for some positive constants $C > c > 0$.

6.2. The complexity of $\mathbb{P}[X_0|X_h]$

Recent efforts are devoted to studying the complexity of inference of X_0 from X_h in the linear regime when $d\theta^2 > 1$ vs. the nonlinear regime where $d\theta^2 < 1$, q is large, and the reconstruction problem is solvable.

Part of the motivation for studying this problem is to identify natural data-generating processes, where inference is possible but requires some nontrivial complexity.

The polynomial degree is one such measure of complexity. Thus we can ask if there is a low degree polynomial of the $d^h q$ indicator variables $(1(x_v = i) : |v| = h, i \in [q])$ that has nonvanishing correlation with X_0 as $h \to \infty$. We can say that:

(1) In the linear regime where $d\theta^2 > 1$, there is a linear function of the variables in X_h that is correlated with X_0 by Theorem 2.1.

(2) In [67] it is shown that for the model (1.2) on the d-ary tree, where $M^e = M$ for all e, there are chains M with $\theta = 0$ for which the reconstruction problem is solvable. The paper [53] shows that for such chains any polynomial of X_h of degree $\le (2^{ch})$ are uncorrelated with X_0, for some positive constant c.

(3) The authors of [53] ask if a similar phenomenon holds through the nonlinear regime. For example, is it true that polynomials of bounded degree have vanishing correlation with X_0 in the regime where $d\theta^2 < 1$?

In an earlier work [64], circuit complexity measures were used to study the inference of X_0 from X_h. The conjectured gap between the linear and nonlinear cases is reflected in the circuit class $\mathbf{TC}^0$ vs. $\mathbf{NC}^1$:

(1) Since the class $\mathbf{TC}^0$ of bounded depth circuits contains majority gates, it can trivially estimate X_0 better than at random when $d\theta^2 > 1$. Moreover, whenever (4.1) holds, $\mathbf{TC}^0$ can estimate X_0 with minimal error.

(2) It is not too hard to show that the computation of BP can always be carried out in $\mathbf{NC}^1$, the class of circuits of logarithmic depth. The paper [64] constructed a chain M with $\theta = 0$ for which estimating X_0 from X_h better than at random is $\mathbf{NC}^1$-complete.

(3) It is conjectured in [64] that estimating better than at random is $\mathbf{NC}^1$-complete when $d\theta^2 < 1$ and the reconstruction problem is solvable.

It is important to note that it is a major open problem to determine if $\mathbf{NC}^1 = \mathbf{TC}^0$.

In an even earlier work, the paper [69] considered a semisupervised version of the phylogenetic problem in the regime $d\theta^2 < 1$ and proved that in this regime it is information-theoretically impossible to classify the unlabeled data for algorithms that ignore correlations between features in the labeled data, while algorithms that do use high-order correlation can classify the data accurately. Moreover, in the regime $d\theta^2 > 1$, high-order correlations are not needed.

ACKNOWLEDGMENTS

Yuval Peres introduced me to the tree broadcast process at the beginning of my graduate studies. I am indebted to Yuval for fascinating discussions and central insights. Much of my work in this area has greatly benefited from collaborations with Mike Steel, Sebastien Roch, Allan Sly, Joe Neeman, Frederic Koehler, and Colin Sandon. Thanks to Sebastien Roch, Colin Sandon, Allan Sly, Nike Sun, Joe Neeman, and André Dixon for feedback on a draft of this survey.

FUNDING

This work was partially supported by NSF grant DMS-2031883, Vannevar Bush Faculty Fellowship ONR-N00014-20-1-2826 and Simons Investigator award (622132).

REFERENCES

[1] E. Abbe and E. Boix-Adserà, An information-percolation bound for spin synchronization on general graphs. *Ann. Appl. Probab.* **30** (2020), no. 3, 1066–1090.

[2] E. Abbe and C. Sandon, Community detection in general stochastic block models: fundamental limits and efficient algorithms for recovery. In *2015 IEEE 56th annual symposium on foundations of computer science*, pp. 670–688, IEEE, 2015.

[3] E. Abbe and C. Sandon, Detection in the stochastic block model with multiple clusters: proof of the achievability conjectures, acyclic bp, and the information-computation gap. 2015, arXiv:1512.09080.

[4] E. Abbe and C. Sandon, Crossing the KS threshold in the stochastic block model with information theory. In *2016 IEEE international symposium on information theory (ISIT)*, pp. 840–844, IEEE, 2016.

[5] E. Abbe and C. Sandon, Proof of the achievability conjectures for the general stochastic block model. *Comm. Pure Appl. Math.* **71** (2018), no. 7, 1334–1406.

[6] N. Alon, I. Benjamini, E. Lubetzky, and S. Sodin, Non-backtracking random walks mix faster. *Commun. Contemp. Math.* **9** (2007), no. 04, 585–603.

[7] K. B. Athreya and P. E. Ney, *Branching processes*. Springer, New York, 1972.

[8] J. Banks, C. Moore, J. Neeman, and P. Netrapalli, Information-theoretic thresholds for community detection in sparse networks. In *Conference on learning theory*, pp. 383–416, PMLR, 2016.

[9] P. Bickel and A. Chen, A nonparametric view of network models and Newman–Girvan and other modularities. *Proc. Natl. Acad. Sci.* **106** (2009), no. 50, 21068–21073.

[10] P. M. Bleher, J. Ruiz, and V. A. Zagrebnov, On the purity of the limiting Gibbs state for the Ising model on the Bethe lattice. *J. Stat. Phys.* **79** (1995), no. 1–2, 473–482.

[11] B. Bollobás, S. Janson, and O. Riordan, The phase transition in inhomogeneous random graphs. *Random Structures Algorithms* **31** (2007), no. 1, 3–122.

[12] C. Bordenave, M. Lelarge, and L. Massoulié, Non-backtracking spectrum of random graphs: community detection and non-regular Ramanujan graphs. In *Foundations of computer science (FOCS), 2015 IEEE 56th annual symposium on*, pp. 1347–1357, IEEE, 2015.

[13] C. Borgs, J. Chayes, E. Mossel, and S. Roch, The Kesten–Stigum reconstruction bound is tight for roughly symmetric binary channels. In *Proceedings of IEEE FOCS 2006*, pp. 518–530, IEEE, 2006.

[14] J. M. Carlson, J. T. Chayes, L. Chayes, J. P. Sethna, and D. J. Thouless, Bethe lattice spin glass: the effects of a ferromagnetic bias and external fields. I. Bifurcation analysis. *J. Stat. Phys.* **61** (1990), no. 5–6, 987–1067.

[15] J. M. Carlson, J. T. Chayes, J. P. Sethna, and D. J. Thouless, Bethe lattice spin glass: the effects of a ferromagnetic bias and external fields. II. Magnetized spin-glass phase and the de Almeida–Thouless line. *J. Stat. Phys.* **61** (1990), no. 5–6, 1069–1084.

[16] J. A. Cavender, Taxonomy with confidence. *Math. Biosci.* **40** (1978), no. 3–4, 271–280.

[17] J. Chang, Full reconstruction of Markov models on evolutionary trees: identifiability and consistency. *Math. Biosci.* **137** (1996), 51–73

[18] J. T. Chayes, L. Chayes, J. P. Sethna, and D. J. Thouless, A mean field spin glass with short-range interactions. *Comm. Math. Phys.* **106** (1986), no. 1, 41–89.

[19] A. Coja-Oghlan, Graph partitioning via adaptive spectral techniques. *Combin. Probab. Comput.* **19** (2010), no. 2, 227–284.

[20] A. Coja-Oghlan, F. Krzakala, W. Perkins, and L. Zdeborová, Information-theoretic thresholds from the cavity method. *Adv. Math.* **333** (2018), 694–795.

[21] A. Coja-Oghlan and K. Panagiotou, The asymptotic k-SAT threshold. *Adv. Math.* **288** (2016), 985–1068.

[22] A. Coja-Oglan and K. Panagiotou, Catching the k-NAESAT threshold. In *Proceedings of the forty-fourth annual ACM symposium on theory of computing*, pp. 899–908, ACM, 2012.

[23] A. Condon and R. Karp, Algorithms for graph partitioning on the planted partition model. *Random Structures Algorithms* **18** (2001), no. 2, 116–140.

[24] C. Darwin, On the origin of species (1859).

[25] C. Daskalakis, E. Mossel, and S. Roch, Evolutionary trees and the Ising model on the Bethe lattice: a proof of Steel's conjecture. *PTRF* **149** (2011), no. 1–2, 149–189.

[26] A. Decelle, F. Krzakala, C. Moore, and L. Zdeborová, Asymptotic analysis of the stochastic block model for modular networks and its algorithmic applications. *Phys. Rev. E* **84** (2011), 066106.

[27] J. Ding, A. Sly, and N. Sun, Proof of the satisfiability conjecture for large k. In *Proceedings of the forty-seventh annual ACM symposium on theory of computing*, pp. 59–68, ACM, 2015.

[28] J. Ding, A. Sly, and N. Sun, Maximum independent sets on random regular graphs. *Acta Math.* **217** (2016), no. 2, 263–340.

[29] J. Ding, A. Sly, and N. Sun, Satisfiability threshold for random regular NAE-SAT. *Comm. Math. Phys.* **341** (2016), no. 2, 435–489.

[30] M. Dyer and A. Frieze, The solution of some random NP-hard problems in polynomial expected time. *J. Algorithms* **10** (1989), no. 4, 451–489.

[31] P. L. Erdős, M. A. Steel, L. A. Székely, and T. A. Warnow, A few logs suffice to build (almost) all trees (part 1). *Random Structures Algorithms* **14** (1999), no. 2, 153–184.

[32] P. L. Erdős, M. A. Steel, L. A. Székely, and T. A. Warnow, A few logs suffice to build (almost) all trees (part 2). *Theoret. Comput. Sci.* **221** (1999), 77–118.

[33] P. Erdős and A. Rényi, On the evolution of random graphs. *Publ. Math. Inst. Hung. Acad. Sci.* **5** (1960), no. 1, 17–60.

[34] W. S. Evans, C. Kenyon, Y. Y. Peres, and L. J. Schulman, Broadcasting on trees and the Ising model. *Ann. Appl. Probab.* **10** (2000), no. 2, 410–433.

[35] W. S. Evans and L. J. Schulman, Signal propagation and noisy circuits. *IEEE Trans. Inf. Theory* **45** (1999), no. 7, 2367–2373.

[36] J. S. Farris, A probability model for inferring evolutionary trees. *Syst. Zool.* **22** (1973), no. 4, 250–256.

[37] J. Felsenstein, Maximum-likelihood estimation of evolutionary trees from continuous characters. *Am. J. Hum. Genet.* **25** (1973), no. 5, 471.

[38] J. Felsenstein, *Inferring phylogenies*. Sinauer, New York, NY, 2004.

[39] J. Friedman, *A proof of Alon's second eigenvalue conjecture and related problems*. Amer. Math. Soc., 2008.

[40] A. Gershchenfeld and A. Montanari, Reconstruction for models on random graphs. In *Foundations of computer science, annual IEEE symposium on*, pp. 194–204, IEEE Computer Society, 2007.

[41] G. Grimmett, The random-cluster model. In *Probability on discrete structures*, pp. 73–123, Encyclopaedia Math. Sci. 110, Springer, Berlin, 2004.

[42] K-i. Hashimoto, Zeta functions of finite graphs and representations of p-adic groups. In *Automorphic forms and geometry of arithmetic varieties*, pp. 211–280, Elsevier, 1989.

[43] Y. Higuchi, Remarks on the limiting Gibbs states on a $(d+1)$-tree. *Publ. Res. Inst. Math. Sci.* **13** (1977), no. 2, 335–348.

[44] J. Hilden, GEN EX—An algebraic approach to pedigree probability calculus. *Clinical Genetics* **1** (1970), no. 5–6, 319–348.

[45] P. Holland, K. Laskey, and S. Leinhardt, Stochastic blockmodels: first steps. *Soc. Netw.* **5** (1983), no. 2, 109–137.

[46] D. Ioffe, Extremality of the disordered state for the Ising model on general trees. In *Trees (Versailles, 1995)*, pp. 3–14, Progr. Probab. 40, Birkhäuser, Basel, 1996.

[47] D. Ioffe, On the extremality of the disordered state for the Ising model on the Bethe lattice. *Lett. Math. Phys.* **37** (1996), no. 2, 137–143.

[48] V. Jain, F. Koehler, J. Liu, and E. Mossel, Accuracy-memory tradeoffs and phase transitions in belief propagation. In *Proceedings of the thirty-second conference on learning theory*, edited by A. Beygelzimer and D. Hsu, pp. 1756–1771, Proc. Mach. Learn. Res. 99, PMLR, Phoenix, USA, 2019.

[49] S. Janson and E. Mossel, Robust reconstruction on trees is determined by the second eigenvalue. *Ann. Probab.* **32** (2004), 2630–2649.

[50] M. Jerrum and G. Sorkin, The Metropolis algorithm for graph bisection. *Discrete Appl. Math.* **82** (1998), no. 1–3, 155–175.

[51] H. Kesten and B. P. Stigum, Additional limit theorems for indecomposable multidimensional Galton–Watson processes. *Ann. Math. Stat.* **37** (1966), 1463–1481.

[52] H. Kesten and B. P. Stigum, Limit theorems for decomposable multi-dimensional Galton–Watson processes. *J. Math. Anal. Appl.* **17** (1967), 309–338.

[53] F. Koehler and E. Mossel, Reconstruction on trees and low-degree polynomials, 2021. arXiv:2109.06915.

[54] F. Krząkała, A. Montanari, F. Ricci-Tersenghi, G. Semerjian, and L. Zdeborová, Gibbs states and the set of solutions of random constraint satisfaction problems. *Proc. Natl. Acad. Sci.* **104** (2007), no. 25, 10318–10323.

[55] F. Krzakala, C. Moore, E. Mossel, J. Neeman, A. Sly Z. L, and P. Zhang, Spectral redemption: clustering sparse networks. *Proc. Natl. Acad. Sci.* **100** (2013), no. 52, 20935–20940.

[56] R. Lyons, The Ising model and percolation on trees and tree-like graphs. *Comm. Math. Phys.* **125** (1989), no. 2, 337–353.

[57] A. Makur, E. Mossel, and Y. Polyanskiy, Broadcasting on random directed acyclic graphs. *IEEE Inf. Theory* **66** (2020), no. 2, 780–812.

[58] E. Maneva, E. Mossel, and M. J. Wainwright, A new look at survey propagation and its generalizations. *J. ACM* **54** (2007), 41.

[59] F. McSherry, Spectral partitioning of random graphs. In *Foundations of computer science, 2001. Proceedings 42nd IEEE symposium on*, pp. 529–537, IEEE, 2001.

[60] M. Mézard and A. Montanari, Reconstruction on trees and the spin glass transition. *J. Stat. Phys.* **124** (2006), 1317–1350.

[61] M. Mézard and A. Montanari, *Information, physics, and computation*. Oxford University Press, USA, 2009.

[62] M. Mézard, G. Parisi, and R. Zecchina, Analytic and algorithmic solution of random satisfiability problems. *Science* **297** (2002), no. 5582, 812–815.

[63] M. Mezard and R. Zecchina, Random k-satisfiability: from an analytic solution to an efficient algorithm. *Phys. Rev. E* **66** (2002).

[64] A. Moitra, E. Mossel, and C. Sandon, Parallels between phase transitions and circuit complexity? In *Conference on learning theory*, pp. 2910–2946, PMLR, 2020.

[65] A. Montanari, R. Restrepo, and P. Tetali, Reconstruction and clustering in random constraint satisfaction problems. *SIAM J. Discrete Math.* **25** (2011), no. 2, 771–808.

[66] E. Mossel, Recursive reconstruction on periodic trees. *Random Structures Algorithms* **13** (1998), no. 1, 81–97.

[67] E. Mossel, Reconstruction on trees: beating the second eigenvalue. *Ann. Appl. Probab.* **11** (2001), no. 1, 285–300.

[68] E. Mossel, Survey: Information flow on trees. In *Graphs, morphisms and statistical physics. DIMACS series in discrete mathematics and theoretical computer science*, edited by J. Nestril and P. Winkler, pp. 155–170, 2004.

[69] E. Mossel, Deep learning and hierarchal generative models. 2019, arXiv:1612.09057.

[70] E. Mossel, J. Neeman, and A. Sly, Reconstruction and estimation in the planted partition model. *Probab. Theory Related Fields* **3–4** (2015), 431–461.

[71] E. Mossel, J. Neeman, and A. Sly, Belief propagation, robust reconstruction, and optimal recovery of block models. *Ann. Appl. Probab.* **26** (2016), no. 4, 2211–2256.

[72] E. Mossel, J. Neeman, and A. Sly, A proof of the block model threshold conjecture. *Combinatorica* **38** (2018), no. 3, 665–708.

[73] E. Mossel and Y. Peres, Information flow on trees. *Ann. Appl. Probab.* **13** (2003), no. 3, 817–844.

[74] E. Mossel and S. Roch, Learning nonsingular phylogenies and hidden Markov models. In *Proceedings of the thirty-seventh annual ACM symposium on theory of computing, Baltimore (STOC'05), MD, USA*, pp. 366–376, ACM, 2005.

[75] E. Mossel, S. Roch, and A. Sly, On the inference of large phylogenies with long branches: How long is too long? *Bull. Math. Biol.* **73** (2011), no. 7, 1627–1644.

[76] E. Mossel and M. Steel, A phase transition for a random cluster model on phylogenetic trees. *Math. Biosci.* **187** (2004), no. 2, 189–203.

[77] J. Neyman, Molecular studies of evolution: a source of novel statistical problems. In *Statistical decision theory and related topics*, edited by S. S. Gupta and J. Yackel, pp. 1–27, Elsevier, 1971.

[78] M. M. G. Parisi, A replica analysis of the travelling salesman problem. *J. Phys.* **47** (1986), 1285–1296.

[79] M. M. G. Parisi, On the solution of the random link matching problem. *J. Phys.* **48** (1987), 1451–1459.

[80] J. Pearl, *Reverend Bayes on inference engines: A distributed hierarchical approach*. Cognitive Systems Laboratory, School of Engineering and Applied Science, 1982.

[81] R. Pemantle and Y. Peres, The critical Ising model on trees, concave recursions and nonlinear capacity. *Ann. Probab.* **38** (2010), no. 1, 184–206.

[82] Y. Polyanskiy and Y. Wu, Application of the information-percolation method to reconstruction problems on graphs. *Math. Stat. Learn.* **2** (2020), no. 1, 1–24.

[83] C. J. Preston, *Gibbs states on countable sets: Gibbs states and Markov random fields*. Cambridge University Press, 1974.

[84] S. Roch and A. Sly, Phase transition in the sample complexity of likelihood-based phylogeny inference. *Probab. Theory Related Fields* **169** (2017), no. 1, 3–62.

[85] K. Rohe, S. Chatterjee, and B. Yu, Spectral clustering and the high-dimensional stochastic blockmodel. *Ann. Statist.* **39** (2011), no. 4, 1878–1915.

[86] C. Semple and M. Steel, *Phylogenetics*. Math. Appl. Ser. 22, Oxford University Press, 2003.

[87] A. Sly, Reconstruction of random colourings. *Comm. Math. Phys.* **288** (2009).

[88] A. Sly, Reconstruction for the Potts model. *Ann. Probab.* **39** (2011), no. 4, 1365–1406.

[89] A. Sly and Y. Zhang, Reconstruction of colourings without freezing. 2016, arXiv:1610.02770.

[90] T. Snijders and K. Nowicki, Estimation and prediction for stochastic blockmodels for graphs with latent block structure. *J. Classification* **14** (1997), no. 1, 75–100.

[91] S. Sodin, Random matrices, nonbacktracking walks, and orthogonal polynomials. *J. Math. Phys.* **48** (2007), 123503.

[92] F. Spitzer, Markov random fields on an infinite tree. *Ann. Probab.* **3** (1975), no. 3, 387–398.

[93] M. Steel, *Phylogeny: Discrete and random processes in evolution*. SIAM, 2016.

[94] M. Steel, My Favourite Conjecture, 2001, http://www.math.canterbury.ac.nz/~mathmas/conjecture.pdf.

[95] J. von Neumann, Probabilistic logics and the synthesis of reliable organisms from unreliable components. In *Automata studies*, pp. 43–98, Ann. of Math. Stud. 34, Princeton University Press, Princeton, NJ, 1956.

ELCHANAN MOSSEL
77 Massachusetts Avenue, Cambridge, MA 02139-4307, USA, elmos@mit.edu

12. PROBABILITY

KPZ LIMIT THEOREMS

JINHO BAIK

ABSTRACT

One-dimensional interacting particle systems, 1+1 random growth models, and two-dimensional directed polymers define 2D height fields. The KPZ universality conjecture posits that an appropriately scaled height function converges to a model-independent universal random field for a large class of models. We survey limit theorems for a few models and discuss changes that arise in different domains. In particular, we present recent results on periodic domains. We also comment on integrable probability models, integrable differential equations, and universality.

MATHEMATICS SUBJECT CLASSIFICATION 2020

Primary 60K35; Secondary 82C22

KEYWORDS

KPZ, TASEP, directed last passage percolation, periodic domain, multi-point distribution, integrable models, integrable differential equations

INTERNATIONAL CONGRESS OF MATHEMATICIANS 2022 JULY 6–14

Proc. Int. Cong. Math. 2022, Vol. 6, pp. 4190–4211
DOI 10.4171/ICM2022/8

Published by EMS Press

1. INTRODUCTION

The KPZ universality is concerned with, among others, one-dimensional interacting particle systems, 1+1 random growth, and two-dimensional directed polymers. These models define height functions $\mathbf{h}(x,t)$, two-dimensional random fields, where x represents the one-dimensional spatial position and t the time. A height function encodes the integrated current for interacting particle systems, the height for random growth models, and the free energy for directed polymer models. See Section 2 for an example. The KPZ universality conjecture is that for a large class of models, the scaled height function

$$\mathbf{h}_T(\gamma,\tau) = \frac{\mathbf{h}(\gamma T^{2/3}, \tau T) - c(T)}{T^{1/3}} \tag{1.1}$$

converges, up to scaling factors, to a model-independent universal 2D random field, which is called the KPZ fixed point. Here $c(T)$ is a nonrandom term determined by the macroscopic limit of the height function. Since the height, position, and time scale as $T^{1/3}$, $T^{2/3}$, and T, respectively, we say that (1.1) is a 1:2:3 scaled height function. We also say that a KPZ limit theorem holds if a 1:2:3 scaled height function converges in any suitable sense for the problem at hand.

Several physics papers [45, 50, 58, 87] conjectured the 1:2:3 scale for various models in the mid-1980s. One of them is the paper [58] of Kardar, Parisi, and Zhang on a nonlinear stochastic partial differential equation, now called the KPZ equation, from which the term KPZ universality is derived. These papers were followed by extensive research in the physics community. However, it remained unknown, even on a conjectural level, what the limit should be.

The situation changed in 1999 with the publication of the paper [6] by Baik, Deift, and Johansson, in which the authors considered the longest increasing subsequence problem of random permutations. This problem is equivalent to the zero-temperature free energy of a directed polymer model. The paper proved that the one-point distribution of an analog of the height function converges in distribution. See Theorem 3.1 below. Moreover, the authors found the limiting distribution explicitly, which turned out to be the Tracy–Widom distribution from random matrix theory. This connection between the KPZ universality and random matrix theory was completely unexpected at that time. Soon after, Johansson [52] proved a similar result for another model, giving yet another example of a KPZ limit theorem.

Exciting developments on KPZ limit theorems followed these results during the next two decades. For example, one-point limit theorems were extended to equal-time, multiposition distributions, multitime distributions, and even to the 2D fields. Results were also generalized to several, mostly isolated, models, and algebraic underpinning of these specific models was studied. The 2D field limit, the KPZ fixed point, was determined, and various properties of the limit were established. Limit theorems were also proved for infinite space, half-infinite space, and recently finite space with periodic boundary condition.

In this article, we give a historical overview of some KPZ limit theorems and present new results on the periodic domain case. We start by introducing the subjects of KPZ universality, interacting particle systems, random growth, and directed polymers, in Section 2, focusing on one particular example. Then, we discuss some limit theorems on infinite spaces in Sections 3–4. After briefly discussing the half-infinite space case in Section 5, we present new results on the periodic case in Section 6. Section 7 compares the formulas of the limiting multipoint distributions for infinite and periodic cases, and Section 8 concerns differential equations associated with distribution functions. We conclude the article with some comments on universality in Section 9.

The research on the KPZ universality has been developing rapidly and extensively over the last two decades. Hence, what is discussed in this article is only a small selection of the activities. The reader may benefit from other excellent survey articles such as [32,34,70] to see other aspects.

2. TASEP, CORNER GROWTH MODEL, AND EXPONENTIAL DLPP

This section discusses one of the most well-studied examples of one-dimensional interacting particle systems, 1+1 random growth, and two-dimensional directed polymers.

2.1. TASEP

The totally asymmetric simple exclusion process (TASEP), introduced by Spitzer [80] in 1970, is a continuous-time Markov process on $\mathbb{Z}$. At any given time, each integer site of $\mathbb{Z}$ is occupied by at most one particle. A particle moves to the adjacent site to its right after a random waiting time, but only if it is empty. The waiting time is exponentially distributed of mean 1, and the clock starts once the neighboring site becomes vacant. All waiting times are independent of each other. Note that all moves are to the right (hence, totally asymmetric), particles can move only one step at a time (simple), and no two particles occupy the same site at the same time (exclusion). Figure 1 is an example of the configuration at a particular time. Black dots denote particles, and white dots mark empty sites. In this configuration, only three particles can move, and they do so independently of each other.

FIGURE 1
TASEP.

The TASEP is an example of interacting particle systems. General systems may allow, for example, left moves in addition to right moves, multirange moves, or several particles at each site.

One particular initial condition we focus on in this article is the step initial condition that the sites in $\mathbb{Z}_- \cup \{0\}$ are occupied, and all sites in $\mathbb{Z}_+$ are empty. The leftmost picture in Figure 2 shows the step initial condition.

2.2. Corner growth model

The configuration space for the TASEP is $\{0, 1\}^{\mathbb{Z}}$ where 1 represents the presence of a particle and 0 an empty site. To each configuration, we can associate a zigzag graph in

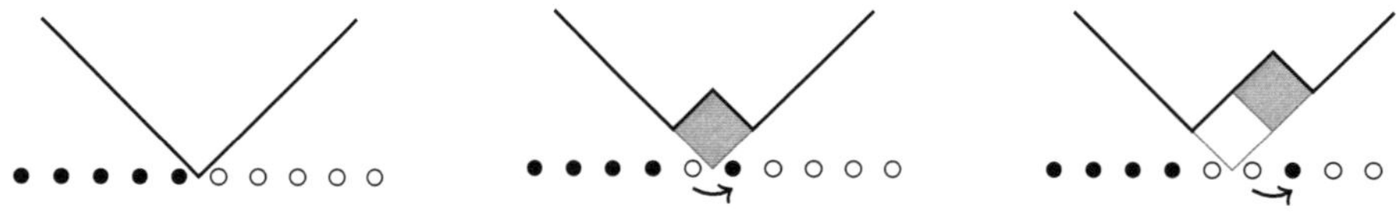

FIGURE 2
Corner growth model.

$\mathbb{R}^2$ as in Figure 2. We assign a particle to a line segment of length $\sqrt{2}$ and slope -1, and an empty site to a line segment of the same length and slope 1. Juxtaposing the line segments, we obtain a zigzag graph as in Figure 2. We call the graph a height function $\mathbf{h} : \mathbb{R} \to \mathbb{R}$ for the configuration of the TASEP, and it is unique up to translations. The leftmost picture in Figure 2 is a translation of $\mathbf{h}(x) = |x|$, and it corresponds to the step initial condition.

The TASEP induces a stochastic evolution of the height function, $\mathbf{h}(x, t)$, in which local valleys (corners) change to local peaks independently with rate 1. The resulting 1+1 random growth process defined by the height function is called the corner growth process. See Figure 5 for a simulation.

2.3. Exponential DLPP

Consider the two-dimensional lattice $\mathbb{Z}_+^2$. Let $(m, n) \in \mathbb{Z}_+^2$. An up/right (i.e., directed) path from $(1, 1)$ to (m, n) is a sequence $p = (p_i)_{i=1}^{m+n-1}$, where $p_i \in \mathbb{Z}_+^2$, $p_1 = (1, 1)$, $p_{m+n-1} = (m, n)$, and $p_{i+1} - p_i \in \{(1, 0), (0, 1)\}$. The thick lines in Figure 3 are an example of a path in which we connected neighboring integer sites for visual aid. Let w_s, $s \in \mathbb{Z}_+^2$, be a collection of independent random variables. The normalized free energy of the directed polymer measure, introduced by Huse and Henley **[50]**, is

$$F(m, n; \beta) = \frac{1}{\beta} \log\Big(\sum_p e^{\beta E(p)}\Big) \quad \text{where } E(p) = \sum_{i=1}^{m+n-1} w_{p_i},$$

the sum is over all directed paths p from $(1, 1)$ to (m, n), and $\beta > 0$ is the inverse temperature. The zero-temperature, $\beta = \infty$, case is called the directed last passage percolation (DLPP). In this case, the normalized free energy becomes

$$L(m, n) = \max_p E(p),$$

which we call the last passage time, interpreting $E(p)$ as the travel time using path p.

For the case when $w_s \geq 0$, the DLPP is related to a random growth model. For $t > 0$, define the subset of $\mathbb{R}^2$ by

$$G_t = \bigcup_{s \in S_t} \big((0, 1]^2 + s\big) \quad \text{where } S_t = \{(m, n) \in \mathbb{Z}^2 : L(m, n) \leq t\}$$

and we set $L(m, n) = 0$ if $m \leq 0$ or $n \leq 0$. See Figure 4. Since $L(m, n)$ is greater than or equal to both $L(m-1, n)$ and $L(m, n-1)$, we see that if $(m, n) \in G_t$, then both points $(m-1, n)$ and $(m, n-1)$ are in G_t. The set G_t grows with time t. If we regard G_t in the first quadrant as a stack of boxes, we can add a new box only at the corners.

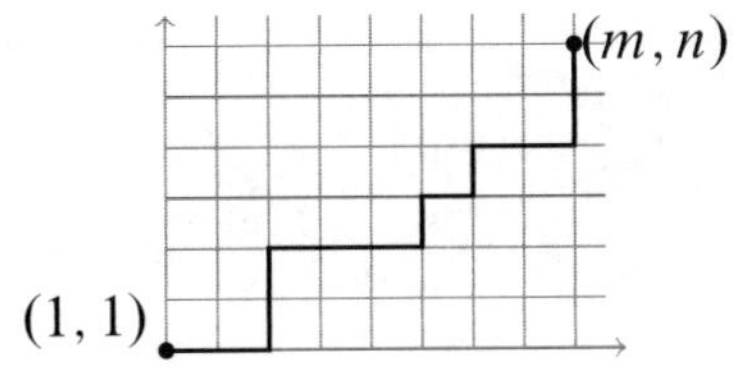

FIGURE 3
Exponential DLPP.

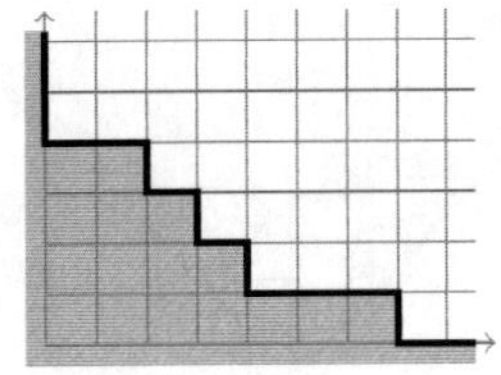

FIGURE 4
An example of G_t.

A special case is the exponential DLPP in which w_s are exponentially distributed with mean 1. In this case, each corner of G_t grows independently of rate 1, i.e., a unit box can be added to each corner independent at rate 1. Thus, the boundary of G_t is a rotation of the height function of the corner growth process. More precisely, for the TASEP with the step initial condition and the exponential DLPP,

$$\mathbf{h}(m-n,t) \geq m+n \quad \text{if and only if} \quad L(m,n) \leq t. \tag{2.1}$$

If the TASEP starts with a different initial condition, we need to consider the exponential DLPP on a subset of $\mathbb{Z}^2$, determined by the initial condition.

2.4. Hydrodynamic limit and KPZ limit

The hydrodynamic limit of TASEP is about $\mathbf{h}(x,t)$ when x and t are proportional. For the step initial condition, $\mathbf{h}(x,0) = |x|$, Rost [75] showed in 1981 that

$$\frac{\mathbf{h}(xT,tT)}{T} \to \bar{\mathbf{h}}(x,t)$$

almost surely as $T \to \infty$, where $\bar{\mathbf{h}}(x,t) = \frac{t^2+x^2}{2t}$ for $|x| \leq t$ and $\bar{\mathbf{h}}(x,t) = |x|$ for $|x| \geq t$. See Figure 6 for the graph. The hydrodynamic limit $\bar{\mathbf{h}}$ is deterministic, and it solves Burger's equation [61,62].

The KPZ limit is about the next term, $\mathbf{h}(xT,tT) - \bar{\mathbf{h}}(x,t)T$. Setting $x=0$ for convenience and following the 1:2:3 scale, the KPZ universality conjecture suggests that

$$\frac{\mathbf{h}(\gamma T^{2/3}, \tau T) - \bar{\mathbf{h}}(0,\tau)T}{T^{1/3}}$$

converges to a 2D random field. If we do not set $x=0$, then we should consider $\mathbf{h}(xT + \gamma T^{2/3}, \tau T) - \bar{\mathbf{h}}(x,\tau)T$ in the numerator. The limiting 2D field, the KPZ fixed point, depends on the initial condition. The step initial condition for the TASEP becomes the so-called narrow wedge initial condition for the KPZ fixed point.

TASEP has interpretations as an interacting particle system, a random growth process, and a last passage percolation model. Each of these interpretations has natural extensions and generalizations. The KPZ universality conjecture is that a large class of models in these generalizations has a universal limit. The exact class is not known, but for random growth models, three key features seem to be the locality of growth, some smoothing mechanism, and lateral growths. For directed last passage percolation, the universality is expected

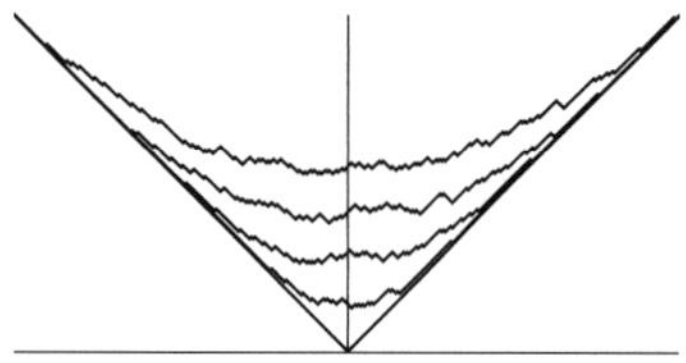

FIGURE 5
Simulation of the corner growth model.

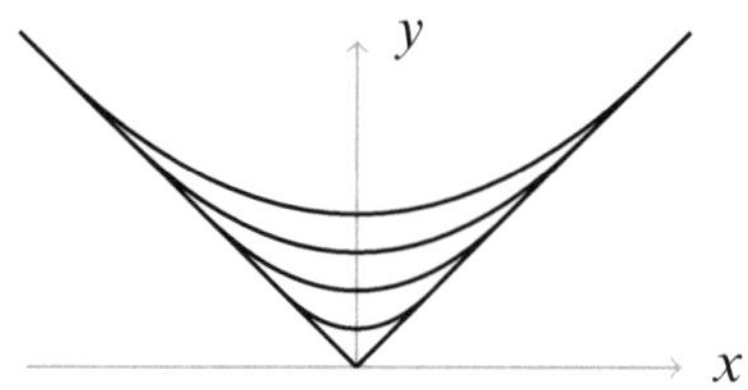

FIGURE 6
$y = \bar{\mathbf{h}}(x, t)$ for a few values of t.

for all random variables w_s with enough moments and without a large atom at the top of the support of the distribution. The last condition is to prevent the situation that there is always a path connecting $(1, 1)$ and (m, n) using only the top value, making the last passage time too concentrated.

3. ONE-POINT DISTRIBUTION

We discuss one-point KPZ limit theorems from [6] and [52] mentioned in the introduction, and extensions to other models.

3.1. Poisson DLPP

Poisson directed last passage percolation is a variation of the exponential DLPP. Consider a realization of a 2D Poisson process in $\mathbb{R}^2_+$. An up/right path p this time is defined as the graph of a continuous piecewise linear function of positive slopes connecting Poisson points, as shown in Figure 7. Let $\mathrm{E}(p)$ denote the number of the Poisson points on p. For $(t, s) \in \mathbb{R}^2_+$, define

$$\mathrm{L}(t, s) = \sup_p \mathrm{E}(p),$$

where the supremum is taken over all up/right paths p from $(0, 0)$ to (t, s). The next theorem follows from [6]. The main theorem of [6] is stated for the case of a fixed number of points, but the paper proves the Poisson points case first, from which the main theorem follows. Since $\mathrm{L}(t, s) \stackrel{d}{=} \mathrm{L}(\sqrt{ts}, \sqrt{ts})$, the next result applies to general points (t, s).

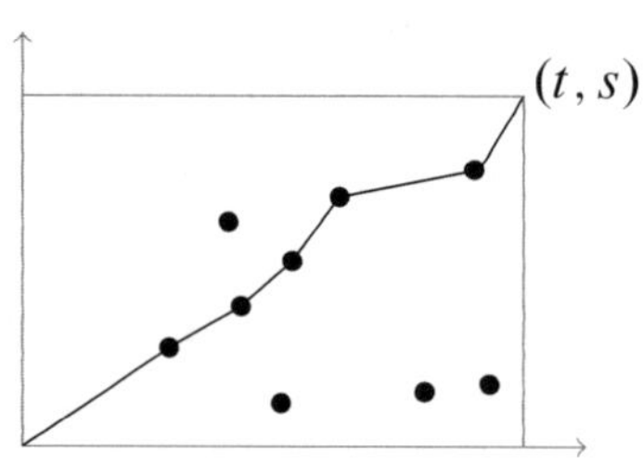

FIGURE 7
Poisson DLPP.

Theorem 3.1 ([6]). *For every $x \in \mathbb{R}$,*

$$\lim_{t\to\infty} \mathrm{P}\left(\frac{\mathrm{L}(t,t)-2t}{t^{1/3}} \le x\right) = \mathrm{F}_{\mathrm{TW}}(x)$$

where F_{TW} is the Tracy–Widom distribution.

The $1/3$-power in $t^{1/3}$ is consistent with the height scale of the KPZ universality; see (2.1). This result shows that the one-point marginal of the KPZ fixed point (for the narrow wedge initial condition) must be distributed as the Tracy–Widom distribution. The Tracy–Widom distribution is the limiting distribution of the largest eigenvalue of random Hermitian matrices such as Gaussian unitary ensemble matrices [84]. The connection of the KPZ fixed point and random matrix theory was surprising and unexpected. See Section 9.3 for more on this connection.

3.2. Longest increasing subsequence

The Poisson DLPP is particularly interesting due to its connection to longest increasing subsequences of random permutations. Note that finitely many points in a rectangle with distinct x and y coordinates can be associated with a permutation by considering the relative orderings of the coordinates. For example, the points in Figure 7 are associated with the permutation $\pi = 475168293$. For this permutation, the subsequence 45689 is an increasing subsequence. Furthermore, it is the longest increasing subsequence, and its length, 5, is equal to the last passage time $L(t,s)$.

Let ℓ_N denote the length of longest increasing subsequences of a uniformly random permutation of size N. Then, $L(t,s)$ has the same distribution as ℓ_N, where N is a Poisson random variable of mean ts. Using this connection, Theorem 3.1 implies, after a de-Poissonization argument, that $\frac{\ell_N - 2\sqrt{N}}{N^{1/6}}$ converges in distribution to the Tracy–Widom distribution.

The problem of determining the large-N behavior of ℓ_N has a long history. The existence of the almost sure limit of $\ell_N/\sqrt{N}$ was proved by Hammersley in [49] using Kingman's subadditive ergodic theorem. The fact that the limit is 2, known as Ulam's problem, was proved independently by two famous papers of Veršik–Kerov [88] and Logan–Shepp [64] in 1977. However, the limiting distribution and the variance (which is of order $N^{2/3}$) remained an open problem until the work [6]. Interested readers are encouraged to consult [3,7,74,81].

3.3. Exponential DLPP

Soon after Theorem 3.1 was proved, Johansson showed that the exponential DLPP model also satisfies a similar limit theorem [52]. We state the result in terms of the height function of the TASEP.

Theorem 3.2 ([52]). *Assume the step initial condition for the TASEP. Then, for every $(\tau, \gamma, \mathsf{h}) \in \mathbb{R}_+ \times \mathbb{R} \times \mathbb{R}$,*

$$\lim_{T\to\infty} \mathrm{P}\left(\frac{\mathbf{h}(\gamma T^{2/3}, 2\tau T) - \tau T}{-T^{1/3}} \le \mathsf{h}\right) = \mathrm{F}_{\mathrm{TW}}\left(\frac{\mathsf{h}}{\tau^{1/3}} + \frac{\gamma^2}{4\tau^{4/3}}\right).$$

Note from either side of the equation that the limit remains unchanged if we rescale

$$(\mathsf{h}, \gamma, \tau) \mapsto (\alpha \mathsf{h}, \alpha^2 \gamma, \alpha^3 \tau) \tag{3.1}$$

for any $\alpha > 0$.

3.4. Integrable models

The above theorems were obtained by explicitly computing the finite-time distribution function and then taking the large limit of the formula. In particular, for the TASEP, the finite-time formula is given by the Fredholm determinant of an operator. After suitable scaling and conjugation, the operator converges to the so-called Airy operator, the Fredholm determinant of which is the Tracy–Widom distribution.

Johansson obtained the finite-time distribution formula for TASEP using a combinatorial interpretation similar to the longest increasing subsequence problem and connecting to the so-called Schur measure [67]. The Schur measure on integer partitions is defined in terms of the Schur function and contains many parameters. The one-point distribution of the TASEP arises by taking a special limit of the parameters.

A different proof computes the transition probabilities of the TASEP explicitly and then takes an appropriate sum over the configuration space to obtain the finite-time distribution. To find the transition probabilities, we solve the Kolmogorov forward equation, which is a linear differential equation with nonconstant coefficients due to the exclusion property of the particles. This equation was solved explicitly in [78] by applying the coordinate Bethe ansatz method from mathematical physics [46, 82], which consists of changing the Kolmogorov equation to a linear differential equation with constant coefficients (the free evolution equation) but with complicated boundary conditions. Taking the sum of transition probabilities over particular configurations is more technical, and this part was done in [73] to rederive the result of Johansson.

Both methods, which are algebraic and exact, are significantly extended to prove a one-point KPZ limit theorem for many other models. The following is the list of some of such integrable (exactly solvable) models. Of course, the list and references are far from exhaustive.

- Interacting particle systems: PushASEP, ASEP, q-TASEP, q-Hahn ASEP [18, 27, 28, 86].
- Random growth models: KPZ equation, stochastic heat equation [4, 24].
- DLPP and directed polymers: O'Connell–Yor semidiscrete polymer, log-gamma polymer [23, 26, 66].

The underlying algebraic structures of these integrable models are generalized greatly by Macdonald processes [22] and the stochastic six-vertex model [25, 35]. They are umbrella models with many parameters whose specializations produce the above models. Though many, these integrable models are still isolated examples. For instance, DLPP with general random variables, other than exponential and geometric random variables, does not seem to be integrable. See Section 9 for some comments for nonintegrable models.

4. MULTIPOINT DISTRIBUTIONS

Prähofer and Spohn [68] and Johansson [54] extended the one-point distribution results of Theorem 3.1 and 3.2 to equal-time, multiposition distributions for the Poisson DLPP and the TASEP with step initial condition, respectively. Their results were further extended to other models and initial conditions by [29–31, 76] in 2005–2008. These results confirmed, in particular, that the spatial correlations are of order $T^{2/3}$ and identified the equal-time slice of the KPZ fixed point for several initial conditions. See also [40] for more recent progress on other more difficult models.

On the other hand, multitime distributions and fully 2D multiposition distributions remained uninvestigated for a while, though some short and long time correlations were studied in [42], confirming that the time scale is T. In a breakthrough paper [65], Mateski, Quastel, and Remenik proved the convergence of the entire 2D height field of the TASEP in 2017. The limiting 2D field, the KPZ fixed point, is constructed as a Markov process with explicit transition probabilities. The result applies to general initial conditions. The authors used the result of [29, 76] on the transition probabilities of the TASEP for general initial conditions and proved that they converge. Dauvergne, Ortmann, and Virág gave an alternative formulation of the KPZ fixed point in terms of a variational formula and proved the field convergence for another model, the Brownian DLPP [36]. See also [89].

In the meantime, Johansson and Rahman [57] and Liu [63] computed the limit of 2D multipoint distributions of the discrete-time TASEP and the continuous-time TASEP, respectively, in 2019. Their results give an explicit formula of multipoint distributions for the KPZ fixed point with the narrow wedge initial condition. See Section 7 for the formula. Two-time distributions were previously computed in [55, 56].

5. HALF-INFINITE SPACE

We discussed so far models on infinite spaces. For example, the TASEP was defined on $\mathbb{Z}$. In this and the following sections, we consider different domains and their effects on the limit.

Consider the TASEP on the half-infinite space $\mathbb{Z}_+ \cup \{0\}$. We introduce a parameter $\alpha > 0$ representing the injection rate at site 0: if the origin is empty, a new particle is injected with rate α. Once injected, particles follow the usual TASEP rule. Suppose that we start with the empty configuration. If we could inject particles freely without being blocked by existing particles in the domain, then the height function at the origin would satisfy $\mathbf{h}(0, T)/T \to 2\alpha$ in probability as $T \to \infty$. However, due to the particles already in the domain, the height grows at a slower rate, and the hydrodynamic limit at the origin turns out to be

$$\frac{\mathbf{h}(0, T)}{T} \to \max\left\{2\alpha(1-\alpha), \frac{1}{2}\right\},$$

showing that the effective injection rate is $\max\{\alpha(1-\alpha), 1/4\}$. The formula changes at $\alpha = 1/2$.

The papers [14, 15] obtained a one-point KPZ limit theorem at the origin for the Poisson PNG and discrete-time TASEP models. The height scales as $T^{1/3}$ for $\alpha \geq 1/2$ and as $T^{1/2}$ for $\alpha < 1/2$. The limiting distribution is a variation of the Tracy–Widom distribution for $\alpha > 1/2$, another variation for $\alpha = 1/2$, and the Gaussian distribution for $\alpha < 1/2$. The result is extended to general positions and equal-time, multiposition distributions in [5,77] for the Poisson PNG and the discrete and continuous-time TASEP. However, generalizations to other models such as directed polymers and other interacting particle systems were missing, even though some algebraic formulas were established. Recently, [17] was able to prove a one-point KPZ limit theorem for the ASEP in which particles can move to the left as well as to the right with asymmetric rates. However, in any of these models, multitime limit theorems are not yet established.

6. RING DOMAIN

Consider the TASEP on the integer ring $\mathbb{Z}_L = \mathbb{Z}/L\mathbb{Z} = \{0, 1, \dots, L-1\}$ where we identify sites L and 0. An equivalent model is the TASEP on $\mathbb{Z}$ that is spatially periodic, which we may call the periodic TASEP. We call L the size of the ring or the period of the periodic TASEP.

The number N of particles in the TASEP on the ring is preserved. We assume the step initial condition shown in Figure 8. For the convenience of presentation, we assume that L is an even integer and $N = L/2$ so that the particle density $\rho = N/L = 1/2$. The initial height function of the periodic TASEP is the bottom curve in Figure 9. The other curves are the hydrodynamic limits, as time t and period L tend to infinity proportionally when $t/L = 0.5n$ for $n = 1, 2, \dots, 6$.

Consider two cases, one that $t \to \infty$ with L fixed and the other that $L \to \infty$ with t fixed. If $t \to \infty$ with L fixed, the periodic corner growth model becomes essentially a one-dimensional growth model, and we expect that the height scales as $t^{1/2}$ and converges to the Gaussian distribution. On the other hand, if $L \to \infty$ with t fixed, then the periodic TASEP becomes the usual TASEP on $\mathbb{Z}$. Thus, the height scales as $t^{1/3}$ if we let $t \to \infty$ after taking

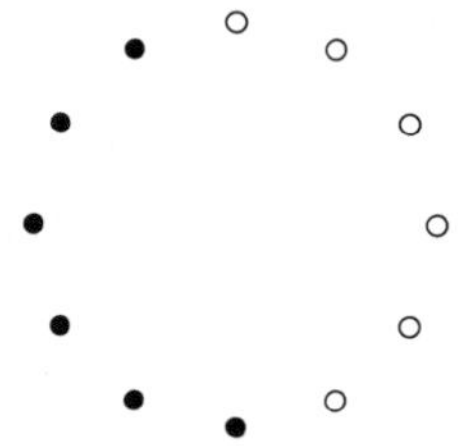

FIGURE 8
Step initial condition on an integer ring.

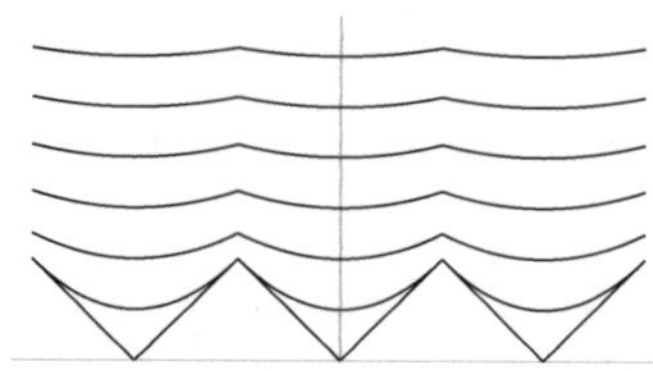

FIGURE 9
Hydrodynamic limits of periodic height function when $t = O(t)$.

$L \to \infty$. An interesting intermediate regime is when $L, t \to \infty$ simultaneously such that

$$t = O(L^{3/2}). \tag{6.1}$$

Since the spatial scale for KPZ limit theorems on infinite spaces is $t^{2/3}$, we expect that the height functions at all positions on the ring of size L are correlated nontrivially. If (6.1) holds, we say that we are in the relaxation time regime.

There were some results on transition probabilities and the spectral gap of the generator in the relaxation time regime, such as [48]. However, KPZ limit theorems were obtained only recently. The physics paper [69], which is not completely rigorous, and the mathematics paper [9] obtained a one-point KPZ limit theorem almost at the same time independently. This result was further extended to 2D multipoint distributions in [10, 11].

Theorem 6.1 ([10]). *Consider the TASEP on a ring of size L with the step initial condition and extend it to the periodic TASEP. Assume that L is even and $\rho = N/L = 1/2$ for the convenience of presentation. Set*

$$T = L^{3/2}.$$

For $i = 1, \dots, m$, let $(\gamma_i, \tau_i, \mathsf{h}_i) \in \mathbb{R} \times \mathbb{R}_+ \times \mathbb{R}$ and assume that $\tau_1 < \dots < \tau_m$. Then,

$$\lim_{T\to\infty} \mathrm{P}\left(\bigcap_{i=1}^{m}\left\{\frac{\mathbf{h}(\gamma_i T^{2/3}, 2\tau_i T) - \tau_i T}{-T^{1/3}} \le \mathsf{h}_i\right\}\right) = F_m^{\mathrm{pKPZ}}(\mathsf{h}; \gamma, \tau)$$

for an m-point distribution function F_m^{pKPZ} described in the next section.

Like the infinite space case, we expect that the height field of the periodic models in the relaxation time regime converges to a universal field, which we may call the periodic KPZ fixed point. The function F_m^{pKPZ} should be the m-point distribution of this conjectured periodic KPZ fixed point with the (periodic) narrow wedge initial condition. It is naturally periodic with respect to $\gamma_i \mapsto \gamma_i + 1$. However, unlike the KPZ fixed point, F_m^{pKPZ} is not invariant under the rescaling (3.1). Indeed, we conjecture that F_m^{pKPZ} interpolates the KPZ fixed point and one-dimensional Brownian motion. Concretely, we expect that

$$\lim_{\varepsilon\to 0} F_m^{\mathrm{pKPZ}}(\mathsf{h}^\varepsilon; \gamma^\varepsilon, \tau^\varepsilon) = F_m(\mathsf{h}; \gamma, \tau) \quad \text{where } (\mathsf{h}_i^\varepsilon, \gamma_i^\varepsilon, \tau_i^\varepsilon) = ((\tau_i\varepsilon)^{1/3}\mathsf{h}_i, (\tau_i\varepsilon)^{2/3}\gamma_i, \tau_i\varepsilon)$$

and F_m is the m-point distribution of the KPZ fixed point, and that

$$\lim_{s\to\infty} F_m^{\mathrm{pKPZ}}(\mathsf{h}^s; \gamma, \tau^s) = G_m(\mathsf{h}; \tau) \quad \text{where } (\mathsf{h}_i^s, \tau^s) = (-s\tau_i + \frac{s^{1/2}\pi^{1/4}}{\sqrt{2}}\mathsf{h}_i, s\tau_i)$$

and G_m is the m-point distribution of a Brownian motion at times $\tau_1, \dots, \tau_m$. These conjectures were proved for $m = 1$ in [12], assuming $\gamma_1 = 0$ for the $\varepsilon \to 0$ case.

Theorem 6.1 is also proved for the discrete-time TASEP on a ring [60]. However, extending the result to other integrable models is yet to be done.

7. FORMULA OF MULTIPOINT DISTRIBUTION FUNCTIONS

7.1. Formula for KPZ fixed point

Let F_m be the m-point distribution of the KPZ fixed point with the narrow wedge initial condition. The result of [63] implies that

$$F_m(\mathsf{h}; \gamma, \tau) = \frac{1}{(2\pi \mathrm{i})^{m-1}} \oint \cdots \oint \frac{\det(1 - \mathsf{K}_\zeta)}{\zeta_1(1-\zeta_1)\cdots\zeta_{m-1}(1-\zeta_{m-1})} \mathrm{d}\zeta_1 \cdots \mathrm{d}\zeta_{m-1} \tag{7.1}$$

where $\zeta = (\zeta_1, \ldots, \zeta_{m-1})$, and the contours are nested circles of radii less than 1 centered at the origin. The operator K_ζ acts on $L^2(\Sigma)$, where Σ is the union of $4m-2$ contours in Figure 10 that extend to infinity with angle $\pi/5$ from the x-axis. The kernel of K_ζ can be written [13] as a simple conjugation of the kernel

$$\mathsf{K}_\zeta^{\mathrm{conj}}(u, v) = \frac{\mathsf{a}(u)^T \mathsf{D}(u)^T \mathsf{D}(v)\mathsf{b}(v)}{u - v}, \qquad u, v \in \Sigma, \tag{7.2}$$

which is zero for $u = v$. The $(m+1) \times (m+1)$ matrix

$$\mathsf{D}(z) = \operatorname{diag}\big(e^{-\frac{1}{3}\tau_1 z^3 + \frac{1}{2}\gamma_1 z^2 + \mathsf{h}_1 z}, \ldots, e^{-\frac{1}{3}\tau_m z^3 + \frac{1}{2}\gamma_m z^2 + \mathsf{h}_m z}, 1\big).$$

The $(m+1) \times 1$ vectors $\mathsf{a}(z)$ and $\mathsf{b}(z)$ are simple and explicit, and they do not depend on $\mathsf{h}_i, \gamma_i, \tau_i$. Note that the exponent $-\frac{1}{3}\tau_i z^3 + \frac{1}{2}\gamma_i z^2 + \mathsf{h}_i z$ in $\mathsf{D}(z)$ is unchanged if we rescale as (3.1) and $z \mapsto \alpha^{-1} z$. This is consistent with the rescaling property of the KPZ fixed point.

7.2. Formula for the periodic case

The formula of [10] for the conjectured m-point distribution of the periodic KPZ fixed point is

$$F_m^{\mathrm{pKPZ}}(\mathsf{h}; \gamma, \tau) = \oint \cdots \oint C(\zeta) \det(1 - \mathsf{K}_\zeta) \mathrm{d}\zeta_1 \cdots \mathrm{d}\zeta_m. \tag{7.3}$$

This time there are m integrals and $C(\zeta)$ is an explicit function expressed in terms of polylog functions. The kernel of the operator K_ζ is of the same form as (7.2) but $\mathsf{a}(z)$ and $\mathsf{b}(z)$ are slightly different. The key change is the space for K_ζ. It is $\ell^2(\mathsf{S})$, where $\mathsf{S} = \mathsf{S}_1 \cup \cdots \cup \mathsf{S}_m$ and S_i is the discrete set of the roots of the equation

$$e^{-s^2/2} = \zeta_i, \tag{7.4}$$

shown in Figure 11. See the following subsection for how this equation arises.

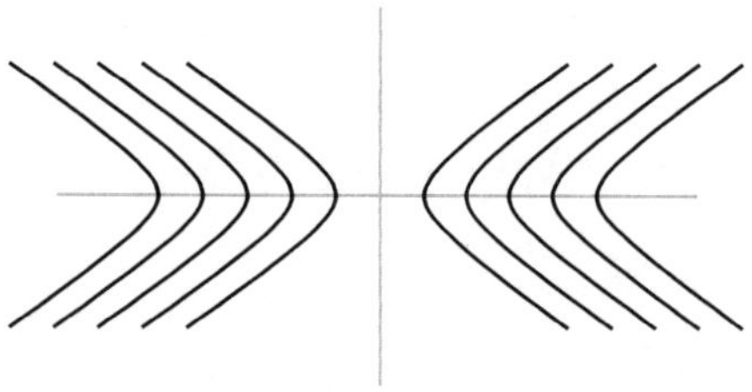

FIGURE 10
The space Σ for KPZ fixed point when $m = 3$.

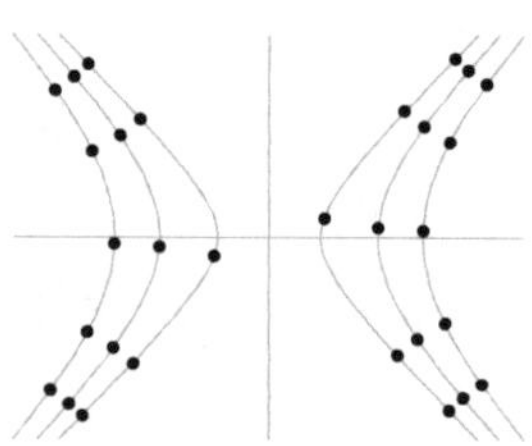

FIGURE 11
The space S for periodic KPZ fixed when $m = 3$.

7.3. Transition probabilities

We mentioned in Section 3.4 that one way of proving a KPZ limit theorem for the TASEP is to compute the transition probabilities explicitly and then take an appropriate sum to find the finite-time distribution functions. The summation part is often more technical, but here we discuss the transition probabilities to see how the equation (7.4) arises.

Suppose that there are only N particles for the TASEP on $\mathbb{Z}$. Let $\mathcal{W}_N = \{(a_1, \dots, a_N) \in \mathbb{Z}^N : a_1 < \cdots < a_N\}$ be the ordered set of the particle locations. Schütz [78] showed that

$$\mathrm{P}_Y(X; t) = \det\left[\frac{1}{2\pi \mathrm{i}} \oint s^{j-i+1}(s+1)^{-x_i+y_j+i-j} e^{ts} \mathrm{d}s\right]_{i,j=1}^{N} \tag{7.5}$$

for X and Y in $\mathcal{W}_N$, where the contour is a circle that encloses the points $s = 0, -1$.

For the periodic TASEP, the particle locations can be represented by the set

$$\mathcal{W}_N^L = \{(a_1, \dots, a_N) \in \mathbb{Z}^N : a_1 < \cdots < a_N < a_1 + L\}.$$

Note that if we consider the TASEP on a ring, this set keeps track of global circulations of the particles. We showed in [9] that

$$\mathrm{P}_Y(X; t) = \oint \det\left[\frac{1}{L} \sum_{w} \frac{w^{j-i+1}(w+1)^{-x_i+y_j+i-j} e^{tw}}{w + N/L}\right]_{i,j=1}^{N} \frac{\mathrm{d}z}{2\pi \mathrm{i} z} \tag{7.6}$$

for $X, Y \in \mathcal{W}_N^L$, where the integral contour for z is any circle enclosing the origin, and the sum inside is over the roots of the equation

$$w^N (w+1)^{L-N} = z. \tag{7.7}$$

See Figure 12.

We now explain equation (7.7). Due to the periodicity, if $A = (a_1, \dots, a_N)$ is in $\mathcal{W}_N(L)$, then $A' = (a_2, \dots, a_N, a_1 + L)$ also represents the same particle configuration of the periodic TASEP. Thus, the transition probability should remain the same if we replace $(x_1, \dots, x_N)$ and $(y_1, \dots, y_N)$ by $(x_2, \dots, x_N, x_1 + L)$ and $(y_2, \dots, y_N, y_1 + L)$,

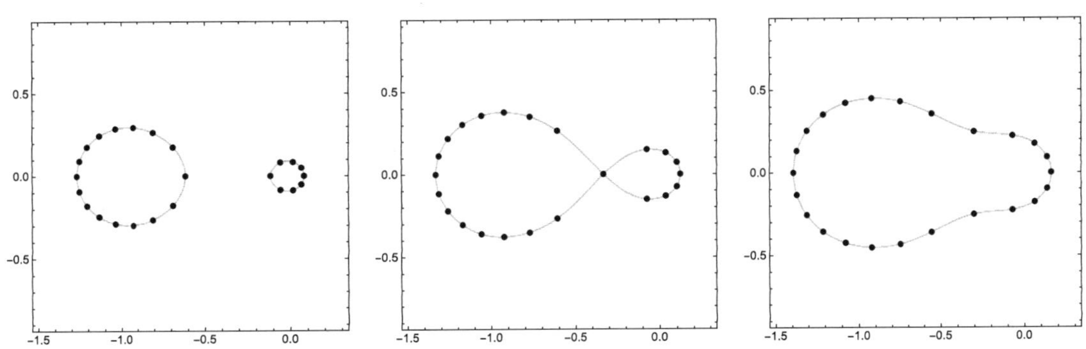

FIGURE 12
Bethe roots when $L = 24$ and $N = 8$ for three values of z.

respectively. We can check directly that the determinant in (7.6) is unchanged thanks to equation (7.7). Equation (7.7) takes care of the labeling ambiguity in the periodic case. Here the variable z could have been any fixed constant, but making it as a free parameter turns out to be the right choice. Since (7.5) and (7.6) were found by solving the Kolmogorov forward equation using the Bethe ansatz method, the roots of (7.7) are called the Bethe roots.

Now, if we set $L = 2N$, $w = -\frac{1}{2} + \frac{s}{2\sqrt{2N}}$, and $z = (-4)^{-N}\zeta$ in (7.7), and let $N \to \infty$, then the equation becomes $e^{-s^2/2} = \zeta$, which is (7.4).

8. INTEGRABLE DIFFERENTIAL EQUATIONS

Distribution functions of the KPZ fixed point have connections to deterministic integrable differential equations. As proved in 1994 **[84]**, the Tracy–Widom distribution is expressible in terms of the Painlevé II equation, one of a family of six special nonlinear ordinary differential equations **[43]**. The papers **[2, 20, 71, 85, 90]** also found differential equations for equal-time, multiposition distribution functions of the KPZ fixed point. We state the following result for multipoint distributions for both infinite and periodic domains.

Define the parameters

$$\mathsf{t}_i = \tau_i/3, \quad \mathsf{y}_i = \gamma_i, \quad \mathsf{x}_i = \mathsf{h}_i,$$

and let

$$\partial_{\mathsf{t}} = \sum_{i=1}^{m} \partial_{\mathsf{t}_i}, \quad \partial_{\mathsf{y}} = \sum_{i=1}^{m} \partial_{\mathsf{y}_i}, \quad \partial_{\mathsf{x}} = \sum_{i=1}^{m} \partial_{\mathsf{x}_i}.$$

Theorem 8.1 (**[12, 13]**). *Let* $\mathsf{K} = \mathsf{K}_\zeta$ *be the operator in either* (7.1) *or* (7.3). *If* $\det(1 - \mathsf{K}) \neq 0$, *which holds for all but at most countably many parameters, then*

$$\partial_{\mathsf{x}}^2 \log \det(1 - \mathsf{K}) = -\mathsf{r}^T \mathsf{p}$$

for complex-valued $m \times 1$ *vector functions* $\mathsf{p}(\mathsf{t}, \mathsf{y}, \mathsf{x})$ *and* $\mathsf{r}(\mathsf{t}, \mathsf{y}, \mathsf{x})$ *which satisfy the equations*

$$\partial_{\mathsf{y}}\mathsf{p} = \frac{1}{2}\partial_{\mathsf{x}}^2\mathsf{p} - \mathsf{p}\mathsf{r}^T\mathsf{p}, \quad \partial_{\mathsf{y}}\mathsf{r} = -\frac{1}{2}\partial_{\mathsf{x}}^2\mathsf{r} + \mathsf{r}\mathsf{p}^T\mathsf{r} \tag{8.1}$$

and

$$\partial_{\mathsf{t}}\mathsf{p} + \partial_{\mathsf{x}}^3\mathsf{p} - 3(\partial_{\mathsf{x}}\mathsf{p})\mathsf{r}^T\mathsf{p} - 3\mathsf{p}\mathsf{r}^T(\partial_{\mathsf{x}}\mathsf{p}) = 0, \quad \partial_{\mathsf{t}}\mathsf{r} + \partial_{\mathsf{x}}^3\mathsf{r} - 3(\partial_{\mathsf{x}}\mathsf{r})\mathsf{p}^T\mathsf{r} - 3\mathsf{r}\mathsf{p}^T(\partial_{\mathsf{x}}\mathsf{r}) = 0. \tag{8.2}$$

Equation (8.2) is a coupled system of vector-valued modified Korteweg–de Vries (mKdV) equations. The scalar mKdV equation is $\partial_{\mathsf{t}} f + \partial_{\mathsf{x}}^3 f - 6(\partial_{\mathsf{x}} f) f^2 = 0$. On the other hand, equation (8.1) forms a coupled system of vector-valued nonlinear forward and backward heat equations. They become vector-valued nonlinear Schrödinger (NLS) equations if we change $\mathsf{y}_i \mapsto i\mathsf{y}_i$. NLS and mKdV equations are two of the most famous integrable partial differential equations **[1]**. The above two systems of equations can be combined to the Kadomtsev–Petviashvili (KP) equation, another integrable differential equation in three variables. Theorem 8.1 was obtained using the fact that the operator is a so-called integrable operator **[37, 51]**.

Integrable differential equations have a long history starting with the work of Gardner, Greene, Kruskal, and Miura in 1967. They found a scattering transform method, an equation-specific nonlinear Fourier transform, to solve the Korteweg–de Vries equation. Like the integrable models in the KPZ universality class, integrable differentiable equations are isolated examples of nonlinear differential equations that can often be solved explicitly and analyzed asymptotically. See, for example, [1, 38]. It is intriguing that integrable probability models are related to integrable differential equations.

9. COMMENTS ON UNIVERSALITY

KPZ limit theorems are proved for many isolated examples of integrable models. In this final section, we discuss a few instances that universality is proved.

9.1. Thin DLPP

The universality should hold for the last passage time $L(n,k)$ as $n,k \to \infty$. It is easy to prove it for thin rectangles. Recall from Section 2.3 that w_s denotes the random variable at site $s \in \mathbb{Z}$, representing the passage time through the site.

Theorem 9.1 ([16, 21]). *Suppose that w_s is an arbitrary random variable which has all moments. Assume that the mean is zero and the variance is one. Then, for every x,*

$$\lim_{n,k\to\infty} \mathrm{P}\left(\frac{L(n,k) - 2\sqrt{nk}}{n^{\frac{1}{2}} k^{-\frac{1}{6}}} \le x \right) = F_{\mathrm{TW}}(x) \quad \text{for } k = [n^a]$$

with any $0 < a < 3/7$.

The restriction $a < 3/7$ is technical. If we assume only finite p moments for $p > 2$, then the result holds for thinner rectangles satisfying $0 < a < \frac{3(p-2)}{7p}$.

For the case of fixed k and large n, a directed path looks like that in Figure 13. Since the sum of w_s on each row converges to a Brownian motion, Donsker's theorem implies that

$$\frac{L(n,k)}{\sqrt{n}} \Rightarrow D_k \quad \text{where } D_k = \sup_{0=t_0 \le t_1 \le \cdots \le t_k = 1} \sum_{i=1}^{k} \left(B_i(t_i) - B_i(t_{i-1}) \right)$$

and $B_i(t)$, $i = 1, \ldots, k$, are independent Brownian motions. On the other hand, it is known [19] that

$$\lim_{k\to\infty} \mathrm{P}\left((D_k - 2\sqrt{k}) k^{1/6} \le x \right) = F_{\mathrm{TW}}(x).$$

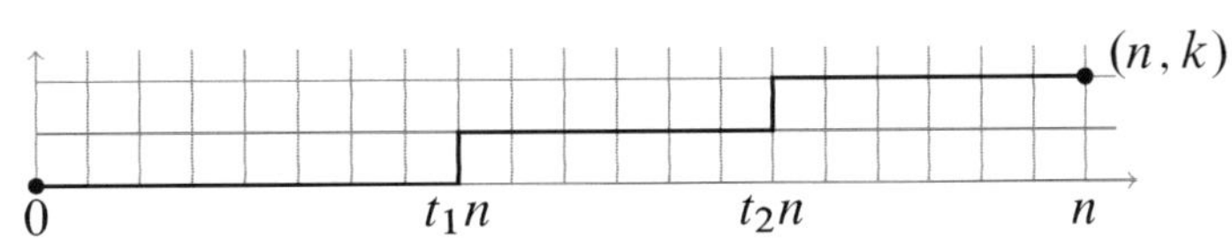

FIGURE 13
Thin DLPP.

This limit is a consequence of an explicit formula of the exponential DLPP by letting $n \to \infty$ first and then taking $k \to \infty$. We prove Theorem 9.1 by showing that if $n, k \to \infty$ but k grows slowly enough, we can take $n \to \infty$ first and $k \to \infty$ later. This argument is achieved by the Skorohod embedding or the Komlós–Major–Tusnády embedding. However, the proof breaks down if $\alpha \geq 3/7$ since we cannot ignore the upward parts of the paths anymore. For $k = O(n)$, limit theorems are proved only for a few examples.

9.2. Interacting particle systems

The particles in the TASEP move only one site to the right. Consider a more general finite-range exclusion process in which a particle at site 0 can potentially move to site v at rate $p(v)$. Assume that $\{v : p(v) > 0\}$ is a finite set generating $\mathbb{Z}$ additively, and $\sum_v vp(v) \neq 0$. In a recent paper [72], Quastel and Sakar proved a KPZ limit theorem for finite-range exclusion processes started from a certain class of initial conditions. They compared the transition probabilities of the general process with those of the TASEP using energy estimates. This work is the first universality result in the $k = O(n)$ regime. It is exciting to see how the method generalizes further.

9.3. TASEP, Coulomb gas, and random matrices

In proving Theorem 3.2, Johansson also proved an unexpected connection of TASEP to Coulomb gas and random matrices [52]. Consider the probability density function on $\mathbb{R}_+^n$ given by

$$p(x_1, \dots, x_n) = c_{n,m} e^{2\sum_{1 \leq i < j \leq n} \log |x_j - x_i| - \sum_{i=1}^n V(x_i)}, \quad V(x) = x - (m-n)\log x, \quad (9.1)$$

where $c_{m,n}$ is the normalization constant. Let $x_{\max} = \max\{x_1, \dots, x_n\}$. Johansson proved that for $m \geq n$, the last passage time $L(m,n)$ of the exponential DLPP has the same distribution as $x_{\max}$. The density function (9.1) is said to define a Coulomb gas with potential V on $\mathbb{R}_+$ since the term $\log |x_i - x_j|$ is the 2D Coulomb potential of two equal charges at x_i and x_j. The function $V(x)$ represents the confining potential.

Let X be an $n \times m$ random matrix with entries that are independent complex normal variables of mean zero and variance $1/2$. The random matrix $W = XX^*$ is called the complex Wishart matrix, and its eigenvalue density function is precisely (9.1) [44]. Thus, $L(m,n)$ has the same distribution as the largest eigenvalue of a complex Wishart matrix. See also [33]. The connection of the exponential DLPP to Coulomb gas and the random matrix is special, and we do not expect to hold for general random variables w_s.

There are universality results for both Coulomb gas and random matrices. The Tracy–Widom limit theorem is proved for the Coulomb gases with a general potential V. The paper [39] showed the limit theorem for generic analytic potentials, and [8] proved for discrete Coulomb gases in which particles are restricted to be only on a discrete set.

Universality is a central question in random matrix theory, and there have been remarkable successes. The largest eigenvalue of a large class of random Hermitian matrices with independent entries converges to the Tracy–Widom distribution [41,59,79,83].

9.4. Universality in many directions

We discussed that the TASEP with the step initial condition is connected to several areas:

- interacting particle system
- 1+1 random growth process
- two-dimensional directed last passage percolation and directed polymer
- Coulomb gas
- random matrix

The TASEP also has interpretations as a random tiling model, and nonintersecting paths [47, 53]. We expect universality results to hold in all of these seven areas. The meaning of universality is different in each area. For example, in random matrix theory, the largest eigenvalue of any random Hermitian matrix with independent and identically distributed entries with $4+\varepsilon$ finite moments converges to the same limit, the Tracy–Widom distribution. On the other hand, the 2D field limit of interacting particle systems depends on the initial condition, a special case of which has the Tracy–Widom distribution as its marginal.

Even though many universality results are proved for Coulomb gases and random matrices, it remains to establish similar results for other areas and develop a general theory that encompasses all of these areas and possibly more.

ACKNOWLEDGMENTS

The author would like to thank Percy Deift, Zhipeng Liu, Andrei Prokhorov, and Guilherme Silva for helpful conversations.

FUNDING

This work was partially supported by the NSF grant DMS-1954790.

REFERENCES

[1] M. J. Ablowitz and P. A. Clarkson, *Solitons, nonlinear evolution equations and inverse scattering*. London Math. Soc. Lecture Note Ser. 149, Cambridge University Press, Cambridge, 1991.

[2] M. Adler and P. van Moerbeke, PDEs for the joint distributions of the Dyson, Airy and sine processes. *Ann. Probab.* **33** (2005), no. 4, 1326–1361.

[3] D. Aldous and P. Diaconis, Longest increasing subsequences: from patience sorting to the Baik–Deift–Johansson theorem. *Bull. Amer. Math. Soc. (N.S.)* **36** (1999), no. 4, 413–432.

[4] G. Amir, I. Corwin, and J. Quastel, Probability distribution of the free energy of the continuum directed random polymer in $1+1$ dimensions. *Comm. Pure Appl. Math.* **64** (2011), no. 4, 466–537.

[5] J. Baik, G. Barraquand, I. Corwin, and T. Suidan, Pfaffian Schur processes and last passage percolation in a half-quadrant. *Ann. Probab.* **46** (2018), no. 6, 3015–3089.

[6] J. Baik, P. Deift, and K. Johansson, On the distribution of the length of the longest increasing subsequence of random permutations. *J. Amer. Math. Soc.* **12** (1999), no. 4, 1119–1178.

[7] J. Baik, P. Deift, and T. Suidan, *Combinatorics and random matrix theory*. Grad. Stud. Math. 172, Am. Math. Soc., Providence, 2016.

[8] J. Baik, T. Kriecherbauer, K. T.-R. McLaughlin, and P. Miller, *Discrete orthogonal polynomials*. Ann. of Math. Stud. 164, Princeton University Press, Princeton, 2007.

[9] J. Baik and Z. Liu, Fluctuations of TASEP on a ring in relaxation time scale. *Comm. Pure Appl. Math.* **71** (2018), no. 4, 747–813.

[10] J. Baik and Z. Liu, Multipoint distribution of periodic TASEP. *J. Amer. Math. Soc.* **32** (2019), no. 3, 609–674.

[11] J. Baik and Z. Liu, Periodic TASEP with general initial conditions. *Probab. Theory Related Fields* **179** (2021), no. 3–4, 1047–1144.

[12] J. Baik, Z. Liu, and G. L. F. Silva, Limiting one-point distribution of periodic TASEP. 2020, arXiv:2008.07024.

[13] J. Baik, A. Prokhorov, and G. L. F. Silva, *Integrable structure of the multipoint distributions of KPZ fixed point*. 2021, in preparation.

[14] J. Baik and E. M. Rains, The asymptotics of monotone subsequences of involutions. *Duke Math. J.* **109** (2001), no. 2, 205–281.

[15] J. Baik and E. M. Rains, Symmetrized random permutations. In *Random matrix models and their applications*, pp. 1–19, Math. Sci. Res. Inst. Publ. 40, Cambridge University Press, Cambridge, 2001.

[16] J. Baik and T. M. Suidan, A GUE central limit theorem and universality of directed first and last passage site percolation. *Int. Math. Res. Not.* **2005** (2005), no. 6, 325–337.

[17] G. Barraquand, A. Borodin, I. Corwin, and M. Wheeler, Stochastic six-vertex model in a half-quadrant and half-line open asymmetric simple exclusion process. *Duke Math. J.* **167** (2018), no. 13, 2457–2529.

[18] G. Barraquand and I. Corwin, The q-Hahn asymmetric exclusion process. *Ann. Appl. Probab.* **26** (2016), no. 4, 2304–2356.

[19] Y. Baryshnikov, GUEs and queues. *Probab. Theory Related Fields* **119** (2001), no. 2, 256–274.

[20] M. Bertola and M. Cafasso, Riemann–Hilbert approach to multi-time processes: the Airy and the Pearcey cases. *Phys. D* **241** (2012), no. 23–24, 2237–2245.

[21] T. Bodineau and J. Martin, A universality property for last-passage percolation paths close to the axis. *Electron. Commun. Probab.* **10** (2005), 105–112.

[22] A. Borodin and I. Corwin, Macdonald processes. *Probab. Theory Related Fields* **158** (2014), no. 1–2, 225–400.

[23] A. Borodin, I. Corwin, and P. Ferrari, Free energy fluctuations for directed polymers in random media in $1 + 1$ dimension. *Comm. Pure Appl. Math.* **67** (2014), no. 7, 1129–1214.

[24] A. Borodin, I. Corwin, P. Ferrari, and B. Vető, Height fluctuations for the stationary KPZ equation. *Math. Phys. Anal. Geom.* **18** (2015), no. 1, Art. 20, 95.

[25] A. Borodin, I. Corwin, and V. Gorin, Stochastic six-vertex model. *Duke Math. J.* **165** (2016), no. 3, 563–624.

[26] A. Borodin, I. Corwin, and D. Remenik, Log-gamma polymer free energy fluctuations via a Fredholm determinant identity. *Comm. Math. Phys.* **324** (2013), no. 1, 215–232.

[27] A. Borodin and P. L. Ferrari, Large time asymptotics of growth models on space-like paths. I. PushASEP. *Electron. J. Probab.* **13** (2008), no. 50, 1380–1418.

[28] A. Borodin and P. L. Ferrari, Anisotropic growth of random surfaces in $2 + 1$ dimensions. *Comm. Math. Phys.* **325** (2014), no. 2, 603–684.

[29] A. Borodin, P. L. Ferrari, M. Prähofer, and T. Sasamoto, Fluctuation properties of the TASEP with periodic initial configuration. *J. Stat. Phys.* **129** (2007), no. 5–6, 1055–1080.

[30] A. Borodin, P. L. Ferrari, and T. Sasamoto, Large time asymptotics of growth models on space-like paths. II. PNG and parallel TASEP. *Comm. Math. Phys.* **283** (2008), no. 2, 417–449.

[31] A. Borodin, P. L. Ferrari, and T. Sasamoto, Transition between Airy_1 and Airy_2 processes and TASEP fluctuations. *Comm. Pure Appl. Math.* **61** (2008), no. 11, 1603–1629.

[32] A. Borodin and V. Gorin, Lectures on integrable probability. In *Probability and statistical physics in St. Petersburg*, pp. 155–214, Proc. Sympos. Pure Math. 91, Am. Math. Soc., Providence, 2016.

[33] A. Borodin and S. Péché, Airy kernel with two sets of parameters in directed percolation and random matrix theory. *J. Stat. Phys.* **132** (2008), no. 2, 275–290.

[34] I. Corwin, The Kardar–Parisi–Zhang equation and universality class. *Random Matrices Theory Appl.* **1** (2012), no. 1, 1130001, 76.

[35] I. Corwin and L. Petrov, Stochastic higher spin vertex models on the line. *Comm. Math. Phys.* **343** (2016), no. 2, 651–700.

[36] D. Dauvergne, J. Ortmann, and B. Virág, The directed landscape. 2018, arXiv:1812.00309.

[37] P. Deift, Integrable operators. In *Differential operators and spectral theory*, pp. 69–84, Amer. Math. Soc. Transl. Ser. 2 189, Am. Math. Soc., Providence, 1999.

[38] P. A. Deift, Three lectures on "Fifty years of KdV: an integrable system". In *Nonlinear dispersive partial differential equations and inverse scattering*, pp. 3–38, Fields Inst. Commun. 83, Springer, New York, [2019] ©2019.

[39] P. Deift, T. Kriecherbauer, K. T.-R. McLaughlin, S. Venakides, and X. Zhou, Uniform asymptotics for polynomials orthogonal with respect to varying exponential weights and applications to universality questions in random matrix theory. *Comm. Pure Appl. Math.* **52** (1999), no. 11, 1335–1425.

[40] E. Dimitrov, Two-point convergence of the stochastic six-vertex model to the Airy process. 2020, arXiv:2006.15934.

[41] L. Erdős, A. Knowles, H.-T. Yau, and J. Yin, Spectral statistics of Erdős–Rényi graphs II: eigenvalue spacing and the extreme eigenvalues. *Comm. Math. Phys.* **314** (2012), no. 3, 587–640.

[42] P. L. Ferrari and H. Spohn, On time correlations for KPZ growth in one dimension. *SIGMA Symmetry Integrability Geom. Methods Appl.* **12** (2016), Paper No. 074, 23.

[43] A. S. Fokas, A. R. Its, A. A. Kapaev, and V. Y. Novokshenov, *Painlevé transcendents*. Math. Surveys Monogr. 128, Am. Math. Soc., Providence, 2006.

[44] P. J. Forrester, *Log-gases and random matrices*. London Math. Soc. Monogr. Ser. 34, Princeton University Press, Princeton, 2010.

[45] D. Forster, D. R. Nelson, and M. J. Stephen, Large-distance and long-time properties of a randomly stirred fluid. *Phys. Rev. A* **16** (1977), 732–749.

[46] M. Gaudin, *The Bethe wavefunction*. Cambridge University Press, New York, 2014.

[47] V. Gorin, *Lectures on random lozenge tilings*. Cambridge Stud. Adv. Math. 193, Cambridge University Press, Cambridge, 2021.

[48] L.-H. Gwa and H. Spohn, Bethe solution for the dynamical-scaling exponent of the noisy Burgers equation. *Phys. Rev. A* **46** (1992), 844–854.

[49] J. M. Hammersley, A few seedlings of research. In *Proceedings of the Sixth Berkeley Symposium on Mathematical Statistics and Probability (Univ. California, Berkeley, Calif., 1970/1971), vol. I: theory of statistics*, pp. 345–394, 1972.

[50] D. A. Huse and C. L. Henley, Pinning and roughening of domain walls in ising systems due to random impurities. *Phys. Rev. Lett.* **54** (1985), 2708–2711.

[51] A. R. Its, A. G. Izergin, V. E. Korepin, and N. A. Slavnov, Differential equations for quantum correlation functions. In *Proceedings of the Conference on Yang–Baxter Equations, Conformal Invariance and Integrability in Statistical Mechanics and Field Theory*, pp. 1003–1037, 1990.

[52] K. Johansson, Shape fluctuations and random matrices. *Comm. Math. Phys.* **209** (2000), no. 2, 437–476.

[53] K. Johansson, Non-intersecting paths, random tilings and random matrices. *Probab. Theory Related Fields* **123** (2002), no. 2, 225–280.

[54] K. Johansson, Discrete polynuclear growth and determinantal processes. *Comm. Math. Phys.* **242** (2003), no. 1–2, 277–329.

[55] K. Johansson, Two time distribution in Brownian directed percolation. *Comm. Math. Phys.* **351** (2017), no. 2, 441–492.

[56] K. Johansson, The two-time distribution in geometric last-passage percolation. *Probab. Theory Related Fields* **175** (2019), no. 3–4, 849–895.

[57] K. Johansson and M. Rahman, Multi-time distribution in discrete polynuclear growth. 2019, arXiv:1906.01053.

[58] M. Kardar, G. Parisi, and Y.-C. Zhang, Dynamic scaling of growing interfaces. *Phys. Rev. Lett.* **56** (1986), 889–892.

[59] J. O. Lee and J. Yin, A necessary and sufficient condition for edge universality of Wigner matrices. *Duke Math. J.* **163** (2014), no. 1, 117–173.

[60] Y. Liao, Multi-point distribution of discrete time periodic TASEP. 2021, arXiv:2011.07726.

[61] T. M. Liggett, *Interacting particle systems*. Grundlehren Math. Wiss. 276, Springer, New York, 1985.

[62] T. M. Liggett, *Stochastic interacting systems: contact, voter and exclusion processes*. Grundlehren Math. Wiss. 324, Springer, Berlin, 1999.

[63] Z. Liu, Multi-time distribution of TASEP. 2019, arXiv:1907.09876.

[64] B. F. Logan and L. A. Shepp, A variational problem for random Young tableaux. *Adv. Math.* **26** (1977), no. 2, 206–222.

[65] K. Matetski, J. Quastel, and D. Remenik, The KPZ fixed point. 2020, arXiv:1701.00018.

[66] N. O'Connell, Directed polymers and the quantum Toda lattice. *Ann. Probab.* **40** (2012), no. 2, 437–458.

[67] A. Okounkov, Infinite wedge and random partitions. *Selecta Math. (N.S.)* **7** (2001), no. 1, 57–81.

[68] M. Prähofer and H. Spohn, Scale invariance of the PNG droplet and the Airy process. *J. Stat. Phys.* **108** (2002), no. 5–6, 1071–1106.

[69] S. Prolhac, Finite-time fluctuations for the totally asymmetric exclusion process. *Phys. Rev. Lett.* **116** (2016), 090601.

[70] J. Quastel, Introduction to KPZ. In *Current developments in mathematics, 2011*, pp. 125–194, Int. Press, Somerville, 2012.

[71] J. Quastel and D. Remenik, KP governs random growth off a one dimensional substrate. 2019, arXiv:1908.10353.

[72] J. Quastel and S. Sarkar, Convergence of exclusion processes and KPZ equation to the KPZ fixed point. 2020, arXiv:2008.06584.

[73] A. Rákos and G. M. Schütz, Current distribution and random matrix ensembles for an integrable asymmetric fragmentation process. *J. Stat. Phys.* **118** (2005), no. 3–4, 511–530.

[74] D. Romik, *The surprising mathematics of longest increasing subsequences*. IMS Textb. 4, Cambridge University Press, New York, 2015.

[75] H. Rost, Nonequilibrium behaviour of a many particle process: density profile and local equilibria. *Z. Wahrsch. Verw. Gebiete* **58** (1981), no. 1, 41–53.

[76] T. Sasamoto, Spatial correlations of the 1D KPZ surface on a flat substrate. *J. Phys. A* **38** (2005), no. 33, L549–L556.

[77] T. Sasamoto and T. Imamura, Fluctuations of the one-dimensional polynuclear growth model in half-space. *J. Stat. Phys.* **115** (2004), no. 3–4, 749–803.

[78] G. M. Schütz, Exact solution of the master equation for the asymmetric exclusion process. *J. Stat. Phys.* **88** (1997), no. 1–2, 427–445.

[79] A. Soshnikov, Universality at the edge of the spectrum in Wigner random matrices. *Comm. Math. Phys.* **207** (1999), no. 3, 697–733.

[80] F. Spitzer, Interaction of Markov processes. *Adv. Math.* **5** (1970), 246–290.

[81] R. P. Stanley, Increasing and decreasing subsequences and their variants. In *International Congress of Mathematicians. Vol. I*, pp. 545–579, Eur. Math. Soc., Zürich, 2007.

[82] B. Sutherland, *Beautiful models*. World Scientific Publishing Co., Inc., River Edge, NJ, 2004.

[83] T. Tao and V. Vu, Random matrices: universality of local eigenvalue statistics up to the edge. *Comm. Math. Phys.* **298** (2010), no. 2, 549–572.

[84] C. A. Tracy and H. Widom, Level-spacing distributions and the Airy kernel. *Comm. Math. Phys.* **159** (1994), no. 1, 151–174.

[85] C. A. Tracy and H. Widom, A system of differential equations for the Airy process. *Electron. Commun. Probab.* **8** (2003), 93–98.

[86] C. A. Tracy and H. Widom, Asymptotics in ASEP with step initial condition. *Comm. Math. Phys.* **290** (2009), no. 1, 129–154.

[87] H. van Beijeren, R. Kutner, and H. Spohn, Excess noise for driven diffusive systems. *Phys. Rev. Lett.* **54** (1985), 2026–2029.

[88] A. M. Veršik and S. V. Kerov, Asymptotic behavior of the Plancherel measure of the symmetric group and the limit form of Young tableaux. *Dokl. Akad. Nauk SSSR* **233** (1977), no. 6, 1024–1027.

[89] B. Virág, The heat and the landscape I. 2020, arXiv:2008.07241.

[90] D. Wang, A PDE for the multi-time joint probability of the Airy process. *Phys. D* **238** (2009), no. 8, 819–833.

JINHO BAIK

Department of Mathematics, University of Michigan, Ann Arbor, MI, 48109, USA, baik@umich.edu

INTRODUCTION TO THE LIOUVILLE QUANTUM GRAVITY METRIC

JIAN DING, JULIEN DUBÉDAT, AND EWAIN GWYNNE

ABSTRACT

Liouville quantum gravity (LQG) is a one-parameter family of models of random fractal surfaces which first appeared in the physics literature in the 1980s. Recent works have constructed a metric (distance function) on an LQG surface. We give an overview of the construction of this metric and discuss some of its most important properties, such as the behavior of geodesics and the KPZ formula. We also discuss some of the main techniques for proving statements about the LQG metric, give examples of their use and discuss some open problems.

MATHEMATICS SUBJECT CLASSIFICATION 2020

Primary 60D05; Secondary 60G60

KEYWORDS

Liouville quantum gravity, Gaussian free field, LQG metric, Liouville first passage percolation, conformal invariance, scaling limits, random planar maps, random geometry, supercritical LQG

INTERNATIONAL CONGRESS OF MATHEMATICIANS 2022 JULY 6–14

Proc. Int. Cong. Math. 2022, Vol. 6, pp. 4212–4244
DOI 10.4171/ICM2022/40

1. INTRODUCTION

Liouville quantum gravity (LQG) is a family of models of random "surfaces," or equivalently random "two-dimensional Riemannian manifolds" which are in some sense canonical. The reason for the quotations is that, as we will see, LQG surfaces are too rough to be Riemannian manifolds in the literal sense. Such surfaces were first studied in the physics literature in the 1980s **[15, 32, 60, 82]**. The purpose of this article is give an overview of the construction of the distance function associated with an LQG surface (Section 2), as well as some of its properties (Section 3), and the main tools used for studying it (Section 4). We also discuss some open problems in Section 5. In the rest of this section, we will give some basic background on the theory of LQG and its motivations.

1.1. Definition of LQG

One can define LQG surfaces with the topology of any orientable surface (disks, spheres, torii, etc.), and all have the same local geometry. We will be primarily interested in the local geometry, so for simplicity we will focus on LQG surfaces with the topology of the whole plane.[1]

To define LQG, we first need to define the Gaussian free field. The whole-plane *Gaussian free field* (GFF) is the centered Gaussian process h with covariance[2]

$$\mathrm{Cov}\big(h(z), h(w)\big) = G(z,w) := \log \frac{\max\{|z|,1\}\max\{|w|,1\}}{|z-w|}, \qquad \forall z, w \in \mathbb{C}.$$

Since $\lim_{w\to z} G(z,w) = \infty$, the GFF is not a function. However, it still makes sense as a generalized function (i.e., a distribution). That is, if $\phi : \mathbb{C} \to \mathbb{R}$ is smooth and compactly supported, then one can define the L^2 inner product $(h, \phi) = \int_{\mathbb{C}} h(z)\phi(z)\, d^2z$ as a random variable. These random variables have covariances

$$\mathrm{Cov}\big((h,\phi),(h,\psi)\big) = \int_{\mathbb{C}\times\mathbb{C}} \phi(z)\psi(w)G(z,w)\, d^2z\, d^2w.$$

The reader can consult **[13, 88, 92]** for more background on the GFF. We have included a simulation of the GFF in Figure 1(left).

More generally, we say that a random generalized function h on $\mathbb{C}$ is a *GFF plus a nice function* if $h = \tilde{h} + f$, where $\tilde{h}$ is the whole-plane GFF and $f : \mathbb{C} \to \mathbb{R}$ is a (possibly random and $\tilde{h}$-dependent) function which is continuous except at finitely many points.

Let $\gamma \in (0, 2]$, which will be the parameter for our LQG surfaces. A *γ-LQG surface* parametrized by $\mathbb{C}$ is the random two-dimensional Riemannian manifold with Riemannian metric tensor

$$e^{\gamma h(z)}\big(dx^2 + dy^2\big), \quad \text{for } z = x + iy, \tag{1.1}$$

where $dx^2 + dy^2$ denotes the Euclidean metric tensor and h is the whole-plane GFF, or, more generally, a whole-plane GFF plus a nice function.

---

1 See **[16, 17, 37, 44, 84]** for constructions of canonical LQG surfaces with various topologies.

2 Our choice of covariance function corresponds to normalizing h so that its average over the unit circle is zero; see, e.g., **[90, SECTION 2.1.1]**.

1.2. Area measure and conformal covariance

The Riemannian metric tensor (1.1) is not well-defined since h is not defined pointwise, so $e^{\gamma h}$ does not make sense literally. However, it is possible to make sense of various objects associated with (1.1) rigorously using regularization procedures. The idea is to consider a collection of continuous functions $\{h_\varepsilon\}_{\varepsilon>0}$ which converge to h in some sense as $\varepsilon \to 0$, define objects associated with the Riemannian metric tensor (1.1) with h_ε in place of h, then take a limit as $\varepsilon \to 0$. In this paper, we will discuss two objects which can be constructed in this way: the LQG area measure (to be discussed just below) and the LQG metric (which is the main focus of the paper). Other examples include the LQG length measure on Schramm–Loewner evolution-type curves [10, 89], Liouville Brownian motion [11, 43], and the correlation functions for the random "fields" $e^{\alpha h}$ for $\alpha \in \mathbb{R}$ [61].

For simplicity, let us restrict attention to the case when h is a whole-plane GFF. A convenient choice of $\{h_\varepsilon\}$ is the convolution of h with the heat kernel. For $t > 0$ and $z \in \mathbb{C}$, we define the heat kernel $p_t(z) := \frac{1}{2\pi t} e^{-|z|^2/2t}$ and set

$$h_\varepsilon^*(z) := (h * p_{\varepsilon^2/2})(z) = \int_{\mathbb{C}} h(w) p_{\varepsilon^2/2}(z-w)\, d^2 w, \quad \forall z \in \mathbb{C}, \tag{1.2}$$

where the integral is interpreted in the sense of distributional pairing.

The easiest nontrivial object associated with (1.1) to construct rigorously is the LQG area measure, or volume form. This is a random measure μ_h on $\mathbb{C}$ which is defined as the a.s. limit, with respect to the vague topology,[3]

$$\mu_h = \lim_{\varepsilon \to 0} \varepsilon^{\gamma^2/2} e^{\gamma h_\varepsilon^*}\, d^2 z, \tag{1.3}$$

where $d^2 z$ denotes Lebesgue measure on $\mathbb{C}$. The reason for the normalizing factor $\varepsilon^{\gamma^2/2}$ is that $\mathbb{E}[e^{\gamma h_\varepsilon^*(z)}] \approx \varepsilon^{-\gamma^2/2}$. The existence of the limit in (1.3) is a special case of the theory of Gaussian multiplicative chaos (GMC) [58, 86]. There are a variety of different ways of approximating μ_h which are all known to converge to the same limit; see [40, 87] for some results in this direction.

The measure μ_h is mutually singular with respect to Lebesgue measure. In fact, it is supported on a dense subset of $\mathbb{C}$ of Hausdorff dimension $2 - \gamma^2/2$; see, e.g., [40, SECTION 3.3]. However, it has no atoms and assigns positive mass to every open subset of $\mathbb{C}$.

The LQG area measure also satisfies a conformal covariance property. Let $U, \tilde{U} \subset \mathbb{C}$ be open and let $f : \tilde{U} \to U$ be a conformal (bijective, holomorphic) map. Let

$$\tilde{h} = h \circ \phi + Q \log|\phi'|, \quad \text{where } Q = \frac{2}{\gamma} + \frac{\gamma}{2}. \tag{1.4}$$

Then $\tilde{h}$ is a random generalized function on $\tilde{U}$ whose law is locally absolutely continuous with respect to the law of h, so $\mu_{\tilde{h}}$ can be defined. It is shown in [40, PROPOSITION 2.1] that a.s.

$$\mu_{\tilde{h}}(X) = \mu_h(\phi(X)), \quad \forall \text{ Borel set } X \subset U. \tag{1.5}$$

---

3 In the case when $\gamma = 2$, there is a log-correction in the scaling factor, see [38, 39, 83].

We can think of the pairs $(U, h|_U)$ and $(\tilde{U}, \tilde{h})$ as representing two different parametrizations of the same LQG surface. The relation (1.5) implies that the LQG area measure is an intrinsic function of the surface, i.e., it does not depend on the choice of parametrization.

The main focus of this article is the *LQG metric*, i.e., the Riemannian distance function associated with the Riemannian metric tensor (1.1). This metric can be constructed via a similar regularization procedure as the measure, but the proof of convergence is much more involved. See Section 2 for details.

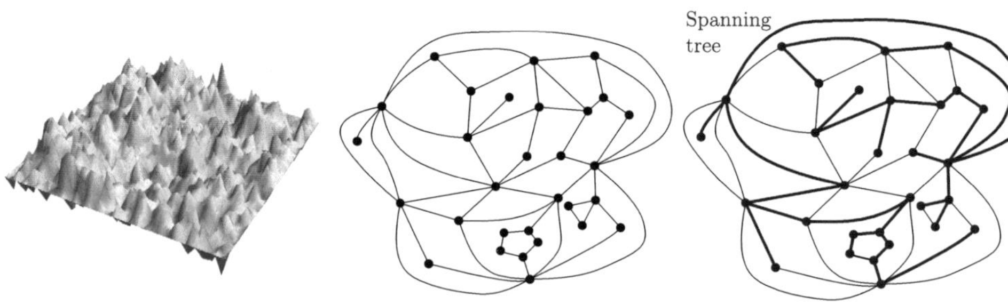

FIGURE 1
(Left) A simulation of the graph of a continuous function which approximates the GFF. (Middle) A planar map. Equivalent representations of the same planar map can be obtained by applying an orientation-preserving homeomorphism from $\mathbb{C}$ to $\mathbb{C}$. (Right) A spanning tree on the planar map.

1.3. Motivation

LQG was first studied by Polyakov [82] in the 1980s in the context of string theory (we discuss Polyakov's motivation in Remark 2.11). LQG is also of interest in conformal field theory since it is closely connected to Liouville conformal field theory, one of the simplest nontrivial conformal field theories. See [90] for an overview of recent mathematical work on Liouville conformal field theory.

One of the most important applications of LQG theory is the so-called Knizhnik–Polyakov–Zamolodchikov (KPZ) formula [60], which gives a relationship between critical exponents for statistical mechanics models in random geometries and deterministic geometries.[4] For example, this formula was used by Duplantier to give nonrigorous predictions for the Brownian intersection exponents [35] (the exponents were predicted earlier by Duplantier and Kwon [36]). These predictions were later verified rigorously by Lawler, Schramm, and Werner in [62–64] using SLE techniques. We discuss the KPZ formula in the context of the LQG metric in Section 3.5.

Another reason to study LQG is that, at least conjecturally, it describes the large-scale behavior of discrete random geometries, such as random planar maps. A *planar map* is a graph embedded in the plane so that no two edges cross, viewed modulo orientation-preserving homeomorphisms of the plane. See Figure 1(middle) for an illustration. There are various interesting types of random planar maps, such as the following:

---

4 The KPZ formula discussed here has no relation with Kardar–Parisi–Zhang equation from [59], except that the initials of the authors for the two papers are the same.

- Uniform planar maps. Consider the (finite) set of planar maps with a specified number $n \in \mathbb{N}$ of edges and choose an element of this set uniformly at random.
- Uniform planar maps with local constraints, such as triangulations (resp. quadrangulations), where each face has exactly 3 (resp. 4) edges.
- Decorated planar maps. Suppose, for example, that we want to sample a uniform pair (M, T) consisting of a planar map M with n edges and a spanning tree T on M (i.e., a subgraph of M which includes every vertex of M and has no cycles). Under this probability measure, the marginal law of M is not uniform; rather, the probability of seeing any particular planar map with n edges is proportional to the number of spanning trees it admits. One can similarly consider planar maps decorated by statistical physics models (such as the Ising model or the FK model) or by various types of orientations on their edges.

It is believed that a large class of different types of planar maps converge to LQG in some sense. The parameter γ depends on the type of planar map under consideration. Uniform planar maps, including maps with local constraints, correspond to $\gamma = \sqrt{8/3}$. This case is sometimes called "pure gravity" in the physics literature. Other values of γ correspond to planar maps decorated by statistical physics models. This case is sometimes called "gravity coupled to matter." For example, the spanning tree-decorated maps discussed above are expected to converge to LQG with $\gamma = \sqrt{2}$.

For this article, the most relevant conjectured mode of convergence of random planar maps toward LQG is the following. View a planar map as a compact metric space, equipped with the graph distance. If we rescale distances in this metric space appropriately, then, as the number of edges tends to ∞, it should converge in the Gromov–Hausdorff sense to an LQG surface equipped with its LQG metric. So far, this type of convergence has only been proven for $\gamma = \sqrt{8/3}$, see Section 2.4. However, weaker connections between random planar maps and γ-LQG have been established rigorously for all $\gamma \in (0, 2)$ using the so-called *mating of trees* theory. See [47] for a survey of this theory.

2. CONSTRUCTION OF THE LQG METRIC

2.1. Liouville first passage percolation

In analogy with the approximation scheme for the LQG measure in (1.3), for a parameter $\xi > 0$, we define

$$D_h^{\varepsilon}(z, w) := \inf_{P: z \to w} \int_0^1 e^{\xi h_{\varepsilon}^*(P(t))} |P'(t)| \, dt, \quad \forall z, w \in \mathbb{C}, \ \forall \varepsilon > 0, \tag{2.1}$$

where the infimum is over all piecewise continuously differentiable paths $P : [0, 1] \to \mathbb{C}$ from z to w. The metrics D_h^{ε} are sometimes referred to as *ε-Liouville first passage percolation* (LFPP).

We want to choose the parameter ξ in a manner depending on γ so that the LFPP metrics (2.1) converge to the distance function associated with the metric tensor (1.1). To determine what ξ should be, we use a heuristic scaling argument. From (1.3), we see that scaling areas by $C > 0$ corresponds to replacing h by $h + \frac{1}{\gamma}\log C$. On the other hand, from (2.1) we see that replacing h by $h + \frac{1}{\gamma}\log C$ scales distances by a factor of $C^{\xi/\gamma}$. Hence ξ/γ is the scaling exponent relating areas and distances. In other words, we want γ/ξ to be the "dimension" of an LQG surface.

It was shown in [22, 30] that there is an exponent $d_\gamma > 2$ which arises in various discrete approximations of LQG and which can be interpreted as the dimension of LQG. For example, d_γ is the ball volume exponent for certain random planar maps [22, THEOREM 1.6]. Once the LQG metric has been constructed, one can show that d_γ is its Hausdorff dimension [54] (see Theorem 3.1). The value of d_γ is not known explicitly except that $d_{\sqrt{8/3}} = 4$. Computing d_γ for general $\gamma \in (0, 2]$ is one of the most important open problems in LQG theory.

The above discussion suggests that one should take

$$\xi = \frac{\gamma}{d_\gamma}. \tag{2.2}$$

It is shown in [22, PROPOSITION 1.7] that ξ is an increasing function of γ, so for $\gamma \in (0, 2]$, ξ takes values in $(0, 2/d_2]$. Estimates for d_γ [22, 53] show that $2/d_2 \approx 0.41$.

The definition of LFPP in (2.1) also makes sense for $\xi > 2/d_2$. In this regime, LFPP metrics do not correspond to γ-LQG with $\gamma \in (0, 2]$. Rather, as we will explain in Section 2.3.2, LFPP for $\xi > 2/d_2$ converges to a metric which is related to LQG with matter central charge in $(1, 25)$, or equivalently $\gamma \in \mathbb{C}$ with $|\gamma| = 2$.

Definition 2.1. We refer to LFPP with $\xi < 2/d_2$, $\xi = 2/d_2$, and $\xi > 2/d_2$ as the *subcritical*, *critical*, and *supercritical* phases, respectively.

Remark 2.2. It is much more difficult to show the convergence of the approximating metrics (2.1) than it is to show the convergence of the approximating measures in (1.3). One intuitive explanation for this is that the infimum in (2.1) introduces a substantial degree of nonlinearity. The minimizing path in (2.1) depends on ε, so one has to keep track of both the location of the minimizing path and its length, whereas for the measure one just has to keep track of the mass of a given set. One can think of the study of LFPP as the study of the extrema of the path-indexed random field whose value on each path is given by the integral in (2.1).

Remark 2.3. The study of LFPP is very different from the study of ordinary first passage percolation (FPP), say on $\mathbb{Z}^2$. In ordinary FPP, the weights of the edges are i.i.d. and the law of the random environment is stationary with respect to spatial translations, neither of which is the case for LFPP (the law of the whole-plane GFF is only translation invariant modulo additive constant). However, for LFPP one has strong independence statements for the field at different Euclidean scales and one can get approximate spatial independence in certain contexts. See Sections 4.2 and 4.3. These independence properties are fundamental tools in the proof of the convergence of LFPP and the study of the limiting metric.

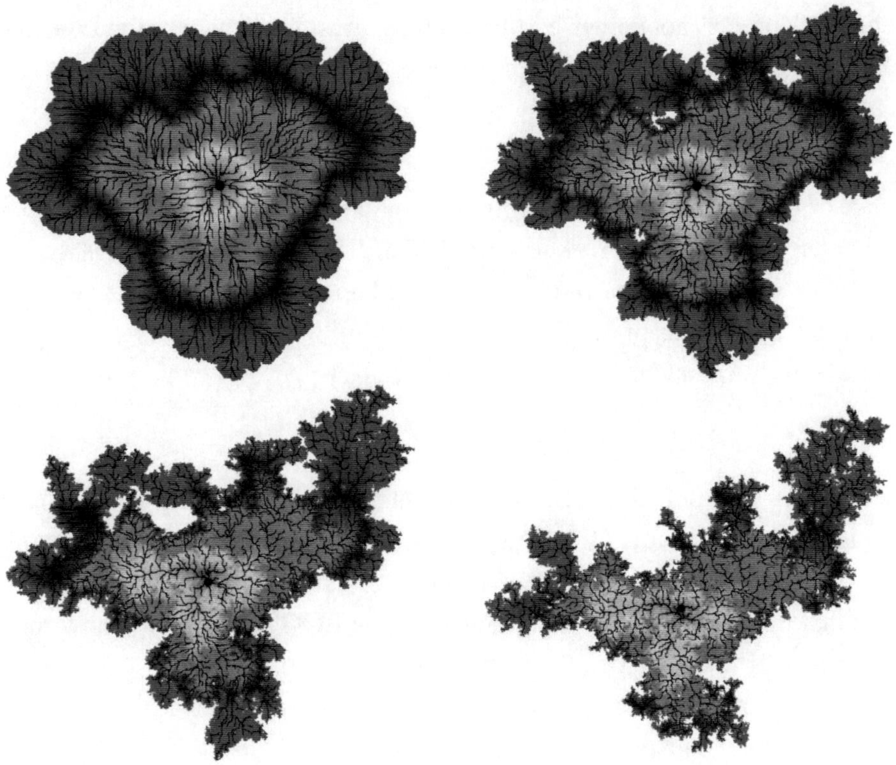

FIGURE 2

Simulation of LFPP metric balls for $\xi = 0.2$ (top left), $\xi = 0.4$ (top right), $\xi = 0.6$ (bottom left), and $\xi = 0.8$ (bottom right). The values $\xi = 0.2, 0.4$ are subcritical and correspond to $\gamma \approx 0.46$ and $\gamma \approx 1.48$, respectively. The values $\xi = 0.6, 0.8$ are supercritical. The colors indicate distance to the center point (marked with a black dot) and the black curves are geodesics from the center point to other points in the ball. These geodesics have a tree-like structure, which is consistent with the confluence of geodesics results discussed in Section 3.3. The pictures are slightly misleading in that the balls depicted do not have enough "holes." Actually, LQG metric balls have infinitely many complementary connected components for all $\xi > 0$, and have empty Euclidean interior for $\xi > 2/d_2$ (Section 3.4). The simulation was produced using LFPP with respect to a discrete GFF on a 1024×1024 subset of $\mathbb{Z}^2$. It is believed that this variant of LFPP falls into the same universality class as the variant in (1.2). The geodesics go from the center of the metric ball to points in the intersection of the metric ball with the grid $20\mathbb{Z}^2$. The code for the simulation was provided by J. Miller.

2.2. Convergence in the subcritical case

2.2.1. Tightness

To extract a nontrivial limit of the metrics D_h^ε, we need to renormalize. We (somewhat arbitrarily) define our normalizing factor by

$$\mathfrak{a}_\varepsilon := \text{median of } \inf\left\{\int_0^1 e^{\xi h_\varepsilon^*(P(t))}\left|P'(t)\right| dt : P \text{ is a left-right crossing of } [0,1]^2\right\}, \quad (2.3)$$

where a left-right crossing of $[0,1]^2$ is a piecewise continuously differentiable path $P : [0,1] \to [0,1]^2$ joining the left and right boundaries of $[0,1]^2$.

The value of $\mathfrak{a}_\varepsilon$ is not known explicitly (in contrast to the case of the LQG measure), but it is shown in [23, PROPOSITION 1.1] that for each $\xi > 0$, there exists $Q = Q(\xi) > 0$ such that

$$\mathfrak{a}_\varepsilon = \varepsilon^{1-\xi Q + o_\varepsilon(1)}, \quad \text{as } \varepsilon \to 0. \tag{2.4}$$

The existence of Q is proven via a subadditivity argument, so the exact relationship between Q and ξ is not known. However, it is known that $Q \in (0, \infty)$ for all $\xi > 0$, Q is a continuous, non-increasing function of ξ, $\lim_{\xi \to 0} Q(\xi) = \infty$, and $\lim_{\xi \to \infty} Q(\xi) = 0$ [23,28]. See also [1, 53] for bounds for Q in terms of ξ.

In the subcritical and critical cases, one has $\xi = \gamma/d_\gamma$ for some $\gamma \in (0, 2]$ and

$$Q(\gamma/d_\gamma) = \frac{2}{\gamma} + \frac{\gamma}{2}. \tag{2.5}$$

In other words, the value of Q for LFPP is the same as the value of Q appearing in the LQG coordinate change formula (1.4). Furthermore, from (2.5) we see that determining the relationship between Q and ξ in the subcritical case is equivalent to computing d_γ.

The first major step in the construction of the LQG metric is to show that the rescaled metrics $\mathfrak{a}_\varepsilon^{-1} D_h^\varepsilon$ are tight, i.e., they admit subsequential limits in distribution. The first paper to prove a version of this was [19], which showed that the metrics $\mathfrak{a}_\varepsilon^{-1} D_h^\varepsilon$ are tight when ξ is smaller than some nonexplicit constant. The proof of this result was simplified in [33]: most importantly, [33] gave a simpler proof of the necessary RSW estimate (for all $\xi > 0$) using a conformal invariance argument. Finally, the tightness for the full subcritical regime $\xi \in (0, 2/d_2)$ was proven in [18].

Theorem 2.4 ([18]). *Assume that $\xi < 2/d_2$. The laws of the metrics $\{\mathfrak{a}_\varepsilon^{-1} D_h^\varepsilon\}_{\varepsilon>0}$ are tight with respect to the topology of uniform convergence on compact subsets of $\mathbb{C} \times \mathbb{C}$. Every possible subsequential limit is a metric on $\mathbb{C}$ which induces the same topology as the Euclidean metric.*

Although the subsequential limit induces the same topology as the Euclidean metric, its geometric properties are very different. See Figure 2 and Section 3.

2.2.2. Uniqueness

The second major step is to show that the subsequential limit is unique. In fact, we want a stronger statement than just the uniqueness of the subsequential limit, since we would like to say that the limiting metric does not depend on the approximation procedure. To this end, the paper [51] established an axiomatic characterization of the LQG metric. To state this characterization, we need some preliminary definitions.

Let $\mathfrak{d}$ be a metric on $\mathbb{C}$. For a path $P : [a, b] \to \mathbb{C}$, we define its $\mathfrak{d}$-length by

$$\operatorname{len}(P; \mathfrak{d}) := \sup_T \sum_{i=1}^{\#T} \mathfrak{d}\big(P(t_i), P(t_{i-1})\big) \tag{2.6}$$

where the supremum is over all partitions $T : a = t_0 < \cdots < t_{\#T} = b$ of $[a, b]$. We say that $\mathfrak{d}$ is a *length metric* if for each $z, w \in \mathbb{C}$, $\mathfrak{d}(z, w)$ is equal to the infimum of the D_h-lengths of all paths joining z and w.

For an open set $U \subset \mathbb{C}$, we define the *internal metric* of $\mathfrak{d}$ on U by

$$\mathfrak{d}(z, w; U) = \inf\{\text{len}(P; \mathfrak{d}) : P \text{ is a path from } z \text{ to } w \text{ in } U\}, \quad \forall z, w \in U. \tag{2.7}$$

We note that $\mathfrak{d}(z, w; U)$ can be strictly larger than the $\mathfrak{d}(z, w)$ since all of the paths from z to w of near-minimal $\mathfrak{d}$-length might exit U.

The following is the axiomatic definition of the LQG metric from [51].

Definition 2.5 (LQG metric). Let $\mathcal{D}'$ be the space of distributions (generalized functions) on $\mathbb{C}$, equipped with the usual weak topology.[5] For $\gamma \in (0, 2)$, a *γ-LQG metric* is a measurable function $h \mapsto D_h$ from $\mathcal{D}'$ to the space of metrics on $\mathbb{C}$ which induce the Euclidean topology with the following properties. Let h be a GFF plus a continuous function on $\mathbb{C}$, i.e., $h = \tilde{h} + f$ where $\tilde{h}$ is a whole-plane GFF and f is a possibly random continuous function. Then the associated metric D_h satisfies the following axioms:

I. *Length space.* Almost surely, D_h is a length metric.

II. *Locality.* Let $U \subset \mathbb{C}$ be a deterministic open set. The D_h-internal metric $D_h(\cdot, \cdot; U)$ is a.s. given by a measurable function of $h|_U$.

III. *Weyl scaling.* Let ξ be as in (2.2). For a continuous function $f : \mathbb{C} \to \mathbb{R}$, define

$$(e^{\xi f} \cdot D_h)(z, w) := \inf_{P : z \to w} \int_0^{\text{len}(P; D_h)} e^{\xi f(P(t))}\, dt, \quad \forall z, w \in \mathbb{C}, \tag{2.8}$$

where the infimum is over all D_h-continuous paths from z to w in $\mathbb{C}$ parametrized by D_h-length. Then a.s. $e^{\xi f} \cdot D_h = D_{h+f}$ for every continuous function $f : \mathbb{C} \to \mathbb{R}$.

IV. *Coordinate change for scaling and translation.* Let $r > 0$ and $z \in \mathbb{C}$. Almost surely,

$$D_h(ru + z, rv + z) = D_{h(r\cdot+z)+Q\log r}(u, v), \quad \forall u, v \in \mathbb{C},$$

where $Q = \frac{2}{\gamma} + \frac{\gamma}{2}$.

The reason why we impose Axioms I. through III. is that we want D_h to be the Riemannian distance function associated to the Riemannian metric tensor (1.1). Axiom IV. is analogous to the conformal coordinate change formula for the LQG area measure (1.5), but restricted to translations and scalings. As in the case of the measure, it can be thought of as saying that the metric D_h is intrinsic to the LQG surface, i.e., it does not depend on the choice of parametrization. The axioms in Definition 2.5 imply a coordinate change formula for general conformal maps, including rotations; see [51, REMARK 1.6] and [50].

The main result of [51] is the following statement, whose proof builds on [18, 34, 48, 49].

5 We do not care about how D_h is defined on any subset of $\mathcal{D}'$ which has measure zero for the law of any random distribution which is a GFF plus a continuous function.

Theorem 2.6 ([51]). *For each $\gamma \in (0,2)$, there exists a γ-LQG metric. This metric is the limit of the rescaled LFPP metrics $\mathfrak{a}_\varepsilon^{-1} D_h^\varepsilon$ in probability with respect to the topology of uniform convergence on compact subsets of $\mathbb{C} \times \mathbb{C}$. Moreover, this metric is unique in the following sense: if D_h and $\tilde{D}_h$ are two γ-LQG metrics, then there is a deterministic constant $C > 0$ such that a.s. $D_h(z,w) = C\tilde{D}_h(z,w)$ for all $z, w \in \mathbb{C}$ whenever h is a whole-plane GFF plus a continuous function.*

Due to Theorem 2.6, we can refer to *the* LQG metric, keeping in mind that this metric is only defined up to a deterministic positive multiplicative constant (the value of this constant is usually unimportant).

Once Theorem 2.6 is established, it is typically easier to prove statements about the LQG metric directly from the axioms, as opposed to going back to the approximation procedure. We explain some of the techniques for doing so in Section 4.

2.2.3. Weak LQG metrics

The existence part of Theorem 2.6, of course, follows from the tightness result in Theorem 2.4, but not as directly as one might expect at first glance. It is relatively easy to check from the definition (2.1) that every possible subsequential limit of the rescaled LFPP metrics $\mathfrak{a}_\varepsilon^{-1} D_h^\varepsilon$ satisfies Axioms I., II., and III. in Definition 2.5. See [34, SECTION 2] for details.

Checking Axiom IV. is much more difficult. The reason is that rescaling space changes the value of ε in (2.1): for $\varepsilon, r > 0$, one has [34, LEMMA 2.6]

$$D_h^\varepsilon(rz, rw) = r D_{h(r\cdot)}^{\varepsilon/r}(z,w), \quad \forall z, w \in \mathbb{C}.$$

So, since we only have subsequential limits of $\mathfrak{a}_\varepsilon^{-1} D_h^\varepsilon$, we cannot deduce that the subsequential limit satisfies an exact spatial scaling property.

To get around this difficulty, we consider a weaker property than Axiom IV. which is sufficient for the proof of uniqueness. To motivate this property, let us consider how Axiom IV. is used in proofs about the LQG metric.

Assume that h is a whole-plane GFF. For $z \in \mathbb{C}$ and $r > 0$, let $h_r(z)$ be the average of h over the circle $\partial B_r(z)$ (see [40, SECTION 3.1] for the definition and basic properties of the circle average process). It is easy to see from the definition of the whole-plane GFF that for any $z \in \mathbb{C}$ and $r > 0$,

$$h(r\cdot + z) - h_r(z) \overset{d}{=} h. \tag{2.9}$$

Furthermore, from Weyl scaling and the LQG coordinate change formula (Axioms III. and IV.), a.s.

$$D_{h(r\cdot+z)-h_r(z)}(u,v) = e^{-\xi h_r(z)} r^{-\xi Q} D_h(ru+z, rv+z), \quad \forall u, v \in \mathbb{C}. \tag{2.10}$$

By (2.9) and (2.10),

$$e^{-\xi h_r(z)} r^{-\xi Q} D_h(r\cdot+z, r\cdot+z) \overset{d}{=} D_h. \tag{2.11}$$

The relation (2.11) allows us to get estimates for D_h which are uniform across different spatial locations and Euclidean scales. However, for many purposes one does not need an exact equality in law in (2.11), but rather just an up-to-constants comparison. This motivates the following definition.

Definition 2.7 (Weak LQG metric). For $\gamma \in (0,2)$, a *weak γ-LQG metric* is a measurable function $h \mapsto D_h$ from $\mathcal{D}'$ to the space of metrics on $\mathbb{C}$ which induce the Euclidean topology, which satisfies Axioms I., II., and III. in Definition 2.5 plus the following further axioms:

VI′. *Translation invariance.* If h is a whole-plane GFF, then for each fixed deterministic $z \in \mathbb{C}$, a.s. $D_{h(\cdot+z)} = D_h(\cdot + z, \cdot + z)$.

V′. *Tightness across scales.* Suppose h is a whole-plane GFF and for $z \in \mathbb{C}$ and $r > 0$ let $h_r(z)$ be the average of h over the circle $\partial B_r(z)$. For each $r > 0$, there is a deterministic constant $\mathfrak{c}_r > 0$ such that the set of laws of the metrics $\mathfrak{c}_r^{-1} e^{-\xi h_r(0)} D_h(r\cdot, r\cdot)$ for $r > 0$ is tight (with respect to the local uniform topology). Furthermore, every subsequential limit of the laws of the metrics $\mathfrak{c}_r^{-1} e^{-\xi h_r(0)} D_h(r\cdot, r\cdot)$ is supported on metrics which induce the Euclidean topology on $\mathbb{C}$.

From (2.11), we see that every strong LQG metric is a weak LQG metric with $\mathfrak{c}_r = r^{\xi Q}$. Furthermore, it is straightforward to check that every subsequential limit of LFPP is a weak LQG metric [34]. In particular, Theorem 2.4 implies that there exists a weak LQG metric for each $\gamma \in (0,2)$. We note that most of the literature requires rather weak a priori bounds for the scaling constants $\mathfrak{c}_r$ in Definition 2.7, but the recent paper [26] shows that these bounds are unnecessary.

It turns out that most statements which can be proven for LQG metrics can also be proven for weak LQG metrics. Using this, [51] established the following statement.

Theorem 2.8 (Uniqueness of weak LQG metrics). *Let $\gamma \in (0,2)$ and let D_h and $\tilde{D}_h$ be two weak γ-LQG metrics which have the* same *values of $\mathfrak{c}_r$ in Definition 2.7. There is a deterministic constant $C > 0$ such that if h is a whole-plane GFF plus a continuous function, then a.s. $D_h = C\tilde{D}_h$.*

Let us now explain why Theorem 2.8 implies Theorem 2.6 (see [51, SECTION 1.4] for more details). If D_h is a weak LQG metric and $b > 0$, then one can establish that $D_{h(b\cdot)+Q\log b}(\cdot/b, \cdot/b)$ is a weak LQG metric with the same scaling constants $\mathfrak{c}_r$ as D_h. From this, one gets that $D_{h(b\cdot)+Q\log b}(\cdot/b, \cdot/b)$ is a deterministic constant multiple of D_h. One can check that the constant has to be 1. This shows that D_h satisfies Axiom IV. in Definition 2.5, i.e., D_h is a strong LQG metric. In particular, D_h is a weak LQG metric with scaling constants $r^{\xi Q}$. This holds for any possible weak LQG metric, so we infer that every weak LQG metric is a strong LQG metric and the weak LQG metric is unique up to constant multiples.

Remark 2.9. There are a few other ways to approximate the LQG metric besides LFPP, which are expected but not proven to give the same object. One possible approximation, called *Liouville graph distance*, is based on the LQG area measure μ_h: for $\varepsilon > 0$ and

$z, w \in \mathbb{C}$, we let $\hat{D}_h^\varepsilon(z, w)$ be the minimal number of Euclidean balls of μ_h-mass ε whose union contains a path from z to w. The tightness of the metrics $\{\hat{D}_h^\varepsilon\}_{\varepsilon>0}$, appropriately rescaled, is proven in [20], but the subsequential limit has not yet been shown to be unique.

Another type of approximation is based on *Liouville Brownian motion*, the "LQG time" parametrization of Brownian motion on an LQG surface [11,43]. Roughly speaking, the idea here is that Liouville Brownian motion conditioned to travel a macroscopic distance in a small time should roughly follow an LQG geodesic. No one has yet established the tightness of any Liouville Brownian motion-based approximation scheme. However, the paper [30] shows that the exponent for the Liouville heat kernel can be expressed in terms of the LQG dimension d_γ, which gives some rigorous connection between Liouville Brownian motion and the LQG metric.

2.3. The supercritical and critical cases

2.3.1. Subsequential limits

Recall that LFPP is related to γ-LQG for $\gamma \in (0, 2)$ in the subcritical case, i.e., when $\xi = \gamma/d_\gamma < 2/d_2 \approx 0.41, \ldots$. In this subsection, we will explain what happens in the supercritical and critical cases, i.e., when $\xi \geq 2/d_2$.

The tightness of supercritical LFPP was established in [23]. Subsequently, it was shown in [27], building on [81], that the subsequential limit is uniquely characterized by a list of axioms analogous to the ones in Definition 2.5 (see [27, SECTION 1.3] for a precise statement). Unlike in the subcritical case, in the supercritical case the limiting metric D_h is not a continuous function on $\mathbb{C} \times \mathbb{C}$, so one cannot work with the uniform topology. However, this metric is lower semicontinuous, i.e., for any $(z, w) \in \mathbb{C} \times \mathbb{C}$ one has

$$D_h(z, w) \leq \liminf_{(z',w') \to (z,w)} D_h(z', w'). \tag{2.12}$$

In [23, SECTION 1.2] the authors describe a metrizable topology on the space of lower semicontinuous functions $\mathbb{C} \times \mathbb{C} \to \mathbb{R} \cup \{\pm\infty\}$, based on the construction of Beer [8]. With this topology in hand, we can state the following generalization of Theorems 2.4 and 2.6.

Theorem 2.10 ([23, 27, 81]). *Let $\xi > 0$. The re-scaled LFPP metrics metrics $\{\mathfrak{a}_\epsilon^{-1} D_h^\epsilon\}_{\epsilon>0}$ converge in probability with respect to the topology on lower semicontinuous functions on $\mathbb{C} \times \mathbb{C}$. The limit D_h is a metric on $\mathbb{C}$, except that it is allowed to take on infinite values. Moreover, D_h is uniquely characterized (up to multiplication by a deterministic positive constant) by a list of axioms similar to the ones in Definition* 2.5.

Let us be more precise about what we mean by allowing the metric to take on infinite values. For $\xi > 2/d_2$, it is shown in [23] that if D_h is as in Theorem 2.10, then a.s. there is an uncountable dense set of *singular points* $z \in \mathbb{C}$ such that

$$D_h(z, w) = \infty, \quad \forall w \in \mathbb{C} \setminus \{z\}. \tag{2.13}$$

However, a.s. each fixed $z \in \mathbb{C}$ is not a singular point (so the singular points have Lebesgue measure zero) and any two nonsingular points lie at a finite D_h-distance from each other. Roughly speaking, if $\{h_r(z) : z \in \mathbb{C}, r > 0\}$ denotes the circle average process of h, then singular points correspond to points in $\mathbb{C}$ for which $\limsup_{r\to 0} h_r(z)/\log r > Q$, where Q is as in (2.4) **[81, PROPOSITION 1.11]**.

Due to the existence of singular points, for $\xi > 2/d_2$, the metric D_h is not continuous with respect to the Euclidean metric on $\mathbb{C} \times \mathbb{C}$, but one can still show that the Euclidean metric is continuous with respect to D_h **[23]**.

In the critical case $\xi = 2/d_2$, which corresponds to $\gamma = 2$, it is shown in **[24]** that D_h induces the Euclidean topology on $\mathbb{C}$. In particular, there are no singular points for $\xi = 2/d_2$. We expect that the rescaled LFPP metrics $\mathfrak{a}_\epsilon^{-1} D_h^\epsilon$ converge uniformly to D_h in this case (not just with respect to the topology on lower semicontinuous functions), but this has not been proven.

2.3.2. Central charge

For $\gamma \in (0, 2]$, the *matter central charge* associated with γ-LQG is

$$\mathbf{c}_{\mathrm{M}} = 25 - 6Q^2 = 25 - 6\left(\frac{2}{\gamma} + \frac{\gamma}{2}\right)^2 \in (-\infty, 1]. \tag{2.14}$$

Note that $\gamma = \sqrt{8/3}$ corresponds to $\mathbf{c}_{\mathrm{M}} = 0$. From physics heuristics, one expects that it should also be possible to define LQG, at least in some sense, in the case when the matter central charge is in $(1, 25)$. However, this regime is much less well understood than the case when $\mathbf{c}_{\mathrm{M}} \in (-\infty, 1]$, even at a physics level of rigor. A major reason for this is that the formula (2.14) shows that $\mathbf{c}_{\mathrm{M}} \in (1, 25)$ corresponds to $\gamma \in \mathbb{C}$ with $|\gamma| = 2$, so various formulas for LQG yield nonphysical complex answers when $\mathbf{c}_{\mathrm{M}} \in (1, 25)$. See **[3, 46]** for further discussion, references, and open problems concerning LQG with $\mathbf{c}_{\mathrm{M}} \in (1, 25)$.

Phase	LFPP parameter	LFPP exponent	Coupling constant	Matter central charge	Topology
Subcritical	$\xi \in (0, 2/d_2)$	$Q > 2$	$\gamma \in (0, 2)$	$\mathbf{c}_{\mathrm{M}} \in (-\infty, 1)$	Bi-Hölder w.r.t. Euclidean
Critical	$\xi = 2/d_2$	$Q = 2$	$\gamma = 2$	$\mathbf{c}_{\mathrm{M}} = 1$	Euclidean topology, not Hölder
Supercritical	$\xi > 2/d_2$	$Q \in (0, 2)$	γ complex, $\lvert\gamma\rvert = 2$.	$\mathbf{c}_{\mathrm{M}} \in (1, 25)$	$\exists$ singular points

FIGURE 3
Table summarizing the phases for the LQG metric.

In light of (2.5) and (2.14), it is natural to define the matter central charge associated with LFPP for $\xi > 2/d_2$ by

$$\mathbf{c}_{\mathrm{M}} = 25 - 6Q(\xi)^2, \tag{2.15}$$

where $Q(\xi)$ is the LFPP distance exponent as in (2.4). One has $Q(\xi) \in (0, 2)$ for $\xi > 2/d_2$, so (2.15) gives $\mathbf{c}_{\mathrm{M}} \in (1, 25)$ for $\xi > 2/d_2$. Hence, the limit of supercritical LFPP can be interpreted as a metric associated with LQG with $\mathbf{c}_{\mathrm{M}} \in (1, 25)$. Since $\xi \mapsto Q(\xi)$ is continuous and non-increasing and $\lim_{\xi \to \infty} Q(\xi) = 0$ [23, PROPOSITION 1.1], there is a $\xi > 2/d_2$ corresponding to each $\mathbf{c}_{\mathrm{M}} \in (1, 25)$.

See Figure 3 for an table summarizing the phases for the LQG metric.

Remark 2.11. From a physics perspective, an LQG surface with matter central charge $\mathbf{c}_{\mathrm{M}}$ represents "two-dimensional gravity coupled to a matter field with central charge $\mathbf{c}_{\mathrm{M}}$." Equivalently, an LQG surface parametrized by a domain U should be a "uniform sample from the space of Riemannian metric tensors g on U, weighted by $(\det \Delta_g)^{-\mathbf{c}_{\mathrm{M}}/2}$, where Δ_g is the Laplace–Beltrami operator." This interpretation is far from being rigorous (e.g., since there is no uniform measure on the space of Riemannian metric tensors), but some partial progress using on regularization procedures has been made in [3].

The central charge also comes up in Polyakov's original motivation for LQG from string theory. If $\mathbf{c}_{\mathrm{M}}$ is an integer, then, roughly speaking, an evolving string in $\mathbb{R}^{\mathbf{c}_{\mathrm{M}}-1}$ traces out a two-dimensional surface embedded in space-time $\mathbb{R}^{\mathbf{c}_{\mathrm{M}}-1} \times \mathbb{R}$, called a *world sheet*. Polyakov wanted to develop a theory of integrals over all possible surfaces embedded in $\mathbb{R}^{\mathbf{c}_{\mathrm{M}}}$ as a string-theoretic generalization of the Feynman path integral (which is an integral over all possible paths). To do this, one needs to define a probability measure on surfaces. It turns out that the "right" measure on surfaces for this purpose is LQG with matter central charge $\mathbf{c}_{\mathrm{M}}$. However, the most relevant case for string theory is $\mathbf{c}_{\mathrm{M}} = 25$, which is outside the range of parameter values for which LQG can be defined probabilistically.

2.4. Alternative construction and planar map connection for $\gamma = \sqrt{8/3}$

In the special case when $\gamma = \sqrt{8/3}$, there is an earlier construction of the $\sqrt{8/3}$-LQG metric due to Miller and Sheffield [76,77,79]. We will comment briefly on the main idea of this construction. See Miller's ICM paper [71] for a more detailed overview.

The idea of the Miller–Sheffield construction is to first construct a candidate for LQG metric balls, then show that these balls are, in fact, the metric balls for a unique metric on $\mathbb{C}$. The candidates for LQG metric balls are generated using a random growth process called *quantum Loewner evolution* (QLE), which is produced by "reshuffling" an SLE_6 curve in a random manner depending on h. Both the construction of this growth process and the proof that one can generate a metric from it rely crucially on special symmetries for $\sqrt{8/3}$-LQG which are established in [37,78], so the construction does not work for any other value of γ.

The Miller–Sheffield metric satisfies the conditions of Definition 2.5, so Theorem 2.6 implies that it agrees with the $\sqrt{8/3}$-LQG metric constructed using LFPP. On the other hand, the construction using QLE gives a number of properties of the $\sqrt{8/3}$-LQG metric which are not apparent from the LFPP construction, for example, various Markov properties for LQG metric balls and the fact that $d_{\sqrt{8/3}} = 4$. These properties can be proven directly using QLE, or can alternatively be deduced from analogous properties of the Brownian map, together with the equivalence between the Brownian map and $\sqrt{8/3}$-LQG discussed just below.

The papers [76,79] also establish a link between the $\sqrt{8/3}$-LQG metric and uniform random planar maps. This link comes by combining two big results:

- Le Gall [66] and Miermont [70] showed independently that certain types of uniform random planar maps (namely, uniform k-angulations for $k = 3$ or k even), equipped with their graph distance, converge in the Gromov–Hausdorff sense to a random metric space called the *Brownian map*. See [67, 68] for a survey of this work.

- Miller and Sheffield showed that there is a certain special variant of the GFF on $\mathbb{C}$ (corresponding to the so-called *quantum sphere*) such that the sphere $\mathbb{C} \cup \{\infty\}$, equipped with the $\sqrt{8/3}$-LQG metric, is isometric to the Brownian map. This is done using the axiomatic characterization of the Brownian map from [74].

Remark 2.12. Building on the aforementioned work (and many additional papers), Holden and Sun [56] showed the rescaled graph distance on uniform triangulations embedded into the plane via the so-called *Cardy embedding* converges to the $\sqrt{8/3}$-LQG metric with respect to a version of the uniform topology. This gives a stronger form of convergence than Gromov–Hausdorff convergence.

3. PROPERTIES OF THE LQG METRIC

In this subsection, we will discuss several properties of the LQG metric which have been established in the literature. Throughout, h denotes a whole-plane GFF and D_h denotes the associated LQG metric with a given parameter $\xi > 0$. We also let Q be as in (2.4) and for $\xi \leq 2/d_2$ we let $\gamma \in (0,2)$ be such that $\xi = \gamma/d_\gamma$, so that $Q = 2/\gamma + \gamma/2$ (2.5). We also let ξ and Q be as above, so that for $\gamma \in (0,2)$ we have $\xi = \gamma/d_\gamma$ and $Q = 2/\gamma + \gamma/2$.

3.1. Dimension

For $\Delta > 0$, the Δ-Hausdorff content of a compact metric space (X, d) is

$$\inf\left\{\sum_{j=1}^{\infty} r_j^{\Delta} : \text{there is a covering of } X \text{ be } d\text{-metric balls with radii } \{r_j\}_{j\in\mathbb{N}}\right\}$$

and the *Hausdorff dimension* of (X, d) is the infimum of the values of Δ for which the Δ-Hausdorff content is zero.

The following theorem follows from the combination of [54, COROLLARY 1.7] and [81, PROPOSITION 1.14].

Theorem 3.1. *In the subcritical case, i.e., when $\gamma \in (0,2)$ and $\xi = \gamma/d_\gamma$, a.s. the Hausdorff dimension of $\mathbb{C}$, equipped with the γ-LQG metric, is equal to d_γ (recall the discussion in Section 2.1). In the supercritical case, i.e., when $\xi > 2/d_2$, the Hausdorff dimension of $\mathbb{C}$, equipped with the LQG metric with parameter ξ, is ∞. ∞.*

As noted above, the value of d_γ is not known except that $d_{\sqrt{8/3}} = 4$, but upper and lower bounds for d_γ have been proven in [1, 22, 53] (see Figure 5). It is shown in

[22, THEOREM 1.2] that $\gamma \mapsto d_\gamma$ is increasing and $\lim_{\gamma\to 0} d_\gamma = 2$. Hence, Theorem 3.1 implies that the LQG metric gets "rougher" as γ increases. We expect that the dimension of $\mathbb{C}$ with respect to the critical ($\gamma = 2$) LQG metric is $d_2 = \lim_{\gamma\to 2} d_\gamma \approx 4.8$, but this has not been proven.

It was shown in [2] that for $\gamma \in (0,2)$, the Minkowski dimension of $(\mathbb{C}, D_h)$ is also equal to d_γ. We expect that in this case, the d_γ-Minkowski content measure for D_h exists and is equal to the γ-LQG area measure μ_h from (1.3). Similarly, the Hausdorff measure associated with D_h, for an appropriate gauge function, should exist and be equal to μ_h. This has been proven for the Brownian map (which is equivalent to $\sqrt{8/3}$-LQG, recall Section 2.4) in [69].

3.2. Quantitative estimates

The optimal Hölder exponents relating D_h and the Euclidean metric can be computed in terms of ξ and Q. For the subcritical (resp. supercritical) case, see [34, THEOREM 1.7] (resp. [81, PROPOSITION 1.10]).

Proposition 3.2 (Hölder continuity). *Let $U \subset \mathbb{C}$ be a bounded open set. Almost surely, for each $\delta > 0$ there is a random $C > 0$ such that*

$$C^{-1}|z-w|^{\xi(Q+2)+\delta} \le D_h(z,w) \le \begin{cases} C|z-w|^{\xi(Q-2)-\delta}, & \xi < 2/d_2, \\ \infty, & \xi \ge 2/d_2. \end{cases}$$

Furthermore, the exponents $\xi(Q+2)$ and $\xi(Q-2)$ are optimal.

In the critical case when $\xi = 2/d_2$, equivalently $Q = 2$, the metric D_h is continuous with respect to the Euclidean metric but not Hölder continuous. Rather, the optimal upper bound for $D_h(z,w)$ is a power of $1/\log(|z-w|^{-1})$ [24].

We also have moment bounds for point-to-point distances, set-to-set distances, and diameters. The following is a compilation of several results from [34,81].

Proposition 3.3 (Moments). *For each distinct $z, w \in \mathbb{C}$, the distance $D_h(z,w)$ has a finite pth moment for all $p \in (-\infty, 2Q/\xi)$. For any two disjoint compact connected sets $K_1, K_2 \subset \mathbb{C}$ which are not singletons, $D_h(K_1, K_2)$ has finite moments of all positive and negative orders. For $\xi < 2/d_2$, for any nonsingleton compact set $K \subset \mathbb{C}$, the D_h-diameter $\sup_{z,w\in K} D_h(z,w)$ has a finite pth moment for all $p \in (-\infty, 4d_\gamma/\gamma^2)$.*

The moment bound for diameters is related to the fact that the LQG area measure has finite moments up to order $4/\gamma^2$ (see, e.g., [86, THEOREM 2.11]).

3.3. Geodesics

Using basic metric space theory, one can show that a.s. for any two points $z, w \in \mathbb{C}$ with $D_h(z,w) < \infty$, there is a D_h-geodesic from z to w, i.e., a path of minimal D_h-length (see, e.g., [14, COROLLARY 2.5.20] for the subcritical case and [81, PROPOSITION 1.12] for the supercritical case). If z and w are fixed, then a.s. this geodesic is unique [72, THEOREM 1.2]. We give a short proof of this fact in Lemma 4.2 below.

It can be shown that the D_h-geodesics started from a specified point have a tree-like structure: two geodesics with the same starting point and different target points stay together for a nontrivial initial time interval. The property is called *confluence of geodesics*, and can be seen in the simulations from Figure 2.

We emphasize that confluence of geodesics is not true for a smooth Riemannian metric (such as the Euclidean metric). Rather, two geodesics for a smooth Riemannian metric with the same starting points and different target points typically intersect only at their starting point.

Confluence of geodesics for the LQG metric was established in the subcritical case ($\xi < 2/d_2$) in [48] and for general $\xi > 0$ in [25]. Let us now state a precise version of this result, which is illustrated in Figure 4. For $s > 0$ and $z \in \mathbb{C}$, let $\mathcal{B}_s(z; D_h)$ be the D_h-metric ball of radius s centered at z.

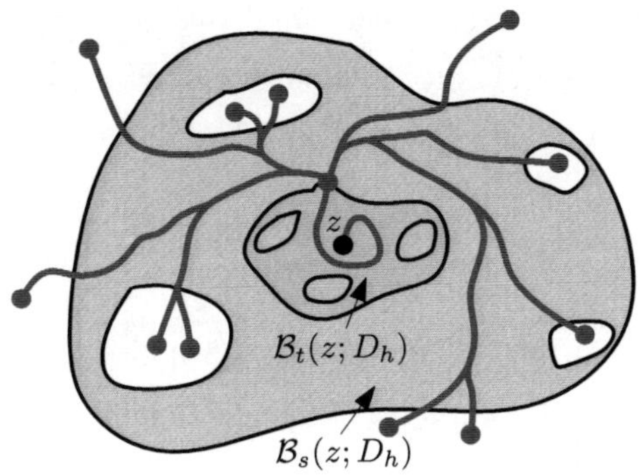

FIGURE 4
Illustration of the statement of Theorem 3.4. The red curves are D_h-geodesics going from z to points outside of the LQG metric ball $\mathcal{B}_s(z; D_h)$. The theorem asserts that these geodesics all coincide until their first exit time from $\mathcal{B}_t(z; D_h)$.

Theorem 3.4 (Confluence of geodesics). *Fix $z \in \mathbb{C}$. Almost surely, for each radius $s > 0$ there exists a radius $t \in (0, s)$ such that any two D_h-geodesics from z to points outside of $\mathcal{B}_s(z; D_h)$ coincide on the time interval* $[0, t]$.

Theorem 3.4 only holds a.s. for a fixed center point $z \in \mathbb{C}$. Almost surely, there is a Lebesgue measure zero set of points in $\mathbb{C}$ where Theorem 3.4 fails. For example, if $P : [0, T] \to \mathbb{C}$ is a D_h-geodesic, then the conclusion of Theorem 3.4 fails for each $z \in P((0, T))$.

Confluence of geodesics is used in the proof of the uniqueness of the γ-LQG metric $\gamma \in (0, 2)$ in [51]. Roughly speaking, confluence is used to establish near-independence for events which depend on small neighborhoods of far-away points on a D_h-geodesic, despite the fact that D_h-geodesics are non-Markovian and do not depend locally on h. See [51] for details. The proof of the uniqueness of the LQG metric for general $\xi > 0$ in [27] does not use confluence of geodesics.

Remark 3.5. Confluence of geodesics was previously established by Le Gall [65] for the Brownian map, which is equivalent to $\sqrt{8/3}$-LQG (see Section 2.4). This result was used in the proof of the uniqueness of the Brownian map in [66,70]. Le Gall's proof was very different from the proof of Theorem 3.4.

Various extensions of the confluence property for $\gamma \in (0,2)$ are proven in [42,55] and for $\gamma = \sqrt{8/3}$ in [73,80].

Little is known about the geometry of a single LQG geodesic. For example, we do not know the Hausdorff dimension of such a geodesic with respect to the Euclidean metric (the dimension with respect to the LQG metric is trivially equal to 1), and we do not have any exact description of its law. The strongest current results in this direction are an upper bound for the Euclidean dimension of an LQG geodesic [54, COROLLARY 1.10], which is not expected to be optimal; and the fact LQG geodesics do not locally look like SLE_κ curves for any value of κ [72]. We do not have a nontrivial lower bound for the Euclidean Hausdorff dimension of an LQG geodesic, but we expect that it is strictly greater than 1 (see [31] for a closely related result for the geodesics for a version of LFPP). Finally, we mention the very recent work [7], which constructs a local limit of the GFF near a typical point of an LQG geodesic.

3.4. Metric balls

From the simulations in Figure 2, one can see that LQG metric balls have a fractal-like geometry. Almost surely, the complement of each LQG metric ball has infinitely many connected components, in both the subcritical and supercritical cases [55,81]. In fact, a.s. "most" points on the boundary of the ball do not lie on any complementary connected component, but rather are accumulation points of arbitrarily small complementary connected components [55, THEOREM 1.14], [25, THEOREM 1.4].

In the subcritical and critical cases, i.e., when $\xi = \gamma/d_\gamma$ for $\gamma \in (0,2]$, the LQG metric induces the same topology as the Euclidean metric so a.s. each closed LQG metric ball is equal to the closure Euclidean interior. In contrast, in the supercritical case a.s. each LQG metric ball has empty Euclidean interior but positive Lebesgue measure. This is a consequence of the fact that the set of singular points from (2.13) is Euclidean-dense but has Lebesgue measure zero.

In the subcritical case, it is shown in [41,55] that a.s. the Hausdorff dimension of the boundary of a γ-LQG metric ball for $\gamma \in (0,2)$ with respect to the Euclidean (resp. LQG) metric is $2 - \xi Q + \xi^2/2$ (resp. $d_\gamma - 1$). We expect that these formulas are also valid for $\gamma = 2$ (equivalently, $\xi = 2/d_2$).

In the supercritical case $\xi > 2/d_2$, the LQG metric D_h does not induce the Euclidean topology, so one has to make a distinction between the boundary with respect to the Euclidean topology or with respect to D_h. The boundary of a closed D_h-metric ball with respect to the Euclidean topology is equal to the ball itself (since the ball is Euclidean closed and has empty Euclidean interior), whereas the boundary with respect to D_h is a proper subset of the ball [25, SECTION 1.2]. It is shown in [81, PROPOSITION 1.14] that for $\xi > 2/d_2$, a.s. the Euclidean boundary of a D_h-metric ball (i.e., the whole D_h-metric ball) is not compact with respect to D_h and has infinite Hausdorff dimension w.r.t. D_h. We expect that the same is true for the D_h-boundary of a D_h-metric ball. The Hausdorff dimension of the Euclidean boundary of a D_h-metric ball with respect to the Euclidean metric is 2 since the metric ball has positive Lebesgue measure. The Hausdorff dimension of the D_h-boundary of a D_h-metric ball with respect to the Euclidean metric has not been computed rigorously.

It is also of interest to consider the boundary of a single complementary connected component of an LQG metric ball. The Hausdorff dimension of such a boundary component with respect to the Euclidean or LQG metric is not known. However, it is known that, even in the supercritical case, each boundary component is a Jordan curve and is compact and finite-dimensional with respect to D_h [25, THEOREM 1.4].

3.5. KPZ formula

The (geometric) Knizhnik–Polyakov–Zamolodchikov (KPZ) formula [60] is a formula which relates the "Euclidean dimension" and the "LQG dimension" of a deterministic set $X \subset \mathbb{C}$, or a random set independent from the GFF h. The first rigorous versions of the KPZ formula appeared in [40, 85]. These papers defined the "LQG dimension" in terms of the LQG area measure. There are several different versions of the KPZ formula in the literature which use different notions of dimension (see, e.g., [4, 6, 9, 12, 45]). Here, we state what is perhaps the most natural version of the KPZ formula, where we compare the Hausdorff dimensions of a set with respect to the LQG metric and the Euclidean metric. We start with the subcritical case, which is [54, THEOREM 1.4].

Theorem 3.6 ([54]). *Let $\gamma \in (0,2)$ and recall that $\xi = \gamma/d_\gamma$ and $Q = 2/\gamma + \gamma/2$. Let $X \subset \mathbb{C}$ be a random Borel set which is independent from the GFF h and let Δ_0 be the Hausdorff dimension of X, equipped the Euclidean metric. Also let Δ_h be the Hausdorff dimension of X, equipped with the γ-LQG metric D_h. Then a.s.*

$$\Delta_h = \xi^{-1}\left(Q - \sqrt{Q^2 - 2\Delta_0}\right). \tag{3.1}$$

Theorem 3.6 does not apply if X is not independent from h. For example, the KPZ formula does not hold for the Hausdorff dimensions of LQG metric ball boundaries with respect to the Euclidean and LQG metrics, as discussed in Section 3.4. However, one has inequalities relating the Hausdorff dimensions of an arbitrary set with respect to the Euclidean and LQG metrics, see [54, THEOREM 1.8].

It is shown in [81, THEOREM 1.15] that the KPZ formula of Theorem 3.6 extends to the case when $\xi \geq 2/d_2$ (modulo some technicalities about the particular notion of "fractal dimension" involved), with the following important caveat. When $\xi > 2/d_2$, we have $Q \in (0,2)$ and the right-hand side of the formula (3.1) is nonreal when $\Delta_0 > Q^2/2$. The extension of the KPZ formula to the supercritical case coincides with (3.1) when $\Delta_0 < Q^2/2$, and gives $\Delta_h = \infty$ when $\Delta_0 > Q^2/2$ (the case when $\Delta_0 = Q^2/2$ is not treated).

4. TOOLS FOR STUDYING THE LQG METRIC

There are a few basic techniques which are the starting point of the majority of the proofs of statements involving the LQG metric. In this subsection, we will discuss a few of the most important such techniques and provide some simple examples of their applications. Throughout, h denotes a whole-plane GFF and D_h denotes an LQG metric in the sense of Definition 2.5. For simplicity, we assume that we are in the subcritical case but our discussion applies in the critical and supercritical cases as well, with only minor modifications.

4.1. Adding a bump function

Suppose that E is an event depending on the LQG metric D_h. For example, maybe we have two points $z, w \in \mathbb{C}$ and E is the event that $D_h(z,w) > 100$, or that the D_h-geodesic from z to w stays in some specified open set. For many choices of E, it is straightforward to show that $\mathbb{P}[E] > 0$ via the following method. Let ϕ be a deterministic smooth, compactly supported function. It is easy to see from basic properties of the GFF that the laws of h and $h + \phi$ are mutually absolutely continuous. See, e.g., [75, PROPOSITION 3.4] for a proof. Using Weyl scaling (Axiom III.), we can choose ϕ so that with high probability, the event E occurs with $h + \phi$ in place of h. The absolute continuity of the laws of $h + \phi$ and h then implies that $\mathbb{P}[E] > 0$. Let us illustrate this idea by showing that an LQG geodesic stays in a specified open set with positive probability.

Lemma 4.1. *Let $z, w \in \mathbb{C}$ and let $U \subset \mathbb{C}$ be a connected open set which contains z and w. With positive probability, every D_h-geodesic from z to w is contained in U.*

Proof. Let $V \subset V' \subset U$ be bounded, connected open sets containing z and w such that $\overline{V} \subset V'$ and $\overline{V}' \subset U$. It is a.s. the case that internal distance $D_h(z,w;V)$ is finite and the distance $D_h(V', \partial U)$ is positive, so we can find $C > 0$ such that

$$\mathbb{P}\big[D_h(z,w;V) \le C, D_h(V', \partial U) > C^{-1}\big] \ge \frac{1}{2}. \tag{4.1}$$

Let ϕ be a smooth, nonnegative bump function which is identically equal to $\frac{2}{\xi}\log C$ on V and is identically equal to zero outside of V'. By Weyl scaling (Axiom III.) and since $\phi \equiv \frac{2}{\xi}\log C$ on V, the $D_{h-\phi}$-internal metric on V is equal to C^{-2} times the D_h-internal metric on V. Furthermore, since $\phi \equiv 0$ outside V', we have $D_h(V', \partial U) = D_{h-\phi}(V', \partial U)$. Therefore, if the event in (4.1) occurs, then

$$D_{h-\phi}(z,w;V) = C^{-2} D_h(z,w;V) \le C^{-1} < D_h(\partial V', \partial U) = D_{h-\phi}(V', \partial U).$$

In particular, $D_{h-\phi}(z,w) < D_{h-\phi}(z, \partial U)$. Therefore, no $D_{h-\phi}$-geodesic from z to w can exit U. This happens with probability at least $1/2$. Since the laws of $h - \phi$ and h are mutually absolutely continuous, the lemma statement follows. ■

In a similar vein, it is sometimes useful to add a *random* bump function to h in order to show that D_h has certain "typical" behavior with probability 1. To be more precise, again let ϕ be a smooth compactly supported bump function and let X be a random variable which is uniform on $[0,1]$, sampled independently from h. Then the laws of h and $h + X\phi$

are mutually absolutely continuous. So, if E is an event depending on D_h, then to show that $\mathbb{P}[E] = 0$ it suffices to show that the probability that E occurs with $h + X\phi$ in place of h is zero. To show this latter statement, it suffices to show that a.s. the Lebesgue measure of the set of $x \in [0, 1]$ such that E occurs with $h + x\phi$ in place of h is zero. Usually, it is possible to show that this set consists of at most a single point. Let us illustrate this technique by proving the uniqueness of D_h-geodesics between typical points.

Lemma 4.2. *Fix distinct points $z, w \in \mathbb{C}$. Almost surely, there is a unique D_h-geodesic from z to w.*

Lemma 4.2 was first established in [72, THEOREM 1.2] via an argument which is similar to, but more complicated than, that we give here. We emphasize that Lemma 4.2 applies only for a fixed pair of points $z, w \in \mathbb{C}$. Almost surely, there are exceptional pairs of points which are joined by multiple D_h-geodesics. See [42, 73, 80] for a discussion of these exceptional pairs of points.

Proof of Lemma 4.2. Let $U, V \subset \mathbb{C}$ be bounded open sets lying at positive distance from z and w such that $\overline{V} \subset U$. Let $E = E(U, V)$ be the event that the following is true: there are distinct D_h-geodesics P, $\tilde{P}$ from z to w such that P is disjoint from U and $\tilde{P}$ enters V. If there is more than one D_h-geodesic from z to w, then $E(U, V)$ must occur for some choice of open sets U, V which we can take to be finite unions of balls with rational centers and radii. Hence it suffices to fix U and V and show that $\mathbb{P}[E] = 0$.

Let $\phi : \mathbb{C} \to [0, 1]$ be a smooth bump function which is identically equal to 1 on a neighborhood of $\overline{V}$ and which vanishes outside of U. For $x \in [0, 1]$, let E_x be the event that E occurs with $h + x\phi$ in place of h. As explained above the lemma statement, it suffices to prove that a.s. the Lebesgue measure of the set of $x \in [0, 1]$ for which E_x occurs is 0. In fact, we will show that a.s. there is at most one values of $x \in [0, 1]$ for which E_x occurs.

For this, it is enough to show that if $0 \le x < y \le 1$ and E_x occurs, then E_y does not occur. To see this, assume that E_x occurs and let P_x and $\tilde{P}_x$ be the $D_{h-x\phi}$-geodesics as in the definition of E_x. By Weyl scaling (Axiom III.) and since ϕ is nonnegative, we have $D_{h+y\phi}(u, v) \ge D_{h+x\phi}(u, v)$ for all $u, v \in \mathbb{C}$. Since P_x does not enter U and ϕ vanishes outside of U, we also have

$$D_{h+y\phi}(z, w) \le \operatorname{len}(P_x; D_{h+y\phi}) = \operatorname{len}(P_x; D_{h+x\phi}) = D_{h+x\phi}(z, w),$$

where here we recall the notation for length with respect to a metric from (2.6). Hence

$$D_{h+y\phi}(z, w) = D_{h+x\phi}(z, w). \tag{4.2}$$

Now suppose that $\tilde{P} : [0, T] \to \mathbb{C}$ is any path from z to w which enters V. We will show that $\tilde{P}$ is not a $D_{h+y\phi}$-geodesic, which implies that E_y does not occur. Indeed, there must be a positive-length interval of times $[a, b]$ such that $P([a, b]) \subset \phi^{-1}(1)$. We therefore have

$$\begin{aligned}
\operatorname{len}(\tilde{P}; D_{h+y\phi}) &= \operatorname{len}(\tilde{P}|_{[0,a]\cup[b,T]}; D_{h+y\phi}) + \operatorname{len}(\tilde{P}|_{[a,b]}; D_{h+y\phi}) \\
&\geq \operatorname{len}(\tilde{P}|_{[0,a]\cup[b,T]}; D_{h+x\phi}) \\
&\qquad + e^{\xi(y-x)} \operatorname{len}(\tilde{P}|_{[a,b]}; D_{h+x\phi}) \quad \text{(by Axiom III.)} \\
&\geq \operatorname{len}(\tilde{P}; D_{h+x\phi}) + (e^{\xi(y-x)} - 1) \operatorname{len}(\tilde{P}|_{[a,b]}; D_{h+x\phi}) \\
&> D_{h+x\phi}(z, w) \quad \text{(by Axiom I.)} \\
&= D_{h+y\phi}(z, w) \quad \text{(by (4.2))}.
\end{aligned}$$

■

Remark 4.3. If ϕ is a deterministic smooth bump function, then the proof of [75, PROPOSITION 3.4] shows that the Radon–Nikodym derivative of the law of $h + \phi$ with respect to the law of h is given by

$$\exp\left((h, \phi)_\nabla - \frac{1}{2}(\phi, \phi)_\nabla \right)$$

where $(f, g)_\nabla := \int_{\mathbb{C}} \nabla f(z) \cdot \nabla g(z) \, d^2 z$ is the Dirichlet inner product. One can use this explicit expression for the Radon–Nikodym derivative, together with arguments of the sort discussed above, to estimate the probabilities of certain rare events for the LQG metric. For example, this is the key idea in the computation of the dimension of a boundary of an LQG metric ball in [41].

4.2. Independence across concentric annuli

Another key tool in the study of the LQG metric is the fact that the restrictions of the GFF to disjoint concentric annuli (viewed modulo additive constant) are nearly independent. In particular, suppose that we have a sequence of events $\{E_{r_k}\}_{k\in\mathbb{N}}$ depending on the restrictions of h to disjoint concentric annuli. If we have a lower bound for $\mathbb{P}[E_{r_k}]$ which is uniform in k, then for $K \in \mathbb{N}$ the number of $k \in \{1, \dots, K\}$ for which E_{r_k} occurs can be compared to a binomial random variable. This leads to the following lemma, which is a special case of [49, LEMMA 3.1].

Lemma 4.4. *Fix $0 < s_1 < s_2 < 1$. Let $z \in \mathbb{C}$ and let $\{r_k\}_{k\in\mathbb{N}}$ be a decreasing sequence of positive real numbers such that $r_{k+1}/r_k \leq s_1$ for each $k \in \mathbb{N}$. Let $\{E_{r_k}\}_{k\in\mathbb{N}}$ be events such that for each $k \in \mathbb{N}$, the event E_{r_k} is a.s. determined by the restriction of $h - h_{r_k}(z)$ to the Euclidean annulus $B_{s_2 r_k}(z) \setminus B_{s_1 r_k}(z)$, where $h_{r_k}(z)$ denotes the circle average.*

(1) *For each $a > 0$, there exist $p = p(a, s_1, s_2) \in (0, 1)$ and $c = c(a, s_1, s_2) > 0$ such that if*

$$\mathbb{P}[E_{r_k}] \geq p, \quad \forall k \in \mathbb{N}, \tag{4.3}$$

then

$$\mathbb{P}[\exists k \in \{1, \dots, K\} \text{ such that } E_{r_k} \text{ occurs}] \geq 1 - ce^{-aK}, \quad \forall K \in \mathbb{N}. \tag{4.4}$$

(2) *For each $p \in (0, 1)$, there exist $a = a(p, s_1, s_2) > 0$ and $c = c(p, s_1, s_2) > 0$ such that if* (4.3) *holds, then* (4.4) *holds.*

We emphasize that the numbers p and c in assertion (1) and the numbers a and c is assertion (2) do *not* depend on z or on $\{r_k\}$ (except via s_1, s_2). The idea of Lemma 4.4 was first used in [72], and the general version stated here was first formulated in [49]. To illustrate the use of Lemma 4.4, we will explain a typical application, namely a polynomial upper bound for the probability that a D_h-geodesic gets near a point.

Lemma 4.5. *For each $\gamma \in (0,2)$, there exist $\alpha = \alpha(\gamma) > 0$ and $c = c(\gamma) > 0$ such that the following is true. For each $z \in \mathbb{C}$ and each $\varepsilon > 0$, the probability that there is a D_h-geodesic between two points in $\mathbb{C} \setminus B_{\varepsilon^{1/2}}(z)$ which enters $B_\varepsilon(z)$ is at most $c\varepsilon^\alpha$.*

Roughly speaking, Lemma 4.5 says that "most" points in $\mathbb{C}$ are not hit by D_h-geodesics except at their endpoints. Lemma 4.5 immediately implies that the Hausdorff dimension of every LQG geodesic with respect to the Euclidean metric is strictly less than 2. Similar (but more complicated) ideas to those in the proof of Lemma 4.5 are used in the proof of confluence of geodesics in [25,48].

Let us now proceed with the proof of Lemma 4.5. The first step is to define the events for which we will apply Lemma 4.4. To lighten notation, we introduce the following terminology.

Definition 4.6. For a Euclidean annulus $A \subset \mathbb{C}$, we define $D_h(\text{across } A)$ to be the D_h-distance between the inner and outer boundaries of A. We define $D_h(\text{around } A)$ to be the infimum of the D_h-lengths of paths in A which separate the inner and outer boundaries of A.

Both $D_h(\text{across } A)$ and $D_h(\text{around } A)$ are determined by the internal metric of D_h on A, so by Axiom II. these quantities are a.s. determined by $h|_A$.

For $z \in \mathbb{C}$ and $r > 0$, let

$$E_r(z) := \{D_h(\text{around } B_{3r}(z) \setminus B_{2r}(z)) < D_h(\text{across } B_{2r}(z) \setminus B_r(z))\}. \tag{4.5}$$

As noted above, Axiom II. implies that $E_r(z)$ is a.s. determined by $h|_{B_{3r}(z)\setminus B_r(z)}$. In fact, adding a constant to h results in scaling D_h-distances by a constant (Axiom III.), so adding a constant to h does not affect whether $E_r(z)$ occurs. Hence $E_r(z)$ is a.s. determined by $(h - h_{4r}(z))|_{B_{3r}(z)\setminus B_r(z)}$.

Lemma 4.7. *There exist $\alpha = \alpha(\gamma) > 0$ and $c = c(\gamma) > 0$ such that for each $z \in \mathbb{C}$ and each $\varepsilon > 0$,*

$$\mathbb{P}[\exists r \in [\varepsilon, \tfrac{1}{4}\varepsilon^{1/2}] \text{ such that } E_r(z) \text{ occurs}] \geq 1 - c\varepsilon^\alpha.$$

Proof. Using a "subtracting a bump function" argument as discussed in Section 4.1, one can show that $p := \mathbb{P}[E_1(0)] > 0$. From (2.11), we see $\mathbb{P}[E_r(z)]$ does not depend on z or r. Hence $\mathbb{P}[E_r(z)] = p$ for each $z \in \mathbb{C}$ and $r > 0$. We now apply Lemma 4.4 with $r_k = 4^{-k}\varepsilon^{1/2}$ and $K = \lfloor \frac{1}{2}\log_4 \varepsilon^{-1} \rfloor$. Then $r_k \in [\varepsilon, \frac{1}{4}\varepsilon^{1/2}]$ for each $k \in \{1, \dots, K\}$, so part (2) of Lemma 4.4 shows that there exists $a = a(\gamma) > 0$ and $c = c(\gamma) > 0$ such that

$$\mathbb{P}[\exists r \in [\varepsilon, \varepsilon^{1/2}] \text{ such that } E_r(z) \text{ occurs}] \geq 1 - cp^{aK}.$$

This last quantity is at least $1 - c\varepsilon^{\alpha}$ for an appropriate $\alpha > 0$ depending on p, a (hence on γ). ■

Proof of Lemma 4.5. By Lemma 4.7, it suffices to show that if there is an $r \in [\varepsilon, \frac{1}{4}\varepsilon^{1/2}]$ such that $E_r(z)$ occurs, then no D_h-geodesic between two points in $\mathbb{C} \setminus B_{\varepsilon^{1/2}}(z)$ can enter $B_\varepsilon(z)$. Indeed, assume that $E_r(z)$ occurs, let $u, v \in \mathbb{C} \setminus B_{\varepsilon^{1/2}}(z)$, and let P be a path from u to v which hits $B_r(z) \supset B_\varepsilon(z)$. We will show that P is not a D_h-geodesic. By the definition (4.5) of $E_r(z)$, there is a path π in $B_{3r}(z) \setminus B_{2r}(z)$ which disconnects the inner and outer boundaries of this annulus and has D_h-length strictly less than $D_h(\text{across } B_{2r}(z) \setminus B_r(z))$. Let σ (resp. τ) be the first (resp. last) time that P hits π. Since P hits $B_r(z)$ and $u, v \notin B_{3r}(z)$, the path P crosses between the inner and outer boundaries of $B_{2r}(z) \setminus B_r(z)$ between times σ and τ. Hence

$$(D_h\text{-length of } P|_{[\sigma,\tau]}) \geq D_h\big(\text{across } B_{2r}(z) \setminus B_r(z)\big). \tag{4.6}$$

But, since $P(\tau), P(\sigma) \in \pi$,

$$\begin{aligned} D_h\big(P(\sigma), P(\tau)\big) \leq (D_h\text{-length of } \pi) &< D_h\big(\text{across } B_{2r}(z) \setminus B_r(z)\big) \\ &\leq \big(D_h\text{-length of } P|_{[\sigma,\tau]}\big). \end{aligned} \tag{4.7}$$

This implies that P is not a D_h-geodesic since it is not the D_h-shortest path from $P(\sigma)$ to $P(\tau)$. ■

4.3. White noise decomposition

A convenient way to approximate the GFF is by convolving the heat kernel with a space-time white noise. To explain this, let W be a space-time white noise on $\mathbb{C} \times [0, \infty)$, i.e., $\{(W, f) : f \in L^2(\mathbb{C} \times [0, \infty))\}$ is a centered Gaussian process with covariances $\mathbb{E}[(W, f)(W, g)] = \int_{\mathbb{C}} \int_0^\infty f(z, s) g(z, s)\, ds\, dz$. For $f \in L^2(\mathbb{C} \times [0, \infty))$ and Borel measurable sets $A \subset \mathbb{C}$ and $I \subset [0, \infty)$, we slightly abuse notation by writing

$$\int_A \int_I f(z, s)\, W(dz, ds) := (W, f \mathbb{1}_{A \times I}).$$

As in (1.2), we denote the heat kernel by $p_t(z) := \frac{1}{2\pi t} e^{-|z|^2/2t}$. Following **[21, SECTION 3]**, we define the centered Gaussian process

$$\hat{h}_t(z) := \sqrt{\pi} \int_{\mathbb{C}} \int_{t^2}^1 p_{s/2}(z - w)\, W(dw, ds), \quad \forall t \in [0, 1], \ \forall z \in \mathbb{C}. \tag{4.8}$$

We write $\hat{h} := \hat{h}_0$. By **[21, LEMMA 3.1]** and Kolmogorov's criterion, each $\hat{h}_t$ for $t \in (0, 1]$ admits a continuous modification. The process $\hat{h}$ does not admit a continuous modification, but makes sense as a distribution: indeed, it is easily checked that its integral against any smooth compactly supported test function is Gaussian with finite variance.

The process $\hat{h}$ is in some ways more convenient to work with than the GFF thanks to the following symmetries, which are immediate from the definition:

- *Rotation/translation/reflection invariance.* The law of $\{\hat{h}_t : t \in [0, 1]\}$ is invariant with respect to rotation, translation, and reflection of the plane.

- *Scale invariance.* For $\delta \in (0,1]$, one has $\{(\hat{h}_{\delta t} - \hat{h}_{\delta})(\delta \cdot) : t \in [0,1]\} \stackrel{d}{=} \{\hat{h}_t : t \in [0,1]\}$.
- *Independent increments.* If $0 \le t_1 \le t_2 \le t_3 \le t_4 \le 1$, then $\hat{h}_{t_2} - \hat{h}_{t_1}$ and $\hat{h}_{t_4} - \hat{h}_{t_3}$ are independent.

One property which $\hat{h}$ does not possess is spatial independence. To get around this, it is sometimes useful to work with a truncated variant of $\hat{h}$ where we only integrate over a ball of finite radius. To this end, we let $\phi : \mathbb{C} \to [0,1]$ be a smooth bump function which is equal to 1 on the ball $B_{1/20}(0)$ and which vanishes outside of $B_{1/10}(0)$. For $t \in [0,1]$, we define

$$\hat{h}_t^{\mathrm{tr}}(z) := \sqrt{\pi} \int_{t^2}^{1} \int_{\mathbb{C}} p_{s/2}(z-w)\phi(z-w)\, W(dw, dt). \tag{4.9}$$

We also set $\hat{h}^{\mathrm{tr}} := \hat{h}_0^{\mathrm{tr}}$. As in the case of $\hat{h}$, it is easily seen from the Kolmogorov continuity criterion that each $\hat{h}_t^{\mathrm{tr}}$ for $t \in (0,1]$ a.s. admits a continuous modification. The process $\hat{h}^{\mathrm{tr}}$ does not admit a continuous modification and is instead viewed as a random distribution.

The key property enjoyed by $\hat{h}^{\mathrm{tr}}$ is spatial independence: if $A, B \subset \mathbb{C}$ with $\mathrm{dist}(A,B) \ge 1/5$, then $\{\hat{h}_t^{\mathrm{tr}}|_A : t \in [0,1]\}$ and $\{\hat{h}_t^{\mathrm{tr}}|_B : t \in [0,1]\}$ are independent. Indeed, this is because $\{\hat{h}_t^{\mathrm{tr}}|_A : t \in [0,1]\}$ and $\{\hat{h}_t^{\mathrm{tr}}|_B : t \in [0,1]\}$ are determined by the restrictions of the white noise W to the disjoint sets $B_{1/10}(A) \times \mathbb{R}_+$ and $B_{1/10}(B) \times \mathbb{R}_+$, respectively. Unlike $\hat{h}$, the distribution $\hat{h}^{\mathrm{tr}}$ does not possess any sort of scale invariance but its law is still invariant with respect to rotations, translations, and reflections of $\mathbb{C}$.

The following lemma, which is proven in the same manner as [22, LEMMA 3.1], tells us that the distributions $\hat{h}$ and $\hat{h}^{\mathrm{tr}}$ and the whole-plane GFF can all be compared up to constant-order additive errors.

Lemma 4.8. *Suppose $U \subset \mathbb{C}$ is a bounded open set. There is a coupling $(h, \hat{h}, \hat{h}^{\mathrm{tr}})$ of a whole-plane GFF normalized so that $h_1(0) = 0$ and the fields from (4.8) and (4.9) such that the following is true. For any $h^1, h^2 \in \{h, \hat{h}, \hat{h}^{\mathrm{tr}}\}$, the distribution $(h^1 - h^2)|_U$ a.s. admits a continuous modification and there are constants $c_0, c_1 > 0$ depending only on U such that for $A > 1$,*

$$\mathbb{P}\Big[\max_{z \in U} \big|(h^1 - h^2)(z)\big| \le A\Big] \ge 1 - c_0 e^{-c_1 A^2}. \tag{4.10}$$

Lemma 4.8 implies that each of $\hat{h}$ and $\hat{h}^{\mathrm{tr}}$ is a GFF plus a continuous function. Hence we can define the LQG metrics $D_{\hat{h}}$ and $D_{\hat{h}^{\mathrm{tr}}}$. The metric $D_{\hat{h}^{\mathrm{tr}}}$ is particularly convenient to work with due to the aforementioned finite range of dependence property of $\hat{h}^{\mathrm{tr}}$. This property allows one to use percolation-style arguments in order to produce large clusters of Euclidean squares where certain "good" events occur. We refer to [21, 22, 30, 52] for examples of this sort of argument.

The white noise decomposition also plays a key role in the proofs of tightness of LFPP in [18, 19, 23, 33]. In fact, these papers first prove the tightness of LFPP defined using the white noise decomposition (4.8) in place of the functions h_ε^*, then transfer to h_ε^* using a comparison lemma which is similar in spirit to Lemma 4.8 (see [18, SECTION 6.1]).

5. OPEN PROBLEMS

Here we highlight some of the most important open problems concerning the LQG metric. Much more substantial lists of open problems can be found in [46, 51].

Problem 5.1. For $\gamma \in (0, 2)$, compute the Hausdorff dimension d_γ of $\mathbb{C}$, equipped with the γ-LQG metric. More generally, for $\xi > 0$ determine the relationship between the parameters Q and ξ of (2.4).

Due to (2.2) and (2.5), computing d_γ for $\gamma \in (0, 2)$ is equivalent to finding the relationship between Q and ξ for $\xi \in (0, 2/d_2)$. As noted above, the only known case is $d_{\sqrt{8/3}} = 4$, equivalently $Q(1/\sqrt{6}) = 5/\sqrt{6}$. One indication of the difficulty of computing Q in terms of ξ is that the relationship between Q and ξ is not universal for LFPP defined using different log-correlated Gaussian fields [29].

Many quantities associated with LQG surfaces and random planar maps can be expressed in terms of d_γ (or ξ and Q), such as the optimal Hölder exponents relating the LQG metric and the Euclidean metric [34], the Hausdorff dimension of the boundary of an LQG metric ball [41], and the ball volume exponent for certain random planar maps [22]. Solving Problem 5.1 would lead to exact formulas for these quantities.

We do not have a guess for the formula relating Q and ξ, nor do we know whether an explicit formula exists. The best-known prediction from the physics literature, due to Watabiki [91], is equivalent to $Q = 1/\xi - \xi$ for $\xi \in (0, 2/d_2)$. The prediction was proven to be false in [21], at least for small values of ξ (equivalently, small values of γ). An alternative proposal, put forward in [22], is that $Q = 1/\xi - 1/\sqrt{6}$ for $\xi \in (0, 2/d_2)$. This formula has not been disproven for any value of $\xi \in (0, 2/d_2)$, but it (like Watabiki's prediction) is inconsistent with the result of [28], which shows that $Q > 0$ for all $\xi > 0$. We expect that both of the above predictions are false for all but finitely many values of ξ.

The best known rigorous bounds relating ξ and Q are obtained in [1, 22, 53]. See Figure 5 for a graph of these bounds.

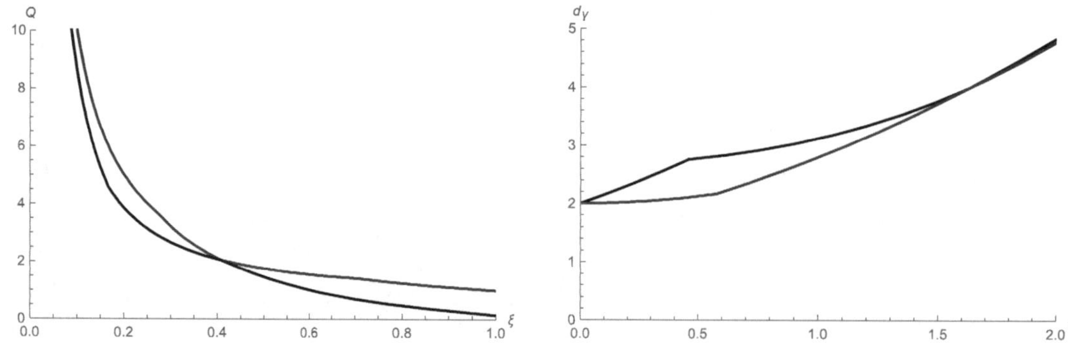

FIGURE 5
(Left) Plot of the best known upper (blue) and lower (red) bounds for Q as a function of ξ. (Right) Plot of the best-known bounds for d_γ as a function of γ.

Our next open problem concerns the relationship between LQG surfaces and random planar maps.

Problem 5.2. Show that, for each $\gamma \in (0, 2]$, appropriate types of random planar maps, equipped with their graph distance (appropriately rescaled), converge in the Gromov–Hausdorff sense to γ-LQG surfaces equipped with the γ-LQG metric.

As discussed in Section 1.3, the value of γ depends on the type of random planar map under consideration. For example, uniform random planar maps correspond to $\gamma = \sqrt{8/3}$, planar maps weighted by the number of spanning trees they admit correspond to $\gamma = \sqrt{2}$, and planar maps weighted by the partition function of the critical Ising model on the map correspond to $\gamma = \sqrt{3}$. So far, Problem 5.2 has only been solved for $\gamma = \sqrt{8/3}$, see Section 2.4.

Problem 5.2 can be made more precise by specifying the scaling factor for the planar maps, as well as the particular types of LQG surfaces one should get in the limit. For concreteness, for $n \in \mathbb{N}$ consider the case of a random planar map M_n with the topology of the sphere, having n total edges. Then M_n, equipped with its graph distance rescaled by n^{-1/d_γ}, should converge in the Gromov–Hausdorff sense to the quantum sphere, a special type of LQG surface which is defined in [16, 37] (the definitions are proven to be equivalent in [5]). Similar statements apply for random planar maps with other topologies, such as the disk, plane, or half-plane.

Finally, we mention a third open problem which has not appeared elsewhere. For $\alpha \in \mathbb{R}$, let $\mathcal{T}_h^\alpha$ be the set of α-thick points of h, i.e., the points $z \in \mathbb{C}$ for which $\limsup_{\varepsilon \to 0} h_\varepsilon(z)/\log \varepsilon^{-1} = \alpha$. Such points exist if and only if $\alpha \in [-2, 2]$ [57]. For a set X, the function which takes α to the Hausdorff dimension of $X \cap \mathcal{T}_h^\alpha$ (with respect to the LQG metric or the Euclidean metric) can be thought of as a sort of "quantum multifractal spectrum" of X.

Problem 5.3. Let $\xi > 0$ and let P be a D_h-geodesic. Is it possible to compute the Hausdorff dimensions of $P \cap T_h^\alpha$ for each $\alpha \in [-2, 2]$ with respect to the D_h (resp. the Euclidean metric)? More weakly, as there a unique value of α which maximizes this dimension? In other words, is there a "typical" thickness for a point on an LQG geodesic?

It is known that the Hausdorff dimensions considered in Problem 5.3 are a.s. equal to deterministic constants, see [55, REMARK 1.12]. The analog of Problem 5.3 for a subcritical LQG metric ball boundary has been solved in [22, 55]. In that case, the maximizing value of α with respect to the Euclidean (resp. LQG) metric is $\alpha = \xi$ (resp. $\alpha = \gamma$). One can also ask the analog of Problem 5.3 with Minkowski dimension instead of Hausdorff dimension. We expect that the answers will be the same.

FUNDING

J. Ding was partially supported by NSF grants DMS-1757479 and DMS-1953848.
J. Dubédat was partially supported by NSF grant DMS-1512853. E. G. was partially supported by a Clay research fellowship.

REFERENCES

[1] M. Ang, Comparison of discrete and continuum Liouville first passage percolation. *Electron. Commun. Probab.* **24** (2019), 64, 12 pp.

[2] M. Ang, H. Falconet, and X. Sun, Volume of metric balls in Liouville quantum gravity. *Electron. J. Probab.* **25** (2020), 160, 50 pp.

[3] M. Ang, M. Park, J. Pfeffer, and S. Sheffield, Brownian loops and the central charge of a Liouville random surface. 2020, arXiv:2005.11845.

[4] J. Aru, KPZ relation does not hold for the level lines and SLE_κ flow lines of the Gaussian free field. *Probab. Theory Related Fields* **163** (2015), no. 3–4, 465–526.

[5] J. Aru, Y. Huang, and X. Sun, Two perspectives of the 2D unit area quantum sphere and their equivalence. *Comm. Math. Phys.* **356** (2017), no. 1, 261–283.

[6] J. Barral, X. Jin, R. Rhodes, and V. Vargas, Gaussian multiplicative chaos and KPZ duality. *Comm. Math. Phys.* **323** (2013), no. 2, 451–485.

[7] R. Basu, M. Bhatia, and S. Ganguly, Environment seen from infinite geodesics in Liouville quantum gravity. 2021, arXiv:2107.12363.

[8] G. Beer, Upper semicontinuous functions and the Stone approximation theorem. *J. Approx. Theory* **34** (1982), no. 1, 1–11.

[9] I. Benjamini and O. Schramm, KPZ in one dimensional random geometry of multiplicative cascades. *Comm. Math. Phys.* **289** (2009), no. 2, 653–662.

[10] S. Benoist, Natural parametrization of SLE: the Gaussian free field point of view. *Electron. J. Probab.* **23** (2018), 103, 16 pp.

[11] N. Berestycki, Diffusion in planar Liouville quantum gravity. *Ann. Inst. Henri Poincaré Probab. Stat.* **51** (2015), no. 3, 947–964.

[12] N. Berestycki, C. Garban, R. Rhodes, and V. Vargas, KPZ formula derived from Liouville heat kernel. *J. Lond. Math. Soc. (2)* **94** (2016), no. 1, 186–208.

[13] N. Berestycki and E. Powell, Gaussian free field, Liouville quantum gravity, and Gaussian multiplicative chaos. https://homepage.univie.ac.at/nathanael.berestycki/Articles/master.pdf

[14] D. Burago, Y. Burago, and S. Ivanov, *A course in metric geometry*. Grad. Stud. Math. 33, American Mathematical Society, Providence, RI, 2001.

[15] F. David, Conformal field theories coupled to 2-D gravity in the conformal gauge. *Modern Phys. Lett. A* **3** (1988), no. 17.

[16] F. David, A. Kupiainen, R. Rhodes, and V. Vargas, Liouville quantum gravity on the Riemann sphere. *Comm. Math. Phys.* **342** (2016), no. 3, 869–907.

[17] F. David, R. Rhodes, and V. Vargas, Liouville quantum gravity on complex tori. *J. Math. Phys.* **57** (2016), no. 2, 022302, 25 pp.

[18] J. Ding, J. Dubédat, A. Dunlap, and H. Falconet, Tightness of Liouville first passage percolation for $\gamma \in (0,2)$. *Publ. Math. Inst. Hautes Études Sci.* **132** (2020), 353–403.

[19] J. Ding and A. Dunlap, Liouville first-passage percolation: Subsequential scaling limits at high temperature. *Ann. Probab.* **47** (2019), no. 2, 690–742.

[20] J. Ding and A. Dunlap, Subsequential scaling limits for Liouville graph distance. *Comm. Math. Phys.* **376** (2020), no. 2, 1499–1572.

[21] J. Ding and S. Goswami, Upper bounds on Liouville first-passage percolation and Watabiki's prediction. *Comm. Pure Appl. Math.* **72** (2019), no. 11, 2331–2384.

[22] J. Ding and E. Gwynne, The fractal dimension of Liouville quantum gravity: universality, monotonicity, and bounds. *Comm. Math. Phys.* **374** (2018), 1877–1934.

[23] J. Ding and E. Gwynne, Tightness of supercritical Liouville first passage percolation. *J. Eur. Math. Soc. (JEMS)* (to appear).

[24] J. Ding and E. Gwynne, The critical Liouville quantum gravity metric induces the Euclidean topology. 2021, arXiv:2108.12067.

[25] J. Ding and E. Gwynne, Regularity and confluence of geodesics for the supercritical Liouville quantum gravity metric. 2021, arXiv:2104.06502.

[26] J. Ding and E. Gwynne, Up-to-constants comparison of Liouville first passage percolation and Liouville quantum gravity. 2021, arXiv:2108.12060.

[27] J. Ding and E. Gwynne, Uniqueness of the critical and supercritical Liouville quantum gravity metrics. 2021, arXiv:2110.00177.

[28] J. Ding, E. Gwynne, and A. Sepúlveda, The distance exponent for Liouville first passage percolation is positive. 2020, arXiv:2005.13570v2.

[29] J. Ding, O. Zeitouni, and F. Zhang, On the Liouville heat kernel for k-coarse MBRW. *Electron. J. Probab.* **23** (2018), 62, 20 pp.

[30] J. Ding, O. Zeitouni, and F. Zhang, Heat kernel for Liouville Brownian motion and Liouville graph distance. *Comm. Math. Phys.* **371** (2019), no. 2, 561–618.

[31] J. Ding and F. Zhang, Liouville first passage percolation: geodesic length exponent is strictly larger than 1 at high temperatures. *Probab. Theory Related Fields* **174** (2019), no. 1–2, 335–367.

[32] J. Distler and H. Kawai, Conformal field theory and 2D quantum gravity. *Nuclear Phys. B* **321** (1989), no. 2.

[33] J. Dubédat and H. Falconet, Liouville metric of star-scale invariant fields: tails and Weyl scaling. *Probab. Theory Related Fields* **176** (2020), no. 1–2, 293–352.

[34] J. Dubédat, H. Falconet, E. Gwynne, J. Pfeffer, and X. Sun, Weak LQG metrics and Liouville first passage percolation. *Probab. Theory Related Fields* **178** (2020), no. 1–2, 369–436.

[35] B. Duplantier, Random walks and quantum gravity in two dimensions. *Phys. Rev. Lett.* **81** (1998), no. 25, 5489–5492.

[36] B. Duplantier and K.-H. Kwon, Conformal invariance and intersections of random walks. *Phys. Rev. Lett.* **61** (1988), 2514–2517.

[37] B. Duplantier, J. Miller, and S. Sheffield, Liouville quantum gravity as a mating of trees. *Astérisque* (to appear).

[38] B. Duplantier, R. Rhodes, S. Sheffield, and V. Vargas, Critical Gaussian multiplicative chaos: convergence of the derivative martingale. *Ann. Probab.* **42** (2014), no. 5, 1769–1808.

[39] B. Duplantier, R. Rhodes, S. Sheffield, and V. Vargas, Renormalization of critical Gaussian multiplicative chaos and KPZ relation. *Comm. Math. Phys.* **330** (2014), no. 1, 283–330.

[40] B. Duplantier and S. Sheffield, Liouville quantum gravity and KPZ. *Invent. Math.* **185** (2011), no. 2, 333–393.

[41] G. Ewain, The dimension of the boundary of a Liouville quantum gravity metric ball. *Comm. Math. Phys.* **378** (2020), no. 1, 625–689.

[42] G. Ewain, Geodesic networks in Liouville quantum gravity surfaces. *Probab. Math. Phys.* (to appear).

[43] C. Garban, R. Rhodes, and V. Vargas, Liouville Brownian motion. *Ann. Probab.* **44** (2016), no. 4, 3076–3110.

[44] C. Guillarmou, R. Rhodes, and V. Vargas, Polyakov's formulation of $2d$ bosonic string theory. *Publ. Math. Inst. Hautes Études Sci.* **130** (2019), 111–185.

[45] E. Gwynne, N. Holden, and J. Miller, An almost sure KPZ relation for SLE and Brownian motion. *Ann. Probab.* **48** (2020), no. 2, 527–573.

[46] E. Gwynne, N. Holden, J. Pfeffer, and G. Remy, Liouville quantum gravity with matter central charge in $(1, 25)$: a probabilistic approach. *Comm. Math. Phys.* **376** (2020), no. 2, 1573–1625.

[47] E. Gwynne, N. Holden, and X. Sun, Mating of trees for random planar maps and Liouville quantum gravity: a survey. 2019, arXiv:1910.04713. To appear in *Panor. Synthèses*.

[48] E. Gwynne and J. Miller, Confluence of geodesics in Liouville quantum gravity for $\gamma \in (0, 2)$. *Ann. Probab.* **48** (2020), no. 4, 1861–1901.

[49] E. Gwynne and J. Miller, Local metrics of the Gaussian free field. *Ann. Inst. Fourier (Grenoble)* **70** (2020), no. 5, 2049–2075.

[50] E. Gwynne and J. Miller, Conformal covariance of the Liouville quantum gravity metric for $\gamma \in (0, 2)$. *Ann. Inst. Henri Poincaré Probab. Stat.* **57** (2021), no. 2.

[51] E. Gwynne and J. Miller, Existence and uniqueness of the Liouville quantum gravity metric for $\gamma \in (0, 2)$. *Invent. Math.* **223** (2021), no. 1, 213–333.

[52] E. Gwynne, J. Miller, and S. Sheffield, The Tutte embedding of the Poisson–Voronoi tessellation of the Brownian disk converges to $\sqrt{8/3}$-Liouville quantum gravity. *Comm. Math. Phys.* **374** (2020), no. 2, 735–784.

[53] E. Gwynne and J. Pfeffer, Bounds for distances and geodesic dimension in Liouville first passage percolation. *Electron. Commun. Probab.* **24** (2019), no. 56, 12 pp.

[54] E. Gwynne and J. Pfeffer, KPZ formulas for the Liouville quantum gravity metric. *Trans. Amer. Math. Soc.* (to appear).

[55] E. Gwynne, J. Pfeffer, and S. Sheffield, Geodesics and metric ball boundaries in Liouville quantum gravity. 2020, arXiv:2010.07889.

[56] N. Holden and X. Sun, Convergence of uniform triangulations under the Cardy embedding. 2019, arXiv:1905.13207. To appear in *Acta Math.*

[57] X. Hu, J. Miller, and Y. Peres, Thick points of the Gaussian free field. *Ann. Probab.* **38** (2010), no. 2, 896–926.

[58] J.-P. Kahane, Sur le chaos multiplicatif. *Ann. Sci. Math. Québec* **9** (1985), no. 2, 105–150.

[59] M. Kardar, G. Parisi, and Y.-C. Zhang, Dynamic scaling of growing interfaces. *Phys. Rev. Lett.* **56** (1986), 889–892.

[60] V. G. Knizhnik, A. M. Polyakov, and A. B. Zamolodchikov, Fractal structure of 2D-quantum gravity. *Modern Phys. Lett. A* **3** (1988), no. 8, 819–826.

[61] A. Kupiainen, R. Rhodes, and V. Vargas, Integrability of Liouville theory: proof of the DOZZ formula. *Ann. of Math. (2)* **191** (2020), no. 1, 81–166.

[62] G. F. Lawler, O. Schramm, and W. Werner, Values of Brownian intersection exponents. I. Half-plane exponents. *Acta Math.* **187** (2001), no. 2, 237–273.

[63] G. F. Lawler, O. Schramm, and W. Werner, Values of Brownian intersection exponents. II. Plane exponents. *Acta Math.* **187** (2001), no. 2, 275–308.

[64] G. F. Lawler, O. Schramm, and W. Werner, Values of Brownian intersection exponents. III. Two-sided exponents. *Ann. Inst. Henri Poincaré Probab. Stat.* **38** (2002), no. 1, 109–123.

[65] J.-F. Le Gall, Geodesics in large planar maps and in the Brownian map. *Acta Math.* **205** (2010), no. 2, 287–360.

[66] J.-F. Le Gall, Uniqueness and universality of the Brownian map. *Ann. Probab.* **41** (2013), no. 4, 2880–2960.

[67] J.-F. Le Gall, Random geometry on the sphere. In *Proceedings of the International Congress of Mathematicians—Seoul 2014. Vol. 1*, pp. 421–442, Kyung Moon Sa, Seoul, 2014.

[68] J.-F. Le Gall, Brownian geometry. *Jpn. J. Math.* **14** (2019), no. 2, 135–174.

[69] J.-F. Le Gall, The volume measure of the Brownian sphere is a Hausdorff measure. 2021, arXiv:2105.05615.

[70] G. Miermont, The Brownian map is the scaling limit of uniform random plane quadrangulations. *Acta Math.* **210** (2013), no. 2, 319–401.

[71] J. Miller, Liouville quantum gravity as a metric space and a scaling limit. In *Proceedings of the International Congress of Mathematicians—Rio de Janeiro 2018. Vol. IV. Invited lectures*, pp. 2945–2971, World Sci. Publ., Hackensack, NJ, 2018.

[72] J. Miller and W. Qian, The geodesics in Liouville quantum gravity are not Schramm–Loewner evolutions. *Probab. Theory Related Fields* **177** (2020), no. 3–4, 677–709.

[73] J. Miller and W. Qian, Geodesics in the Brownian map: strong confluence and geometric structure. 2020, arXiv:2008.02242.

[74] J. Miller and S. Sheffield, An axiomatic characterization of the Brownian map. *J. Éc. Polytech.* **8** (2021), 609–731.

[75] J. Miller and S. Sheffield, Imaginary geometry I: interacting SLEs. *Probab. Theory Related Fields* **164** (2016), no. 3–4, 553–705.

[76] J. Miller and S. Sheffield, Liouville quantum gravity and the Brownian map II: geodesics and continuity of the embedding. *Ann. Probab.* **49** (2021), no. 6, 2732–2829.

[77] J. Miller and S. Sheffield, Liouville quantum gravity and the Brownian map III: the conformal structure is determined. *Probab. Theory Related Fields* **179** (2021), no. 3–4, 1183–1211.

[78] J. Miller and S. Sheffield, Liouville quantum gravity spheres as matings of finite-diameter trees. *Ann. Inst. Henri Poincaré Probab. Stat.* **55** (2019), no. 3, 1712–1750.

[79] J. Miller and S. Sheffield, Liouville quantum gravity and the Brownian map I: the QLE$(8/3, 0)$ metric. *Invent. Math.* **219** (2020), no. 1, 75–152.

[80] Omer Angel, B. Kolesnik, and G. Miermont, Stability of geodesics in the Brownian map. *Ann. Probab.* **45** (2017), no. 5, 3451–3479.

[81] J. Pfeffer, Weak Liouville quantum gravity metrics with matter central charge $\mathbf{c} \in (-\infty, 25)$. 2021, arXiv:2104.04020.

[82] A. M. Polyakov, Quantum geometry of bosonic strings. *Phys. Lett. B* **103** (1981), no. 3, 207–210.

[83] E. Powell, Critical Gaussian chaos: convergence and uniqueness in the derivative normalisation. *Electron. J. Probab.* **23** (2018), 31, 26 pp.

[84] G. Remy, Liouville quantum gravity on the annulus. *J. Math. Phys.* **59** (2018), no. 8, 082303, 26 pp.

[85] R. Rhodes and V. Vargas, KPZ formula for log-infinitely divisible multifractal random measures. *ESAIM Probab. Stat.* **15** (2011), 358–371.

[86] R. Rhodes and V. Vargas, Gaussian multiplicative chaos and applications: a review. *Probab. Surv.* **11** (2014), 315–392.

[87] A. Shamov, On Gaussian multiplicative chaos. *J. Funct. Anal.* **270** (2016), no. 9, 3224–3261.

[88] S. Sheffield, Gaussian free fields for mathematicians. *Probab. Theory Related Fields* **139** (2007), no. 3–4, 521–541.

[89] S. Sheffield, Conformal weldings of random surfaces: SLE and the quantum gravity zipper. *Ann. Probab.* **44** (2016), no. 5, 3474–3545.

[90] V. Vargas, Lecture notes on Liouville theory and the DOZZ formula. 2017, arXiv:1712.00829.

[91] Y. Watabiki, Analytic study of fractal structure of quantized surface in two-dimensional quantum gravity. *Progr. Theoret. Phys. Suppl.* **114** (1993), 1–17. Quantum gravity (Kyoto, 1992).

[92] W. Werner and E. Powell, Lecture notes on the Gaussian free field. 2020, arXiv:2004.04720.

JIAN DING

The Wharton School, University of Pennsylvania, 3733 Spruce Street, Philadelphia, PA 19104, USA, dingjian@wharton.upenn.edu

JULIEN DUBÉDAT

Department of Mathematics, Columbia University, Room 509, MC 4406, 2990 Broadway, New York, NY 10027, USA, dubedat@math.columbia.edu

EWAIN GWYNNE

Department of Mathematics, Eckhart Hall, 5734 S University Ave, Chicago, IL, 60637, USA, ewain@uchicago.edu

This Page Intentionally Left Blank

ANALYSIS OF HIGH-DIMENSIONAL DISTRIBUTIONS USING PATHWISE METHODS

RONEN ELDAN

ABSTRACT

The goal of this note is to present an emerging method in the analysis of high-dimensional distributions, which exhibits applications to several mathematical fields, such as functional analysis, convex and discrete geometry, combinatorics, and mathematical physics. The method is based on pathwise analysis: One constructs a stochastic process, driven by Brownian motion, associated with the high-dimensional distribution in hand. Quantities of interest related to the distribution, such as covariance, entropy, and spectral gap, are then expressed via corresponding properties of the stochastic process, such as quadratic variation, making the former tractable through the analysis the latter. We focus on one particular manifestation of this approach, the *Stochastic Localization* process. We review several results which can be obtained using Stochastic Localization and outline the main steps towards their proofs. By doing so, we try to demonstrate some of the ideas and advantages of the pathwise approach. We focus on two types of results relevant to high-dimensional distributions: The first one has to do with dimension-free concentration bounds, manifested by functional inequalities which have no explicit dependence on the dimension. Our main focus in this respect will be on the Kannan–Lovász–Simonovits conjecture, concerning the isoperimetry of high-dimensional log-concave measures. Additionally, we discuss concentration inequalities for Ising models and expansion bounds for complex-analytic sets. The second type of results concern the decomposition of a high-dimensional measure into mixtures of measures attaining a simple structure, with applications to mean-field approximations.

MATHEMATICS SUBJECT CLASSIFICATION 2020

Primary 60-02; Secondary 60H30, 60G17, 60G44, 52A23

KEYWORDS

Concentration of measure, Isoperimetric inequalities, High dimensional phenomena, Mean-field models, Gaussian space

ICM INTERNATIONAL CONGRESS OF MATHEMATICIANS 2022 JULY 6–14

Proc. Int. Cong. Math. 2022, Vol. 6, pp. 4246–4270
DOI 10.4171/ICM2022/61

1. INTRODUCTION

This note is concerned with probability measures on high-dimensional spaces. The intuition derived from low-dimensional examples in various fields such as topology and partial differential equations may suggest that an attempt to understand the behavior of high-dimensional objects is futile, since a system's behavior quickly becomes complex and intractable as the dimension increases.

Nevertheless, a recently emerging theory of "high-dimensional phenomena" reveals that some important classes of distributions turn out to be surprisingly well-behaved (some introductory books on this theory are [2, 32, 43]). We focus on one particular facet of this theory which concerns *dimension-free* phenomena: It is often the case that the behavior of objects of interest is dictated by their marginals onto a fixed number of directions. This is manifested, for example, in the fact that several important functional inequalities have no explicit dependence on the dimension.

An exemplary illustration of this phenomenon is given by the *Gaussian isoperimetric inequality*. Consider the space $\mathbb{R}^n$ equipped with the standard Gaussian measure whose density is

$$\frac{d\gamma_n}{dx} := (2\pi)^{-n/2} \exp\big(-|x|^2/2\big),$$

which we refer to as the Gaussian space. A subset $H \subset \mathbb{R}^n$ is called a *half-space* if it has the form $H = \{x : \langle x, v\rangle \leq b\}$ for some $v \in \mathbb{R}^n$ and $b \in \mathbb{R}$. For $A \subset \mathbb{R}^n$ and $\varepsilon > 0$, we define the ε-extension of A by $A_\varepsilon := \{x \in \mathbb{R}^n : \exists y \in A, \|y - x\|_2 \leq \varepsilon\}$. Moreover, define

$$\Phi(t) := \frac{1}{\sqrt{2\pi}} \int_{-\infty}^{t} e^{-x^2/2} dx,$$

the normal cumulative distribution function. The Gaussian isoperimetric inequality reads,

Theorem 1.1 (Borell, Sudakov–Tsirelson [8,42]). *If $A \subset \mathbb{R}^n$ is a measurable set and $H \subset \mathbb{R}^n$ is a half-space satisfying $\gamma_n(A) = \gamma_n(H)$ then for all $\varepsilon > 0$ we have*

$$\gamma_n(A_\varepsilon) \geq \gamma_n(H_\varepsilon) = \Phi\big(\Phi^{-1}\big(\gamma_n(A)\big) + \varepsilon\big).$$

This theorem highlights an important metaproperty of Gaussian space: The extremizers of functional and geometric inequalities are one-dimensional objects, in the sense that they only depend on one direction. This is, for example, the case with the logarithmic-Sobolev inequality, Ehrhad's inequality, and Talagrand's transportation–entropy inequality (see [31, 32] for details). A recent breakthrough by Milman and Neeman [39] shows that the k-set analog of the isoperimetric inequality is saturated by partitions which only depend on $k-1$ directions.

Is it reasonable to look for larger classes of measures which are Gaussian-like in the sense that they obey similar principles? Product measures are one natural candidate: By considering the harmonics, it is clear that several inequalities, such as the Poincaré inequality, will be saturated by one-dimensional functions. The central limit theorem ensures us that product distributions are Gaussian-like in the sense that, under mild conditions, marginals onto "typical" directions are close to a Gaussian.

In recent years, the class of measures which satisfy a convexity property, called *log-concave* measures, arose as another promising candidate. One remarkable result which supports this is Klartag's central limit theorem for convex sets [27], which asserts that typical one-dimensional marginals of such measures have an approximately normal law. In this note, we discuss the aspects of *isoperimetry* and *concentration of measure* in this class, in search of a counterpart to Theorem 1.1.

Log-concave measures and the Kannan–Lovász–Simonovits conjecture. A measure ν on $\mathbb{R}^n$ is *log-concave* if its density with respect to the Lebesgue measure is of the form $d\nu = e^{-V}dx$ where $V : \mathbb{R}^n \to \mathbb{R} \cup \{\infty\}$ is convex. This class captures, for example, the Gaussian measure, as well as the uniform measure on a convex set.

For a set $A \subset \mathbb{R}^n$, define the surface area measure of A with respect to ν by

$$\nu^+(\partial A) = \limsup_{\varepsilon \to 0+} \frac{1}{\varepsilon} \nu(A_\varepsilon \setminus A). \tag{1.1}$$

(recalling that $A_\varepsilon = \{x \in \mathbb{R}^n : \exists y \in A, \|y - x\|_2 \leq \varepsilon\}$). In analogy with the Gaussian isoperimetric inequality, we would like to obtain a lower bound on $\nu^+(\partial A)$ in terms of $\nu(A)$. This gives rise to the definition

$$\psi_\nu := \inf_{A \subset \mathbb{R}^n} \frac{\nu^+(\partial A)}{\nu(A)(1-\nu(A))},$$

known as the *Cheeger constant* of the measure ν. Up to universal constants, ψ_ν^2 is equivalent to the Neumann spectral gap of ν, see [38]. It plays a central role in the theory of *concentration of measure* phenomena; a lower bound on ψ_ν implies, for example, that a Lipschitz function is typically close to its mean (see [32] and Section 2 below).

Since the problem is not scale invariant, it is hopeless to find a lower bound on ψ_ν that holds uniformly over all log-concave measures. Indeed, by considering the push-forward through the map $x \to \lambda x$ and replacing A by λA, the Cheeger constant scales as $\frac{1}{\lambda}$. We therefore need to assume that the measure is normalized in some way. A natural way to do so is to require that $\mathrm{Cov}(\nu) = \mathrm{Id}$ where $\mathrm{Cov}(\nu)$ is the covariance matrix of ν, defined by $\mathrm{Cov}(\nu)_{i,j} := \mathbb{E}_{X\sim\nu}[X_i X_j]$. A centered measure ν satisfying $\mathrm{Cov}(\nu) = \mathrm{Id}$ is called *isotropic*. It turns out that, as a consequence of the Brunn–Minkowski inequality, this normalization essentially corresponds to the fact that half-spaces satisfy an isoperimetric inequality.

Fact 1.2 (see [30, SECTION 2]). *Let ν by a log-concave measure in $\mathbb{R}^n$. Consider the quantity*

$$\alpha_\nu := \inf_{\substack{H \subset \mathbb{R}^n \\ \text{half-space}}} \frac{\nu^+(\partial H)}{\nu(H)(1-\nu(H))}$$

(the difference from ψ_ν being that the infimum is only taken over half-spaces). Then,

$$\frac{1}{3}\alpha_\nu \leq \|\mathrm{Cov}(\nu)\|_{\mathrm{OP}}^{-1/2} \leq 3\alpha_\nu.$$

We are now ready to state the Kannan–Lovász–Simonivits conjecture.

Conjecture 1.3 (KLS conjecture, [25]). *There exists a universal constant $c > 0$ such that any isotropic, log-concave measure ν on $\mathbb{R}^n$ satisfies $\psi_\nu \geq c$.*

In light of Fact 1.2, the KLS conjecture equivalently asserts that for any log-concave measure,

$$c\alpha_\nu \le \psi_\nu \le \alpha_\nu$$

for a universal constant $c > 0$. In words, up to a constant independent of the measure or the dimension, the isoperimetric minimizer of any (not necessarily isotropic) log-concave measure is a half-space, hence the analogy to Theorem 1.1.

This conjecture has a wide array of implications in high-dimensional convex geometry and computational geometry, see [1, 33] for extensive reviews. Here, we only mention what is perhaps the most important of implications, a conjecture due to Bougain, known as the *hyperplane conjecture* or the *slicing problem*.

Conjecture 1.4. *There is a universal constant $c > 0$ such that, for every n and every convex $K \subset \mathbb{R}^n$ of unit volume, there exists an affine hyperplane H such that*

$$\mathrm{Vol}_{n-1}(K \cap H) > c. \tag{1.2}$$

For a survey on the hyperplane conjecture and other related problems, see [30]. Denote by $\psi_n = \inf_\nu \psi_\nu$ where the infimum is over all isotropic log-concave measures ν on $\mathbb{R}^n$ (so that the KLS conjecture states that $\psi_n \ge c$ for a universal constant $c > 0$), and by L_n the largest (possibly dimension-dependent) constant which can replace the constant c on the right-hand side of (1.2). It was shown by Klartag and the author [20], that $\psi_n \lesssim L_n$. In particular, Bourgain's hyperplane conjecture is implied by the KLS conjecture (see also [5]).

Let us briefly review some of the history around the KLS and hyperplane conjectures. In their original work, Kannan, Lovász, and Simonovits showed that $\psi_n \gtrsim n^{-1/2}$. The exponent $1/2$ was improved in several consecutive works, by Klartag [27] (relying on Bobkov [7]), by the author [16] (relying on Guédon–Milman [24]), and by Lee and Vempala [34], obtaining $\psi_n \gtrsim n^{-1/4}$. Regarding the hyperplane conjecture, Bourgain [12] showed that $L_n \gtrsim n^{-1/4}\log(n)^{-1}$. Until recently, the only improvement of this bound was $L_n \gtrsim n^{-1/4}$, due to Klartag [26].

A recent breakthrough by Chen makes a very significant improvement upon these bounds, nearly proving both conjectures.

Theorem 1.5 (Chen, [15]). *One has $\psi_n = n^{-o(1)}$. As a corollary, $L_n = n^{-o(1)}$.*

The pathwise approach and the Stochastic Localization scheme. Chen's proof is based on the so-called *Stochastic Localization* scheme, introduced in [16] and described in detail below. This scheme is one example of a more general metatechnique which we call pathwise analysis. In recent years, this metatechnique was proven useful in obtaining a variety of results that have to do with the analysis of high-dimensional distributions. The goal of this note is to highlight the main ideas behind it and review several applications thereof.

The use of ideas from diffusion and heat-flow to concentration inequalities dates back at least to the 1960s and to the seminal works of Nelson and Gross, which introduced the hypercontractivity property of heat semigroups and derived the log-Sobolev inequality for Gaussian space, respectively. In the following decades, heat flow (or semigroup) techniques were realized to be a very powerful tool in proving concentration inequalities. These are, for example, the main ingredients in the celebrated Bakry–Emery theory [3]. These ideas rely on differentiation formulas for the heat semigroup which can alternatively be obtained via pathwise integration along the corresponding diffusion process.

The pathwise approach takes one more step and inspects the behavior of the process along a single path; it turns out that, when averaging over paths, quite a bit of information is lost, which can otherwise be revealed by using stochastic calculus. For example, bounds on the spectral gap and mixing times of diffusion processes can be obtained by coupling the paths of two diffusion processes. Some examples of works which manage to prove new bounds by direct analysis of the diffusion process are [10,11,13,36]. In this note we focus on a seemingly new type of pathwise proofs where, rather than considering the path of a diffusion process, one constructs an *evolution on the space of measures*, driven by Brownian motion, associated with a given distribution.

Structure of the paper. In what follows, in order to give an initial glimpse into pathwise techniques, in Section 2 we begin with a warm-up where we prove a concentration inequality for Lipschitz functions on Gaussian space using stochastic calculus. Then, in Section 3 we prove a generalization of the Gaussian isoperimetric inequality, due to Borell. In Section 4 we introduce the Stochastic Localization process and discuss the main ideas used in obtaining bounds for the KLS conjecture. Finally, in Section 5 we outline several other applications of Stochastic Localization towards (i) expansion bounds for complex-analytic sets, (ii) concentration inequalities for Ising models, and (iii) structure theorems which represent measures on the discrete hypercube as mixtures of product-like components.

2. A FIRST TASTE OF PATHWISE ANALYSIS: CONCENTRATION OF LIPSCHITZ FUNCTIONS IN GAUSSIAN SPACE

A useful property of Gaussian space, due to Maurey and Pisier, is the fact that Lipchitz functions have a sub-Gaussian tail:

Fact 2.1. *For any 1-Lipschitz function $f : \mathbb{R}^n \to \mathbb{R}$, we have*

$$\gamma_n\left(\left\{x : \left|f(x) - \int f d\gamma_n\right| > \alpha\right\}\right) \le 4\big(1 - \Phi(\alpha)\big) \le 2e^{-\alpha^2/2}, \quad \forall \alpha > 0.$$

This type of behavior is often referred to as *concentration of measure*. Put forth by V. Milman, such bounds have far-reaching applications and the behavior of this type is a cornerstone in the theory of high-dimensional phenomena (see, e.g., [41]). In this warm-up section, we provide a proof of this fact which will highlight some of the advantages of pathwise analysis. We assume that the reader has some familiarity with basic concepts in stochastic calculus.

Throughout the section, we fix a measurable function $f : \mathbb{R}^n \to \mathbb{R}$. Let $(B_t)_{t\ge 0}$ be a standard Brownian motion on $\mathbb{R}^n$; recall that B_1 has law γ_n. Consider the Doob martingale

$$M_t := \mathbb{E}\big[f(B_1) \mid B_t\big].$$

For a function $g : \mathbb{R}^n \to \mathbb{R}$, define $P_t[g](x) := \int_{\mathbb{R}^n} g(x + \sqrt{t}y)\gamma_n(dy)$. Since the law of B_1, conditioned on B_t, is $\mathcal{N}(B_t, (1-t)\mathrm{Id})$, we have

$$M_t = P_{1-t}[f](B_t). \tag{2.1}$$

Recall that, given a stochastic process $(X_s)_{s\geq 0}$, its *quadratic variation* is defined as

$$[X]_t := \lim_{\|\mathcal{P}=(t_0=0,t_1,\ldots,t_n=t)\|\to 0} \sum_{k=1}^{n} (X_{t_k} - X_{t_{k-1}})^2,$$

where $\|\mathcal{P}\|$ denotes the mesh of the partition. Ito's isometry tells us that

$$\mathrm{Var}_{\gamma_n}[f] = \mathrm{Var}\big[f(B_1)\big] = \mathbb{E}[M]_1$$

(this is just the continuous version of the fact that for a discrete time martingale $X_0, X_1, \ldots$, one has that $\mathrm{Var}[X_t] = \sum_{i=1}^{t} \mathbb{E}(X_i - X_{i-1})^2$).

In order to obtain a bound on $[M]_t$, using Itô's formula, we calculate

$$\begin{aligned} dM_t &\overset{(2.1)}{=} d\big(P_{1-t}[f](B_t)\big) \\ &= \big\langle \nabla P_{1-t}[f](B_t), dB_t\big\rangle + \frac{\partial}{\partial t} P_{1-t}[f](B_t)dt + \frac{1}{2}\Delta P_{1-t}[f](B_t)dt \\ &= \big\langle \nabla P_{1-t}[f](B_t), dB_t\big\rangle, \end{aligned}$$

where we used the identity $\frac{d}{ds} P_s[f] = \frac{1}{2}\Delta P_s[f]$. It follows that

$$\frac{d}{dt}[M]_t = \big\|\nabla P_{1-t}[f](B_t)\big\|_2^2 = \|V_t\|_2^2, \tag{2.2}$$

where $V_t := \nabla P_{1-t}[f](B_t)$. Since the operators ∇ and P_t commute, we also have

$$V_t = \mathbb{E}\big[\nabla f(B_1) \mid B_t\big],$$

which teaches us that V_t is a martingale. By the convexity of $\|\cdot\|_2^2$, we have that $\|V_t\|^2$ is a submartingale. We conclude that

$$\mathrm{Var}_{\gamma_n}[f] = \mathbb{E}[M]_1 = \int_0^1 \mathbb{E}\|V_s\|_2^2 ds \leq \mathbb{E}\|V_1\|_2^2 = \mathbb{E}_{\gamma_n}\|\nabla f\|_2^2.$$

This is precisely the Poincaré inequality. Alternatively, it can be easily proven using spectral methods. Moreover, instead of using Brownian motion, we could essentially repeat the argument directly using the semigroup P_t. Writing

$$\int f^2 d\gamma_n - \left(\int f d\gamma_n\right)^2 = P_1[f^2](0) - P_1[f](0)^2 = \int_0^1 \left(\frac{d}{dt} P_t\big[(P_{1-t}[f])^2\big](0)\right)dt,$$

a simple calculation using integration by parts gives that

$$\frac{d}{dt} P_t\big[(P_{1-t}[f])^2\big](0) = P_t\big[\|\nabla P_{1-t}[f]\|_2^2\big](0),$$

and an application of Jensen's inequality yields the Poincaré inequality.

Thus, in what we have seen so far, the "pathwise" aspect merely provides different viewpoint on a proof that can be carried out via elementary calculus. To see where it has a real advantage, let us now assume that the function f is 1-Lipschitz. Under this assumption, with the help of Jensen's inequality, we learn that

$$\|V_t\|_2^2 = \left\|\nabla P_{1-t}[f](B_t)\right\|_2^2 \le P_{1-t}\left[\|\nabla f\|_2^2\right](B_t) \le 1, \quad \forall 0 \le t \le 1.$$

By equation (2.2), we have that $[M]_1 \le 1$ almost surely. Since M_1 has the same law as the push-forward of γ_n under f, Fact 2.1 follows as an immediate corollary of the following:

Proposition 2.2. *Let $(M_t)_{0 \le t \le 1}$ be a martingale satisfying $[M]_1 \le 1$ almost surely. Then,*

$$\mathbb{P}\left(|M_1 - M_0| > \alpha\right) \le 4\left(1 - \Phi(\alpha)\right), \quad \forall \alpha > 0. \tag{2.3}$$

The key to the proof of this proposition is the Dambis/Dubins–Schwartz theorem which, roughly speaking, asserts that every continuous martingale can be represented as a time-changed Brownian motion. More formally, if M_t is a continuous martingale adapted to a filtration $\mathcal{F}_t$, then one can define a process $(W_t)_{t \ge 0}$ and a filtration $(\tilde{F}_t)_t$ over the same underlying probability space, such that:

(i) W_t is a Brownian motion with respect to the filtration $\tilde{F}_t$.

(ii) One has $M_t - M_0 = W_{[M]_t}$ and $\mathcal{F}_t = \tilde{\mathcal{F}}_{[M]_t}$ for all $t \ge 0$.

Next, we claim that $\tau := [M]_1$ is an $\tilde{\mathcal{F}}_t$-stopping time. Indeed, the claim that for all t, the event $\{\tau \le t\}$ is $\tilde{\mathcal{F}}_t$-measurable is equivalent to the claim that for all t the event $\{\tau \le [M]_t\}$ is $\mathcal{F}_t$-measurable, which is evident. Note that, by assumption, we have $\tau \le 1$ almost surely.

We finally conclude the following: There exists a Brownian motion W_t adapted to a filtration $\tilde{\mathcal{F}}_t$ and an $\tilde{\mathcal{F}}_t$-stopping time τ such that $\tau \le 1$ almost surely and such that W_τ is equal in law to $M_1 - M_0$. At this point we can write

$$\mathbb{P}(M_1 - M_0 \ge \alpha) = \mathbb{P}(W_\tau \ge \alpha) \le \mathbb{P}\left(\exists t \in [0,1] \text{ such that } W_t \ge \alpha\right).$$

The proof of Proposition 2.2 is now concluded via the following "reflection principle."

Fact 2.3. *Let W_t be a standard Brownian motion. Then, for all $\alpha > 0$, we have*

$$\mathbb{P}\left(\exists t \in [0,1] \text{ such that } W_t \ge \alpha\right) = 2\mathbb{P}(W_1 \ge \alpha) = 2\left(1 - \Phi(\alpha)\right).$$

Proof. (sketch) Consider the stopping time $\tau = \inf\{t; W_t = \alpha\}$. Since, conditioned on $\tau \le 1$, we have that $W_1 - W_\tau$ has a symmetric law, we have $\mathbb{P}(W_1 \ge W_\tau | \tau \le 1) = \frac{1}{2}$. ■

Fact 2.1 may be alternatively proven by combining the Gaussian isoperimetric inequality and the coarea formula, or by a direct coupling argument (see [41, THEOREM 2.2]). Nevertheless, the above proof highlights the advantage in considering the martingale M_t in a "path-by-path" manner, and the reason that this approach can reveal dimension-free phenomena: The process V_t extracts the "important" directions, in which the function f varies, and the law of $f(B_1)$ eventually only depends on the behavior of the *one-dimensional* process M_t. The reduction of the analysis of an n-dimensional function, or measure, to the behavior one-dimensional process will be a recurring motif later on.

3. THE GAUSSIAN ISOPERIMETRIC INEQUALITY AND NOISE-SENSITIVITY

As a next step towards demonstrating the pathwise technique, we provide a proof of the Gaussian isoperimetric inequality, Theorem 1.1. In fact, we prove a stronger statement, known as Borel's noise stability inequality [9].

Let B_t be a standard Brownian motion in $\mathbb{R}^n$, adapted to a filtration $\mathcal{F}_t$. Define $Z_t := \int_0^t e^{-s/2} dB_s$. Observe that $Z_\infty := \lim_{t\to\infty} Z_t$ has the law γ_n, since $\int_0^\infty e^{-t} dt = 1$. For all measurable $A \subset \mathbb{R}^n$ and $t > 0$, define

$$\mathrm{Sens}_t(A) := \mathbb{E}\big[\mathbb{P}(Z_\infty \in A \mid Z_t)\mathbb{P}(Z_\infty \notin A \mid Z_t)\big],$$

referred to as the t-noise sensitivity of A. From an analytic point of view, this quantity can be understood as the rate at which heat escapes the set A under the heat flow on Gaussian space, defined by the Ornstein–Uhlenbeck operator $\mathcal{L} = \Delta - x \cdot \nabla$.

A standard argument shows that noise-sensitivity is related to isoperimetry by

$$\gamma_n^+(\partial A) = \lim_{t\to 0} \frac{\mathrm{Sens}_t(A)}{\sqrt{t}}, \tag{3.1}$$

which holds, for example, under the assumption that A has finite perimeter.

Theorem 3.1 (Borell [9]). *If $A \subset \mathbb{R}^n$ is a measurable set and $H \subset \mathbb{R}^n$ is a half-space satisfying $\gamma_n(A) = \gamma_n(H)$, then for all $t \geq 0$,*

$$\mathrm{Sens}_t(A) \geq \mathrm{Sens}_t(H).$$

This theorem has far-reaching applications in statistics and theoretical computer science which we do not discuss here, but we refer the reader to [17,40] and references therein. Combined with equation (3.1), Theorem 1.1 follows as a corollary.

Towards proving Theorem 3.1, define for a set $A \subset \mathbb{R}^n$,

$$b(A) := \int_A x \gamma_n(dx),$$

the Gaussian first-moment of A. Moreover, we define for $s \in \mathbb{R}$,

$$q(s) = \int_{\Phi^{-1}(s)}^{\infty} t \gamma_1(dt).$$

Evidently, if $H \subset \mathbb{R}^n$ is a half-space then one has $\|b(H)\|_2 = q(\gamma_n(H))$. At the center of our proof lies the following simple fact.

Fact 3.2 (Level-1 inequality). *For any measurable $A \subset \mathbb{R}^n$,*

$$\big\|b(A)\big\|_2 \leq q\big(\gamma_n(A)\big), \tag{3.2}$$

with equality when A is a half-space.

This fact is referred to as the *level-1 inequality* since it characterizes the sets which maximize the L_2-energy on the first-order Hermite expansion. It constitutes the only inequality in the proof to come.

Proof. Set $\theta = \frac{b(A)}{\|b(A)\|_2}$. Let H be a half-space of the form $H = \{x; \langle x, \theta \rangle \geq \alpha\}$ with α chosen so that $\gamma_n(H) = \gamma_n(A)$. Note that, by definition,

$$q(\gamma_n(A)) = \left\| \int_H x \gamma_n(dx) \right\|_2 = \int_H \langle x, \theta \rangle \gamma_n(dx),$$

so we only need to show that

$$\int_H \langle x, \theta \rangle \gamma_n(dx) \geq \int_A \langle x, \theta \rangle \gamma_n(dx).$$

Since $\gamma_n(A) = \gamma_n(H)$, we may subtract α from both integrands, thus the above is equivalent to

$$\int_{\mathbb{R}^n} (\langle x, \theta \rangle - \alpha)(\mathbf{1}_{\langle x, \theta \rangle \geq \alpha} - \mathbf{1}_{x \in A}) \gamma_n(dx) \geq 0,$$

which is evident. ■

Define μ_t to be the law of Z_∞ conditioned on Z_t, which easily checked to be $\mathcal{N}(Z_t, e^{-t/2}\mathrm{Id})$, or in other words,

$$\mu_t(dx) = (2\pi)^{-n/2} e^{nt/2} \exp\left(-\frac{1}{2} e^t |x - Z_t|^2\right) dx.$$

Set $M_t := \mu_t(A) = \mathbb{P}(Z_\infty \in A \mid Z_t)$. Note that, by definition,

$$\mathrm{Sens}_t(A) = \mathbb{E}\big[M_t(1 - M_t)\big] = M_0(1 - M_0) - \mathrm{Var}[M_t]. \tag{3.3}$$

Consider a half-space H satisfying $\gamma_n(H) = \gamma_n(A)$ and, analogously, define $N_t = \mu_t(H)$. Equation (3.3) tells us that the statement of Theorem 3.1 is equivalent to the assertion that

$$\mathrm{Var}[M_t] \leq \mathrm{Var}[N_t], \quad \forall t > 0. \tag{3.4}$$

In order to compare the variances of the two processes, we first calculate the corresponding quadratic variations. Using Itô's formula, we write

$$\begin{aligned} dM_t &= d \int_A (2\pi)^{-n/2} e^{nt/2} \exp\left(-\frac{1}{2} e^t \|x - Z_t\|_2^2\right) dx \\ &= e^t \int_A \langle x - Z_t, dZ_t \rangle \mu_t(dx) \\ &= \big\langle b(A_t), dB_t \big\rangle, \end{aligned} \tag{3.5}$$

where $A_t := e^{t/2}(A - Z_t)$. Observing that $M_t = \mu_t(A) = \gamma_n(A_t)$, with the help of (3.2), we arrive at the inequality

$$\frac{d}{dt}[M]_t = \big\|b(A_t)\big\|_2^2 \leq q(M_t)^2. \tag{3.6}$$

Defining $H_t := e^{t/2}(H - Z_t)$, a similar calculation shows

$$\frac{d}{dt}[N]_t = \big\|b(H_t)\big\|_2^2 = q(N_t)^2. \tag{3.7}$$

On an intuitive level, equations (3.6) and (3.7) tell us that, in a certain sense, the martingale N_t is moving faster than M_t. Naively, we might hope that the above implies that $\mathbb{E}\frac{d}{dt}[N]_t \geq \mathbb{E}\frac{d}{dt}[M]_t$ for all t, which would conclude (3.4). This is not true, however.

Observe that $\mathbb{E}[N]_\infty = \mathrm{Var}[N_\infty] = \mathrm{Var}[M_\infty] = \mathbb{E}[M]_\infty$. The following lemma extracts the power of the pathwise approach. We can couple the two processes in a way that gives us the desired domination.

Lemma 3.3. *Let $v : \mathbb{R} \to [0, \infty)$ be a continuous function. Let $(M_t)_{t=0}^{\infty}$, $(N_t)_{t=0}^{\infty}$ be two continuous real-valued martingales such that $M_0 = N_0$, and such that*

$$\frac{d}{dt}[N]_t = v(N_t) \quad \text{and} \quad \frac{d}{dt}[M]_t \le v(M_t), \tag{3.8}$$

almost surely, for all $t \ge 0$. Then for all $t \ge 0$, one has

$$\mathrm{Var}[M_t] \le \mathrm{Var}[N_t].$$

Since equations (3.6) and (3.7) verify (3.8), an application of the above lemma yields (3.4), which concludes the proof of Theorem 3.1. It therefore only remains to prove this lemma.

Proof of Lemma 3.3 (sketch). Without loss of generality, assume $M_0 = N_0 = 0$. By the Dambis/Dubins–Schwartz theorem, there exist standard Brownian motions $B_t, \tilde{B}_t$ such that $N_t = B_{[N]_t}$ and $M_t = \tilde{B}_{[M]_t}$. By a standard disintegration theorem, the processes maybe defined on the same probability space in a way that $B_t = \tilde{B}_t$. In other words, there exist two martingales X_t, Y_t and a standard Brownian motion B_t, defined over the same probability space, such that X_t, Y_t have the same laws as N_t, M_t, respectively, and such that

$$X_t = B_{[X]_t} \text{ and } Y_t = B_{[Y]_t}, \quad \forall t \ge 0.$$

Let τ_X, τ_Y be the inverse functions of $[X]_t, [Y]_t$, respectively. Then the last display implies $X_{\tau_X(T)} = Y_{\tau_Y(T)} = B_T$, and by formula (3.8) we have

$$\frac{d}{dt}[X]_t|_{t=\tau_X(T)} = v(B_T) \ge \frac{d}{dt}[Y]_t|_{t=\tau_Y(T)}, \quad \forall T \ge 0,$$

which implies that $[X]_t \ge [Y]_t$ for all $t \ge 0$. By Itô's isometry, the lemma follows. ■

4. STOCHASTIC LOCALIZATION AND THE KLS CONJECTURE

In this section we introduce the main technique discussed in this note, the Stochastic Localization process, and demonstrate how it can be used to produce lower bounds on the Cheeger constant of a log-concave measure.

4.1. Construction of the process and basic properties

Let B_t be a standard Brownian motion in $\mathbb{R}^n$, adapted to a filtration $\mathcal{F}_t$. As in the previous section, define $Z_t := \int_0^t e^{-s/2} dB_s$ and let μ_t be defined as the law of Z_∞ conditioned on Z_t. The measure-valued process (μ_t) interpolates between the standard Gaussian measure, at time 0, and a Dirac measure at time ∞. A key formula in the previous section was (3.5), which can be restated as follows: Setting $p_t(x) := \frac{\mu_t(dx)}{dx}$, we have

$$\forall x \in \mathbb{R}^n, \quad dp_t(x) = e^{t/2} p_t(x) \left\langle x - \int x p_t(x) dx, dB_t \right\rangle. \tag{4.1}$$

Now, let ν be an arbitrary probability measure on $\mathbb{R}^n$. We would like to consider a similar evolution with ν taking the place of the Gaussian measure. Suppose that $(C_t)_{t\geq 0}$ is a stochastic process adapted to $\mathcal{F}_t$, such that for all t, C_t is an $n \times n$ positive semidefinite matrix. Inspired by (4.1), consider the system of stochastic differential equations

$$\forall x \in \mathbb{R}^n, \quad F_0(x) = 1, \quad dF_t(x) = F_t(x)\langle x - a_t, C_t dB_t\rangle, \tag{4.2}$$

where

$$a_t := \int x F_t(x)\nu(dx).$$

We can now define a measure-valued process, $(\nu_t)_{t\geq 0}$, by $\nu_t(dx) = F_t(x)\nu(dx)$. Note that $\nu_0 = \nu$. The choice $\nu = \gamma_n$ and $C_t = e^{t/2}\mathrm{Id}$ recovers the evolution defined by (4.1), so the process ν_t can be thought of as a generalization of μ_t.

We remark that the system (4.2) is an infinite system of stochastic differential equations, but as we will see below, it may instead be written as a finite system. Its existence and uniqueness is proven in [16]. Informally, we can think of equation (4.2) as

$$F_{t+dt}(x) = F_t(x)\big(1 + \langle x - a_t, \mathcal{N}(0, C_t^2 dt)\rangle\big),$$

so that process can be understood as a continuous version the following iterative procedure: Start with a density on $\mathbb{R}^n$, and at each iteration multiply this density by a linear function, which is equal to 1 at the center of mass of ν_t, and whose gradient is distributed according to an infinitesimal Gaussian.

Before we continue, let us point out several basic properties of this process. First, using Itô's formula, we calculate

$$d \log F_t(x) = \frac{dF_t(x)}{F_t(X)} - \frac{d[F(x)]_t}{2F_t(x)^2} \overset{(4.2)}{=} \langle x - a_t, C_t dB_t\rangle - \frac{1}{2}\big\|C_t(x - a_t)\big\|_2^2 dt.$$

Consequently, the measure ν_t attains the form

$$\nu_t(dx) = \exp\left(z_t + \langle v_t, x\rangle - \frac{1}{2}\langle G_t x, x\rangle\right)\nu(dx), \tag{4.3}$$

with $G_t := \int_0^t C_s^2 ds$, where $v_t \in \mathbb{R}^n$ is an Itô process adapted to $\mathcal{F}_t$ and z_t is a normalizing constant. In particular, if we choose $C_t = \mathrm{Id}$ for all t, we have

$$\nu_t(dx) = \exp\left(z_t + \langle v_t, x\rangle - \frac{t}{2}\|x\|_2^2\right)\nu(dx). \tag{4.4}$$

Next, we calculate

$$d\nu_t(\mathbb{R}^n) = \left\langle \int_{\mathbb{R}^n} C_t(x - a_t)F_t(x)\nu(dx), dB_t\right\rangle = 0.$$

Equation (4.3) shows that $F_t(x)$ is positive for all x and t, so we conclude that ν_t is almost surely a probability measure for all t and that a_t is its center of mass. Finally, it is evident from (4.2) that $F_t(x)$ is a martingale for every x, which immediately gives the following.

Fact 4.1. *For every measurable $W \subset \mathbb{R}^n$, the process $\nu_t(W)$ is a martingale.*

4.2. Isoperimetry for log-concave measures using Stochastic Localization

Fix a log-concave measure ν on $\mathbb{R}^n$ and a measurable set $A \subset \mathbb{R}^n$. We would like to use the process constructed above in order to produce a lower bound on $\nu^+(\partial A)$. Consider the process $(\nu_t)_t$ defined in (4.2) with the choice $C_t = \mathrm{Id}$.

Applying Fact 4.1 to the set $A_\varepsilon \setminus A$ gives

$$\nu^+(\partial A) = \mathbb{E}\nu_t^+(\partial A). \tag{4.5}$$

To continue the analogy with Section 3, we consider the martingale $M_t := \nu_t(A)$. Recall that the proof of the Gaussian isoperimetric inequality amounted to obtaining an upper bound on $\frac{d}{dt}[M]_t$, in terms of M_t, for all $t \geq 0$. In our case, we will only be able to establish such a bound for small enough values of t. This will be complemented by the fact that for large t, the measure ν_t satisfies an isoperimetric inequality, a consequence of the following:

Theorem 4.2 (Bakry–Ledoux [4]; see also [33, THEOREM 25]). *Let μ be a probability measure on $\mathbb{R}^n$ whose density is of the form*

$$d\mu(x) = e^{-V(x)-\frac{\alpha}{2}\|x\|_2^2}dx \tag{4.6}$$

where $V(x) : \mathbb{R}^n \to \mathbb{R} \cup \{\infty\}$ is convex and $\alpha > 0$. Then for all $A \subset \mathbb{R}^n$, we have

$$\mu^+(\partial A) \geq \sqrt{\alpha}\mu(A)\big(1 - \mu(A)\big). \tag{4.7}$$

We sketch an alternative proof of this theorem in Section 5.1. In light of (4.4), we may apply the theorem to the measure ν_t with $\alpha = t$, which yields

$$\begin{aligned} \nu^+(\partial A) &\overset{(4.5)}{=} \mathbb{E}\nu_t^+(\partial A) \\ &\overset{(4.4)+(4.7)}{\geq} \mathbb{E}\big[\sqrt{t}\nu_t(A)\big(1 - \nu_t(A)\big)\big] \\ &= \sqrt{t}\big(M_0(1 - M_0) - \mathrm{Var}[M_t]\big). \end{aligned} \tag{4.8}$$

As in Section 3, our goal is once again to bound from above the quantity $\mathrm{Var}[M_t] = \mathbb{E}[M]_t$. To this end, we calculate

$$dM_t = d\int_A F_t(x)\nu(dx) \overset{(4.2)}{=} \int_A \langle x - a_t, dB_t\rangle \nu_t(dx), \tag{4.9}$$

implying that

$$\frac{d}{dt}[M]_t = \left\| \int_A (x - a_t)\nu_t(dx) \right\|_2^2. \tag{4.10}$$

The right-hand side can be bounded with the help of the following simple lemma.

Lemma 4.3. *For every probability measure μ on $\mathbb{R}^n$ and every measurable $A \subset \mathbb{R}^n$,*

$$\left\| \int_A \left(x - \int_{\mathbb{R}^n} x\mu(dx)\right)\mu(dx) \right\|_2^2 \leq \|\mathrm{Cov}(\mu)\|_{\mathrm{OP}}. \tag{4.11}$$

Proof. Define $\theta := \frac{\int_A x\mu(dx)}{\|\int_A x\mu(dx)\|_2}$ (if the denominator vanishes, there is nothing to show). Also, without loss of generality assume $\int_{\mathbb{R}^n} x\mu(dx) = 0$. Then we have

$$\left\| \int_A x\mu(dx) \right\|_2^2 = \left(\int_{\mathbb{R}^n} \langle x, \theta \rangle \mu(dx) \right)^2$$
$$\leq \int_{\mathbb{R}^n} \langle x, \theta \rangle^2 \mu(dx) = \langle \theta, \mathrm{Cov}(\mu)\theta \rangle \leq \left\| \mathrm{Cov}(\mu) \right\|_{\mathrm{OP}}. \qquad \blacksquare$$

In the process μ_t considered in Section 3 (which corresponds to ν_t, only with the measure ν replaced by the Gaussian measure), the matrix $\mathrm{Cov}(\mu_t)$ was deterministic. The crucial difference here is that we have to account for $\|\mathrm{Cov}(\nu_t)\|_{\mathrm{OP}}$.

Combining (4.10) and (4.11), we have

$$\mathrm{Var}[M_t] = \mathbb{E}[M]_t \leq \int_0^t \left\| \mathrm{Cov}(\nu_s) \right\|_{\mathrm{OP}} ds.$$

Together with (4.8), the state of events can be concluded by the following proposition.

Proposition 4.4. *Let ν by a log-concave measure on $\mathbb{R}^n$. Construct the process $(\nu_t)_{t=0}^{\infty}$ using equation (4.2). Suppose that for some $t, \alpha > 0$, one has*

$$\mathbb{E}\left[\int_0^t \left\| \mathrm{Cov}(\nu_s) \right\|_{\mathrm{OP}} ds \right] \leq \alpha. \tag{4.12}$$

Then, for every $A \subset \mathbb{R}^n$ such that $\nu(A)(1-\nu(A)) \geq 2\alpha$, we have the Cheeger-type inequality

$$\nu^+(\partial A) \geq \frac{1}{2}\sqrt{t}\nu(A)\big(1-\nu(A)\big).$$

The condition $\nu(A)(1-\nu(A)) \geq 2\alpha$ is not crucial; one can show that it may, in fact, be ignored (see [38, THEOREM 1.8]), so that a bound of the form (4.12) implies $\psi_\nu \geq \frac{1}{2}\sqrt{t}$. Our goal will therefore be to establish (4.12).

Note that condition (4.12) does not involve the set A at all. In that sense, we have managed to reduce a statement with a quantifier "for every $A \subset \mathbb{R}^n$," to a bound which involves only the measure ν. We now need to produce an upper bound on $\|\mathrm{Cov}(\nu_s)\|_{\mathrm{OP}}$. The process $\mathrm{Cov}(\nu_s)$ becomes tractable thanks to a "moment-generating" property described in the next subsection.

Before we proceed, let us note the trade-off between two conflicting goals, namely controlling from above the variance of $\nu_t(A)$ (which corresponds to taking t small enough), and controlling from below the "uniform concavity" term in $\log \frac{d\nu_t}{d\nu}$ (which corresponds to taking t large enough). It is reasonable to expect that a clever choice of the matrix C_t could be fruitful: For a general choice of C_t, equation (4.9) becomes

$$d\nu_t(A) = \langle C_t b_t(A), dB_t \rangle, \quad \text{where } b_t(A) := \int_A (x - a_t)\nu_t(dx). \tag{4.13}$$

Equations (4.3) and (4.13) suggest that the choice of the driving matrix C_t allows a more intricate control of this trade-off. On the one hand, by taking C_t to be small in the direction of b_t, we gain more control of the variance of $\nu_t(A)$, but, on the other hand, we would like the matrix $\int_0^t C_s^2 ds$ to be large. In the context of the KLS conjecture, it is not known if this strategy can produce better bounds, however, in Section 5 we give several examples which crucially rely on a careful choice of C_t.

4.2.1. Stochastic Localization as a moment-generating process

Recall that a_t is the center of mass of ν_t. A calculation shows that

$$\begin{aligned} da_t &= d\int_{\mathbb{R}^n} x\nu_t(dx) \\ &\overset{(4.2)}{=} \int_{\mathbb{R}^n} x\langle x - a_t, C_t dB_t\rangle F_t(x)\nu(dx) \\ &= \left(\int_{\mathbb{R}^n} x \otimes (x - a_t)\nu_t(dx)\right) C_t dB_t = \mathrm{Cov}(\nu_t) C_t dB_t. \end{aligned} \tag{4.14}$$

In words, the time-differential of the first cumulant of ν_t is equal to its second cumulant multiplied by the generating increment $C_t dB_t$. A calculation of similar spirit gives

$$d\,\mathrm{Cov}(\nu_t) = \mathcal{M}^{(3)}[\nu_t] C_t dB_t - \mathrm{Cov}(\nu_t) C_t^2 \,\mathrm{Cov}(\nu_t) dt, \tag{4.15}$$

where

$$\mathcal{M}^{(k)}[\nu_t] := \int (x - a_t)^{\otimes k} \nu_t(dx)$$

is the kth moment tensor of ν_t. In general, the time differential of $\mathcal{M}^{(k)}[\nu_t]$ will involve the term $\mathcal{M}^{(k+1)}[\nu_t] C_t dB_t$.

This property is reminiscent of the *logarithmic Laplace* transform, where derivatives with respect to the space parameter correspond to cumulants of a tilted measure. This fact has far-reaching applications in asymptotic geometric analysis, notably it has been used in several works of Klartag, in particular in his breakthrough on the slicing problem [26].

4.2.2. Obtaining a bound for the KLS conjecture

We now give an overview of the next steps needed to obtain a bound for the KLS conjecture. Going back to Proposition 4.4, such a bound is reduced to obtaining upper bounds on the growth of $\|\mathrm{Cov}(\nu_t)\|_{\mathrm{OP}}$. According to equation (4.15), the expression for the differential of $\mathrm{Cov}(\nu_t)$ involves the process $\mathcal{M}^{(3)}[\nu_t]$.

First let us consider a simple (but somewhat wasteful) way to obtain a bound for $\|\mathrm{Cov}(\nu_t)\|_{\mathrm{OP}}$, based on the fact that

$$\left\|\mathrm{Cov}(\nu_t)\right\|_{\mathrm{OP}} \le \mathrm{Tr}\big(\mathrm{Cov}(\nu_t)^2\big).$$

Equation (4.15) combined with Itô's formula gives

$$\frac{d}{dt}\mathbb{E}\,\mathrm{Tr}\big(\mathrm{Cov}(\nu_t)^2\big) \le \mathbb{E}\left\|\mathcal{M}^{(3)}[\nu_t]\right\|_{\mathrm{HS}}^2. \tag{4.16}$$

The quantity on the right-hand side involves third moments of the measure ν_t. On a conceptual level, at this point, the state of events is that we have the implications:

$$\text{Upper bound on } \mathcal{M}^{(3)}(\nu_t) \Rightarrow \text{Upper bound on } \left\|\mathrm{Cov}(\nu_t)\right\|_{\mathrm{OP}} \Rightarrow \text{Lower bound on } \psi_\nu. \tag{4.17}$$

One way to continue from here would be to look for bounds on $\mathcal{M}^{(3)}(\mu)$ in terms of $\mathrm{Cov}(\mu)$ which hold universally over all log-concave measures μ on $\mathbb{R}^n$. This would imply that the rate of growth of $\mathrm{Cov}(\nu_t)$ is bounded by $\mathrm{Cov}(\nu_t)$ itself. Following this route, the work [16] used a priori estimates on the "thin-shell" constant, defined as

$$\sigma_n := \sup_{\mu} \mathrm{Var}_{X\sim\mu} \|X\|_2,$$

where the supremum runs over all isotropic, log-concave probability measures μ on $\mathbb{R}^n$. Upper bounds on σ_n imply the type of bounds on $\mathcal{M}^{(3)}[\nu_t]$ which, when plugged into the implications above, give a reduction, up to a logarithmic factor, from the KLS conjecture to a weaker conjecture called the *variance conjecture* stating that $\sigma_n = O(1)$ (see [1, 16]).

Later on, Lee and Vempala ([35, LEMMA 33]) realized that, when taking the driving matrix C_t to be the identity, one could instead use the bound

$$\mathbb{E}\big\|\mathcal{M}^{(3)}[\mu]\big\|_{\mathrm{HS}}^2 \lesssim \mathrm{Tr}\big(\mathrm{Cov}(\mu)^2\big)^{3/2} \tag{4.18}$$

which holds uniformly for all log-concave measures. Together with equation (4.16) and an application of Gronwall's inequality, this gives that $\mathbb{E}\,\mathrm{Tr}(\mathrm{Cov}(\nu_t)^2) = O(n)$ for all $t \le \frac{1}{\sqrt{n}}$. Plugging this into Proposition 4.4 yields the bound $\psi_\nu \gtrsim n^{-1/4}$.

Let us now briefly discuss the additional steps needed to produce Chen's bound, $\psi_\nu = n^{-o(1)}$. First of all, by rather direct arguments, one can reverse implication (4.17) in the sense that

$$\begin{matrix}\text{Lower bound on } \psi_\mu \\ \text{for all } \mu \text{ isotropic, log-concave}\end{matrix} \quad\Rightarrow\quad \begin{matrix}\text{Improved upper bounds on } \mathcal{M}^{(3)}[\mu] \\ \text{in terms of } \mathrm{Cov}(\mu).\end{matrix} \tag{4.19}$$

In other words, if we have a priori bounds for the KLS conjecture we can improve, in some sense, on (4.18). Lee and Vempala speculated that the implications (4.17) and (4.19) can be chained in a way that "bootstraps" an a priori bound on the KLS constant yield a better bound, however, they were not able to successfully implement this strategy.

Chen added another important ingredient to the mix: In light of equation (4.4), we know that for large enough t, the measure ν_t is not only log-concave, but is t-uniformly log-concave in the sense of (4.6). According to Theorem 4.2, it has concentration properties which do not a priori hold for log-concave measures. The main strategy is then to split the interval $[0, t]$, in Proposition 4.4, into two intervals: In the first interval, the "bootstrap" bound on ψ_n is used, whereas in the second, he manages to leverage on the uniform log-concavity of ν_t in order to find a version of the implication (4.19) which gives a yet stronger lower bound on ψ_n, thereby closing the implication circle of (4.17) and (4.19).

5. DECOMPOSITION OF MEASURES AND FURTHER APPLICATIONS OF STOCHASTIC LOCALIZATION

In this section we describe several additional applications of the Stochastic Localization process. Common to these applications is that they rely on the fact that this process gives rise to a decomposition scheme which expresses a given measure as a *mixture*. We fix a measure ν on $\mathbb{R}^n$ and use the same notation as in Section 4.1.

Recall that, according to Fact 4.1, for every fixed measurable set $A \subset \mathbb{R}^n$, the process $\nu_t(A)$ is a martingale, which implies in particular that $\nu = \mathbb{E}\nu_t$. More generally, the optional stopping theorem implies that, for every $\mathcal{F}_t$-stopping time τ, one has $\nu = \mathbb{E}\nu_\tau$. Therefore, every such stopping time induces a *decomposition* of the measure ν, in the sense that it can be viewed as a mixture whose components are the measures ν_τ.

To summarize, every stopping time τ may be associated with a probability measure $m = m_\tau$ on an abstract index set $\mathcal{I}$, and every $\alpha \in \mathcal{I}$ may be associated with a probability measure ν_α on $\mathbb{R}^n$, so that

$$\nu(W) = \int_{\mathcal{I}} \nu_\alpha(W) m(d\alpha), \quad \forall W \subset \mathbb{R}^n \text{ measurable.} \tag{5.1}$$

Above, the random measure ν_α with $\alpha \sim m$ has the same distribution as ν_τ where ν_t is defined by equations (4.2). In the next sections, we review several applications of this decomposition.

5.1. Needle decompositions

At the heart of the argument found in the original paper of Kannan, Lovász, and Simonovits **[25]** lies a procedure that takes the uniform measure on a convex set in $\mathbb{R}^n$ and represents it as a decomposition into measures whose support is contained in a one-dimensional affine subspace (referred to as "needles"). This was done using an iterative scheme which repeatedly cuts the set via hyperplane bisections which preserve the relative volume of the set A. This type of scheme, referred to as "localization" generalizes an earlier lemma by Lovász and Simonovits **[37]**, and is based on ideas going back to Gromov and Milman **[23]**. Klartag **[28]** gives a somewhat canonical construction which generalizes this concept to Riemannian manifolds (where needles are supported on geodesics).

We will not discuss the aforementioned localization schemes in detail here. Instead, we describe an alternative way to obtain a "needle decomposition" (hence, a decomposition into measures with one-dimensional support) for a prescribed measure on $\mathbb{R}^n$, using the (generalized) stochastic localization equations. We first demonstrate this by outlining an alternative proof of the Gaussian isoperimetric inequality, as well as to Theorem 4.2.

Take $\nu = \gamma_n$ and consider the process generated by (4.2). The main idea is the following one: In view of equation (4.13), by choosing $C_t = \mathrm{Proj}_{b_t^\perp}$, we have that $\nu_t(A)$ remains constant along the process. By doing so, we obtain a decomposition of γ_n into measures which satisfy $\nu_\alpha(A) = \gamma_n(A)$. Next, we will argue that as $t \to \infty$, we obtain a decomposition into measures with one-dimensional support.

Denote $G_t = \int_0^t C_s^2 ds$. Since C_t is a projection matrix of codimension 1, we have $\mathrm{Tr}(G_t) = (n-1)t$ and $\|G_t\|_{\mathrm{OP}} \le t$. This implies that all but one of the eigenvalues of G_t are at least $t/2$. According to (4.3), the measure ν_t is a Gaussian measure whose covariance matrix $\mathrm{Cov}(\nu_t)$ converges, as $t \to \infty$, to a matrix M of rank at most 1. For all $A \subset \mathbb{R}^n$, define $\nu_\infty(A) = \lim_{t\to\infty} \nu_t(A)$ (the limit exists by the martingale convergence theorem). It is straightforward to show that ν_∞ is σ-additive and therefore a probability measure, and in fact, it is a Gaussian measure whose covariance is $\lim_{t\to\infty} \mathrm{Cov}(\nu_t)$. Moreover, since $\nu(A) = \mathbb{E}\nu_t(A)$ for all t and $A \subset \mathbb{R}^n$, by taking limits we have that $\nu(A) = \mathbb{E}\nu_\infty(A)$.

In light of equation (5.1) (taking $\tau = \infty$), we arrive at the following lemma.

Lemma 5.1. *For every measurable $A \subset \mathbb{R}^n$, there exist a probability measure m on an index set $\mathcal{J}$ and, for every $\alpha \in \mathcal{J}$, a probability measure ν_α on $\mathbb{R}^n$ such that*

$$\gamma_n(W) = \int_{\mathcal{J}} \nu_\alpha(W) m(d\alpha), \quad \forall W \subset \mathbb{R}^n \text{ measurable.} \tag{5.2}$$

Moreover, for every $\alpha \in \mathcal{J}$, the measure ν_α has a Gaussian law with covariance matrix C_α such that (i) $\operatorname{rank}(C_\alpha) = 1$, (ii) $\|C_\alpha\|_{\mathrm{OP}} \leq 1$, *and* (iii) $\nu_\alpha(A) = \gamma_n(A)$.

We now use this decomposition to show that the n-dimensional Gaussian measure "inherits" the isoperimetric properties of the one-dimensional Gaussian measure. Indeed, assuming the bound

$$\gamma_1^+(\partial W) \geq \mathrm{I}\big(\gamma_1(W)\big), \quad \forall W \subset \mathbb{R} \text{ measurable}, \tag{5.3}$$

for some function $\mathrm{I} : [0,1] \to [0,\infty)$, and given any measurable set $A \subset \mathbb{R}^n$, we can find a decomposition of γ_n as in (5.2) such that every ν_α is a one-dimensional Gaussian measure of variance at most 1 and $\nu_\alpha(A) = \gamma_n(A)$. We get that

$$\gamma_n(A_\varepsilon \setminus A) \overset{(5.2)}{=} \int_{\mathcal{J}} \nu_\alpha(A_\varepsilon \setminus A) m(d\alpha) \geq \int_{\mathcal{J}} \nu_\alpha\big(\big(A \cap \operatorname{Supp}(\nu_\alpha)\big)_\varepsilon \setminus A\big) m(d\alpha).$$

By taking limits,

$$\gamma_n^+(\partial A) \geq \int_{\mathcal{J}} \nu_\alpha^+(\partial A) m(d\alpha) \overset{(5.3)}{\geq} \int_{\mathcal{J}} \mathrm{I}\big(\nu_\alpha(A)\big) m(d\alpha) = \mathrm{I}\big(\gamma_n(A)\big).$$

We have therefore reduced the proof of the Gaussian isoperimetric inequality in dimension n to the same inequality in dimension 1.

If ν has density of the form $d\nu = \exp(-V(x) - \alpha|x|^2)dx$ with $\alpha > 0$ and $V : \mathbb{R}^n \to \mathbb{R}$ convex, then the same procedure gives rise to a decomposition into one-dimensional needles whose potential exhibits uniform convexity of a similar form. Thus an analogous argument gives a reduction of Theorem 4.2 to the one-dimensional case of the theorem (which has an elementary proof that we omit due to space considerations).

Next, we discuss a needle decomposition obtained by Stochastic Localization, in a different setting, where the role of convexity is replaced by complex-analyticity.

5.1.1. A waist inequality for complex-analytic functions

In [29], Klartag uses a decomposition of the Gaussian measure γ_n, via Stochastic Localization, to prove several expansion inequalities for complex-analytic sets. For example, he obtains the following bound.

Theorem 5.2 (Klartag [29]). *Let $f : \mathbb{C}^n \to \mathbb{C}^k$ be a holomorphic function such that $f(0) = 0$. Write $Z = f^{-1}(0)$. Then one has*

$$\gamma_n(Z_\varepsilon) \geq \gamma_k\big(\{x \in \mathbb{C}^k : \|x\|_2 \leq \varepsilon\}\big), \quad \forall \varepsilon > 0, \tag{5.4}$$

where Z_ε is the ε-extension of Z and γ_m is the complex standard Gaussian measure on $\mathbb{C}^m$.

The above may be thought of in context of Gromov's waist inequality [22], according to which, every continuous function $f : \mathbb{R}^n \to \mathbb{R}^k$ has a level set $Z = f^{-1}(a)$ which satisfies (5.4). The key to the proof is to find a decomposition of γ_n of the form

$$\gamma_n = \int_{\mathcal{I}} \nu_\alpha m(d\alpha)$$

such that:

(i) The measures ν_α are Gaussian measures with covariance matrix of rank at most k and operator norm bounded by 1.

(ii) The center of mass of each ν_α lies on Z.

Such a decomposition effectively reduces the proof of the theorem to the trivial case $k = n$.

We give a high-level sketch of ideas used to obtain such a decomposition. Consider the Stochastic Localization process of equation (4.2) taking the background measure ν to be the Gaussian measure γ_n. Our goal is to find a control matrix C_t so that the two properties above hold. In order to obtain property (ii), the idea is to make sure that the $a_t \in Z$ for all $t \geq 0$ (the center of mass of ν_t). The evolution of a_t obeys the equation (as in (4.14))

$$da_t = \mathrm{Cov}(\nu_t) C_t dB_t,$$

where now B_t is a Brownian motion in $\mathbb{C}^n$ and C_t is an $n \times n$ Hermitian matrix. We want to make sure that $f(a_t)$ remains constant. The key observation is that, due to the fact that f is holomorphic, there will be no quadratic variation terms in the formula for $df(a_t)$, and we have that

$$df_i(a_t) = \nabla f_i(a_t)^T \,\mathrm{Cov}(\nu_t) C_t dB_t, \quad \forall 1 \leq i \leq k.$$

For each t, by dimension considerations, we can find a projection matrix C_t of rank $n - k$ such that $df(a_t) = 0$. With this choice of driving-matrix, all but k eigenvalues of the matrix $G_t = \int_0^t C_s^2 ds$ must converge to infinity as $t \to \infty$ and, in light of (4.3), we get that $\mathrm{Cov}(\nu_t)$ tends to a matrix of rank k, as required by property (i).

5.2. Measures on the discrete hypercube

Up to this point, we were focused on absolutely continuous measures on $\mathbb{R}^n$ (or $\mathbb{C}^n$). In this section, we discuss applications of Stochastic Localization to discrete measures, where there is no natural notion of convexity and heat-flow techniques typically do not apply.

5.2.1. Concentration for Ising models via decomposition into low-rank systems

An *Ising model* is a measure ν on the discrete hypercube $\{-1, 1\}^n$ whose potential is a quadratic function or, in other words, its density is of the form

$$\nu(\{x\}) = Z_\nu^{-1} \exp(\langle x, Jx \rangle + \langle h, x \rangle), \quad \forall x \in \{-1, 1\}^n \tag{5.5}$$

for some $n \times n$ symmetric matrix J (called an interaction matrix) and some $h \in \mathbb{R}^n$ (an "external field"), and where Z_ν is a normalization constant. An important question in statistical mechanics is to characterize the pairs (J, h) for which the model is in *high temperature*.

One interpretation of high temperature is that $\sqrt{\mathrm{Var}\langle X, Y\rangle} \ll n$ where X, Y are independent vectors with law ν.

It is a common belief that for most cases of interest, measures in the high-temperature regime will admit stronger forms of concentration. For example, it is expected that the so-called Glauber dynamics admits a polynomially-large spectral gap in the high-temperature regime, which implies the existence of a polynomial-time sampling algorithm for ν, see [21] for definitions and background.

In what follows, we outline a way to obtain a concentration inequality for high-temperature Ising models using Stochastic Localization. In order to keep things simple and avoid encumbering the reader with definitions, we will derive a weaker form of concentration than what the method allows. A function $\varphi : \mathbb{R}^n \to \mathbb{R}$ is 1-Hamming–Lipschitz (1-Lipschitz in short) if $|\varphi(x) - \varphi(y)| \le \|x - y\|_1$ for all $x, y \in \{-1, 1\}^n$. We will show the following.

Theorem 5.3. *For every ν of the form (5.5) such that $\|J\|_{\mathrm{OP}} \le 1/2$ and every 1-Lipschitz test function $\varphi : \{-1, 1\}^n \to \mathbb{R}$,*

$$\mathrm{Var}_\nu[\varphi] \le \frac{n}{\frac{1}{2} - \|J\|_{\mathrm{OP}}}.$$

The above bound was first obtained as a corollary of a result by Bauerschmidt and Bodineau [6]. A modification of the argument below produces a stronger bound which also establishes polynomial mixing of the Glauber dynamics, see [21]. We now outline the proof.

Without loss of generality, we may assume that J is positive semidefinite (we can always add a multiple of the identity without changing the distribution). Given an Ising model ν and a test function $\varphi : \{-1, 1\}^n \to \mathbb{R}$, consider the Stochastic Localization equations (4.2), with the matrix C_t to be defined later on. Define

$$b_t := \int_{\{-1,1\}^n} \varphi(x)(x - a_t)\nu_t(dx)$$

so that, by (4.13), we have

$$d \int \varphi(x)\nu_t(dx) = \langle C_t b_t, dB_t\rangle.$$

Set $J_t := J - \frac{1}{2}\int_0^t C_s^2 ds$. Equation (4.3) implies that ν_t is an Ising model with interaction matrix J_t. The idea now is to choose C_t to be the orthogonal projection on the intersection $\mathrm{Im}(J_t) \cap b_t^{\perp}$. By continuity, the matrix J_t is decreasing in the positive definite sense, but remains positive semidefinite. Since $C_t b_t = 0$, we have, using (4.13), that

$$\mathbb{E} \int \varphi d\nu_t = \int \varphi d\nu, \quad \forall t \ge 0.$$

By dimension considerations, C_t is nonzero as long as $\dim(\mathrm{Im}(J_t)) > 1$. By running the process until J_t is of rank at most 1 and using the decomposition (5.1), we arrive at the "needle decomposition" theorem formulated below. For every $u, h \in \mathbb{R}^n$, define

$$\nu_{u,h}(\{x\}) = Z_{u,h}^{-1} \exp(\langle x, u\rangle^2 + \langle h, x\rangle), \tag{5.6}$$

with $Z_{u,h}$ being a constant normalizing $\nu_{u,h}$ to be a probability measure.

Theorem 5.4. *Let ν be an Ising measure on $\{-1,1\}^n$ of the form (5.5) with J positive semidefinite, and let $\varphi : \{-1,1\}^n \to \mathbb{R}$. There exists a probability measure m on $\mathbb{R}^n \times \mathbb{R}^n$ such that ν admits the decomposition*

$$\nu = \int_{\mathbb{R}^n \times \mathbb{R}^n} \nu_{u,h} dm(u,h) \tag{5.7}$$

and such that m-almost surely the pair (u,h) satisfies $\int \varphi d\nu_{u,h} = \int \varphi d\nu$ and $\|u\|_2 \le \|J\|_{\mathrm{OP}}$.

This decomposition theorem allows us to show that an Ising measure inherits the concentration properties satisfied by rank-one Ising models whose interaction matrix has a corresponding norm. For models of rank-one, we rely on the following fact.

Fact 5.5 (see [21]). *For all $u, h \in \mathbb{R}^n$ such that $|u| < \frac{1}{2}$ and for all 1-Lipschitz $\varphi : \mathbb{R}^n \to \mathbb{R}$, we have*

$$\mathrm{Var}_{\nu_{u,h}}[\varphi] \le \frac{n}{1/2 - |u|}. \tag{5.8}$$

Now, given an Ising model ν with positive semidefinite interaction matrix J of norm at most $1/2$ and given a 1-Lipschitz test function φ, use Theorem 5.4 to find a measure m corresponding to ν, φ. We have, by the law of total variance,

$$\mathrm{Var}_\nu[\varphi] \stackrel{(5.7)}{=} \int_{\mathbb{R}^n \times \mathbb{R}^n} \mathrm{Var}_{\nu_{u,h}}[\varphi] dm(u,h) \stackrel{(5.8)}{\le} \int_{\mathbb{R}^n \times \mathbb{R}^n} \frac{n}{1/2 - |u|} dm(u,h) \le \frac{n}{1/2 - \|J\|_{\mathrm{OP}}}.$$

5.2.2. Entropy-efficient decomposition of discrete measures

In the previous subsections we saw how the Stochastic Localization process allows us to decompose a measure into well-behaved "needles." We now present a family of related applications which has proven useful in the context of interacting particle systems, random graphs, and large deviation theory.

We begin the discussion with a simple example referred to as the Curie–Weiss model: Fix $\beta > 0$ and consider the measure ν on $\{-1,1\}^n$, defined by

$$\nu(\{x\}) = Z_\beta^{-1} \exp\left(\frac{\beta}{n} \sum_{i \neq j} x_i x_j\right),$$

with Z_β a normalizing constant. Let $X \sim \nu$. It is well known that this measure exhibits the following phase transition: If $\beta < 1/2$, then $\mathrm{Cov}(X_1, X_2) \to 0$ as $n \to \infty$, whereas if $\beta > 1/2$ then $\mathrm{Cov}(X_1, X_2)$ is bounded away from 0 as $n \to \infty$ (and hence, also $\mathrm{Var}[\sum_i X_i] = \Omega(n^2)$). On the other hand, in the latter case, there exist two measures $\nu^\pm$ such that $\nu = \frac{1}{2}(\nu^+ + \nu^-)$ and such that $\nu^\pm$ are approximate product measures in the sense that $\|\mathrm{Cov}(\nu^\pm)\|_{\mathrm{OP}} = O(1)$ and, in fact, in a much stronger sense discussed later on.

This simple, yet somewhat prototypical example motivates the question of finding sufficient conditions on a measure ν on $\{-1,1\}^n$ under which it can be expressed as a decomposition $\nu = \sum_{i=1}^N \nu_i$ where the measures ν_i attain a simple form, and N is not too large. Here, we consider a more general form of decomposition where our goal is to express ν as

$$\nu = \int_{\mathcal{I}} \nu_\alpha m(d\alpha)$$

such that the ν_α's have a simple form. In this context, it is natural to replace the requirement that N is not too large by an upper bound on the *entropic-deficit* of the decomposition, defined as

$$\mathrm{Ent}[\nu] - \int_{\mathcal{I}} \mathrm{Ent}[\nu_\alpha] m(d\alpha),$$

where, for a measure μ on $\{-1,1\}^n$, we define $\mathrm{Ent}(\mu) := -\int_{\{-1,1\}^n} \log(\mu(\{x\}))\mu(dx)$.

Stochastic Localization is a useful tool in obtaining decompositions of this sort, via equation (5.1). The key is to analyze the evolution of the processes $\mathrm{Cov}(\nu_t)$ and $\mathrm{Ent}[\nu_t]$, which turn out to be quite tractable. As an initial idea of how it can be done, observe that choosing $C_t = \mathrm{Id}$ and taking expectations on both sides of equation (4.15), we have that

$$\frac{d}{dt}\mathbb{E}\,\mathrm{Cov}(\nu_t) = -\mathbb{E}\big[\mathrm{Cov}(\nu_t)^2\big].$$

One may interpret the last display as follows: The localization process "shrinks," in expectation, the large directions of the covariance matrix. Let us now outline an argument which builds on this intuition.

Fix a measure ν on $\{-1,1\}^n$ and consider the process ν_t obtained by running the process of equation (4.2) with the initial condition $\nu_0 = \nu$. For every t, take C_t to be the projection onto the span of the top eigenvector of $\mathrm{Cov}(\nu_t)$. Using (4.15), we have that

$$d\,\mathrm{Tr}\big(\mathrm{Cov}(\nu_t)\big) = -\big\|\mathrm{Cov}(\nu_t)\big\|_{\mathrm{OP}}^2 dt + \text{martingale term}.$$

A straightforward calculation using Itô's formula yields that

$$\begin{aligned} d\,\mathrm{Ent}(\nu_t) &= -\mathrm{Tr}\big(C_t\,\mathrm{Cov}(\nu_t)\big)dt + \text{martingale term} \\ &= -\big\|\mathrm{Cov}(\nu_t)\big\|_{\mathrm{OP}} dt + \text{martingale term}. \end{aligned}$$

By comparing that last two displays, we see that as long as $\|\mathrm{Cov}(\nu_t)\|_{\mathrm{OP}}$ is large, the trace of the covariance matrix of ν_t decays, in expectation, much faster than its entropy. Now fix $\lambda > 0$ and consider the stopping time

$$\tau := \min\big\{t;\ \big\|\mathrm{Cov}(\nu_t)\big\|_{\mathrm{OP}} \le \lambda\big\}.$$

By the above, we have that $\lambda\,\mathrm{Ent}(\nu_t) - \mathrm{Tr}(\mathrm{Cov}(\nu_t))$ is a submartingale up to the stopping time τ. Using the optional stopping theorem, we have that

$$\mathbb{E}\big[\mathrm{Ent}[\nu] - \mathrm{Ent}[\nu_\tau]\big] \le \frac{1}{\lambda}\mathbb{E}\big(\mathrm{Tr}\big(\mathrm{Cov}(\nu)\big) - \mathrm{Tr}\big(\mathrm{Cov}(\nu_\tau)\big)\big) \le \frac{\mathrm{Tr}(\mathrm{Cov}(\nu))}{\lambda} \le \frac{n}{\lambda}.$$

Using the decomposition (5.1), we arrive at the following theorem.

Theorem 5.6. *Let ν be a measure on $\{-1,1\}^n$. Then for every $\lambda \ge 1$, there exist a probability measure m on an index set $\mathcal{I}$ and a family of probability measures $\{\nu_\theta\}_{\theta\in\mathcal{I}}$ on $\{-1,1\}^n$ such that the measure ν admits the decomposition*

$$\nu(W) = \int_{\mathcal{I}} \nu_\theta(W) dm(\theta), \quad \forall W \subset \mathbb{R}^n \text{ measurable}, \tag{5.9}$$

such that

$$\big\|\mathrm{Cov}(\nu_\alpha)\big\|_{\mathrm{OP}} \le \lambda, \quad \forall \alpha \in \mathcal{I}$$

and

$$\mathrm{Ent}[\nu] - \int_{\mathcal{I}} \mathrm{Ent}[\nu_\alpha] m(d\alpha) \leq \frac{n}{\lambda}.$$

A related argument can also produce bounds on the Frobenius norm of $\mathrm{Cov}(\nu_\alpha)$. We refer the reader to [19] for other inequalities of this form, as well as application to mean-field approximation, which we do not discuss here.

The measures ν_α given by the above theorem are close to product measures in a rather weak sense, and one may consider stronger notions of approximating a product measure. A particularly useful notion is defined in terms of the transportation distance to a product measure. For probability measures μ_1, μ_2 on $\{-1,1\}^n$, we define

$$\mathrm{W}(\mu_1, \mu_2) = \sup_{\|\varphi\|_{\mathrm{Lip}} \leq 1} \left(\int_{\{-1,1\}^n} \varphi d\mu_1 - \int_{\{-1,1\}^n} \varphi d\mu_2 \right),$$

where $\|\cdot\|_{\mathrm{Lip}}$ denotes the Hamming–Lipschitz norm. This quantity is referred to as the (Wasserstein) transportation distance with respect to the Hamming metric. Given a probability measure μ on $\{-1,1\}^n$, let $\xi(\mu)$ be the unique product measure having the same center of mass of μ. Consider the quantity

$$\mathcal{P}(\mu) := \mathrm{W}\big(\mu, \xi(\mu)\big)$$

which quantifies how close μ is to a product measure. What conditions on a measure ν on $\{-1,1\}^n$ ensure that it admits a decomposition of the form (5.9) such that both the entropic deficit and $\mathcal{P}(\mu)$ are nontrivially small (say, both are $o(n)$)? The work [18] establishes this under a condition inspired by an earlier work of Chatterjee and Dembo [14] and referred to as *low complexity*. For a measure ν on $\{-1,1\}^n$, denote by f_ν its density with respect to the uniform measure. Define the *complexity* of ν by

$$\mathcal{D}(\nu) := \mathbb{E}_{\Gamma \sim \mathcal{N}(0,\mathrm{Id})} \sup_{x \in \{-1,1\}^n} \langle \nabla \log f_\nu, \Gamma \rangle$$

(which can be understood as the Gaussian-width of the gradient of its potential). The following decomposition theorem can be obtained via Stochastic Localization.

Theorem 5.7 ([18]). *For every measure ν on $\{-1,1\}^n$ and every $\varepsilon > 0$, there exists a decomposition of the form* (5.9) *such that its entropic deficit satisfies*

$$\mathrm{Ent}[\nu] - \int_{\mathcal{I}} \mathrm{Ent}[\nu_\alpha] m(d\alpha) \leq \varepsilon n$$

and such that

$$\int_{\mathcal{I}} \mathcal{P}(\nu_\alpha) m(d\alpha) \lesssim \sqrt{\frac{n \mathcal{D}(\nu)}{\varepsilon}}.$$

Note that, as long as $\mathcal{D}(\nu) = o(n)$, we may obtain a decomposition with entropic deficit $o(n)$, such that $\mathcal{P}(\nu_\alpha) = o(n)$ for all but an $o(1)$ fraction of α's (with respect to m). This type of structure theorem has several applications, in particular to the emerging field of *nonlinear large deviations* pioneered by Chatterjee and Dembo in [14] and to mean-field approximations. We refer the reader to [18] for more details.

FUNDING

The author is supported by a European Research Council starting Grant (ERC StG), project number 803084.

REFERENCES

[1] D. Alonso-Gutiérrez and J. Bastero, *Approaching the Kannan–Lovász–Simonovits and variance conjectures*. Lecture Notes in Math. 2131, Springer, Cham, 2015.

[2] S. Artstein-Avidan, A. Giannopoulos, and V. D. Milman, *Asymptotic geometric analysis. Part I*. Math. Surveys Monogr. 202, American Mathematical Society, Providence, RI, 2015.

[3] D. Bakry and M. Emery, Diffusions hypercontractives. In *Séminaire de probabilités, XIX, 1983/1984*, pp. 177–206, Lecture Notes in Math. 1123, Springer, Berlin, 1985.

[4] D. Bakry and M. Ledoux, Lévy–Gromov isoperimetric inequality for an infinite dimensional diffusion generator. *Invent. Math.* **123** (1995), 259–281.

[5] K. Ball and V. H. Nguyen, Entropy jumps for isotropic log-concave random vectors and spectral gap. *Studia Math.* **213** (2012), no. 1, 81–96.

[6] R. Bauerschmidt and T. Bodineau, A very simple proof of the LSI for high temperature spin systems. *J. Funct. Anal.* **276** (2019), no. 8, 2582–2588.

[7] S. Bobkov, On isoperimetric constants for log-concave probability distributions. In *Geometric aspects of functional analysis. Israel Seminar 2004–2005*, pp. 81–88, Lecture Notes in Math. 1910, Springer, 2007.

[8] C. Borell, The Brunn–Minkowski inequality in Gauss space. *Invent. Math.* **30** (1975), no. 2, 207–216.

[9] C. Borell, Geometric bounds on the Ornstein–Uhlenbeck velocity process. *Z. Wahrsch. Verw. Gebiete* **70** (1985), no. 1, 1–13.

[10] C. Borell, Diffusion equations and geometric inequalities. *Potential Anal.* **12** (2000), no. 1, 49–71.

[11] C. Borell, Isoperimetry, log-concavity, and elasticity of option prices. In *New directions in mathematical finance*, edited by P. Wilmott and H. Rasmussen, pp. 73–91, Wiley, 2002.

[12] J. Bourgain, On the distribution of polynomials on high-dimensional convex sets. In *Geometric Aspects of Functional Analysis (1989–1990)*, pp. 127–137, Lecture Notes in Math. 1469, Springer, 1991.

[13] P. Cattiaux, A pathwise approach of some classical inequalities. *Potential Anal.* **20** (2004), no. 4, 361–394.

[14] S. Chatterjee and A. Dembo, Nonlinear large deviations. *Adv. Math.* **299** (2019), 396–450.

[15] Y. Chen, An almost constant lower bound of the isoperimetric coefficient in the KLS conjecture. *Geom. Funct. Anal.* **31** (2021), no. 1, 34–61.

[16] R. Eldan, Thin shell implies spectral gap up to polylog via a Stochastic Localization scheme. *Geom. Funct. Anal.* **23** (2012), no. 2, 532–569.

[17] R. Eldan, A two-sided estimate for the Gaussian noise stability deficit. *Invent. Math.* **201** (2015), no. 2, 561–624.

[18] R. Eldan, Gaussian-width gradient complexity, reverse log-Sobolev inequalities and nonlinear large deviations. *Geom. Funct. Anal.* **28** (2018), no. 6, 1548–1596.

[19] R. Eldan, Taming correlations through entropy-efficient measure decompositions with applications to mean-field approximation. *Probab. Theory Related Fields* **176** (2020), no. 3–4, 737–755.

[20] R. Eldan and B. Klartag, Approximately gaussian marginals and the hyperplane conjecture. In *Proc. of a workshop on "Concentration, Functional Inequalities and Isoperimetry"*, pp. 55–68, Contemp. Math. 545, Amer. Math. Soc., 2011.

[21] R. Eldan, F. Koehler, and O. Zeitouni, A spectral condition for spectral gap: fast mixing in high-temperature Ising models. 2020, arXiv:2007.08200.

[22] M. Gromov, Isoperimetry of waists and concentration of maps. *Geom. Funct. Anal.* **13** (2003), no. 1, 178–215.

[23] M. Gromov and V. D. Milman, Generalization of the spherical isoperimetric inequality to uniformly convex Banach spaces. *Compos. Math.* **62** (1987), 263–282.

[24] O. Guedon and E. Milman, Interpolating thin-shell and sharp large-deviation estimates for isotropic log-concave measures. *Geom. Funct. Anal.* **21** (2011), no. 5, 1043–1068.

[25] R. Kannan, L. Lovász, and M. Simonovits, Isoperimetric problems for convex bodies and a localization lemma. *Discrete Comput. Geom.* **13** (1995), 541–559.

[26] B. Klartag, On convex perturbations with a bounded isotropic constant. *Geom. Funct. Anal.* **16** (2006), no. 6, 1274–1290.

[27] B. Klartag, A central limit theorem for convex sets. *Invent. Math.* **168** (2007), 91–131.

[28] B. Klartag, Needle decompositions in Riemannian geometry. *Mem. Amer. Math. Soc.* **249** (2017), no. 1180.

[29] B. Klartag, Eldan's Stochastic Localization and tubular neighborhoods of complex-analytic sets. *J. Geom. Anal.* **28** (2018), no. 3, 2008–2027.

[30] B. Klartag and V. D. Milman, The slicing problem by Bourgain. In *Analysis at Large, A Collection of Articles in Memory of Jean Bourgain*. Springer (to appear).

[31] M. Ledoux, Isoperimetry and Gaussian analysis. In *Lectures on probability theory and statistics*, pp. 165–294, Lecture Notes in Math., 1648. Springer, Berlin, Heidelberg, 1996.

[32] M. Ledoux, *The concentration of measure phenomenon*. Math. Surveys Monogr. 89, American Mathematical Society, Providence, RI, 2001.

[33] Y. T. Lee and S. Vempala, The Kannan–Lovász–Simonovits conjecture. In *Current developments in mathematics 2017*, pp. 1–36, Int. Press, Somerville, MA, 2017.

[34] Y. T. Lee and S. S. Vempala, Eldan's Stochastic Localization and the KLS hyperplane conjecture: an improved lower bound for expansion. In *2017 IEEE 58th annual symposium on foundations of computing (FOCS)*, pp. 998–1007, ACM, New York, 2017.

[35] Y. T. Lee and S. S. Vempala, Eldan's stochastic localization and the KLS conjecture: isoperimetry, concentration and mixing. 2018, arXiv:1612.01507.

[36] J. Lehec, Representation formula for the entropy and functional inequalities. *Ann. Inst. Henri Poincaré Probab. Stat.* **49** (2013), no. 3, 885–899.

[37] L. Lovász and M. Simonovits, Random walks in a convex body and an improved volume algorithm. *Random Struct. Algebra* **4** (1993), 359–412.

[38] E. Milman, On the role of convexity in isoperimetry, spectral-gap and concentration. *Invent. Math.* **177** (2009), no. 1, 1–43.

[39] E. Milman and J. Neeman, The Gaussian multi-bubble conjecture. 2018, arXiv:1805.10961.

[40] E. Mossel and J. Neeman, Robust optimality of Gaussian noise stability. *J. Eur. Math. Soc. (JEMS)* **17** (2015), no. 2, 433–482.

[41] G. Pisier, Probabilistic methods in the geometry of Banach spaces. In *Probability and analysis (Varenna, 1985)*, pp. 167–241, Lecture Notes in Math. 1206, Springer, Berlin, 1986.

[42] V. N. Sudakov and B. S. Tsirel'son, Extremal properties of half-spaces for spherically invariant measures. *J. Sov. Math.* **9** (1978), no. 1, 9–18.

[43] R. Vershynin, *High-dimensional probability. An introduction with applications in data science. With a foreword by Sara van de Geer*. Camb. Ser. Stat. Probab. Math. 47, Cambridge University Press, Cambridge, 2018.

RONEN ELDAN

Department of Mathematics, Weizmann Institute of Science, Rehovot 76100, Israel, ronen.eldan@weizmann.ac.il

This Page Intentionally Left Blank

NATURAL SELECTION IN SPATIALLY STRUCTURED POPULATIONS

ALISON ETHERIDGE

ABSTRACT

Mathematical models play a fundamental role in theoretical population genetics and, in turn, population genetics provides a wealth of mathematical challenges. Here we illustrate this by using mathematical caricatures of the evolution of genetic types in a spatially distributed population to demonstrate the complex interplay between spatial structure, natural selection, and so-called random genetic drift (the randomness due to reproduction in a finite population). In particular, we highlight the role that the shape of the domain inhabited by the population can play in mediating the interplay between the different forces of evolution acting upon it.

MATHEMATICS SUBJECT CLASSIFICATION 2020

Primary 92D15; Secondary 60J70, 60H30, 35K57, 35B30

KEYWORDS

Natural selection, spatial structure, genetic drift, blocking

ICM INTERNATIONAL CONGRESS OF MATHEMATICIANS 2022 JULY 6–14

Proc. Int. Cong. Math. 2022, Vol. 6, pp. 4272–4300
DOI 10.4171/ICM2022/195

Published by EMS Press

1. INTRODUCTION

Theoretical population genetics is concerned with understanding genetic differences within and between populations. It finds its origins at the beginning of the twentieth century in *the modern evolutionary synthesis*, in which Darwin's theory of evolution through natural selection and Mendel's laws of genetic inheritance were integrated. This work, pioneered by Fisher, Haldane, and Wright, provided a unified mathematical framework within which to discuss possible causes of evolution. As a result some consensus emerged about which forces influence evolution, but questions such as their relative importance remained unresolved.

The intervening century has seen a rich interplay between population genetics and the mathematical sciences: mathematical modeling has been employed to explore concepts such as adaptation, speciation, and population structure, and in the process questions arising from population genetics have stimulated the development of elegant new mathematical models and techniques, often of much wider applicability.

In population genetics, mathematical models are used both as a basis for statistical inference and as a means to validate or dismiss concepts. The purpose of the model is then not to provide detailed predictions of the fate of a particular biological population, but rather to use caricatures of the forces of evolution, and the ways in which they interact, to gain some insight into the evolutionary process. Our aim here is to illustrate this approach through models that attempt to capture some features of the interactions between natural selection, spatial structure, and the randomness due to reproduction in a finite population (so-called *genetic drift*). Rather than giving detailed proofs, for which we refer to the original papers, we shall provide informal arguments and draw out some of the lessons learned.

We shall try to minimize the use of biological jargon, but it is convenient to fix some terminology. The term *locus* is used to refer in a general way to a location on the genome. For our purposes, it will correspond to a region that codes for a gene, and it will be passed on as a single unit from parent to offspring. Genes can occur in different forms, called *alleles*, and we shall make the simplifying assumption that the gene in which we are interested has just two alleles, denoted a, A. Evolution is fueled by mutation, the source of the genetic diversity on which natural selection acts, but we shall assume that any new mutations arising at the locus of interest are neutral, that is, do not affect fitness. Moreover, since genes are organized onto chromosomes, different genetic loci do not evolve independently of one another. However, our models will neglect this *genetic structure* and suppose that (relative) fitness is determined entirely by the alleles at the locus of interest. Although crude, such single locus models exhibit a surprisingly rich variety of behaviors.

There is a huge literature devoted to understanding the interaction between natural selection and genetic drift. In particular, in the absence of spatial structure, it is well understood that in larger populations, not only is a beneficial mutation more likely to establish and sweep to fixation (that is, increase in frequency until it is carried by every individual in the population), but it is also more likely that deleterious mutations will be expunged. Genetic drift, which drives random fluctuations in the proportions of the different alleles, is stronger in a smaller population, and this increases the chance that a deleterious mutation is fixed just by chance [27].

The interaction between natural selection and spatial structure (ignoring genetic drift) is often investigated through reaction–diffusion equations. This was initiated by Fisher [22], who studied traveling wave solutions to the equation

$$\frac{\partial u}{\partial t} = m\frac{\partial^2 u}{\partial x^2} + su(1-u), \quad x \in \mathbb{R},\ t \geq 0, \tag{1.1}$$

as a model of the spread of a favorable allele through a one-dimensional population. Here $u(t,x) \in [0,1]$ models the proportion of the alleles carried by the individuals at location x at time t that are of the fitter type. In [28], Kolmogorov, Petrovsky, and Piscounov considered the analogous equation in two spatial dimensions and with a general reaction term $F(u)$ in place of $su(1-u)$, although they focused on solutions that are independent of y (and thus essentially one-dimensional). Motivated by the discussion of Fisher [21], they specialized to a reaction term of the form $\alpha u(1-u)^2$ for their application to population genetics. Equation (1.1), with $\partial^2 u/\partial x^2$ replaced by Δu in dimensions $\mathrm{d} \geq 2$, is often referred to as the Fisher–KPP equation.

A special case of a result of Skorokhod [36] expresses the distribution of a branching Brownian motion in terms of the solution to (1.1). Conversely, in the particular case of a Heaviside initial condition, this allows one to express the solution to (1.1) in terms of the distribution of the rightmost particle in a binary branching Brownian motion at time t. This is often referred to as McKean's representation [31], and underpins the remarkable work of Bramson [9], which provides much of our understanding of the traveling wave to which the solution started from a Heaviside initial condition converges. More recently, these results have been considerably extended, with a particular focus on adding a stochastic term to (1.1) to capture the effect of random genetic drift (see, for example, [33] and the references therein), resulting in a stochastic PDE:

$$du = \left(m\frac{\partial^2 u}{\partial x^2} + su(1-u)\right)dt + \sqrt{\frac{1}{\rho}u(1-u)}W(dt,dx), \tag{1.2}$$

with W a space-time white noise and ρ a measure of local population size. (The form of this so-called *Wright–Fisher noise* term will be motivated in Section 2.1.) The vast majority of this work is restricted to one spatial dimension. In the biologically natural setting of two spatial dimensions, although equation (1.1) generalizes in a natural way, the obvious generalization of equation (1.2) has no solution. In Section 6 we shall describe one way to circumvent this, and provide a mathematical model through which we can explore the interaction of natural selection and genetic drift in a population distributed across a spatial continuum (of any dimension). Depending on the dispersal mechanism and the local population density, an individual may be competing with its own close (and equally fit) relatives, limiting the effect of natural selection. We shall see that if the local population density is bounded, *the dimension of the space in which the population lives is important.*

Natural selection can take many forms. While equation (1.1) models the spread of an allele which is always favorable to the individual carrying it, in much of what follows we shall be interested in populations in which individuals carry two copies of the gene and those carrying *different* alleles are selectively disadvantaged. As we explain in Section 2.1, this form of selection can be captured by replacing the reaction term in (1.1) to obtain

$$\frac{\partial u}{\partial t} = m\frac{\partial^2 u}{\partial x^2} + su(1-u)(2u-1), \quad x \in \mathbb{R},\ t \geq 0. \tag{1.3}$$

Inference from genetic data typically involves using differences between the DNA sequences of a sample of individuals from the population to reconstruct information about *genealogical* ancestors of those individuals. This can then be compared to the predictions of mathematical models under different hypotheses about the forces of evolution acting on the population. The neutral mutation rate therefore dictates the scales over which we can glean meaningful information, and, since it is very small, this leads us to consider very large spatial and temporal scales. With this in mind, we apply a diffusive scaling, corresponding to modeling proportions of different alleles over spatial regions of diameter $\mathcal{O}(1/\varepsilon)$ at times of $\mathcal{O}(1/\varepsilon^2)$, to obtain

$$\frac{\partial u^\varepsilon}{\partial t} = \Delta u^\varepsilon + \frac{1}{\varepsilon^2}u^\varepsilon(1-u^\varepsilon)(2u^\varepsilon - 1), \tag{1.4}$$

where we have set $m = 1, s = 1$. In a sense made precise in Theorem 2.4, for suitable initial conditions, as $\varepsilon \to 0$, u^ε converges to the indicator function of a set whose boundary evolves according to mean curvature flow (see Definition 2.1). We emphasize that although our main interest is in two spatial dimensions (where mean curvature flow is simply curvature flow), our mathematical results are valid in arbitrary spatial dimension $\mathbb{d}$.

More generally (see Section 5.1), if there is a fitness difference between individuals carrying two a alleles and those carrying two A alleles, we consider the equation

$$\frac{\partial u}{\partial t} = m\Delta u + su(1-u)\big(2u - (1-\gamma)\big), \quad x \in \mathbb{R}^{\mathbb{d}},\ t \geq 0. \tag{1.5}$$

As we shall explain, in order to obtain a nontrivial limit under the diffusive scaling, we also scale $\gamma = \varepsilon\nu$. The limit is then the indicator function of a set whose boundary evolves according to a mixture of "constant flow" of rate $-\nu$ and mean curvature flow (for as long as this flow is defined).

Whereas mean curvature flow has no nontrivial fixed point, the spherical shell of radius $(\mathbb{d}-1)/\nu$ (whose interior is completely occupied by the favored type) is fixed by this mixture of curvature and constant flow. In this scenario, the two components of the selection acting on the population work against one another and at this critical radius are finely balanced; for any larger radius constant flow dominates and the circle expands without bound; for a smaller radius, mean curvature flow wins out, and the circle shrinks to a point. This behavior is in sharp contrast to the situation in one spatial dimension, and it is natural then to ask about other domains; for example, what is the fate of an expanding population that must pass through an isthmus? In Section 5.2, we shall see examples of domains for which the effect of curvature flow leads to "blocking" of the expansion of the range of the selectively favored type (but in a way which will result in a stable nontrivial steady state). *The geometry of the domain in which the population lives is important.*

The main mathematical tool that we use is a representation of the solution to (1.4) in terms of a ternary branching Brownian motion which we explain in Section 3. Although reminiscent of the Skorokhod/McKean representations of the solution to the Fisher–KPP equation, it differs in using the entire tree structure of the branching process. Our approach can be seen as an adaptation of that of de Masi et al. in [13], and similar ideas have also been exploited in [29]. For us, it provides an intuitive and flexible representation of the solutions to equations like (1.3), (1.4), and (1.5), that is readily adapted to the framework of Section 6, allowing us to incorporate the effects of genetic drift.

The rest of this article is laid out as follows. In Section 2, we motivate (1.3) from a biological perspective and give a more precise statement about its limiting behavior as $\varepsilon \to 0$. In Section 3, we present the probabilistic representation of the solution to (1.4) and use it to provide some intuition for the emergence of mean curvature flow. In Section 4, we replace the Laplacian in (1.4) by a *fractional* Laplacian, in order to capture the corresponding behaviour in populations with long-range dispersal. In Section 5, we turn to the situation modeled by (1.5) in which there is a fitness difference between type aa and type AA individuals. In particular, we shall consider what happens when the population no longer occupies the whole Euclidean space, and we provide conditions on the geometry of its range under which the expansion of the region occupied by the fitter type is, or is not, blocked. Finally, in Section 6, we extend our models to incorporate genetic drift and explore the extent to which it breaks down the effect of natural selection. In particular, we shall see how the impact of genetic drift depends on both the local population density and the spatial dimension.

2. HYBRID ZONES AND CURVATURE FLOW

A hybrid zone is a narrow geographic region where two genetically distinct populations are found close together and hybridize to produce offspring of mixed ancestry. Hybrid zones are ubiquitous in nature; see, for example, [5] and [6] for an extensive catalogue and discussion. They can be maintained by a variety of mechanisms. For example, consider two populations, each of which is adapted to a different set of environmental conditions. If hybrids are less well adapted to those conditions, then an abrupt change in the environment could result in a hybrid zone. In that case the hybrid zone will not move.

The situation that we shall be trying to caricature is one in which the hybrid zone is maintained by a balance between dispersal and selection against hybrids. For instance, this might arise if two populations regain contact after a period of geographic isolation such as that imposed by the last glacial maximum (c. 18,000 years ago) when many species were forced into isolated refugia. Because they are not dependent on changes in local environmental conditions, hybrid zones maintained by this mechanism can move from place to place. In [3], Barton presented a theoretical study of the dynamics of hybrid zones. In the interests of space, we do not attempt to examine all of the influences on the motion of the zone considered by [3]; instead we present our mathematical approach and illustrate its application in three contrasting settings before adapting it to include genetic drift in Section 6.

2.1. Modeling selection against heterozygosity

As advertised in the introduction, we are going to focus on the case in which the hybrid zone is maintained by selection acting on a single genetic locus. We suppose that the gene at that locus has two alleles, denoted a and A, and that each individual carries two copies of the gene. One population consists of aa individuals, the other of AA individuals. Although it is possible to obtain equations like (1.1)–(1.5) as scaling limits of a variety of individual based models (see, for example, [12, 23, 34]), it is generally highly technical and so instead we shall motivate the models using an argument commonly found in the biological literature.

Our first aim is to understand the form of the reaction term in (1.3), and so we begin with the case in which the population is infinitely large, and has no spatial structure. We assume Hardy–Weinberg equilibrium; that is, if the proportion of a-alleles across the whole population is $\bar{u}$, then the proportions of individuals of types aa, aA, and AA are given by

aa	aA	AA
$\bar{u}^2$	$2\bar{u}(1-\bar{u})$	$(1-\bar{u})^2$

This is expected to be a reasonable approximation if selection is not too strong (which we shall assume here). To model selection against hybrids, we assume that the three types have *relative fitnesses*

aa	aA	AA
1	$1-s_0$	1

We define relative fitness implicitly by explaining its effect. During reproduction, each individual produces a large (effectively infinite) number of germ cells, each of which carries a copy of all the genetic material of the parent (for our purposes this is just two copies of the gene in which we are interested). The germ cells then split into gametes (each containing one copy of the gene). All the gametes are put into a pool, and each individual in the next generation, independently, is created by fusing two gametes sampled at random from that pool.

The relative fitnesses above are reflected in each *heterozygote* (aA) individual producing $(1-s_0)$ times as many germ cells as a *homozygote* (aa or AA) individual. The proportion of type a gametes in the pool is then

$$\begin{aligned}\bar{u}^* &= \frac{(\bar{u}^2+\bar{u}(1-\bar{u})(1-s_0))}{(\bar{u}^2+2\bar{u}(1-\bar{u})(1-s_0)+(1-\bar{u})^2)}\\ &= \frac{\bar{u}^2+\bar{u}(1-\bar{u})(1-s_0)}{1-2s_0\bar{u}(1-\bar{u})}\\ &= (1-s_0)\bar{u}+s_0(3\bar{u}^2-2\bar{u}^3)+\mathcal{O}(s_0^2)\\ &= \bar{u}+s_0\bar{u}(1-\bar{u})(2\bar{u}-1)+\mathcal{O}(s_0^2).\end{aligned}$$

In particular,

$$\bar{u}^*-\bar{u}=s_0\bar{u}(1-\bar{u})(2\bar{u}-1)+\mathcal{O}(s_0^2).$$

In an infinite population, the proportions of alleles among offspring will exactly follow those in the pool of gametes, and if $s_0 = s/M$ (where M is large), measuring time in units of M generations, this suggests the approximation

$$\frac{d\bar{u}}{dt} = s\bar{u}(1-\bar{u})(2\bar{u}-1)$$

for the dynamics of the proportion of a-alleles. The Laplacian term in (1.3) is then added to capture dispersal of offspring.

In a finite population, we must account for the randomness inherent in drawing a finite sample from the pool of gametes. We assume that the population size N is large and fixed. The number of a-alleles among offspring is $\mathtt{Bin}(2N, \bar{u}^*)$, and so the *proportion* of a-alleles has mean $\bar{u}^*$ and variance $\frac{1}{2N}\bar{u}^*(1-\bar{u}^*)$. Notice in particular, that we can expect the effects of the fluctuations to be relevant over timescales of $\mathcal{O}(N)$ generations. As before we suppose that $s_0 = s/M$, and measure time in units of M generations. If $M/N = \mathcal{O}(1)$, the dynamics of the proportion of a-alleles can then be approximated by the Wright–Fisher diffusion

$$du = su(1-u)(2u-1)dt + \sqrt{\frac{M}{2N}u(1-u)}dB_t, \tag{2.1}$$

where B is a one-dimensional Brownian motion.

Replacing the reaction term in equation (1.2) by $su(1-u)(2u-1)$, in one spatial dimension we obtain what can be thought of as a spatial analogue of (2.1) in which offspring sample gametes from a pool generated by adult individuals at the location at which they were born. The Wright–Fisher noise is supposed to capture the randomness inherent in the sampling.

2.2. (Mean) curvature flow

We are primarily interested in two spatial dimensions, when mean curvature flow reduces to curvature flow, but our results are valid for all $\mathrm{d} \geq 2$. Recall that a function is said to be a smooth embedding if it is a diffeomorphism onto its image (which we shall implicitly assume is a subset of $\mathbb{R}^{\mathrm{d}}$).

Definition 2.1 ((Mean) curvature flow). Let S^1 denote the unit circle in $\mathbb{R}^2$. Let $\mathbf{\Gamma} = (\mathbf{\Gamma}_t(\cdot))_t$ be a family of smooth embeddings, indexed by $t \in [0, \mathscr{T})$, where, for each t, $\mathbf{\Gamma}_t : S^1 \to \mathbb{R}^2$. Let $\boldsymbol{n} = \boldsymbol{n}_t(u)$ denote the unit (inward) normal vector to $\mathbf{\Gamma}_t$ at u and let $\kappa_t(u)$ denote the curvature of $\mathbf{\Gamma}_t$ at u. We say that $\mathbf{\Gamma}$ is a *curvature flow* if

$$\frac{\partial \mathbf{\Gamma}_t(u)}{\partial t} = \kappa_t(u)\boldsymbol{n}_t(u) \tag{2.2}$$

for all t, u.

In higher dimensions, we replace S^1 by $S^{\mathrm{d}-1}$, $\mathbb{R}^2$ by $\mathbb{R}^{\mathrm{d}}$, and κ_t by the *mean* curvature of $\mathbf{\Gamma}_t$ to obtain *mean curvature flow*.

Remark 2.2. Perhaps the easiest way to visualize the curvature at a point P on a differentiable curve in $\mathbb{R}^2$ is as the reciprocal of the radius of the *osculating circle* at P which (if it exists) is the circle that best approximates the curve at the point P.

The curvature tells us how quickly the tangent to the curve changes as we traverse the curve. To make this concrete, first parametrize the curve in terms of its arc length: $\Gamma(s) =$

$(x(s), y(s))$ with $x'(s)^2 + y'(s)^2 = 1$. The tangent vector to the curve at $(x(s), y(s))$, $\boldsymbol{T}(s) = (x'(s), y'(s))$, has norm one, and the unit normal is $\boldsymbol{n}(s) = (-y'(s), x'(s))$. If the curve is twice differentiable, then $\boldsymbol{T}'(s) = \kappa(s)\boldsymbol{n}(s)$, where $\kappa(s)$ is the (signed) curvature at the point. For example, for a circle of radius R, $(x(s), y(s)) = (R\cos(s/R), R\sin(s/R))$, and $\kappa(s) \equiv 1/R$.

The circle is, of course, a rare example for which arc length is easy to calculate, but by an application of the chain rule, this allows one to calculation κ in terms of an arbitrary parametrization

$$\kappa = \frac{\det(\Gamma', \Gamma'')}{\|\Gamma'\|^3}.$$

In the biologically relevant case of two dimensions, curvature flow is sometimes called the *curve-shortening* flow and its behavior is well understood. The flow has a finite lifetime $\mathscr{T}$. For example, if $\boldsymbol{\Gamma}_0$ is a circle of radius R_0, then $\boldsymbol{\Gamma}_t$ will be a circle with radius R_t satisfying $dR/dt = -1/R$, so the curve shrinks to a point in time $R_0^2/2$. In fact, this behavior is generic in $\mathrm{d} = 2$: in [24], it was shown that if $\boldsymbol{\Gamma}_0$ is convex then so is $\boldsymbol{\Gamma}_t$ for all $t < \mathscr{T}$, and that as $t \uparrow \mathscr{T}$ the asymptotic "shape" of $\boldsymbol{\Gamma}_t$ is a circle; [26] showed that any smoothly embedded closed curve becomes convex at a time $\tau < \mathscr{T}$.

2.3. The motion of hybrid zones

To state a result about the behavior of the solution to (1.4) as $\varepsilon \to 0$, we shall need some regularity assumptions on the initial condition.

Assumptions 2.3 (Assumptions on $u^\varepsilon(0, x)$). Let $u^\varepsilon(0, x) = p(x)$ where p takes values in $[0, 1]$ and set

$$\Gamma = \left\{x \in \mathbb{R}^{\mathrm{d}} : p(x) = \frac{1}{2}\right\}.$$

We suppose that Γ is a smooth hypersurface which is the boundary of an open set which is topologically equivalent to a sphere. (When $\mathrm{d} = 2$, this just says Γ is a smooth curve, topologically equivalent to a circle.) We further assume:

($\mathscr{C}1$) Γ is C^α for some $\alpha > 3$;

($\mathscr{C}2$) for x outside Γ, $p(x) < \frac{1}{2}$; for x inside Γ, $p(x) > \frac{1}{2}$;

($\mathscr{C}3$) there exists $r, \mu > 0$ such that, for all $x \in \mathbb{R}^2$, $|p(x) - \frac{1}{2}| \geq \mu(\mathtt{dist}(x, \Gamma) \wedge r)$.

Condition ($\mathscr{C}1$) guarantees that mean curvature flow $(\boldsymbol{\Gamma}_t(\cdot))_t$ started from Γ exists up to some time $\mathscr{T} > 0$. The second condition is just a convention; the third is to ensure that the slope of p near Γ is not too small, and that p is bounded away from $1/2$ for points that are not close to Γ.

We write $d(x, t)$ for the signed distance from x to $\boldsymbol{\Gamma}_t$, chosen to be positive inside $\boldsymbol{\Gamma}_t$ and negative outside. As sets, $\boldsymbol{\Gamma}_t = \{x \in \mathbb{R}^{\mathrm{d}} : d(x, t) = 0\}$.

Theorem 2.4 (Special case of Chen [11, THEOREM 3]). *Let $u^\varepsilon(t,x)$ solve (1.4) with $u^\varepsilon(0,x) = p(x)$ satisfying Assumptions 2.3. Fix $T^* \in (0,\mathscr{T})$ and let $k \in \mathbb{N}$. There exists $\varepsilon_{\mathrm{d}}(k) > 0$, and $a_{\mathrm{d}}(k), c_{\mathrm{d}}(k) \in (0,\infty)$ such that for all $\varepsilon \in (0,\varepsilon_{\mathrm{d}})$ and t satisfying $a_{\mathrm{d}}\varepsilon^2|\log\varepsilon| \le t \le T^*$,*

(1) *for x such that $d(x,t) \ge c_{\mathrm{d}}\varepsilon|\log\varepsilon|$, we have $u^\varepsilon(t,x) \ge 1-\varepsilon^k$;*

(2) *for x such that $d(x,t) \le -c_{\mathrm{d}}\varepsilon|\log\varepsilon|$, we have $u^\varepsilon(t,x) \le \varepsilon^k$.*

3. A PROBABILISTIC REPRESENTATION OF SOLUTIONS TO (1.4)

In [14] an analogue of Theorem 2.4 for a model which incorporates (weak) genetic drift is proved. A large part of that paper is devoted to providing a new proof of Theorem 2.4, for which we now explain the key ideas. It is based on a probabilistic representation of the solution to (1.4), and is readily adapted to a host of other situations, some of which we describe in Sections 4, 5, and 6.

For compatibility with the literature on partial differential equations, we shall suppose that all Brownian motions run at rate 2 (and so have infinitesimal generator Δ rather than $\frac{1}{2}\Delta$). The representation is in terms of a ternary branching Brownian motion in which:

(1) each individual has an independent exponentially distributed lifetime with mean ε^2 at the end of which it is replaced, at the location where it died, by three offspring; and

(2) during its lifetime, each individual follows an independent Brownian motion.

We shall only ever be interested in this process started from a single individual at time 0.

Whereas Skorokhod's representation of the solution to the Fisher–KPP equation in [36] is just in terms of the locations of the individuals in a (binary) branching Brownian motion at time t, our representation of the solution to (1.4) will also require the structure of the tree relating the individuals in the ternary branching Brownian motion. The simplest way to encode that information is to use Ulam–Harris notation. Each individual is labeled by an element of $\mathscr{U} = \bigcup_{m=0}^{\infty}\{1,2,3\}^m$. The original ancestor is labeled $\emptyset$. The offspring of an individual with label $\bar{\imath} = (i_1,\dots,i_m)$ receive the labels $(\bar{\imath},1)$, $(\bar{\imath},2)$, and $(\bar{\imath},3)$. Thus, for example, $(1,3)$ is the label of the third child of the first child of the original ancestor.

We shall use $\boldsymbol{W}(t) =$ to denote *historical ternary branching Brownian motion*, that is, the tree of Brownian paths traced out by ternary branching Brownian motion up until time t, and $\mathcal{T}(\boldsymbol{W}(t))$ for the corresponding ternary tree, obtained by ignoring the spatial positions of individuals.

Definition 3.1 (Majority voting in (historical) branching Brownian motion). For a fixed function $p : \mathbb{R}^{\mathrm{d}} \to [0,1]$, define a voting procedure on $\boldsymbol{W}(t)$ as follows. We write $\{W_i(t)\}_{i=1}^{N_t}$ for the spatial locations of the random number $N(t)$ of individuals alive at time t. We call these individuals the leaves.

(1) Each leaf, independently, votes a with probability $p(W_i(t))$; otherwise it votes A.

(2) At each branch point in $\mathcal{T}(\boldsymbol{W}(t))$, the vote of the parent particle $\bar{\iota}$ is the majority vote of the votes of its three children $(\bar{\iota}, 1)$, $(\bar{\iota}, 2)$, and $(\bar{\iota}, 3)$.

This defines an iterative voting procedure, which runs inwards from the leaves of $\boldsymbol{W}(t)$ to the root. We define $\mathbb{V}_p(\boldsymbol{W}(t))$ to be the vote associated to the root.

This majority voting procedure is illustrated in Figure 1.

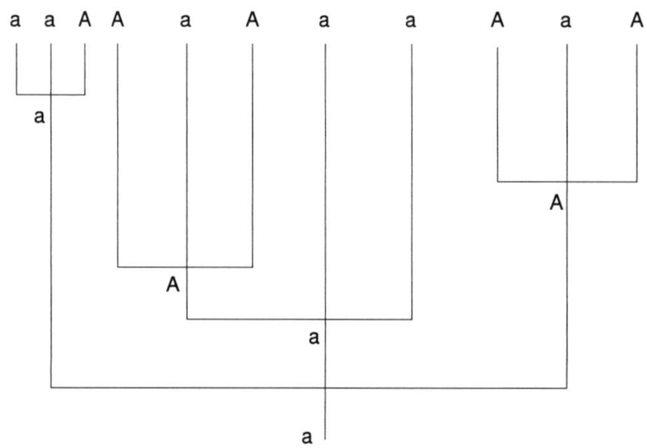

FIGURE 1
Majority voting on a ternary tree. Starting at the leaves, we move back to the root. At each branch point, the parent adopts the majority view of its children. In this example, the vote at the root is a.

Lemma 3.2 (Majority voting and the Allen–Cahn equation). *Let $\boldsymbol{W}(t) =$ be a historical ternary branching Brownian motion with branching rate $1/\varepsilon^2$, and let $p : \mathbb{R}^{\mathrm{d}} \to [0, 1]$. The function*

$$u^\varepsilon(t, x) = \mathbb{P}_x^\varepsilon\big[\mathbb{V}_p\big(\boldsymbol{W}(t)\big) = a\big] \tag{3.1}$$

solves equation (1.4) with $u^\varepsilon(0, x) = p(x)$.

The subscript x on the right hand side of equation (3.1) indicates that $\boldsymbol{W}$ starts from a single individual at x at time zero. The proof of Lemma 3.2 proceeds in a standard way by partitioning over whether or not the ancestor in the branching Brownian motion dies in the first δt of time, and thus calculating

$$\lim_{\delta t \to 0} \frac{(u^\varepsilon(t + \delta t, x) - u^\varepsilon(t, x))}{\delta t}. \tag{3.2}$$

To understand why majority voting gives rise to the desired nonlinearity, consider what happens if the ancestor *does* die in the first δt of time, which happens with probability $\delta t/\varepsilon^2$. Each offspring, independently, votes a with the same probability, u say. The probability that the majority of their 3 votes is a is $u^3 + 3u^2(1 - u) = u(1 - u)(2u - 1) + u$. Assuming some continuity so that we can take $u \approx u^\varepsilon(t,x)$ as $\delta t \to 0$, we see that the contribution to (3.2) from the event {the ancestor died before time δt} is $u^\varepsilon(1-u^\varepsilon)(2u^\varepsilon - 1)/\varepsilon^2$ as required. With probability $1 - \delta t/\varepsilon^2$ the ancestor *did not* die before δt, and it followed a Brownian motion which is at position $W_{\delta t}$ at time δt. Using the Markov property at time δt, the Laplacian term arises from this event as $\lim_{\delta t \to 0}(\mathbb{E}_x[u^\varepsilon(t, W_{\delta t})] - u(t,x))/\delta t$.

With the representation (3.1), the conclusion of Theorem 2.4 can be written:

(1) for x with $d(x,t) \geq c_{\mathrm{d}}\varepsilon|\log \varepsilon|$, $\mathbb{P}_x^\varepsilon[\mathbb{V}_p(\boldsymbol{W}(t)) = a] \geq 1 - \varepsilon^k$;

(2) for x with $d(x,t) \leq -c_{\mathrm{d}}\varepsilon|\log \varepsilon|$, $\mathbb{P}_x^\varepsilon[\mathbb{V}_p(\boldsymbol{W}(t)) = a] \leq \varepsilon^k$.

The intuition behind the proof of these statements is very simple. First observe that majority voting increases bias: if $p < \frac{1}{2}$, $p^3 + 3p^2(1-p) < p$; if $p > \frac{1}{2}$, $p^3 + 3p^2(1-p) > p$. Since the branching rate is $1/\varepsilon^2$, our branching Brownian motion sees many rounds of majority voting in a very short space of time, and so a small bias in votes at the leaves of the tree translates into a large bias at the root. As a result, a narrow interface will be generated across which there is a rapid transition from $\mathbb{P}_x^\varepsilon[\mathbb{V}_p(\boldsymbol{W}(t)) = a]$ being close to zero, to it being close to one. Suppose that in fact this transition is sharp, and the solution to equation (1.4) is the indicator function of a region bounded by a surface Γ. Taking this solution as the new initial condition, after a small time h, we once again expect that the solution is close to a sharp interface whose position, Γ_h, marks the transition from a voting bias in favor of type a, to one in favor of type A. That is, $\Gamma_h \approx \{x \in \mathbb{R}^{\mathrm{d}} : T_h \mathbf{1}_\Gamma(x) = 1/2\}$ where T denotes the heat semigroup. If we replace the solution at time h by $\mathbf{1}_{\Gamma_h}$ and repeat this process, we are actually performing the Merriman–Bence–Osher (MBO) algorithm for simulating mean curvature flow **[32]**. To gain some intuition for the role of mean curvature flow, consider the special case in which Γ is a sphere of radius R in $\mathbb{R}^{\mathrm{d}}$. Then we are approximating Γ_h by $\{x \in \mathbb{R}^{\mathrm{d}} : \mathbb{P}_x[\|W_h\| > R] = 1/2\}$, where W is d-dimensional Brownian motion. Since the radial part of a d-dimensional Brownian motion is a d-dimensional Bessel process, while it is close to Γ, $\|W\|$ will be distributed as a one-dimensional Brownian motion B with drift close to $(\mathrm{d}-1)/R$ (remembering that our Brownian motions all run at rate two), and so we are approximating Γ_h by the set of points for which $\mathbb{P}_{\|x\|}[B_h + h(\mathrm{d}-1)/R > R] = 1/2$; in other words, by symmetry of B, by $\{x : \|x\| = R - h(\mathrm{d}-1)/R\}$. The mean curvature of Γ is $(\mathrm{d}-1)/R$, and so for small h, Γ_h is close to the surface obtained by evolving Γ according to mean curvature flow for time h.

This intuitive picture is close to the structure of the rigorous proof which has two main ingredients: an analysis of the one-dimensional solution, started from a Heaviside initial condition; and coupling (close to the interface) of $d(W_s, t-s)$ with a one-dimensional Brownian motion.

4. LONG-RANGE DISPERSAL

The Laplacian in equation (1.4) reflects an assumption that offspring remain close to their parents. However, for many organisms this may fail; see, for example, **[10]** for a discussion of long range seed dispersal in plants. To incorporate this into equation (1.5), we replace the Laplacian by a fractional Laplacian,

$$\frac{\partial v}{\partial t} = (-\Delta)^{\frac{\alpha}{2}} v + \mathrm{s}v(1-v)(2v-1), \tag{4.1}$$

where (for smooth functions f which decay sufficiently fast)

$$(-\Delta)^{\frac{\alpha}{2}} f(x) := C_\alpha \lim_{\delta\to 0} \int_{\mathbb{R}^{\mathrm{d}}\setminus B_\delta(x)} \frac{f(y)-f(x)}{\|y-x\|^{\mathrm{d}+\alpha}} dy. \tag{4.2}$$

Here $C_\alpha := 2^\alpha \Gamma(\frac{\alpha}{2}+\frac{\mathrm{d}}{2})/(\pi^{\mathrm{d}/2}|\Gamma(-\frac{\alpha}{2})|)$, where Γ is the Gamma function, and $B_\delta(x)$ is the ball of radius δ about x.

The operator (4.2) is the generator of a symmetric α-stable process, and the probabilistic representation of solutions to (4.1) follows by substituting a branching α-stable process for the branching Brownian motion in Section 3. If we are to recover an analogue of Theorem 2.4, we expect to need to consider scales over which the spatial motion along each branch is close to a Brownian motion. We appeal to a decomposition often used in the numerical simulation of symmetric stable processes, see, for example, [2]. If we run an α-stable process at rate $I(\varepsilon)^{\alpha-2}$ (where $I(\varepsilon)\to 0$ as $\varepsilon\to 0$), then the process obtained by censoring jumps of size greater than $I(\varepsilon)$ can be approximated by Brownian motion. To show that the *uncensored* process along a branch of our ternary tree is close to a Brownian motion, we need to control the number of jumps of size at least $I(\varepsilon)$ before an exponential time with mean ε^2. Since we have time-changed the stable process by $I(\varepsilon)^{\alpha-2}$, the rate of such jumps is $\mathcal{O}(I(\varepsilon)^{-2})$, and so in order that branches on which we see jumps of size more than $I(\varepsilon)$ be rare, we take $I(\varepsilon)/\varepsilon\to\infty$.

With this in mind, set $\sigma^2 = 2C_\alpha/(2-\alpha)$ and rescale time and space by $t\mapsto\varepsilon^2 t$, $x\mapsto \varepsilon^{2/\alpha} I(\varepsilon)^{1-2/\alpha} x$. When $\alpha=2$, we recover the diffusive scaling. Equation (4.1) becomes

$$\frac{\partial v^\varepsilon}{\partial t} = \frac{\sigma^{-2}}{I(\varepsilon)^{2-\alpha}} (-\Delta)^{\frac{\alpha}{2}} v^\varepsilon + \frac{1}{\varepsilon^2} v^\varepsilon(1-v^\varepsilon)(2v^\varepsilon-1), \quad v^\varepsilon(0,x)=p(x). \tag{4.3}$$

In [7], an analogue of Theorem 2.4 is proved for functions $I:\mathbb{R}_+\to\mathbb{R}_+$ satisfying:

($\mathscr{A}1$) $\lim_{\varepsilon\to 0} I(\varepsilon)|\log(\varepsilon)|^k = 0\ \forall k\in\mathbb{N}$.

($\mathscr{A}2$) $\lim_{\varepsilon\to 0} \frac{\varepsilon^2|\log(\varepsilon)|}{I(\varepsilon)^2} = 0$.

($\mathscr{A}3$) $\lim_{\varepsilon\to 0} H(\varepsilon) := I(\varepsilon)^2|\log(\varepsilon)|\varepsilon^{\frac{2\mathrm{d}}{\alpha}-\mathrm{d}-1} + \frac{I(\varepsilon)^{2\alpha}}{\varepsilon^2}|\log(\varepsilon)|^\alpha = 0$.

Note that Assumptions ($\mathscr{A}2$) and ($\mathscr{A}3$) are incompatible as soon as $\alpha\le 1$.

Theorem 4.1 ([7, THEOREM 1.5]). *Let $\alpha\in(1,2)$ and suppose that $I(\varepsilon)$ satisfies Assumptions ($\mathscr{A}1$)–($\mathscr{A}3$) above. Suppose v^ε solves equation* (4.3) *with initial condition p satisfying Assumptions* 2.3. *Let $\mathscr{T}$ and $d(x,t)$ be as in Section* 2.3, *and fix $T^*\in(0,\mathscr{T})$. Then there exists $\varepsilon_{\mathrm{d}}(\alpha,I)$, $a_{\mathrm{d}}(\alpha,I)$, $c_{\mathrm{d}}(\alpha,I)$, $M(\alpha,I)>0$ such that, for $\varepsilon\in(0,\varepsilon_{\mathrm{d}})$ and $a_{\mathrm{d}}\varepsilon^2|\log\varepsilon|\le t\le T^*$,*

(1) *for x with $d(x,t)\ge c_{\mathrm{d}} I(\varepsilon)|\log\varepsilon|$, we have $v^\varepsilon(t,x)\ge 1-\frac{\varepsilon^2}{I(\varepsilon)^2} - M(H(\varepsilon)+I(\varepsilon)^{\alpha-1})$;*

(2) *for* x *with* $d(x,t) \le -c_{\mathrm{d}} I(\varepsilon)|\log\varepsilon|$, *we have* $v^\varepsilon(t,x) \le \frac{\varepsilon^2}{I(\varepsilon)^2} + M(H(\varepsilon) + I(\varepsilon)^{\alpha-1})$.

For example, $I(\varepsilon) = \varepsilon|\log(\varepsilon)|$ fulfills Assumptions $(\mathscr{A}1)$–$(\mathscr{A}3)$, and the "error" $\varepsilon^2/I(\varepsilon)^2 + M(H(\varepsilon) + I(\varepsilon)^{\alpha-1})$ is of order $1/(\log\varepsilon)^2$. There are two competing effects: we want to take $I(\varepsilon)$ as small as possible if the approximation of the small jumps of the stable process by a Brownian motion is to be good; on the other hand, we need $I(\varepsilon)$ to be large if branches along which we see a jump of size more than $I(\varepsilon)$ are to be rare. In contrast to the Brownian case, these cannot be balanced to obtain an error of order ε^k for arbitrary k.

5. ASYMMETRY AND BLOCKING

So far we have worked exclusively on the whole of Euclidean space. In this section we see that, in some scenarios, the geometry of the domain can be important.

5.1. An asymmetric reaction: homozygotes of different fitnesses

In our justification of equation (1.3) in Section 2, we assumed that both homozygotes were equally fit. It is natural to ask what happens if that is not the case? Suppose, for example, that we take relative fitnesses

aa	aA	AA
$1+\gamma_1 s_1$	$1-s_1$	1

,

where γ_1 is assumed small. Mimicking our previous approach, and setting $(2+\gamma_1)s_1/2 = \mathbf{s}/M$, $2/(2+\gamma_1) = 1-\gamma$, we recover equation (1.5). The one-dimensional equation

$$\frac{\partial u}{\partial t} = m\frac{\partial^2 u}{\partial x^2} + \mathbf{s}u(1-u)\big(2u-(1-\gamma)\big)$$

has a traveling wave solution of the form

$$u(x,t) = \left(1+\exp\left(-\sqrt{\frac{\mathbf{s}}{m}}(x+\gamma\sqrt{m\mathbf{s}}t)\right)\right)^{-1}, \tag{5.1}$$

connecting 0 at $-\infty$ to 1 at ∞, and with wave speed $\gamma\sqrt{m\mathbf{s}}$. In particular, if we scale m and/or $\mathbf{s}$, then we may also have to scale γ in order to obtain a finite wavespeed. With this in mind, **[25]** considers the equation

$$\frac{\partial u^\varepsilon}{\partial t} = \varepsilon^{1-\ell}\Delta u^\varepsilon + \frac{1}{\varepsilon^{1+\ell}}u^\varepsilon(1-u^\varepsilon)\big(2w^\varepsilon-(1-\gamma_\varepsilon)\big), \quad x\in\mathbb{R}^{\mathrm{d}},\ t>0, \tag{5.2}$$

where $\gamma_\varepsilon = \nu\varepsilon^{\tilde\ell}$ for some nonnegative ν and $\tilde\ell$, with the additional condition that $\nu<1$ when $\tilde\ell = 0$, and $\ell = \min(\tilde\ell, 1)$.

Notice that with these parameters, the one-dimensional wave has speed of $\mathcal{O}(1)$ if $\tilde\ell \le 1$ and tending to zero as $\varepsilon^{\tilde\ell-1}$ if $\tilde\ell > 1$. We define

$$\nu_\varepsilon = \begin{cases} \nu & \text{if } \tilde\ell \le 1, \\ \gamma_\varepsilon/\varepsilon & \text{if } \tilde\ell \in (1,2], \\ 0 & \text{if } \tilde\ell > 2. \end{cases} \tag{5.3}$$

Set $u^\varepsilon(x,0) = p(x)$, take $\Gamma = \{x \in \mathbb{R}^d : p(x) = (1+\gamma_\varepsilon)/2\}$, and modify Assumptions 2.3 in the obvious way (by replacing $1/2$ by $(1+\gamma_\varepsilon)/2$).

Theorem 5.1 (Restatement of [25, THEOREM 2.4]). *Let u^ε solve equation (5.2) with initial condition p satisfying Assumptions 2.3 (modified as described above), and let*

$$\frac{\partial \widetilde{\mathbf{\Gamma}}(s)}{\partial t} = \big(-\nu_\varepsilon + \kappa_t(s)\big)\mathbf{n}_t(s), \tag{5.4}$$

until the time $\mathscr{T}$ at which $\widetilde{\mathbf{\Gamma}}$ develops a singularity. Write $\tilde{d}$ for the signed distance to $\widetilde{\mathbf{\Gamma}}$ (chosen to be positive inside $\widetilde{\mathbf{\Gamma}}$). Fix $T^* \in (0,\mathscr{T})$ and $k \in \mathbb{N}$. There exists $\varepsilon_{\mathrm{d}}(k) > 0$, and $a_{\mathrm{d}}(k), c_{\mathrm{d}}(k) \in (0,\infty)$ such that for all $\varepsilon \in (0,\varepsilon_{\mathrm{d}})$ and t satisfying $a_{\mathrm{d}}\varepsilon^{1+\ell}|\log\varepsilon| \le t \le T^*$,

1. *for x such that $\tilde{d}(x,t) \ge c_{\mathrm{d}}\varepsilon|\log\varepsilon|$, we have $u^\varepsilon(t,x) \ge 1-\varepsilon^k$;*
2. *for x such that $\tilde{d}(x,t) \le -c_{\mathrm{d}}\varepsilon|\log\varepsilon|$, we have $u^\varepsilon(t,x) \le \varepsilon^k$.*

Remark 5.2. When $\tilde{l} = 1$, Theorem 5.1 is a special case of Theorem 1.3 of [1] in which more general "slightly unbalanced" bistable nonlinearities are considered.

For $\tilde{\ell} \le 2$, ν_ε in (5.3) and (5.4) corresponds to the one-dimensional wavespeed derived above. For $\tilde{\ell} > 2$, the wavespeed converges to zero sufficiently quickly as $\varepsilon \to 0$ that it is not necessary to include the corresponding small contribution from the constant flow in (5.4).

We shall focus on equation (5.2) with $\tilde{\ell} = 1$,

$$\frac{\partial u^\varepsilon}{\partial t} = \Delta u^\varepsilon + \frac{1}{\varepsilon^2}u^\varepsilon(1-u^\varepsilon)\big(2u^\varepsilon - (1-\varepsilon\nu)\big) \quad x \in \mathbb{R}^d,\ t > 0. \tag{5.5}$$

The approach of [25] is to extend the probabilistic representation to the asymmetric case.

Lemma 5.3. *Let $\widetilde{\boldsymbol{W}}(t)$ be a historical ternary branching Brownian motion with branching rate $(1+\varepsilon\nu)/\varepsilon^2$, and let $p : \mathbb{R}^d \to [0,1]$. Define a voting procedure on $\widetilde{\boldsymbol{W}}(t)$ as follows:*

1. *Each leaf, independently votes a with probability $p(\widetilde{W}_i(t))$, otherwise it votes A;*
2. *at a branch point, the parental vote is the majority vote of the children* unless *precisely one offspring vote is a, in which case the parent votes a with probability $2\varepsilon\nu/(3+3\varepsilon\nu)$.*

Write $\widetilde{\mathbb{V}}_p(\widetilde{\boldsymbol{W}}(t))$ for the vote associated with the root. Then

$$u^\varepsilon(t,x) = \mathbb{P}_x^\varepsilon\big[\widetilde{\mathbb{V}}_p\big(\widetilde{\boldsymbol{W}}(t)\big) = a\big]$$

solves equation (5.5) with $u^\varepsilon(0,x) = p(x)$.

The proof of Theorem 5.1 closely follows the probabilistic proof of Theorem 2.4 in [14], except that the signed distance $\tilde{d}(W_s, \widetilde{\Gamma}_{t-s})$ is coupled to a one-dimensional Brownian motion with drift ν.

Remark 5.4 (Other voting schemes). The probabilistic representation above is far from unique. For example, it might seem more natural to write the reaction term in (5.5) as $\frac{1}{\varepsilon^2}(u^\varepsilon(1-u^\varepsilon)(2u^\varepsilon-1)+\varepsilon\nu u^\varepsilon(1-u^\varepsilon))$, and express the solution in terms of a branching Brownian motion with a mixture of binary branching at rate ν/ε, with the rule that the parent votes a unless both offspring vote A, and ternary branching at rate $1/\varepsilon^2$ with the majority voting rule. However, it turns out to be much more convenient to base the proof on a ternary tree. To obtain the voting mechanism above, we rewrite the quadratic term $u(1-u)$ as the sum of two cubic terms using that $1 = u + (1-u)$.

Voting schemes are very general. In [35], O'Dowd showed that if $P(u)$ is any polynomial with $P(0) \geq 0$ and $P(1) \leq 0$ (or vice versa), then the solution to

$$\frac{\partial u}{\partial t} = \Delta u + P(u)$$

can be represented in terms of a historical n-ary branching Brownian motion (where n is the degree of P) and a rule for assigning votes to a parent according to the votes of its offspring.

5.2. Geometry matters: blocking

We now turn our attention to solutions to (5.5) on domains $\Omega \subseteq \mathbb{R}^{\mathrm{d}}$ with reflecting boundary conditions. We focus on the fate of the favored allele as it tries to expand through a semiinfinite domain. We shall consider "cylindrical" domains of the form

$$\Omega = \{(x_1, x') : x_1 \in \mathbb{R},\ x' \in \phi(x_1) \subseteq \mathbb{R}^{\mathrm{d}-1}\}. \tag{5.6}$$

We shall always take the initial condition $u^\varepsilon(0,x) = \mathbf{1}_{x_1 \geq 0}$.

Theorem 5.5 ([8], Theorems 1.4, 1.5, 1.6, 1.7, paraphrased). *Let u be the solution to equation* (1.5) *on Ω with normal reflection on the boundary and initial condition $u(x,0) = \mathbf{1}_{x_1 \geq 0}$. Depending on the geometry of the domain Ω we have one of three possible asymptotic behaviors of the solution of equation* (5.5)*:*

(1) *there can be* complete invasion, *that is, $u(x,t) \to 1$ as $t \to \infty$ for every $x \in \Omega$;*

(2) *there can be* blocking *of the solution, meaning that $u(x,t) \to u_\infty(x)$ as $t \to \infty$, with $u_\infty(x) \to 0$ as $x_1 \to -\infty$;*

(3) *there can be* axial partial propagation, *meaning that $u(x,t) \to u_\infty(x)$ as $t \to \infty$, with $\inf_{x \in \mathbb{R} \times B_R} u_\infty(x) > c > 0$ for some $R > 0$, where B_R is the ball of radius R centered at 0 in $\mathbb{R}^{\mathrm{d}-1}$.*

Which behavior is observed depends on the geometry of the domain Ω. For example, there will be complete invasion if Ω is decreasing as x_1 decreases; axial partial propagation if it contains a straight cylinder of sufficiently large cross-section; and there can be blocking if there is an abrupt change in the geometry.

The results of [18] concerning the behavior of solutions to (5.5) complement those of [8]. (The addition of the parameter ε, which is not present in the work of [8], prevents direct

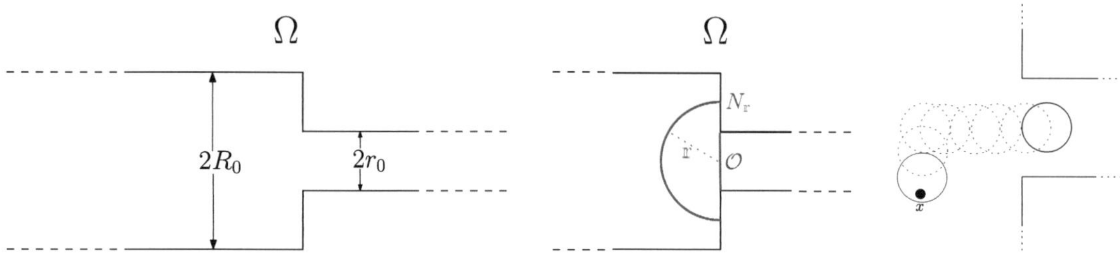

FIGURE 2
Left to right: (a) the domain Ω of Theorems 5.6 and 5.7; (b) the opening $\mathcal{O}$ and the hemispherical shell $N_{\mathbb{r}}$ used in the proof of Theorem 5.6; (c) an illustration of "chaining" used in the proof of Theorem 5.7. Image taken from **[18]**.

comparison.) As in the previous sections, they are based on the probabilistic representation of solutions. Following **[8]**, we begin with the very special form of Ω depicted in Figure 2.

Theorem 5.6 ([18, THEOREM 1.6]). *Let u^ε denote the solution to equation (5.5) on the domain Ω in Figure 2, with reflecting boundary condition and $u^\varepsilon(0, x) = \mathbf{1}_{x_1 \geq 0}$. Suppose $r_0 < \frac{\mathrm{d}-1}{\nu} \wedge R_0$. Define $N_{\mathbb{r}} = \{x \in \Omega : \|x\| = \mathbb{r},\ x_1 < 0\}$, where $\frac{\mathrm{d}-1}{\nu} \wedge R_0 > \mathbb{r} > r_0$, and let $\hat{d}(x)$ be the signed (Euclidean) distance of any point $x \in \Omega$ to $N_{\mathbb{r}}$ (chosen to be negative as $x_1 \to -\infty$). Let $k \in \mathbb{N}$. Then there is $\hat{\varepsilon}(k) > 0$ and $M(k) > 0$ such that for all $\varepsilon \in (0, \hat{\varepsilon})$, and all $t \geq 0$,*

$$\text{for } x = (x_1, \ldots, x_{\mathrm{d}}) \in \Omega \text{ such that } \hat{d}(x) \leq -M(k)\varepsilon\big|\log(\varepsilon)\big|, \text{ we have } u^\varepsilon(x, t) \leq \varepsilon^k.$$

In other words, if the aperture r_0 is too small, then, for sufficiently small ε, blocking occurs. With the machinery of Section 5.1 in place, the proof is straightforward. First we check that the solution to (5.5) on Ω is monotone in the initial condition, which allows us to compare with the solution started from an initial condition p which dominates $\mathbf{1}_{x_1 \geq 0}$, is radially symmetric in the left half plane, satisfies $p(x) = (1 - \gamma_\varepsilon)/2$ on the hemispherical shell $N_{\mathbb{r}}$, and fulfills the analogue of conditions ($\mathscr{C}2$) and ($\mathscr{C}3$) from Assumptions 2.3 (with Γ replaced by $N_{\mathbb{r}}$ and $1/2$ by $(1 - \gamma_\varepsilon)/2$). For this initial condition, it is straightforward to adapt the proof for the whole Euclidean space from **[25]**, and indeed things are simplified considerably by the radial symmetry.

The converse of Theorem 5.6 is also true in the following sense.

Theorem 5.7 ([18, THEOREM 1.7]). *Let u^ε be as in Theorem 5.6. Suppose $r_0 > \frac{\mathrm{d}-1}{\nu}$, then for all $x \in \Omega$ and $\delta > 0$ there is $\hat{t} := \hat{t}(x_1, \delta, R_0, r_0) > 0$ and $\hat{\varepsilon}$ such that, for all $\varepsilon \in (0, \hat{\varepsilon})$ and $t \geq \hat{t}$, we have $u^\varepsilon(t, x) \geq 1 - \delta$.*

Again the proof exploits monotonicity in the initial condition. The solution dominates one started from $(1 - \varepsilon)$ times the indicator of a ball of radius $r > (\mathrm{d} - 1)/\nu$, with center sitting on the x_1-axis and contained in $\Omega \cap \{x : x_1 \geq 0\}$. This time, adapting the arguments for the solution on $\mathbb{R}^{\mathrm{d}}$ tells us that at a later time that solution dominates $(1 - \varepsilon)$ times the indicator of a ball with larger radius r', but the same center, strictly contained within Ω. We now start the process again, taking as initial condition $1 - \varepsilon$ times the indicator of a ball of radius r, and with center shifted a distance $r' - r$. Continuing in this way, we can find a chain of balls connecting any point $x \in \Omega$ to the original ball. This process of "chaining" is illustrated in Figure 2. It mirrors the use of the "sliding ball" assumption to prove complete propagation in **[8]**.

Together, Theorems 5.6 and 5.7 say that there is a sharp transition at the critical radius $(\mathbb{d} - 1)/\nu$. As described in Section 1, this is the radius of the shell at which the constant and curvature flow exactly balance on the whole of $\mathbb{R}^{\mathbb{d}}$. However, in that case a small perturbation results in complete invasion or extinction of the favored type, here a stable interface will be maintained.

The domain Ω of Theorem 5.6 is very special. However, the crucial step was to be able to cover the opening $\mathcal{O}$ illustrated in Figure 2 by a hemispherical shell of less than the critical radius $(\mathbb{d} - 1)/\nu$, and orthogonal to the boundary of the domain where they intersect. The same result will follow (from essentially the same argument) for any domain which can be "blocked" by a portion of such a shell in this way. As a first step, consider the domain $\widetilde{\Omega}$, which opens out as a truncated cone, and the shell of radius $\mathbb{r}$ shown in Figure 3. We can choose $\mathbb{r} < (\mathbb{d} - 1)/\nu$ precisely when $r_0 < (\mathbb{d} - 1)\sin\alpha/\nu$.

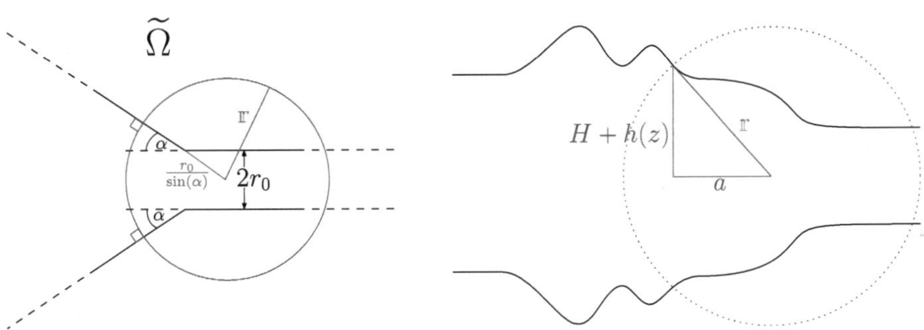

FIGURE 3
(Left) The domain $\widetilde{\Omega}$. (See text below Theorem 5.7.) (Right) An example of a domain from Theorem 5.8. Condition (5.7) that guarantees that we can insert a portion of a spherical shell as shown with radius less than $(\mathbb{d} - 1)/\nu$ can be read off from that for $\widetilde{\Omega}$ on setting $r_0 = H + h(z)$ and $\sin\alpha = h'(z)/\sqrt{1 + h'(z)^2}$. Image taken from **[18]**.

The intuition behind blocking is that if the domain opens out too rapidly, then offspring of favored individuals are "spread too thin" and selection against hybrids will rapidly eliminate their descendants. Our approach has been to seek a shell of sufficiently small radius that intercepts the boundary of the domain orthogonally, but if the domain is opening even faster, in the sense that expanding the shell radially one stays within the domain, at least for a short time, this effect will be further amplified. This is the meaning of the condition (5.7) in the following theorem.

Theorem 5.8 (**[18, THEOREM 1.9]**). *Suppose that u^{ε} solves* (5.5) *where $\Omega \subseteq \mathbb{R}^{\mathbb{d}}$ is defined as in* (5.6) *with*

$$\phi(x_1) = \left\{ \|x'\| \leq H + h(-x_1) \right\},$$

and h a nonnegative C^1 function. Suppose that

$$\inf_{z>0}\left\{H + h(z) - \left(\frac{\mathrm{d}-1}{\nu}\right)\frac{h'(z)}{\sqrt{1+h'(z)^2}}\right\} < 0. \tag{5.7}$$

Fix $k \in \mathbb{N}$. There exist $x_0 < 0$, $\hat{\varepsilon}(k) > 0$ and $M(k) > 0$ such that for all $\varepsilon \in (0, \hat{\varepsilon})$ and $t \geq 0$,

for $x = (x_1, \dots, x_{\mathrm{d}}) \in \Omega$ such that $x_1 \leq x_0 - M(k)\varepsilon\left|\log(\varepsilon)\right|$ we have $u^\varepsilon(x,t) \leq \varepsilon^k$.

Condition (5.7) can be understood from the condition $r_0 < (\mathrm{d}-1)\sin\alpha/\nu$ on $\widetilde{\Omega}$ on setting $r_0 = H + h(z)$ and $\sin\alpha = h'(z)/\sqrt{1+h'(z)^2}$.

Conversely, if the domain does not open up sufficiently fast, we have invasion.

Theorem 5.9 ([18, THEOREM 1.10]). *Suppose that u^ε solves (5.5) where $\Omega \subseteq \mathbb{R}^{\mathrm{d}}$ is defined as in (5.6) with*

$$\phi(x_1) = \left\{\left\|x'\right\| \leq H + h(-x_1)\right\},$$

and h a nonnegative C^1 function. Suppose that

$$\inf_{z>0}\left\{H + h(z) - \left(\frac{\mathrm{d}-1}{\nu}\right)\frac{h'(z)}{\sqrt{1+h'(z)^2}}\right\} > 0.$$

Then for all $x \in \Omega$ and $\delta > 0$ there is $\hat{t} := \hat{t}(x_1, \delta) > 0$ and $\hat{\varepsilon}$ such that, for all $\varepsilon \in (0, \hat{\varepsilon})$ and $t \geq \hat{t}$, we have $u^\varepsilon(t,x) \geq 1 - \delta$.

These results (valid for any $\mathrm{d} \geq 2$) are somewhat analogous to those of [30], which consider a plane curve evolving according to equation (5.4) in a two-dimensional cylinder with a periodic saw-toothed boundary. The authors say that such a curve is a *periodic* traveling wave with effective speed δ/T_δ if $\widetilde{\Gamma}_{t+T_\delta}(s) = \widetilde{\Gamma}_t(s) + \delta$ for some $\delta > 0$. Setting $h_\delta(x) = \delta h_1(x/\delta)$ and letting $\delta \to 0$ leads to the homogenization limit of the wave, with speed $c_0 = \lim_{\delta\to 0} c_\delta$. They show that $c_0 > 0$ for $\nu H > \sin\alpha$ with α determined by $\tan\alpha = \max_x h'(x)$, but that the wave is blocked for small enough δ if $\nu H < \sin\alpha$.

6. ADDING NOISE

In Section 2, we motivated the noise appearing in equation (1.2) as a means of taking account of the randomness due to resampling inherent in reproduction in a finite population. Although in $\mathrm{d} = 1$, where the equations are well-posed, quite a lot is known about the solutions to stochastic reaction–diffusion equations like (1.2), in $\mathrm{d} \geq 2$ such equations have no solution. On the other hand, we have seen in Section 5.2 that populations may behave quite differently in $\mathrm{d} = 1$ and $\mathrm{d} = 2$ and so it may be misleading to only consider the one-dimensional equation.

The Spatial Λ-Fleming–Viot process was introduced in [4,17] as an alternative way to capture the effect of genetic drift in models for proportions of different allelic types in populations evolving in a spatial continuum. Although originally introduced for selectively neutral populations, it can be thought of as providing a framework for modeling, which can readily be adapted to incorporate a wealth of biologically relevant features, including natural selection.

First, we define a very special version of the Spatial Λ-Fleming–Viot process for a neutral population evolving in $\mathbb{R}^d$. As usual, we are most interested in $d = 2$. At each time t, the random function $\{w_t(x) : x \in \mathbb{R}^d\}$ will model the proportion of a-alleles at spatial position x at time t. Strictly speaking, the process is only defined up to a Lebesgue-null set. The identification

$$\int_{\mathbb{R}^d} \{w_t(x) f(x,a) + (1 - w_t(x)) f(x,A)\} dx = \int_{\mathbb{R}^d \times \{a,A\}} f(x,\kappa) M(dx, d\kappa)$$

provides a one-to-one correspondence between its state space and the space $\mathcal{M}_\lambda$ of measures on $\mathbb{R}^d \times \{a, A\}$ with "spatial marginal" Lebesgue measure, which we endow with the topology of vague convergence. We abuse notation and also denote the state space of the process $(w_t)_{t \in \mathbb{R}_+}$ by $\mathcal{M}_\lambda$.

Definition 6.1 (A neutral Spatial Λ-Fleming–Viot process (SLFV)). Fix $u \in (0, 1]$ and $r > 0$. Let Π be a Poisson Point Process on $\mathbb{R}_+ \times \mathbb{R}^d$ with intensity measure $dt \otimes dx$. The *Spatial Λ-Fleming–Viot process* driven by Π, with *event radius* r and *impact parameter* u, is the $\mathcal{M}_\lambda$-valued process $(w_t)_{t \geq 0}$ with dynamics given as follows.

If $(t, x) \in \Pi$, a reproduction event occurs at time t within the closed ball $B(x, r)$ of radius r centered on x:

(1) Choose a parental location z uniformly at random in $B(x, r)$, and a parental type, α_0, according to $w_{t-}(z)$; that is $\alpha_0 = a$ with probability $w_{t-}(z)$ and $\alpha_0 = A$ with probability $1 - w_{t-}(z)$.

(2) For every $y \in B(x, r)$, set $w_t(y) = (1 - u) w_{t-}(y) + u \mathbf{1}_{\{\alpha_0 = a\}}$.

Remark 6.2. Suppose that a reproduction event affects the ball $B(x, r)$ in which the proportion of a-alleles immediately before the event is w, and write w^* for the proportion of a-alleles immediately after the event. Then

$$\mathbb{E}\big[w^* - w\big] = 0, \quad \text{and} \quad \mathtt{var}(w^* - w) = u^2 w(1 - w).$$

This can be compared to the (Wright–Fisher) sampling noise in Section 2.1.

This is a very special case of the SLFV, even for a neutral population. More generally, one can take both r and u to be random. See [19] for a construction of the process under very much more general conditions.

Instead of sampling a parental location, and then a parental type, we could equally have just sampled types independently and uniformly at random according to the proportions in the region affected by the event. The two-step description is convenient as we wish to trace the ancestry of a sample from the population. Things are made particularly simple as the Poisson process Π that dictates reproduction events is reversible (with the same distribution). We write $\overleftarrow{\Pi}$ for the time-reversed process.

Definition 6.3 (SLFV dual). The process $(\mathcal{P}_t)_{t \geq 0}$ is the $\bigcup_{l \geq 1} (\mathbb{R}^d)^l$-valued Markov process with dynamics defined as follows.

The process starts from a finite collection of points $\xi_1(0), \ldots, \xi_{N(0)} \in \mathbb{R}^d$. We write $\mathcal{P}_t = (\xi_1(t), \ldots, \xi_{N(t)}(t))$, where the random number $N(t) \in \mathbb{N}$ is the number of individuals alive at time t, and $\{\xi_i(t)\}_{i=1}^{N(t)}$ are their locations. For each $(t, x) \in \overleftarrow{\Pi}$:

(1) for each $\xi_i(t-) \in B(x, r)$, independently mark the corresponding individual with probability u;

(2) if at least one individual is marked, all marked individuals coalesce into a single individual, whose location is chosen uniformly in $B(x, r)$.

If no individual is marked, then nothing happens.

One can write down a formal duality between this "backwards in time" ancestral process and the SLFV. It requires a little care because the SLFV is only defined up to a Lebesgue null set. However, informally, suppose that we know $\{w_0(x) : x \in \mathbb{R}^d\}$ and that we would like to find the type of an individual sampled from the point z at time t. Starting the dual from a single individual with $\xi_1(0) = z$, $\xi_1(t)$ is the location of the ancestor of the sampled individual at time 0, and its type is determined by sampling according to $w_0(\xi_1(t))$.

Each ancestral lineage evolves in a series of jumps. By translation invariance, its distribution is determined by the rate at which an ancestral lineage jumps from 0 to $x \in \mathbb{R}^d$. For such a jump to occur, three things must happen: first, an event has to fall that covers both 0 and x; second, the lineage has to be among the offspring of the event; third, x has to be chosen as the location of the parent. Writing $L_r(x) = |B_r(0) \cap B_r(x)|$ for the volume of the region in $\mathbb{R}^d$ of possible centers for balls of radius r that cover both 0 and x, and V_1 for the volume of a unit ball in $\mathbb{R}^d$, we see that a single ancestral lineage evolves in a series of jumps with intensity

$$dt \otimes L_r(x)\, u \frac{1}{V_1 r^d}\, dx. \tag{6.1}$$

In particular, under our assumptions, the motion of a lineage is a spatially and temporally homogeneous continuous time random walk in $\mathbb{R}^d$, with uniformly bounded jumps taking place at a rate proportional to u.

Note that ancestral lineages evolve independently (only) if they are far enough apart that they cannot be covered by the same event.

6.1. Adding (genic) selection to the SLFV

There are many ways in which to add selection to the SLFV. Perhaps the simplest is to weight the choice of parental type during a reproduction event. For example, we might weight A alleles by a factor $1 - s$ for some small parameter s. Mimicking our approach in Section 2, if the proportion of a-alleles in $B(x, r)$ immediately before a reproduction event is w, then the chance of choosing a type a parent is

$$w^* = \frac{w}{1 - s(1 - w)} = w + sw(1 - w) + \mathcal{O}(s^2).$$

We rewrite this as

$$w^* = (1-s)w + s\big(1-(1-w)^2\big) + \mathcal{O}(s^2), \tag{6.2}$$

and incorporate (weak) selection into the SLFV as follows:

Definition 6.4 (A Spatial Λ-Fleming–Viot process with genic selection (SLFVGS)). Fix u, r, and Π as in Definition 6.1 and $s \in (0,1)$. The *Spatial Λ-Fleming–Viot process with genic selection (SLFVGS)* driven by Π, with event radius r, impact parameter u, and *selection coefficient* s, is the $\mathcal{M}_\lambda$-valued process $(w_t)_{t\geq 0}$ with dynamics given as follows.

If $(t,x) \in \Pi$, with probability $1-s$, a neutral reproduction event occurs as described in Definition 6.1. With the complementary probability s the event is *selective*, in which case:

(1) Choose two "potential" parental locations $z_1, z_2 \in \mathbb{R}^{\mathrm{d}}$ independently and uniformly at random from $B(x,r)$. Sample types α_1, α_2, according to $w_{t-}(z_1)$, $w_{t-}(z_2)$, respectively.

(2) For every $y \in B(x,r)$, set $w_t(y) = (1-u)w_{t-}(y) + u(1-\mathbf{1}_{\{\alpha_1=A=\alpha_2\}})$.

Once again we define a dual process.

Definition 6.5 (Dual to SLFVGS). The process $(\mathcal{P}_t)_{t\geq 0}$ is the $\bigcup_{l\geq 1}(\mathbb{R}^{\mathrm{d}})^l$-valued Markov process with dynamics defined as follows.

For each $(t,x) \in \overleftarrow{\Pi}$, the corresponding event is neutral with probability $1-s$, in which case proceed as in Definition 6.3. With the complementary probability s, the event is selective, in which case:

(1) for each $\xi_i(t-) \in B(x,r)$, independently mark the corresponding individual with probability u;

(2) if at least one individual is marked, all of the marked individuals are replaced by *two* offspring, whose locations are drawn independently and uniformly in $B(x,r)$.

In both cases, if no individual is marked, then nothing happens.

Remark 6.6. From the perspective of the SLFVGS, it would be more natural to call the individuals created during a selective event in the dual process "parents" (or "potential parents"), as they are situated at the locations from which the parental alleles are sampled. We choose to call them offspring in order to emphasize that the dual process plays the role for the SLFVGS that branching Brownian motion plays for equation (1.1).

This time, to determine the type of an individual sampled from the population at time t, construct the dual as in Definition 6.5 and assign a type to each of the individuals alive at time t by sampling (independently) according to $w_0(\xi_i(t))$. The individual that we sampled is of the unfavored type A if and only if all of the individuals in $\mathcal{P}_t$ are assigned type A. If there were no coalescence, this would parallel the McKean/Skorokhod representation for the Fisher–KPP equation (with Brownian motion replaced by the random walk of ancestral lineages); genetic drift appears as coalescence. It is natural to ask what happens if we scale the SLFVGS in such a way that the random walk followed by an ancestral lineage converges to Brownian motion.

Theorem 6.7 (Informal restatement of [20, THEOREM 1.11]). *Consider the process of Definition* 6.1. *Take β, γ, $\delta > 0$, and let the impact and selection coefficients be $u_n = u/n^\gamma$, and $s_n = s/n^\delta$ (for some positive constants u, s). Define the scaled process $w^{(n)}(t,x) = w(nt, n^\beta x)$. Suppose that*

$$1 - \gamma = 2\beta, \quad \text{and} \quad 1 - \delta - \gamma = 0.$$

Then:

(1) *If $\mathrm{d} \geq 2$ and $\beta \leq \gamma$, or $\mathrm{d} = 1$ and $\beta < \gamma$, $w^{(n)}$ converges weakly to a (weak) solution of the Fisher–KPP equation.*

(2) *If $\beta = \gamma = 1/3$, $\delta = 2/3$, and $\mathrm{d} = 1$, as $n \to \infty$, $w^{(n)}$ converges weakly to the solution of the stochastic Fisher–KPP equation* (1.2).

This result is most easily understood through the dual process of branching and coalescing lineages. Recalling (6.1), in the scaled process that is dual to $w^{(n)}$, a lineage jumps a distance of $\mathcal{O}(1/n^\beta)$ at rate proportional to $nu_n = n^{1-\gamma}$. To obtain a nontrivial limit, we choose $1 - \gamma = 2\beta$, corresponding to the diffusive scaling.

Now suppose that a selective event covers a lineage. With probability $1/n^\gamma$ the lineage is an offspring of the event, in which case two lineages are created at separation $\mathcal{O}(1/n^\beta)$. They may almost immediately coalesce, but with positive probability they will move apart to a distance at which they cannot be covered by the same event. In the limit as $n \to \infty$, we will only "see" the branching event, if the lineages move apart to distance $\mathcal{O}(1)$ before coalescing. By comparison with simple random walk, we expect that the number of times that they will come back to a separation less than $2r/n^\beta$ (and so have a chance to coalesce) before reaching a separation of $\mathcal{O}(1)$ is $\mathcal{O}(n^\beta)$ in $\mathrm{d} = 1$, $\mathcal{O}(\log n)$ in $\mathrm{d} = 2$, and $\mathcal{O}(1)$ in $\mathrm{d} \geq 3$. Now consider how many times they come back together before they coalesce. When they are overlapped by the same event, given that one of them is an offspring, the chance that the second lineage is also an offspring, and so they coalesce, is $\mathcal{O}(1/n^\gamma)$, from which we deduce that they must come back together $\mathcal{O}(n^\gamma)$ times before coalescence.

Combining the above, in $\mathrm{d} \geq 2$, as $n \to \infty$, the chance that they escape to a separation of $\mathcal{O}(1)$ before coalescing is $\mathcal{O}(1)$. Since selective events happen at rate $ns_n u_n = \mathcal{O}(n^{1-\delta-\gamma})$, we take $1 - \delta - \gamma = 0$ in order that branching of ancestral lineages has rate of $\mathcal{O}(1)$. In $\mathrm{d} = 1$, if $\beta < \gamma$ the chance of coalescing before separating is also asymptotically negligible. In all these cases, as $n \to \infty$ the dual process converges to a branching Brownian motion, corresponding to the forwards-in-time process converging to a weak solution to the Fisher–KPP equation. If $\mathrm{d} = 1$ and $\beta = \gamma$, which combined with our other conditions requires $\beta = \gamma = 1/3$ and $\delta = 2/3$, there is a positive chance of lineages separating to $\mathcal{O}(1)$, but they also coalesce in finite time, reflected by the Wright–Fisher noise in (1.2).

In the argument above we took $u_n \to 0$, corresponding to the local population density tending to infinity. In [15, 16] scaling limits of the SLFVGS are considered in which the impact u is fixed. The diffusive scaling then requires us to set $w^{(n)}(t, x) = w(nt, \sqrt{n}x)$. This time, when lineages are covered by the same event, they have a strictly positive chance of coalescing. Reproducing the argument above, since lineages will coalesce after coming together only a finite number of times, most branches will rapidly be lost to coalescence. In order to see *any* lineages separate to $\mathcal{O}(1)$ requires $s_n = \mathcal{O}(1/\sqrt{n})$ in $\mathrm{d} = 1$, $\mathcal{O}(\log n/n)$ in $\mathrm{d} = 2$, and $\mathcal{O}(1/n)$ in $\mathrm{d} \geq 3$. In contrast to the setting of [20], the local population density remains bounded as we pass to the limit and in low dimensions we see the effect of individuals competing with their own close relatives. Recall that one motivation for taking a scaling limit is that we use neutral mutations to infer information about genetic ancestry. This result says that if local population density is bounded, if selection is to be detected, the selection coefficient must be much larger in one spatial dimension than in two, and in turn larger in two dimensions than in a population without spatial structure. In particular, when local population density is bounded, *spatial dimension is important in limiting the effect of selection*.

6.2. The effect of genetic drift on blocking

In order to investigate the effect of genetic drift on the blocking that we saw in Section 5.2, we adapt the SLFV to incorporate the selection mechanism of Section 5.1. There is not yet any accepted way in which to incorporate boundary conditions into the SLFV. An obvious approach that can be applied to simple domains (including for example the domain Ω of Figure 2) based on "reflected sampling" (essentially mimicking Lord Kelvin's method of images for the heat equation) is used in [18]. The important consequence of that choice is that scaled ancestral lineages will converge to reflected Brownian motions. For brevity we shall only describe the adaptation of the SLFV on the whole Euclidean space.

Definition 6.8 (A Spatial Λ-Fleming process with (asymmetric) selection against heterozygotes (SLFVSH)). Fix r, u and Π as in Definition 6.1. Fix $\gamma \in (0, 1]$ and $\boldsymbol{s} \in (0, 1/(1+\gamma))$. In the *Spatial Λ-Fleming–Viot process with selection against heterozygosity (SLFVSH)*, if $(t, x) \in \Pi$, with probability $1 - (1+\gamma)\boldsymbol{s}$ a neutral reproduction event occurs as described in Definition 6.1. With the complementary probability $(1+\gamma)\boldsymbol{s}$ the event is selective, in which case:

(1) Choose *three* "potential" parental locations $z_1, z_2, z_3 \in \mathbb{R}^{\mathrm{d}}$ independently and uniformly at random from $B(x, r)$. Sample types $\alpha_1, \alpha_2, \alpha_3$, according to $w_{t-}(z_1), w_{t-}(z_2), w_{t-}(z_3)$, respectively. Let $\hat{\alpha}$ denote the most common allelic type in $\alpha_1, \alpha_2, \alpha_3$, except that if precisely one of $\alpha_1, \alpha_2, \alpha_3$, is a, with probability $\frac{2\gamma}{3+3\gamma}$ set $\hat{\alpha} = a$.

(2) For every $y \in B(x, r)$, set $w_t(y) = (1-u)w_{t-}(y) + u\mathbf{1}_{\{\hat{\alpha}=a\}}$.

The dual process mirrors the process $(\mathcal{P}_t)_{t\geq 0}$ of Definition 6.5, except that this time, in a selective event, if at least one individual is marked then all marked individuals are replaced by *three* offspring. Just as for the deterministic setting of Lemma 5.3, the duality relation that we exploit is between the SLFVSH and the *historical process* of branching and coalescing lineages, $\Xi(t) := (\mathcal{P}_s)_{0\le s\le t}$, and rests on a voting scheme:

(1) Each leaf of $\Xi(t)$ independently votes a with probability $p(\xi_i(t))$, and A otherwise;

(2) at each neutral event in $\overleftarrow{\Pi}$, all marked individuals adopt the vote of the offspring;

(3) at each selective event in $\overleftarrow{\Pi}$, all marked individuals adopt the majority vote of the three offspring, unless precisely one vote is a, in which case they all vote a with probability $\frac{2\gamma}{3+3\gamma}$, otherwise they vote A.

This defines an iterative voting procedure, which runs inwards from the "leaves" of $\Xi(t)$ to the ancestral individual $\emptyset$ situated at the point x. The special case of majority voting, corresponding to $\gamma = 0$, is illustrated in Figure 4.

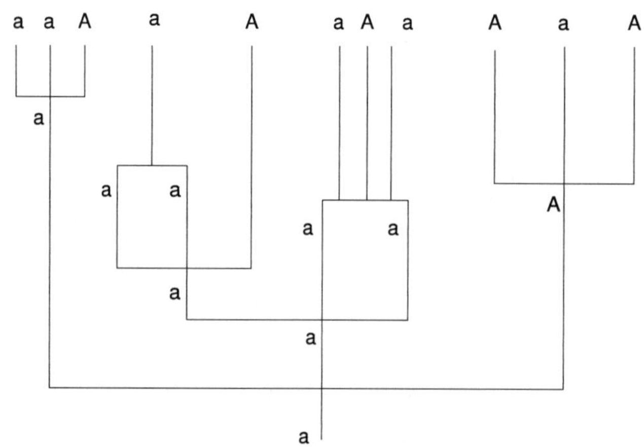

FIGURE 4
Example of majority voting on the dual to the SLFV with selection against heterozygosity. This corresponds to the duality when both homozygotes are equally fit.

Lemma 6.9. *With the voting procedure described above, define $\widetilde{\mathbb{V}}_p(\Xi(t))$ to be the vote associated to the root $\emptyset$. Write $\mathbb{P}_x$ for the law of Ξ when $\mathcal{P}_0$ is the single point x, and $\mathbb{E}_x$ for the corresponding expectation. Then*

$$\mathbb{E}_p\big[w_t(x)\big] = \mathbb{P}_x\big[\widetilde{\mathbb{V}}_p\big(\Xi(t)\big) = a\big].$$

To understand the influence of the genetic drift on blocking we consider two different scalings of the SLFVSH. In both cases we shall be taking a sequence $\varepsilon_n \to 0$ as $n \to \infty$. Our results require that ancestral lineages converge to Brownian motion sufficiently quickly, compared to the rate at which $\varepsilon_n \to 0$, which is the purpose of the following assumption.

Assumption 6.10. The sequence $\{\varepsilon_n\}_{n\in\mathbb{N}}$ is such that $\varepsilon_n \to 0$ and $(\log n)^{1/2}\varepsilon_n \to \infty$ as $n \to \infty$.

Weak noise/selection ratio

Our first scaling is what we shall call the *weak noise/selection ratio* regime. In this regime, selection overwhelms genetic drift. It mirrors that explored in [14] and is also considered in [25]. For each $n \in \mathbb{N}$, and some $\beta \in (0, 1/4)$, set $w^{(n)}(t,x) = w(nt, n^{\beta}x)$. Let $\nu > 0$. We denote by u_n the impact parameter, and by s_n and γ_n the selection parameters at the nth stage of the scaling. They will be given by

$$u_n = \frac{u}{n^{1-2\beta}}, \quad s_n = \frac{1}{\varepsilon_n^2 n^{2\beta}}, \quad \gamma_n = \nu\varepsilon_n. \tag{6.3}$$

Adapting the proof of Theorem 1.11 in [20], and arguments in Section 3 of [14], one can show that under this scaling, for large n, the SLFVSH will be close to the solution of equation (5.5).

Strong noise/selection ratio

We shall refer to our second scaling as the *strong noise/selection ratio* regime. In this regime, genetic drift overcomes selection. We take a sequence of impact parameters $(u_n)_{n\in\mathbb{N}} \subseteq (0,1)$. Consider $\beta \in (0, 1/2)$ and let $\hat{u}_n := u_n n^{1-2\beta}$. This time, we scale time by $n/\hat{u}_n$ and space by n^{β}: $w^{(n)}(t,x) = w(nt/\hat{u}_n, n^{\beta}x)$. We consider a sequence of selection coefficients, $(s_n)_{n\in\mathbb{N}} \subseteq (0,1)$, satisfying one of the following conditions:

$$\begin{cases} s_n n^{2\beta} \to 0, & \liminf_{n\to\infty} u_n \log n < \infty \text{ or } \mathrm{d} \geq 3, \\ \dfrac{s_n n^{2\beta}}{u_n \log n} \to 0, & \liminf u_n \log n = \infty \text{ and } \mathrm{d} = 2. \end{cases} \tag{6.4}$$

The first case includes some choices of impact that were allowed in the first (weak noise/selection ratio) regime; it is the strength of drift *relative to selection* that matters. In this regime, we can take the parameters $(\gamma_n)_{n\in\mathbb{N}}$ that dictate the asymmetry in our selection to be any sequence in $(0,1)$.

Remark 6.11. The rationale behind these scalings is that (at least if $u = 1$ in (6.3)) we can choose parameters in such a way that the scaled models only differ in the strength of the genetic drift (which can be thought of as the reciprocal of the impact). To see this, consider a single ancestral lineage: in the first regime, the rate at which it jumps is proportional to $nu_n = n^{2\beta}$; in the second regime, it is proportional to $nu_n/\hat{u}_n = n^{2\beta}$ (with the same constant of proportionality). In both cases we take the same spatial scaling, so the motion of ancestral lineages is the same. The rate at which a lineage "branches" as a result of being covered by a selective event in the first regime is proportional to $nu_n s_n = n^{2\beta}s_n$. In the second regime, it is the same, $nu_n s_n/\hat{u}_n = n^{2\beta}s_n$, so if we choose the same coefficients s_n, the "branching rate" is the same in both regimes. From the perspective of the dual process, the only difference between the two scalings will then be in the probability of coalescence (determined by u_n).

Theorem 6.12 ([18, SPECIAL CASE OF THEOREM 1.19]). *Let $\rho_* = (\mathrm{d}-1)/\nu$ and suppose $r_0 < \rho_*$. Let $(w^{(n)}(t,\cdot))_{t\geq 0}$ be the scaled SLFVSH defined above on the domain Ω of Figure 2, with initial condition $w^{(n)}(0,x) = \mathbf{1}_{x_1 \geq 0}$.*

(1) *Under the weak noise/selection ratio regime, for any* $k \in \mathbb{N}$, *there exist* $n_*(k) < \infty$, *and* $a_*(k), d_*(k) \in (0, \infty)$ *such that for all* $n \geq n_*$ *and all* $t > 0$,

$$\text{for almost every } x \text{ such that } x_1 \leq -d_* \varepsilon_n |\log \varepsilon_n|, \quad \mathbb{E}\big[w^{(n)}(t,x)\big] \leq \varepsilon_n^k.$$

(2) *Under the strong noise/selection ratio regime, a sharp interface does not develop as* n *goes to infinity. Instead, there is* $\sigma^2 > 0$ *such that for every* $\varepsilon > 0$ *and* $t \geq 0$, *there are a reflected Brownian motion* $(W_t)_{t \geq 0}$, *and* n_* *such that for all* $n \geq n_*$,

$$\big|\mathbb{E}_{w_0}\big[w^{(n)}(t,x)\big] - \mathbb{P}_x\big[W(\sigma^2 t) \geq 0\big]\big| \leq \varepsilon.$$

More generally, one can show that in the strong noise/selection ratio regime for $x \neq y$, $w^{(n)}(t,x)$ and $w^{(n)}(t,y)$ decorrelate as $n \to \infty$.

The first statement says that in the weak noise/selection ratio regime the SLFVSH behaves approximately as the deterministic equation (5.5). The key step in the proof is to couple the dual process to a system of branching random walks in which there is no coalescence. The proof then follows the same pattern as the deterministic result with the extra twist that one must control the error arising from approximating the random walks by Brownian motions.

In the strong noise/selection ratio regime, as one can convince oneself using the argument outlined in the case of bounded neighborhood size in Section 6.1, the genetic drift is strong enough to counter the effects of selection and it breaks down the interface. We see coexistence of the populations throughout the domain. Perhaps counterintuitively, the favored type expands its range further when the population density is lower.

7. CONCLUSION

There is a vast body of literature that seeks to understand the interactions between natural selection, spatial structure, and genetic drift. Mathematics has provided a powerful tool; a great deal has been learned from apparently crude caricatures of the ways in which these forces interact with one another. However, as with any mathematical models, one must be cognisant of the assumptions and simplifications that are being made. In the examples presented here, we have aimed to draw out the importance of not neglecting the dimension and geometry of the domain in which a population is evolving, and of taking account of the randomness inherent in reproduction in a finite population.

ACKNOWLEDGMENTS

My thanks to Nick Barton for introducing me to mathematical population genetics and to my graduate students, without whom very few of the results presented here would have been proved.

REFERENCES

[1] M. Alfaro, D. Hilhorst, and H. Matano, The singular limit of the Allen–Cahn equation and the FitzHugh-Nagumo system. *J. Differential Equations* **245** (2008), 505–565.

[2] S. Asmussen and J. Rosiński, Approximations of small jumps of Lévy processes with a view towards simulation. *J. Appl. Probab.* **38** (2001), 482–493.

[3] N. H. Barton, The dynamics of hybrid zones. *Heredity* **43** (1979), no. 3, 341–359.

[4] N. H. Barton, A. M. Etheridge, and A. Véber, A new model for evolution in a spatial continuum. *Electron. J. Probab.* **15** (2010), 162–216.

[5] N. H. Barton and G. M. Hewitt, Analysis of hybrid zones. *Annu. Rev. Eol. Syst.* **16** (1985), 113–148.

[6] N. H. Barton and G. M. Hewitt, Adaptation, speciation and hybrid zones. *Nature* **341** (1989), 497–503.

[7] K. Becker, A. Etheridge, and I. Letter, Branching stable processes and the fractional Allen–Cahn equation. 2022, in preparation.

[8] H. Berestycki, J. Bouhours, and G. Chapuisat, Blocking and propagation in cylinders with varying cross section. *Calc. Var. Partial Differ. Equ.* **55** (2016), no. 44.

[9] M. Bramson, Convergence of solutions of the Kolmogorov equation to travelling waves. *Mem. Amer. Math. Soc.* **44** (1983), no. 285.

[10] M. L. Cain, B. G. Milligan, and A. E. Strand, Long-distance seed dispersal in plant populations. *Am. J. Bot.* **87** (2000), no. 9, 1217–1227.

[11] X. Chen, Generation and propagation of interfaces for reaction diffusion equations. *J. Differential Equations* **96** (1992), 116–141.

[12] J. T. Cox, R. Durrett, and E. A. Perkins, Voter model perturbations and reaction diffusion equations. *Astérisque* **349** (2013).

[13] A. de Masi, P. A. Ferrari, and J. L. Lebowitz, Reaction–diffusion equations for interacting particle systems. *J. Stat. Phys.* **44** (1986), no. 3/4, 589–644.

[14] A. Etheridge, N. Freeman, and S. Penington, Branching Brownian motion, mean curvature flow and the motion of hybrid zones. *Electron. J. Probab.* **22** (2017), no. 103, 1–40.

[15] A. Etheridge, N. Freeman, S. Penington, and D. Straulino, Branching Brownian motion and selection in the spatial Λ-Fleming–Viot process. *Ann. Appl. Probab.* **27** (2017), 2605–2645.

[16] A. Etheridge, N. Freeman, and D. Straulino, The Brownian net and selection in the spatial Λ-Fleming–Viot process. *Electron. J. Probab.* **22** (2017), 1–36.

[17] A. M. Etheridge, Drift, draft and structure: some mathematical models of evolution. *Banach Center Publ.* **80** (2008), 121–144.

[18] A. M. Etheridge, M. D. Gooding, and I. Letter, On the effects of a wide opening in the domain of the (stochastic) Allen–Cahn equation and the motion of hybrid zones. 2022, arXiv:2204.00316.

[19] A. M. Etheridge and T. G. Kurtz, Genealogical constructions of population models. *Ann. Probab.* **47** (2019), no. 4, 1827–1910.

[20] A. M. Etheridge, A. Véber, and F. Yu, Rescaling limits of the spatial Lambda-Fleming–Viot process with selection. *Electron. J. Probab.* **25** (2020), no. 120, 1–89.

[21] R. A. Fisher, *The genetical theory of natural selection*. Oxford University Press, 1930.

[22] R. A. Fisher, The wave of advance of advantageous genes. *Annu. Eugen.* **7** (1937), 355–369.

[23] F. Flandoli and R. Huang, The KPP equation as a scaling limit of locally interacting Brownian particles. *J. Differential Equations* **303** (2021), 608–644.

[24] M. Gage and R. Hamilton, The heat equation shrinking convex plane curves. *J. Differential Geom.* **23** (1986), 417–491.

[25] M. D. Gooding, *Long term behaviour of spatial population models with heterozygous or asymmetric homozygous selection*. PhD thesis, University of Oxford, 2018.

[26] M. A. Grayson, The heat equation shrinks embedded plane curves to round points. *J. Differential Geom.* **26** (1987), 285–314–491.

[27] M. Kimura, T. Maruyama, and J. F. Crow, The mutation load in small populations. *Genetics* **48** (1963), no. 10, 1303–1312.

[28] A. Kolomogorov, I. Petrovsky, and N. Piscounov, Étude de l'équation de la diffusion avec croissance de la quantité de matière et son application à un problème biologique. *Moscow Univ. Math. Bull.* **1** (1937), 1–25.

[29] T. Mach, A. Sturm, and J. Swart, Recursive tree processes and the mean-field limit of stochastic flows. *Electron. J. Probab.* **25** (2020), 1–63.

[30] H. Matano, K. I. Nakamura, and B. Lou, Periodic traveling waves in a two-dimensional cylinder with saw-toothed boundary and their homogenization limit. *Netw. Heterog. Media* **1** (2006), no. 4, 537–568.

[31] H. P. McKean, Application of Brownian motion to the equation of Kolmogorov–Petrovski–Piskunov. *Comm. Pure Appl. Math.* **28** (1975), 323–331.

[32] B. Merriman, J. K. Bence, and S. J. Osher, Motion of multiple junctions: A level set approach. *J. Comput. Phys.* **112** (1994), no. 2, 334–363.

[33] C. Mueller, L. Mytnik, and L. Ryzhik, The speed of a random front for stochastic reaction–diffusion equations with strong noise. *Comm. Math. Phys.* **384** (2021), 699–732.

[34] C. Mueller and R. Tribe, Stochastic p.d.e.'s arising from the long range contact and long range voter processes. *Probab. Theory Related Fields* **102** (1994), 519–546.

[35] Z. O'Dowd, *Branching Brownian motion and partial differential equations*. University of Oxford MMath Dissertation, 2019.

[36] A. V. Skorokhod, Branching diffusion processes. *Theory Probab. Appl.* **9** (1964), 492–497.

ALISON ETHERIDGE
Department of Statistics, University of Oxford, OX1 3LB, UK, etheridg@stats.ox.ac.uk

This Page Intentionally Left Blank

HYDRODYNAMIC LIMIT AND STOCHASTIC PDES RELATED TO INTERFACE MOTION

TADAHISA FUNAKI

ABSTRACT

The hydrodynamic limit gives a link between microscopic and macroscopic systems via a space–time scaling. Its notable feature is the averaging effect due to the local ergodicity under the local equilibria. In this article, as the microscopic system, we consider several types of interacting particle systems, in which particles perform random walks with interaction. We derive, under the hydrodynamic limit or its nonlinear fluctuation limit, three different objects: the motion by mean curvature, Stefan free boundary problem, and coupled KPZ equation. These are all related to the interface motion. The Boltzmann–Gibbs principle plays a fundamental role. We discuss the coupled KPZ equation from the aspect of singular SPDEs and renormalizations. Ginzburg–Landau $\nabla\phi$-interface model, stochastic motion by mean curvature, and stochastic eight-vertex model are also briefly discussed.

MATHEMATICS SUBJECT CLASSIFICATION 2020

Primary 60K35; Secondary 82C22, 60H15, 35R60, 35B40, 35K57

KEYWORDS

Hydrodynamic limit, interacting particle system, stochastic PDE, motion by mean curvature, Stefan problem, KPZ equation, Boltzmann–Gibbs principle

ICM INTERNATIONAL CONGRESS OF MATHEMATICIANS 2022 JULY 6–14

Proc. Int. Cong. Math. 2022, Vol. 6, pp. 4302–4325
DOI 10.4171/ICM2022/26

1. INTRODUCTION

The hydrodynamic limit is a scaling limit in space and time for interacting systems at the microscopic level, and leads to macroscopic evolutional rules usually prescribed by nonlinear PDEs, via an averaging effect due to the local ergodicity in local equilibria, cf. [17,44,48]. It is formulated as a law of large numbers. Its fluctuation limit is also studied, and we obtain linear or nonlinear stochastic PDEs (SPDEs) in the limit.

In this review article, we discuss the derivation of three different objects from interacting particle systems: the motion by mean curvature (MMC, Section 2.1), Stefan free boundary problem (Section 3) and coupled Kardar–Parisi–Zhang (KPZ) equation (Section 4.2). We also discuss the coupled KPZ equation from the aspect of singular SPDEs (Section 4.1). This is an ill-posed equation in a classical sense and requires renormalizations.

We consider particle systems, in which each particle moves performing a random walk and interacting with other particles on the d-dimensional discrete torus $\mathbb{T}_N^d = \{1, 2, \dots, N\}^d$ (with periodic boundary) with large N. Specifically, we consider a zero-range process, in which several particles may occupy each site of $\mathbb{T}_N^d$ and interact only at the same site, or Kawasaki dynamics sometimes called exclusion process, in which particles obey the hard-core exclusion rule so that at most one particle can occupy each site. To derive Stefan problem or coupled KPZ equation, we consider multiple types of particles. In addition, we introduce the Glauber mechanism, which governs the creation and annihilation of particles. More precisely, we consider both creation and annihilation for MMC problem, annihilation only for Stefan problem, and neither creation nor annihilation for the coupled KPZ problem.

Our problems have a common feature that relates to the interface motion. The system leading to MMC exhibits a phase separation to sparse and dense regions of particles and, macroscopically, the interface is created to separate these two phases and evolves under the MMC, while, in that leading to the Stefan problem, we observe the segregation of different species. Scalar KPZ equation was originally introduced as an equation for a growing interface. Technically, the so-called Boltzmann–Gibbs principle plays a fundamental role.

2. MOTION BY MEAN CURVATURE

2.1. From particle systems

Here, to illustrate the idea and the results, we take Glauber–zero-range process as a microscopic model based on El Kettani et al. [7,8]. Instead of zero-range process, one can take simple Kawasaki dynamics (independent random walks with exclusion rule, [26,43]) or Kawasaki dynamics with speed change (see [21]).

Properly tuning the Glauber part, the system exhibits phase separation and one can derive the MMC as a macroscopic evolutional rule for the phase separation surface. Our method is a combination of the techniques of the hydrodynamic limit, based on the relative entropy method (Proposition 2.3) and Boltzmann–Gibbs principle (Theorem 2.6), and the

PDE technique called the sharp interface limit (Proposition 2.7) and the discrete Schauder estimate (Proposition 2.5).

2.1.1. Glauber–zero-range process and hydrodynamic limit with fixed K

Glauber–zero-range process on $\mathbb{T}_N^d$ is the Markov process $\eta^N(t) = \{\eta_x^N(t)\}_{x\in\mathbb{T}_N^d}$ on the configuration space $\mathcal{X}_N = \mathbb{Z}_+^{\mathbb{T}_N^d}$ ($\mathcal{X}_N = \{0,1\}^{\mathbb{T}_N^d}$ in Kawasaki case) with the generator given by $L_N = N^2 L_Z + K L_G$ with $K > 0$, where

$$(L_Z f)(\eta) = \sum_{x\in\mathbb{T}_N^d} \sum_{e\in\mathbb{Z}^d:|e|=1} g(\eta_x)\{f(\eta^{x,x+e}) - f(\eta)\},$$

$$(L_G f)(\eta) = \sum_{x\in\mathbb{T}_N^d} \sum_{\pm} c_x^{\pm}(\eta)\{f(\eta^{x,\pm}) - f(\eta)\},$$

for $\eta = \{\eta_x\}_{x\in\mathbb{T}_N^d} \in \mathcal{X}_N$ and functions f on $\mathcal{X}_N$. Here, $\eta_x \in \mathbb{Z}_+ = \{0,1,2,\dots\}$ denotes the number of particles at x, $\eta^{x,y}$ is η after one particle jumps from x to y, $\eta^{x,+}$ is η after one particle is created at x, and $\eta^{x,-}$ is η after one particle is annihilated at x.

The flip rates of the Glauber part are shift-invariant, that is, $c_x^{\pm}(\eta) = c^{\pm}(\tau_x\eta)$ with the creation and annihilation rates $c^{\pm}(\eta)(= c_0^{\pm}(\eta))$ of a particle at $x = 0$ and the spatial shift τ_x acting on $\mathcal{X}_N$. We assume $c^-(\eta) = 0$ if $\eta_0 = 0$. The jump rate $g(k)$, $k \in \mathbb{Z}_+$, of the zero-range part is bounded from above and below by linear functions of k. In particular, $g(0) = 0$.

The invariant measures or equilibrium states, being shift-invariant in space, of the zero-range process, that is, the leading part of our dynamics, are superpositions of the product measures $\bar\nu_\varphi$ on $\mathcal{X}_N$ (or on $\mathcal{X} = \mathbb{Z}_+^{\mathbb{Z}^d}$) with one-site marginal distribution given by

$$\bar\nu_\varphi(k) = \frac{1}{Z_\varphi}\frac{\varphi^k}{g(k)!}, \qquad Z_\varphi = \sum_{k=0}^{\infty} \frac{\varphi^k}{g(k)!},$$

with parameter $\varphi \ge 0$ called fugacity, where $g(k)! = \prod_{i=1}^k g(i)$, $k \ge 1$, and $g(0)! = 1$. We denote $\nu_\rho := \bar\nu_{\varphi(\rho)}$ by changing the parameter with its mean $\rho \ge 0$. In fact, ρ and $\varphi = \varphi(\rho)$ are related by $\rho = \varphi(\log Z_\varphi)' = E^{\bar\nu_\varphi}[k] \equiv E^{\bar\nu_\varphi}[\eta_0]$, and $\varphi = E^{\bar\nu_\varphi}[g(k)]$ holds.

The macroscopic empirical measure (density field of particles) on $\mathbb{T}^d$ $(= [0,1)^d$ with periodic boundary), which is the macroscopic region corresponding to microscopic $\mathbb{T}_N^d$, associated with the configuration $\eta \in \mathcal{X}_N$, is defined by

$$\alpha^N(dv;\eta) = \frac{1}{N^d}\sum_{x\in\mathbb{T}_N^d} \eta_x \delta_{\frac{x}{N}}(dv), \quad v \in \mathbb{T}^d, \tag{2.1}$$

or equivalently, for a test function $G \in C^\infty(\mathbb{T}^d)$,

$$\langle \alpha^N(\cdot;\eta), G\rangle = \frac{1}{N^d}\sum_{x\in\mathbb{T}_N^d} \eta_x G\left(\frac{x}{N}\right). \tag{2.2}$$

Thus, the scaling from micro to macro is given by $\frac{1}{N}$ in space, $\frac{1}{N^d}$ in mass, as well as N^2 (for the zero-range part) and K (for the Glauber part) in time. Our problem is to study the limit as $N \to \infty$.

For a fixed K, one can expect that the hydrodynamic limit holds, that is,

$$\alpha^N\big(dv;\eta^N(t)\big) \to \rho(t,v)dv \quad \text{as } N \to \infty$$

holds in probability multiplying a test function G on $\mathbb{T}^d$, if this holds at $t=0$, where $\rho(t,v)$ is a unique weak solution of the reaction–diffusion equation with a nonlinear diffusion term

$$\partial_t \rho = \Delta\varphi(\rho) + Kf(\rho), \quad v \in \mathbb{T}^d, \tag{2.3}$$

with initial value $\rho(0)$ and

$$f(\rho) = E^{\nu_\rho}\big[c^+(\eta) - c^-(\eta)\big]. \tag{2.4}$$

Recall that $\varphi(\rho) = E^{\nu_\rho}[g]$ and Δ is the Laplacian on $\mathbb{T}^d$. This was shown for Glauber–Kawasaki dynamics in **[5]**, and the result for the zero-range process without Glauber part (i.e., when $K=0$) is found in **[44]**. See Section 4.2.2 for some related heuristic arguments to derive (2.3).

2.1.2. Mesoscopic Glauber perturbation and derivation of MMC

We consider the Glauber–zero-range process $\eta^N(t)$, that is, the $\mathcal{X}_N$-valued process with generator $L_N = N^2 L_Z + KL_G$, now with $K = K(N) \to \infty$. One can construct flip rates $c^\pm(\eta)$ of the Glauber part in such a way that the corresponding f determined by (2.4) is bistable, that is, f has exactly three zeros $0 < \alpha_1 < \alpha_* < \alpha_2 < \infty$ and $f'(\alpha_1) < 0$, $f'(\alpha_2) < 0$ hold, and satisfies the φ-balance condition $\int_{\alpha_1}^{\alpha_2} f(\rho)\varphi'(\rho)d\rho = 0$. We actually take $c_x^+(\eta) = \frac{\hat{c}^+(\tau_x\eta)}{g(\eta_x+1)}$ and $c_x^-(\eta) = \hat{c}^-(\tau_x\eta)1_{\{\eta_x\ge 1\}}$, where $\hat{c}^\pm(\eta)$ are nonnegative local functions on $\mathcal{X} = \mathbb{Z}_+^{\mathbb{Z}^d}$ (regarded as those on $\mathcal{X}_N$), which do not depend on η_0. Microscopically, there are two phases: sparse phase (with density α_1 of particles) and dense phase (density α_2). Macroscopically, these two phases are separated by an interface Γ_t in $\mathbb{T}^d$. The creation and annihilation mechanism at the microscopic level forces the macroscopic density to one of those two stable phases.

For a function $u = \{u(\frac{x}{N})\}_{x\in\mathbb{T}_N^d}$, we define the local equilibrium state ν_u as the product measure on $\mathcal{X}_N$ defined by $\nu_u(d\eta) = \prod_{x\in\mathbb{T}_N^d} \nu_{u(\frac{x}{N})}(d\eta_x)$. For two probability measures μ and ν, the relative entropy of μ with respect to ν is defined by

$$H(\mu|\nu) := \int \frac{d\mu}{d\nu}\log\frac{d\mu}{d\nu}\cdot d\nu.$$

For the initial distribution μ_0^N of $\eta^N(0)$, we assume $H(\mu_0^N|\nu_0^N) = O(N^{d-\varepsilon_0})$ with some $\varepsilon_0 > 0$, where $\nu_0^N = \nu_{u^N(0)}$ for some $u^N(0) = \{u^N(0,\frac{x}{N})\}_{x\in\mathbb{T}_N^d}$ which satisfies

- $u^N(0,\frac{x}{N}) = u_0(\frac{x}{N})$, $x \in \mathbb{T}_N^d$, with some $u_0 \in C^5(\mathbb{T}^d)$ such that $u_0 > 0$;
- $\Gamma_0 := \{v \in \mathbb{T}^d; u_0(v) = \alpha_*\}$ is a $(d-1)$-dimensional $C^{5+\theta}$-hypersurface, $\theta > 0$, without boundary in $\mathbb{T}^d$ and ∇u_0 is nondegenerate in the normal direction to Γ_0.

Theorem 2.1 (**[7]**). *We assume $d \ge 2$, the above conditions, and that $K(N)$ diverges to ∞ satisfying $1 \le K(N) \le \delta_0(\log N)^{\frac{\sigma}{2}}$ with small enough $\delta_0 > 0$ and the Hölder exponent $\sigma \in$*

(0, 1) *determined by Nash estimate, see Proposition 2.5. Let* $\alpha^N(t,dv) := \alpha^N(dv;\eta^N(t))$ *be the macroscopic empirical measure associated with* $\eta^N(t)$. *Then, we have for* $t \in (0,T]$,

$$\alpha^N(t) \to \chi_{\Gamma_t} := \begin{cases} \alpha_1, & \text{on one side of } \Gamma_t, \\ \alpha_2, & \text{on the other side of } \Gamma_t, \end{cases} \tag{2.5}$$

in probability, where the hypersurface Γ_t *in* $\mathbb{T}^d$ *moves according to the MMC,* $V = \lambda_0 \kappa$.

Here, κ *is the mean curvature of* Γ_t *multiplied by* $d-1$ *and* V *is the normal velocity of* Γ_t *from the* α_1*-side to* α_2*-side. The sides of* Γ_t *are determined continuously from* Γ_0. *We assume* Γ_t *is* $C^{5+\theta}$ *for* $t \le T$.

The constant λ_0 *is determined by the homogenization effect from the nonlinear Laplacian and given by*

$$\lambda_0 = \frac{\int_{\alpha_1}^{\alpha_2} \varphi'(u)\sqrt{W(u)}du}{\int_{\alpha_1}^{\alpha_2} \sqrt{W(u)}du}$$

with the potential defined by $W(u) = \int_u^{\alpha_2} f(s)\varphi'(s)ds$, $u > 0$. *Note that* $\lambda_0 = 1$ *if* $g(k) = k$ *so that* φ *is linear,* $\varphi(u) = u$.

2.1.3. Proof of Theorem 2.1

(a) Probabilistic part. Let μ_t^N be the distribution of $\eta^N(t)$ on $\mathcal{X}_N$. Let $u^N(t) = \{u^N(t,\frac{x}{N})\}_{x\in\mathbb{T}_N^d}$ be the solution of the quasilinear discrete PDE (ODE):

$$\partial_t u^N\left(t,\frac{x}{N}\right) = \Delta^N \varphi\left(u^N\left(t,\frac{x}{N}\right)\right) + Kf\left(u^N\left(t,\frac{x}{N}\right)\right), \tag{2.6}$$

with initial value $u^N(0)$, where Δ^N is the discrete Laplacian defined by

$$\Delta^N \psi\left(\frac{x}{N}\right) = N^2 \sum_{y\in\mathbb{T}_N^d:|y-x|=1} \left(\psi\left(\frac{y}{N}\right) - \psi\left(\frac{x}{N}\right)\right), \tag{2.7}$$

for $\psi = \{\psi(\frac{x}{N})\}_{x\in\mathbb{T}_N^d}$. Note that (2.6) is a discretized version of (2.3). Let $\nu_t^N = \nu_{u^N(t)}$ be the local equilibrium state on $\mathcal{X}_N$ with mean density $\{u^N(t,\frac{x}{N})\}_{x\in\mathbb{T}_N^d}$.

The main estimate in the probabilistic part is the following:

Theorem 2.2. *Under the condition* $H(\mu_0^N|\nu_0^N) = O(N^{d-\varepsilon_0})$ *for some* $\varepsilon_0 > 0$, *if* $1 \le K(N) \le \delta_0(\log N)^{\frac{\sigma}{2}}$ *with small enough* $\delta_0 > 0$, *then we have* $H(\mu_t^N|\nu_t^N) = o(N^d)$ *as* $N \to \infty$.

Once this is shown, one can show that $\alpha^N(t)$ is close to $u^N(t)$ in the sense that

$$\lim_{N\to\infty} \mu_t^N(\mathcal{A}_{N,t}) = 0, \tag{2.8}$$

for the event $\mathcal{A}_{N,t} := \{\eta \in \mathcal{X}_N;\ |\langle \alpha^N, G\rangle - \langle u^N(t,\cdot), G\rangle| > \delta\}$ and every $\delta > 0$ and $G \in C^\infty(\mathbb{T})$. Indeed, we may combine the entropy inequality, $\mu(A) \le \frac{\log 2 + H(\mu|\nu)}{\log(1+1/\nu(A))}$, and the large deviation estimate for the product measure ν_t^N, $\nu_t^N(\mathcal{A}_{N,t}) \le e^{-CN^d}$ for some $C = C_{\delta,G} > 0$.

The proof of Theorem 2.2 is divided into five steps.

(1) The time derivative of the relative entropy.

Proposition 2.3 ([33, 40, 51]). *Let m be a reference measure on $\mathcal{X}_N$ with full support and set $\psi_t^N = \frac{d\nu_t^N}{dm}$. Then, we have*

$$\partial_t H(\mu_t^N|\nu_t^N) \le -N^2 \mathfrak{D}\left(\sqrt{\frac{d\mu_t^N}{d\nu_t^N}}; \nu_t^N\right) + \int_{\mathcal{X}_N} \{L_N^{*,\nu_t^N} 1 - \partial_t \log \psi_t^N\} d\mu_t^N,$$

where $L^{*,\nu}$ denotes the adjoint of L on $L^2(\nu)$ in general, and $\mathfrak{D}(f;\nu) \ge 0$ is the Dirichlet form associated with L_Z, which may be dropped since we actually do not use it.

(2) Computation of $L_N^{*,\nu}1$ and $\partial_t \log \psi_t^N$. We write $\sum_x$ for $\sum_{x\in\mathbb{T}_N^d}$ for simplicity.

Lemma 2.4. *Let $\nu = \nu_u$ and $\varphi(\frac{x}{N}) = \varphi(u(\frac{x}{N}))$ in the following first two equalities. Then,*

$$N^2 L_Z^{*,\nu} 1 = \sum_x \frac{(\Delta^N \varphi)(\frac{x}{N})}{\varphi(\frac{x}{N})} \left\{ g(\eta_x) - \varphi\left(\frac{x}{N}\right)\right\},$$

$$L_G^{*,\nu} 1 = \sum_x \left\{ \hat{c}^+(\tau_x\eta)\left(\frac{1(\eta_x \ge 1)}{\varphi(\frac{x}{N})} - \frac{1}{g(\eta_x+1)}\right) + \hat{c}^-(\tau_x\eta)\left(\frac{\varphi(\frac{x}{N})}{g(\eta_x+1)} - 1(\eta_x \ge 1)\right)\right\},$$

$$\partial_t \log \psi_t^N = \sum_x \frac{\partial_t \varphi(u^N(t,\frac{x}{N}))}{\varphi(u^N(t,\frac{x}{N}))} \left(\eta_x - u^N\left(t, \frac{x}{N}\right)\right).$$

(3) Schauder estimate. To bound the prefactor appearing in $N^2 L_Z^{*,\nu}1$, we need

Proposition 2.5 (Schauder estimate for quasilinear discrete PDEs, [24]). *If $\sup_N \|u^N(0)\|_{C_N^4} < \infty$ (which holds under our assumption), the solution of (2.6) has the bound*

$$\|u^N(t)\|_{C_N^2} \le CK^{\frac{2}{\sigma}},$$

where $\|u\|_{C_N^k} = \sum_{i=0}^k \max_{x;e_1,\dots,e_i} |\nabla_{e_1}^N \cdots \nabla_{e_i}^N u(\frac{x}{N})|$, $\sigma \in (0,1)$ is the Hölder exponent obtained in Nash estimate and $\nabla_e^N u(\frac{x}{N}) = N(u(\frac{x+e}{N}) - u(\frac{x}{N}))$ is the discrete derivative of u in the direction $e \in \mathbb{Z}^d$, $|e| = 1$.

(4) First-order Boltzmann–Gibbs principle. For a local function $h = h(\eta)$ on $\mathcal{X}$ (i.e., h depends only on finitely many $\{\eta_x\}$) growing at most linearly in η, we set $\tilde{h}(\rho) = E^{\nu_\rho}[h]$, $\rho \ge 0$, and

$$f_x(\eta) = h(\tau_x\eta) - \tilde{h}(u_x) - \tilde{h}'(u_x)(\eta_x - u_x),$$

where we write $u_x = u^N(t, \frac{x}{N})$ for simplicity. Roughly saying, one can replace h by the first-order Taylor expansion of its equilibrium average.

Theorem 2.6 (Boltzmann–Gibbs principle). *Let $\{a_{t,x}\}_{t\ge 0, x\in\mathbb{T}_N^d}$ be nonrandom coefficients satisfying $|a_{t,x}| \le M$. Then, there exist $\varepsilon_1, C > 0$ such that*

$$E\left|\int_0^T \sum_x a_{t,x} f_x(\eta^N(t)) dt\right| \le CMKN^{d-\varepsilon_1} + CM\int_0^T H(\mu_t^N|\nu_t^N) dt.$$

For the proof, we apply truncation, entropy inequality, estimate on the exponential moment under ν_t^N, Feynman–Kac formula, Raleigh estimate, and equivalence of ensembles.

First, take $h = g(\eta_0) - \varphi(u_x)$ and $a_{t,x} = \frac{\Delta^N \varphi(u_x)}{\varphi(u_x)}$. Then, $\tilde{h}(\rho) = \varphi(\rho) - \varphi(u_x)$ and, noting $\tilde{h}(u_x) = 0$, by Theorem 2.6, $N^2 L_Z^{*,\nu_t^N} 1$ is replaced by

$$\sum_x \frac{\Delta^N \varphi(u_x)}{\varphi(u_x)} \varphi'(u_x)(\eta_x - u_x).$$

We use Proposition 2.5 to bound $|a_{t,x}|$ by $M = CK^{\frac{2}{\sigma}}$. Note that $\varphi(u_x) \geq c$ holds for some $c > 0$ by the maximum principle, the property of f, and the assumption on $u^N(0)$.

Next, take the function inside curly braces in $L_G^{*,\nu} 1$ in Lemma 2.4 replacing $\tau_x \eta$ and η_x, respectively, by η and η_0 as h and $a_{t,x} = K$. Noting $E^{\nu_\beta}[\frac{1}{g(\eta_0+1)}] = \frac{1}{\varphi(\beta)} E^{\nu_\beta}[1(\eta_0 \geq 1)]$ and $\tilde{h}(u_x) = 0$, by Theorem 2.6, $KL_G^{*,\nu_t^N} 1$ is replaced by

$$K \sum_x E^{\nu_{u_x}}[c^+ - c^-] \frac{\varphi'(u_x)}{\varphi(u_x)} (\eta_x - u_x).$$

Summarizing these and noting $\partial_t \varphi(u_x) = \varphi'(u_x) \partial_t u_x$ for $\varphi(u_x) = \varphi(u^N(t, \frac{x}{N}))$, $L_N^{*,\nu_t^N} 1 - \partial_t \log \psi_t^N$ is replaced, with an error given by Theorem 2.6, by

$$\sum_x \frac{\varphi'(u_x)}{\varphi(u_x)} (\Delta^N \varphi(u_x) + Kf(u_x) - \partial_t u_x)(\eta_x - u_x).$$

This vanishes if $u_x = u^N(t, \frac{x}{N})$ is the solution of (2.6).

(5) Completion of the proof. Finally, since $K = K(N) \leq \delta_0 (\log N)^{\frac{\sigma}{2}}$, we obtain

$$\partial_t H(\mu_t^N | \nu_t^N) \leq CK^{\frac{2}{\sigma}} H(\mu_t^N | \nu_t^N) + O(N^{d-\varepsilon_2}),$$

for $0 < \varepsilon_2 < \varepsilon_1$ in integrated form in t. Gronwall's inequality shows

$$H(\mu_t^N | \nu_t^N) \leq (H(\mu_0^N | \nu_0^N) + tO(N^{d-\varepsilon_2})) e^{CK^{\frac{2}{\sigma}} t}.$$

Note that $e^{CK^{\frac{2}{\sigma}} t} \leq N^{C\delta_0^{\frac{2}{\sigma}} t}$ from $K \leq \delta_0 (\log N)^{\frac{\sigma}{2}}$. Thus, taking $\delta_0 > 0$ small enough, Theorem 2.2 is shown.

(b) PDE part. The following proposition is a purely PDE result, which establishes the sharp interface limit for the solution $u^N(t)$ of (2.6) and leads to the MMC. Theorem 2.1 follows from (2.8) and this proposition.

Proposition 2.7. *Under our assumption, $u^N(t)$ converges to χ_{Γ_t} as $N \to \infty$, where χ_{Γ_t} is defined in* (2.5) *and the hypersurface Γ_t in $\mathbb{T}^d$ moves according to the MMC, $V = \lambda_0 \kappa$.*

The proof of Proposition 2.7 relies on the comparison theorem for the discrete PDE (2.6) due to the nondecreasingness of φ and consists of two parts: generation of interface and propagation of interface. In a short time, the reaction term $Kf(u^N)$ is dominant and the solution $u^N(t, \frac{x}{N})$ is pushed to one of the two stable points, α_1 and α_2, of f within the time $t = \frac{c}{K} \log K$, $c > 0$. This is called the generation of interface.

Once the interface is created, we can construct super- and subsolutions to (2.6) based on the traveling (standing) wave solution U_0: $\varphi(U_0)'' + f(U_0) = 0$ on $\mathbb{R}$ combing with the second-order term U_1 in the asymptotic expansion in K of the continuous PDE (2.3). By sandwiching the solution of (2.6) within super- and subsolutions, and studying the asymptotic behavior of these solutions as $K = K(N) \to \infty$, we obtain Proposition 2.7.

2.2. Other approaches

2.2.1. Ginzburg–Landau interface model

The Ginzburg–Landau $\nabla\phi$-interface model is an evolutional model of height functions of discretized interface. After characterizing all (tempered and shift-invariant) invariant measures of $\nabla\phi$-dynamics on $\mathbb{Z}^d$ as nonlinear version of massless Gaussian lattice free fields with long correlations, an anisotropic motion by mean curvature was derived under the hydrodynamic limit, see Funaki and Spohn [25] and Funaki [14]. Funaki [13] derived a PDE with an obstacle described by an evolutionary variational inequality from the Ginzburg–Landau interface model on a wall.

2.2.2. SPDE approach to stochastic MMC

An approach to stochastic MMC from SPDEs is also known. The sharp interface limit of the time-dependent Ginzburg–Landau model of nonconservative type, or equivalently the stochastic Allen–Cahn equation, was studied in [10, 12, 28]. In one dimension with space–time Gaussian white noise, the limit motion of a phase separation point is described by a stochastic differential equation [10]. In higher dimensions, stochastic MMC was derived in the limit in [12, 28]. Chapter 4 of [16] gives a survey of related results. Physical background of these SPDEs is found in [38].

3. STEFAN PROBLEM

3.1. From two-component Glauber–Kawasaki dynamics

De Masi et al. [6] derived a system of diffusion equations with Stefan free boundary condition from two-component simple Kawasaki dynamics with relatively large mesoscopic annihilation effect when different types of particles meet. The annihilation effect leads to segregation in a competition–diffusion system at the macroscopic level.

We consider two-component Glauber–Kawasaki dynamics on $\mathbb{T}_N^d$, that is, the Markov process $(\eta_1^N(t), \eta_2^N(t)) = \{\eta_{1,x}^N(t), \eta_{2,x}^N(t)\}_{x\in\mathbb{T}_N^d}$ on $\mathcal{X}_N \times \mathcal{X}_N$ with generator $L_N = N^2 L_K^{(2)} + K L_G$, where $\mathcal{X}_N = \{0,1\}^{\mathbb{T}_N^d}$ in the present setting and

$$
\begin{aligned}
\big(L_K^{(2)} f\big)(\eta_1, \eta_2) &= d_1\big(L_K f(\cdot, \eta_2)\big)(\eta_1) + d_2\big(L_K f(\eta_1, \cdot)\big)(\eta_2), \\
(L_K f)(\eta) &= \frac{1}{2} \sum_{x,y\in\mathbb{T}_N^d : |x-y|=1} \big\{f(\eta^{x,y}) - f(\eta)\big\}, \quad \eta \in \mathcal{X}_N, \\
(L_G f)(\eta_1, \eta_2) &= \sum_{x\in\mathbb{T}_N^d} \eta_{1,x}\eta_{2,x}\big\{f\big(\eta_1^x, \eta_2^x\big) - f(\eta_1, \eta_2)\big\},
\end{aligned}
$$

for $(\eta_1, \eta_2) \in \mathcal{X}_N \times \mathcal{X}_N$ and functions f on $\mathcal{X}_N \times \mathcal{X}_N$ (on $\mathcal{X}_N$ in the second line). Here, $d_1, d_2 > 0$, $\eta^{x,y}$ is the configuration $\eta = \{\eta_x\}_{x\in\mathbb{T}_N^d}$ with η_x and η_y exchanged, and η^x is η after a flip $\eta_x \leftrightarrow 1 - \eta_x$ which happens at x. In particular, when two particles of different types meet, both of them disappear with high probability as $K = K(N) \to \infty$.

The macroscopic empirical measure $\alpha^N(dv; \eta)$ is defined for $\eta \in \mathcal{X}_N$ as in (2.1), and we set $\alpha_i^N(t, dv) = \alpha^N(dv; \eta_i^N(t))$ for $i = 1, 2$ and $t \geq 0$.

Theorem 3.1 ([6]). *Let $1 \le K = K(N) \le \delta_0(\log N)^{1/2}$ with small enough $\delta_0 > 0$, $K(N) \to \infty$, and assume proper conditions on the initial distribution of $(\eta_1^N(0), \eta_2^N(0))$ including those on the relative entropy as in Theorem 2.1. We additionally assume $e^{-c_1 K} \le u_i^N(0, \frac{x}{N}) \le c_2$ with $c_1 > 0$, $0 < c_2 < 1$ and their convergence to some $u_i(0, v)$ as $N \to \infty$. Then, $\alpha_i^N(t, dv)$ converges to $u_i(t, v)dv$ as $N \to \infty$ in probability for $i = 1, 2$. In the limit, $u_1(t, v)u_2(t, v) = 0$ a.e. holds and $w(t, v) := u_1(t, v) - u_2(t, v)$ is the unique weak solution of the equation*

$$\partial_t w = \Delta D(w), \tag{3.1}$$

where $D(w) = d_1 w$ for $w \ge 0$ and $= d_2 w$ for $w < 0$; cf. (3.4) *for a formulation of the weak solution.*

The last sentence in the theorem means that u_i are solutions of diffusion equations $\partial_t u_i = d_i \Delta u_i$ on the regions $\{u_i > 0\}$ for $i = 1, 2$, and satisfy two-phase Stefan free boundary condition (cf., PDE literature [4] and references therein):

$$d_1 \partial_{\mathbf{n}} u_1 + d_2 \partial_{\mathbf{n}} u_2 = 0$$

at the free boundary $\Gamma_t := \{v \in \mathbb{T}^d; u_1 = u_2 = 0\}$, where $\mathbf{n}$ is the unit normal vector at Γ_t directed to the region $\{v \in \mathbb{T}^d; u_1 > 0\}$. Some extension of this theorem is given in [37].

For the proof, we apply again the relative entropy method to show that the microscopic system is close to $u_i^N(t, \frac{x}{N})$, which is determined as the solution of the system of the discretized hydrodynamic equation

$$\partial_t u_i^N\left(t, \frac{x}{N}\right) = d_i \Delta^N u_i^N\left(t, \frac{x}{N}\right) - K u_1^N\left(t, \frac{x}{N}\right) u_2^N\left(t, \frac{x}{N}\right), \quad i = 1, 2. \tag{3.2}$$

Then, we show convergence of the solution of (3.2) to that of the free boundary problem. Indeed, one can show two estimates, $\int_0^T \int_{\mathbb{T}^d} u_1^N(t, v) u_2^N(t, v) dt dv \le \frac{1}{K}$ and $\int_0^T \int_{\mathbb{T}^d} |\nabla^N u_i^N(t, v)|^2 dt dv \le \frac{1}{2d_i}$ for $i = 1, 2$, where $u_i^N(t, v)$ are the extensions on $\mathbb{T}^d$ of $u_i^N(t, \frac{x}{N})$ as step functions. These two estimates show the relative compactness of $\{u_i^N(t, v)\}_N$ in $L^2([0, T] \times \mathbb{T}^d)$. Take any limit $\{u_i\}$ of $\{u_i^N\}_N$. Then, one can show that $u_1 u_2 = 0$ a.e. from the first estimate and also that $w = u_1 - u_2$ is the weak solution of (3.1). Therefore, the uniqueness of the weak solution of (3.1) completes the proof.

3.2. From two-component Kawasaki dynamics with speed change and annihilation

Funaki [11] studied the derivation of the Stefan free boundary problems from Kawasaki dynamics with a speed change having two types of particles called water/ice (W/I). When W hits I, W is instantaneously killed, while I disappears after receiving ℓ hits of W. This models the effect of latent heat. We obtain a one-phase Stefan problem when I is immobile, and a two-phase Stefan problem when I is mobile. This model appears by first letting $K \to \infty$ for that discussed in Section 3.1 and, indeed, the two-phase Stefan problem obtained in the limit is the same (if we take $c^{\pm}(\eta) \equiv d_1, d_2$ and $\ell = 1$). The derivation of the two-phase Stefan problem in the repelling case is also possible in one dimension.

3.2.1. One-phase Stefan problem (Immobile Ice)

To record the number of hits by W particles, we label I particles by $-\ell, \dots, -1$ ($\ell \in \mathbb{N}$) and regard as different microscopic states for I. The I particles melt and disappear after they are hit ℓ times by W particles. Thus the configuration space is $\mathcal{X}_N := \{-\ell, \dots, 1\}^{\mathbb{T}_N^d}$. For $\eta = \{\eta_x\}_{x \in \mathbb{T}_N^d} \in \mathcal{X}_N$, $\eta_x = 1$ and 0 mean that the site x is occupied by a W particle or is vacant, respectively. The jump rate $c_{x,y}(\xi)$ of W particles is defined for $\xi \in \mathcal{X}^+ := \{0,1\}^{\mathbb{Z}^d}$, the I-disregarded configuration space, and $x, y \in \mathbb{T}_N^d$, $|x-y| = 1$.

Then, the generator of our model is given by

$$L_N f(\eta) = \sum_{x,y \in \mathbb{T}_N^d : |x-y|=1} c_{x,y}(\eta^+)\{f(\eta^{x,y}) - f(\eta)\}, \tag{3.3}$$

for functions f on $\mathcal{X}_N$, where $\eta^+ := \eta \vee 0$ denotes the I-disregarded configuration, while $\eta^{x,y}$ denotes the configuration after a W particle jumps from x to y, changing if $\eta_x = 1$ and $\eta_y^+ = 0$, and $\eta^{x,y} = \eta$ (remaining unchanged) otherwise.

We assume that the jump rate $c_{x,y}$ satisfies "symmetry, spatial homogeneity, locality, positivity" and the "detailed balance condition" with respect to the local specification of a certain extreme canonical Gibbs measure ν_ρ, which exists uniquely for each density $\rho \in [0,1]$ and has the uniform mixing property. In addition, we assume the "gradient condition": There exist local functions $\{h_i\}_{1 \le i \le d}$ of ξ such that the currents have the forms

$$c_{0,e_i}(\xi)(\xi_{e_i} - \xi_0) = h_i(\tau_{e_i}\xi) - h_i(\xi), \quad 1 \le i \le d, \ \xi \in \mathcal{X}^+,$$

where $e_i \in \mathbb{Z}^d$, $|e_i| = 1$, stands for the unit vector in the direction i. We also assume that the equilibrium means $P^+(\rho) := E^{\nu_\rho}[h_i]$, $\rho \in [0,1]$, are independent of i. One can see that $P^+(\rho)$ is nondecreasing and continuous in ρ.

Consider the macroscopic empirical measure $\alpha^N(t, dv) = \alpha^N(dv; \eta^N(t))$ of $\eta^N(t) = \{\eta_x^N(t)\}_{x \in \mathbb{T}_N^d} \in \mathcal{X}_N$, generated by $N^2 L_N$, defined similarly as in (2.1). We assume that $\alpha^N(0) \to a(0,v)dv$ in probability as $N \to \infty$, where $a(0,v) \in [-\ell, 1]$.

Theorem 3.2 ([11]). *For every $t > 0$, $\alpha^N(t)$ converges to $a(t,v)dv$ in probability. The limit density $a(t,v) \in [-\ell, 1]$ is a unique solution of the equation:*

$$\langle a(t), G\rangle = \langle a(0), G\rangle + \int_0^t \langle P(a(s)), \Delta G\rangle \, ds, \tag{3.4}$$

for every $G \in C^\infty(\mathbb{T}^d)$, where $\langle a, G\rangle = \int_{\mathbb{T}^d} a(v)G(v)\, dv$. The function P on $[-\ell, 1]$ is defined by $P(a) = P^+(a)$ for $a \in [0,1]$ and $= P^+(0)$ for $a \in [-\ell, 0]$.

Equation (3.4) is the weak (or enthalpy) formulation of the following one-phase Stefan problem for the density $u(t,v) \in [0,1]$ of W:

$$\begin{aligned} \partial_t u &= \Delta P^+(u) && \text{on } \mathcal{L}(t), \\ u(t,v) = 0 \text{ and } \ell V &= -\partial_{\mathbf{n}} P^+(u) && \text{at } \Sigma(t) := \partial\mathcal{L}(t), \end{aligned}$$

where $\mathcal{L}(t) := \{v \in \mathbb{T}^d; u(t,v) > 0\}$, $\mathbf{n}$ denotes the unit normal vector at $\Sigma(t)$ directed toward $\mathcal{L}(t)$, and V is the velocity of $\Sigma(t)$ in the direction $\mathbf{n}$. The speeds of loosing masses of W and I at $\Sigma(t)$ are given by $\partial_{\mathbf{n}} P^+(u)$ and $-V$, respectively. Since the loosing speed for W is ℓ times faster than that for I, we have the last Stefan free boundary condition.

3.2.2. Two-phase Stefan problem (Mobile Ice)

We make I particles with label $-\ell$ active. They perform Kawasaki dynamics with jump rates different from those for W particles. The particles with labels $-\ell+1,\dots,-1$ remain immobile and are regarded as those in intermediate states between I and W. One can determine the dynamics by properly introducing jump rates $c^+_{x,y}(\xi)$ and $c^-_{x,y}(\xi)$, $\xi \in \mathcal{X}^+$ of W/I particles, both of which satisfy the conditions in Section 3.2.1 with different extreme canonical Gibbs measures ν^+_ρ, ν^-_ρ and functions $\{h^+_i\}$, $\{h^-_i\}$, respectively. We write $P^+(\rho) = E^{\nu^+_\rho}[h^+_i]$ and $P^-(\rho) = E^{\nu^-_\rho}[h^-_i]$, $\rho \in [0,1]$.

In this setting, one can derive the following two-phase Stefan problem, written in strong form, for the density $u_1(t,v) \in [0,1]$ of W and $u_2(t,v) \in [0,1]$ of I:

$$\partial_t u_1 = \Delta P^+(u_1) \quad \text{on } \mathcal{L}_1(t), \quad \partial_t u_2 = \Delta P^-(u_2) \quad \text{on } \mathcal{L}_2(t),$$
$$u_1 = u_2 = 0 \quad \text{and} \quad (\ell-1)V = -\partial_{\mathbf{n}} P^+(u_1) - \partial_{\mathbf{n}} P^-(u_2) \quad \text{at } \Sigma(t),$$

where $\Sigma(t) := \partial\mathcal{L}_1(t) = \partial\mathcal{L}_2(t)$, $\mathbf{n}$ and V are the same as above and $\mathcal{L}_i(t) = \{u_i > 0\}$.

4. KPZ EQUATION

4.1. KPZ equation as singular SPDE

4.1.1. Scalar KPZ equation

The KPZ (Kardar, Parisi, and Zhang [42]) equation describes the motion of a growing interface with random fluctuation. It is an equation for the height function $h(t,v)$ of a curve (interface) in the plane:

$$\partial_t h = \frac{1}{2}\partial_v^2 h + \frac{1}{2}(\partial_v h)^2 + \dot{W}(t,v), \quad v \in \mathbb{R} \text{ or } \mathbb{T}. \tag{4.1}$$

We consider it in one-dimension on the whole line $\mathbb{R}$ or on a finite interval $\mathbb{T} = [0,1)$ under the periodic boundary condition; $\dot{W}(t,v)$ is a space–time Gaussian white noise with mean 0 and covariance structure

$$E\big[\dot{W}(t,v_1)\dot{W}(s,v_2)\big] = \delta(t-s)\delta(v_1-v_2). \tag{4.2}$$

This means that the noise is independent for different (t,v), and $\dot{W}(t,v)$ is realized only as a generalized function (distribution), cf. [16].

The KPZ equation attracts a lot of attention from viewpoints of integrable probability [1, 41, 46, 47], singular ill-posed SPDEs [31, 34, 35], and microscopic interacting particle systems [3].

Equation (4.1) is ill-posed in a classical sense due to the conflict between nonlinearity and roughness of the noise. It is known that the linear SPDE

$$\partial_t h = \frac{1}{2}\partial_v^2 h + \dot{W}(t,v), \quad v \in \mathbb{R} \text{ or } \mathbb{T}, \tag{4.3}$$

obtained by dropping the nonlinear term has a continuous solution which is α-Hölder continuous in v for every $\alpha < \frac{1}{2}$. Therefore, one can imagine that the nonlinear term $(\partial_v h)^2$

in (4.1) is undefinable in the usual sense. Actually, it requires a renormalization. The following renormalized KPZ equation with compensator $\delta_v(v)$ $(= +\infty)$ would have a meaning

$$\partial_t h = \frac{1}{2}\partial_v^2 h + \frac{1}{2}\{(\partial_v h)^2 - \delta_v(v)\} + \dot{W}(t,v). \tag{4.4}$$

To see (4.4) heuristically, consider the linear stochastic heat equation for $Z = Z(t,v)$:

$$\partial_t Z = \frac{1}{2}\partial_v^2 Z + Z\dot{W}(t,v), \quad v \in \mathbb{R} \text{ or } \mathbb{T}, \tag{4.5}$$

with a multiplicative noise defined in Itô's sense. It is known that (4.5) is well-posed in the mild or generalized functions' sense and the strong comparison principle holds: $Z(t,v) > 0$ for all $t > 0$ if it holds at $t = 0$. In particular, we can define the so-called Cole–Hopf solution

$$h_{\mathrm{CH}}(t,v) := \log Z(t,v). \tag{4.6}$$

Then, applying Itô's formula for (4.6) and noting $dW(t,v_1)dW(t,v_2) = \delta(v_1 - v_2)dt$ which follows from (4.2), we obtain the renormalized KPZ equation (4.4) for h_{CH} from (4.5). Note that $-\frac{1}{2}\delta_v(v)$ arises as an Itô correction term. To give a meaning to (4.4), we need to introduce approximations, see Section 4.1.3.

4.1.2. Coupled KPZ equation

One can extend the KPZ equation (4.1) to the equation for the system with n-components $h(t,v) = (h^i(t,v))_{i=1}^n$ on $\mathbb{T}$ (or $\mathbb{R}$):

$$\partial_t h^i = \frac{1}{2}\partial_v^2 h^i + \frac{1}{2}\Gamma^i_{jk}\partial_v h^j \partial_v h^k + \dot{W}^i(t,v), \quad 1 \le i \le n. \tag{4.7}$$

We use Einstein's convention to omit the sum over j and k for the second term, and $(\dot{W}^i(t,v))_{i=1}^n$ is a family of n independent space–time Gaussian white noises.

The coupling constants Γ^i_{jk} always satisfy the bilinear condition, $\Gamma^i_{jk} = \Gamma^i_{kj}$ for all i,j,k, and we sometimes assume the trilinear condition (T), namely $\Gamma^i_{jk} = \Gamma^i_{kj} = \Gamma^k_{ji}$ for all i,j,k.

The coupled KPZ equation is ill-posed. We need to introduce approximations with smooth noises and renormalizations. Equation (4.7) appears in the study of nonlinear fluctuating hydrodynamics [49,50] for a system with n-conserved quantities by taking second-order terms into account. We will discuss this for the interacting particle system in Section 4.2.3.

We also consider the coupled KPZ equation with constant drifts c^i as in [49,50]:

$$\partial_t h^i = \frac{1}{2}\partial_v^2 h^i + \frac{1}{2}\Gamma^i_{jk}\partial_v h^j \partial_v h^k + c^i\partial_v h^i + \dot{W}^i(t,v), \quad 1 \le i \le n. \tag{4.8}$$

We may assume $c^i = 0$ and reduce to (4.7) (with new space–time Gaussian white noises) by considering $\tilde{h}^i(t,v) := h^i(t, v - c^i t)$.

4.1.3. Two approximations, local and global well-posedness and invariant measure

We now discuss the approximations for (4.4) in the framework of the coupled KPZ equation. Indeed, one can introduce two types of approximations: one is simple, the other is suitable to find invariant measures. We replace the noise by a smeared one obtained by convoluting with the nonnegative symmetric kernel $\eta^\varepsilon(v) := \frac{1}{\varepsilon}\eta(\frac{v}{\varepsilon})$, which converges to $\delta_0(v)$ as $\varepsilon \downarrow 0$.

The first approximation is simple in the sense that we replace only the noise and introduce the renormalizations. It is given as follows. For $h^i = h^{\varepsilon,i}$, $\varepsilon > 0$,

$$\partial_t h^i = \frac{1}{2}\partial_v^2 h^i + \frac{1}{2}\Gamma^i_{jk}\left(\partial_v h^j \partial_v h^k - c^\varepsilon \delta^{jk} - B^{\varepsilon,jk}\right) + \dot{W}^i * \eta^\varepsilon(t,v), \tag{4.9}$$

where δ^{jk} is Kronecker's δ, $c^\varepsilon = \frac{1}{\varepsilon}\|\eta\|^2_{L^2(\mathbb{R})} - 1$ (on $\mathbb{T}$, while -1 is unnecessary on $\mathbb{R}$), and $B^{\varepsilon,jk}$ ($= O(\log\frac{1}{\varepsilon})$ in general) are renormalization factors. The renormalizations $B^{\varepsilon,jk}$ and $\tilde{B}^{\varepsilon,jk}$ introduced in (4.10) below are unnecessary under condition (T), especially, in the scalar-valued case, see Theorem 4.2(1).

The second approximation, which is suitable to find the invariant measure, is given as follows. For $\tilde{h}^i = \tilde{h}^{\varepsilon,i}$, $\varepsilon > 0$,

$$\partial_t \tilde{h}^i = \frac{1}{2}\partial_v^2 \tilde{h}^i + \frac{1}{2}\Gamma^i_{jk}\left(\partial_v \tilde{h}^j \partial_v \tilde{h}^k - c^\varepsilon \delta^{jk} - \tilde{B}^{\varepsilon,jk}\right) * \eta_2^\varepsilon + \dot{W}^i * \eta^\varepsilon(t,v), \tag{4.10}$$

with renormalization factors c^ε as above and $\tilde{B}^{\varepsilon,jk}(= O(\log\frac{1}{\varepsilon}))$, where $\eta_2^\varepsilon = \eta^\varepsilon * \eta^\varepsilon$. This approximation for the scalar KPZ equation was introduced in Funaki and Quastel **[23]** and the idea behind (4.10) is the fluctuation–dissipation relation. Renormalization factor c^ε comes from the second-order terms in the related Wiener–Itô chaos expansion, while $B^{\varepsilon,jk}$ and $\tilde{B}^{\varepsilon,jk}$ are from the fourth-order terms. For the solution of (4.10) (with $\tilde{B} = 0$), **[15]** showed on $\mathbb{R}$, under condition (T), the infinitesimal invariance of the distribution of $B * \eta^\varepsilon(v)$, $v \in \mathbb{R}$, where B is the $\mathbb{R}^n$-valued two-sided Brownian motion, cf. Theorem 4.2(2).

When $n = 1$ and $\Gamma = 1$, the solution of (4.9) with $B^\varepsilon = 0$ converges as $\varepsilon \downarrow 0$ to the Cole–Hopf solution h_{CH} of the KPZ equation, while **[23]** showed on $\mathbb{R}$ that the solution of (4.10) with $\tilde{B}^\varepsilon = 0$ converges to $h_{\mathrm{CH}} + \frac{1}{24}t$, see also **[39]**. This was shown based on the Boltzmann–Gibbs principle, which follows by Kipnis–Varadhan estimate (cf. **[44, 45]**), see (4.19). This estimate is sometimes called Itô–Tanaka trick.

The method of **[23]** is based on the Cole–Hopf transform, but it is not available for the coupled equation in general. Instead, the paracontrolled calculus due to Gubinelli et al. **[31]** is applicable. Funaki and Hoshino **[19]** showed the following three results on $\mathbb{T}$.

First is the convergence of h^ε and $\tilde{h}^\varepsilon$ and local-in-time well-posedness of coupled KPZ equation (4.7). Let $\mathcal{C}^\alpha = (\mathcal{B}^\alpha_{\infty,\infty}(\mathbb{T}))^n$, $\alpha \in \mathbb{R}$ be an $\mathbb{R}^n$-valued Hölder–Besov space on $\mathbb{T}$.

Theorem 4.1 (**[19]**). (1) *Assume $h_0 \in \mathcal{C}^\delta$, $\delta \in (0, \frac{1}{2})$, then a unique solution h^ε of (4.9) exists up to survival time $T^\varepsilon \in (0, \infty]$. With a proper choice of $B^{\varepsilon,jk}$, there exists $T_{sur} > 0$ such that $T_{sur} \le \liminf_{\varepsilon\downarrow 0} T^\varepsilon$ holds, and h^ε converges in probability as $\varepsilon \downarrow 0$ to some h in $C([0,T],\mathcal{C}^\delta) \cap C((0,T],\mathcal{C}^\alpha)$ for every $\alpha < \frac{1}{2}$ and $0 < T < T_{sur}$.*

(2) *Similar result holds for the solution $\tilde{h}^\varepsilon$ of (4.10) with some limit $\tilde{h}$. Under proper choices of $B^{\varepsilon,jk}$ and $\tilde{B}^{\varepsilon,jk}$, we can actually make $h = \tilde{h}$.*

Second, under the trilinear condition (T), we have the invariance of Wiener measure on $\mathbb{T}$ in the following sense and can explicitly compute the difference of two limits.

Theorem 4.2 ([19]). *Assume the trilinear condition* (T). *Then,*

(1) $B^{\varepsilon,jk}$, $\tilde{B}^{\varepsilon,jk} = O(1)$ *as* $\varepsilon \downarrow 0$ *so that the solutions of* (4.9) *with* $B^{\varepsilon,jk} = 0$ *and* (4.10) *with* $\tilde{B}^{\varepsilon,jk} = 0$ *converge. In the limit, we have* $\tilde{h}^i(t,v) = h^i(t,v) + c^i t$, $1 \le i \le n$, *where*

$$c^i = \frac{1}{24} \sum_{j,k,k_1,k_2} \Gamma^i_{jk} \Gamma^j_{k_1k_2} \Gamma^k_{k_1k_2}.$$

(2) *Moreover, the distribution of* $\{\partial_v B\}_{v\in\mathbb{T}}$, *where* B *is the* $\mathbb{R}^n$*-valued periodic Brownian motion on* $\mathbb{T}$, *is the unique invariant (probability) measure for the tilt process* $u = \partial_v h$. *Or, one can say that the periodic Wiener measure on the quotient space* $\mathcal{C}^\alpha/\sim$, $\alpha < \frac{1}{2}$, *where "*$\sim$*" is defined by* $h \sim h + c$ *for* $c \in \mathbb{R}$, *is invariant for* h. *The uniqueness of invariant measures does not hold on* $\mathbb{R}$. *Wiener measures with constant drifts are all invariant on* $\mathbb{R}$, *see* [23].

Third is the global-in-time well-posedness (existence and uniqueness of solutions) of (4.7) in paracontrolled sense under (T). Indeed, assuming (T), we take the initial value $h(0)$ as $h(0,0) = 0$ and $u_0 := \partial_v h(0) \overset{\text{law}}{=} \{\partial_v B\}_{v\in\mathbb{T}}$ (i.e., stationary). Then, one can show the uniform bound for $u = \partial_v h$, namely that $E[\sup_{t\in[0,T]} \|u(t;u_0)\|^p_{\mathcal{C}^{\alpha-1}}] < \infty$ for every $T > 0$, $p \ge 1$, $\alpha < \frac{1}{2}$. This implies the global-in-time existence of the solution for a.a.-u_0. Combing this with the strong Feller property of $\partial_v h$ for h in (4.7) on $\mathcal{C}^{\alpha-1}$, $\alpha \in (0, \frac{1}{2})$ shown by [36], we obtain the global existence for $u = \partial_v h$ for all given u_0.

For $u \equiv u^\varepsilon = (u^i)_i = (\partial_v h^i)_i$ for h^i in (4.9), we have

$$\sum_{i,j,k} \Gamma^i_{jk} \int_{\mathbb{T}} u^i \partial_v (u^j u^k) dv = 0$$

under (T). This shows an a priori estimate and global well-posedness for (4.9) at least if $h(0) \in H^1(\mathbb{T})$. Therefore, Theorem 4.1(1) holds globally in time if $h(0) \in H^1(\mathbb{T})$.

The example given in [9] with $n = 2$ does not satisfy (T), but the logarithmic renormalization term is unnecessary, and one can show the existence of an invariant measure. The role of the trilinear condition (T) is discussed further in [18].

4.1.4. Proof of Theorems 4.1 and 4.2

We think of the Ornstein–Uhlenbeck part (as in (4.3) for the scalar-valued case) as the leading term and of the nonlinear term as its perturbation. This leads to an expansion of the equation. We introduce finitely many driving terms $\mathbb{H}$, which involve renormalizations, and show that, once these terms are determined, the rest of the equation is solvable in the framework of the paracontrolled calculus. In particular, we can show the local-in-time solvability and continuity of the solutions in $\mathbb{H}$.

4.2. From interacting particle systems

Bernardin et al. [2] derived the coupled KPZ equation (4.8) with drifts from a microscopic interacting particle system called multispecies zero-range process with weak asymmetry (WA). The derivation of scalar KPZ(–Burgers) equation from particle systems was studied in [3] (WA simple exclusion process), [29] (WA exclusion process with speed change), and [30] (WA zero-range process).

4.2.1. n-species zero-range process on $\mathbb{T}_N$

To derive an n-component system in the limit, we need to consider a system with n-conserved quantities at the microscopic level. We consider n-species zero-range process on $\mathbb{T}_N = \{1, 2, \dots, N\}$ with periodic boundary condition, namely, particles of n-types, which perform random walks on $\mathbb{T}_N$ and interact only at the same sites. The corresponding macroscopic space is $\mathbb{T} = [0, 1)$. Compared to the model discussed in Section 2.1.1, our system has multiple species, but is limited in one-dimension and is without Glauber part. A configuration of particles is denoted by $\boldsymbol{\eta} = (\eta^i)_{i=1}^n = (\{\eta_x^i\}_{x\in\mathbb{T}_N})_{i=1}^n \in \mathcal{X}_N^n$, where $\mathcal{X}_N = \mathbb{Z}_+^{\mathbb{T}_N}$ is the configuration space of single-species particles and $\eta_x^i \in \mathbb{Z}_+$ denotes the number of particles of the ith species at $x \in \mathbb{T}_N$.

We introduce a weak asymmetry (WA) in jump rates. Once a jump happens, the probabilities of a jump of the ith particles to the right or the left are given by $p_i^N(\pm 1) = \frac{1}{2} \pm c^{i,N}$ (+ for right, − for left) with small $c^{i,N}$. As we will see, $c^{i,N} = \frac{c^i}{N}$, i.e., $O(\frac{1}{N})$ for the hydrodynamic limit and linear fluctuation (see Section 4.2.2), while $c^{i,N} = \frac{c}{\sqrt{N}} + \frac{c^i}{N}$, i.e., $O(\frac{1}{\sqrt{N}})$ for KPZ nonlinear fluctuation (see Section 4.2.3). Note that the constant c in leading order is common for all i.

We consider the Markov process $\boldsymbol{\eta}^N(t) = \{\eta_x^{N,i}(t)\}_{x,i}$ on $\mathcal{X}_N^n$ with the generator

$$L_N f(\boldsymbol{\eta}) = N^2 \sum_{x\in\mathbb{T}_N, 1\le i\le n, e=\pm 1} p_i^N(e) g_i(\boldsymbol{\eta}_x)\{f(\boldsymbol{\eta}^{x,x+e;i}) - f(\boldsymbol{\eta})\},$$

for functions f on $\mathcal{X}_N^n$, where $\boldsymbol{\eta}_x = (\eta_x^i)_{i=1}^n$ and $\boldsymbol{\eta}^{x,y;i}$ stands for the configuration $\boldsymbol{\eta}$ after one ith particle jumps from x to y (which is possible only when $\eta_x^i \ge 1$). The diffusive time change N^2 is introduced. The jump rate g_i of the ith particles has the zero-range property, that is, it is a function on $\mathbb{Z}_+^n$, which is the configuration space at a single site, so that $g_i = g_i(\mathbf{k})$ for $\mathbf{k} = (k_i)_{i=1}^n \in \mathbb{Z}_+^n$. In particular, interaction occurs only at the same sites. We assume that the jump rates $\{g_i(\mathbf{k})\}_{1\le i\le n, \mathbf{k}\in\mathbb{Z}_+^n}$ satisfy the conditions of "nondegeneracy, linear growth, nontriviality of Dom_Z (defined below)" and the "detailed balance condition" with respect to product measures, $\frac{g_i(\mathbf{k})}{g_i(\mathbf{k}_{j,-})} = \frac{g_j(\mathbf{k})}{g_j(\mathbf{k}_{i,-})}$ for all $i \ne j$ and $\mathbf{k} \in \mathbb{Z}_+^n$ with $k_i, k_j \ge 1$, where $\mathbf{k}_{j,-} = (k_1, \dots, k_{j-1}, k_j - 1, k_{j+1}, \dots, k_n)$, see [2] for details.

The invariant measures, or equilibrium states of $\boldsymbol{\eta}^N(t)$, are superpositions of the product measures $\{\bar{\nu}_{\boldsymbol{\varphi}} := p_{\boldsymbol{\varphi}}^{\otimes\mathbb{T}_N}\}$ with one-site marginal

$$p_{\boldsymbol{\varphi}}(\mathbf{k}) = \frac{1}{Z_{\boldsymbol{\varphi}}} \frac{\boldsymbol{\varphi}^{\mathbf{k}}}{\mathbf{g}(\mathbf{k})!}, \quad Z_{\boldsymbol{\varphi}} = \sum_{\mathbf{k}\in\mathbb{Z}_+^n} \frac{\boldsymbol{\varphi}^{\mathbf{k}}}{\mathbf{g}(\mathbf{k})!}.$$

Here $\boldsymbol{\varphi} = (\varphi_i)_{i=1}^n$ are nonnegative parameters, called fugacity, $\boldsymbol{\varphi}^{\mathbf{k}} = \varphi_1^{k_1} \cdots \varphi_n^{k_n}$, and

$$\mathbf{g}(\mathbf{k})! = \prod_{\ell=1}^{|\mathbf{k}|} g_{i(\ell)}(\mathbf{k}_\ell),$$

with $|\mathbf{k}| = k_1 + \cdots + k_n$, is a product along an increasing path $\mathbf{k}_0 = \mathbf{0} \to \cdots \to \mathbf{k}_\ell \to \cdots \to \mathbf{k}_{|\mathbf{k}|} = \mathbf{k}$ connecting $\mathbf{0}$ and $\mathbf{k}$ in $\mathbb{Z}_+^n$ such that $|\mathbf{k}_\ell| = \ell$, $0 \le \ell \le |\mathbf{k}|$, where $i(\ell)$ is the coordinate increased by 1 from $\mathbf{k}_{\ell-1}$ to $\mathbf{k}_\ell$. We set $\mathrm{Dom}_Z := \{\boldsymbol{\varphi} \in (0,\infty)^n; Z_{\boldsymbol{\varphi}} < \infty\}$. Note that, by the detailed balance condition, $\mathbf{g}(\mathbf{k})!$ does not depend on the choice of the increasing path $\{\mathbf{k}_\ell\}$, so is well-defined.

As in Section 2.1.1, we change the parameter from fugacity $\boldsymbol{\varphi}$ to density $\mathbf{a} = (a^i)_{i=1}^n$ of particles. Namely, define the map $R : \boldsymbol{\varphi} \mapsto \mathbf{a} = (a^i(\boldsymbol{\varphi}))_{i=1}^n$ by

$$a^i \equiv a^i(\boldsymbol{\varphi}) := E^{\bar{\nu}_{\boldsymbol{\varphi}}}\big[\eta_0^i\big], \quad 1 \le i \le n, \tag{4.11}$$

which is defined on $\mathrm{Dom}_R := \{\boldsymbol{\varphi} \in \mathrm{Dom}_Z; a^i(\boldsymbol{\varphi}) < \infty, 1 \le i \le n\}$ and denote $\nu_{\mathbf{a}} := \bar{\nu}_{\boldsymbol{\varphi}}$. The correspondence $\boldsymbol{\varphi} \leftrightarrow \mathbf{a}$ is one-to-one. We accordingly have a family of invariant measures $\{\nu_{\mathbf{a}}\}_{\mathbf{a}}$ parametrized by density $\mathbf{a} \in [0,\infty)^n$.

4.2.2. Hydrodynamic limit and linear fluctuation

We discuss the hydrodynamic limit (LLN) and the equilibrium linear fluctuation problem (CLT).

Hydrodynamic limit. Recall that the weak asymmetry is $O(\frac{1}{N})$, i.e., $p_i^N(\pm 1) = \frac{1}{2} \pm \frac{c^i}{N}$ and c^i may be different for different species. We consider an $\mathbb{R}^n$-valued macroscopic empirical measure $X_t^N = (X_t^{N,i})_{i=1}^n$ on $\mathbb{T}$ defined as in (2.1) taking $\eta_x^{N,i}(t)$ for η_x:

$$X_t^{N,i}(dv) = \frac{1}{N} \sum_{x \in \mathbb{T}_N} \eta_x^{N,i}(t) \delta_{\frac{x}{N}}(dv), \quad v \in \mathbb{T},\ 1 \le i \le n.$$

One can show that, multiplied by a test function $G \in C^\infty(\mathbb{T})$ as in (2.2), X_t^N converges as $N \to \infty$ to $\mathbf{a}(t,v)dv = (a^i(t,v)dv)_{i=1}^n$ in probability and the limit density $a^i(t,v)$ is the solution of the system of nonlinear PDEs:

$$\partial_t a^i = \frac{1}{2}\partial_v^2 \varphi_i(\mathbf{a}) - 2c^i \partial_v \varphi_i(\mathbf{a}), \quad v \in \mathbb{T},\ 1 \le i \le n, \tag{4.12}$$

where $\varphi_i(\mathbf{a}) := E^{\nu_{\mathbf{a}}}[g_i(\boldsymbol{\eta}_0)]$. This is a multispecies version of (2.3) with $K = 0$ and weak asymmetry. The diffusion matrix is parabolic in the sense that $\sum_{ij} \frac{\partial \varphi_i}{\partial a^j} \xi_i \xi_j \ge 0$ for any $(\xi_i) \in \mathbb{R}^n$.

The hydrodynamic equation (4.12) can be heuristically derived as follows. By Dynkin's formula, we have

$$\langle X_t^{N,i}, G\rangle = \langle X_0^{N,i}, G\rangle + \int_0^t L_N X_s^{N,i}(G)ds + M_t^{N,i}(G) \tag{4.13}$$

and

$$L_N X^{N,i}(G) = \frac{1}{2N} \sum_{x \in \mathbb{T}_N} g_i(\boldsymbol{\eta}_x) \Delta^N G\Big(\frac{x}{N}\Big)$$

$$+\frac{c^i}{N}\sum_{x\in\mathbb{T}_N} g_i(\boldsymbol{\eta}_x)\Big\{\nabla^N G\Big(\frac{x}{N}\Big)+\nabla^N G\Big(\frac{x-1}{N}\Big)\Big\},$$

where $\nabla^N G(\frac{x}{N})=N(G(\frac{x+1}{N})-G(\frac{x}{N}))$ and we recall (2.7) for Δ^N. For martingale terms, $\lim_{N\to\infty}E[M_t^{N,i}(G)^2]=0$ hold. By local ergodicity in local equilibria, one can replace $g_i(\boldsymbol{\eta}_x)$ by its local average $\varphi_i(\mathbf{a}(t,\frac{x}{N}))$ and obtain the weak form of (4.12) for $a^i(t,v)$ in the limit.

Linear fluctuation. Keeping the weak asymmetry the same as above, we discuss equilibrium fluctuation so that we assume $\boldsymbol{\eta}^N(0)\overset{\text{law}}{=}\nu_{\mathbf{a}_0}$ for any fixed $\mathbf{a}_0=(a_0^i)_{i=1}^n\in(0,\infty)^n$. Consider the fluctuation field $Y_t^N=(Y_t^{N,i})_{i=1}^n$ around $\mathbf{a}_0$ defined by

$$Y_t^{N,i}(dv)=\frac{1}{\sqrt{N}}\sum_{x\in\mathbb{T}_N}(\eta_x^{N,i}(t)-a_0^i)\delta_{\frac{x}{N}}(dv),\quad v\in\mathbb{T},\ 1\le i\le n. \tag{4.14}$$

The limit $Y_t=(Y_t^i)_{i=1}^n$ is the solution of linear SPDE (Ornstein–Uhlenbeck process)

$$\partial_t Y^i=\frac{1}{2}\sum_{j=1}^n\partial_{a^j}\varphi_i(\mathbf{a}_0)\partial_v^2Y^j-2c^i\sum_{j=1}^n\partial_{a^j}\varphi_i(\mathbf{a}_0)\partial_vY^j+\sqrt{\varphi_i(\mathbf{a}_0)}\partial_v\dot{W}^i, \tag{4.15}$$

where $\dot{W}=(\dot{W}^i(t,v))_{i=1}^n$ is a family of n independent space–time Gaussian white noises. The coefficient $\partial_{a^j}\varphi_i(\mathbf{a}_0)$ arises as a linearization of $\varphi_i(\mathbf{a})$ in equation (4.12) around $\mathbf{a}_0$,

$$\begin{aligned}\varphi_i(\mathbf{a})&=\varphi_i(\mathbf{a}_0)+\sum_{j=1}^n\partial_{a^j}\varphi_i(\mathbf{a}_0)(a^j-a_0^j)+\cdots\\&\cong\varphi_i(\mathbf{a}_0)+\frac{1}{\sqrt{N}}\sum_{j=1}^n\partial_{a^j}\varphi_i(\mathbf{a}_0)\cdot Y^j+\cdots.\end{aligned}$$

To make this replacement rigorous, we need to establish the first-order Boltzmann–Gibbs principle. The limit noise $(\sqrt{\varphi_i(\mathbf{a}_0)}\partial_v\dot{W}^i)_i$ is obtained by computing quadratic and cross-variations of the martingale terms $\tilde{M}_t^{N,i}(G)=\sqrt{N}M_t^{N,i}(G)$ of $\langle Y_t^{N,i},G\rangle$. Indeed, we have

$$\begin{aligned}\frac{d}{dt}\langle\tilde{M}^{N,i}(G)\rangle_t&=N\Big(L_N\langle\eta^{N,i}(t),G\rangle^2-2\langle\eta^{N,i}(t),G\rangle L_N\langle\eta^{N,i}(t),G\rangle\Big)\\&=\frac{1}{N}\sum_{x\in\mathbb{T}_N}g_i(\boldsymbol{\eta}_x^N(t))\Big(\nabla^NG\Big(\frac{x}{N}\Big)\Big)^2+O\Big(\frac{1}{N}\Big)\to\varphi_i(\mathbf{a}_0)\|G'\|_{L^2(\mathbb{T})}^2,\end{aligned}$$

as $N\to\infty$, since $\boldsymbol{\eta}^N(t)\overset{\text{law}}{=}\nu_{\mathbf{a}_0}$ for all $t\ge 0$, while $\langle\tilde{M}^{N,i}(G_1),\tilde{M}^{N,j}(G_2)\rangle_t=0$ for $i\ne j$.

See Section 2.2 of **[16]** for nonequilibrium fluctuation. The class of models having Ornstein–Uhlenbeck scaling limit is sometimes called Edwards–Wilkinson university class.

4.2.3. Nonlinear fluctuation leading to coupled KPZ–Burgers equation

Now the weak asymmetry is $O(\frac{1}{\sqrt{N}})$, i.e., $p_i^N(\pm 1)=\frac{1}{2}\pm(\frac{c}{\sqrt{N}}+\frac{c^i}{N})$, which is larger than before. Note that the leading constant c is common to have the common moving frame. Compared to Section 4.2.2, c^i are replaced by $c\sqrt{N} + c^i$ so that equation (4.12) for the density of the ith particles will look like

$$\partial_t a^i = \frac{1}{2}\partial_v^2 \varphi_i(\mathbf{a}) - 2(c\sqrt{N} + c^i)\partial_v \varphi_i(\mathbf{a}) + \frac{1}{\sqrt{N}}(\text{noise}). \tag{4.16}$$

We consider the fluctuation field under equilibrium, i.e., $\boldsymbol{\eta}^N(0) \stackrel{\text{law}}{=} \nu_{\mathbf{a}_0}$ for some $\mathbf{a}_0$. This time, $\mathbf{a}_0$ should be chosen properly. To cancel the diverging factor $2c\sqrt{N}$ in (4.16), we introduce the moving frame with speed $2c\lambda\sqrt{N}$ at the macroscopic level with a suitably chosen $\lambda = \lambda(\mathbf{a}_0)$ to the fluctuation field so that (4.14) is modified to become

$$Y_t^{N,i}(dv) = \frac{1}{\sqrt{N}} \sum_{x \in \mathbb{T}_N} (\eta_x^{N,i}(t) - a_0^i)\delta_{\frac{x}{N} - 2c\lambda\sqrt{N}t}(dv), \quad v \in \mathbb{T},\ 1 \le i \le n. \tag{4.17}$$

The frame should have common speed for all i and this gives the restriction on the choice of $\mathbf{a}_0$. Indeed, we need to assume the frame condition [FC], namely $\partial_{a^j}\varphi_i(\mathbf{a}_0) = -\lambda\delta^{ij}$ for $\mathbf{a}_0$ and λ. Then, our main result for the nonlinear fluctuation is stated as follows.

Theorem 4.3 ([2]). *Assume the frame condition* [FC]. *Then,* $Y_t^N = (Y_t^{N,i})_{i=1}^n$ *converges to* $Y_t = (Y_t^i)_{i=1}^n$ *in law on the Skorohod space* $D([0,T], \mathcal{S}'(\mathbb{T})^n)$ *with dual of* $\mathcal{S}(\mathbb{T}) = C^\infty(\mathbb{T})$. *The limit* Y_t *is the unique stationary martingale solution of the coupled KPZ–Burgers equation*

$$\begin{aligned} \partial_t Y^i =& \frac{1}{2}\partial_{a^i}\varphi_i(\mathbf{a}_0)\partial_v^2 Y^i + \Gamma_{jk}^i(\mathbf{a}_0)\partial_v(Y^j Y^k) \\ &- 2c^i \partial_{a^i}\varphi_i(\mathbf{a}_0)\partial_v Y^i + \sqrt{\varphi_i(\mathbf{a}_0)}\partial_v \dot{W}^i, \quad v \in \mathbb{T}, \end{aligned} \tag{4.18}$$

where $(\dot{W}^i)_{i=1}^n$ *is the same as in* (4.15) *and* $\Gamma_{jk}^i(\mathbf{a}_0) = -c\partial_{a^j}\partial_{a^k}\varphi_i(\mathbf{a}_0)$. *We use Einstein's convention for the second term. Observe that if* $c = 0$ *and under* [FC] *then* (4.18) *is the same as* (4.15) *for the linear fluctuation.*

The reason to have the limit noises in (4.18) is the same as (4.15); note that they have the same distribution under the shift by moving frame. We give a heuristic reason to have the nonlinear drift term in the limit. The main idea is the combination of the averaging effect due to ergodicity under $\nu_{\mathbf{a}_0}$ and Taylor expansion, now up to the second-order terms. Noting $a^i = a_0^i + \frac{1}{\sqrt{N}}Y^i + \cdots$ and the moving frame in (4.17), in (4.16) we will have

$$\partial_t a^i = \frac{1}{\sqrt{N}}\partial_t Y^i + 2c\lambda\sqrt{N}\partial_v a^i + \cdots = \frac{1}{\sqrt{N}}\partial_t Y^i + 2c\lambda\partial_v Y^i + \cdots,$$

$$\varphi_i(\mathbf{a}) = \varphi_i(\mathbf{a}_0) + \frac{1}{\sqrt{N}}\sum_{j=1}^n \partial_{a^j}\varphi_i(\mathbf{a}_0)\cdot Y^j + \frac{1}{2N}\sum_{j,k=1}^n \partial_{a^j}\partial_{a^k}\varphi_i(\mathbf{a}_0)\cdot Y^j Y^k + \cdots.$$

We put these expansions into (4.16) and multiply it by $\sqrt{N}$. Then, noting $\partial_v\varphi_i(\mathbf{a}_0) = 0$ and observing that the diverging terms of order $O(\sqrt{N})$ exactly cancel by the frame condition [FC], we will obtain (4.18).

More precisely, in Dynkin's formula (4.13), by the averaging effect, we replace

$$g_i(\eta_x) \cong E^{\nu_{\mathbf{a}_0 + \frac{1}{\sqrt{N}}Y_t(\frac{x}{N} - 2c\lambda\sqrt{N}t)}}[g_i(\eta_x)] = \varphi_i\left(\mathbf{a}_0 + \frac{1}{\sqrt{N}}Y_t\left(\frac{x}{N} - 2c\lambda\sqrt{N}t\right)\right)$$

and then make the above expansion. This procedure will be made rigorous by the second-order Boltzmann–Gibbs principle. We give a slightly more detailed outline of the proof.

Proof of Theorem 4.3. For the proof, we need to establish the second-order Boltzmann–Gibbs principle (Theorem 4.4), which is a combination of averaging due to ergodicity and Taylor expansion. It guarantees the replacement under space–time average of a function $f = f(\boldsymbol{\eta})$, whose ensemble average and its first derivatives vanish at $\mathbf{a}_0$, by quadratic function of $\eta^i - a^i$, in equilibrium $\nu_{\mathbf{a}_0}$. For the identification of the limit, we use the uniqueness of stationary coupled KPZ–Burgers martingale solutions obtained in [32]. In the limit SPDE, the drift term with c^i can be killed by the spatial shift as noted below (4.8) (at the level of KPZ equation) so that we assume $c^i = 0$ for simplicity. We also show the tightness of $\{Y_t^N\}_N$ in the uniform topology in $D([0,T], \mathcal{S}'(\mathbb{T})^n)$.

For $\zeta = (\zeta_x)$, the sample average of ζ around x in size $\ell \geq 1$ is defined by $\zeta_x^{(\ell)} := \frac{1}{2\ell+1}\sum_{|y|\leq \ell} \zeta_{x+y}$. The ensemble average of f is denoted by $\langle f\rangle(\mathbf{a}) = E^{\nu_{\mathbf{a}}}[f]$.

Theorem 4.4. *Let $f = f(\boldsymbol{\eta}) \in L^5(\nu_{\mathbf{a}_0})$ be a local function supported on sites $|y| \leq \ell_0$ such that $\langle f\rangle(\mathbf{a}_0) = 0$ and $\partial_{a^i}\langle f\rangle(\mathbf{a}_0) = 0$ for all i. Then, there exists $C = C(\ell_0) > 0$ such that for $T > 0$, $\ell \geq \ell_0$, and $\psi : \mathbb{T}_N \to \mathbb{R}$, we have*

$$E^{\nu_{\mathbf{a}_0}}\Bigg[\sup_{0\leq t\leq T}\Bigg(\int_0^t ds \sum_{x\in\mathbb{T}_N} \psi_{x-[cs]}\Bigg(f\big(\tau_x \boldsymbol{\eta}^N(s)\big) - \frac{1}{2}\sum_{j,k=1}^{n} \partial_{a^j}\partial_{a^k}\langle f\rangle(\mathbf{a}_0) \times \bigg\{\big((\eta^{N,j})_x^{(\ell)}(s) - a_0^j\big)\big((\eta^{N,k})_x^{(\ell)}(s) - a_0^k\big) - \frac{V_{jk}(\mathbf{a}_0)}{2\ell+1}\bigg\}\Bigg)\Bigg)^2\Bigg]$$
$$\leq C\|f\|^2_{L^5(\nu_{\mathbf{a}_0})}\bigg(\frac{T\ell}{N}\|\psi\|^2_{L^2(\mathbb{T}_N)} + \frac{T^2N^2}{\ell^3}\|\psi\|^2_{L^1(\mathbb{T}_N)}\bigg),$$

where $(V_{jk}(\mathbf{a}_0)) = \mathrm{cov}(\nu_{\mathbf{a}_0})$ and $\|\psi\|_{L^p(\mathbb{T}_N)} = (\frac{1}{N}\sum_{x\in\mathbb{T}_N} |\psi_x|^p)^{1/p}$, $p \geq 1$.

Proof. We apply Kipnis–Varadhan estimate to reduce the dynamic problem to a static one (bound by H^{-1}-norm): roughly for a mean-zero function F,

$$E^{\nu_{\mathbf{a}_0}}\bigg[\sup_{0\leq t\leq T}\bigg(\int_0^t F\big(\boldsymbol{\eta}(s)\big)ds\bigg)^2\bigg] \leq CTE^{\nu_{\mathbf{a}_0}}\big[F\cdot\big(-L_N^{\mathrm{sym}}\big)^{-1}F\big], \tag{4.19}$$

where L_N^{sym} is the symmetric part of the generator L_N of $\boldsymbol{\eta}(t)$. To estimate the H^{-1}-norm by L^2-norm, we apply the spectral gap estimate of the operator $-L_N^{\mathrm{sym}}$, but this works only on a bounded region and depends on the size of this region. Let $L_{\mathbf{k},\ell}^{\mathrm{sym}}$ be the symmetrized generator on $\Lambda_\ell = \{x; |x| \leq \ell\}$ with particles numbered $\mathbf{k}$ on Λ_ℓ, and let $W(\mathbf{k},\ell)$ be the inverse of the spectral gap of $-L_{\mathbf{k},\ell}^{\mathrm{sym}}$. Then, one can show $E^{\nu_{\mathbf{a}}}[W(\mathbf{k},\ell)^2] \leq C\ell^4$. We need some assumption on $(g_i)_{i=1}^n$ to show this. So, we need to confine ourselves to a bounded region of size ℓ by conditioning, that is, under the canonical ensemble.

Then we give static estimates. More precisely, we give a decay estimate for the canonical average as $\ell \to \infty$ to get grandcanonical average called the equivalence of ensembles (shown by applying the local CLT, see the first "$\sim$" below) and also Taylor expansion (see the second "$\sim$"). To give some feeling, for $y \in \mathbb{R}^n$, we present

$$E^{\nu_{\mathbf{a}_0}}\big[f(\eta)|\eta^{(\ell)} = y\big] = \frac{E^{\nu_{\mathbf{a}_0}}[f(\eta)\cdot 1_{\{\eta^{(\ell)}=y\}}]}{\nu_{\mathbf{a}_0}(\eta^{(\ell)} = y)} \sim E^{\nu_y}\big[f(\eta)\big]$$
$$\sim \langle f\rangle(\mathbf{a}_0) + \nabla_{\mathbf{a}}\langle f\rangle(\mathbf{a}_0)\cdot(y-\mathbf{a}_0) + \frac{1}{2}\big(y-\mathbf{a}_0, D^2_{\mathbf{a}}\langle f\rangle(\mathbf{a}_0)(y-\mathbf{a}_0)\big) + \cdots.$$

This leads to (the static version of) Theorem 4.4 by taking $y = \eta^{(\ell)}$. ■

To characterize the limit, we apply the martingale problem approach called energy solution, especially, its uniqueness. The authors of [32] established the uniqueness of stationary energy solutions satisfying Yaglom reversibility, that is, the time reversed process has the negative nonlinear drift compared to the original process.

Indeed, the coupled KPZ–Burgers equation for $Y^i = \partial_v h^i$ for h satisfying (4.7) in canonical form is written as

$$\partial_t Y^i = \frac{1}{2}\partial_v^2 Y^i + \frac{1}{2}\Gamma^i_{jk}\partial_v(Y^j Y^k) + \partial_v \dot{W}^i(t,v), \quad v \in \mathbb{T},\ 1 \le i \le n. \tag{4.20}$$

Its formal generator is given by $\mathcal{L} = \mathcal{L}_0 + \mathcal{A}$, where

$$\mathcal{L}_0\Phi(Y) = \frac{1}{2}\sum_i\Big(\int_{\mathbb{T}} \partial_v^2 D^2_{Y^i(v)}\Phi\,dv + \int_{\mathbb{T}} \partial_v^2 Y^i(v)\cdot D_{Y^i(v)}\Phi\,dv\Big),$$
$$\mathcal{A}\Phi(Y) = \frac{1}{2}\sum_{i,j,k}\Gamma^i_{jk}\int_{\mathbb{T}} \partial_v\big(Y^j(v)Y^k(v)\big)D_{Y^i(v)}\Phi\,dv,$$

for $\Phi = \Phi(Y)$ and D, D^2 denoting the Fréchet derivatives. The authors of [32] gave the precise definition of $\mathcal{L}$ and its domain $\mathcal{D}(\mathcal{L})$. Then they showed that the Kolmogorov backward equation $\partial_t\Psi = \mathcal{L}\Psi$ is solvable in the paracontrolled sense in $\Psi = \Psi(t,Y) \in \mathcal{D}(\mathcal{L})$ for a wide class of initial values $\Psi(0) = \Psi_0$. In particular, this shows the uniqueness for the $(\mathcal{L},\mathcal{D}(\mathcal{L}))$-martingale problem as follows: $\Phi(t,Y_t) - \Phi(0,Y_0) - \int_0^t(\partial_s\Phi + \mathcal{L}\Phi)(s,Y_s)ds$ is a martingale for $\Phi(t,\cdot) \in \mathcal{D}(\mathcal{L})$. Take $\Phi(t,Y) = \Psi(T-t,Y)$, $t \in [0,T]$, with the solution Ψ of the Kolmogorov equation. Then, $\Psi(T-t,Y_t) - \Psi(T,Y_0)$ is a martingale. Take $t = T$, and we have $E_{Y_0}[\Psi_0(Y_T)] = \Psi(T,Y_0)$. This shows the uniqueness.

They further showed that the stationary solution of the cylinder function martingale problem, that is, instead of $\Phi \in \mathcal{D}(\mathcal{L})$ the martingale property holds for tame functions $\Phi(Y) = f(\langle Y,\psi_1\rangle,\ldots,\langle Y,\psi_n\rangle)$ satisfying Kipnis–Varadhan estimate, is a solution of the $(\mathcal{L},\mathcal{D}(\mathcal{L}))$-martingale problem.

The proof of Theorem 4.3 is completed by combining all these arguments. ■

The coupling constants $\Gamma^i_{jk}(\mathbf{a}_0)$ in our coupled KPZ–Burgers equation (4.18) satisfy the trilinear condition (T) after rewriting it in a canonical form (4.20) by a proper change of the time and magnitude taking $c^i = 0$. The scaling limit of Y^N under the product measure $\nu_{\mathbf{a}_0}$ is the "spatial Gaussian white noise" (at Burgers' level), so that this is consistent in view of Theorem 4.2(2).

4.3. Related results

4.3.1. Stochastic eight-vertex model

Funaki et al. [22] introduced the stochastic eight-vertex model motivated by the eight-vertex model in statistical mechanics. It is a totally asymmetric discrete time particle system on $\mathbb{Z}$ with jumps to one of the consecutive vacant sites on the right. Moreover, a Glauber-type mechanism, that is, creation of pair of particles and annihilation of colliding two particles, is allowed. A new type of KPZ–Burgers equation is obtained in the scaling limit for a properly defined fluctuation field:

$$\partial_t Y = \frac{\nu_*}{2}\partial_v^2 Y - \frac{\kappa_*}{2}\partial_v Y^2 - \frac{\lambda_*}{2} Y + \sqrt{D_1}\partial_v \dot{W}^1(t,v) + \sqrt{D_2}\dot{W}^2(t,v), \quad v \in \mathbb{R}, \tag{4.21}$$

with some constants $\nu_*, \kappa_*, \lambda_*, D_1, D_2 > 0$ satisfying the Einstein relation and two independent space–time Gaussian white noises $\dot{W}^1$ and $\dot{W}^2$.

4.3.2. Related singular quasilinear SPDE

Under the hydrodynamic limit for a zero-range process on $\mathbb{T}_N$ in a random environment, first for smeared one and then removing it, we obtain the singular quasilinear SPDE for the limit density $u = u(t,v)$ of particles:

$$\partial_t u = \partial_v^2 \varphi(u) + \partial_v\{\varphi(u)\dot{W}(v)\}, \quad v \in \mathbb{T}, \tag{4.22}$$

where $\dot{W}(v)$ is the spatial Gaussian white noise; compare with (4.12) taking $c^i = c^i(v)$ and moving it in the inside of ∂_v. In Funaki and Xie [27], after proving the global-in-time well-posedness in paracontrolled sense for (4.22) (i.e., $u \in C([0,\infty), \mathcal{C}^{\alpha-1})$, $\alpha \in (\frac{13}{9}, \frac{3}{2})$), the asymptotic behavior of the solution $u(t)$ as $t \to \infty$ was studied at least when $W(v)$ is nearly periodic. The equation has the conserved quantity $m = \int_{\mathbb{T}} u(t,v)dv$ and the limit as $t \to \infty$ is uniquely determined for each $m \in \mathbb{R}$. In Funaki et al. [20], a more general SPDE with second $\varphi(u)$ replaced by another function $\chi(u)$ was studied.

ACKNOWLEDGMENTS

The author thanks Sunder Sethuraman for recent fruitful collaborations.

FUNDING

This research was supported in part by JSPS KAKENHI, Grant-in-Aid for Scientific Research (A) 18H03672.

REFERENCES

[1] G. Amir, I. Corwin, and J. Quastel, Probability distribution of the free energy of the continuum directed random polymer in 1 + 1 dimensions. *Comm. Pure Appl. Math.* **64** (2011), 466–537.

[2] C. Bernardin, T. Funaki, and S. Sethuraman, Derivation of coupled KPZ–Burgers equation from multi-species zero-range processes. *Ann. Appl. Probab.* **31** (2021), 1966–2017.

[3] L. Bertini and G. Giacomin, Stochastic Burgers and KPZ equations from particle systems. *Comm. Math. Phys.* **183** (1997), 571–607.

[4] E. C. M. Crooks, E. N. Dancer, D. Hilhorst, M. Mimura, and H. Ninomiya, Spatial segregation limit of a competition–diffusion system with Dirichlet boundary conditions. *Nonlinear Anal. Real World Appl.* **5** (2004), 645–665.

[5] A. De Masi, P. Ferrari, and J. Lebowitz, Reaction–diffusion equations for interacting particle systems. *J. Stat. Phys.* **44** (1986), 589–644.

[6] A. De Masi, T. Funaki, E. Presutti, and M. E. Vares, Fast-reaction limit for Glauber–Kawasaki dynamics with two components. *ALEA Lat. Am. J. Probab. Math. Stat.* **16** (2019), 957–976.

[7] P. El Kettani, T. Funaki, D. Hilhorst, H. Park, and S. Sethuraman, Mean curvature interface limit from Glauber+Zero-range interacting particles. 2021, arXiv:2004.05276v2.

[8] P. El Kettani, T. Funaki, D. Hilhorst, H. Park, and S. Sethuraman, Singular limit of an Allen-Cahn equation with nonlinear diffusion. 2021, arXiv:2112.13081.

[9] D. Ertaş and M. Kardar, Dynamic roughening of directed lines. *Phys. Rev. Lett.* **69** (1992), 929–932.

[10] T. Funaki, The scaling limit for a stochastic PDE and the separation of phases. *Probab. Theory Related Fields* **102** (1995), 221–288.

[11] T. Funaki, Free boundary problem from stochastic lattice gas model. *Ann. Inst. Henri Poincaré B, Probab. Stat.* **35** (1999), 573–603.

[12] T. Funaki, Singular limit for stochastic reaction–diffusion equation and generation of random interfaces. *Acta Math. Sin. (Engl. Ser.)* **15** (1999), 407–438.

[13] T. Funaki, Hydrodynamic limit for $\nabla\phi$ interface model on a wall. *Probab. Theory Related Fields* **126** (2003), 155–183.

[14] T. Funaki, Stochastic Interface Models. In *Lectures on Probability Theory and Statistics, Ecole d'Eté de Probabilités de Saint-Flour XXXIII – 2003*, edited by J. Picard, pp. 103–274, Lecture Notes in Math. 1869, Springer, 2005.

[15] T. Funaki, Infinitesimal invariance for the coupled KPZ equations. In *Séminaire de Probabilités XLVII*, pp. 37–47, Lecture Notes in Math. 2137, Springer, 2015.

[16] T. Funaki, *Lectures on Random Interfaces*. SpringerBriefs Probab. Math. Statist., Springer, 2016, xii+138 pp.

[17] T. Funaki, Hydrodynamic limit for exclusion processes. *Commun. Math. Stat.* **6** (2018), 417–480.

[18] T. Funaki, Invariant measures in coupled KPZ equations. In *Stochastic Dynamics Out of Equilibrium*, Institut H. Poincaré (2017), pp. 560–568, Springer, 2019.

[19] T. Funaki and M. Hoshino, A coupled KPZ equation, its two types of approximations and existence of global solutions. *J. Funct. Anal.* **273** (2017), 1165–1204.

[20] T. Funaki, M. Hoshino, S. Sethuraman, and B. Xie, Asymptotics of PDE in random environment by paracontrolled calculus. *Ann. Inst. Henri Poincaré B, Probab. Stat.* **57** (2021), 1702–1735.

[21] T. Funaki, P. van Meurs, S. Sethuraman, and K. Tsunoda, Motion by mean curvature from Glauber–Kawasaki dynamics with speed change. 2022, preprint.

[22] T. Funaki, Y. Nishijima, and H. Suda, Stochastic eight-vertex model, its invariant measures and KPZ limit. *J. Stat. Phys.* **184** (2021), no. 11, 1–30.

[23] T. Funaki and J. Quastel, KPZ equation, its renormalization and invariant measures. *Stoch. Partial Differ. Equ. Anal. Comput.* **3** (2015), 159–220.

[24] T. Funaki and S. Sethuraman, Schauder estimate for quasilinear discrete PDEs of parabolic type. 2021, arXiv:2112.13973.

[25] T. Funaki and H. Spohn, Motion by mean curvature from the Ginzburg–Landau $\nabla\phi$ interface model. *Comm. Math. Phys.* **185** (1997), 1–36.

[26] T. Funaki and K. Tsunoda, Motion by mean curvature from Glauber–Kawasaki dynamics. *J. Stat. Phys.* **177** (2019), 183–208.

[27] T. Funaki and B. Xie, Global solvability and convergence to stationary solutions in singular quasilinear stochastic PDEs. 2021, arXiv:2106.01102.

[28] T. Funaki and S. Yokoyama, Sharp interface limit for stochastically perturbed mass conserving Allen–Cahn equation. *Ann. Probab.* **47** (2019), 560–612.

[29] P. Gonçalves and M. Jara, Nonlinear fluctuations of weakly asymmetric interacting particle systems. *Arch. Ration. Mech. Anal.* **212** (2014), 597–644.

[30] P. Gonçalves, M. Jara, and S. Sethuraman, A stochastic Burgers equation from a class of microscopic interactions. *Ann. Probab.* **43** (2015), 286–338.

[31] M. Gubinelli, P. Imkeller, and N. Perkowski, Paracontrolled distributions and singular PDEs. *Forum Math. Pi* **3** (2015), no. e6, 1–75.

[32] M. Gubinelli and N. Perkowski, The infinitesimal generator of the stochastic Burgers equation. *Probab. Theory Related Fields* **178** (2020), 1067–1124.

[33] M. Z. Guo, G. C. Papanicolaou, and S. R. S. Varadhan, Nonlinear diffusion limit for a system with nearest neighbor interactions. *Comm. Math. Phys.* **118** (1988), 31–59.

[34] M. Hairer, Solving the KPZ equation. *Ann. of Math.* **178** (2013), 559–664.

[35] M. Hairer, A theory of regularity structures. *Invent. Math.* **198** (2014), 269–504.

[36] M. Hairer and J. Mattingly, The strong Feller property for singular stochastic PDEs. *Ann. Inst. Henri Poincaré B, Probab. Stat.* **54** (2018), 1314–1340.

[37] K. Hayashi, Spatial-segregation limit for exclusion processes with two components under unbalanced reaction. *Electron. J. Probab.* **26** (2021), no. 51, 1–36.

[38] P. C. Hohenberg and B. I. Halperin, Theory of dynamic critical phenomena. *Rev. Modern Phys.* **49** (1977), 435–475.

[39] M. Hoshino, Paracontrolled calculus and Funaki–Quastel approximation for the KPZ equation. *Stochastic Process. Appl.* **128** (2018), 1238–1293.

[40] M. Jara and O. Menezes, Non-equilibrium fluctuations of interacting particle systems. 2018, arXiv:1810.09526.

[41] K. Johansson, Shape fluctuations and random matrices. *Comm. Math. Phys.* **209** (2000), 437–476.

[42] M. Kardar, G. Parisi, and Y.-C. Zhang, Dynamic scaling of growing interfaces. *Phys. Rev. Lett.* **56** (1986), 889–892.

[43] M. A. Katsoulakis and P. E. Souganidis, Interacting particle systems and generalized evolution of fronts. *Arch. Ration. Mech. Anal.* **127** (1994), 133–157.

[44] C. Kipnis and C. Landim, *Scaling Limits of Interacting Particle Systems*. Springer, 1999.

[45] T. Komorowski, C. Landim, and S. Olla, *Fluctuations in Markov Processes: Time Symmetry and Martingale Approximation*. Springer, 2012.

[46] J. Quastel and H. Spohn, The one-dimensional KPZ equation and its universality class. *J. Stat. Phys.* **160** (2015), 965–984.

[47] T. Sasamoto and H. Spohn, One-dimensional KPZ equation: An exact solution and its universality. *Phys. Rev. Lett.* **104** (2010), 230602 (4 pages).

[48] H. Spohn, *Large Scale Dynamics of Interacting Particles*. Springer, 1991.

[49] H. Spohn, Nonlinear fluctuating hydrodynamics for anharmonic chains. *J. Stat. Phys.* **154** (2014), 1191–1227.

[50] H. Spohn and G. Stolz, Nonlinear fluctuating hydrodynamics in one dimension: the case of two conserved fields. *J. Stat. Phys.* **160** (2015), 861–884.

[51] H.-T. Yau, Relative entropy and hydrodynamics of Ginzburg–Landau models. *Lett. Math. Phys.* **22** (1991), 63–80.

TADAHISA FUNAKI

Department of Mathematics, Waseda University, Okubo, Tokyo 169-8555, Japan, and Graduate School of Mathematical Sciences, University of Tokyo, Komaba, Tokyo 153-8914, Japan, funaki@ms.u-tokyo.ac.jp

ON THE UNIVERSALITY FROM INTERACTING PARTICLE SYSTEMS

PATRÍCIA GONÇALVES

Dedicated to my mother.

ABSTRACT

In these notes, we review recent results for the limiting behavior of equilibrium fluctuations of interacting particle systems with one or several conserved quantities. Two main classes of models are considered. First, the weakly asymmetric simple exclusion process, a model with one conservation law, and whose fluctuations cross from the Edwards–Wilkinson (EW) universality class to the Kardar–Parisi–Zhang (KPZ) universality class. Second, we consider a class of Hamiltonian systems perturbed by a noise and conserving two quantities. In the case of an exponential potential, the transition occurs from diffusion to fractional $\frac{5}{3}$ behavior, while for a harmonic potential the fluctuations cross from diffusive to fractional $\frac{3}{2}$ behavior. We review two different methods which rigorously prove some of the aforementioned results.

MATHEMATICS SUBJECT CLASSIFICATION 2020

Primary 60K35; Secondary 60G52, 60H15, 60F17, 35R11

KEYWORDS

Interacting particle systems, fluctuating hydrodynamics, stochastic burgers equation, Fractional diffusion, Lévy process, weakly asymmetric systems, universality classes

ICM INTERNATIONAL CONGRESS OF MATHEMATICIANS 2022 JULY 6–14

Proc. Int. Cong. Math. 2022, Vol. 6, pp. 4326–4348
DOI 10.4171/ICM2022/111

1. INTRODUCTION

Over the last years, there has been much progress in understanding the emergence of universality at the level of the macroscopic equations that rule the space-time evolution of the conserved quantities of 1-d interacting particle systems (IPS). To be concrete, we describe the simplest dynamics of an IPS, namely the exclusion process. In this process, particles evolve as 1-d continuous-time random walks, with the constraint that does not allow more than a particle per site at any given time. This means that after an exponential clock of parameter one, a particle jumps from x to y according to a transition probability $p(y-x)$. The number of particles in the system is conserved and one of the questions that one might ask is about starting from some configuration, how to figure out what is the typical configuration at any given time. A fundamental question in the IPS literature, known as *hydrodynamic limit*, is to describe the evolution of the distribution of the conserved quantities as a function of space and time in the thermodynamic limit, i.e., when the size of the system is taken to infinity. The hydrodynamic limit is nothing but a law of large numbers for the empirical measure associated with the conserved quantities of the system, in a suitable time-scale [20]. The limit is given by a partial differential equation (PDE) which can be parabolic, hyperbolic, or even of a fractional form.

Our focus on this article is to describe the limiting laws that appear when one looks at the deviations of the system from the hydrodynamical profile, therefore we are in the central limit theorem scaling. As expected, contrarily to the deterministic solution obtained in the law of large numbers, in this case, the fluctuations are described by some stochastic partial differential equation (SPDE), and the challenge is to derive and characterize it.

The explanation and characterization of anomalous behavior in 1-d nonequilibrium systems are challenging even when the interactions are on a finite size window. A way to characterize the behavior of systems that exhibit an anomalous behavior is by studying the dynamical structure function, describing the time-dependent fluctuations of the conserved quantities of the system. For systems with one conservation law, two universality classes can be obtained: the (Gaussian) universality class, whose scale-invariance is $1:2:4$; and the superdiffusive KPZ universality class whose scale invariance is $1:2:3$. The latter is conjectured to be the universal law for the fluctuations of models with some smoothing mechanism, slope-dependent growth speed, and short-range randomness, while the former should be universal for models without slope-dependence and therefore with Gaussian fluctuations.

For systems with more conservation laws, the situation is much more complicated and many other universality classes exist. To give some concrete examples, we start by considering the weakly asymmetric simple exclusion process (WASEP). By tuning the asymmetry, one can observe different limiting equations depending on whether the symmetry or the asymmetry is the dominant dynamics. In the case of nearest-neighbor jumps and with an asymmetry of order $O(\frac{1}{n^\kappa})$, where n is the scaling parameter, the crossover goes from a diffusive behavior (corresponding to the phase where the symmetry dominates, that is, $\kappa > \frac{1}{2}$) to a behavior given in terms of the stochastic Burgers equation (corresponding to the phase where both the symmetry and the asymmetry have the same impact, that is, for

$\kappa = \frac{1}{2}$). In the strong asymmetric regime, recent results show that the limiting behavior should be given in terms of the KPZ fixed point [22]. This has been rigorously proved only for totally asymmetric jumps and the current field, but the same behavior should be true for partially asymmetric jumps and even in the whole phase where the asymmetry dominates (corresponding to $\kappa \in [0, \frac{1}{2})$), see the upper dashed line in Figure 1.

The particle system described above has a unique conservation law—the number of particles—and therefore its analysis is much simpler when compared to systems with more conservation laws and whose hydrodynamic limit consists of a system of PDEs. Let us now describe the type of systems with several conservation laws that we focus on, namely the chains of oscillators. They consist of Hamiltonian systems that are perturbed by a conservative noise. In [7] it was introduced and studied from a numerical point of view, a class of Hamiltonian systems that present strong analogies with the standard chains of oscillators. These models, denoted by $(r(t), p(t))_{t\geqslant 0}$, can be described by considering two nonnegative potentials V and U and the equations of motion are given on x by

$$dp_x = \big(V'(r_{x+1}) - V'(r_x)\big)dt \quad \text{and} \quad dr_x = \big(U'(p_x) - U'(p_{x-1})\big)dt,$$

where p_x denotes the momentum of the particle x, q_x is its position, and $r_x = q_x - q_{x-1}$ is the deformation of the lattice at x. If we assume that $V = U$ and by mapping $\eta_{2x-1} = r_x$ and $\eta_{2x} = p_x$, the dynamics above can be rewritten as

$$d\eta_x(t) = \big(V'(\eta_{x+1}) - V'(\eta_{x-1})\big)dt.$$

With respect to the variables η, the energy of the system corresponds to $\sum_x V(\eta_x)$. To make the model mathematically tractable, the dynamics just described, which is purely deterministic, is perturbed by adding a noise that exchanges η_x with η_{x+1} at random exponentially distributed times, and this is done independently for each bond $\{x, x+1\}$. These models have two conserved quantities, the energy $\sum_x V(\eta_x)$ and the volume $\sum_x \eta_x$ and the analysis of their asymptotic behavior is much more intricate than for the case of models with just one conserved quantity as we described above. Examples of the Hamiltonian systems introduced above are the models of [1,3–6]. In those articles, it was studied the fluctuations of the conserved quantities, namely, the energy and the volume, starting the system from the invariant measure, which is of product form. By tuning the strength of the Hamiltonian dynamics by a factor $\frac{1}{n^\kappa}$ one can analyze the crossover fluctuations. The main problem when studying these Hamiltonian systems is that depending on the chosen potential, the volume and the energy can be linearly transported in the system, each one having its own velocity and living on its own time scale. This is the main problem in general when dealing with systems with more than one conservation law. However, there are cases in which the situation simplifies since the conserved quantities have the same velocity and they live on the same time scale, e.g., in [2], where multicomponent coupled equations have been obtained as scaling limits of the empirical measures of the conserved quantities for the multispecies zero-range process.

When dealing with systems with only one conservation law, there are not many doubts about the field that one should consider, *the field of the conserved quantity*. Nevertheless, in presence of more than one conservation law, since any linear combination of the conserved quantities is again conserved, the possibility on the choice of the fluctuation fields is wider.

In [26], with a focus on anharmonic chains of oscillators, it was developed the nonlinear fluctuating hydrodynamics theory for the equilibrium time-correlations of the conserved quantities of that model, see also [28] for previous results on the anomalous transport in 1-d Hamiltonian systems with an emphasis on the KPZ behavior. In those articles, there are analytical predictions, based on a mode-coupling approximation, for the form of the fluctuations of the conserved quantities. Depending on the value of the coupling constants many other universality classes pop up, besides the Gaussian and the KPZ, already seen in systems with only one conservation law. At the same time in [23], by analyzing coupled single-lane asymmetric simple exclusion processes, the authors obtained numerically some universality classes with several dynamical exponents for the two conserved quantities of the system. All the possible combinations of limits in that model are summarized in Table 1 of [23]. We apply in Sections 3.2 and 3.5 the strategy developed in [23, 26] to compute the fluctuation fields for models of chains of oscillators with two different potentials, the exponential and the harmonic, respectively. We also describe in Section 3.3 an alternative way to compute these fields, based on the action of the generator on the conserved quantities of the system. Once the proper choice of the fluctuation fields is done, the next question is related to the predictions on the form of the fluctuations for those quantities. This means that one has to write down the equations for the time evolution of those fields and one has to close those equations in terms of those fields. But since each field evolves in a certain time scale (which is not necessarily the same for both), then one has to analyze the leading terms in the expansion of the equations in such a way that one can recover the limiting SPDE, and this can be a hard task. The strategy to do this, is to write down the instantaneous current of the system in terms of the field of the conserved quantities and this can be done by the so-called Boltzmann–Gibbs principle. This principle was introduced in [8] for systems with one conserved quantity and it states that any local field of the dynamics can be replaced (in a proper topology) by the fluctuation field of the conserved quantity. When one is looking at the fluctuation field of the conserved quantity of the WASEP described above, the aforementioned Boltzmann–Gibbs principle is sufficient to recover the SPDE satisfied by the limiting field in the regime where the symmetry dominates (i.e., $\kappa > \frac{1}{2}$), but when the asymmetry has the same impact as the symmetry, then the Boltzmann–Gibbs principle of [8] does not give any information about the limiting field. For that purpose a second-order Boltzmann–Gibbs principle has been derived in [12, 14, 15] which allows replacing any local field of the dynamics by the square of the fluctuation field of the conserved quantity of the system. By using this principle, it becomes simple to close the equation for the field of the conserved quantity and to recognize the SPDE satisfied by the limiting field. The proof for the second-order Boltzmann–Gibbs principle of [15] does not impose strong conditions on the underlying microscopic dynamics and allows obtaining the crossover fluctuations from a diffusive behavior to a behavior given by the stochastic Burgers equation, as described above for the WASEP, and, more generally, for any system which has an instantaneous current that can be written as a sum of polynomial functions. Therefore the application of this principle even to the Hamiltonian systems described above, allows getting information on the form of the fluctuations. But below the critical case, $\kappa = \frac{1}{2}$, nothing rigorous is known apart from the strong asymmetric regime and for a specific choice of the jump rates. We believe that the extension of this result to Hamiltonian systems will open a new way to obtain the fluctuations of the energy and the volume and to establish the precise dependence on the strength of the perturbing noise, at the level of the crossover from different SPDEs.

The description of the universality classes from Gaussian to KPZ is for one component systems and in the case of multicomponent systems, as the Hamiltonian systems just mentioned, the scenario is much less understood. In the last years, this problem has attracted a lot of interest in both the physics and the mathematics communities, since up to very recently, there was no theoretical explanation and the numerical simulations were too controversial [10,21].

As mentioned above, in [23,26,27] with the so-called nonlinear fluctuating hydrodynamics theory, which has been developed during the last years, the authors proposed a rich and complex phase diagram of the universality classes for the aforementioned Hamiltonian systems. The richness of the diagram is explained by the nontrivial nonlinear couplings, occurring at different time scales, between the conserved quantities. This means, as mentioned above, that each conserved quantity will have its own velocity in a certain time scale and its own limiting SPDE ruling its fluctuations. This results have been, proved rigorously in the context of harmonic chains perturbed by a conservative noise (see [4,5,18]) and also for a 1-d infinite chain of coupled charged harmonic oscillators with a magnetic field [24]. The predictions in [26,27] are done starting from the macroscopic equation, which is assumed to be an Euler equation given by $\partial_t \varrho + \partial_x \langle j \rangle_\varrho = 0$, where $\varrho = (\varrho_1, \varrho_2, \varrho_3)$ is the vector whose ith ($i = 1, 2, 3$) component represents one of the conserved quantities of the system, j is the vector whose ith component represents the instantaneous current of the system for one of the conserved quantities and $\langle \cdot \rangle_\varrho$ represents the average with respect to the invariant state with parameter ϱ. By linearizing the equation, one arrives at $\partial_t \varrho + A \partial_x \varrho = 0$, and by adding a noise term and a dissipating term, one gets $\partial_t \varrho + \partial_x [A\varrho - D\partial_x \varrho + BW] = 0$, where W is an n-d white noise and A and D are matrices. For the dynamical correlation function given by $S(i,t) = \langle j_i(t), j_0(0) \rangle_\varrho$, one has $\sum_i S(i,t) = ACt$, where C is another matrix related to A and B. By taking certain ansatz for the matrices, one can predict many universality classes (only in the strong asymmetric regime). Besides the predictions not being mathematically rigorous, they bring up a new insight to approach the problem from the mathematical point of view: in order to study the fluctuations of systems with multicomponent conserved quantities, one has to look at a proper linear combination of the fields of the conserved quantities.

In [26,27] the authors give very detailed predictions about the correct time scales that one should see a nontrivial behavior for each one of the conserved quantities, and more than that, they also predicted what are the limiting processes that one is searching for. They did it for the models of chains of oscillators, but these models should have the same asymptotic behavior as the dynamics introduced above with only two conserved quantities. According to their predictions, one can get conserved quantities with a Gaussian behavior, or a KPZ behavior, or a fractional behavior given in terms of Lévy processes with a certain exponent that depends on the dynamics.

We highlight that the list of universality classes is not exhausted by those described above, as the EW [11], the KPZ, or those given by Lévy processes. Recently in [9], the authors analyzed a temperature-dependent model (that for zero temperature gives the classical ballistic deposition model) and the ∞-temperature version is a random interface, that does not belong to any of the universality classes mentioned above. Its scaling limit is given by the Brownian Castle, a renormalization fixed point, whose scale-invariance is given by $1:1:2$, distinct from both the EW or the KPZ classes.

Here we report two ways of rigorously obtaining some of the universality classes mentioned above. We present the methods for the WASEP and also for the model of chains of oscillators as described above for two different potentials. With this, we establish the existence of crossover lines, by tuning the parameter κ, in the phase diagram connecting some universality classes.

2. EXCLUSION PROCESS: A PROTOTYPE MODEL WITH ONE CONSERVATION LAW

We start by explaining in detail a model whose dynamics conserves one quantity, namely, the density of particles. Our prototype model is the exclusion process which we denote by $\eta(t)$. We consider the process evolving on the discrete torus $\mathbb{T}_n = \{0, 1, \ldots, n-1\}$ and its dynamics can be described as follows. Each particle waits an exponential time of parameter 1 and then it jumps to a site according to a certain probability transition rate $p(\cdot)$. The exclusion rule dictates that the jump of a particle is performed if and only if the destination site is empty, otherwise nothing happens and the particle waits a new random time. The space state of this process is $\{0,1\}^{\mathbb{T}_n}$ and a configuration is denoted by $\eta = \{\eta_x \in \{0,1\} : x \in \mathbb{T}_n\}$. We denote the jump rate from the site x to the site y by $p(x,y) = p(y-x)$, and note that $p(\cdot)$ only depends on the size of the jump and not on the exact location where the jump is performed. When jumps are allowed only to nearest-neighbor sites the process is simple, so that $p(z) = 0$ if $|z| > 1$. We make the following choice $p(-1) = 1 - p(1)$ and $p(1) = p + \frac{E}{n^\kappa}$, where p, E, and κ are constants. If $E = 0$ (no dependence on κ) and $p = \frac{1}{2}$, we get the symmetric simple exclusion process (SSEP); if $E = 0$ and $p \neq \frac{1}{2}$, we get the asymmetric simple exclusion process (ASEP), and if $E \neq 0$ and $p = \frac{1}{2}$, the process is the WASEP. Observe that the parameter κ rules the strength of the asymmetry, that is, the higher the value of κ, the weaker the asymmetry. Its infinitesimal generator is given on functions $f : \{0,1\}^{\mathbb{T}_n} \to \mathbb{R}$ by

$$\mathscr{L}^{ex} f(\eta) = \sum_{x \in \mathbb{T}_n} \{p(1)\eta_x(1-\eta_{x+1}) + p(-1)\eta_{x+1}(1-\eta_x)\}\big(f(\eta^{x,x+1}) - f(\eta)\big), \quad (2.1)$$

where $\eta^{x,x+1}$ is the configuration obtained from the configuration η by swapping the occupation variables η_x and η_{x+1}:

$$\eta_y^{x,x+1} = \eta_{x+1}\mathbf{1}_{y=x} + \eta_x \mathbf{1}_{y=x+1} + \eta_y \mathbf{1}_{y \neq x,x+1}. \quad (2.2)$$

The system is speeded up in the time scale tn^a, where $a > 0$ is a constant. A simple computation shows that $\mathscr{L}^{ex}(\eta_x) = j_{x-1,x}(\eta) - j_{x,x+1}(\eta)$ where

$$j_{x,x+1}(\eta) = \eta_x(1-\eta_{x+1})\Big(p + \frac{E}{n^\kappa}\Big) - \eta_{x+1}(1-\eta_x)\Big(1 - p - \frac{E}{n^\kappa}\Big). \tag{2.3}$$

The system conserves one quantity—the *number of particles*, $\sum_{x\in\mathbb{T}_n}\eta_x$. The invariant measures are denoted by ν_ϱ and are, in fact, Bernoulli product measures of parameter ϱ:

$$\nu_\varrho(d\eta) = \prod_{x\in\mathbb{T}_n} \varrho^{\eta_x}(1-\varrho)^{1-\eta_x} \tag{2.4}$$

for $\varrho \in (0,1)$. In the case $E = 0$ and $p = \frac{1}{2}$, these measures are also reversible.

2.1. Hydrodynamic limit

The trajectories of the process live on the Skorohod space $\mathcal{D}([0,T],\{0,1\}^{\mathbb{T}_n})$ and $\mathbb{P}_{\mu_n}$ is the probability measure on that space induced by an initial measure μ_n and by the process $\{\eta(tn^a)\}_{t\geqslant 0}$. The expectation with respect to $\mathbb{P}_{\mu_n}$ is denoted by $\mathbb{E}_{\mu_n}$. We define the empirical measure associated to the density by

$$\pi^n(\eta, du) := \frac{1}{n}\sum_{x\in\mathbb{T}_n} \eta_x \delta_{\frac{x}{n}}(du),$$

where $\delta_{\frac{x}{n}}$ is a Dirac mass on $\frac{x}{n} \in \mathbb{T}$ and $\pi^n_t(\eta, du) := \pi^n(\eta(tn^a), du)$. The statement of the hydrodynamic limit can be rigorously stated as follows. We assume that the process starts from a probability measure μ_n for which a law of large numbers holds, i.e., the sequence of random measures $\pi^n_0(\eta, du)$ converges, in probability with respect to μ_n and when n is taken to infinity, to the deterministic measure $\varrho(0,u)du$, where the density $\varrho(0,u)$ is a measurable function. The claim in the hydrodynamic limit is that under the previous assumption, the same result holds at any time t, that is, the random measure $\pi^n_t(\eta, du)$ converges, in probability with respect to the distribution of the process at time t and when n is taken to infinity, to a deterministic measure $\varrho(t,u)du$. The function $\varrho(t,u)$ is the solution (usually in a weak sense) of a PDE, which is called, the *hydrodynamic equation* of the system. For the exclusion processes introduced above, one can get as hydrodynamic equations: for the SSEP and by rescaling time diffusively $a = 2$, the heat equation given by

$$\partial_t \varrho(t,u) = \frac{1}{2}\Delta\varrho(t,u).$$

For the WASEP with $\kappa = 1$ (resp. for the ASEP) and by rescaling time diffusively $a = 2$ (resp., in the hyperbolic scale $a = 1$), the viscous Burgers equation (resp., the inviscid Burgers equation),

$$\partial_t \varrho(t,u) = \frac{1}{2}\Delta\varrho(t,u) + (1-2E)\nabla F\big(\varrho(t,u)\big),$$
$$\partial_t \varrho(t,u) = (1-2E)\nabla F\big(\varrho(t,u)\big),$$

where $F(\varrho) = \varrho(1-\varrho)$. The last result is a Law of Large Numbers for the unique conserved quantity of the system, i.e., the density. Now the natural question that comes next is related to the fluctuations around the obtained hydrodynamical profile. Moreover, we could ask if there are equations that can be obtained for different dynamics which share common grounds. If so, what are their form and how do they relate? In these notes, we will explain two different methods that allow obtaining some answers in this direction.

2.2. Fluctuations

We consider the system starting from the invariant measure, which, for the model under investigation, is of product form and homogeneous, see (2.4). We define the empirical field associated to the conserved quantity—the *density fluctuation field*—which is the linear functional acting on functions f as

$$\mathcal{Y}_t^n(f) = \frac{1}{\sqrt{n}} \sum_{x\in\mathbb{T}_n} f\left(\frac{x}{n}\right)\overline{\eta}_x(tn^a),$$

where $\overline{\eta}_x := \eta_x(tn^a) - \varrho$. The last identity is obtained by first integrating f with respect to the density empirical measure $\pi_t^n(\eta, du)$, then removing its mean (with respect to the invariant state) and then multiplying it by $\sqrt{n}$. The question that arises now is to understand the limit in distribution, as $n \to +\infty$, of $\mathcal{Y}_t^n$ that we denote by $\mathcal{Y}_t$. For the exclusion processes introduced above, one can get for the SSEP and rescaling time diffusively $a = 2$, the Ornstein–Uhlenbeck (OU) equation given by

$$d\mathcal{Y}_t = \frac{1}{2}\Delta\mathcal{Y}_t dt + \sqrt{F(\varrho)}\nabla\mathcal{W}_t.$$

For the WASEP with $\kappa > \frac{1}{2}$ and rescaling time diffusive $a = 2$, one can get exactly the same OU equation as in the symmetric case, while for $\kappa = \frac{1}{2}$ and still rescaling time diffusively $a = 2$, the KPZ equation (introduced in [19]) or its companion, namely the stochastic Burgers (SB) equation, respectively, for the height fluctuation field and the density fluctuation field,

$$d\mathcal{h}_t = \frac{1}{2}\Delta\mathcal{h}_t dt + 4E(\nabla\mathcal{h}_t)^2 dt + \sqrt{F(\varrho)}\mathcal{W}_t,$$
$$d\mathcal{Y}_t = \frac{1}{2}\Delta\mathcal{Y}_t dt + 4E\nabla\mathcal{Y}_t^2 dt + \sqrt{F(\varrho)}\nabla\mathcal{W}_t.$$

Above $\mathcal{W}_t$ is a space-time white-noise. Last results were proved in [14] and [13]. For the case $E = 0$, $p \neq 1/2$ and taking the system in the hyperbolic time scale $a = 1$, one can get

$$d\mathcal{Y}_t = (1 - 2\varrho)(1 - 2p)\nabla\mathcal{Y}_t dt.$$

Observe that in the last equation if we consider $\varrho = \frac{1}{2}$, we get a trivial evolution for the density field. The same result would be obtained if, instead of choosing $\varrho = \frac{1}{2}$, we redefine the field in a frame with the velocity $(1 - 2\varrho)n^{a-1}$. To simplify the presentation we consider $\varrho = \frac{1}{2}$ in what follows. Therefore, to get a nontrivial behavior, one has to speed up the time and in that case, and for the choice $a = \frac{3}{2}$, the limiting field should be given in terms of the KPZ fixed point, see [22]. In [12] it was proved that, up to the time scale $a = \frac{4}{3}$, there is no evolution of the field; beyond that time scale, the limit of this field is not known yet, but it should be given in terms of the KPZ fixed point. The results in [12] applied to the WASEP show that below the line $a = \frac{4}{3}(\kappa + 1)$ there is no time evolution of the limiting field, but the trivial evolution should go up to the line $a = \kappa + \frac{3}{2}$. Last results are summarized in Figure 1.

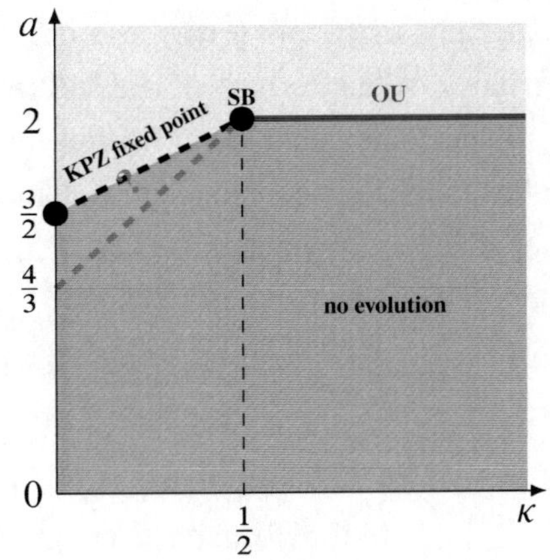

FIGURE 1
Density fluctuations.

The starting point to prove the latter results (that we now restrict to WASEP with $\varrho = \frac{1}{2}$) is to use Dynkin's formula, so that for $f \in C^2(\mathbb{T})$, where $\mathbb{T}$ denotes the one-dimensional torus,

$$\mathcal{M}_t^n(f) = \mathcal{Y}_t^n(f) - \mathcal{Y}_0^n(f) - \int_0^t (\partial_s + n^a \mathcal{L}^{ex})\mathcal{Y}_s^n(f)ds, \tag{2.5}$$

is a martingale with respect to the natural filtration of the process. We say that it is the martingale associated to the field $\mathcal{Y}_t^n(f)$. A simple computation shows that

$$\mathcal{M}_t^n(f) = \mathcal{Y}_t^n(f) - \mathcal{Y}_0^n(f) - \frac{n^a}{2n^2}\mathcal{S}_t^n(f) - \frac{En^a}{n^{\frac{3}{2}+\kappa}}\mathcal{A}_t^n(f).$$

Above, the contribution of the symmetric and asymmetric parts of the dynamics are respectively given by

$$\mathcal{S}_t^n(f) = \int_0^t \mathcal{Y}_s^n(\Delta_n f)ds \quad \text{and} \quad \mathcal{A}_t^n(f) = \int_0^t \sum_{x\in\mathbb{T}_n} \nabla_n f\left(\frac{x}{n}\right)\overline{\eta}_x(sn^a)\overline{\eta}_{x+1}(sn^a)ds,$$

where Δ_n and ∇_n, respectively, denote the discrete Laplacian and discrete derivative

$$\Delta_n f\left(\frac{x}{n}\right) = n^2\left\{f\left(\frac{x+1}{n}\right) - 2f\left(\frac{x}{n}\right) + f\left(\frac{x-1}{n}\right)\right\} \quad \text{and}$$
$$\nabla_n f\left(\frac{x}{n}\right) = n\left\{f\left(\frac{x+1}{n}\right) - f\left(\frac{x}{n}\right)\right\}.$$

A simple computation shows that the quadratic variation is given by

$$\langle \mathcal{M}^n(f)\rangle_t = \int_0^t \frac{n^a}{n^3}\left(\frac{1}{2} + \frac{E}{n^\kappa}\right)\sum_{x\in\mathbb{T}_n}\left(\nabla_n f\left(\frac{x}{n}\right)\right)^2 \left(\eta_x(sn^a) - \eta_{x+1}(sn^a)\right)^2 ds,$$

so that if $a = 2$ we get $\lim_{n\to\infty} \mathbb{E}_{\nu_\varrho}[(\mathcal{M}_t^n(f))^2] = tF(\varrho)\int(\nabla f(u))^2du$, while for $a < 2$ it vanishes. To close the equations for the density fluctuation field, one just has to analyze the integral terms in the martingale above. The term coming from the symmetric part of the dynamics is simple since it is already written in terms of the fluctuation field $\mathcal{Y}_s^n$. A simple computation shows that the variance of that term is of order $O(n^{2(a-2)})$, so that it converges if $a = 2$, but for $a < 2$ it vanishes. The most complicated term is that coming from the asymmetric part of the dynamics, namely, the term $\mathcal{A}_t^n(f)$. Our tool to analyze the variance of this term is given in terms of a $\mathcal{H}_{-1}$-norm estimate, stated in Theorem 4 of [3] as

$$\mathbb{E}_{\nu_\varrho}\left[\left(\int_0^t \sum_{x\in\mathbb{T}_n} \nabla_n f\left(\frac{x}{n}\right)\overline{\eta}_x(sn^a)\overline{\eta}_{x+1}(sn^a)\,ds\right)^2\right] \leqslant \frac{C_f t n}{\sqrt{n^a}}. \tag{2.6}$$

From the latter result, the variance of the term involving $\mathcal{A}_t^n(f)$ is of order $O(n^{\frac{3}{2}a-2(\kappa+1)})$, so that it vanishes when $a < \frac{4}{3}(\kappa+1)$. In the diffusive time scale, that term vanishes for any $\kappa > \frac{1}{2}$, while for $\kappa = \frac{1}{2}$ we can use the second-order Boltzmann–Gibbs principle proved in [14,15]. This principle states that for any $t \in [0,T]$, any positive integer n, and any $\varepsilon \in (0,1)$, it holds

$$\mathbb{E}_{\nu_\varrho}\left[\left(\mathcal{A}_t^n(f) - \int_0^t \frac{1}{n}\sum_{x\in\mathbb{T}_n} \nabla_n f\left(\frac{x}{n}\right)\left(\mathcal{Y}_s^n\left(\iota_\epsilon\left(\frac{x}{n}\right)\right)\right)^2 ds\right)^2\right] \leqslant C_f T\left(\varepsilon + \frac{t}{\varepsilon^2 n}\right), \tag{2.7}$$

where for $u \in (0,1)$ and $y \in (0,1)$ we have $\iota_\epsilon(u)(y) = \mathbf{1}_{(u,u+\epsilon]}(y)$. From the latter results we see that for $a < \inf(\frac{4}{3}(\kappa+1), 2)$ we have $\mathcal{Y}_t^n(f) = \mathcal{Y}_0^n(f)$ plus terms that vanish in the $\mathbb{L}^2(\mathbb{P}_{\nu_\varrho})$-norm as $n \to +\infty$, so that the limit field has a trivial evolution. We note that in fact last result should be true for $a < \inf(\frac{3}{2}+\kappa, 2)$ but this has not been proved, yet. When $a = 2$ and $\kappa > \frac{1}{2}$, we see that

$$\mathcal{M}_t^n(f) = \mathcal{Y}_t^n(f) - \mathcal{Y}_0^n(f) - \int_0^t \mathcal{Y}_s^n(\Delta_n f)ds$$

plus terms that vanish in the $\mathbb{L}^2(\mathbb{P}_{\nu_\varrho})$-norm as $n \to +\infty$, so that we get the OU equation; while for $\kappa = \frac{1}{2}$, we get

$$\begin{aligned}\mathcal{M}_t^n(f) &= \mathcal{Y}_t^n(f) - \mathcal{Y}_0^n(f) - \int_0^t \mathcal{Y}_s^n(\Delta_n f)ds \\ &\quad + \int_0^t \frac{1}{n}\sum_{x\in\mathbb{T}_n} \nabla_n f\left(\frac{x}{n}\right)\left(\mathcal{Y}_s^n\left(\iota_\epsilon\left(\frac{x}{n}\right)\right)\right)^2 ds\end{aligned}$$

plus terms that vanish in the $\mathbb{L}^2(\mathbb{P}_{\nu_\varrho})$-norm as $n \to +\infty$, so that we get the SB equation. In fact, with a little more effort one can get an energy solution to the SB as introduced in [14] and [15], for which the uniqueness as been proved in [17] by using paracontrolled calculus, see [16]. In the next subsection we analyze the same problem for a model with two conservation laws.

3. A PROTOTYPE MODEL WITH TWO CONSERVATION LAWS

Fix once and for all a positive real parameter $b > 0$. We start with the 1-d potential $V_b : \mathbb{R} \to [0,+\infty)$ defined by $V_b(u) = e^{-bu} - 1 + bu$ and we consider the Markov process $\{\eta_x(t)\}_{t\geqslant 0}$ with state space $\Omega_n := \mathbb{R}^{\mathbb{T}_n}$, whose infinitesimal generator is denoted by $\mathcal{L}$ and is given by $\mathscr{L} = \alpha_n \mathscr{A}_b + \gamma \mathcal{S}$, where $\gamma > 0$, $\alpha_n = \alpha n^{-\kappa}$, $\alpha \in \mathbb{R}$, and $\kappa > 0$. The operators $\mathscr{A}_b$ and $\mathcal{S}$ act on differentiable functions $f : \Omega_n \to \mathbb{R}$ as follows:

$$(\mathscr{A}_b f)(\eta) = \sum_{x \in \mathbb{T}_n} \big(V_b'(\eta_{x+1}) - V_b'(\eta_{x-1})\big)(\partial_{\eta_x} f)(\eta)$$

and

$$(\mathcal{S} f)(\eta) = \sum_{x \in \mathbb{T}_n} \big(f(\eta^{x,x+1}) - f(\eta)\big). \tag{3.1}$$

The system under investigation is a Hamiltonian system that is perturbed by a stochastic noise generated by $\mathcal{S}$. Above, the configuration $\eta^{x,x+1}$ is given in (2.2). The interested reader can find more details on these models in **[3, 7, 27]**.

Observe that all the objects defined above should be indexed on the scaling parameter n, but in order to simplify notation we will omit the dependence on it. The parameter $\alpha_n = \alpha n^{-\kappa}$ regulates the intensity of the asymmetry in the system in terms of the scaling parameter n. The role of the parameter γ is to regulate the intensity of the stochastic noise. The system will be speed up in the time scale tn^a with $a > 0$. From the expression of the potential $V_b(u)$, we see that if $\xi_x = e^{-b\eta_x}$ then $V_b(\eta_x) = \xi_x - 1 + b\eta_x$. A simple computation shows that

$$\mathscr{L}\big(V_b(\eta_x)\big) = j^e_{x-1,x}(\eta) - j^e_{x,x+1}(\eta), \quad \mathscr{L}(\eta_x) = j^v_{x-1,x}(\eta) - j^v_{x,x+1}(\eta), \tag{3.2}$$

where

$$\begin{aligned} j^e_{x,x+1}(\eta) &= -\alpha_n b^2 \xi_x \xi_{x+1} + \alpha_n b^2(\xi_x + \xi_{x+1}) - \gamma \nabla\big(V_b(\eta_x)\big), \\ j^v_{x,x+1}(\eta) &= \alpha_n b(\xi_x + \xi_{x+1}) - \gamma \nabla \eta_x, \end{aligned} \tag{3.3}$$

and $\nabla \tau_x f(\eta) = \tau_{x+1} f(\eta) - \tau_x f(\eta)$, where $\tau_x f(\eta) = f(\tau_x \eta)$. The same is true for ξ_x, that is, $\mathscr{L}(\xi_x) = j^\xi_{x-1,x}(\eta) - j^\xi_{x,x+1}(\eta)$, where

$$j^\xi_{x,x+1}(\eta) = -\alpha_n b^2 \xi_x \xi_{x+1} - \gamma \nabla \xi_x. \tag{3.4}$$

We have two conserved quantities, namely *energy* and *volume*, $\sum_{x \in \mathbb{T}_n} V_b(\eta_x)$ and $\sum_{x \in \mathbb{T}_n} \eta_x$, see **[7]** where it is proved that, in some sense, they are the only conserved quantities. Observe that any linear combination (plus constants) of energy and volume is also conserved, as the quantity $\sum_{x \in \mathbb{T}_n} \xi_x$ which will be very relevant in what follows. The invariant measures of the process are denoted by $\mu_{\bar\beta,\bar\lambda}$ and are explicitly given by

$$\mu_{\bar\beta,\bar\lambda}(d\eta) = \prod_{x \in \mathbb{T}_n} \bar Z^{-1}(\bar\beta, \bar\lambda) \exp\{-\bar\beta e^{-b\eta_x} - \bar\lambda \eta_x\} d\eta_x, \tag{3.5}$$

for $\bar\beta, \bar\lambda > 0$, where $\bar Z(\bar\beta, \bar\lambda) = \Gamma(\bar\lambda/b)/(b\bar\beta^{\bar\lambda/b})$ is the normalization constant. Let us denote by $E_{\mu_{\bar\beta,\bar\lambda}}$ the expectation with respect to $\mu_{\bar\beta,\bar\lambda}$. We denote by $e := e(\bar\beta, \bar\lambda)$, $v := v(\bar\beta, \bar\lambda)$, and $\rho = \rho(\bar\beta, \bar\lambda)$ the averages of the quantities $V_b(\eta_x)$, η_x, and ξ_x with respect to $\mu_{\bar\beta,\bar\lambda}$, that is,

$$e = E_{\mu_{\bar\beta,\bar\lambda}}\big[V_b(\eta_x)\big], \quad v = E_{\mu_{\bar\beta,\bar\lambda}}[\eta_x], \quad \text{and} \quad \rho = E_{\mu_{\bar\beta,\bar\lambda}}[\xi_x]. \tag{3.6}$$

With the notations that we have just introduced, we see that

$$
\begin{aligned}
E_{\mu_{\bar\beta,\bar\lambda}}\big[j^e_{x,x+1}\big] &= -\alpha_n b^2(e-bv)^2+\alpha_n b^2 = -\alpha_n b^2\rho(1-2\rho),\\
E_{\mu_{\bar\beta,\bar\lambda}}\big[j^v_{x,x+1}\big] &= 2\alpha_n b(1+e-bv) = 2\alpha_n b\rho,\\
E_{\mu_{\bar\beta,\bar\lambda}}\big[j^\xi_{x,x+1}\big] &= -\alpha_n b^2\rho^2.
\end{aligned} \tag{3.7}
$$

Note that $\rho=\frac{\bar\lambda}{b\bar\beta}$ and $\tau^2=\frac{\bar\lambda}{b\bar\beta^2}$.

3.1. Hydrodynamic limits

Now we describe the space-time evolution of the relevant quantities of the system. Therefore, for any configuration $\eta\in\Omega_n$, we define the empirical measures associated to the energy and the volume as $\pi^{n,e}(\eta,du)$ and $\pi^{n,v}(\eta,du)$ in $\mathbb{R}$ by

$$
\pi^{n,e}(\eta,du)=\frac{1}{n}\sum_{x\in\mathbb{T}_n}V_b(\eta_x)\delta_{\frac{x}{n}}(du)\quad\text{and}\quad \pi^{n,v}(\eta,du)=\frac{1}{n}\sum_{x\in\mathbb{T}_n}\eta_x\delta_{\frac{x}{n}}(du),
$$

and let $\pi^{n,\cdot}_t(\eta,du):=\pi^{n,\cdot}(\eta(tn^a),du)$. In [7], for $a=1$ and in the strong asymmetric regime, if $e_0:\mathbb{R}\to\mathbb{R}$ and $v_0:\mathbb{R}\to\mathbb{R}$ are measurable functions and if $\pi^{n,e}_0(\eta,du)\to^w_{n\to+\infty}e_0(u)du$ and $\pi^{n,v}_0(\eta,du)\to^w_{n\to+\infty}v_0(u)du$, where the convergence is in the weak sense and with respect to an initial measure μ_n, then the same result is true for any $t\in[0,T]$ (before the appearance of shocks), namely $\pi^{n,e}_t(\eta,du)\to^w_{n\to+\infty}e(t,u)du$ and $\pi^{n,v}_t(\eta,du)\to^w_{n\to+\infty}v(t,u)du$, where

$$
\begin{cases}
\partial_t e(t,u)-\alpha b^2\partial_u\big((e(t,u)-bv(t,u))^2\big)=0,\\
\partial_t v(t,u)+2\alpha b\partial_u\big(e(t,u)-bv(t,u)\big)=0,
\end{cases} \tag{3.8}
$$

with initial conditions e_0 and v_0, respectively.

Our interest in these notes is to go beyond the hydrodynamic limit and ask about the form of the fluctuations around the hydrodynamical profile. The study of nonequilibrium fluctuations is usually very intricate, since it requires knowledge on the correlations of the system, therefore we restrict ourselves to the equilibrium scenario. So from now on, we assume that our Markov process starts from the invariant measure $\mu_{\bar\beta,\bar\lambda}$.

For systems with only one conservation law, there is no ambiguity in the choice of the fields that one should look at. When systems have more than one conserved quantity, and whose evolution is coupled, as is our case here, we have to be careful when we define those fields. In the next section, we apply the mode-coupling theory explained in detail in [27] and compute the correct fields that one should look at. We also explain in Section 3.3 another way of computing those fields just by employing Dynkin's formula.

3.2. Predictions from mode coupling theory

The system under investigation has two conserved quantities and it is possible to have an estimate on the scaling exponent and the form of the liming fluctuations by using the nonlinear fluctuating hydrodynamics theory. This is a macroscopic theory that requires only the knowledge of the hydrodynamic equations (3.8) in the hyperbolic time scale, we refer to [23, 26, 27]. Nevertheless, observe that this theory has only been developed in the strong asymmetric regime corresponding to $\kappa = 0$. Let us fix the quantities ξ_x and η_x, and recall (3.7). Assume $\rho \neq 0$. Consider the column flux matrix

$$j = \begin{pmatrix} E_{\mu_{\bar\beta,\bar\lambda}}[j^{\xi}_{x,x+1}] \\ E_{\mu_{\bar\beta,\bar\lambda}}[j^{v}_{x,x+1}] \end{pmatrix} = \begin{pmatrix} -\alpha_n b^2 \rho^2 \\ 2\alpha_n b \rho \end{pmatrix}. \tag{3.9}$$

The Jacobian is thus given by

$$J = \begin{pmatrix} \partial_\rho E_{\mu_{\bar\beta,\bar\lambda}}[j^{\xi}_{x,x+1}] & \partial_v E_{\mu_{\bar\beta,\bar\lambda}}[j^{\xi}_{x,x+1}] \\ \partial_\rho E_{\mu_{\bar\beta,\bar\lambda}}[j^{v}_{x,x+1}] & \partial_v E_{\mu_{\bar\beta,\bar\lambda}}[j^{v}_{x,x+1}] \end{pmatrix} = \begin{pmatrix} -2\alpha_n b^2 \rho & 0 \\ 2\alpha_n b & 0 \end{pmatrix}. \tag{3.10}$$

Now observe that the eigenvalues of last matrix are given by $v_1 := -2\alpha_n b^2 \rho$ and $v_2 = 0$. This corresponds to the velocity that we should take in order to see the evolution of the fields. The corresponding eigenvectors are given by $\tau_1 = \left(\begin{smallmatrix} 1 \\ -\frac{1}{b\rho} \end{smallmatrix}\right)$ and $\tau_2 = \left(\begin{smallmatrix} 0 \\ \mathrm{c} \end{smallmatrix}\right)$, where c is a constant. To obtain the linear combination of the fields that one should look at, we need to find the matrix R that diagonalizes J, that is, $RJR^{-1} = \left(\begin{smallmatrix} v_1 & 0 \\ 0 & v_2 \end{smallmatrix}\right)$. Observe that R^{-1} is the matrix whose columns are the eigenvectors of J so that

$$R^{-1} = \begin{pmatrix} 1 & 0 \\ -\frac{1}{b\rho} & \mathrm{c} \end{pmatrix} \quad \text{and} \quad R = \frac{1}{\mathrm{c}} \begin{pmatrix} \mathrm{c} & 0 \\ \frac{1}{b\rho} & 1 \end{pmatrix}. \tag{3.11}$$

The free constant c is determined by the equation $RKR^{-1} = \mathbb{I}$, where $\mathbb{I}$ denotes the identity matrix. The matrix K is a symmetric matrix and it is called the compressibility matrix. According to the nonlinear fluctuating hydrodynamic theory, the quantities that we should look at are given by the identity $(\mathcal{U}_1, \mathcal{U}_2) = R(\overline{\xi}_x, \overline{\eta}_x)$, which gives $\mathcal{U}_1 = \overline{\xi}_x$ and $\mathcal{U}_2 = \frac{1}{\mathrm{c}} \frac{\overline{\xi}_x}{b\rho} + \overline{\eta}_x$. Therefore, the quantities $\mathcal{U}_1$ and $\mathcal{U}_2$ are the conserved quantities that we should look at and on a frame with velocity v_1 and v_2, respectively. Now let us see the predictions on the form of the fluctuations for each one of these quantities. Given the two entries of the matrix (3.10), we look now at the corresponding Hessians:

$$\mathcal{H}^1 = -2\alpha_n b^2 \begin{pmatrix} 1 & 0 \\ 0 & 0 \end{pmatrix} \quad \text{and} \quad \mathcal{H}^2 = \begin{pmatrix} 0 & 0 \\ 0 & 0 \end{pmatrix}. \tag{3.12}$$

The coupling constants, which are determined by the above matrices, are given on $i \in \{1, 2\}$ by $G^i = \frac{1}{2} \sum_{j=1}^{2} R_{i,j} [(R^{-1})^\dagger \mathcal{H}^j R^{-1}]$ where $R_{i,j}$ is the entry of the matrix R. A simple computation shows that

$$[(R^{-1})^\dagger \mathcal{H}^1 R^{-1}] = -\alpha_n b^2 \begin{pmatrix} 1 & 0 \\ 0 & 0 \end{pmatrix},$$

and from this we get

$$G^1 = -\alpha_n b^2 \begin{pmatrix} 1 & 0 \\ 0 & 0 \end{pmatrix} \quad \text{and} \quad G^2 = -\frac{\alpha_n b \mathrm{c}}{\rho} \begin{pmatrix} 1 & 0 \\ 0 & 0 \end{pmatrix}.$$

From Section 2.2 of [27], we obtain for the strong asymmetric regime ($\kappa = 0$), since $G^1_{1,1} = 1$, $G^2_{2,2} = 0$, and $G^2_{1,1} = 1$, that the equilibrium fluctuations of the quantity $\mathcal{U}_1$ should be in the KPZ universality class, while the fluctuations of the quantity $\mathcal{U}_2$ should be described by a Lévy process with exponent $\frac{5}{3}$.

3.3. Choice of the fluctuations fields

Suppose the system starts from the invariant measure $\mu_{\bar{\beta},\bar{\lambda}}$ as defined in (3.5). Recall also (3.6). In the same spirit as we defined above, we present now an alternative way to obtain the fields that we need to look at. Let us define on $f \in C^2(\mathbb{T})$ the quantities:

$$\mathcal{X}_t^n(f) = \frac{1}{\sqrt{n}} \sum_{x\in\mathbb{T}_n} f\Big(\frac{x}{n}\Big)\big(\xi_x(tn^a) - \rho\big) \quad \text{and}$$

$$\mathcal{V}_t^n(f) = \frac{1}{\sqrt{n}} \sum_{x\in\mathbb{T}_n} f\Big(\frac{x}{n}\Big)\big(\eta_x(tn^a) - v\big).$$

At this point it is important to recall (3.3) and (3.4) and to center them with respect to the invariant measure $\mu_{\bar{\beta},\bar{\lambda}}$. We see that the centered currents become

$$j^{\xi}_{x,x+1}(\eta) = -\alpha_n b^2 \bar{\xi}_x \bar{\xi}_{x+1} - \alpha_n b^2 \rho \bar{\xi}_x - \alpha_n b^2 \rho \bar{\xi}_{x+1} - \alpha_n b^2 \rho^2 - \gamma \nabla \bar{\xi}_x, \tag{3.13}$$

$$j^{v}_{x,x+1}(\eta) = \alpha_n b(\bar{\xi}_x + \bar{\xi}_{x+1} + 2\rho) - \gamma \nabla \bar{\eta}_x. \tag{3.14}$$

Our starting point is Dynkin's formula, which allows us to associate the martingales to each field $\mathcal{X}_t^n(f)$ and $\mathcal{V}_t^n(f)$ as in (2.5). The important terms to analyze are the time integrals. Observe that since our test functions are time-independent the contributions from the terms ∂_s are null. Let us now check the contribution from the action of the generator. We start with the field $\mathcal{X}_t^n$. Note that from (3.13), since $\sum_{x\in\mathbb{T}_n} \nabla_n f(\frac{x}{n}) = 0$, and from a summation by parts, we have that

$$\begin{aligned} n^a \mathcal{L} \mathcal{X}_s^n(f) = {} & \frac{\gamma}{n^{2-a}} \mathcal{X}_s^n(\Delta_n f) + \frac{\alpha_n b^2 \rho}{n^{2-a}} \mathcal{X}_s^n(\Delta_n f) + \frac{2b^2 \rho \alpha_n}{n^{1-a}} \mathcal{X}_s^n(\nabla_n f) \\ & - \frac{b^2 \alpha_n}{n^{3/2-a}} \sum_{x\in\mathbb{T}_n} \nabla_n f\Big(\frac{x}{n}\Big) \bar{\xi}_x(sn^a) \bar{\xi}_{x+1}(sn^a). \end{aligned} \tag{3.15}$$

At this point we stop with the computations for the ξ field and observe that by similar computations we obtain for the volume field

$$n^a \mathcal{L} \mathcal{V}_s^n(f) = \frac{\gamma}{n^{2-a}} \mathcal{V}_s^n(\Delta_n f) + \frac{2b\alpha_n}{n^{1-a}} \mathcal{X}_s^n(\nabla_n f) - \frac{b\alpha_n}{n^{2-a}} \mathcal{X}_s^n(\Delta_n f). \tag{3.16}$$

As one can see from the expansions above, the evolution of the $\mathcal{X}_t^n$ field is independent of the evolution of the $\mathcal{V}_t^n$ field but for the volume field, this is not the case. Let us now explain how to get a linear combination of the fields that one should focus on to have both fields drifting at the same velocity $\mathfrak{v}_n$.

Let us redefine the fields above, by considering a test function which is time dependent and given by a translation with a velocity $\mathfrak{v}_n$ and let $\mathcal{Z}_t^{n,\mathfrak{u}}(f)$ be the field corresponding to the variable $\bar{\zeta}_x^{\mathfrak{u}} = \bar{\xi}_x + \mathfrak{u}\bar{\eta}_x$, that is,

$$\mathcal{Z}_t^{n,\mathfrak{u}}(f) = \mathcal{X}_t^n(T^-_{\mathfrak{v}_n t} f) + \mathfrak{u}\mathcal{V}_t^n(T^-_{\mathfrak{v}_n t} f),$$

where $u \in \mathbb{R}$ and $T^-_{\mathfrak{v}_n t} f(\frac{x}{n}) = f(\frac{x}{n} - \mathfrak{v}_v t)$. A simple computation based on (3.15) and (3.16) shows that

$$
\begin{aligned}
n^a \mathcal{L}\mathfrak{T}^{n,\mathfrak{u}}_s(f) &= \frac{\gamma}{n^{2-a}} \mathfrak{T}^{n,\mathfrak{u}}_s(\Delta_n f) \\
&+ \frac{n^a \alpha_n b}{n}(-2b\rho + 2\mathfrak{u})\mathfrak{X}^n_s(\nabla_n T^-_{\mathfrak{v}_n s} f) - \frac{\alpha_n(\mathfrak{u}b - b^2\rho)}{n^{2-a}}\mathfrak{X}^n_s(\Delta_n T^-_{\mathfrak{v}_n s} f) \\
&- \frac{b^2\alpha_n}{n^{3/2-a}} \sum_{x \in \mathbb{T}_n} \nabla_n T^-_{\mathfrak{v}_n s} f\left(\frac{x}{n}\right)\bar{\xi}_x(sn^a)\bar{\xi}_{x+1}(sn^a).
\end{aligned}
$$

Observe also that since now the test functions are time-dependent, we get a contribution from the term $\partial_s \mathfrak{T}^{n,\mathfrak{u}}_s(f)$ given by $\partial_s \mathfrak{T}^{n,\mathfrak{u}}_s(f) = \mathfrak{v}_n \mathfrak{T}^{n,\mathfrak{u}}_s(\nabla f)$. To find the velocity $\mathfrak{v}_n$, we only look at the degree-one terms in the last display. Observe that both the rightmost term in the second line of the latter display and the rightmost term on the first line of the same display have a smaller variance when compared to the leftmost term on the second line of that display. Therefore the latter is the term that one has to get rid of. By combining that term with the contribution from $\partial_s \mathfrak{T}^{n,\mathfrak{u}}_s(f)$, we see that to find the constants $\mathfrak{u}$ and $\mathfrak{v}_n$ we have to solve the system of equations:

$$
\begin{cases}
\frac{n^a \alpha_n b}{n}(-2b\rho + 2\mathfrak{u}) = \mathfrak{v}_n, \\
0 = \mathfrak{v}_n \mathfrak{u}.
\end{cases}
\tag{3.17}
$$

The latter system is obtained by equating the coefficients in front of the quantities $\bar{\xi}_x$ and $\bar{\eta}_x$ in the expression

$$
\frac{n^a \alpha_n b}{n}(-2b\rho + 2\mathfrak{u})\mathfrak{X}^n_s(\nabla_n T^-_{\mathfrak{v}_n s} f) + \mathfrak{v}_n \mathfrak{T}^{n,\mathfrak{u}}_s(\nabla f).
$$

The system (3.17) has two solutions:

(I) $\mathfrak{u} = 0$, which gives $\mathfrak{v}_n = -bn^{a-1}\alpha_n \mathfrak{v}$, where $\mathfrak{v} = 2b\rho$. In this case, we should consider the field associated to the quantity $\mathcal{U}_1 = \bar{\xi}_x$ in a moving time-dependent frame.

(II) $\mathfrak{v}_n = 0$, which gives $\mathfrak{u} = b\rho$. In this case, we should consider the field associated to the quantity $\mathcal{U}_2 = \bar{\xi}_x + b\rho\bar{\eta}_x$ and with no velocity since $\mathfrak{v}_n = 0$.

Observe that these results match the predictions from the previous subsection. Now that the fields are fixed, let us see what we can say in these two cases.

3.4. Limiting equations

From the computations of the last subsections, we know exactly the linear combination of the conserved quantities that we should look at and the corresponding velocities. Now we explain what we can rigorously prove for each one of them.

3.4.1. Case (I)

In this case $\mathfrak{v}_n = -2b^2\rho n^{a-1}\alpha_n$. Since $\mathfrak{u} = 0$, we have that

$$
\mathfrak{T}^{n,0}_t(f) = \mathfrak{X}^n_t(T^-_{\mathfrak{v}_n t} f) = \frac{1}{\sqrt{n}} \sum_{x \in \mathbb{T}_n} (T^-_{\mathfrak{v}_n t} f)\left(\frac{x}{n}\right)\bar{\xi}_x(tn^a).
$$

Therefore,

$$(\partial_s + n^a \mathscr{L})\mathfrak{X}^{n,0}_s(f) = \frac{\gamma}{n^{2-a}}\mathfrak{X}^{n,0}_s(\Delta_n f) + \frac{\alpha_n b^2 \rho}{n^{2-a}}\mathfrak{X}^{n,0}_s(\Delta_n f)$$
$$- 2b^2\rho n^{a-1}\alpha_n \mathfrak{X}^{n,0}_s(\nabla_n f) - v_n \mathfrak{X}^{n,0}_s(\nabla f)$$
$$- \frac{b^2\alpha_n}{n^{3/2-a}} \sum_{x\in\mathbb{T}_n} \nabla_n T^{-}_{v_n s} f\left(\frac{x}{n}\right)\bar{\xi}_x(sn^a)\bar{\xi}_{x+1}(sn^a). \quad (3.18)$$

By a Taylor expansion on f and the choice of v_n, the second line above vanishes as $n \to +\infty$. From (2.6), we see that the term in the last line of (3.18) has a variance of order $O(\alpha_n^2 n^{3a/2-2})$ so that for $a < \frac{4}{3}(\kappa+1)$ the $\mathbb{L}^2(\mathbb{P}_{\mu_{\bar\beta,\bar\lambda}})$-norm of that term vanishes as $n \to +\infty$. Moreover, if $a < 2$ (resp. $a < 2+\kappa$), the $\mathbb{L}^2(\mathbb{P}_{\mu_{\bar\beta,\bar\lambda}})$-norm of the first (resp. second) term on the right-hand side of the first line in (3.18) vanishes as $n \to +\infty$. In the case $a = 2$ and $\kappa = 1/2$, we can treat the term in the last line of (3.18) by using (2.7). From that result, the last line of (3.18) can be written, for n sufficiently big and ε sufficiently small, as

$$b^2\alpha \int_0^t \frac{1}{n}\sum_{x\in\mathbb{T}_n} \nabla_n f\left(\frac{x}{n}\right)\left(\mathfrak{X}^{n,0}_s\left(\iota_\epsilon\left(\frac{x}{n}\right)\right)\right)^2 ds. \quad (3.19)$$

Let us denote the martingale associated to $\mathfrak{X}^{n,0}_t(f)$ by $\mathscr{M}^{n,0}_t(f)$. We note that its quadratic variation is given by

$$\langle \mathscr{M}^{n,0}(f)\rangle_t = \int_0^t \{n^a \mathscr{L}(\mathfrak{X}^{n,0}_s(f))^2 - 2\mathfrak{X}^{n,0}_s(f) n^a \mathscr{L}\mathfrak{X}^{n,0}_s(f)\}\, ds$$
$$= \gamma \int_0^t \frac{n^a}{n}\sum_{x\in\mathbb{T}_n}\left(f\left(\frac{x+1}{n}\right) - f\left(\frac{x}{n}\right)\right)^2 (\xi_{x+1}(sn^a) - \xi_x(sn^a))^2 ds.$$
(3.20)

From simple computations, we see that if $a = 2$ and $\kappa > \frac{3}{4}a - 1$ then

$$\mathscr{M}^{n,0}_t(f) = \mathfrak{X}^{n,0}_t(f) - \mathfrak{X}^{n,0}_0(f) - \gamma\int_0^t \mathfrak{X}^{n,0}_s(\Delta_n f) ds$$

plus a term that vanishes in $\mathbb{L}^2(\mathbb{P}_{\mu_{\bar\beta,\bar\lambda}})$ as $n \to +\infty$. Moreover, the quadratic variation of the martingale satisfies

$$\lim_{n\to+\infty} \mathbb{E}_{\mu_{\bar\beta,\bar\lambda}}[\langle \mathscr{M}^{n,0}(f)\rangle_t] = 2t\gamma\tau^2\|\nabla f\|_0^2,$$

where τ^2 is the variance of ξ_x with respect to $\mu_{\bar\beta,\bar\lambda}$ and $\|f\|_0^2$ denotes the $\mathbb{L}^2$-norm of f. Then $(\mathfrak{X}^{n,0}_t)_n$ converges to the solution of the OU equation

$$d\mathfrak{X}^0_t = \gamma\Delta\mathfrak{X}^0_t dt + \sqrt{2\gamma\tau^2}\nabla\mathscr{W}_t.$$

Now, for $a < 2$ and $\kappa > \frac{3}{4}a - 1$, we have $\mathscr{M}^{n,0}_t(f) = \mathfrak{X}^{n,0}_t(f) - \mathfrak{X}^{n,0}_0(f)$ plus a term that vanishes in $\mathbb{L}^2(\mathbb{P}_{\mu_{\bar\beta,\bar\lambda}})$ as $n \to +\infty$. Moreover, the quadratic variation of the martingale satisfies

$$\lim_{n\to+\infty} \mathbb{E}_{\mu_{\bar\beta,\bar\lambda}}[\langle \mathscr{M}^{n,0}(f)\rangle_t] = 0. \quad (3.21)$$

Then $\mathcal{Z}_t^0$ has a trivial evolution given by $\mathcal{Z}_t^0 = \mathcal{Z}_0^0$, so that $d\mathcal{Z}_t^0 = 0$. If $a = 2$ and $\kappa = \frac{3}{4}a - 1 = \frac{1}{2}$, then

$$\mathcal{M}_t^{n,0}(f) = \mathcal{Z}_t^{n,0}(f) - \mathcal{Z}_0^{n,0}(f) - \gamma \int_0^t \mathcal{Z}_s^{n,0}(\Delta_n f) ds$$
$$+ b^2\alpha \int_0^t \frac{1}{n} \sum_{x\in\mathbb{T}_n} \nabla_n f\left(\frac{x}{n}\right)\left(\mathcal{Z}_s^{n,0}\left(\iota_\epsilon\left(\frac{x}{n}\right)\right)\right)^2 ds.$$

Moreover, the quadratic variation of the martingale satisfies (3.21), so that $\mathcal{Z}_s^0$ is solution of the stochastic Burgers equation

$$d\mathcal{Z}_t^0 = \gamma\Delta\mathcal{Z}_t^0 dt + b^2\alpha\nabla(\mathcal{Z}_t^0)^2 dt + \sqrt{2\gamma\tau^2}\nabla\mathcal{W}_t.$$

The evolution of this quantity should be as described in Figure 1.

3.4.2. Case (II)

In this case $\mathfrak{u} = b\rho$ and $\mathfrak{v}_n = 0$. Then

$$n^a\mathcal{L}\mathcal{Z}_s^{n,\mathfrak{u}}(f) = \frac{\gamma}{n^{2-a}}\mathcal{Z}_s^{n,\mathfrak{u}}(\Delta_n f) - \frac{b^2\alpha_n}{n^{3/2-a}} \sum_{x\in\mathbb{T}_n} \nabla_n f\left(\frac{x}{n}\right)\bar{\xi}_x(sn^a)\bar{\xi}_{x+1}(sn^a).$$

Doing the same analysis as above, we see that the first term on the right-hand side of the latter display vanishes, as $n \to +\infty$, if $a < 2$ and the last term has a variance of order $O(\alpha_n^2 n^{3a/2-2})$, so that for $a < \frac{4}{3}(\kappa + 1)$ the $\mathbb{L}^2(\mathbb{P}_{\mu_{\bar{\beta},\bar{\lambda}}})$-norm of that term vanishes as $n \to +\infty$. This means that for this quantity we can show that its behavior is diffusive if $\kappa > \frac{1}{2}$ and $a = 2$, and trivial if $a < \inf(\frac{4}{3}(\kappa + 1), 2)$.

We note that in **[1]** the second quantity that was analyzed was the joint field for both quantities $\bar{\xi}_x$ and $\bar{\eta}_x$, and all the limiting behavior was derived rigorously. Nevertheless, as we have seen above, the second quantity that one should look at is $\bar{\zeta}_x^{b\rho} = \bar{\xi}_x + b\rho\bar{\eta}_x$, and only partial results are proved. According to the nonlinear fluctuating hydrodynamics theory developed in **[23, 26, 28]**, one should get for $\kappa = 0$ a fractional behavior given by a Lévy $\frac{5}{3}$ and this should persist up to $\kappa < 1/3$, and for $\kappa > \frac{1}{2}$ one should see a diffusive behavior **[25]**. The predictions from mode-coupling theory in the weak asymmetric regime are a bit controversial so that the regime $\kappa \in [\frac{1}{3}, \frac{1}{2}]$ is still unclear **[25]**. In the next section, we analyze one potential for which we can prove rigorously all possible limits.

3.5. The harmonic case

Let us now consider the same model as in the beginning of this section but with the potential $V(x) = \frac{x^2}{2}$ so that $\mathcal{L} = \alpha_n\mathcal{A} + \gamma\mathcal{S}$, where $\gamma > 0$, $\alpha_n = \alpha n^{-\kappa}$, $\alpha \in \mathbb{R}$, $\kappa > 0$,

$$(\mathcal{A}f)(\eta) = \sum_{x\in\mathbb{T}_n} (\eta_{x+1} - \eta_{x-1})(\partial_{\eta_x} f)(\eta),$$

and the operator $\mathcal{S}$ is defined in (3.1). The translation invariant stationary measures $\mu_{v,\beta}$ are explicitly given by the product of Gaussian measures

$$\mu_{v,\beta}(d\eta) = \prod_{x\in\mathbb{T}_n} \left(\frac{\beta}{2\pi}\right)^{1/2} \exp\left\{-\frac{\beta}{2}(\eta_x - v)^2\right\} d\eta_x.$$

In this case the system also conserves two quantities, the energy $\sum_x \eta_x^2$ and the volume $\sum_x \eta_x$. Note that the average with respect to $\mu_{v,\beta}$ of η_x and η_x^2 is equal to v and $v^2 + \frac{1}{\beta}$, respectively. If we repeat the computations of Section 3.2 applied to this potential, we see that

$$j^e_{x,x+1}(\eta) = -2\alpha_n \eta_x \eta_{x+1} - \gamma \nabla \eta_x^2, \tag{3.22}$$

$$j^v_{x,x+1}(\eta) = \alpha_n(\eta_x + \eta_{x+1}) - \gamma \nabla \eta_x. \tag{3.23}$$

In this case the Jacobian matrix is given by

$$J = \begin{pmatrix} 0 & -4\alpha_n v \\ 0 & 2\alpha_n \end{pmatrix}$$

with eigenvalues $v_1 = 2\alpha_n$ and $v_2 = 0$, and the corresponding eigenvectors $\tau_1 = \binom{-2v\mathfrak{d}}{\mathfrak{d}}$ and $\tau_2 = \binom{\mathfrak{e}}{0}$, where $\mathfrak{d}$ and $\mathfrak{e}$ are constants. Moreover,

$$R^{-1} = \begin{pmatrix} -2v\mathfrak{d} & \mathfrak{e} \\ \mathfrak{d} & 0 \end{pmatrix} \quad \text{and} \quad R = \begin{pmatrix} 0 & \frac{1}{\mathfrak{d}} \\ \frac{1}{\mathfrak{e}} & \frac{2v}{\mathfrak{e}} \end{pmatrix}.$$

From this, we see that the quantities that we should analyze are

$$(\mathcal{U}_1, \mathcal{U}_2) = R(\overline{\xi}_x, \overline{\eta}_x),$$

with $\mathcal{U}_1 = \frac{\overline{\eta}_x}{\mathfrak{d}}$ and $\mathcal{U}_2 = \frac{2v\overline{\eta}_x}{\mathfrak{e}} + \frac{\overline{\eta_x^2}}{\mathfrak{e}}$. Note that for $v = 0$ we simply get $(\mathcal{U}_1, \mathcal{U}_2)$ as the volume and energy. By computing the Hessian matrices associated with the currents, we see that the predictions tell us that in the strong asymmetric regime ($\kappa = 0$) we should have $\mathcal{U}_1$ diffusive and $\mathcal{U}_2$ Lévy with exponent $\frac{3}{2}$. In the case $v = 0$ last result was proved in **[4]**. For the volume, i.e. the quantity $\mathcal{U}_1$, when we take the fluctuation field with velocity zero, we get a process that is linearly transported in time, see the line in red colour in Figure 2, while if we take it with the velocity v_1 we get an OU without drift, see the line in green colour in Figure 2 below. For $\mathcal{U}_2$ with velocity $v = 0$, i.e. the energy (recall that $v_2 = 0$) we have the results summarized in Figure 3.

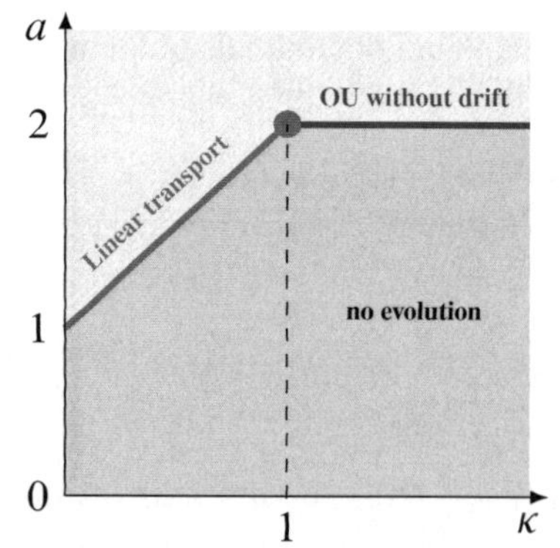

FIGURE 2
$\mathcal{U}_1$ fluctuations.

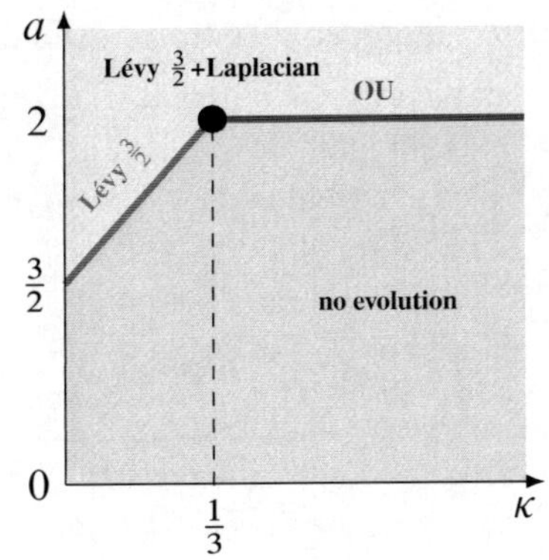

FIGURE 3
$\mathcal{U}_2$ fluctuations.

In Figure 2, the line in red colour corresponds to $a = \kappa + 1$; while in Figure 3 the line where we see the Lévy process with exponent $\frac{3}{2}$ is given by $a = \frac{3}{2}(\kappa + 1)$. Last results were proved in [4,5]. When the volume is taken with velocity v_1 then the line in green colour reaches the vertical line corresponding to $\kappa = 0$. Let us comment a bit on the proof of this result. We focus on the energy and note that $v_2 = 0$, so that the associated fluctuation field is given by

$$\mathscr{E}_t^n(f) = \frac{1}{\sqrt{n}} \sum_{x \in \mathbb{T}_n} f\left(\frac{x}{n}\right) \overline{\eta_x^2}(tn^a).$$

Moreover, by assuming that $v = 0$, we get the following action of the generator:

$$(\partial_s + n^a \mathscr{L})\mathscr{E}_s^n(f) = \gamma n^{a-2} \mathscr{E}_s^n(\Delta_n f) - 2\alpha n^{a-\kappa-3/2} \sum_{x \in \mathbb{T}_n} \nabla_n f\left(\frac{x}{n}\right) \eta_x(sn^a)\eta_{x+1}(sn^a). \tag{3.24}$$

From (2.6) we know that the second term on the right-hand side of the last display vanishes, as $n \to +\infty$, for $a < \frac{4}{3}(\kappa + 1)$. Let us now explain how to prove that, in fact, the last term vanishes for $a < \frac{3}{2}(\kappa + 1)$, giving rise to the rosy area in Figure 3 above, and on the line $a = \frac{3}{2}(\kappa + 1)$ and for $\kappa \in [0, \frac{1}{3})$ we get the fractional behavior given by the Lévy process with exponent $\frac{3}{2}$. At this point we need to use the deterministic part of the dynamics given by the Hamiltonian $\mathscr{A}$. In order to do that, the idea is to rewrite the rightmost term of the last display in terms of the correlation field of the volume η_x, that we denote by $\mathbb{Q}_t^n$. This fields acts on functions $h : \mathbb{T}^2 \to \mathbb{R}$ as follows:

$$\mathbb{Q}_t^n(h) = \frac{1}{n} \sum_{x \neq y} h\left(\frac{x}{n}, \frac{y}{n}\right) \eta_x(tn^a)\eta_y(tn^a),$$

and since $v = 0$, the field is centered. Note that the definition of $\mathbb{Q}_t^n$ does not depend on the value of the function h on the diagonal $x = y$ because, when $x = y$, from $\eta_x\eta_y$ we would recover the energy η_x^2. With this notation, we can rewrite (3.24) as

$$(\partial_s + n^a \mathscr{L})\mathscr{E}_s^n(f) = \gamma n^{a-2} \mathscr{E}_s^n(\Delta_n f) - 2\alpha n^{a-\kappa-3/2} \mathbb{Q}_s^n(\nabla_n f \otimes \delta),$$

where $\nabla_n f \otimes \delta : \mathbb{T}_n^2 \to \mathbb{R}$ is a discrete approximation of the distribution $f'(x) \otimes \delta(x = y)$ and $\delta(x = y)$ is the δ of Dirac at the line $x = y$ and it is given by

$$(\nabla_n f \otimes \delta)\Big(\frac{x}{n}, \frac{y}{n}\Big) = \frac{n}{2} \nabla_n f\Big(\frac{x}{n}\Big) \mathbf{1}_{y=x+1} + \frac{n}{2} \nabla_n f\Big(\frac{x-1}{n}\Big) \mathbf{1}_{y=x-1}. \tag{3.25}$$

This means that for the energy we get

$$\mathcal{M}_t^{n,e}(f) = \mathcal{E}_t^n(f) - \mathcal{E}_0^n(f) - \int_0^t \gamma n^{a-2} \mathcal{E}_s^n(\Delta_n f) - 2\alpha n^{a-\kappa-3/2} \mathbb{Q}_s^n(\nabla_n f \otimes \delta) ds. \tag{3.26}$$

In order to close the equation for the energy field, we need to understand the behavior of the correlation field. A long computation (for details we refer the reader to Appendix A of [5]) shows that

$$\begin{aligned}\mathcal{M}_t^{n,q}(h) &= \mathbb{Q}_t^n(h) - \mathbb{Q}_0^n(h) \\ &\quad - \int_0^t \mathbb{Q}_s^n(\gamma n^{a-2} \Delta_n h + \alpha n^{a-\kappa-1} A_n h) - 2\alpha n^{a-\kappa-3/2} \mathcal{E}_s^n(\mathcal{D}_n h) \\ &\quad + 2\mathbb{Q}_s^n(n^{a-2} \tilde{\mathcal{D}}_n h) ds,\end{aligned}$$

where the operator $\Delta_n h$ is a discrete approximation of the 2-d Laplacian of h given by

$$\begin{aligned}\Delta_n h\Big(\frac{x}{n}, \frac{y}{n}\Big) &= n^2\Big(h\Big(\frac{x+1}{n}, \frac{y}{n}\Big) + h\Big(\frac{x-1}{n}, \frac{y}{n}\Big) + h\Big(\frac{x}{n}, \frac{y+1}{n}\Big) + h\Big(\frac{x}{n}, \frac{y-1}{n}\Big) \\ &\quad - 4h\Big(\frac{x}{n}, \frac{y}{n}\Big)\Big).\end{aligned}$$

Above $A_n h$ is a discrete approximation of the directional derivative $(-2, -2) \cdot \nabla h$ given by

$$A_n h\Big(\frac{x}{n}, \frac{y}{n}\Big) = n\Big(h\Big(\frac{x}{n}, \frac{y-1}{n}\Big) + h\Big(\frac{x-1}{n}, \frac{y}{n}\Big) - h\Big(\frac{x}{n}, \frac{y+1}{n}\Big) - h\Big(\frac{x+1}{n}, \frac{y}{n}\Big)\Big),$$

the operator $\mathcal{D}_n h$ is a discrete approximation of the directional derivative of h along the diagonal $x = y$ and it is given by $\mathcal{D}_n h(\frac{x}{n}) = n(h(\frac{x}{n}, \frac{x+1}{n}) - h(\frac{x-1}{n}, \frac{x}{n}))$, and the operator $\tilde{\mathcal{D}}_n$ is defined as follows:

$$\begin{aligned}\tilde{\mathcal{D}}_n h\Big(\frac{x}{n}, \frac{y}{n}\Big) &= n^2\Big(\tilde{\mathcal{E}}_n h\Big(\frac{x}{n}\Big) - \frac{1-\kappa}{2} \tilde{\mathcal{F}}_n h\Big(\frac{x}{n}\Big)\Big) \mathbf{1}_{y=x+1} \\ &\quad + n^2\Big(\tilde{\mathcal{E}}_n h\Big(\frac{y}{n}\Big) - \frac{1-\kappa}{2} \tilde{\mathcal{F}}_n h\Big(\frac{y}{n}\Big)\Big) \mathbf{1}_{y=x-1},\end{aligned}$$

with $\tilde{\mathcal{E}}_n h(\frac{x}{n}) = h(\frac{x}{n}, \frac{x+1}{n}) - h(\frac{x}{n}, \frac{x}{n})$ and $\tilde{\mathcal{F}}_n h(\frac{x}{n}) = h(\frac{x+1}{n}, \frac{x+1}{n}) - h(\frac{x}{n}, \frac{x}{n})$. Now we need to link the two equations above. To do so, we take h_n, as the symmetric function, solution of the Poisson equation

$$\gamma \Delta_n h\Big(\frac{x}{n}, \frac{y}{n}\Big) + \alpha n^{1-\kappa} \mathcal{A}_n h\Big(\frac{x}{n}, \frac{y}{n}\Big) = 2\alpha n^{1/2-\kappa} \nabla_n f \otimes \delta\Big(\frac{x}{n}, \frac{y}{n}\Big). \tag{3.27}$$

We get, by neglecting the martingales (since it can be shown that they vanish in $\mathbb{L}^2(\mathbb{P}_{\mu_{v,\beta}})$ as $n \to +\infty$ if $a < 2$), that

$$\begin{aligned}\mathcal{E}_t^n(f) - \mathcal{E}_0^n(f) &= \int_0^t \mathcal{E}_s^n(\gamma n^{a-2} \Delta_n f - 2\gamma n^{a-\kappa-3/2} \mathcal{D}_n h_n)\, ds \\ &\quad + \mathbb{Q}_0^n(h_n) - \mathbb{Q}_t^n(h_n) + 2\int_0^t \mathbb{Q}_s^n(n^{a-2} \tilde{\mathcal{D}}_n h_n)\, ds.\end{aligned} \tag{3.28}$$

Now we need to analyze each term at the right-hand side of the last display. By Fourier estimates, one can show that the discrete $\mathbb{L}^2$-norm of h_n vanishes as $n \to +\infty$, and from this and the Cauchy–Schwarz inequality we get that the $\mathbb{L}^2(\mathbb{P}_{\mu_{v,\beta}})$-norm of the second and third terms on the right-hand side of last display vanish, as $n \to \infty$. From long computations, one can also show the next result whose proof can be seen in [5].

Lemma 1. *Let h_n be the solution of the Poisson equation given in (3.27), $a = \inf(\frac{3}{2}(1+\kappa), 2)$ and $\kappa \in (0,1)$. For any $t > 0$, we have that*

$$\lim_{n\to\infty} \mathbb{E}_{\mu_{v,\beta}}\left[\left(\int_0^t \mathbb{Q}_s^n(n^{a-2}\tilde{\mathfrak{D}}_n h_n)\,ds\right)^2\right] = 0.$$

Finally, the remaining term has a nontrivial contribution to the limit which is given by the next lemma, whose proof can be found in [5].

Lemma 2. *If $a = \inf(\frac{3}{2}(1+\kappa), 2)$ and $\kappa \in (0,+\infty)$, then*

$$\lim_{n\to\infty} \frac{1}{n}\sum_{x\in\mathbb{Z}} \left| \{\gamma n^{a-2}\Delta_n f - 2\alpha n^{a-\kappa-3/2}\mathfrak{D}_n h_n\}\left(\frac{x}{n}\right) - \mathbb{L}_{\alpha,\kappa} f\left(\frac{x}{n}\right)\right|^2 = 0, \qquad (3.29)$$

where $\mathbb{L}_{\alpha,\kappa} = \mathbf{1}_{\kappa\geq 1/3}\Delta + \alpha^{3/2}\mathbf{1}_{\kappa\leq 1/3}\mathcal{L}$, with $\mathcal{L} = -\frac{1}{\sqrt{2}}\{(-\Delta)^{3/4} - \nabla(-\Delta)^{1/4}\}$.

From last results, for $a < 2$ the limiting field satisfies $\mathscr{E}_t(f) - \mathscr{E}_0(f) = \int_0^t \mathscr{E}_s(\mathcal{L}f)\,ds$. For $a = 2$, we recover the latter identity plus the contribution from the martingale, which in the diffusive time scale does not vanish anymore. This proves the results for the energy in Figure 3. We observe that the previous method allowed us to extend the Gaussianity of the limit field for $\kappa > \frac{1}{3}$, for which the second-order Boltzmann–Gibbs principle would give the same result but only for $\kappa > \frac{1}{2}$. In this sense, this method allows us to reach areas of the phase diagram that we could not reach with the previous method. Nevertheless, it relies on the specific form of the dynamics and the fact that the equation for the quadratic field only involves terms of the energy field and the quadratic field itself, and this is not the case for the majority of the dynamics, some other mixtures of fields of the conserved quantities might appear. We note, however, that in [6], by perturbing the harmonic potential weakly by a quartic potential, the result obtained above for harmonic case persists up to some small critical value of the anharmonicity. There is still work to do in this direction, and we believe that one should analyze the action of the generator in those mixtures of the fields and keep track of the relevant quantities that give a nontrivial contribution to the limit. This study could give a way to prove rigorously the results predicted by mode-coupling theory for many other models.

ACKNOWLEDGMENTS

I thank Gunter Schütz for valuable discussions on mode-coupling theory. I also thank Cédric Bernardin and Pedro Cardoso for helping me to improve the reading of this article.

FUNDING

This project received funding from the European Research Council under the European Union's Horizon 2020 research and innovative program (grant agreement no. 715734).

REFERENCES

[1] R. Ahmed, C. Bernardin, P. Gonçalves, and M. Simon, A microscopic derivation of coupled SPDE's with a KPZ flavor. *Ann. Inst. Henri Poincaré Probab. Stat.* (to appear).

[2] C. Bernardin, T. Funaki, and S. Sethuraman, Derivation of coupled KPZ–Burgers equation from multi-species zero-range processes. *Ann. Appl. Probab.* **31** (2021), no. 4, 1966–2017.

[3] C. Bernardin and P. Gonçalves, Anomalous fluctuations for a perturbed Hamiltonian system with exponential interactions. *Comm. Math. Phys.* **325** (2014), 291–332.

[4] C. Bernardin, P. Gonçalves, and M. Jara, 3/4-Fractional superdiffusion in a system of harmonic oscillators perturbed by a conservative noise. *Arch. Ration. Mech. Anal.* **220** (2016), no. 2, 505–542.

[5] C. Bernardin, P. Gonçalves, and M. Jara, Weakly harmonic oscillators perturbed by a conserving noise. *Ann. Appl. Probab.* **28** (2018), no. 3, 1315–1355.

[6] C. Bernardin, P. Gonçalves, M. Jara, and M. Simon, Nonlinear perturbation of a noisy Hamiltonian lattice field model: universality persistence. *Comm. Math. Phys.* **361** (2018), no. 2, 605–659.

[7] C. Bernardin and G. Stoltz, Anomalous diffusion for a class of systems with two conserved quantities. *Nonlinearity* **25** (2012), no. 4, 1099–1133.

[8] T. Brox and H. Rost, Equilibrium fluctuations of stochastic particle systems: the role of conserved quantities. *Ann. Probab.* **12** (1984), no. 3, 742–759.

[9] G. Canizzaro and M. Hairer, The Brownian castle. 2021, arXiv:2010.02766.

[10] A. Dhar, Heat transport in low-dimensional systems. *Adv. Phys.* **57** (2008), no. 5, 457–537.

[11] S. Edwards and D. Wilkinson, The surface statistics of a granular aggregate. *Proc. R. Soc. Lond. Ser. A, Math. Phys. Sci.* **381** (1980), 17–31.

[12] P. Gonçalves, Central limit theorem for a tagged particle in asymmetric simple exclusion. *Stochastic Process. Appl.* **118** (2008), 474–502.

[13] P. Gonçalves and M. Jara, Crossover to the KPZ equation. *Ann. Henri Poincaré* **13** (2012), no. 4, 813–826.

[14] P. Gonçalves and M. Jara, Nonlinear fluctuations of weakly asymmetric interacting particle systems. *Arch. Ration. Mech. Anal.* **212** (2014), no. 2, 597–644.

[15] P. Gonçalves, M. Jara, and M. Simon, Second order Boltzmann–Gibbs principle for polynomial functions and applications. *J. Stat. Phys.* **166** (2017), no. 1, 90–113.

[16] M. Gubinelli, P. Imkeller, and N. Perkowski, Paracontrolled distributions and singular PDEs. *Forum Math. Pi* **3** (2015).

[17] M. Gubinelli and N. Perkowski, Energy solutions of KPZ are unique. *J. Amer. Math. Soc.* **31** (2018), no. 2, 427–471.

[18] M. Jara, T. Komorowski, and S. Olla, Superdiffusion of energy in a chain of harmonic oscillators with noise. *Comm. Math. Phys.* **339** (2015), no. 2, 407–453.

[19] M. Kardar, G. Parisi, and Y. C. Zhang, Dynamic scaling of growing interfaces. *Phys. Rev. Lett.* **56** (1986), no. 9, 889–892.

[20] C. Kipnis and C. Landim, *Scaling limits of interacting particle systems*. Fundam. Principles Math. Sci. 320, Springer, Berlin, 1999.

[21] S. Lepri, R. Livi, and A. Politi, Thermal conduction in classical low-dimensional lattices. *Phys. Rep.* **377** (2003), no. 1, 1–80.

[22] K. Matetski, J. Quastel, and D. Remenik, The KPZ fixed point. *Acta Math.* **227** (2021), no. 1, 115–203.

[23] V. Popkov, J. Schmidt, and G. Schütz, Universality classes in two-component driven diffusive systems. *J. Stat. Phys.* **160** (2015), no. 4, 835–860.

[24] K. Saito, M. Sasada, and H. Suda, 5/6-superdiffusion of energy for coupled charged harmonic oscillators in a magnetic field. *Comm. Math. Phys.* **372** (2019), 151–182.

[25] G. Schütz, Private communication.

[26] H. Spohn, Nonlinear fluctuating hydrodynamics for anharmonic chains. *J. Stat. Phys.* **154** (2014), no. 5, 1191–1227.

[27] H. Spohn and G. Stoltz, Nonlinear fluctuating hydrodynamics in one dimension: the case of two conserved fields. *J. Stat. Phys.* **160** (2015), 861–884.

[28] H. van Beijeren, Exact results for anomalous transport in one-dimensional Hamiltonian systems. *Phys. Rev. Lett.* **108** (2012), 180601.

PATRÍCIA GONÇALVES

Center for Mathematical Analysis, Geometry and Dynamical Systems, Instituto Superior Técnico, Universidade de Lisboa, Av. Rovisco Pais, 1049-001 Lisboa, Portugal, pgoncalves@tecnico.ulisboa.pt

This Page Intentionally Left Blank

MIXING TIME AND CUTOFF FOR ONE-DIMENSIONAL PARTICLE SYSTEMS

HUBERT LACOIN

ABSTRACT

We survey recent results concerning the total-variation mixing time of the simple exclusion process on the segment (symmetric and asymmetric) and a continuum analog, the simple random walk on the simplex with an emphasis on cutoff results. A Markov chain is said to exhibit cutoff if on a certain time scale, the distance to equilibrium drops abruptly from 1 to 0. We also review a couple of techniques used to obtain these results by exposing and commenting some elements of proof.

MATHEMATICS SUBJECT CLASSIFICATION 2020

Primary 60J10; Secondary 37L60, 82C20

KEYWORDS

Markov chains, mixing time, particle systems, cutoff profile

ICM INTERNATIONAL CONGRESS OF MATHEMATICIANS 2022 JULY 6–14

Proc. Int. Cong. Math. 2022, Vol. 6, pp. 4350–4374
DOI 10.4171/ICM2022/41

1. A SHORT INTRODUCTION TO MARKOV CHAINS

1.1. Definition of a Markov chain

A stochastic process $(X_t)_{t\geq 0}$ indexed by $\mathbb{R}_+$ with values in a state-space Ω is said to be a *Markov process* if at each time $t \geq 0$, the distribution of the future $(X_{t+u})_{u\geq 0}$, conditioned on the past $(X_s)_{s\in[0,t]}$ is only determined by its present state X_t. This is equivalent to saying that for every bounded measurable function $F : \Omega^{\mathbb{R}_+} \to \mathbb{R}$, there exists $G : \Omega \to \mathbb{R}$ such that

$$\mathbb{E}\big[F\big[(X_{t+s})_{s\geq 0}\big] \mid (X_u)_{u\in[0,t]}\big] = G(X_t). \tag{1.1}$$

The assumption (1.1) can be interpreted as the *absence of memory* of the process and is called the *Markov property* (we refer to [41, CHAPTER III] for an introduction to Markov processes). *Markov chains* are Markov processes which are right continuous for the discrete topology on Ω, meaning that (X_t) always remains for some time in its current state before always jumping from it

$$\forall t \geq 0, \quad \inf\{s, X_{t+s} \neq X_t\} > 0.$$

Remark 1.1. The name Markov chains also (and perhaps more frequently) refers to discrete-time Markov processes, that is, processes indexed by $\mathbb{Z}_+$ rather than $\mathbb{R}_+$; see, for instance, [39]. Let us mention that all the continuous-time Markov chains mentioned in this paper are equivalent to discrete-time Markov chains in the sense that they can be obtained by composing a discrete-time Markov chain with a homogeneous Poisson process on $\mathbb{R}$, even when the considered state-space is infinite. In particular, they are càdlàg and do not display accumulation of jumps (a phenomenon called *explosion* see [45, CHAPTER 4]). We study these processes in continuous time rather than discrete mostly for practical and aesthetic reasons, but the results remain valid for the discrete-time version of the chains (and the adaptation of the proof from one setup to another is straightforward; see, for instance, [25, APPENDIX B]). While some references we refer to, such as [46], mention only the discrete-time version of the chains, we always transpose the cited results in the continuous-time setup for a better presentation.

1.2. Markov semigroup, generator, invariant measures, and reversibility

The distribution of a Markov chain $(X_t)_{t\geq 0}$ is determined by two inputs:

(A) The distribution of its initial condition X_0, which is a probability distribution on Ω, which we denote by μ.

(B) The rules of evolution of the future given the present, that is, the mapping $(\Omega^{\mathbb{R}_+} \to \mathbb{R}) \to (\Omega \to \mathbb{R})$ that associates G to F in equation (1.1). It can be encoded in an operator acting on functions defined on Ω, the *generator* of the Markov chain.

Since, in the present paper, we are interested in statements which are valid for every initial distribution μ, when introducing examples of Markov chains, we are going to specify only their generator.

1.2.1. Finite state-space case

Let us start by defining the generator of a Markov chain in the simpler case when the state-space is finite (the reader can find in [31, CHAPTER 20] a more substantial introduction and proofs of the results mentioned in this section). An important intermediate step is the definition of a Markov semigroup $(P_t)_{t\geq 0}$ associated with the Markov chain. It is a sequence of $\Omega \times \Omega$ matrices that satisfy the semigroup property $P_{s+t} = P_s P_t$ (where matrix multiplication is considered) and such that for every $x, y \in \Omega$ and $s, t \geq 0$, when $\mathbb{P}[X_s = x] > 0$,

$$\mathbb{P}[X_{t+s} = y \mid X_s = x] = P_t(x, y). \tag{1.2}$$

Note that $(P_t)_{t\geq 0}$ jointly with the initial distribution fully determines the finite-dimensional distributions of the process since the iteration of (1.2) yields

$$\begin{aligned} &\mathbb{P}[X_0 = x_0, X_{t_1} = x_1, X_{t_1+t_2} = x_2, \dots, X_{\sum_{i=1}^k t_i} = x_k] \\ &= \mathbb{P}[X_0 = x_0] P_{t_1}(x_0, x_1) P_{t_2}(x_1, x_2) \cdots P_{s_k}(t_{k-1}, t_k). \end{aligned} \tag{1.3}$$

The semigroup property, together with our assumption that (X_t) is càdlàg, implies that there exists an $\Omega \times \Omega$ matrix $\mathcal{L}$ – the generator of the Markov chain – such that for all $t \geq 0$,

$$\forall t > 0, \quad P_t = e^{\mathcal{L}t} := \sum_{k=1}^{\infty} \frac{s^k}{k!} \mathcal{L}^k.$$

Note that when we have for $x, y \in \Omega$, $x \neq y$,

$$\begin{cases} \mathcal{L}(x, y) = \lim_{t\to 0} \frac{1}{t} P_t(x, y), \\ -\mathcal{L}(x, x) = \lim_{t\to 0} \frac{1}{t}(1 - P_t(x, x)), \end{cases} \tag{1.4}$$

then $\mathcal{L}(x, y)$ represents the rate at which our Markov chain jumps from x to y, while $-\mathcal{L}(x, x)$ corresponds to the rate at which the chain jumps away from x. In practice, when introducing the generator of a Markov chain, we simply write its action (by left multiplication) on $\mathbb{R}$-valued functions on Ω. That is,

$$\mathcal{L} f(x) := \sum_{y\in\Omega} \mathcal{L}(x, y) f(y) = \sum_{y\in\Omega\setminus\{x\}} \mathcal{L}(x, y)\big[f(y) - f(x)\big].$$

We focus on the case of *irreducible* Markov chains, that is, we assume that every state of Ω can be reached from any other state with a finite number of jumps. Formally, for each x, y, there exist $k \geq 1$ and a sequence $x_0, x_1, \dots, x_k$ with $x_0 = x$ and $x_k = y$ such that

$$\forall i \in [\![1, k]\!], \quad \mathcal{L}(x_{i-1}, x_i) > 0.$$

This condition immediately implies that $P_s(x, y) > 0$ for every $x, y \in \Omega$. If $\mathcal{L}$ is irreducible, and $\mathbb{P}_\mu$ denotes the law of the Markov chain with generator $\mathcal{L}$ and initial distribution μ, then there exists a unique probability π on Ω such that $\mathbb{P}_\pi(X_t = x) = \pi(x)$ for every π. Such a probability is called the *invariant distribution* of the Markov chain. Considering π as a (line) vector on Ω, this is equivalent to either of the two relations below

$$\begin{cases} \forall t > 0, \quad \pi P_t = \pi, \\ \pi \mathcal{L} = 0. \end{cases} \tag{1.5}$$

The convergence theorem for irreducible finite state-space Markov chains states (see, for instance, [31, THEOREMS 4.9 AND 20.1]) that the invariant probability measures π is also the limit distribution for X_t when $t \to \infty$. More precisely, for any probability μ on Ω, we have

$$\lim_{t\to\infty} \mathbb{P}_\mu(X_t = x) = \pi(x). \tag{1.6}$$

We want to investigate the quantitative aspect of this convergence. For the Markov chains in this paper, the stationary measure satisfies the so-called *detailed balance* condition

$$\forall x, y \in \Omega, \quad \pi(x)\mathcal{L}(x,y) = \pi(y)\mathcal{L}(y,x), \tag{1.7}$$

where we use the notation

$$[\![a,b]\!] := [a,b] \cap \mathbb{Z}. \tag{1.8}$$

It can be easily checked that (1.7) implies (1.5), but there are irreducible Markov chains for which the stationary probability does not satisfy (1.7). Markov chains for which the stationary measure satisfies (1.7) are called *reversible*.

1.2.2. Continuum state-space case

When our state-space is a continuum, the above description of the generator as a matrix cannot be used. In that case the semigroup associated to the Markov chain $(P_t)_{t\geq 0}$ is a sequence of probability kernels such that for every bounded measurable function f on Ω, and every s and t, we have[1]

$$\mathbb{E}\big[f(X_{t+s}) \mid X_s\big] = P_t f(X_s) \quad \text{with} \quad P_t f(x) := \int_\Omega f(y) P_t(x, \mathrm{d}y). \tag{1.9}$$

Informally, $P_t(x, A)$ is the probability that $X_{s+t} \in A$ given $X_s = x$. In analogy with (1.3), the semigroup $(P_t)_{t\geq 0}$, jointly with the initial distribution, determines fully the finite-dimensional distributions of $(X_t)_{t\geq 0}$. The generator of the Markov chain $\mathcal{L}$ can be defined in analogy with (1.4) by

$$\mathcal{L}f := \lim_{t\to 0} \frac{P_t f - f}{t}. \tag{1.10}$$

For a general Markov processes, the limit on the right-hand side in (1.10) may not exist for every bounded measurable f; the set of functions for which the limit (1.10) does exist is called the *domain* of the generator. In this paper, however, we are going to consider only Markov chains with uniformly bounded jump rates, so we will not have to worry about this. Conditions for the existence and uniqueness of a stationary probability distribution and for a convergence such as that in (1.6) in continuous state-space are very far from being as nice as in the finite case (see, for instance, [41, CHAPTER 3]). In this survey, we consider only chains for which the stationary measure exists and is unique. They also satisfy the continuum counterpart of (1.7), that is, the operator $\mathcal{L}$ is self-adjoint in $L_2(\pi)$.

---

1 Strictly speaking, the relation (1.9) does not uniquely define $(P_t)_{t\geq 0}$, since one can modify $P_t(x,\cdot)$ for on a set of xs which is visited with probability zero but is is not a relevant issue for our discussion.

1.3. Total variation distance and mixing time

In order to quantify the convergence to equilibrium (1.6), we need a notion of distance on the set $M_1(\Omega)$ of probability measures on Ω, equipped with a σ-algebra (which is simply the power set $\mathcal{P}(\Omega)$ when Ω is finite). We consider the *total variation* distance, which quantifies how well two variables with different distributions can be coupled. Given $\alpha, \beta \in M_1(\Omega)$, the total variation distance between α and β is defined by

$$\|\alpha - \beta\|_{\mathrm{TV}} := \sup_{A \subset \Omega} |\alpha(A) - \beta(A)|,$$

where the supremum is taken over measurable sets. The following equivalent characterizations of the total variation distance helps to better grasp the notion. It is a sort of L_1 distance which measures how well two random variables can be coupled.

Proposition 1.2. *If Ω is finite or countable then we have*

$$\|\alpha - \beta\|_{\mathrm{TV}} := \frac{1}{2} \sum_{x \in \Omega} |\alpha(x) - \beta(y)|$$

If ν is a measure on Ω such that both α and β are absolutely continuous with respect to ν then

$$\|\alpha - \beta\|_{\mathrm{TV}} := \frac{1}{2} \int_{\Omega} \left| \frac{\mathrm{d}\alpha}{\mathrm{d}\nu} - \frac{\mathrm{d}\beta}{\mathrm{d}\nu} \right| \nu(\mathrm{d}x).$$

We have

$$\|\alpha - \beta\|_{\mathrm{TV}} := \min_{\substack{X_1 \sim \alpha \\ X_2 \sim \beta}} \mathbf{P}[X_1 = X_2]$$

where the minimum is taken over the set of all probability distribution $\mathbf{P}$ *on* $\Omega \times \Omega$ *which have marginal laws* α *and* β.

The total variation distance to equilibrium of the Markov chain with generator $\mathcal{L}$ and stationary measure π at time t is given by

$$d(t) := \sup_{\mu \in M_1(\Omega)} \left\| \mathbb{P}_\mu(X_t \in \cdot) - \pi \right\|_{\mathrm{TV}},$$

where $\mathbb{P}_\mu$ is the law of the Markov chain with generator $\mathcal{L}$ and initial measure μ. A standard coupling argument is sufficient to show that $d(t)$ is nondecreasing as a function of t. Given $\varepsilon \in (0,1)$, the mixing time associated to the threshold ε, or ε-mixing time of the Markov chain X_t, is given by

$$T_{\mathrm{mix}}(\varepsilon) := \inf\{t > 0 : d(t) \le \varepsilon\} = \sup\{t > 0 : d(t) > \varepsilon\}.$$

It indicates how long it takes, for a Markov chain starting from an arbitrary initial condition, to get close to its equilibrium measure. Note that when Ω is finite and the chain is irreducible, (1.6) guarantees that $\lim_{t \to \infty} d(t) = 0$ so that $T_{\mathrm{mix}}(\varepsilon) < \infty$ for all ε. For chains with a continuum state space, it is relevant to study the mixing time in the form defined above only if there is a unique stationary probability measure $\lim_{t \to \infty} d(t) = 0$.

Remark 1.3. In the case when $d(t) \nrightarrow 0$, some relevant variant of the mixing time can be defined by considering a restriction on the initial condition, for instance, by restricting x to a compact subset of Ω; see, e.g., [5, 13].

1.4. Organization of the paper

The main object of this paper is to survey some results and methods concerning the mixing time of some Markovian one-dimensional particle systems (with discrete and continuum state-space). In Section 2 we introduce these processes. In Section 3 we expose some results obtained with coauthors in the past decade, and propose a short survey of related research. In Section 4 we review a couple of pivotal ideas, which first appeared in [46] (in a slightly different form) and show how they can be combined to obtain (nonoptimal) upper bounds on the mixing time. In Section 5, we discuss the technical refinements that are required to improve these bounds to get optimal results.

Remark 1.4. In both Sections 4 and 5, we have made the choice to focus exclusively on upper-bound estimates for the mixing time. For the theorems presented in this survey – and in most instances of mixing-time problems – this is generally thought to be the hardest part of the results.

Some comments on notation. In the remainder of the paper, we always use the letter π (with superscripts and subscripts to underline the dependence on parameters) to denote the equilibrium measure of each of the considered Markov chain, so that the meaning of, say, π_N or $\pi_{N,k}$ will depend on the context. When several Markov chains with different initial distributions are considered, we may use a superscript to underline the initial distribution (for instance, (X_t^π) denotes a Markov chain starting from the stationary distribution). If the initial distribution is a Dirac mass δ_x with $x \in \Omega$, we write X_t^x rather than $X_t^{\delta_x}$.

2. ONE-DIMENSIONAL PARTICLE SYSTEMS AND INTERFACE MODELS

The Markov chains introduced in this section model the motion of particles in a one-dimensional space. In each instance, we do not introduce a single chain but rather a sequence of chains, which are indexed by one or two parameters, which correspond to the size of the system and/or the number of particles. We want to understand the evolution of the mixing time when these parameters diverge to infinity.

2.1. The interchange process on a segment

The symmetric interchange process on a segment. For $N \geq 2$, we let $\mathcal{S}_N$ denote the symmetric group, that is, the set of permutations on N elements. For $i \neq j$, we let $\tau_{i,j}$ denote the transposition which exchanges the position of i and j. We define the (symmetric) interchange process on the segment $[\![1, N]\!]$ (recall (1.8)) as the Markov chain on $\mathcal{S}_N$ with generator

$$\mathcal{L}^{(N)} f(\sigma) := \frac{1}{2} \sum_{i=1}^{N-1} \big[f(\sigma \circ \tau_{i,i+1}) - f(\sigma) \big].$$

It takes little effort to check that the Markov chain described above is irreducible, and that the uniform probability on $\mathcal{S}_N$ satisfies the detailed balanced condition (1.7). A more intuitive description of the process, which we denote by (σ_t), can be obtained using equation (1.4): it jumps away from its current stat with rate $(N-1)/2$ (that is, the times between consecutive jumps are IID exponential variables of mean $2/(N-1)$), and when it jumps, it chooses uniformly among the permutations obtained by composing on the right with a transposition of the form $\tau_{i,i+1}$ for $i \in [\![1, N]\!]$, or in other words, it interchanges the value of two randomly chosen consecutive coordinates.

An alternative description is that σ_t is *updated* with a rate $N-1$ (which is twice the previous rate). At an update time t, one coordinate $i \in [\![1, N-1]\!]$ is chosen uniformly at random, and σ_t is resampled by choosing uniformly at random in the set $\Theta(i, \sigma_{t-})$, where σ_{t-} is used to denote the left limit at t and

$$\Theta(i,\sigma) := \{\sigma' \in \mathcal{S}_N : \forall j \in [\![1, N]\!] \setminus \{i, i+1\}, \sigma'(j) = \sigma(j)\} = \{\sigma, \sigma \circ \tau_{i,i+1}\}.$$

Note that with this description, at each update, the value of σ_t remains unchanged with probability $1/2$. This second description might seem initially less natural than the first, but it turns out to be more convenient to construct monotone couplings, see Section 4.1.

Remark 2.1. The process described above is one of many examples of random walks on $\mathcal{S}_N$. This family of processes has attracted attention since the origin of the study of mixing times, due to the connection it has with the problem of card shuffling (see [31, CHAPTER 8] and the references therein). The symmetric interchange process, which we have considered here on the segment can be generalized: the study of the mixing properties for the interchange process on an arbitrary graph has been an active field of research; see, for instance, [7, 18, 35] and the references therein.

The biased interchange process. We consider a variant of the process which induces a bias towards more "ordered" permutations, that is, favors moves which drive the chain "closer" to the identity permutation. The set $\Theta(i,\sigma)$ is composed of two elements. We let $\sigma^{(i,+)}$ be the element of $\Theta(i,\sigma)$ such that $\sigma^{(i,+)}(i) < \sigma^{(i,+)}(i+1)$ and let $\sigma^{(i,-)}$ denote the element of $\Theta(i,\sigma)$ such that $\sigma^{(i,-)}(i) > \sigma^{(i,-)}(i+1)$ (intuitively, $\sigma^{(i,+)}$ is the permutation which is more ordered). Letting $p \in (1/2, 1)$ ($p = 1/2$ corresponds to the symmetric case considered above, the case $p \in (0, 1/2)$ is equivalent to $p \in (1/2, 1)$ after reverting the order of the coordinates) and setting $q := 1 - p$, we define the generator of the biased interchange process of the segment

$$\mathcal{L}_N^{(p)} f(\sigma) := \sum_{i=1}^{N-1} p\big[f(\sigma^{(i,+)}) - f(\sigma)\big] + q\big[f(\sigma^{(i,-)}) - f(\sigma)\big].$$

The introduction of a bias drastically modifies the stationary distribution. We let $D(\sigma)$ denote the minimal number of transpositions of type $\tau_{i,i+1}$ which we need to compose to obtain σ – it corresponds to the distance between σ and the identity permutation in the Cayley graph generated by the nearest-neighbor transpositions $(\tau_{i,i+1})_{i=1}^{N-1}$ (see [33, SECTION 3.4] for an introduction to Cayley graphs). We have

$$D(\sigma) = \sum_{1 \le i < j \le N} \mathbf{1}_{\{\sigma(i) > \sigma(j)\}}.$$

Setting $\lambda := p/q$ ($\lambda > 1$), the probability measure $\pi_N^{(p)}$ defined by

$$\pi_N^{(p)}(\sigma) = \frac{\lambda^{-D(\sigma)}}{\sum_{\sigma' \in \mathcal{S}_N} \lambda^{-D(\sigma')}}$$

satisfies the detailed balance condition for $\mathcal{L}_N^{(p)}$. As $\lambda > 1$, the measure $\pi_N^{(p)}$ concentrates most of its mass in a small neighborhood of the identity (more precisely, $D(\sigma)$ is typically of order N under $\pi_N^{(p)}$, while it is of order N^2 under the uniform measure).

2.2. The exclusion process on the segment

This Markov chain models the evolution of particles diffusing on a segment and subject to *exclusion*: each site can host at most one particle. Let N denote the length of the segment. A particle configuration is encoded by a sequence of 0 and 1 on the segment, ones and zeros respectively indicating the presence/absence of particle at a site. The space of configurations with a fixed number of particles k is defined by

$$\Omega_{N,k} := \left\{ \xi, [\![1, N]\!] \to \{0,1\} : \sum_{i=1}^{N} \xi(i) = k \right\}.$$

Given $\xi \in \Omega_{N,k}$ and distinct $i, j \in [\![1, N]\!]$, we set $\xi^{(i,j)} = \xi \circ \tau_{i,j}$ and define the generator of the *Symmetric Simple Exclusion Process* (or SSEP) to be

$$\mathcal{L}_{N,k} f(\xi) := \frac{1}{2} \sum_{i=1}^{N-1} \left[f(\xi \circ \tau_{i,i+1}) - f(\xi) \right].$$

An intuitive way to describe the above Markov chain is to say that each particle particle performs an independent, continuous-time nearest neighbor random walk with jump rate $1/2$ to the left and to the right, but that any jump which would result in either a particle moving out of the segment (that is, a jump to the site 0 or $N+1$) or two particles occupying the same site (that is, a jump of a particle to an already occupied site) are canceled (see Figure 1). The uniform probability on $\Omega_{N,k}$ satisfies the detailed balance condition (1.7).

Given $p \in (1/2, 1)$, we can also define the *Asymmetric Simple Exclusion Process* (or ASEP) which is a similar process on $\Omega_{N,k}$, but where the particles perform a random walk with respective jump rates p and q to the right and to the left. The corresponding generator is

$$\begin{aligned} \mathcal{L}_{N,k}^{(p)} f(\xi) := & \sum_{i=1}^{N-1} p \mathbf{1}_{\{\xi(i) > \xi(i+1)\}} \left[f(\xi \circ \tau_{i,i+1}) - f(\xi) \right] \\ & + \sum_{i=1}^{N-1} q \mathbf{1}_{\{\xi(i) < \xi(i+1)\}} \left[f(\xi \circ \tau_{i,i+1}) - f(\xi) \right]. \end{aligned} \tag{2.1}$$

Here also the introduction of the bias yields a modification of the stationary probability. The probability which satisfies the detailed balance condition is given by (recall that $\lambda = p/q$)

$$\pi_{N,k}^{(p)}(\xi) := \frac{\lambda^{-A(\xi)}}{\sum_{\xi' \in \Omega_{N,k}} \lambda^{-A(\xi')}}$$

where $A(\xi) := \sum_{i=1}^{N} (N-i)\xi(i) - \frac{k(k-1)}{2}$ denotes the (minimal) number of particle moves that separates ξ from the configuration $\mathbf{1}_{[\![N-k+1,N]\!]}$ with all particles packed to the right of the segment. Note that $A(\xi)$ is typically of order 1 under $\pi_{N,k}^{(p)}$ whereas it is of order N^2 under the uniform measure.

2.3. The corner-flip dynamics

We consider a Markov chain that models the motion of an interface, which is subject only to local moves. The one-dimensional interface is the graph of a one-dimensional nearest-neighbor path which belongs to the state space

$$\begin{aligned}\Xi_{N,k} := \big\{\zeta,\ [\![0,N]\!] \to \mathbb{Z} : \zeta(0) = 0, \zeta(N) = N-2k,\\ \forall i \in [\![0,N-1]\!], |\zeta(i+1) - \zeta(i)| = 1\big\}.\end{aligned} \tag{2.2}$$

We introduce a Markov chain on $\Xi_{N,k}$ that only changes the coordinates of ζ one at a time. Given $\zeta \in \Xi_N$ and $i \in [\![1,N-1]\!]$, we introduce $\zeta^{(i)}$ to be the element of Ω_N for which only the coordinate at i has been changed (see Figure 1):

$$\begin{cases} \zeta^{(i)}(j) := \zeta(j) & \text{if } j \neq i,\\ \zeta^{(i)}(i) := \zeta(i) - 2 & \text{if } \zeta(i+1) = \zeta(i-1) := \zeta(i) - 1,\\ \zeta^{(i)}(i) := \zeta(i) + 2 & \text{if } \zeta(i+1) = \zeta(i-1) := \zeta(i) + 1,\\ \zeta^{(i)}(i) := \zeta(i) & \text{if } |\zeta(i+1) - \zeta(i-1)| = 2.\end{cases}$$

The generator of the symmetric corner flip dynamics is given by

$$\mathfrak{L}_{N,k} := \frac{1}{2}\sum_{i=1}^{N-1}\big[f(\zeta^{(i)}) - f(\zeta)\big].$$

A way to visualize this dynamics is to say that each "corner" displayed by the the graph of ζ is flipped with rate $1/2$. The uniform measure on $\Xi_{N,k}$ satisfies the detailed balance condition (1.7).

We can also define an asymmetric version of the dynamics which favors flipping the corners in one direction. Given $\zeta \in \Xi_N$ and $i \in [\![i,i+1]\!]$, we define $\zeta^{(i,\pm)}$ to be respectively the "highest" and "lowest" path in the set $\{\zeta^{(i)}, \zeta\}$ (the set is possibly a singleton, so that we may have $\zeta^{(i,+)} = \zeta^{(i,-)}$)

$$\begin{cases} \zeta^{(i,\pm)}(j) := \zeta(j) & \text{if } j \neq i,\\ \zeta^{(i,+)}(i) := \max\big(\zeta(i), \zeta^{(i)}(i)\big),\\ \zeta^{(i,-)}(i) := \min\big(\zeta(i), \zeta^{(i)}(i)\big),\end{cases}$$

and define the generator of the asymmetric corner-flip dynamics as

$$\mathfrak{L}_{N,k}^{(p)} := \sum_{i=1}^{N-1} p\big[f(\zeta^{(i,+)}) - f(\zeta)\big] + q\big[f(\zeta^{(i,-)}) - f(\zeta)\big].$$

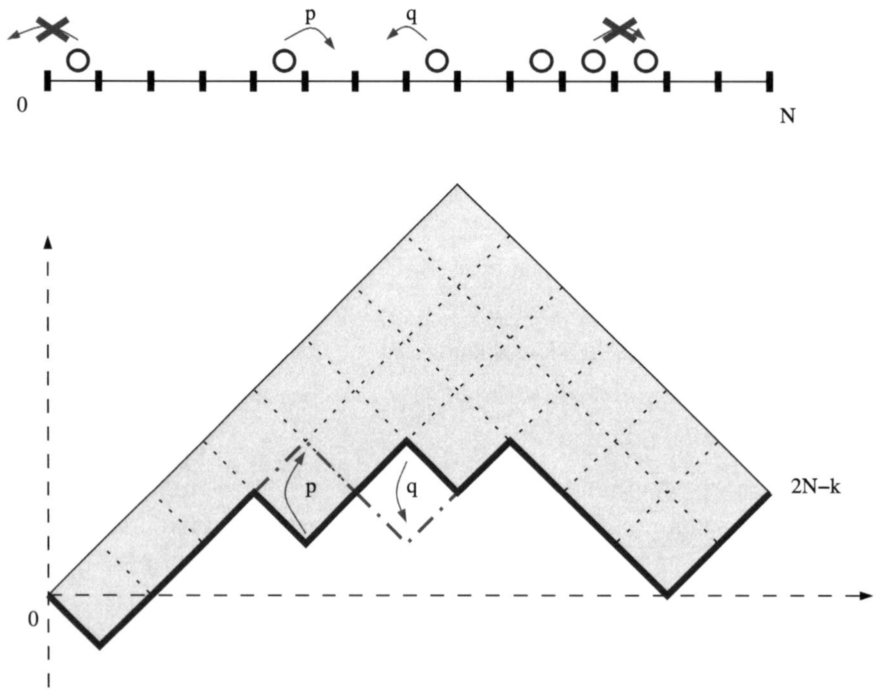

FIGURE 1
Graphical representation of the exclusion process and of the corner-flip dynamics with $k = 6$ and $N = 14$. Particles represented by circles, jump to the right with rate p and to the left with rate $q = 1 - p$ ($p = 1/2$ in the symmetric case), jumps are canceled if a particle tries to jump to an already occupied site. After applying the transformation h given in (2.5), we obtain the corner-flip dynamics, where each downward pointing corner on our interface is flipped up with rate p and each upward pointing corner is flipped down with rate q. The quantity $A(\zeta)$ which is the number of up-flips that need to be performed to reach the maximal configuration $\wedge$ (represented as a thin solid line of the figure) is equal to 22.

For the asymmetric corner flip, the reversible measure $\pi_{N,k}^{(p)}$ is defined by

$$\pi_{N,k}^{(p)}(\zeta) := \frac{\lambda^{-A(\zeta)}}{\sum_{\zeta' \in \Xi_{N,k}} \lambda^{-A(\zeta')}},$$

where $A(\zeta)$ denotes the halved geometric area lying between ζ and the highest path in $\Omega_{N,k}$ defined by

$$\wedge(i) = \min\bigl(i, 2(N-k) - i\bigr), \tag{2.3}$$

that is,

$$A(\zeta) := \frac{1}{2} \sum_{i=1}^{N-1} \bigl(\wedge(i) - \zeta(i)\bigr).$$

2.4. Random walk on the simplex

Let us finally consider a Markov chain for which the state-space is a continuum. We let $\mathfrak{X}_N$ denote the $(N-1)$-dimensional simplex defined by

$$\mathfrak{X}_N := \bigl\{(x_1, \ldots, x_{N-1}) \in \mathbb{R}^{N-1} : 0 \le x_1 \le \cdots \le x_{N-1} \le N\bigr\}.$$

We introduce a dynamics which is a continuum analog of the symmetric exclusion, the coordinates $x_1, \dots, x_{N-1}$ can be thought as the positions of $N-1$ particles on the segment $[0, N]$. The generator of the dynamics is given by

$$\mathrm{L}_N f(x) = \sum_{i=1}^{N-1} \int_0^1 \big[f(x^{(u,i)}) - f(x)\big] \mathrm{d}u,$$

for $f : \mathfrak{X}_N \to \mathbb{R}$ bounded and measurable, where $x^{(u,i)} \in \mathfrak{X}_N$ is defined by

$$\begin{cases} x_j^{(u,i)} := x_j & \text{for } j \neq i, \\ x_i^{(u,i)} := u x_{i+1} + (1-u) x_{i-1}, \end{cases}$$

with the convention that $x_0 = 0$ and $x_N = N$. In words, at rate one, each coordinate is resampled uniformly on its possible range of values, which is the segment $[x_{i-1}, x_{i+1}]$ (see Figure 2).

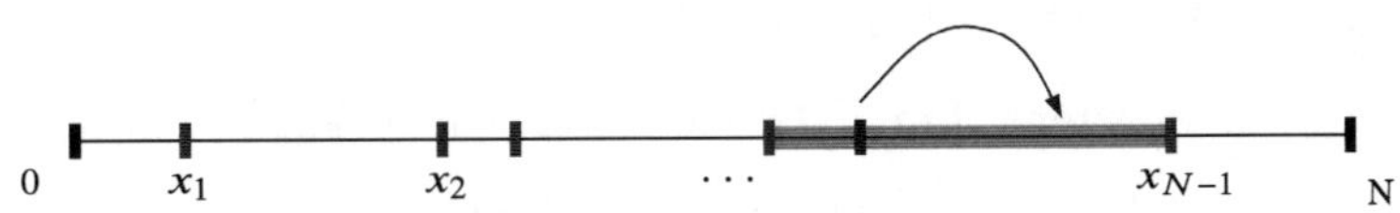

FIGURE 2

Graphical representation of the random walk on the simplex ($N = 7$). When the position of a particle is updated (x_5 on the picture), it is resampled uniformly in the interval delimited by the neighboring particles (that is, $[x_4, x_6]$), with the convention that $x_0 = 0$ and $x_N = N$.

The uniform probability on $\mathfrak{X}_N$, π_N, defined by

$$\pi_N(\mathrm{d}x) := \frac{(N-1)!}{N^{N-1}} \mathbf{1}_{\{x \in \mathfrak{X}_N\}} \mathrm{d}x_1 \cdots \mathrm{d}x_{N-1},$$

is stationary for L_N, and the generator is self-adjoint in $L^2(\pi_N)$.

Let us present an explicit construction of the Markov chain $(\mathbf{X}_t^x)_{t \geq 0}$ with initial distribution δ_x using auxiliary random variables. Such a construction is referred to as *a graphical construction* and turns out to be very convenient to work with (see, for instance, Section 4.1 below). It follows the following steps:

(i) To each coordinate $i \in [\![1, N-1]\!]$, we associate an independent rate-1 Poisson clock process $(\mathcal{T}_n^{(i)})_{n \geq 1}$ (the increments of $\mathcal{T}^{(i)}$ are i.i.d. exponential variables of mean 1) and a sequence of uniform random variables on $[0, 1]$, $(U_n^{(i)})_{n \geq 1}$.

(ii) We set $\mathbf{X}^x(0) = x$. The process is càdlàg and $(\mathbf{X}^x(t))_{t \geq 0}$ remains constant on each open interval of the set $(0, \infty) \setminus \{\mathcal{T}_n^{(i)}, n \geq 1, i \in [\![1, N-1]\!]\}$.

(iii) At time $t = \mathcal{T}_n^{(i)}$, we determine $\mathbf{X}^x(t)$ from $\mathbf{X}^x(t_-)$ (the left limit at t) by setting

$$X_i(t) := U_n^{(i)} X_{i+1}(t_-) + \big(1 - U_n^{(i)}\big) X_{i-1}(t).$$

The other coordinates are unchanged, $X_j(t) = X_j(t_-)$ for $j \neq i$.

The reader can check that, for any bounded measurable function f and every $x \in \mathcal{X}_N$,

$$\lim_{t\to 0} \frac{\mathbb{E}[f(\mathbf{X}^x(t))] - f(x)}{t} = \mathrm{L}_N f(x). \tag{2.4}$$

2.5. Correspondences

The Markov chains presented in Sections 2.1, 2.2, and 2.3 are very much related to each other. Let us first describe the correspondence between the particle system and discrete interfaces. Let us consider $\zeta : \Omega_{N,k} \to \Xi_{N,k}$ defined by

$$h(\xi)(x) := \sum_{y=1}^{x} (1 - 2\xi(x)). \tag{2.5}$$

It is immediate to check that $h(\xi) \in \Xi_{N,k}$ for every $\xi \in \Omega_{N,k}$ and that h is a bijection (we have $h^{-1}(\zeta)(x) = \frac{1+\zeta(x-1)-\zeta(x)}{2}$). Furthermore, we have $\mathfrak{L}^{(p)}_{N,k} \circ h = h \circ \mathcal{L}^{(p)}_{N,k}$ and, as a consequence, if $(\eta_t)_{t\geq 0}$ is a Markov chain with generator $\mathcal{L}^{(p)}_{N,k}$, then its image $h(\eta_t)_{t\geq 0}$ is a Markov chain on $\Xi_{N,k}$ with generator $\mathfrak{L}^{(p)}_{N,k}$ (this is, of course, also true in the symmetric case, when $p = 1/2$).

The corner-flip representation of the exclusion process can be convenient for reasoning since it allows for a better visual representation of an order relation which is conserved by the dynamics (see Section 4.1).

Another useful – although not bijective – correspondence is that between the interchange and exclusion processes. Given $k \in [\![1, N-1]\!]$, we define $\xi^{(k)} : \mathcal{S}_N \to \Omega_{N,k}$ as

$$\xi^{(k)}(\sigma) = \mathbf{1}_{[\![N-k+1,N]\!]} \circ \sigma.$$

Since $\mathcal{L}^{(p)}_{N,k} \circ \xi^{(k)} = \xi^{(k)} \circ \mathcal{L}^{(p)}_{N}$, if $(\sigma_t)_{t\geq 0}$ is a Markov chain on $\mathcal{S}_N$ with generator $\mathcal{L}^{(p)}_N$ then $(\xi^{(k)}(\sigma_t))_{t\geq 0}$ is a Markov chain on $\Omega_{N,k}$ with generator $\mathcal{L}^{(p)}_{N,k}$. The whole sequence of projections $(\xi^{(k)}(\sigma))_{k=1}^{N-1}$ allows recovering σ since we have

$$\sigma(i) = N - k + 1 \quad \Leftrightarrow \quad \xi^{(k)}(i) - \xi^{(k-1)}(i) = 1. \tag{2.6}$$

3. REVIEW OF MIXING-TIME RESULTS FOR ONE-DIMENSIONAL PARTICLE SYSTEMS

3.1. The cutoff phenomenon

Let us now survey a few results concerning the mixing time of the Markov chains introduced in the previous section. For all of these processes, an asymptotic equivalent to the mixing time $T^{(N)}_{\mathrm{mix}}(\varepsilon)$ (or $T^{(N,K)}_{\mathrm{mix}}(\varepsilon)$ if we have several parameters) is obtained in the limit when the parameter N (or N and k) tends to infinity. A striking common feature of all these results is that, in the asymptotic equivalent of $T^{(N)}_{\mathrm{mix}}(\varepsilon)$, there is no dependence on ε ($T^{(N)}_{\mathrm{mix}}(\varepsilon)$ depends on ε but this dependence only appear in higher order terms). In particular, we have for any $\varepsilon > 0$,

$$\lim_{N\to\infty} \frac{T^{(N)}_{\mathrm{mix}}(\varepsilon)}{T^{(N)}_{\mathrm{mix}}(1-\varepsilon)} = 1.$$

This mean that on a certain time scale, the distance to equilibrium drops abruptly from 1 to 0. This phenomenon is known as a cutoff and is believed to hold for a wide class of Markov chains (we refer to [12] and [31, CHAPTER 18], and the references therein for a historical introduction to cutoff). Cutoff is delicate to prove most of the time. For many Markov chains, while a short argument allows identifying the mixing time up to a constant multiplicative factor (cf. Section 4), much more effort is usually needed to obtain asymptotically matching upper and lower bounds.

3.2. Mixing time results

The SSEP and the interchange process. We group the results concerning the exclusion and interchange processes as their proofs share a lot of common ideas. We start with the symmetric case. The lower bounds in the result below have been proved by Wilson in [46] while the upper bounds have been obtained by the author in [25].

Theorem 3.1. *For any sequence k_N such that* $\lim \min(k_N, N - k_N) = \infty$, *the mixing time of the symmetric exclusion process on* $[\![1, N]\!]$ *with k_N particles satisfies, for any* $\varepsilon \in (0, 1)$,

$$\lim_{N\to\infty} \frac{T_{\mathrm{mix}}^{\mathrm{SSEP},(N,k_N)}(\varepsilon)}{N^2 \log[\min(k_N, N - k_N)]} = \frac{1}{\pi^2}. \tag{3.1}$$

For the interchange process on $[\![1, N]\!]$, *we have, for any* $\varepsilon \in (0, 1)$,

$$\lim_{N\to\infty} \frac{T_{\mathrm{mix}}^{\mathrm{IP},(N)}(\varepsilon)}{N^2 \log N} = \frac{1}{\pi^2}.$$

In view of the correspondence discussed in Section 2.5, the first part of the result, that is, (3.1), is also valid for the corner-flip dynamics introduced in Section 2.3.

The random walk on the simplex. The process is in a sense very similar to the simple exclusion process with a positive density of particles. However, the methods developed in [25, 46] – and, more generally, many of the techniques concerning upper bounds for the mixing time – rely on the fact that the state-space is discrete. The following, proved in [4], is one of a few cutoff results that have been proved for a Markov chain evolving in a continuum (see [19] for another example).

Theorem 3.2. *For the random walk on the simplex* $\mathfrak{X}_N$, *we have, for any* $\varepsilon \in (0, 1)$,

$$\lim_{N\to\infty} \frac{T_{\mathrm{mix}}^{\mathrm{RW},(N)}(\varepsilon)}{N^2 \log N} = \frac{1}{\pi^2}.$$

Upper and lower bounds of the right order - that is $N^2 \log N$ - but without the right constant factor has been proved prior to the above theorem in [38] (see also [37] for similar results in a periodic setting).

The ASEP and the biased interchange process. The introduction of a bias has the effect of making the system mix considerably faster: a time of order N is required for mixing instead of $N^{2+o(1)}$ in the symmetric case (this was proved in [2]). In [22] jointly with C. Labbé, we were able to identify the sharp asymptotics of the mixing time, proving cutoff both for the ASEP and for the biased interchange process.

Theorem 3.3. *For any $p \in (1/2, 1]$ and any sequence (k_N) such that*

$$\forall N \geq 2, k_N \in [\![1, N-1]\!] \quad \text{and} \quad \lim_{N\to\infty} \frac{k_N}{N} = \alpha \in [0, 1],$$

the mixing time of the asymmetric exclusion process with k_N particles satisfies, for every $\varepsilon \in (0, 1)$,

$$\lim_{N\to\infty} \frac{T_{\mathrm{mix}}^{\mathrm{ASEP},(p,N,k_N)}(\varepsilon)}{N} = \frac{(\sqrt{\alpha} + \sqrt{1-\alpha})^2}{2p-1}.$$

For the biased interchange process on a segment, we have, for every $\varepsilon \in (0, 1)$,

$$\lim_{N\to\infty} \frac{T_{\mathrm{mix}}^{\mathrm{BIP},(p,N)}(\varepsilon)}{N} = \frac{2}{2p-1}.$$

Note that the expression for the mixing time in the above result diverges when p tends to $1/2$. In [23,30] the crossover regime between the symmetric and asymmetric case is investigated. The right order of magnitude for the mixing time is established in [30], while [23] proves cutoff results.

3.3. Review of related works

Cutoff window and profile. The results above concern the first-order asymptotics of the mixing time. However, one can aim for results with a finer precision. For instance, one can try to estimate the order of magnitude of $T_{\mathrm{mix}}^{(N)}(\varepsilon) - T_{\mathrm{mix}}^{(N)}(1-\varepsilon)$ (say, for a fixed $\varepsilon \in (0, 1/2)$, this could theoretically depend on the value of ε, but in practice it does not for most chains), a quantity called the width of *the cutoff window*. One can further refine the picture and look for the limit of the distance to equilibrium $d^{(N)}(t)$ after recentering the picture at $t = T_{\mathrm{mix}}^{(N)}(1/2)$ and rescaling time by the cutoff window width. This is called *the cutoff profile*, and is the finest degree of description of convergence to equilibrium. For the SSEP on the circle – which is the closest cousin for the exclusion SSEP on the segment – the cutoff window of order N^2 and the profile have been identified in [24,26]. In the asymmetric case, the cutoff window of order $N^{1/3}$ and the profile have been identified in [3] (in the case where the density of particles is positive).

The exclusion process with an open boundary condition. We have considered the above dynamics where the number of particles is conserved. It is possible to consider the case of open boundaries, where particles can enter and exit the segment on the left and on the right. In that case, the equilibrium and dynamical behavior of the system depends a lot on the value chosen for the exit and entrance rate of the particles at the left and right boundary. Mixing-time results for the exclusion of the segment with a variety of boundary conditions are proved in [15], where several open questions and conjectures are also displayed. One of these conjectures is solved in [43], where it is shown that in the maximal current phase, for the totally asymmetric exclusion process (TASEP), the mixing time in that case is of order $N^{3/2}$. A similar result is predicted to hold for the asymmetric exclusion process on the circle, and the corresponding lower bound on the mixing time can be deduced from the results in [1].

The exclusion process in a random environment. Another variant of the process has been considered where the bias that each particle feels depends on the site at which it lies. That is, p varies with i. In [29,42] the case of i.i.d. random biases has been considered. This is a multiparticle version of the the classical Random Walk in a Random Environment (RWRE) (see, e.g., [20,44] for seminal references and [14] for a study of the mixing time of RWRE). The works [29,42] show that the presence of inhomogeneities in the environment can slow down the convergence to equilibrium.

The exclusion process in higher dimensions. The symmetric exclusion process on a higher-dimensional rectangle or torus has also been investigated. Proving result beyond dimension one turns out to be more difficult since monotonicity (in the sense of Section 4.1), which is a tool of crucial importance, cannot be used. It has been shown in [34] that the exclusion process in that case continues with a mixing time of order $N^2 \log k$ (see also [36,48] for earlier functional inequalities, which implies that the mixing time is of order $N^2 \log N$ when there is a density of particles).

More general interfaces. The mixing of one-dimensional interfaces has been studied well beyond the case of the corner-flip dynamics. In [6,8,10,27,28,47], the case of interfaces interacting with a substrate has been considered. The references [8,9,11,17] investigate the mixing time of higher-dimensional interfaces. In [5], interfaces with real-valued height functions are considered beyond the case of the random walk on the simplex. Let us finally mention [13] which proves a cutoff for Gaussian interfaces (the lattice free field) in arbitrary dimension.

4. A FEW TECHNICAL TOOLS USED TO PROVE THESE RESULTS

We review of a few key ingredients used in the proof of the results presented in the previous section. More precisely, to illustrate these techniques, we present a proof of nonoptimal results concerning the mixing time of the simple exclusion process on the segment (symmetric and asymmetric), or rather, its corner-flip representation. Although the presentation slightly differs, the argument found below is in spirit very similar to that found in [46, SECTION 3]. The reasoning can be applied without much change to the interchange process (see Remark 4.7) but, for clarity and conciseness, we limit the exposition of details to the case of the exclusion process. We discuss in Section 5 which additional ideas are needed to improve on this nonoptimal result.

4.1. Order preservation

Let $\leq$ be a partial order relation on Ω. Given $\alpha, \beta \in M_1(\Omega)$, we say that α is stochastically dominated by β (for the order $\leq$), and write $\alpha \preccurlyeq \beta$, if one can construct – on the same probability space – a pair of Ω-valued variables Z_α and Z_β with respective distributions α and β such that we have $Z_\alpha \leq Z_\beta$ with probability one. A Markov chain with generator $\mathcal{L}$ is said to be *order preserving* or *attractive* if its semigroup preserves stochastic ordering, that is, for any $t > 0$,

$$\alpha \preccurlyeq \beta \Rightarrow \alpha P_t \preccurlyeq \beta P_t.$$

An equivalent way of saying this is that the dynamic is order preserving if, for any $x, y \in \Omega$ such that $x \preccurlyeq y$, one can couple two Markov chains $(X_t^x)_{t\geq 0}$ and $(X_t^y)_{t\geq 0}$ with respective initial conditions x and y in such a way that

$$\forall t \geq 0, \quad X_t^x \leq X_t^y.$$

Order preservation for the corner-flip dynamics. We define $\leq$ on $\Xi_{N,k}$ to simply be the coordinatewise order, that is,

$$\zeta \leq \zeta' \quad \Leftrightarrow \quad \forall i \in [\![1, 2N-1]\!], \quad \zeta(i) \leq \zeta'(i). \tag{4.1}$$

To show that the corner-flip dynamics on $\Xi_{N,k}$ is order preserving, we use a construction which is similar to that presented above in equation (2.4), using clock processes $(\mathcal{T}_n^{(i)})_{i\in[\![1,N-1]\!],n\geq 0}$ (independent Poisson processes with mean-1 interarrival law) and accessory variables $(U_n^{(i)})_{i\in[\![1,N-1]\!],n\geq 0}$ which are i.i.d. uniform variables on the interval $[0,1)$. The clock processes $(\mathcal{T}^{(i)})_{n\geq 0}$ determine when the updates of coordinate i are performed, and the variables $U_n^{(i)}$ are used to determine whether the corner should be flipped up or down. Given $\zeta \in \Xi_{N,k}$, we construct (h_t^ζ) as the unique càdlàg process which satisfies:

(i) $h_0^\zeta = \zeta$,

(ii) $(h_t^\zeta)_{t\geq 0}$ remains constant on the intervals of $\mathbb{R}_+ \setminus (\mathcal{T}_n^{(i)})_{i\in[\![1,N-1]\!],n\geq 0}$;

(iii) If $t = \mathcal{T}_n^{(i)}$ and $h_{t_-}^\zeta = \xi$, then

 (A) if $U_n^{(i)} \in [1-p, 1)$ set $h_t = \xi^{(i,+)}$,

 (B) if $U_n^{(i)} \in [0, 1-p)$ set $h_t = \xi^{(i,-)}$.

Since the sets $\{\mathcal{T}_n^{(i)}\}_{i\in[\![1,N-1]\!],n\geq 0}$ display no accumulation points, (h_t^ζ) can be constructed by performing the updates sequentially. We can use this construction (using the same $\mathcal{T}$ and U) to obtain a collection of processes (h_t^ζ), $\zeta \in \Xi_{N,k}$, constructed on the same probability space, such that

$$\zeta \leq \zeta' \quad \Rightarrow \quad \forall t \geq 0, \quad h_t^\zeta \leq h_t^{\zeta'}. \tag{4.2}$$

The validity of (4.2) follows from the fact that each update is order preserving, which holds true because, for any fixed i, the applications $\zeta \mapsto \zeta^{(i,\pm)}$ are order preserving. A coupling such as that presented above, where chains starting from all initial conditions are constructed on a common probability space, is called a *grand coupling*. This type of construction using an auxiliary variable is called the *graphical construction* and is quite common for interacting particle or spin systems (see, for instance, [32, CHAPTER III.6]).

Remark 4.1. For the interchange process, we can use the order which corresponds to (4.1) after applying the correspondences of Section 2.5, and a similar construction allows obtaining a monotone grand coupling. An analogous construction also provides a monotone grand coupling for the random walk on the simplex.

4.2. Connection with the discrete heat equation

Let us expose first how the evolution of the mean of simple observables – the height function in the symmetric case, the exponential of the height in the asymmetric case – can be described by a simple system of linear equations.

4.2.1. The symmetric case

Given $\zeta \in \Xi_{N,k}$, we define $u^{\zeta}(t,\cdot)$ to be the recentered mean height of the interface at time t for the corner-flip dynamics with initial condition ζ,

$$u^{\zeta}(t,i) := \mathbb{E}\big[h_t^{\zeta}(i)\big]. \tag{4.3}$$

For a real-valued function f defined on $[\![1, N-1]\!]$, we define $\Delta_D f$ (Δ_D being the discrete Laplace operator with Dirichlet boundary condition) by

$$\Delta_D f(i) := f(i+1) + f(i-1) - 2f(i) \quad \text{for } i \in [\![1, N-1]\!], \tag{4.4}$$

with the convention that $f(0) = 0$ and $f(N) = N - 2k$. The function u^{ζ} is the unique solution of the following system of differential equations that can be considered as a partial differential equation where the space variable is discrete (Δ_D acts on the second variable)

$$\partial_t u(t,i) = \frac{1}{2}\Delta_D u(t,i), \quad \forall i \in [\![1, N-1]\!]. \tag{4.5}$$

Setting $U^{(i)}(\zeta) := \zeta(i)$, equation (4.5) is deduced from the identity (that can be checked from the definition of the generator), namely

$$\mathfrak{L}_{N,k} U^{(i)}(\zeta, i) = \frac{1}{2}\Delta_D \zeta(i).$$

More precisely, (4.5) is obtained by combining (1.10), the Markov property, the above identity, and the fact that Δ_D, being an affine transformation, commutes with the expectation, as follows:

$$\partial_t u(t,i) = \mathbb{E}\big[\mathfrak{L}_{N,k} U^{(i)}(h_t^{\zeta})\big] = \mathbb{E}\big[\Delta_D h_t^{\zeta}(i)\big] = \Delta_D\big(\mathbb{E}\big[h_t^{\zeta}\big]\big)(i) = \Delta_D u(t,i). \tag{4.6}$$

The fact that u^{ζ} does not satisfy the zero boundary condition is not a problem since, in computations, we consider the difference $u^{\zeta} - u^{\zeta'}$ which displays the zero boundary condition. The Dirichlet Laplacian with the zero boundary condition $\Delta_D^{(0)}$ is a linear operator that can easily be diagonalized. The family $(\overline{\sin}^{(j)})_{j=1}^{N-1}$ defined by $\overline{\sin}^{(j)}(i) := \sin(\frac{ij\pi}{N})$ forms a base of eigenvectors of $\Delta_D^{(0)}$ in $\mathbb{R}^N$, and we have

$$\Delta_D^{(0)}\overline{\sin}^{(j)} = -2\gamma_N^{(j)}\overline{\sin}^{(j)} \quad \text{where } \gamma_N^{(j)} = 1 - \cos\left(\frac{j\pi}{N}\right). \tag{4.7}$$

Using Parceval's inequality, we obtain the following contractive estimates, which we use to bound the mixing time.

Lemma 4.2. *If $u : [0,\infty) \times [\![1, N-1]\!]$ satisfies $\partial_t u = \Delta_D^{(0)} u$, then we have, for any $t \geq 0$,*

$$\sum_{i=1}^{N-1} u(t,i)^2 \leq e^{-2\gamma_1^{(N)} t} \sum_{i=1}^{N-1} u(0,i)^2.$$

4.2.2. The asymmetric case

When $p \neq 1/2$, the quantity $\mathfrak{L}_{N,k}^{(p)} U^{(i)}(\zeta)$ cannot be expressed as a linear combination of $U^{(j)}(\zeta)$, $j \in [\![1, N]\!]$ so that there is no way to recover a linear system analogous to (4.5) for the averaged heights.

However, we can obtain something similar for the evolution of an averaged quantity related to the heights. The key idea which can be traced back to [16] (where it is used to derive hydrodynamic limits) is to apply the so-called discrete Cole–Hopf transform. We consider exponentials of heights rather the than heights themselves. Recalling that $\lambda = p/q$, we define

$$V(\zeta, i) := \lambda^{\frac{1}{2}\zeta(i)} \quad \text{and} \quad v^{\zeta}(t, i) := \mathbb{E}^{(p)}\big[V\big(h_t^{\zeta}\big)(i)\big].$$

Setting $\varrho := (\sqrt{p} - \sqrt{q})^2$, it can be checked from the definition of the generator that for every ζ and $i \in [\![1, N-1]\!]$, we have

$$\mathfrak{L}_{N,k}^{(p)} V(\zeta, i) = \sqrt{pq}\Delta_D V(\zeta, i) - \varrho V(\zeta, i), \tag{4.8}$$

where this time Δ_D denotes the Dirichlet Laplacian defined as in (4.4) but with the boundary condition $f(0) = 1$ and $f(N) = \lambda^{\frac{N}{2}-k}$ (we refer to [22, SECTION 3.3] for details on the computation leading to (4.8)). In (4.8) note that $\mathfrak{L}_{N,k}^{(p)}$ acts on the first coordinate while Δ_D acts on the second. As in (4.6), we obtain from (4.8) that v^{ζ} satisfies

$$\partial_t v(t, i) := (\sqrt{pq}\Delta_D - \varrho)v(t, i), \quad \forall i \in [\![1, N-1]\!]. \tag{4.9}$$

Again, the nonzero boundary condition for Δ_D here is of no importance since in practice we are going to consider the difference $v^{\zeta} - v^{\zeta'}$. As in the symmetric case, the diagonalization of the operator with the zero boundary condition $\Delta_D^{(0)}$ yields the following estimate.

Lemma 4.3. *If v satisfies $\partial_t v = \sqrt{pq}\Delta_D^{(0)} v - \varrho v$ then we have*

$$\sum_{i=1}^{N-1} v(t, i)^2 \le e^{-2(\gamma_1^{(N)}+\varrho)t} \sum_{i=1}^{N-1} v(0, i)^2.$$

4.3. Using the heat equations to obtain bounds on the mixing time

Let $(h_t^{(1)})$ and $(h_t^{(2)})$ be two *ordered* corner flip dynamics, that is, such that $h_t^{(1)} \le h_t^{(2)}$ for all t. Using only Lemmas 4.2, 4.3, and order preservation, we can control the coupling time of $(h_t^{(1)})$ and $(h_t^{(2)})$ defined by

$$\tau := \inf\{t > 0 : h_t^{(1)} = h_t^{(2)}\}. \tag{4.10}$$

Proposition 4.4. *If $(h_t^{(1)})$ and $(h_t^{(2)})$ are two ordered symmetric corner flip dynamics then, for any $t > 0$, we have*

$$\mathbb{P}[\tau > t] \le k(N-1)e^{-\gamma_1^{(N)} t}.$$

If $(h_t^{(1)})$ and $(h_t^{(2)})$ are two ordered asymmetric corner flip dynamics with parameter p, then we have

$$\mathbb{P}[\tau > t] \le k(N-1)\lambda^{N/2-1} e^{-\varrho t}.$$

From these coupling estimates, we can derive upper estimates on the mixing time.

Corollary 4.5. *We have*

$$T_{\text{mix}}^{\text{SSEP},(N,k_N)} \le \frac{1}{\gamma_1^{(N)}} \log\left(\frac{2k(N-1)}{\varepsilon}\right),$$

$$T_{\text{mix}}^{\text{ASEP},(p,N,k_N)} \le \frac{1}{\varrho}\left[\left(\frac{N}{2} - 1\right)\log\lambda + \log\left(\frac{2k(N-1)}{\varepsilon}\right)\right].$$

Remark 4.6. Replacing $\gamma_1^{(N)}$ by an asymptotic equivalent $(\frac{\pi^2}{2N^2})$, we find that the upper bound on the SSEP mixing time is $\frac{2N^2}{\pi^2}(\log N + \log k)(1 + o(1))$ which is, in the best case, a factor of 4 away from the estimate given in Theorem 3.1. For the ASEP, our upper bound is asymptotically equivalent to $\frac{\log \lambda}{2\varrho} N$. Since we have, for every $p \in (1/2, 1)$,

$$\frac{\log \lambda}{2\varrho} > \frac{2}{2p-1} = \max_{\alpha \in [0,1]} \frac{(\sqrt{\alpha} + \sqrt{1-\alpha})^2}{2p-1},$$

in this case again the estimate is not sharp. The reason why the bounds in Corollary 4.5 are not sharp is further discussed in Section 5.

Proof of Corollary 4.5. Using the correspondence of Section 2.5, we can reason with the corner flip dynamics since it has the same mixing time. In order to prove an upper bound on the mixing time, one must bound from above the distance between $\mathbb{P}[h_t^\zeta \in \cdot]$ and the stationary measure π for an arbitrary $\zeta \in \Xi_{N,k}$. In order to transform this into a coupling problem, note that $\pi = \mathbb{P}[h_t^\pi \in \cdot]$ where, with some abuse of notation, we let h_t^π denote a Markov chain with initial condition π.

Let us consider now three different dynamics, $h_t^\zeta, h_t^\wedge$, and h_t^π, with respective initial conditions ζ, $\wedge$ (defined in (2.3)), and stationary distribution. They are constructed on the same probability space and coupled in such a way that, for all $t \geq 0$ (Section 4.1 gives such a coupling),

$$h_t^\pi \leq h_t^\wedge \quad \text{and} \quad h_t^\zeta \leq h_t^\wedge.$$

Using (1.2), stationarity, and union bound, we have

$$\|\mathbb{P}[h_t^\zeta \in \cdot] - \pi\|_{\mathrm{TV}} \leq \mathbb{P}[h_t^\zeta \neq h_t^\pi] \leq \mathbb{P}[h_t^\zeta \neq h_t^\wedge] + \mathbb{P}[h_t^\pi \neq h_t^\wedge] = \mathbb{P}[\tau_1 > t] + \mathbb{P}[\tau_2 > t],$$

where we have set

$$\tau_1 := \inf\{t : h_t^\zeta \neq h_t^\wedge\} \quad \text{and} \quad \tau_2 := \inf\{t : h_t^\pi \neq h_t^\wedge\}. \tag{4.11}$$

The tail distributions of τ_1 and τ_2 can be estimated using Proposition 4.4, and we obtain (let us now for the first time highlight the difference in p)

$$\begin{cases} \|\mathbb{P}[h_t^\zeta \in \cdot] - \pi\|_{\mathrm{TV}} \leq 2(N-1)ke^{-\gamma_1^{(N)}t} & \text{in the symmetric case,} \\ \|\mathbb{P}[h_t^\zeta \in \cdot] - \pi\|_{\mathrm{TV}} \leq 2(N-1)k\lambda^{N/2-1}e^{-\varrho t} & \text{in the asymmetric case.} \end{cases} \tag{4.12}$$

The reader can then check that the value of t which makes the right-hand side in (4.12) equal to ε is the claimed upper bound on the mixing time. ■

Proof of Proposition 4.4. Let us start with the symmetric case. We set

$$h_t^{(1,2)}(i) := h_t^{(2)}(i) - h_t^{(1)}(i) \quad \text{and} \quad u^{(1,2)}(t,i) := \mathbb{E}[(h_t^{(2)} - h_t^{(1)})(i)].$$

Since $h_t^{(1)} \leq h_t^{(2)}$, we have $h_t^{(1,2)}(i) \geq 0$ for all i and, if $h_t^{(1)} \neq h_t^{(2)}$, the inequality must be strict for at least one value of i. Since the minimal discrepancy between two values of $\zeta(i)$ is 2, this implies that

$$\mathbb{P}[\tau > t] = \mathbb{P}[h_t^{(1)} \neq h_t^{(2)}] = \mathbb{P}\left[\sum_{i=1}^{N-1} h_t^{(1,2)}(i) \geq 2\right] \leq \frac{1}{2}\sum_{n=1}^{N-1} u^{(1,2)}(t,i). \tag{4.13}$$

Combining Cauchy–Schwarz inequality with Lemma 4.2 – from (4.5) we know that $u^{(1,2)}$ satisfies the assumption – we have

$$\sum_{i=1}^{N-1} u^{(1,2)}(t,i) \le \left((N-1)\sum_{i=1}^{N-1} u^{(1,2)}(t,i)^2\right)^{1/2} \le e^{-\gamma_1^{(N)}t}\left((N-1)\sum_{i=1}^{N-1} u^{(1,2)}(0,i)^2\right)^{1/2},$$

and we can conclude using the fact $u^{(1,2)}(0,i) \le 2k$ since $2k$ is a bound for the maximal height difference between two elements in $\Xi_{N,k}$.

For the asymmetric case, we apply the reasoning to the exponential of the heights

$$W(\zeta) := \sum_{i=1}^{N-1} V(\zeta,i) \quad \text{and} \quad W_t^{(1,2)} := W(h_t^{(2)}) - W(h_t^{(1)}). \tag{4.14}$$

Note that, since $\zeta(i) \ge -k$ for all ζ and i, the minimal positive value of $W_t^{(1,2)}$ is given by

$$\delta_{\min} := \min_{\substack{\zeta' \ge \zeta \\ \zeta' \ne \zeta}} W(\zeta') - W(\zeta) = (\lambda - 1)\lambda^{-k/2}.$$

Repeating the reasoning in (4.13) in the asymmetric case, we obtain that

$$\mathbb{P}[\tau > t] \le \frac{\mathbb{E}[W_t^{(1,2)}]}{\delta_{\min}}.$$

Now from (4.9), $v^{(1,2)}(t,i) := \mathbb{E}[V(h_t^{(2)},i) - V(h_t^{(1)},i)]$ satisfies the assumptions of Lemma 4.3. We obtain, using Cauchy–Schwarz inequality, that

$$\mathbb{E}[W_t^{(1,2)}] \le \sum_{i=1}^{N-1} v^{(1,2)}(t,i) \le e^{-\varrho t}\left((N-1)\sum_{i=1}^{N-1} v^{(1,2)}(0,i)^2\right)^{1/2}.$$

Now considering that the maximal possible height difference is $2k$ and that the maximal possible value of $\zeta(i)$ is always smaller than $N-k$, we have, for every $i \in [\![1, N-1]\!]$,

$$v^{(1,2)}(0,i) \le \max_{\zeta' \in \Xi_{N,k}} \lambda^{\frac{\zeta'(i)}{2}}(1 - \lambda^{-k}) \le k(\lambda - 1)\lambda^{\frac{N-k}{2}-1}.$$

Setting $\delta_{\max} := (\lambda - 1)\lambda^{\frac{N-k}{2}-1}$, we obtain that $\sum_{i=1}^{N-1} v^{(1,2)}(0,i)^2 \le \delta_{\max}^2 (N-1)k^2$ so that

$$\mathbb{P}[\tau > t] \le \frac{\delta_{\max}}{\delta_{\min}}(N-1)ke^{-\varrho t},$$

which is the desired result. ■

Remark 4.7. Note that the argument exposed in this section can also be used without changes for the interchange process. Indeed, the correspondences exposed in Section 2.5 allow us to associate, to the dynamics σ_t, $N-1$ corner-flip dynamics $(h_t^{(k)})$, $k = 1, \dots, N$, defined by

$$h_t^{(k)} = h \circ \xi^{(k)} \circ \sigma_t,$$

where the transformations h and $\xi^{(k)}$ are those of Section 2.5. The observation (2.6) guarantees that two dynamics $\sigma_t^{(1)}$ and $\sigma_t^{(2)}$ are coupled when all the corresponding corner-flip dynamics are coupled, so that the analog of Proposition 4.4 is valid for the interchange process on the segment, with the factor $k(N-1)$ replaced by $(N-1)^3$. The reader can refer to [46, SECTION 3] and [22, SECTION 3.4] for more details in the symmetric and asymmetric cases, respectively.

5. SHORTCOMINGS AND POSSIBLE IMPROVEMENTS OF THE REASONING ABOVE

5.1. Symmetric dynamics

As mentioned in Remark 4.6, the upper-bound on the SSEP mixing time is suboptimal, off by a factor of 4 in the case when k and $N-k$ are of order N. There are two separate reasons for which the method does not yield an optimal result, each being accountable for a multiplicative factor of 2. To illustrate this, let us mention [46, SECTION 8], where it is proved that for the monotone coupling inherited from the graphical construction (described in Section 4.1), the coupling time τ_1 in (4.11) is of order $\frac{2}{\pi^2}N^2 \log k$. This results shows that not only the method above is off by a factor of 2 to estimate the coupling time, but also, compared to Theorem 3.1, that this coupling time itself does allow for a sharp estimate on the mixing time. This means that in order to improve the bound on the mixing time, we have to design a monotone coupling that makes the value of the coupling time τ as small as possible.

This becomes particularly obvious when the random walk on the simplex is considered (recall Section 2.4). If one considers the monotone grand coupling based on the graphical construction presented in Section 2.4, then trajectories starting with different initial conditions *never* coalesce ($\tau = \infty$ almost surely). Hence for this model, there can be no equivalent of Proposition 4.4: any nontrivial estimate of τ must rely on specific features of the coupling beyond monotonicity.

In [4,5,23–26], refinements have been performed in order to obtain optimal estimates on the mixing time. This first one is the introduction of a coupling that is aimed at minimizing the coalescence time. The basic idea for the discrete model is to make the corner-flips performed by $h_t^{(1)}$ and $h_t^{(2)}$ less synchronized while preserving monotonicity so that the quantity

$$A(t) := \sum_{i=1}^{N-1} \left(h_t^{(2)} - h_t^{(1)}\right),$$

which is an integer-valued supermartingale, hits zero faster. Roughly speaking, this is achieved by having, at any given time, independent corner flips for coordinates at which $h_t^{(2)}(x) > h_t^{(1)}(x)$, and synchronized corner flips for coordinates at which $h_t^{(2)}(x) = h_t^{(1)}(x)$ (the couplings used in the continuous setup in [4,5] are based on an analogous intuition). The second key improvement is to use diffusion estimates in order to estimate the time when $A(t)$ hits 0, instead of relying on Markov's inequality. For the corner-flip dynamics, $A(t)$ is a time-changed random walk on $\mathbb{Z}_+$, and the hitting time of 0 can be precisely estimated if one has some control over its jump rate (see [24–26]). This idea was considerably improved in [4,5,23] where we need to estimate the hitting time of zero of a supermartingale which is not integer-valued. The improvement comes from reasoning in terms of martingale brackets instead of jump rate.

5.2. Asymmetric dynamics

Remark 4.6 also underlines that the result of the previous section is also suboptimal in the asymmetric case. The reason for this is that the quantity $W_t^{(2,1)}$ considered in (4.14) is typically much smaller than its average (by a factor which is exponential in N). Since this quantity has very wild fluctuation, it is not possible to apply to it the same technique as in the symmetric case. The proof of Theorem 3.2 presented in [22] relies on two key ingredients:

(A) Hydrodynamic limits;

(B) The control of particle speed when the density is vanishing.

Hydrodynamic limits are an extensively studied topic for particle systems (see [21]). The hydrodynamic limit of a system is the limit obtained for the evolution of the particle density after rescaling time and space. It usually takes the form of the solution to partial differential equation. In the case of the asymmetric exclusion process, is has been established (see [40] where the result is proved in a much broader context) that the hydrodynamic limit – after rescaling time and space by N – is the solution of the equation

$$\partial_t \rho = (2p-1)\partial_x\big[\rho(1-\rho)\big]. \tag{5.1}$$

More precisely, for the exclusion on the segment, we have to consider some specific notion of a solution and boundary conditions (see [22, SECTION 5] for details). In this context, given any initial condition ρ_0, which satisfies

$$\forall x \in [0,1],\ 0 \leq \rho_0(x) \leq 1 \quad \text{and} \quad \int_{[0,1]} \rho(x) = \alpha,$$

(5.1) has a unique solution which stabilizes to the fixed point $\mathbf{1}_{[1-\alpha,1]}$ after a time $\frac{(\sqrt{\alpha}+\sqrt{1-\alpha})^2}{2p-1}$, indicating that at time $\frac{(\sqrt{\alpha}+\sqrt{1-\alpha})^2 N}{2p-1}$ the system is macroscopically at equilibrium.

What remains to check afterwards is whether around that time the system is also at equilibrium in the total variation sense, which is *a priori* a much finer statement. The important point is to verify that the position of the leftmost particle and rightmost empty site match the indication given by the macroscopic profile (that is, are both $(1-\alpha)N + o(N)$), and this is where the point (B) comes into play (we refer to [22, SECTION 6] for more details).

Once we have proved that both the density of particles and the position of the leftmost particle/rightmost empty site have reached their equilibrium, we still have not proved that the system is at equilibrium. However, this information implies that with the notation of Section 4.3, when $t = t_{\alpha,N} := \frac{(\sqrt{\alpha}+\sqrt{1-\alpha})^2}{2p-1}$, we have $W_t^{(1,2)} = \exp(o(N))\delta_{\min}$. Hence we can use, as a third step of our reasoning, the contraction estimate of Lemma 4.3 to show that a coupling must occur shortly after time $t_{\alpha,N}$.

ACKNOWLEDGMENTS

The author is gratefull to P. Caputo and C. Labbé for feedback on the manuscript. The author acknowledges support from a productivity grant from CNPq and from a JCNE grant from FAPERj.

REFERENCES

[1] J. Baik and Z. Liu, Fluctuations of TASEP on a ring in relaxation time scale. *Comm. Pure Appl. Math.* **71** (2018), no. 4, 747–813.

[2] I. Benjamini, N. Berger, C. Hoffman, and E. Mossel, Mixing times of the biased card shuffling and the asymmetric exclusion process. *Trans. Amer. Math. Soc.* **357** (2005), no. 8, 3013–3029 (electronic).

[3] A. M. Bufetov and P. Nejjar, Cutoff profile of ASEP on a segment. 2020, arXiv:2012.14924.

[4] P. Caputo, C. Labbé, and H. Lacoin, Mixing time of the adjacent walk on the simplex. *Ann. Probab.* **48** (2020), no. 5, 2449–2493.

[5] P. Caputo, C. Labbé, and H. Lacoin, Spectral gap and cutoff phenomenon for the Gibbs sampler of $\nabla\varphi$ interfaces with convex potential. *Ann. Inst. Henri Poincaré Probab. Stat.* (to appear).

[6] P. Caputo, H. Lacoin, F. Martinelli, F. Simenhaus, and F. L. Toninelli, Polymer dynamics in the depinned phase: metastability with logarithmic barriers. *Probab. Theory Related Fields* **153** (2012), no. 3–4, 587–641.

[7] P. Caputo, T. M. Liggett, and T. Richthammer, Proof of Aldous' spectral gap conjecture. *J. Amer. Math. Soc.* **23** (2010), no. 3, 831–851.

[8] P. Caputo, E. Lubetzky, F. Martinelli, A. Sly, and F. L. Toninelli, Dynamics of $(2+1)$-dimensional SOS surfaces above a wall: slow mixing induced by entropic repulsion. *Ann. Probab.* **42** (2014), no. 4, 1516–1589.

[9] P. Caputo, F. Martinelli, F. Simenhaus, and F. L. Toninelli, "Zero" temperature stochastic 3D Ising model and dimer covering fluctuations: a first step towards interface mean curvature motion. *Comm. Pure Appl. Math.* **64** (2011), no. 6, 778–831.

[10] P. Caputo, F. Martinelli, and F. L. Toninelli, On the approach to equilibrium for a polymer with adsorption and repulsion. *Electron. J. Probab.* **13** (2008), no. 10, 213–258.

[11] P. Caputo, F. Martinelli, and F. L. Toninelli, Mixing times of monotone surfaces and SOS interfaces: a mean curvature approach. *Comm. Math. Phys.* **311** (2012), no. 1, 157–189.

[12] P. Diaconis, The cutoff phenomenon in finite Markov chains. *Proc. Natl. Acad. Sci. USA* **93** (1996), no. 4, 1659–1664.

[13] S. Ganguly and R. Gheissari, Cutoff for the Glauber dynamics of the lattice free field. 2021, arXiv:2108.07791.

[14] N. Gantert and T. Kochler, Cutoff and mixing time for transient random walks in random environments. *ALEA Lat. Am. J. Probab. Math. Stat.* **10** (2013), no. 1, 449–484.

[15] N. Gantert, E. Nestoridi, and D. Schmid, Mixing times for the simple exclusion process with open boundaries. 2020, arXiv:2003.03781.

[16] J. Gärtner, Convergence towards Burger's equation and propagation of chaos for weakly asymmetric exclusion processes. *Stochastic Process. Appl.* **27** (1988), 233–260.

[17] S. Greenberg, D. Randall, and A. P. Streib, Sampling biased monotonic surfaces using exponential metrics. *Combin. Probab. Comput.* **29** (2020), no. 5, 672–697.

[18] J. Hermon and J. Salez, The interchange process on high-dimensional products. *Ann. Appl. Probab.* **31** (2021), no. 1, 84–98.

[19] B. Hough and Y. Jiang, Cut-off phenomenon in the uniform plane Kac walk. *Ann. Probab.* **45** (2017), no. 4, 2248–2308.

[20] H. Kesten, M. V. Kozlov, and F. Spitzer, A limit law for random walk in a random environment. *Compos. Math.* **30** (1975), 145–168.

[21] C. Kipnis and C. Landim, *Scaling limits of interacting particle systems*. Grundlehren Math. Wiss. 320, Springer, Berlin, 1999.

[22] C. Labbé and H. Lacoin, Cutoff phenomenon for the asymmetric simple exclusion process and the biased card shuffling. *Ann. Probab.* **47** (2019), no. 3, 1541–1586.

[23] C. Labbé and H. Lacoin, Mixing time and cutoff for the weakly asymmetric simple exclusion process. *Ann. Appl. Probab.* **30** (2020), no. 4, 1847–1883.

[24] H. Lacoin, The cutoff profile for the simple exclusion process on the circle. *Ann. Probab.* **44** (2016), no. 5, 3399–3430.

[25] H. Lacoin, Mixing time and cutoff for the adjacent transposition shuffle and the simple exclusion. *Ann. Probab.* **44** (2016), no. 2, 1426–1487.

[26] H. Lacoin, The simple exclusion process on the circle has a diffusive cutoff window. *Ann. Inst. Henri Poincaré Probab. Stat.* **53** (2017), no. 3, 1402–1437.

[27] H. Lacoin and A. Teixeira, A mathematical perspective on metastable wetting. *Electron. J. Probab.* **20** (2015), no. 17, 23.

[28] H. Lacoin and S. Yang, Metastability for expanding bubbles on a sticky substrate. 2020, arXiv:2007.07832.

[29] H. Lacoin and S. Yang, Mixing time for the asymmetric simple exclusion process in a random environment. 2021, arXiv:2102.02606.

[30] D. A. Levin and Y. Peres, Mixing of the exclusion process with small bias. *J. Stat. Phys.* **165** (2016), no. 6, 1036–1050.

[31] D. A. Levin and Y. Peres, *Markov chains and mixing times*. 2nd edn., MBK, American Mathematical Society, 2017.

[32] T. M. Liggett, *Interacting particle systems*. Classics Math., Springer, 2005.

[33] R. Lyons and Y. Peres, *Probability on trees and networks*. Camb. Ser. Stat. Probab. Math. 42, Cambridge University Press, New York, 2016.

[34] B. Morris, The mixing time for simple exclusion. *Ann. Appl. Probab.* **16** (2006), no. 2, 615–635.

[35] R. I. Oliveira, Mixing of the symmetric exclusion processes in terms of the corresponding single-particle random walk. *Ann. Probab.* **41** (2013), no. 2, 871–913.

[36] J. Quastel, Diffusion of color in the simple exclusion process. *Comm. Pure Appl. Math.* **45** (1992), no. 6, 623–679.

[37] D. Randall and P. Winkler, Mixing points on a circle. In *Approximation, randomization and combinatorial optimization. algorithms and techniques*, pp. 426–435, Springer, 2005.

[38] D. Randall and P. Winkler, Mixing points on an interval. In *Proceedings of the second workshop on analytic algorithms and combinatorics, vancouver, 2005*, pp. 216–221, 2005.

[39] D. Revuz, *Markov chains*. 2nd edn., N.-Holl. Math. Libr. 11, North-Holland Publishing Co., Amsterdam, 1984.

[40] F. Rezakhanlou, Hydrodynamic limit for attractive particle systems on $\mathbf{Z}^d$. *Comm. Math. Phys.* **140** (1991), no. 3, 417–448.

[41] L. C. G. Rogers and D. Williams, *Diffusions, Markov processes, and martingales*. 2nd edn., Cambridge Math. Lib. 1, Cambridge University Press, 2000.

[42] D. Schmid, Mixing times for the simple exclusion process in ballistic random environment. *Electron. J. Probab.* **24** (2019), Paper No. 22, 25.

[43] D. Schmid, Mixing times for the TASEP in the maximal current phase. 2021, arXiv:2104.12745.

[44] F. Solomon, Random walks in a random environment. *Ann. Probab.* **3** (1975), 1–31.

[45] D. W. Stroock, *An introduction to Markov processes*. 2nd edn., Grad. Texts in Math. 230, Springer, Berlin–Heidelberg, 2014.

[46] D. B. Wilson, Mixing times of Lozenge tiling and card shuffling Markov chains. *Ann. Appl. Probab.* **14** (2004), no. 1, 274–325.

[47] S. Yang, Cutoff for polymer pinning dynamics in the repulsive phase. *Ann. Inst. Henri Poincaré Probab. Stat.* **57** (2021), no. 3, 1306–1335.

[48] H.-T. Yau, Logarithmic Sobolev inequality for generalized simple exclusion processes. *Probab. Theory Related Fields* **109** (1997), no. 4, 507–538.

HUBERT LACOIN

IMPA – Estrada Dona Castorina, 110 Rio de Janeiro – 22460-320 – RJ, Brazil, lacoin@impa.br

This Page Intentionally Left Blank

ULTRAMETRICITY IN SPIN GLASSES

DMITRY PANCHENKO

ABSTRACT

Ultrametricity of the Gibbs measure is a fundamental feature of the Parisi solution of the Sherrington–Kirkpatrick model of spin glasses. We will start by describing one origin of ultrametricity in a way that requires no special knowledge, and after that review some background and discuss some applications.

MATHEMATICS SUBJECT CLASSIFICATION 2020

Primary 82B44; Secondary 60K35

KEYWORDS

Spin glasses, Parisi ansatz, ultrametricity

ICM INTERNATIONAL CONGRESS OF MATHEMATICIANS 2022 JULY 6–14

Proc. Int. Cong. Math. 2022, Vol. 6, pp. 4376–4392
DOI 10.4171/ICM2022/69

1. INTRODUCTION

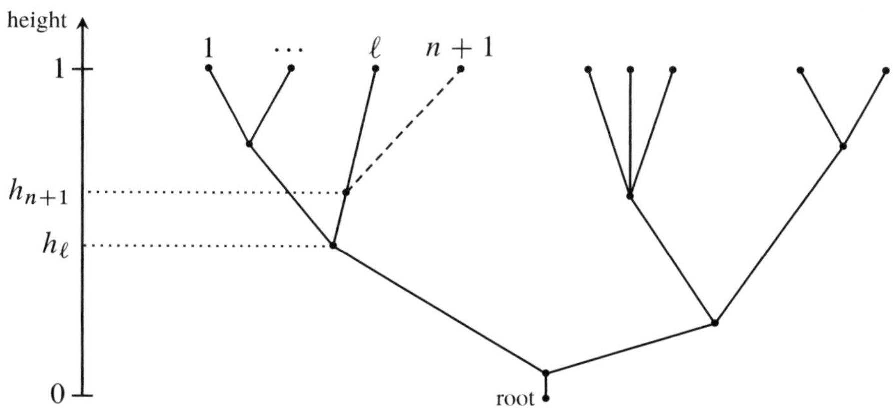

FIGURE 1
New leaf $n+1$ is attached to the path from a randomly chosen leaf $\ell \in \{1, \dots, n\}$ to the root. Because $h_{n+1} := \max(t_{n+1}, h_\ell) \geq h_\ell$, it is attached at or above the height h_ℓ where the leaf ℓ was originally attached to the tree.

Given a probability distribution ζ on $[0, 1]$, let us generate a tree with the root at height 0 and countably many leaves at height 1 using the following simple sequential process (see Figure 1). The height of any point on the tree refers to its coordinate on the vertical axis (labeled "height" in the figure). In this process, h_ℓ will represent the height at which leaf ℓ was attached to the tree (during its turn). Proceed as follows:

(1) Attach leaf 1 by a new branch from the root and set $h_1 = 0$.

(2) For $n+1 = 2, 3, \dots$, repeat the following steps:

(2a) Pick a leaf $\ell \in \{1, \dots, n\}$ uniformly at random.

(2b) Generate a new random variable t_{n+1} from the distribution ζ.

(2c) Attach the leaf $n+1$ to the path from the chosen leaf ℓ to the root at the height $h_{n+1} := \max(t_{n+1}, h_\ell)$, by adding a new branch.

For any two leaves $\ell, \ell' \geq 1$, let $R_{\ell,\ell'}$ be the height at which the paths from these leaves to the root meet. We will call $R := (R_{\ell,\ell'})_{\ell,\ell' \geq 1}$ the *overlap array*, since these quantities measure how much the paths overlap. We will denote by $R^n := (R_{\ell,\ell'})_{\ell,\ell' \leq n}$ the $n \times n$ block of overlaps corresponding to the first n leaves. The array R satisfies the following three properties, which we list from the most obvious to less obvious:

(a) The conditional distribution of the overlap $R_{1,n+1}$ given R^n is equal to

$$\mathcal{L}(R_{1,n+1} \mid R^n) = \frac{1}{n}\zeta + \frac{1}{n}\sum_{\ell=2}^{n} \delta_{R_{1,\ell}}. \tag{1.1}$$

(b) The array R is nonnegative definite, with diagonal elements $R_{\ell,\ell} = 1$ and off-diagonal elements $R_{\ell,\ell'} \in [0, 1]$.

(c) The array $(R_{\ell,\ell'})$ is *weakly exchangeable*, which means that

$$(R_{\ell,\ell'})_{\ell,\ell' \leq n} \stackrel{d}{=} (R_{\pi(\ell),\pi(\ell')})_{\ell,\ell' \leq n}$$

for any $n \geq 2$ and any permutation π of $\{1, \ldots, n\}$.

Property (a) holds because, if in step (2a) we picked $\ell = 1$ then there would be no constraint on where new branch is attached and so $R_{1,n+1} = t_{n+1}$ had distribution ζ, and if we picked $\ell \geq 2$ then $R_{1,n+1} = R_{1,\ell}$.

Property (b) can be seen in different ways, but one way to see it is to notice that, because of the tree structure, for any $q \in [0, 1]$, the relation $\ell \sim_q \ell'$ on the leaves defined by

$$\ell \sim_q \ell' \iff R_{\ell,\ell'} \geq q \tag{1.2}$$

is an equivalence relation and, therefore, the array $(\mathrm{I}(R_{\ell,\ell'} \geq q))_{\ell,\ell' \geq 1}$ is block-diagonal with entries of each block all equal to 1 and, thus, nonnegative definite. Using that $R_{\ell,\ell'} = \int_0^1 \mathrm{I}(R_{\ell,\ell'} \geq q)\, dq$, we see that the array R is also nonnegative definite. Another way to express that (1.2) is an equivalence relation is to say that

$$R_{\ell_2,\ell_3} \geq \min(R_{\ell_1,\ell_2}, R_{\ell_1,\ell_3}) \tag{1.3}$$

for any three leaves ℓ_1, ℓ_2, and ℓ_3. If property (b) holds then the array R satisfying (1.3) is called an *ultrametric* array, because a subset $\{h_\ell : \ell \geq 1\}$ of the unit sphere in a Hilbert space such that $R_{\ell,\ell'} = h_\ell \cdot h_{\ell'}$ will form an ultrametric set satisfying

$$\|\sigma_{\ell_2} - \sigma_{\ell_3}\| \leq \max(\|\sigma_{\ell_1} - \sigma_{\ell_2}\|, \|\sigma_{\ell_1} - \sigma_{\ell_3}\|).$$

One can also embed the entire tree isometrically into a unit ball of a Hilbert space, with the overlap equal to the scalar product in this embedding.

Property (c) is not obvious and requires some calculation, but it is not difficult and we will leave it as an exercise. The basic idea is that one can compute the probability of observing a finite tree in a particular configuration by "unwinding" how this tree was formed starting from clusters of closest leaves (those with the largest overlaps), and ignoring the order, because we will end up with the factor $1/n!$ no matter what the order was. So, we have this symmetry in distribution, although it does not appear immediately obvious from the construction. This also means that (1.1) also holds with indices permuted.

Notice that the properties (a), (b), and (c) do not explicitly refer to the tree structure in Figure 1. However, it turns out that these properties do imply that such a tree structure must be present, even if it is a priori not given.

Theorem 1.1 ([102]). *If the array $(R_{\ell,\ell'})_{\ell,\ell' \geq 1}$ satisfies properties (a), (b), and (c) then* (1.3) *holds and the array can be generated as in Figure* 1 *with $\zeta = \mathcal{L}(R_{1,2})$.*

The distributional identities (1.1) in property (a) are called the *Ghirlanda–Guerra identities* [71] and, in all intended applications, properties (b) and (c) are, essentially, built into the construction. So, in words, Theorem 1.1 says that the Ghirlanda–Guerra identities (1.1) imply ultrametricity (1.3). As we will discuss below, the result is useful because the Ghirlanda–Guerra identities often appear rather naturally.

Below we will discuss the origin of this result in the setting of the so-called spin glass models from statistical physics. Before we begin, let us mention that the construction of the tree in Figure 1 is called the *Goldschmidt–Martin algorithm* [72] for generating a sample from the *Ruelle Probability Cascades* [115] corresponding to the overlap distribution ζ or, equivalently, for constructing the *Bolthausen–Sznitman coalescent* [34] (see also [98]). Nonnegative definite random arrays satisfying property (c) are called *Gram–de Finetti arrays*, and the analogue of de Finetti's representation for such arrays is called the *Dovbysh–Sudakov representation* [66] (see, e.g., [103, SECTION 1.5]).

2. SOME BACKGROUND

The name "spin glass" refers to certain dilute magnetic alloys (for example, dilute solutions of manganese in copper, or other magnetic atoms in nonmagnetic metals), and it seems to have been coined by Philip Anderson and Wai-Chao Kok (according to [7]). An entertaining account of a part of the history of spin glasses in physics up to 1990 can be found in [5–11]. Here I will only mention a few fundamental results related to the Sherrington–Kirkpatrick model [117]. In this model, given integer $N \geq 1$, one considers a Gaussian process $H_N(\sigma)$ indexed by $\sigma \in \Sigma_N := \{-1, +1\}^N$,

$$H_N(\sigma) = \frac{1}{\sqrt{N}} \sum_{i,j=1}^{N} g_{ij}\sigma_i\sigma_j,$$

where the coefficients g_{ij} are i.i.d. standard Gaussian random variables. This process is called the *Hamiltonian* of the model. It is sometimes viewed as the energy function of a random optimization problem of assigning students to two dorms, called dean's problem. Let i, j be indices corresponding to N students, g_{ij} be an interaction parameter describing how much student i likes or dislikes student j, and σ_i be the label of one of two dorms $\{-1, +1\}$ that student i is assigned to. If we write $\sigma_i\sigma_j = 2\mathrm{I}(\sigma_i = \sigma_j) - 1$, we can see that maximizing $H_N(\sigma)$ over all possible assignments σ is equivalent to maximizing the so-called *comfort function* $\sum_{i\neq j} g_{ij}\mathrm{I}(\sigma_i = \sigma_j)$, which is the sum of interactions within the same dorms. It is not difficult to check that $\max_\sigma H_N(\sigma)$ is of order $\mathcal{O}(N)$, and so one may try to compute the exact limit of

$$\frac{1}{N} \max_{\sigma\in\Sigma_N} H_N(\sigma)$$

as $N \to \infty$. Related to this maximum is the *free energy*

$$F_N = F_N(\beta) := \frac{1}{N} \log \sum_{\sigma\in\Sigma_N} \exp \beta H_N(\sigma),$$

where $\beta > 0$ is called the *inverse temperature* parameter. Free energy can be viewed as a "smooth approximation" of the maximum, because

$$\frac{F_N(\beta)}{\beta} \le \frac{1}{N} \max_{\sigma \in \Sigma_N} H_N(\sigma) \le \frac{F_N(\beta)}{\beta} + \frac{\log 2}{\beta},$$

which can be seen by bounding the sum from below by the largest term, or replacing all terms by the largest one. This means that $F_N(\beta)/\beta$ is a good approximation of the maximum when β is large, so one can try to compute the limit of the free energy for any fixed β first. This turns out to be closely related to understanding the geometric and probabilistic structure of the *Gibbs measure* of the model,

$$G_N(\sigma) = \frac{\exp \beta H_N(\sigma)}{\sum_{\rho \in \Sigma_N} \exp \beta H_N(\rho)}.$$

The formula for the limit of $F_N(\beta)$ was proposed by David Sherrington and Scott Kirkpatrick in **[117]** based on the so-called replica formalism: using the formula $\log x = \lim_{n \downarrow 0} n^{-1}(x^n - 1)$, interchanging limits $N \to \infty$ and $n \to 0$, computing the $N \to \infty$ limit for integer $n \geq 0$, and hoping that the formula survives in the $n \to 0$ limit. They observed that the formula they obtained exhibited "unphysical behavior" at low temperature, which meant that it could only be correct for small enough values of β. Several years later, Giorgio Parisi **[112,113]** proposed another formula, now called the *Parisi formula*, which seemed to pass all the consistency checks. It stated that (almost surely)

$$\lim_{N \to \infty} F_N(\beta) = \inf_{\zeta} \left(\Phi(0,0) - \beta^2 \int_0^1 t\zeta(t)\, dt \right),$$

where the infimum is taken over all probability distributions $\zeta \in \Pr[0,1]$ on $[0,1]$ with $\zeta(t) := \zeta([0,t])$ being a cumulative distribution function, and $\Phi(t,x) : [0,1] \times \mathbb{R} \to \mathbb{R}$ the solution of

$$\Phi_t = -\beta^2 \big(\Phi_{xx} + \zeta(t)(\Phi_x)^2\big), \quad \Phi(1,x) := \log 2 \cosh(x).$$

Parisi's calculation started along the same lines of the replica formalism, but it required breaking from conventional wisdom in more than one way, as well as making some creative choices along the way. One of these choices was the *ultrametric parametrization* of the replica matrix that comes up in the calculation. Here this replica matrix appeared in a purely algebraic way, and it was only later given a physical meaning in another paper of Parisi **[114]**, as the matrix of *overlaps*

$$R_{\ell,\ell'} = \frac{1}{N} \sum_{i=1}^{N} \sigma_i^{\ell} \sigma_i^{\ell'} \tag{2.1}$$

of an i.i.d. sample $(\sigma^{\ell})_{\ell \geq 1}$ from the Gibbs measure G_N. The overlap array R discussed in the introduction and in Theorem 1.1 arises as a limit (in distribution) of the array (2.1) and, in the context of spin glass models, the purpose of Theorem 1.1 is to understand the distribution of this array in the thermodynamic limit $N \to \infty$. The reason why the overlaps (2.1) appear in the computation of the free energy is simple. The Hamiltonian is a Gaussian process, so its distribution is determined by its covariance, which in this case happens to be a function of the overlap,

$$\mathbb{E}H_N(\sigma^1)H_N(\sigma^2) = N(R_{1,2})^2. \tag{2.2}$$

Following the work of Parisi, there was a tremendous activity in physics using, extending, and analyzing these ideas. A classic summary of spin glasses at the end of the 1980s is the book of Mézard, Parisi, and Virasoro "Spin Glass Theory and Beyond" [91]. Some key developments appeared in a series of papers by Marc Mézard et al. [88–90, 92] in the mid-1980s, where, in particular, an algebraic choice of ultrametric parametrization in Parisi's replica calculation was expressed in terms of the familiar ultrametric geometry—in this case, the geometry of the support of the Gibbs measure in the thermodynamic limit $N \to \infty$. A very important role was also played by the study of toy models of spin glasses by Bernard Derrida et al.—the random energy model, REM, in [59, 60], and the generalized random energy model, GREM, in [61, 62]. It was shown in [58, 63, 90] that various statistics of the Gibbs sample in these toy models coincide with those in the SK model. When David Ruelle [115] gave an explicit description of the Gibbs measure in the GREM in terms of a certain family of Poisson processes, this meant that one now had an explicitly defined object conjecturally describing the Gibbs measure in the SK model. This object is now called the *Ruelle probability cascades* (RPC). For an explicit description, we will refer to Chapter 2 in [103], but Figure 1 above describes how to generate a sample from RPC corresponding to the parameter $\zeta \in \Pr[0, 1]$.

The fact that the limit of the free energy actually exists was proved by Guerra and Toninelli in [75]. The Parisi formula was proved by Michel Talagrand in a celebrated paper [126], following a discovery by Francesco Guerra [74] of an ingenious interpolation that showed that the Parisi formula is an upper bound on the limit of the free energy.

3. THE GHIRLANDA–GUERRA IDENTITIES

Talagrand's proof of the Parisi formula found a way around the ultrametricity (1.3) that played such an important role in the physics literature, but there is another approach based on the above Theorem 1.1 and the Aizenman–Sims–Starr scheme [4] (see [103, 104]). If we look at the array of overlaps (2.1), by definition, it satisfies properties (b) and (c) in Theorem 1.1, except for $R_{\ell,\ell'} \in [0, 1]$. The original SK Hamiltonian $H_N(\sigma)$ is symmetric under $\sigma \to -\sigma$, so the distribution of the overlaps is symmetric. However, there are various ways to break this symmetry in a way that enforces $R_{\ell,\ell'} \geq 0$ in the limit $N \to \infty$ without affecting the free energy much and, as a result, we can pretend that properties (b) and (c) always hold. This means that property (a)—the so-called Ghirlanda–Guerra identities—is really at the heart of Theorem 1.1. In fact, these identities also imply that $R_{\ell,\ell'} \geq 0$ (which is known as Talagrand's positivity principle), but, of course, their main role is to ensure that the ultrametricity of the overlap array in (1.3) holds.

So where are the Ghirlanda–Guerra identities (1.1) coming from? If we denote by $\langle \cdot \rangle$ the average with respect to the Gibbs measure G_N and by $\mathbb{E}$ the average with respect to the *Gaussian disorder* (g_{ij}) then, roughly speaking, the form (1.1) is simply another way to express the concentration of the Hamiltonian,

$$\mathbb{E}\left\langle \left| \frac{H_N(\sigma)}{N} - \mathbb{E}\left\langle \frac{H_N(\sigma)}{N} \right\rangle \right| \right\rangle \to 0, \tag{3.1}$$

by testing this concentration against a test function and then integrating by parts using the formula for covariance of H_N in (2.2). Precise details are a bit more complicated, but the main question becomes: Where is the concentration (3.1) coming from? The particular statement (3.1) is not easy to prove (see Section 3.7 in [103]), but, for example, the same statement on average over the inverse temperature parameter β is rather straightforward and follows readily from the convexity of $F_N(\beta)$ and its concentration around the expectation $\mathbb{E}F_N(\beta)$, as was demonstrated by Guerra in [73] and generalized by Ghirlanda and Guerra in [71]. Once the overall idea became clear, there was a lot of room to tweak this approach and make it applicable in a variety of situations.

For example, given some Hamiltonian $H_N(\sigma)$ with maximum of order $\mathcal{O}(N)$, one can add a smaller-order Gaussian perturbation term (with covariance given by a function of the overlap similarly to (2.2)) in such a way that the Ghirlanda–Guerra identities hold in the limit. The only requirement from the model is that the free energy satisfies some mild concentration assumptions. Hence, the Ghirlanda–Guerra identities become a property of the perturbation and not the model itself. Since the perturbation is of smaller order, its presence does not affect the limit of the free energy. Such perturbative approach even allows considering more general overlaps, leading to further applications as will be discussed in the next section.

Closely related to the Ghirlanda–Guerra identities is the so-called *Aizenman–Contucci stochastic stability* [3,56]. The first approach to ultrametricity using this stochastic stability was developed by Louis-Pierre Arguin and Michael Aizenman in [12], which inspired the line of research [99,101] that lead to Theorem 1.1 in [102]. The general idea of forcing nontrivial properties on a model using perturbations has emerged as one of the most important ideas on the mathematical side of spin glasses. As the physicists like to say, you can learn a lot about the system by observing how it reacts to small perturbations, and mathematicians like to think that in a small neighborhood of a system you might be able to find another one with better properties.

4. SYNCHRONIZATION MECHANISM

In this section, we will describe how Theorem 1.1 can be combined with the Ghirlanda–Guerra identities for more general overlaps to study various generalizations of the SK model. For illustration purposes, we will use the following two examples.

Example 4.1 (Nonhomogeneous SK model). In the language of dean's problem, suppose that students are divided into two groups, Girls $= \{1, \ldots, N_1\}$ and Boys $= \{N_1 + 1, \ldots, N\}$, where $N_1/N \to \lambda \in (0, 1)$, and suppose that the variance of the Gaussian interactions depends on the group membership,

$$\mathrm{Var}(g_{ij}) = \sigma^2_{S,S'}, \quad \text{if } i \in S, j \in S' \text{ for } S, S' \in \{\text{Girls, Boys}\}. \tag{4.1}$$

Otherwise, the Hamiltonian is the same as in the SK model. In this case, the computation of the free energy involves understanding the joints distribution of two types of overlaps,

$$R^S_{\ell,\ell'} = \frac{1}{N}\sum_{i\in S}\sigma_i^\ell \sigma_i^{\ell'} \quad \text{for } S \in \{\text{Girls, Boys}\},$$

over the entire array $\ell, \ell' \geq 1$.

Example 4.2 (Potts SK model). Here we consider $K \geq 2$ dorms, so assignments σ belong to $\{1, \dots, K\}^N$, and suppose that the dorm sizes are fixed,

$$\lim_{N\to\infty} \frac{|\{i : \sigma_i = k\}|}{N} = p_k \in (0,1) \quad \text{for all } k \leq K,$$

where $\sum_{k\leq K} p_k = 1$. Now it is more natural to define the Hamiltonian as

$$H_N(\sigma) = \frac{1}{\sqrt{N}} \sum_{i,j=1}^N g_{ij} \mathrm{I}(\sigma_i = \sigma_j).$$

In this case, the computation of the free energy involves understanding the joints distribution over all $\ell, \ell' \geq 1$ of the matrix of overlaps

$$R(\sigma^\ell, \sigma^{\ell'}) = \big(R_{x,y}(\sigma^\ell, \sigma^{\ell'})\big)_{x,y\in\text{Dorms}}, \tag{4.2}$$

where, for $x, y \in \text{Dorms} := \{1, \dots, K\}$,

$$R_{x,y}(\sigma^\ell, \sigma^{\ell'}) = \frac{1}{N}\sum_{i=1}^N \mathrm{I}(\sigma_i^\ell = x)\mathrm{I}(\sigma_i^{\ell'} = y),$$

which is the proportion of students assigned to dorm x in the assignment σ^ℓ and dorm y in another assignment $\sigma^{\ell'}$.

To study joint distributions of more than one type of overlap, a *synchronization mechanism* was developed in [106,110,111], which we will now describe. Let Σ_N be the space of assignments, for example, $\Sigma_N = \{-1, +1\}^N$ in the SK model and nonhomogeneous SK model, and $\Sigma_N = \{1, \dots, K\}^N$ in the Potts model. Let H be a Hilbert space and

$$\Phi_N : \Sigma_N \to H$$

be such that $\|\Phi_N(\sigma)\|_H = \text{const}$, where for simplicity we will assume that this constant in independent of N. We will call

$$R_{\ell,\ell'} = \Phi_N(\sigma^\ell) \cdot \Phi_N(\sigma^{\ell'})$$

a *generalized overlap*. Since we can define a Gaussian process with the covariance equal to this generalized overlap (or its powers), the perturbation approach we described in the previous section allows us to force any such generalized overlap to satisfy the Ghirlanda–Guerra identities and, by way of Theorem 1.1, satisfy ultrametricity in the limit $N \to \infty$.

Moreover, if we consider two generalized overlaps $R_{\ell,\ell'}$ and $Q_{\ell,\ell'}$ then $R^n_{\ell,\ell'} Q^m_{\ell,\ell'}$ will also be a generalized overlap for any integer $n, m \geq 0$, and the perturbative approach allows us to simultaneously force all of them be ultrametric in the limit. The ultrametricity puts strong constraints on how this can happen and, in fact, implies that $R_{\ell,\ell'}$ and $Q_{\ell,\ell'}$ have to be synchronized in the following sense (Theorem 4 in [106]):

$$R_{\ell,\ell'} = f(R_{\ell,\ell'} + Q_{\ell,\ell'}), \quad Q_{\ell,\ell'} = g(R_{\ell,\ell'} + Q_{\ell,\ell'}), \tag{4.3}$$

for some deterministic 1-Lipschitz functions f and g, which depend only on the distribution of the array (R, Q).

In the case of nonhomogeneous SK model, this means that both overlaps $R^S_{\ell,\ell'}$ for $S \in \{\text{Girls}, \text{Boys}\}$ are determined by their sum, which is just the usual overlap $R_{\ell,\ell'}$. When the matrix $\Sigma = (\sigma^2_{S,S'})_{S,S' \in \{\text{Girls},\text{Boys}\}}$ of variances in (4.1) is nonnegative definite, this allows computing the Parisi-type formula for the free energy [28, 106]. When Σ is not nonnegative definite, the upper bound via Guerra's interpolation [74] is missing and the problem is still open, but the synchronization mechanism plays a crucial role in another promising approach to this problem developed in a series of papers [2, 29, 94–97].

In the case of the Potts SK model, the quadratic form

$$\sum_{x,y \in \text{Dorms}} R_{x,y}(\sigma^\ell, \sigma^{\ell'})^m \lambda_x \lambda_y$$

is a generalized overlap for any $m \in \mathbb{N}$ and $\lambda \in \mathbb{R}^K$, and, again, one can force all of them to be synchronized in the sense of equation (4.3). This yields even more surprising constraints on the overlap matrix (4.2) in the limit $N \to \infty$. Namely, in this case, one can show that

$$R(\sigma^\ell, \sigma^{\ell'}) = \Phi\big(\text{tr}\big(R(\sigma^\ell, \sigma^{\ell'})\big)\big)$$

for some deterministic function $\Phi : \mathbb{R} \to \text{SPD}$ (where SPD is the set of symmetric nonnegative definite matrices) that depends only on the distribution of the array R. Moreover, Φ is Lipschitz elementwise, and nondecreasing in SPD, i.e., $\Phi(a) - \Phi(b) \in \text{SPD}$ for any $a \geq b$. The reason such constraints are surprising is that a priori the matrix $R(\sigma^\ell, \sigma^{\ell'})$ in (4.2) is not even symmetric. However, synchronization of the generalized overlaps above enforces such strong symmetries and, in particular, all the overlaps are determined by the trace $\text{tr}(R(\sigma^\ell, \sigma^{\ell'}))$. This yields a Parisi-type formula for the free energy [110, 111].

Besides the above two examples, some more general results include showing that the upper bounds of Talagrand [128, 129] for multiple systems with overlap constraints are sharp. These results and the synchronization mechanism itself were used in a variety of applications (see, e.g., [1, 31, 44, 57, 65, 77, 83]).

5. OTHER APPLICATIONS OF ULTRAMETRICITY

Theorem 1.1 or the ideas in its proof were used in a number of other places. We will not describe them in detail, but will at least mention briefly. One application is to show the so-called *chaos in temperature* for generic mixed p-spin models [108]. The phenomenon of chaos in temperature in spin glass models was first studied in the physics literature by Fisher and Huse [70] and Bray and Moore [36], and its states that, if we change the inverse temperature parameter β even a little, configurations sampled from the Gibbs measure will become uncorrelated with those sampled at the original temperature. In other words, the Gibbs measure is chaotic under small changes in temperature.

Thouless–Anderson–Palmer approach [130] to computing the free energy in the SK model is a landmark work in the physics literature and its ideas play an important role also in connection to the Parisi solution [91]. In [50,51], the so-called *generalized TAP free energy* was studied, extending the ideas of Eliran Subag from the setting of spherical models [121] to the original SK and related models. Theorem 1.1 played a role there in computing the TAP correction, similarly to the approach in [104].

The Ghirlanda–Guerra identities and Theorem 1.1 also played a key role in the series of papers [105,107,109] that studied the so-called Mézard–Parisi ansatz [87] in the setting of diluted spin glass models. The main problem here (the so-called reproducibility hypothesis) is still open, but quite general special case was proved in [109].

6. SOME RELATED WORK

Finally, we will briefly summarize some related work in spin glasses. It is not really feasible to give a detailed overview, so we will only list some important results. Talagrand first discovered in [124] that the Ghirlanda–Guerra identities hold in the setting of the simplest Ruelle probability cascades (Poisson–Dirichlet processes) and, following a similar idea, the Ghirlanda–Guerra identities in the setting of the general Ruelle probability cascades were proved by Bovier and Kurkova in [35], where the Gibbs measure of the GREM was studied rigorously. The Ghirlanda–Guerra identities in a strong sense (not on average over β) were derived in [18,38,100]. The ultrametricity of the overlap array in the thermodynamic limit can be translated to a similar description for finite-size systems in some approximate sense, which was done in [41,76]. The properties of the Parisi formula were studied in [15,16,19,23,78,80,127,131], and the Parisi formula at zero temperature (for the maximum of the Hamiltonian) was obtained and studied in [17,45,52,79]. Various results related to chaos in temperature and chaos in disorder appeared in [32,37,39,40,43,47,48,69]. A tiny sample of results related to diluted spin glass models is [54,55,64]. Spherical analogues of the SK and related models were studied in [24–27,42,52,79,81–86,119,122,125]. The complexity of critical points in spherical models was analyzed in great detail in [13,14,20,118,123]. Various results related to the TAP approach and optimization can be found in [21,22,30,33,46,49,53,67,68,93,116,120].

FUNDING

This work was partially supported by NSERC and Simons Fellowship.

REFERENCES

[1] A. Adhikari and C. Brennecke, Free energy of the quantum Sherrington–Kirkpatrick spin-glass model with transverse field. *J. Math. Phys.* **61** (2020), 083302.

[2] E. Agliari, A. Barra, R. Burioni, and A. Di Biasio, Notes on the p-spin glass studied via Hamilton–Jacobi and smooth-cavity techniques. *J. Math. Phys.* **53** (2012), no. 6, 063304.

[3] M. Aizenman and P. Contucci, On the stability of the quenched state in mean-field spin-glass models. *J. Stat. Phys.* **92** (1998), no. 5–6, 765–783.

[4] M. Aizenman, R. Sims, and S. L. Starr, An extended variational principle for the SK spin-glass model. *Phys. Rev. B* **68** (2003), 214403.

[5] P. W. Anderson, Spin glass I: a scaling law rescued. *Phys. Today* **41** (1988), no. 1, 9–11.

[6] P. W. Anderson, Spin glass II: is there a phase transition? *Phys. Today* **41** (1988), no. 3, 9–11.

[7] P. W. Anderson, Spin glass III: theory raises its head. *Phys. Today* **41** (1988), no. 6, 9–11.

[8] P. W. Anderson, Spin glass IV: glimmerings of trouble. *Phys. Today* **41** (1988), no. 9, 9–11.

[9] P. W. Anderson, Spin glass V: real power brought to bear. *Phys. Today* **42** (1989), no. 7, 9–11.

[10] P. W. Anderson, Spin glass VI: spin glass as cornucopia. *Phys. Today* **42** (1989), no. 9, 9–11.

[11] P. W. Anderson, Spin glass VII: spin glass as paradigm. *Phys. Today* **43** (1990), no. 3, 9–11.

[12] L.-P. Arguin and M. Aizenman, On the structure of quasi-stationary competing particles systems. *Ann. Probab.* **37** (2009), no. 3, 1080–1113.

[13] A. Auffinger and G. Ben, Arous, Complexity of random smooth functions on the high dimensional sphere. *Ann. Probab.* **41** (2013), no. 6, 4214–4247.

[14] A. Auffinger, G. Ben Arous, and J. Cerný, Random matrices and complexity of spin glasses. *Comm. Pure Appl. Math.* **66** (2013), no. 2, 165–201.

[15] A. Auffinger and W.-K. Chen, On properties of Parisi measures. *Probab. Theory Related Fields* **161** (2015), no. 3, 817–850.

[16] A. Auffinger and W.-K. Chen, The Parisi formula has a unique minimizer. *Comm. Math. Phys.* **335** (2015), no. 3, 1429–1444.

[17] A. Auffinger and W.-K. Chen, Parisi formula for the ground state energy in the mixed p-spin model. *Ann. Probab.* **45** (2017), 4617–4631.

[18] A. Auffinger and W.-K. Chen, On concentration properties of disordered Hamiltonians. *Proc. Amer. Math. Soc.* **146** (2018), 1807–1815.

[19] A. Auffinger, W.-K. Chen, and Q. Zeng, The SK model is infinite step replica symmetry breaking at zero temperature. *Comm. Pure Appl. Math.* **73** (2020), no. 5, 921–943.

[20] A. Auffinger and J. Gold, The number of saddles of the spherical p-spin model. 2020, arXiv:2007.09269.

[21] A. Auffinger and A. Jagannath, On spin distributions for generic p-spin models. *J. Stat. Phys.* **174** (2019), 316–332.

[22] A. Auffinger and A. Jagannath, Thouless–Anderson–Palmer equations for generic p-spin glasses. *Ann. Probab.* **47** (2019), no. 4, 2230–2256.

[23] A. Auffinger and Q. Zeng, Existence of two-step replica symmetry breaking for the spherical mixed spin glass at zero temperature. *Comm. Math. Phys.* **370** (2019), no. 1, 377–402.

[24] J. Baik, E. Collins-Woodfin, P. Le Doussal, and H. Wu, Spherical spin glass model with external field. *J. Stat. Phys.* **183** (2021), no. 2.

[25] J. Baik and J. O. Lee, Fluctuations of the free energy of the spherical Sherrington–Kirkpatrick model. *J. Stat. Phys.* **165** (2016), no. 2, 185–224.

[26] J. Baik and J. O. Lee, Free energy of bipartite spherical Sherrington–Kirkpatrick model. *Ann. Inst. Henri Poincaré Probab. Stat.* **56** (2020), no. 4, 2897–2934.

[27] J. Baik, J. O. Lee, and H. Wu, Ferromagnetic to paramagnetic transition in spherical spin glass. *J. Stat. Phys.* **173** (2018), no. 5, 1484–1522.

[28] A. Barra, P. Contucci, E. Mingione, and D. Tantari, Multi-species mean-field spin-glasses. Rigorous results. *Ann. Henri Poincaré* **16** (2015), 691–708.

[29] A. Barra, A. Di Biasio, and F. Guerra, Replica symmetry breaking in mean-field spin glasses through the Hamilton–Jacobi technique. *J. Stat. Mech. Theory Exp.* (2010), P09006.

[30] D. Belius and N. Kistler, The TAP–Plefka variational principle for the spherical SK model. *Comm. Math. Phys.* **367** (2019), no. 3, 991–1017.

[31] G. Ben Arous and A. Jagannath, Spectral gap estimates in mean field spin glasses. *Comm. Math. Phys.* **361** (2018), no. 1, 1–52.

[32] G. Ben Arous, E. Subag, and O. Zeitouni, Geometry and temperature chaos in mixed spherical spin glasses at low temperature – the perturbative regime. 2018, arXiv:1804.10573.

[33] E. Bolthausen, An iterative construction of solutions of the TAP equations for the Sherrington–Kirkpatrick model. *Comm. Math. Phys.* **325** (2014), no. 1, 333–366.

[34] E. Bolthausen and A.-S. Sznitman, On Ruelle's probability cascades and an abstract cavity method. *Comm. Math. Phys.* **197** (1998), no. 2, 247–276.

[35] A. Bovier and I. Kurkova, Derrida's generalized random energy models. I. Models with finitely many hierarchies. *Ann. Inst. Henri Poincaré Probab. Stat.* **40** (2004), no. 4, 439–480.

[36] A. J. Bray and M. A. Moore, Chaotic nature of the spin-glass phase. *Phys. Rev. Lett.* **58** (1987), no. 1, 5760.

[37] S. Chatterjee, Disorder, chaos, and multiple valleys in spin glasses. 2008, arXiv:0907.3381.

[38] S. Chatterjee, The Ghirlanda–Guerra identities without averaging. 2009, arXiv:0911.4520.

[39] S. Chatterjee, *Superconcentration and related topics*. Springer Monogr. Math., Springer, Berlin–Heidelberg, 2014.

[40] S. Chatterjee, Chaos, concentration, and multiple valleys. 2018, arXiv:0810.4221.

[41] S. Chatterjee and L. Sloman, Average Gromov hyperbolicity and the Parisi ansatz. *Adv. Math.* **376** (2021), 107417.

[42] W.-K. Chen, The Aizenman–Sims–Starr scheme and Parisi formula for mixed p-spin spherical models. *Electron. J. Probab.* **18** (2013), no. 94, 1–14.

[43] W.-K. Chen, Chaos in the mixed even-spin models. *Comm. Math. Phys.* **328** (2014), no. 3, 867–901.

[44] W.-K. Chen, Phase transition in the spiked random tensor with Rademacher prior. *Ann. Statist.* **47** (2019), no. 5, 2734–2756.

[45] W.-K. Chen, M. Handschy, and G. Lerman, On the energy landscape of the mixed even p-spin model. *Probab. Theory Related Fields* **171** (2018), no. 1–2, 53–95.

[46] W.-K. Chen and W.-K. Lam, Universality of approximate message passing algorithms. *Electron. J. Probab.* **26** (2021), no. 36, 1–44.

[47] W.-K. Chen and D. Panchenko, Temperature chaos in some spherical mixed p-spin models. *J. Stat. Phys.* **166** (2017), no. 5, 1151–1162.

[48] W.-K. Chen and D. Panchenko, Disorder chaos in some diluted spin glass models. *Ann. Appl. Probab.* **28** (2018), no. 3, 1356–1378.

[49] W.-K. Chen and D. Panchenko, On the TAP free energy in the mixed p-spin models. *Comm. Math. Phys.* **362** (2018), no. 1, 219–252.

[50] W.-K. Chen, D. Panchenko, and E. Subag, The generalized TAP free energy. 2018, arXiv:1812.05066.

[51] W.-K. Chen, D. Panchenko, and E. Subag, The generalized TAP free energy II. *Comm. Math. Phys.* **381** (2021), no. 1, 257–291.

[52] W.-K. Chen and A. Sen, Parisi formula, disorder chaos and fluctuation for the ground state energy in the spherical mixed p-spin models. *Comm. Math. Phys.* **350** (2017), no. 1, 129–173.

[53] W.-K. Chen and S. Tang, On convergence of the cavity and Bolthausen's TAP iterations to the local magnetization. 2020, arXiv:2011.00495.

[54] A. Coja-Oghlan and K. Panagiotou, Going after the k-SAT threshold. In *Proc. 45th STOC*, pp. 705–714, ACM, 2013.

[55] A. Coja-Oghlan and K. Panagiotou, The asymptotic k-SAT threshold. *Adv. Math.* **288** (2016), 985–1068.

[56] P. Contucci and C. Giardinà, Spin-glass stochastic stability: a rigorous proof. *Ann. Henri Poincaré* **6** (2005), no. 5, 915–923.

[57] P. Contucci and E. Mingione, A multi-scale spin-glass mean-field model. *Comm. Math. Phys.* **268** (2018), no. 3, 1323–1344.

[58] C. de Dominicis and H. Hilhorst, Random (free) energies in spin glasses. *J. Phys. Lett.* **46** (1985), L909–L914.

[59] B. Derrida, Random-energy model: limit of a family of disordered models. *Phys. Rev. Lett.* **45** (1980), no. 2, 79–82.

[60] B. Derrida, Random-energy model: an exactly solvable model of disordered systems. *Phys. Rev. B (3)* **24** (1981), no. 5, 2613–2626.

[61] B. Derrida, A generalization of the random energy model that includes correlations between the energies. *J. Phys. Lett.* **46** (1985), 401–407.

[62] B. Derrida and E. Gardner, Solution of the generalised random energy model. *J. Phys. C* **19** (1986), 2253–2274.

[63] B. Derrida and G. Toulouse, Sample to sample fluctuations in the random energy model. *J. Phys. Lett.* **46** (1985), L223–L228.

[64] J. Ding, A. Sly, and N. Sun, Proof of the satisfiability conjecture for large k. In *STOC'15*, pp. 59–68, ACM, 2015.

[65] T. Dominguez, The ℓ^p-Gaussian–Grothendieck problem with vector spins. Preprint, 2021.

[66] L. N. Dovbysh and V. N. Sudakov, Gram–de Finetti matrices. *Zap. Nauchn. Sem. Leningrad. Otdel. Mat. Inst. Steklov.* **119** (1982), 77–86.

[67] A. El Alaoui and A. Montanari, Algorithmic thresholds in mean field spin glasses. 2020, arXiv:2009.11481.

[68] A. El Alaoui, A. Montanari, and M. Sellke, Optimization of mean-field spin glasses. 2020, arXiv:2001.00904.

[69] R. Eldan, A simple approach to chaos for p-spin models. *J. Stat. Phys.* **181** (2020), 1266–1276.

[70] D. S. Fisher and D. A. Huse, Ordered phase of short-range Ising spin glasses. *Phys. Rev. Lett.* **56** (1986), no. 15, 16011604.

[71] S. Ghirlanda and F. Guerra, General properties of overlap probability distributions in disordered spin systems. Towards Parisi ultrametricity. *J. Phys. A* **31** (1998), no. 46, 9149–9155.

[72] C. Goldschmidt and J. B. Martin, Random recursive trees and the Bolthausen–Sznitman coalescent. *Electron. J. Probab.* **10** (2005), no. 21, 718–745.

[73] F. Guerra, About the overlap distribution in mean field spin glass models. *Internat. J. Modern Phys. B* **10** (1996), no. 13–14, 1675–1684.

[74] F. Guerra, Broken replica symmetry bounds in the mean field spin glass model. *Comm. Math. Phys.* **233** (2003), no. 1, 1–12.

[75] F. Guerra and F. L. Toninelli, The thermodynamic limit in mean field spin glass models. *Comm. Math. Phys.* **230** (2002), no. 1, 71–79.

[76] A. Jagannath, Approximate ultrametricity for random measures and applications to spin glasses. *Comm. Pure Appl. Math.* **70** (2017), no. 4, 611–664.

[77] A. Jagannath, J. Ko, and S. Sen, A connection between MAX κ-CUT and the inhomogeneous Potts spin glass in the large degree limit. *Ann. Appl. Probab.* **28** (2018), no. 3, 1536–1572.

[78] A. Jagannath and I. Tobasco, A dynamic programming approach to the Parisi functional. *Proc. Amer. Math. Soc.* **144** (2016), 3135–3150.

[79] A. Jagannath and I. Tobasco, Low temperature asymptotics of spherical mean field spin glasses. *Comm. Math. Phys.* **352** (2017), no. 3, 979–1017.

[80] A. Jagannath and I. Tobasco, Some properties of the phase diagram for mixed p-spin glasses. *Probab. Theory Related Fields* **167** (2017), no. 3–4, 615–672.

[81] P. Kivimae, Critical fluctuations for the spherical Sherrington–Kirkpatrick model in an external field. 2019, arXiv:1908.07512.

[82] J. Ko, The Crisanti–Sommers formula for spherical spin glasses with vector spins. 2019, arXiv:1911.04355.

[83] J. Ko, Free energy of multiple systems of spherical spin glasses with constrained overlaps. *Electron. J. Probab.* **25** (2020), no. 28, 1–34.

[84] B. Landon, Free energy fluctuations of the 2-spin spherical SK model at critical temperature. 2020, arXiv:2010.06691.

[85] B. Landon and P. Sosoe, Fluctuations of the overlap at low temperature in the 2-spin spherical SK model. 2019, arXiv:1905.03317.

[86] B. Landon and P. Sosoe, Fluctuations of the 2-spin SSK model with magnetic field. 2020, arXiv:2009.12514.

[87] M. Mézard and G. Parisi, The Bethe lattice spin glass revisited. *Eur. Phys. J. B* **20** (2001), no. 2, 217–233.

[88] M. Mézard, G. Parisi, N. Sourlas, G. Toulouse, and M. A. Virasoro, On the nature of the spin-glass phase. *Phys. Rev. Lett.* **52** (1984), 1156.

[89] M. Mézard, G. Parisi, N. Sourlas, G. Toulouse, and M. A. Virasoro, Replica symmetry breaking and the nature of the spin-glass phase. *J. Phys.* **45** (1984), 843.

[90] M. Mézard, G. Parisi, and M. A. Virasoro, Random free energies in spin glasses. *J. Phys. Lett.* **46** (1985), L217–L222.

[91] M. Mézard, G. Parisi, and M. A. Virasoro, *Spin glass theory and beyond*. Lecture Notes in Phys. 9, World Scientific, Teaneck, NJ, 1987.

[92] M. Mézard and M. A. Virasoro, The microstructure of ultrametricity. *J. Phys.* **46** (1985), 1293–1307.

[93] A. Montanari, Optimization of the Sherrington–Kirkpatrick Hamiltonian. *SIAM J. Comput.* **FOCS19-38** (2021).

[94] J.-C. Mourrat, Parisi's formula is a Hamilton–Jacobi equation in Wasserstein space. 2019, arXiv:1906.08471.

[95] J.-C. Mourrat, Free energy upper bound for mean-field vector spin glasses. 2020, arXiv:2010.09114.

[96] J.-C. Mourrat, Nonconvex interactions in mean-field spin glasses. *Probab. Math. Phys.* **2** (2021), no. 2, 281–339.

[97] J.-C. Mourrat and D. Panchenko, Extending the Parisi formula along a Hamilton–Jacobi equation. *Electron. J. Probab.* **25** (2020), no. 23, 1–17.

[98] J. Neveu, *A continuous-state branching process in relation with the GREM model of spin glass theory*. Rapport interne no. 267 Ecole Polytechnique, 1992.

[99] D. Panchenko, A connection between Ghirlanda–Guerra identities and ultrametricity. *Ann. Probab.* **38** (2010), no. 1, 327–347.

[100] D. Panchenko, The Ghirlanda–Guerra identities for mixed p-spin model. *C. R. Acad. Sci. Paris, Ser. I* **348** (2010), 189–192.

[101] D. Panchenko, Ghirlanda–Guerra identities and ultrametricity: an elementary proof in the discrete case. *C. R. Acad. Sci. Paris, Ser. I* **349** (2011), no. 13–14, 813–816.

[102] D. Panchenko, The Parisi ultrametricity conjecture. *Ann. of Math. (2)* **177** (2013), no. 1, 383–393.

[103] D. Panchenko, *The Sherrington–Kirkpatrick model.* Springer Monogr. Math., Springer, New York, 2013.

[104] D. Panchenko, The Parisi formula for mixed p-spin models. *Ann. Probab.* **42** (2014), no. 3, 946–958.

[105] D. Panchenko, Structure of 1-RSB asymptotic Gibbs measures in the diluted p-spin models. *J. Stat. Phys.* **155** (2014), no. 1, 1–22.

[106] D. Panchenko, Free energy in the multi-species Sherrington–Kirkpatrick model. *Ann. Probab.* **43** (2015), no. 6, 3494–3513.

[107] D. Panchenko, Hierarchical exchangeability of pure states in mean field spin glass models. *Probab. Theory Related Fields* **161** (2015), no. 3, 619–650.

[108] D. Panchenko, Chaos in temperature in generic $2p$-spin models. *Comm. Math. Phys.* **346** (2016), no. 2, 703–739.

[109] D. Panchenko, Structure of finite-RSB asymptotic Gibbs measures in the diluted spin glass models. *J. Stat. Phys.* **162** (2016), no. 1, 1–42.

[110] D. Panchenko, Free energy in the mixed p-spin models with vector spins. *Ann. Probab.* **46** (2018), no. 2, 865–896.

[111] D. Panchenko, Free energy in the Potts spin glass. *Ann. Probab.* **46** (2018), no. 2, 829–864.

[112] G. Parisi, Infinite number of order parameters for spin-glasses. *Phys. Rev. Lett.* **43** (1979), 1754–1756.

[113] G. Parisi, A sequence of approximate solutions to the S-K model for spin glasses. *J. Phys. A* **13** (1980), L-115.

[114] G. Parisi, Order parameter for spin glasses. *Phys. Rev. Lett.* **50** (1983), 1946.

[115] D. Ruelle, A mathematical reformulation of Derrida's REM and GREM. *Comm. Math. Phys.* **108** (1987), no. 2, 225–239.

[116] M. Sellke, Optimizing mean field spin glasses with external field. 2021, arXiv:2105.03506.

[117] D. Sherrington and S. Kirkpatrick, Solvable model of a spin glass. *Phys. Rev. Lett.* **35** (1975), 1792–1796.

[118] E. Subag, The complexity of spherical p-spin models – a second moment approach. *Ann. Probab.* **45** (2017), no. 5, 3385–3450.

[119] E. Subag, The geometry of the Gibbs measure of pure spherical spin glasses. *Invent. Math.* **210** (2017), no. 1, 135–209.

[120] E. Subag, Following the ground-states of full-RSB spherical spin glasses. 2018, arXiv:1812.04588.

[121] E. Subag, Free energy landscapes in spherical spin glasses. 2018, arXiv:1804.10576.

[122] E. Subag, The free energy of spherical pure p-spin models – computation from the TAP approach. 2021, arXiv:2101.04352.

[123] E. Subag and O. Zeitouni, The extremal process of critical points of the pure p-spin spherical spin glass model. *Probab. Theory Related Fields* **168** (2017), no. 3–4, 773–820.

[124] M. Talagrand, *Spin glasses: a challenge for mathematicians*. Ergeb. Math. Grenzgeb. (3)/Ser. Mod. Surv. Math. 43, Springer, 2003.

[125] M. Talagrand, Free energy of the spherical mean field model. *Probab. Theory Related Fields* **134** (2006), no. 3, 339–382.

[126] M. Talagrand, The Parisi formula. *Ann. of Math. (2)* **163** (2006), no. 1, 221–263.

[127] M. Talagrand, Parisi measures. *J. Funct. Anal.* **231** (2006), no. 2, 269–286.

[128] M. Talagrand, Large deviations, Guerra's and A.S.S. schemes, and the Parisi hypothesis. *J. Stat. Phys.* **126** (2007), no. 4–5, 837–894.

[129] M. Talagrand, *Mean-field models for spin glasses. Volumes I and II*. Ergeb. Math. Grenzgeb. (3)/Ser. Mod. Surv. Math., Springer, 2011.

[130] D. J. Thouless, P. W. Anderson, and R. G. Palmer, Solution of 'solvable model of a spin glass'. *Phys. Mag.* **35** (1977), no. 3, 593–601.

[131] F. L. Toninelli, About the Almeida–Thouless transition line in the Sherrington–Kirkpatrick mean-field spin glass model. *Europhys. Lett.* **60** (2002), no. 5, 764.

DMITRY PANCHENKO

Department of Mathematics, University of Toronto, Toronto, Canada, panchenk@math.toronto.edu

This Page Intentionally Left Blank

INTERACTING STOCHASTIC PROCESSES ON SPARSE RANDOM GRAPHS

KAVITA RAMANAN

ABSTRACT

Large ensembles of stochastically evolving interacting particles describe phenomena in diverse fields including statistical physics, neuroscience, biology, and engineeering. In such systems, the infinitesimal evolution of each particle depends only on its own state (or history) and the states (or histories) of neighboring particles with respect to an underlying, possibly random, interaction graph. While these high-dimensional processes are typically too complex to be amenable to exact analysis, their dynamics are quite well understood when the interaction graph is the complete graph. In this case, classical theorems show that in the limit as the number of particles goes to infinity, the dynamics of the empirical measure and the law of a typical particle coincide and can be characterized in terms of a much more tractable dynamical system of reduced dimension called the mean-field limit. In contrast, until recently not much was known about corresponding convergence results in the complementary case when the interaction graph is sparse (i.e., with uniformly bounded average degree). This article provides a brief survey of classical work and then describes recent progress on the sparse regime that relies on a combination of techniques from random graph theory, Markov random fields, and stochastic analysis. The article concludes by discussing ramifications for applications and posing several open problems.

MATHEMATICS SUBJECT CLASSIFICATION 2020

Primary 60K35; Secondary 60G60, 60J60, 60J27, 60J80, 60F15

KEYWORDS

Interacting particle systems, Markov random fields, mean-field limits, nonlinear processes, sparse random graphs, local convergence, Erdős–Rényi graphs, Galton–Watson processes

ICM INTERNATIONAL CONGRESS OF MATHEMATICIANS 2022 JULY 6–14

Proc. Int. Cong. Math. 2022, Vol. 6, pp. 4394–4425
DOI 10.4171/ICM2022/188

1. INTRODUCTION

1.1. Background

A recurring theme in probability theory is the emergence of deterministic (or more predictable) behavior when there is an aggregation of many random elements. A classical result is the strong law of large numbers established by Kolmogorov in 1933 [43]. This states that given a sequence of random variables $(X_i)_{i\in\mathbb{N}}$ that are independent and identically distributed (i.i.d.) and have finite mean (equivalently, $(X_i)_{i\in\mathbb{N}}$ is distributed according to some product probability measure $\otimes^{\mathbb{N}}\nu$, where ν is a probability measure on the Borel sets of $\mathbb{R}$ that satisfies $\int_{\mathbb{R}}|x|\nu(dx)<\infty$), then with probability one,

$$S_n := \frac{1}{n}\sum_{i=1}^{n} X_i \to \mathbb{E}[X_1] = \int_{\mathbb{R}} x\nu(dx), \quad \text{as } n\to\infty. \tag{1.1}$$

In a similar spirit, the Glivenko–Cantelli theorem, also established in 1933 [10, 35], provides information on the asymptotic behavior of empirical measures of i.i.d. random variables. Specifically, it shows that with probability one,

$$\mu_n := \frac{1}{n}\sum_{i=1}^{n}\delta_{X_i} \to \mathbb{E}[\delta_{X_1}] = \mathscr{L}(X_1) = \nu, \quad \text{as } n\to\infty, \tag{1.2}$$

where δ_x represents the Dirac delta measure at x and $\mathscr{L}(Y)$ denotes the law or distribution of a random variable Y. The convergence in (1.2) is in the so-called Kolmogorov distance, which in particular implies weak convergence, that is, for every bounded, continuous function f on $\mathbb{R}$, $\int_{\mathbb{R}} f(x)\mu_n(dx)\to\int_{\mathbb{R}} f(x)\nu(dx)$.

Similar results also hold when the random variables are not independent, but exhibit some form of weak dependence. For instance, consider a triangular array of (dependent) random variables $(X_i^n, i=1,\ldots,n)_{n\in\mathbb{N}}$ that have a common mean, finite variances, and exhibit *asymptotic correlation decay* in the sense that there exist positive real numbers $\{f_{n,k}, k=1,\ldots,n\}_{n\in\mathbb{N}}$ such that $\sup_{k,n\in\mathbb{N}} f_{n,k}<\infty$,

$$\left|\mathrm{Cov}\left(X_i^n, X_j^n\right)\right| \le f_{n,|i-j|} \quad \text{and} \quad \lim_{n\to\infty}\frac{1}{n}\sum_{i=1}^{n} f_{n,k} = 0, \tag{1.3}$$

where $\mathrm{Cov}(X_i^n, X_j^n)$ represents the covariance of X_i^n and X_j^n. Then it follows from Chebyshev's inequality [74] that the normalized partial sum $S_n = \frac{1}{n}\sum_{i=1}^{n} X_i^n$ satisfies

$$\mathbb{P}\left(\left|S_n - \mathbb{E}\left[X_1^1\right]\right| > \varepsilon\right) \to 0, \quad \forall\varepsilon>0.$$

On the other hand, in many interesting cases one wants to analyze large collections of strongly dependent random elements. Such an analysis is often facilitated by graphical model representations, which capture *conditional independence* properties of the random elements via a graph. A specific class of graphical models that will be important for the present discussion is a Markov random field (MRF) (a precise definition is given in Section 4.1). The theory of MRFs and associated Gibbs measures goes back to the late 1960s with the pioneering works of Dobrushin [25, 26] and Lanford and Ruelle [51], who were motivated by models in statistical physics involving static interacting random elements. In this case a key question is efficient computation or analytical characterization of marginal distributions of the high-dimensional ensemble of random elements.

1.2. Questions of interest

This article focuses on the dynamics of large ensembles of stochastic processes whose interactions are governed by an underlying graph $G = (V, E)$. Here V represents a finite or countably infinite vertex set and E is a subset of unordered pairs of distinct vertices in V that represent the (undirected) edges of the graph. The graph G is always assumed to be simple (i.e., each pair in E is comprised of two distinct vertices) and locally finite, that is, for each $v \in V$, the size of its neighborhood $\partial_G(v) := \{u \in V : uv \in E\}$ is finite. The notation $u \sim v$ will often also be used to indicate $uv \in E$. Given the graph G and an initial condition $\xi = (\xi_v)_{v \in V}$, we are interested in a collection of stochastic processes $X^{G,\xi} = (X_v^{G,\xi}(t), t \geq 0)_{v \in V}$ indexed by the vertices of G, that satisfies $X_v^{G,\xi}(0) = \xi_v$ for $v \in V$, and whose interaction structure is governed by the graph G. Specifically, for each $v \in V$, the infinitesimal evolution of $X_v^{G,\xi}$ at any time only depends on its own state (or history) and the states (or histories) of neighboring particles in G at that time. Note that this includes both the case when $X^{G,\xi}$ is Markovian, where the infinitesimal evolution depends only on the current states of particles, as well as non-Markovian evolutions, where the infinitesimal evolution of a particle can depend on its own history and the histories of particles in its neighbrhood. For conciseness, we will restrict our discussion to two types of dynamics: interacting diffusions, which are described in Section 2.1, and interacting jump processes, which are described in Section 2.3. Given such (Markovian or non-Markovian) interacting processes on a large finite graph, quantities of interest include the following:

A. The macroscopic behavior of the system as captured by the (global) empirical measure process, defined by

$$\mu^{G,\xi}(t) = \frac{1}{|V|} \sum_{v \in V} \delta_{X_v^{G,\xi}(t)}, \quad t \geq 0. \tag{1.4}$$

Note that for each $t > 0$, $\mu^{G,\xi}(t)$ is a random probability measure on the state space that encodes the fractions of particles taking values in different (measurable) subsets of the state space.

B. The microscopic behavior, in particular the marginal dynamics of a "typical particle." By this we mean the dynamics of $X_o^{G,\xi}$, where the vertex o, referred to as the root, is assumed to be chosen uniformly at random from the finite vertex set V. An important question here is to ascertain how the dynamics depends on the graph topology?

Due to the complexity and high dimensionality of the dynamics, these quantities are typically not amenable to exact analysis or efficient computation. The goal instead is to identify more tractable approximations of reduced dimension that can be rigorously justified by limit theorems, as the number of particles goes to infinity. A desirable goal is to obtain an *autonomous characterization* of the limiting marginal dynamics of a typical particle and evolution of the empirical measure, which does not refer to the full particle system dynamics.

In Section 2 we review the well understood case when $G = K_n$, the n-clique or the complete graph on n vertices, in which all pairs of distinct vertices are connected by an

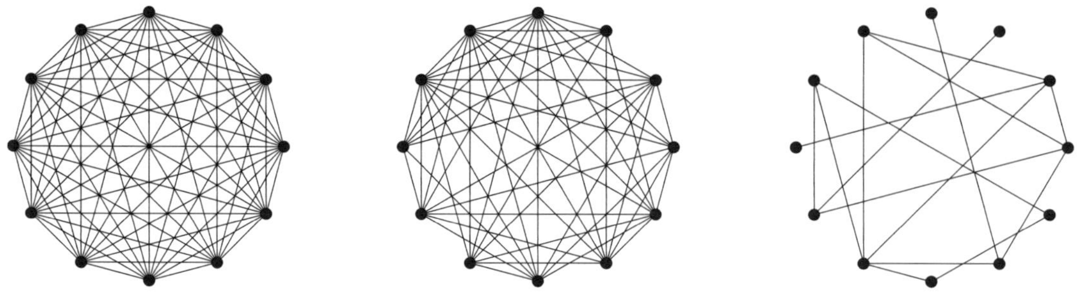

(a) Complete graph (b) Dense E–R graph (c) Sparse E–R graph

FIGURE 1

edge; see Figure 1(a). For suitable initial conditions ξ, convergence results for $\mu^{K_n,\xi}$ and $X_o^{K_n,\xi}$ in the context of interacting diffusion models go back more than half a century to the seminal works of McKean [57, 58], and fall under the rubric of mean-field limits. As briefly described in Section 2, under broad conditions, the limits, as $n \to \infty$, of both $\mu^{K_n,\xi}$ and the law of $X_o^{K_n,\xi}$ exist and coincide, and are described by a certain nonlinear stochastic process. More recent work has also considered interacting processes on certain dense random graph sequences. A generic example of a random graph is the so-called Erdős–Rényi graph $G(n, p_n)$, which is a graph on n vertices in which each pair of vertices has an edge with probability $p_n \in (0, 1)$ independently of all other edges; see Figures 1(b) and 1(c) for realizations of $G(n, p_n)$ with $n = 12$ and $p_n = 0.8$ and $p_n = 0.25$, respectively. Motivated by the study of synchronization phenomena, the work [19] considers suitably scaled pairwise interacting diffusions on "dense" Erdős–Rényi graph $G(n, p_n)$ sequences with divergent average degree ($np_n \to \infty$), and shows that the law of $X_o^{G_n,\xi}$ converges to the same mean-field limit as in the complete graph case. The key idea is that in this regime, particles are only weakly interacting and become asymptotically independent, and thus the empirical measure behaves as in the i.i.d. case (1.2) described in Section 1.1.

The main focus of this article is on the complementary setting of interacting stochastic processes on sequences of *sparse* (possibly random) graphs, where the (average) degrees of vertices are uniformly bounded as $n \to \infty$. A typical example is the Erdős–Rényi graph $G(n, p_n)$ sequence when $np_n \to c \in (0, 1)$. There has been extensive analysis of various interacting stochastic processes on deterministic sparse graphs, originating with the work of Spitzer [71], followed by significant analysis of several Markovian models including the contact process, exclusion process, and voter model. These were first studied on the d-dimensional lattice (see the monographs [27, 41, 52, 53]) and then on d-regular trees (e.g., [65,72]). More recent work has also considered processes on sparse random interaction graphs (see, e.g., [6, 12, 19, 32, 38, 61] for an incomplete list), but none of these latter works appear to address the main question listed above of autonomous characterization of the marginal dynamics of a typical particle. In fact, for interacting diffusions on the sequence of sparse

Erdős–Rényi graphs $G_n = G(n, c/n)$, with $c \in (0, \infty)$, obtaining such a characterization has remained an important open question (e.g., see [19, P. 9]).

The sparse regime is more challenging because particles have strong interactions, neighboring particles remain correlated in the limit as $n \to \infty$, and the topology of the graph has a strong influence. Sections 3 and 4 describe recent progress that in particular provides a resolution of the open question in [19]. The article concludes in Section 5 with generalizations and open questions. The work on both mean-field models and various other aspects of interacting particle systems is so extensive that it will be impossible to be representative in this short article. Instead, I hope to just provide enough pointers for the reader to get a flavor of the classical results and set the context for more recent results. Monographs covering various aspects of interacting particles systems include [7, 20, 34, 41, 44, 52, 53, 73].

2. CLASSICAL MEAN-FIELD RESULTS FOR INTERACTING STOCHASTIC PROCESSES

Given a (simple, locally finite, undirected) graph $G = (V, E)$ and $v \in V$, we use $\mathrm{cl}_G(v) := \{v\} \cup \partial_G(v)$ to denote the closure of v in G. Note that $|\mathrm{cl}_G(v)|$ is always finite, where $|A|$ denotes the cardinality of a set A. We now describe the dynamics of locally interacting diffusions and interacting jump processes. Rather than provide the most general setting, we make simplifying assumptions whenever convenient to illustrate the key issues.

2.1. Interacting diffusions

Given an initial condition $\xi = (\xi_v)_{v \in V} \in \mathbb{R}^V$ with $\mathbb{E}[\xi_v^2] < \infty$ for every $v \in V$, consider the collection $X^{G,\xi} = \{X_v^{G,\xi}\}_{v \in V}$ of diffusive particles, indexed by the nodes of the graph G, that evolve according to the following coupled system of stochastic differential equations (SDEs):

$$dX_v^{G,\xi}(t) = b\big(t, X_v^{G,\xi}(t), \mu_v^{G,\xi}(t)\big)\, dt + dW_v(t), \quad X_v^{G,\xi}(0) = \xi_v, \quad t > 0, v \in V, \tag{2.1}$$

where $(W_v)_{v \in V}$ are i.i.d. standard Brownian motions independent of $(\xi_v)_{v \in V}$, and for any vertex that is not isolated, $\mu_v^{G,\xi}(t)$ represents the *local empirical measure* of a neighborhood of v at time $t \geq 0$,

$$\mu_v^{G,\xi}(t) = \frac{1}{|\partial_G(v)|} \sum_{u \in \partial_v} \delta_{X_u^{G,\xi}(t)},$$

and b is a drift coefficient that is sufficiently regular to ensure that the SDE (2.1) has a unique weak solution. (When v is isolated, the precise definition of $\mu_v^{G,\xi}$ is not so important; it can be set equal to an arbitrary quantity.)

A special case of interest is when b has linear dependence on the measure term, say, of the form

$$b(t, x, \nu) = \int_{\mathbb{R}} \beta(t, x, y)\nu(dy), \tag{2.2}$$

for some interaction potential $\beta : \mathbb{R}_+ \times \mathbb{R}^2 \to \mathbb{R}$ that is symmetric in the last two variables. In this case, system (2.1) reduces to the following system of pairwise interacting diffusions:

$$dX_v^{G,\xi}(t) = \frac{1}{|\partial_G(v)|} \sum_{u\sim v} \beta\big(t, X_v^{G,\xi}(t), X_u^{G,\xi}(t)\big)dt + dW_v(t), \quad v \in V, t > 0, \tag{2.3}$$

which models phenomena in different fields, including statistical physics and neuroscience [21,56,68]. The trajectories of each particle lie in the space $\mathcal{C}$ of continuous real-valued functions on $[0, \infty)$, which we endow with the topology of uniform convergence on compact sets.

2.2. Mean-field limits and nonlinear diffusion processes

Now consider the SDE (2.3) with $G = K_n$, the complete graph, and assume without loss of generality that G has vertex set $\{1, \dots, n\}$. We present a sufficient condition on the drift under which one can establish a standard mean-field result. Given any $p \geq 1$ and Polish space E, the Wasserstein-p metric on E is defined as follows:

$$\mathcal{W}_{E,p}(\nu, \tilde{\nu}) := \inf_{\pi}\bigg(\int_{E\times E} d^p(x, y)\pi(dx, dy)\bigg)^{1/p}, \tag{2.4}$$

where the infimum is over all couplings π of ν and $\tilde{\nu}$, namely probability measures π on E^2 with first and second marginals ν and $\tilde{\nu}$, respectively. Let $\mathcal{P}^p(E)$ be the space of probability measures on E equipped with the Wasserstein-p metric $\mathcal{W}_{E,p}$.

Assumption 2.1. Suppose that b is bounded and for every $t > 0$, the map $\mathbb{R} \times \mathcal{P}^2(\mathbb{R}) \ni (x, \nu) \mapsto b(t, x, \nu) \in \mathbb{R}$ is Lipschitz continuous, uniformly with respect to t in compact subsets of $\mathbb{R}_+$.

Note that Assumption 2.1 is satisfied when the drift b is of the form (2.2), where the interaction potential β is such that $\mathbb{R}^2 \ni (x, y) \to \beta(t, x, y)$ is Lipschitz continuous and bounded, uniformly with respect to t in compact subsets of $\mathbb{R}_+$.

Theorem 2.2. *Suppose Assumption* 2.1 *holds, and there exists $\mu_o \in \mathcal{P}^2(\mathbb{R})$ such that the initial conditions $(\xi_i^n)_{i=1,\dots,n}$, $n \in \mathbb{N}$, satisfy*

$$\mathbb{E}\bigg[\mathcal{W}_{\mathbb{R},2}\bigg(\frac{1}{n}\sum_{i=1}^{n}\delta_{\xi_i^n}, \mu_o\bigg)\bigg] \to 0 \quad \text{as } n \to \infty. \tag{2.5}$$

Then there is a unique strong solution to the SDE

$$dX_o(t) = b\big(t, B(t), \mu(t)\big)\,dt + dB(t), \quad \mu(t) = \mathcal{L}\big(X_o(t)\big), \quad t > 0, \tag{2.6}$$

with $\mathcal{L}(X_o(0)) = \mu_0$. Moreover, if for each $n \in \mathbb{N}$, $X^n := X^{K_n,\xi^n}$ is the unique solution to the SDE (2.1)*, then the global empirical measure $\mu^n := \mu^{K^n,\xi^n}$ defined in* (1.4) *satisfies*

$$\lim_{n\to\infty} \mathbb{E}\Big[\sup_{s\in[0,t]} \mathcal{W}_{\mathbb{R},2}\big(\mu^n(s), \mu(s)\big)\Big] = 0, \quad \forall t > 0. \tag{2.7}$$

Furthermore, for any $k \in \mathbb{N}$ and $t > 0$, the law of $(X_1^n(t), \ldots, X_k^n(t))$ converges weakly to the product $(\mu(t))^{\otimes k}$, that is, for all bounded continuous functions $f_i : \mathbb{R} \to \mathbb{R}$, $i = 1, \ldots, k$,

$$\lim_{n\to\infty} \mathbb{E}\big[f_1\big(X_1^n(t)\big) \ldots f_k\big(X_k^n(t)\big)\big] = \prod_{i=1}^{k} \int_{\mathbb{R}} f_i(x)\mu(t)(dx). \tag{2.8}$$

If there were no interaction, $b \equiv 0$, then the theorem would simply be a (functional) strong law of large numbers result. However, even when $b \not\equiv 0$, the particles are only weakly interacting because the symmetry of the interaction ensures that the influence of any particle on the drift of another particle is $O(1/n)$, which vanishes in the limit. The property (2.8) that any finite subset of random variables from $\{X_i^n(t), i = 1, \ldots, n\}_{n\in\mathbb{N}}$ are asymptotically independent is referred to as *chaoticity*, and is well known to be equivalent to the convergence of $\mu^n(t)$ to a deterministic law **[73, PROPOSITION 2.2]**. Now, (2.5) implies that the initial conditions are chaotic. Thus Theorem 2.2 asserts that the dynamics are such that this chaoticity also holds for positive times $t > 0$, a phenomenon referred to as *propagation of chaos*. In turn, this leads to an *autonomous* description of the limiting marginal process X_o, which is a Markov process whose infinitesimal evolution at any time t also depends on its own law $\mu(t)$ at that time. As a result, the forward Kolmogorov equation (or master equation), which is the partial differential equation (PDE) describing the evolution of the marginal law μ, is nonlinear. Consequently, such a process is referred to as a *nonlinear* Markov process. When the drift has the form (2.2), under suitable conditions it can be shown that the law $\mu(t)$ is absolutely continuous with respect to Lebesgue measure and that its density satisfies the granular media equation **[58]**. Thus, PDE techniques can be useful for studying nonlinear Markov processes (see, e.g., **[4]**).

There are many different approaches to establishing mean-field limits, including PDE analysis, fixed point arguments, martingale techniques and stochastic coupling constructions. First, one needs to establishing well-posedness of the nonlinear SDE (2.6). An analytical approach to this problem entails proving uniqueness of the nonlinear PDE describing the evolution of the marginal law. Another, more probabilistic, approach is to first consider the mapping that takes any continuous measure flow $t \mapsto \nu(t) \in \mathcal{P}^2(\mathbb{R})$ to the measure flow $t \mapsto \mathcal{L}(X^\nu(t))$, where X^ν is the unique solution to the SDE in (2.6) when μ is replaced with ν. Observing that the flow $t \mapsto \mathcal{L}(X_o(t))$ must be a fixed point for this mapping, well-posedness is equivalent to uniqueness of the fixed point of this mapping. The latter can be established by showing the mapping is a contraction by exploiting the Lipschitz continuity of the drift. Given well-posedness, the coupling approach to proving convergence proceeds by first defining $\bar{X}^n$ to be the n-dimensional process whose every coordinate is an independent copy of the nonlinear process X_o. Then one couples this process with the original process X^n so that they are both driven by the same Brownian motions. Using Itô's formula, the Lipschitz condition on the drift and standard estimates, one can then show that the $\mathcal{W}_{\mathbb{R},2}$ distance between the empirical measures of X^n and $\bar{X}^n$ vanishes as $n \to \infty$. Since the strong law of large numbers ensures that the empirical measure of the latter converges to the law of X_o, which is equal to μ, this concludes the proof. An alternative approach to proving convergence is to first use the generator of the Markov process X^n to identify martingales involving the empirical measure process μ^n, next show that the sequence $\{\mu^n\}$ is relatively compact (or tight), then characterize any subsequential limit satisfies what is known as a nonlinear martingale problem, and finally establish well-posedness of the latter [33, 62].

Remark 2.3. One can consider more general dynamics where both the drift and diffusion coefficients are functions of the current state and the empirical measure process, as well as non-Markovian versions that depend on the history of the process.

2.3. Interacting jump processes and their mean-field limits

2.3.1. Description of dynamics

We will also be interested in interacting pure jump processes, which describe models in statistical physics, engineering, epidemiology and the dynamics of opinion formation [7, 53]. For concreteness, consider the voter model [53] that aims to capture opinion dynamics, in which each particle takes values in the state space $\mathcal{X} = \{0, 1\}$ that represents two possible opinions. The allowed transitions or jump directions of a particle lie in the set $\mathcal{J} = \{1, -1\}$. The rate at which any particle changes its opinion is equal to the fraction of its neighbors with the opposite opinion. Note that the dependence of the rate on the neighboring states is symmetric. More generally, when the state of the system is $(x_v)_{v\in V}$, the jump rate of a particle at v could be a more complicated symmetric functional of the neighboring states $(x_u)_{u\sim v}$ and also depend on time t, in addition to its own state x_v. This symmetric dependence on neighboring states is most succinctly captured by saying the rate is a functional of the *unnormalized* empirical measure $\theta_v = \sum_{u\sim v} \delta_{x_u}$ of the neighboring states. Note that θ_v lies in the space $\mathcal{M}(\mathcal{X})$ of locally finite nonnegative integer-valued measures on $\mathcal{X}$.

In the general setup, we consider a finite state space $\mathcal{X}$, a subset $\mathcal{J} \subset \{i - j : i, j \in \mathcal{X}\}$ of possible jump directions, and a collection of jump rate functions $\bar{r}_j : \mathbb{R}_+ \times \mathcal{X} \times \mathcal{M}(\mathcal{X}) \to \mathbb{R}_+$, $j \in \mathcal{J}$. Given a (simple) finite graph $G = (V, E)$ and initial condition $\xi = (\xi_v)_{v\in V} \in \mathcal{X}^V$, the $\mathcal{X}^V$-valued process representing the configuration of the associated IPS evolves according to the following system of (jump) SDEs:

$$X_v^{G,\xi}(t) = \xi_v + \sum_{j\in\mathcal{J}} j \int_{(0,t]\times\mathbb{R}_+} \mathbb{I}_{\{r\le\bar{r}_j(s, X_v^{G,\xi}(s-), \theta_v^{G,\xi}(s-))\}} N_v(ds, dr), \quad t \ge 0, v \in V, \tag{2.9}$$

where $(N_v)_{v\in V}$, are i.i.d. Poisson random measures on $\mathbb{R}_+^2$ with intensity measure Leb^2, where Leb represents Lebesgue measure on $\mathbb{R}$, and for each $s \ge 0$, $\theta_v^{G,\xi}(s)$ is the random (unnormalized) empirical measure corresponding to the states of the particles in the neighborhood of v at time s:

$$\theta_v^{G,\xi}(s) := \sum_{u\sim v} \delta_{X_u^{G,\xi}(s)}, \quad v \in V, s \ge 0. \tag{2.10}$$

The SDE (2.9) captures a simple evolution. For any $j \in \mathcal{J}$ and time t, the particle at a node v makes a transition from its state $X_v^{G,\xi}(t-)$ to $X_v^{G,\xi}(t-) + j$ at a rate $\bar{r}_j(t, x, \theta_v^{G,\xi}(t-))$ that depends on the current time, the state of the node just prior to the current time, and symmetrically on the states of neighboring nodes just prior to the current time, as encoded by $\theta_v^{G,\xi}(t-)$. Use of the unnormalized measure $\theta_v^{G,\xi}(t)$ instead of the empirical measure allows one to capture a broader class of models in which jump rates depend on the number of neighboring nodes in particular states (and not just their fractions), as is the case for models like the contact process [53]. Note that the trajectory of each particle lies in the càdlàg space $\mathcal{D}$ of right continuous $\mathcal{X}$-valued functions on $[0,\infty)$ that have finite left limits on $(0,\infty)$.

The solution $X^{G,\xi}$ to the jump SDE (2.9) is a Markov jump process and so its law can also be characterized via the associated *infinitesimal generator* [52]: for functions $f : \mathcal{X}^V \mapsto \mathbb{R}$,

$$\begin{aligned}\mathcal{A}_t f(x) &= \lim_{h\downarrow 0}\frac{\mathbb{E}[f(X^{G,\xi}_{t+h}) - f(X^{G,\xi}_t)|X^{G,\xi}_t = x]}{h}\\ &= \sum_{j\in\mathcal{J},v\in V}\bar{r}_j\Big(t,x_v,\sum_{u\sim v}\delta_{x_u}\Big)\big[f(x+je_v)-f(x)\big], \quad t>0, x\in\mathcal{X}^V,\end{aligned}$$

where $e_v \in \{0,1\}^V$ is the vector with 1 in the vth coordinate and 0 elsewhere. However, the jump SDE representation in (2.9) is more convenient for generalizations to non-Markovian processes (see [29]). Furthermore, the jump SDE formulation is also better suited to describing the form of limiting marginal dynamics on sparse graphs, as described in Section 4.4.

2.3.2. Mean-field limits and nonlinear jump processes

Mean-field results analogous to Theorem 2.2 also hold in the jump setting under the following regularity assumption on the jump rate functions:

Assumption 2.4. For each $j \in \mathcal{J}$, the jump rate function takes the form $\bar{r}_j(t,x,\theta) = \hat{r}_j(t,x,\frac{1}{\theta(\mathcal{X})}\theta)$ when $\theta(\mathcal{X}) \neq 0$, and $\bar{r}_j(t,x,\theta) = 0$ otherwise, where the function $\hat{r}_j : \mathbb{R}\times\mathcal{X}\times\mathcal{P}^1(\mathcal{X}) \mapsto \mathbb{R}_+$ is such that $\mathcal{P}^1(\mathcal{X}) \ni \nu \mapsto \hat{r}(t,x,\nu)$ is Lipschitz continuous, uniformly for $x \in \mathcal{X}$ and t in compact subsets of $\mathbb{R}_+$.

Assumption 2.4 reflects the fact that in the mean-field setting, the dependence of the jump rates on the neighboring particles must be a sufficiently regular function of the usual (normalized) empirical measure. The following result is established in [62, THEOREM 2]; see also [44].

Theorem 2.5. *Suppose Assumption 2.4 holds and the initial conditions are chaotic, that is,* $\frac{1}{n}\sum_{i=1}^n \delta_{\xi_i^n}$ *converges in the total variation metric to a deterministic limit* μ_0*, then* $\mu^{K_n,\xi^n}(t)$ *converges weakly to* $\mathcal{L}(X_o(t))$ *where* $X_o(t) = X^{\mu_0}(t)$, $t \geq 0$, *is the unique solution to the following nonlinear jump SDE:* $\mathcal{L}(X_o(0)) = \mu_0$*, and for* $t \geq 0$,

$$\begin{aligned}X_o(t) &= X_o(0) + \sum_{j\in\mathcal{J}} j\int_{(0,t]\times\mathbb{R}_+}\mathbb{I}_{\{r\leq\hat{r}_j(s,X_o(s-),\mu(s-))\}}N(ds,dr),\\ \mu(t) &= \mathcal{L}\big(X_o(t)\big),\end{aligned} \tag{2.11}$$

where N *is a Poisson process on* $\mathbb{R}^2_+$ *with intensity* Leb^2, *independent of* $X_o(0)$. *Furthermore, for any* $k\in\mathbb{N}$, *the law of* $(X_1^{K_n,\xi},\ldots,X_k^{K_n,\xi})$ *on* $\mathcal{D}^k$ *converges weakly to* $(\mathcal{L}(X_o))^{\otimes_k}$.

Just as the evolution of the law of the mean-field diffusion limit in (2.6) can be characterized by a nonlinear PDE, the evolution of the law of the nonlinear jump process X_o

in (2.11) can be characterized as the unique solution to its forward equation, which is now a nonlinear integrodifferential equation.

Remark 2.6. Theorems 2.2 and 2.5 are meant to only provide a flavor of mean-field results. While a survey of mean-field limits is not the current focus, it is worth mentioning that in both the diffusive and jump process settings, one can obtain mean-field limits under weaker assumptions and for much more general dynamics where the diffusion coefficient is also a function of the current state and empirical measure process, as well as non-Markovian versions where the drift coefficient or jump rates depend on the history of the process (see, e.g., [59] for propagation of chaos results on interacting non-Markovian jump diffusions and [3] for a large deviations analysis of non-Markovian weakly interacting diffusions).

2.4. Limitations of mean-field approximations

The mean-field limit theorems established in Theorems 2.2 and 2.5 indicate that the law of the nonlinear Markov processes X_o in (2.6) and (2.11), respectively, can be used to approximate quantities of interest for interacting diffusions or jump processes on finite graphs. In particular, consider the voter model described in Section 2.3.1. Its jump rates take the explicit form

$$\bar{r}_1(t,x,\theta) = \frac{\mathbb{I}_{\{x=0\}}}{|\theta(\mathcal{X})|}\int_{\mathcal{X}} y\theta(dy), \quad \bar{r}_{-1}(t,x,\theta) = \frac{\mathbb{I}_{\{x=1\}}}{|\theta(\mathcal{X})|}\int_{\mathcal{X}} (1-y)\theta(dy).$$

In this case, one would expect that the dynamics of X_o in (2.11) with these reates could provide an approximation for the probability of agreement of any two neighboring particles in the voter model on a sufficiently large complete graph. However, for lack of a better alternative, mean-field approximations are used even for dynamics on other graphs. While these approximations may do reasonably well on dense graphs (where vertices have high degrees) [19], they can be very inaccurate on sparse graphs. Figure 2 plots the evolution of the probabil-

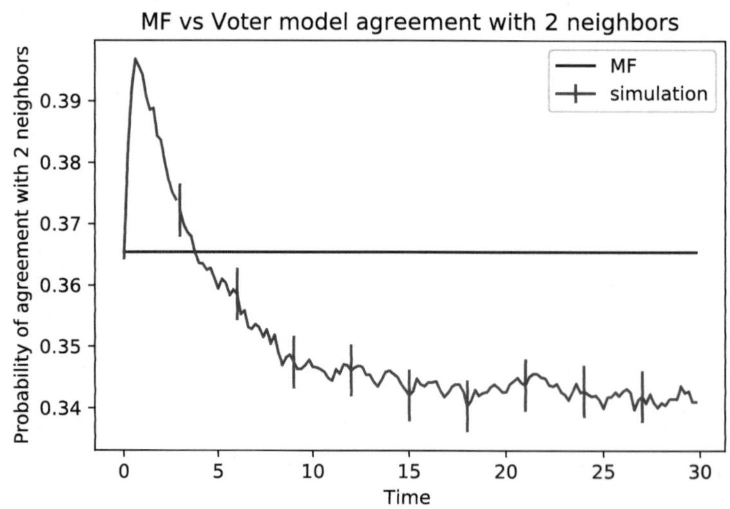

FIGURE 2
Simulations and mean-field (MF) approximations for the voter model on a 3-tree.

ity that the state of the root agrees with precisely two of its neighbors for the voter model on a rooted 3-regular tree with 9 generations, given at time zero, each particle independently has an opinion 1 with probability 0.3. The vertical bars in Figure 2 provide confidence intervals for the simulation. The mean-field approximation assumes neighboring vertices are independent, and thus performs poorly. Ad hoc refinements of the mean-field approximation that take into account correlations also remain inaccurate in this setting. This strongly motivates the development of a convergence theory for the empirical distribution and marginal dynamics on sparse graph sequences that could lead to more principled approximations.

3. INTERACTING PROCESSES ON SPARSE GRAPHS: HYDRODYNAMIC LIMITS

We now turn to interacting processes $(X^{G_n,\xi^n})_{n\in\mathbb{N}}$ on sparse graph sequences $(G_n)_{n\in\mathbb{N}}$ with initial conditions $(\xi^n)_{n\in\mathbb{N}}$. Assume each G_n is finite and o_n is a vertex chosen uniformly at random from the vertices of G_n. Unlike in the case of the complete graph (or even dense graph sequences), the degree of a vertex remains bounded and so neighboring vertices do not become asymptotically independent. Thus, the number of neighbors becomes important and so it is clear that one cannot expect $(X_{o_n}^{G_n,\xi^n})_{n\in\mathbb{N}}$ to have a limit just by sending the number of vertices n to infinity, without imposing any additional consistency requirements on the graphs in the sequence. This leads to the following questions:

Q1. For what graph sequences $(G_n)_{n\in\mathbb{N}}$ would one expect $(X_{o_n}^{G_n,\xi^n})_{n\in\mathbb{N}}$ to have a limit?

Q2. For such sequences, will $(\mu^{G_n,\xi^n})_{n\in\mathbb{N}}$ converge to a deterministic limit?

Q3. When $(\mu^{G_n,\xi^n})_{n\in\mathbb{N}}$ converges to a deterministic limit, will this limit always coincide with the limit law of $X_{o_n}^{G_n,\xi^n}$?

Q4. Is there an autonomous reduced-dimension description of the limit of the marginal $X_{o_n}^{G_n,\xi^n}$ whenever this limit exists?

In light of the first question above, we review a natural notion of convergence of sparse graphs called local convergence in Section 3.1. This notion was used to study asymptotic properties of static models (Gibbs measures) of discrete-valued marked random graphs in [20].

3.1. Local convergence of sparse graph sequences

Given a graph $G=(V,E)$ and two vertices $u,v\in V$, a path of length n between u and v is a sequence $u=u_0,u_1,\dots,u_n=v$ such that $u_{i-1}\sim u_i$ for every $i=1,\dots,n$. A graph is said to be connected if there exists a finite path between any two vertices and the graph distance between two vertices is the minimum length of a path between them. A rooted graph (G,o) is a graph $G=(V,E)$ with a special vertex $o\in V$, referred to as the root. A useful notion of convergent sequences of (connected) rooted sparse graphs is that of *local convergence*, which was introduced by Benjamini and Schramm [5]. Other references on local convergence include [1, 8]. We first introduce some terminology that is required to define local convergence. An *isomorphism* from one rooted graph (G_1, o_1) to another (G_2, o_2) is a bijection φ from the vertex set of G_1 to that of G_2 such that $\varphi(o_1) = o_2$ and such that (u, v) is an edge in G_1 if and only if $(\varphi(u), \varphi(v))$ is an edge in G_2. Two rooted graphs are said to be *isomorphic* if there exists an isomorphism between them. Let $\mathcal{G}_*$ denote the set of isomorphism classes of connected rooted graphs. We will also need to consider convergence of graphs that carry "marks" representing the initial condition or trajectory of the state dynamics at that vertex. With that in mind, given a Polish space $\mathcal{S}$, we define a *$\mathcal{S}$-marked rooted graph* to be a tuple (G, x, o), where (G, o) is a rooted graph and $x = (x_v)_{v \in G} \in \mathcal{S}^G$ is a vector of marks, indexed by the vertices of G. We say that two marked rooted graphs (G_1, x^1, o_1) and (G_2, x^2, o_2) are *isomorphic* if there exists an isomorphism φ from the rooted graph (G_1, o_1) to the rooted graph (G_2, o_2) that maps the marks of (G_1, o_1) to the marks of (G_2, o_2) (i.e., for which $x^2_{\varphi(v)} = x^1_v$ for all $v \in G$). Let $\mathcal{G}_*[\mathcal{S}]$ denote the set of isomorphism classes of $\mathcal{S}$-marked rooted graphs.

We now define the topologies of local convergence on the spaces $\mathcal{G}_*$ and $\mathcal{G}_*[\mathcal{S}]$. For $r \in \mathbb{N}$ and $(G, o) \in \mathcal{G}_*$, let $B_r(G, o)$ denote the induced subgraph of G (rooted at o) containing those vertices with (graph) distance at most r from the root o. The distance between (G_1, o_1) and (G_2, o_2) in $\mathcal{G}_*$ is defined to be $1/(1 + \bar{r})$, where $\bar{r}$ is the supremum over $r \in \mathbb{N}_0$ such that $B_r(G_1, o_1)$ and $B_r(G_2, o_2)$ are isomorphic, where we interpret $B_0(G_i, o_i) = \{o_i\}$. Now, let d denote a metric that induces the Polish topology on $\mathcal{S}$. We then metrize $\mathcal{G}_*[\mathcal{S}]$ by similarly defining the distance between two $\mathcal{S}$-marked graphs (G_i, x^i, o_i), $i = 1, 2$, to be $1/(1 + \bar{r})$, where now $\bar{r}$ is the supremum over $r \in \mathbb{N}_0$ such that there exists an isomorphism φ from $B_r(G_1, o_1)$ to $B_r(G_2, o_2)$ for which $d(x^1_v, x^2_{\varphi(v)}) \le 1/r$ for all $v \in B_r(G_1, o_1)$. Under the respective topologies, $\mathcal{G}_*$ and $\mathcal{G}_*[\mathcal{S}]$ are Polish spaces (see [8, LEMMA 3.4] or [46, APPENDIX A]). For any Polish space $\mathcal{S}$, let $C_b(\mathcal{S})$ denote the space of bounded continuous functions on S.

We will always assume the spaces $\mathcal{G}_*$ and $\mathcal{G}_*[\mathcal{S}]$ are equipped with their Borel σ-algebras. One can then talk about weak convergence or convergence in distribution of random graphs and random marked graphs as random elements in $\mathcal{G}_*$ or $\mathcal{G}_*[\mathcal{S}]$. Specifically, a sequence of random $\mathcal{G}_*$-valued random elements $\{(G_n, o_n)\}$ is said to converge in distribution in the local weak sense to a $\mathcal{G}_*$-valued limit (G, o) if for every bounded continuous function $f : \mathcal{G}_* \to \mathbb{R}$, $\mathbb{E}[f(G_n, o_n)] \to \mathbb{E}[f(G, o)]$. Likewise, convergence in distribution in the local weak sense of (isomorphism classes of) random $\mathcal{S}$-marked graphs is equivalent to weak convergence on the space $\mathcal{G}_*[\mathcal{S}]$.

Remark 3.1. Figure 3 illustrates two generic examples of locally convergent graph sequences. Let G_n be the n-cycle, which is the connected graph on n vertices where every vertex has degree 2, along with the root o_n chosen uniformly at random from the n vertices. Then (G_n, o_n) converges weakly in $\mathcal{G}_*$ to a infinite line graph rooted at some fixed vertex; see Figure 3(a). A less trivial example is illustrated in Figure 3(b). Given $c \in (0, \infty)$, the sequence of Erdős–Rényi graphs $G(n, c/n)$ converges in distribution in the local weak

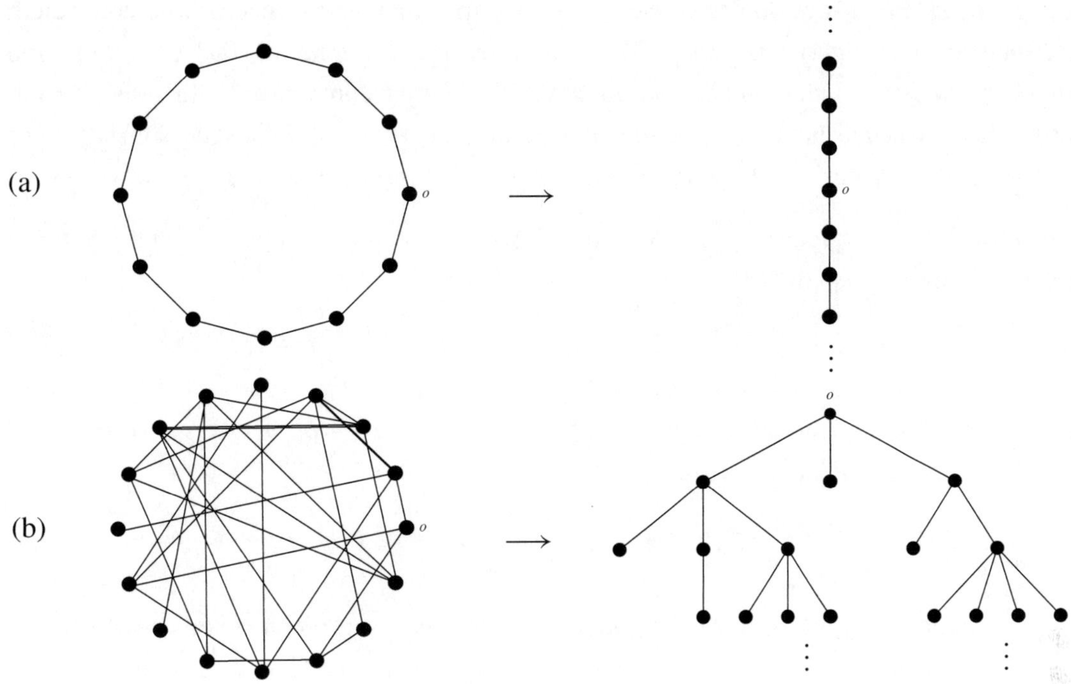

FIGURE 3
Local convergence: (a) cycle to infinite line; (b) Erdős–Rényi graph to a UGW tree.

sense to the Galton–Watson (GW) tree with offspring distribution given by the Poisson(c) distribution. The latter is an example of a unimodular Galton–Watson (UGW) tree, which is defined as follows. Given a probability distribution ρ on $\mathbb{N} \cup \{0\}$ that has finite nonzero first moment, that is, satisfies $0 < \sum_{k \in \mathbb{N}} k\rho(k) < \infty$, the random tree UGW($\rho$) has a root whose neighborhood size is distributed according to ρ. The neighbors of the vertices are referred to as the offspring of the root and form the first generation of the tree. Recursively, for $n \geq 1$, each vertex in the nth generation of the tree has an independent random number of offspring (equivalently, neighbors that are further away from the root than itself) with distribution $\hat{\rho}$

$$\hat{\rho}(k) = \frac{(k+1)\rho(k+1)}{\sum_{n \in \mathbb{N}} n\rho(n)}, \quad k \in \mathbb{N} \cup \{0\}. \tag{3.1}$$

The $(n+1)$th generation of the tree is comprised of all offspring of vertices in the nth generation. It is easy to verify that if ρ is a Poisson distribution, then $\hat{\rho} = \rho$. Hence, a Galton–Watson tree with a Poisson(c) offspring distribution is in fact a UGW (Poisson(c)) distribution. Another special case is the κ-regular tree, for $\kappa \geq 2$, which is given by $\mathbb{T}_\kappa :=$ UGW (δ_κ). UGW trees are in a sense canonical objects since they arise as local weak limits of many sparse random graph sequences including Erdős–Rényi graphs, configuration models and preferential attachment graphs; see **[46, SECTION 2.2.4]** for further discussion of these examples.

To extend this notion of convergence to graphs that are not necessarily connected, given an (unrooted) graph $G = (V, E)$ and a vertex $v \in V$, define $\mathsf{C}_v(G) \in \mathcal{G}_*$ to be the isomorphism class of the connected component of G that contains v, with v as its root. Furthermore, when G is finite, we let o denote a random vertex of G chosen uniformly from the set V, in which case $\mathsf{C}_o(G)$ denotes the connected component of that random vertex.

Definition 3.2. A sequence of finite (random) graphs $\{G_n\}$ is said to *converge in distribution in the local weak sense* to G if

$$\lim_{n\to\infty} \mathbb{E}\left[\frac{1}{|G_n|}\sum_{v\in G_n} f\big(\mathsf{C}_v(G_n)\big)\right] = \mathbb{E}\big[f(G)\big], \quad \forall f \in C_b(\mathcal{G}_*). \tag{3.2}$$

A sequence of finite (random) graphs $\{G_n\}$ is said to *converge in probability in the local weak sense* to G if for every $\varepsilon > 0$,

$$\lim_{n\to\infty} \mathbb{P}\left(\left|\frac{1}{|G_n|}\sum_{v\in G_n} f\big(\mathsf{C}_v(G_n)\big) - \mathbb{E}\big[f(G)\big]\right| > \varepsilon\right) \to 0, \quad \forall f \in C_b(\mathcal{G}_*). \tag{3.3}$$

Analogously, given a marked (unroooted, not necessarily connected) graph (G, x), $\mathsf{C}_v((G, x))$ denotes the connected component of G containing v, with v as its root and with the corresponding marks. The notions of convergence in distribution and in probability in the local weak sense for marked graphs are defined in an exactly analogous fashion as Definition 3.2, with $\mathsf{C}_v((G_n, x^n))$, $\mathsf{C}_v((G,x))$ and (G, x) in place of $\mathsf{C}_v(G_n)$, $\mathsf{C}_v(G)$ and G, respectively. For both unmarked and marked graphs, convergence in probability clearly implies convergence in distribution. We will use the same notation for graphs and their isomorphism classes and often omit the root from the notation and simply refer to $G \in \mathcal{G}_*$ rather than $(G, o) \in \mathcal{G}_*$.

Remark 3.3. Given any sequence of random graphs $\{G_n\}_{n\in\mathbb{N}}$ that converges (either in distribution or in probability) in the local weak sense to a limit graph G, if $x^n = (x^n_v)_{v\in G_n}$ are i.i.d. marks on some Polish space $\mathcal{S}$ with the same distribution irrespective of n, then it is easy to show that the marked graph sequence $\{(G_n, x^n)\}_{n\in\mathbb{N}}$ also converges (in the same local weak sense as the unmarked counterparts) to (G, x), where $x = (x_v)_{v\in G}$ is i.i.d. with the same distribution. In fact, as shown in [46, PROPOSITION 2.16], convergence of the marked graph sequence holds for the larger class of possibly dependent marks distributed according to a Gibbs measure on the graph with respect to a fixed pairwise interaction functional.

3.2. Hydrodynamic limits

3.2.1. Interacting Diffusion processes

We now address Q1–Q3 raised at the beginning of Section 3. The first result below states that if a sequence of graphs marked with initial conditions converges (either in probability or in distribution) in the local weak sense to a limit graph, then the graphs marked with the trajectories that solve the corresponding SDE also converge to the limit graph in the same sense. The result also characterizes the limit of the global empirical measure under suitable conditions.

Theorem 3.4. *Suppose Assumption 2.1 holds, and the sequence $\{(G_n, \xi^n)\}_{n\in\mathbb{N}}$ of (not necessarily connected, finite) random marked graphs converges in distribution in the local weak sense to a $\mathcal{G}_*[B_r(\mathbb{R})]$-valued limit (G, ξ) for some $r > 0$. Also, for each $n \in \mathbb{N}$, let X^{G_n,ξ^n} be the solution to the SDE (2.3) with initial data (G_n, ξ^n) and let $X^{G,\xi}$ be the unique weak solution to the SDE (2.3) on the limit graph (G, ξ). Then $\{(G_n, X^{G_n,\xi^n})\}_{n\in\mathbb{N}}$ converges in distribution in the local weak sense to the $\mathcal{G}_*[\mathcal{C}]$-valued element $(G, X^{G,\xi})$. In particular, $\{X^{G_n,\xi^n}_{o_n}\}_{n\in\mathbb{N}}$ converges weakly to $X^{G,\xi}_o$. Moreover, if $\{(G_n, \xi^n)\}_{n\in\mathbb{N}}$ converges in probability in the local weak sense to (G, ξ), then $\{(G_n, X^{G_n,\xi^n})\}_{n\in\mathbb{N}}$ also converges in probability in the local weak sense to $(G, X^{G,\xi})$ and additionally, $\{\mu^{G_n,\xi^n}\}_{n\in\mathbb{N}}$ converges weakly to the law of $X^{G,\xi}_o$.*

This result follows from [46, THEOREMS 3.3 AND 3.7], which establish this result for more general, possibly non-Markovian diffusive dynamics. A version of the first assertion of the above theorem was also established for a slightly different class of interacting diffusions in [64]. Theorem 3.4 can be seen as establishing continuity in the local weak topology of the dynamics with respect to the initial data, comprising the graph marked with initial conditions. It also provides conditions under which the empirical measure can be shown to have a deterministic limit (equivalently, hydrodynamic limit) that additionally coincides with the limit law of the root particle, thus answering in the affirmative Q2 and Q3 at the beginning of Section 3. As discussed earlier, on complete graphs the analogous phenomena holds due to asymptotic independence of the trajectories of any two particles. In contrast, in the sparse regime, neighboring particles remain dependent in the limit. Instead, the proof relies on showing that the trajectories on finite neighborhoods of two independent vertices, both chosen uniformly at random from the graph become asymptotically independent in the limit. The latter property relies on a certain correlation decay property of the dynamics in the spirit of (1.3); see [46, LEMMA 5.2] for details.

However, it should be emphasized that in the sparse regime the deterministic hydrodynamic limit result holds *only* when the initial data converges in the stronger sense of convergence *in probability* in the local weak sense. Indeed, as shown in [46, THEOREMS 3.9 AND 6.4], the limiting empirical measure can be stochastic when the initial data only converges in distribution in the local weak sense. In particular, fix $c \in (0, \infty)$ and suppose $\tilde{G}_n$ is the graph obtained by taking the *connected component* of a vertex chosen uniformly at random from the Erdős–Rényi graph $G(n, c/n)$, and setting the root to be that chosen vertex. Also, let the initial conditions $\xi^n = \{\xi^n_v\}_{v\in G_n}$ be i.i.d. with common distribution γ, and let $\tilde{\xi}^n$ denote the restriction of the initial conditions to $\tilde{G}_n$. Then both $\{(G_n, \xi^n)\}_{n\in\mathbb{N}}$ and $\{(\tilde{G}_n, \tilde{\xi}^n)\}_{n\in\mathbb{N}}$ converge in distribution in the local weak sense to (G, ξ), where $\xi = (\xi_v)_{v\in V}$ is i.i.d. with distribution γ, and $G = \mathcal{T} :=$ UGW(Poiss(c)). However, whereas $\mu^{G_n,\xi}$ converges weakly to $\mathcal{L}(X^{\mathcal{T},\xi}_o)$, the law of the dynamics at the root vertex in $\mathcal{T}$, $\mu^{\tilde{G}_n,\tilde{\xi}^n}$ converges weakly to the following *random limit* $\tilde{\mu}^{\mathcal{T},\xi}$ given by

$$\tilde{\mu}^{\mathcal{T},\xi} := \begin{cases} \mu^{\mathcal{T},\xi} & \text{on the event } |\mathcal{T}| < \infty, \\ \mathrm{Law}\big(X^{\mathcal{T},\xi}_o \mid |\mathcal{T}| = \infty\big) & \text{on the event } |\mathcal{T}| = \infty. \end{cases}$$

This limit is truly stochastic because, as is well known from the elementary theory of branching processes [2], there is always a positive probability for the UGW tree $\mathcal{T}$ to be finite (and there is a positive probability of $\mathcal{T}$ being infinite only when $c > 1$). Furthermore, there also exist examples that show that even when the limiting empirical measure is deterministic, it need not coincide with the law of the root particle in the limit. For example, this can occur if the graph itself is not homogeneous and the root is not chosen uniformly at random from the vertices of the graph (see, e.g., [46, SECTION 3.6]). The above discussion shows that the existence and nature of the hydrodynamic limit is far more subtle in the sparse regime than in the case of complete or dense graphs, even for diffusive dynamics.

3.2.2. Interacting jump processes

In the setting of jump processes, additional subtleties arise. Whereas in the diffusion setting, Assumption 2.1 ensures that the drift of a particle at a node v experiences only an $O(1/|\partial v|)$ effect when there is a perturbation in the state of a neighboring particle, an analogous assumption would be too stringent to cover most jump models of interest on sparse graphs. In the latter case, the effect of a neighboring particle on the jump intensity at a vertex either remains constant or $O(1)$ as in the voter model, or grows with the degree of the vertex in many other models, including the contact process (see [53]). It is precisely to accommodate such a dependence that the jump intensity $\bar{r}_j$ in (2.9) is expressed as a function of the *unnormalized* sum of the Dirac masses at neighboring states introduced in (2.10), rather than the normalized empirical measure. For hydrodynamic limits on sparse graphs, it will suffice to impose the following mild assumption on the rates.

Assumption 3.5. For every $T > 0$, suppose there exist constants $C_{k,T}$, $k \in \mathbb{N}$, such that $k \mapsto C_{k,T}$ is nondecreasing and for every $j \in \mathcal{J}$, $\sup_{x \in \mathcal{X}, t \in [0,T]} \bar{r}_j(t, x, \theta) \leq C_{\theta(\mathcal{X}),T}$.

Note that in the jump SDE (2.9) describing the dynamics, the third argument θ of $\bar{r}_j$ is equal to the unnormalized empirical measure of the states of the neighbors of a vertex, as defined in (2.10). Thus, $\theta(\mathcal{X})$ irepresents the degree of the vertex and Assumption 3.5 allows the uniform bound on the jump rates at a vertex to grow with the degree of a vertex.

We now state an analog of Theorem 3.4 for jump diffusions. Recall the definition of a UGW tree given in Remark 3.1.

Theorem 3.6. *Suppose Assumption 3.5 holds, and the sequence $\{(G_n, \xi^n)\}_{n \in \mathbb{N}}$ of (not necessarily connected, finite) random rooted marked graphs converges in probability in the local weak sense to a limit (G, ξ), where G is a UGW(ρ) tree with ρ having finite, strictly positive first and second moments. For each $n \in \mathbb{N}$, let X^{G_n, ξ^n} be the solution to the jump SDE (2.9) with initial data (G_n, ξ^n), and let $X^{G,\xi}$ be the unique strong solution to the SDE (2.9) on the limit graph (G, ξ). Then $\{(G_n, X^{G_n,\xi^n})\}_{n \in \mathbb{N}}$ converges in probability in the local weak sense to the $\mathcal{G}_*[\mathcal{D}]$-valued element $(G, X^{G,\xi})$. Furthermore, $\{\mu^{G_n,\xi^n}\}_{n \in \mathbb{N}}$ converges weakly to the law of $X_o^{G,\xi}$.*

This theorem follows from more general results established in [29, THEOREM 4.8 AND COROLLARY 5.16]. As in the diffusion case, the proof of the theorem involves establishing continuity properties of the dynamics with respect to the graph and initial condition as well as a correlation decay property. However, the proofs of these properties are considerably more involved than in the diffusion case due to the weaker conditions imposed on the jump intensities in Assumption 3.5. For one, in the jump setting even well-posedness of the particle system on an infinite random graph of unbounded degree is not automatic. As shown in [29, APPENDIX B], there exist examples of simple jump particle systems with uniformly bounded jump rate functions that can have multiple solutions on certain graphs with exponential growth. To quote Liggett [54], "Given an intuitive description of the behavior of the particles, it is often not clear whether or not there exists a ... process which corresponds to that description. Therefore it is important to find conditions under which infinite particle systems exist." On finite graphs, there are only finitely number of jumps in any bounded interval, and the process remains constant between jumps. Thus, one can simply order the jumps and define the process recursively. The problem in the infinite graph setting is that one cannot always identify a "first" jump. In [54], Liggett used an analytical construction invoking the theory of semi-groups to establish a general existence theorem for the law of Markovian particle systems on infinite graphs with quite general (not necessarily finite-range) interactions. In the context of nearest-neighbor interactions on lattices, an alternative, probabilistic approach was used to establish well-posedness of Markovian interacting particle systems in Harris [36, 37]. However, both approaches seem to only apply to graphs with finite maximal degree. On the other hand, graphs of particular interest like the UGW(Poisson(c)) tree discussed above (see Remark 3.1) have unbounded degrees. Under Assumption 3.5, well-posedness of (possibly non-Markovian) interacting jump processes was established in [29, THEOREM 4.3] for a large class of possibly random graphs that satisfy a certain "finite dissociability" property almost surely, and this property was shown to hold for UGW trees in [29, COROLLARY 5.16]. The proof of Theorem 3.6 then follows on combining this well-posedness result with continuity properties of the dynamics (with respect to the initial data) and a correlation decay property (established in [29, PROPOSITION 6.8] and [29, THEOREM 4.9], respectively).

4. MARGINAL DYNAMICS ON TREES

The hydrodynamic limit result reduces the characterization of the limit law of $X_{o_n}^{G_n,\xi^n}$ to the understanding of the marginal law $X_o^{G,\xi}$ of the root dynamics on the *infinite* limit graph G. Given that local weak limits of many random graphs are trees (see Remark 3.1), we focus here on understanding marginal dynamics on (random) trees. The evolution of the root $X_{o_n}^{G_n,\xi^n}$ in (2.1) or (2.9) is driven by the *local* neighborhood empirical measure $\mu_{o_n}^{G_n,\xi^n}$ or its unnormalized counterpart $\theta_{o_n}^{G_n,\xi^n}$. In the complete (or sufficiently dense) graph case, in the limit as the number of particles goes to infinity, the local empirical measure coincides with the global empirical measure. Thus, in this case the hydrodynamic limit yields an autonomous characterization of the limit marginal dynamics. In contrast, when the graph sequence G_n is sparse, neighboring vertices remain strongly correlated, the local neighborhood empirical measure remains stochastic and thus the hydrodynamic limit results in Section 3 are not adequate to provide an autonomous characterization of the marginal dynamics.

Instead, we adopt a different perspective, which is better suited to the analysis of large collections of dependent random elements. As mentioned in Section 1.1, as a first step we try to identify the conditional independence structures in such random variables. To this end, we identify a certain Markov random field (MRF) property for the trajectories of $X^{G,\xi} = \{X_v^{G,\xi}, v \in G\}$ in Section 4.1 below. We then describe how to exploit this property, along with filtering results from stochastic analysis and symmetry properties of the graph, to identify an autonomously defined "local equation" satisfied by marginal dynamics on cl_o, the *root and its neighborhood.* We do this first for diffusions on the line in Section 4.2, then for diffusions on UGW trees in Section 4.3, and finally for jump processes in Section 4.4. Unlike in the mean-field case, consideration of the marginal at the root *and* its neighborhood (rather than just at the root), appears necessary in order to obtain an autonomous characterization. This is also necessary in order to capture correlations between neighboring vertices, which do not vanish in the sparse regime. Further discussion of the local equation is given in Sections 4.2-4.4. But it is worth noting here that since the graphs we consider are locally finite, the neighorhood of the root is (almost surely) finite. Thus, the local equation describes the evolution of an (almost surely) finite number of interacting particles. When combined with the convergence results of Theorems 3.4 and 3.6, this finite-dimensional interacting process serves as an approximation for the marginals of (possibly non-Markovian) interacting processes with an arbitrarily large number of particles, and thus constitutes a significant dimension reduction.

4.1. A Markov random field property

We first introduce the definition of an MRF.

Definition 4.1. Fix a measurable space $\mathcal{Y}$, and a (possibly infinite, but locally finite) graph $G = (V, E)$. A random element $Y = (Y_v)_{v\in V}$ is said to be a *(first-order) MRF* (abbreviated as MRF) on $\mathcal{Y}^V$ with respect to G if for every *finite* set $A \subset V$, Y_A is conditionally independent of $Y_{(A\cup\partial A)^c}$ given $Y_{\partial A}$, which we denote as

$$Y_A \perp\!\!\!\perp Y_{(A\cup\partial A)^c} | Y_{\partial A}. \tag{4.1}$$

On the other hand, Y is said to be a *(first-order) global MRF* if (4.1) holds for all $A \subset V$, possibly infinite. Furthermore Y is said to be a *(first-order) semi-global MRF* (abbreviated as SGMRF) if (4.1) holds for all $A \subset V$ such that ∂A is finite. Furthermore, $Y = (Y_v)_{v\in V}$ is said to be a *second-order MRF* or 2-MRF (respectively, second-order SGMRF or 2-SGMRF) with respect to G if it is an MRF (respectively, SGMRF) with respect to the square graph $G^2 = (V, E^2)$, where E^2 contains E as well as vertex pairs that are a distance two apart in G. In all cases, we will say $\nu \in \mathcal{P}(\mathcal{Y}^V)$ exhibits a certain MRF property whenever some $\mathcal{Y}^V$)-valued random element Y with law ν satisfies that MRF property.

The SGMRF property, introduced in **[29]**, can be viewed as a generalization of tree-indexed Markov chains **[34, CHAPTER 12]** to general graphs, and is clearly strictly stronger than the MRF property. For any $t > 0$, and collections of paths $x = (x_v)_{v\in V}$ (either in $\mathcal{C}$ or $\mathcal{D}$), let

$$x_v[t] := \big(x_v(s)\big)_{s\in[0,t]} \quad \text{and} \quad x_v[t) := \big(x_v(s)\big)_{s\in[0,t)} \tag{4.2}$$

represent the trajectory of x_v in the intervals $[0,t]$ and $[0,t)$, respectively, and for any subset $A \subset V$, let $x_A[t] := (x_v[t])_{v\in A}$ and $x_A[t) := (x_v[t))_{v\in A}$. Certain MRF properties are preserved under the evolution of interacting processes, in a sense made precise in the following theorem.

Theorem 4.2. *Let $G = (V, E)$ be a (deterministic) graph with uniformly bounded degree or the almost sure realization of a UGW tree. Suppose the $\mathbb{R}^V$-valued element $\xi = (\xi_v)_{v\in V}$ forms a 2-MRF (or 2-SGMRF) with respect to G, Assumption 2.1 holds and $X^{G,\xi}$ is the unique solution to the diffusive SDE (2.3). Then the $\mathcal{C}$-valued trajectories $X^{G,\xi} = (X_v^{G,\xi})_{v\in V}$ also form a 2-MRF (respectively, 2-SGMRF) with respect to G. On the other hand, suppose Assumption 3.5 holds, $\xi = (\xi_v)_{v\in V}$ is a $\mathcal{X}^V$-valued random element that forms a 2-MRF (or 2-SGMRF) with respect to G, and $X^{G,\xi}$ is the unique solution to the jump SDE (2.9). Then $X^{G,\xi} = (X_v^{G,\xi})_{v\in V}$ also forms a 2-MRF (respectively, 2-MRF) on G in $\mathcal{D}$. In both cases, the same assertions also hold with $X^{G,\xi}$ replaced with $X^{G,\xi}[t]$ or $X^{G,\xi}[t)$ for any $t \geq 0$.*

The discussion in Section 4.2 provides insight into why only the second-order, and not in general the first-order, MRF property is preserved by the dynamics. The preservation of the 2-MRF property for diffusions for graphs with bounded degree follows from [47, THEOREM 2.7]. The proof proceeds by first establishing the result on finite graphs by appealing to Girsanov's theorem and the Gibbs–Markov theorem [34, THEOREM 2.30] (also often referred to as the Hammersley–Clifford theorem), and then suitably approximating infinite systems by a sequence of finite systems. The proof of preservation of both the 2-MRF and 2-SGMRF properties for jump processes in [30, THEOREM 3.7] follows a rather different approach. It exploits a certain duality between marginals of the interacting system and nonexplosive point processes to directly establish an infinite-dimensional Girsanov theorem, obviating the need for any approximation arguments. This approach also allows more general initial conditions that can incorporate infinite histories, which is required to characterize solutions to the local equation described in Section 4.4 as flows on a suitable path space [31]. The result for diffusions in [47] can be generalized in a similar fashion using the approach developed in [30].

Prior work on such questions has largely focused on interacting diffusions, specifically characterizing them as Gibbs measures on path space in order to construct weak solutions to infinite-dimensional SDEs. Deuschel [24] initiated this perspective for diffusions with drifts of gradient type. Although not explicitly stated, the 2MRF property is implicit in his proof of existence of the weak solution, which relies on estimates of Dobrushin's contraction coefficient that crucially require additional smoothness and boundedness properties of the drift. Cattiaux, Roelly, and Zessin [11] relaxed the boundedness condition to allow Markovian, Malliavin differential drifts, using a variational characterization and an integration-by-parts formula. Subsequent works [14,60] used a cluster expansion method that applies to systems obtained as small perturbations of non-interacting systems. Dereudre and

Roelly [22] established Gibbsian properties of paths of interacting one-dimensional diffusions on $\mathbb{Z}^m$ with (possibly history-dependent) drift having sublinear growth using specific entropy, but this crucially requires shift-invariant initial conditions. In another direction, several other works have considered the MRF (or Gibbsian) nature of marginals rather than of paths, both in the diffusion and jump process contexts [42,45,68,69,75], but preservation of this property holds in general only for sufficiently small time horizons or interaction strengths. Furthermore, none of this work seems to have considered the SGMRF property, which is crucial for the derivation of the local equation, as elaborated in Sections 4.2-4.4 below.

4.2. Outline of derivation of the local equation for diffusions on the line

We now describe how the 2-SGMRF property of Theorem 4.2 can be used to obtain an autonomous characterization of the marginal dynamics of the root neighborhood on the 2-regular tree $\mathbb{T}_2$. For simplicity, we identify $\mathbb{T}_2$ with $\mathbb{Z}$ and identify the root o with 0. Additionally, rather than consider the general form in (2.1) with $G = \mathbb{Z}$, we focus on the special case of pairwise interacting diffusions in (2.3), but without time dependence in the drift:

$$dX_v(t) = \frac{1}{2}\big[\beta\big(X_v(t), X_{v+1}(t)\big) + \beta\big(X_v(t), X_{v-1}(t)\big)\big]dt + dW_v(t), \quad v \in \mathbb{Z}, \tag{4.3}$$

where we have dropped the superscripts denoting graph dependence for notational conciseness. We also assume that $(X_v(0))_{v\in\mathbb{Z}}$ is a shift-invariant 2-SGMRF.

Given the above setup, our goal is to understand the marginal dynamics $X_{\{-1,0,1\}} = (X_{-1}, X_0, X_1)$ of the root *and its neighborhood*. The characterization of this marginal via the local equation entails four key ingredients, which we elaborate upon below.

(i) *A mimicking theorem.* By (4.3), we can rewrite the dynamics of the marginal $X_{\{-1,0,1\}}$ of interest as follows:

$$dX_v(t) = b_v(t, X)dt + dW_v(t), \quad v \in \{-1, 0, 1\}, \tag{4.4}$$

where for $v \in \{-1, 0, 1\}$ and $X = (X_v)_{v\in\mathbb{Z}}$,

$$b_v(t, X) := \frac{1}{2}\big[\beta\big(X_v(t), X_{v+1}(t)\big) + \beta\big(X_v(t), X_{v-1}(t)\big)\big], \quad v \in \{-1, 0, 1\}. \tag{4.5}$$

Let $(\Omega, \mathcal{F}, \mathbb{F}, \mathbb{P})$ denote the filtered space that supports the $\mathbb{F}$-adapted process X. For the root node, observe that at time t, the drift b_0 of X_0 depends on X only through $X_{\{-1,0,1\}}(t)$. However, at time t the drift b_1 of X_1 depends on $X_2(t)$ and likewise the drift b_{-1} of node -1 depends on $X_{-2}(t)$. Since 2 and -2 do not lie in the closure $\{-1, 0, 1\}$ of the root, the system of equations (4.4) is not autonomous since its drift at time t depends on random elements beyond $X_{\{-1,0,1\}}(t)$. Nevertheless, (4.4) and (4.5) together show that $X_{\{-1,0,1\}}$ is what is known as an Itô process, which means that its drift (b_{-1}, b_0, b_1) is $\mathbb{F}$-progressively measurable (as a consequence of (4.5), the continuity of β and the fact that X is an $\mathbb{F}$-adapted continuous process). Therefore, one can appeal to a "mimicking" theorem for Itô processes from filtering theory (see [55] or [48, APPENDIX A]), which allows one to express $X_{\{-1,0,1\}}$ as the solution to an SDE whose drift at time t is a functional only of the past of $X_{\{-1,0,1\}}$ up to time t, rather than an arbitrary $\mathbb{F}$-adapted process. Then the mimicking theorem allows one to conclude that (by extending the probability space if necessary) there exist independent Brownian motions $(\widetilde{W}_{-1}, \widetilde{W}_0, \widetilde{W}_1)$ on the extended probability space such that $X = (X_{-1}, X_0, X_1)$ satisfies

$$X_v(t) = \tilde{b}_v(t, X)dt + d\widetilde{W}_v(t), \quad v \in \{-1, 0, 1\}, \tag{4.6}$$

where for $v \in \{-1,0,1\}$, $\tilde{b}_v : [0,\infty) \times \mathcal{C}^{\{-1,0,1\}} \mapsto \mathbb{R}^d$ is a progressively measurable version of the conditional expectation:

$$\tilde{b}_v(t, x) := \frac{1}{2}\mathbb{E}\big[\beta\big(X_v(t), X_{v+1}(t)\big) + \beta\big(X_v(t), X_{v-1}(t)\big) \,\big|\, X_{\{-1,0,1\}}[t] = x[t]\big]. \tag{4.7}$$

Recall from (4.2) that $x[t] = (x(s))_{s\in[0,t]}$. Clearly, the conditioning does not alter the drift coefficient for the root particle, which remains the same as in the original system (4.3):

$$\tilde{b}_0(t, x) = \frac{1}{2}\big[\beta\big(x_0(t), x_1(t)\big) + \beta\big(x_0(t), x_{-1}(t)\big)\big]. \tag{4.8}$$

However, $\tilde{b}_1$ and $\tilde{b}_{-1}$ will be altered by the conditioning. We now see how the MRF property can be used to simplify the expression for $\tilde{b}_1$ and $\tilde{b}_{-1}$.

(ii) *Markov random field structure.* To compute the drifts $\tilde{b}_1$ and $\tilde{b}_{-1}$ and provide a self-contained description of the law of the dynamics of $X_{\{-1,0,1\}}$, one needs to be able to express the conditional law of $X_2(t)$ and $X_{-2}(t)$ given the past $X_{\{-1,0,1\}}[t]$ in terms of the (joint) law of $X_{\{-1,0,1\}}$ or preferably, in terms of the law of $X_{\{-1,0,1\}}[t]$ to get a nonanticipative description of the dynamics. To this end, we invoke the property from Theorem 4.2 that the trajectories $(X_i[t])_{i\in\mathbb{Z}}$ up to time t form a 2-SGMRF or second-order Markov chain in $\mathbb{Z}$:

$$\big(X_j[t]\big)_{j<i} \perp\!\!\!\perp \big(X_j[t]\big)_{j>i+1} \,\big|\, \big(X_i[t], X_{i+1}[t]\big), \quad \forall i \in \mathbb{Z}. \tag{4.9}$$

Before we use this property, let us consider the corresponding first-order property: namely to ask whether we should in fact expect that for every $t > 0$,

$$\big(X_j[t]\big)_{j<i} \perp\!\!\!\perp \big(X_j[t]\big)_{j>i} \,\big|\, X_i[t], \quad \forall i \in \mathbb{Z}.$$

One may attempt to bolster this hypothesis by reasoning that conditioned on $X_i[t] = \psi$, $X_{i-1}[t]$ and $X_{i+1}[t]$ become decoupled and satisfy the following SDE: for $s \in [0,t]$,

$$dX_{i-1}(s) = \frac{1}{2}\big[\beta\big(X_{i-1}(s), \psi(s)\big) + \beta\big(X_{i-1}(s), X_{i-2}(s)\big)\big]ds + dW_{i-1}(s),$$
$$dX_{i+1}(s) = \frac{1}{2}\big[\beta\big(X_{i+1}(s), X_{i+2}(s)\big) + \beta\big(X_{i+1}(t), \psi(s)\big)\big]ds + dW_{i+1}(s),$$

where W_{i-1} and W_{i+1} are independent Brownian motions. However, a more careful inspection would reveal that such a reasoning is fallacious because the evolution of X_i, and thus the random element $X_i[t]$, directly depends on the states $(X_{i-1}(s), X_{i+1}(s))_{s\in[0,t]}$, which are in turn dependent on the Brownian motions W_{i-1} and W_{i+1}. Thus, conditioning on $X_i[t] = \psi$ causes the driving Brownian motions $W_{i-1}[t]$ and $W_{i+1}[t]$ to become correlated. Thus, under this conditioning, X_{i-1} and X_{i+1} are not independent and do not follow the above SDE driven by independent Brownian motions on $[0,t]$. However, (4.9) shows that by conditioning on both $X_i[t]$ and $X_{i+1}[t]$, the driving noise processes W_{i-1} and W_{i+2} remain decoupled (i.e., independent), although the conditioning does alter the distributions of W_{i-1} and W_{i+2}; they are no longer Brownian motions or even martingales.

Returning to the simplification of the expression for the drifts $\tilde{b}_1$ and $\tilde{b}_2$ given in (4.7), note that since the relation in (4.9) implies that $X_2(t)$ is independent of $X_{-1}[t]$ when conditioned on $X_{\{0,1\}}[t]$, the drift $\tilde{b}_1$ in (4.7) can be rewritten as

$$\begin{aligned}\tilde{b}_1(t,x) &:= \frac{1}{2}\mathbb{E}\big[\beta\big(X_1(t),X_0(t)\big)+\beta\big(X_1(t),X_2(t)\big)\,\big|\,X_{\{-1,0,1\}}[t]=x[t]\big]\\ &= \frac{1}{2}\beta\big(x_1(t),x_0(t)\big)+\frac{1}{2}\mathbb{E}\big[\beta\big(X_1(t),X_2(t)\big)\,\big|\,X_{\{0,1\}}[t]=x_{\{0,1\}}[t]\big]. \qquad (4.10)\end{aligned}$$

By the same reasoning, an analogous expression holds for $\tilde{b}_{-1}$.

(iii) *Symmetry considerations.* Despite the simplification of the last section, the second term on the right-hand side of (4.10) still involves X_2, and thus has not been written purely in terms of $X_{-1,0,1}[t]$ and its law. However, it can be rewritten in this form by exploiting the shift-variance of the particle system on $\mathbb{Z}$ (since this is true of both the initial condition and the dynamics). More precisely, the fact that (X_0, X_1, X_2) has the same distribution as (X_{-1}, X_0, X_1) allows us to conclude that

$$\tilde{b}_1(t,x)=\frac{1}{2}\beta\big(x_1(t),x_0(t)\big)+\frac{1}{2}\mathbb{E}\big[\beta\big(X_0(t),X_1(t)\big)\,\big|\,X_{\{-1,0\}}[t]=x_{\{0,1\}}[t]\big].$$

Along with the analogous expression for $\tilde{b}_{-1}$, and equations (4.6), (4.7), and (4.8), this shows that

$$\begin{aligned}dX_0(t) &= \frac{1}{2}\big[\beta\big(X_0(t),X_1(t)\big)dt+\beta\big(X_0(t),X_{-1}(t)\big)\big]dt+d\widetilde{W}_0(t),\\ dX_k(t) &= \frac{1}{2}\big[\beta\big(X_k(t),X_0(t)\big)+\tilde{\gamma}_t(X_k,X_0)\big]dt+d\widetilde{W}_k(t), \quad k\in\{-1,1\},\end{aligned} \qquad (4.11)$$

where $\widetilde{W}_{-1}$, $\widetilde{W}_0$ and $\widetilde{W}_1$ are independent d-dimensional Brownian motions, and

$$\tilde{\gamma}_t(x,y):=\mathbb{E}\big[\beta\big(X_0(t),X_1(t)\big)\,\big|\,(X_0,X_{-1})[t]=(x,y)[t]\big], \quad (x,y)\in\mathcal{C}^2. \qquad (4.12)$$

Modulo some additional technical (measurability and integrability) conditions, this identifies the form of the *local equation* satisfied by the $\{-1,0,1\}$ marginal dynamics on the line (see [48, DEFINITION 3.5 WITH $\kappa=2$] for a complete definition).

Observe that even though the original system (4.3) describes a (linear) Markov process, its marginal $X_{\{-1,0,1\}}$, characterized by the local equation system (4.11)– (4.12), is a *nonlinear*, describes a nonlinear *non-Markovian* process since the drift functional $\tilde{\gamma}_t$ depends on both the history of $X_{\{-1,0,1\}}[t]$ up to time t and its law. However, the structure of $\tilde{\gamma}_t$ ensures that the coupled system (4.11) no longer depends on values of the process X_v for $v\notin\{-1,0,1\}$, and is thus autonomously defined.

(iv) *Uniqueness of solutions to the local equation.* The above argument shows that the law of the marginal solves the local equation system (4.11) and (4.12). To complete the autonomous *characterization* of the marginal it only remains to show that the law of the marginal of $X_{-1,0,1}$ is the *unique* solution to the local equation system (or rather, its complete specification as stated in [48, DEFINITION 3.5 WITH $\kappa=2$]). The methods described in Section 2.2 to prove well-posedness of nonlinear Markov processes (which characterize the limit marginal law of a node in the complete graph case) all run into difficulties here due to the path-dependence and, more importantly, the nonlinearity occurring through dependence on conditional laws, which are less regular. Nevertheless, it is possible to prove well-posedness using other approaches, entailing relative entropy estimates and symmetry properties, or via a correspondence with the infinite particle system; further details can be found in **[48, SECTIONS 4.3.1 AND 4.2]**.

The local equation on the root neighborhood of a κ-regular tree $\mathbb{T}_\kappa$, with $\kappa \geq 3$, can be derived in a manner similar to the case of $\mathbb{T}_2$, once again invoking the mimicking theorem and Theorem 4.2, but now exploiting the additional "rotational and transational" symmetries arising from the automorphism groups of $\mathbb{T}_\kappa$, in place of just the translational and reflection symmetries of $\mathbb{T}_2$. However, the full expression of the local equation is omitted as it is a special case of the UGW tree discussed in the next section, whose analysis is more subtle.

4.3. Local equations for diffusions on unimodular Galton–Watson trees

Let ρ be a probability distribution on $\mathbb{N} \cup \{0\}$ satisfying $\sum_{k \in \mathbb{N}} k\rho_k < \infty$, and let $\mathcal{T}$ be a UGW(ρ) tree as in Remark 3.1. Again, we would like to describe the marginal dynamics of the particle process on the root node and its (random) neighborhood. As elucidated in the last section, the main ingredients in the derivation of the local equation on the line are a mimicking theorem, a conditional independence property, symmetry considerations and, finally, well-posedness of the local equation. The mimicking theorem can be applied without change also on $\mathcal{T}$. However, the MRF property in Theorem 4.2 applies to deterministic graphs and is thus not sufficient. Instead, one needs an *annealed version*, that is, one that also takes into account the random structure of the tree $\mathcal{T}$. For any $t > 0$, one has to show that (on the event the root is not isolated) for any child k of the root, conditioned on the trajectories $(X_o^{\mathcal{T}}[t], X_\kappa[t])$, the trajectories $X_{\mathcal{T}_k}^{\mathcal{T}}[t]$ of particles on the subtree $\mathcal{T}_k$ rooted at k are independent of the trajectories $X_{\mathrm{cl}_{\mathcal{T}}(o)}[t]$ on the root and its neighborhood, and, moreover, that the conditional law of $X_{\mathcal{T}_k}^{\mathcal{T}}[t]$ given $X_{\{o,k\}}^{\mathcal{T}}[t]$ does not depend on k (see **[48, PROPOSITION 3.17]**). In the case when $\rho = \delta_\kappa$, this would follow from Theorem 4.2 and homogeneity of the dynamics, but in the general case, one also has to account for the randomness of $\mathcal{T}_k$ and the root neighborhood.

Furthermore, the symmetry properties are now considerably more subtle – the appropriate notion here being *unimodularity*. Unimodularity can be viewed as the analog on an infinite graph of the property on the finite graph that the root is uniformly distributed on the graph. Since the latter statement is not well defined on an infinite graph, this is instead phrased in terms of a "mass transport principle" that on finite graphs is equivalent to having a uniformly distributed root. The definition of unimodularity involves the space $\mathcal{G}_{**}$ of isomorphism classes of doubly rooted graphs, which is defined as follows, in a fashion analogous to $\mathcal{G}_*$. A *doubly rooted graph* (G, o, o') is a rooted graph (G, o) with an additional distinguished vertex o' (which may equal o). Two doubly rooted graphs (G_i, o_i, o_i') are isomorphic if there is an isomorphism from (G_1, o_1) to (G_2, o_2) which also maps o_1' to o_2'. We write $\mathcal{G}_{**}$ for the set of isomorphism classes of doubly rooted graphs. A double rooted marked graph is defined in the obvious way, and $\mathcal{G}_{**}[\mathcal{S}]$ denotes the set of isomorphism classes of doubly rooted marked graphs. The space $\mathcal{G}_{**}[\mathcal{S}]$ is equipped with its Borel σ-algebra.

Definition 4.3. For a metric space $\mathcal{S}$, a $\mathcal{G}_*[\mathcal{S}]$-valued random variable (G, o, S) is said to be *unimodular* if the following *mass-transport principle* holds: for every (nonnegative) bounded Borel measurable function $F : \mathcal{G}_{**}[\mathcal{S}] \to \mathbb{R}_+$,

$$\mathbb{E}\Bigg[\sum_{o' \in G} F(G, S, o, o')\Bigg] = \mathbb{E}\Bigg[\sum_{o' \in G} F(G, S, o', o)\Bigg].$$

Combining all these properties it was shown in **[48]** that the marginal of the particle system $X^{\mathcal{T}}$ on the closure $\bar{\mathcal{T}}$ of the root neighborhood of the UGW (ρ) tree $\mathcal{T}$ can be characterized by a local equation that has a similar form to (4.11), except that $\tilde{\gamma}_t$ is now a reweighted version of the conditional expectation of the drift that takes into account the structure of the tree: on the event that the root is not isolated,

$$\gamma_t(X_o, X_1) = 2\frac{\mathbb{E}[\alpha_t \beta(X_o, X_{\partial_{\bar{\mathbb{T}}}(o)}) \mid X_o[t], X_1[t]]}{\mathbb{E}[\alpha_t \mid X_o[t], X_1[t]]},$$

where $\alpha_t := |\partial_{\bar{\mathcal{T}}}(o)|/(1 + \hat{C}_1)$, with $\hat{C}_1$ being a random variable distributed according to $\hat{\rho}$ (representing the number of offspring of a child of the root) that is independent of the root neighborhood structure, the initial conditions and the driving Brownian motions, and the factor of 2 in the expression arises just to compensate for the $1/2$ that arises in (4.11). This extra weighting by α_t arises due to the unimodularity property and significantly complicates the proof of well-posedness of the local equation.

4.4. Marginal dynamics for pure jump processes

As in the last section, let $\mathcal{T} = (V, E)$ be the UGW(ρ) tree, and given initial conditions $\xi = (\xi)_{v \in V}$, let $X = X^{\mathcal{T}, \xi}$ be the solution of the jump SDE (2.9) with $G = \mathcal{T}$, and also denote $\theta_v = \theta_v^{G,\xi}$. We are once again interested in obtaining an autonomous characterization of the marginal law of X on the root and its neighborhood in terms of a corresponding local equation. The derivation of this local equation follows the same broad outline that was used in the case of diffusions, although the justification of each step require substantially different arguments. For simplicity, we flesh out a few details in the special case when $\mathcal{T}$ is the rooted κ-regular tree $\mathbb{T}_\kappa$ for $\kappa \geq 2$. Let $\tilde{\mathbb{T}}_\kappa$ denote the subtree of $\mathbb{T}_\kappa$ consisting of the root and its neighborhood. Then, on verifying certain technical conditions, one can first appeal to filtering results for point processes (see, e.g., **[15]**) to establish an analogous mimicking theorem for jump processes. Specifically, we use the latter result to argue that X restricted to $\tilde{\mathbb{T}}_\kappa$ can be expressed (on a possibly extended probability space) as the solution to the following functional jump SDE: for $t \geq 0$,

$$X_v(t) = X_v(0) + \sum_{j \in \mathcal{J}} j \int_{(0,t] \times \mathbb{R}_+} \mathbb{I}_{\{r \leq \tilde{r}_j(s, X)\}} \tilde{N}_v(ds, dr), \quad v \in \tilde{\mathbb{T}}_\kappa, \tag{4.13}$$

where $(\tilde{N}_v)_{v \in V}$, are i.i.d. Poisson random measures on $\mathbb{R}_+^2$ with intensity measure Leb^2, and $\tilde{r}_j : \mathbb{R}_+ \times \mathcal{D}^{\kappa+1} \to \mathbb{R}_+$ is a predictable version of the conditional expectation

$$\tilde{r}_j^v(t, x) = \mathbb{E}\left[\bar{r}_j^v(t, X_v(t-), \theta_v(t-)) | X_{\tilde{\mathbb{T}}_\kappa}[t) = x[t)\right],$$

for a.s. every x.

Combining this with the 2-SGMRF property for X established in Theorem 4.2 for interacting jump processes with $G = \mathbb{T}_\kappa$, and invoking the symmetries of the law of X with respect to the automorphisms of $\mathbb{T}_\kappa$, the equation (4.13) can be further simplified into an autonomous local equation of the following form (see [31]):

$$X_o(t) = X_o(0) + \sum_{j\in\mathcal{J}} \int_{[0,t]\times\mathbb{R}_+} \mathbb{I}_{\{r\le\bar{r}_j(s,X_o(s-),\theta_o(s-))\}} \tilde{N}_0(ds,dr),$$

$$X_k(t) = X_k(0) + \sum_{j\in\mathcal{J}} \int_{[0,t]\times\mathbb{R}_+} \mathbb{I}_{\{r\le\tilde{\gamma}_j(s,X_k,X_o)\}} \tilde{N}_k(ds,dr), \quad k = 1,\dots,\kappa, \tag{4.14}$$

where recall $\theta_o(s-) = \sum_{k=1}^{\kappa} \delta_{X_k(s-)}$ and $\tilde{\gamma} : \mathbb{R}_+ \times \mathcal{D}^2 \mapsto \mathbb{R}_+$ is defined by

$$\tilde{\gamma}_j(t,x,y) := \mathbb{E}[\bar{r}_j(t, X_o(t), \theta_o(t)) \mid (X_o, X_1)[t] = (x,y)[t]]. \tag{4.15}$$

Here, once again, we have omitted various measurability and other technical conditions required to define a solution to the local equation, referring the reader to [31] for full details. As in the case of diffusions, it is evident from the local equation the marginal dynamics is non-Markovian even when the original dynamics in (2.9) is, and it is also nonlinear in the sense that the evolution of the process depends on its own law. The local equations identify precisely the nature of this nonlinear non-Markovian dynamics. As in the diffusion case, it is also possible to define analogous local equations describing marginal dynamics on the UGW tree, which rely on a more involved (annealed) SGMRF property and a more complicated proof of well-posedness of the local equation.

4.5. Generalizations and approximations

For both diffusive and jump dyncamis, one could also consider non-Markovian interacting processes. For example, consider solutions to the SDE (2.1) in which the drift b at the vertex v is replaced by a suitably regular nonanticipative functional $F_v : \mathbb{R}_+ \times \mathcal{C}^V \to \mathbb{R}$. For example, consider $F_v(t,x) = b(t, x_v(t-\tau), \mu_v(t-\tau))$, with $\mu_v(s) = \frac{1}{\partial v}\sum_{u\sim v} \delta_{x_u(s)}$ for some $\tau > 0$. Or likewise, consider solutions to the jump SDE (2.9) in which the jump rate $\bar{r}_j^v$ at vertex v is replaced with a suitable predictable functional of the paths such as $F_{j,v}(t,x) = \bar{r}_j(t, x_v(t-\tau), \theta_v(t-\tau))$, where $\theta_v(s) = \sum_{u\sim v} \delta_{x_u(s)}$. It is not too difficult to see from the discussions of the derivation of the local equation given in Sections 4.2-4.4 that even in the non-Markovian setting, under suitable regularity conditions, the marginal law of the process on the root and its neighborhood could be characterized by an analogous local equation. Indeed, the frameworks in both [46–48] and [29–31] allow for quite general non-Markovian dynamics. Furthermore, one can also allow more general initial conditions that are not necessarily i.i.d. but form a 2-SGMRF and (in the non-Markovian setting) incorporate histories of the process up to time 0. The latter is useful for studying flow properties of the local equation dynamics and gaining insight into stationary measures for the local equations [31]. Furthermore, the framework in [29] can also be used to handle interacting particle systems on directed graphs (see [29, REMARK 2.2]).

In the case when the full system is non-Markovian, the local equation yields a significant dimension reduction parallel to that achieved in mean-field limits, since it approximates the marginal of a non-Markovian system on an arbitrarily large high-dimensional (random) graph by a nonlinear non-Markovian process of a fixed finite (average) dimension. On the other hand, when one seeks to use the local equations to approximate the marginal of a Markovian system, since the local equation is still non-Markovian. Thus, in terms of computing or simulating the process, there is a tradeoff between the size of the time interval one is interested in and the size (or number of particles) in the original Markovian system. It is thus natural to ask if any further principled approximations are possible in that case to make the local equations more analytically and computationally tractable even over long time intervals.

Recall from Figure 2 that for the voter model on the 3-regular tree $\mathbb{T}_3$ (truncated after 9 generations) the mean-field approximation for the probability of agreement of the root with precisely two of its neighbors was rather inaccurate. Figure 4 plots, for the same tree and parameters, an *ad hoc* Markovianization of the local equation, wherein $\tilde{\gamma}_j$ in (4.15) is replaced with a modified state-dependent version $\bar{\gamma}_j : \mathbb{R}_+ \times \mathcal{X}^2 \to \mathbb{R}_+$ given by

$$\bar{\gamma}_j(t, x, y) := \mathbb{E}[\bar{r}_j(t, X_o(t), \theta_o(t)) \mid (X_o, X_1)[t-) = (x, y)[t-)].$$

The good agreement of the simulation with the Markovian version of the local equation in Figure 4 suggests that such a Markovian local equation may serve as a good approximation for several models. This motivates a more rigorous investigation of the accuracty of the Markovian local equation for various classes of models and derivation of rigorous error bounds between the laws of the solution to the Markovianized local equation and the original local equation.

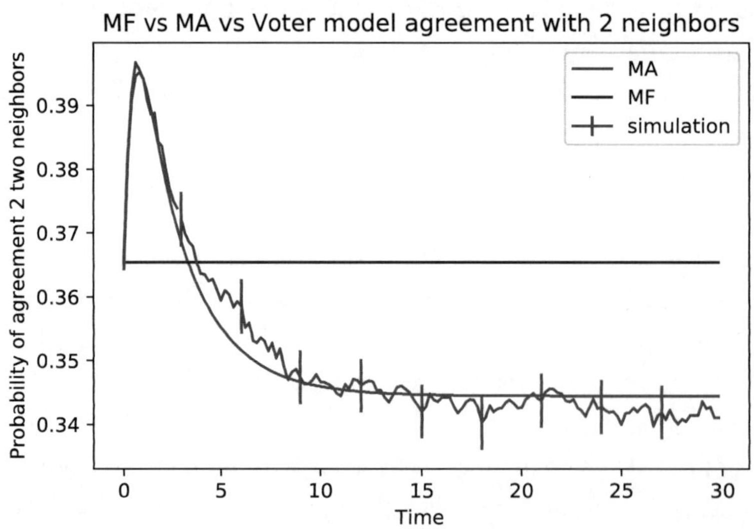

FIGURE 4
Comparing simulations, the mean-field approximation and a Markovian version of the local equation

5. OPEN QUESTIONS

This article describes the first reduced dimension characterization of marginal dynamics on sparse random graphs, thereby resolving an open question raised in [19] in the context of interacting diffusions. A plethora of open questions remain, related to the structure of the solution to the local equation such as ergodic properties, as well as theoretical guarantees for developing more computationally tractable principled approximations, and also applications (see, for example, [67]). A few open questions are listed below.

A. Long-time behavior and invariant measures for the local equations

The results thus far have focused on transient dynamics over finite time intervals. There are several open questions about equilibrium behavior and long-time behavior.

Q1. Can one can establish general conditions for existence and uniqueness of stationary or invariant measures for the local equation?

Q2. Can one identify which of these stationary measures correspond to a stationary measure of the evolution on the corresponding infinite graph? Can the local equation be used to identify phase transitions (i.e., identify parameters for existence of multiple stationary distributions on random trees)?

Q3. Can one study ergodic properties and use the local equation to sample from marginal stationary distributions on random graphs?

Recent work [31,50] has analyzed some stationarity properties for interacting systems indexed by a regular tree. The work [50] studies limits of systems of diffusions with gradient drift that have an explicit Gibbs measure (i.e., 1-MRF) as the unique invariant measure on any finite graph, and relates the stationary distribution of the full system to that of a modified local equation. Continuous-state MRFs on infinite trees can also be studied via recursions (see, e.g., [28,70]). On the other hand, the work [31] studies jump processes with possibly nonreversible dynamics.

B. Refined convergence results

A natural question is whether one can obtain more refined convergence results, that provide concentration results and rates of convergence, as well as a characterization of fluctuation and large deviations from the hydrodynamic limit. Large deviations principles also provide an alternative way of characterizing hydrodynamic limits.

Q4. Can one establish large deviation principles and concentration results for interacting particle systems on sparse random graphs?

Such results have been obtained for weakly interacting particle systems on complete and dense graphs (see, e.g., [3,9,13,16,18,63,66]).

C. Analytic characterizations

In the mean-field setting, the corresponding nonlinear process and its stationary distribution can also be described by nonlinear PDEs (in the diffusive case) or nonlinear integrodifferential equations (in the jump case).

Q5. Is it possible to develop a corresponding theory for these new types of path-dependent nonlinear equations that involve conditional laws? Also, can one determine when the marginal laws are absolutely continuous with respect to Lebesgue measure?

D. From interacting particle systems to games

Mean-field approximations have been used to study not only interacting particle systems but also games where where strategic agents control their dynamics to maximize an objective function. When the dynamics and objective functions are symmetric, a limit problem called the mean-field game has shown to provide tractable approximations to Nash equilibria in finite-agent games, which are notoriously hard to compute (see [17] for surveys on different aspects of mean-field games).

Q6. Can one establish limit theorems for Nash equilibria of games with a large number of agents in which the interaction network of agents is sparse rather than the complete graph? While there have been several recent results looking at mean-field games on networks with nodes whose degrees diverge to infinity, there are only a few works studying this on graphs with uniformly bounded degree (see [49] for the study of linear–quadratic games and the works [23,39,40] for games on directed graphs).

ACKNOWLEDGMENTS

I am grateful to my collaborators D. Lacker, A. Ganguly and R. Wu for various joint works on this subject. I would also like to thank A. Ganguly for generating Figures 1, 2, and 4, and thank J. Cocomello and K. Hu for comments on the exposition.

FUNDING

This work was partially supported by the Vannevar Bush Faculty Fellowship ONR-N0014-21-1-2887, Army Research Office grant 911NF2010133 and the National Science Foundation grant DMS-1954351.

REFERENCES

[1] D. Aldous and J. Steele, The objective method: probabilistic combinatorial optimization and local weak convergence. In *Probability on discrete structures*, pp. 1–72, Springer, 2004.

[2] K. Athreya and P. Ney, *Branching processes*. Grundlehren Math. Wiss. 196, Springer, 1972.

[3] R. Baldasso, A. Pereira, and G. Reis, Large deviations for interacting diffusions with path-dependent McKean–Vlasov limit. *Ann. Appl. Probab.* **32** (2022), no. 1, 665–695.

[4] V. Barbu and M. Röckner, From nonlinear Fokker–Planck equations to solutions of distribution dependent SDE. *Ann. Probab.* **48** (2020), no. 4, 1902–1920.

[5] I. Benjamini and O. Schramm, Recurrence of distributional limits of finite planar graphs. *Electron. J. Probab.* **6** (2001).

[6] S. Bhamidi, D. Nam, O. Nguyen, and A. Sly, Survival and extinction of epidemics on random graphs with general degree. *Ann. Probab.* **49** (2021), no. 1, 1–39.

[7] P. Biane and R. Durrett, Ten lectures on particle systems. In *Lectures on probability theory*, edited by P. Bernard, pp. 97–201, Éc. Été Probab. St.-Flour XXIII, Springer, 1993.

[8] C. Bordenave, Lecture notes on random graphs and probabilistic combinatorial optimization. https://www.math.univ-toulouse.fr/~bordenave/coursRG.pdf, 2016.

[9] A. Budhiraja, P. Dupuis, and M. Fischer, Large deviation properties of weakly interacting processes via weak convergence methods. *Ann. Probab.* (2012), 74–102.

[10] F. Cantelli, Sulla determinazione empirica dellee leggi di probabilità. *G. Ist. Ital. Attuari* **4** (1933), 421–424.

[11] P. Cattiaux, S. Roelly, and H. Zessin, Une approche Gibbsienne des diffusions Browniennes infini-dimensionnelles. *Probab. Theory Related Fields* **104** (1996), 147–179.

[12] S. Chatterjee and R. Durrett, Contact processes on random graphs with power law degree distributions have critical value 0. *Ann. Probab.* **37** (2009), no. 6, 2332–2356.

[13] F. Coppini, H. Dietert, and G. Giacomin, A law of large numbers and large deviations for interacting diffusions on Erdős–Rényi graphs. *Stoch. Dyn.* **20** (2020), no. 2.

[14] P. Dai Pra and S. Roelly, An existence result for infinite-dimensional Brownian diffusions with non-regular and non-Markovian drift. *Markov Process. Related Fields* **10** (2004), no. 1, 113–136.

[15] D. Daley and D. Vere-Jones, *An Introduction to the Theory of Point Processes: Volume II: General Theory and Structure*. Probability and its Applications. Springer, 2012.

[16] D. Dawson and J. Gärtner, Large deviations from the McKean–Vlasov limit for weakly interacting diffusions. *Stochastics* **20** (1987), no. 4, 247–308.

[17] F. Delarue (ed.), *Mean-field games*. Proc. Sympos. Appl. Math. 78, American Mathematical Society, 2021.

[18] F. Delarue, D. Lacker, and K. Ramanan, From the master equation to mean field game limit theory: large deviations and concentration of measure. *Ann. Probab.* **48** (2020), no. 1, 211–263.

[19] S. Delattre, G. Giacomin, and E. Luçon, A note on dynamical models on random graphs and Fokker–Planck equations. *J. Stat. Phys.* **165** (2016), no. 4, 785–798.

[20] A. Dembo and A. Montanari, Gibbs measures and phase transitions on sparse random graphs. *Braz. J. Probab. Stat.* **24** (2010), no. 2, 137–211.

[21] D. Dereudre, Interacting Brownian particles and Gibbs fields on pathspaces. *ESAIM Probab. Stat.* **7** (2003), 251–277.

[22] D. Dereudre and S. Rœlly, Path-dependent infinite-dimensional SDE with non-regular drift: an existence result. *Ann. Inst. Henri Poincaré Probab. Stat.* **53** (2017), no. 2, 641–657.

[23] N. Detering, J.-P. Fouque, and T. Ichiba, Directed chain stochastic differential equations. *Stochastic Process. Appl.* **130** (2020), no. 4, 2519–2551.

[24] J. Deuschel, Infinite-dimensional diffusion processes as Gibbs measures on $C[0,1]^{Z^d}$. *Probab. Theory Related Fields* **76** (1987), 325–340.

[25] R. L. Dobrušin, Description of a random field by means of conditional probabilities and conditions for its regularity. *Teor. Veroyatn. Primen.* **13** (1968), 201–229.

[26] R. L. Dobrušin, Gibbsian random fields for lattice systems with pairwise interactions. *Funktsional. Anal. i Prilozhen.* **2** (1968), no. 4, 31–43.

[27] R. Durret, T. Liggett, F. Spitzer, and A.-S. Sznitman, *Interacting particle systems at Saint-Flour*. Probab. St.-Flour, Springer, 2012.

[28] D. Gamarnik and K. Ramanan, Gibbs measures for continuous hardcore models. *Ann. Probab.* **47** (2019), no. 4, 1949–1981.

[29] A. Ganguly and K. Ramanan, Hydrodynamic limits of non-Markovian interacting particle systems on sparse graphs. 2022, arXiv:2205.01587v1.

[30] A. Ganguly and K. Ramanan, Interacting jump processes preserve semi-global Markov random fields. 2022, arXiv:2210.09253v1.

[31] A. Ganguly and K. Ramanan, Marginal dynamics of interacting particle systems on regular trees: stationarity and Markovian approximations. 2022, preprint.

[32] N. Gantert and D. Schmid, The speed of the tagged particle in the exclusion process on Galton–Watson trees. *Electron. J. Probab.* **25** (2020), 1–27.

[33] J. Gärtner, On the McKean–Vlasov limit for interacting diffusions. *Math. Nachr.* **137** (1988), 197–248.

[34] H.-O. Georgii, *Gibbs measures and phase transitions*. 2nd edn., Stud. Math., De Gruyter, Berlin/New York, 2011.

[35] V. Glivenko, Sulla determinazione empirica delle leggi di probabilità. *G. Ist. Ital. Attuari* **4** (1933), 92–99.

[36] T. E. Harris, Nearest-neighbor Markov interaction processes on multidimensional lattices. *Adv. Math.* **9** (1972), 66–89.

[37] T. E. Harris, Contact interactions on a lattice. *Ann. Probab.* **2** (1974), no. 6, 969–988.

[38] X. Huang and R. Durrett, The contact process on random graphs and Galton Watson trees. *ALEA Lat. Am. J. Probab. Math. Stat.* **17** (2020), no. 1, 159–182.

[39] T. Ichiba, Y. Feng, and J.-P. Fouque, Linear-quadratic stochastic differential games on directed chain networks. *J. Math. Stat. Sci.* **7** (2021), 25–67.

[40] T. Ichiba, Y. Feng, and J.-P. Fouque, Linear-quadratic stochastic differential games on random directed networks. *J. Math. Stat. Sci.* **7** (2021), 79–108.

[41] C. Kipnis and C. Landim, *Scaling limits of interacting particle systems*. Grundlehren Math. Wiss. 320, Springer, Berlin, 1999.

[42] S. Kissel and C. Külske, Dynamical Gibbs–non-Gibbs transitions in lattice Widom–Rowlinson models with hard-core and soft-core interactions. *J. Stat. Phys.* **178** (2020), no. 3, 725–762.

[43] A. Kolmogorov, *Grundbegriffe der Wahrscheinlichkeitsrchnung*. Springer, 1933.

[44] N. Kolokoltsov, *Nonlinear Markov processes and kinetic equations*. Cambridge Tracts in Math. 182, Cambridge University Press, 2010.

[45] C. Külske, Gibbs-Non Gibbs transitions in different geometries: the Widom–Rowlinson model under stochastic spin-flip dynamics. In *Statistical mechanics of classical and disordered systems*, edited by V. Gayrard, L.-P. Arguin, N. Kistler, and I. Kourkova, pp. 3–19, Springer, 2019.

[46] D. Lacker, K. Ramanan, and R. Wu, Local weak convergence for sparse networks of interacting processes. *Ann. Appl. Probab.* **33** (2023), no. 2, 843–888.

[47] D. Lacker, K. Ramanan, and R. Wu, Locally interacting diffusions as Markov random fields on path space. *Stochastic Process. Appl.* **140** (2021), 81–114.

[48] D. Lacker, K. Ramanan, and R. Wu, Marginal dynamics of interacting diffusions on unimodular Galton–Watson trees. 2021, arXiv:1904.02585v3, to appear in *Probab. Theor. Related Fields*.

[49] D. Lacker and A. Soret, A case study on stochastic games on large graphs in mean field and sparse regimes. *Math. Oper. Res.* (2021).

[50] D. Lacker and J. Zhang, Stationary solutions and local equations for interacting diffusions on regular trees. 2021, arXiv:2111.05416v2.

[51] O. E. Lanford, III and D. Ruelle, Observables at infinity and states with short range correlations in statistical mechanics. *Comm. Math. Phys.* **13** (1969), 194–215.

[52] T. Liggett, *Interacting particle systems*. 1st edn., Springer, New York, 1985.

[53] T. Liggett, *Stochastic interacting systems: contact, voter and exclusion processes*. Grundlehren Math. Wiss. 324, Springer, 1999.

[54] T. M. Liggett, Existence theorems for infinite particle systems. *Trans. Amer. Math. Soc.* **165** (1972), 471–481.

[55] R. Liptser and A. Shiryaev, *Statistics of random processes: I. General theory. 1.* Springer, 2001.

[56] E. Luçon and W. Stannat, Mean field limit for disordered diffusions with singular interactions. *Ann. Appl. Probab.* **24** (2014), no. 5, 1946–1993.

[57] H. McKean, Propagation of chaos for a class of non-linear parabolic equations. In *Stochastic differential equations*, pp. 41–57, Lect. Ser. Differ. Equ., Session 7, Catholic Univ, 1967.

[58] H. P. McKean, Jr., A class of Markov processes associated with nonlinear parabolic equations. *Proc. Natl. Acad. Sci. USA* **56** (1966), 1907–1911.

[59] S. Mehri, M. Scheutzow, W. Stannat, and B. Z. Zangeneh, Propagation of chaos for stochastic spatially structured neuronal networks with delay driven by jump diffusions. *Ann. Appl. Probab.* **30** (2020), no. 1, 175–207.

[60] R. Minlos, S. Roelly, and H. Zessin, Gibbs states on space-time. *Potential Anal.* **13** (2000), 367–408.

[61] D. Nam, O. Nguyen, and A. Sly, Critical value asymptotics for the contact process on random graphs. 2019, arXiv:1910.13958v1.

[62] K. Oelschlager, A martingale approach to the law of large numbers for weakly interacting stochastic processes. *Ann. Probab.* (1984), 458–479.

[63] R. Oliveira and G. Reis, Interacting diffusions on random graphs with diverging degrees: hydrodynamics and large deviations. *J. Stat. Phys.* **176** (2019), 1057–1087.

[64] R. Oliveira, G. Reis, and L. Stolerman, Interacting diffusions on sparse graphs: hydrodynamics from local weak limits. *Electron. J. Probab.* **25** (2020), no. 110.

[65] R. Pemantle, The contact process on trees. *Ann. Probab.* **20** (1992), no. 4, 2089–2116.

[66] K. Ramanan, Refined convergence results for interacting diffusions and mean-field games. In *Mean-field games*, pp. 105–161, Proc. Sympos. Appl. Math. 78, American Mathematical Society, 2021.

[67] K. Ramanan, Beyond mean-field limits for the analysis of large-scale networks. *Queueing Syst.* **100** (2022), no. 3–4.

[68] F. Redig, S. Roelly, and W. Ruszel, Short-time Gibbsianness for infinite-dimensional diffusions with space-time interaction. *J. Stat. Phys.* **138** (2010), 1124–1144.

[69] S. Roelly and W. Ruszel, Propagation of Gibbsianness for infinite-dimensional diffusions with space-time interaction. *Markov Process. Related Fields* **20** (2014), 653–574.

[70] U. Rozikov, *Gibbs measures on Cayley trees*. World Scientific, 2013.

[71] F. Spitzer, Interaction of Markov processes. *Adv. Math.* **5** (1970), 246–290.

[72] A. Stacey, The contact process on finite homogeneous trees. *Probab. Theory Related Fields* **121** (2001), no. 4, 551–576.

[73] A. Sznitman, *Topics in propagation of chaos*. Springer, 1991.

[74] P. Tchebichef, Des valeurs moyennes. *J. Math. Pures Appl.* **2** (1867), no. 12, 177–184.

[75] A. Van Enter, et al., Possible loss and recovery of Gibbsianness during the stochastic evolution of Gibbs measures. *Comm. Math. Phys.* **226** (2002), no. 1, 101–130.

KAVITA RAMANAN

Box F, Brown University, Providence, RI 02912 USA, Kavita_Ramanan@brown.edu

INTEGRABLE FLUCTUATIONS IN THE KPZ UNIVERSALITY CLASS

DANIEL REMENIK

ABSTRACT

The KPZ fixed point is a scaling-invariant Markov process which arises as the universal scaling limit of a broad class of models of random interface growth in one dimension, the one-dimensional KPZ universality class. In this survey we review the construction of the KPZ fixed point and some of the history that led to it, in particular through the exact solution of the totally asymmetric simple exclusion process, a special solvable model in the class. We also explain how the construction reveals the KPZ fixed point as a stochastic integrable system, and how from this it follows that its finite-dimensional distributions satisfy a classical integrable dispersive PDE, the Kadomtsev–Petviashvili (KP) equation.

MATHEMATICS SUBJECT CLASSIFICATION 2020

Primary 60K35; Secondary 82C21, 82B23

KEYWORDS

KPZ fixed point, KPZ universality class, random growth, interacting particle systems

ICM INTERNATIONAL CONGRESS OF MATHEMATICIANS 2022 JULY 6–14

Proc. Int. Cong. Math. 2022, Vol. 6, pp. 4426–4450
DOI 10.4171/ICM2022/3

Published by EMS Press

1. THE KPZ UNIVERSALITY CLASS

The subject of this survey are the universal fluctuations of a large collection of models known as the one-dimensional *Kardar–Parisi–Zhang (KPZ) universality class*. This class includes many physical and probabilistic models of one-dimensional random growth, as well as several other models, including directed polymers in a random potential, some interacting particle systems, stochastic reaction–diffusion equations, and random stirred fluids, all of which can be represented in terms of the evolution of a one-dimensional interface. *Universality* here refers to the idea that the long time, large-scale fluctuations of all the models in the class share a common description, in the form of common scaling exponents and a common scaling limit, which are independent of the microscopic description of each model.

While the belief in KPZ universality originates in statistical physics, much of the progress in its understanding, and in particular in the description of the universal KPZ scaling limits, has been achieved in the mathematical literature, through the study of some particular models which present a striking degree of exact solvability, and borrowing methods from algebraic combinatorics, representation theory, mathematical physics, and integrable systems. These *integrable probabilistic systems* comprise a sprawling subject, to which we cannot do justice in this article; we refer the interested reader instead to the reviews [11, 12] on this topic, as well as to [23, 41] for more physical perspectives. Our main focus will be to describe part of the work in the field which in recent years has led to a very complete description of the universal scaling limit of KPZ models and its connection with integrable systems and random matrix theory.

Two examples. We begin by introducing a simple model which is not in the KPZ universality class. Suppose that blocks of unit height fall at each site of $\mathbb{Z}$ at rate 1 (i.e., at the times of a rate-1 Poisson process). If the tower of blocks at each site grows independently of the others, the height $h(t,x)$ at time $t \geq 0$ at each site $x \in \mathbb{Z}$ can be described, by the classical central limit theorem (applied to the Poisson distribution), as $h(t,x) \approx t + t^{1/2}\xi$ with ξ a standard normal random variable. In other words, the height grows linearly with time, with Gaussian fluctuations of size $t^{1/2}$. But since sites in this model, sometimes called *random deposition*, are independent, $h(t,x)$ presents no interesting spatial structure. In order to obtain a more interesting one-dimensional interface, one can add a *relaxation* mechanism to the model as follows: when a block falls over site x, it lands on either x or any of its two nearest neighbors, whichever has the lowest height (choosing, say, uniformly in case of ties). To first order, $h(t,x)$ still grows like t, but fluctuations are now of order $t^{1/4}$: the relaxation mechanism has the effect of smoothing the interface, which now presents nontrivial correlations on a spatial scale of order $t^{1/2}$ (see [6] for a discussion). This model belongs to what physicists call the *Edwards–Wilkinson universality class* [56], which still has Gaussian fluctuations.

A very different picture arises in the *ballistic deposition* model, first introduced in [54] as a model for colloidal aggregates. In this model the falling blocks are sticky, and they attach to the side of the first neighboring block they come in contact with, see Figure 1. The interface $h(t,x)$, defined as the location of the highest block above x at time t, now has overhangs. Seppäläinen [48] proved that the height still grows linearly; growth has to

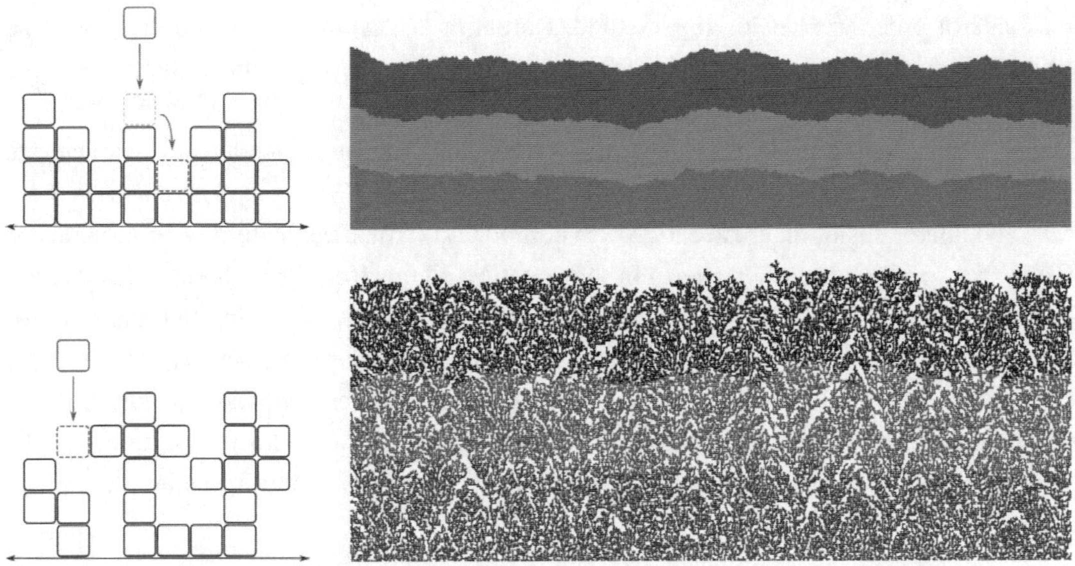

FIGURE 1
Random deposition with relaxation (top) and ballistic deposition (bottom), simulations on the right (different shadings depict snapshots at different times).

be faster than for the previous models (as the aggregate grows, it is left with holes inside), but the exact rate remains unknown. Fluctuations, on the other hand, are expected in this case to be of order $t^{1/3}$: the ballistic mechanism produces a rougher interface than random deposition with relaxation. In the same way, the lateral growth of the interface makes for a longer range of spatial correlations, expected in this case to be of order $t^{2/3}$. These two scaling exponents are one of the hallmarks of the KPZ universality class, but for this model they remain out of reach of rigorous analysis.

Ballistic deposition is representative both of the main features of models in the class and of the difficulty in analyzing them. This is why progress in the field has had to take place mostly through the analysis of some specific models with a very special structure. In the next section we will describe one of the main examples among these special models, TASEP. Now we introduce the model that gives the KPZ universality class its name: the (one-dimensional) *Kardar–Parisi–Zhang equation*, which is the nonlinear stochastic PDE

$$\partial_t h = \lambda(\partial_x h)^2 + \nu\partial_x^2 h + \sigma\xi, \tag{1.1}$$

where ξ is space-time white noise and λ, ν, and σ are physical parameters. This equation was introduced in 1986 [26] by the physicists Kardar, Parisi, and Zhang, and was conceived as the simplest (and has become the canonical) continuum model for random interface growth which incorporates the physical features of models such as ballistic deposition. Using physical arguments (based on nonrigorous dynamic renormalization group methods), Forster, Nelson, and Stephen [21] had predicted that the closely related stochastic Burgers equation (essentially the equation satisfied by $\partial_x h(t,x)$, which can be thought of as a much simplified model of a random stirred fluid) had fluctuations of order $t^{1/3}$ with nontrivial correlations on a scale of order $t^{2/3}$. Using this method, Kardar et al. [26] then predicted that the same should hold for the KPZ equation and for a large class of models, a fortiori identified as the KPZ universality class.

The right-hand side of (1.1) identifies the main elements which loosely characterize a model in the class: local dynamics, short range randomness, a smoothing mechanism (here $\partial_x^2 h$), and a lateral, slope-dependent growth component (which can naturally be modeled as $F(\partial_x h)$ for some F; the $(\partial_x h)^2$ term in the equation comes from keeping only the second-order term in the expansion $F(u) = F(0) + F'(0)u + \frac{1}{2}F''(0)u^2$, noting that the first two terms can be removed by a change of variables in the equation). The crucial feature here is the nonlinear lateral growth term (note how it becomes macroscopically apparent in the ballistic deposition simulation in Figure 1). In fact, setting $\lambda = 0$ yields the (simpler, linear) *additive stochastic heat equation*, which identifies the Edwards–Wilkinson class mentioned in the previous example.

In terms of solvability, the KPZ equation lies halfway between currently intractable models such as ballistic deposition and integrable models such as TASEP (this is separate from the delicate issue of well-posedness of (1.1), which we will not discuss, see [22]).

The KPZ universality conjecture. The Kardar–Parisi–Zhang paper marked the beginning of a long period of intense research interest, and has been one of the main drivers for advances the field, both in the physics and in the mathematics literature. Numerical simulations and experiments confirmed the KPZ scaling prediction for many different systems and later on, as results on the distribution of the fluctuations of special KPZ models were first obtained, the following picture began to emerge: if $h(t,x)$ is the height function describing the evolution of the interface associated to a model in the KPZ class, then (here $\stackrel{(d)}{=}$ denotes equality in distribution)

$$\lim_{t\to\infty} t^{-1/3}\big(h(c_1 t, c_2 t^{2/3}x) - c_3 t\big) \stackrel{(d)}{=} \mathcal{A}(x) \tag{1.2}$$

for a universal limiting process $(\mathcal{A}(x))_{x\in\mathbb{R}}$ which depends only on the initial data of the model (more precisely, on the limit $\lim_{\varepsilon\to 0} \varepsilon^{1/2} h(0, c_2\varepsilon^{-1}x)$). The scaling on the left-hand side reflects the KPZ prediction: after subtracting a linear term ($c_3 t$) which represents the first-order (deterministic) linear growth of a typical KPZ interface, we obtain a random variable which fluctuates at the order of $t^{1/3}$, so we need to multiply by $t^{-1/3}$ to obtain a meaningful limit, while nontrivial correlations for two spatial points are observed when they are at distance order $t^{2/3}$ (i.e., at shorter scales the height function at the two points looks the same as $t \to \infty$, while at longer scales they become independent), so the spatial variable x has to be observed at that scale to see a nontrivial spatial process. The constants c_1, c_2, c_3 are model dependent; they are used to provide a common normalization.

As we will see in Section 2, the description (1.2) emerged at first only partially: it was initially restricted only to one-point distributions (i.e., fixed x instead of the whole spatial process $\mathcal{A}(x)$) and, crucially, only to some very special choices of initial data. Remarkably, it was realized that in those special cases the fluctuations arising in KPZ models were connected to random matrix theory, although to some extent the connection remained mysterious.

1:2:3 scaling and the KPZ fixed point. On the other hand, (1.2) does not provide a full description, since it loses all information about the temporal evolution of the interface. To recover it, it is convenient to introduce a parameter $\varepsilon > 0$, rescale the variables (t, x) as $(\varepsilon^{-3/2}t, \varepsilon^{-1}x)$, as well as the height h (after subtracting the first order linear growth term) as $\varepsilon^{1/2}h$, and take $\varepsilon \to 0$ (instead of $t \to \infty$). This is usually referred to as the *1:2:3 KPZ scaling* (reflecting the ratios of the exponents associated to the size of fluctuations, space, and time). The KPZ universality conjecture, first expressed in this form in [14], then asserts that for any model in the class,

$$\lim_{\varepsilon\to 0} \varepsilon^{1/2}\big(h(c_1\varepsilon^{-3/2}t, c_2\varepsilon^{-1}x) - c_3\varepsilon^{-3/2}t\big) \stackrel{(\mathrm{d})}{=} \mathfrak{h}(t,x) \tag{1.3}$$

for a universal process $(\mathfrak{h}(t,x))_{t\geq 0, x\in\mathbb{R}}$ which, again, should only depend on the initial data $\mathfrak{h}_0(x) := \lim_{\varepsilon\to 0} \varepsilon^{1/2}h(0, c_2\varepsilon^{-1}x)$ prescribed for the model. Taking $t = 1$ in this limit recovers the spatial processes prescribed in (1.2).

The limiting process $\mathfrak{h}(t,x)$ appearing on the right-hand side of (1.3) is known as the *KPZ fixed point*. Since many of the models which the process should arise as a limit of are Markovian, one expects it to be a Markov process (taking values in a suitable space of real valued curves). The name of the process comes from the fact that, by its definition as a limit of 1:2:3 rescaled models, it should be invariant under such rescaling: if $\mathfrak{h}(t,x;\mathfrak{h}_0)$ denotes the KPZ fixed point with initial data $\mathfrak{h}(0,x) = \mathfrak{h}_0$, then for $\alpha > 0$ one expects, writing $\mathfrak{h}_0^{(\alpha)}(x) = \alpha^{-1}\mathfrak{h}_0(\alpha^2 x)$, that

$$\alpha\mathfrak{h}(\alpha^{-3}t, \alpha^{-2}x; \mathfrak{h}_0^{(\alpha)}) \stackrel{(\mathrm{d})}{=} \mathfrak{h}(t,x;\mathfrak{h}_0). \tag{1.4}$$

The rough picture one should have in mind is of $\mathfrak{h}(t,x)$ as an attracting fixed point, under the renormalization map defined by the left-hand side of (1.4) with $\alpha \to 0$, in some (loosely defined) space of models. As such, one can think of this fixed point alternatively as *defining* the KPZ universality class (as the family of models which lie in its domain of attraction). In other words, if the KPZ fixed point can be constructed explicitly, then (1.3) can be used to turn the vague characterization of membership in the KPZ universality class described above into a concrete definition.

It is worth stressing that the KPZ fixed point should not be confused with the KPZ equation, which is just one (albeit very special) member of the class; in fact, the KPZ equation is not invariant under the KPZ 1:2:3 scaling (1.4) (which sends the parameters (λ, ν, σ) to $(\lambda, \alpha\nu, \alpha^{1/2}\sigma)$).

Much (though certainly not all) of the progress in the field during the last 20 years can be understood as an effort to describe the KPZ fixed point, understand its properties, and explore its connections with objects coming from random matrix theory and integrable systems. The purpose of this review is to describe one part of this story, which in particular leads to the construction of the KPZ fixed point and its description as a stochastic integrable system. There is a priori no reason to believe that any of this should be possible; as we will see, what comes to our rescue is the remarkable exact solvability of some special discrete models in the class, which can be used to access the KPZ fixed point in the limit.

One of the main players in this story is the totally asymmetric simple exclusion process. We turn to it in the next section.

2. TASEP WITH SPECIAL INITIAL DATA

The (one-dimensional, continuous time) *totally asymmetric simple exclusion process (TASEP)* is an interacting particle system made out of particles at positions $\cdots < X_t(2) < X_t(1) < X_t(0) < X_t(-1) < X_t(-2) < \cdots$ on $\mathbb{Z} \cup \{-\infty, \infty\}$ performing totally asymmetric nearest neighbor random walks with exclusion: each particle independently attempts jumps to the neighboring site to the right at rate 1, the jump being allowed only if that site is unoccupied. Placing particles at $\pm\infty$ allows for systems with a rightmost and/or leftmost particle with no change of notation (such particles play no role in the dynamics). Since its introduction (in a more general form) in 1970 by F. Spitzer [49], TASEP has become one of the basic and most heavily studied out-of-equilibrium models in probability and statistical physics.

In order to associate an interface to the TASEP particle system, we let $X_t^{-1}(u) = \min\{k \in \mathbb{Z} : X_t(k) \leq u\}$ and define the *TASEP height function* as

$$h(t,x) = -2\big(X_t^{-1}(x-1) - X_0^{-1}(-1)\big) - x, \quad t \geq 0,\ x \in \mathbb{Z}.$$

In words, this fixes $h(0,0) = 0$ and constructs the height function by moving up from x to $x+1$ whenever there is a particle at x and down from x to $x-1$ if the site is empty. By interpolating piecewise linearly (and shifting x by $1/2$, which makes no difference), we can picture $h(t,x)$ as a continuous function made out of line segments of slope $+1$ above every particle and -1 above every hole. The dynamics of this height function is that every local maximum $\wedge$ becomes a local minimum $\vee$ at rate 1, as in the figure; in this guise the model is sometimes known as the *corner growth model* (for special initial data) or *restricted solid-on-solid model.*

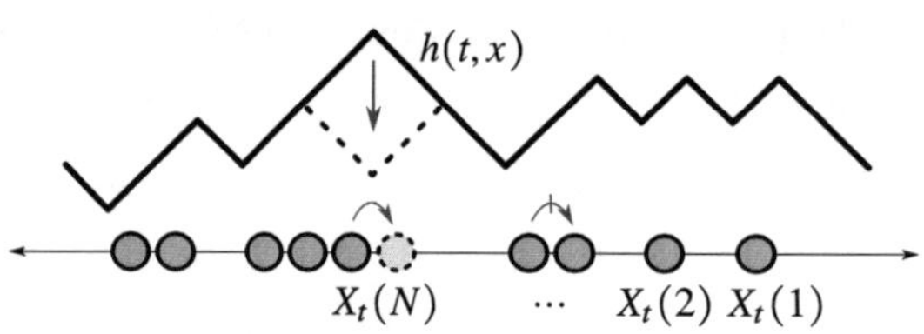

To see how the features of a KPZ model described in the introduction arise in TASEP, think of writing the evolution of the height function as a stochastic equation involving a family of independent Poisson processes at each site. Such an equation can be rewritten roughly as $\mathrm{d}h(t,x) = -2\mathbf{1}_{\wedge}\mathrm{d}t + \mathrm{d}M_t(x)$ with M_t a martingale which provides the random forcing, and where the drift term (which says that the height function goes down by two at rate 1 at sites where we see a local maximum) contains the smoothing and lateral growth mechanisms, as can be seen by rewriting it as

$$-2\mathbf{1}_{\wedge} = \frac{1}{2}\Big[(\nabla^- h)(\nabla^+ h) - 1 + \frac{1}{2}\nabla^+\nabla^- h\Big]$$

with $\nabla^{\pm}$ the forward/backward discrete difference operators

$$\nabla^{+} f(x) = f(x+1) - f(x), \quad \nabla^{-} f(x) = f(x) - f(x-1) \tag{2.1}$$

(in fact, using this decomposition, it can be shown that if particles now jump to the right at rate $p \in [0,1]$ and to the left at rate $q = 1-p$, then the associated height function converges, in the *weakly asymmetric limit* corresponding to $p - q = \varepsilon^{1/2}$ with $\varepsilon \to 0$ under diffusive scaling, to a solution of the KPZ equation (1.1), see [7]).

Wedge initial data. The simplest possible choice of (infinite) TASEP initial condition is the *packed*, or *step*, initial data where particles are initially placed at every negative integer site (i.e., $X_0(i) = -i$, $i \geq 1$). For the TASEP height function, it translates into the *wedge* initial condition $h(0,x) = -|x|$. The (essentially) first result about the limiting fluctuations for a KPZ model was proved by Johansson in 1999 (to be more precise, a version of this result was proved a couple of months earlier by Baik, Deift, and Johansson in their seminal paper [4] for Poissonian last passage percolation):

Theorem 2.1 ([24]). *For the TASEP height function with wedge initial data $h(0,x) = -|x|$, one has*

$$\lim_{t\to\infty} \mathbb{P}\left(\frac{h(2t, 2t^{2/3}x) + t}{t^{1/3}} \leq r\right) = F_{\mathrm{GUE}}(r - x^2), \tag{2.2}$$

where F_{GUE} is the Tracy–Widom GUE distribution [50].

The parabolic shift appearing on the right-hand side of (2.2) reflects the curvature of the (deterministic, first order) hydrodynamic limit for the model in this case, which states [42] that $\lim_{\kappa\to\infty} \kappa^{-1} h(\kappa t, \kappa x) = -t + x^2/2t$ for $|x| \leq t$. What is remarkable in this result, and was in fact very surprising, is the nature of the distribution of the limiting fluctuations: they coincide with the asymptotic fluctuations of the largest eigenvalue of a matrix from the *Gaussian Unitary Ensemble (GUE)*, i.e., a Hermitian random matrix with properly scaled (complex) Gaussian entries.

In terms of the KPZ universality conjecture (1.3), this result can be reinterpreted as the first computation of a marginal of the KPZ fixed point $\mathfrak{h}(t,x)$, for $t = 1$, fixed $x \in \mathbb{R}$ and initial data $\mathfrak{h}(0,x) = 0$ for $x = 0$ and $-\infty$ everywhere else. We will denote this choice of initial data (which may look singular, but is natural and, as we will see, is in fact the simplest possible initial condition for the KPZ fixed point) by $\mathfrak{d}_0$; it is known as a *narrow wedge*, as it arises from wedge initial data becoming increasingly narrower in the scaling limit.

The proof of Theorem 2.1 in [24] is based on the analysis of TASEP as a determinantal point process, using as a basic tool the Robinson–Schensted–Knuth (RSK) correspondence from algebraic combinatorics. The eigenvalues of an $N \times N$ GUE random matrix are also determinantal, and in fact it was later understood [9, 55] (see also [33]) that using these and related tools, these eigenvalues and the location of the first N TASEP particles (with step initial data) can be realized as projections of a larger process. We will describe shortly a different proof.

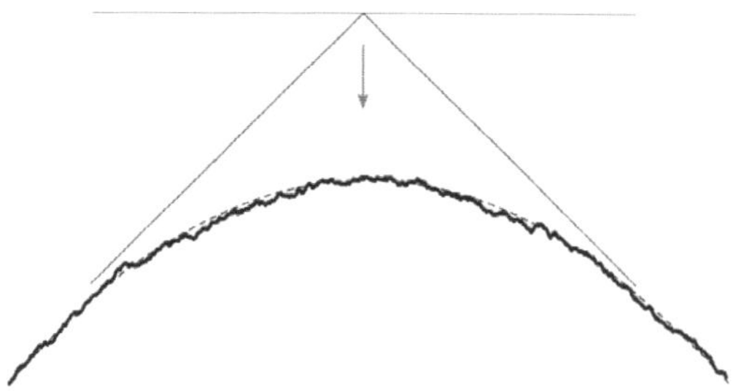

FIGURE 2
A simulation of the KPZ fixed point with narrow wedge initial data $\mathfrak{h}(1, x; \mathfrak{d}_0) \overset{(d)}{=} \mathcal{A}_2(x) - x^2$ as the limit of the TASEP height function.

The next step was to extend Theorem 2.1 to the full (fixed time) spatial process:

Theorem 2.2. *The rescaled TASEP height function $t^{-1/3}(h(2t, 2t^{2/3}x) + t)$ with wedge initial data converges in distribution as $t \to \infty$, uniformly in x on compact sets, to $\mathcal{A}_2(x) - x^2$, where $\mathcal{A}_2$ is the* Airy$_2$ process.

The Airy$_2$ process was introduced by Prähofer and Spohn in 2001 **[35]** as the limit (at the level of finite-dimensional distributions) of the closely related polynuclear growth (PNG) model; a version of the result quoted here was proved by Johansson **[25]** in 2003 (for a related discrete time model, see the coming discussion about LPP). Comparing with Theorem 2.1, we see that $\mathcal{A}_2(x)$ has to be stationary, with Tracy–Widom GUE marginals at each x. The process itself is in fact closely related to random matrices. In particular, it arises as the scaling limit of the top path of GUE Dyson Brownian motion, the eigenvalue process associated to a GUE matrix whose entries evolve as independent (complex) Brownian motions. The process is defined through its finite dimensional distributions, which are given by a Fredholm determinant formula, see (4.5) below.

In terms of the KPZ fixed point, Theorem 2.2 now tells us that, as a process in x,

$$\mathfrak{h}(1, x; \mathfrak{d}_0) \overset{(d)}{=} \mathcal{A}_2(x) - x^2 \tag{2.3}$$

(where $\mathfrak{d}_0$ is the narrow wedge initial data introduced after Theorem 2.1). See Figure 2.

Last passage percolation. We make a brief detour now to introduce another model in the KPZ universality class, *last passage percolation (LPP)*. We focus on the discrete case; there are similar models in other settings (e.g., Poisson LPP in the continuous case, Brownian LPP in the semidiscrete case). Consider a family $\{w_{i,j}\}_{i,j \in \mathbb{Z}}$ of i.i.d. random variables and define the *point-to-point last passage time*

$$L[(m_1, n_1) \to (m_2, n_2)] = \max_{\pi \in \Pi:(m_1,n_1)\to(m_2,n_2)} \sum_i w_{\pi_i},$$

where the max is taken over the set of all paths connecting (m_1, n_1) to (m_2, n_2) which take unit steps up or right; if this set is empty, we take the max to be $-\infty$. We can associate a growing cluster to this model by considering the set of points with passage times less than t, i.e.,

$$\mathscr{C}_0(t) = \{(m,n) \in \mathbb{Z}^2 : L[(0,0) \to (m,n)] \le t\}.$$

If the $w_{i,j}$'s are exponentials with parameter 1 then the model can be mapped to TASEP with step initial condition by interpreting each $w_{i,j}$ as the waiting time that it takes particle j to jump from site $i-j$ to site $i-j+1$ (counted from the instant when that jump first becomes possible); in fact, it is not too hard to check that, after a rotation by $-3\pi/4$ (and a slight shift), the boundary of $\mathscr{C}_0(t)$ encodes the TASEP height function $h(t,\cdot)$. In view of Theorem 2.2, the LPP fluctuations are thus governed by the Airy$_2$ process. Johansson's result in [25] was for the case when the $w_{i,j}$'s have a geometric distribution, which maps to a discrete time version of TASEP; it gives

$$c_1 N^{-1/3}\big(L[(0,0) \to (N + c_2 N^{2/3}x, N - c_2 N^{2/3}x)] - c_3 N\big) \xrightarrow[N\to\infty]{} \mathcal{A}_2(x) - x^2$$

(note that, as stated, this LPP result is not quite equivalent to Theorem 2.2 even if one lets the weights be exponential). By universality one expects the same to hold for general choices of weights, but the problem is completely open.

To map LPP to TASEP with general initial data, one can consider paths which start from any point in a given curve instead of just at the origin. Another, perhaps more natural, way of changing the LPP initial data, is to let paths start at any point in the antidiagonal line $\{(\ell,-\ell)\}_{\ell\in\mathbb{Z}}$ and add an extra reward $g(\ell)$ (the *boundary condition*) there, i.e., to set

$$L_g[(m,n)] = \sup_{\ell\in\mathbb{Z}}\big(L[(\ell,-\ell) \to (m,n)] + g(\ell)\big). \tag{2.4}$$

In the scaling limit, g will now become the initial data for the KPZ fixed point.

Periodic initial data. Coming back to TASEP, the next case which could be solved corresponds to *periodic* initial data $X_0(i) = -2i$, $i \in \mathbb{Z}$, which at the level of the TASEP height function translates into the (asymptotically) *flat* initial condition of the form $\wedge\!\!\wedge\!\!\wedge\!\!\wedge$. This case was first solved for Poissonian LPP with zero boundary condition [5], which translated into the context of TASEP suggested that for periodic initial data,

$$\lim_{t\to\infty} \mathbb{P}\left(\frac{h(2t, 2t^{2/3}x) + t}{t^{1/3}} \le r\right) = F_{\mathrm{GOE}}(4^{1/3}r), \tag{2.5}$$

where F_{GOE} is the Tracy–Widom GOE distribution [51], the analog of F_{GUE} for the *Gaussian Orthogonal Ensemble (GOE)*, i.e., symmetric random matrices with properly scaled (real) Gaussian entries. This was later confirmed, and extended to the full spatial process:

Theorem 2.3 ([10, 45]). *The rescaled TASEP height function $t^{-1/3}(h(2t, 2t^{2/3}x) + t)$ with flat initial data converges in distribution as $t \to \infty$ to the* Airy$_1$ process $\mathcal{A}_1(x)$.

The Airy$_1$ process is the analog of the Airy$_2$ process for flat initial data. It is stationary, and has Tracy–Widom GOE marginals. In terms of the KPZ fixed point, Theorem 2.3 now tells us that, as a process in x,

$$\mathfrak{h}(1, x; 0) \overset{(\mathrm{d})}{=} \mathcal{A}_1(x). \tag{2.6}$$

Transition probabilities. We turn now to a sketch of the proof of Theorems 2.2 and 2.3. It is based on Schütz's 1997 solution [47] of TASEP with N particles, which shows that the transition probabilities of $(X_t(1),\dots,X_t(N))$ have the following determinantal form:

$$\mathbb{P}_{X_0}\big(X_t(1)=x_1,\dots,X_t(N)=x_N\big)=\det\big(F_{j-i}\big(t,x_i-X_0(j)\big)\big)_{1\le i,j\le N}, \tag{2.7}$$

where X_0 in the subscript denotes the initial data of the process and where

$$F_n(t,x)=\frac{1}{2\pi}\oint_{\Gamma_{0,1}}\mathrm{d}w\,\frac{(1-w)^{-n}}{w^{x-n+1}}e^{t(w-1)}, \tag{2.8}$$

with $\Gamma_{0,1}$ any positively oriented simple loop which includes $w=0$ and $w=1$. The derivation uses a method known in physics as the coordinate Bethe ansatz to find a solution of Kolmogorov forward (or master) equation of the process. The key ingredients in the derivation become apparent after rewriting the functions F_n, $n\in\mathbb{Z}$, as

$$F_n(t,x)=(\nabla^+)^n e^{-t\nabla^-}\delta_0(x) \tag{2.9}$$

where $\nabla^\pm$ are the discrete difference operators from (2.1) (with the inverse of ∇^+ defined through $(\nabla^+)^{-1}f(x)=\sum_{y>x}f(y)$) and $\delta_i(y)=\mathbf{1}_{y=i}$. The operator $e^{-t\nabla^-}$ is simply the transition semigroup of a Poisson process with jumps to the left at rate 1 (and has a kernel acting by convolution with precisely the right-hand side of (2.8) with $n=0$, which is where (2.9) comes from); this factor encodes the dynamics of a free (i.e., not subject to exclusion) TASEP particle. The factor $(\nabla^+)^n$, on the other hand, encodes the exclusion restriction: very roughly put, in a situation where one particle tries to jump on top of another one, this factor produces two identical rows in the determinant obtained by using (2.7) on the right-hand side of the Kolmogorov equation, and hence terms corresponding to those transitions will not contribute.

In principle, (2.8) contains all the information one needs in order to compute a limit like (1.3) for TASEP, at least for initial data which has a rightmost particle $X_t(1)$. In fact, computing the distribution of the TASEP height function at a given (finite) set of locations is equivalent to computing, for some given indices $n_1<\cdots<n_m$,

$$\mathbb{P}_{X_0}\big(X_t(n_1)>a_1,\dots,X_t(n_m)>a_m\big) \tag{2.10}$$

and the evolution of $(X_t(i))_{i=1,\dots,n_m}$ is independent of the particles to their left, so we may restrict to a system with a finite number N of particles. However, (2.8) is not by itself conducive to asymptotic analysis, for which we need to sum over the positions of the other $N-m$ particles and then take N, which is also the dimension of the determinant, to infinity.

Biorthogonalization. This difficulty was overcome in [10, 45], where the authors were able to show that the right-hand side of (2.7) can be expressed as a marginal of a (signed) determinantal point process on a larger space of Gelfand–Tsetlin patterns (i.e., triangular arrays of integers with interlaced consecutive levels). This allowed them to use techniques from random matrix theory (more precisely, a version of the Eynard–Mehta Theorem [19]) to derive an explicit Fredholm determinant formula for (2.10). We will not describe the derivation, and content ourselves with stating a version of their result.

To do so, we need to introduce some notation. Fix an initial condition $(X_0(n))_{n\geq 1}$ for the particle system (which is right-finite, i.e., with a rightmost particle). Define

$$Q(x,y) = 2^{y-x}\mathbf{1}_{x>y}, \quad Q^{-1}(x,y) = 2^{y-x}\nabla^+(x,y) = 2^{y-x}(\mathbf{1}_{x=y-1} - \mathbf{1}_{x=y});$$

Q is invertible as an operator acting on $\ell^2(\mathbb{Z})$, with inverse given by Q^{-1} as defined above ($\nabla^+(x,y)$ is similarly just the kernel of ∇^+). Next, for $n \geq 0$ and $k < n$ let

$$\Psi^n_k(x) = 2^{X_0(n-k)-x}F_{-k}\big(t, x - X_0(n-k)\big) = Q^{-k}e^{-\frac{1}{2}t\nabla^-}\delta_{X_0(n-k)}(x).$$

The powers of 2 which we have introduced should be thought of as a convenient normalization, the crucial point being that Q is the transition matrix of a random walk with strictly negative Geom$[\frac{1}{2}]$ steps; Ψ^n_k can be thought of (cf. (2.9)) as coming from applying repeatedly the forward difference operator to the Poisson weight $w_{t/2}(x) = e^{-t/2}(t/2)^x/x!\mathbf{1}_{x\geq 0}$, shifted by the initial data $X_0(n-k)$. One checks directly then, using the classical recurrence equations satisfied by the *Charlier polynomials* $C_k(x,t)$ (i.e., the family of discrete orthogonal polynomials with respect to the Poisson weight $w_t(x)$) that

$$\Psi^n_k(x) = 2^{X_0(n-k)-x}f_k\big(x + k - X_0(n-k)\big) \quad \text{with } f_k(x) = C_k(x,t)w_{t/2}(x).$$

Note that the functions Ψ^n_k only depend on n through a shift by the TASEP initial data. Next for $n \geq 0$ define $\{\Phi^n_k(x)\}_{k=0,\dots,n-1}$ as the (unique) solution of the following *biorthogonalization problem*:

> Given the family of *shifted Charlier functions* $\{\Psi^n_k\}_{k=0,\dots,n-1}$, find a family of functions $\{\Phi^n_k\}_{k=0,\dots,n-1}$ on $\mathbb{Z}$ so that
>
> (i) the two families are biorthogonal, i.e., $\sum_{x\in\mathbb{Z}}\Psi^n_k(x)\Phi^n_\ell(x) = \mathbf{1}_{k=\ell}$;
>
> (ii) $2^{-x}\Phi^n_k(x)$ is a polynomial of degree k.

Finally, for a fixed vector $a \in \mathbb{R}^m$ and indices $n_1 < \cdots < n_m$ let

$$\chi_a(n_j, x) = \mathbf{1}_{x>a_j} \quad \text{and} \quad \bar{\chi}_a(n_j, x) = \mathbf{1}_{x\leq a_j}, \tag{2.11}$$

which we also regard as multiplication operators (actually, projections) acting on the space $\ell^2(\{n_1,\dots,n_m\}\times\mathbb{Z})$, and later also on $L^2(\{n_1,\dots,n_m\}\times\mathbb{R})$. If a is a scalar, we write similarly $\chi_a(x) = 1 - \bar{\chi}_a(x) = \mathbf{1}_{x>a}$.

Theorem 2.4 ([10,45]). *Consider TASEP with initial data $(X_0(n))_{n\geq 1}$ and let $n_1,\dots,n_m$ be distinct positive integers. Then for $t > 0$ we have*

$$\mathbb{P}\big(X_t(n_j) > a_j,\ j = 1,\dots,m\big) = \det(I - \bar{\chi}_a K_t \bar{\chi}_a)_{\ell^2(\{n_1,\dots,n_m\}\times\mathbb{Z})}, \tag{2.12}$$

where

$$K_t(n_i, x_i; n_j, x_j) = -Q^{n_j-n_i}(x_i,x_j)\mathbf{1}_{n_i<n_j} + Q^{n_j-n_i}K^{(n_i)}_t(x_i,x_j) \tag{2.13}$$

with

$$K^{(n)}_t(x,y) = \sum_{k=1}^{n}\Psi^n_{n-k}(x)\Phi^n_{n-k}(y). \tag{2.14}$$

The determinant in (2.12) is the *Fredholm determinant*: for an integral operator A acting on $L^2(X,\mu)$ with kernel $A(x,y)$,

$$\det(I+A)=\sum_{n\geq 0}\frac{1}{n!}\int_{X^n}\mathrm{d}\mu(x_1)\cdots\mathrm{d}\mu(x_n)\det\big[A(x_i,x_j)\big]_{i,j=1}^{n}.$$

Note that the result holds for any choice of right-finite initial data X_0. The point is that, if we can solve the above biorthogonalization problem for X_0, then we have an explicit formula for the TASEP multipoint distributions which, at least in principle, is amenable to asymptotic analysis (in fact, the size of the determinant is now fixed, and computing the scaling limit will now involve only calculating suitable limits of the kernel K_t).

The challenge is then to solve the above biorthogonalization problem. One sees immediately why the choice of step/wedge initial data is the simplest in this setting. In fact, in this case $X_0(i)=-i$, so $\Psi^n_k(x)=2^{k-n-x}f_k(x+n)$ and hence by definition the biorthogonalization problem is solved by the Charlier polynomials themselves, $\Phi^n_k(x)=c_k2^{x+n-k}C_k(x+n,t)$ for a suitable normalization constant c_k. This leads to a relatively simple form for K_t in (2.13) as the *(extended) Charlier kernel* (related to what in random matrix theory would be called the *Charlier ensemble*), from which the TASEP limit can be extracted essentially by classical orthogonal polynomial asymptotics, leading to a kernel in terms of Airy functions (this explains the name of the Airy$_2$ process in Theorem 2.2), see (4.5). To prove Theorem 2.3, one first considers the half-periodic initial condition $X_0(i)=-2i$, $i\geq 1$, in which case the biorthogonalization was solved (in 2005 **[45]**) essentially by linear algebra (the answer **[10]** is $\Phi^n_k(x)=c'_k2^x\sum_{\ell=0}^{n-1}\frac{t^\ell}{(2k-\ell)(\ell-1)!}\binom{2k-\ell}{k-\ell}C_\ell(x,t)$); the full periodic case is recovered by focusing on particles far to the left and taking a suitable limit. But for more general choices of X_0 the method stalled for about a decade. (A third choice of initial data, namely a product measure, could be analyzed **[20]** by other methods which use crucially its stationarity for the TASEP evolution; also mixed versions of the three initial conditions could be handled, with one choice on the positive integers and another on the negative integers).

3. GENERAL SOLUTION OF TASEP

As the reader can probably guess by now, what we are aiming for is to construct the KPZ fixed point as the scaling limit (1.3) of the TASEP height function. Two obstacles lie in our way: we only know how to compute the limit for two special choices of initial data, and we can only do it for fixed time t (we chose $t=1$ above, but other choices of fixed t follow in the same way by adjusting the scaling). To some extent, however, the two obstacles are the same: in fact, TASEP is a Markov process and we thus expect the KPZ fixed point to also be Markovian, so in order to define its temporal evolution it should be enough to characterize its transition probabilities from an arbitrary initial condition in a suitable space.

Biorthogonalization solution. The general solution of the biorthogonalization problem for TASEP appeared in **[28]**, and leads to a representation for the kernel K_t from which asymptotics can be performed naturally. Two main ingredients were used in its derivation. The first one is the time reversal invariance satisfied by TASEP.

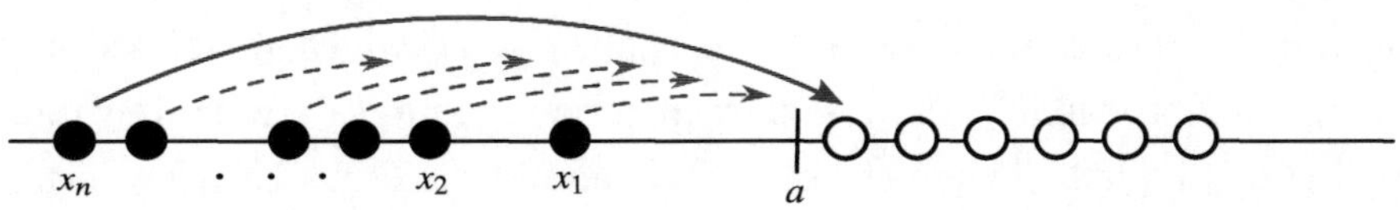

Suppose we start with particles at locations $x_1 > \cdots > x_n$. By the exclusion condition, the probability that $X_t(n) > a$ is the same as the probability that $X_t(i) > a + n - i$ for each $i = 1, \ldots, n$. But, by symmetry, this is the same as starting TASEP at $(a+1, \ldots, a+n)$, running it backwards, and computing the probability that $X_t(i) \le x_{n+1-i}$ for each i. Using now simple reflection and shift invariance properties of the TASEP dynamics, we deduce that

$$\mathbb{P}_{(x_1,\ldots,x_n)}\big(X_t(n) > a\big) = \mathbb{P}_{(-1,\ldots,-n)}\big(X_t(i) > a - x_{n+1-i} - 1,\ i = 1, \ldots, n\big). \tag{3.1}$$

We have thus turned the one-point distribution of TASEP with arbitrary initial data into the multipoint distribution of TASEP with step initial data, which as we explained in the last section can be computed explicitly.

The second ingredient is a *path integral* version of the extended kernel formula (2.12), which reads as follows (recall the definition of χ_a for scalar a after (2.11)):

$$\begin{aligned}\mathbb{P}\big(X_t(n_j) > a_j,\ j = 1, \ldots, m\big)\\ = \det\big(I - K_t^{(n_m)}(I - Q^{n_1-n_m}\chi_{a_1} Q^{n_2-n_1}\chi_{a_2} \cdots Q^{n_m-n_{m-1}}\chi_{a_m})\big)_{\ell^2(\mathbb{Z})},\end{aligned} \tag{3.2}$$

where $K_t^{(n)} = K_t(n, \cdot; n, \cdot)$. A formula of this type was first derived in **[35]** for the Airy$_2$ process and later extended to the Airy$_1$ process in **[38]**, and to a very wide class of processes in **[8]**. To see how this helps, observe that the factor $\chi_{a_1} Q^{n_2-n_1}\chi_{a_2} \cdots Q^{n_m-n_{m-1}}\chi_{a_m}(x,y)$ inside the determinant in (3.2) is nothing but the probability that a random walk with geometric steps goes from x at time n_1 to y at time n_m, staying above a_1 at time n_1, above a_2 at time n_2, etc. On the other hand, through (3.1) this formula computes the distribution of $X_t(n)$ with general initial data. From this we see that the Φ_k^n's should be related to the probability that the geometric random walk hits the curve prescribed by the a_i's, which together with the fact that the factor $K_t^{(n)}$ appearing in (3.2) is explicit (it is the one-point kernel for step initial data) allows one to try to guess the form of the functions. Theorem 2.4 is then set up perfectly, because one can simply check (in fact, in just a few lines) that the guess gives the right answer, which is as follows: $\Phi_k^n(x) = (e^{\frac{t}{2}\nabla^-})^* h_k^n(0,x)$, where $h_k^n(\ell,x)$ is the unique solution to the initial–boundary value problem for the discrete backwards heat equation

$$\begin{cases} (Q^*)^{-1} h_k^n(\ell,x) = h_k^n(\ell+1,x), & \ell < k,\ x \in \mathbb{Z},\\ h_k^n(k,x) = 2^{x-X_0(n-k)}, & x \in \mathbb{Z},\\ h_k^n(\ell, X_0(n-\ell)) = 0, & \ell < k. \end{cases}$$

For $x < X_0(n-k)$, $h_k^n(0,x)$ is simply the probability, starting from x, that the (reversed) random walk first goes above the curve $(X_0(n-\ell+1))_{\ell=1,\ldots,n}$ at time $\ell = k+1$.

Explicit formula. In order to obtain a usable formula for the TASEP kernel K_t, we need to evaluate the sum in (2.14) using the solution for the Φ_k^n's. With a bit of work, this leads to a formula which is explicitly given in terms of random walk hitting times.

In order to state it, we introduce a kernel $Q^n_{\mathrm{epi}(X_0)}$ which is defined as follows: $Q^n_{\mathrm{epi}(X_0)}(x,y)$ is the probability, starting at x, that the random walk with transition matrix Q hits the strict epigraph of (i.e the region strictly above) the curve $(X_0(\ell+1))_{\ell=0,\dots,n-1}$ and ends at y at time n. One can check that, for fixed x, the mapping $y \mapsto 2^{-y} Q^n_{\mathrm{epi}(X_0)}(x,y)$ defines a polynomial for $y \le X_0(n)$.

Theorem 3.1 ([28]). *The kernel $K_t^{(n)}$ in (2.14) can be written as follows:*

$$K_t^{(n)} = e^{\frac{t}{2}\nabla^-} Q^{-n}\, \overline{Q}^n_{\mathrm{epi}(X_0)} e^{-\frac{t}{2}\nabla^-}, \tag{3.3}$$

where $\overline{Q}^n_{\mathrm{epi}(X_0)}(x,y)$ equals 2^y times the polynomial extension from $y \le X_0(n)$ to all $y \in \mathbb{Z}$ of the kernel $2^{-y} Q^n_{\mathrm{epi}(X_0)}(x,y)$.

The polynomial extension in the formula comes from using the above representation of the functions $h_k^n(0,x)$ as hitting probabilities for x below the curve; since $2^{-x}\Phi_k^n(x)$ has to be a polynomial, one can recover it everywhere through this extension. The operators are relatively simple, however, and the polynomial extension can be computed explicitly.

Other processes. The scheme which we have described works for a more general class of particle systems with determinantal transition functions of the form (2.7). It has been applied, in particular, for PushASEP [32], one-sided reflected Brownian motions [31], and several discrete time versions of TASEP [29].

4. THE KPZ FIXED POINT

Recapitulating, what we would like to do now is to extract the limit in (1.3), with h the TASEP height function, using the explicit formula supplied by Theorems 2.4 and 3.1 (this involves a simple translation from the particle system to the height function). In view of the scaling in Theorems 2.2 and 2.3, we take $c_1 = c_2 = 2$ and $c_3 = -1$. Let us briefly sketch how the limit arises. Consider the factor $e^{\frac{t}{2}\nabla}Q^{-n}$ in (3.3). Using the scaling from (1.3), it becomes approximately $e^{\varepsilon^{-3/2}t[-\nabla^- + \frac{1}{2}\log(I+2\nabla^+)]}$ (ignoring lower-order terms). After suitably scaling the variables inside the kernel, the limit is computed on the scaled lattice $\varepsilon^{1/2}\mathbb{Z}$, so $\nabla^\pm \sim \varepsilon^{1/2}$ and therefore $-\nabla^- + \frac{1}{2}\log(I+2\nabla^+) = -\nabla^- + \nabla^+ - (\nabla^+)^2 + \frac{4}{3}(\nabla^+)^3 + \mathcal{O}(\varepsilon^2) \sim \frac{1}{3}\varepsilon^{3/2}\partial^3$ after a simple Taylor expansion, where ∂ is the derivative operator. Similarly (or by the central limit theorem), we have $Q^{\varepsilon^{-1}x} \sim e^{x\partial^2}$. This tells us that, as $\varepsilon \to 0$, $e^{\frac{t}{2}\nabla}Q^{-n}$ becomes

$$\mathbf{S}_{t,x} := e^{\frac{1}{3}t\partial^3 + x\partial^2}. \tag{4.1}$$

At first sight, this operator may appear to be problematic because the heat kernel $e^{x\partial^2}$ is ill-defined for $x<0$, but, in fact, $\mathbf{S}_{t,x}$ makes sense for all $t \neq 0$ as an integral operator on a suitable domain with integral kernel (here Ai is the Airy function)

$$\mathbf{S}_{t,x}(u,v) = t^{-1/3} e^{-\frac{2x^3}{3t^2} - (u-v)\frac{x}{t}} \operatorname{Ai}\big(t^{-1/3}(v-u) + t^{-4/3}x^2\big)$$

and it satisfies the group property $\mathbf{S}_{t,x}\mathbf{S}_{s,y} = \mathbf{S}_{t+s,x+y}$ as long as t, s and $t+s$ are all nonzero. The convergence to $\mathbf{S}_{t,x}$ which we have sketched can be proved by using a contour integral formula (similar to (2.8)) for $e^{\frac{t}{2}\nabla^-}Q^{-n}$. A similar argument works for the other factor, $\overline{Q}^n_{\mathrm{epi}(X_0)}e^{-\frac{t}{2}\nabla^-}$ (using a similar contour integral formula). The scaling under which we are working, on the other hand, has the effect of rescaling the random walk inside this last kernel diffusively, so in the limit the random walk hitting times become Brownian hitting times.

Brownian scattering operator. The above sketch explains how the ingredients which will make up the formula for the KPZ fixed point arise. The actual proof of the limit, in a suitably strong sense and with appropriate estimates, involves some heavy asymptotic analysis. The final result, in its most appealing form after some postprocessing, involves a kernel which we introduce next.

The natural class of initial data for our (continuum) random growth models is UC, the space of upper-semicontinuous functions $\mathfrak{h}\colon \mathbb{R} \to [-\infty, \infty)$ satisfying $\mathfrak{h}(x) \le a + b|x|$ for some $a, b > 0$ and $\mathfrak{h} \not\equiv -\infty$, which we endow with the topology of local Hausdorff convergence (with $[-\infty,\infty)$ compactified at $-\infty$). The linear bound which we are assuming on the initial data is not quite optimal, but it ensures that $\mathfrak{h}(t,x)$ is defined for all $t > 0$. Note that the narrow wedge $\mathfrak{d}_0$ is in UC. Given $\mathfrak{h} \in$ UC and $\ell_1 < \ell_2$, let

$$\mathbf{P}^{\text{No hit }\mathfrak{h}}_{\ell_1,\ell_2}(u_1,u_2)du_2 = \mathbb{P}_{\mathbf{B}(\ell_1)=u_1}\big(\mathbf{B}(y) > \mathfrak{h}(y) \text{ on } [\ell_1,\ell_2],\ \mathbf{B}(\ell_2) \in du_2\big)$$

with $\mathbf{B}$ a Brownian motion with diffusion coefficient 2, and define $\mathbf{P}^{\text{Hit }\mathfrak{h}}_{\ell_1,\ell_2} = \mathbf{I} - \mathbf{P}^{\text{No hit }\mathfrak{h}}_{\ell_1,\ell_2}$. The ($t$-dependent) *Brownian scattering operator* associated to $\mathfrak{h}$ is

$$\mathbf{K}^{\mathrm{hypo}(\mathfrak{h})}_t = \lim_{\substack{\ell_1\to-\infty\\ \ell_2\to\infty}} e^{-\frac{1}{3}t\partial^3+\ell_1\partial^2}\mathbf{P}^{\text{Hit }\mathfrak{h}}_{\ell_1,\ell_2}e^{\frac{1}{3}t\partial^3-\ell_2\partial^2}. \tag{4.2}$$

In words, the Brownian scattering operator computes a sort of asymptotic "transition density" for a Brownian motion in the whole line, killed if it does not hit hypo($\mathfrak{h}$), the hypograph of (i.e., the region below) $\mathfrak{h}$. The fact that the right-hand side of (4.2) makes sense is far from obvious; it was first proved (for a more restricted class of $\mathfrak{h}$) in [39]. A more explicit formula can be given in terms of the operators $\mathbf{S}_{t,x}$ and the law of the hitting time by a Brownian motion of hypo($\mathfrak{h}$).

The Brownian scattering operator $\mathbf{K}^{\mathrm{hypo}(\mathfrak{h})}_t$ plays a key role in the construction of the KPZ fixed point, with the function $\mathfrak{h}$ as the initial data for $\mathfrak{h}(t,x)$. One may wonder about whether the limit on the right-hand side of (4.2) contains all the necessary information about $\mathfrak{h}$. It does: for any $t > 0$, $\mathfrak{h} \mapsto \mathbf{K}^{\mathrm{hypo}(\mathfrak{h})}_t$ is invertible, and moreover continuous, as a mapping from UC to a suitable space of trace class operators (the invertibility is obtained essentially directly from (4.3) below).

Definition of the KPZ fixed point. We are finally ready to define our main object of interest: the *KPZ fixed point* is the (unique) Markov process taking values in UC whose transition probabilities satisfy (here $\mathfrak{h}_0$ in the subscript denotes the initial condition)

$$\mathbb{P}_{\mathfrak{h}_0}\big(\mathfrak{h}(t,x_1) \le r_1, \dots, \mathfrak{h}(t,x_m) \le r_m\big) = \det(\mathbf{I} - \chi_r\mathbf{K}^{\mathrm{hypo}(\mathfrak{h}_0)}_{t,\mathrm{ext}}\chi_r)_{L^2(\{x_1,\dots,x_m\}\times\mathbb{R})} \tag{4.3}$$

for any $r = (r_1, \dots, r_m) \in \mathbb{R}^m$, with $\mathbf{K}^{\mathrm{hypo}(\mathfrak{h}_0)}_{t,\mathrm{ext}}$ the *extended Brownian scattering operator*

$$\mathbf{K}^{\mathrm{hypo}(\mathfrak{h}_0)}_{t,\mathrm{ext}}(x_i, u_i; x_j, u_j) = -e^{(x_j - x_i)\partial^2}(u_i, u_j)\mathbf{1}_{x_i < x_j} + e^{-x_i\partial^2}\mathbf{K}^{\mathrm{hypo}(\mathfrak{h}_0)}_{t} e^{x_j\partial^2}(u_i, u_j) \quad (4.4)$$

and where, we recall, χ_r was defined in (2.11). Two (related) statements are implicit in this definition: that the right-hand side of (4.3) defines uniquely a probability measure on UC, and that it in fact defines the transition kernel of a Markov process. The first one holds because the events in the probability on the left-hand side generate the Borel σ-algebra on UC. The second one is proved based on the fact, stated next, that $\mathfrak{h}(t, x)$ arises as the scaling limit of TASEP, which is Markovian, plus a compactness argument which allows one to show that the property is preserved in the limit.

Theorem 4.1 ([28]). *Let the 1:2:3 rescaled TASEP height function be defined as*

$$\mathfrak{h}^{\varepsilon}(t, x) = \varepsilon^{1/2}\big[h(2\varepsilon^{-3/2}t, 2\varepsilon^{-1}x) + 2\varepsilon^{-1}x\big].$$

Fix $\mathfrak{h}_0 \in$ UC and assume that $\mathfrak{h}^{\varepsilon}(0, \cdot) \to \mathfrak{h}_0$ in distribution in UC. Then $\mathfrak{h}^{\varepsilon}(t, x) \to \mathfrak{h}(t, x)$ as $\varepsilon \to 0$, in distribution in UC as a process in t, x, where $\mathfrak{h}(t, x)$ is the KPZ fixed point started at $\mathfrak{h}_0$, i.e., the UC-valued Markov process defined through (4.3).

Properties of the KPZ fixed point. The formula for the KPZ fixed point transition probabilities (4.3) looks perhaps too complicated to be of any use, but in fact it can be used to derive many of the conjectured properties of the KPZ fixed point (as well as some surprising ones, as we will see in the next section), including the following (some of which we state vaguely, see [28] for the details):

- $\mathfrak{h}(t, x)$ is 1:2:3 scaling invariant, i.e., it satisfies (1.4).
- $\mathfrak{h}(t, x)$ is invariant under spatial shifts, reflections and affine translations.
- (Skew time reversibility) $\mathbb{P}_{\mathfrak{g}}(\mathfrak{h}(\mathbf{t}, \mathbf{x}) \le -\mathfrak{f}(\mathbf{x})) = \mathbb{P}_{\mathfrak{f}}(\mathfrak{h}(\mathbf{t}, \mathbf{x}) \le -\mathfrak{g}(\mathbf{x}))$ for any $\mathfrak{f}, \mathfrak{g} \in$ UC.
- $\mathfrak{h}(t, x)$ is Hölder-$\frac{1}{2}$ in x, and Hölder-$\frac{1}{3}$ in time.
- (Brownian invariance and ergodicity) If $\mathbf{B}$ is a two-sided Brownian motion with diffusion coefficient 2 then for any $t > 0$, the process $x \longmapsto \mathfrak{h}(t, x; \mathbf{B}) - \mathfrak{h}(t, 0; \mathbf{B})$ has the same distribution as $\mathbf{B}$. Moreover, for any initial condition and any fixed $t > 0$, the finite-dimensional distributions of $\mathfrak{h}(t, x) - \mathfrak{h}(t, 0)$ are locally Brownian and, under some conditions, they converge as $t \to \infty$ to those of $\mathbf{B}$.

Formula (4.3) can also be used to recover the Airy$_1$ and Airy$_2$ processes which had already been derived for special initial data (see (2.3) and (2.6)) because the Brownian hitting probabilities in (4.2) are explicit in those cases. For example, for narrow wedge initial data $\mathfrak{d}_0$, the only way to hit hypo($\mathfrak{d}_0$) is for the Brownian path to pass below the origin at time 0, so one trivially has, for $\ell_1 < 0 < \ell_2$, $\mathbf{P}^{\mathrm{Hit}\,\mathfrak{d}_0}_{\ell_1,\ell_2} = e^{-\ell_1\partial^2}\bar{\chi}_0 e^{\ell_2\partial^2}$ and therefore $\mathbf{K}^{\mathrm{hypo}(\mathfrak{d}_0)}_t = e^{-\frac{t}{3}\partial^3}\bar{\chi}_0 e^{\frac{t}{3}\partial^3}$ which, using (4.1) and setting $t = 1$, leads directly to the known formula for the Airy$_2$ process:

$$\mathbb{P}\big(\mathcal{A}_2(x_1) \le r_1, \ldots, \mathcal{A}_2(x_m) \le r_m\big) = \det(\mathbf{I} - \chi_r \mathbf{K}^{\mathrm{Ai}}_{\mathrm{ext}} \chi_r)_{L^2(\{x_1,\ldots,x_m\}\times\mathbb{R})} \tag{4.5}$$

with $\mathbf{K}^{\mathrm{Ai}}_{\mathrm{ext}}$ defined by the right-hand side of (4.4) with $\mathbf{K}^{\mathrm{hypo}(\mathfrak{h}_0)}_t(u, v)$ replaced by the *Airy kernel* $\int_0^\infty d\lambda\, \mathrm{Ai}(x+\lambda)\,\mathrm{Ai}(y+\lambda)$. For flat initial data, the calculation involves the hitting probability by a Brownian motion of a straight line, which can be computed using the reflection principle.

Variational formula. An alternative description of the KPZ fixed point is through a variational (Hopf–Lax type) formula involving a nontrivial input noise called the *Airy sheet* $\mathcal{A}(x, y)$ (which is natural, for instance, from the point of view of LPP with boundary conditions, see (2.4)): for the KPZ fixed point starting from $\mathfrak{h}(0, x) = \mathfrak{h}_0(x)$,

$$\mathfrak{h}(t, x) \overset{(d)}{=} \sup_{y\in\mathbb{R}} \Big\{ t^{1/3}\mathcal{A}(t^{-2/3}x, t^{-2/3}y) - \frac{1}{t}(x-y)^2 + \mathfrak{h}_0(y) \Big\}. \tag{4.6}$$

The Airy sheet $\mathcal{A}(x, y)$ can be thought of as $\mathfrak{h}(1, x)$ starting from a narrow wedge at y at time 0, and it therefore involves coupling different initial conditions. The construction of the KPZ fixed point from TASEP described above leads to the Airy sheet through subsequential limits: TASEP can be constructed with coupled initial conditions but, as far as is known, one loses access to explicit formulas, and hence the distribution of the Airy sheet is unknown. This led to a problem in that it was unclear that (4.6) even involved a unique object on the right-hand side. This problem was overcome in [15], where the Airy sheet (and, more generally, its space-time version known as the *directed landscape*) was constructed directly in terms of an LPP problem on the Airy line ensemble, and where the authors showed that it is the scaling limit of Brownian LPP, putting the variational formula (4.6) on a solid footing ([31] confirmed that, as expected, both constructions define the same object).

The methods used in [15] are very different from those presented here (they use heavily, in particular, a version of the RSK correspondence), and provide an alternative approach to the study of the KPZ fixed point. As an example, they have been used to prove a strong version of the local Brownian property mentioned above: the KPZ fixed point $\mathfrak{h}(t, x)$ is absolutely continuous (in x) with respect to a Brownian motion on compact intervals [44].

Convergence for other models. We have constructed the KPZ fixed point as a scaling limit of TASEP. With the description of this universal limit at hand, an important and natural problem that follows is to show that it is too the limit of other models conjectured to be in the class. Since the methods we described in Section 3 are applicable to a wider class of determinantal interacting particle systems, it is natural to expect that convergence can be proved for those, too. This has been done for the model of one-sided reflected Brownian motions [31], and can also be done for the variants of TASEP covered in [29] (although in this last case the details have not yet been worked out). The methods from [15], on the other hand, have been extended in [16] to show convergence of several (exactly solvable) LPP models.

Recently in [43,53] the convergence was extended to the KPZ equation, asymmetric exclusion processes and Brownian last passage percolation. These major results required new ideas, since those processes are not determinantal; however, the methods still rely on integrability to a certain extent, and the convergence to the KPZ fixed point for more general models (e.g., ballistic deposition or LPP with general weights) remains wide open.

5. INTEGRABILITY OF THE KPZ FIXED POINT TRANSITION PROBABILITIES

The description of the KPZ fixed point given in the last section leaves open the question as to whether it satisfies some sort of a stochastic equation. The variational formula (4.6) gives a partial answer; however, the distribution of the Airy sheet and the functional of the Airy line ensemble from which it arises are not explicit. In this sense, (4.6) is not satisfying as a universal scaling invariant equation. But we can do something else.

Stochastic integrability. Recall the definition (4.2) of the Brownian scattering operator, the main building block in the KPZ fixed point formulas. Notice that, at least formally, we can write $\mathbf{K}_t^{\mathrm{hypo}(\mathfrak{h}_0)} = e^{-\frac{t}{3}\partial^3}\mathbf{K}_0^{\mathrm{hypo}(\mathfrak{h}_0)}e^{\frac{t}{3}\partial^3}$. Remarkably, the dependence of $\mathbf{K}_t^{\mathrm{hypo}(\mathfrak{h}_0)}$ on t is completely decoupled from the dependence on the initial data. Moreover, the time evolution is linear: at the level of the extended Brownian scattering operator (4.4), it satisfies the Lax equation

$$\partial_t \mathbf{K}_{t,\mathrm{ext}}^{\mathrm{hypo}(\mathfrak{h}_0)} = \left[-\frac{1}{3}\partial^3, \mathbf{K}_{t,\mathrm{ext}}^{\mathrm{hypo}(\mathfrak{h}_0)}\right], \tag{5.1}$$

where $[A, B] = AB - BA$; in other words, the equation reads

$$\partial_t \mathbf{K}_{t,\mathrm{ext}}^{\mathrm{hypo}(\mathfrak{h}_0)}(x_i, u_i; x_j, u_j) = -\frac{1}{3}(\partial_{u_i}^3 + \partial_{u_j}^3)\mathbf{K}_{t,\mathrm{ext}}^{\mathrm{hypo}(\mathfrak{h}_0)}(x_i, u_i; x_j, u_j).$$

The dynamics is thus trivial at the level of the kernels, and the KPZ fixed point finite-dimensional distributions are recovered by projecting down via the Fredholm determinant (4.3). This provides a representation for the temporal evolution of the KPZ fixed point under which the flow is linearized; from this perspective, one might say that this presents the KPZ fixed point as a *stochastic integrable system* (cf. [17]). Note that, in view of Theorems 2.4 and 3.1, TASEP is integrable in the same sense; this is separate from (though, of course, not unrelated to) other aspects of TASEP's exact solvability such as the coordinate Bethe ansatz leading to (2.8), the algebraic Bethe ansatz leading to its diagonalization [36], or its relation with the Schur process (see, e.g., [11]).

Note that an equation of the same nature can be written for the dependence of the Brownian scattering operator on the spatial variables: one has

$$(\partial_{x_1} + \cdots + \partial_{x_m})\mathbf{K}_{t,\mathrm{ext}}^{\mathrm{hypo}(\mathfrak{h}_0)}(x_i, u_i; x_j, u_j) = (\partial_{u_j}^2 - \partial_{u_j}^2)\mathbf{K}_{t,\mathrm{ext}}^{\mathrm{hypo}(\mathfrak{h}_0)}(x_i, u_i; x_j, u_j). \tag{5.2}$$

We have stated this identity in terms of the differential operator $\partial_{x_1} + \cdots + \partial_{x_m}$ in order to write a simple formula, but this will actually be consequential below.

Kadomtsev–Petviashvili equation. For fixed $\mathfrak{h}_0 \in \mathrm{UC}$ and $m \in \mathbb{N}$, let

$$F(t, x_1, \ldots, x_m, r_1, \ldots, r_m) = \mathbb{P}_{\mathfrak{h}_0}\big(\mathfrak{h}(t, x_1) \le r_1, \ldots, \mathfrak{h}(t, x_m) \le r_m\big)$$

denote the m-point distribution of the KPZ fixed point. We will see now that the stochastic integrability of $\mathbf{K}_t^{\text{hypo}(\mathfrak{h}_0)}$ leads to a description of F in terms of a classical dispersive PDE. Shifting the variables inside the Fredholm determinant in (4.3), F can be written as

$$F(t, x_1, \ldots, x_m, r_1, \ldots, r_m) = \det(\mathbf{I} - \mathbf{K}) \tag{5.3}$$

where the determinant is on the m-fold direct sum of $L^2([0,\infty))$ (which we have identified with $L^2(\{x_1, \ldots, x_m\} \times [0,\infty))$) and

$$\mathbf{K}_{ij}(u_i, u_j) = \mathbf{K}_{t,\text{ext}}^{\text{hypo}(\mathfrak{h}_0)}(u_i + r_i, u_j + r_j).$$

Next we introduce an $m \times m$ matrix-valued function Q defined in terms of $\mathbf{K}$ as follows:

$$Q(t, x_1, \ldots, x_m, r_1, \ldots, r_m) = (\mathbf{I} - \mathbf{K})^{-1}\mathbf{K}(0,0)$$

(see [40] for the fact that the right hand side is well-defined). Note that each entry of Q depends on t and each of the x_i's and r_i's, but we omit this from the notation. Finally, let

$$\mathcal{D}_r = \partial_{r_1} + \cdots + \partial_{r_n}, \quad \mathcal{D}_x = \partial_{x_1} + \cdots + \partial_{x_n}. \tag{5.4}$$

Theorem 5.1 ([40]). *Fix an initial condition $\mathfrak{h}_0 \in \mathrm{UC}$ for the KPZ fixed point and define F and Q as above. Then*

$$\mathcal{D}_r \log F = \operatorname{tr} Q,$$

while Q and its derivative $q = \mathcal{D}_r Q$ solve the matrix Kadomtsev–Petviashvili (KP) equation

$$\partial_t q + \frac{1}{2}\mathcal{D}_r q^2 + \frac{1}{12}\mathcal{D}_r^3 q + \frac{1}{4}\mathcal{D}_x^2 Q + \frac{1}{2}[q, \mathcal{D}_x Q] = 0. \tag{5.5}$$

In particular, for the one point marginals of the KPZ fixed point $F(t,x,r) = \mathbb{P}_{\mathfrak{h}_0}(\mathfrak{h}(t,x) \le r)$, $\phi = \partial_r^2 \log F$ satisfies the scalar KP-II equation

$$\partial_t \phi + \frac{1}{2}\partial_r \phi^2 + \frac{1}{12}\partial_r^3 \phi + \frac{1}{4}\partial_r^{-1}\partial_x^2 \phi = 0. \tag{5.6}$$

The KP equation (5.6) was originally derived from studies of long waves in shallow water [1], and plays the role of a natural two-dimensional extension of the Korteweg–de Vries (KdV) equation. In fact, when ϕ is independent of x, as is the case for flat initial data $\mathfrak{h}_0 \equiv 0$ in our setting (though other, nondeterministic choices are possible [27]), it reduces to KdV,

$$\partial_t \phi + \frac{1}{2}\partial_r \phi^2 + \frac{1}{12}\partial_r^3 \phi = 0. \tag{5.7}$$

KP (as well as its matrix version) is completely integrable and plays an important role in the Sato theory as the first equation in the KP hierarchy [30]. However, none of the previous physical derivations of KP seem to be related to our problem. The proof of Theorem 5.1, sketched below, is essentially by algebra, and we have no physical intuition yet as to why it is true. (As a separate note, we mention that [37] derived essentially concurrently a connection with superpositions of KP solitons for KPZ fluctuations with special initial data in a finite volume setting).

In the one-point case (5.6), the initial data in our setting is $\phi(0,x,r) = 0$ for $r \ge \mathfrak{h}_0(x)$, $\phi(0,x,r) = -\infty$ for $r < \mathfrak{h}_0(x)$. The formal $-\infty$ can be replaced by a suitable decay condition as $t \searrow 0$ but, in any case, this type of initial data is very far from what known well-posedness schemes for KP can handle, and uniqueness for the solutions of (5.6) arising from KPZ growth remains open. The initial data for the matrix version (5.5) is more delicate, see [40].

What should be most striking about the statement of Theorem 5.1 is that the finite-dimensional distributions satisfy a closed equation at all. In retrospect, one realizes that a PDE for the evolution of the one-point distributions follows, in the special case of narrow wedge and flat initial data, from scaling considerations and (2.2), (2.5) (see below). But for general initial data $\mathfrak{h}_0 \in \mathrm{UC}$, this is way outside the scope of what had been expected. And in any case, even knowing that the one-point distributions satisfy a closed equation, in general one does not necessarily expect there to be multipoint equations (moreover, one would tipically hope at best that the multipoint distributions satisfy a hierarchy of equations, linking the m-point distribution to the k-point distributions for $k < m$); an exception is the multipoint distributions of the Airy_2 process, for which PDEs were expected to be satisfied, some answers had been derived in [2, 52]. It is also not clear why the multipoint equation should be written in terms of derivatives with respect to the variables $r_1 + \cdots + r_m$ and $x_1 + \cdots + x_m$ in (5.4).

Let us very briefly sketch how Theorem 5.1 is proved. For simplicity, we restrict to the one-point distribution $F(t, x, r)$ (in which case the proof amounts essentially to a rediscovery of an argument which had been employed before in an abstract setting; see, e.g., [34]). Letting $\Phi(t, x, r) = \partial_r \log(F(t, x, r))$, we have, using (5.3),

$$\Phi(t, x, r) = \partial_r \log\big(\det(\mathbf{I} - \mathbf{K})\big) = -\operatorname{tr}\big((\mathbf{I} - \mathbf{K})^{-1} \partial_r \mathbf{K}\big) = (\mathbf{I} - \mathbf{K})^{-1} \mathbf{K}(0, 0);$$

the second equality is standard, while the third follows from a simple computation of the trace of $(\mathbf{I} - \mathbf{K})^{-1} \partial_r \mathbf{K}$ as $\int_0^\infty dx\, (\mathbf{I} - \mathbf{K})^{-1} \partial_r \mathbf{K}(x, x)$ and the crucial fact that, by definition,

$$\partial_r \mathbf{K}(u, v) = (\partial_u + \partial_v)\mathbf{K}(u, v). \tag{5.8}$$

We can now compute derivatives of Φ directly in terms of the derivatives of $\mathbf{K}$ with respect to each parameter, using the last identity together with the evolution equations (5.1) and (5.2); putting them together leads, after a couple of pages of computations, to (5.6) (the main difficulty is the nonlinear term in the KP equation, but this can be handled through a suitable integration by parts formula).

Tracy–Widom distributions. One of the most striking aspects of the study of the KPZ universality class is its connection with random matrix theory. The most prominent instance of this connection is provided by (2.2) and (2.5), which state that the one point distribution of the KPZ fixed point with narrow wedge and flat initial data are distributed, respectively, as Tracy–Widom GUE and GOE random variables. A partial explanation for this in the GUE/narrow wedge case is the fact, which we mentioned in Section 2, that some special KPZ models (with very particular initial conditions, e.g., step for TASEP) can be coupled to models which are naturally of random matrix type; in the GOE/flat case, such a connection appears not to have been fully uncovered. In any case, the correct picture one should have in mind seems to be of KPZ and random matrix theory as two separate (though related) domains which intersect, most prominently via the central role played by the Tracy–Widom distributions on both sides.

This bears the natural question as to what makes the Tracy–Widom distributions so special. One difficulty is that the Tracy–Widom distributions themselves seem to lack any meaningful invariance. In the KPZ setting, however, Theorem 5.1 makes such an invariance apparent. The crucial point is that the KP equation (5.6) is invariant (as it has to be, in view of (1.4)) under the 1:2:3 rescaling

$$\phi(t,x,r) \mapsto \alpha^{-2}\phi(\alpha^{-3}t,\alpha^{-2}x,\alpha^{-1}r), \quad \mathfrak{h}_0(x) \mapsto \alpha^{-1}\mathfrak{h}_0(\alpha^2 x). \tag{5.9}$$

As we explain next, the Tracy–Widom distributions then appear in the context of the KPZ universality class as special self-similar solutions of KP (the key being that both narrow wedge and flat initial are invariant under the rescaling in (5.9)).

Consider first the narrow wedge case. From (2.3) and the 1:2:3 scaling invariance (1.4), we have $\mathfrak{h}(t,x) + x^2/t \overset{(d)}{=} t^{1/3}\mathcal{A}_2(t^{-2/3}x)$ where $\mathcal{A}_2$ is the Airy$_2$ process, which is stationary. In view of (5.9), it is then natural to look for a self-similar solution of the form $\phi^{\mathrm{nw}}(t,x,r) = t^{-2/3}\psi^{\mathrm{nw}}(t^{-1/3}r + t^{-4/3}x^2)$. This turns (5.6) into the ODE

$$(\psi^{\mathrm{nw}})''' + 12\psi^{\mathrm{nw}}(\psi^{\mathrm{nw}})' - 4r(\psi^{\mathrm{nw}})' - 2\psi^{\mathrm{nw}} = 0. \tag{5.10}$$

The transformation $\psi^{\mathrm{nw}} = -u^2$ takes (5.10) into the Painlevé II equation:

$$u'' = ru + 2u^3. \tag{5.11}$$

Known tail estimates for the Tracy–Widom GUE distribution (i.e. the one-point marginal of the Airy$_2$ process) imply that, as $r \to -\infty$, one has $\phi^{\mathrm{nw}}(t,x,r) \sim -(\frac{r}{2t} + \frac{x^2}{2t^2})$. This picks out the Hastings–McLeod solution of (5.11), $u(r) \sim -\operatorname{Ai}(r)$ as $r \to \infty$, and thus we get

$$F(t,x,r) = \exp\left\{-\int_{\hat{r}}^{\infty} \mathrm{d}s\,(s-\hat{r})u^2(s)\right\} = F_{\mathrm{GUE}}(t^{-1/3}r + t^{-4/3}x^2)$$

with $\hat{r} = \frac{r}{t^{1/3}} + \frac{x^2}{t^{4/3}}$, the last equality being the Painlevé II representation for F_{GUE} famously derived by Tracy and Widom [50].

In the flat case, $\mathfrak{h}_0 = 0$, there is no dependence on x, so we need to look for self-similar solutions of KdV (5.7), of the form $\phi^{\mathrm{fl}}(t,r) = (t/4)^{-2/3}\psi^{\mathrm{fl}}((t/4)^{-1/3}r)$ (the factor of $1/4$ is for convenience), leading to $(\psi^{\mathrm{fl}})''' + 12(\psi^{\mathrm{fl}})'\psi^{\mathrm{fl}} - r(\psi^{\mathrm{fl}})' - 2\psi^{\mathrm{fl}} = 0$. Miura's transform $\psi^{\mathrm{fl}} = \frac{1}{2}(u' - u^2)$ brings this to Painlevé II (5.11), with the same asymptotics as $r \to -\infty$, and we recover, writing $\hat{r} = 4^{1/3}t^{-1/3}r$ (the second equality comes from [51])

$$F(t,x,r) = \exp\left\{-\frac{1}{2}\int_{\hat{r}}^{\infty} \mathrm{d}s\,u(s)\right\} F_{\mathrm{GUE}}(\hat{r})^{1/2} = F_{\mathrm{GOE}}(4^{1/3}t^{-1/3}r).$$

6. KP IN SPECIAL SOLUTIONS OF THE KPZ EQUATION

The proof of Theorem 5.1 which we sketched above does not make any use of the particular form of the Brownian scattering operator other than the fact that it satisfies the differential equations (5.1), (5.2), and (5.8) (together with some technical conditions). The method thus applies in general to Fredholm determinants of kernels which satisfy the same identities.

Surprisingly, it turns out that, in the one-point case, the method is applicable to some special solutions of the KPZ equation (1.1), where for convenience we fix the scaling $\lambda = \nu = \frac{1}{4}$ and $\sigma = 1$. The simplest case is, again, the narrow wedge solution h_{nw} of (1.1), by which what is meant (see [7]) is that $h_{\mathrm{nw}} = \log(Z)$ with Z the fundamental solution of the stochastic heat equation with multiplicative noise (i.e., $\partial_t Z = \frac{1}{4}\partial_x^2 Z + \xi Z$, $Z(0,x) = \delta_0(x)$): in the early 2010s a formula was obtained [3,13,18,46] for the KPZ generating function

$$G_{\mathrm{nw}}(t,x,r) = \mathbb{E}\big[\exp\big\{-e^{h_{\mathrm{nw}}(t,x)+\frac{t}{12}-r}\big\}\big]$$

which can be rewritten as $\det(\mathbf{I} - \mathbf{K})$ with

$$\mathbf{K}(u,v) = \int_{-\infty}^{\infty} d\lambda\, t^{-2/3} \frac{e^{(v-u)x/t}}{1+e^{\lambda}} \times \mathrm{Ai}\big(t^{-1/3}(u+r-\lambda) + t^{-4/3}x^2\big)\,\mathrm{Ai}\big(t^{-1/3}(v+r-\lambda) + t^{-4/3}x^2\big).$$

This kernel satisfies the necessary equations, and from this ones gets [40] that, remarkably,

$$\phi_{\mathrm{nw}} := \partial_r^2 \log G_{\mathrm{nw}} \quad \text{solves the KP equation (5.6).} \tag{6.1}$$

Similar derivations are available for the KPZ equation with half-Brownian/spiked, two-sided Brownian and stationary initial data [40, 57]. What is common to these cases is that there are special solvable models which converge to the KPZ equation and for which it has been possible to derive explicit formulas under these choices of initial data. But this remains out of reach for general initial conditions (and for multipoint distributions), and at this point it is not known whether a statement such as (6.1) holds in any greater generality.

ACKNOWLEDGMENTS

Much of the work reported in this note comes from joint works with Konstantin Matetski and Jeremy Quastel. The author is also grateful to Joaquín Fontbona and Jeremy Quastel for comments on a draft of this article.

FUNDING

This work was partially supported by Centro de Modelamiento Matemático (CMM) Basal Funds AFB170001, ACE210010 and FB210005 from ANID-Chile, and by Fondecyt Grant 1201914.

REFERENCES

[1] M. J. Ablowitz and H. Segur, On the evolution of packets of water waves. *J. Fluid Mech.* **92** (1979), no. 4, 691–715.

[2] M. Adler and P. van Moerbeke, PDEs for the joint distributions of the Dyson, Airy and sine processes. *Ann. Probab.* **33** (2005), no. 4, 1326–1361.

[3] G. Amir, I. Corwin, and J. Quastel, Probability distribution of the free energy of the continuum directed random polymer in 1 + 1 dimensions. *Comm. Pure Appl. Math.* **64** (2011), no. 4, 466–537.

[4] J. Baik, P. Deift, and K. Johansson, On the distribution of the length of the longest increasing subsequence of random permutations. *J. Amer. Math. Soc.* **12** (1999), no. 4, 1119–1178.

[5] J. Baik and E. M. Rains, Symmetrized random permutations. In *Random matrix models and their applications*, pp. 1–19, Math. Sci. Res. Inst. Publ. 40, Cambridge University Press, Cambridge, 2001.

[6] A.-L. Barabási and H. E. Stanley, *Fractal concepts in surface growth*. Cambridge University Press, Cambridge, 1995.

[7] L. Bertini and G. Giacomin, Stochastic Burgers and KPZ equations from particle systems. *Comm. Math. Phys.* **183** (1997), no. 3, 571–607.

[8] A. Borodin, I. Corwin, and D. Remenik, Multiplicative functionals on ensembles of non-intersecting paths. *Ann. Inst. Henri Poincaré Probab. Stat.* **51** (2015), no. 1, 28–58.

[9] A. Borodin and P. L. Ferrari, Anisotropic growth of random surfaces in 2 + 1 dimensions. *Comm. Math. Phys.* **325** (2014), no. 2, 603–684.

[10] A. Borodin, P. L. Ferrari, M. Prähofer, and T. Sasamoto, Fluctuation properties of the TASEP with periodic initial configuration. *J. Stat. Phys.* **129** (2007), no. 5–6, 1055–1080.

[11] A. Borodin and V. Gorin, Lectures on integrable probability. In *Probability and statistical physics in St. Petersburg*, pp. 155–214, Proc. Sympos. Pure Math. 91, Am. Math. Soc., Providence, 2016.

[12] A. Borodin and L. Petrov, Integrable probability: from representation theory to Macdonald processes. *Probab. Surv.* **11** (2014), 1–58.

[13] P. Calabrese, P. L. Doussal, and A. Rosso, Free-energy distribution of the directed polymer at high temperature. *Europhys. Lett.* **90** (2010), no. 2, 20002.

[14] I. Corwin, J. Quastel, and D. Remenik, Renormalization fixed point of the KPZ universality class. *J. Stat. Phys.* **160** (2015), no. 4, 815–834.

[15] D. Dauvergne, J. Ortmann, and B. Virág, The directed landscape. *Acta Math.* (to appear), arXiv:1812.00309.

[16] D. Dauvergne and B. Virág, The scaling limit of the longest increasing subsequence. 2021, arXiv:2104.08210.

[17] P. A. Deift, Three lectures on "Fifty years of KdV: an integrable system". In *Nonlinear dispersive partial differential equations and inverse scattering*, pp. 3–38, Fields Inst. Commun. 83, Springer, New York, 2019.

[18] V. Dotsenko, Bethe ansatz derivation of the Tracy–Widom distribution for one-dimensional directed polymers. *Europhys. Lett.* **90** (2010), no. 2, 20003.

[19] B. Eynard and M. L. Mehta, Matrices coupled in a chain. I. Eigenvalue correlations. *J. Phys. A* **31** (1998), no. 19, 4449–4456.

[20] P. L. Ferrari and H. Spohn, Scaling limit for the space-time covariance of the stationary totally asymmetric simple exclusion process. *Comm. Math. Phys.* **265** (2006), no. 1, 1–44.

[21] D. Forster, D. R. Nelson, and M. J. Stephen, Large-distance and long-time properties of a randomly stirred fluid. *Phys. Rev. A* **16** (1977), no. 2, 732–749.

[22] M. Hairer, Solving the KPZ equation. *Ann. of Math. (2)* **178** (2013), no. 2, 559–664.

[23] T. Halpin-Healy and K. A. Takeuchi, A KPZ cocktail—shaken, not stirred: toasting 30 years of kinetically roughened surfaces. *J. Stat. Phys.* **160** (2015), no. 4, 794–814.

[24] K. Johansson, Shape fluctuations and random matrices. *Comm. Math. Phys.* **209** (2000), no. 2, 437–476.

[25] K. Johansson, Discrete polynuclear growth and determinantal processes. *Comm. Math. Phys.* **242** (2003), no. 1–2, 277–329.

[26] M. Kardar, G. Parisi, and Y.-C. Zhang, Dynamical scaling of growing interfaces. *Phys. Rev. Lett.* **56** (1986), no. 9, 889–892.

[27] K. Liechty, G. B. Nguyen, and D. Remenik, Airy process with wanderers, KPZ fluctuations, and a deformation of the Tracy–Widom GOE distribution. *Ann. Inst. Henri Poincaré Probab. Stat.* (to appear), arXiv:2009.07781.

[28] K. Matetski, J. Quastel, and D. Remenik, The KPZ fixed point. *Acta Math.* **227** (2021), no. 1, 115–203.

[29] K. Matetski and D. Remenik, TASEP and generalizations: method for exact solution. 2021, arXiv:2107.07984.

[30] T. Miwa, M. Jimbo, and E. Date, *Solitons*. Cambridge Tracts in Math. 135, Cambridge University Press, Cambridge, 2000.

[31] M. Nica, J. Quastel, and D. Remenik, One-sided reflected Brownian motions and the KPZ fixed point. *Forum Math. Sigma* **8** (2020), Paper No. e63, 16.

[32] M. Nica, J. Quastel, and D. Remenik, Solution of the Kolmogorov equation for TASEP. *Ann. Probab.* **48** (2020), no. 5, 2344–2358.

[33] N. O'Connell and M. Yor, A representation for non-colliding random walks. *Electron. Commun. Probab.* **7** (2002), 1–12.

[34] C. Pöppe, General determinants and the τ function for the Kadomtsev–Petviashvili hierarchy. *Inverse Probl.* **5** (1989), no. 4, 613–630.

[35] M. Prähofer and H. Spohn, Scale invariance of the PNG droplet and the Airy process. *J. Stat. Phys.* **108** (2002), no. 5–6, 1071–1106.

[36] S. Prolhac, Spectrum of the totally asymmetric simple exclusion process on a periodic lattice-first excited states. *J. Phys. A* **47** (2014), no. 37, 375001.

[37] S. Prolhac, Riemann surfaces for KPZ with periodic boundaries. *SciPost Phys.* **8** (2020).

[38] J. Quastel and D. Remenik, Local behavior and hitting probabilities of the Airy_1 process. *Probab. Theory Related Fields* **157** (2013), no. 3–4, 605–634.

[39] J. Quastel and D. Remenik, How flat is flat in random interface growth? *Trans. Amer. Math. Soc.* **371** (2019), no. 9, 6047–6085.

[40] J. Quastel and D. Remenik, KP governs random growth off a one-dimensional substrate. *Forum Math. Pi* (to appear), arXiv:1908.10353.

[41] J. Quastel and H. Spohn, The one-dimensional KPZ equation and its universality class. *J. Stat. Phys.* **160** (2015), no. 4, 965–984.

[42] H. Rost, Nonequilibrium behaviour of a many particle process: density profile and local equilibria. *Z. Wahrsch. Verw. Gebiete* **58** (1981), no. 1, 41–53.

[43] S. Sarkar and J. Quastel, Convergence of exclusion processes and KPZ equation to the KPZ fixed point. 2020, arXiv:2008.06584.

[44] S. Sarkar and B. Virág, Brownian absolute continuity of the KPZ fixed point with arbitrary initial condition. *Ann. Probab.* **49** (2021), no. 4, 1718–1737.

[45] T. Sasamoto, Spatial correlations of the 1D KPZ surface on a flat substrate. *J. Phys. A* **38** (2005), no. 33, L549.

[46] T. Sasamoto and H. Spohn, Exact height distributions for the KPZ equation with narrow wedge initial condition. *Nuclear Phys. B* **834** (2010), no. 3, 523–542.

[47] G. M. Schütz, Exact solution of the master equation for the asymmetric exclusion process. *J. Stat. Phys.* **88** (1997), no. 1–2, 427–445.

[48] T. Seppäläinen, Strong law of large numbers for the interface in ballistic deposition. *Ann. Inst. Henri Poincaré Probab. Stat.* **36** (2000), no. 6, 691–736.

[49] F. Spitzer, Interaction of Markov processes. *Adv. Math.* **5** (1970), 246–290.

[50] C. A. Tracy and H. Widom, Level-spacing distributions and the Airy kernel. *Comm. Math. Phys.* **159** (1994), no. 1, 151–174.

[51] C. A. Tracy and H. Widom, On orthogonal and symplectic matrix ensembles. *Comm. Math. Phys.* **177** (1996), no. 3, 727–754.

[52] C. A. Tracy and H. Widom, A system of differential equations for the Airy process. *Electron. Commun. Probab.* **8** (2003), 93–98.

[53] B. Virág, The heat and the landscape I. 2020, arXiv:2008.07241.

[54] M. J. Vold, A numerical approach to the problem of sediment volume. *J. Colloid Sci.* **14** (1959), no. 2, 168–174.

[55] J. Warren, Dyson's Brownian motions, intertwining and interlacing. *Electron. J. Probab.* **12** (2007), no. 19, 573–590.

[56] D. R. Wilkinson and S. F. Edwards, The surface statistics of a granular aggregate. *Proc. R. Soc. Lond. Ser. A, Math. Phys. Sci.* **381** (1982), no. 1780, 17–31.

[57] X. Zhang, A system of PDEs for the Baik-Rains distribution. 2020, arXiv:2010.09779.

DANIEL REMENIK

Departamento de Ingeniería Matemática and Centro de Modelamiento Matemático (IRL-CNRS 2807), Universidad de Chile, Santiago, Chile, dremenik@dim.uchile.cl

This Page Intentionally Left Blank

HEAT KERNEL ESTIMATES ON HARNACK MANIFOLDS AND BEYOND

LAURENT SALOFF-COSTE

ABSTRACT

On a Riemannian manifold, M, the heat kernel is a smooth function on $(0, +\infty) \times M \times M$, $(t, x, y) \mapsto p(t, x, y)$, and the shape of this function depends on the properties of M. This article pays particular attention to the long-time, large-scale behavior of the heat kernel and its relation to the global geometry of M. When does the heat kernel look like a bell curve? If it does not, what does it look like and why? To answer such questions, one needs tools to obtain sharp two-sided estimates for the heat kernel in terms of the time variable $t > 0$ and basic geometric quantities depending on $x, y \in M$. Under what assumptions on M, can one hope to obtain such bounds?

MATHEMATICS SUBJECT CLASSIFICATION 2020

Primary 58J35; Secondary 60J60, 58J65, 60J46

KEYWORDS

Heat kernel, manifolds, Brownian motion

INTERNATIONAL CONGRESS OF MATHEMATICIANS 2022 JULY 6–14

Proc. Int. Cong. Math. 2022, Vol. 6, pp. 4452–4473
DOI 10.4171/ICM2022/11

1. INTRODUCTION

Over the last 50 years, the heat kernel has become the subject of many studies in many different settings and for many purposes. Several fields of mathematics have obvious good reasons to pay particular attention to the heat kernel. In partial differential equations, it is the fundamental solution of the most basic parabolic equation which is the model for all evolution equations. In probability theory, it is the density of the distribution of Brownian motion at a given time. In mathematical physics, beyond its original role in the theory of heat, it leads to the notion of "abstract Wiener space," a building block in quantum field theory. But interest in the heat kernel goes well beyond these natural areas. It has been called ubiquitous and a universal gadget by mathematicians interested in topology (index theorems) or number theory (trace formulae). On a Riemannian manifold M, the heat kernel is a smooth function p on $(0, +\infty) \times M \times M$, $(t, x, y) \mapsto p(t, x, y)$. When M is the real line, $y \mapsto p(t, x, y)$ is a scaled version of the bell curve. See Figure 1.

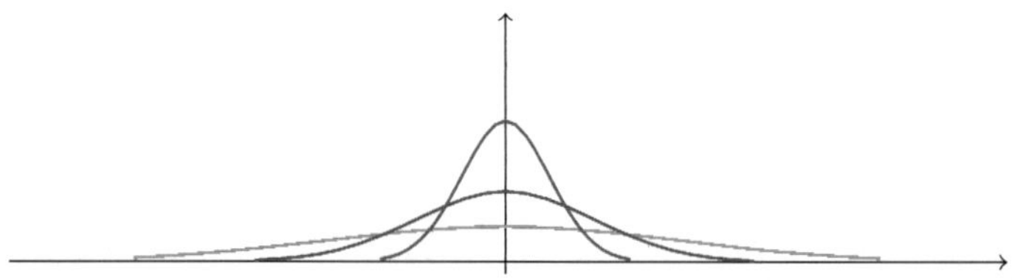

FIGURE 1
The bell curve as a model for the heat kernel: the heat kernel on the real line for $x = 0$ and three values of t: $t = 1/16, 1/4, 1$; $p(t, 0, y) = (4\pi t)^{-1/2} \exp(-|y|^2/4t)$.

This article pays particular attention to the long-time, large-scale behavior of the heat kernel and its meaning in global geometry. The goal is to develop tools to obtain sharp two-sided estimates for the heat kernel in terms of time, $t > 0$, and basic geometric quantities depending on $x, y \in M$. Because the heat kernel is the density function of the probability distribution of Brownian motion on M started at x at time t, if one can obtain sharp two-sided bounds on p valid for all $t > 0$, $x, y \in M$ and uniform over all manifolds M in a certain class $\mathcal{M}$, then one can say that, in that class $\mathcal{M}$, the geometry controls the behavior of Brownian motion in a precise sense. Such bounds have many further implications concerning spectral theory, potential theory, and global analysis.

2. EXISTENCE

When does the heat kernel exist? What is needed to define it uniquely? Even in the basic setting of Riemannian manifolds, these questions require some attention, and the answers involve the use of some significant machinery. It is useful to proceed by stages: first, define and prove the existence of an abstract weaker version of the heat kernel; then extract from this weaker version a proper heat kernel. For instance, one could prove existence in the sense of distribution theory and then prove that the constructed distribution is, in fact, a smooth function. Instead, we appeal here to semigroup theory so that our first step is to define the heat semigroup $(P_t)_{t>0}$, from which we later intend to extract the heat kernel itself. What is needed for this purpose is a reasonable underlying space M equipped with a measure μ, the Hilbert space $L^2(M,\mu)$, and a Dirichlet form, $(\mathcal{E},\mathcal{D}(\mathcal{E}))$, that is, a densely defined closed nonnegative bilinear form on $L^2(M,\mu)$ with one additional property, the Markovian property. Namely, one requires that, for any $u\in\mathcal{D}(\mathcal{E})$, it holds that $|u|\in\mathcal{D}(\mathcal{E})$ and $\mathcal{E}(|u|,|u|)\le\mathcal{E}(u,u)$. See [9] (the Markovian property is akin to restricting ourselves to a positivity-preserving semigroup, something related to versions of the maximum principle). Functional analysis associates to the data $(M,\mu,(\mathcal{E},\mathcal{D}(\mathcal{E})))$ a self-adjoint semigroup of operators

$$P_t : L^2(M,\mu)\to L^2(M,\mu),\quad f\mapsto P_t f,$$

which solves the initial value problem

$$\begin{cases}\partial_t u=\Delta u,\\ u(0,\cdot)=f,\end{cases}$$

in the sense that $u(t,x)=P_tf(x)$ is the only solution of this problem when $f\in L^2(M,\mu)$. Here, Δ is the operator extracted from $\mathcal{E}$ in the same way that a symmetric matrix can be associated with any given positive-definite bilinear form on a finite Euclidean space. Namely, $\Delta u\in L^2(M,\mu)$ is such that $\phi\mapsto\int_M(\Delta u)\phi d\mu=\mathcal{E}(u,\phi)$ for all $\phi\in\mathcal{D}(\mathcal{E})$ whenever $u\in\mathcal{D}(\mathcal{E})$ has the property that $|\mathcal{E}(u,\phi)|\le C_u\|\phi\|_{L^2(M,\mu)}$. This densely-defined linear operator is called the infinitesimal generator of the semigroup $(P_t)_{t>0}$, and it can also be obtained using the formula

$$\Delta v=\lim_{t\to 0_+}t^{-1}(P_tv-v)$$

when this limit exists in $L^2(M,\mu)$. Moreover, $P_t=e^{t\Delta}$ where one can think of the right-hand side as defined by using the spectral theory applied to the self-adjoint operator Δ. In the context of a Riemannian manifold M equipped with its Riemannian measure μ, the classical choice is

$$\mathcal{E}(f,f)=\int_M|\nabla f|^2d\mu$$

with the domain equal to the closure of smooth compactly supported functions for the norm $(\|f\|_2^2+\mathcal{E}(f,f))^{1/2}$. Using integration by parts, the infinitesimal generator of the associated semigroup is, indeed, the Laplacian, $\Delta f=\operatorname{div}(\nabla f)$ (in the case $M=\mathbb{R}^n$, the Euclidian space, $\Delta=\sum_1^n\partial_i^2$ where ∂_i is the partial derivative in the direction of the ith basis unit vector).

This simple construction provides us with a *transition kernel* $p(t,x,dy)$, which, for each t and x, is a nonnegative measure of finite total mass at most 1 in y. This is sometimes referred to as the heat kernel measure (at time t and centered at $x\in M$), and its definition is simply that $P_tf(x)=\int_M f(y)p(t,x,dy)$ for all $f\in L^2(M,d\mu)$. The question of the existence of the heat kernel (as a function) becomes the question of the absolute continuity of the measure $p(t,x,dy)$ with respect to the base measure μ. If absolute continuity holds then, abusing notation somewhat, $p(t,x,y)$ is defined by

$$p(t,x,dy)=p(t,x,y)d\mu(y).$$

Hence, to prove the existence of the heat kernel as a measurable locally bounded function, it suffices to prove bounds of the type $\sup_{x\in U}\{|P_t f(x)|\} \le C(t,U,V)\|f\|_1$, $f \in \mathcal{C}_c(V)$, for pairs (U,V) of open relatively compact sets that cover $M \times M$. Here $\mathcal{C}_c(V)$ is the space of continuous functions with compact support in V. On a smooth Riemannian manifold, the parabolic nature of the heat equation and local PDE theory provide such bounds, as well as the smoothness of the heat kernel. Properties of this type are known under the name of "ultracontractivity," and they can often be proved via the use of functional inequalities such as Sobolev or Nash inequalities. For instance, on a Riemannian manifold M, for any fixed $\nu > 0$, the Nash inequality ([24])

$$\|f\|_2^{2+4/\nu} \le C_1\|\nabla f\|_2^2\|f\|_1^{4/\nu}, \quad f \in \mathcal{C}_c^\infty(M), \tag{2.1}$$

is equivalent to the ultracontractivity inequality

$$\|P_t f\|_\infty \le C_2 t^{-\nu/2}\|f\|_1, \quad t > 0, f \in L^1(M,\mu),$$

which, in turn, is equivalent to

$$\sup_{x,y\in M}\{p(t,x,y)\} \le C_2 t^{-\nu/2}, \quad t > 0. \tag{2.2}$$

Even in the case of Riemannian manifolds where, thanks to local PDE theory, the existence and smoothness of the heat kernel are not in question, this Nash inequality technique gives access to a more quantitative control of the heat kernel. Because $\sup_{x,y}\{p_t(x,y)\} = \sup_x\{p(t,x,x)\}$, bounds of type (2.2), possibly with different functions of t on the right-hand side, are often called "on-diagonal heat kernel upper-bounds." They capture the decay of the heat kernel as time tends to infinity; see, e.g., [5]. In fact, with a little more work, the same set of ideas leads to the fact that (2.1) implies a Gaussian upper-bound involving the Riemannian distance between two points x, y, that is,

$$\forall t > 0,\ x,y \in M, \quad p(t,x,y) \le C_\varepsilon t^{-\nu/2}\exp\left(-\frac{d(x,y)^2}{4(1+\varepsilon)t}\right) \tag{2.3}$$

for any small $\varepsilon > 0$ (see, e.g., [27, CHAPTER 4, SECTION 2] and the references therein). For comparison, in our notation, the heat kernel in $\mathbb{R}^n$ is

$$\frac{1}{(4\pi t)^{n/2}}\exp\left(-\frac{|x-y|^2}{4t}\right).$$

The essentially universal nature of the Gaussian factor, $\exp(-d^2/4t)$, is somewhat surprising and very useful in practice: it makes the heat kernel behave almost like a compactly supported function, and this facilitates many manipulations.

3. THE GEOMETRY OF NICE DIRICHLET SPACES

The Nash inequality easily makes sense on a Dirichlet space $(M,\mu,(\mathcal{E},\mathcal{D}(\mathcal{E})))$ simply by interpreting $\|\nabla f\|_2^2$ as $\mathcal{E}(f,f)$ for $f \in \mathcal{D}(\mathcal{E}) \cap L^1(M,\mu)$. It then implies (2.2) and, in particular, the existence of a bounded heat kernel for all $t > 0$. Dirichlet forms can be local or nonlocal (the associated Markov process has continuous paths in the first case and includes jumps in the second). Under some additional relatively mild assumptions, it is possible to extract from a (strictly) local Dirichlet form $(\mathcal{E}, \mathcal{D}(\mathcal{E}))$ a measure-valued bilinear form defined on $\mathcal{D}(\mathcal{E}) \times \mathcal{D}(\mathcal{E})$, $(f, g) \mapsto d\Gamma(f, g)$, such that $\mathcal{E}(f, g) = \int_M d\Gamma(f, g)$. For a given function $f \in \mathcal{D}(\mathcal{E})$, $d\Gamma(f, f)$ may or may not be absolutely continuous with respect to $d\mu$. When it is, we write $d\Gamma(f, f) = \Gamma(f, f)d\mu$. This is called the "carré du champ," and it is a substitute for the classical $|\nabla f|^2$. It extends naturally to a local version $\mathcal{D}_{\mathrm{loc}}(\mathcal{E})$ of $\mathcal{D}(\mathcal{E})$. For arbitrary points $x, y \in M$, set

$$d(x, y) = \sup\{|f(x) - f(y)| : f \in \mathcal{C}(M) \cap \mathcal{D}_{\mathrm{loc}}(\mathcal{E}), d\Gamma(f, f) \leq d\mu\}.$$

This symmetric function of x and y can vanish or take the value $+\infty$. When it is finite, continuous, and defines the topology of M, it provides a good notion of distance called the "intrinsic distance" of the given Dirichlet form; see, e.g., **[22, 28]**. In the Riemannian setting, the intrinsic distance is simply the Riemannian distance. To see a different (non-Riemannian), yet classical set of examples, let $M = G$ be a unimodular Lie group equipped with its Haar measure μ, and a family $\{X_1, \ldots, X_k\}$ of left-invariant vector fields which generates the Lie algebra $\mathfrak{g}$ of G (i.e., these fields together with all their iterated Lie brackets span $\mathfrak{g}$, linearly). Set

$$\mathcal{E}(f, f) = \int_G \sum_1^k |X_i f|^2 d\mu$$

for f in the closure of $\mathcal{C}_c^\infty(G)$ for the norm $(\int_G |f|^2 d\mu + \int_G \sum_1^k |X_i f|^2 d\mu)^{1/2}$. In this case, the distance d is the sub-Riemannian distance associated with the family $\{X_1, \ldots, X_k\}$ and this example serves as a model for the development of sub-Riemannian geometry and the analysis of the related subelliptic Laplacians (here, $\Delta = \sum_1^k X_i^2$, because G is unimodular).

In general, when the intrinsic distance d is continuous and defines the topology of M, it is possible again to obtain (2.3) from (2.1) (e.g., **[28]**). Recently, based in part on the notions and techniques described here, Carron and Tewodrose proved the following rigidity result **[4]**. Consider a σ-compact complete metric space (M, d), equipped with a positive Radon measure μ and with a Dirichlet form $(\mathcal{E}, \mathcal{D}(\mathcal{E}))$. Assume that the associated heat semigroup admits a heat kernel p which satisfies, for all $(t, x, y) \in (0, +\infty) \times M \times M$,

$$p(t, x, y) = \frac{1}{(4\pi t)^{\alpha/2}} \exp\left(-\frac{d(x, y)^2}{4t}\right).$$

Then α is an integer, M is $\mathbb{R}^\alpha$, d is the Euclidean metric on $\mathbb{R}^\alpha$, and μ is the α-dimensional Hausdorff measure. The Dirichlet form $(\mathcal{E}, \mathcal{D}(\mathcal{E}))$ is the usual Euclidean Dirichlet form.

On the other hand, there are many interesting examples that are definitively not Euclidean and whose heat kernel satisfies

$$\frac{c_1}{t^{\alpha/2}} \exp\left(-C_1 \frac{d(x, y)^2}{t}\right) \leq p(t, x, y) \leq \frac{C_2}{t^{\alpha/2}} \exp\left(-c_2 \frac{d(x, y)^2}{t}\right).$$

These examples include uniformly elliptic operators in $\mathbb{R}^n$, for which

$$\mathcal{E}(f, g) = \int_{\mathbb{R}^n} \sum_{i,j=1}^n a_{ij}(x) \nabla f(x) \cdot \nabla g(x) dx,$$

where the coefficients a_{ij} are bounded, measurable, and satisfy $\sum_{ij} a_{ij}(x)\xi_i\xi_j \geq \varepsilon\|\xi\|_2^2$ for all $x, \xi \in \mathbb{R}^n$ and some $\varepsilon > 0$. In this case $\alpha = n$ (see [1, 2, 25]). Another example is the Heisenberg group H of 3×3 matrices

$$\begin{pmatrix} 1 & x & z \\ 0 & 1 & y \\ 0 & 0 & 1 \end{pmatrix}, \quad x, y, z \in \mathbb{R},$$

equipped with the Dirichlet form $\mathcal{E}(f, f) = \int_H (|Xf|^2 + |Yf|^2)d\mu$ where X (resp. Y) is the left-invariant vector field equal to $\partial/\partial x$ (resp. $\partial/\partial y$) at the identity. In this case $\alpha = 4$. In all these examples the correct interpretation of the on-diagonal factor, $t^{-\alpha/2}$, is that it is $1/V(x, \sqrt{t})$ where $V(x, r) = \mu(\{z \in M : d(x, z) < r\})$, the volume of the ball of radius r and center x. This leads us to consider the following hypothetical two-sided Gaussian bound:

$$\frac{c_1}{V(x, \sqrt{t})} \exp\left(-C_1 \frac{d(x, y)^2}{t}\right) \leq p(t, x, y) \leq \frac{C_2}{V(x, \sqrt{t})} \exp\left(-c_2 \frac{d(x, y)^2}{t}\right). \tag{3.1}$$

When (3.1) holds on a Riemannian manifold, one can answer many questions. For instance, Brownian motion on such a manifold is transient if and only if $\int^{+\infty} \frac{ds}{V(x, \sqrt{s})} < +\infty$. If this integral is finite then the Green function $G(x, y)$, i.e., the function such that

$$\Delta^{-1} f(x) = \int_M G(x, y) f(y) d\mu(y), \quad f \in \mathcal{C}_c^\infty(M),$$

satisfies

$$c \int_{d(x,y)^2}^{+\infty} \frac{ds}{V(x, \sqrt{s})} \leq G(x, y) \leq C \int_{d(x,y)^2}^{+\infty} \frac{ds}{V(x, \sqrt{s})}.$$

By integrating (3.1) with respect to y over the ball $B(x, 2\sqrt{t})$ and noting that the integral of the heat kernel is at most 1, one easily sees that (3.1) implies that the manifold M must be doubling in the following sense.

Definition 3.1. A metric measure space is called doubling if there exists a constant D such that, for all $x \in M$ and $r > 0$, $V(x, 2r) = \mu(B(x, 2r)) \leq DV(x, r) = D\mu(B(x, r))$.

In the next section, we answer the following question: Which Riemannian manifolds satisfy (3.1)?

4. HARNACK MANIFOLDS AND DIRICHLET SPACES

On a smooth manifold with boundary, a function u is harmonic in an open ball B if it is smooth in B, satisfies $\Delta u = 0$, and has vanishing normal derivative on $\delta_M \cap B$. Similarly, a function u is a solution of the heat equation in a time-space cylinder $Q = (a, b) \times B$ if it is smooth there, satisfies $(\partial_t - \Delta)u = 0$ in Q, and has vanishing normal derivative on $\delta_M \cap B$. When dealing with more general contexts, including local (regular) Dirichlet spaces, the appropriate notion of a *weak solution* must be used instead; see, e.g., [20, 29] for details.

The elliptic Harnack inequality is one of the most well-known inequalities in analysis and goes back to the nineteenth century. In $\mathbb{R}^n$, it states that there is a constant C_n

such that any nonnegative function u, harmonic in a ball $B = B(x,r)$, satisfies $\sup_{\frac{1}{2}B}\{u\} \le C_n \inf_{\frac{1}{2}B}\{u\}$ where $\frac{1}{2}B = B(x,r/2)$. A Riemannian manifold (or Dirichlet space as above which admits a good intrinsic distance) satisfies the elliptic Harnack inequality when there is a constant C_M such that any nonnegative function u which is harmonic in a ball $B = B(x,r)$ satisfies

$$\sup_{\frac{1}{2}B}\{u\} \le C_M \inf_{\frac{1}{2}B}\{u\}.$$

Until recently, there was no clear characterization of this property in geometric terms, but this problem is resolved beautifully in [3], to which the reader is referred.

The importance and usefulness of the parabolic Harnack inequality only became apparent in the second half of the twentieth century in the work of Nash [24], Moser [23], and many others after them. Consider a time-space cylinder $Q = (s - r^2, s) \times B(x,r)$ and the smaller separated subcylinders $Q_- = (s - 3r^2/4, s - r^2/2) \times B(x, r/2)$ and $Q_+ = (s - r^2/4, s] \times B(x, r/2)$. We say that M satisfies the parabolic Harnack inequality (at all scales and locations) if there is a constant C_M such that any nonnegative solution u of the heat equation in Q satisfies

$$\sup_{Q_-}\{u\} \le C_M \inf_{Q_+}\{u\}. \tag{4.1}$$

In what follows we will make constant use of the following definition.

Definition 4.1. We say that a Riemannian manifold M is Harnack with constant C if it satisfies the parabolic Harnack inequality (4.1), at all scales and locations, with a constant $C_M \le C$.

Given a precompact open subset $\Omega \subseteq M$, set

$$\lambda(\Omega) = \inf\left\{\frac{\int_\Omega |\nabla f|^2 d\mu}{\int_\Omega |f - f_\Omega|^2 d\mu} : f \in W^1(\Omega), f - f_\Omega \ne 0\right\}.$$

Here, f_Ω is the average value of f on Ω and $W^1(\Omega)$ is the set of all L^2-functions in Ω whose gradient in Ω (in the sense of distributions) can be represented as an L^2-vector field in Ω. The language of Dirichlet spaces allows us to view $\lambda(\Omega)$ as the lowest *positive* eigenvalue of the Neumann-Laplacian in Ω (even if Ω does not have a smooth boundary).

Definition 4.2. We say that a Riemannian manifold M satisfies the Poincaré inequality at all scales and locations, with a constant at most P, if

$$\forall x \in M,\ r > 0, \quad \lambda\big(B(x,r)\big) \ge 1/\big(Pr^2\big). \tag{4.2}$$

Definition 4.3. Fix $\kappa \ge 1$. We say that a Riemannian manifold M satisfies the weak Poincaré inequality with parameter κ at all scales and locations, with a constant at most P, if

$$\forall x \in M, r > 0, \forall f \in \mathcal{C}_b^\infty\big(B(x,\kappa r)\big), \quad \int_{B(x,r)} |f - f_{B(x,r)}|^2 d\mu \le Pr^2 \int_{B(x,\kappa r)} |\nabla f|^2 d\mu.$$

With these definitions we can answer the question posed at the end of the previous section: Which Riemannian manifolds satisfy the two-sided Gaussian heat kernel estimate (3.1)?

Theorem 4.4. *A complete Riemannian manifold M is Harnack if and only if it satisfies* (3.1). *These properties are also equivalent to the fact that M is doubling and satisfies the Poincaré inequality* (4.2) *at all scales and locations. Finally, doubling and the weak Poincaré inequality with a fixed parameter $\kappa \geq 1$ are enough to imply that M is Harnack.*

This theorem is essentially taken from the independent works [18] and [26], which both used ideas developed in the earlier works by various other authors. The proof in [26,27] uses the well-know techniques of Nash and Moser, as well as ideas developed by D. Jerison and S. Kusuoka and D. Stroock. K. T. Sturm extended this theorem in an important and useful way to the context of local Dirichlet spaces admitting a good intrinsic distance [28,29]. In this abstract context, the well-known fact that (4.1) implies the Hölder continuity of the (weak) solutions of the heat equation provides an important method to prove the continuity of the heat kernel. Each property used in Theorem 4.4, the parabolic Harnack inequality (4.1), the two-sided-Gaussian bound (3.1), and the conjunction of doubling and the Poincaré inequality, comes with a small set of fundamental constants, the constant C_M in (4.1), the constants c_1, C_1, c_2, C_2 in (3.1), and the doubling constant D and Poincaré constants κ, P. In each case, the constants of a given property can be controlled solely in terms of the constants of one of the other equivalent properties. For instance, fix large, positive, reals D and P. There is a constant $C = C(D, P)$ such that any complete manifold satisfying doubling with constant at most D and the Poincaré inequality (4.2) with constant at most P also satisfies the parabolic Harnack inequality (4.1) with constant at most C.

This brings us to the following conjecture due to Maria Gordina, Nate Eldredge, and the author.

Conjecture 4.5 ([8]). *Given a compact Lie group G, there exists a constant $H(G)$ such that all left-invariant Riemannian metrics on G are Harnack with constant at most $H(G)$.*

Because of (3.1), what this would mean is that, on a given compact Lie group, all left-invariant diffusion processes are uniformly controlled by their own geometry.

A simple argument going back to the work of N. Varopoulos shows that a left-invariant Riemannian metric on a unimodular Lie group G which is doubling also satisfies the Poincaré inequality. It follows that the conjecture above can be stated in the following much simpler form.

Conjecture 4.6 ([8]). *Given a compact Lie group G, there exists a constant $D(G)$ such that all left-invariant Riemannian metrics on G are doubling with constant at most $D(G)$.*

The best evidence for these conjectures is that they hold for abelian Lie groups (compact or not), for nilpotent Lie groups (not compact, if not abelian), and for SU(2) (see [8]). The case of $U(2)$ and $\mathrm{SU}(2) \times A$ where A is an abelian Lie group is in preparation by the same authors (it is surprisingly more involved than the case of SU(2)).

These conjectures can be compared with the following theorem which follows from [21] (this theorem covers the case of abelian groups because their left-invariant Riemannian metrics have 0 curvature).

Theorem 4.7. *Fix a dimension n. There is a constant C_n such that all complete Riemannian manifolds with nonnegative Ricci curvature are Harnack with constant at most C_n.*

Repeating something said above, one of the consequence of this theorem is that, on manifolds with nonnegative Ricci curvature and dimension at most n, the behavior of Brownian motion is controlled by the geometry of the manifold, uniformly over all such manifolds.

For a typical compact Lie group G in Conjectures 4.5–4.6, there is no finite common lower bound on the Ricci curvature of all left-invariant Riemannian metrics. So, Theorem 4.7 does not help much in settling these conjectures. In fact, although we stated these conjectures for left-invariant Riemannian metrics, they automatically extend to (e.g., include) left-invariant sub-Riemannian geometries because the desired property is uniform over all Riemannian metrics on G. They further extend to left-invariant structures on analytic subgroups of the compact group G, showing that G cannot contain an analytic subgroup of exponential volume growth (the fact that a compact Lie group G cannot contain an analytic subgroup of exponential volume growth is indeed known; it can be viewed as supporting evidence for the conjectures).

Remark 4.8. Most results concerning Harnack inequalities (elliptic or parabolic) in the context of Riemannian geometry are based on Ricci curvature lower bounds in the spirit of the famous works of S. T. Yau, Cheng and Yau, and Li and Yau. This technique leads to "gradient Harnack inequalities" which imply inequalities of the type (4.1). One can strengthen Conjecture 4.5 by asking for a uniform *parabolic Harnack gradient inequality* over all left-invariant Riemannian metrics on G. This stronger conjecture is open even for SU(2).

5. NON-HARNACK MANIFOLDS

In the remaining sections, we will focus on examples of non-Harnack manifolds. Finding ways to obtain sharp heat kernel estimates for manifolds to which Theorem 4.4 DOES NOT apply is a major challenge. Techniques exist that provide good on-diagonal estimates in various particular situations, and it is known that the Gaussian factor of the type $\exp(-cd(x,y)^2/t)$ is somewhat universal (although not always sharp at large scales and large time). One can phrase this challenge more precisely by asking for upper/lower bounds for the heat kernel $p(t,x,y)$ in terms of some explicit functions $(t,x,y) \mapsto g_{\mathbf{c}}(t,x,y)$ which are expressed in terms of t and basic geometric quantities, including the volume functions $V(x,r)$, $V(y,r)$, $r > 0$, the distance function $d(x,y)$, and perhaps other similar quantities. Here, $\mathbf{c}$ represents a positive constant (more generally, finite set of positive constants) that enters the definition of the function g and may be different in upper and lower bounds. In case when one knows or expects that $\int_M p(t,x,y)dy = 1$ (i.e., heat diffusion on M is conservative), it is highly desirable that the functions $g_{\mathbf{c}}$ used to estimate p satisfy $\varepsilon_{\mathbf{c}} \leq \int_M g_{\mathbf{c}}(t,x,y)dy \leq \varepsilon_{\mathbf{c}}^{-1}$, for some $\varepsilon_{\mathbf{c}} > 0$.

This challenge has many facets. Here, we will focus only on one of them. Namely, we are going to focus on manifolds which lack basic homogeneity but can be decomposed into simpler pieces that are Harnack. The simplest basic example of such is the catenoid which, roughly speaking, is the connected sum of two planes. The catenoid is doubling but does not satisfy the Poincaré inequality at all scales and locations; see the next section.

First, we briefly describe explicitly a different type of challenging example. The Lie group Sol is the model for one of the eight "geometries" that are the building blocks of manifolds in dimension 3 (Perelman's theorem, formerly Thurston conjecture; the Heisenberg group mentioned earlier is another one of these eight). We can describe Sol as the matrix group

$$\begin{pmatrix} e^x & 0 & y \\ 0 & e^{-x} & z \\ 0 & 0 & 1 \end{pmatrix}, \quad x, y, z \in \mathbb{R}.$$

We equip Sol with the left-invariant metric associated with the orthonormal basis of $\mathbb{R}^3$ viewed as the tangent space of Sol at the identity, id. The rough on-diagonal behavior of the heat kernel $p(t, \mathrm{id}, \mathrm{id})$ is described by the function $\exp(-t^{1/3})$ for large t, but no off-diagonal estimate of the type described above is known. The simplest way to see that Theorem 4.4 does not apply (positively) to this example is to note that the volume of large balls grows exponentially fast with the radius. Hence the doubling condition fails. In a similar spirit, sharp off-diagonal estimates of the heat kernel of the universal cover of a compact manifold whose fundamental group is a finitely generated solvable group of exponential volume growth is a challenge that goes well beyond existing techniques.

6. MANIFOLDS MADE OF NICE PIECES AND RECONSTRUCTION: BASIC EXAMPLES

Consider the connected sum of two Euclidean spaces, $M = \mathbb{R}^n \# \mathbb{R}^n$, $n \geq 2$, equipped with a Riemannian metric, that is, the Euclidean metric away from the central collar gluing the two copies of $\mathbb{R}^n$ together. This manifold is made of two very nice Harnack pieces, the two Euclidean spaces. It is doubling at all scales and locations, but the Poincaré inequality fails to hold for balls of large radius centered at the collar. In fact, the second-lowest Neumann eigenvalue for such a large central ball is of order $1/(r^2 \log r)$ when $n = 2$ and $1/r^n$ when $n > 2$ (for the Poincaré inequality to hold, we need $1/r^2$). Write M as the disjoint union of a compact part K (the collar), and the two disconnected ends E_1, E_2, both equal to $\mathbb{R}^n \setminus \mathbf{B}$, where $\mathbf{B}$ is a ball centered at the origin in $\mathbb{R}^n$, of radius large enough so that the metric of M on each E_i is Euclidean. For $x, y \in M$, consider the following geometric quantities:

- $|x| = \sup_{z \in K}\{d(x, z)\}$;
- $d_+(x, y)$, the infimum of the lengths of smooth curves joining x to y in M having a *nonempty intersection* with K;
- $d_\emptyset(x, y)$, the infimum of the lengths of smooth curves joining x to y in M having a *empty intersection* with K.

For $n > 2$, set

$$g_{M,c}(t,x,y) = \frac{1}{ct^{n/2}}\left(\frac{1}{|x|^{n-2}} + \frac{1}{|y|^{n-2}}\right)\exp\left(-c\frac{d_+(x,y)^2}{t}\right) + \frac{1}{ct^{n/2}}\exp\left(-c\frac{d_\emptyset(x,y)^2}{t}\right).$$

It is proved in [18] that there are constants c_1, c_2 such that, for all $(t,x,y) \in (0,+\infty) \times M \times M$, the heat kernel of M, $p_M(t,x,y)$, satisfies

$$g_{M,c_1}(t,x,y) \le p_M(t,x,y) \le g_{M,c_2}(t,x,y). \tag{6.1}$$

More complicated formulae of the same type apply when $M = M_1\#\cdots\#M_k$ with each $M_i = \mathbb{R}^{n_i} \times \mathbb{S}^{N-n_i}$ for $n_i \ge 3$. In particular, in this case, for any fixed point $o \in M$, there are constants c_1, C_1 such that, for all $t > 1$, $c_1 t^{-\min\{n_i\}/2} \le p_M(t,o,o) \le C_1 t^{-\min\{n_i\}/2}$.

The case of the connected sum of two planes, $\mathbb{R}^2\#\mathbb{R}^2$, is different because Brownian motion on $\mathbb{R}^2$ is recurrent (an open ball is visited with probability 1 from any starting point; equivalently, there is no positive Green's function). It is proved in [13, 18] (some technical elementary manipulations are required to turn the results of [13, 18] into the statement given here) that the heat kernel for $M = \mathbb{R}^2\#\mathbb{R}^2$ satisfies (6.1) with

$$g_{\mathbb{R}^2\#\mathbb{R}^2,c}(t,x,y) = \frac{1}{ct}\left(\frac{\log(e(1+t/|x|^2))}{\log(1+t+|x|^2)} + \frac{\log(e(1+t/|y|^2))}{\log(1+t+|y|^2)}\right)\exp\left(-c\frac{d_+(x,y)^2}{t}\right) + \frac{1}{ct}\exp\left(-c\frac{d_\emptyset(x,y)^2}{t}\right).$$

In this formula, the second term, $\frac{1}{ct}\exp(-cd_\emptyset(x,y)^2/t)$, is 0 if x, y are in different planes, and it always dominates if x, y are in the same plane.

7. MANIFOLDS WITH FINITELY MANY NONPARABOLIC HARNACK ENDS

Our goal now is to generalize as much as possible the results described above in model cases. Consider the connected sum $M = M_1\#\cdots\#M_k$ of k complete noncompact weighted Riemannian manifolds with boundary. So M may have a nonempty "boundary" $\delta M \subset M$ along which it is modeled locally by the half-space $\mathbb{R}^n_+$, and M equipped with its Riemannian distance is a complete metric space. The weight σ is a positive smooth function (in fact, continuity is more than enough). The heat equation on this weighted manifold, and the heat kernel p_M, are associated with the Dirichlet space

$$\left(M, \mu, \int_M |\nabla f|^2 d\mu, W_0^1(M, d\mu)\right)$$

where dx is the Riemannian measure, $\mu(dx) = \sigma(x)dx$, and $W_0^1(M)$ is the closure of smooth compactly supported functions on M under the norm $(\int_M(|f|^2 + |\nabla f|^2)d\mu)^{1/2}$. By definition, we can write M as the disjoint union $M = K \cup E_1 \cup \cdots \cup E_k$ where K is compact and $\overline{E_i}$ are smooth manifolds with boundary isometric to $M_i \setminus K_i$ for some compact K_i in M_i. Each E_i inherits a weight $\sigma_i = \sigma|_{E_i}$. In more classical terms, the Laplacian of a smooth function f on (M,μ) is $\frac{1}{\sigma}\text{div}(\sigma\nabla f)(x)$ at point $x\in M\setminus\delta M$, and the heat equation is taken with Neumann boundary condition along δM. Although we informally refer to the M_i or the E_i, $1\le i\le k$, as the "ends" of M, it is not necessarily the case in the setting described above that they represent the full list of the topological ends of M as any one of them could possibly split if a very large ball is removed.

In this section, we make two fundamental assumptions:

(HE) Each weighted manifold M_i is Harnack at all scales and locations.

(NPE) Each weighted manifold M_i, $1\le i\le k$, is nonparabolic.

Regarding (NPE), note that the dichotomy parabolic/nonparabolic (nonparabolic means the "existence of a positive Green function") is identical to the dichotomy recurrent/transient (recurrence means an open ball is visited with probability 1). Moreover, a manifold satisfying (HE) is nonparabolic if and only if $\int_1^\infty \frac{ds}{V(x,\sqrt{s})}<+\infty$. See [11] for a comprehensive review. Under the two assumptions (HE)–(NPE), each E_i is indeed a representative of an end of M in the classical sense because nonparabolic Harnack manifolds can only have one end [15].

For any $x,y\in M$, define $|x|, d_+(x,y), d_\emptyset(x,y)$ in terms of the compact set K as before. Also, set

$$i_x=\begin{cases} i & \text{if } x\in E_i, 1\le i\le k,\\ 0 & \text{if } x\in K.\end{cases}$$

For $x\in E_i$, set $V_i(x,r)=\mu(B(x,r)\cap E_i)$ and $V_i(r)=\mu(B(o,r)\cap(K\cup E_i))$ where o is a fixed central point in K. Set

$$V_0(r)=\min\{V_i(r):1\le i\le k\}$$

and

$$H_{(M,\mu)}(t,x)=H(t,x)=\min\left\{1,\frac{|x|^2}{V_{i_x}(|x|)}+\left(\int_{|x|^2}^t\frac{ds}{V_{i_x}(\sqrt{s})}\right)_+\right\}. \tag{7.1}$$

To understand the behavior of $H(x,t)$, note that (HE) and (NPE) imply $\int^{+\infty}\frac{ds}{V_{i_x}(\sqrt{s})}<+\infty$. Whenever $V_{i_x}(r)/V_{i_x}(s)\ge c(r/s)^\alpha$ with $\alpha>2$, the integral $\int_{|x|^2}^t\frac{ds}{V_{i_x}(\sqrt{s})}$ is dominated by the term $\frac{|x|^2}{V_{i_x}(|x|)}$. This integral becomes relevant when the end containing x, E_{i_x}, is only barely nonparabolic, for instance, if $V_{i_x}(r)$ grows like $r^2(\log r)^2$.

Theorem 7.1 ([18]). *Assuming that* (HE) *and* (NPE) *are satisfied, the heat kernel* $p_{(M,\mu)}(t,x,y)$ *satisfies the two-sided estimate* (6.1) *with*

$$\begin{aligned} g_{M,c}(t,x,y)=&\left(\frac{H(x,t)H(y,t)}{cV_0(\sqrt{t})}+\frac{H(y,t)}{cV_{i_x}(\sqrt{t})}+\frac{H(x,t)}{cV_{i_y}(\sqrt{t})}\right)\exp\left(-c\frac{d_+(x,y)^2}{t}\right)\\ &+\frac{1}{cV_{i_x}(x,\sqrt{t})}\exp\left(-c\frac{d_\emptyset(x,y)^2}{t}\right).\end{aligned} \tag{7.2}$$

Note that the last term comes into play only when x and y are in the same end. For $t \in (0,1)$, it is possible to show that

$$\frac{1}{c_1 V(x,\sqrt{t})} \exp\left(-c_1 \frac{d(x,y)^2}{t}\right) \le g_{M,c}(t,x,y) \le \frac{1}{c_2 V(x,\sqrt{t})} \exp\left(-c_2 \frac{d(x,y)^2}{t}\right),$$

as expected, though this takes a bit of technical work.

For fixed x_0, y_0 and large t, $p_M(t,x_0,y_0)$ behaves as $\frac{1}{V_0(\sqrt{t})}$, that is, it is controlled by the volume growth of the smallest end at scale $\sqrt{t}$ (it is possible that the "the smallest end" changes depending on the scale at which the question is asked).

8. NONPARABOLIC MANIFOLDS WITH FINITELY MANY HARNACK ENDS

In this section, the manifold $M = M_1 \# \cdots \# M_k$ is a complete noncompact weighed Riemannian manifold with boundary as before, and we continue to make assumption (HE) that each end is a Harnack manifold. The manifold M is nonparabolic if and only if at least one of the ends M_i, $1 \le i \le k$, is nonparabolic. So, we may weaken assumption (NPE) to

(NP) At least one of the weighted manifolds M_i, $1 \le i \le k$, is nonparabolic.

However, under these circumstances, we need to make a further assumption in order to obtain sharp heat kernel estimates. Namely, we assume the following:

(RCA*) All the ends M_i, $1 \le i \le k$, that are parabolic must satisfy the relatively connected annulus condition (RCA): There exists a constant $A > 1$ such that, for any $R > A^2$ and any two points $x, y \in E$ with $|x| = |y| = R$, there is a continuous curve connecting x to y in $\{z : R/A \le |z| \le AR\}$.

In words, we assume that, in any parabolic end E_{i_0} of M, two points at a distance about R from the central part K can be connected without going too far toward infinity (no further than AR) and without coming back too close to the central part K (no closer than R/A). This condition is key to obtaining the results below. Note that again, under these assumptions, $E_1, \dots, E_k$ are indeed representative of the ends of M in the classical sense as any one of them is a manifold with only one end in the classical sense: the nonparabolic ones because they are Harnack, and the parabolic ones because of condition (RCA).

Each E_i is an incomplete manifold with boundary such that $\delta E_i = \delta M \cap E_i$ and $\partial E_i = \overline{E_i} \setminus E_i \subset K$. A *harmonic profile* for E_i is a function u_i which is positive in E_i, vanishes along ∂E_i, and is harmonic in E_i (this includes the condition that u_i has vanishing normal derivative along δE_i). It is known that such a function exists, is continuous on $\overline{E_i}$, and is unique up to multiplication by a positive real (recall hypothesis (HE)). Moreover, there is a constant $c > 0$ such that, for all $x \in E_i$ with $|x|$ large enough (e.g., $|x| \ge 2(1 + \mathrm{diam}(K))$),

$$c \int_1^{|x|^2} \frac{ds}{V_i(\sqrt{s})} \le u_i(x) \le c^{-1} \int_1^{|x|^2} \frac{ds}{V_i(\sqrt{s})}.$$

This fact (see [15,18,20]) depends crucially upon the hypotheses (HE) and (RCA*). It implies that the harmonic profile of a nonparabolic end E_i is bounded and bounded away from 0 in E_i away from K while the harmonic profile of a parabolic end tends to infinity at infinity.

A *harmonic profile* for $M = K \cup E_1 \cup \cdots \cup E_k$ is a positive harmonic function h on M (this implies it has vanishing normal derivative along δM) which, in each E_i, behaves as u_i at infinity. Again, it is known that such a function exists under assumption (NP), see [18,30].

We use this positive harmonic function h on M to consider the new weighted manifold, (M, μ_{h^2}), where

$$\mu_{h^2}(dx) = h^2(x)\mu(dx) = h^2(x)\sigma(x)dx,$$

whose Dirichlet form is

$$\int_M |\nabla f|^2 d\mu_{h^2} = \int_M |\nabla f|^2 h^2 d\mu.$$

Because h is harmonic,

$$\int_M |\nabla f|^2 h^2 d\mu = \int_M |\nabla (hf)|^2 d\mu, \quad f \in \mathscr{C}_c^\infty(M).$$

This means that the heat equation associated with

$$\left(M, \mu_{h^2}, \int_M |\nabla f|^2 d\mu_{h^2}, W_0^1(M, \mu_{h^2})\right)$$

is $\partial_t u - \frac{1}{h}\Delta_\sigma(hu) = 0$, and the associated heat kernel $p_{(M,\mu_{h^2})}$ is given by

$$p_{(M,\mu_{h^2})} = \frac{1}{h(x)h(y)} p_{(M,\mu)}(t,x,y). \tag{8.1}$$

In probability theory, the use of this relation is often referred to as the "Doob transform" technique after Joseph Doob. For us, its significance is that, assuming we know the profile h, it is possible to turn estimates of $p_{(M,\mu_{h^2})}$ into estimates of $p_{(M,\mu)}$. This is useful because of the following theorem.

Theorem 8.1. *Assume that* (M, μ) *satisfies* (HE), (NP), *and* (RCA*). *Then* (M, μ_{h^2}) *satisfies* (HE) *and* (NPE).

What this theorem says is that the *weighted* Riemannian manifold $(M, h^2\mu)$,

$$M = M_1 \# \cdots \# M_k = K \cup E_1 \cup \cdots \cup E_k,$$

is a connected sum of Harnack weighted manifolds, and, moreover, each of them is nonparabolic. The proof that each $(\overline{E_i}, \mu_{h^2}|_{E_i})$ is a Harnack manifold proceeds by showing that doubling and Poincaré inequalities hold at all scales and locations; see [17,20].

Theorems 7.1–8.1 and (8.1) lead to a sharp two-sided estimate for the heat kernel of the nonparabolic weighted manifold (M, μ), $\mu(dx) = \sigma(x)dx$ in terms of the functions

$$g_{(M,\mu),c}(t,x,y) = h(x)h(y)g_{(M,\mu_{h^2}),c}(t,x,y), \tag{8.2}$$

where $g_{(M,\mu_{h^2}),c}$ is given by (7.2).

Theorem 8.2. *Assume that* (M,μ) *satisfies* (HE), (NP), *and* (RCA*). *Then, there exist* c_1, c_2 *such that, for all* $t > 0$, $x, y \in M$, *we have*

$$g_{(M,\mu),c_1}(t,x,y) \le p_{(M,\mu)}(t,x,y) \le g_{(M,\mu),c_2}(t,x,y) \tag{8.3}$$

where $g_{(M,\mu),c}$ *is given by* (8.2).

It takes quite a bit of work to unpack this statement. How explicit the obtained estimates are depends very much on our ability to understand the function h, the harmonic profile of M. One key point to notice is that everything depends on the volume growth function of each end and our ability to compute, for $r > 1$, quantities such as

$$r \mapsto 1 + \int_1^{r^2} \frac{ds}{V_i(\sqrt{s})},$$

which controls u_i (hence h in E_i), and

$$(r,t) \mapsto \int_{r^2}^{t} \frac{ds}{(1 + \int_1^s \frac{d\tau}{V_i(\sqrt{\tau})})^2 V_i(\sqrt{s})}, \qquad r^2 < t,$$

which are needed to control the function $H_{(M,\mu_{h^2})}$ in E_i. These computations are the trickiest for ends that are near the threshold separating parabolic from nonparabolic ends, e.g., when $V_i(r)$ grows as r^2 up to a slowly-varying factor (think of $r^2[\log(1+r)]^\alpha$ with $\alpha \in \mathbb{R}$). It is worth noting that the result holds without restriction on the behavior of the volume growth through the parabolic/nonparabolic threshold, as long as (M,μ) itself is nonparabolic. The simplest general result concerns the long-time behavior of $p_{(M,\mu)}(t,x_0,y_0)$ for fixed $x_0, y_0 \in M$ which is that

$$\min_{1\le i\le k}\left\{\frac{c}{(1+\int_1^t \frac{ds}{V_i(\sqrt{s})})^2 V_i(\sqrt{t})}\right\} \le p_{(M,\mu)}(t,x_0,y_0) \le \min_{1\le i\le k}\left\{\frac{C}{(1+\int_1^t \frac{ds}{V_i(\sqrt{s})})^2 V_i(\sqrt{t})}\right\}.$$

Example 8.3. To illustrate what this says, consider the case when $M = K \cup E_1 \cup E_2 \cup E_3$ is a solid 3-dimensional body with 3 ends that can be described as follows:

- E_1 is a half-cylinder of radius 1 around the the bottom part of the z-axis,
$$E_1 = \{(x,y,z) : x^2 + y^2 \le 1, z < -1\};$$
- E_2 is essentially a solid planar slab around the xy-plane,
$$E_2 = \{(x,y,z) : x^2 + y^2 > 2, -1 \le 2z \le 1\};$$
- E_3 is essentially a solid half-cone of revolution of positive aperture around the positive z-axis, say
$$E_3 = \{(x,y,z) : x^2 + y^2 \le z, z > 1\};$$
- K is a compact set joining these ends smoothly together, and the measure on M is Lebesgue measure (i.e., $\sigma \equiv 1$).

In this description, $\overline{E_i}$ are smooth manifolds with corners but this can easily be fixed. The attentive reader will note that the enumeration of the ends corresponds precisely to their volume growth, with $V_1(r)$ growing linearly, $V_2(r)$ growing quadratically, and $V_3(r)$ growing as r^3. These ends are all Harnack, satisfy (RCA), and E_3 is nonparabolic so that M is nonparabolic. The other two ends are parabolic. The harmonic profile h of M satisfies $c|x| \le h(x) \le c^{-1}|x|$ in E_1, $c\log(1+|x|) \le h(x) \le c^{-1}\log(1+|x|)$ in E_2, and $c \le h(x) \le c^{-1}$ in E_3. It follows that $V_{i,h^2}(r)$ grows as r^3 in both E_1 and E_3, while $V_2(r)$ grows as $r^2\log^2 r$. For $o = (0,0,0) \in M$ and $t > 1$, these computations give

$$\frac{c}{t\log^2(1+t)} \le p_{(M,\mu)}(t,o,o) \le \frac{C}{t\log^2(1+t)}.$$

Now, consider the following two questions (see Figure 2):

(a) At time $t > 1$, where is $p_M(t,o,x)$ approximately the largest?

(b) Can we find balls $B_t = B(x_t, \frac{1}{4}\sqrt{t})$ with $|x_t| \le 4\sqrt{t}$ such that

$$\lim_{t\to+\infty} p_M(t,o,B_t) = \lim_{t\to+\infty} \int_{B(x_t,\frac{1}{4}\sqrt{t})} p_M(t,o,y)dy = 0?$$

Such balls contain an unusually small amount of heat given their sizes and locations.

The answer to the first question is that the heat kernel $p_M(t,o,x)$ is the largest when x is relatively close to $(0,0,-\sqrt{t})$, down in the cylinder E_1 where its approximate value is $1/t$. For comparison, note that $p_M(t,o,(0,0,+\sqrt{t}))$ is of the order of $1/t^{3/2}$. However, the ball $B((0,0,-\sqrt{t}),\frac{1}{4}\sqrt{t})$ has small volume, of order $\sqrt{t}$, so that $p_M(t,o,B((0,0,-\sqrt{t}),\frac{1}{4}\sqrt{t}))$ is approximately equal to $1/\sqrt{t}$. In the slab around the xy plane, $p_M(t,o,B((\sqrt{t},\sqrt{t},0),\frac{1}{4}\sqrt{t}))$ is approximately $1/\log t$. However, in the largest end, E_3, where the heat kernel is the smallest, $p_M(t,o,B((0,0,\sqrt{t}),\frac{1}{4}\sqrt{t}))$ is approximately 1; see Figure 2. In terms of heat diffusion, the heat kernel describes punctual temperature, and the integral over a ball is the caloric content. The caloric content of balls of a given radius

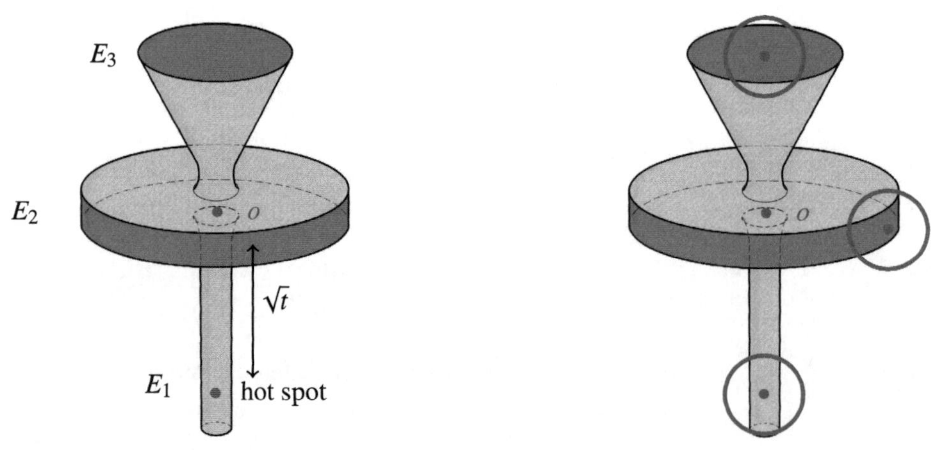

FIGURE 2
The solid body of Example 8.3: the "hot spot" and balls of interest (scale is $\sqrt{t}$).

depends on both the local temperature and the local volume growth. These computations illustrate the detailed information provided by Theorem 8.2.

9. PARABOLIC MANIFOLDS WITH FINITELY MANY HARNACK ENDS

It turns out that the case when the weighted manifold $M = M_1\#\cdots\#M_k$ is parabolic (i.e., Brownian motion is recurrent) is harder, and the treatment remains incomplete despite good results presented in [13] and a forthcoming companion paper. To give an idea of what is expected, let us consider the very simple case when each M_i is a surface of revolution in $\mathbb{R}^3$ associated with the rotation of the graph of

$$\phi : [0, +\infty) \to \big[0, +\infty), x \mapsto z = \phi_i(x) \text{ with } \phi_i(x) = x^{\alpha_i} \text{ for } x > 2 \text{ and } \alpha_i \in (0, 1\big].$$

The behavior of ϕ near 0 is $\sqrt{x}$ so that the surface M_i is smooth. These smooth surfaces in $\mathbb{R}^3$ are equipped with their natural Riemannian metric and measure. Each such surface is Harnack and (RCA) and its heat kernel satisfies

$$\frac{c}{V_i((s,\phi_i(s)),\sqrt{t})} \le p_{M_i}\big(t, (s,\phi_i(s)), (s,\phi_i(s))\big) \le \frac{C}{V_i((s,\phi_i(s)),\sqrt{t})}$$

with $V_i((s,\phi_i(s)), r)$ approximately equal to

$$\begin{cases} r^2 & \text{if } 0 \le r < \max\{1, s^{\alpha_i}\}, \\ s^{\alpha_i} r & \text{if } \max\{1, s^{\alpha_i}\} \le r \le s, \\ r^{1+\alpha_i} & \text{if } \max\{1, s\} \le r. \end{cases}$$

Reference [13] gives sharp global two-sided estimates for $M = M_1\#\cdots\#M_k$ as above. Here are some highlights:

- If for all $i \in \{1, \dots, k\}$, $\alpha_i = \alpha \in (0, 1)$, then $M = M_1\#\cdots\#M_k$ is Harnack.
- If $k \ge 2$ and for all $i \in \{1, \dots, k\}$, $\alpha_i = 1$, then M is doubling but does not satisfy the Poincaré inequality in large balls centered at a fixed point o in M. For large t and a fixed point $o \in M$, $p_M(t, o, o)$ is approximately equal to $1/t$, whereas if x, y are in different ends, at distance $\sqrt{t}$ from o, then $p_M(t, x, y)$ is approximately equal to $1/(t \log^2(1 + t))$.
- In all cases, $p_M(t, o, o)$ is approximately equal to $1/\max_{1\le i\le k}\{V_i(\sqrt{t})\}$ where $V_i(r) = \begin{cases} r^2 & \text{if } r \in (0, 1), \\ r^{1+\alpha_i} & \text{if } r \ge 1. \end{cases}$

The simplest and most important thing to note is that $p_M(t, o, o)$ is now controlled by the volume of the largest end whereas, in the case when each end is nonparabolic (i.e., Section 7), $p_M(t, o, o)$ is controlled by the volume of the smallest end. The first observation of this phenomena in a simplified model case which appeared in [7]. It is also worth stressing that the following problem remains open (see [14] for additional details on what is known).

Problem 9.1. Prove a sharp two-sided heat kernel estimate for $M = M_1\#\cdots\#M_k$ under the assumption that each M_i is Harnack, parabolic, and satisfies (RCA).

10. MIXED BOUNDARY CONDITIONS ON HARNACK MANIFOLDS

Although we did not insist much on this aspect, the results discussed in Sections 7, 8 and 9 depend in a significant way on our ability to derive sharp heat kernel estimates with Dirichlet boundary condition on domains obtained from a Harnack manifold by deleting a compact set with nonempty interior and smooth boundary; see [15, 16, 18, 19]. One is then naturally led to consider the problem of heat kernel estimates for manifolds with boundary and mixed boundary conditions (Neumann and Dirichlet). This requires new types of hypotheses. Assume that M is a complete, weighted Riemannian manifold with boundary δM, and that Ω is an open subset of M such that $M \setminus \Omega \subseteq \delta M$. To simplify the presentation, assume that $\partial\Omega = M \setminus \Omega$ has finitely many connected components which are manifolds with boundary. Our new object of interest here is the minimal heat kernel of Ω, $p_\Omega(t, x, y)$, which is the heat kernel of the Dirichlet form $(L^2(\Omega, \mu), \int_\Omega |\nabla f|^2 d\mu, W_0^1(\Omega))$ where the domain $W_0^1(\Omega)$ is the closure of smooth, compactly-supported functions in Ω for the norm $(\int_\Omega (|f|^2 + |\nabla f|^2) d\mu)^{1/2}$. The corresponding heat equation has Dirichlet boundary condition along $\partial\Omega$ and Neumann boundary condition along the rest of the original boundary of M, δM. In [20], sharp heat kernel estimates are derived under the condition that (1) (M, μ) is Harnack, and (2) Ω is a uniform domain in M. Before we describe what *uniform* means, observe that the distance between two points x, y in Ω, $d_\Omega(x, y)$, is the same as the distance between x and y in M, $d_M(x, y)$.

To say that Ω is uniform in M with constant C is to say that, for any pair of points $x, y \in \Omega$, there is a rectifiable curve parametrized by arc length, $\gamma_{xy} : [0, T_{xy}] \to \Omega$, joining x to y, of length $T_{x,y} \le C d_\Omega(x, y)$, and satisfying $d(\gamma_{xy}(s), M \setminus \Omega) \ge C^{-1} \min\{s, T_{xy} - s\}$, for all $s \in [0, T_{xy}]$. In words, the curve γ_{xy} is roughly of optimal length and, when moving away from x (or y) along γ_{xy}, one also moves away from the boundary in a roughly linear fashion. For instance, the open upper-half plane in the closed upper-half plane is uniform, but the open infinite strip $\Omega = \{(x, y) : -1 < y < 1, x \in \mathbb{R}\}$ is NOT uniform in its closure because one cannot escape from being close to the boundary; see Figure 3.

When Ω is uniform in M, it admits a harmonic profile h_Ω, which is positive harmonic in Ω, vanishing continuously along $\partial\Omega$ (this function has vanishing normal derivative along $\delta M \setminus \partial\Omega$).

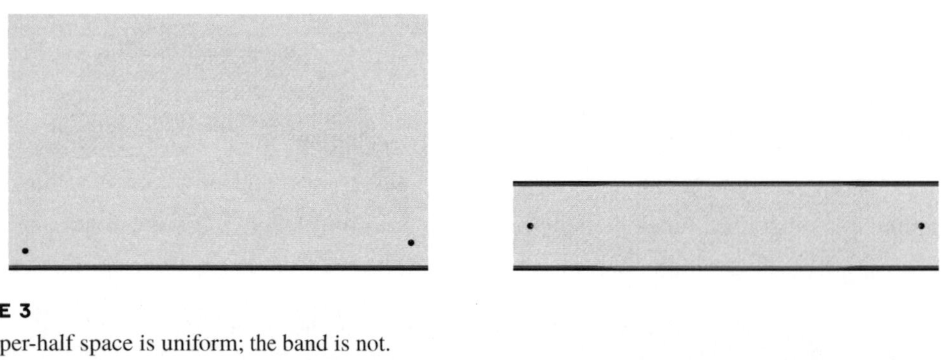

FIGURE 3
The upper-half space is uniform; the band is not.

Theorem 10.1 (See [20, THEOREM 5.11]). *Referring to the setting outlined above, assume that (M, μ) is Harnack and that Ω is uniform in M with harmonic profile h_Ω. Then there are constants c_1, c_2 such that*

$$g_{\Omega,c_1}(t,x,y) \leq p_\Omega(t,x,y) \leq g_{\Omega,c_2}(t,x,y), \tag{10.1}$$

where

$$g_{\Omega,c}(t,x,y) = \frac{h_\Omega(x)h_\Omega(y)}{cV_{h_\Omega^2}(x,\sqrt{t})}\exp\left(-c\frac{d_M(x,y)^2}{t}\right).$$

When reading this theorem, recall that

$$d_\Omega(x,y) = d_M(x,y) \quad \text{and} \quad V_{h_\Omega^2}(x,r) = \int_{B_M(x,r)} h_\Omega^2(y)\mu(dy).$$

The lack of symmetry between x and y is intentional; because p_Ω is symmetric, a symmetric estimate can easily be derived from that stated here. This is a satisfactory and useful result from a theoretical viewpoint, but detailed applications require estimating the profile h_Ω, which is a difficult problem.

11. MIXED BOUNDARY CONDITIONS ON MANIFOLDS WITH ENDS

The techniques used in the previous section to study uniform domains in complete Riemannian manifolds with boundary can be implemented together with the techniques of Section 7 to study the minimal heat kernel p_Ω of a domain Ω in a complete Riemannian manifold $M = K \cup E_1 \cup \cdots \cup E_k$ with boundary when $\partial\Omega \subset \delta M$. A simplistic, yet interesting example is depicted in Figure 4. Can you guess the behavior of $p_\Omega(t,o,o)$ in Figure 4? The answer is $1/(t \log^2 t)$ because, far from o, one of the three cones is free of Dirichlet boundary condition. If each cone had at least one of its sides contained in $\partial\Omega$, the behavior would depend in an explicit way on the apertures of the cones and whether each cone has one or two sides contained in $\partial\Omega$. See [6] for a detailed discussion and general results in this direction, including complete two-sided heat kernel estimates for such mixed boundary problems.

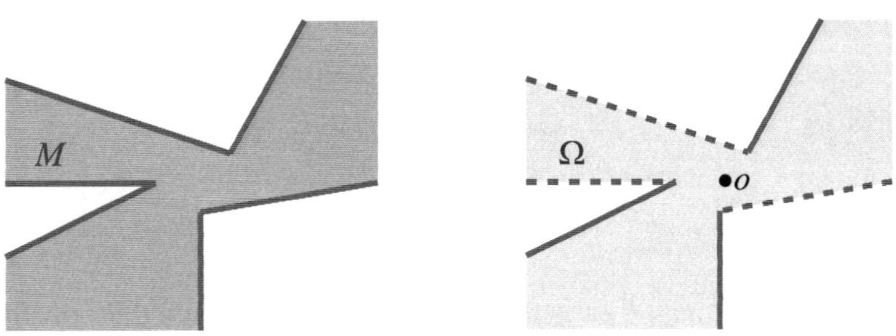

FIGURE 4
Sketch (corners should be rounded) of M (left, thick boundary lines are part of M) and Ω (right) with "Dirichlet boundary" $\partial\Omega \subseteq \delta M$ (dashed) not part of Ω.

12. ATTACHMENTS ALONG NONCOMPACT SUBMANIFOLDS

The basic ideas implemented in the study of connected sums above can be described informally in greater generality as follows. Given a (metrically complete) manifold M, identify large chunks, hopefully, finitely many, $M_1, \dots, M_k$, which, taken by themselves, are Harnack manifolds. Each chunk has attachment boundaries along which they are attached to each other to form the manifold M. Call Att_i the attachment boundary of M_i and set $\Omega_i = M_i \setminus \mathrm{Att}_i$. If each Ω_i is uniform in M_i, $1 \le i \le k$, not only can we have good estimates for the heat kernel p_{M_i} of M_i (because M_i is Harnack), but we can also estimate the minimal heat kernel p_{Ω_i} of Ω_i (this heat kernel satisfies the Dirichlet boundary condition along Att_i). For the next step, it may be necessary to make further assumptions about the manifolds M_i and their open subsets Ω_i (see, for instance, conditions (NP) and (RCA) above). Now, find a way to use the known information regarding the different large chunks M_i to reconstruct and estimate the heat kernel p_M of M. Sections 7–9 above describe how these ideas apply successfully to connected sums (i.e., compact attachments). The article [12] is, so far, the lone published attempt to carry out this approach when two large chunks are glued along a noncompact attachment boundary. Emily Dautenhahn and the author are working on applying these ideas beyond the cases treated in [12].

The results of Sections 7–12 should ultimately be developed in the more abstract context of Dirichlet spaces so as to include ends that satisfy the Harnack inequalities that appear in the context of fractals (see [3] and the references therein). This would allow for the treatment of a larger class of Riemannian manifolds, as the geometry of a Riemannian manifold at infinity can mimic that of a fractal object.

ACKNOWLEDGMENTS

A. Grigor'yan and the author have shared their interests in heat kernel estimates and their applications to potential theory, probability theory, and geometry ever since they met, a few years after having proved, independently, some version of Theorem 4.4. This article is partly the result of this long lasting collaboration and friendship. Thanks to T. Coulhon, N. Varopoulos, and to all my colleagues, collaborators, and students who helped me make progress in the study of heat kernels over the years.

FUNDING

The author's research is partially supported by a grant from the National Science Foundation (USA), DMS-2054593.

REFERENCES

[1] D. G. Aronson, Bounds for the fundamental solution of a parabolic equation. *Bull. Amer. Math. Soc.* **73** (1967), 890–896.

[2] D. G. Aronson and J. Serrin, Local behavior of solutions of quasilinear parabolic equations. *Arch. Ration. Mech. Anal.* **25** (1967), 81–122.

[3] M. Barlow and M. Murugan, Stability of the elliptic Harnack inequality. *Ann. of Math. (2)* **187** (2018), no. 3, 777–823.

[4] G. Carron and D. Tewodrose, A rigidity result for metric measure spaces with Euclidean heat kernel. *J. Éc. Polytech. Math.* **9** (2022), 101–154.

[5] T. Coulhon, Ultracontractivity and Nash type inequalities. *J. Funct. Anal.* **141** (1996), no. 2, 510–539.

[6] E. Dautenhahn and L. Saloff-Coste, Heat kernel estimates on manifolds with ends with mixed boundary condition. 2021, arXiv:2108.05790.

[7] E. B. Davies, Non-Gaussian aspects of heat kernel behaviour. *J. Lond. Math. Soc.* **55** (1997), 105–125.

[8] N. Eldredge, M. Gordina, and L. Saloff-Coste, Left-invariant geometries on SU(2) are uniformly doubling. *Geom. Funct. Anal.* **28** (2018), no. 5, 132–1367.

[9] M. Fukushima, Y. Oshima, and M. Takeda, *Dirichlet forms and symmetric Markov processes*. Second revised and extended edn., de Gruyter Stud. Math. 19, Walter de Gruyter & Co., Berlin, 2011.

[10] A. Grigor'yan, The heat equation on non-compact Riemannian manifolds. *Mat. Sb.* **182** (1991), 55–87. Engl. Transl. *Math. USSR Sb.* **72** (1992), 47–77.

[11] A. Grigor'yan, Analytic and geometric background of recurrence and non-explosion of the Brownian motion on Riemannian manifolds. *Bull. Amer. Math. Soc. (N.S.)* **36** (1999), no. 2, 135–249.

[12] A. Grigor'yan and S. Ishiwata, Heat kernel estimates on a connected sum of two copies of $\mathbb{R}^n$ along a surface of revolution. *Glob. Stoch. Anal.* **2** (2012), no. 1, 29–65.

[13] A. Grigor'yan, S. Ishiwata, and L. Saloff-Coste, Heat kernel estimates on connected sums of parabolic manifolds. *J. Math. Pures et Appl.* **113** (2018), 155–194.

[14] A. Grigor'yan, S. Ishiwata, and L. Saloff-Coste, Geometric analysis on manifolds with ends. In *Analysis and partial differential equations on manifolds, fractals and graphs*, edited by A. Grigor'yan and Y. Sun, pp. 325–344, De Gruyter, Berlin, Boston. 2021. DOI 10.1515/9783110700763-011.

[15] A. Grigor'yan and L. Saloff-Coste, Dirichlet heat kernel in the exterior of a compact set. *Comm. Pure Appl. Math.* **55** (2002), 93–133.

[16] A. Grigor'yan and L. Saloff-Coste, Hitting probabilities for Brownian motion on Riemannian manifolds. *J. Math. Pures et Appl.* **81** (2002), 115–142.

[17] A. Grigor'yan and L. Saloff-Coste, Stability results for Harnack inequalities. *Ann. Inst. Fourier* **55** (2005), 825–890.

[18] A. Grigor'yan and L. Saloff-Coste, Heat kernel on manifolds with ends. *Ann. Inst. Fourier* **59** (2009), 1917–1997.

[19] A. Grigor'yan and L. Saloff-Coste, Surgery of the Faber–Krahn inequality and applications to heat kernel bounds. *Nonlinear Anal.* **131** (2016), 243–272.

[20] P. Gyrya and L. Saloff-Coste, *Neumann and Dirichlet heat kernels in inner uniform domains*. Astérisque 336, Société Mathématique de France, 2011.

[21] P. Li and S.-T. Yau, On the Schrödinger equation and the eigenvalue problem. *Comm. Math. Phys.* **88** (1983), 309–318.

[22] U. Mosco, Composite media and asymptotic Dirichlet forms. *J. Funct. Anal.* **123** (1994), no. 2, 368–421.

[23] J. Moser, A Harnack inequality for parabolic differential equations. *Comm. Pure Appl. Math.* **16** (1964), 101–134. Correction in **20** (1967), 231–236.

[24] J. Nash, Continuity of solutions of parabolic and elliptic equations. *Amer. J. Math.* **80** (1958), 931–954.

[25] F. Porper and S. Eidel'man, Two-sided estimates of fundamental solutions of second-order parabolic equations and some applications. *Russian Math. Surveys* **39** (1984), 119–178.

[26] L. Saloff-Coste, A note on Poincaré, Sobolev and Harnack inequalities. *Duke Math. J.* **65**, *Int. Math. Res. Not. IMRN* **2** (1992), 27–38.

[27] L. Saloff-Coste, *Aspects of Sobolev-type inequalities*. London Math. Soc. Lecture Note Ser. 289, Cambridge University Press, Cambridge, 2002.

[28] K. T. Sturm, Analysis on local Dirichlet spaces. II. Upper Gaussian estimates for the fundamental solutions of parabolic equations. *Osaka J. Math.* **32** (1995), no. 2, 275–312.

[29] K. T. Sturm, Analysis on local Dirichlet spaces. III. The parabolic Harnack inequality. *J. Math. Pures Appl. (9)* **75** (1996), no. 3, 273–297.

[30] C.-J. Sung, L.-F. Tam, and J. Wang, Spaces of harmonic functions. *J. Lond. Math. Soc. (2)* **3** (2000), 789–806.

LAURENT SALOFF-COSTE

567 Malott Hall, Cornell University, Ithaca NY, 14850, USA, lps2@cornell.edu

This Page Intentionally Left Blank

13. COMBINATORICS

SPECIAL LECTURE

PROBABILITY THEORY FOR RANDOM GROUPS ARISING IN NUMBER THEORY

MELANIE MATCHETT WOOD

ABSTRACT

We consider the probability theory, and in particular the moment problem and universality theorems, for random groups of the sort that arise or are conjectured to arise in number theory, and in related situations in topology and combinatorics. The distributions of random groups that are discussed include those conjectured in the Cohen–Lenstra–Martinet heuristics to be the distributions of class groups of random number fields, as well as distributions of nonabelian generalizations, and those conjectured to be the distributions of Selmer groups of random elliptic curves. For these sorts of distributions on finite and profinite groups, we survey what is known about the moment problem and universality, give a few new results including new applications, and suggest open problems.

MATHEMATICS SUBJECT CLASSIFICATION 2020

Primary 60E05; Secondary 60B99, 60F05, 15B52, 11R29, 11R45

KEYWORDS

Random groups, moment problem, universality, Cohen-Lenstra heuristics, arithmetic statistics, random matrices

ICM INTERNATIONAL CONGRESS OF MATHEMATICIANS 2022 JULY 6–14

Proc. Int. Cong. Math. 2022, Vol. 6, pp. 4476–4508
DOI 10.4171/ICM2022/145

1. INTRODUCTION

In this paper we will discuss the probability theory of random groups that arise in number theory and related areas, and the applications of that probability theory to other fields. We focus on the moment problem and on universality results for these random groups. While our focus is on the probability theory, we use potential applications in number theory, as well as topology and combinatorics, to motivate the kind of random groups on which we focus our probabilistic study.

One of the first motivating examples is the Cohen–Lenstra distribution on finite abelian p-groups. Let p be a prime and let X_{CL} be a random finite abelian p-group such that

$$\operatorname{Prob}(X_{\mathrm{CL}} \simeq A) = \frac{\prod_{i \geq 1}(1 - p^{-i})}{|\operatorname{Aut}(A)|}$$

for each finite abelian p-group A. For p an odd prime, let C_B be the Sylow p-subgroup of the class group of a uniform random imaginary quadratic field K with $|\operatorname{Disc} K| \leq B$. Then Cohen and Lenstra [15] conjectured that for each finite abelian p-group A,

$$\lim_{B \to \infty} \operatorname{Prob}(C_B \simeq A) = \operatorname{Prob}(X_{\mathrm{CL}} \simeq A), \tag{1.1}$$

i.e., that the C_B converge (in distribution) to X_{CL}. This X_{CL} is our starting example of a random group whose probability theory we wish to understand. Throughout the paper, we will consider more examples, including those related to generalizations of C_B such as when quadratic extensions are replaced by higher degree extensions or when the base field $\mathbb{Q}$ is replaced by another number field or $\mathbb{F}_q(t)$. We will consider nonabelian analogs where we consider $\operatorname{Gal}(K^{un}/K)$, the Galois group of the maximal unramified extension of K, in place of the class group. We will mention connections to analogous random groups arising in other fields, such as $\pi_1(M)$ or $H_1(M)$ for a random 3-manifold M, or the Jacobian (also known as sandpile group) of a random graph.

With these examples in mind, we first discuss the moment problem. Given a random variable X of a certain type, based on the type of random variable, we choose certain real-valued functions $f_0, f_1, \ldots$ and call the averages $\mathbb{E}(f_k(X))$ the *moments* of X. When X is real valued, we usually take $f_k(X) = X^k$, but when X is a random group we usually take $f_k(X)$ to be the number $\# \operatorname{Sur}(X, G_k)$ of surjective homomorphisms from X to a group G_k. The moment problem asks when the distribution of the random variable is determined uniquely from these moments. This is very useful in applications because the moments are usually easier to access than a distribution itself, and we will discuss many applications to class groups and their generalizations.

Next, we discuss universality questions in the sense of the central limit theorem. We ask when and how can we build a random group from many independent inputs such that in a limit the random group is insensitive to the distribution of the random inputs. When this happens, the output distribution is, of course, a natural one (as in the normal distribution in the Central Limit Theorem), and it tells us that such distributions are likely to arise in nature. This can help provide further motivation and context for conjectures in number theory. As the theory develops, we expect there will be further applications of these universality results to other fields.

For both topics, we will review what is known for random abelian and nonabelian groups, mention many applications, prove a few new results and applications, and suggest open problems.

1.1. Notation and conventions

We use $\mathbb{E}$ to denote the expectation of a real-valued random variable.

For a finite set S, we use $\#S$ or $|S|$ to denote the size of the set.

We write $\mathbb{F}_q$ for the finite field with q elements.

For a set of primes P, a P-group is a group whose order is a product of powers of primes in P, and a pro-P group is a profinite group all of whose continuous finite quotients are P-groups. The pro-P completion of a group is the inverse limit of all of its P-group quotients (and is a pro-P group).

We use Hom, resp. Sur, to denote homomorphisms, resp. surjective homomorphisms, always in the category of whatever the objects are in, e.g., for profinite groups we take continuous homomorphisms, and for R-modules we take R-module homomorphisms. Sometimes we use a subscript, e.g., $\mathrm{Sur}_R(A, B)$, as a reminder of the category. We use Aut to denote automorphisms with the same caveats.

When we take a random finite group, it is always with the discrete σ-algebra on the set of finite groups.

For random variables $Y, X_0, \dots$ with respect to a Borel σ-algebra, we say the X_n weakly converge in distribution to Y if for every open set U we have $\liminf \mathrm{Prob}(X_n \in U) \geq \mathrm{Prob}(Y \in U)$. By the Portmanteau theorem, this is equivalent to many other conditions. In this paper, in the topologies we consider, every open set is a countable disjoint union of basic open sets (used to define the topology), and each basic open set is also closed. In these settings, weak convergence in distribution is equivalent to having, for each basic open set U, $\lim_{n\to\infty} \mathrm{Prob}(X_n \in U) = \mathrm{Prob}(Y \in U)$ (see [33, PROOF OF THEOREM 1.1]).

For an abelian group A, we have $\wedge^2 A$ is the quotient of $A \otimes A$ by the subgroup generated by elements of the form $a \otimes a$ (and for an R-module A, we define $\wedge^2_R A$ similarly with the tensor product over R) and $\mathrm{Sym}^2 A$ is the quotient of $A \otimes A$ by the subgroup generated by elements of the form $a \otimes b - b \otimes a$.

For a group (resp. profinite group) G and elements $g_1, \dots \in G$, we write $\langle g_1, \dots \rangle$ for the normal subgroup (resp. closed normal subgroup) generated by $g_1, \dots$.

For a function $f(x)$, we write $f(x) = O(g(x))$ to mean that there exists a constant C such that for all x such that $f(x)$ is defined, we have $|f(x)| \leq Cg(x)$.

For a group G with an action of a group Γ, we write G^Γ for the invariants, i.e., elements of G that are fixed by every element of Γ.

In the distributions of interest from number theory, there will usually be a random number field, or random elliptic curve, or some such object behind the scenes. In these situations, there are a countable number of objects of interest (such as imaginary quadratic number fields), and we consider some enumeration of them such that there are a finite number up to some bound B, and then we take a uniform random object up to bound B, and consider the limit of these distributions as $B \to \infty$. We do not wish to suggest that the uniform distribution is the only distribution on a finite set. Indeed, the entire point of Section 3 is based on the fact that there are many nonuniform distributions. Even beyond the question of the distribution on the objects up to bound B, there is still a question of which enumeration one takes and this can have interesting and important effects (e.g., see [3, 45]). However, since we are using the examples from number theory mainly as motivation, in this paper we will usually be very brief or not mention at all how exactly we take the random number theoretic objects.

2. THE MOMENT PROBLEM

In probability, one often detects the distribution of a random variable by its moments, i.e., the averages of certain functions of the random variable. Most classically, the moments of a random variable $X \in \mathbb{R}$ are the averages $\mathbb{E}(X^k)$, indexed by natural numbers k, and the (mixed) moments of a random variable $(X_1, \ldots, X_n) \in \mathbb{R}^n$ are the averages $\mathbb{E}(X_1^{k_1} \cdots X_n^{k_n})$, indexed by n-tuples of natural numbers $(k_1, \ldots, k_n)$.

The moment problem asks whether moments determine a unique distribution, and results on the moment problem, such as the following, are foundational in probability theory.

Theorem 2.1 (Carleman's condition). *Let X be a random real number such that $M_k = \mathbb{E}(X^k)$ is finite for all integers $k \geq 0$. Then if*

$$\sum_{k \geq 1} M_{2k}^{-\frac{1}{2k}} = \infty, \tag{2.2}$$

then there is a unique distribution for a random real number Y such that $\mathbb{E}(Y^k) = M_k$ for all $k \geq 0$. In particular, if $M_k = O(e^k)$, then (2.2) *holds.*

This kind of uniqueness result is useful in a situation when we have a conjectural distribution, know its moments, and then can prove some random variable is distributed as conjectured by showing it has those moments. In other situations, we have an unknown distribution, compute its moments, and then recognize those as moments of a well-known distribution, and can use a uniqueness result to show our distribution matches the well-known one.

In many applications we have not a single random variable, but rather a sequence of random variables, and we seek their limiting distribution. For this, we require a uniqueness theorem that is *robust*, in the sense that we can prove that a sequence of random variables whose moments converge to certain values must converge in distribution to a certain limit.

Now we will clarify some, slightly informal, language to talk about different aspects of the moment problem. Suppose we are considering random variables taking values in some set, and a sequence of real-valued functions $f_0, f_1, f_2, \ldots$ on that set whose averages give the moments of the random variables. We say we have *uniqueness* in the moment problem for moments $M_k \in \mathbb{R}$, if the following holds: for any two random variables X, Y under consideration, if for all k we have $\mathbb{E}(f_k(X)) = \mathbb{E}(f_k(Y)) = M_k$, then X and Y have the same distribution. We say we have *robust uniqueness* in the moment problem for moments $M_k \in \mathbb{R}$, if the following holds: for any sequence of random variables $Y, X_1, X_2, X_3, \ldots$ under consideration, if for all k we have $\lim_{n\to\infty} \mathbb{E}(f_k(X_n)) = \mathbb{E}(f_k(Y)) = M_k$, then the X_n weakly converge in distribution to Y. We have *existence* in the moment problem for moments $M_k \in \mathbb{R}$, if we know there exists a random variable X with $\mathbb{E}(f_k(X)) = M_k$ for all k, and we have *construction* in the moment problem for moments $M_k \in \mathbb{R}$ if we have existence and can moreover explicitly describe X by giving useful formulas for its distribution on enough subsets to generate the underlying σ-algebra.

2.1. Robust uniqueness for random abelian groups

For example, Fouvry and Klüners [21] determined the distribution of the 4-ranks $(2C_B)[2]$, where C_B is the 2-Sylow subgroup of the class group of a random imaginary quadratic field as in Section 1, as $B \to \infty$. Fouvry and Klüners proved the following result (which covers all aspects of the moment problem for certain average values).

Theorem 2.3 ([22, THEOREM 1]). *If p is a prime and $X_1, X_2, \ldots$ are random finite-dimensional $\mathbb{F}_p$-vector spaces such that for every integer $k \geq 0$, we have*

$$\lim_{n\to\infty} \mathbb{E}\big(\#\operatorname{Sur}\big(X_n, \mathbb{F}_p^k\big)\big) = 1,$$

then for each integer $r \geq 0$, we have

$$\lim_{n\to\infty} \operatorname{Prob}\big(X_n \simeq \mathbb{F}_p^r\big) = p^{-r^2} \frac{\prod_{j=r+1}^{\infty}(1-p^{-j})}{\prod_{j=1}^{r}(1-p^{-j})}.$$

The distribution and averages in Theorem 2.3 are known to occur as $\operatorname{Prob}(X_{\mathrm{CL}}/pX_{\mathrm{CL}} \simeq \mathbb{F}_p^r)$ and $\mathbb{E}(\#\operatorname{Sur}(X_{\mathrm{CL}}/pX_{\mathrm{CL}}, \mathbb{F}_p^k))$, respectively, for the random group X_{CL} introduced in Section 1 (see [15, THEOREM 6.3, COROLLARY 6.5]), so there does exist a random variable with these averages and its distribution can be explicitly described. In the paper [21], Fouvry and Klüners determined the averages $\mathbb{E}(\#\operatorname{Sur}((2C_B)[2], \mathbb{F}_2^k))$, and then applied Theorem 2.3 to determine the distribution of 4-ranks of class groups of imaginary quadratic fields (and did the analogous work for class groups of real quadratic fields).

Fouvry and Klüners actually write $\prod_{0\leq i<k}(p^{\mathrm{rk}_p(X_n)}-p^i)$, and we have interpreted that as the number of surjective homomorphisms $\#\operatorname{Sur}(X_n, \mathbb{F}_p^k)$. In [22], Fouvry and Klüners translate the knowledge of the averages of $\prod_{0\leq i<k}(p^{\mathrm{rk}_p(X)}-p^i)$ for all k to the knowledge of the averages of $p^{\mathrm{rk}_p(X)k} = \#\operatorname{Hom}(X, \mathbb{F}_p^k)$ for all k (which can be done by a finite sum over the subgroups of $\mathbb{F}_p^k$). These latter averages are the classical moments of the random number $p^{\mathrm{rk}_p(X)} = |X|$. When our random groups get more complicated (and in particular nonabelian), we will not be able to capture the entire data of our groups so simply in a number, or even a sequence of numbers, but the functions $\#\operatorname{Sur}(-, G)$ or $\#\operatorname{Hom}(-, G)$ will continue to be important and convenient functions whose averages we will call the moments (or Sur-moments, Hom-moments) of a random group. (See [13, SECTION 3.3] for a discussion about the fact that the Hom-moments for finite abelian p-groups are classical mixed moments of certain numerical invariants of the groups.) The relationship between the Hom-moments and the Sur-moments is analogous to the relationship of the moments $\mathbb{E}(X^k)$ and the factorial

moments $\mathbb{E}(X(X-1)\cdots(X-k+1))$ of a random real number—knowledge of either kind of moments for $k \leq m$ easily gives knowledge of the other kind for $k \leq m$, and the choice of which to use mainly depends which is more convenient for the problem at hand.

Fouvry and Klüners's proof of the robust uniqueness part of Theorem 2.3 actually works whenever

$$\mathbb{E}(\#\operatorname{Hom}(X, \mathbb{F}_p^k)) = \mathbb{E}(|X|^k) = O(p^{k^2/2})$$

(see [22, PROPOSITION 3]), echoing the refrain that moments that do not grow too quickly determine a distribution. (Note that in this generality we are not claiming existence of a distribution, but only uniqueness.) Such moments are too large to use Carleman's condition to conclude the distribution of $|X|$ as a real number, and indeed there are different distributions of real numbers that give the same moments with this order of growth (e.g., various distributions that have the same moments as the log-normal distribution). However, in our setting, of course, $|X|$ is restrained to be a power of p.

For a random cyclic cubic field K, with class group Cl_K with 3-torsion $\mathrm{Cl}_K[3]$, Klys [28] found the asymptotic moments of $\mathrm{Cl}_K[3]/\,\mathrm{Cl}_K[3]^{\mathrm{Gal}(K/\mathbb{Q})}$, and then applied the more general form of Theorem 2.3 to determine the limiting distribution of $\mathrm{Cl}_K[3]/\,\mathrm{Cl}_K[3]^{\mathrm{Gal}(K/\mathbb{Q})}$.

Ellenberg, Venkatesh, and Westerland prove the following.

Theorem 2.4 ([19, PROPOSITION 8.3]). *If for each $n \geq 0$, we have a random abelian p-groups X_n such that for every abelian p-group A we have*

$$\lim_{n\to\infty} \mathbb{E}(\#\operatorname{Sur}(X_n, A)) = 1,$$

then the X_n weakly converge in distribution to X_{CL}, i.e., for every abelian p-group B, we have

$$\lim_{n\to\infty} \operatorname{Prob}(X_n \simeq B) = \operatorname{Prob}(X_{\mathrm{CL}} \simeq B) = \frac{\prod_{i\geq 1}(1-p^{-i})}{|\operatorname{Aut}(B)|}.$$

Ellenberg, Venkatesh, and Westerland use Theorem 2.4, along with a determination of certain limiting moments of class groups of imaginary quadratic extensions of $\mathbb{F}_q(t)$, to prove that in a limit where the discriminant goes to infinity and then q goes to infinity, that the ℓ-Sylow subgroups of these class groups are as predicted by the Cohen–Lenstra heuristics for any odd prime ℓ, as long as $\ell \nmid q-1$ [19, THEOREM 1.2]. The work of Ellenberg, Venkatesh, and Westerland also particularly pioneered the idea that it is useful to consider these averages of surjection counts to be moments.

If we would like to consider more general finite abelian groups, and also distributions that have other moments, we have the following theorem by the author. (The cited results are stated with stronger bounds on the M_A, but one can see that all that is used in the proof is the hypotheses below.)

Theorem 2.5 (see [44, THEOREM 8.3, PROOF OF COROLLARY 9.2]). *Let P be a finite set of primes, and let $\mathcal{A}$ be the set of finite abelian P-groups. Let $M_A \in \mathbb{R}$ for each $A \in \mathcal{A}$ such that $M_A = O(|\wedge^2 A|)$. Let $Y, X_1, X_2, \ldots$ be random groups in $\mathcal{A}$. If for every $A \in \mathcal{A}$, we have*

$$\lim_{n\to\infty} \mathbb{E}(\#\operatorname{Sur}(X_n, A)) = \mathbb{E}(\#\operatorname{Sur}(Y, A)) = M_A,$$

then the X_n weakly converge in distribution to Y, i.e., for every $B \in \mathcal{A}$,

$$\lim_{n\to\infty} \operatorname{Prob}(X_n \simeq B) = \operatorname{Prob}(Y \simeq B).$$

When $A = \mathbb{F}_p^k$, we have $|\wedge^2 A| = p^{k(k-1)/2}$, so we see a similar upper bound to that of Fouvry and Klüners. Theorem 2.5 was applied in [44] to determine the limiting distribution of the Jacobians (also known as sandpile groups) of Erdős–Rényi random graphs, and by Mészáros [37] to determine the limiting distribution of the Jacobians of random regular graphs. Mészáros's result then had the striking corollary that the adjacency matrix of a random regular graph is invertible with high probability, answering a long-standing open question that is not a priori about random groups at all.

If we consider a random finite abelian group X, without any condition on primes dividing its order, we have a uniqueness result by W. Wang and the author as a corollary of Theorem 2.5.

Corollary 2.6 ([43, THEOREM 6.13]). *Let $M_A \in \mathbb{R}$ for each finite abelian group A such that $M_A = O(|\wedge^2 A|)$. Let X, Y be random finite abelian groups. If for every finite abelian group A, we have*

$$\mathbb{E}\big(\#\operatorname{Sur}(X, A)\big) = \mathbb{E}\big(\#\operatorname{Sur}(Y, A)\big) = M_A,$$

then X and Y have the same distribution, i.e., for every finite abelian group B,

$$\operatorname{Prob}(X \simeq B) = \operatorname{Prob}(Y \simeq B).$$

Proof. For a finite abelian group C, let C_p denote its Sylow p-subgroup. We have

$$\operatorname{Prob}(X \simeq A) = \lim_{z\to\infty} \operatorname{Prob}\Big(\prod_{p\le z} X_p \simeq \prod_{p\le z} A_p\Big).$$

Then we can apply Theorem 2.5 with P the set of primes at most z to conclude the corollary. ■

However, for general finite abelian groups, robustness no longer holds (as it is possible the limit in n cannot be exchanged with the limit in z). As in [43, EXAMPLE 6.14], we can consider a random finite abelian group X, e.g., such that

$$\operatorname{Prob}(X \simeq A) = \frac{\zeta(2)^{-1}\zeta(3)^{-1}\zeta(4)^{-1}\cdots}{|A||\operatorname{Aut} A|},$$

where ζ is the Riemann zeta function and we can also write $\zeta(2)^{-1}\zeta(3)^{-1}\zeta(4)^{-1}\cdots$ as a product over primes $\prod_p \prod_{i\ge 2}(1-p^{-i})$. (There is a random group with this distribution–see e.g., [46, PROPOSITION 2.1], and it is the limiting distribution predicted by Gerth's extension [25] of the Cohen–Lenstra heuristics for $2\operatorname{Cl}_K$, where K is a random real quadratic field.) Then consider the random groups $X \times \mathbb{Z}/p\mathbb{Z}$ for each prime p. For any finite abelian group A, we have $\lim_{p\to\infty} \mathbb{E}(\#\operatorname{Sur}(X \times \mathbb{Z}/p\mathbb{Z}, A)) = \mathbb{E}(\#\operatorname{Sur}(X, A))$ since for p large enough $p \nmid |A|$. Yet the limiting distribution of the $X \times \mathbb{Z}/p\mathbb{Z}$ is the zero distribution, i.e., for each A we have $\lim_{p\to\infty} \operatorname{Prob}(X \times \mathbb{Z}/p\mathbb{Z} \simeq A) = 0$. This is in stark contrast to the situation for random real numbers [7, THEOREM 30.2], where whenever the moments determine a unique distribution, they do so robustly.

2.2. When uniqueness fails

Another important example of distributions arising in number theory are those predicted by Poonen and Rains **[39]** as the asymptotic distributions of p-Selmer groups of random elliptic curves. We consider two different random $\mathbb{F}_p$ vector spaces, with distributions given as follows:

$$\begin{aligned} \operatorname{Prob}\bigl(X_{\mathrm{odd}} \simeq \mathbb{F}_p^k\bigr) &= \begin{cases} p^{-(k^2-k)/2} \frac{\prod_{j=0}^{\infty}(1-p^{-2j-1})}{\prod_{j=1}^{k}(1-p^{-j})} & k \text{ odd}, \\ 0 & k \text{ even}, \end{cases} \\ \operatorname{Prob}\bigl(X_{\mathrm{even}} \simeq \mathbb{F}_p^k\bigr) &= \begin{cases} p^{-(k^2-k)/2} \frac{\prod_{j=0}^{\infty}(1-p^{-2j-1})}{\prod_{j=1}^{k}(1-p^{-j})} & k \text{ even}, \\ 0 & k \text{ odd}. \end{cases} \end{aligned} \tag{2.7}$$

Poonen and Rains **[39]** conjecture that these are the limiting distributions of p-Selmer group of elliptic curves over $\mathbb{Q}$ of odd and even parity, respectively, and note **[39, PROPOSITION 2.22(C)]** that these distributions have the same moments, even though they are quite different distributions, supported on entirely disjoint sets of groups. Indeed, there moments are as follows, and we see that these cases are just beyond the bounds of the uniqueness results mentioned above.

Theorem 2.8. *For each $k \geq 0$, we have*

$$\mathbb{E}\bigl(\#\operatorname{Sur}\bigl(X_{\mathrm{odd}}, \mathbb{F}_p^k\bigr)\bigr) = \mathbb{E}\bigl(\#\operatorname{Sur}\bigl(X_{\mathrm{even}}, \mathbb{F}_p^k\bigr)\bigr) = p^{(k^2+k)/2}, \quad \textit{and}$$

$$\mathbb{E}\bigl(\#\operatorname{Hom}\bigl(X_{\mathrm{odd}}, \mathbb{F}_p^k\bigr)\bigr) = \mathbb{E}\bigl(\#\operatorname{Hom}\bigl(X_{\mathrm{even}}, \mathbb{F}_p^k\bigr)\bigr) = p^{(k^2+k)/2} \prod_{j=1}^{k}\bigl(1 + p^{-j}\bigr).$$

Proof sketch. The Hom-moments are shown in **[39, PROPOSITION 2.22(C)]**. The Sur-moments can be found, in principle, by applying Möbius inversion to the Hom-moments. However, the following argument is perhaps more practical. The distributions of X_{odd} and X_{even} occur as the limiting distribution of cokernels of uniform random $n \times n$ alternating matrices over $\mathbb{F}_p$ (where n is odd or even, respectively). It is a general feature that for various computations it can be helpful, even for a known distribution, to recognize it as the limit of natural distributions. We can see the claimed limit by counting exactly how many alternating matrices over $\mathbb{F}_p$ have corank k for each k as in **[31, PROPOSITION 3.8]** (see also **[5, THEOREM 1.10]**). Then, one can make a simple argument to compute the limiting moments of these random cokernels as in **[13, THEOREM 11]** (which does the analogous thing for symmetric matrices), and use the explicit formulas for the distribution of the random cokernels for each k and n along with the dominated convergence theorem, as in **[13, THEOREM 10]**, to deduce that the limiting moments of the random cokernels agree with the moments of X_{odd} and X_{even}. ■

However, in a setting as we have described, we could also use the additional information that we are looking for a distribution supported only on groups of even rank (or odd rank), along with the moments, to determine a distribution.

One important motivation for the conjectures of Poonen and Rains was the result of Heath-Brown **[26]** determining the limiting distribution of 2-Selmer groups of a random quadratic twist of the congruent number curve. Heath-Brown showed that the limiting distribution for the quotient of the 2-Selmer group by the $\mathbb{F}_2^2$ coming from the 2-torsion points on the curve is the X_{odd} distribution for twists $D \equiv 5, 7 \pmod 8$ (when the Selmer rank is odd), and the X_{even} distribution for twists $D \equiv 1, 3 \pmod 8$ (when the Selmer rank is even). Heath-Brown determined these distributions by first determining the moments and then proving a robust uniqueness result for the moment problem. Heath-Brown pointed out that it was surprising that these different distributions had the same moment, and proved the following robust uniqueness result, taking into account the parity.

Theorem 2.9 ([26, LEMMA 18, PROOF OF THEOREM 2]). *Let $M_0, M_2, \dots$ be nonnegative real numbers such that $M_k = O(2^{k(k+1)/2})$. Let $Y, X_1, X_2, \dots$ be random even dimensional $\mathbb{F}_2$-vector spaces. Then if for every even $k \geq 0$, we have*

$$\lim_{n\to\infty} \mathbb{E}\big(\#\operatorname{Hom}(X_n, \mathbb{F}_2^k)\big) = \mathbb{E}\big(\#\operatorname{Hom}(Y, \mathbb{F}_2^k)\big) = M_k,$$

then the X_n weakly converge in distribution to Y, i.e., for every even r, we have

$$\lim_{n\to\infty} \operatorname{Prob}\big(X_n \simeq \mathbb{F}_2^r\big) = \operatorname{Prob}\big(Y \simeq \mathbb{F}_2^r\big).$$

The statement also holds if we replace "even" with "odd."

Feng, Landesman, and Rains [20] face a similar issue (in a slightly different context, where the random groups have fixed finite support of a given parity, but they only know half the moments) and use knowledge of the parity along with moments to determine the distribution of n-Selmer groups of elliptic curves of fixed height over $\mathbb{F}_q(t)$ as $q \to \infty$.

Given the two distributions of X_{odd} and X_{even} on $\mathbb{F}_p$-vector spaces given in (2.7), one natural question is what are all the distributions on $\mathbb{F}_p$-vector spaces with those same moments. We will now show that these (plus their linear combinations) are the only such distributions.

Theorem 2.10. *Given nonnegative reals $M_{-1}, M_0, M_1, \dots,$ and $p > 1$, and $b < 3$, such that $M_k = O(p^{\frac{k^2+bk}{2}})$, there is at most one simultaneous solution $(x_s)_s$ to*

$$\sum_{s=0}^{\infty} (-1)^s x_s = M_{-1} \quad \text{and}$$

$$\sum_{s=0}^{\infty} x_s p^{sk} = M_k, \quad k = 0, 1, \dots,$$

such that $x_s \geq 0$ for all s.

We note that this proof strategy is in the style of the earliest work on this problem, and not the more recent work, but it will also let us see some of the main features of the moment problem.

Proof. We modify the method from [26, LEMMA 18]. First, assuming we have a nonnegative solution, we can bound x_s using the $k = s$ equation to obtain

$$x_s = O\big(p^{\frac{-s^2+bs}{2}}\big).$$

From this it follows that for any $N \geq 0$ and $k \leq N-2$,

$$\sum_{s \geq N} x_s p^{sk} = O\Big(\sum_{s \geq N} p^{\frac{-s^2+bs+2ks}{2}}\Big) = O\big(p^{\frac{-N^2+bN+2kN}{2}}\big),$$

where we allow the constant in the O to depend on p.

We take some positive integer N, and we truncate the system to write

$$\sum_{s=0}^{N-1} x_s p^{sk} = M'_k$$

for $k = -1, 0, 1, \ldots, N-2$ (except for $k = -1$ we replace p^{sk} with $(-1)^s$). Let V be the $N \times N$ matrix whose i, j coefficient is $p^{(i-2)(j-1)}$ for $i \geq 2$ and $(-1)^{j-1}$ for $i = 1$. Let x be the vector with entries $x_0, \ldots, x_{N-1}$ and M' the vector with entries $M'_{-1}, \ldots, M'_{N-2}$. Then $Vx = M'$. (All of these implicitly depend on N.) We will just give the first row of V^{-1} explicitly. Since V is Vandermonde, we have $\det V = \prod_{0 \leq i < j \leq N-2}(p^j - p^i)\prod_{i=0}^{N-2}(p^i + 1)$. Note that the $(i, 1)$ minor of V is also Vandermonde (after dividing out a factor from each row) on the same elements, except for p^{i-2}, (or -1 when $i = 1$). So we have

$$(V^{-1})_{1,j} = \frac{\pm p^{\frac{(N-2)(N-1)}{2} - (j-2)}}{(p^{j-2}+1)\prod_{\substack{0 \leq i \leq N-2 \\ i \neq j-2}}(p^{j-2} - p^i)} \tag{2.11}$$

for $j > 1$, and

$$(V^{-1})_{1,1} = \frac{\pm p^{\frac{(N-2)(N-1)}{2}}}{\prod_{i=0}^{N-2}(p^i+1)},$$

and in all cases

$$(V^{-1})_{1,j} = O\big(p^{\frac{-j^2+j}{2}}\big).$$

So

$$\begin{aligned} x_0 &= \sum_{j=1}^{N} (V^{-1})_{1,j} M'_{j-2} \\ &= \sum_{j=1}^{N} (V^{-1})_{1,j} M_{j-2} + O\Big(\sum_{j=1}^{N} p^{\frac{-j^2+j}{2}} \big|M_{j-2} - M'_{j-2}\big|\Big) \\ &= \sum_{j=1}^{N} (V^{-1})_{1,j} M_{j-2} + O\big(p^{\frac{(b-3)N}{2}}\big), \end{aligned}$$

meaning that x_0 must be $\lim_{N \to \infty} \sum_{j=1}^{N} (V^{-1})_{1,j} M_{j-2}$ (where the matrix V implicitly depends on N).

Once x_0 is determined, we notice that our equations imply

$$\sum_{s=1}^{\infty} (-1)^{s-1} x_s = -(M_{-1} - x_0) \quad \text{and}$$

$$\sum_{s=1}^{\infty} x_s p^{(s-1)k} = (M_k - x_0)p^{-k},$$

and we have a new system whose constants are still $O(p^{\frac{k^2+bk}{2}})$, and thus we can apply to same reasoning to deduce $x_1, \dots,$ each have at most 1 possible value. ∎

Corollary 2.12. *If $\mu_{\text{odd}}, \mu_{\text{even}}$ are the distributions of $X_{\text{odd}}, X_{\text{even}}$, then any random $\mathbb{F}_p$-vector space X such that for all k,*

$$\mathbb{E}\big(\#\operatorname{Hom}\big(X, \mathbb{F}_p^k\big)\big) = p^{(k^2+k)/2} \prod_{j=1}^{k} (1 + p^{-j})$$

has distribution $\lambda\mu_{\text{odd}} + (1-\lambda)\mu_{\text{even}}$ for some $0 \le \lambda \le 1$.

Proof. Clearly, $\lambda\mu_{\text{odd}} + (1-\lambda)\mu_{\text{even}}$ give distributions with these same moments, and they each assign a different probability to the group being odd rank. Let λ be the probability that X has odd rank. We apply Theorem 2.10 with $x_s = \operatorname{Prob}(X \simeq \mathbb{F}_p^s)$, and $M_{-1} = 1 - 2\lambda$, and $M_k = p^{(k^2+k)/2} \prod_{j=1}^{k} (1 + p^{-j})$, and find that there are unique values x_s satisfying the equations, which proves the corollary. ∎

Open Problem 2.13. Besides the parity of the rank, are there other natural moments that we can consider for random finite $\mathbb{F}_p$-vector spaces, or finite abelian groups more generally, so that with the additional moments we can strengthen uniqueness results to allow for larger growing moments?

In forthcoming work of Nguyen and the author, we prove a generalization of the robust uniqueness result of Theorem 2.9 for random finite abelian groups whose orders are supported on a finite set of primes, with a parity condition on the group.

Theorem 2.14 (Nguyen–Wood, forthcoming). *Let P be a finite set of primes, and let $\mathcal{A}$ be the set of finite abelian P-groups. Let $M_A \in \mathbb{R}$ for each $A \in \mathcal{A}$ such that $M_A = O(|\operatorname{Sym}^2 A|)$. Let a be an integer and $Y, X_1, X_2, \dots$ be random groups in $\mathcal{A}$, either*

(1) *all supported on groups of the form $G \times G$, or*

(2) *all supported on groups of the form $\mathbb{Z}/a\mathbb{Z} \times G \times G$, for G with $aG = 0$.*

If for every $A \in \mathcal{A}$, we have

$$\lim_{n\to\infty} \mathbb{E}\big(\#\operatorname{Sur}(X_n, A)\big) = \mathbb{E}\big(\#\operatorname{Sur}(Y, G)\big) = M_A,$$

then the X_n weakly converge in distribution to Y, i.e., for every $B \in \mathcal{A}$,

$$\lim_{n\to\infty} \operatorname{Prob}(X_n \simeq B) = \operatorname{Prob}(Y \simeq B).$$

2.3. Random finite abelian groups with additional structure

The class groups of Galois fields are not just abelian groups, but are also $\mathbb{Z}[G]$-modules, where G is the Galois group. Let $\mathbb{Z}[G]' = \mathbb{Z}[G, |G|^{-1}]$. Given a number field k and a finite group G, the Cohen–Lenstra–Martinet heuristics [15, 16] give a distribution on $\mathbb{Z}[G]'$-modules, and conjecture that a random G-extension of k has class group who prime-to-$|G|$ part is according to their distribution. Thus for potential number theoretic applications, one would like robust uniqueness for the moment problem for random finite $\mathbb{Z}[G]'$-modules. W. Wang and the author have given such a robust uniqueness result (the stated results are only for particular moments that occur in the Cohen–Lenstra–Martinet heuristics, but the proof works without change for the result given here).

Theorem 2.15 (See [43, THEOREM 6.11]). *Let G be a finite group. Let P be a finite set of primes, none dividing $|G|$, and let $\mathcal{A}$ be the set of finite P-group $\mathbb{Z}[G]'$-modules. Let $M_A \in \mathbb{R}$ for each $A \in \mathcal{A}$ such that $M_A = O(|\wedge^2_{\mathbb{Z}[G]'} A|)$. Let $Y, X_1, X_2, \ldots$ be random $\mathbb{Z}[G]'$-modules in $\mathcal{A}$. If for every $A \in \mathcal{A}$, we have*

$$\lim_{n\to\infty} \mathbb{E}\big(\#\operatorname{Sur}_G(X_n, A)\big) = \mathbb{E}\big(\#\operatorname{Sur}_G(Y, A)\big) = M_A,$$

then the X_n weakly converge in distribution to Y, i.e., for every $B \in \mathcal{A}$,

$$\lim_{n\to\infty} \operatorname{Prob}(X_n \simeq B) = \operatorname{Prob}(Y \simeq B).$$

Theorem 2.15 can be applied to work of Liu, Zureick-Brown, and the author [34], to prove, for every finite group G, a function field analog of the Cohen–Lenstra–Martinet heuristics for G-extensions over $\mathbb{F}_q(t)$, as $q \to \infty$, as we will see below. Wang and the author [43, THEOREM 6.2] have found the moments of the Cohen–Lenstra–Martinet distributions on $\mathbb{Z}[G]'$-modules. In [34], we count and compare components of various Hurwitz schemes to estimate the moments of the class groups of random G-extensions of $\mathbb{F}_q(t)$, and notice those moments, in the limit where $q \to \infty$ and then the degree n of the (reduced) branch locus of the cover (i.e., the size of the radical of the discriminant) goes to infinity, match those predicted by Cohen–Lenstra–Martinet. Theorem 2.15 then tells us that the limiting distribution of these class groups, when q and n both go to ∞, and q is sufficiently large in terms of n, is as predicted by the Cohen–Lenstra–Martinet heuristics. (Some caveats: these results are only in the case of extensions split completely over infinity, are only about the part of the class group that is prime to $|G|$, and q must be taken so that $q-1$ is relatively prime to all the primes in P, and q is prime to $|G|$ and the primes in P. So these results do not see the part of the class group that is affected by roots of unity in $\mathbb{F}_q(t)$ [24, 36].) Precisely, we have the following.

Theorem 2.16 (Corollary of [34, COROLLARY 1.5] and [43, THEOREMS 6.2 AND 6.11]). *Let G be a finite group and P be a finite set of primes that are relatively prime to $|G|$. Let B be a finite abelian P-group $\mathbb{Z}[G]$-module, and $B^G = 0$.*

Let $K_{q,n}$ be a uniform random Galois G-extension K of $\mathbb{F}_q(t)$, split completely over ∞, with the norm of the radical of its discriminant $K/\mathbb{F}_q(t)$ at most q^n. Let $X_{q,n}$ be the product of the Sylow p-subgroups of the class group of $K_{q,n}$ (more precisely, of its ring of integers over $\mathbb{F}_q[t]$) for $p \in P$.

Then if q_n is a sequence of prime powers growing sufficiently fast in n, such that for all n we have that q_n is relatively prime to $|G|$ and all the primes in P and $q_n - 1$ is relatively prime to all the primes in P, then

$$\lim_{n\to\infty} \operatorname{Prob}(X_{q_n,n} \simeq B) = \frac{c}{|B||\operatorname{Aut}_G(B)|},$$

where c is a constant depending on G and P such that the limiting probabilities above sum, over B, to 1.

Proof. By **[34, COROLLARY 1.5]**, for every finite abelian P-group $\mathbb{Z}[G]$-module H with $H^G = 0$, and every $\epsilon > 0$, there is an N_ϵ, such that for $n \geq N_\epsilon$, we have

$$\left| \lim_{\substack{q\to\infty \\ (q,|G|)=1 \\ (q(q-1),p)=1 \text{ for } p \in P}} \mathbb{E}\big(\#\operatorname{Sur}_G(X_{q,n}, H)\big) - |H|^{-1} \right| \leq \epsilon/2.$$

For $n \geq N_\epsilon$, we choose a $Q_{n,\epsilon}$ such that for $q \geq Q_{n,\epsilon}$ (satisfying the conditions above) we have

$$\left|\mathbb{E}\big(\#\operatorname{Sur}_G(X_{q,n}, H)\big) - |H|^{-1}\right| \leq \epsilon.$$

So, if for each n, we consider the smallest ϵ such that $n \geq N_\epsilon$, and then take $q_n \geq Q_{n,\epsilon}$, we have

$$\lim_{n\to\infty} \mathbb{E}\big(\#\operatorname{Sur}_G(X_{q_n,n}, H)\big) = |H|^{-1}.$$

Since $\operatorname{Cl}\mathcal{O}_K$ is trivial and $(|X_{q,n}|, |G|) = 1$, we have $X_{q,n}^G = 0$ **[16, COROLLARY 7.7]**, so if H is such that $H^G \neq 0$, we have $\#\operatorname{Sur}_G(X_{q,n}, H) = 0$. By **[43, THEOREM 6.2]**, we have that these are also the moments of the random $\mathbb{Z}[G]$-module Y such that for any finite abelian P-group $\mathbb{Z}[G]$-module B with $B^G = 0$ (on which Y is supported)

$$\operatorname{Prob}(Y \simeq B) = \frac{c}{|B||\operatorname{Aut}_G(B)|},$$

where c is a constant depending only on P and G. Thus by Theorem 2.15 we conclude the theorem. ■

As described by Wang and the author **[43, SECTIONS 7–8]**, the class groups of non-Galois fields, away from certain bad primes, are also modules for a certain maximal order $\mathfrak{o}$ in a semisimple algebra depending on the Galois group G of the Galois closure over $\mathbb{Q}$ and over the field itself, and moreover are determined (as modules) from the class group of the Galois closure. The algebra $\mathfrak{o}$ can be nontrivial even when the non-Galois field has no automorphism. We can thus show that the Cohen–Lenstra–Martinet heuristics imply conjectures for the distribution of class groups of non-Galois fields. For the part of the class group prime to $|G|$, analogous results to Theorem 2.16 for the non-Galois case then follow formally from Theorem 2.16 and the results in **[43]**. However, for non-Galois extensions, the "bad" primes avoided by the conjectures are not always all primes dividing $|G|$. So at certain "good" primes p dividing $|G|$, we have shown in **[43, THEOREM 8.14]** that the Cohen–Lenstra–Martinet heuristics imply a conjectural distribution on the Sylow p-subgroups of class groups of non-Galois extensions (with Galois closure of group G) as well. See **[43, THEOREM 8.14]** for the relevant notion of good primes. Here we mention a few examples of good primes: 2 for S_3 cubic extensions, 3 for A_4 and S_4 quartic extensions, 2 for quintic D_5 or A_5 extensions. The moment calculations and the unique robustness of the moment problem results in **[43]** include the situations for all good primes for non-Galois extensions, as they are more generally for distributions of modules over maximal orders in semisimple algebras.

In particular, the robust uniqueness result in **[43, THEOREM 6.11]** is a version of Theorem 2.5 in which $\mathbb{Z}[G]'$ is replaced by a maximal order in a semisimple algebra. Sawin **[41, THEOREM 1.3]** has proven a version of Theorem 2.5, in which $\mathbb{Z}[G]'$ is replaced by any associative algebra R such that there are only finitely many isomorphism classes of finite simple R-modules, and Ext^1_R between any two finite R-modules is finite, but one requires the stronger assumption that $M_A = O(|A|^{O(1)})$.

As another example of additional structure, for the Sylow p-subgroups of class groups of quadratic extensions of $\mathbb{F}_q(t)$, Lipnowski, Sawin, and Tsimerman find that these groups have additional structure when $p^n \mid q-1$ **[32]** (where $q-1$ crucially is the number of roots of unity in $\mathbb{F}_q(t)$). This structure involves two pairings and a compatibility relation, and they call a group with such structure a p^n-Bilinearly Enhanced Group. In **[32, SECTION 8]**, they define moments for these enhanced groups and address the uniqueness and robustness aspects of the moments problem in this context. They then apply their moment problem result, along with the homological stability results of Ellenberg, Venkatesh, and Westerland **[19]**, to give a limiting distribution of Sylow p-subgroups of class groups of quadratic extensions of $\mathbb{F}_q(t)$, along with this extra structure.

2.4. Random nonabelian groups

One can also consider random nonabelian groups. A natural such group arising in number theory is $\mathrm{Gal}(K^{un}/K)$, the Galois group of the maximal unramified extension of some random number field K. We have that $\mathrm{Gal}(K^{un}/K) = \pi_1^{\acute{e}t}(\mathrm{Spec}\, \mathcal{O}_K)$ and this group has abelianization Cl_K. The maximal pro-p quotient $G_p(K)$ of $\mathrm{Gal}(K^{un}/K)$ is the p-class tower group of K, the Galois group of K^p, the p-class tower of K.

Boston, Bush, and Hajir **[9,10]**, inspired by the Cohen–Lenstra heuristics, developed heuristics predicting the distribution of $G_p(K)$ for K a random imaginary (respectively, real) quadratic field and p an odd prime. Boston and the author **[11]** found the moments of the conjectural distribution of Boston–Bush–Hajir for imaginary quadratic fields, and prove robust uniqueness for the moment problem for these moments.

Now, as we are considering random profinite groups, the set of isomorphism classes of groups under consideration is uncountable, and we need to be more precise about the measure theory. For a quadratic field K, note that $G_p(K)$ has an action of $\mathbb{Z}/2\mathbb{Z} = \mathrm{Gal}(K/\mathbb{Q})$, by lifting elements to $\mathrm{Gal}(K^p/\mathbb{Q})$ and conjugating. In general, this would only be an outer action, but since p is odd, by the Schur–Zassenhaus theorem we can find a splitting of $\mathrm{Gal}(K^p/\mathbb{Q}) \to \mathrm{Gal}(K/\mathbb{Q})$, and the resulting action of $\mathrm{Gal}(K/\mathbb{Q})$ on $G_p(K)$ does not depend, up to isomorphism, on the choice of splitting. Let $\mathcal{G}_p$ be the set of isomorphism classes of finitely generated pro-p groups with a continuous action of $\mathbb{Z}/2\mathbb{Z}$ (i.e., where morphisms must be equivariant for the $\mathbb{Z}/2\mathbb{Z}$ action). A pro-p group has a canonical lower p-central series defined by $P_0(G) := G$, and for $n \geq 0$, we define $P_{n+1}(G)$ to be the closed subgroup generated by the commutators $[G, P_n(G)]$ and $P_n(G)^p$. A finitely generated pro-p group G then has canonical finite quotients $Q_n(G) := G/P_n(G)$. We let Ω be the σ-algebra on $\mathcal{G}_p$ generated by the sets

$$\{G \mid Q_c(G) \simeq P\},$$

as P ranges over p-groups. We consider all random variables valued in $\mathcal{G}_p$ to be for the σ-algebra Ω. (See [11, SECTION 3] for more details.) With these preliminaries, we can state the uniqueness result of Boston and the author.

Theorem 2.17 ([11, THEOREMS 1.3 AND 1.4]). *Let p be an odd prime. There is a random $X_{\mathrm{BBH}} \in \mathcal{G}_p$ whose distribution is the predicted distribution of Boston–Bush–Hajir for $G_p(K)$ for imaginary quadratic K. For all finite $P \in \mathcal{G}_p$, we have*

$$\mathbb{E}\big(\#\operatorname{Sur}_{\mathbb{Z}/2\mathbb{Z}}(X_{\mathrm{BBH}}, P)\big) = 1.$$

If we have a random $X \in \mathcal{G}_p$ such that, for all finite $P \in \mathcal{G}_p$, we have

$$\mathbb{E}\big(\#\operatorname{Sur}_{\mathbb{Z}/2\mathbb{Z}}(X, P)\big) = 1,$$

then X has the same distribution as X_{BBH}.

The argument in [11] actually shows the following more general uniqueness result.

Theorem 2.18 (see [11, LEMMA 4.7, PROOF OF THEOREM 4.9]). *Let p be a prime and $M_P \in \mathbb{R}$ for each finite $P \in \mathcal{G}_p$. Let $\mathcal{G}_p^c$ the image of $\mathcal{G}_p$ under Q_c. Suppose that for each $c \geq 0$ and each $P \in \mathcal{G}_p^c$, we have*

$$\sum_{Q \in \mathcal{G}_p^c} \frac{M_Q |\operatorname{Sur}_{\mathbb{Z}/2\mathbb{Z}}(Q, P)|}{M_P |\operatorname{Aut}_{\mathbb{Z}/2\mathbb{Z}}(Q)|} < 2. \tag{2.19}$$

If we have random $X, Y \in \mathcal{G}_p$ such that, for all finite $P \in \mathcal{G}_p$, we have

$$\mathbb{E}\big(\#\operatorname{Sur}_{\mathbb{Z}/2\mathbb{Z}}(X, P)\big) = \mathbb{E}\big(\#\operatorname{Sur}_{\mathbb{Z}/2\mathbb{Z}}(Y, P)\big) = M_P,$$

then X and Y have the same distribution.

The challenge in applying Theorem 2.18 is that it is not at all clear how one can evaluate the sum in (2.19). Note that (2.19) is a sum of quite a different flavor than if we were considering abelian groups. In particular, we do not have any convenient enumeration of all finite p-groups, and so evaluating this sum seems to involve a rather difficult group theory problem. In [11], we prove that (2.19) holds when p is odd and all M_P are 1, but by a round-about argument that uses the construction of X_{BBH}.

In [11], we analyze components of certain Hurwitz schemes to prove that in a certain function field analog some of the moments of $G_p(K)$ for quadratic $K/\mathbb{F}_q(t)$ (ramified at infinity) agree with the conjectures of Boston, Bush, and Hajir. In our result [11, THEOREM 1.5], we let the degree of the discriminant go to infinity, and then let q go to infinity, and as in Theorem 2.16 we require that $(q, 2p) = 1$ and $(q-1, p) = 1$. This result involves the generally more difficult limit of letting q go to infinity after the bound on the discriminant, as in the theorem of [19], and we also use the theorem of Ellenberg, Venkatesh, and Westerland [19] on the homological stability of Hurwitz spaces in the proof.

While Theorem 2.17 certainly helps contextualize the result of [11] on function field moments, it does not immediately apply because Theorem 2.17 proves only uniqueness and not robust uniqueness, which would be required in our desired applications, as they involve limits of distributions. In the nonabelian setting, Sawin recently proved a robust uniqueness result however that can be applied.

We will now explain what is required for this robust uniqueness result for nonabelian profinite groups. Fix a finite group Γ, and consider the set $\mathcal{G}$ of isomorphism classes of profinite groups with a continuous action of Γ, finitely many surjections to any finite group, and all continuous finite quotients of order relatively prime to $|\Gamma|$. We will define a topology on $\mathcal{G}$, introduced by Liu, Zureick-Brown, and the author **[34]** (based on **[33]**), and our σ-algebra Ω will be the Borel σ-algebra for that topology. As we used $Q_c(G)$ above, we would like our topology to filter our profinite groups by certain canonical finite quotients. We will make such a canonical finite quotient for any finite set $\mathcal{C}$ of finite groups with an action of Γ (we call these Γ-*groups*). Let $\bar{\mathcal{C}}$ be the closure of $\mathcal{C}$ under taking Γ-equivariant subgroups, products, and quotients. Let $G^{\mathcal{C}}$ be the inverse limit of all quotients of G that are in $\bar{\mathcal{C}}$. Then these $G^{\mathcal{C}}$ (indexed by finite sets $\mathcal{C}$ of finite groups) are the canonical quotients we will use. We then use the topology on $\mathcal{G}$ whose open sets are generated by

$$\{G \mid G^{\mathcal{C}} \simeq H\},$$

where H ranges over all finite Γ-groups. Then Sawin's robust uniqueness result can be stated as follows.

Theorem 2.20 (**[41, THEOREM 1.2]**). *Let Γ be a finite group and $\mathcal{C}$ be a finite set of finite Γ-groups whose orders are relatively prime to $|\Gamma|$. For every finite Γ-group H, let $M_H \in \mathbb{R}$ such that $M_H = O(|H|^{O(1)})$. Let $Y, X_1, X_2, \ldots$ be random groups in $\mathcal{G}$. Assume that for every finite Γ-group H with $H^{\mathcal{C}} = H$, we have*

$$\lim_{n\to\infty} \mathbb{E}\big(\#\operatorname{Sur}_\Gamma(X_n, H)\big) = \mathbb{E}\big(\#\operatorname{Sur}_\Gamma(Y, H)\big).$$

Then for every finite group H with an action of Γ,

$$\lim_{n\to\infty} \operatorname{Prob}(X_n^{\mathcal{C}} \simeq H) = \operatorname{Prob}(Y^{\mathcal{C}} \simeq H). \tag{2.21}$$

Corollary 2.22. *Let Γ be a finite group. For every finite Γ-group H, let $M_H \in \mathbb{R}$ such that $M_H = O(|H|^{O(1)})$. Let $Y, X_1, X_2, \ldots$ be random groups in $\mathcal{G}$. Assume that for every finite Γ-group H, we have*

$$\lim_{n\to\infty} \mathbb{E}\big(\#\operatorname{Sur}_\Gamma(X_n, H)\big) = \mathbb{E}\big(\#\operatorname{Sur}_\Gamma(Y, H)\big).$$

Then the distributions of the X_i weakly converge to the distribution of Y.

Sawin proved Theorem 2.20 in order to apply it to results of Liu, Zureick-Brown, and the author **[34]**. We discussed above that the moments of the class groups of random Γ-extensions $K/\mathbb{F}_q(t)$ were found in the paper **[34]** (as $q \to \infty$), but this paper found, more generally, the moments of $\operatorname{Gal}(K^\#/K)$, where $K^\#$ is the maximal unramified extension of K that is prime to $|\Gamma|$, prime to $q(q-1)$, and split completely at infinity **[34, THEOREM 1.4]**. Moreover, the paper constructed a distribution on random groups with these moments **[34, THEOREMS 1.2 AND 6.2]**. Sawin applied his result **[41, THEOREM 1.1]** to conclude that (in a limit where $q \to \infty$ fast enough compared to n, similar to Theorem 2.16) the random profinite groups $\mathrm{Gal}(K^{\#}/K)$ converge in distribution to the group constructed in [34].

For quadratic extensions $K/\mathbb{F}_q(t)$, we can apply the work of Liu, Zureick-Brown, and the author [34], the homological stability result of Ellenberg, Venkatesh, and Westerland [19], and Sawin's result Theorem 2.20, and find the limiting distribution of the maximal unramified odd extension of K when $q, n \to \infty$ in any way. Let X be a random profinite group with an action of $\mathbb{Z}/2\mathbb{Z}$ with distribution μ_1 from [34, SECTION 4] (with $\Gamma = \mathbb{Z}/2\mathbb{Z}$). The measure of this distribution on basic opens is given explicitly in [34, EQUATION (4.14)]. Let $\mathcal{F}_m$ be the free odd profinite group on m generators, with a $\mathbb{Z}/2\mathbb{Z} = \langle\sigma\rangle$ action inverting each of the generators, and let y_i be independent random elements of $\mathcal{F}_m$ from Haar measure. Then in [34, SECTION 3], it is shown that $\mathcal{F}_m/\langle y_1^{-1}\sigma(y_1), \dots, y_{m+1}^{-1}\sigma(y_{m+1})\rangle$ converge in distribution to X, as $m \to \infty$. Let X_P be the pro-P completion (i.e., the inverse limit of all the finite P-group quotients) of X.

Theorem 2.23. *Let P be a finite set of odd primes. Let $K_{q,n}$ be a uniform random quadratic extension K of $\mathbb{F}_q(t)$, split completely over ∞, with $\mathrm{Nm}\,\mathrm{Disc}\, K/\mathbb{F}_q(t) \leq q^n$. Let K^P be the maximal unramified extension of K, split completely at infinity, all of whose finite subextensions have degree a product of primes in P. Let $X_{q,n} = \mathrm{Gal}(K_{q,n}^P/K_{q,n})$.*

Then as $q, n \to \infty$ in any way such that q is odd, relatively prime to the primes in P, and $q-1$ is relatively prime to the primes in P, then

$$X_{q,n} \text{ converge in distribution to } X_P.$$

Proof. Let $\Gamma = \mathbb{Z}/2\mathbb{Z}$. We follow [34, PROOF OF THEOREM 1.4], but will use the homological stability result of Ellenberg, Venkatesh, and Westerland [19]. Let H be a finite P-group with an action of Γ, such that the coinvariants H_Γ are trivial (note this is equivalent to the admissibility condition in [34], given the condition on P).

Let q be a prime power relatively prime to 2 and all the primes in P, and let $q-1$ be relatively prime to all the primes in P. Let $E_\Gamma(n,q)$ be the set of quadratic extensions $K/\mathbb{F}_q(t)$, split completely at infinity, with $\mathrm{Nm}\,\mathrm{Disc}\, K/\mathbb{F}_q(t) = q^n$. Note n must be even for there to exist such a K (e.g., by the Riemann–Hurwitz formula). Let $G = H \rtimes \Gamma$. Let c be the set of elements of G of order 2, and note by the Schur–Zassenhaus Theorem this is a single conjugacy class of G. Then there are Hurwtiz schemes $\mathsf{Hur}^n_{G,c}$, $\mathsf{Hur}^n_{\Gamma,\Gamma\setminus\{1\}}$ constructed in [34], such that by [34, LEMMA 10.2]

$$[H : H^\Gamma] \sum_{K \in E_\Gamma(n,q)} \#\mathrm{Sur}_\Gamma\big(\mathrm{Gal}(K^P/K), H\big) = \#\mathsf{Hur}^n_{G,c}(\mathbb{F}_q)$$

and

$$\#E_\Gamma(n,q) = \#\mathsf{Hur}^n_{\Gamma,\Gamma\setminus\{1\}}(\mathbb{F}_q).$$

For n sufficiently large given G, by [34, THEOREM 10.4], we have that $\#\mathsf{Hur}^n_{G,c}$ and $\#\mathsf{Hur}^n_{\Gamma,\Gamma\setminus\{1\}}$ have the same number, z_n, of Frobenius fixed components over $\bar{\mathbb{F}}_q$. Moreover, z_n is positive for even n because we know $\mathbb{F}_q(t)$ has quadratic extensions split completely at infinity and so $\#\mathrm{Hur}^n_{\Gamma,\Gamma\setminus\{1\}}$ has $\mathbb{F}_q$-points. By the Grothendieck–Lefschetz trace formula, we have

$$\left|\#\mathrm{Hur}^n_{G,c}(\mathbb{F}_q) - z_n q^n\right| \leq \sum_{j=0}^{2n-1} q^{j/2} \dim H^j_{c,\acute{e}t}\big((\mathrm{Hur}^n_{G,c})_{\bar{\mathbb{F}}_q}, \mathbb{Q}_\ell\big),$$

for some ℓ (**[34, LEMMA 10.3]** tells us $(\mathrm{Hur}^n_{G,c})_{\bar{\mathbb{F}}_q}$ is smooth and n-dimensional). By **[34, LEMMA 10.3]**, we then have

$$\left|\#\mathrm{Hur}^n_{G,c}(\mathbb{F}_q) - z_n q^n\right| \leq \sum_{j=0}^{2n-1} q^{j/2} \dim H^{2n-j}\big((\mathrm{Hur}^n_{G,c})_{\mathbb{C}}, \mathbb{Q}\big).$$

By **[19, THEOREM 6.1, PROPOSITION 2.5]** (their $\mathrm{CHur}^c_{G,n}$ is the topological space of the analytic topology of our $(\mathrm{Hur}^n_{G,c})_{\mathbb{C}}$ by **[34, SECTION 11.3]**, and we can easily check their nonsplitting condition is satisfied here), there exist constants C and D, depending on G, such that $\dim H^k((\mathrm{Hur}^n_{G,c})_{\mathbb{C}}, \mathbb{Q}) \leq CD^k$. Thus we have

$$\left|\#\mathrm{Hur}^n_{G,c}(\mathbb{F}_q) - z_n q^n\right| \leq \sum_{j=0}^{2n-1} q^{j/2} C D^{2n-j}.$$

For $q \geq D^4$, we have

$$\left|\#\mathrm{Hur}^n_{G,c}(\mathbb{F}_q) - z_n q^n\right| \leq \sum_{j=0}^{2n-1} C q^{n/2+j/4} \leq \frac{2Cq^{n-1/4}}{1-q^{-1/4}}.$$

By the same argument, we have the same inequalities for $\mathrm{Hur}^n_{\Gamma,\Gamma\setminus\{1\}}$ Summing over even $n \leq N$, we conclude that if $q, n \to \infty$ in any way, we have

$$\frac{1}{\#E_\Gamma(n,q)} \sum_{K \in E_\Gamma(n,q)} \#\operatorname{Sur}_\Gamma\big(\mathrm{Gal}(K^P/K), H\big) \to \big[H : H^\Gamma\big]^{-1}.$$

By **[34, THEOREM 6.2]**, we see these are exactly the moments of the random Γ-group X_P described above. Thus applying Theorem 2.20, we conclude the result. ■

The methods of the paper **[34]** can find the moments of the maximal unramified extension of a random Γ extension $K/\mathbb{F}_q(t)$ even when we allow parts not prime to $q-1$, but the obstruction to proceeding is that there is no candidate conjectural random group with those moments. This brings us to the first case in this story when there was not an already known conjectural distribution that one was trying to show some distributions from number theory converged to. So we naturally turn to the existence and construction aspects of the moment problem.

All of the questions on moment problems for random groups discussed above have been reducible to questions of a countable list of linear equations in a countable number of variables, and whether they have a unique solution. The equations and variables are parametrized by groups, and the coefficients are given by group theoretic quantities (numbers of surjective homomorphisms). In Theorem 2.10, we made these equations quite explicit, and inverted the implicit infinite matrices by truncating them to finite matrices that we could explicitly invert. This is an approach that works well when the groups involved are $\mathbb{F}_p$-vector spaces, but it becomes less and less tractable as the groups get more complicated. For finite abelian groups, one relies on the classification of the groups and the ability to write a formula for the number of surjections from one to another. For nonabelian groups, there is no reasonable formulaic parametrization of the groups and their numbers of surjections. Theorem 2.20 is proved by a localization process that reduces the question to one only involving a smaller list of groups that can be classified and for which the number of surjections can be simply expressed.

All of these proofs of uniqueness, at least in principle, give some expression for the (only possible) solutions to these systems of equations. What then remains of the existence question? (1) The solutions must be nonnegative in order to describe a measure. (2) The determined values must further be shown to satisfy the equations. (3) In some cases, the solutions must be compatible in order to describe a measure.

We elaborate a bit on what these remaining problems are like. First we consider (1). In Theorem 2.10, we find an expression for x_0, the probability of the trivial group, as

$$\lim_{N\to\infty}\sum_{j=1}^{N}(V^{-1})_{1,j}M_{j-2},$$

where the M_j are the given moments, and the coefficients of the inverse matrix are given explicitly in (2.11). The other x_i are given similarly, with modified values of M_j. It is not clear whether one should expect a simple criterion for whether these values are nonnegative, but it seems conceivable that for a particular nice family of M_j of interest that one could, with work, prove the values of the x_i that are determined are indeed positive. Addressing (2), one could hope to prove for sufficiently bounded moments that these determined values satisfied the equations. We cannot see problem (3) above when the random groups are just $\mathbb{F}_p$-vector spaces, but even in the case of finite abelian p-groups, some approaches prove that the distribution on groups mod p is determined, and then that the distribution on groups mod p^2 is determined, etc. One can see this feature explicitly in the statement of Theorem 2.20. So, in such cases, to prove existence, one would have to check that the determined values were compatible and could be pieced together into a probability distribution.

The *construction* problem, which we have described above as giving *useful* formulas for the distribution, now turns on what useful means. The formulas for the distributions that arise from the uniqueness proofs above are generally infinite sums. One might not expect to solve this for general moments, but perhaps only for specific moments that arise in particular problems. We propose as one test of usefulness—can one detect if the distribution assigns value 0 to any particular basic open set? Note that the distributions on finite abelian groups we have seen above in Theorems 2.3 and 2.4 and in (2.7) all have this property. The distributions on non-abelian groups we have discussed, including those of Boston, Bush, and Hajir, and Liu, Zureick-Brown, and the author also have this property (see [11, LEMMA 4.8], [34, THEOREM 4.12]). Other tests for usefulness may come from the features of the desired application, but we emphasize that there can be a significant gap between having a formula for a distribution as an infinite sum, and being able to use that formula in practice to answer questions about the distribution.

We mention briefly forthcoming work of Sawin and the author on the moment problem for profinite groups. This work will strengthen Theorem 2.20 so that larger growing moments M_H are allowed, up to the point where the statement is no longer true (e.g., because of the example (2.7)). We also prove a general existence result addressing the problems (2) and (3) mentioned above. Our first applications are to problems where moments are known but the distribution is not known. The first of these applications is mentioned above, and is for the distribution of class groups or their nonabelian analogs, or order not prime to roots of unity in the base field $\mathbb{F}_q(t)$. The second is to the distribution of the profinite completion of random 3-manifolds (from random Heegaard splittings), as introduced by Dunfield and Thurston [18]. In these applications, we also solve the construction problem, e.g., we can describe explicitly the support of the limiting distribution, and we can use our formulas for the limiting distribution to answer open questions about the distributions from number theory and topology. Moreover, the 3-manifold application requires addressing situations where uniqueness does not actually hold, and we recover uniqueness with additional parity hypotheses, such as in Theorems 2.9, 2.10, and 2.14 above.

3. UNIVERSALITY

A central concept in probability theory is that of *universality*, which describes the ubiquitous phenomena that many input independent distributions can be combined to make an output distribution, and as the number of input distributions goes to infinity, the output distribution becomes quite insensitive to the input distributions. The first and most well-known example is the Central Limit Theorem.

Theorem 3.1 (Central Limit Theorem). *Let $X_1, X_2, \ldots$ be independent, identically distributed random real numbers with finite mean $\mu = \mathbb{E}(X_i)$ and finite variance σ^2. Then as $n \to \infty$,*

$$\sqrt{n}\left(\frac{X_1 + \cdots + X_n}{n} - \mu\right)$$

converge in distribution to the normal distribution with mean 0 *and variance σ^2.*

Here the X_i are the input distributions, and their normalized sum is the output distribution, and we see that the output, asymptotically, only depends on the variance of the input distributions. The Central Limit Theorem is the tip of the iceberg, and probability theory is filled with further examples of this kind of phenomenon.

Here we discuss a somewhat newer line of inquiry, namely universality for random groups. In this case, the output distribution should be a random group, and the random group is somehow built out of the input distributions. One natural way to obtain such a random group is to start with a fixed random group F and take the quotient by random elements of F that we call relators. If F is a free abelian group, $F = \mathbb{Z}^n$, and we collect m random relators as the columns of a matrix M, then the quotient of F by our relators is the cokernel cok M (by definition of the cokernel). This shows that questions about random abelian groups built in this way can be rephrased as questions about cokernels of random integral matrices.

3.1. Random finite abelian groups

The simplest sort of groups to consider, as in our discussion above on the moment problem, are $\mathbb{F}_p$-vector spaces. Let $F = \mathbb{F}_p^n$. If M is an $n \times m$ matrix with coefficients in $\mathbb{F}_p$, then the quotient of F by the columns of M, i.e., $\operatorname{cok} M$, has rank equal to $n - \operatorname{rank} M = \operatorname{corank} M$. Hence we translate questions about random $\mathbb{F}_p$-vector spaces into questions about ranks of random matrices over $\mathbb{F}_p$. We note here that determining the rank distribution of random matrices over $\mathbb{F}_p$ is a simple exercise if the matrices are uniformly distributed. The entire interest here is when the matrix coefficients (still independent) are drawn from a wide range of distributions, and in particular if there is a resulting universality in the distribution of the ranks. There is a long history of work on this question. Kozlov [30] showed a universality result for the ranks over $\mathbb{F}_2$, and Kovalenko and Levitskaja [29] showed a version over $\mathbb{F}_p$. Both works require that the matrix entries take all possible values with positive probability. Charlap, Rees, and Robbins [12] only determined the probability that a square matrix is invertible, but allowed more general matrix entries. Balakin [2], Blömer, Karp, and Welzl [8], and Cooper [17] determined the ranks for sparser matrices, with entries uniformly distributed over nonzero values. The most general result we know is the following result of Nguyen and the author.

Theorem 3.2 (Corollary of [38, THEOREM 4.1]). *Let p be a prime. Let u be a nonnegative integer and α_n a function of integers n such that for any constant $\Delta > 0$, for n sufficiently large we have $\alpha_n \geq \Delta(\log n)/n$. For every positive integer n, let M_n be a random $n \times (n+u)$ matrix with independent entries $\xi_{i,j,n} \in \mathbb{F}_p$ that satisfy*

$$\max_{a \in \mathbb{F}_p} \operatorname{Prob}(\xi_{i,j,n} = a) \leq 1 - \alpha_n$$

for every i, j, n. Then for every $r \geq 0$,

$$\lim_{n\to\infty} \operatorname{Prob}(\operatorname{cok} M_n \simeq \mathbb{F}_p^r) = \lim_{n\to\infty} \operatorname{Prob}(\operatorname{rank} M_n = n - r) = p^{-r(r+u)} \frac{\prod_{j=r+u+1}^{\infty}(1-p^{-j})}{\prod_{j=1}^{r}(1-p^{-j})}.$$

We see that there are separate universality classes for different u, i.e., different numbers of relations compared to the number of generators, but for fixed u a wide range of entry distributions all give random groups in the same universality class. Note that Theorem 3.2 does not require the matrix entries to be identically distributed. It also allows the matrices to be quite sparse. If $\operatorname{Prob}(\xi_{i,j,n} = 0) = 1 - (\log n)/n$, the matrix would have a row of all zeroes with (asymptotically) positive probability, and this crosses a threshold for the behavior of the random matrix, similar to the well-known threshold for the behavior of random graphs and sparse random matrices in other contexts.

Open Problem 3.3. Lower the bound on α_n in Theorem 3.2, as close to the $(\log n)/n$ threshold as possible (and similarly for Theorems 3.4 and 3.6 below).

We next consider finite abelian p-groups, and now $F = \mathbb{Z}_p^n$ (and $\mathbb{Z}_p$ are the p-adic integers). If we form a random group by taking $n + u$ random relators, then the group is $\operatorname{cok} M$, where M is the matrix whose columns are the relations. Indeed, Theorem 3.2 is actually a corollary of the following.

Theorem 3.4 ([38, THEOREM 4.1]). *Let p be a prime. Let u be a nonnegative integer and α_n a function of integers n such that for any constant $\Delta > 0$, for n sufficiently large we have $\alpha_n \geq \Delta(\log n)/n$. For every positive integer n, let M_n be a $n \times (n+u)$ matrix with independent entries $\xi_{i,j,n} \in \mathbb{Z}_p$ that satisfy*

$$\max_{a \in \mathbb{F}_p} \operatorname{Prob}\big(\xi_{i,j,n} \equiv a \ (\mathrm{mod}\ p)\big) \leq 1 - \alpha_n$$

for every i, j, n. Then for every abelian p-group A, we have

$$\lim_{n\to\infty} \mathbb{P}\big(\operatorname{cok}(M_n) \simeq A\big) = \frac{1}{|A|^u |\operatorname{Aut}(A)|} \prod_{k=1}^{\infty} (1 - p^{-k-u}).$$

The proof of Theorem 3.4 builds heavily on the method in [47], but extends the statement to include the sparse regime.

Proof of Theorem 3.2. The probabilities in Theorem 3.4 sum over A to 1 to give a probability distribution for each u [47, LEMMA 3.2]. Thus it follows from Fatou's Lemma that we can simply add up the probabilities from Theorem 3.4 for groups of rank r to obtain the limiting probabilities in Theorem 3.2. This is done in [15, COROLLARY 6.5]. ■

When $u = 0$, the distribution in Theorem 3.4 is the Cohen–Lenstra distribution of X_{CL} we have mentioned above, and when $u = 1$ it is the distribution conjectured by Cohen and Lenstra [15] for the Sylow p-subgroups of class groups of real random quadratic fields (for p odd). Let us now put these class groups in the context of random matrices, following Venkatesh and Ellenberg [42, SECTION 4.1]. Let $K = \mathbb{Q}(\sqrt{D})$ for some negative (resp. positive) square-free integer D, and S be any finite set of primes of K that generate $\mathrm{Cl}(K)$. We write $\mathcal{O}_S^*$ for the S-units in the integers $\mathcal{O}_K$, and I_K^S for the abelian group of fractional ideals generated by the elements of S. Then

$$\mathrm{Cl}(K) = \operatorname{cok}(\mathcal{O}_S^* \to I_K^S), \tag{3.5}$$

where the map takes α to the ideal (α). So the Sylow p-subgroup of $\mathrm{Cl}(K)$ is $\operatorname{cok}(\mathcal{O}_S^* \otimes_{\mathbb{Z}} \mathbb{Z}_p \to I_K^S \otimes_{\mathbb{Z}} \mathbb{Z}_p)$. Since I_K^S and $\mathcal{O}_S^*$ are both abelian groups of rank $|S|$ (resp. of ranks $|S|$ and $|S|+1$), we have written the Sylow p-subgroup of $\mathrm{Cl}(K)$ as a cokernel of a p-adic $n \times n$ matrix R_D (resp. $n \times (n+1)$ matrix). One can now view the Cohen–Lenstra conjecture for class groups of quadratic fields as asking whether universality of Theorem 3.4 extends to the random matrix R_D for random D. This point of view was a motivation for the paper [47].

Now we consider random finite abelian groups more generally. For a finite set P of primes, considering finite abelian P-groups turns out to be only notationally more challenging than considering abelian p-groups, and indeed [38, THEOREM 4.1] is proven in this slightly more general context. However, considering all primes at once is quite a bit more of a challenge, because there will always be primes large compared to n. Nguyen and the author develop a method to handle large primes (compared to n) and we prove the following.

Theorem 3.6 ([38, THEOREM 1.1]). *For integers $n, u \geq 0$, let $M_{n\times(n+u)}$ be an integral $n \times (n+u)$ matrix with entries i.i.d. copies of a random integer ξ_n, with*

$$\limsup_{p \text{ prime}} \max_{a\in\mathbb{F}_p} \operatorname{Prob}(\xi_n \equiv a \pmod p) \leq 1 - n^{-1+\epsilon}$$

and $|\xi_n| \leq n^T$ for any fixed parameters $0 < \epsilon < 1$ and $T > 0$ not depending on n. For any fixed finite abelian group A and $u \geq 0$,

$$\lim_{n\to\infty} \mathbb{P}\left(\operatorname{cok}(M_{n\times(n+u)}) \simeq A\right) = \frac{1}{|A|^u|\operatorname{Aut}(A)|} \prod_{k=u+1}^{\infty} \zeta(k)^{-1}, \tag{3.7}$$

where $\zeta(s)$ is the Riemann zeta function.

Note this theorem has nice corollaries like the probability that a random map as in Theorem 3.6 (for $u = 1$) from $\mathbb{Z}^{n+1} \to \mathbb{Z}^n$ is surjective is $\prod_{k=2}^{\infty} \zeta(k)^{-1} \approx 0.4358$. As in the proof of Theorem 3.2, one can obtain other probabilities as corollaries, such as (for $u \geq 1$) the probability that $\operatorname{cok}(M_{n\times(n+u)})$ is cyclic. However, when $u = 0$, the probabilities in Theorem 3.6 are all 0 (from the $\zeta(1)^{-1}$ term), so this theorem tells us little about the distribution of random abelian groups from n generators and n random relations. In [38, THEOREM 1.2] we do find the probability that $\operatorname{cok}(M_{n\times n})$ is cyclic, and in [38, THEOREM 2.4] more generally give the probability that $\operatorname{cok}(M_{n\times n})$ is any set of groups $\{A \times C \mid C \text{ cyclic}, p \nmid |C| \text{ for } 1 < p < Y\}$. However, we are not able to distinguish a factor of $\mathbb{Z}/p\mathbb{Z}$ for large p from one of $\mathbb{Z}/p^2\mathbb{Z}$, for example.

Open Problem 3.8. Find

$$\lim_{n\to\infty} \operatorname{Prob}(|\operatorname{cok} M_{n\times n}| \text{ is square-free}) = \lim_{n\to\infty} \operatorname{Prob}(|\det M_{n\times n}| \text{ is square-free}).$$

Note that finding the probability that a polynomial takes square-free values on even the nicest distributions of integers is difficult and generally open, but there has been some progress for certain discriminant polynomials by Bhargava [4] and Bhargava, Shankar, and Wang [6]

Open Problem 3.9. Extend Theorem 3.6 to nonidentical entries.

The first connection of the Cohen–Lenstra heuristics to random matrices came from work of Friedman and Washinton [23]. They considered the analog of the Cohen–Lenstra conjectures for quadratic extensions of $\mathbb{F}_q(t)$. In this case one can also describe the Sylow p-subgroup of the class group of K (or more precisely of the Pic^0) as the cokernel of a certain random $2g \times 2g$ random matrix $I - F$ over $\mathbb{Z}_p$, where I is the identity matrix, and F describes the action of Frobenius on the p-adic Tate module of the curve corresponding to K [23, PROPOSITION 2]. (Here p is *not* the characteristic of $\mathbb{F}_q$.) Friedman and Washington showed that the cokernels of random matrices from the (additive) Haar measure on $n \times n$ matrices over $\mathbb{Z}_p$, as $n \to \infty$, approach the Cohen–Lenstra distribution. However, the matrix F above is not just any matrix; since it acts on the Weil pairing by scaling the pairing by q, it lies in a generalized symplectic coset $\operatorname{GSp}_{2g}^q(\mathbb{Z}_p)$ ($\operatorname{GSp}_{2g}^q(\mathbb{Z}_p)$ is the coset of matrices M such that $M^tJM = qJ$, where J is an invertible alternating matrix, and in particular the $M \in \mathrm{GSp}_{2g}^q(\mathbb{Z}_p)$ are invertible). Friedman and Washington prove that the cokernels of random matrices $I - M$, where M is random from the (multiplicative) Haar measure on $\mathrm{GL}_{2g}(\mathbb{Z}_p)$, as $g \to \infty$, approach the Cohen–Lenstra distribution [23, SECTION 4]. Eventually, it was understood that this also holds for $I - M$, where M is random from the Haar induced measure on $\mathrm{GSp}_{2g}^q(\mathbb{Z}_p)$ and $\gcd(q-1, p) = 1$. (This is not clearly stated in the literature, but follows from work of Achter [1] and Ellenberg and Venkatesh [19] in a very round about way, as outlined by Garton [24, P.153].)

We can view these results as additional examples of random matrices in the universality class of Theorem 3.4, even though the matrices do not have independent entries, and also come from very special distributions. Another example that would fit into this category is Mészáros's theorem [37, THEOREM 1] that says that the Laplacians of uniform random d-regular directed graphs, for any $d \geq 3$, also have these limiting cokernel distributions. It is a very interesting problem to extend this universality to matrices with dependent entries but for broader classes of random matrices, where the degrees of freedom in choosing the distribution of random matrices is large. As an example, in [38, THEOREM 1.6], we extend universality to Laplacians of random matrices with independent entries (so matrices whose off-diagonal entries are independent and whose columns sum to 0), which includes Laplacians of directed Erdős–Rényi random graphs. However, this is a very special kind of dependency among entries for which the methods are well-suited.

Open Problem 3.10. Extend Theorem 3.4 to more classes of matrices with dependent entries.

Open Problem 3.11. Give a unified proof that multiple special classes of random matrices are in the universality class of Theorem 3.4.

3.2. Random finite abelian groups with additional structure

Of course, if the entries of the random matrices have too much dependence in some particular way, their cokernels may land in another universality class. For example, for symmetric matrices the author has proved the following.

Theorem 3.12 ([44]). *Let p be a prime and $0 < \alpha < 1$. For every positive integer n, let M_n be a symmetric random $n \times n$ matrix with independent entries $\xi_{i,j,n} \in \mathbb{Z}_p$ for $i \geq j$ that satisfy*

$$\max_{a \in \mathbb{F}_p} \mathrm{Prob}\big(\xi_{i,j,n} \equiv a \pmod{p}\big) \leq 1 - \alpha$$

for every i, j, n. Then for every abelian P-group A, we have

$$\lim_{n \to \infty} \mathbb{P}\big(\mathrm{cok}(M_n)_P \simeq A\big) = \frac{\#\{\text{symmetric, bilinear, perfect } \phi : A \times A \to \mathbb{C}^*\}}{|A||\mathrm{Aut}(A)|} \prod_{k=0}^{\infty} \big(1 - p^{-2k-1}\big).$$

(Note the number of pairings can be described explicitly in terms of the partition corresponding to the group A [44, EQUATION (2)].)

Proof. Theorem 6.1 in [44] gives the moments, and then Theorem 2.5 shows they determine a unique distribution, and [13, THEOREM 2] gives formulas for the distribution when M_n is taken from Haar measure, as in [44, COROLLARY 9.2]. ■

Nguyen and the author have forthcoming work in which we extend Theorem 3.12 to integer matrices (and all primes), analogous our results on $n \times n$ matrices over $\mathbb{Z}$ described above (including obtaining the probability that the cokernel is cyclic).

One way of understanding why some random groups are in a different universality class is that the groups may be naturally coming with further structure than just group structure. For example, the cokernel of a symmetric matrix over the integers (or $\mathbb{Z}_p$) [13, SECTION 1.1] comes with a natural symmetric bilinear pairing. Clancy, Leake, and Payne [14] suggested that for random graphs, the cokernels of the graph Laplacian, along with their symmetric pairing, should be distributed proportionally to $|A|^{-1}|\mathrm{Aut}(A, \langle \cdot, \cdot \rangle)|^{-1}$. If we sum these expressions over isomorphism classes of pairings for a fixed group, we exactly obtain the probabilities for groups in Theorem 3.12 (see [44, COROLLARY 9.2]). This reflects an important part of the philosophy of the Cohen–Lenstra–Martinet heuristics—that the natural distributions on algebraic objects must take into account all of the structure of the objects. For example, when considering class groups of Galois number fields with Galois group G, we consider the class group not just as a group but rather as a G-module, and the predicted probabilities for a particular G-module involve the number of automorphisms of the G-module (as a G-module). Since the distributions that arise from universality theorems are certainly natural, we would expect them to share this sensitivity to extra structure, and thus it makes sense that cokernels of symmetric matrices, since as such they have natural symmetric pairings, should be distributed in a distribution that sees those pairings.

Open Problem 3.13. Prove that the cokernels of random symmetric matrices as in Theorem 3.12, along with their pairings, are distributed as suggested by Clancy, Leake, and Payne [14, SECTION 4]. One might naturally use moments of groups with pairings, and the corresponding moment problem, as in [32, SECTION 8].

There are a few other classes of random groups that we know in this universality class. The result [44, THEOREM 1.1] extends Theorem 3.12 to cokernels of Erdős–Rényi random graph Laplacians, also known as sandpile groups or Jacobians of the graphs. Mészáros [37, THEOREM 1.2] extends Theorem 3.12 to sandpile groups of d-regular graphs for $d \geq 3$ (unless d is even and $p = 2$, in which case a different distribution arises, likely reflecting further structure of the pairing). Dunfield and Thurston [18, SECTION 8.7] show that the homology $H_1(M, \mathbb{F}_p)$ for a 3-manifold from a random Heegaard splitting of genus g as $g \to \infty$ approaches the universal distribution of Theorem 3.12, or more precisely the pushforward of that distribution to elementary abelian p-groups under the map $A \mapsto A/pA$. (See [44, COROLLARY 9.4] to see that this is indeed the pushforward.) Forthcoming work of Sawin and the author finds the distribution more generally of $H_1(M, \mathbb{Z}_p)$, along with the torsion linking pairing, of these random 3-manifolds, and finds that it is in the natural distribution suggested by Clancy, Leake, and Payne [14, SECTION 4]. So the presence of the symmetric pairing from the torsion linking pairing explains why the homology of random 3-manifolds appears in this universality class.

Open Problem 3.14. Prove that the sandpile groups of Erdős–Rényi random graphs (or uniform random d-regular graphs) along with their pairings, are distributed as suggested by Clancy, Leake, and Payne [14, SECTION 4].

There are, however, many more algebraic structures that are important in arithmetic statistics and other fields whose universality classes should be studied, such as random abelian groups with an action of a group, or random modules.

Open Problem 3.15. Prove an analog of Theorem 3.4 for $\mathbb{Z}_p[G]$-modules for a finite group G (with $p \nmid |G|$). More generally (as would be related to the Cohen–Lenstra–Martinet heuristics for non-Galois fields, see [43, SECTIONS 7–8]), prove an analog of Theorem 3.4 for random $\mathfrak{o}$-modules, where $\mathfrak{o}$ is a maximal order (over $\mathbb{Z}_p$) in a semisimple $\mathbb{Q}_p$-algebra.

Note that the reduction of Problem 3.15 mod p is a question about matrices over finite fields. So part of solving the above will include generalizing Theorem 3.2 from $\mathbb{F}_p$ to general finite fields $\mathbb{F}_q$. In this more general case, the requirement that the $\xi_{i,j,n}$ are not concentrated at a single point is not sufficient, and must be replaced with something like $\xi_{i,j,n}$ not concentrated on a translate of a subfield. Kahn and Komlós [27] have shown universality of the singularity probability of a random $n \times n$ matrix over $\mathbb{F}_q$ under such a condition.

Open Problem 3.16. For $\mathfrak{o}$ a maximal order (over $\mathbb{Z}_p$) in a semisimple $\mathbb{Q}_p$-algebra, with an order two automorphism σ, such as $\mathfrak{o}$ being the ring of integers of the unramified quadratic extension of $\mathbb{Q}_p$, or $\mathfrak{o} = \mathbb{Z}_p \times \mathbb{Z}_p$, prove an analog of Theorem 3.12 for random σ-Hermitian matrices (i.e. M such that $\sigma(M) = M^t$).

3.3. Random nonabelian groups

We now turn to universality questions for nonabelian random groups, which are largely unstudied, but we expect contain much potential. One naturally starts with a free group (or free profinite group) F_n and takes the quotient by independent random relations in some way that involves many independent choices for each relations. As $n \to \infty$, one hopes that the limiting distribution is somewhat insensitive to the distribution from which the relations are chosen. The first stumbling block when considering such questions is that it is less clear how to take a random relation built up from many independent choices. When the relation was in $\mathbb{F}_p^n$ or $\mathbb{Z}_p^n$, we could just take each coordinate independently. However, if F_n is the free group (or free profinite group) on n generators, there are not analogous coordinates in F_n. In the case of random nilpotent groups, one might consider using Mal'cev coordinates. Another way to characterize the probability measures on $\mathbb{Z}_p^n$ from which we drew relations above, e.g., in Theorem 3.4, is that they are not concentrated at a point in any finite simple quotient, so it may be interesting to consider the nonabelian version of that condition. While it is not so clear what the parameters for the universality class should be, one has a natural target for the universal distribution from a result of Liu and the author on the quotient of the free group by random relations. Let $\mathcal{G}$ be the set of isomorphism classes of profinite groups with finitely many surjections to any finite group. (This is $\mathcal{G}$ from Section 2.4 with $\Gamma = 1$, and we consider the same topology on it as defined there.)

Theorem 3.17 ([33, THEOREM 1.1]). *For every integer u, there is a random group X_u in $\mathcal{G}$ whose measure is described explicitly on each basic open* [33, EQUATION (3.2)]. *If F_n is the free profinite group on n generators, and r_i are independent random elements of F_n drawn from Haar measure, then as $n \to \infty$ the quotients*

$$F_n/\langle r_1, \ldots, r_{n+u}\rangle \quad \textit{weakly converge in distribution to } X_u.$$

As in [33, SECTION 14], one can consider usual (not profinite) free group F_n and take random relations obtained from a random walk on F_n. However, as the length of the random walk goes to infinity, these relation become equidistributed with respect to Haar measure, and so this is not really a new example for the universality class.

Open Problem 3.18. Find some more general hypotheses for a distribution on F_n from which one can draw independent relations so that Theorem 3.17 still holds.

While it would be nice to have hypotheses that allow a wide range of distributions, i.e., a universality theorem, it would even be interesting to find other specific random groups converging to the distributions X_u. We give one example here, which is a nonabelian analog of the result of Friedman and Washington on cokernels of $I - M$, where M is random from the Haar measure on $\mathrm{GL}_n(\mathbb{Z}_p)$.

Theorem 3.19. *Let F_n be the free profinite group on n generators, and let $\mathrm{Aut}(F_n)$ be the group of (continuous) automorphisms of F_n, which is a profinite group* [40, COROLLARY 4.4.4]. *Let $I \in \mathrm{Aut}(F_n)$ be the identity and let α_n be a random element of $\mathrm{Aut}(F_n)$ with respect to Haar measure. Then, as $n \to \infty$,*

$$F_n/\big\langle \alpha_n(x)x^{-1} \mid x \in F_n\big\rangle \quad \textit{weakly converge in distribution to } X_0,$$

(where X_0 is defined as in Theorem 3.17).

We will compute the moments of these random groups, and then apply Corollary 2.22 from Sawin's result on the moment problem. To do that, we first need the moments of X_0. While it is an easy to see that for independent Haar relations r_i, we have

$$\lim_{n\to\infty} \mathbb{E}\big(\#\operatorname{Sur}\big(F_n/\langle r_1, \ldots, r_{n+u}\rangle, A\big)\big) = 1,$$

it does require some argument to interchange the limit in n and the expectation and obtain these same moments for X_0.

Lemma 3.20. *Let X_0 be defined as in Theorem 3.17. Then for any finite group H, we have*

$$\mathbb{E}\big(\operatorname{Sur}(X_0, H)\big) = 1.$$

Proof. We follow the strategy of [34, THEOREM 6.2] adapted to our situation. Let F_n be the free profinite group on n generators and let Z_n be the random profinite group $F_n/\langle r_1, \ldots, r_n\rangle$, where the r_i are random elements of F_n from Haar measure. For any positive integer ℓ, let $\mathcal{C}_\ell$ be the set of finite groups of order at most ℓ. We consider the following function defined for any positive integer ℓ and any finite group G of level with $G^{\mathcal{C}_\ell} \simeq G$,

$$f_n(G,\ell) = \mathbb{E}\big(|\mathrm{Sur}(Z_n, H)| \times \mathbb{1}_{{Z_n}^{\mathcal{C}_\ell} \simeq G}\big),$$

where $\mathbb{1}_{{Z_n}^{\mathcal{C}_\ell} \simeq G}$ is the indicator function of ${Z_n}^{\mathcal{C}_\ell} \simeq G$. We let $\pi_{F_n}\colon F_n \to (F_n)^{\mathcal{C}_\ell}$ and $\pi_H :$ $H \to H^{\mathcal{C}_\ell}$ be the natural quotient maps. Each $\phi \in \mathrm{Sur}(F_n, H)$ induces a map $\overline{\phi} \in \mathrm{Sur}((F_n)^{\mathcal{C}_\ell}, H^{\mathcal{C}_\ell})$. By the definition of random group Z_n, we have

$$\begin{aligned} &\mathbb{E}\big(|\mathrm{Sur}(Z_n, H)| \times \mathbb{1}_{{Z_n}^{\mathcal{C}_\ell} \simeq G}\big) \\ &\quad = \sum_{\phi \in \mathrm{Sur}(F_n,H)} \mathrm{Prob}\big(r_1, \dots, r_n \in \ker\phi \text{ and } (F_n)^{\mathcal{C}_\ell}/\langle \pi_{F_n}(r_1), \dots, \pi_{F_n}(r_n)\rangle \simeq G\big). \end{aligned} \tag{3.21}$$

Given $\phi \in \mathrm{Sur}_\Gamma(F_n, H)$ and $y_1, \dots, y_n \in \ker\overline{\phi}$, we have that

$$\mathrm{Prob}\big(r_1, \dots, r_n \in \ker\phi \mid \pi_{F_n}(r_i) = y_i \text{ for all } i\big) = \frac{|H^{\mathcal{C}_\ell}|^n}{|H|^n}.$$

This follows from the straightforward calculation that

$$\big|\pi_{F_n}^{-1}(y_i) \cap \ker\phi\big| = |F_n|\,|H^{\mathcal{C}_\ell}|/\big(|F_n^{\mathcal{C}_\ell}|\,|H|\big).$$

Then, summing over choices of $y_i \in \ker\overline{\phi}$ such that $(F_n)^{\mathcal{C}_\ell}/\langle y_1, \dots, y_n\rangle \simeq G$, we have

$$\begin{aligned} &\mathrm{Prob}\left(\begin{matrix} r_1, \dots, r_n \in \ker\phi \text{ and} \\ (F_n)^{\mathcal{C}_\ell}/\langle \pi_{F_n}(r_1), \dots, \pi_{F_n}(r_n)\rangle \simeq G \end{matrix} \;\middle|\; \begin{matrix} \pi_{F_n}(r_i) \in \ker\overline{\phi} \text{ for all } i, \text{ and} \\ (F_n)^{\mathcal{C}_\ell}/\langle \pi_{F_n}(r_1), \dots, \pi_{F_n}(r_n)\rangle \simeq G \end{matrix}\right) \\ &\quad = \frac{|H^{\mathcal{C}_\ell}|^n}{|H|^n}. \end{aligned}$$

Thus (3.21) is equal to

$$\begin{aligned} &\frac{|H^{\mathcal{C}_\ell}|^n}{|H|^n} \sum_{\phi \in \mathrm{Sur}(F_n,H)} \mathrm{Prob}\left(\begin{matrix} \pi_{F_n}(r_i) \in \ker\overline{\phi} \text{ for all } i, \text{ and} \\ (F_n)^{\mathcal{C}_\ell}/\langle \pi_{F_n}(r_1), \dots, \pi_{F_n}(r_n)\rangle \simeq G \end{matrix}\right) \\ &\quad = \frac{|H^{\mathcal{C}_\ell}|^n}{|H|^n} \sum_{\phi \in \mathrm{Sur}(F_n,H)} \frac{\#\left\{(\tau, \pi) \;\middle|\; \begin{matrix} \tau \in \mathrm{Sur}((F_n)^{\mathcal{C}_\ell}, G) \\ \pi \in \mathrm{Sur}(G, H^{\mathcal{C}_\ell}) \\ \text{and } \pi \circ \tau = \overline{\phi} \end{matrix}\right\}}{|\mathrm{Aut}(G)|\,|G|^n} P_{0,n}(U_{\mathcal{C}_\ell,G}), \end{aligned} \tag{3.22}$$

where $P_{0,n}(U_{\mathcal{C}_\ell,G})$ is defined in **[33]** just before Lemma 9.5, and is the probability that n independent uniform random elements in the kernel of $(F_n)^{\mathcal{C}_\ell} \to G$ generate that kernel as a normal subgroup (as worked out in the proof of **[33, Theorem 8.1]**). (To explain the above equality a bit more: if $(F_n)^{\mathcal{C}_\ell}/\langle \pi_{F_n}(r_1), \dots, \pi_{F_n}(r_n)\rangle \simeq G$ then there is a choice of $\tau \in \mathrm{Sur}((F_n)^{\mathcal{C}_\ell}, G)$ inducing that isomorphism, whose $\mathrm{Aut}(G)$ orbit is unique, and $\bar{\phi}$ must factor through τ since $\pi_{F_n}(r_i) \in \ker\overline{\phi}$. Given a τ, the probability that the relations are in $\ker\tau$ and generate it as a normal subgroup is $|G|^{-n} P_{0,n}(U_{\mathcal{C}_\ell,G})$.)

On the other hand, let $\overline{\phi} \in \mathrm{Sur}((F_n)^{\mathcal{C}_\ell}, H^{\mathcal{C}_\ell})$. Then the composition map $\rho := \overline{\phi} \circ \pi_{F_n}$ is a surjection $F_n \to H^{\mathcal{C}_\ell}$. The number of $\phi \in \mathrm{Sur}(F_n, A)$ such that ϕ induces $\overline{\phi}$ we denote by $\mathrm{Sur}(\rho, \pi_H)$. It is easy to see that

$$\frac{|\mathrm{Sur}((F_n)^{\mathcal{C}_\ell}, G)|}{|G|^n} \le 1 \quad \text{and} \quad \lim_{n\to\infty} \frac{|\mathrm{Sur}((F_n)^{\mathcal{C}_\ell}, G)|}{|G|^n} = 1,$$

and similarly

$$\frac{|\mathrm{Sur}(\rho, \pi_H)|}{|H|^n |H^{\mathcal{C}_\ell}|^{-n}} \le 1, \quad \lim_{n\to\infty} \frac{|\mathrm{Sur}(\rho, \pi_H)|}{|H|^n |H^{\mathcal{C}_\ell}|^{-n}} = 1.$$

Then by (3.22), we obtain that $f_n(G,\ell) = g_n(G,\ell) P_{0,n}(U_{\mathcal{C}_\ell,G})$ where

$$g_n(G,\ell) = \frac{|H^{\mathcal{C}_\ell}|^n |\mathrm{Sur}(\rho, \pi_H)| |\mathrm{Sur}((F_n)^{\mathcal{C}_\ell}, G)| |\mathrm{Sur}(G, H^{\mathcal{C}_\ell})|}{|H|^n |G|^n |\mathrm{Aut}(G)|} \quad \text{and}$$

$$g(G,\ell) := \lim_{n\to\infty} g_n(G,\ell) = \frac{|\mathrm{Sur}(G, H^{\mathcal{C}_\ell})|}{|\mathrm{Aut}(G)|}.$$

Now we apply **[34, LEMMA 5.10]**, where condition (1) holds by definition, (2) from the above, and (3) follows from the definition of $f_n(G,\ell)$. This allows us to conclude, for every ℓ,

$$\sum_{\substack{G \\ G^{\mathcal{C}_\ell} \simeq G}} \lim_{n\to\infty} f_n(G,\ell) = \lim_{n\to\infty} f_n(\text{trivial group}, 1) = \lim_{n\to\infty} \mathbb{E}\big(|\mathrm{Sur}(X_n, H)|\big) = 1. \quad (3.23)$$

When ℓ is sufficiently large such that $H^{\mathcal{C}_\ell} \simeq H$,

$$\lim_{n\to\infty} f_n(G,\ell) = \lim_{n\to\infty} |\mathrm{Sur}(G,H)| \, \mathrm{Prob}\big((Z_n)^{\mathcal{C}_\ell} \simeq G\big) = |\mathrm{Sur}(G,H)| \, \mathrm{Prob}\big((X_0)^{\mathcal{C}_\ell} \simeq G\big),$$

where the last equality is by Theorem 3.17. Hence (3.23) gives the desired result in the lemma. ■

Proof of Theorem 3.19. We compute the moments of $F_n/\langle \alpha_n(x)x^{-1} \mid x \in F_n\rangle$. Consider a fixed finite group H. If H can be generated by n elements, then there are some number of surjections $\phi : F_n \to H$. Those surjections that factor through the quotient

$$F_n/\langle \alpha_n(x)x^{-1} \mid x \in F_n\rangle$$

are exactly those ϕ such that $\phi\alpha = \phi$. So

$$\mathbb{E}\big(\#\mathrm{Sur}\big(F_n/\langle \alpha_n(x)x^{-1} \mid x \in F_n\rangle, H\big)\big) = \sum_{\phi \in \mathrm{Sur}(F_n, H)} \mathrm{Prob}(\phi\alpha = \alpha).$$

The action of $\mathrm{Aut}(F_n)$ on $\mathrm{Sur}(F_n, H)$ is transitive **[35, PROPOSITION 2.2]**, and factors through a finite group. So $\mathrm{Prob}(\phi\alpha = \alpha) = |\mathrm{Sur}(F_n, H)|^{-1}$. Thus, as long as H can be generated by n elements, we have

$$\mathbb{E}\big(\#\mathrm{Sur}\big(F_n/\langle \alpha_n(x)x^{-1} \mid x \in F_n\rangle, H\big)\big) = 1.$$

Thus we can use Theorem 2.20 and Lemma 3.20 to conclude the theorem. ■

Of course, if any kind of universality result can be proven for nonabelian random groups, it would then be interesting to extend the methods to particular nonabelian groups with additional structure that are arising in number theory and topology. So far the applications of these sort of universality methods for random groups have largely been in combinatorics. We expect that as the methods become developed, there will be further applications, including in number theory and topology.

ACKNOWLEDGMENTS

Much of this paper is based on the author's work with coauthors Nigel Boston, Hoi Nguyen, Yuan Liu, Will Sawin, Weitong Wang, and David Zureick-Brown, and we thank them for all the insights gained from these collaborations. The outlook in this paper also benefited from useful conversations with Jordan Ellenberg, Aaron Landesman, Sam Payne, Bjorn Poonen, and Philip Matchett Wood. The author would like to thank Yuan Liu, Hoi Nguyen, and Will Sawin for useful feedback on an earlier draft of this paper.

FUNDING

This work was done with the support of a Packard Fellowship for Science and Engineering, National Science Foundation Waterman Award DMS-2140043, and National Science Foundation grant DMS-2052036.

REFERENCES

[1] J. D. Achter, Results of Cohen–Lenstra type for quadratic function fields. In *Computational arithmetic geometry*, pp. 1–7, Contemp. Math. 463, Amer. Math. Soc., Providence, RI, 2008.

[2] G. V. Balakin, The distribution of the rank of random matrices over a finite field. *Akad. Nauk SSSR. Teor. Veroâtn. Primen.* **13** (1968), 631–641.

[3] A. Bartel and H. W. Lenstra, On class groups of random number fields. *Proc. Lond. Math. Soc.* **121** (2020), no. 4, 927–953.

[4] M. Bhargava, The geometric sieve and the density of squarefree values of invariant polynomials. 2014, arXiv:1402.0031.

[5] M. Bhargava, D. M. Kane, H. W. Jr. Lenstra, B. Poonen, and E. Rains, Modeling the distribution of ranks, Selmer groups, and Shafarevich–Tate groups of elliptic curves. *Camb. J. Math.* **3** (2015), no. 3, 275–321.

[6] M. Bhargava, A. Shankar, and X. Wang, Squarefree values of polynomial discriminants I. 2016, arXiv:1611.09806.

[7] P. Billingsley, *Probability and measure. Second edn.* Wiley Ser. Prob. Math. Stat., Prob. Math. Stat., John Wiley & Sons, Inc., New York, 1986.

[8] J. Blömer, R. Karp, and E. Welzl, The rank of sparse random matrices over finite fields. *Random Structures Algorithms* **10** (1997), no. 4, 407–419.

[9] N. Boston, M. R. Bush, and F. Hajir, Heuristics for p-class towers of imaginary quadratic fields. *Math. Ann.* **368** (2017), no. 1–2, 633–669.

[10] N. Boston, M. R. Bush, and F. Hajir, Heuristics for p-class towers of real quadratic fields. *J. Inst. Math. Jussieu* **20** (2021), no. 4, 1429–1452.

[11] N. Boston and M. M. Wood, Non-abelian Cohen–Lenstra heuristics over function fields. *Compos. Math.* **153** (2017), no. 7, 1372–1390.

[12] L. S. Charlap, H. D. Rees, and D. P. Robbins, The asymptotic probability that a random biased matrix is invertible. *Discrete Math.* **82** (1990), no. 2, 153–163.

[13] J. Clancy, N. Kaplan, T. Leake, S. Payne, and M. M. Wood, On a Cohen–Lenstra heuristic for Jacobians of random graphs. *J. Algebraic Combin.* (2015), 1–23.

[14] J. Clancy, T. Leake, and S. Payne, A note on Jacobians, Tutte polynomials, and two-variable zeta functions of graphs. *Exp. Math.* **24** (2015), no. 1, 1–7.

[15] H. Cohen and H. W. Jr. Lenstra, Heuristics on class groups of number fields. In *Number theory, Noordwijkerhout 1983 (Noordwijkerhout, 1983)*, pp. 33–62, Lecture Notes in Math. 1068, Springer, Berlin, 1984.

[16] H. Cohen and J. Martinet, étude heuristique des groupes de classes des corps de nombres. *J. Reine Angew. Math.* **404** (1990), 39–76.

[17] C. Cooper, On the distribution of rank of a random matrix over a finite field. *Random Structures Algorithms* **17** (2000), no. 3–4, 197–212.

[18] N. M. Dunfield and W. P. Thurston, Finite covers of random 3-manifolds. *Invent. Math.* **166** (2006), no. 3, 457–521.

[19] J. S. Ellenberg, A. Venkatesh, and C. Westerland, Homological stability for Hurwitz spaces and the Cohen–Lenstra conjecture over function fields. *Ann. of Math. (2)* **183** (2016), no. 3, 729–786.

[20] T. Feng, A. Landesman, and E. Rains, The geometric distribution of Selmer groups of elliptic curves over function fields. 2020, arXiv:2003.07517.

[21] É. Fouvry and J. Klüners, Cohen–Lenstra heuristics of quadratic number fields. In *Algorithmic number theory*, edited by F. Hess, S. Pauli, and M. Pohst, pp. 40–55, Lecture Notes in Comput. Sci. 4076, Springer, Berlin–Heidelberg, 2006.

[22] É. Fouvry and J. Klüners, On the 4-rank of class groups of quadratic number fields. *Invent. Math.* **167** (2006), no. 3, 455–513.

[23] E. Friedman and L. C. Washington, On the distribution of divisor class groups of curves over a finite field. In *Théorie des nombres (Quebec, PQ, 1987)*, pp. 227–239, de Gruyter, Berlin, 1989.

[24] D. Garton, Random matrices, the Cohen–Lenstra heuristics, and roots of unity. *Algebra Number Theory* **9** (2015), no. 1, 149–171.

[25] F. Gerth III, Extension of conjectures of Cohen and Lenstra. *Expo. Math.* **5** (1987), no. 2, 181–184.

[26] D. R. Heath-Brown, The size of Selmer groups for the congruent number problem. II. *Invent. Math.* **118** (1994), no. 2, 331–370.

[27] J. Kahn and J. Komlós, Singularity Probabilities for Random Matrices over Finite Fields. *Combin. Probab. Comput.* **10** (2001), no. 02, 137–157.

[28] J. Klys, The distribution of p-torsion in degree p cyclic fields. *Algebra Number Theory* **14** (2020), no. 4, 815–854.

[29] I. N. Kovalenko and A. A. Levitskaja, Limiting behavior of the number of solutions of a system of random linear equations over a finite field and a finite ring. *Dokl. Akad. Nauk SSSR* **221** (1975), no. 4, 778–781.

[30] M. V. Kozlov, On the rank of matrices with random Boolean elements. *Sov. Math., Dokl.* **7** (1966), 1048–1051.

[31] J. B. Lewis, R. I. Liu, A. H. Morales, G. Panova, S. V. Sam, and Y. X. Zhang, Matrices with restricted entries and q-analogues of permutations. *J. Comb.* **2** (2011), no. 3, 355–395.

[32] M. Lipnowski, W. Sawin, and J. Tsimerman, Cohen–Lenstra heuristics and bilinear pairings in the presence of roots of unity. 2020, arXiv:2007.12533.

[33] Y. Liu and M. M. Wood, The free group on n generators modulo $n + u$ random relations as n goes to infinity. *J. Reine Angew. Math.* **2020** (2020), no. 762, 123–166.

[34] Y. Liu, M. M. Wood, and D. Zureick-Brown, A predicted distribution for Galois groups of maximal unramified extensions. 2019, arXiv:1907.05002.

[35] A. Lubotzky, Pro-finite presentations. *J. Algebra* **242** (2001), no. 2, 672–690.

[36] G. Malle, Cohen–Lenstra heuristic and roots of unity. *J. Number Theory* **128** (2008), no. 10, 2823–2835.

[37] A. Mészáros, The distribution of sandpile groups of random regular graphs. *Trans. Amer. Math. Soc.* **373** (2020), no. 9, 6529–6594.

[38] H. H. Nguyen and M. M. Wood, Random integral matrices: universality of surjectivity and the cokernel. *Invent. Math.* (2021). DOI 10.1007/s00222-021-01082-w.

[39] B. Poonen and E. Rains, Random maximal isotropic subspaces and Selmer groups. *J. Amer. Math. Soc.* **25** (2012), no. 1, 245–269.

[40] L. Ribes and P. Zalesskii, *Profinite groups. Second edn., Ergebnisse Der Mathematik Und Ihrer Grenzgebiete. 3. Folge*. Ergeb. Math. Grenzgeb. (3) 40, Springer, Berlin, 2010.

[41] W. Sawin, Identifying measures on non-abelian groups and modules by their moments via reduction to a local problem. 2020, arXiv:2006.04934.

[42] A. Venkatesh and J. S. Ellenberg, Statistics of number fields and function fields. In *Proceedings of the International Congress of Mathematicians. Volume II*, pp. 383–402, Hindustan Book Agency, New Delhi, 2010.

[43] W. Wang and M. Wood, Moments and interpretations of the Cohen–Lenstra–Martinet heuristics. *Comment. Math. Helv.* **96** (2021), no. 2, 339–387.

[44] M. Wood, The distribution of sandpile groups of random graphs. *J. Amer. Math. Soc.* **30** (2017), no. 4, 915–958.

[45] M. M. Wood, On the probabilities of local behaviors in abelian field extensions. *Compos. Math.* **146** (2010), no. 1, 102–128.

[46] M. M. Wood, Cohen–Lenstra heuristics and local conditions. *Res. Number Theory* **4** (2018), no. 4, 41.

[47] M. M. Wood, Random integral matrices and the Cohen–Lenstra heuristics. *Amer. J. Math.* **141** (2019), no. 2, 383–398.

MELANIE MATCHETT WOOD

Department of Mathematics, Harvard University, Science Center Room 325, 1 Oxford Street, Cambridge, MA 02138, USA, mmwood@math.harvard.edu

13. COMBINATORICS

THE GEOMETRY OF GEOMETRIES: MATROID THEORY, OLD AND NEW

FEDERICO ARDILA-MANTILLA

ABSTRACT

The theory of *matroids* or *combinatorial geometries* originated in linear algebra and graph theory, and has deep connections with many other areas, including field theory, matching theory, submodular optimization, Lie combinatorics, and total positivity. Matroids capture the combinatorial essence that these different settings share.

In recent years, the (classical, polyhedral, algebraic, and tropical) geometric roots of the field have grown much deeper, bearing new fruits. We survey some recent successes, stemming from three geometric models of a matroid: the matroid polytope, the Bergman fan, and the conormal fan.

MATHEMATICS SUBJECT CLASSIFICATION 2020

Primary 52B40; Secondary 05B35, 14C17, 14T15, 16T30, 52B45

KEYWORDS

Combinatorial geometry, combinatorial Hodge theory, generalized permutahedron, Hopf algebra, Lefschetz fan, log-concavity, matroid, polyhedral valuation, submodular function, tropical geometry, Tutte polynomial.

ICM INTERNATIONAL CONGRESS OF MATHEMATICIANS 2022 JULY 6–14

Proc. Int. Cong. Math. 2022, Vol. 6, pp. 4510–4541
DOI 10.4171/ICM2022/144

Published by EMS Press

1. INTRODUCTION

There are natural notions of *independence* in linear algebra, graph theory, field theory, matching theory, routing theory, rigidity theory, model theory, and many other areas. When one seeks to understand the pleasing similarities between these different contexts, one is led to the powerful theory of *matroids* or *combinatorial geometries*. These intriguing, multifaceted objects turn out to also play a fundamental role in Lie combinatorics, tropical geometry, total positivity, and other settings.

The geometric approach to matroid theory has recently led to the solution of long-standing questions, and to the discovery of deep, fascinating interactions between combinatorics, algebra, and geometry. This survey is a selection of some recent achievements of the theory.

Section 2 reviews basic definitions, and Section 3 discusses key invariants of a matroid, like the characteristic and Tutte polynomials, and the finite field method to compute them. Section 4 focuses on the *matroid polytope*, starting from its parallel origins in combinatorial optimization and in the geometry of the Grassmannian. We discuss its connections to root systems, generalized permutahedra, the theory of matroid subdivisions and valuative invariants, and the role played by Hopf algebraic methods in these developments. Section 5 concerns the *Bergman fan* of a matroid. We discuss its central role in tropical geometry, its Hodge-theoretic properties, their role in the log-concavity of matroid f-vectors, and the theory of Chern–Schwartz–MacPherson cycles of matroids. Section 6 discusses the *conormal fan* of a matroid, its Hodge-theoretic properties, their role in the log-concavity of matroid h-vectors, and a Lagrangian geometric interpretation of CSM cycles. It also discusses two polytopes that play an important role in this theory and have elegant combinatorial properties: the bipermutahedron and harmonic polytope. Finally, Section 7 offers some closing remarks.

2. MATROIDS

Matroids were defined independently in the 1930s by Nakasawa [83] and Whitney [113]. We choose one of many equivalent definitions. A *matroid* $M = (E, \mathcal{I})$ consists of a finite set E and a collection $\mathcal{I}$ of subsets of E, called the *independent sets*, such that

(I1) $\emptyset \in \mathcal{I}$.

(I2) If $J \in \mathcal{I}$ and $I \subseteq J$ then $I \in \mathcal{I}$.

(I3) If $I, J \in \mathcal{I}$ and $|I| < |J|$ then there exists $j \in J - I$ such that $I \cup j \in \mathcal{I}$.

A matroid where every set of size at most 2 is independent is called a *simple matroid* or a *combinatorial geometry*.

Thanks to (I2), it is enough to list the collection $\mathcal{B}$ of maximal independent sets; these are called the *bases* of M. By (I3), they have the same size $r = r(M)$, which we call the *rank* of M. Our running example will be the matroid with

$$E = abcde, \quad \mathcal{B} = \{abc, abd, abe, acd, ace\}, \tag{2.1}$$

omitting brackets for easier readability. Throughout the paper we let $n = |E|$ and $r = r(M)$.

The two most important motivating examples are graphical and linear matroids. Figure 1 shows how (2.1) is a member of both families.

Graphical matroids. Let E be the set of edges of a graph G and $\mathcal{I}$ be the collection of forests of G, that is, the subsets of E containing no cycle.

Linear or realizable matroids. Let $\mathbb{F}$ be a field.

(a) (Vector configurations) Let E be a finite set of vectors in a vector space over $\mathbb{F}$, and let $\mathcal{I}$ be the collection of linearly independent subsets of E.

(b) (Subspaces) Let $V = \mathbb{F}^E$ be a finite-dimensional vector space and $U \subseteq V$ be a subspace. Let $\mathcal{I}$ be the collection of subsets $I \subseteq E$ such that U intersects the coordinate subspace $V_I = \{\mathbf{v} \in V : v_i = 0 \text{ for } i \in I\}$ transversally, that is, $\dim(U \cap V_I) = \dim U - |I|$.

The latter two constructions are equivalent: for a matrix A, the matroid of the set of columns of A equals the matroid of the rowspace of A.

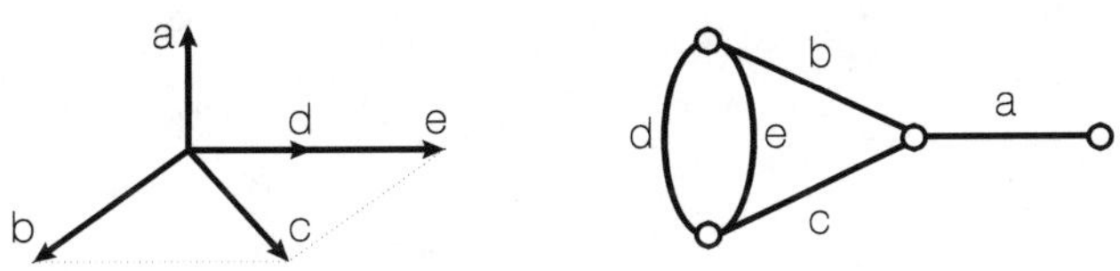

FIGURE 1
The matroid (2.1) with bases $\mathcal{B} = \{abc, abd, abe, acd, ace\}$ is linear and graphical.

Matroids arise naturally in many important settings, e.g., the study of algebraic dependences in a field extension, the combinatorics of root systems of semisimple Lie algebras, the perfect matchings in a bipartite graph, the nonintersecting paths in a directed graph, and total positivity of matrices, to name a few **[8, 87]**. Many of the matroids in natural applications are linear, but most matroids are not realizable over any field **[84]**, and including them leads to a much more powerful and robust theory of matroids and their geometry.

There are several natural operations on matroids. For $S \subseteq E$, the *restriction* $M|S$ (or *deletion* $M\backslash(E - S)$) and the *contraction* M/S are matroids on the ground sets S and $E - S$, respectively, with independent sets

$$\mathcal{I}|S := \{I \subseteq S : I \in \mathcal{I}\},$$
$$\mathcal{I}/S := \{I \subseteq E - S : I \cup I_S \in \mathcal{I}\},$$

for any maximal independent subset I_S of S; the latter is independent of the choice of I_S. When M is a linear matroid in a vector space V, $M|S$ and M/S are the linear matroids on the ground sets S and $E - S$ that M determines on the vector spaces $\operatorname{span}(S)$ and $V/\operatorname{span}(S)$, respectively. If $S \subseteq T$, we write

$$M[S,T] := (M|T)/S.$$

The *direct sum* $M_1 \oplus M_2$ of $M_1 = (E_1, \mathcal{I}_1)$ and $M_2 = (E_2, \mathcal{I}_2)$ with $E_1 \cap E_2 = \emptyset$ is the matroid on $E_1 \cup E_2$ with independent sets $\mathcal{I}_1 \oplus \mathcal{I}_2 = \{I_1 \cup I_2 : I_1 \in \mathcal{I}_1, I_2 \in \mathcal{I}_2\}$. Every matroid decomposes uniquely as a direct sum of its *connected components*.

Finally, the matroid $M^\perp$ *dual* or *orthogonal* to M is the matroid on E with bases

$$\mathcal{B}^\perp := \{E - B : B \in \mathcal{B}\}.$$

Remarkably, this simple notion simultaneously generalizes orthogonal complements and dual graphs. If M is the matroid of a subspace $U \subseteq V$ of an inner product space, then $M^\perp$ is the matroid of its orthogonal complement $U^\perp \subseteq V$. If M is the matroid for a planar graph G, drawn on the plane without edge intersections, then $M^\perp$ is the matroid for the dual graph $G^\perp$, whose vertices and edges correspond to the faces and edges of G, respectively, as shown in Figure 2.

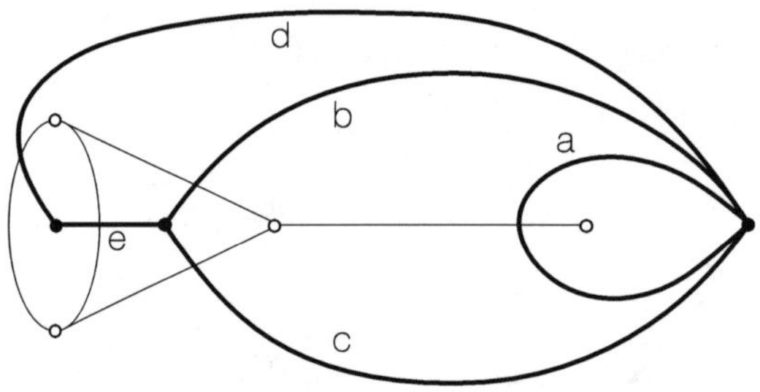

FIGURE 2
The dual matroid $\mathcal{B}^\perp = \{bd, be, cd, ce, de\}$ is realized by the graph dual to Figure 1(b).

An element a is a *loop* of M if $\{a\}$ is dependent. A *coloop* of M is a loop of $M^\perp$.

3. ENUMERATIVE INVARIANTS

Two matroids $M_1 = (E_1, \mathcal{I}_1)$ and $M_2 = (E_2, \mathcal{I}_2)$ are *isomorphic* if there is a *relabeling* bijection $\phi : E_1 \to E_2$ that maps $\mathcal{I}_1$ to $\mathcal{I}_2$. A *matroid invariant* is a function f on matroids such that $f(M_1) = f(M_2)$ whenever M_1 and M_2 are isomorphic. In 1964, Rota introduced the first foundational example [91], which we now define.

The characteristic polynomial. We define the *rank function* $r : 2^E \to \mathbb{Z}$ of a matroid M by

$$r(A) := \text{largest size of an independent subset of } A,$$

for $A \subseteq E$. When M is the matroid of a vector configuration A, $r(A) = \dim \operatorname{span}(A)$. The *characteristic polynomial* of a loopless matroid M is

$$\chi_M(q) := \sum_{A \subseteq E} (-1)^{|A|} q^{r-r(A)}.$$

For the example in (2.1), we have $\chi_M(q) = q^3 - 4q^2 + 5q - 2$. The characteristic polynomial of a matroid is one of its most fundamental invariants. For graphical and linear matroids, it has the following interpretations [38, 86, 115].

1. Graphs. If M is the matroid of a connected graph G, then $q\chi_M(q)$ is the *chromatic polynomial* of G; it counts the *proper colorings* of the vertices of G with q given colors, where no two neighboring vertices have the same color.

2. Hyperplane arrangements. Suppose M is the matroid of a set of nonzero vectors $v_1, \dots, v_n$ spanning $\mathbb{F}^d$. Consider the arrangement $\mathcal{A}$ of hyperplanes $H_i = \{x \in \mathbb{F}^d : v_i \cdot x = 0\}$ for $1 \le i \le n$, and its complement $V(\mathcal{A}) = \mathbb{F}^d \setminus (H_1 \cup \cdots \cup H_n)$. Depending on the underlying field, $\chi_M(q)$ stores different information about $V(\mathcal{A})$:

(a) ($\mathbb{F} = \mathbb{R}$) The complement $V(\mathcal{A})$ consists of $(-1)^d \chi_M(-1)$ regions.

(b) ($\mathbb{F} = \mathbb{C}$) The Betti numbers of the complement $V(\mathcal{A})$ are the coefficients of $(-q)^d \chi_M(-1/q)$.

(c) ($\mathbb{F} = \mathbb{F}_q$) The complement $V(\mathcal{A})$ consists of $\chi_M(q)$ points. For a significantly stronger result on the ℓ-adic étale cohomology of the arrangement, see [30].

Two related invariants that arise in several contexts are the *Möbius* and *beta invariants* $\mu(M) = \chi_M(0)$ and $\beta(M) = (-1)^{r-1}\chi'_M(1)$, where $\chi'_M(x)$ is the derivative of $\chi_M(x)$. If $|E| \ge 2$ then $\beta(M) = \beta(M^\perp)$.

The independence and broken circuit complex and their f- and h-vectors. Let $<$ be a linear order on E. A *circuit* is a minimal dependent set. A *broken circuit* is a set of the form $C - \{\min C\}$ where C is a circuit. An *nbc set* is a subset of E not containing a broken circuit.

We consider two simplicial complexes associated to a matroid M: the *independence complex* and *broken circuit complex*:

$$\mathcal{I}(M) := \{\text{independent sets of } M\}, \quad \mathcal{BC}_<(M) := \{\text{nbc sets of } M\}.$$

The *f-vector* of a simplicial complex Δ of dimension $d-1$ counts the number $f_k(\Delta)$ of faces of Δ of size k for each $0 \le k \le d$. The *h-vector* of M stores this information more compactly; it is given by

$$\sum_{k=0}^{d} f_k (q-1)^{d-k} = \sum_{k=0}^{d} h_k q^{d-k}.$$

The simplicial complexes $\mathcal{I}(M)$ and $\mathcal{BC}_<(M)$ are $(r-1)$-dimensional. For the example in (2.1),

$$f(\mathcal{I}(M)) = (1,5,9,5), \quad f(\mathcal{BC}_<(M)) = (1,4,5,2),$$
$$h(\mathcal{I}(M)) = (1,2,2,0), \quad h(\mathcal{BC}_<(M)) = (1,1,0,0).$$

Topologically, $\mathcal{I}(M)$ and $\mathcal{BC}_<(M)$ are wedges of $\mu(M^\perp)$ and $\beta(M)$ spheres of dimension $r-1$, respectively **[29]**. Up to alternating signs, the coefficients of $\chi_M(q)$ and $\chi_M(q+1)$ give the f-vector and h-vector of $\mathcal{BC}_<(M)$.

The Tutte polynomial. The invariant that appears most often in geometric, algebraic, and enumerative questions related to matroids is the *Tutte polynomial* **[108]**:

$$T_M(x,y) := \sum_{A\subseteq M} (x-1)^{r-r(A)}(y-1)^{|A|-r(A)}. \tag{3.1}$$

For the example in (2.1), we have $T_M(x,y) = x^3 + x^2y + x^2 + xy^2 + xy$.

The ubiquity of this polynomial is explained by the following universality property. If a function f : Matroids $\to \mathbb{F}$ satisfies a *deletion–contraction* of the following form:

$$f(M) = \begin{cases} af(M\backslash e) + bf(M/e) & \text{if } e \text{ is neither a loop nor a coloop,} \\ f(M\backslash e)f(L) & \text{if } e = L \text{ is a loop,} \\ f(M/e)f(C) & \text{if } e = C \text{ is a coloop,} \end{cases} \tag{3.2}$$

then it is an evaluation of the Tutte polynomial, namely $f(M) = a^{n-r}b^r T_M\left(\frac{f(C)}{b}, \frac{f(L)}{a}\right)$. The Tutte polynomial also behaves very nicely under duality; we have $T_{M^\perp}(x,y) = T_M(y,x)$.

Many natural enumerative, algebraic, geometric, and topological quantities in numerous settings satisfy deletion–contraction recurrences, and hence are given by the Tutte polynomial; see **[8, SECTION 7.7]** and **[10]** for examples. In particular,

$$\chi_M(q) = (-1)^r T_M(1-q,0), \quad \mu(M) = T_M(1,0), \quad \beta(M) = [x^1y^0]T_M(x,y),$$

where the last equality holds for $|E| \geq 2$, and

$$\sum_i f_i(\mathcal{I}(M))x^{r-i} = T_M(1+x,1), \quad \sum_i f_i(\mathcal{BC}_<(M))x^{r-i} = T_M(1+x,0),$$
$$\sum_i h_i(\mathcal{I}(M))x^{r-i} = T_M(x,1), \quad \sum_i h_i(\mathcal{BC}_<(M))x^{r-i} = T_M(x,0).$$

In particular, though $\mathcal{BC}_<(M)$ depends delicately on the order $<$, its f- and h-vector do not.

3.1. Computing Tutte polynomials: the finite field method

Given how many quantities are given by the Tutte polynomial, it should not be a surprise that computing Tutte polynomials is extremely difficult (#P-complete **[111]**) for general matroids. Nevertheless, we introduced a *finite field method* **[6,7]**, building on **[24,38]**, that has been effective for computing $T_M(x,y)$ in some special cases of interest. This method is similar in spirit to Weil's philosophy **[110]** of learning about a complex projective variety

$X(\mathbb{C})$ defined by integer polynomials from its reductions $X(\mathbb{F}_q)$ to various finite fields $\mathbb{F}_q$; in that setting we lose access to the complex geometry but we gain the ability to count.

To compute the Tutte polynomial of M with this method, we need a linear realization of M as a set of vectors in $\mathbb{Q}^d$; most examples of interest have one. For any power q of a large enough prime number, this gives a linear realization of M in $\mathbb{F}_q^d$; let $\mathcal{A}_q$ be the corresponding hyperplane arrangement. One then needs to count the points in $\mathbb{F}_q^d$ according to the number of hyperplanes of $\mathcal{A}_q$ that they lie on. If one is able to do this for enough values of q, one can obtain the Tutte polynomial $T_M(x, y)$ from that enumeration.

Theorem 1 (Finite field method, [7]). *Let M be a matroid of rank r realized by a set of vectors in $\mathbb{Q}^d$. Let q be a power of a large enough prime and $\mathcal{A}_q$ be the induced hyperplane arrangement in $\mathbb{F}_q^d$. Then*

$$\sum_{p \in \mathbb{F}_q^d} t^{h(p)} = q^{d-r}(t-1)^r T_M\left(1 + \frac{q}{t-1}, t\right),$$

where $h(p)$ is the number of hyperplanes of $\mathcal{A}_q$ containing p.

An equivalent result in the context of coding theory was obtained earlier by Greene [60]. This finite field method is successful for root systems, arguably the most important vector configurations. The Tutte polynomial of the four families of *classical root systems*

$$\begin{aligned} A_{n-1}^{+} &= \{\mathsf{e}_i - \mathsf{e}_j : i < j\} \\ B(C)_n^{+}, &= \{\mathsf{e}_i \pm \mathsf{e}_j : i < j\} \cup \{(2)\mathsf{e}_i\}, \\ D_n^{+} &= \{\mathsf{e}_i \pm \mathsf{e}_j : i < j\} \subset \mathbb{R}^n, \end{aligned}$$

where $\{\mathsf{e}_1, \ldots, \mathsf{e}_n\}$ is the standard basis of $\mathbb{R}^n$, are given by the coefficient of Z^n in the series

$$\begin{aligned} T_A &= F(Z, Y)^X, \\ T_{B(C)} &= F(2Z, Y)^{(X-1)/2} F(YZ, Y^2), \\ T_D &= F(2Z, Y)^{(X-1)/2} F(Z, Y^2), \end{aligned}$$

where $F(\alpha, \beta) = \sum_{n \geq 0} \alpha^n \beta^{\binom{n}{2}}/n!$ is the *deformed exponential function* [7]. Formulas for the exceptional root systems and the complex reflection groups are given in [45,59] and [89].

The characteristic polynomial is particularly elegant: if $\Phi^+(\mathfrak{g})$ is the set of positive roots of a semisimple Lie algebra $\mathfrak{g}$ and $e_1, \ldots, e_n$ are its *exponents*, then

$$\chi_{\Phi^+(\mathfrak{g})}(q) = (q - e_1) \cdots (q - e_n).$$

This is one of several examples where a characteristic polynomial surprisingly factors into linear factors; [94] outlines three conceptual explanations for this phenomenon.

The *arithmetic Tutte polynomial* $M_A(x, y)$ of a vector configuration A in a lattice also keeps track of arithmetic properties of A. This polynomial is related to the lattice point enumeration of zonotopes [39,103], to complements of toric arrangements [44,49,81], and to Dahmen–Micchelli and De Concini–Procesi–Vergne modules [40,46]. There is also a finite field method for computing $M_A(x, y)$ [14,33] that can be used successfully for root systems, and describes the volume and Ehrhart theory of Coxeter permutahedra [11,14,45].

4. GEOMETRIC MODEL 1: MATROID POLYTOPES

A crucial insight on the geometry of matroids came from two seemingly unrelated places: combinatorial optimization and algebraic geometry. From both points of view, it is natural to model a matroid in terms of the following polytope.

Definition 2 ([48]). The *matroid polytope* of a matroid M on E is

$$P_M := \operatorname{conv}\{\mathrm{e}_B : B \text{ is a basis of } M\} \subset \mathbb{R}^E,$$

where $\{\mathrm{e}_i : i \in E\}$ is the standard basis of $\mathbb{R}^E$ and $\mathrm{e}_B = \mathrm{e}_{b_1} + \cdots + \mathrm{e}_{b_r}$ for $B = \{b_1, \ldots, b_r\}$.

Figure 3 shows this polytope for the example in (2.1). It is a three-dimensional polytope in $\mathbb{R}^5$.

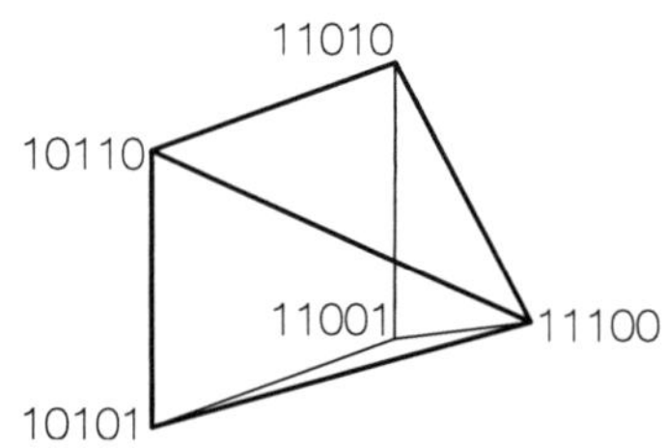

FIGURE 3
The matroid polytope for the matroid in (2.1). The vertices correspond to the bases.

4.1. Algebraic geometry

The intimate relation between matroids and the geometry of the Grassmannian is well studied and mutually beneficial to both fields. Let us describe it briefly.

Instead of studying the r-dimensional subspaces of $\mathbb{C}^n$ one at a time, it is often useful to study them all at once. They can be conveniently organized into the *Grassmannian* $\mathrm{Gr}(r, n)$; each point of $\mathrm{Gr}(r, n)$ represents an r-subspace of $\mathbb{C}^n$.

A choice of a coordinate system on $\mathbb{C}^n$ gives rise to the *Plücker embedding*

$$\mathrm{Gr}(r,n) \overset{p}{\hookrightarrow} \mathbb{CP}^{\binom{n}{r}-1},$$
$$V \mapsto \big(\det(A_B) : B \text{ is an } r\text{-subset of } [n]\big)$$

defined as follows. For an r-subspace $V \subset \mathbb{C}^n$, choose an $r \times n$ matrix A with $V = \operatorname{rowspan}(A)$. For each r-subset B of $[n]$, let $p_B(V) := \det(A_B)$ be the determinant of the $r \times r$ submatrix A_B of A whose columns are given by the subset B. Different choices of A lead to the same *Plücker vector* $p(V)$ in projective space $\mathbb{CP}^{\binom{n}{r}-1}$. The map p provides a realization of the Grassmannian as a smooth projective variety.

Let $\mathbb{C}^* = \mathbb{C} \setminus \{0\}$. The torus $\mathbb{T} = (\mathbb{C}^*)^n/\mathbb{C}^*$ acts on the Grassmannian $\mathrm{Gr}(r,n)$ by stretching the n coordinate axes of $\mathbb{C}^n$, modulo simultaneous stretching. Symplectic geometry then gives a *moment map* $\mu : \mathrm{Gr}(r,n) \to \mathbb{R}^n$, which in this setting is given by

$$\mu(V) = \frac{\sum_B |\det(A_B)|^2 \mathrm{e}_B}{\sum_B |\det(A_B)|^2},$$

where B ranges over the r-subsets of $[n]$.

Now consider the orbit $\mathbb{T} \cdot V$ of the r-subspace $V \in \mathrm{Gr}(r,n)$ as the torus $\mathbb{T}$ acts on it, and the toric variety $X_V := \overline{\mathbb{T} \cdot V}$. Gelfand, Goresky, MacPherson, and Serganova [58] proved that the moment map takes this toric variety to the matroid polytope of $M(V)$,

$$\mu(\overline{\mathbb{T} \cdot V}) = P_{M(V)}. \tag{4.1}$$

Thus matroid polytopes arise naturally in this algebro-geometric setting as well.

As a sample application, the degree of the closure of a torus orbit in the Grassmannian $X_V = \overline{\mathbb{T} \cdot V} \subset \mathbb{CP}^{\binom{n}{r}-1}$ is then given by the volume of the matroid polytope $P_{M(V)}$. Ardila, Benedetti, and Doker [12] used this to find a purely combinatorial formula for it.

The projective coordinate ring of the toric variety X_V is isomorphic to the subalgebra of the polynomial ring $\mathbb{C}[t_e : e \in E]$ generated by the monomials $t_B = \prod_{b \in B} t_b$ for the bases B of $M(V)$. White [104, 112] conjectured that its defining toric ideal is generated by quadratic binomials. For the current state-of-the-art on this conjecture, see [72].

4.2. A geometric characterization of matroids

In most contexts where polytopes arise, it is advantageous if their faces can be described combinatorially. The vertices and edges often play an especially important role. For example, in geometry, they control the GKM presentation of the equivariant cohomology and K-theory of the Grassmannian [56, 70]. In optimization, they are crucial to various algorithms for linear programming.

Matroid polytopes can be described entirely by their vertices and edges, as shown in the following beautiful combinatorial characterization.

Theorem 3 ([58]). *A collection $\mathcal{B}$ of subsets of E is the set of bases of a matroid if and only if every edge of the polytope*

$$P_{\mathcal{B}} := \mathrm{conv}\{\mathrm{e}_B : B \in \mathcal{B}\} \subset \mathbb{R}^E$$

is a translate of $\mathrm{e}_i - \mathrm{e}_j$ for some i, j in E.

Therefore, one could *define* a matroid to be a lattice subpolytope of the cube $[0,1]^n$ that only uses these vectors as edges. Even if one is led to this family of polytopes through the geometry of subspaces as in (4.1), one finds that nonlinear matroids are equally natural from the polytopal point of view. *Matroid theory provides the correct level of generality.*

Positively oriented matroids. *Oriented matroids* are an abstraction of linear algebra over $\mathbb{R}$, abstracting linear dependence relations and their sign patterns. *Positively oriented matroids* are those where every basis is positively oriented. Their matroid polytopes are precisely the polytopes whose edges are translates of $\mathrm{e}_i - \mathrm{e}_j$ and whose facet directions are $\overline{\mathrm{e}}_i - \overline{\mathrm{e}}_j$ for $i, j \in E$, where $\overline{\mathrm{e}}_i := \mathrm{e}_1 + \cdots + \mathrm{e}_i$. Ardila, Rincón, and Williams [22] used this characterization to prove da Silva's 1987 conjecture [42] that every positively oriented matroid is realizable.

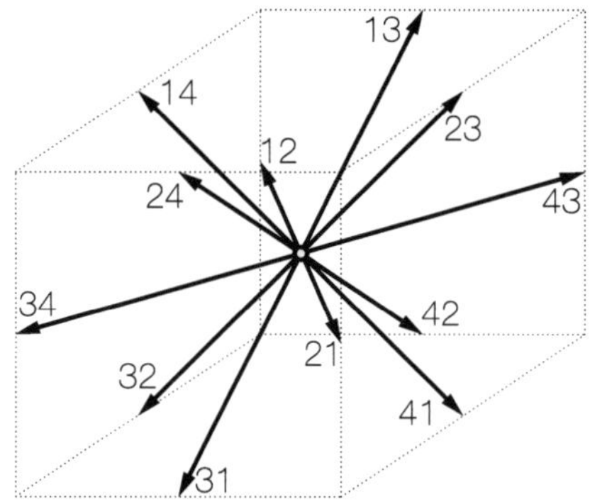

FIGURE 4

The root system $A_3 = \{e_i - e_j : 1 \leq i, j \leq 4\}$, where $e_i - e_j$ is denoted ij. Root systems play an essential role in matroid theory, as demonstrated by Theorem 3.

Coxeter matroids. Theorem 3 shows that in matroid theory, a central role is played by one of the most important vector configurations in mathematics, the root system for the special linear group SL_n,

$$A_{n-1} = \{e_i - e_j : 1 \leq i, j \leq n\} \subset \mathbb{R}^n,$$

shown in Figure 4 for $n = 4$. It is then natural to extend this construction to other semisimple Lie groups. The resulting theory of *Coxeter matroids* **[32]**, introduced by Gelfand and Serganova, starts with a generalization of (4.1) and includes many other interesting results, but Borovik, Gelfand, and White's 2002 assessment still applies today:

> *"the focal point of the theory: the relations between Coxeter matroids and the geometry of flag varieties [...] will need a few more years to settle in a definite form."* **[32]**

The enumerative combinatorics of Coxeter matroids, and its potential applications outside of matroid theory, are ripe for further exploration as well.

4.3. Combinatorial optimization and generalized permutahedra

Matroid theory also benefits from its close connection to submodular optimization, as discovered by Edmonds **[48]** in 1970. This connection begins with a simple, but fundamental, observation: Matroids are precisely the simplicial complexes for which the greedy algorithm finds the facets of minimum weight, as we now explain.

For any *weight function* $w : E \to \mathbb{R}$ on the elements of a matroid $M = (E, \mathcal{I})$, let the weight of a basis B be $w(B) = \sum_{b \in B} w(b)$. Then the bases of minimum weight of M are exactly those that can be obtained by applying the following greedy algorithm:

> *Start with $B = \emptyset$. Then, at each step, add to B any element $e \notin B$ of minimum weight $w(e)$ such that $B \cup e \in \mathcal{I}$. Stop when B is maximal in $\mathcal{I}$.*

Conversely, a simplicial complex $\mathcal{I}$ constitutes the independent sets of a matroid if and only if this greedy algorithm works for any weight function w.

We may rewrite this greedy property as follows **[19, 32]**. Let $\mathcal{S}(w) := \{\emptyset = S_0 \subsetneq S_1 \subsetneq \cdots \subsetneq S_k \subsetneq S_{k+1} = E\}$ be the flag of subsets of E such that $w(s_i)$ is constant for $s_i \in S_i - S_{i-1}$ and $w(S_1) < w(S_2 - S_1) < \cdots < w(S_k - S_{k-1}) < w(E - S_k)$. Then the w-minimum bases of M are the bases of the *w-minimum matroid*

$$M_w := \bigoplus_{i=0}^{k} M[S_i, S_{i+1}]. \tag{4.2}$$

It is useful to restate this geometrically as well. The bases of the matroid M are the vertices of the matroid polytope P_M. The w-minimum bases of M are the vertices of the w-minimum face $(P_M)_w = \{x \in P_M : w(x) \leq w(y) \text{ for all } y \in P_M\}$ of P_M, and that face is itself the matroid polytope of M_w, that is, $(P_M)_w = P_{M_w}$.

Now consider the *braid fan* $\mathcal{A}_E$ in $\mathbb{R}^E$ cut out by the hyperplanes $x_i = x_j$ for $i \neq j$ in E. This is the normal fan of the permutahedron Π_E, whose vertices are the $n!$ permutations of $[n]$. The faces of $\mathcal{A}_E$ are in bijection with the flags of subsets of E: the open face $\sigma_\mathcal{S}$ consists of those $w \in \mathbb{R}^E$ such that $\mathcal{S}(w) = \mathcal{S}$. Then (4.2) shows that for any weight function w in a fixed open face $\sigma_\mathcal{S}$, the matroid M_w depends only on $\mathcal{S}$. This means that the braid fan $\mathcal{A}_E$ refines the normal fan of the matroid polytope P_M.

Generalized permutahedra and submodularity. A *generalized permutahedron*[1] is a polytope P in $\mathbb{R}^E$ satisfying the following three equivalent conditions **[48, 88]**:

- The braid fan $\mathcal{A}_E$ is a refinement of the normal fan of P.
- The edges of P are parallel to roots $e_i - e_j$ for $i, j \in E$.
- $P = P(z) := \{x \in \mathbb{R}^E : \sum_{i=1}^n x_i = z(E), \sum_{i \in I} x_i \leq z(I)\}$ for a (unique) *submodular function* z on E: a function $z : 2^E \to \mathbb{R}$ with $z(A \cup B) + z(A \cap B) \leq z(A) + z(B)$ for $A, B \subseteq E$.

Allowing $z : 2^E \to \mathbb{R} \cup \{\infty\}$, we get *extended generalized permutahedra*. Figure 5 shows some three-dimensional examples.

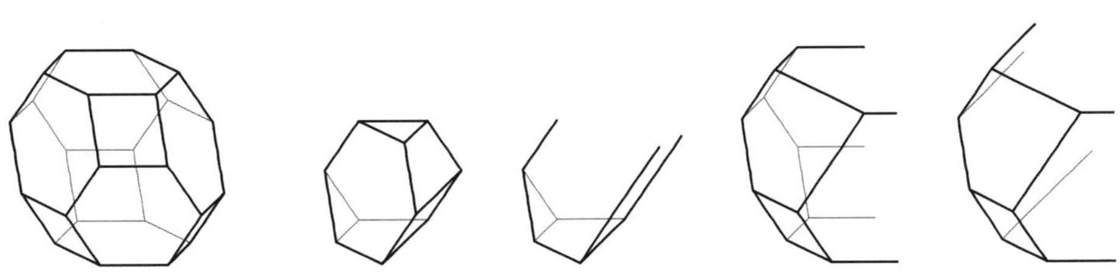

FIGURE 5
The standard 3-permutahedron and four other extended generalized permutahedra.

1 Generalized permutahedra are the translates of the base polytopes of *polymatroids*, of **[48]**.

The rank function of a matroid M is one of the prototypical examples of a submodular function; the corresponding generalized permutahedron is the matroid polytope P_M. This explains why matroid theory informs and benefits greatly from the theory of submodular optimization.

Submodular functions arise in many contexts, partly because submodularity is equivalent to a natural *diminishing returns property* that we now describe. If z measures some quantifiable benefit $z(A)$ associated to each subset $A \subseteq E$, then the contraction $[z/S](e) = z(S \cup e) - z(S)$ measures the *marginal return* of adding e to $S \not\ni e$. A function $z : 2^E \to \mathbb{R}$ is submodular if and only if $[z/S](e) \geq [z/T](e)$ for all $S \subseteq T \subseteq E - e$, that is, if the marginal return $[z/S](e)$ diminishes as we add elements to $S \not\ni e$.

4.4. Matroid subdivisions

A *matroid subdivision* is a polyhedral subdivision $\mathcal{P}$ of a matroid polytope P_M where every polytope $P \in \mathcal{P}$ is itself a matroid polytope. Equivalently, by Theorem 3, it is a subdivision of P_M that only uses the edges of P_M. Let $\mathcal{P}^{\text{int}}$ be the set of interior faces of $\mathcal{P}$; these are the polytopes in $\mathcal{P}$ that are not on the boundary of P_M. In the most important case, $M = U_{d,n}$ is the *uniform matroid* where every d-tuple of $[n]$ is a basis, and P_M is the hypersimplex $\Delta(d,n)$.

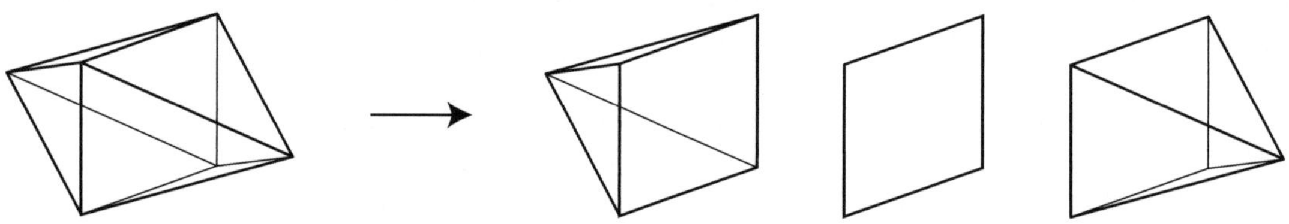

FIGURE 6
The interior faces of a matroid subdivision of the uniform matroid $U_{2,4}$.

Matroid subdivisions were first studied by Lafforgue in his 2003 work on surgery on Grassmannians [71]. These subdivisions also arose in algebraic geometry [61, 68, 71], in tropical geometry [97], and in the theory of valuated matroids in optimization [47, 82].

Lafforgue gave an intriguing application of matroid subdivisions: if a matroid polytope P_M has no nontrivial matroid subdivisions, then the matroid M has (up to trivial transformations) only finitely many realizations over a fixed field $\mathbb{F}$. This is in stark contrast with Mnëv's Universality Theorem, which roughly states that every singularity type appears in the space of realizations of some oriented matroid M. This theorem was used by Vakil to construct several families of moduli spaces with arbitrarily bad singularities [109].

The following conjecture of Speyer [97] has led to many interesting developments.

Conjecture 4. *If M is a matroid on $[n]$ of rank d, then a matroid subdivision of M has at most $\frac{(n-c-1)!}{(d-c)!(n-d-c)!(c-1)!}$ interior faces of dimension $n-c$ for each $1 \leq c \leq \min\{d, n-d\}$.*

For example, the subdivision of Figure 6 has two interior faces of dimension three and one of dimension two, achieving equality in Conjecture 4.

4.5. Matroid valuations

Matroid valuations are ways of measuring matroids that behave well under subdivision. Concretely, let Mat be the family of matroids and A be an abelian group. A function $f : \mathsf{Mat} \to A$ is a *matroid valuation* if for any matroid subdivision $\mathcal{P}$ of a matroid polytope P_M we have the inclusion–exclusion relation

$$f(M) = \sum_{P_N \in \mathcal{P}^{\text{int}}} (-1)^{\dim P_M - \dim P_N} f(N). \tag{4.3}$$

The volume, number of lattice points, and Ehrhart polynomial (given by $\text{Ehr}_P(t) = |tP \cap \mathbb{Z}^d|$ for $t \in \mathbb{N}$) are natural ways of measuring a polytope, and an Euler characteristic computation shows that they are matroid valuations. More interestingly, matroids can also be measured using seemingly unrelated combinatorial and algebro-geometric invariants that, unexpectedly, also satisfy (4.3). These valuations include, among many others:

- the Tutte polynomial of a matroid **[18, 97]**,
- the Chern–Schwartz–MacPherson cycles of a matroid **[74]**,
- the Kazhdan–Lusztig polynomial of a matroid **[23, 51]**,
- the motivic zeta function of a matroid **[23, 65]**,
- the Speyer polynomial of a matroid **[98]**,
- the volume polynomial of the Chow ring of a matroid **[52]**.

Ardila, Fink, and Rincón **[18]** gave a general geometric technique to construct many valuations of matroids, including valuative invariants. Derksen and Fink **[43]** constructed the universal valuative invariant of matroids. They used a slighly different definition of valuation; the equivalence of these two definitions is proved in **[18]**.

As a sample application of matroid valuations, we sketch Speyer's proof of Conjecture 4 for $c = 1$. Since $T_M(x, y)$ is a valuation, so is the *beta invariant* $\beta(M) = [x^1 y^0]T_M(x, y)$. The deletion–contraction recursion gives $\beta(U_{d,n}) = \binom{n-2}{d-1}$, and one can show that $\beta(N) = 0$ if N is not connected (or, equivalently, if P_N is not full-dimensional) and $\beta(N) \geq 1$ otherwise. Thus, for a matroid subdivision $\mathcal{P}$ of $U_{d,n}$,

$$\binom{n-2}{d-1} = \beta(U_{d,n}) = \sum_{\substack{P_N \in \mathcal{P} \\ P_N \text{ facet}}} \beta(N) \geq (\text{number of facets of } \mathcal{P}).$$

Similarly, Speyer **[98]** constructed a polynomial invariant $g_M(t)$ motivated by the K-theory of the Grassmannian, and he used it to prove Conjecture 4 for matroid subdivisions whose matroids are realizable over a field of characteristic 0. His proof relies on the nonnegativity of $g_M(t)$, which is only known for matroids realizable in characteristic 0, for which the coefficients of this polynomial have a geometric interpretation. The nonnegativity of $g_M(t)$ for all M, which would prove Conjecture 4 in full generality, remains open.

The numerous examples of this section raise a natural question.

Question 1. Why are many natural functions of matroids also matroid valuations? How might we find others?

One answer is given by the Derksen–Fink invariant, which is the universal matroid valuation. However, in practice it is often not clear why a conjectural valuation is an evaluation of this invariant. Next we offer a different answer to Question 1, coming from Hopf algebras.

4.6. Hopf algebras and valuations

In 1978, Joni and Rota [66] showed that many combinatorial families have natural *merging* and *breaking* operations that give them the structure of a Hopf algebra, with many useful consequences. For matroids, there is a pleasant surprise: the geometric point of view plays a central role in the Hopf algebra, and connects it with the theory of valuations.

The Hopf algebra of matroids. Joni–Rota [66] and Schmitt [95] defined the *Hopf algebra of matroids* $\mathbf{M}$ as the span of the set of matroids modulo isomorphism, with the product $\cdot : \mathbf{M} \otimes \mathbf{M} \to \mathbf{M}$ and coproduct $\Delta : \mathbf{M} \to \mathbf{M} \otimes \mathbf{M}$ given by:

$$M \cdot N := M \oplus N, \quad \Delta(M) := \sum_{A \subseteq E} (M|A) \otimes (M/A).$$

A Hopf algebra has an *antipode map* $S : \mathbf{M} \to \mathbf{M}$, which is the Hopf-theoretic analog of the inverse map $g \mapsto g^{-1}$ in groups. Takeuchi [107] gave a general formula for the antipode of any connected, graded Hopf algebra; it is an alternating sum with a superexponential number of terms, that is generally not tractable. A central problem for a Hopf algebra of interest H is to use the structure of H to find an explicit, cancelation-free formula for the antipode S.

The optimal formula for the antipode of matroids was discovered by Aguiar and Ardila [2]. The key new insight is that, although they arose in optimization and geometry, *matroid polytopes are also fundamental in the Hopf algebraic structure of matroids.*

Theorem 5 ([2]). *The antipode of the Hopf algebra of matroids* $\mathbf{M}$ *is*

$$S(M) = \sum_{P_N \text{ face of } P_M} (-1)^{c(N)} N$$

for any matroid M, where $c(N)$ denotes the number of connected components of N.

The indicator Hopf algebra of matroids. Theorem 5 makes it very tempting to replace each matroid M with the indicator function $\mathbb{1}_M : \mathbb{R}^E \to \mathbb{R}$ given by $\mathbb{1}_M(p) = 1$ if p is in the matroid polytope P_M and $\mathbb{1}_M(p) = 0$ otherwise. This would give the formula

$$S(P_M) = (-1)^{c(M)} \operatorname{int}(P_M),$$

suggesting connections with the *Euler map* of McMullen's polytope algebra [77] and with *Ehrhart reciprocity* for lattice polytopes [50].

Ardila and Sanchez made this precise, constructing the *indicator Hopf algebra of matroids* $\mathbb{I}(\mathbf{M})$. Its component of degree n is spanned by the indicator functions of matroid polytopes on $[n]$. We have

$$\mathbb{I}(\mathbf{M}) \cong \mathbf{M}/\mathrm{ie}(\mathbf{M}) \quad \text{for } \mathrm{ie}(\mathbf{M}) := \mathrm{span}\left\{ P_M - \sum_{P_N \in \mathcal{P}^{\mathrm{int}}} (-1)^{\mathrm{codim} P_N} P_N : \mathcal{P} \text{ subdivides } P_M \right\}.$$

The subspace $\mathrm{ie}(\mathbf{M})$ is a Hopf ideal of $\mathbf{M}$, so the quotient $\mathbb{I}(\mathbf{M})$ is indeed a Hopf algebra.

Matroid valuations are precisely the functions on matroids that descend to the vector space $\mathbb{I}(\mathbf{M})$. The Hopf algebraic structure on $\mathbb{I}(\mathbf{M})$ then provides a straightforward, unifying framework to discover and prove many known and new matroid valuations, including all those discussed in Section 4.5. This offers one possible answer to Question 1.

Generalized permutahedra and universality. The constructions and theorems of this section hold more generally for the Hopf algebras $\mathbf{GP}^{(+)}$ of (extended) generalized permutahedra and their indicator functions. The Hopf algebras of symmetric functions, Faá di Bruno, matroids, graphs, posets, and many others can be realized as subalgebras of $\mathbf{GP}^+$, with many useful consequences. The *character theory* of Hopf algebras and Theorem 5 give a unified explanation of various *combinatorial reciprocity theorems*: instances where the same polynomial $p(x)$ gives the number $p(n)$ of A-structures of size n and the number $|p(-n)|$ of B-structures of size n for two *reciprocal* combinatorial families A and B. They also explain why the coefficients of the multiplicative and compositional inverses of power series are given by the face structure of the permutahedron and associahedron, respectively [2].

This raises another natural question:

Question 2. Why are many Hopf algebras in combinatorics related to generalized permutahedra and their characters?

As a partial answer to this question, we offer two universality results.

To define $\mathbf{GP}$, the key fact is that for any generalized permutahedron $P \subset \mathbb{R}^E$ and any subset $S \subseteq E$, the e_A-maximal face of P decomposes as $P_A = (P|A) \times (P/A)$ for generalized permutahedra $P|A \subset \mathbb{R}^A$ and $P/A \subset \mathbb{R}^{E-A}$. Define a product and coproduct by

$$P \cdot Q := P \times Q, \quad \Delta(P) := \sum_{A \subseteq E} (P|A) \otimes (P/A). \tag{4.4}$$

Aguiar and Ardila [2] proved that generalized permutahedra are the maximal family of polytopes for which (4.4) defines a Hopf algebra. The antipode is analogous to Theorem 5 [2].

In a different, and more general direction, we have the following universality theorem. A *generalized polynomial character* on a Hopf algebra $\mathbf{H}$ is a multiplicative function from $\mathbf{H}$ to the ring of generalized polynomials, which can have any real numbers as exponents. For example, the *canonical character* on $\mathbf{GP}^+$ is $\beta(P) = t^{r(P)}$ where P lies on the hyperplane $\sum_{i \in E} x_i = r(P)$. Ardila and Sanchez [23] proved that the indicator Hopf algebra $(\mathbb{I}(\mathbf{GP}^+), \beta)$ is the terminal Hopf algebra with a generalized polynomial character. *Any Hopf algebra with a generalized polynomial character factors through* $\mathbb{I}(\mathbf{GP}^+)$. This partially explains the ubiquity of these polytopes in combinatorial Hopf algebras.

These results are closely related to Derksen and Fink's universal valuative invariant for generalized permutahedra [43]. Generalizing to the setting of finite root systems, Ardila, Castillo, Eur, and Postnikov described *generalized Coxeter permutahedra* [13] and Eur, Sanchez, and Supina computed their universal valuation [53]. It would be interesting to construct a Coxeter–Hopf-theoretic framework where generalized Coxeter permutahedra, Coxeter matroids, and other related objects fit naturally.

5. GEOMETRIC MODEL 2: BERGMAN FANS

We now introduce a second geometric model of matroids, coming from tropical geometry. It relies on the *flats* of M; these are the subsets $F \subseteq E$ such that $r(F \cup e) > r(F)$ for all $e \notin F$. We say a flat F is *proper* if it does not have rank 0 or r. The *lattice of flats* of M, denoted L_M, is the set of flats, partially ordered by inclusion.

When M is the matroid of a vector configuration E in a vector space V, the flats of M correspond to the subspaces of V spanned by subsets of E, as illustrated in Figure 7. In this section we assume that the matroid M has no loops.

Ardila and Klivans introduced the following polyhedral rendering of a matroid:

Definition 6 ([19]). The *Bergman fan* or *matroid fan* Σ_M of a matroid M on E is the polyhedral fan in $\mathbb{R}^E/\langle \mathrm{e}_E \rangle$ consisting of the cones

$$\sigma_{\mathcal{F}} := \mathrm{cone}\{\mathrm{e}_F : F \in \mathcal{F}\}$$

for each flag $\mathcal{F} = \{F_1 \subsetneq \cdots \subsetneq F_l\}$ of proper flats of M. Here $\mathrm{e}_F := \mathrm{e}_{f_1} + \cdots + \mathrm{e}_{f_k}$ for $F = \{f_1, \ldots, f_k\}$.

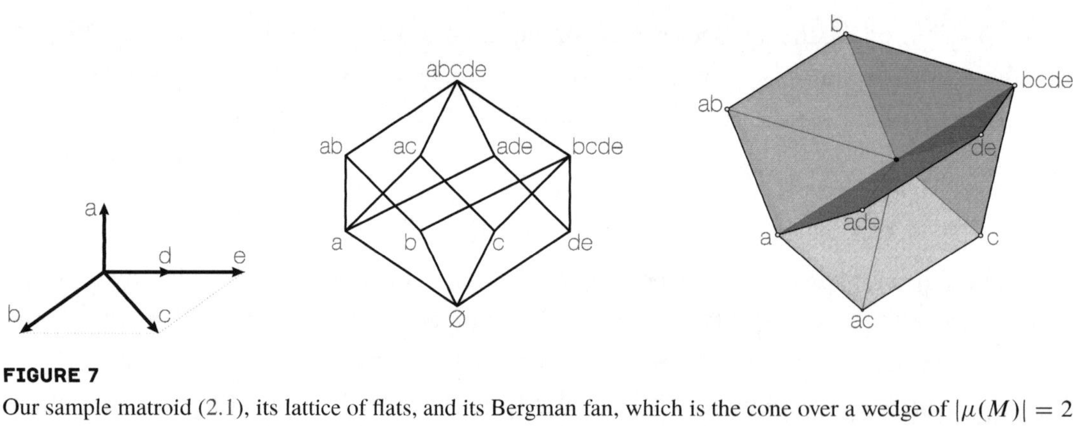

FIGURE 7
Our sample matroid (2.1), its lattice of flats, and its Bergman fan, which is the cone over a wedge of $|\mu(M)| = 2$ circles.

Let us discuss the tropical geometric origin of this fan, and some of its applications.

5.1. Tropical geometry

Tropical geometry is a powerful technique designed to answer questions in algebraic geometry by translating them into polyhedral questions that can be approached combinatorially.[2] In one of its manifestations, tropical geometry sends a complex algebraic variety $V \subset (\mathbb{C}^*)^n$ to its amoebas $\mathcal{A}_t(V)$, whose limit as t approaches 0 is a piecewise linear space Trop V called the *tropicalization* or *logarithmic limit set* of V:

$$\mathcal{A}_t(V) := \left\{(\log_{\frac{1}{t}}(|z_1|, \dots, \log_{\frac{1}{t}}(|z_n|)) : (z_1, \dots, z_n) \in V\right\}, \quad \operatorname{Trop} V := \lim_{t \to 0} \mathcal{A}_t(V).$$

For an introduction and a more precise discussion, see [3, 4, 75, 80]. An important early success of the theory was Mikhalkin's 2005 tropical computation [79] of the *Gromov–Witten invariants of* $\mathbb{CP}^2$, which count the plane curves of degree d and genus g passing through $3d + 1 - g$ general points. Since then, many new results in classical algebraic geometry have been obtained through tropical techniques.

The tropical approach requires two steps. Firstly, one needs to recognize what features of a geometric object V can be recovered from its tropicalization Trop V, which only captures part of the behavior of V at infinity. Secondly, one needs to realize that Trop V may be simpler than V, but it is still usually very intricate.

Additionally, to develop a robust theory, one is led to define *tropical varieties* that are not necessarily tropicalizations of algebro-geometric objects, but are equally important tropically. Understanding their structure is the source of very interesting combinatorial problems.

An important development towards the algebraic foundations of tropical geometry was Sturmfels's description [105] of Trop V in terms of the Gröbner fan of its ideal $I(V)$:

$$\operatorname{Trop} V = \left\{w \in \mathbb{R}^n : \text{the } w\text{-initial ideal of } I(V) \text{ has no monomials}\right\}.$$

This led him to define the *tropical variety of a matroid* M on E to be

$$\operatorname{Trop} M := \left\{w \in \mathbb{R}^E : \text{the } w\text{-minimum matroid } M_w \text{ has no loops}\right\}.$$

If $M = M(V)$ is the matroid of a linear subspace $V \subset (\mathbb{C}^*)^n$ then Trop M = Trop V. If M is not linear, Trop M is not the tropicalization of a variety.

Bergman [26] conjectured and Bieri and Groves [27] showed that the tropicalization of an irreducible variety in $(\mathbb{C}^*)^n$ is pure and connected. Sturmfels [106] conjectured that Trop M should have these same properties, even if M is not a linear matroid. Ardila and Klivans [19] first introduced the Bergman fan with the goal of settling this conjecture.

Theorem 7. *The Bergman fan Σ_M of a matroid M is a triangulation of the tropical space* Trop M. *Therefore* Trop M *is a cone over a wedge of $|\mu(M)|$ spheres of dimension $r - 2$, where $\mu(M)$ is the Möbius number of the matroid.*

The first statement relies on the fact, explained in Section 4.3, that the matroid M_w only depends on the face of the braid fan $\mathcal{A}_E$ containing w. Intersecting Trop M with the

---

2 This mathematician from the tropics finds the name "tropical geometry" questionable.

braid fan induces the triangulation Σ_M. The second statement then follows from Björner's result [29] that the order complex of the lattice of flats of M is a wedge of spheres.

Total positivity and oriented matroids. Motivated by the theory of total positivity, Speyer and Williams [99] introduced the positive part $\mathrm{Trop}^+ V$ of the tropicalization of an affine variety V. Analogously, Ardila, Klivans, and Williams [20] studied the *positive part of the tropical variety of an acyclic oriented matroid* M, which is $\mathrm{Trop}^+ M = \{w \in \mathbb{R}^E :$ the w-minimum matroid M_w is acyclic$\}$. They showed that the *Las Vergnas face lattice* of M gives a triangulation of $\mathrm{Trop}^+ M$; and hence [31] $\mathrm{Trop}^+ M$ is a cone over a sphere.

Furthermore, Ardila, Reiner, and Williams [21] constructed $|\mu(M)|$ reorientations M^ε of M that decompose $\mathrm{Trop}(M)$ explicitly as a wedge of the corresponding spheres $\mathrm{Trop}^+ M^\varepsilon$ [21]. When M is the matroid of a root system Δ, each sphere $\mathrm{Trop}^+ M^\varepsilon$ is dual to the *graph associahedron* of the Dynkin diagram of Δ [21].

There are also very interesting connections between the positive part of the tropical Grassmannian, matroid theory, and cluster algebras; see [114] for a survey.

5.2. A tropical characterization of matroids

A *tropical fan* is a subset $X \subseteq \mathbb{R}^n$ that has the structure of a pure, integral polyhedral fan $\mathcal{X}$ with weights $w : \{\text{facets of } \mathcal{X}\} \to \mathbb{N}$ satisfying the *balancing condition*:

$$\sum_{\text{facets } \sigma \supset \tau} w(\sigma) v_{\sigma/\tau} = 0 \bmod \mathrm{span}(\tau) \quad \text{for any face } \tau \text{ of codimension 1,} \tag{5.1}$$

where $v_{\sigma/\tau} \in \mathbb{Z}^n$ is the primitive generator of the ray σ/τ in $\mathbb{Z}^n/\mathbb{Z}\tau$. Examples include $\mathrm{Trop}\, V$ for any subvariety $V \subset (\mathbb{C}^*)^n$ and $\mathrm{Trop}\, M$ for any matroid M on $[n]$.

In analogy with the classical setting, the *degree* of a tropical fan X is obtained by counting the intersection points of X with $\mathrm{Trop}\, V$ for a generic linear subspace V of codimension $\dim X$, with certain multiplicities. For precise definitions, see [4,80].[3] Fink [55] gave the following remarkable characterization:

Theorem 8 ([55]). *The tropical fans of degree* 1 *in* $\mathbb{R}^E$ *are precisely the tropical varieties of the matroids on* E.

Thus Bergman fans of matroids can be thought of as the tropical analogs to linear subspaces. In fact, one could *define* a matroid on E to be a tropical fan of degree 1 in $\mathbb{R}^E$. Notice that, although the matroid fan $\mathrm{Trop}\, M$ only arises via tropicalization when M is a linear matroid, one should really consider the matroid fans of nonrealizable matroids as well; they are equally natural from the tropical point of view. Again, *matroid theory provides the correct level of generality.*

Theorems 7 and 8 explain two important roles that matroids play in tropical geometry. On the one hand, they offer a useful testing ground, providing hints for the kinds of general results that may be possible, and the sorts of difficulties that one should expect. On

---

3 One sometimes allows bounded faces in a tropical variety; Fink works in this setting.

the other hand, they are fundamental building blocks; for instance, in analogy with the classical definition of a manifold, a *tropical manifold* is an abstract tropical variety that locally has the structure of a Bergman fan of a matroid [80]; Figure 8 shows an example. Clarifying the foundations of the theory of tropical manifolds is an important project; we expect it will continue to shape and benefit from the further development of the geometry of matroids.

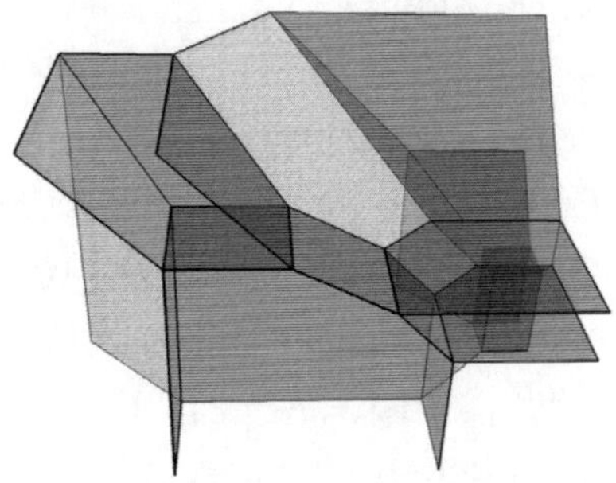

FIGURE 8
A tropical manifold is a tropical variety that is a matroid fan locally. (Picture: Johannes Rau)

Another interesting direction is the tropical geometry of Coxeter matroids and homogenous spaces; for some initial efforts, see [25,34,90]. As the next section will illustrate, this could have interesting enumerative applications.

5.3. The Chow ring, combinatorial Hodge theory, and log-concavity

We say that a sequence $a_0, a_1, \dots, a_r$ of nonnegative integers is:

- *unimodal* if $a_0 \leq a_1 \leq \cdots \leq a_{m-1} \leq a_m \geq a_{m+1} \geq \cdots \geq a_r$ for some $0 \leq m \leq r$,
- *log-concave* if $a_{i-1}a_{i+1} \leq a_i^2$ for all $1 \leq i \leq r-1$, and
- *flawless* if $a_i \leq a_{s-i}$ for all $1 \leq i \leq \frac{s}{2}$, where s is the largest index with $a_s \neq 0$.

Many sequences in combinatorics have these properties, but proving them often requires a fundamentally new construction or connection to algebra or geometry, and gives rise to unforeseen structural results about the objects of interest.

In 1970, Rota [92] first raised such questions in the context of matroid theory, and suggested the Alexandrov–Fenchel inequalities in convex geometry as a possible approach. In the early 1980s, Stanley [101,102] systematically used the hard Lefschetz theorem and the representation theory of Lie algebras to prove similar combinatorial inequalities. In recent years, building on these techniques, a *combinatorial Hodge theory of matroids* has led to the solution of several long-standing open problems in matroid theory.

The *Chow ring* of the Bergman fan Σ_M is

$$A^*(\Sigma_M) := \mathbb{R}[x_F : F \text{ proper flat of } M]/(I_M + J_M),$$

where

$$I_M := \langle x_{F_1} x_{F_2} : F_1 \subsetneq F_2 \text{ and } F_1 \supsetneq F_2 \rangle, \quad J_M := \Big\langle \sum_{F \ni i} x_F - \sum_{F \ni j} x_F : i, j \in E \Big\rangle.$$

The work of Brion [35] implies that $A^*(\Sigma_M)$ is isomorphic to the Chow ring of the toric variety associated to Σ_M. The work of Billera [28] implies that $A^*(\Sigma_M)$ is also isomorphic to the algebra of continuous piecewise polynomial functions on Σ_M, modulo the restrictions of global linear functions to Σ_M. When studying this ring, it is often useful to keep in mind both its algebraic presentation and its interpretation in terms of piecewise polynomial functions.

When M is linear over $\mathbb{C}$, Feichtner and Yuzvinsky [54] proved that $A^*(\Sigma_M)$ is the Chow ring of De Concini and Procesi's *wonderful compactification* of the complement of a hyperplane arrangement. Surprisingly, for any matroid M, the Chow ring $A^*(\Sigma_M)$ has many of the properties of the cohomology ring of a smooth projective variety.

We say a fan Σ is *Lefschetz* if its Chow ring $A^*(\Sigma)$ satisfies Poincaré duality, the hard Lefschetz theorem, and the Hodge–Riemann relations, and the star of any face in Σ also has these properties. For a precise definition, see [16]. Huh [62], Huh and Katz [64], and Adiprasito, Huh, and Katz [1] developed the first steps in the Hodge theory of matroids:

Theorem 9 ([1]). *The Bergman fan of a matroid is Lefschetz.*

The inspiration for this theorem is geometric, coming from the Grothendieck standard conjectures on algebraic cycles. The statement and proof are combinatorial.

Instead of giving a complete definition of Lefschetz fans here, we focus on a comparatively small but powerful consequence. The Chow ring $A^*(\Sigma_M)$ is graded of degree $r-1$, and there is an isomorphism $\deg : A^{r-1} \to \mathbb{R}$ characterized by the property that $\deg(F_1 \cdots F_{r-1}) = 1$ for any complete flag $F_1 \subsetneq \cdots \subsetneq F_{r-1}$ of proper flats.

Consider the *ample cone* $K(\Sigma_M) \subset A^1(\Sigma_M)$ given by the piecewise linear functions on the Bergman fan Σ_M that are strictly convex around every cone. In Brion's presentation, $K(\Sigma_M) = \{\sum_{F \text{ flat}} c_F x_F \text{ for } c : 2^E \to \mathbb{R} \text{ strictly submodular}\}$. The Hodge–Riemann relations imply that for any ample classes $L_1, \ldots, L_{r-3}, a, b \in K(\Sigma_M)$, if we write $L = L_1 \cdots L_{r-3}$,

$$\deg(La^2)\deg(Lb^2) \leq \deg(Lab)^2. \tag{5.2}$$

By continuity, this property also holds for *nef classes*, i.e., classes in the closure $\overline{K}(\Sigma_M)$.

Combining these ingredients, Adiprasito, Huh, and Katz [1] considered the elements

$$\alpha := \alpha_i = \sum_{F \ni i} x_F, \quad \beta := \beta_i = \sum_{F \not\ni i} x_F$$

of the Chow ring $A^*(\Sigma_M)$, which are independent of i and lie in the nef cone $\overline{K}(\Sigma_M)$. An algebraic combinatorial computation in $A^*(\Sigma_M)$ shows that

$$\deg(\alpha^k \beta^{r-1-k}) = (-1)^{r-1-k}\bigg(\text{coefficient of } q^k \text{ in } \frac{\chi_M(q)}{q-1}\bigg). \tag{5.3}$$

As k varies, this sequence of degrees is log-concave by (5.2). In turn, by elementary arguments [36, 67, 73], this implies the following theorems, which were conjectured by Rota, Heron, Mason, and Welsh in the 1970s and 1980s.

Theorem 10 ([1]). *For any matroid M, the following sequences, defined in Section 3, are unimodal, log-concave, and flawless:*

- *the f-vector $f(\mathcal{I}(M))$ of the independence complex of M, and*
- *the f-vector $f(\mathcal{BC}_<(M))$ of the broken circuit complex of M.*

The latter is the sequence of absolute values of the coefficients of $\chi_M(q)$.

5.4. Chern–Schwartz–MacPherson cycles of matroids

Chern–Schwartz–MacPherson (CSM) cycles generalize the Chern class of a tangent bundle to the setting of possibly singular or noncompact complex algebraic varieties. When $\mathcal{A}$ is a complex hyperplane arrangement, López de Medrano, Rincón, and Shaw [74] computed the CSM class of the complement $\mathbb{C}^E \setminus \mathcal{A}$ of $\mathcal{A}$ in its wonderful compactification $W_{\mathcal{A}}$ in terms of the matroid $M = M(\mathcal{A})$.

A k-dimensional *Minkowski weight* on a fan Σ is a choice of weights $w(\sigma)$ for each k-dimensional face σ of Σ satisfying the balancing condition (5.1) for every $(k-1)$-dimensional face τ of Σ. We write $\mathrm{MW}_k(\Sigma)$ for the additive group of k-dimensional Minkowski weights and $\mathrm{MW}(\Sigma) = \bigoplus_{k\geq 0} \mathrm{MW}_k(\Sigma)$. This is dual to the Chow ring of Σ in the following sense. Fulton and Sturmfels [57] showed that $\mathrm{MW}_k(\Sigma) \cong \mathrm{Hom}(A^k(\Sigma), \mathbb{R})$. The product in $A(\Sigma)$ then gives $\mathrm{MW}(\Sigma)$ the structure of an $A(\Sigma)$-module, and $\mathrm{MW}(\Sigma) \cong \mathrm{Hom}(A(\Sigma), \mathbb{R})$ as modules. For details, see, for example, [16, SECTION 3.1].

The *kth CSM cycle of a matroid M* is the k-skeleton of the Bergman fan Σ_M with weights

$$w(\sigma_{\mathcal{F}}) := (-1)^{r-k} \prod_{i=0}^{k} \beta\big(M[F_i, F_{i+1}]\big) \quad \text{for } \mathcal{F} = \{\emptyset = F_0 \subset F_1 \subset \cdots \subset F_k \subset F_{k+1} \subset E\},$$

where $\beta(M[F_i, F_{i+1}])$ is the beta invariant of the minor $M[F_i, F_{i+1}]$ [74]. For any matroid, the fan above satisfies the balancing condition (5.1), giving a Minkowski weight.

When M is the matroid of a complex hyperplane arrangement $\mathcal{A}$, the (geometric) CSM class of the wonderful compactification $W_{\mathcal{A}}$ is given by the (combinatorial) CSM cycles of M. The above construction makes sense for arbitrary matroids, and further, it defines the *CSM cycles of tropical manifolds*.

As shown in [5, 74], the tropical degrees of the CSM cycles of a matroid M are the entries of the h-vector of the broken circuit complex of M:

$$\deg\big(\mathrm{csm}_k(M)\big) = \text{coefficient of } q^k \text{ in } \chi_M(q+1). \tag{5.4}$$

6. GEOMETRIC MODEL 3: CONORMAL FANS

Motivated by Lagrangian geometry, Ardila, Denham, and Huh [16] introduced a third polyhedral model that enriches the geometry of matroids and leads to stronger inequalities for matroid invariants. In this section we assume that the matroid M has no loops or coloops.

A *biflag* $(\mathcal{F}, \mathcal{G})$ of M consists of flags $\mathcal{F} = \{F_1 \subseteq \cdots \subseteq F_l\}$ and $\mathcal{G} = \{G_1 \supseteq \cdots \supseteq G_l\}$ of nonempty flats of M and $M^\perp$, respectively, such that

$$\bigcap_{i=1}^{l} (F_i \cup G_i) = E, \quad \bigcup_{i=1}^{l} (F_i \cap G_i) \neq E.$$

All maximal biflags have length $n-2$.

Definition 11 ([16]). The *conormal fan* $\Sigma_{M,M^\perp}$ of a matroid M is the polyhedral fan in $\mathbb{R}^E/\langle \mathsf{e}_E \rangle \times \mathbb{R}^E/\langle \mathsf{f}_E \rangle$ consisting of the cones

$$\sigma_{\mathcal{F},\mathcal{G}} := \operatorname{cone}\{\mathsf{e}_{F_i} + \mathsf{f}_{G_i} : 1 \leq i \leq l\} \quad \text{for each biflag } (\mathcal{F}, \mathcal{G}).$$

Here $\{\mathsf{e}_i : i \in E\}$ and $\{\mathsf{f}_i : i \in E\}$ are the standard bases for two copies of $\mathbb{R}^E$.

This is a simplicial fan whose support is the product $\operatorname{Trop} M \times \operatorname{Trop} M^\perp$.

6.1. The conormal Chow ring and combinatorial Hodge theory

A *biflat* (F, G) of M is a biflag of length 1. It consists of flats $F, G \neq \emptyset$ of $M, M^\perp$, respectively, not both equal to E, such that $F \cup G = E$. Consider the polynomial ring with a variable $x_{F,G}$ for each biflat (F, G). For a set $(\mathcal{F}, \mathcal{G})$ of distinct biflats $(F_1, G_1), \ldots, (F_l, G_l)$ write $x_{\mathcal{F},\mathcal{G}} = x_{F_1,G_1} \cdots x_{F_l,G_l}$. For $i \in E$, let

$$\gamma_i := \sum_{\substack{F \ni i \\ F \neq E}} x_{F,G}, \quad \gamma_i' := \sum_{\substack{G \ni i \\ G \neq E}} x_{F,G}, \quad \delta_i := \sum_{F \cap G \ni i} x_{F,G}.$$

The *Chow ring of the conormal fan* of M is

$$A^*(\Sigma_{M,M^\perp}) := \mathbb{R}[x_{F,G}]/(I_{M,M^\perp} + J_{M,M^\perp}),$$

where $I_{M,M^\perp} = \langle x_{\mathcal{F},\mathcal{G}} : (\mathcal{F}, \mathcal{G}) \text{ is not a biflag}\rangle$, $J_{M,M^\perp} = \langle \gamma_i - \gamma_j, \gamma_i' - \gamma_j' : i, j \in E\rangle$. The elements $\gamma := \gamma_i, \gamma' := \gamma_i'$, and $\delta := \delta_i$ of $A^1(\Sigma_{M,M^\perp})$ are independent of i.

Ardila, Denham, and Huh [16] showed the conormal analog of Theorem 9.

Theorem 12 ([16]). *The conormal fan of a matroid is Lefschetz.*

Since $|\Sigma_{M,M^\perp}| = \operatorname{Trop} M \times \operatorname{Trop} M^\perp = |\Sigma_M| \times |\Sigma_{M^\perp}|$ and the product of Lefschetz fans is Lefschetz, the key step in the proof of Theorem 12 is the general result that the Lefschetz property of a simplicial fan Σ depends only on the support $|\Sigma|$:

Theorem 13 ([16]). *If two simplicial fans Σ_1 and Σ_2 have the same support $|\Sigma_1| = |\Sigma_2|$, then Σ_1 is Lefschetz if and only if Σ_2 is Lefschetz.*

This Chow ring $A^*(\Sigma_{M,M^\perp})$ has degree $n-2$, and there is a unique isomorphism $\deg : A^{n-2} \to \mathbb{R}$ characterized by the property that $\deg(x_{\mathcal{F},\mathcal{G}}) = 1$ for any maximal biflag $(\mathcal{F}, \mathcal{G})$ of M. The log-concavity inequality (5.2) holds in the ample cone $K(\Sigma_{M,M^\perp})$ of the conormal fan, and hence in the nef cone $\overline{K}(\Sigma_{M,M^\perp})$ as well.

6.2. Lagrangian interpretation of CSM classes

We return to the Chern–Schwartz–MacPherson classes of Section 5.4. Schwartz's and MacPherson's constructions [76, 96] of csm for a complex algebraic variety X are rather subtle. Sabbah [93] later observed that CSM classes can be interpreted more simply as "shadows" of the characteristic cycles in the cotangent bundle T^*X.

Similarly, the CSM cycles of a matroid M are combinatorially intricate fans supported on the Bergman fan Σ_M. We prove that they are "shadows" of much simpler cycles of the conormal fan $\Sigma_{M,M^\perp}$. There is a natural projection map $\pi : \Sigma_{M,M^\perp} \to \Sigma_M$ which gives a pushforward map $\pi_* : \mathrm{MW}_k(\Sigma_{M,M^\perp}) \to \mathrm{MW}_k(\Sigma_M)$. We have:

Theorem 14 ([16]). *If M has no loops and no coloops, we have*

$$\mathrm{csm}_k(M) = (-1)^{r-k}\pi_*(\delta^{n-k-1} \cap 1_{M,M^\perp}) \quad \textit{for } 0 \le k \le r,$$

where $1_{M,M^\perp}$ is the top-dimensional constant Minkowski weight 1 on the conormal fan.

6.3. Unimodality, log-concavity, and flawlessness

Applying the projection formula to Theorem 14 and (5.4), we then express the h-vector of the broken circuit complex of M in the intersection theory of the conormal fan

$$\deg(\gamma^k \delta^{n-k-1}) = (-1)^{r-k}\big(\text{coefficient of } q^k \text{ in } \chi_M(q+1)\big). \tag{6.1}$$

We give an alternative proof of this identity that does not rely on CSM classes in [15], through a careful study of the *Lagrangian combinatorics of matroids*.

The classes γ and δ are nef, so the log-concavity inequalities (5.2) apply to the sequence of degrees in (6.1). This implies the following strengthening of Theorem 10, parts of which were originally conjectured by Brylawski, Dawson, and Colbourn in the early 1980s [36, 37, 41] and left open in Huh's 2018 ICM paper [63].

Theorem 15 ([16]). *For any matroid M, the following sequences, defined in Section 3, are unimodal, log-concave, and flawless:*

- *the h-vector $h(\mathcal{I}(M))$ of the independence complex of M, and*
- *the h-vector $h(\mathcal{BC}_<(M))$ of the broken circuit complex of M.*

The most difficult part of Theorem 15 is the log-concavity of $h(\mathcal{BC}_<(M))$ (6.1). The remaining parts follow from it by elementary arguments.

In 1977, Stanley [100] conjectured that $h(\mathcal{I}(M))$ is the f-vector of a *pure multicomplex*: a set of monomials such that if $m' \in X$ and $m|m'$ then $m \in X$, and the maximal monomials in X have the same degree. This conjecture has been proved in rather different ways for various families of matroids, e.g., [78, 85], but remains wide open in general.

6.4. Lagrangian combinatorics, bipermutahedra, and harmonic polytopes

A subtle technical issue in the proofs of Theorems 10 and 15 leads to some combinatorial constructions of independent interest. For a Lefschetz fan Σ, the log-concavity inequalities (5.2) hold inside the open cone $K(\Sigma)$, corresponding to piecewise linear functions on Σ that are *strictly* convex around every cone. In light of (5.3) and (6.1), we wish to apply these inequalities to the classes α, β in $A^1(\Sigma_M)$ and γ, δ in $A^1(\Sigma_{M,M^\perp})$, which are *weakly* convex locally, that is, they lie in the closed cone $\overline{K}(\Sigma)$. Continuity will guarantee that the log-concavity inequalities in $K(\Sigma)$ will still hold in the closure $\overline{K}(\Sigma)$ *if* the open cone $K(\Sigma)$ is nonempty. This is not a trivial condition.

A complete polyhedral fan Σ is *projective* if it is the normal fan $\Sigma = \mathcal{N}(P)$ of a polytope P; each such polytope produces a strictly convex piecewise linear function on the fan, namely, $w \mapsto \max_{p \in P} w(p)$. Under this correspondence, we can think of $K(\Sigma)$ as the space of polytopes whose normal fan is Σ. An incomplete fan Σ is *quasiprojective* if it is a subfan of a projective fan $\mathcal{N}(P)$; this guarantees that $K(\Sigma) \neq \emptyset$ as well.

By construction, the Bergman fan Σ_M of every matroid M on E is a subfan of the braid arrangement Σ_E; this is the normal fan of the permutahedron Π_E, which is simple. This guarantees $K(\Sigma_M) \neq \emptyset$, as required in the proof of Theorem 10. Similarly, we had to construct a simple polytope, called the *bipermutahedron* $\Pi_{E,E}$, whose normal fan contains the conormal fan $\Sigma_{M,M^\perp}$ of any matroid M on E. This guarantees $K(\Sigma_{M,M^\perp}) \neq \emptyset$, as required in the proof of Theorem 15. Let us briefly discuss this polytope.

The harmonic polytope. A *bisubset* of E, denoted $S|T \sqsubset E$, consists of subsets $S, T \neq \emptyset$ of E, not both equal to E, with $S \cup T = E$. Ardila and Escobar [17] studied the *harmonic polytope* $H_{E,E}$, given by

$$\sum_{e \in E} x_e = \sum_{e \in E} y_e = \frac{n(n+1)}{2} + 1,$$

$$\sum_{s \in S} x_s + \sum_{t \in T} y_t \geq \frac{|S|(|S|+1) + |T|(|T|+1)}{2} + 1 \text{ for } S|T \sqsubset E.$$

It has $3^n - 3$ facets and $(n!)^2(1 + \frac{1}{2} + \cdots + \frac{1}{n})$ vertices. Its volume is a weighted sum of the degrees of the toric varieties associated to all connected bipartite graphs with n edges.

The harmonic polytope is the minimal polytope whose normal fan contains all conormal fans as subfans: every such polytope contains $H_{E,E}$ as a Minkowski summand. Thus the harmonic polytope arises very naturally in this setting, but it has the disadvantage of not being simple, which makes it difficult to work with its Chow ring.

The bipermutahedron. The *bipermutahedron* $\Pi_{E,E} \subset \mathbb{R}^E \times \mathbb{R}^E$ is given by

$$\sum_{e \in E} x_e = \sum_{e \in [n]} y_e = 0,$$

$$\sum_{s \in S} x_s + \sum_{t \in T} y_t \geq -(|S| + |S - T|)(|T| + |T - S|) \text{ for } S|T \sqsubset E.$$

It was constructed by Ardila, Denham, and Huh [16] and further studied in [9]. This is the most elegant simple polytope that we know which has the harmonic polytope as a Minkowski summand, so its normal fan contains all matroid conormal fans. Its normal fan is simplicial, so we can use Brion and Billera's descriptions of its Chow ring. Finally, it has an elegant combinatorial structure that allows us to prove Theorem 14 and (6.1) in [16].

The bipermutahedron also has $3^n - 3$ facets, and it has $(2n)!/2^n$ vertices corresponding to the *bipermutations of* E. It is combinatorially isomorphic to a unimodular triangulation of the product of n unit triangles, and Ehrhart theory then gives a simple formula for its h-vector. In analogy with the permutahedron, this h-vector counts the bipermutations according to their number of descents, and it is also log-concave [9].

7. GEOMETRY AND GEOMETRIES: THE FUTURE

We have centered our discussion on three related geometric models of a combinatorial geometry and their consequences within and outside of matroid theory. There are certainly many other such models—some already known, some yet to be discovered. What these constructions share is their deep connection with or analogy to natural geometric constructions associated to vector configurations or subspaces of a vector space. This situation reminds us of a 50-year old prophetic remark of Bose, relayed by Kelly and Rota:

> *"We combinatorialists have much to gain from the study of algebraic geometry, if not by its applications to our field, at least by the analogies between the two subjects."* [69]

The recurring theme of developing discrete versions of geometric techniques is not born from a wish to avoid algebraic geometry; quite the opposite. Our goal is to develop the necessary tools to solve combinatorial and geometric problems when the current algebro-geometric technology is not sufficient. The applications of this program are not only combinatorial. As Gelfand, Goresky, MacPherson, and Serganova wrote in 1987,

> *"We believe that combinatorial methods will play an increasing role in the future of geometry and topology."* [58]

Today, these predictions ring true more than ever.

ACKNOWLEDGMENTS

I warmly thank Gian-Carlo Rota, Richard Stanley, and Bernd Sturmfels for their very beautiful and very different lessons on combinatorial geometries, and my wonderful coauthors and students for the great fun we've had understanding matroids together. I am also grateful to Carolina Benedetti, Dusty Ross, Felipe Rincón, Graham Denham, Johannes Rau, June Huh, Kris Shaw, Laura Escobar, Mariel Supina, Mario Sanchez, May-Li Khoe, and Raúl Penaguião for their valuable feedback on a draft of this manuscript.

FUNDING

This work was partially supported by United States National Science Foundation grant DMS-1855610 and Simons Fellowship 613384.

REFERENCES

[1] K. Adiprasito, J. Huh, and E. Katz, Hodge theory for combinatorial geometries. *Ann. of Math. (2)* **188** (2018), no. 2, 381–452.

[2] M. Aguiar and F. Ardila, Hopf monoids and generalized permutahedra. 2017, arXiv:1709.07504. To appear in *Mem. Amer. Math. Soc.*

[3] D. Alessandrini, Logarithmic limit sets of real semi-algebraic sets. *Adv. Geom.* **13** (2013), no. 1, 155–190.

[4] L. Allermann and J. Rau, First steps in tropical intersection theory. *Math. Z.* **264** (2010), no. 3, 633–670.

[5] P. Aluffi, Grothendieck classes and Chern classes of hyperplane arrangements. *Int. Math. Res. Not. IMRN* **8** (2013), 1873–1900.

[6] F. Ardila, *Enumerative and algebraic aspects of matroids and hyperplane arrangements*. Ph.D. thesis, Massachusetts Institute of Technology, 2003.

[7] F. Ardila, Computing the Tutte polynomial of a hyperplane arrangement. *Pacific J. Math.* **230** (2007), no. 1, 1–26.

[8] F. Ardila, Algebraic and geometric methods in enumerative combinatorics. In *Handbook of enumerative combinatorics*, pp. 3–172, Discrete Math. Appl. (Boca Raton), CRC Press, Boca Raton, FL, 2015.

[9] F. Ardila, The bipermutahedron. 2020, arXiv:2008.02295.

[10] F. Ardila, Tutte polynomials of hyperplane arrangements and the finite field method. To appear in *Handbook of the Tutte polynomial and related topics*. CRC Press, Boca Raton, FL.

[11] F. Ardila, M. Beck, and J. McWhirter, The arithmetic of Coxeter permutahedra. *Rev. Acad. Colombiana Cienc. Exact. Fis. Natur.* **44** (2020), no. 173, 1152–1166.

[12] F. Ardila, C. Benedetti, and J. Doker, Matroid polytopes and their volumes. *Discrete Comput. Geom.* **43** (2010), no. 4, 841–854.

[13] F. Ardila, F. Castillo, C. Eur, and A. Postnikov, Coxeter submodular functions and deformations of Coxeter permutahedra. *Adv. Math.* **365** (2020), 107039, 36.

[14] F. Ardila, F. Castillo, and M. Henley, The arithmetic Tutte polynomials of the classical root systems. *Int. Math. Res. Not.* **2015** (2014), no. 12, 3830–3877.

[15] F. Ardila, G. Denham, and J. Huh, Lagrangian combinatorics of matroids. 2021, arXiv:2109.11565.

[16] F. Ardila, G. Denham, and J. Huh, Lagrangian geometry of matroids. 2020, arXiv:2004.13116.

[17] F. Ardila and L. Escobar, The harmonic polytope. *Selecta Math. (N.S.)* **27** (2021), no. 5, 91, 31 pp.

[18] F. Ardila, A. Fink, and F. Rincón, Valuations for matroid polytope subdivisions. *Canad. J. Math.* **62** (2010), no. 6, 1228–1245.

[19] F. Ardila and C. J. Klivans, The Bergman complex of a matroid and phylogenetic trees. *J. Combin. Theory Ser. B* **96** (2006), no. 1, 38–49.

[20] F. Ardila, C. Klivans, and L. Williams, The positive Bergman complex of an oriented matroid. *European J. Combin.* **27** (2006), no. 4, 577–591.

[21] F. Ardila, V. Reiner, and L. Williams, Bergman complexes, Coxeter arrangements, and graph associahedra. *Sém. Lothar. Combin.* **54A** (2005/2007), B54Aj, 25 pp.

[22] F. Ardila, F. Rincón, and L. K. Williams, Positively oriented matroids are realizable. *J. Eur. Math. Soc. (JEMS)* **19** (2017), no. 3, 815–833.

[23] F. Ardila and M. Sanchez, Valuations and the Hopf monoid of generalized permutahedra. *Int. Math. Res. Not.* (2022), rnab355.

[24] C. A. Athanasiadis, Characteristic polynomials of subspace arrangements and finite fields. *Adv. Math.* **122** (1996), no. 2, 193–233.

[25] G. Balla and J. A. Olarte, The tropical symplectic Grassmannian. *Int. Math. Res. Not.* (2021), rnab267.

[26] G. M. Bergman, The logarithmic limit-set of an algebraic variety. *Trans. Amer. Math. Soc.* **157** (1971), 459–469.

[27] R. Bieri and J. R. J. Groves, The geometry of the set of characters induced by valuations. *J. Reine Angew. Math.* **347** (1984), 168–195.

[28] L. J. Billera, The algebra of continuous piecewise polynomials. *Adv. Math.* **76** (1989), no. 2, 170–183.

[29] A. Björner, The homology and shellability of matroids and geometric lattices. In *Matroid applications*, pp. 226–283, Encyclopedia Math. Appl. 40, Cambridge Univ. Press, Cambridge, 1992.

[30] A. Björner and T. Ekedahl, Subspace arrangements over finite fields: cohomological and enumerative aspects. *Adv. Math.* **129** (1997), no. 2, 159–187.

[31] A. Björner, M. Las Vergnas, B. Sturmfels, N. White, and G. M. Ziegler, *Oriented matroids. Second edn*. Encyclopedia Math. Appl. 46, Cambridge University Press, Cambridge, 1999.

[32] A. V. Borovik, I. M. Gelfand, and N. White, *Coxeter matroids*. Progr. Math. 216, Birkhäuser Boston, Inc., Boston, MA, 2003.

[33] P. Brändén and L. Moci, The multivariate arithmetic Tutte polynomial. *Trans. Amer. Math. Soc.* **366** (2014), no. 10, 5523–5540.

[34] M. Brandt, C. Eur, and L. Zhang, Tropical flag varieties. *Adv. Math.* **384** (2021), 107695.

[35] M. Brion, Piecewise polynomial functions, convex polytopes and enumerative geometry. In *Parameter spaces (Warsaw, 1994)*, pp. 25–44, Banach Center Publ. 36, Polish Acad. Sci. Inst. Math., Warsaw, 1996.

[36] T. Brylawski, A combinatorial model for series–parallel networks. *Trans. Amer. Math. Soc.* **154** (1971), 1–22.

[37] C. J. Colbourn, *The combinatorics of network reliability*. Internat. Ser. Monogr. Comput. Sci., The Clarendon Press, Oxford University Press, New York, 1987.

[38] H. H. Crapo and G.-C. Rota, *On the foundations of combinatorial theory: Combinatorial geometries. Preliminary edn*. The MIT Press, Cambridge, Mass.-London, 1970.

[39] M. D'Adderio and L. Moci, Ehrhart polynomial and arithmetic Tutte polynomial. *European J. Combin.* **33** (2012), no. 7, 1479–1483.

[40] W. Dahmen and C. A. Micchelli, On the local linear independence of translates of a box spline. *Studia Math.* **82** (1985), no. 3, 243–263.

[41] J. E. Dawson, A collection of sets related to the Tutte polynomial of a matroid. In *Graph theory, Singapore 1983*, pp. 193–204, Lecture Notes in Math. 1073, Springer, Berlin, 1984.

[42] I. P. F. Da Silva, *Quelques propriétés des matroides orientés*. Ph.D. thesis, Université Pierre-et-Marie-Curie [Paris VI], 1987.

[43] H. Derksen and A. Fink, Valuative invariants for polymatroids. *Adv. Math.* **225** (2010), no. 4, 1840–1892.

[44] C. De Concini and C. Procesi, On the geometry of toric arrangements. *Transform. Groups* **10** (2005), no. 3–4, 387–422.

[45] C. De Concini and C. Procesi, The zonotope of a root system. *Transform. Groups* **13** (2008), no. 3–4, 507–526.

[46] C. De Concini, C. Procesi, and M. Vergne, Vector partition functions and index of transversally elliptic operators. *Transform. Groups* **15** (2010), no. 4, 775–811.

[47] A. W. M. Dress and W. Wenzel, Valuated matroids. *Adv. Math.* **93** (1992), no. 2, 214–250.

[48] J. Edmonds, Submodular functions, matroids, and certain polyhedra. In *Combinatorial structures and their applications* (Proc. Calgary Internat. Conf., Calgary, Alta., 1969) pp. 69–87, Gordon and Breach, New York, 1970.

[49] R. Ehrenborg, M. Readdy, and M. Slone, Affine and toric hyperplane arrangements. *Discrete Comput. Geom.* **41** (2009), no. 4, 481–512.

[50] E. Ehrhart, Sur les polyèdres rationnels homothétiques à n dimensions. *C. R. Acad. Sci. Paris* **254** (1962), 616–618.

[51] B. Elias, N. Proudfoot, and M. Wakefield, The Kazhdan–Lusztig polynomial of a matroid. *Adv. Math.* **299** (2016), 36–70.

[52] C. Eur, Divisors on matroids and their volumes. *J. Combin. Theory Ser. A* **169** (2020), 105135, 31.

[53] C. Eur, M. Sanchez, and M. Supina, The universal valuation of Coxeter matroids. *Bull. Lond. Math. Soc.* **53** (2021), no. 3, 798–819.

[54] E. M. Feichtner and S. Yuzvinsky, Chow rings of toric varieties defined by atomic lattices. *Invent. Math.* **155** (2004), no. 3, 515–536.

[55] A. Fink, Tropical cycles and chow polytopes. *Beitr. Algebra Geom.* **54** (2013), no. 1, 13–40.

[56] A. Fink and D. E. Speyer, K-classes for matroids and equivariant localization. *Duke Math. J.* **161** (2012), no. 14, 2699–2723.

[57] W. Fulton and B. Sturmfels, Intersection theory on toric varieties. *Topology* **36** (1997), no. 2, 335–353.

[58] I. M. Gel'fand, R. M. Goresky, R. D. MacPherson, and V. V. Serganova, Combinatorial geometries, convex polyhedra, and Schubert cells. *Adv. Math.* **63** (1987), no. 3, 301–316.

[59] T. W. Geldon, *Computing the Tutte polynomial of hyperplane arrangements*. Ph.D. thesis, University of Texas, Austin, 2009.

[60] C. Greene, Weight enumeration and the geometry of linear codes. *Stud. Appl. Math.* **55** (1976), no. 2, 119–128.

[61] P. Hacking, S. Keel, and J. Tevelev, Compactification of the moduli space of hyperplane arrangements. *J. Algebraic Geom.* **15** (2006), no. 4, 657–680.

[62] J. Huh, Milnor numbers of projective hypersurfaces and the chromatic polynomial of graphs. *J. Amer. Math. Soc.* **25** (2012), no. 3, 907–927.

[63] J. Huh, Combinatorial applications of the Hodge–Riemann relations. In *Proceedings of the International Congress of Mathematicians—Rio de Janeiro 2018. Vol. IV. Invited lectures*, pp. 3093–3111, World Sci. Publ., Hackensack, NJ, 2018.

[64] J. Huh and E. Katz, Log-concavity of characteristic polynomials and the Bergman fan of matroids. *Math. Ann.* **354** (2012), no. 3, 1103–1116.

[65] D. Jensen, M. Kutler, and J. Usatine, The motivic zeta functions of a matroid. *J. Lond. Math. Soc. (2)* **103** (2021), no. 2, 604–632.

[66] S. A. Joni and G.-C. Rota, Coalgebras and bialgebras in combinatorics. In *Umbral calculus and Hopf algebras (Norman, OK, 1978)*, pp. 1–47, Contemp. Math. 6, Amer. Math. Soc., Providence, R.I., 1982.

[67] M. Juhnke-Kubitzke and D. V. Le, Flawlessness of h-vectors of broken circuit complexes. *Int. Math. Res. Not. IMRN* **5** (2018), 1347–1367.

[68] M. M. Kapranov, Chow quotients of Grassmannians. I. In *I. M. Gel'fand Seminar*, pp. 29–110, Adv. Sov. Math. 16, Amer. Math. Soc., Providence, RI, 1993.

[69] D. Kelly and G.-C. Rota, Some problems in combinatorial geometry. In *A survey of combinatorial theory*, pp. 309–312, Elsevier, 1973.

[70] A. Knutson and T. Tao, Puzzles and (equivariant) cohomology of Grassmannians. *Duke Math. J.* **119** (2003), no. 2, 221–260.

[71] L. Lafforgue, *Chirurgie des Grassmanniennes*. CRM Monogr. Ser. 19, American Mathematical Society, Providence, RI, 2003.

[72] M. Lasoń, On the toric ideals of matroids of a fixed rank. *Selecta Math. (N.S.)* **27** (2021), no. 2, 18, 17 pp.

[73] M. Lenz, The f-vector of a representable-matroid complex is log-concave. *Adv. in Appl. Math.* **51** (2013), no. 5, 543–545.

[74] L. López de Medrano, F. Rincón, and K. Shaw, Chern–Schwartz–MacPherson cycles of matroids. *Proc. Lond. Math. Soc. (3)* **120** (2020), no. 1, 1–27.

[75] D. Maclagan and B. Sturmfels, *Introduction to tropical geometry*. Grad. Stud. Math. 161, American Mathematical Society, Providence, RI, 2015.

[76] R. D. MacPherson, Chern classes for singular algebraic varieties. *Ann. of Math. (2)* **100** (1974), 423–432.

[77] P. McMullen, The polytope algebra. *Adv. Math.* **78** (1989), no. 1, 76–130.

[78] C. Merino, The chip firing game and matroid complexes. In *Discrete models: combinatorics, computation, and geometry (Paris, 2001)*, pp. 245–255 (electronic), Discrete Math. Theor. Comput. Sci. Proc., AA, Maison Inform. Math. Discrèt. (MIMD), Paris, 2001.

[79] G. Mikhalkin, Enumerative tropical algebraic geometry in $\mathbb{R}^2$. *J. Amer. Math. Soc.* **18** (2005), no. 2, 313–377.

[80] G. Mikhalkin and J. Rau, Tropical geometry. Nov. 16, 2018. Available at author's website: https://math.uniandes.edu.co/~j.rau/downloads/main.pdf.

[81] L. Moci, A Tutte polynomial for toric arrangements. *Trans. Amer. Math. Soc.* **364** (2012), no. 2, 1067–1088.

[82] K. Murota, Convexity and Steinitz's exchange property. *Adv. Math.* **124** (1996), no. 2, 272–311.

[83] T. Nakasawa, Zur Axiomatik der linearen Abhängigkeit. I. *Sci. Rep. Tokyo Bunrika Daigaku, Sect. A* **2** (1935), no. 43, 235–255.

[84] P. Nelson, Almost all matroids are nonrepresentable. *Bull. Lond. Math. Soc.* **50** (2018), no. 2, 245–248.

[85] S. Oh, Generalized permutohedra, h-vectors of cotransversal matroids and pure O-sequences. *Electron. J. Combin.* **20** (2013), no. 3, 14, 14 pp.

[86] P. Orlik and L. Solomon, Combinatorics and topology of complements of hyperplanes. *Invent. Math.* **56** (1980), no. 2, 167–189.

[87] J. Oxley, *Matroid theory. Second edn.* Oxf. Grad. Texts Math. 21, Oxford University Press, Oxford, 2011.

[88] A. Postnikov, Permutohedra, associahedra, and beyond. *Int. Math. Res. Not. IMRN* **6** (2009), 1026–1106.

[89] H. Randriamaro, The Tutte polynomial of symmetric hyperplane arrangements. *J. Knot Theory Ramifications* **29** (2020), no. 3, 2050004, 19 pp.

[90] F. Rincón, Isotropical linear spaces and valuated delta-matroids. *J. Combin. Theory Ser. A* **119** (2012), no. 1, 14–32.

[91] G.-C. Rota, On the foundations of combinatorial theory. I. Theory of Möbius functions. *Z. Wahrsch. Verw. Gebiete* **2** (1964), 340–368.

[92] G.-C. Rota, Combinatorial theory, old and new. In *Actes du Congrès International des Mathématiciens (Nice, 1970), Tome 3*, pp. 229–233, Gauthier-Villars, Paris, 1971.

[93] C. Sabbah, Quelques remarques sur la géométrie des espaces conormaux. *Astérisque* **130** (1985), 161–192.

[94] B. E. Sagan, Why the characteristic polynomial factors. *Bull. Amer. Math. Soc. (N.S.)* **36** (1999), no. 2, 113–133.

[95] W. R. Schmitt, Antipodes and incidence coalgebras. *J. Combin. Theory Ser. A* **46** (1987), no. 2, 264–290.

[96] M.-H. Schwartz, Classes caractéristiques définies par une stratification d'une variété analytique complexe. I. *C. R. Acad. Sci. Paris* **260** (1965), 3262–3264.

[97] D. E. Speyer, Tropical linear spaces. *SIAM J. Discrete Math.* **22** (2008), no. 4, 1527–1558.

[98] D. E. Speyer, A matroid invariant via the K-theory of the Grassmannian. *Adv. Math.* **221** (2009), no. 3, 882–913.

[99] D. Speyer and L. Williams, The tropical totally positive Grassmannian. *J. Algebraic Combin.* **22** (2005), no. 2, 189–210.

[100] R. P. Stanley, Cohen–Macaulay complexes. In *Higher combinatorics (Proc. NATO Advanced Study Inst., Berlin, 1976)*, pp. 51–62, NATO Adv. Stud. Inst. Ser., Ser. C, Math. Phys. Sci. 31, Reidel, Dordrecht, 1977.

[101] R. P. Stanley, Unimodal sequences arising from Lie algebras. In *Combinatorics, representation theory and statistical methods in groups*, pp. 127–136, Lect. Notes Pure Appl. Math. 57, Dekker, New York, 1980.

[102] R. P. Stanley, Combinatorial applications of the hard Lefschetz theorem. In *Proceedings of the International Congress of Mathematicians, Vol. 1, 2 (Warsaw, 1983)*, pp. 447–453, PWN, Warsaw, 1984.

[103] R. P. Stanley, A zonotope associated with graphical degree sequences. In *Applied geometry and discrete mathematics*, pp. 555–570, DIMACS Ser. Discrete Math. Theoret. Comput. Sci. 4, Amer. Math. Soc., Providence, RI, 1991.

[104] B. Sturmfels, Equations defining toric varieties. In *Algebraic geometry—Santa Cruz 1995*, pp. 437–449, Proc. Sympos. Pure Math. 62, Amer. Math. Soc., Providence, RI, 1997.

[105] B. Sturmfels, *Solving systems of polynomial equations*. CBMS Reg. Conf. Ser. Math. 97, Published for the Conference Board of the Mathematical Sciences, Washington, DC; by the American Mathematical Society, Providence, RI, 2002.

[106] B. Sturmfels, Personal communication. 2003.

[107] M. Takeuchi, Free Hopf algebras generated by coalgebras. *J. Math. Soc. Japan* **23** (1971), 561–582.

[108] W. T. Tutte, On dichromatic polynominals. *J. Combin. Theory* **2** (1967), 301–320.

[109] R. Vakil, Murphy's law in algebraic geometry: badly-behaved deformation spaces. *Invent. Math.* **164** (2006), no. 3, 569–590.

[110] A. Weil, Numbers of solutions of equations in finite fields. *Bull. Amer. Math. Soc.* **55** (1949), 497–508.

[111] D. J. A. Welsh, *Complexity: knots, colourings and counting*. London Math. Soc. Lecture Note Ser. 186, Cambridge University Press, Cambridge, 1993.

[112] N. L. White, A unique exchange property for bases. *Linear Algebra Appl.* **31** (1980), 81–91.

[113] H. Whitney, On the abstract properties of linear dependence. *Amer. J. Math.* **57** (1935), no. 3, 509–533.

[114] L. Williams, The positive Grassmannian, the amplituhedron, and cluster algebras. In *Proceedings of the international congress of mathematicians (St. Petersburg)*, 2022.

[115] T. Zaslavsky, Facing up to arrangements: face-count formulas for partitions of space by hyperplanes. *Mem. Amer. Math. Soc.* **1** (1975), no. 1, 154, vii+102 pp.

FEDERICO ARDILA-MANTILLA

San Francisco State University, San Francisco, CA, USA, and Universidad de Los Andes, Bogotá, Colombia, federico@sfsu.edu

GRAPH AND HYPERGRAPH PACKING

JULIA BÖTTCHER

ABSTRACT

Packing problems in combinatorics concern the edge disjoint embedding of a family of guest (hyper)graphs into a given host (hyper)graph. Questions of this type are intimately connected to the field of design theory, and have a variety of significant applications. The area has seen important progress in the last two decades, with a number of powerful new methods developed. Here, I will survey some major results contributing to this progress, alongside background, and some ideas concerning the methods involved.

MATHEMATICS SUBJECT CLASSIFICATION 2020

Primary 05B40; Secondary 05C51, 05B30, 05C78

KEYWORDS

Graph packing, graph decomposition

ICM INTERNATIONAL CONGRESS OF MATHEMATICIANS 2022 JULY 6–14

Proc. Int. Cong. Math. 2022, Vol. 6, pp. 4542–4566
DOI 10.4171/ICM2022/136

Published by EMS Press

1. INTRODUCTION AND BACKGROUND

Assume that we want to test the efficacy of n drugs. One challenge when setting up an experiment for this is that the efficacy may vary with different characteristics of the individuals a drug is used by, such as age, ethnicity, sex or existing medical conditions; and we want to control for such variances. One method applied to address this in statistical experiments is *blocking*: Individuals are grouped into blocks of similar characteristics, so that within one block we can directly compare outcomes. For simplicity, let us assume that each of these blocks has size q, which we shall think of as relatively small, allowing for fine-grained control. Let us also assume that the number n of drugs tested is large compared to q. In this scenario, if we want to gain an overall picture of the (pairwise) relative efficacy of the drugs, and each individual is given one of the drugs, we also want the property that each pair of drugs is tested on two individuals from the same block. The most efficient way of guaranteeing this is to require additionally that each pair of drugs appears only in one block. Mathematically, what this is asking for is a certain type of combinatorial (block) design (for a definition see Section 3).

Various generalizations are natural: We could ask for r-wise comparisons instead of pairwise comparisons; we could ask that each set of r drugs is contained in exactly λ instead of only one block; we could allow blocks of different sizes; or we could ask that for each block there is always a collection of other blocks, disjoint among themselves and the chosen block, that partition the set of administered drugs (this is called a resolvable design; a different example of this is given below). All these and many others have been considered in what is known as design theory (for a comprehensive overview of the area, see [16]). These designs are also special types of so-called (hyper)graph packings (we formally introduce packings in Section 2). Two of the most fundamental questions concerning these mathematical objects are: For which parameters do these packings exist? And how can they be constructed? Combinatorics has recently seen particularly rich progress concerning these questions, alongside the development of powerful new methods. In this survey I will outline some of the most important of these results and methods.

The connection of designs to statistical experiments was formalized by Fisher in the first half of the 20th century (see, e.g., [26]). Historically, however, questions related to designs appeared already earlier; the following description is based on [66]. In the 1830s, motivated by the study of certain plane cubic curves, Plücker discovered a particular design, the so-called 9-point affine plane. Later extensions of his work by Fano were important for the development of projective geometry. Soon after, designs featured in recreational mathematics. The 1844 edition of the *Lady's and Gentleman's Diary*, according to its title page a journal "designed principally for the amusement and instruction of students in mathematics: comprising many useful and entertaining particulars, interesting to all persons engaged in that delightful pursuit", presented the following prize problem, posed by the editor, Revd. Woolhouse:

- Determine the number of combinations that can be made out of n symbols, p symbols in each; with this limitation, that no combination of q symbols, which may appear in any one of them shall be repeated in any other.

This asks for a partial design maximizing the number of blocks. In the 1850 edition of the *Lady's and Gentleman's Diary*, Revd. Kirkman posed what is nowadays known as "Kirkman's schoolgirls problem":

- Fifteen young ladies in a school walk out three abreast for seven days in succession: it is required to arrange them daily, so that no two shall walk twice abreast.

This asks for a resolvable design with $n = 15$, block size $q = 3$, and with $r = 2$. A quest for solutions, generalizations, along with rediscoveries, and discussions over priority ensued. See [66] for a detailed historical account and the mathematicians involved in these early developments. I will return to designs in Section 3.

Another recreationally motivated packing problem was posed by Ringel in 1967 (see [32]) and had the well-being of the mathematical community in mind: At Oberwolfach meetings, each meal the participants are assigned a seat at one of the possibly differently sized tables. Is a succession of assignments possible for these meals, such that no participant sits next to another participant twice? This asks for a packing of so-called cycle factors in the complete graph on n vertices, where n is the number of meeting participants, and each cycle represents one of the tables. I will return to this in Section 4.

These are just some examples. There is a wealth of other prominent problems that can be formulated as graph or hypergraph packing problems, including the search for Latin squares, or orthogonal Latin squares. Applications also arise in the construction of certain codes, or in information security. (See, for example, [40] for more details.) In this survey though I will not explore these further, but concentrate on (mainly recent) mathematical progress concerning packings instead.

Organization. It is not my goal in this contribution to exhaustively survey the vast amount of results that have so far been obtained concerning designs and packings. Instead, I aim to highlight some important recent progress and to discuss some newly developed techniques that made this progress possible. The remainder is organized as follows. In Section 2 some notation and basic definitions are introduced. In Section 3 we discuss recent breakthrough results concerning designs. Sections 4 and 5 concentrate on packing results for cycles, and trees, respectively. Section 6 considers packing problems for more general classes of graphs, while Section 7 briefly mentions related results for hypergraphs. In Section 8, finally, some important open problems are collected.

2. NOTATION AND BASIC DEFINITIONS

We denote the set $\{1, \dots, n\}$ by $[n]$. A *graph* $G = (V, E)$ consists of a set of vertices V (which is always finite here) and a set of edges $E \subseteq \binom{V}{2}$, where each edge contains two different elements of V. An *r-uniform hypergraph* $H = (V, E)$ generalizes this notion in that it allows r vertices in each edge, that is, it requires $E \subseteq \binom{V}{r}$. We write $V(H)$ and $E(H)$ for the vertices and edges of H, respectively, and $e(H)$ for $|E(H)|$. For a vertex u and a set of vertices S in a graph G, the *neighborhood* $N_G(u)$ of u is the set of all vertices v such that uv is an edge, the *degree* of v is $|N_G(v)|$, and $N_G(S) = \bigcap_{u \in S} N_G(u)$ is the *common neighborhood* of S. Similarly, for a set U of $r-1$ vertices in an r-uniform hypergraph H, the neighborhood $N_H(U)$ of U is the set of all vertices v such that Uv is an edge. For a (hyper)graph H and a vertex set S, we write $H \setminus S$ for the sub(hyper)graph of H induced on vertex set $V(H) \setminus S$. A sub(hyper)graph of H is called *spanning* if it uses all vertices of H.

A *complete graph* K_n on n vertices is a graph in which all pairs of vertices form an edge. Analogously, the complete r-uniform hypergraph $K_n^{(r)}$ contains all r-sets as edges. A *path* in a graph is a sequence of different vertices $v_1, \dots, v_\ell$ such that $v_i v_{i+1}$ is an edge for every $i \in [\ell - 1]$; a *cycle* in a graph is a path $v_1, \dots, v_\ell$ plus the edge $v_\ell v_1$. A *tree* on n vertices is a graph with $n-1$ edges without cycles. A graph is r*-regular* if each vertex has degree r. A *cycle factor* of a graph H on n vertices is a 2-regular subgraph of H on n vertices. For a (hyper)graph F, an F*-factor* in a (hyper)graph H on n vertices is a spanning sub(hyper)graph of H consisting of vertex disjoint copies of F. A *Hamilton cycle* in a graph is a cycle using all the vertices. In an r-uniform hypergraph H, a *tight Hamilton cycle* is given by an ordering $v_1, \dots, v_n$ of the vertices of H such that $v_i, \dots, v_{i+r-1}$ forms an edge for each $i \in [n]$, where indices are taken modulo n. (The Hamilton cycle is then formed by the involved edges.)

Definition 2.1 (Packing, decomposition). For a collection $G_1, \dots, G_t$ of (hyper)graphs, and another (hyper)graph H, the family $(G_1, \dots, G_t)$ is said to *pack into* H if there are edge-disjoint copies of $G_1, \dots, G_t$ in H. The packing is called *perfect* if it uses exactly all the edges of H once, that is, if $\sum_{i \in [t]} e(G_i) = e(H)$. It is called *almost-perfect* if $\sum_{i \in [t]} e(G_i) = (1 - o(1))e(H)$. A perfect packing of $(G_1, \dots, G_t)$ is also called a *decomposition* of H into $(G_1, \dots, G_t)$. We occasionally refer to H as the *host (hyper)graph* and to the G_i as the *guest (hyper)graphs* of the packing.

More generally, given a natural number λ, a λ*-fold packing* of $(G_1, \dots, G_t)$ into H allows edges of H to be used more than once but not more than λ times, where we require, however, that the embeddings use distinct subgraphs of H. A packing (in the sense above) thus is a 1-fold packing. A λ-fold packing is *perfect* if it uses each edge of H exactly λ times. A perfect λ-fold packing of $(G_1, \dots, G_t)$ into H is also called a λ*-fold decomposition* of H into $(G_1, \dots, G_t)$.

3. DESIGNS

Given a set X of size n and a family S of distinct subsets of X, each of size q, we say that S is a *design* with parameters (n, q, r, λ) if every subset Y of X of size r is contained in exactly λ members of S. For example, it is easy to see that for $X = [7]$ the family $S = \{123, 145, 167, 246, 257, 356, 347\}$, where we denote the subset $\{s_1, s_2, s_3\}$ of S

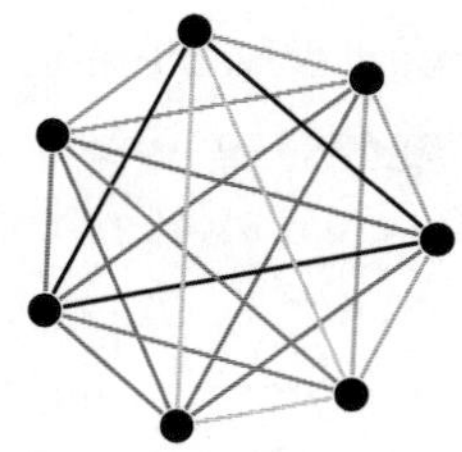

FIGURE 1
A design with parameters (7, 3, 2, 1) as a packing of triangles in K_7.

by $s_1 s_2 s_3$, is a design with parameters $(7, 3, 2, 1)$. In the language of hypergraphs, the problem of finding a design with parameters (n, q, r, λ) translates to the problem of finding a perfect λ-fold packing of complete r-uniform hypergraphs $K_q^{(r)}$ on q vertices in a complete r-uniform hypergraph $K_n^{(r)}$ on n vertices. A design is *resolvable* if (in the language of hypergraphs) it is also a λ-fold packing of $K_q^{(r)}$-factors. For example, the design with parameters $(7, 3, 2, 1)$ corresponds to a packing of triangles in the complete graph K_7 which uses each edge exactly once (see Figure 1). This cannot be a resolvable design since 7 is not divisible by 3.

What are necessary conditions for a design with certain parameters to exist? Firstly, it is clear that, for example, no design with parameters $(5, 3, 2, 1)$ can exist because the complete graph K_5 on 5 vertices has $\binom{5}{2} = 10$ edges, which is not divisible by 3. On the other hand, for parameters $(6, 3, 2, 1)$ the number of edges in K_6 is $\binom{6}{2} = 15$, hence divisible by 3; but each vertex of K_6 is contained in 5 edges, which cannot all edge-disjointly be covered by triangles because each triangle would use 2 edges at the vertex. The conditions resulting from simple obstacles like this are called divisibility conditions. In full generality they are as follows.

Definition 3.1 (Divisibility conditions). The parameters (n, q, r, λ) satisfy the *divisibility conditions*, if $\binom{q-i}{r-i}$ divides $\lambda\binom{n-i}{r-i}$ for every $0 \le i \le r-1$.

It is clear that these conditions are necessary because any set I of i vertices in the complete r-uniform hypergraph on n vertices is contained in $\binom{n-i}{r-i}$ edges, each of which we need to cover λ times; and for this we can use, from any copy of the complete r-uniform hypergraph using the vertices of I, the $\binom{q-i}{r-i}$ edges touching I. The existence conjecture for designs states that these conditions are also sufficient, apart from some small counterexamples.

Conjecture 3.2 (existence conjecture for designs). *Given q, r, and λ, there is n_0 so that for each $n \ge n_0$, if (n, q, r, λ) satisfy the divisibility conditions, then there is a design with these parameters.*

This conjecture and related problems inspired much work. The case of packing triangles in a complete graph, that is, $r = 2$, $q = 3$, and $\lambda = 1$ was already solved by Kirkman (see **[66]**). It took much longer to solve the graph case for all q: In a celebrated series of papers, Wilson settled the problem for $r = 2$ in the 1970s **[67, 68, 70]**. Ray-Chaudhuri and

Wilson [59] established the existence of resolvable designs in the graph case. In the 1980s Teirlinck [65] proved that nontrivial designs exist for all r (and some q and λ). Already before that, natural variations of the problem were considered. Graver and Jurkat [31] and independently Wilson [69] proved that the divisibility conditions are sufficient for so-called integral designs, where we allow the assignment of arbitrary integer weights to q-sets (instead of just weights 0 and 1) and these have to add up to λ on each r-set. Rödl [61], on the other hand, established the existence of almost-designs: families of subsets of size q that cover all but a small fraction of the r-sets of the ground set. The following theorem makes this precise for the case $\lambda = 1$.

Theorem 3.3 (Rödl [61]). *Given $r, q \in \mathbb{N}$ with $1 \leq r \leq q$ and given $\gamma > 0$, there is n_0 such that for each $n \geq n_0$, there is a partition of the edges of $K_n^{(r)}$ into edge-disjoint copies of $K_q^{(r)}$ and a leftover set of size at most γn^r.*

This result did not only represent important progress, but fundamentally influenced Combinatorics through the novel technique its proof introduced: the so-called *Rödl nibble*, which has been used to resolve a multitude of other important problems.

Let us briefly sketch the basic idea for the proof of Theorem 3.3 in the special case $r = 2$ and $q = 3$. In this case we want to pack roughly $\frac{1}{6}(1 - \gamma')n^2$ triangles in the complete graph H on n vertices. We approach this by embedding triangles *randomly*. Now, it is clear that we cannot simply randomly throw in all the triangles at once since this would lead to lots of overlaps of triangles on edges. However, if instead we randomly throw in only a small constant proportion, say $\alpha_1 n^2$ triangles, then the following will be true. Among the $\alpha_1 n^2$ randomly embedded triangles in expectation only a small proportion, of order $\alpha_1^2 n^2$, will overlap. So, assuming we have a typical outcome of random choices, we can simply discard all overlapping triangles, and the remainder will still be a packing of more than $\frac{1}{2}\alpha_1 n^2$ triangles. Rödl's idea now was to iterate this procedure in the following way. We remove from H all edges that have been used in the packing just obtained. One can then show that what remains of H has good quasirandomness properties. This is why another random embedding round of triangles will be successful: We embed $\alpha_2 n^2$ triangles randomly into H, with high probability not many of these will overlap, which we can again discard. We can then again delete all used edges from H and proceed to the next round, and so on, until almost all edges are used (at which point the error hidden in the quasirandomness condition gets out of control). The choice of the constants α_i here is somewhat delicate but we shall not discuss this here further (see, e.g., [6, CHAPTER 4] for more details).

Returning to perfect packings, Kuperberg, Lovett, and Peled [51] proved the existence of nontrivial designs for a large range of parameters (but with λ comparatively large) before, in a celebrated breakthrough, Keevash [36] resolved the existence conjecture. The result Keevash obtains is stronger in that it allows more generally for packings in all hypergraphs with certain quasirandomness properties.

An n-vertex r-uniform hypergraph H is called (ε, a)-*typical*, if every set $A \subseteq \binom{V(H)}{r-1}$, that is, of subsets of $V(H)$ of size $r-1$, such that $|A| \leq a$ satisfies $|\bigcap_{U \in A} N_H(U)| = (1 \pm \varepsilon)d^{|A|}n$, where $d = |E(H)|/\binom{n}{r}$ is the *density* of H. The mandated sizes of common neighborhoods in this definition is what we would expect to see in a random graph of density d. The divisibility conditions are adapted to this setting of an incomplete host hypergraph in the obvious way: We say that H is *$K_q^{(r)}$-divisible* if for each $0 \leq i \leq r$ and every set I of i vertices in H we have that $\binom{q-i}{r-i}$ divides $|\{e \in E(H) : I \subseteq e\}|$. Keevash's result then reads as follows.

Theorem 3.4 (Keevash [36]). *Given $q > r \geq 1$ and $\lambda \geq 1$, there exist $\varepsilon_0, \alpha > 0$ and $s, n_0 \in \mathbb{N}$ so that the following holds. Let H be a $K_q^{(r)}$-divisible (ε, s)-typical r-uniform hypergraph on $n \geq n_0$ vertices with $e(H) = d\binom{n}{r}$ edges such that $d \geq n^{-\alpha}$ and $\varepsilon \leq \varepsilon_0 d^{s^2}$. Then H has a λ-fold decomposition into $K_q^{(r)}$-copies.*

Keevash's proof of this result combines the nibble method with a powerful new approach, which he calls *randomized algebraic construction*. In its underlying philosophy, this in turn can be seen as inspired by the so-called *absorbing method*, an important contemporary technique in combinatorics pioneered by Krivelevich [48] and Rödl, Ruciński, and Szemerédi [62]. Roughly, the idea of the absorbing method is as follows: We set aside at the start a clever structure, which we call the absorber. We then obtain an almost-solution to our problem (often with the help of certain random processes, or greedy methods), leaving a small leftover. We then use the absorber to incorporate the leftover into the almost-solution, obtaining a full solution. In Keevash's proof the absorber is constructed by using edge-disjoint $K_q^{(r)}$-copies satisfying certain algebraic relations (whose definition uses some randomness); showing that this construction can absorb any leftover is complicated and requires a sequence of intricate steps. See [39] for a more detailed outline of this approach, and [37] for a detailed discussion in the special case of triangle decompositions.

Using a different influential variation of the absorbing method called *iterative absorption*, Glock, Kühn, Lo, and Osthus [30] provided a different proof of this result and more: They establish the existence of hypergraph F-designs, that is, perfect packings of copies of an arbitrary fixed r-uniform hypergraph F into the complete r-uniform hypergraph on n vertices (where n is large compared to the number of vertices in F) when suitable divisibility conditions are satisfied. For graphs, this result is due to Wilson [71].

Given F and $0 \leq i \leq r-1$, we define d_i to be the greatest common divisor of all $|\{e \in E(F) : I \subseteq e\}|$ such that I is a set of i vertices in F. We say that an r-uniform hypergraph H is *F-divisible* if for each $0 \leq i \leq r$ and every set I of i vertices in H we have that d_i divides $|\{e \in E(H) : I \subseteq e\}|$.

Theorem 3.5 (Glock, Kühn, Lo, and Osthus [30]). *Given $q > r \geq 1$, and $\varepsilon, d > 0$ such that $\varepsilon \leq 0.9(d/2)^s/(\tilde{q}^r 4^{\tilde{q}})$, where $\tilde{q} := 2q \cdot q!$ and $s := 2^r\binom{\tilde{q}+r}{r}$, there are $n_0 \in \mathbb{N}$ and $\gamma > 0$ such that the following holds. Let F be any r-uniform hypergraph on q vertices and H be an F-divisible (ε, s)-typical r-uniform hypergraph on $n \geq n_0$ vertices with $e(H) = d\binom{n}{r}$ edges, and let $\lambda \leq \gamma n$. Then H has a λ-fold decomposition into F-copies.*

The basic idea of iterative absorption is to repeatedly apply an absorbing-type technique, making the leftover more and more structured in every round, until it is structured enough so that it can be absorbed entirely. In the case of [30] "structured enough" means that before the last absorption step, the leftover will be contained in a constant-sized part of H, so that only a constant number of different possibilities for the leftover remain. For these constant number of possibilities, one can prepare at the very start by setting aside for each of them a suitable absorber. The details are involved and require lots of complex ideas, also building on previous work in **[12,13,29,45,50]**. An excellent exposition of the application of iterative absorption for obtaining triangle decompositions can be found in **[11]**.

Keevash **[38]** greatly extended Theorem 3.4 to a decomposition result allowing a partite setting. This means that the host hypergraph as well as the guest hypergraphs come with a partition and we require that vertices of part i in a guest hypergraph gets embedded into part i of the host hypergraph. This immediately allows the construction of resolvable designs via the following observation: Assume we want to obtain a resolvable $K_q^{(r)}$-decomposition of H, a hypergraph on n vertices. Then we construct an auxiliary hypergraph $\tilde{H}$ whose vertices come in two parts, one containing the vertices of H, and the other containing $(r-1)n/q$ new vertices $(u_{i,j})_{i\in[n/q],j\in[r-1]}$. The edges of $\tilde{H}$ are the edges of H and additionally for each vertex $v \in V(H)$ and each $i \in [n/q]$ we add the edge $\{v, u_{i,1}, \dots, u_{i,r-1}\}$. Also, let G be the graph obtained from $K_q^{(r)}$, which forms one part of G, by adding $r-1$ new vertices $y_1, \dots, y_{r-1}$, which form the second part, and adding all edges $\{x, y_1, \dots, y_{r-1}\}$ with $x \in V(K_q^{(r)})$. Then, a decomposition of $\tilde{H}$ into copies of G which respects the partition automatically gives a resolvable $K_q^{(r)}$-decomposition of H.

In fact, the main result in **[38]** is even more general in that it allows scenarios where the edges of the hypergraphs are colored, where multihypergraphs and ordered edges are allowed. This result is powerful and general, and stating it is complex. To convey an idea of what can be handled, let us look at the special case of graphs, which is handled in Theorem 3.10 and taken from **[2]**. A similar formulation of a special case is given in **[41, THEOREM 3.4]**. See also **[40]** for more general special cases.

Theorem 3.10 handles *partially directed multigraphs*, that is, $(V, E \dot{\cup} D)$, where V is the vertex set, E is a set of undirected edges, and D is a set of directed edges, with multiedges in E and D allowed, antiparallel directed edges allowed, but no loops. For an integer D, a partially directed multigraph is $[D]$-*edge-colored* if each edge (directed or undirected) is assigned a color from $[D]$. We are interested in obtaining decompositions in this colored setting: Let $\mathcal{G}$ be a family of $[D]$-edge-colored partially directed multigraphs on vertex set $[q]$, and let H be a $[D]$-edge-colored partially directed multigraph on vertex set $[n]$. We say that H has a $\mathcal{G}$-*decomposition* if the edges of G can be partitioned into copies of partially directed multigraphs from $\mathcal{H}$ that preserve the coloring. We further operate in a partite setting, where we allow edges inside the parts. The restrictions on the host graph H and the guest graphs G are as follows.

Definition 3.6 (Compatible partite partially directed multigraphs). Let $D, q, n \in \mathbb{N}$, let $\mathcal{P} = \{P_1, \dots, P_t\}$ be a partition of $[q]$ and $\mathcal{P}' = \{P'_1, \dots, P'_t\}$ be a partition of $[n]$. Let $\mathcal{G}$ be a family of partially directed multigraphs on $[q]$, and H be a partially directed multigraph on $[n]$, with $\mathcal{G}$ and H all $[D]$-edge-colored. Further, for each color $d \in [D]$, assume we are given a pair $(i, j) \in [t]^2$, which call the *color location* of d, and each color is specified as being either a *directed color* or an *undirected color*.

In this case, we say that H and $\mathcal{G}$ are *(n, q)-compatible partite partially directed multigraphs* with partitions $\mathcal{P}$ and $\mathcal{P}'$ if the following hold for each color d and its color location (i, j):

(i) If d is a directed (undirected) color, then all edges in H and in $\mathcal{G}$ of color d are directed (undirected).

(ii) In H all edges of color d start in P'_i and end in P'_j, and in all $G \in \mathcal{G}$ all edges of color d start in P_i and end in P_j.

(iii) For each $G \in \mathcal{G}$, there are no parallel (directed or undirected) or antiparallel edges in G. (In H parallel and antiparallel edges are allowed.)

Keevash's result requires a number of very general conditions under which the desired decompositions exist, which we define next in our specific setup. We start with the divisibility conditions. For a color d and an edge-colored partially directed multigraph F, we let $e_d(F)$ denote the number of edges in F colored d, and for a vertex v in F we let $\deg_{F,d}(x)$, $\deg^{\text{out}}_{F,d}(x)$, and $\deg^{\text{out}}_{F,d}(x)$, respectively, denote the number of undirected edges of color d incident to v in D, the number of directed edges of color d leaving v in F, and the number of directed edges of color d entering v in F, respectively.

Definition 3.7 (Partite divisibility conditions). Let $\mathcal{G}$, H, and $\mathcal{P}$, $\mathcal{P}'$ be as in Definition 3.6. We say that $(H, \mathcal{P}')$ is *$(\mathcal{G}, \mathcal{P})$-divisible* if the following hold:

(i) There are integers $(m_G)_{G\in\mathcal{G}}$ such that for each $d \in [D]$ we have

$$e_d(H) = \sum_{G\in\mathcal{G}} m_G \cdot e_d(G).$$

(ii) For each $i \in [t]$ and every vertex $v \in P'_i$, there are integers $(m_{G,x})_{G\in\mathcal{G},x\in P_i}$ such that for each undirected color $d \in [D]$ we have

$$\deg_{H,d}(v) = \sum_{G\in\mathcal{G},x\in P_i} m_{G,x} \cdot \deg_{G,d}(x),$$

and for each directed color $d \in [D]$ we have

$$\deg^{\text{out}}_{H,d}(v) = \sum_{G\in\mathcal{G},x\in P_i} m_{G,x} \cdot \deg^{\text{out}}_{G,d}(x) \quad \text{and}$$

$$\deg^{\text{in}}_{H,d}(v) = \sum_{G\in\mathcal{G},x\in P_i} m_{G,x} \cdot \deg^{\text{in}}_{G,d}(x).$$

Further, the following regularity condition is required, which can be seen as a robust fractional decomposition requirement mandating that suitably weighted copies of the guest graphs G in the host graph H are distributed regularly on edges of H. We denote the fact that a colored subgraph G' of H is a copy of some $G \in \mathcal{G}$ (with colors preserved) by writing $G' \sim_H \mathcal{G}$.

Definition 3.8 (Regularity condition). Let $\mathcal{G}$, H, $\mathcal{P}$, $\mathcal{P}'$, q, and n be as in Definition 3.6 and let $c, \omega > 0$ be reals. We say that $(H, \mathcal{P}')$ is *$(\mathcal{G}, \mathcal{P}, c, \omega)$-regular*, if there are weights $(w_{G'})_{G' \sim_H \mathcal{G}}$ with $w_{G'} \in [\omega \cdot n^{2-q}, \frac{1}{\omega} \cdot n^{2-q}]$ such that for each edge $e \in E(H)$ we have

$$\sum_{G' \sim_H \mathcal{G} : e \in E(G')} w_{G'} = (1 \pm c).$$

Finally, Keevash's result uses the following condition that considers any guest graph vertex $x \in V(G_i)$ and requires that for every choice of linear-sized sets A_y for each other vertex y in G_i, where A_y is chosen in the part of $\mathcal{P}'$ where y should be embedded, the common neighborhood of these sets A_y in the part of $\mathcal{P}'$ where x should be embedded is of linear size. We first need to make precise what we mean by common neighborhood in this setting. For any two vertices y, x in some guest graph G with $x \in P_i$ and a set A of vertices in the host graph H, we define the neighborhood of A in H mandated by yx as

$$N_H^{(y,x,G)}(A) = \begin{cases} N_{H,d}(A) \cap P_i' & \text{if } yx \in E \text{ is undirected,} \\ N_{H,d}^{\text{out}}(A) \cap P_i' & \text{if } yx \in E \text{ is directed towards } x, \\ N_{H,d}^{\text{in}}(A) \cap P_i' & \text{if } yx \in E \text{ is directed towards } y, \\ P_i' & \text{if there is no edge with endpoints } x \text{ and } y, \end{cases}$$

where d is the unique color of the (directed or undirected) edge with endpoints x and y (if it exists), $N_{H,d}(A)$ is the set of common neighbors of A in H of color d, and $N_{H,d}^{\text{out}}(A)$, $N_{H,d}^{\text{in}}(A)$ are defined analogously.

Definition 3.9 (Vertex extendibility). Let $\mathcal{G}$, H, $\mathcal{P}$, $\mathcal{P}'$, q, and n be as in Definition 3.6, let h be an integer, and $\omega > 0$ be a real. We say that $(H, \mathcal{P}')$ is *$(\mathcal{G}, P, \omega, h)$-vertex extendable*, if the following holds for every guest graph $G \in \mathcal{G}$ and each of its vertices $x \in [q]$. For every choice of pairwise disjoint vertex sets $(A_y)_{y \in [q] \setminus \{x\}}$ in H, one for each vertex in G other than x, of size $|A_y| \leq h$ and with $A_y \subseteq P_j'$ whenever $y \in P_j$, we have

$$\left| \bigcap_{y \in [q] \setminus \{x\}} N_H^{(y,x,H)}(A_y) \right| \geq \omega n.$$

The following is a special case of Keevash's result [40, THEOREM 19] (see also [2] for more detailed explanations).

Theorem 3.10. *Given $q, D \in \mathbb{N}$, and $\sigma > 0$, there exist $\omega_0 > 0$, $n_0 \in \mathbb{N}$, such that for $q' = \max\{q, 8 + \log_2(1/\sigma)\}$, $h' = 2^{50q'^3}$, $\delta = 2^{-10^3 q'^5}$ the following holds for every $n > n_0$ and $\omega \in (n^{-\delta}, \omega_0)$.*

Let H and $\mathcal{G} = (G_i)_{i \in [m]}$ be $[D]$-edge-colored and (n, q)-compatible partite partially directed multigraphs with partitions $\mathcal{P} = (P_i)_{i \in [t]}$ and $\mathcal{P}' = (P_i')_{i \in [t]}$, respectively, such that $|P_i'| \geq \sigma n$ for each $i \in [t]$. If $(H, \mathcal{P}')$ is $(\mathcal{G}, \mathcal{P})$-divisible, $(\mathcal{G}, \mathcal{P}, \omega^{h^{20}}, \omega)$-regular, and $(\mathcal{G}, \mathcal{P}, \sqrt[q'h]{\omega}, h)$-vertex-extendable, then H has a $\mathcal{G}$-decomposition.

Let us illustrate the power of this result with a simple example (that will be useful for packing spanning trees in Section 5). Assume we are given a partially directed multigraph H

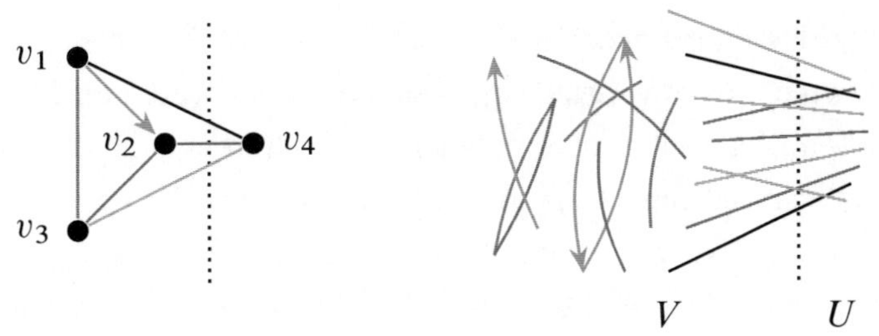

FIGURE 2
A diamond and an illustration of a colored partially directed multigraph with partition $U\dot{\cup}V$ that can be decomposed into diamond-copies with the help of Theorem 3.10.

with a partition $V(H) = V\dot{\cup}U$ such that U is much smaller than V. There are edges of 6 different colors in H; edges of the first color (let us call it green) are directed, the others are undirected; edges of the first three colors (green and, say, blue and red) run within V and edges of the other three colors (say, grey, black, and purple) run between V and U. Our goal is to pack colored *diamonds* in H such that all green, blue, red, and grey edges are used in H. Here, our diamonds D have four vertices $v_1, \dots, v_4$, a directed green edge v_1v_2, and undirected edges v_1v_3 of color blue, v_2v_3 of color red, v_1v_4 of color black, v_2v_4 of color purple, and v_3v_4 of color grey. See Figure 2 for an illustration. In such a packing we do not need to use all black and purple edges of H (which makes the conditions we shall need in order to obtain such a packing easier). In order to apply Theorem 3.10 in this setting, we thus let $\mathcal{G}$ have three elements: D, a graph containing a single black edge, and a graph containing a single purple edge. We can then apply Theorem 3.10, if appropriate divisibility conditions for the edges of the different colors are satisfied, and the colored edges in H are suitably quasirandomly distributed: The quasirandomness conditions will imply the more general regularity and extendibility conditions. For more details, see **[2, SECTION 4]**.

Another direction of generalization that received considerable attention concerns packings of small graphs into graphs with sufficiently high minimum degrees. Here, the most famous (and still open) problem is a conjecture of Nash-Williams concerning triangle packings, to which I will return in Section 8. Refer to the survey **[49]** for more details on progress concerning problems of this type.

4. PACKING CYCLES

Let us now turn to a famous problem concerning the packing of cycle factors into the complete graph that we already mentioned in the introduction.

Problem 4.1 (Oberwolfach problem **[32]**). Let C be any 2-regular graph on an odd number n of vertices. Can we perfectly pack copies of C into K_n?

Here, n has to be odd so that each vertex in K_n has an even number of neighbors. If such a perfect packing can be obtained, this means that n workshop participants can be placed around the tables represented by the cycles in C for $\frac{n-1}{2}$ meals so that no two sit next to each other twice. Several variations of this problem were considered, for example, the so-called Hamilton–Waterloo problem, where two different 2-regular graphs C_1 and C_2 have to be used a prescribed number of times in the packing, representing a meeting that takes place at the two close venues Hamilton and Waterloo. The Oberwolfach problem and its variants have inspired a vast number of research papers, but only recently was the problem solved for large n, independently by Glock, Joos, Kim, Kühn, Osthus [28] and Keevash, Staden [41]. (These papers are also good references for further background and previous results.) The results obtained by both groups are more general (in particular they also both solve the Hamilton–Waterloo problem), but in different ways.

The result in [28] allows for more general classes of graph to be packed into complete graphs. A graph on n vertices is called *μ-separable* if it contains a set S of at most μn vertices such that in $H \setminus S$ each component has at most μn vertices.

Theorem 4.2 (Glock, Joos, Kim, Kühn, Osthus [28]). *For all $\Delta \in \mathbb{N}$ and $\alpha > 0$, there are $\mu > 0$ and $n_0 \in \mathbb{N}$ such that for all $n \geq n_0$ the following holds. If $\mathcal{C}$ and $\mathcal{G}$ are families of n vertex graphs containing together at most $\binom{n}{2}$ edges such that*

(i) *$\mathcal{C}$ consists of at least αn copies of a 2-regular n-vertex graph,*

(ii) *each graph in $\mathcal{G}$ is a μ-separable r-regular n-vertex graph with $r \leq \Delta$,*

then $\mathcal{C} \cup \mathcal{G}$ pack into K_n.

In their proof, Glock, Joos, Kim, Kühn, and Osthus use results on hypergraph matchings by Alon and Yuster [7], as well as an approximate decomposition result by Condon, Kim, Kühn, and Osthus [17] (see also Section 6), and a special case of the colored partite designs results of [38], alongside many new ideas. Here I will not describe more in detail how a reduction to these design results can be obtained for this packing problem; but in the next section I provide more details of a reduction of this type in a different setting.

Theorem 4.2 allows for the packing of a linear fraction of copies of the same cycle factor and any collection of separable regular graphs. In particular, it allows the perfect packing of any family of cycle factors in which one cycle factor appears linearly often. The result in [41], on the other hand, allows the perfect packing of any family of cycle factors (but not more generally separable graphs), also in host graphs which are not necessarily complete. Recall that an n-vertex graph H is called (ε, s)-typical, if every set $S \subseteq V(H)$ of at most s vertices has $(1 \pm \varepsilon)d^{|S|}n$ common neighbors in H, where $d = |E(H)|/\binom{n}{2}$.

Theorem 4.3 (Keevash, Staden [41]). *For all $\delta > 0$ there are $\varepsilon > 0$, s, and n_0 such that the following holds for any $n \geq n_0$ and $r \geq \delta n$. If $\mathcal{C}$ is a family of 2-regular n-vertex graphs with $|\mathcal{C}| = r$, and H is (ε, s)-typical, then $\mathcal{C}$ packs into H.*

For proving this result, Keevash and Staden also use the colored partite designs results of [38] as a tool. Here, a direct reduction is only possible if all cycles in $\mathcal{C}$ have lengths bounded by some constant. For nonconstant cycle lengths a more intricate reduction to colored partite designs is needed, embedding constant-length paths and connecting them into larger cycles; this requires a lot of extra work and ideas.

Packing, more generally, cycles in regular host graphs was also considered; see [18] for results concerning the packing of Hamilton cycles, and the survey [49] for more background and related results.

5. PACKING TREES

In the area of tree packings, there are two influential conjectures, which are remarkable for their elegant statements. The first of these was formulated by Ringel [60] in 1968.

Conjecture 5.1 (Ringel's conjecture [60]). *For each $n \in \mathbb{N}$ and for each tree T on $n+1$ vertices, we have that $2n+1$ copies of T pack into the complete graph K_{2n+1}.*

The second of these conjectures is attributed to Gyárfás (see [33]) and from 1978.

Conjecture 5.2 (Gyárfás's tree packing conjecture). *For each $n \in \mathbb{N}$ and for each family of trees $(T_s)_{s\in[n]}$ such that T_s has s vertices for each $s \in [n]$, we have that $(T_s)_{s\in[n]}$ packs into the complete graph K_n.*

These conjectures have in common that they ask for perfect packings, because trees on s vertices have $s-1$ edges and we have $(2n+1)(n+1) = \binom{2n+1}{2}$ and $\sum_{s=1}^{n}(s-1) = \binom{n}{2}$. However, they differ in that Ringel's conjecture concerns the packing of copies of the same tree, which has roughly half the number of vertices of the host graph, and the tree packing conjecture concerns the packing of different trees, some of which have essentially the same number of vertices as the host graph. Thus, it is not surprising that Gyárfás's tree packing conjecture turned out more challenging than Ringel's conjecture.

Both these conjectures inspired much work, and we shall turn to recent progress shortly, concentrating on the highlights. But first I will mention connections to some other well-studied combinatorial objects. Indeed, when it turned out that Ringel's conjecture was difficult, even for special classes of trees, then it was suggested that some symmetry might help in attacking the problem. To this end, Rosa [63] introduced a notion which by now we call graceful labelings. A *graceful labeling* of a graph H is an injective mapping $f : V(H) \to \{1, \dots, e(H)+1\}$ such that the induced edge labels $|f(x) - f(y)|$ for $xy \in E(H)$ are distinct. It is easy to see that if H has a graceful labeling, then there is a packing of k copies of H into the complete graph on k vertices as long as $k \geq 2e(H)+1$: Given a graceful labeling f of H consider the embeddings $f_i : V(H) \to V(K_k)$ with $V(K_k) = [k]$ and $f_i(x) = f(x) + i$ for $0 \leq i < k$, where $k \geq 2e(H)+1$ guarantees that these embeddings are edge disjoint. This means that f_0 embeds H as mandated by the labeling, and then we take translates (or "rotations") of this embedding to pack all other copies of H. Consequently, the following conjecture, which is attributed to Kotzig (see [63]), implies Ringel's conjecture (Conjecture 5.1).

Conjecture 5.3 (Graceful labeling conjecture). *Every tree has a graceful labelling.*

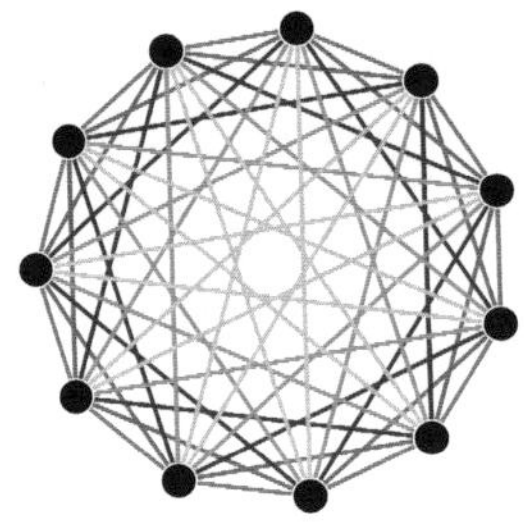

FIGURE 3
The near distance coloring of K_{11}.

This conjecture is still open, but Adamaszek, Allen, Grosu, and Hladký [1] proved an approximate version for almost all trees: They showed that every n-vertex tree with maximum degree at most $cn/\log n$ has a labeling satisfying the gracefulness property which uses labels from $[(1+\varepsilon)n]$, where $\varepsilon > 0$, $c > 0$ is small compared to ε, and n is sufficiently large.

What proved more important for the resolution of Ringel's conjecture (for large n) is the following connection to rainbow copies in edge-colored graphs. A *rainbow* copy of a graph G in an edge-colored graph H is a copy of G in H whose edges have pairwise distinct colors. The *near distance coloring* of the complete graph K_k on vertex set $[k]$ assigns to each edge $uv \in E(K_k)$ the smallest number d as color such that $u + d = v$ or $v + d = u$ modulo k (see Figure 3 for an example). Again, it is easy to see that a rainbow copy of H in K_k under the near distance coloring gives a packing of k (uncolored) copies of H in K_k (uncolored) by taking the rainbow copy and considering translates as before. Using this formulation, Montgomery, Pokrovskiy, and Sudakov [57] proved that Ringel's conjecture holds for large n. (Preliminary results were obtained in [15, 34], but these results work more generally also for Gyárfás's conjecture, so they are listed below.)

Theorem 5.4 (Montgomery, Pokrovskiy, Sudakov [57]). *For every sufficiently large n and every tree T on $n+1$ vertices, there is a rainbow copy of T in the near-distance coloring of K_{2n+1}.*

For proving this result, Montgomery, Pokrovskiy, and Sudakov distinguish different cases, depending on whether the tree under consideration has many nonneighboring leaves, many bare paths, or most vertices in the tree are leaves with many neighboring leaves. Here, a bare path is a path in the tree whose internal vertices have no neighbors outside the path. These cases are exhaustive because trees have average degree smaller than 2. In the first two cases, completing the rainbow copy relies on a version of absorption introduced in [54]. Techniques from Montgomery, Pokrovskiy, and Sudakov's earlier papers [55, 56] are also used.

Keevash and Staden [42] prove the following generalization of Ringel's conjecture to quasirandom graphs.

Theorem 5.5 (Keevash, Staden [42]). *There exists $s \in \mathbb{N}$ such that for all $p > 0$ there are $\varepsilon > 0$ and $n_0 \in \mathbb{N}$ such that for all $n \geq n_0$ with $pn \in \mathbb{N}$ the following holds. Let T be a tree on $pn + 1$ vertices. Let H be a graph on $2n + 1$ vertices with $pn(2n + 1)$ edges which is (ε, s)-typical. Then H can be decomposed into $2n + 1$ copies of T.*

Their proof distinguishes similar cases as the previous result: almost all vertices belong to large stars; many leaves are in small stars; and many vertices are in disjoint bare paths. They also apply a recent result on pseudorandom hypergraph matchings by Ehard, Glock, and Joos [21], their methods developed for proving Theorem 4.3, a result of Barát, Gyarfás, and Sárközy [10] on rainbow matchings in bipartite multigraphs, alongside many new ideas.

As indicated before, for Gyárfás's conjecture (Conjecture 5.2) less is known. An almost perfect packing version for trees with constant maximum degree was obtained in [15], applying a version of the Rödl nibble. Ferber and Samotij [25] considered trees with maximum degree up to $cn/\log n$, proving these can be almost-perfectly packed into random host graphs (in fact, they have more general results for sparse random host graphs). Getting perfect packing results turned out much harder. Joos, Kim, Kühn, and Osthus [34] proved that Gyárfás's conjecture holds for families of trees with constant maximum degree if n is sufficiently large. Their proof uses an array of important tools developed previously: Szemerédi's regularity lemma, robust expanders, random walks, iterative absorption (we refer to [34] for more details) and a blow-up lemma for approximate decompositions (that we shall return to in Section 6).

In [2,3] it is shown that Gyárfás's conjecture holds when n is sufficiently large for all families of trees with maximum degree $cn/\log n$ for some universal constant $c > 0$. In fact, the following more general result is obtained, which applies to quasirandom host graphs, and also implies a version of Gyárfás's conjecture for families of different trees (of the same size) under the same maximum degree restriction.

Theorem 5.6. *For each $\delta, d > 0$ there exist $c, \varepsilon > 0$ and $n_0, s \in \mathbb{N}$ such that for each $n \geq n_0$ in any (ε, s)-typical graph H on n vertices with at least dn^2 edges we can pack any family $(T_t)_{t\in[N]}$ of trees satisfying*

(i) $\sum_{t\in[N]} e(T_t) \leq e(H)$ *and* $\Delta(T_t) \leq \frac{cn}{\log n}$ *for all* $t \in [N]$,

(ii) $\delta n \leq v(T_t) \leq (1 - \delta)n$ *for all* $1 \leq t \leq (\frac{1}{2} + \delta)n$ *and* $v(T_t) \leq n$ *for all* $(\frac{1}{2} + \delta)n < t \leq N$.

Let me briefly sketch some proof ideas used for obtaining this result. Similarly to the approaches above, we distinguish two cases: Either a linear number of the nonspanning trees we are given contain a linear number of leaves, or a linear number of them contain a linear number of disjoint bare paths of length 11. In both cases we first remove these leaves/paths and obtain an almost perfect packing of what remains using a random packing process which we outline in the subsequent section, with a leftover graph $\tilde{H}$. We show that this random process preserves many nice properties, which we shall need to complete the packing, for example, that $\tilde{H}$ is quasirandom. In the first case (which is treated in [3]), it then remains to pack the omitted leaves. We obtain this by another random process, in one round randomly mapping leaves "dangling" at one host graph vertex to remaining host graph edges, moving on to the next host graph vertex in the next round, an so on. Here, a leaf "dangles" at a host graph vertex v if its neighbor has been embedded to v. For being successful in this process, we further need to use a random orientation of the host graph and exploit superregularity properties of certain auxiliary graphs representing all choices we have for embedding leaves "dangling" at v. More precisely, our random process will select a random perfect matching in this auxiliary graph, and then we need to show that this does not too negatively affect the auxiliary graphs for the remaining host graph vertices v'. Some of these ideas are inspired by methods from [34].

In the second case (which is treated in [2]), it remains to pack the remaining bare paths, and also a small number of leaves that we also omitted when obtaining the almost perfect packing. These leaves were omitted for the following reason: All vertices in bare paths have degree 2, which creates some obvious parity restrictions when we want to pack the bare paths. We now first embed the omitted leaves in such a way that we obtain the necessary parity of edges remaining at each host graph vertex. Each of the bare paths that now remain has 11 vertices. We will then, in a sequence of carefully tailored intermediate stages, embed some of these paths completely and some of these paths partly, until we arrive at the following scenario, where only a set of bare paths of length 3 remain to be packed which are paths in a subset $(T_t)_{t\in \mathcal{S}}$ of our trees. Our goal is to apply Keevash's partite and colored designs result (in the form of Theorem 3.10) to pack these. In the scenario we obtain, the remainder H^* of the host graph has an even number of vertices which are partitioned into two sets $V_{\boxminus} = \{\boxminus_i : i \in [\ell]\}$ and $V_{\boxplus} = \{\boxplus_i : i \in [\ell]\}$. We call each pair $\{\boxminus_i, \boxplus_i\}$ a *terminal pair*, and each of the remaining paths x, x', y', y is *anchored* at some terminal pair, that is x is embedded to some $\boxminus_i$ and y is embedded to the corresponding $\boxplus_i$, but the edges xy, yz, zw still need to be embedded. When these will be embedded, then we insist that x' is also embedded in $V_{\boxminus}$ and y' is also embedded in $V_{\boxplus}$, which implies that in our intermediate stages mentioned above we will need to guarantee that $H^*[V_{\boxplus}]$ and $H^*[V_{\boxplus}]$ have the same number of edges.

We shall apply the designs result, Theorem 3.10, on the following auxiliary partially directed colored multigraph, which we call chest and which has vertex parts $V = \{1, \dots, |V_{\boxminus}|\}$ and $U = \mathcal{S}$. The chest has the following colored edges running within V: A blue undirected edge ij for each edge $\boxminus_i\boxminus_j \in E(H^*)$, a red undirected edge ij for each edge $\boxplus_i\boxplus_j \in E(H^*)$, and a green directed edge from i to j for each edge $\boxminus_i\boxplus_j \in E(H^*)$. This means that the embedding of a remaining path x, x', y', y to vertices $\boxminus_i, \boxminus_j, \boxplus_k, \boxplus_i$ in H^* corresponds to a triangle i, j, k in the chest, in which ij is blue, jk is green and directed to k, and ki is red. The chest further has the following edges running from V to U: a grey undirected edge ti for each remaining path in T_t anchored at $\{\boxminus_i, \boxplus_i\}$, a black undirected edge ti for each for each terminal $\boxminus_i$ that does not host any vertex of T_t yet (embedded in previous stages), and a purple undirected edge ti for each terminal $\boxplus_i$ that does not host any vertex of T_t, yet. The reason for inserting these edges is that we need to

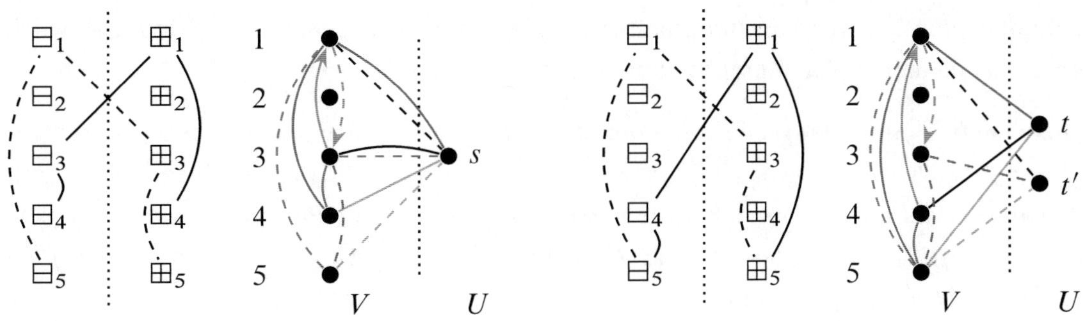

FIGURE 4

The two left-hand pictures show an example of two remaining paths from the same tree T_t embedded in H^* to the paths $\boxminus_4 \boxminus_3 \boxplus_1 \boxplus_4$ and $\boxminus_5 \boxminus_1 \boxplus_3 \boxplus_5$ and the corresponding diamonds in the chest. The two right-hand pictures show an example of two remaining paths from different trees T_t and $T_{t'}$ embedded in H^* to the paths $\boxminus_5 \boxminus_4 \boxplus_1 \boxplus_5$ and $\boxminus_5 \boxminus_1 \boxplus_3 \boxplus_5$, respectively, and the corresponding diamonds in the chest. The solid and dashed lines in this picture are only used to distinguish the two paths/diamonds.

guarantee that the remaining paths are embedded vertex disjointly if they come from the same tree and that they also do not use vertices previously used for this tree; in addition we need that for a given terminal pair exactly the trees T_t for which some remaining bare path is anchored at this terminal pair receive exactly one path with these endpoints in the packing (this is guaranteed by the grey edges). A packing of the remaining paths in H^* which satisfies these properties corresponds precisely to a packing of diamonds (recall the definition from Section 3) in the chest using all red, green, blue, and grey edges. See Figure 4 for an illustration. Consequently, our strategy is to use the intermediate stages for ensuring the necessary divisibility, regularity, and extendibility properties of the chest so that we can then apply Theorem 3.10 to complete our perfect packing. (For more details, see [2, SECTION 3].)

6. PACKING MORE GENERAL GRAPH CLASSES

The results on designs in Section 3 concern packings of graphs that are small in comparison to the host graph. The results on cycle and tree packings discussed in the previous two sections concern the packing of graphs from very special classes of graphs. It is natural to ask for which other types of guest graphs one can hope for analogous packing results.

Early progress was made by Messuti, Rödl, and Schacht [52], who used the results of Ray-Chaudhury and Wilson [59] on resolvable graph decompositions to obtain almost perfect packings of almost spanning graphs from minor-closed families of graphs with bounded maximum degree. I omit the definition of minors here; suffice it to say that graphs embeddable on a fixed surface are a special case. Ferber, Lee, and Mousset [24] generalized this result, moving from almost spanning graphs to spanning graphs.

In [5] more generally families of D-degenerate graphs are considered. A graph is called *D-degenerate* if there is an ordering of its vertices such that each vertex has at most D

neighbors preceding it. Many important classes of graphs are degenerate, such as trees, which are 1-degenerate, or planar graphs, which are 5-degenerate.

Theorem 6.1 ([5]). *For every $D \in \mathbb{N}$ and $\eta > 0$, there are $n_0 \in \mathbb{N}$ and $c, \varepsilon > 0$ such that for each $n \geq n_0$ the following holds. Suppose that $(G_s)_{s\in\mathcal{S}}$ is a family of D-degenerate graphs, each on at most n vertices and of maximum degree $\frac{cn}{\log n}$, whose total number of edges is at most $(p-\eta)\binom{n}{2}$, and suppose that H is an $(\varepsilon, 2D+3)$-typical n-vertex graph with $p\binom{n}{2}$ edges. Then $(G_s)_{s\in\mathcal{S}}$ packs into H.*

The techniques developed for proving this result form the starting point for the results on tree packings obtained in [2, 3] discussed in the previous section. Accordingly, the results of [2,3] also more generally apply to certain classes of D-degenerate guest graphs with many leaves and many bare paths, respectively. The details are more complex, and we omit them here.

In the proof of this result we use the following natural random packing process. We embed the guest graphs G_s one after the other. When constructing an embedding ϕ_s of G_s, we proceed vertex by vertex, following a degeneracy order $x_1, x_2, \ldots$. When embedding x_i, we consider all previously embedded neighbors $y_1, \ldots, y_\ell$ of x_i, of which there are at most D. It is clear that we need to embed x_i into the set X_i that is given by the common neighborhood in the host graph of the $\phi_s(y_j)$ with $j \in [\ell]$ minus the set U_i of vertices used already for earlier vertices of G_s, that is, $U_i = \{\phi(x_j) : j < i\}$. We choose a random vertex in X_i as $\phi(x_i)$. Then we delete from the host graph all edges used for embedding x_i, that is, all edges $\phi_s(y_j)\phi_s(x_i)$ with $j \in [\ell]$. This process will of course only have a chance of succeeding if all of our guest graphs are a bit smaller than the host graph H. To obtain this setting, we omit a small (linear) number of vertices from each guest graph G_s that is too large before running the random process. We also set aside a small random proportion of the edges of H, which we use after running the random packing process to pack the omitted vertices greedily. While the described random packing process is easy, analyzing it is not: In order to prove Theorem 6.1, we show that the sets X_i always stay as large as expected because the random process preserves pseudorandomness of the (changing) host graph, as well as a suitably random distribution of the sets U_i, among other nice properties. This then allows us to complete the packing.

Kim, Kühn, Osthus, and Tyomkyn [44] provide an important general purpose tool for obtaining almost-perfect decompositions, namely a blow-up lemma for decompositions. The blow-up lemma [46] is an integral part of the powerful *regularity method*, complementing Szemerédi's celebrated regularity lemma [64]. For simplicity, we only state the bipartite version here, which displays the essence of the setup; the generalization of this setup to more general partite graphs is standard. For a bipartite graph H with partition (V_1, V_2) and sets $V_1' \subseteq V_1$, $V_2' \subseteq V_2$, we let $d_H(V_1', V_2') = e_H(V_1', V_2')/(|V_1'||V_2'|)$ be the *density* of (V_1', V_2'). We say that H is (ε, d)*-regular* if $d_H(V_1', V_2') = d \pm \varepsilon$ for each $V_1' \subseteq V_1$, $V_2' \subseteq V_2$ with $|V_1'| \geq \varepsilon|V_1|$, $|V_2'| \geq \varepsilon|V_2|$. Further, H is (ε, d)-superregular, if it is (ε, d)-regular and each vertex in V_1 (respectively V_2) has $(d \pm \varepsilon)|V_2|$ (respectively $(d \pm \varepsilon)|V_1|$) neighbors in V_2 (respectively V_1).

Theorem 6.2. *For every $\alpha > 0$, there are $\varepsilon > 0$ and $n_0 \in \mathbb{N}$ such that the following holds for all $n \geq n_0$ and $d \geq \alpha$. Suppose H is a bipartite graph with partition classes of size n, which is (ε, d)-superregular. Suppose $(G_s)_{s\in\mathcal{S}}$ with $|\mathcal{S}| \leq \alpha^{-1}n$ is a family of bipartite graphs with partition classes of site n, with maximum degree $\Delta(G_s) \leq \alpha^{-1}$ for each $s \in \mathcal{S}$, whose total number of edges is at most $(1-\alpha)dn^2$. Then there is a packing of $(G_s)_{s\in\mathcal{S}}$ into H.*

Ehard and Joos [23] provide a simplified proof of this result, and are also able to obtain a generalization, yielding packings with stronger quasirandomness properties. Applications of this result are manifold. It has been applied as a tool (among other techniques) in [34] and [43] for obtaining packings of trees, in [17] for obtaining a packing version of the so-called bandwidth theorem, and in [47] for decompositions in more general graphs.

7. PACKING LARGE HYPERGRAPHS

Analogues for hypergraph packings have been considered for various results discussed in the preceding sections. In particular, Keevash's result on decompositions in the coloured and partite setting discussed in Section 3 does more generally allow k-partite hypergraphs (under suitable conditions). In the same section we also already mentioned the results on general hypergraph F-designs by Glock, Kühn, Lo, and Osthus [30]. Here, I will just briefly mention some important further developments.

Almost-perfect decompositions of regular hypergraphs satisfying certain quasirandomness properties into Hamilton cycles (of different types) were obtained by Bal and Frieze [9]. Packings of more general tight cycle factors in hypergraphs with large co-degrees were considered by Joos, Kühn, and Schülke [35]. Results on almost-perfect decompositions quasirandom hypergraphs into arbitrary families of hypergraphs of bounded maximum degree are proved by Ehard and Joos [22]. Turning to hypergraphs with larger degrees, in [4] the almost-perfect packing result for D-degenerate graphs of [5] and the perfect packing result for D-degenerate graphs with many leaves of [3] are generalized to hypergraphs.

I close this section by remarking that graph and hypergraph packings are intimately related to the problem of finding perfect matchings in certain hypergraphs, highlighting the importance of new results in this direction, such as those by Ehard, Glock, and Joos [21], for the area. For example, consider the problem of finding an F-factor of $K_n^{(r)}$ for some r-regular hypergraph F. This is equivalent to finding a perfect matching in the $e(F)$-uniform hypergraph with vertex set $E(K_n^{(r)})$ which has an edge $\tilde{F} = \{e_1, \ldots, e_{e(F)}\}$ whenever $\tilde{F}$ is a copy of F (not necessarily induced) in $K_n^{(r)}$. More details on connections between packings, hypergraph matchings, and also rainbow subgraphs can be found in [21].

8. SOME OPEN PROBLEMS

I close with a small collection of what I consider some important open problems in the area. Firstly, it remains to resolve Gyárfás's tree packing conjecture in full

Problem 8.1. Solve the tree packing conjecture (Conjecture 5.2) in full.

Similarly, the graceful labeling conjecture also remains open.

Problem 8.2. Solve the graceful labeling conjecture (Conjecture 5.3) in full.

Another famous conjecture that I only mentioned in passing so far concerns triangle packings.

Conjecture 8.3 (Nash-Williams [58]). *For large n every K_3-divisible graph H on n vertices with $\delta(H) \geq \frac{3}{n}/4$ has a K_3-decomposition.*

By a result of Barber, Kühn, Lo, and Osthus [13], the approximate version of this conjecture follows from a fractional version. Recent progress on fractional triangle decompositions was made by Dross [20] and Delcourt and Postle [19], but it remains open to show that a fractional triangle decomposition exists in n-vertex graphs of minimum degree $3n/4$. Decomposition problems for other graphs F than the triangle were considered in [12, 29, 53].

For perfectly packing cycle factors in a graph H, we clearly cannot impose any nontrivial minimum degree condition; instead we need the host graph H to be regular. The following conjecture from [28] concerns packings of cycle factors in sufficiently dense regular graphs.

Conjecture 8.4 (Glock, Joos, Kim, Kühn, Osthus [28]). *Any large n-vertex r-regular graph H with even $r \geq \frac{3}{4}n + o(n)$ has a decomposition into G-copies for any 2-regular graph G on n vertices.*

For more general classes of graphs than cycles, trees and F-factors, little is known so far when it comes to perfect packings. The following conjecture from [28] concerns packings of arbitrary regular graphs into the complete graph.

Conjecture 8.5 (Glock, Joos, Kim, Kühn, Osthus [28]). *For all $\Delta \in \mathbb{N}$, there exists an $n_0 \in \mathbb{N}$ so that for $n \geq n_0$ any family $(G_i)_{i \in [t]}$ of n-vertex graphs such that G_i is r_i-regular with $r_i \leq \Delta$ and $\sum_{i \in [t]} r_i = n - 1$ packs into K_n.*

Similarly, for hypergraphs many problems remain. In particular, showing that $K_n^{(r)}$ has a decomposition into tight Hamilton cycles (under appropriate divisibility conditions) is still open. This was conjectured by Bailey and Stevens [8]. Glock, Kühn, and Osthus propose the following more general conjecture.

Conjecture 8.6 (Glock, Kühn, and Osthus [49]). *For fixed k and large n, every vertex disjoint union G of tight k-uniform cycles, each of length at least $2k - 1$, with in total n vertices, decomposes $K_n^{(k)}$ if k divides $\binom{n-1}{k-1}$.*

Another direction that has not yet seen much progress is that of packing problems in sparse graphs. As mentioned earlier, tree packing problems in this context were considered in [25]. Packings of Hamilton cycles in sparse random graphs were considered in [14, 27, 45]. It would be interesting to obtain similar results for other families of guest graphs.

Problem 8.7. For which families of graphs and probabilities p is the following true? Given a family $(G_i)_{i\in[t]}$ of graphs on at most $(1-\varepsilon)n$ vertices with in total at most $(1-\varepsilon)p\binom{n}{2}$ edges, we can pack $(G_i)_{i\in[t]}$ into the random graph $G(n,p)$.

REFERENCES

[1] A. Adamaszek, P. Allen, C. Grosu, and J. Hladký, Almost all trees are almost graceful. *Random Structures Algorithms* **56** (2020), no. 4, 948–987.

[2] P. Allen, J. Böttcher, D. Clemens, J. Hladký, D. Piguet, and A. Taraz, The tree packing conjecture for trees of almost linear maximum degree. 2021, arXiv:2106.11720.

[3] P. Allen, J. Böttcher, D. Clemens, and A. Taraz, Perfectly packing graphs with bounded degeneracy and many leaves. *Israel J. Math.* (accepted).

[4] P. Allen, J. Böttcher, and A. Dankovics, Packing degenerate hypergraphs, manuscript, 70 pp.

[5] P. Allen, J. Böttcher, J. Hladký, and D. Piguet, Packing degenerate graphs. *Adv. Math.* **354** (2019), 106739, 58 pp.

[6] N. Alon and J. H. Spencer, *The probabilistic method. Fourth edn.* Wiley Ser. Discrete Math. Optim., John Wiley & Sons, Inc., Hoboken, NJ, 2016.

[7] N. Alon and R. Yuster, On a hypergraph matching problem. *Graphs Combin.* **21** (2005), no. 4, 377–384.

[8] R. F. Bailey and B. Stevens, Hamiltonian decompositions of complete k-uniform hypergraphs. *Discrete Math.* **310** (2010), no. 22, 3088–3095.

[9] D. Bal and A. Frieze, Packing tight Hamilton cycles in uniform hypergraphs. *SIAM J. Discrete Math.* **26** (2012), no. 2, 435–451.

[10] J. Barát, A. Gyárfás, and G. N. Sárközy, Rainbow matchings in bipartite multigraphs. *Period. Math. Hungar.* **74** (2017), no. 1, 108–111.

[11] B. Barber, S. Glock, D. Kühn, A. Lo, R. Montgomery, and D. Osthus, Minimalist designs. *Random Structures Algorithms* **57** (2020), no. 1, 47–63.

[12] B. Barber, D. Kühn, A. Lo, R. Montgomery, and D. Osthus, Fractional clique decompositions of dense graphs and hypergraphs. *J. Combin. Theory Ser. B* **127** (2017), 148–186.

[13] B. Barber, D. Kühn, A. Lo, and D. Osthus, Edge-decompositions of graphs with high minimum degree. *Adv. Math.* **288** (2016), 337–385.

[14] B. Bollobás and A. M. Frieze, On matchings and Hamiltonian cycles in random graphs. In *Random graphs '83 (Poznań, 1983)*, pp. 23–46, North-Holl. Math. Stud. 118, North-Holland, Amsterdam, 1985.

[15] J. Böttcher, J. Hladký, D. Piguet, and A. Taraz, An approximate version of the tree packing conjecture. *Israel J. Math.* **211** (2016), no. 1, 391–446.

[16] C. J. Colbourn and J. H. Dinitz (eds.), *Handbook of combinatorial designs. Second edn.* Chapman & Hall/CRC, Boca Raton, FL, 2007.

[17] P. Condon, J. Kim, D. Kühn, and D. Osthus, A bandwidth theorem for approximate decompositions. *Proc. Lond. Math. Soc.* **118** (2019), no. 6, 1393–1449.

[18] B. Csaba, D. Kühn, A. Lo, D. Osthus, and A. Treglown, Proof of the 1-factorization and Hamilton decomposition conjectures. *Mem. Amer. Math. Soc.* **244** (2016), no. 1154, v+164 pp.

[19] M. Delcourt and L. Postle, Progress towards Nash-Williams' conjecture on triangle decompositions. *J. Combin. Theory Ser. B* **146** (2021), 382–416.

[20] F. Dross, Fractional triangle decompositions in graphs with large minimum degree. *SIAM J. Discrete Math.* **30** (2016), no. 1, 36–42.

[21] S. Ehard, S. Glock, and F. Joos, Pseudorandom hypergraph matchings. *Combin. Probab. Comput.* **29** (2020), no. 6, 868–885.

[22] S. Ehard and F. Joos, Decompositions of quasirandom hypergraphs into hypergraphs of bounded degree. 2020, arXiv:2011.05359.

[23] S. Ehard and F. Joos, A short proof of the blow-up lemma for approximate decompositions. *Combinatorica* (to appear).

[24] A. Ferber, C. Lee, and F. Mousset, Packing spanning graphs from separable families. *Israel J. Math.* **219** (2017), no. 2, 959–982.

[25] A. Ferber and W. Samotij, Packing trees of unbounded degrees in random graphs. *J. Lond. Math. Soc.* **99** (2019), no. 3, 653–677.

[26] R. A. Fisher, *Statistical methods, experimental design, and scientific inference*. Oxford University Press, 1990.

[27] A. Frieze and M. Krivelevich, On packing Hamilton cycles in ϵ-regular graphs. *J. Combin. Theory Ser. B* **94** (2005), no. 1, 159–172.

[28] S. Glock, F. Joos, J. Kim, D. Kühn, and D. Osthus, Resolution of the Oberwolfach problem. *J. Eur. Math. Soc. (JEMS)* **23** (2021), 2511–2547.

[29] S. Glock, D. Kühn, A. Lo, R. Montgomery, and D. Osthus, On the decomposition threshold of a given graph. *J. Combin. Theory Ser. B* **139** (2019), 47–127.

[30] S. Glock, D. Kühn, A. Lo, and D. Osthus, The existence of designs via iterative absorption: hypergraph F-designs for arbitrary F. *Mem. Amer. Math. Soc.* (to appear).

[31] J. E. Graver and W. B. Jurkat, The module structure of integral designs. *J. Combin. Theory Ser. A* **15** (1973), 75–90.

[32] R. K. Guy, Unsolved combinatorial problems. In *Combinatorial mathematics and its applications(Proc. Conf., Oxford, 1969)*, pp. 121–127, Academic Press, Oxford, 1971.

[33] A. Gyárfás and J. Lehel, Packing trees of different order into K_n. In *Combinatorics (Proc. Fifth Hungarian Colloq., Keszthely, 1976)*, pp. 463–469, Colloq. Math. Soc. János Bolyai 18, North-Holland, Amsterdam, 1978.

[34] F. Joos, J. Kim, D. Kühn, and D. Osthus, Optimal packings of bounded degree trees. *J. Eur. Math. Soc. (JEMS)* **21** (2019), no. 12, 3573–3647.

[35] F. Joos, M. Kühn, and B. Schülke, Decomposing hypergraphs into cycle factors. 2021, arXiv:2104.06333.

[36] P. Keevash, The existence of designs. 2014, arXiv:1401.3665.

[37] P. Keevash, Counting designs. *J. Eur. Math. Soc. (JEMS)* **20** (2018), no. 4, 903–927.

[38] P. Keevash, The existence of designs II. 2018, arXiv:1802.05900.

[39] P. Keevash, Hypergraph matchings and designs. In *Proceedings of the International Congress of Mathematicians—Rio de Janeiro 2018. Vol. IV. Invited lectures*, pp. 3113–3135, World Sci. Publ., Hackensack, NJ, 2018.

[40] P. Keevash, Coloured and directed designs. In *Building bridges II*, pp. 279–315, Bolyai Soc. Math. Stud. 28, Springer, Berlin, 2019.

[41] P. Keevash and K. Staden, The generalised Oberwolfach problem. 2020, arXiv:2004.09937.

[42] P. Keevash and K. Staden, Ringel's tree packing conjecture in quasirandom graphs. 2020, arXiv:2004.09947.

[43] J. Kim, Y. Kim, and H. Liu, Tree decompositions of graphs without large bipartite holes. *Random Structures Algorithms* **57** (2020), no. 1, 150–168.

[44] J. Kim, D. Kühn, D. Osthus, and M. Tyomkyn, A blow-up lemma for approximate decompositions. *Trans. Amer. Math. Soc.* **371** (2019), no. 7, 4655–4742.

[45] F. Knox, D. Kühn, and D. Osthus, Edge-disjoint Hamilton cycles in random graphs. *Random Structures Algorithms* **46** (2015), no. 3, 397–445.

[46] J. Komlós, G. N. Sárközy, and E. Szemerédi, Blow-up lemma. *Combinatorica* **17** (1997), no. 1, 109–123.

[47] D. Král', B. Lidický, T. L. Martins, and Y. Pehova, Decomposing graphs into edges and triangles. *Combin. Probab. Comput.* **28** (2019), no. 3, 465–472.

[48] M. Krivelevich, Triangle factors in random graphs. *Combin. Probab. Comput.* **6** (1997), no. 3, 337–347.

[49] D. Kühn, S. Glock, and D. Osthus, Extremal aspects of graph and hypergraph decomposition problems. In *Surveys in Combinatorics 2021*, London Math. Soc. Lecture Note Ser. **470**, pp. 235–265.

[50] D. Kühn and D. Osthus, Hamilton decompositions of regular expanders: a proof of Kelly's conjecture for large tournaments. *Adv. Math.* **237** (2013), 62–146.

[51] G. Kuperberg, S. Lovett, and R. Peled, Probabilistic existence of regular combinatorial structures. *Geom. Funct. Anal.* **27** (2017), no. 4, 919–972.

[52] S. Messuti, V. Rödl, and M. Schacht, Packing minor-closed families of graphs into complete graphs. *J. Combin. Theory Ser. B* **119** (2016), 245–265.

[53] R. Montgomery, Fractional clique decompositions of dense partite graphs. *Combin. Probab. Comput.* **26** (2017), no. 6, 911–943.

[54] R. Montgomery, Spanning trees in random graphs. *Adv. Math.* **356** (2019), 106793, 92 pp.

[55] R. Montgomery, A. Pokrovskiy, and B. Sudakov, Decompositions into spanning rainbow structures. *Proc. Lond. Math. Soc. (3)* **119** (2019), no. 4, 899–959.

[56] R. Montgomery, A. Pokrovskiy, and B. Sudakov, Embedding rainbow trees with applications to graph labelling and decomposition. *J. Eur. Math. Soc. (JEMS)* **22** (2020), no. 10, 3101–3132.

[57] R. Montgomery, A. Pokrovskiy, and B. Sudakov, A proof of Ringel's conjecture. 2020, arXiv:2001.02665.

[58] C. S. J. A. Nash-Williams, Edge-disjoint Hamiltonian circuits in graphs with vertices of large valency. In *Studies in pure mathematics (presented to Richard Rado)*, pp. 157–183, Academic Press, London, 1971.

[59] D. K. Ray-Chaudhuri and R. M. Wilson, The existence of resolvable block designs. In *Survey of combinatorial theory (Proc. Internat. Sympos., Colorado State Univ., Fort Collins, CO, 1971)*, pp. 361–375. North-Holland, Amsterdam, 1973.

[60] G. Ringel, Problem 25. In *Theory of graphs and its applications (Proc. Int. Symp. Smolenice 1963)*, pp. 85–90. Publ. House Czech. Acad. Sci., Prague, 1963.

[61] V. Rödl, On a packing and covering problem. *European J. Combin.* **6** (1985), no. 1, 69–78.

[62] V. Rödl, A. Ruciński, and E. Szemerédi, A Dirac-type theorem for 3-uniform hypergraphs. *Combin. Probab. Comput.* **15** (2006), no. 1–2, 229–251.

[63] A. Rosa, On certain valuations of the vertices of a graph. In *Theory of Graphs (Internat. Sympos., Rome, 1966)*, pp. 349–355, Gordon and Breach, New York; Dunod, Paris, 1967.

[64] E. Szemerédi, Regular partitions of graphs. In *Problèmes combinatoires et théorie des graphes (Colloq. Internat. CNRS, Univ. Orsay, Orsay, 1976)*, pp. 399–401, Colloq. Int. Cent. Natl. Rech. Sci. 260, CNRS, Paris, 1978.

[65] L. Teirlinck, Nontrivial t-designs without repeated blocks exist for all t. *Discrete Math.* **65** (1987), no. 3, 301–311.

[66] R. Wilson, The early history of block designs. *Rend. Sem. Mat. Messina Ser. II* **9** (2003), 267–276.

[67] R. M. Wilson, An existence theory for pairwise balanced designs. I. Composition theorems and morphisms. *J. Combin. Theory Ser. A* **13** (1972), 220–245.

[68] R. M. Wilson, An existence theory for pairwise balanced designs. II. The structure of PBD-closed sets and the existence conjectures. *J. Combin. Theory Ser. A* **13** (1972), 246–273.

[69] R. M. Wilson, The necessary conditions for t-designs are sufficient for something. *Util. Math.* **4** (1973), 207–215.

[70] R. M. Wilson, An existence theory for pairwise balanced designs. III. Proof of the existence conjectures. *J. Combin. Theory Ser. A* **18** (1975), 71–79.

[71] R. M. Wilson, Decompositions of complete graphs into subgraphs isomorphic to a given graph. In *Proceedings of the Fifth British Combinatorial Conference (Univ. Aberdeen, Aberdeen, 1975)*, pp. 647–659. Congressus Numerantium, No. XV, Utilitas Math., Winnipeg, Man., 1976.

JULIA BÖTTCHER
Department of Mathematics, London School of Economics and Political Science, Houghton Street, London WC2A 2AE, UK, j.boettcher@lse.ac.uk

This Page Intentionally Left Blank

KKL'S INFLUENCE ON ME

EHUD FRIEDGUT

Dedicated to my dear friends and mentors K., K., and L.

ABSTRACT

In 1988 Kahn, Kalai, and Linial published their landmark paper in which they proved a lower bound on the maximal influence of variables on a Boolean function. Their use of Fourier analysis to solve the question, and especially their introduction of a hypercontractive inequality (due to Bonami, Beckner, and Gross), has shaped the field of study of Boolean functions and has had a great influence on combinatorics and theoretical computer science.

In this paper I survey how my own work has been influenced by their approach, via a collection of various problems that I have approached throughout the years.

MATHEMATICS SUBJECT CLASSIFICATION 2020

Primary 0502; Secondary 06E30

KEYWORDS

Boolean functions, hypercontractivity

INTERNATIONAL CONGRESS OF MATHEMATICIANS 2022 JULY 6–14

Proc. Int. Cong. Math. 2022, Vol. 6, pp. 4568–4581
DOI 10.4171/ICM2022/78

1. INTRODUCTION

In 1988 Kahn, Kalai, and Linial published their landmark paper [34] in which they proved that if $f : \{0,1\}^n \to \{0,1\}$ is a function with expectation α then there exists a coordinate i for which the (discrete) derivative in the ith direction, f_i, is substantial: $\|f_i\|_2^2 \geq c\alpha(1-\alpha)\log(n)/n$, for some global constant c, which does not depend on n. Stated in this language this result might well be categorized by the reader as a theorem in analysis. However, the range of the function f, the set $\{0,1\}$, is indicative of the fact that the motivation for this problem comes from a combinatorial and computer-science-theoretical angle. Indeed, what made this paper so influential is the use of discrete Fourier analysis, and especially the introduction of the hypercontractive inequality of Beckner [4], Bonami [6], and Gross [32] (henceforth the BBG inequality), to the end of proving a conjecture that has a strong combinatorial flavor, and arose in the context of theoretical computer science. In the third of a century since, this approach has had a huge impact on the study of Boolean functions, and their applications to combinatorics. Seeing that a substantial part of combinatorics deals with sets and their subsets, it is no surprise that the Boolean functions that indicate these subsets are such a useful language for dealing with these problems, and that the analytical tools applied to these functions are so fruitful.

When I approached Gil Kalai in 1992, and requested a research project, with the intention of it potentially growing to be a PhD thesis, he offered me several papers to read: a paper on the diameter of polytopes, a paper dealing with Cohen–Macaulay rings, and the duo of papers [34] and its sequel where Bourgain and Katznelson joined to produce [7]. I was intrigued by the latter two papers, and have been working on related problems ever since, to this very day. In this paper I want to present some samples of my related work. By doing this, I hope to give a glimpse into some of the interesting problems, notions, and techniques that I have encountered over the years during my work in the field. Of course, this sample is in no way representative of the progress of the field of Boolean functions, and the rich collection of results produced in it during the past decades; it is heavily skewed by focusing on my own work. For a nice reference for many of the foundational and important results in the field (at least up to 2014), I recommend Ryan O'Donnell's book on Boolean functions [39].

Following this introduction, this paper contains two main sections: in the first, I present a set of problems, and in the second, I present the tools and ideas used to solve each of them.

1.1. Basic terminology and definitions

Any function $f : \{0,1\}^n \to \mathbb{R}$ has a unique Fourier expansion

$$f = \sum_{S \subset \{1,2,\ldots,n\}} \hat{f}(S)\chi_S$$

where the functions χ_S are the characters of $\mathbb{Z}_2^n$,

$$\chi_S(x) = (-1)^{\sum_{i \in S} x_i}.$$

The *degree* of f is the maximal $|S|$ for which $\hat{f}(S)$ is nonzero.

We often consider the product measure μ_p on $\{0,1\}^n$, where

$$\mu_p(A) = \sum_{x \in A} p^{\sum x_i}(1-p)^{\sum(1-x_i)}.$$

When $p = 1/2$, this is the uniform measure. For other p, the measure induces the inner product

$$\langle f, g\rangle = E_x[f(x)g(x)],$$

and we expand functions according to an orthogonal basis with respect to this inner product. We refer to the elements of this basis also as characters (although this is a misnomer, as there is no group involved). So now $f = \sum_{S \subset \{1,2,\ldots,n\}} \hat{f}(S)\chi_S$ where χ_S is defined by

$$\chi_S(x) = \left(-\sqrt{(1-p)/p}\right)^{\sum_{i\in S} x_i}\left(\sqrt{p/(1-p)}\right)^{\sum_{i\in S}(1-x_i)}.$$

When considering the distribution of the Fourier coefficients of a function f, we refer to the *Fourier weight* of f on a set $A \subset \{0,1\}^n$, meaning $\sum_{S\in A} \hat{f}^2(S)$.

Finally, we need the important notion of *influence* (appearing also in the title of this paper): given a Boolean function $f : \{0,1\}^n \to \{0,1\}$, the influence of the ith coordinate on f is the probability that $f(x) \neq f(y)$, where x is chosen uniformly at random, and y is obtained by flipping the ith coordinate of x.

2. PROBLEMS

In this section I describe the parting point of several papers, i.e., the problems that they set out to solve.

2.1. Juntas rule

For a subset S of the discrete cube $\{0,1\}^n$, the *edge boundary* of S, denoted by $\partial_e(S)$ is the set of all pairs (x, y) with $x \in S$, $y \notin S$, and such that x and y differ in a single coordinate (they form an edge in the Hamming graph on $\{0,1\}^n$). Finding the minimum size of $\partial_e(S)$ given the size of S is a classical isoperimetric problem, first solved by Harper [33]. It is quite easy to prove that if $|S| = 2^{n-1}$ the unique minimizers of $\partial_e(S)$ are the subcubes of codimension 1, for which $|\partial_e(S)| = 2^{n-1}$. These are sometimes known as *dictators*, as their characteristic functions are dictated by a single coordinate. The question is what can be said if the edge boundary is within a multiplicative constant of the minimum:

Question 1. Given $S \subset \{0,1\}^n$, with, say, $|S| = 2^{n-1}$ and $|\partial_e(S)| = c \cdot 2^{n-1}$, is it true that S may be approximated by *a junta*, a set whose characteristic function is determined by a small set of coordinates, whose size depends only on the constant c and on the precision of the approximation?

2.2. Almost-dictator functions

A Boolean function on $\{0,1\}^n$ of the form $f(x) = x_i$ is called a dictator; $f(x) = 1 - x_i$ is called an antidictator. It is easy to show that dictators, antidictators, and the constant functions 0 and 1 are the only Boolean functions of degree 1. A question that arose quite naturally in the work of Kalai on social choice was whether this is robust:

Question 2. Is it true that a Boolean function which has almost all of its Fourier weight on the first two levels (empty set and singletons) is close to a dictator, antidictator, or constant? In other words, if $f : \{0,1\}^n \to \{0,1\}$ and

$$\sum_{|S|>1} \hat{f}^2(S) < \varepsilon,$$

is there a Boolean function g of degree 1 such that $\|f - g\|_2^2$ is small as a function of ε?

2.3. Thresholds for every graph property

The very first problem I worked on during my PhD studies was the following question, proposed by Gil Kalai and Nati Linial.

Question 3. Given a property of graphs on n vertices, how wide can the threshold interval be for the appearance of the property in $G(n,p)$?

For example, for a given n, if k is the most likely size of the maximal clique in $G(n,1/2)$, then for $\varepsilon = O(1/\log^2(n))$ there is an interval of length ε in $(0,1)$, such that for all p in that interval the probability that the largest clique in $G(n,p)$ is of size at least k is between 0.01 and 0.99. Fixing these parameters (0.01 and 0.99), are there other graph properties for which the threshold interval for p is much larger?

2.4. Sharp/coarse thresholds – global/local properties

The previous question is interesting mainly when the threshold interval is bounded away from 0 and 1. However, for many interesting graph properties, the critical probability for the appearance of the property in the random graph $G(n,p)$ is p which is a function of n, tending to zero as n grows. If p^* is such that a given monotone graph property appears in $G(n,p^*)$ with probability 1/2, then a result of Bollobás and Thomason [5] tells us that the threshold length is at most of order p^*. The following is a natural, if ambitious, question.

Question 4. Which monotone graph properties have a *sharp threshold*? That is, for which properties is the length of the threshold interval $o(p^*)$?

For example, the appearance of a triangle in $G(n,p)$ typically occurs for $p = \Theta(1/n)$, but for any constant c the probability that $G(n,c/n)$ contains a triangle is bounded away from 0 and 1. In contrast, the critical probability for connectivity is $p = \log(n)/n$, whereas the threshold is of width only $\Theta(1/n)$.

2.5. Maximizing copies of one (hyper)graph in another

Question 5. Given a fixed graph (or uniform hypergraph) H, and an integer m, what is the maximal number of copies of H one can have as subgraphs of a graph with m edges? (And what in the world does this possibly have to do with hypercontractive inequalities?)

It turns out that the answer is of the form $\Theta(m^{\pi^*(H)})$. The main challenge is to understand the function π^* (although finding the precise constants is also a challenging question.) In his MSc thesis, Noga Alon [2] provided an algorithm that computes $\pi^*(H)$. It was later recognized that this is the fractional covering number of H. In a joint paper with Jeff Kahn [26], we proved this for all uniform hypergraphs, using an entropy approach. Prior to this, a chance encounter with Noga led me to consider whether this result is related to the Bonami–Beckner–Gross hypercontractive inequality. I was trying to use a special case of the graph-theoretic inequality in order to prove a special case of the BBG inequality, and Noga, upon hearing this, first introduced me to his old(est) result. Studying this led me to the realization that the opposite implication is also a possibility, and led me to wonder whether one can prove the graph theorem using hypercontractivity.

2.6. The traffic light problem

Consider the following combinatorial problem:

Question 6. Assume that at a given road junction there are n three-position switches that control the red–amber–green position of the traffic light. You are told that, whenever you change the position of all the switches, the color of the light changes. Is it true that the light is, in fact, controlled by a single switch?

This problem was solved by Greenwell and Lovász [31] in 1974, using straightforward combinatorial arguments. In 2003, with Alon, Dinur, and Sudakov [3] we tackled this problem and generalizations thereof using spectral analysis, and also used the existing techniques from the study of Boolean functions to prove robustness results, i.e., what can be said if the hypothesis holds only for, say, 99.99% of the switch-configurations?

2.7. Subsets of independent sets in product graphs

The problem in the previous section gives rise to characterizing the maximal independent sets in the graph $(K_3)^{\otimes n}$, and showing that they are, in fact, cylinders of codimension 1, i.e., dictators. These sets have measure $1/3$ (according to the uniform measure on the graph). But what about independent sets whose size differs by a fixed multiplicative constant from the maximum?

Question 7. What can be said about independent sets in $K_3^{\otimes n}$ whose measure is, say, $1/100$? Are such sets essentially described by a small (constant) number of coordinates?

The hope to completely describe, or even approximate such sets, using a bounded number of coordinates is, of course, too far-reaching, as a random subset of a maximal independent set is also independent, and cannot be characterized by a small number of coordinates. Could it be, however, that any large independent set is (essentially) *contained* in an independent set determined by few coordinates (a junta)?

2.8. t-intersecting subsets of the cube

The Erdős–Ko–Rado theorem [18] is a (or perhaps the) fundamental theorem in extremal combinatorics. The ground set it considers is all subsets of size k of $\{1, 2, \ldots, n\}$ for some $k \leq n/2$. It then bounds the size of a maximal intersecting family of such subsets. The "complete intersection theorem" of Ahlswede and Khachatarian [1] expands this to families of subsets, where every two of them have an intersection of size at least t, for some integer $t > 1$.

It is quite natural to consider a "smoothed" version of this question. For $p \in (0, 1/2]$, consider the product measure $\mu = \mu_p$ on the discrete cube $\{0, 1\}^n$, and ask what the maximal measure of a family of vectors in the cube is if every two of them have nondisjoint support, or a common support of size at least t. The fact that if $A \subseteq \{0, 1\}^n$ is intersecting, then $\mu_p(A) \leq p$ was proven in many papers, e.g., [20] and [22] to mention just a few.

Question 8. Does this fact have a Fourier-theoretic proof? Does it have a robust version? What about t-intersecting families?

2.9. Triangle-intersecting families of graphs

There are many beautiful questions that generalize the original question of Erdős–Ko–Rado regarding intersecting families. One of my favorites was posed by Miklos Simonovits and Vera Sós, and I was intrigued by it from the moment I heard Vera introduce it in an open problems session in Oberwolfach. The question arises if we impose some structure on the ground set from which the intersecting subsets are taken. Their question is the following:

Question 9. Given a family of subgraphs of the complete graph on n vertices, if the intersection of every two members of the family contains a triangle, how large can the family be? Is it true that the maximum is attained by taking all graphs containing a fixed triangle?

2.10. Intersecting families of permutations

Another nice generalization of EKR was considered by Deza and Frankl in [11]. A family of permutations in S_n is called intersecting if for every two permutations in the family, σ, τ, there exists i such that $\sigma(i) = \tau(i)$. The family is t-intersecting for an integer $t \geq 1$ if every two permutations in the family agree on t points. Deza and Frankl made the observation that an intersecting family has size at most $(n-1)!$, and conjectured that this is achieved only by dictatorships, namely families of the form $\{\tau \in S_n : \tau(i) = j\}$ for some i and j. This conjecture, made in 1977, was finally proved in 2003 by Cameron and Ku in [9].

The more general conjecture made by Deza and Frankl was that t-intersecting families are of size at most $(n-t)!$, with the unique maximizers being families of the form

$$\{\tau \in S_n : \tau(i_r) = j_r, r = 1, \ldots, t\}$$

for some $i_1, \ldots, i_t$ and $j_1, \ldots, j_t$. Here even the conjectured upper bound of $(n-t)!$, which was very simple in the case $t = 1$, turned out to be difficult for $t > 1$.

Question 10. Prove that the families mentioned above are the unique maximal size t-intersecting families of permutations.

3. APPROACHES AND TECHNIQUES

In this section I give a bird's-eye view of the ways in which the questions presented in the previous section were resolved.

3.1. Juntas rule

In [23] I prove that, indeed, if the edge boundary of $S \subset \{0,1\}^n$ is small, then S is close to a set that is determined by a small number of coordinates. Here are the main steps and ideas of the proof:

(1) Let f be the characteristic function of S. The size of the edge boundary of S is equal (after normalizing) to the sum of the influences of the variables on f. Thus the sum of the influences in this case is small (bounded).

(2) Partition the variables according to their influence into two sets, J – those with large influence, and those with small influence, where the cutoff is appropriately chosen. The variables in J will form the junta. Since the sum of influences is bounded there cannot be too many variables in J.

(3) Let $f^{(1)} = \sum_{|S|<K} \hat{f}(S)\chi_S$ be a truncated version of f, where the cutoff point K is appropriately chosen. It is easy to show that $f^{(1)}$ is close to f (the fact that the sum of influences is small implies that there is not much Fourier weight on large sets.)

(4) Next, discard all coefficients of characters χ_S where S is a set not contained in the junta J, i.e.,

$$f^{(2)} = \sum_{|S|<K, S\subseteq J} \hat{f}(S)\chi_S.$$

Clearly, $f^{(2)}$ depends only on the junta variables.

(5) The following step is the heart of the argument – one wishes to show that $\|f - f^{(2)}\|_2^2$ is small, i.e., that the Fourier weight on sets containing variables with small influence is small. Note that this is not immediate since, although they each have small influence individually, it is not clear that their collective influence on f is small. To this end, the hypercontractive estimate of Bonami–Beckner–Gross is applied à-la KKL. By comparing different norms of the derivatives f_i, where i is a coordinate with small influence, one can show that the total Fourier weight on sets involving these variables is small.

(6) Using a rounding procedure, replace $f^{(2)}$ by a Boolean function that also depends only on the junta variables, and is a good approximation of f.

3.2. Almost-dictator functions

In [28], together with Kalai and Naor, we proved that, indeed, Boolean functions with almost all their Fourier weight concentrated on the first two levels are close to either a constant function, or a dictator, or an antidictator. In the paper we supply two different proofs. One proof uses a Berry–Eseen-type theorem of König, Schütt, and Tomczak-Jaegermann, [37], to show that there is at most one singleton $\{i\}$ such that $\hat{f}(\{i\})$ is large. The other proof uses the Bonami–Beckner–Gross hypercontractive inequality to show that once one truncates all Fourier coefficients above the first level, one is still left with a very well-behaved function, a fact that can only be explained by it being close to a degree-1 Boolean function. In particular, if $f^{(1)} = \sum_{|S|\le 1} \hat{f}(S)\chi_S$, we measure the "Booleanity" of $f^{(1)}$ by bounding the variance of $(f^{(1)})^2 - f^{(1)}$. Ultimately, this enables us to conclude that if for a Boolean function f one has $\sum_{|S|>1} \hat{f}^2(S) < \varepsilon$ then there exists a degree-1 Boolean function g such that $\|f - g\|_2^2 < O(\varepsilon)$.

3.3. Thresholds for every graph property

In [27] Kalai and I proved that for every monotone graph property the threshold length is of order at most $O(1/\log(n))$. The proof is embarrassingly simple: we simply observe that this is true for every weakly symmetric function – a function where there is a transitive group action on the variables of the function. By KKL and symmetry, if there are m coordinates, then all m influences are of order at least $\log(m)/m$, and their sum is of order at least $\log(m)$. Then, the Russo–Margulis lemma relates the sum of the influences to the slope of $\mu(p)$, the probability that $G(n, p)$ possesses the property in question, as a function of p. This slope determines the length of the threshold interval.

The paper [27], published in 1996, contains a conjecture that is still open, and, if true (as is widely believed), would have many interesting consequences. It is known as the *Entropy-Influence Conjecture*.

Conjecture 1. *There exists a constant c such that for every Boolean function* $f : \{0,1\}^n \to \{0,1\}$ *it holds that*

$$\sum \hat{f}^2(S) \log\bigl(1/|\hat{f}(S)|\bigr) \le c \sum \hat{f}^2(S)|S|.$$

Another interesting question raised in [27] is that of the dependence of the maximal possible threshold interval length on the structure of the orbits of the symmetry group of the property. In particular, we suspected that the maximal interval length for graph properties should be $O(1/\log^2(n))$, and not $1/\log(n)$. This was finally (almost) proved by Bourgain and Kalai in [8] where they gave a bound of $O(1/\log^{2-\varepsilon}(n))$, for every ε. Recently Kelman, Kindler, Lifshitz, Minzer, and Safra [36] improved this to $O(\frac{\log\log(n)}{\log^2(n)})$.

3.4. Sharp/coarse thresholds – global/local properties

In my PhD thesis [24], I settled this problem by proving that any monotone graph property which does not have a sharp threshold for appearance in $G(n, p)$ must be close to a "local" property: one of containing a graph from a fixed list of graphs (e.g., the property of containing a triangle or a cycle of length 4). The main steps of the proof are:

(1) Given a graph property with a coarse threshold, consider f, the characteristic function of the graphs containing this property, and its "skew-Fourier" expansion.

(2) Using the information that the threshold interval is large, prove that there exists a list of graphs S such that almost all the Fourier weight of f is concentrated on basis-functions indexed by graphs from this list. This is the main difficulty in the proof.

(3) For a function f, with Fourier expansion as above, partition the space of graphs according to the subgraph count of graphs from the list S, and show that on each part f is close to being constant.

(4) Show that any monotone function that behaves as described on the different parts of the space is a "local" function, i.e., its behavior on the different parts of the space is consistent.

3.5. Maximizing copies of one (hyper)graph in another

In [26] Jeff Kahn and I proved that $\pi^*(H)$ is equal to the fractional covering number of H for any hypergraph H. The fractional covering number of a graph is the solution to a certain linear programming problem, and we used the strong duality theorem of linear programming to prove the (equal) upper and lower bounds using the problem and its dual. The challenging part was the upper bound, for which we provided two different proofs. The first used an information-theoretic approach, via Shearer's entropy lemma. The other proof, which actually quite surprised us, was by relating the subgraph count to two different norms of a certain low degree polynomial function, and then using the BBG hypercontractive inequality to bound one norm in terms of the other, deducing the required inequality.

3.6. The traffic light problem

In [3] we proved for a large family of graphs (which includes complete graphs, and in particular triangles, which are the basis for the traffic light problem) that all maximal independent sets in tensor powers of these graphs are cylinders of codimension 1. Furthermore, we proved a stability result, showing that independent sets of size close to the maximal must also be close in structure to one of the extremal cylinders. The idea of the proof was to take the proof of the Hoffman bound on the size of independent sets in the base graph, and to "tensor" it. We built an orthonormal basis for the space of functions on the powers of the base graph, that consisted of tensor powers of the eigenvectors of the base graph. We then proceeded to use spectral analysis in a manner that is completely analogous to the Fourier analysis on $\{0,1\}^n$, including proving robustness results by using a variant of [28].

3.7. Subsets of independent sets in product graphs

In [12] Dinur, Regev, and I prove that, indeed, any large independent set in $K_3^{\otimes n}$ (and in many other product graphs, and also in Kneser graphs) is essentially contained in a set determined by few coordinates, a junta. Later, in [29] a paper with Regev, we complement this by proving that the junta itself is also an independent set (whereas in [12] we only proved that it is sparse.) The main tool in [12] is *noise*. The noise operator is the object which is the focus of the BBG hypercontractive inequality. Recall that the inequality studies a noisy version of a function f, i.e., $Tf(x) = E[f(y)]$, where y is a randomly perturbed version of x. By applying noise to an independent set contained in a junta, one is able to recover the junta, and the hypercontractivity serves to control this procedure.

In [29] we complement this result by showing that the junta we have recovered (which we proved in [12] to span very few edges) can be made independent by removing a small set of vertices. To this end, we prove an edge removal lemma, and prove it in the spirit of Fox's proof of the graph removal lemma [21]. We show that for the graphs in question (e.g., $K_3^{\otimes n}$ and also Kneser graphs) any sparse set can be made independent by removing a small set of vertices.

3.8. t-intersecting subsets of the cube

In [25] I manage to apply a Fourier approach to prove the claim that for $p \leq 1/2$ and an intersecting family $I \subseteq \{0,1\}^n$ we have $\mu_p(I) \leq p$. The idea of the proof comes from encoding the problem via a weighted graph, and applying (the proof of) Hoffman's bound to this graph. It turns out that the t-intersecting subsets' problem is more intricate. First of all, the natural extremal candidate of a t-umvirate, the AND of t bits, is extremal only for $p \leq 1/(t+1)$. For larger values of p, other examples take over (such as "three out of four" for 2-intersecting families). Still, for $p \leq 1/(t+1)$ it is possible to give a Fourier proof, and a robustness theorem.

Since the heart of the proof for the 1-intersecting case relies on the product structure of the cube, it uses the deep mathematical fact that if $x, y \in \{0,1\}$ then $x \cdot y = 1$ if and only if $x = y = 1$. This suffices to "penalize" two sets that intersect. However, when we want to penalize sets that have intersection of size at least, say, 2, we do not want to disqualify two sets that have one joint element, only "warn" them. To turn this into a scheme that is amenable to products, we need an element $X \neq 0$ such that $X^2 = 0$. The solution is to work over a ring of polynomials in the formal variable X, modulo the relation $X^2 = 0$. This adds various complications, as the eigenvalues of the resulting matrices are no longer real numbers, rather they are ring elements, hence it no longer makes sense to speak of the minimal eigenvalue. These difficulties can be handled, and eventually one can produce a set of t linear inequalities involving the Fourier distribution of the characteristic function of the t-intersecting family. Taking an appropriate linear combination of the inequalities, one may prove that the Fourier transform of the function sits on the first $t+1$ levels, and from there the road to proving the uniqueness and stability results is not too rough.

I would like to mention that Ryan O'Donnell [38] also discovered a simpler, more elementary way to deduce the inequalities regarding the Fourier coefficients of the functions that are used in this proof.

3.9. Triangle-intersecting families of graphs

In [14] we proved that, indeed, the largest triangle-intersecting families of subgraphs of K_n are those consisting of all graphs containing a given, fixed, triangle.

The difference between the 1-intersecting and t-intersecting for $t > 1$ in the previous subsection is that in the former case one gets an inequality regarding the Fourier coefficients of the characteristic function of the family in question, and in the latter case one gets a system of t inequalities. These inequalities are generated by the various constraints on the intersections of the family elements. In the problem now at hand, the constraints we used were of the form that for any bipartite graph B, and any two graphs G and H in the family, the intersection of G and H is not contained in B (as it contains a triangle). The potentially unbounded set of inequalities arising from these constraints is formidable, yet it offers a wealth of possibilities. By taking various appropriate linear combinations of these inequalities, we were able to prove that the characteristic function of the family in question has its Fourier coefficients concentrated precisely on the characters we expect.

3.10. Intersecting families of permutations

This problem was settled in [17], written with David Ellis and Haran Pilpel. The idea of using weighted graphs in order to prove a Hoffman-type bound is a recurring theme in solving the problems we have seen so far. This usually leads to expanding the functions in question in terms of a nice orthonormal basis, often the characters of a group, or the "skew" version, especially if the graph in question is a Cayley graph. In the case of the permutation version of EKR, this approach leads us to study the representation theory of S_n, and the (non-Abelian) Fourier expansion of functions. It turns out that, after carefully choosing the weights on the edges of the graph that represents the problem, one can deduce that the Fourier transform of a t-intersecting family is concentrated on representations indexed by partitions where the first part is of length at most $n - t$. This enables us to prove that the natural examples of extremal t-intersecting families, cosets of stabilizers of t points, are indeed maximal and unique. In later papers with Ellis and Filmus ([15] and [16]), we proved results that imply stability versions of [17], that t-intersecting families in S_n whose size is close to the maximum must be close in structure to the true maximizers. The results of [15] and [16] are analogs of known results regarding the Fourier transform of Boolean functions on $\{0, 1\}^n$. In [15] we prove an analog of the FKN theorem: Boolean functions on S_n whose Fourier transform is mainly concentrated on the trivial representation and the representation indexed by $(n - 1, 1)$ must be close to a dictator. In [16] we study functions on S_n whose Fourier transform is concentrated mostly on the representations indexed by by partitions of n with first part of size at least $n - t$, and proving that they are "junta-like." This is an analog of several similar theorems regarding Boolean functions on $\{0, 1\}^n$ whose Fourier transform is mostly concentrated on low levels (almost-low-degree functions), see, e.g., [35] and its references for the latest results in this vein.

It turns out, however, that there was a gap in the proof in [17], one of the final lemmas in the proof was false, thus, for $t > 1$ our proof of uniqueness of the extremal examples is incorrect. This was noticed by Filmus [19]. Luckily, this has been circumvented, both in a 2011 paper by Ellis [13] who used a clever bootstrapping approach, and in a recent paper [10].

REFERENCES

[1] R. Ahlswede and L. H. Khachatrian, The complete intersection theorem for systems of finite sets. *European J. Combin.* **18** (1997), 125–136.

[2] N. Alon, On the number of subgraphs of prescribed type of graphs with a given number of edges. *Israel J. Math.* **38** (1981), 116–130.

[3] N. Alon, I. Dinur, E. Friedgut, and B. Sudakov, Graph products, Fourier analysis and spectral techniques. *Geom. Funct. Anal.* **14** (2004), no. 5, 913–940.

[4] W. Beckner, Inequalities in Fourier analysis. *Ann. of Math.* **102** (1975), 159–182.

[5] B. Bollobás and A. Thomason, Threshold functions. *Combinatorica* **7** (1986), 35–38.

[6] A. Bonami, Etude des coefficients Fourier des fonctiones de $L^p(G)$. *Ann. Inst. Fourier (Grenoble)* **20** (1970), no. 2, 335–402.

[7] J. Bourgain, J. Kahn, G. Kalai, Y. Katznelson, and N. Linial, The influence of variables in product spaces. *Israel J. Math.* **77** (1992), 55–64.

[8] J. Bourgain and G. Kalai, Influences of variables and threshold intervals under group symmetries. *Geom. Funct. Anal.* **7** (1997), 438–46.

[9] P. Cameron and C. Y. Ku, Intersecting families of permutations. *European J. Combin.* **24** (2003), 881–890.

[10] N. Dafni, Y. Filmus, N. Lifshitz, N. Lindzey, and M. Vinyals, Complexity measures on symmetric group and beyond. In *12th Innovations in Theoretical Computer Science Conference (ITCS 2021)*, pp. 87:1–87:5, Schloss Dagstuhl - Leibniz-Zentrum für Informatik, 2021.

[11] M. Deza and P. Frankl, On the maximum number of permutations with given maximal or minimal distance. *J. Combin. Theory Ser. A* **22** (1977), 352–360.

[12] I. Dinur, E. Friedgut, and O. Regev, Independent sets in graph powers are almost contained in juntas. *Geom. Funct. Anal.* **18** (2008), no. 1, 77–97.

[13] D. Ellis, Stability for t-intersecting families of permutations. *J. Combin. Theory Ser. A* **118** (2011), 208–227.

[14] D. Ellis, Y. Filmus, and E. Friedgut, Triangle-intersecting families of graphs. *J. Eur. Math. Soc. (JEMS)* **14** (2012), 841–885.

[15] D. Ellis, Y. Filmus, and E. Friedgut, A stability result for balanced dictatorships in S_n. *Random Structures Algorithms* **46** (2015), no. 3, 494–530.

[16] D. Ellis, Y. Filmus, and E. Friedgut, Low-degree Boolean functions on S_n, with an application to isoperimetry. *Forum Math. Sigma* **5** (2017).

[17] D. Ellis, E. Friedgut, and H. Pilpel, Intersecting families of permutations. *J. Amer. Math. Soc.* **24** (2011), 649–682.

[18] P. Erdős, C. Ko, and R. Rado, Intersection theorems for systems of finite sets. *Quart. J. Math. Oxford (Ser. 2)* **12** (1961), 313–320.

[19] Y. Filmus, A comment on Intersecting Families of Permutations. 2017, arXiv:1706.10146.

[20] P. C. Fishburn, P. Frankl, D. Freed, J. C. Lagarias, and A. M. Odlyzko, Probabilities for intersecting systems and random subsets of finite sets. *SIAM J. Algebr. Discrete Methods* **7** (1986), no. 1, 73–79.

[21] J. Fox, A new proof of the graph removal lemma. *Ann. of Math. (2)* **174** (2011), no. 1, 561–579.

[22] P. Frankl and N. Tokushige, Weighted multiply intersecting families. *Studia Sci. Math. Hungar.* **40** (2003), no. 3, 287–291.

[23] E. Friedgut, Boolean functions with low average sensitivity depend on few coordinates. *Combinatorica* **18** (1998), no. 1, 27–36.

[24] E. Friedgut, Sharp thresholds of graph properties, and the k-sat problem. *J. Amer. Math. Soc.* **12** (1999), no. 4, 1017–1054.

[25] E. Friedgut, On the measure of intersecting families, uniqueness and stability. *Combinatorica* **28** (2008), 503–528.

[26] E. Friedgut and J. Kahn, On the number of copies of one hypergraph in another. *Israel J. Math.* **105** (1998), 251–256.

[27] E. Friedgut and G. Kalai, Every monotone graph property has a sharp threshold. *Proc. Amer. Math. Soc.* **124** (1996), 2993–3002.

[28] E. Friedgut, G. Kalai, and A. Naor, Boolean functions whose Fourier transform is concentrated on the first two levels and neutral social choice. *Adv. in Appl. Math.* **29** (2002), 427–437.

[29] E. Friedgut and O. Regev, Kneser graphs are like Swiss cheese. *Discrete Anal.* **2** (2018), DOI 10.19086/da.3103.

[30] C. Godsil and K. Meagher, A new proof of the Erdős–Ko–Rado theorem for intersecting families of permutations. *European J. Combin.* **30** (2009), no. 2, 404–414.

[31] D. Greenwell and L. Lovász, Applications of product colorings. *Acta Math. Acad. Sci. Hung.* **25** (1974), no. 3–4, 335–340.

[32] L. Gross, Logarithmic Sobolev inequalities. *Amer. J. Math.* **97** (1975), 1061–1083.

[33] L. H. Harper, Optimal assignments of numbers to vertices. *SIAM J. Appl. Math.* **12** (1964), 13–135.

[34] J. Kahn, G. Kalai, and N. Linial, The influence of variables on Boolean functions. In *Proc. 29-th Ann. Symp. on Foundations of Comp. Sci*, pp. 68–80, Computer Society Press, 1988.

[35] N. Keller and O. Klein, A structure theorem for almost low-degree functions on the slice. *Israel J. Math.* **240** (2020), 179–221.

[36] E. Kelman, G. Kindler, N. Lifshitz, D. Minzer, and S. Safra, Towards a proof of the Fourier-entropy conjecture? 2019, arXiv:1911.10579.

[37] H. König, C. Schütt, and N. Tomczak-Jaegermann, Projection constants of symmetric spaces and variants of Khintchine's inequality. *J. Reine Angew. Math.* **511** (1999), 1–42.

[38] R. O'Donnell, private communication.

[39] R. O'Donnell, *Analysis of Boolean functions*. Cambridge University Press, Cambridge, 2014.

EHUD FRIEDGUT

Weizmann Institute of Science, Rehovot, Israel, ehud.friedgut@weizmann.ac.il

SCHUBERT CALCULUS AND QUIVER VARIETIES

ALLEN KNUTSON

ABSTRACT

The Littlewood–Richardson rule (1934) is a combinatorial (and, in particular, manifestly positive) way to compute the structure constants of two a priori unrelated rings-with-basis: the representation ring of $\mathrm{GL}_k(\mathbb{C})$, and the cohomology ring of the Grassmannian $\mathrm{Gr}(k, \mathbb{C}^n)$. We recall a wealth of generalizations of the latter ring (changing the space, the cohomology theory, or the basis), all of which have non-manifestly-positive rules for computation, nowadays called their *Schubert calculus.* Until this century very few of these structure constants had combinatorial rules for their calculation, although many of the structure constants have been proven (ineffectively) to be nonnegative.

In recent years the formal similarity of one of these rules (the Knutson–Tao "puzzle" rule for equivariant cohomology) to quantum integrable systems has been traced to the geometry of *quiver varieties,* a class among which one finds the cotangent bundles to Grassmannians. This allowed for the discovery and proof of rules for many heretofore unsolved Schubert calculus problems, and new connections to representation theory.

MATHEMATICS SUBJECT CLASSIFICATION 2020

Primary 14M15; Secondary 16T25, 16G20

KEYWORDS

Schubert calculus, Knutson–Tao puzzles, quiver varieties, flag manifolds, R-matrices, stable envelopes, Chern–Schwartz–MacPherson classes

INTERNATIONAL CONGRESS OF MATHEMATICIANS 2022 JULY 6–14

Proc. Int. Cong. Math. 2022, Vol. 6, pp. 4582–4605
DOI 10.4171/ICM2022/118

Published by EMS Press

FIGURE 1
Donald Knutson, left, at the author's first ICM (see also Figure 2).

1. LITTLEWOOD–RICHARDSON COEFFICIENTS

1.1. From intersection theory on Grassmannians

Given a compact oriented manifold M (so, one enjoying Poincaré duality) and a Morse function, one obtains a decomposition of M into cells. With luck[1] the Morse function is "perfect," meaning that the cellular homology chain maps vanish, and the cells therefore give a basis of homology and (using the Poincaré duality) cohomology. The product $[X] \cdot [Y]$ in the cohomology ring[2] can be interpreted using the intersection $[X \cap Y]$, assuming that the cell closures X and Y have been moved to be transverse.

In this ring-with-basis, there is no reason to expect the structure constants to be non-negative. For example, if M is the blowup $\widetilde{\mathbb{CP}^2}$ of $\mathbb{CP}^2$ at a point, and E is the exceptional divisor, then $[E] \cdot [E]$ is *minus* the class of a point. (From this one can infer that the two-sphere E cannot be perturbed to some E' inside the real four-manifold $\widetilde{\mathbb{CP}^2}$ while *staying complex,* as the complex intersection $E \cap E'$ would then have the right orientation to be a positive number of points.)

---

1 Of course, this is situation is very special—for example, it can only hold when the homology has no torsion.

2 It is worth noting that this homology/cohomology technology was invented exactly to answer the 15th question Hilbert proposed at the 1900 ICM **[29]**, about putting Schubert's calculus on a rigorous footing.

Sometimes there is a cheap source of such perturbations. If M is a *homogeneous space* for a complex Lie group G, then [34] shows that for $X, Y \subseteq M$ complex subvarieties and $g \in G$ a generically chosen group element, the subvarieties X and $g \cdot Y$ are transverse. From this and a certain duality property of the cohomology basis (to be defined below, in the case M compact), one finds that the structure constants are nonnegative.

We now focus on the first such M of real interest: the Grassmannian $\mathrm{Gr}(k, \mathbb{C}^n)$ of k-planes in n-space. One can overparametrize this manifold using the "row span" map taking a full-rank $k \times n$ matrix R to its row span, a k-plane. Since this map is invariant under row operations, we can use Gaussian elimination to restrict the domain to full-rank $k \times n$ matrices in reduced row-echelon form. There are now $\binom{n}{k}$ cases, according to where the k pivots occur, and each case gives a complex cell; in all this gives the *Bruhat decomposition* of $\mathrm{Gr}(k, \mathbb{C}^n)$ into complex cells. (There is also a Morse theory picture [1].)

The cells X_λ° are naturally indexed by partitions $(\lambda_1 \geq \cdots \geq \lambda_k \geq 0)$ as follows: erase the pivot columns, and count the 0s in each row, from bottom to top. That is to say, the zeros form a "French partition" inside the smaller matrix:

$$\begin{bmatrix} 1 & * & 0 & 0 & * & 0 & * \\ 0 & 0 & 1 & 0 & * & 0 & * \\ 0 & 0 & 0 & 1 & * & 0 & * \\ 0 & 0 & 0 & 0 & 0 & 1 & * \end{bmatrix} \mapsto \begin{bmatrix} * & * & * \\ 0 & * & * \\ 0 & * & * \\ 0 & 0 & * \end{bmatrix} \mapsto \lambda = (2, 1, 1, 0). \qquad (*)$$

The closures $\{X_\lambda := \overline{X_\lambda^\circ}\}$ of these cells are the *Schubert varieties* in the Grassmannian, and we denote the Poincaré duals of their homology classes by $\{[X_\lambda]\}$, the *Schubert classes*.

Though we will not pursue the following viewpoint further here, it is worth recalling the reasons for the general interest in moduli spaces and especially in their cohomology, where the Grassmannian (the "moduli space of k-dimensional subspaces of $\mathbb{C}^n$") is the most basic example. Whenever one has a family $F \to X$ of some kind of mathematical object $\mathcal{O}$, one may hope to interpret it as the pullback of a *universal family* $\mathcal{F} \to M(\mathcal{O})$ along a "classifying map" $X \to M(\mathcal{O})$. What would be even better is if this recipe $\mathrm{Map}(X, M(\mathcal{O})) \to$ {isomorphism classes of $\mathcal{O}$-families over X} were bijective. Assuming both, and applying H^* to the classifying map gives us

$$\{\text{isomorphism classes of } \mathcal{O}\text{-families over } X\} \to \mathrm{Map}\big(H^*\big(M(\mathcal{O})\big) \to H^*(X)\big)$$

In the case $\mathcal{O} = \{k\text{-planes in } \mathbb{C}^\infty\}$ so $F \to X$ is a k-dimensional vector bundle, the cohomology ring $H^*(M(\mathcal{O}))$ is a polynomial ring in k generators $c_1, \ldots, c_k$, and their images in $H^*(X)$ are the Chern classes of F. The images of the Schubert classes arise as the classes of "degeneracy loci" of generic bundle maps from F to a (flagged) trivial bundle $\mathbb{C}^n \times X$. For more of this viewpoint on Schubert classes, see, e.g., [24].

1.2. From representation theory

The representation ring $\mathrm{Rep}(G)$ of a group G has a natural $\mathbb{Z}$-basis consisting of the finite-dimensional irreducible representations, and the multiplication in this basis has a positivity property: the expansion of (the semisimplification of) the tensor product of two irreps is an $\mathbb{N}$-combination of irreps. When G is a complex Lie group (or even Kac–Moody group), there has been a great deal of work on combinatorial interpretations of these structure constants, with a reasonably complete answer given by the work of Littelmann [46].

In the specific case $G = \mathrm{GL}_k(\mathbb{C})$, the basis is naturally indexed by dominant weights $\{\lambda \in \mathbb{Z}^k : \lambda_1 \geq \lambda_2 \geq \cdots \geq \lambda_k\}$, and many authors restrict to the subcase $\lambda_k \geq 0$ of "polynomial representations" V_λ when considering the **Littlewood–Richardson coefficients**

$$c^{\nu}_{\lambda\mu} := \dim \mathrm{Hom}_{\mathrm{GL}_k(\mathbb{C})}(V_\nu, V_\lambda \otimes V_\mu).$$

Effectively, this subcase is the representation theory of the Lie *monoid* $(\mathrm{End}(\mathbb{C}^k), \bullet)$ rather than of the group $\mathrm{GL}_k(\mathbb{C})$ sitting densely within. If one is willing to stray this far from groups, it is natural to consider the representation theory of the entire category **Vec** (of finite-dimensional complex vector spaces), i.e., functors **Vec** $\to$ **Vec**, where the irreps are the "Schur functors" such as $V \mapsto \mathrm{Alt}^m V$. (For technical reasons, one usually restricts to representations that are finite direct sums of Schur functors.)

This Rep(**Vec**) picture lets one observe a nice stability: for $\lambda, \mu, \nu \in \mathbb{N}^k$ and $k' \geq k$, if we construct $\lambda', \mu', \nu' \in \mathbb{N}^{k'}$ by concatenating $k' - k$ zeros at the end, then $c^{\nu}_{\lambda\mu} = c^{\nu'}_{\lambda'\mu'}$.

Theorem 1. *The linear map* $\mathrm{Rep}(\mathrm{End}(\mathbb{C}^k)) \twoheadrightarrow H^*(\mathrm{Gr}(k, \mathbb{C}^n))$ *taking*

$$[V_\lambda] \mapsto \begin{cases} [X_\lambda] & \text{if } \lambda_1 \leq n-k, \\ 0 & \text{otherwise} \end{cases}$$

is a ring homorphism. In particular, the structure constants in the Schubert basis of the Grassmannian are again Littlewood–Richardson coefficients.

The original proof [45] of Theorem 1 is rather indirect—essentially, one checks that the "Pieri rule" (which governs multiplication in the case $\mu_1 = 1$) holds in both cases. These Pieri classes generate the ring, and associativity takes care of the rest.

Since then, there have been a number of more satisfying linkages drawn between the two rings-with-bases. In [41, §8] Kostant, developing further a proof by Horrocks [38] (see also [14]), approaches $H^*(\mathrm{Gr}(k, \mathbb{C}^n)) \cong H^*(\mathrm{GL}_n(\mathbb{C})/P_{k,n-k})$ using de Rham cohomology and $U(n)$-invariant forms. The resulting Lie algebra cohomology differential vanishes because the radical $\mathrm{Rad}(P_{k,n-k})$ of the parabolic subgroup $P_{k,n-k}$ is abelian, and the representation theory points very directly to the Schubert basis. Compact homogeneous spaces G/P with $\mathrm{Rad}(P)$ abelian are called **cominuscule** and will appear again in Section 2.5.

In the cunningly titled paper "The connection between Schubert calculus and representation theory" [58], a natural map $\mathrm{Rep(Vec)} \to H^*(\mathrm{Gr}(k, \mathbb{C}^n))$ is constructed, applying a Schur functor to the tautological bundle over the Grassmannian and (à la Chern–Weil theory) using the $U(n)$-invariant Hermitian connection on that bundle to build a cohomology class. Unfortunately, while the *map* is natural and visibly multiplicative, the proof that it takes basis elements to basis elements (or zero) again amounts to observing that both rings enjoy the Pieri rule.

A tighter connection appears in [5] (see also [6]), in which Belkale uses a point in the transverse triple intersection $X_\lambda \cap (g \cdot X_\mu) \cap (g' \cdot X_\nu)$ to define a vector in

$$(V_\lambda \otimes V_\mu \otimes V_\nu)^{\mathrm{SL}(\mathbb{C}^k)},$$

and shows that the resulting vectors are linearly independent. (This proves only an inequality between the intersection-theoretical vs. the representation-theoretical numbers. A related approach in [49] establishes the equality.) This is perhaps the most satisfying (or "most categorical") in that it works directly with the vector space rather than just its dimension. The same is true in a more general statement about quivers in [20].

1.3. From group theory

Given a partition λ and a prime p, one can construct a finite abelian p-group $\Gamma_\lambda := \prod_i (\mathbb{Z}/p^{\lambda_i})$. The number of short exact sequences $0 \to \Gamma_\lambda \to \Gamma_\nu \to \Gamma_\mu \to 0$ turns out to be a polynomial in p, with leading coefficient $c^\nu_{\lambda\mu}$. We refer the reader to [22] for more on this source of LR coefficients.

1.4. Combinatorial approaches

Before going further, we draw a distinction here between the computation of Schubert *classes* vs. their products, the Schubert *calculus*. Every one of the rings-with-bases we will consider here and in Section 2 has a known presentation with generators and relations, and (with greater difficulty in some cases than others) a known system of polynomial representatives for the desired basis elements. These polynomials themselves often have interesting positivity properties, giving statements such as "The Schubert polynomials of Lascoux–Schützenberger have nonnegative coefficients." However, such positivity results (or even combinatorial formulæ) do not directly give positivity results about the multiplication. As such we will not focus further here on the (very interesting) questions of constructing these representatives.

We cannot emphasize strongly enough that the name of the game is to give *manifestly nonnegative formulæ for the (known to be nonnegative) structure constants.* We mention three reasons to seek such formulæ, even where nonpositive formulæ are readily available.

(1) For applications (including real-world engineering applications), it is more important to know that some structure constant c is *positive*, than it is to know its actual value. This is much more easily studied with a noncancelative formula. The same is true for another problem of frequent interest, determining when $c = 1$.

(2) Alternating sum formulæ tend to be *much* less efficient computationally.

(3) When positive integers appear, they suggest that there may be a possibility for categorification, in which each coefficient is promoted to a vector space of that dimension. A combinatorial rule for the coefficient then suggests an indexing of a basis for the vector space.

There are several rules of the form "$c^{\nu}_{\lambda\mu}$ is the number of Young tableaux of a certain shape and content satisfying several conditions [semistandard/ballot/Yamanouchi/reverse lattice word]" all of which go by the name "Littlewood–Richardson rules." The history of these rules is somewhat convoluted—in particular, the original proof had an error not corrected for decades—and we refer the interested reader to **[22, 62]** for it. The most concise modern proofs seem to be **[56]** (based on a sign-reversing involution) and **[12]** (which uses the associativity argument of **[45]**).

For these same numbers there are a wealth of other rules counting combinatorial objects, e.g., the "pictures" of **[65]** (which allows for a generalization involving skew Schur functions), the "cartons" of **[59]** which manifest an S_3-symmetry in the problem, the "Mondrian tableaux" of **[17]**—but we focus now on the *puzzles* that we introduced in **[36, 37]** which have so far admitted the most generalizations.

We will need to index Schubert classes on $\mathrm{Gr}(k, \mathbb{C}^n)$ not by partitions as in $(*)$, but by binary strings where the 0s indicate the pivot columns. In the example from $(*)$, the string would be 0100101, and more generally has content $0^k 1^{n-k}$. One typical cohomology calculation is $S_{101}^2 = S_{110} \in H^*(\mathrm{Gr}(1,3))$, which says that two lines in the projective plane intersect in a point.

A $c^{\nu}_{\lambda\mu}$ **puzzle** will be an equilateral triangle of edge-length n, with the northwest, northeast, and south sides of this Δ labeled by λ, μ, ν all written left-to-right.

Theorem 2. **[36, 37]** *There exists a finite set of "puzzle pieces" (unit triangles with edge labels, oriented either as Δ or ∇) such that the number of $c^{\nu}_{\lambda\mu}$ puzzles assembled from them is the Littlewood–Richardson coefficient, for all k, n and λ, μ, ν.*

Once we grant this oracular statement, it is extremely easy to reverse-engineer the pieces (which is why I did not include them in the statement above). On the point $\mathrm{Gr}(1, \mathbb{C}^1)$ there is a unique Schubert class $S_0 = 1$, hence $S_0^2 = S_0$, so there should exist a unique puzzle with boundary 0 on all three sides. We have found our first piece, the 0–0–0 Δ-piece. The same argument on the point $\mathrm{Gr}(0, \mathbb{C}^1)$ lets us also discover a Δ-piece with 1 on each side. The similar formula $S_{00}^2 = S_{00}$ in $H^*(\mathrm{Gr}(2, \mathbb{C}^2))$ is almost as easy; the 0–0–0 Δ-piece fits nicely into the three corners, forcing us to invent a 0–0–0 ∇-piece to go in the middle. Again, the corresponding calculation on $H^*(\mathrm{Gr}(0, \mathbb{C}^2))$ suggests we admit a 1–1–1 ∇-piece as well.

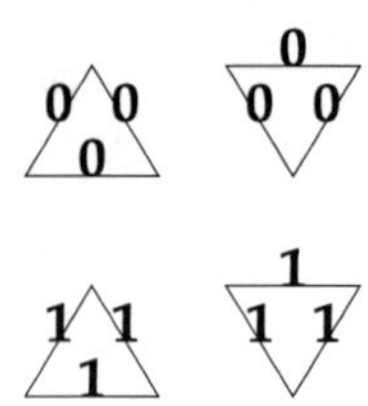

A new phenomenon enters when we consider $H^*(\mathrm{Gr}(1, 2))$, where we need to find a puzzle computing $S_{01}^2 = S_{01}$. In the southwest and southeast corners we can place 0–0–0 and 1–1–1 Δ-pieces. If we try to put either the label 0 or 1 on the remaining edge, we run into problems (overcounting $c^{\nu}_{\lambda\mu}$ somewhere down the line), so we invent a new label "10" to go on this edge. The two rotations of this puzzle give the $S_{01}S_{10} = S_{10}S_{01} = S_{10}$ computations, so all six rotations of the 1–0–10 piece should be admitted.

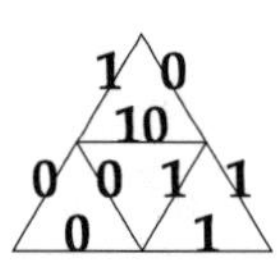

It is then an easily checked experimental fact (for $n \leq 10$, say) that no more labels or pieces are needed: these three (up to rotation) pieces are already giving the right count for every $c^{\nu}_{\lambda\mu}$! Of course, this is not a proof, and the first proof of Theorem 2 was a bit unsatisfying—just a reduction to another, known, rule for LR coefficients. A much more concrete link between the Grassmannian geometry and combinatorics was first laid out in [61], and connected to puzzles in [35, 40].

These puzzles enjoy six symmetries: rotation by multiples of 120°, left–right reflection composed with the label swap $0 \leftrightarrow 1$, and composites thereof. In fact, the LR coefficients have these symmetries and more, once one observes

$$c^{\nu}_{\lambda\mu} = \int_{\mathrm{Gr}(k,\mathbb{C}^n)} S_\lambda S_\mu S_{\nu \text{ reversed}} = \int_{\mathrm{Gr}(n-k,\mathbb{C}^n)} S_{\lambda^*} S_{\mu^*} S_{\nu^* \text{ reversed}}$$

where λ^* means "reverse and swap $0 \leftrightarrow 1$." The first equality comes from the dual-basis statement $\int_{\mathrm{Gr}(k,\mathbb{C}^n)} S_\lambda S_{\nu \text{ reversed}} = \delta_{\lambda\nu}$, the second from Grassmannian duality. It is rather hard to directly see that the puzzle rule defines a commutative product, which is manifest in the carton rule from [60].

2. SEVERAL INDEPENDENT AXES OF GENERALIZATION

In Sections 2.1–2.6 below we present various mutually compatible axes of generalization "**KTFQGC**" of the basic problem (that being Schubert calculus in $H^*(\mathrm{Gr}(k,\mathbb{C}^n))$), and comment afterward on the combinations thereof. While we have endeavored to report the state-of-the-art (as concerns combinatorial rules), for reasons of space we have not included a complete timeline of earlier results.

2.1. *K*-theory [K]

The K-theory of the Grassmannian is again a free $\mathbb{Z}$-module of dimension $\binom{n}{k}$, but there are two big differences between it and the cohomology: it is naturally filtered rather than graded (with $H^*(\mathrm{Gr}(k,\mathbb{C}^n))$ as the associated graded ring), and there are *two* natural bases for it, dual to one another under a natural pairing. The more commonly studied basis $\{G_\lambda\}$ consists of (Klasses of the) structure sheaves of the Schubert varieties. The other basis $\{G^*_\lambda\}$ consists of the "ideal sheaves", functions on a Schubert variety vanishing on all smaller Schubert varieties.

The first proof that the multiplicative structure constants of the $\{G_\lambda\}$ basis are positive (up to a sign convention) came hand in hand with a formula for their computation, in terms of Buch's "set-valued semistandard Young tableaux" [8]. A (more widely applicable, but noneffective) geometric argument for this positivity appeared afterward in [7].

It is easy to guess puzzle pieces for these two products, from the computations $(G^*_{01})^2 = G^*_{01} - G^*_{10}$ and $G^2_{0101} = G_{0110} + G_{1001} - G_{1010}$. For the $\{G_\lambda\}$ multiplication, one introduces a Δ-piece with labels 10–10–10 (announced in [61]). For the $\{G^*_\lambda\}$ multiplication, one introduces the 10–10–10 ∇-piece [64]. In both cases, the sign convention requires that each 10–10–10 piece contribute a factor of -1, but this does not lead to any cancelation.

2.2. T-equivariant cohomology [T]

The invertible diagonal matrices $T \leq \mathrm{GL}_n(\mathbb{C})$ act on $\mathrm{Gr}(k, \mathbb{C}^n)$ preserving each of the Schubert varieties X_λ. Hence, each X_λ defines an element of the T-equivariant cohomology ring $H_T^*(\mathrm{Gr}(k, \mathbb{C}^n))$. These classes (again called $\{S_\lambda\}$) form a basis not over $\mathbb{Z} = H^*(pt)$, but over the base ring $H_T^*(pt) \cong \mathbb{Z}[y_1, \ldots, y_n]$, and hence the structure constants live in this polynomial ring. The coefficients were shown (again, geometrically and more generally but ineffectively) in [27] to lie in $\mathbb{N}[y_1 - y_2, \ldots, y_{n-1} - y_n]$.

The first two computations $S_{10}^2 = (y_1 - y_2)S_{10}$ and $S_{010}^2 = S_{100} + (y_2 - y_3)S_{010}$ suggest a vertically[3] rhomboidal **equivariant piece** whose **fugacity** $\mathrm{fug}(\Diamond) = y_i - y_j$ depends on the piece's location in the puzzle (namely, the $\Diamond$ is in the ith SW/NE diagonal and the jth NW/SE diagonal). See the examples below.

Theorem 3 ([36]). *Define an **equivariant puzzle** P to be one in which this additional piece is allowed, and define its **fugacity** $\mathrm{fug}(P)$ to be the product of the fugacities of the equivariant pieces. Then the equivariant structure constant $c_{\lambda\mu}^{\nu}$ is the sum of the fugacities of the equivariant puzzles with boundary λ, μ, ν as in Theorem 2.*

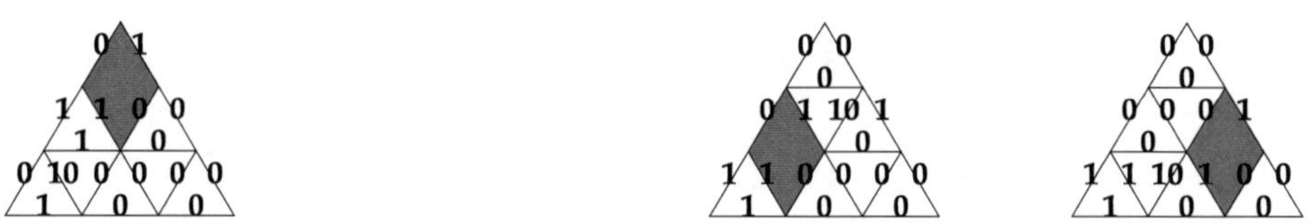

$$S_{010}\, S_{100} = (y_1 - y_3)S_{100} \quad = \quad S_{100}\, S_{010} = \quad (y_1 - y_2)S_{100} \quad + \quad (y_2 - y_3)S_{100}.$$

(In particular, $i < j$ for any equivariant piece, so this formula verifies the Grassmannian case of Graham's positivity theorem [27].) The proof proceeds from the "most equivariant case" $c_{\lambda\lambda}^{\lambda}$, with an induction based on a recursive formula for the $\{c_{\lambda\mu}^{\nu}\}$. The recursion involves a denominator that vanishes if one passes to the nonequivariant case $y_i \equiv 0$, so does not allow for a directly nonequivariant proof of Theorem 2.

We take a moment to foreshadow the framework that will come in Sections 3–4.

The matching requirement for adjacent puzzle labels can be interpreted nicely in terms of matrix multiplication (where the formula $(AB)_{iq} = \sum_{j=p} A_{ij} B_{pq}$ requires a similar matching). Following [66], we introduce a 9×9 matrix $R(a, b)$ whose 3^2 columns are labeled by the possible rhombus tops, and rows by the 3^2 possible rhombus bottoms. A matrix entry is 1 if the resulting labeled rhombus can be filled by two triangular puzzle pieces; is

---

3 In fact, the search was also directed by the need to break the $\mathbb{Z}/3$-symmetry, whose derivation depended on the dual-basis equation $\int_{\mathrm{Gr}(k,\mathbb{C}^n)} S_\lambda S_\nu = \delta_{\lambda,\nu \text{ backwards}}$. This equation makes sense, but does not hold, in equivariant cohomology. The $\mathbb{Z}/3$-symmetry of the K-theory puzzles comes from a different source, $G_\lambda^* = G_\lambda[\mathcal{O}(1)]$ for all λ, a property of "minuscule" flag manifolds (which includes Grassmannians); see Section 2.5.

$a-b$ in the one case that it can be filled by an equivariant piece; and is 0 otherwise:

	$1\wedge 1$	$1\wedge 0$	$1\wedge 10$	$0\wedge 1$	$0\wedge 0$	$0\wedge 10$	$10\wedge 1$	$10\wedge 0$	$10\wedge 10$
$1\vee 1$	1	0	0	0	0	1	0	0	0
$1\vee 0$	0	0	0	$a-b$	0	0	0	0	0
$1\vee 10$	0	0	0	0	1	0	1	0	0
$0\vee 1$	0	1	0	0	0	0	0	0	0
$0\vee 0$	0	0	0	0	1	0	1	0	0
$0\vee 10$	0	0	0	0	0	0	0	0	0
$10\vee 1$	0	0	0	0	0	0	0	0	0
$10\vee 0$	1	0	0	0	0	1	0	0	0
$10\vee 10$	0	0	0	0	0	0	0	0	0

There is a corresponding 3×9 matrix U, with rows labeled $\underline{1}, \underline{0}, \underline{10}$, controlled by the triangular pieces taken alone. With R and U, one reinterprets the sum in Theorem 3 as a *matrix entry* inside the $3^n \times (3^n \cdot 3^n)$ matrix

$$U^{\otimes n}\left(\prod_{i=1}^{n-1}\prod_{j=i+1}^{n}\left(I_{3^{2j-i-2}} \otimes R(y_{j-i}, y_j) \otimes I_{3^{2n-2j+i}}\right)\right).$$

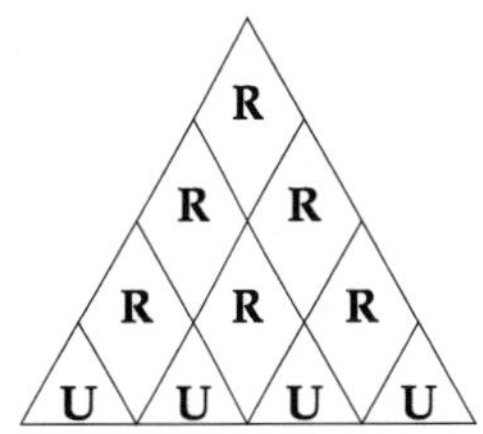

The key observation in [66] is that this R satisfies the "rational Yang–Baxter equation"

$$\left(R(a,b)\otimes I_3\right)\left(I_3\otimes R(a,c)\right)\left(R(b,c)\otimes I_3\right) = \left(I_3\otimes R(b,c)\right)\left(R(a,c)\otimes I_3\right)\left(I_3\otimes R(a,b)\right)$$

whose relevance will be explained in Section 3.

There is also a "trigonometric Yang–Baxter equation" whose R-matrix entries depend on a/b instead of $a-b$. This becomes relevant in creating puzzle formulæ in equivariant K-theory, as in [38,39,64], foreshadowed in [54].

2.3. Larger flag manifolds [F]

Define the **d-step flag manifold** $\mathrm{Fl}(n_1, n_2, \ldots, n_d; \mathbb{C}^n)$ to be the space of chains $(V_1 \leq V_2 \leq \cdots \leq V_d : \dim(V_i) = n_i)$, so Grassmannians are 1-step flag manifolds. (Of course, one can reduce to the case (n_i) strictly increasing, in which case $d = n$ is the maximal situation, but it will be mildly convenient not to.) The row-span and Gaussian elimination technique from Section 1.1 lead again to a cell decomposition with strata now labeled by strings with content $0^{n_1} 1^{n_2-n_1} \ldots d^{n-n_d}$.

In 1999 I followed the oracle of Theorem 2 to $d > 1$, and came up with a set of edge labels and puzzle pieces that looked promising for general d. But it was already wrong for $d = 3$ and $n = 5$, so I (prematurely) abandoned it, without even pursuing $d = 2$. Buch observed experimentally that my incorrect rule was consistently *undercounting* at $d = 3$, and he suggested some additional puzzle labels, though $d \geq 4$ remained seemingly out of reach. My 2-step puzzle conjecture was proven in [10], again by an associativity check, and in [38] we proved Buch's modified 3-step conjecture (plus a 151-piece extension to K-theory) by techniques to be recalled in Section 4. Another combinatorial rule for 2-step appears in [17], where it is suggested that the techniques involved should extend to higher d (see the survey [19] for a $d = 3$ example).

The "Schur times Schubert" subproblem, in which one of the two classes is pulled back from the Grassmannian, is easily shown to be equivalent to the problem of expanding the class of a "positroid variety" into Schubert classes. Positroid varieties have become of much interest in the physical theory of scattering amplitudes [4].

2.4. Quantum cohomology [Q]

On a compact oriented manifold, where cohomology classes can be thought of homologically, the structure constants of multiplication can be computed as the finite number of points $p \in X \cap Y \cap Z$ in a transverse triple intersection. Physicists introduced in the 1990s the "quantum cohomology" of an almost complex manifold M, in which one instead counts almost complex maps $\gamma : \mathbb{CP}^1 \to M$ such that $\gamma(0) \in X$, $\gamma(1) \in Y$, and $\gamma(\infty) \in Z$. To define the (small) quantum product for most honest complex manifolds, one must deform the complex structure to generic almost complex (the 1990s solution), or involve an "obstruction bundle" over the moduli space of maps (the 21st-century solution), but for Grassmannians these niceties are unnecessary [23].

It turns out that for generic enough $\gamma : \mathbb{CP}^1 \to \mathrm{Gr}(k, \mathbb{C}^n)$ (and as proved in [11], the γs we want to count *will* be generic enough), the two nested subspaces $\bigcap_{z \in \mathbb{CP}^1} \gamma(z)$, $\sum_{z \in \mathbb{CP}^1} \gamma(z)$ will have dimensions $k - \deg(\gamma)$, $k + \deg(\gamma)$, respectively (where $\deg(\gamma)$ is defined by $\gamma_*([\mathbb{CP}^1]) = \deg(\gamma)[X_{1\ldots1010\ldots0}]$). With a little more work, it is shown in [11] that the degree m structure constants in $QH(\mathrm{Gr}(k, \mathbb{C}^n))$ can be calculated as structure constants in *ordinary* $H^*(\mathrm{Fl}(k - m, k + m; \mathbb{C}^n))$. (This sparked particular interest in the 2-step case, causing Buch–Kresch–Purbhoo–Tamvakis to revisit and eventually prove my 2-step conjecture from Section 2.3.)

2.5. Other Lie groups [G]

Grassmannians $\mathrm{Gr}(k, \mathbb{C}^n)$ are minimal homogeneous spaces for $\mathrm{GL}_n(\mathbb{C})$, with the property that their point stabilizers have abelian unipotent radical. Such **cominuscule flag manifolds** are very rare—each connected simply-connected group G has only $|Z(G)| - 1$ of them up to conjugacy (and even fewer up to isomorphism, such as $\mathrm{Gr}(k, \mathbb{C}^n) \cong \mathrm{Gr}(n - k, \mathbb{C}^n)$). For example, there are two E_6 cominuscule flag manifolds, but they are isomorphic.

In [60] is given a uniform rule for Schubert calculus in the ordinary cohomology of cominuscule flag manifolds. The proof of its validity, however, is case-by-case.

There is a Langlands dual notion, of "minuscule" flag manifold, which is one whose minimal equivariant embedding $G/P \hookrightarrow \mathbb{P}V$ is into a G-representation V with only extremal weights. This tight control on the homogeneous coordinate ring is felicitous for K-theory considerations (see, e.g., [13]).

2.6. Cotangent Schubert calculus [C]

Although a complex manifold M and its cotangent bundle T^*M are homotopic, hence bear the same cohomology ring, they may have different natural bases. One source for the latter is the *characteristic cycle* $\mathrm{cc}(\mathcal{F})$ of a $\mathcal{D}_M$-module $\mathcal{F}$. This is a Lagrangian cycle inside T^*M, and is invariant under the $\mathbb{C}^\times$-action dil dilating the cotangent fibers, hence defines a class

$$[\mathrm{cc}(\mathcal{F})] \in H^*_{\mathrm{dil}}(T^*M) \cong H^*_{\mathrm{dil}}(M) \cong H^*_{\mathrm{dil}}(pt) \otimes H^*(M) \cong H^*(M)[\hbar]$$

(where $\hbar$ is the standard generator of $H^*_{\mathrm{dil}}(pt)$). When $\iota : A \hookrightarrow M$ is the inclusion of a locally closed submanifold and $\mathcal{F} = \iota_*(\mathcal{O}_A)$ is the sheaf of distributions supported on A, this class $[\mathrm{cc}(\mathcal{F})]$ is essentially the **Chern–Schwarz–MacPherson class** of the submanifold [26], up to a sign.

If we invert $\hbar$ (or set it to -1, as is conventional for CSM classes) then the classes associated to the ($\mathcal{D}$-modules of distributions on the) Bruhat cells are again a basis, now of $H^*(\mathrm{Gr}(k, \mathbb{C}^n))[\hbar^{\pm}]$. There is some work on their structure constants of multiplication [16], but it has been more fruitful to consider multiplying the **Segre–Schwarz–MacPherson classes** $\{\mathrm{SSM}_\lambda := [\mathrm{cc}(\iota_*(\mathcal{O}_{X^\circ_\lambda}))] \,/\, e(T^* \mathrm{Gr}(k, \mathbb{C}^n))\}$. The necessity of introducing this denominator is hinted at in Section 4.

Theorem 4 ([39]). *The product of SSM classes on* $\mathrm{Gr}(k, \mathbb{C}^n)$ *can be computed using puzzles as before, within which one now allows* both *the* Δ *and* ∇ *10–10–10 pieces.*

As a consequence, the Euler characteristic of a transverse triple intersection $X^\circ_\lambda \cap (g \cdot X^\circ_\mu) \cap (h \cdot X^\circ_\nu)$ *is* $(-1)^{\text{its dimension}}$ *times the number of such puzzles, where* λ, μ, ν *are all written clockwise on the puzzle boundary. That dimension also predicts the number of 10–10–10 pieces in every puzzle.*

"Cotangent" is the newest adjective in the subject and is getting a lot of attention, e.g., [2, 57]. The involvement of $\mathcal{D}_{G/P}$-modules is especially exciting because of the Beilinson–Bernstein localization theorem, which relates them to representations of $U\mathfrak{g}$. The representations that are relevant here are the parabolic Verma modules of central character 0.

In addition, it appears that a proper understanding of Schubert calculus in *elliptic* cohomology (the next step beyond K-theory, in a sense) requires passage to the cotangent bundle [42].

2.7. Mixing and matching

The theorems in Sections 2.1–2.6 may make the subject sound closed, but each of the 2^6 combinations of **KTFQGC** is its own problem, and most are unsolved. By time of writing, the maximal positively solved problems (in the sense of having a manifestly noncancelative combinatorial rule for all products) are

- **KG** for minuscule flag manifolds [13],
- **KTFC** for $d = 2$ and **KFC** for $d = 3$ [39],

- **QT** via the connection to 2-step [9–11],
- **KQT** for projective space [15].

There are many partial results concerning multiplication by special classes, as well as non-effective positivity results such as [3, 48].

2.8. A few other generalizations

The cohomology of a space M bears a ring structure exactly because M has a canonical map $M \to M \times M$, the diagonal inclusion; the multiplication then comes from that pullback. If more generally we have a map $F/P \to G/Q$ of generalized flag manifolds, we can consider the map $H^*(G/Q) \to H^*(F/P)$ in the bases of Schubert classes.

Theorem 5 ([28]). *Let $\iota : \mathrm{SpGr}(k, \mathbb{C}^{2n}) \hookrightarrow \mathrm{Gr}(k, \mathbb{C}^{2n})$ be the inclusion of the Grassmannian of isotropic k-planes with respect to a symplectic form. Then the pullback in T^n-equivariant cohomology can be computed using puzzles with 10-labels allowed on the bottom, and that are invariant under flipping left–right while exchanging $0 \leftrightarrow 1$. (Another, nonequivariant, rule appears in [18].)*

The *affine Grassmannian* Gr_G is a homogeneous space for the affine Lie group, so the study of its cohomology is covered by case **G** above. But since Gr_G is homotopic to a group, its *homology* also bears a ring structure. Fascinatingly, this ring is tightly connected to the quantum cohomology of the corresponding finite-dimensional *full* flag manifold [33, 43, 44] (itself very far from having a positive rule).

There is also a ring structure on the K-homology $\bigoplus_{a,b} K_\bullet(\mathrm{Gr}(a, \mathbb{C}^{a+b}))$ induced by the "direct sum" map, computed with new puzzle pieces in [55]. Finally, the "separated descents" pullback along the inclusion $\mathrm{Fl}(\mathbb{C}^n) \hookrightarrow \mathrm{Fl}(1, \dots, k; \mathbb{C}^n) \times \mathrm{Fl}(k, \dots, n; \mathbb{C}^n)$ is computed in cohomology in [31] and will be given a **KTC** puzzle rule elsewhere.

3. QUIVER VARIETIES, STABLE ENVELOPES, AND STABLE BASES

We switch gears to define a very different family of varieties, following [25,47,50–52].

We comment briefly on the 20-year journey we took from puzzles to these *quiver varieties*. P. Zinn-Justin [66] reproved the equivariant puzzle rule from [36], along the way showing that one could build an "R-matrix" (meaning, a solution to the "Yang–Baxter equation") from the equivariant pieces; see Section 2.2. Through this he was able to replace much of the bespoke combinatorial arguments we had used in [36] with standard tricks from the theory of quantum integrable systems. Further puzzle results were obtained by this algebraic technique in [38, 64]—in particular, *discovering* and proving the rule for K(3-step), which requires 151 new puzzle pieces—but the deeper relation of this algebra to flag manifolds was not clear.

Solutions to the Yang–Baxter equation typically come from commutors of representations of quantized affine algebras (see, e.g., [21, §13] and [32]). Nakajima constructed some such representations on the K-theory of quiver schemes (theorem 7 below), and Maulik–

Okounkov interpreted the R-matrices directly on the quiver varieties [47,53]. In the remainder of this paper we recall the constructions from [39, §7], in which we geometrically reinterpret the algebraic results of [38] and extend to cotangent Schubert calculus in the bargain.

3.1. Quiver varieties

Consider a directed graph (Γ_0, Γ_1) with some of the vertices Γ_0 called "gauged" and the others called "framed", which we will generally indicate by $\boxed{\text{framing}}$ them. To simplify notation (while not ruling out any of the cases of most interest here), we assume that there is at most one edge connecting any two vertices.

To each labeling $d : \Gamma_0 \to \mathbb{N}$ of the vertices, called a **dimension vector**, we construct a **quiver variety** $\mathcal{M}(\Gamma, d)$ in four steps:

(1) Consider the vector space $\prod_{(t\to h)\in\Gamma_1}(\mathrm{Hom}(\mathbb{C}^{d(t)}, \mathbb{C}^{d(h)}) \times \mathrm{Hom}(\mathbb{C}^{d(h)}, \mathbb{C}^{d(t)}))$, where a typical element is a tuple $(M_{ab} \in \mathrm{Hom}(\mathbb{C}^{d(a)}, \mathbb{C}^{d(b)}))$.

(2) Impose the closed "complex moment map" condition that at each gauged vertex v, $\sum_{(v\to w)\in\Gamma_1} M_{vw}M_{wv}$ equals $\sum_{(w\to v)\in\Gamma_1} M_{wv}M_{vw}$ plus a scalar.

(3) Impose the open "stability" condition that for any $\vec{w} \neq \vec{0}$ at a gauged vertex v_0, there exists an undirected path $v_0 \leftrightarrow v_1 \leftrightarrow \cdots \leftrightarrow v_m$ with v_m framed, such that $M_{v_{m-1},v_m} \cdots M_{v_1,v_0}\vec{w} \neq \vec{0}$. (There are other stability conditions one might use but we will not need them.)

(4) Divide by the action of the group of basis transformations at the gauge vertices.

It is more traditional to fix the scalars used in the moment map condition, especially to 0, but for convenience of exposition we work with this enlarged point of view. While quiver varieties (as defined here) are naturally Poisson varieties, with symplectic leaves given by fixing those scalars, we will focus attention on a certain circle action that does *not* preserve the Poisson structure. It is induced from the scaling action on half the original variables, $\prod_{(t\to h)\in\Gamma_1}\mathrm{Hom}(\mathbb{C}^{d(t)}, \mathbb{C}^{d(h)})$, and we call it the **dilation** action dil.

The following is a folklore observation, which we include in part to recall the Grothendieck–Springer deformation of $T^*\,\mathrm{Fl}(n_1, \dots, n_d; \mathbb{C}^n)$:

Theorem 6 (see, e.g., [50]).

$$\mathcal{M}\begin{pmatrix} \boxed{n} & & \\ \uparrow & & \\ n_d \leftarrow & \cdots & \leftarrow n_1 \end{pmatrix}$$

is isomorphic to the Grothendieck–Springer deformation of the cotangent bundle $T^*\,\mathrm{Fl}(n_1, \dots, n_d; \mathbb{C}^n)$, where the deformation parameters are the d scalars used at the gauge vertices. The dilation action on the cotangent bundle is given by scaling the cotangent vector.

Proof sketch (the morphism in one direction). For convenience, index the vertices by their dimensions. The gauge-invariant functions we will use are the endomorphism

$M_{n_d,n} M_{n,n_d} \circlearrowright \mathbb{C}^n$ and the nested subspaces $V_i := im(M_{n,n_d} M_{n_{d-1},n_d} \cdots M_{n_{i+1},n_i}) \le \mathbb{C}^n$. The stability condition ensures that $\dim(V_i) = n_i$, giving us the flag, and the moment map condition implies that $(X - \varepsilon_i)V_i \le V_{i+1}$ where ε_i is the scalar at vertex i. This resulting space $\{(X, V_\bullet \in \mathrm{Fl}(n_1, \ldots, n_d; \mathbb{C}^n), \vec{\varepsilon}) : (X - \varepsilon_i)V_i \le V_{i-1}\}$ is the **Grothendieck–Springer family**, whose central fiber $\vec{\varepsilon} = \vec{0}$ is the Springer resolution of the closure of a nilpotent orbit. ■

It will also be convenient to fix only the dimension vector $\boxed{d}$ on the framed vertices, and define the **quiver scheme** $\mathcal{M}(\Gamma, \boxed{d})$ as the disjoint union of all $\mathcal{M}(\Gamma, e)$ where e agrees with $\boxed{d}$ on the framed vertices. (That may seem like a lot of components, but because of the stability condition these $\mathcal{M}(\Gamma, e)$ are frequently empty, such as when the (n_i) in the example of Theorem 6 are not weakly increasing.) One way these enter geometric representation theory is as follows:

Theorem 7 ([51]). *Assume Γ's gauge vertices form an ADE quiver, corresponding to a simple Lie algebra $\mathfrak{g}$. Then there is a family of natural actions of the quantized loop algebra $U_q(g[z^\pm])$ on the finite-dimensional vector space $K(\mathcal{M}(\Gamma, \boxed{d}))$. The decomposition $K(\mathcal{M}(\Gamma, \boxed{d})) = \bigoplus_e K(\mathcal{M}(\Gamma, e))$ is into the weight spaces.*

Moreover, if $d = d_1 + d_2$, then $K(\mathcal{M}(\Gamma, \boxed{d})) \cong K(\mathcal{M}(\Gamma, \boxed{d_1})) \otimes K(\mathcal{M}(\Gamma, \boxed{d_2}))$ generically in this family.

The parameters on the family can be interpreted geometrically as follows. Pick a basis of each $\boxed{\text{framed}}$ space, and let T be the torus that acts by scaling these basis elements. Then $U_q(g[z^\pm])$ acts on $K_T(\mathcal{M}(\Gamma, \boxed{d}))$, and the base of the family above is the space $T \cong \operatorname{Spec} K_T(pt)$ of equivariant parameters.

Applying this to the A_d example in Theorem 6 when $\boxed{n} = 1$, the nonempty quiver varieties are the $d+1$ points $T^* \mathrm{Fl}(0, \ldots, 0, 1, \ldots, 1; \mathbb{C})$. Their total K-theory gives the standard representation $\mathbb{C}^{d+1}$ of $U_q(\mathfrak{sl}_{d+1}(\mathbb{C}[z^\pm]))$. For general $\boxed{n}$, we get the rep $(\mathbb{C}^{d+1})^{\otimes n}$, with a basis consisting of length n strings in $0, 1, \ldots, d$, compatible with the indexing from Section 2.3.

One can degenerate the algebra to a Yangian (essentially) and act instead on $H(\mathcal{M}(\Gamma, \boxed{d}))$ [63], with the benefit that one can extract $H_{\mathrm{top}}(\mathcal{M}(\Gamma, \boxed{d}))$; this latter space bears a less-natural *irreducible* action of $U\mathfrak{g}$ itself [50] that was found first. In the example above, we recover the irrep $\mathrm{Sym}^n(\mathbb{C}^{d+1})$ of $U\mathfrak{sl}_{d+1}$, whose highest weight is $n\omega_1$; this reflects the fact that we attached the $\boxed{n}$ to the first vertex of A_d.

3.2. Stable envelopes and bases

Fix Γ and d. A circle action S on the framed vector spaces induces a circle action on $\mathcal{M}(\Gamma, d)$. Loosely following [47, THEOREM 3.7.4], we define the **stable envelope**

$$\operatorname{env}(S) := \overline{\left\{\left(p \in \mathcal{M}(\Gamma, d)^S, q \in \mathcal{M}(\Gamma, d)\right) : \lim_{z \to 0} S(z) \cdot q \text{ exists and is } p\right\}}$$
$$\subseteq \mathcal{M}(\Gamma, d)^S \times \mathcal{M}(\Gamma, d)$$

which we regard as providing a *correspondence,* not an actual map, $\mathcal{M}(\Gamma, d)^S \to \mathcal{M}(\Gamma, d)$. Sidestepping some compactness issues (discussed, in e.g., **[39, §7]**), the envelope induces an isomorphism $\tilde{H}^*_{\text{dil}}(\mathcal{M}(\Gamma, d)^S) \to \tilde{H}^*_{\text{dil}}(\mathcal{M}(\Gamma, d))$ where dil is the dilation action, and the tilde indicates that (as in Section 2.6) we have inverted the $\hbar$ in $H^*_{\text{dil}}(pt) = \mathbb{Z}[\hbar]$. This is of particular interest when $\mathcal{M}(\Gamma, d)^S$ is a finite set, in which case we can use the isomorphism to carry the obvious basis of $\tilde{H}^*_{\text{dil}}(\mathcal{M}(\Gamma, d)^S)$ to a **stable basis** of the target. This basis depends on S, though the action $S'(z) \cdot m := S(z^N) \cdot m$ for fixed $N > 0$ leads to the same envelope and basis.

Lemma 1 (Special case and restatement of **[47, LEMMA 3.6.1]**). *Fix Γ, d and let A_1, A_2 be two commuting circle actions on the framed spaces. Then for $N \gg 1$, the triangle*

$$\begin{array}{ccc} \mathcal{M}(\Gamma, d)^{A_1, A_2} & \xrightarrow{\text{env}(A_{1+})} & \mathcal{M}(\Gamma, d) \\ \text{env}(A_2) \searrow & & \nearrow \text{env}(A_1) \\ & \mathcal{M}(\Gamma, d)^{A_1} & \end{array}$$

commutes (in the sense of convolutions of correspondences), where $A_{1+}(z) = A_1(z^N)A_2(z)$.

In particular, if $\mathcal{M}(\Gamma, d)^{A_1, A_2}$ is finite, this implies that the correspondence $\text{env}(A_1)$ takes stable basis elements to stable basis elements.

3.3. Comparison of stable bases

There is another important application of Lemma 1. Assume that A, A' are commuting **regular** circle actions on $M := \mathcal{M}(\Gamma, d)$ in the sense that they each have isolated fixed points—necessarily the *same* set of fixed points, as each of A, A' acts (trivially) on the fixed points of the other. How can we compute the change-of-basis matrix between the two stable bases?

Let $\langle A, A' \rangle$ be the 2-torus they generate, and $\Lambda := \text{Hom}(\mathbb{C}^\times, \langle A, A' \rangle) \cong \mathbb{Z}^2$ be its coweight lattice. Within this plane Λ, the subset $\{S \in \Lambda : M^S \textit{ not } \text{finite}\}$ is easily shown to be the union of a finite number of lines through the origin. These lines cut the plane into sectors, and within each sector the associated stable basis is constant.

Draw a path from A to A' inside this coweight lattice. It may pass through many sectors, giving us many stable bases along the way, and give thereby a factorization of the change-of-basis matrix as a product. Wherever the path crosses a wall $C_+ \cap C_-$ between two chambers C_+, C_-, we can pick coweights S_1, S_2 where S_1 lies in the interior of the wall, and for all $N \gg 0$ we have $S_1^N S_2 \in C_+$, $S_1^N S_2^{-1} \in C_-$. Then we get a diagram of correspondences

$$\begin{array}{ccccc} M^{\langle A, A' \rangle} & \xrightarrow{\text{env}(S_2)} & M^{S_1} & \xleftarrow{\text{env}(S_2^{-1})} & M^{\langle A, A' \rangle} \\ \text{env}(S_1^N S_2) \searrow & & \downarrow \text{env}(S_1) & & \swarrow \text{env}(S_1^N S_2^{-1}) \\ & & M & & \end{array}$$

whose triangles commute (by Lemma 1), and whose diagonal arrows induce the stable bases from chambers C_+, C_-.

The change-of-basis matrix R across the wall $C_+ \cap C_-$ amounts to following the induced map on cohomology from the northwest $M^{\langle A,A'\rangle}$, down to the M, followed by the inverse of the map from the northeast $M^{\langle A,A'\rangle}$. Because the triangles commute we can instead work across the top line, from the northwest $M^{\langle A,A'\rangle}$, to M^{S_1}, followed by the inverse of the map from the northeast $M^{\langle A,A'\rangle}$. Typically, M^{S_1} is highly disconnected, from which we infer that the change-of-basis matrix R is very sparse.

In the example of the next section, each component of M^{S_1} is either a point or $T^*\mathbb{CP}^1$, so (up to reordering of rows and columns) R is a direct sum of blocks of size 1 or 2.

4. A FACTORIZATION OF CORRESPONDENCES GIVES THE PUZZLE RULE

This material is ahistorically drawn from **[38, 39]**. In this section, we specialize the deformation parameters in our quiver varieties to 0, and work with the action of $U_{\exp(\hbar)}(\mathfrak{g}[z])$ on cohomology rather than $U_q(\mathfrak{g}[z^{\pm}])$ on K-theory.

We have a big clue: in **[66]** the puzzle pieces for $H_T^*(\mathrm{Gr}(k, \mathbb{C}^n))$ are utilized to construct an R-matrix (see Section 2.2), which by the work of Drinfel'd and Jimbo referenced above suggests we seek a quantum group representation on $\mathbb{C}^3$ (where $3 = \#\{0, 1, 10\}$ is the number of edge labels). There is an obvious guess: $\mathfrak{sl}_3 \circlearrowright \mathbb{C}^3$ or, more precisely, $U_{\exp(\hbar)}(\mathfrak{sl}_3[z]) \circlearrowright \mathbb{C}^3(y)$ where y is the "evaluation parameter" of the representation (arising in Theorem 7 as an equivariant parameter).

The R-matrix[4] $\mathbb{C}^3(a) \otimes \mathbb{C}^3(b) \to \mathbb{C}^3(b) \otimes \mathbb{C}^3(a)$ gives[5] the rhombi, but we will also need triangular pieces $U : \mathbb{C}^3(a) \otimes \mathbb{C}^3(b) \to V(c)$, where V is again 3-dimensional. The $\mathfrak{sl}_3$-equivariance allows for only one possibility: V must be $\mathrm{Alt}^2\, \mathbb{C}^3$. This asymmetry suggests a reformulation of the puzzle pieces:

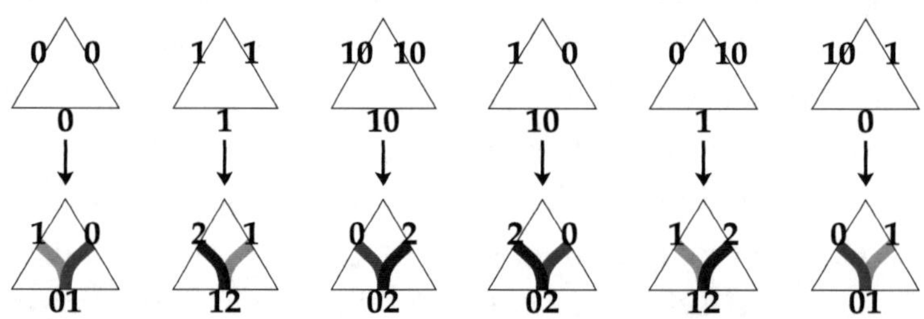

4 Technically, in the quantum integrable systems literature this is the "$\check{R}$-matrix."

5 Actually the R-matrix from Section 2.2 is only a degenerate limit of the one provided by the representation theory—for example, in the nondegenerate R-matrix there are nonzero entries corresponding to fillings that use 10–10–10 pieces. One of the purposes of **[39]** was to provide a cohomological question that would be answered by the richer puzzles constructed from the nondegenerate R-matrix. This turned out to be the Cotangent story.

The t-equivariance, or weight conservation, of the map U becomes the statement that the $0, 1, 2$-pipes propagate all the way to the boundary of a puzzle. See these:

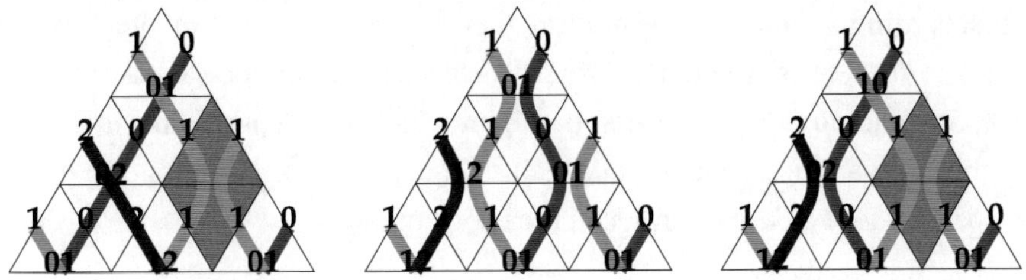

Under this new labeling, the puzzles have content $0^0 1^k 2^{n-k}$ on the northwest side, $0^k 1^{n-k} 2^0$ on the northeast side, and $(01)^k (02)^0 (12)^{n-k}$ on the south side. Those contents pick out certain weight spaces of $(\mathbb{C}^3)^{\otimes n}$, $(\mathbb{C}^3)^{\otimes n}$, and $(\mathrm{Alt}^2\,\mathbb{C}^3)^{\otimes n}$, suggesting that we look at the following A_2 quiver varieties,

$$\begin{array}{ccccccc} / & & \backslash & & \wedge & & — \\ \mathcal{M}\begin{pmatrix} \boxed{n} & \\ k & 0 \end{pmatrix} & \times & \mathcal{M}\begin{pmatrix} \boxed{n} & \\ n & k \end{pmatrix} & \to & \mathcal{M}\begin{pmatrix} \boxed{2n} & \\ n+k & k \end{pmatrix} & \to & \mathcal{M}\begin{pmatrix} & \boxed{n} \\ k & k \end{pmatrix} \\ T^*\,\mathrm{Fl}(0,k;\mathbb{C}^n) & \times & T^*\,\mathrm{Fl}(k,n;\mathbb{C}^n) & & T^*\,\mathrm{Fl}(k,n+k;\mathbb{C}^{2n}) & & T^*\,\mathrm{Fl}(k,k;\mathbb{C}^n) \end{array}$$

(The middle dimension vector is the sum of the two on the left. The two arrows are yet to be discussed.)

This is looking good: the first space is $T^*(\mathrm{Gr}(k,\mathbb{C}^n)^2)$, the last is $T^*(\mathrm{Gr}(k,\mathbb{C}^n))$, and multiplication (our goal) is pullback along the diagonal inclusion $\mathrm{Gr}(k,\mathbb{C}^n) \hookrightarrow \mathrm{Gr}(k,\mathbb{C}^n)^2$. There is no natural *map* between their cotangent bundles, but there is a natural correspondence[6]—the conormal bundle to the graph of the diagonal inclusion. (It does not *quite* induce the multiplication map on $H^*(\mathrm{Gr}(k,\mathbb{C}^n))$; as explained in **[39, LEMMA 11]**, this will be why we need introduce the denominator in our SSM classes.)

This suggests we attach correspondences to the two arrows in the diagram above, so as to make their composite the conormal bundle to the graph of the diagonal inclusion (or, more correctly, to its transpose). We have a good choice for the first arrow: a certain component of the stable envelope for the circle $S(z) := \mathrm{diag}(z^n, 1^n) \in \mathrm{GL}_{2n}(\mathbb{C})$. As an effect of working with the "leftmost" component in some sense—the gauge dimensions $k, 0$ in the left factor are smallest possible—the closure step in the definition of stable envelope may be skipped. Identifying each cotangent bundle with its Springer description (as in Theorem 6), this component of the envelope is the correspondence

$$\left\{ \begin{array}{l} (((A,W),(D,V)),(X,(W',V'))) : \\ \quad X = \begin{bmatrix} A & 0 \\ * & D \end{bmatrix}, \begin{array}{l} W' = W \oplus 0 \\ V' = \mathbb{C}^n \oplus V \end{array} \end{array} \right\} \subseteq T^*\,\mathrm{Gr}(k,\mathbb{C}^n)^2 \times T^*\,\mathrm{Fl}(k,n+k;\mathbb{C}^n).$$

6 This is an example of a *Lagrangian* correspondence and, more specially, is a "conical Lagrangian correspondence" as it is invariant under the dilation action on the cotangent bundle. Although it is intriguing, we will not make any real use of this additional geometry.

The second correspondence is subtler in that it must break symmetry—the middle variety has a $\mathrm{GL}_{2n}(\mathbb{C})$-action, whereas the third has only a $\mathrm{GL}_n(\mathbb{C})$-action. We now draw inspiration from the $U_{\exp(\hbar)}(\mathfrak{sl}_3[z])$-equivariance of the maps $\mathbb{C}^3(y_i) \otimes \mathbb{C}^3(z_i) \to \mathrm{Alt}^2\,\mathbb{C}^3(c)$: according to the representation theory, these maps can only exist if $y_i = \hbar/2 + c = \hbar + z_i$ (otherwise the tensor product is either irreducible, or has only $\mathrm{Sym}^2(\mathbb{C}^3)$ as a quotient). Specializing the evaluation parameters is equivalent (thanks to Theorem 7) to specializing the equivariant parameters, which is equivalent to shrinking the group action. That clue helped suggest a correspondence that does the job:

$$\left\{ \begin{array}{ll} (X, (W', V')), (((A', W''))) : & \\ X = \begin{bmatrix} A & * \\ I_n & D \end{bmatrix}, & \begin{array}{c} W'' = W'/(0 \oplus \mathbb{C}^n) \\ W'' = V' \cap (0 \oplus \mathbb{C}^n) \\ A' = A + D \end{array} \end{array} \right\} \subseteq T^* \mathrm{Fl}(k, n+k; \mathbb{C}^n) \times T^* \mathrm{Gr}(k, \mathbb{C}^n).$$

(Because of the I_n, the space of such X is only invariant under $\{[\begin{smallmatrix} M & 0 \\ 0 & M \end{smallmatrix}] : M \in \mathrm{GL}_n(\mathbb{C})\}$, which is how the symmetry is broken.)

Theorem 8 ([39, §7.6.2]). *The composite of these two correspondences is the conormal bundle to the transpose of the graph of the diagonal inclusion.*

The standard stable basis on $T^* \mathrm{Fl}(k, n+k; \mathbb{C}^n)$ does not interact well with the second correspondence; as will be explained in a moment, we need to change to the stable basis based on the Weyl chamber corresponding to the riffle–shuffle permutation $1\ 3\ 5\ \ldots\ 2n-1\ 2\ 4\ \ldots\ 2n$, of length $\binom{n}{2}$. Changing from one basis to another, as explained in Section 3.3, involves passing through $\binom{n}{2}$ walls and thereby composing $\binom{n}{2}$ many very sparse change-of-basis matrices. Puzzlewise this amounts to filling in the $\binom{n}{2}$ vertical rhombi.

The benefit of working in this second stable basis on $T^* \mathrm{Fl}(k, n+k; \mathbb{C}^n)$ is that under the map induced by the second arrow, each basis element maps either to 0 on $T^* \mathrm{Gr}(k, \mathbb{C}^n)$, or to a (fixed, rational-function) multiple of a basis element. The corresponding puzzle calculation amounts to filling in the n triangles at the bottom (if possible). Together, modulo an $\hbar \to \infty$ limit to be discussed in a moment, this is the calculation we did at the end of Section 2.2.

Theorem 9 ([39]). *The structure constants for multiplying equivariant SSM classes on $T^* \mathrm{Gr}(k, \mathbb{C}^n)$ can be computed with the puzzle pieces from Theorem 4, plus two new equivariant pieces, where the fugacities are derived from the entries of the R-matrix for $\mathbb{C}^3 \otimes \mathbb{C}^3$.*

The Schubert classes arise as a limit of the SSM classes, $S_\lambda = \lim_{\hbar \to \infty} \hbar^{\ell(\lambda)}\, \mathrm{SSM}_\lambda$. If one distributes powers of $\hbar$ carefully among the fugacities (essentially, conjugating the R-matrix with a diagonal matrix), one can then derive Theorem 3 as a limit of Theorem 4.

There is an analogous basis in K-theory, the "motivic Segre classes," and a similar theorem holds but we do not yet have a fully geometric proof. One should be available through upgrading the correspondences, from cycles to instead sheaves supported on those cycles.

4.1. $d = 2, 3, 4, \geq 5$

There are 8 labels in the $d = 2$ puzzle rule, so the R-matrix technology suggests we look for a group G with 8-dimensional representations $V_{r,g,b}$ and an intertwiner $V_r \otimes V_g \to V_b$. This (plus some extra weight-conservation considerations, spelled out in [38, §2.3–§2.6]) suggests $G = \mathrm{Spin}(8)$ acting on its three minuscule representations. The 3-fold symmetry of puzzles is then based on D_4's triality! It is worth noting that unlike at $d = 1$, the intermediate quiver variety is not a cotangent bundle (proof: its middle homology is not 1-dimensional). We are properly in quiver variety territory here.

A new phenomenon arises at $d = 3$ (based on the 27-dim reps of E_6): there is no way to distribute the powers of $\hbar$ so as to regularize the limit $\hbar \to \infty$ of the puzzle piece fugacities. One can just barely sidestep this, but only through giving up T-equivariance.

Another new phenomenon arises at $d = 4$: the representations involved (each the $(\mathfrak{e}_8 \oplus \mathbb{C})$-rep of $U_q(E_8[z^{\pm}])$) have a weight space of dimension > 1. Without a canonical choice of basis and dual basis, we can no longer guarantee that the dot products and fugacities all come out simultaneously positive. So there *is* a puzzle rule, but it is not positive.

At $d \geq 5$ there is still a natural choice of Cartan matrix and representation [38, §2], but both the Lie algebra and its representations are infinite-dimensional, and the rule will undoubtedly suffer the same lack of positivity as at $d = 4$.

The appearance of the Dynkin diagrams A_2, D_4, E_6, E_8 at $d = 1, 2, 3, 4$ is very suggestive of a connection to cluster algebras, as these are the types of the finite-type cluster varieties $\mathrm{Gr}(3, n)$ for $n = 5, 6, 7, 8$. (For $n \geq 9$ they, like our Lie algebras, are infinite type.)

5. FUTURE DIRECTIONS

Obviously there is a great deal of work left to do on the 2^6 problems from Section 2. In my personal estimation, the problems likely to have the most impact on/interaction with other fields are those around

- **C**otangent, for the connection to representation theory,
- **Q**uantum, for the connection to 2-d mirror symmetry,
- elliptic cohomology, for the connection to 3-d mirror symmetry, and
- H_* (affine Grassmannian), for the connection to the geometric Satake correspondence.

That admitted, my heart tells me to continue pursuing **F**lag manifolds.

FIGURE 2
The author (center) en route to the Nice ICM (see Figure 1) .

FUNDING

This work was partially supported by NSF grant 1953948.

REFERENCES

[1] E. Akyıldız, Bruhat decomposition via $\mathbb{G}_m$-action. *Bull. Acad. Pol. Sci., Sér. Sci. Math.* **28** (1980), no. 11–12, 541–547.

[2] P. Aluffi, L. C. Mihalcea, J. Schürmann, and C. Su, Shadows of characteristic cycles, Verma modules, and positivity of Chern–Schwartz–MacPherson classes of Schubert cells. 2017, arXiv:1709.08697.

[3] D. Anderson, S. Griffeth, and E. Miller, Positivity and Kleiman transversality in equivariant K-theory of homogeneous spaces. *J. Eur. Math. Soc. (JEMS)* **13** (2011), no. 1, 57–84. arXiv:0808.2785.

[4] N. Arkani-Hamed, J. L. Bourjaily, F. Cachazo, A. B. Goncharov, A. Postnikov, and J. Trnka, *Grassmannian geometry of scattering amplitudes*. Cambridge University Press, Cambridge, 2016, ix+194 pp. arXiv:arxiv:1212.5605.

[5] P. Belkale, Invariant theory of GL(n) and intersection theory of Grassmannians. *Int. Math. Res. Not.* **69** (2004), 3709–3721.

[6] P. Belkale, The tangent space to an enumerative problem. In *Proceedings of the International Congress of Mathematicians. Volume II*, pp. 405–426, Hindustan Book Agency, New Delhi, 2010.

[7] M. Brion, Positivity in the Grothendieck group of complex flag varieties. *J. Algebra (special volume in honor of Claudio Procesi)* **258** (2002), 137–159. arXiv:math/0105254.

[8] A. S. Buch, A Littlewood–Richardson rule for the K-theory of Grassmannians. *Acta Math.* **189** (2002), 37–78. arXiv:math.AG/0004137.

[9] A. S. Buch, Mutations of puzzles and equivariant cohomology of two-step flag varieties. *Ann. of Math. (2)* **182** (2015), no. 1, 173–220. arXiv:1401.3065.

[10] A. S. Buch, A. Kresch, K. Purbhoo, and H. Tamvakis, The puzzle conjecture for the cohomology of two-step flag manifolds. *J. Algebraic Combin.* **44** (2016), 973–1007. arXiv:1401.1725.

[11] A. S. Buch, A. Kresch, and H. Tamvakis, Gromov–Witten invariants on Grassmannians. *J. Amer. Math. Soc.* **16** (2003), 901–915. arXiv:math.AG/0306388.

[12] A. S. Buch, A. Kresch, and H. Tamvakis, Littlewood–Richardson rules for Grassmannians. *Adv. Math.* **185** (2004), 80–90. arXiv:math/0306391.

[13] A. S. Buch and M. J. Samuel, K-theory of minuscule varieties. *J. Reine Angew. Math.* **719** (2016), 133–171. arXiv:1306.5419.

[14] J. B. Carrell, Chern classes of the Grassmannians and Schubert calculus. *Topology* **17** (1978), no. 2, 177–182.

[15] S. Chung, *Cominuscule flag varieties and their quantum K-theory*. Ph.D. thesis, Rutgers University, 2017.

[16] V. Collins, *Crystal branching for non-Levi subgroups and a puzzle formula for the equivariant cohomology of the cotangent bundle on projective space*. Ph.D. thesis, Cornell University, 2016.

[17] İ. Coşkun, A Littlewood–Richardson rule for two-step flag varieties. *Invent. Math.* **176** (2009), no. 2, 325–395.

[18] İ. Coşkun, Restriction varieties and geometric branching rules. *Adv. Math.* **228** (2011), no. 4, 2441–2502.

[19] İ. Coşkun and R. Vakil, Geometric positivity in the cohomology of homogeneous spaces and generalized Schubert calculus. In *Algebraic geometry–Seattle 2005. Part 1*, pp. 77–124, Proc. Sympos. Pure Math. 80, Amer. Math. Soc., Providence, RI, 2009, arXiv:math/0610538.

[20] H. Derksen, A. Schofield, and J. Weyman, On the number of subrepresentations of a general quiver representation. *J. Lond. Math. Soc. (2)* **76** (2007), no. 1, 135–147. arXiv:math/0507393.

[21] V. G. Drinfel'd, Quantum groups. In *Proceedings of the International Congress of Mathematicians, Vol. 1, 2 (Berkeley, Calif., 1986)*, pp. 798–820, Amer. Math. Soc., Providence, RI, 1987.

[22] W. Fulton, Eigenvalues, invariant factors, highest weights, and Schubert calculus. *Bull. Amer. Math. Soc. (N.S.)* **37** (2000), no. 3, 209–249. arXiv:math/9908012.

[23] W. Fulton and R. Pandharipande, Notes on stable maps and quantum cohomology. In *Algebraic geometry–Santa Cruz 1995. Part 2*, pp. 45–96, Proc. Sympos. Pure Math. 62, Amer. Math. Soc., Providence, RI, 1997, arXiv:alg-geom/9608011.

[24] W. Fulton and P. Pragacz, *Schubert varieties and degeneracy loci. Appendix J by the authors in collaboration with I. Ciocan-Fontanine*. Lecture Notes in Math. 1689, Springer, Berlin, 1998.

[25] V. Ginzburg, Lectures on Nakajima's quiver varieties, The summer school "Geometric methods in representation theory", Grenoble, June 16–July 4. 2008, arXiv:0905.0686.

[26] V. Ginzburg (as Victor Ginsburg), Characteristic varieties and vanishing cycles. *Invent. Math.* **84** (1986), no. 2, 327–402.

[27] W. Graham, Positivity in equivariant Schubert calculus. *Duke Math. J.* **109** (2001), no. 3, 599–614. arXiv:math/9908172.

[28] I. Halacheva, A. Knutson, and P. Zinn-Justin, Restricting Schubert classes to symplectic Grassmannians using self-dual puzzles. *Sém. Lothar. Combin.* **82B** (2020), 83, 12 pp. arXiv:1811.07581.

[29] D. Hilbert, Mathematical problems. Lecture delivered before the International Congress of Mathematicians at Paris in 1900. Translated from the German by M. Winston Neson. *Math. Today (Southend-on-Sea)* **36** (2000), no. 1, 14–17.

[30] G. Horrocks, On the relation of S-functions to Schubert varieties. *Proc. Lond. Math. Soc. (3)* **7** (1957), 265–280.

[31] D. Huang, Schubert products for permutations with separated descents. 2021, arXiv:2105.01591.

[32] M. Jimbo, Solvable lattice models and quantum groups. In *Proceedings of the International Congress of Mathematicians, Vol. I, II (Kyoto, 1990)*, pp. 1343–1352, Math. Soc. Japan, Tokyo, 1991.

[33] S. Kato, Loop structure on equivariant K-theory of semi-infinite flag manifolds. arXiv:1805.01718.

[34] S. Kleiman, The transversality of a general translate. *Compos. Math.* **28** (1974), no. 3, 287–297.

[35] A. Knutson, Schubert calculus and shifting of interval positroid varieties. 2014, arXiv:1408.1261.

[36] A. Knutson and T. Tao, Puzzles and (equivariant) cohomology of Grassmannians. *Duke Math. J.* **119** (2003), no. 2, 221–260. arXiv:math.AT/0112150.

[37] A. Knutson, T. Tao, and C. Woodward, The honeycomb model of $\mathrm{GL}(n)$ tensor products II: Puzzles determine facets of the Littlewood–Richardson cone. *J. Amer. Math. Soc.* **17** (2004), no. 1, 19–48. arXiv:math.CO/0107011.

[38] A. Knutson and P. Zinn-Justin, Schubert puzzles and integrability I: invariant trilinear forms. 2020, arXiv:1706.10019v6.

[39] A. Knutson and P. Zinn-Justin, Schubert puzzles and integrability II: multiplying motivic Segre classes. 2021, arXiv:2102.00563.

[40] A. Knutson, Schubert calculus and puzzles. In *Schubert calculus–Osaka 2012*, pp. 185–209, Adv. Stud. Pure Math. 71, Math. Soc. Japan, Tokyo, 2016. https://pi.math.cornell.edu/~allenk/puzzlenotes.pdf.

[41] B. Kostant, Lie algebra cohomology and the generalized Borel–Weil theorem. *Ann. of Math. (2)* **74** (1961), 329–387.

[42] S. Kumar, R. Rimányi, and A. Weber, Elliptic classes of Schubert varieties. *Math. Ann.* **378** (2020), no. 1–2, 703–728. arXiv:1910.02313.

[43] T. Lam, C. Li, L. C. Mihalcea, and M. Shimozono, A conjectural Peterson isomorphism in K-theory. *J. Algebra* **513** (2018), 326–343. arXiv:1705.03435.

[44] T. Lam and M. Shimozono, Quantum cohomology of G/P and homology of affine Grassmannian. *Acta Math.* **204** (2010), no. 1, 49–90. arXiv:0705.1386.

[45] L. Lesieur, Les problémes d'intersection sur une variété de Grassmann. *C. R. Acad. Sci. Paris* **225** (1947), 916–917.

[46] P. Littelmann, The path model for representations of symmetrizable Kac–Moody algebras. In *Proceedings of the International Congress of Mathematicians, Vol. 1, 2 (Zürich, 1994)*, pp. 298–308, Birkhäuser, Basel, 1995.

[47] D. Maulik and A. Okounkov, Quantum groups and quantum cohomology. *Astérisque* **408** (2019), ix+209 pp. arXiv:1211.1287.

[48] L. C. Mihalcea, Positivity in equivariant quantum Schubert calculus. *Amer. J. Math.* **128** (2006), no. 3, 787–803. arXiv:math/0407239.

[49] E. Mukhin, V. Tarasov, and A. Varchenko, Schubert calculus and representations of the general linear group. *J. Amer. Math. Soc.* **22** (2009), no. 4, 909–940. arXiv:0711.4079.

[50] H. Nakajima, Instantons on ALE spaces, quiver varieties, and Kac–Moody algebras. *Duke Math. J.* **76** (1994), no. 2, 365–416.

[51] H. Nakajima, Quiver varieties and finite-dimensional representations of quantum affine algebras. *J. Amer. Math. Soc.* **14** (2001), no. 1, 145–238. arXiv:math/9912158.

[52] H. Nakajima, Geometric construction of representations of affine algebras. In *Proceedings of the International Congress of Mathematicians, Vol. I (Beijing, 2002)*, pp. 423–438, Higher Ed. Press, Beijing, 2002.

[53] A. Okounkov, On the crossroads of enumerative geometry and geometric representation theory. In *Proceedings of the International Congress of Mathematicians—Rio de Janeiro 2018. Vol. I. Plenary lectures*, pp. 839–867, World Sci. Publ., Hackensack, NJ, 2018.

[54] O. Pechenik and A. Yong, Equivariant K-theory of Grassmannians II: the Knutson–Vakil conjecture. *Compos. Math.* **153** (2017), no. 4, 667–677. arXiv:1508.00446.

[55] P. Pylyavskyy and J. Yang, Puzzles in K-homology of Grassmannians. *Pacific J. Math.* **303** (2019), no. 2, 703–727. arXiv:1801.07667.

[56] J. R. Stembridge, A concise proof of the Littlewood–Richardson rule. *Electron. J. Combin.* **9** (2002), no. 1, note 5, 4 pp.

[57] C. Su and C. Zhong, Stable bases of the springer resolution and representation theory. Schubert calculus and its applications in combinatorics and representation theory. In *Springer Proc. Math. Stat.*, pp. 195–221, 332, Springer, Singapore, 2020, arXiv:1904.06613.

[58] H. Tamvakis, The connection between representation theory and Schubert calculus. *Enseign. Math. (2)* **50** (2004), no. 3–4, 267–286. arXiv:math/0306414.

[59] H. Thomas and A. Yong, An S_3-symmetric Littlewood–Richardson rule. *Math. Res. Lett.* **15** (2008), no. 5, 1027–1037. arXiv:0704.0817.

[60] H. Thomas and A. Yong, A combinatorial rule for (co)minuscule Schubert calculus. *Adv. Math.* **222** (2009), no. 2, 596–620. arXiv:math/0608276.

[61] R. Vakil, A geometric Littlewood–Richardson rule. Appendix A written with A. Knutson. *Ann. of Math. (2)* **164** (2006), no. 2, 371–421. arXiv:math/0302294.

[62] M. A. A. van Leeuwen, The Littlewood–Richardson rule, and related combinatorics. In *Interaction of combinatorics and representation theory*, pp. 95–145, MSJ Mem. 11, Math. Soc. Japan, Tokyo, 2001, arXiv:math/9908099.

[63] M. Varagnolo, Quiver varieties and Yangians. *Lett. Math. Phys.* **53** (2000), no. 4, 273–283. arXiv:math/0005277.

[64] M. Wheeler and P. Zinn-Justin, Littlewood–Richardson coefficients for Grothendieck polynomials from integrability. *J. Reine Angew. Math.* **757** (2019), 159–195. arXiv:1607.02396.

[65] A. V. Zelevinsky, A generalization of the Littlewood–Richardson rule and the Robinson–Schensted–Knuth correspondence. *J. Algebra* **69** (1981), no. 1, 82–94.

[66] P. Zinn-Justin, Littlewood–Richardson coefficients and integrable tilings. *Electron. J. Combin.* **16** (2009), no. 1, research paper 12, 33 pp. arXiv:0809.2392.

ALLEN KNUTSON

Department of Mathematics, Cornell University, Ithaca, New York, USA,
allenk@math.cornell.edu

RECENT PROGRESS TOWARDS HADWIGER'S CONJECTURE

SERGEY NORIN

ABSTRACT

In 1943 Hadwiger conjectured that every graph with no K_t minor is $(t-1)$-colorable for every $t \geq 1$. Hadwiger's conjecture generalizes the Four Color Theorem and is among most studied problems in graph theory.

In this paper we survey the ideas behind recent progress towards this conjecture, which, in particular, allowed for the first asymptotic improvement since 1980s on the number of colors sufficient to color every graph with no K_t minor.

MATHEMATICS SUBJECT CLASSIFICATION 2020

Primary 05C15; Secondary 05C83

KEYWORDS

Graph minors, graph coloring, Hadwiger's conjecture, extremal function

INTERNATIONAL CONGRESS OF MATHEMATICIANS 2022 JULY 6–14

Proc. Int. Cong. Math. 2022, Vol. 6, pp. 4606–4620
DOI 10.4171/ICM2022/102

Published by EMS Press

1. INTRODUCTION

In 1852 Francis Guthrie (see, e.g., [33]) conjectured that every planar graph is four colorable. This Four Color Conjecture was the central driving force behind many of the developments in graph theory for over a hundred years. Eventually, it was proved in 1976 by Appel and Haken [2,3] and became the Four Color Theorem. Appel and Haken's proof is one of the first and most well-known examples of computer assisted proofs. To date there are no known proofs of the Four Color theorem that can be reasonably considered to be human readable, and a deeper reason behind it remains elusive.

If true, the following famous conjecture made by Hadwiger [17] in 1943 points to such a reason. Its statement eliminates the topological component present in the Four Color Theorem's statement and instead involves graph minors.

Given graphs H and G, we say that G *has an* H *minor* or H *is a minor of* G if a graph isomorphic to H can be obtained from a subgraph of G by contracting edges. We denote the complete graph on t vertices by K_t.

Conjecture 1.1 (Hadwiger's conjecture [17]). *For every integer $t \geq 1$, every graph with no K_t minor is $(t-1)$-colorable.*

We refer the reader to a comprehensive survey by Seymour [51] for the detailed history of the conjecture, and only present the background necessary to motivate the discussion of recent progress on two particular weakenings of the conjecture that we focus on here.

Hadwiger [17] and Dirac [10] independently showed that Conjecture 1.1 holds for $t \leq 4$. As the class of planar graphs is closed under taking minors and the complete graph K_5 is not planar, the case $t = 5$ of Hadwiger's conjecture implies the Four Color Theorem. In fact, Wagner already shown in 1937 that this case is equivalent to the Four Color Theorem. Robertson, Seymour, and Thomas [49] went one step further and proved Hadwiger's conjecture for $t = 6$, also by reducing it to the Four Color Theorem. Settling the conjecture exactly for $t \geq 7$ appears to be extremely challenging, in part due to the aforementioned absence of a transparent proof of the Four Color Theorem.

Another notable challenging case of Hadwiger's conjecture is the case of graphs with no independent set of size three. If G is such a graph on n vertices then properly coloring G requires at least $n/2$ colors, and so Hadwiger's conjecture implies that G has a $K_{\lceil n/2 \rceil}$ minor. This is still open. In fact, as mentioned in [51], it is not known whether there exists any $c > 1/3$ such that every graph G as above has a K_t minor for some $t \geq cn$.

The following natural weakening of Hadwiger's conjecture, which has been considered by several researchers, sidesteps the above challenges.

Conjecture 1.2 (Linear Hadwiger's conjecture [22,23,47]). *There exists $C > 0$ such that for every integer $t \geq 1$, every graph with no K_t minor is Ct-colorable.*

Until recently the best bound on the number of colors needed to color the graphs with no K_t minor was $O(t\sqrt{\log t})$, obtained independently by Kostochka [26,27] and Thomason [54] in the 1980s. The only improvement since then [24,55,59] and until the last two years since then has been in the constant factor. In the last two years, however, using in part the methods we survey here this bound has been improved, first, by Postle, Song, and I [40] to $O(t(\log t)^{\beta})$ for every $\beta > 1/4$, then by Postle [46] to $O(t(\log \log t)^6)$ and, very recently by Delcourt and Postle [9] to $O(t \log \log t)$. We sketch the proof of the first of these bounds here.

Investigation of another series of weakening of Hadwiger's conjecture has been proposed more recently by Seymour.

Conjecture 1.3 (H-Hadwiger's conjecture [50, 51]). *For every graph H on t vertices, every graph with no H minor is $(t-1)$-colorable.*

Note that the bound on the number of colors in Conjecture 1.3 is tight for every graph H on t vertices, as K_{t-1} has no H minor and requires $t-1$ colors to properly color. Until recently, Conjecture 1.3 was only verified for a few very structured families of graphs H. As noted by Seymour [50], Conjecture 1.3 holds if H is a tree, and Kostochka [27] proved that Conjecture 1.3 holds for $H = K_{s,t}$ which is a sufficiently unbalanced complete bipartite graph, i.e., $t \geq C(s \log s)^3$ for some constant C. Using the methods surveyed in this paper, Turcotte and I [43] recently proved Conjecture 1.3 for a fairly large class of structurally sparse bipartite graphs H, and we present the sketch of our arguments in this survey.

We overview the main tools behind the above mentioned recent progress towards Conjectures 1.2 and 1.3, which mainly relies on the interplay between the very basic parameters of the graph G with no H minor, namely

- $\mathsf{v}(G)$—the number of vertices of G,
- $\mathsf{e}(G)$—the number of edges of G,
- $\mathsf{d}(G) = \mathsf{e}(G)/\mathsf{v}(G)$—the *density* of G,
- $\chi(G)$—the *chromatic number* of G, that is the minimum positive integer r such that G is r-colorable,
- $\kappa(G)$—the *connectivity* of G, the maximum positive integer $k < \mathsf{v}(G)$ such that G remains connected after deleting any set of fewer than k vertices.

The rest of the paper is structured as follows. In Section 2 we present the basic tools relating connectivity, density, and chromatic number of graphs with no K_t minor (and more generally, for graphs in classes closed under taking minors). In Section 3 we survey known bounds on density of graphs with no H minors. In Section 4, we present the crucial tool behind the recent progress—the density increment theorem, which is used to locate small dense subgraphs in large graphs without a dense minor. Using this theorem, one can build the minors of the graph under consideration by combining smaller pieces found in the dense subgraphs. This procedure is described in Section 5. Finally, in Section 6 we sketch how the presented tools are combined to obtain the results of [40] and [43] mentioned above.

2. BASIC DEPENDENCIES

For both Conjectures 1.2 and 1.3, it is enough to investigate *contraction critical* graphs G, that is, graphs G such that $\chi(H) < \chi(G)$ for every minor H of G unless H is isomorphic to G.

The basic relationship between density and the chromatic number of contraction critical graphs is given by the following standard *degeneracy* argument. Let G be a contraction critical graph, and let $t = \chi(G)$. If $\deg(v) \le t - 2$ for some $v \in V(G)$, then we have $\chi(G \setminus v) \le t - 1$ and any $(t-1)$-coloring of $G \setminus v$ can be extended to v, a contradiction. Thus every vertex of G has degree at least $t - 1$ and so

$$\mathsf{d}(G) \le \frac{\chi(G) - 1}{2}, \tag{2.1}$$

by averaging.

The following harder theorem of Kawarabayshi [22] guarantees that the connectivity of every contraction critical graph is also linear in the chromatic number.

Theorem 2.1 ([22]). *If G is a contraction critical graph, then*

$$\kappa(G) \ge \frac{2}{27}\chi(G).$$

In the more technical arguments, one works with subgraphs of contraction critical graphs, which are not by themselves contraction critical. To be useful for building the minors, we need these subgraphs to be highly-connected. The following classical result of Mader allows us to gain connectivity without losing to much density.

Theorem 2.2 ([36]). *Every graph G contains a subgraph G' such that $\kappa(G') \ge \mathsf{d}(G)/2$.*

We frequently want to additionally guarantee that by passing to the highly-connected subgraph or minor, we do not reduce the chromatic number excessively. This is possible due to a recent theorem of Girao and Narayanan [16].

Theorem 2.3 ([16]). *For every positive integer k, every graph G with $\chi(G) \ge 7k$ contains a subgraph G' such that $\kappa(G') \ge k$ and $\chi(G') \ge \chi(G) - 3k$.*

Finally, for small graphs G, we have another tool, the following classical bound due to Duchet and Meyniel [12], on the independence number of graphs with no K_t minor.

The set $X \subseteq V(G)$ is *independent* in G if no pair of vertices of X are adjacent. The *independence number* $\alpha(G)$ of a graph G is the maximum size of an independent set in G.

Theorem 2.4 ([12]). *For every $t \ge 2$, every graph G with no K_t minor has an independent set of size at least $\frac{\mathsf{v}(G)}{2(t-1)}$.*

Theorem 2.4 implies that every graph with no K_t minor contains a t-colorable subgraph on a constant proportion of vertices. Woodall [60] proved the following stronger result, which as observed by Seymour [51] also follows from the proof of Theorem 2.4 in [12].

Theorem 2.5 ([60]). *Let G be a graph with no K_t minor. Then there exists $X \subseteq V(G)$ with $|X| \ge \frac{\mathsf{v}(G)}{2}$ such that $\chi(G[X]) \le t - 1$.*

Theorem 2.5 straightforwardly implies the following bound on the chromatic number of graphs with no K_t minor.

Corollary 2.6. *Let G be a graph with no K_t minor. Then*

$$\chi(G) \leq \left(\log_2\left(\frac{\mathsf{v}(G)}{t}\right) + 2\right)t.$$

3. DENSITY

Until recently the best bounds on the chromatic number of graph with no K_t minor for large t relied exclusively on the degeneracy bound (2.1). To determine the optimum bounds that can be obtained in this manner towards Conjecture 1.3, we investigate the maximum density of graphs G with no H minor for a fixed H.

More formally, following [38], for a graph H with $\mathsf{v}(H) \geq 2$, we define the *extremal function* $c(H)$ of H as the supremum of $\mathsf{d}(G)$ taken over all nonnull graphs G not containing H as a minor.

Mader [34] proved that $c(H)$ is finite for every graph H. The exact value has been determined for various small graphs H. For example, if K_t is the complete graph on $t \leq 9$ vertices, then $c(K_t) = t - 2$ (see [11,21,35,52]); and if P is the Petersen graph, then $c(P) = 5$ (see [20]). We primarily focus on asymptotic results for classes of graphs H.

The asymptotic behavior of $c(K_t)$ was studied in [26, 27, 54], and was determined precisely by Thomason [55], who showed that

$$c(K_t) = \big(\lambda + o(1)\big)t\sqrt{\log t}, \tag{3.1}$$

where

$$\lambda = \max_{\alpha>0} \frac{1-e^{-\alpha}}{2\sqrt{\alpha}} = 0.319\ldots$$

Improving on results of [38,48], Thomason and Wales [56] recently extended the upper bound from (3.1) to general graphs, by showing that for every graph H,

$$c(H) \leq \big(\lambda + o_{\mathsf{d}(H)}(1)\big)\,\mathsf{v}(H)\sqrt{\log \mathsf{d}(H)}. \tag{3.2}$$

The inequality (3.2) is tight in many regimes. Myers and Thomason [38] showed that it is tight (up to the choice of the error term) for almost all graphs with n vertices and $n^{1+\varepsilon}$ edges for every fixed $\varepsilon > 0$, and for all regular graphs with these parameters. They also gave an explicit asymptotic formula for $c(H)$ for all such polynomially dense graphs.

Reed, Thomason, Wood, and I [41] recently showed that (3.2) is also tight for almost all regular graphs of constant density, that is, for almost all d-regular graphs H,

$$c(H) \geq \big(\lambda - o_d(1)\big)\,\mathsf{v}(H)\sqrt{\log d}. \tag{3.3}$$

However, for several concrete sparse families, the extremal function behaves qualitatively differently:

- Chudnovsky, Reed, and Seymour [7] proved that $c(K_{2,t}) = \frac{t+1}{2}$ for all $t \geq 2$;
- Kostochka and Prince [28] proved that $c(K_{3,t}) = \frac{t+3}{2}$ for all $t \geq 6300$;
- More generally, Myers [37] considered the asymptotic behavior of $c(K_{s,t})$ for fixed s and t and conjectured that $c(K_{s,t}) \leq c_s t$ for some constant independent on t. Kühn and Osthus [32] and Kostochka and Prince [25] independently proved this conjecture by showing that $c(K_{s,t}) = (\frac{1}{2} + o_s(1))t$.
- Csóka et al. [8] proved that if H is a disjoint union of cycles, then

$$c(H) \leq \frac{\mathsf{v}(H) + \mathrm{comp}(H)}{2} - 1,$$

which is tight whenever every component of H is an odd cycle.

All of the above families are structurally sparse and the extremal function is linear in the number of vertices. (In fact, $c(H) < (1 + o(1))\,\mathsf{v}(H)$ for all these graphs.)

This property generalizes to the large and well-studied class of sparse graph families defined as follows. A graph family is *monotone* if it is closed under taking subgraphs. A *separation* of a graph G is a pair (A_1, A_2) of subsets of $V(G)$ such that $G = G[A_1] \cup G[A_2]$ and $A_1 \setminus A_2 \neq \emptyset$ and $A_2 \setminus A_1 \neq \emptyset$. A separation (A_1, A_2) has *order* $|A_1 \cap A_2|$. A separation (A_1, A_2) is *balanced* if $|A_1|, |A_2| \geq \frac{\mathsf{v}(G)}{3}$. A graph family $\mathcal{F}$ admits *strongly sublinear separators* (written $\mathcal{F}$ is *s.s.s.*, for brevity) if $\mathcal{F}$ is monotone, and there exist $\beta < 1$ and $c > 0$ such that every graph $G \in \mathcal{F}$ has a balanced separation of order at most $c\,\mathsf{v}(G)^{\beta}$. For example, every *proper minor-closed family* (a family that is closed under isomorphisms and taking minors, and does not include all graphs) is s.s.s., as proved by [1] with $\beta = \frac{1}{2}$. More generally, every family with polynomial expansion is s.s.s. [13].

Before formally stating the general asymptotic bound on the extremal function of graphs in s.s.s. graph families, we describe two natural lower bounds on $c(H)$. First, since H is not a minor of $K_{\mathsf{v}(H)-1}$,

$$c(H) \geq \mathsf{d}(K_{\mathsf{v}(H)-1}) = \frac{\mathsf{v}(H)}{2} - 1. \tag{3.4}$$

A *vertex cover* of H is a set $S \subseteq V(H)$ such that $H - S$ has no edges. Let $\tau(H)$ be the minimum size of a vertex cover of H. For the second bound, observe that $\tau(H) \leq \tau(G)$ whenever H is a minor of G. It follows that H is not a minor of the complete bipartite graph $K_{\tau(H)-1,n}$ for any n and

$$c(H) \geq \lim_{n\to\infty} \mathsf{d}(K_{\tau(H)-1,n}) = \tau(H) - 1. \tag{3.5}$$

Hendrey, Wood, and I [19] have recently shown that the lower bounds (3.4) and (3.5) are asymptotically tight for 4-colorable graphs in s.s.s. families, strengthening the result of Haslegrave, Kim, and Liu [18] for bipartite graphs. The resulting density theorem below is one of the main tools used in the recent progress towards Conjecture 1.3 discussed below.

Theorem 3.1. *For every s.s.s. family $\mathcal{F}$ and for every $H \in \mathcal{F}$ with $\chi(H) \leq 4$,*

$$c(H) = \big(1 + o_{\mathcal{F}}(1)\big) \cdot \max\left(\frac{\mathsf{v}(H)}{2}, \tau(H)\right), \tag{3.6}$$

where the error term $o_{\mathcal{F}}(1)$ depends on $\mathcal{F}$ and satisfies $o_{\mathcal{F}}(1) \to 0$ as $\mathsf{v}(H) \to \infty$.

Finally, we mention the following tight bound on density of unbalanced bipartite graph without a K_t minor, due to Postle and I. It is used to supplement the density increment arguments presented in the next section

Theorem 3.2 ([39]). *There exists $C > 0$ such that, for every $t \geq 3$ and every bipartite graph G with bipartition (A, B) and no K_t minor, we have*

$$\mathsf{e}(G) \leq Ct\sqrt{\log t}\sqrt{|A||B|} + (t-2)\mathsf{v}(G). \tag{3.7}$$

4. DENSITY INCREMENT

Perhaps the most important new ingredient in the recent progress towards Conjectures 1.2 and 1.3 is a density increment argument, which informally says that every graph either contains a substantially denser minor, or a small subgraph with density not much smaller than that of the whole graph.

Let us proceed by giving a more detailed motivation. By (3.1) there exists $D = O(t\sqrt{\log t})$ such that every graph G with density $\mathsf{d}(G) \geq D$ has a K_t minor. For a graph G with smaller density, one might still hope to guarantee a K_t minor by finding a minor H of G with $\mathsf{d}(H) \geq D$. Thus we are interested, for given d and D, in properties of graphs G of density $\mathsf{d}(G) = d$ and no minor of density D.

One family of obstructions are graphs G which simply do not have enough edges. As every graph of density D has at least D^2 edges, if G has a minor of density D, we must have $D^2 \leq \mathsf{e}(G) = d \cdot \mathsf{v}(G)$. It follows that all the graphs G with $\mathsf{v}(G) < D^2/d$ are among the obstructions to our approach. One can obtain further obstructions by taking disjoint union of such graphs, and, more generally, by gluing smaller obstructions along small sets in a "tree-like fashion." However, the graphs obtained in this way contain a subgraph with at most D^2/d vertices and density close to d.

A series of density increment results culminating in the following result by Wang [57] shows that a similar subgraph can be found in every obstruction.

Theorem 4.1 ([57]). *There exists $C > 0$ satisfying the following. Let $D > 0$ be real, G be a graph with $\mathsf{d}(G) \geq C$, and let $s = D/\mathsf{d}(G)$. Then G contains at least one of the following:*

(i) *a minor J with $\mathsf{d}(J) \geq D$, or*

(ii) *a subgraph H with $\mathsf{v}(H) \leq g(s)D^2/\mathsf{d}(G)$ and $\mathsf{d}(H) \geq \mathsf{d}(G)/g(s)$,*

where $g(s) = C(1 + \log s)^5$.

The first variant of Theorem 4.1 was proved by Song and I [42] with $g(s) = Cs^{\alpha}$ for a particular constant α. The magnitude of $g(s)$ was subsequently improved by Postle, first in [44] to $g(s) = o(s^\delta)$ for every $\delta > 0$, then in [45] in $g(s) = C(1 + \log s)^6$, and, finally, by Wang [57] to the bound stated in Theorem 4.1.

A similar theorem with narrower scope of application, but stronger bounds, was very recently obtained, using Theorem 3.2, by Delcourt and Postle [9].

Theorem 4.2 ([9]). *There exists $C > 0$ satisfying the following. Let $t \geq 1$ be an integer and let G be a graph with $\mathsf{d}(G) \geq Ct$. Then G contains at least one of the following:*

(i) *a K_t minor, or*

(ii) *a subgraph H with $\mathsf{v}(H) \leq Ct \log^3 t$ and $\mathsf{d}(H) \geq Ct$,*

We finish this section with an example of application of Theorem 4.1 due to Postle, Song, and I [40].

For a pair of graphs G and H, we say that G is *H-free* if no subgraph of G is isomorphic to H. The next theorem due to Kühn and Osthus [31] shows that H-free graphs have exceptionally dense minors for every complete bipartite graph H.

Theorem 4.3 ([31]). *For every integer $s \geq 2$, every $K_{s,s}$-free graph G has a minor J with*

$$\mathsf{d}(J) \geq (\mathsf{d}(G))^{1+\frac{1}{2(s-1)}-o_{\mathsf{d}(G)}(1)}. \tag{4.1}$$

Krivelevich and Sudakov [29] tightened (4.1) to $\mathsf{d}(J) \geq c_s(\mathsf{d}(G))^{1+\frac{1}{s-1}}$ for some $c_s > 0$ independent of $\mathsf{d}(G)$. They also proved the following, strengthening a result of Kühn and Osthus [30].

Theorem 4.4 ([29]). *For every integer $k \geq 2$, there exists $c_k > 0$ such that every C_{2k}-free G has a minor J with*

$$\mathsf{d}(J) \geq c_k(\mathsf{d}(G))^{\frac{k+1}{2}}.$$

The exponents appearing in Theorems 4.3 and 4.4 cannot be improved, subject to well known conjectures on the Turán numbers of $K_{s,s}$ and C_{2k}, which we mention below.

In this section we use Theorem 4.1 to extend Theorems 4.3 and 4.4 to general bipartite graphs. Stating our result requires a couple of definitions. The *Turán number* $\mathrm{ex}(n, H)$ of a graph H with $\mathsf{e}(H) \neq 0$ is the maximum number of edges in an H-free graph G with $\mathsf{v}(G) = n$. The *Turán exponent* $\gamma(H)$ of a graph H with $\mathsf{e}(H) \geq 2$ is defined as

$$\gamma(H) := \limsup_{n\to\infty} \frac{\log \mathrm{ex}(n, H)}{\log n}.$$

Many fundamental questions about Turán exponents of bipartite graphs remain open. In particular, a famous conjecture of Erdős and Simonovits (see [15, CONJECTURE 1.6]) states that $\gamma(H)$ is rational for every graph H, and that $\lim_{n\to\infty} \mathrm{ex}(n, H)/n^{\gamma(H)}$ exists and is positive. We refer the reader to a comprehensive survey by Füredi and Simonovits [15] for further background.

Theorem 4.1 implies an essentially tight analogue of Theorems 4.3 and 4.4 for H-free graphs G for general bipartite H.

Theorem 4.5 ([48]). *For every bipartite graph H with $\gamma(H) > 1$, every H-free graph G has a minor J with*

$$\mathsf{d}(J) \geq \left(\mathsf{d}(G)\right)^{\frac{\gamma(H)}{2(\gamma(H)-1)} - o_{\mathsf{d}(G)}(1)}.$$

5. BUILDING THE MINORS

As mentioned earlier, we build the required minors from pieces. Describing how the minors combine together is more convenient in the language of models. A *model μ of a graph H in a graph G* assigns to every vertex of $v \in V(H)$ a set $\mu(v)$ of vertices of G such that

- $\mu(u) \cap \mu(v) = \emptyset$ for every pair of distinct $u, v \in V(H)$,
- the subgraph $G[\mu(v)]$ of G induced by $\mu(v)$ is connected for every $v \in V(H)$, and
- for every edge $uv \in E(H)$ there exist $u' \in \mu(u)$ and $v' \in \mu(v)$ such that $u'v' \in E(G)$.

It is well known and not hard to see that G has an H minor if and only if there exists a model of H in G.

Given an injection $\phi : V(H) \to V(G)$, we say that a model μ of H in G is *ϕ-rooted* if $\phi(v) \in \mu(v)$ for every $v \in V(H)$. Finally, we say that G is *H-linked* if $\mathsf{v}(G) \geq \mathsf{v}(H)$ and for every injection $\phi : V(H) \to V(G)$ there exists a ϕ-rooted model of H in G. Thus every H-linked graph has an H minor, but the converse does not hold.

The case when H is a matching of size k, i.e., $H = kK_2$ is of particular interest. Note that a graph G is kK_2-linked, if and only if $\mathsf{v}(G) \geq 2k$ and, for every collection of distinct $s_1, s_2, \ldots, s_k, t_1, t_2, \ldots, t_k \in V(G)$, there exist pairwise vertex disjoint paths $P_1, \ldots, P_k$ such that P_i has ends s_i and t_i for every $i \in [k]$. We will write *k-linked* instead of kK_2-linked for brevity. (Our definition coincides with the standard definition of k-linked graphs.)

The following theorem of Thomas and Wollan [53], improving an earlier result of Bollobás and Thomason [4], ensures that connectivity linear in k is sufficient to guarantee that the graph is k-linked.

Theorem 5.1 ([53]). *For every integer $k \geq 1$, every graph G with $\kappa(G) \geq 10k$ is k-linked.*

Connectivity linear in t is certainly insufficient to guarantee that a graph is K_t-linked, but interestingly, as observed by Delcourt and Postle [9], connectivity linear in t together with a slightly larger complete minor is sufficient.

Lemma 5.2 ([9]). *For every integer $t \geq 1$, every graph G with $\kappa(G) \geq t$ that has a $K_{\lceil 5t/2 \rceil}$ minor is K_t-linked.*

We are further interested in a more general setting, where we need to find a rooted model of a disjoint union H' of a given graph H and a matching of size k. If a graph G is

H'-linked for such an H' then we write that G is $(H+k)$*-linked* for brevity. The following theorem can be easily derived from the results of Wollan **[58]**.

Theorem 5.3 (**[58]**). *There exists $C > 0$ satisfying the following. Let H and G be graphs and let $k \geq 0$ be an integer. If*

$$\kappa(G) \geq C\big(c(H) + k\big)$$

then G is $(H+k)$-linked.

Our final tool describes the conditions under which we can glue a larger minor from small pieces in a highly connected graph. Let $H_1, H_2, \ldots, H_s$ be graphs and let $H = H_1 \cup H_2 \cup \cdots \cup H_s$. Then we say that $\{H_i\}_{i\in[s]}$ is *a decomposition* of H with *excess* $(\sum_{i\in[s]} \mathsf{v}(H_i)) - \mathsf{v}(H)$.

Theorem 5.4 (**[43]**). *There exists $C > 1$ satisfying the following. Let $\{H_i\}_{i\in[s]}$ is a decomposition of a graph H with excess k, let G be a graph and let $G_1, \ldots, G_s$ be pairwise vertex disjoint subgraphs of G. If*

- *G_i is $(H_i + k)$-linked for every $i \in [s]$, and*
- *G is k-linked,*

then G has an H minor.

6. BRINGING IT ALL TOGETHER

Having introduced the necessary toolkit in the preceding section, let us describe how combining them we can progress forward.

We start by sketching a proof of the following theorem by Postle and I **[39]**.

Theorem 6.1 (**[39]**). *For every $\beta > \frac{1}{4}$, if G is a graph with $\kappa(G) = \Omega(t(\log t)^\beta)$ no K_t minor then $\mathsf{v}(G) = O(t(\log t)^{7/4})$.*

Note that using Theorem 2.1 and Corollary 2.6, we immediately obtain from Theorem 6.1 the following bound on the chromatic number of graphs with no K_t minors, originally proved in **[40]**.

Theorem 6.2 (**[40]**). *For every $\beta > \frac{1}{4}$, if G is a graph with no K_t minor then $\chi(G) = O(t(\log t)^\beta)$.*

Proof sketch of Theorem 6.1. Note that there exists a decomposition of K_t into $O((\log t)^{1/4})$ complete subgraphs $H_1, H_2, \ldots, H_s$ of excess $k = O(t(\log t)^{1/4})$ such that $s = O((\log t)^{1/2})$ and $\mathsf{v}(H_i) = O(t/(\log t)^{1/4})$ for every $i \in [s]$.

Assume that $\mathsf{v}(G) = \Omega(t(\log t)^{7/4})$. By Theorem 5.4, it suffices to find vertex disjoint subgraphs $G_1, \ldots, G_s$ of G such that G_i is $(H_i + k)$-linked for every $i \in [s]$, as G is k-linked by Theorem 5.1 and the lower bound on $\kappa(G)$. By (3.1), we have $c(H_i) = O(t(\log t)^{1/4})$ and so, by Theorem 5.3, it suffices to guarantee that

$\kappa(G_i) \geq Ct(\log t)^{1/4})$ for some C' independent on t. By Theorem 2.3, we can further relax this condition to $\mathsf{d}(G_i) \geq Ct(\log t)^{1/4})$ (possibly changing C).

Finally, to find the required G_i's, select the maximum collection $G_1, \dots, G_{s'}$ of vertex disjoint subgraphs of G such that $\mathsf{d}(G_i) \geq Ct(\log t)^{1/4}$ and $\mathsf{v}(G_i) \leq t(\log t)^{3/4}$. Assume for a contradiction that $s' < s$. Let $A = \bigcup_{i\in[s']}(V(G_i))$, then $|A| = Ct(\log t)^{5/4}$. Let $B = V(G) - A$. Then $|B| = (1 - o(1))\mathsf{v}(G)$. By Theorem 3.2, there are $O(t^2(\log t)^{9/8}\sqrt{\mathsf{v}(G)})$ edges joining A and B, and so as the minimum degree of G is $\Omega(t(\log t)^{\beta})$, and

$$t^2(\log t)^{9/8}\sqrt{\mathsf{v}(G)} = o\big(t(\log t)^{\beta}\mathsf{v}(G)\big),$$

the average degree of the subgraph $G[B]$ of G induced by B is still $\Omega(t(\log t)^{\beta})$. As $G[B]$ has no K_t minor, applying Theorem 4.1 with $D = c(K_t)$ to $G[B]$, we conclude that G contains a subgraph $G_{s'+1}$ with $\mathsf{v}(G_{s'+1}) = O(t(\log t)^{1-\beta}(\log\log t)^5)$ and $\mathsf{d}(G_{s'+1}) = \Omega(t(\log t)^{\beta}/(\log\log t)^5)$, contradicting the choice of s'. ■

Secondly, let us outline the proof of the following recent theorem due to Turcotte and I **[43]**.

Theorem 6.3 (**[43]**). *For every s.s.s. graph family $\mathcal{F}$ and every positive integer Δ, there exists N such that for every bipartite graph $H \in \mathcal{F}$ with $\Delta(H) \leq \Delta$ and $\mathsf{v}(H) \geq N$, every graph G with $\chi(G) \geq \mathsf{v}(H)$ has an H minor. (That is, Conjecture 1.3 holds for H.)*

To prepare for the proof of Theorem 6.3, we need to introduce a couple of final tools from the literature. The first is the well-known "bandwidth theorem" of Böttcher, Schachts, and Taraz **[6]**, which using the results of **[5]** can be adapted to our setting, to imply the following.

Theorem 6.4 (**[5,6]**). *For every s.s.s. graph family $\mathcal{F}$, every positive integer Δ, every $\gamma > 0$, and for every bipartite graph $H \in \mathcal{F}$ with $\Delta(H) \leq \Delta$ and $\mathsf{v}(H) \geq N$, if G is a graph such that $\mathsf{v}(G) \geq \mathsf{v}(H)$ and $\deg(v) \geq (1+\gamma)\frac{\mathsf{v}(G)}{2}$ for every $v \in V(G)$ then G contains a subgraph isomorphic to H.*

The second is a fairly straightforward lemma, present, in particular, in **[14]**.

Lemma 6.5 (**[14]**). *Let $\mathcal{F}$ be s.s.s. graph family. Then for every $\varepsilon > 0$ there exists C such that every graph $G \in \mathcal{F}$ admit a decomposition into subgraphs on at most C vertices with excess at most $\varepsilon\mathsf{v}(G)$.*

We are now ready to sketch the proof of Theorem 6.3 using our toolkit.

Proof sketch of Theorem 6.3. By Theorem 2.1, we may assume that $\kappa(G) \geq 2/27 \cdot \mathsf{v}(H)$. By the argument in Section 2, we may further assume $\deg(v) \geq \mathsf{v}(H) - 1$ for every $v \in V(G)$. In particular, $\mathsf{d}(G) \geq (\mathsf{v}(H) - 1)/2$.

If $\mathsf{v}(G) \leq 3/2 \cdot \mathsf{v}(H)$ then, assuming N is chosen to be appropriately large, G contains a subgraph isomorphic to H by Theorem 6.4.

Assume next that $\mathsf{v}(G) \leq K\dot{\mathsf{v}}(H)$ for some sufficiently large K dependent only on the constant in the preceding theorems. By Theorem 3.1, $c(H) = (1 + o(1))\mathsf{v}(H)/2$. Thus by Theorem 4.1 applied with $D = c(H)$, G contains a subgraph G_1 with $\mathsf{v}(G_1) \leq C\mathsf{v}(H)$ and $\mathsf{d}(G_1) \geq \mathsf{v}(H)/C$ for some constant C. In fact, if K is sufficiently large compared to k and C, we can find vertex disjoint subgraphs $G_1, \dots, G_k$ of G with the same properties, using a variant of the argument used for a similar purpose in the proof sketch of Theorem 6.1 above. By Lemma 6.5, for any $\varepsilon > 0$ there exists a decomposition of H with excess $k' \leq \varepsilon\mathsf{v}(H)$ into subgraphs $H_1, \dots, H_k$ such that $(1 - \varepsilon)\mathsf{v}(H)/k \geq \mathsf{v}(H_i) \leq (1 + \varepsilon)\mathsf{v}(H)/k$, as long as $\mathsf{v}(H)$ is large enough as a function of k and ε. Applying Theorem 3.1 to H_i this time, we have $c(H_i) \leq \mathsf{v}(H)/k$, and so G_i is $(H_i + k')$-linked for every $i \in [k]$ by Theorem 5.3, as long as ε is sufficiently small and k sufficiently large compared to C. By Theorem 5.4, it now follows that H is a minor of G as desired.

It remains to consider the case $3/2 \cdot \mathsf{v}(H) \leq \mathsf{v}(G) \leq K \cdot \mathsf{v}(H)$. This regime is somewhat complicated and we do not present all the details. We consider a decomposition $H_1, \dots, H_k$ of H with excess at most $\varepsilon\mathsf{v}(H)$ such that $\mathsf{v}(H_i) \leq C$ for every $i \in [k]$ and excess at most $\varepsilon\mathsf{v}(H)$, where k is no longer constant, but C is. We reserve a randomly chosen $Z \subseteq V(G)$ with $\varepsilon\mathsf{v}(H) \ll |Z| \ll \mathsf{v}(G)$ for future use. Next, we choose l maximum such that the disjoint union of graphs $H_1, H_2, \dots, H_l$ is isomorphic to a subgraph G' of $G \setminus Z$. If $l < k$, by further choosing G' such that $G[V(G')]$ is as sparse as possible, we guarantee that $G \setminus V(G') \setminus Z$ is dense enough to contain a subgraph isomorphic to H_{l+1} by Theorem 4.5, which is a contradiction, implying $l = k$. The subgraphs $H_1, H_2, \dots, H_k$ can now be linked together to obtain the H minor by using Z. ■

FUNDING

This work was partially supported by an NSERC Discovery grant.

REFERENCES

[1] N. Alon, P. Seymour, and R. Thomas, A separator theorem for nonplanar graphs. *J. Amer. Math. Soc.* **3** (1990), no. 4, 801–808.

[2] K. Appel and W. Haken, Every planar map is four colorable. I. Discharging. *Illinois J. Math.* **21** (1977), no. 3, 429–490.

[3] K. Appel, W. Haken, and J. Koch, Every planar map is four colorable. II. Reducibility. *Illinois J. Math.* **21** (1977), no. 3, 491–567.

[4] B. Bollobás and A. Thomason, Highly linked graphs. *Combinatorica* **16** (1996), no. 3, 313–320.

[5] J. Böttcher, K. P. Pruessmann, A. Taraz, and A. Würfl, Bandwidth, expansion, treewidth, separators and universality for bounded-degree graphs. *European J. Combin.* **31** (2010), no. 5, 1217–1227.

[6] J. Böttcher, M. Schacht, and A. Taraz, Proof of the bandwidth conjecture of Bollobás and Komlós. *Math. Ann.* **343** (2009), 175–205.

[7] M. Chudnovsky, B. Reed, and P. Seymour, The edge-density for $K_{2,t}$ minors. *J. Combin. Theory Ser. B* **101** (2011), no. 1, 18–46.

[8] E. Csóka, I. Lo, S. Norin, H. Wu, and L. Yepremyan, The extremal function for disconnected minors. *J. Combin. Theory Ser. B* **121** (2017), 162–174.

[9] M. Delcourt and L. Postle, Reducing linear Hadwiger's conjecture to coloring small graphs. 2021, arXiv:2108.01633.

[10] G. A. Dirac, A property of 4-chromatic graphs and some remarks on critical graphs. *J. Lond. Math. Soc.* **27** (1952), 85–92.

[11] G. A. Dirac, Homomorphism theorems for graphs. *Math. Ann.* **153** (1964), 69–80.

[12] P. Duchet and H. Meyniel, On Hadwiger's number and the stability number. In *In North-Holland Mathematics Studies*, pp. 71–73, 62, Elsevier, 1982.

[13] Z. Dvorák and S. Norin, Strongly sublinear separators and polynomial expansion. *SIAM J. Discrete Math.* **30** (2016), no. 2, 1095–1101.

[14] D. Eppstein, Densities of minor-closed graph families. *Electron. J. Combin.* **17** (2010), no. 1, 21, research paper 136.

[15] Z. Füredi and M. Simonovits, The history of degenerate (bipartite) extremal graph problems. In *Erdős centennial*, pp. 169–264, Bolyai Soc. Math. Stud. 25, János Bolyai Math. Soc., Budapest, 2013.

[16] A. Girão and B. Narayanan, Subgraphs of large connectivity and chromatic number. 2020, arXiv:2004.00533.

[17] H. Hadwiger, Über eine Klassifikation der Streckenkomplexe. *Vierteljahrsschr. Nat.forsch. Ges. Zür.* **88** (1943), 133–142.

[18] J. Haslegrave, J. Kim, and H. Liu, Extremal density for sparse minors and subdivisions. To appear in *Int. Math. Res. Not.*

[19] K. Hendrey, S. Norin, and D. R. Wood, Extremal functions for sparse minors. 2021, arXiv:2107.08658.

[20] K. Hendrey and D. R. Wood, The extremal function for Petersen minors. *J. Combin. Theory Ser. B* **131** (2018), 220–253.

[21] L. K. Jørgensen, Contractions to K_8. *J. Graph Theory* **18** (1994), no. 5, 431–448.

[22] K-i. Kawarabayashi, On the connectivity of minimum and minimal counterexamples to Hadwiger's Conjecture. *J. Combin. Theory Ser. B* **97** (2007), no. 1, 144–150.

[23] K.-i. Kawarabayashi and B. Mohar, Some recent progress and applications in graph minor theory. *Graphs Combin.* **23** (2007), no. 1, 1–46.

[24] T. Kelly and L. Postle, A local epsilon version of Reed's conjecture. *J. Combin. Theory Ser. B* **141** (2020), 181–222.

[25] A. Kostochka and N. Prince, On $K_{s,t}$-minors in graphs with given average degree. *Discrete Math.* **308** (2008), no. 19, 4435–4445.

[26] A. V. Kostochka, The minimum Hadwiger number for graphs with a given mean degree of vertices. *Metody Diskretn. Anal.* **38** (1982), 37–58.

[27] A. V. Kostochka, Lower bound of the Hadwiger number of graphs by their average degree. *Combinatorica* **4** (1984), no. 4, 307–316.

[28] A. V. Kostochka and N. Prince, Dense graphs have $K_{3,t}$ minors. *Discrete Math.* **310** (2010), no. 20, 2637–2654.

[29] M. Krivelevich and B. Sudakov, Minors in expanding graphs. *Geom. Funct. Anal.* **19** (2009), no. 1, 294–331.

[30] D. Kühn and D. Osthus, Minors in graphs of large girth. *Random Structures Algorithms* **22** (2003), no. 2, 213–225.

[31] D. Kühn and D. Osthus, Complete minors in $K_{s,s}$-free graphs. *Combinatorica* **25** (2005), no. 1, 49–64.

[32] D. Kühn and D. Osthus, Forcing unbalanced complete bipartite minors. *European J. Combin.* **26** (2005), no. 1, 75–81.

[33] D. MacKenzie, *Mechanizing proof: computing, risk, and trust.* Inside Technol., MIT Press, 2001.

[34] W. Mader, Homomorphieeigenschaften und mittlere Kantendichte von Graphen. *Math. Ann.* **174** (1967), 265–268.

[35] W. Mader, Homomorphiesätze für Graphen. *Math. Ann.* **178** (1968), 154–168.

[36] W. Mader, Existenz n-fach zusammenhängender Teilgraphen in Graphen genügend grosser Kantendichte. *Abh. Math. Semin. Univ. Hambg.* **37** (1972), 86–97.

[37] J. S. Myers, The extremal function for unbalanced bipartite minors. *Discrete Math.* **271** (2003), no. 1–3, 209–222.

[38] J. S. Myers and A. Thomason, The extremal function for noncomplete minors. *Combinatorica* **25** (2005), no. 6, 725–753.

[39] S. Norin and L. Postle, Connectivity and choosability of graphs with no K_t minor. 2020, arXiv:2004.10367.

[40] S. Norin, L. Postle, and Z.-X. Song, Breaking the degeneracy barrier for coloring graphs with no K_t minor. 2020, arXiv:1910.09378.

[41] S. Norin, B. Reed, A. Thomason, and D. R. Wood, A lower bound on the average degree forcing a minor. *Electron. J. Combin.* **27** (2020), no. 2, Paper 2.4, 9 pp.

[42] S. Norin and Z.-X. Song, Breaking the degeneracy barrier for coloring graphs with no K_t minor. 2019.

[43] S. Norin and J. Turcotte, Chromatic number of graphs excluding a sparse bipartite minor (in preparation).

[44] L. Postle, Halfway to Hadwiger's conjecture. 2019, arXiv:1911.01491.

[45] L. Postle, An even better density increment theorem and its application to Hadwiger's conjecture. 2020, arXiv:2006.14945.

[46] L. Postle, Further progress towards Hadwiger's conjecture. 2020, arXiv:2006.11798.

[47] B. Reed and P. Seymour, Fractional colouring and Hadwiger's conjecture. *J. Combin. Theory Ser. B* **74** (1998), no. 2, 147–152.

[48] B. Reed and D. R. Wood, Forcing a sparse minor. *Combin. Probab. Comput.* (2015), 1–23.

[49] N. Robertson, P. Seymour, and R. Thomas, Hadwiger's conjecture for K_6-free graphs. *Combinatorica* **13** (1993), no. 3, 279–361.

[50] P. Seymour, Open problem presented at BIRS workshop: Geometric and structural graph theory.

[51] P. Seymour, Hadwiger's conjecture. In *Open problems in mathematics*, pp. 417–437, Springer, 2016.

[52] Z.-X. Song and R. Thomas, The extremal function for K_9 minors. *J. Combin. Theory Ser. B* **96** (2006), no. 2, 240–252.

[53] R. Thomas and P. Wollan, An improved linear edge bound for graph linkages. *European J. Combin.* **26** (2005), no. 3–4, 309–324.

[54] A. Thomason, An extremal function for contractions of graphs. *Math. Proc. Cambridge Philos. Soc.* **95** (1984), no. 2, 261–265.

[55] A. Thomason, The extremal function for complete minors. *J. Combin. Theory Ser. B* **81** (2001), no. 2, 318–338.

[56] A. Thomason and M. Wales, On the extremal function for graph minors. 2019, arXiv:1907.11626.

[57] Y. Wang, Improved bound for Hadwiger's conjecture. 2021, arXiv:2108.09230.

[58] P. Wollan, Extremal functions for rooted minors. *J. Graph Theory* **58** (2008), 159–178.

[59] D. R. Wood, A note on Hadwiger's conjecture. 2013, arXiv:1304.6510.

[60] D. R. Woodall, Subcontraction-equivalence and Hadwiger's conjecture. *J. Graph Theory* **11** (1987), no. 2, 197–204.

SERGEY NORIN

Department of Mathematics and Statistics, McGill University, Montreal, QC, Canada, sergey.norin@math.mcgill.ca

This Page Intentionally Left Blank

FACE NUMBERS: THE UPPER BOUND SIDE OF THE STORY

ISABELLA NOVIK

ABSTRACT

We survey the theory of face numbers of simplicial complexes through the lens of upper bound type results and neighborliness. We focus on the classes of polytopes, simplicial spheres, and simplicial manifolds, along with the classes of centrally symmetric polytopes and centrally symmetric simplicial spheres. Along the way, we sketch some of the ideas and methods used in the proofs. We also highlight some of the many open problems in the field.

MATHEMATICS SUBJECT CLASSIFICATION 2020

Primary 05E45; Secondary 05E40, 13F55, 52B05, 52B12, 57Q15

KEYWORDS

f-numbers, h-numbers, simplicial polytopes, simplicial spheres, face ring, centrally symmetric, neighborly, stacked, upper bound problem, moment curve

ICM INTERNATIONAL CONGRESS OF MATHEMATICIANS 2022 JULY 6–14

Proc. Int. Cong. Math. 2022, Vol. 6, pp. 4622–4644
DOI 10.4171/ICM2022/43

Published by EMS Press

1. INTRODUCTION

The focus of this survey is simplicial polytopes and more general simplicial complexes with nice topological properties such as triangulations of spheres and manifolds. Simplicial complexes have played an important role in topology since its early days, as topological spaces were often studied through their triangulations. Due to their discrete nature, simplicial complexes have also been studied by combinatorialists and discrete geometers. The theory of simplicial complexes has always been inseparable from the theory of polytopes. Although polytopes were studied since antiquity, the field has become extremely active since the last half of the 20th century, partly due to rapid developments in optimization and statistics.

There are many excellent surveys on combinatorics of polytopes and simplicial complexes; see, for instance, [12, 44]. Here we concentrate on polytopes and simplicial complexes with and without symmetry, with a particular emphasis on the upper bound type results on their face numbers and the related notion of neighborliness. Among the questions we discuss are: What is the largest number of i-faces that a simplicial sphere of dimension $d-1$ with n vertices can have? How many combinatorially distinct neighborly spheres of dimension $d-1$ with n vertices are there? How neighborly can a centrally symmetric d-polytope be? How different is the answer for centrally symmetric $(d-1)$-spheres? Along the way, we mention some of the algebraic, combinatorial, analytical, and topological tools that have been developed and used over the last 50 years and have brought the field to its current state. The interplay between these various methods is a really fascinating part of the story. We also discuss some of the many open problems in the field. We are only able to touch on a limited number of topics and the choice of these topics is rather subjective. Yet we hope this paper provides the reader with a view into the beautiful theory of face numbers.

2. SOME BASICS

A convex *d-dimensional polytope* (a *d-polytope*, for short) is the convex hull of a finite set of points in $\mathbb{R}^d$ that affinely span $\mathbb{R}^d$. A *supporting hyperplane* of a polytope P is a hyperplane H in $\mathbb{R}^d$ that intersects P nontrivially and such that all points of P lie on the same (closed) side of H. A *proper face* of P is the intersection of P with a supporting hyperplane. The empty set and P itself are the *improper faces* of P. A polytope P is *simplicial* if all of its proper faces are simplices. Faces of dimension 0, 1, and $d-1$ are called *vertices, edges*, and *facets*, respectively; faces of dimension i are called *i-faces*.

A *simplicial complex* Δ on a (finite) vertex set $V = V(\Delta)$ is a collection of subsets of V that is closed under inclusion. The elements $F \in \Delta$ are called *faces*. The *dimension of a face* $F \in \Delta$ is $\dim(F) = |F| - 1$ and the *dimension of* Δ is $\dim(\Delta) = \max\{\dim(F) \mid F \in \Delta\}$. As in the case of polytopes, an i-dimensional face is abbreviated as an *i-face*. A $(d-1)$-dimensional simplicial complex is *pure* if all of its maximal faces (with respect to inclusion) are $(d-1)$-faces; in this case the $(d-1)$-faces are called *facets* and the $(d-2)$-faces are called *ridges*. Unless Δ is the void complex $\emptyset$, the empty set is a face of Δ.

Although simplicial complexes are defined as purely combinatorial objects, each simplicial complex Δ admits a geometric realization $\|\Delta\|$ that contains a geometric i-simplex for each i-face of Δ. We typically do not distinguish between the combinatorial object Δ and the geometric object $\|\Delta\|$ and often say that Δ has certain geometric or topological properties in addition to certain combinatorial properties. For instance, we say that Δ is a *simplicial sphere* (respectively a *simplicial manifold*) if $\|\Delta\|$ is homeomorphic to a sphere (a compact topological manifold without boundary, respectively). Each simplicial d-polytope P gives rise to a simplicial $(d-1)$-sphere, namely the *boundary complex* of P, denoted ∂P.

For a $(d-1)$-dimensional simplicial complex Δ, we denote by $f_i(\Delta)$ the number of i-faces of Δ, and by $f(\Delta) = (f_{-1}(\Delta), f_0(\Delta), \ldots, f_{d-1}(\Delta))$ the *f-vector* of Δ. For reasons that will become apparent below, it is often more natural to study a certain invertible integer transformation of the f-vector called the *h-vector* of Δ, $h(\Delta) = (h_0(\Delta), h_1(\Delta), \ldots, h_d(\Delta))$; it is defined by the following polynomial relation:

$$\sum_{j=0}^{d} h_j(\Delta) \cdot t^{d-j} = \sum_{j=0}^{d} f_{j-1}(\Delta) \cdot (t-1)^{d-j}.$$

For instance, $h_d = f_{d-1} - f_{d-2} + \cdots + (-1)^{d-1} f_0 + (-1)^d = (-1)^{d-1}\tilde{\chi}(\Delta)$ is, up to a sign, the *reduced Euler characteristic* of Δ. Abusing notation, for a simplicial polytope P, we write $f(P)$ and $h(P)$ instead of $f(\partial P)$ and $h(\partial P)$, respectively.

3. THE CYCLIC POLYTOPE AND MCMULLEN'S UPPER BOUND THEOREM

Our story begins with an amazing object—the *cyclic polytope*. This polytope, $C_d(n)$, is the convex hull of $n \geq d+1$ distinct points on the dth *moment curve* $M(t) = M_d(t) = (t, t^2, t^3, \ldots, t^d)$, that is, $C_d(n) = \mathrm{conv}(M_d(t_1), \ldots, M_d(t_n))$, where $t_1 < t_2 < \cdots < t_n$ are real numbers. It is a d-dimensional simplicial polytope with n vertices whose combinatorial type is independent of the choice of $t_1, t_2, \ldots, t_n$. The most remarkable property of $C_d(n)$ is that it is *$\lfloor d/2 \rfloor$-neighborly*, meaning that every $k \leq d/2$ vertices of $C_d(n)$ form the vertex set of a face. (No d-polytope except a d-simplex can be $(\lfloor d/2 \rfloor + 1)$-neighborly.)

To see that $C_d(n)$ is $\lfloor d/2 \rfloor$-neighborly, consider any integer $0 < k \leq \lfloor d/2 \rfloor$ and a k-subset $I = \{i_1, \ldots, i_k\}$ of $[n] = \{1, 2, \ldots, n\}$. Let $P(t) = (t - t_1 + 1)^{d-2k}(t - t_{i_1})^2(t - t_{i_2})^2 \cdots (t - t_{i_k})^2$. Observe that $P(t)$ is a polynomial of degree d, and so it can be written as $P(t) = \gamma_d t^d + \gamma_{d-1} t^{d-1} + \cdots + \gamma_1 t + \gamma_0$. Observe also that $P(t_i) = 0$ for all $i \in I$ while $P(t_i) > 0$ for all $i \in [n] \setminus I$. It follows that the hyperplane

$$H = \{\vec{x} = (x_1, \ldots, x_d) \mid (\gamma_1, \ldots, \gamma_d) \cdot \vec{x} = -\gamma_0\}$$

is a supporting hyperplane of $C_d(n)$ (here $\vec{a} \cdot \vec{x}$ denotes the dot product) and that $H \cap C_d(n) = \mathrm{conv}(M_d(t_i) \mid i \in I)$. Thus $\{M(t_i) \mid i \in I\}$ is the vertex set of a face.

That the combinatorial type of $C_d(n)$ is independent of the choice of $t_1 < \cdots < t_n$ is a consequence of *Gale's evenness condition* [26]. This result asserts that for a subset $I = \{i_1, \ldots, i_d\} \subset [n]$, $\{M(t_{i_1}), M(t_{i_2}), \ldots, M(t_{i_d})\}$ is the vertex set of a facet of $C_d(n)$ if and only if for all $i, j \in [n] \setminus I$, the number of elements $\ell \in I$ that lie between i and j is even.

The cyclic polytope was discovered and rediscovered by many people, including Carathéodory [19, 20], Gale [26], Motzkin [65], and others; see [34, CHAPTER 7] for historic remarks and additional references. The importance of the cyclic polytope is that by virtue of its neighborliness, the face numbers $f_{i-1}(C_d(n))$ for $i \leq \lfloor d/2 \rfloor$ are equal to $\binom{n}{i}$, which is the maximum possible number of $(i-1)$-faces that any $(d-1)$-dimensional simplicial complex with n vertices can have. This led Motzkin [65] to propose the following *Upper Bound Conjecture* (UBC, for short): *in the class of all d-polytopes with n vertices, the cyclic polytope simultaneously maximizes all the face numbers.*

The motivation for the UBC partly comes from optimization: stated in a dual form, it posits that among all d-polytopes defined by n linear constraints, the polar of the cyclic polytope has the largest number of vertices.

By a standard trick of "pulling vertices," to prove the UBC for all polytopes, it suffices to prove it for simplicial polytopes. One advantage of working with a simplicial polytope P is that the first half of the f-vector of P determines the entire f-vector. This important fact is known as the Dehn–Sommerville relations [47]. More specifically, when $i \geq \lfloor d/2 \rfloor$, $f_i(P)$ can be written as a linear combination of $f_j(P)$ for $j < i$. This result and the fact that the cyclic polytope is $\lfloor d/2 \rfloor$-neighborly make the UBC even more plausible. Unfortunately, the main difficulty in trying to derive the UBC for the upper half of the face numbers from the lower half is that the linear combinations $f_i = \sum_{j=-1}^{i-1} a_{ij} f_j$ contain positive and negative coefficients, which makes it very hard to control the magnitude of the sums.

After many partial results and premature announcements, Motzkin's conjecture was finally proved by McMullen [58]. McMullen's insight was to work with the h-numbers instead of the f-numbers. At this point we should note that stated in terms of the h-numbers, the Dehn–Sommerville relations for simplicial polytopes take on the following elegant form: $h_i = h_{d-i}$ for all $0 \leq i \leq d$. (The $h_0 = h_d$ instance of this result is the Euler relation.)

McMullen used shellability of polytopes (established by Brugesser and Mani [17]) and the Dehn–Sommerville relations to prove that, in the class of simplicial polytopes, the cyclic polytope simultaneously maximizes not only the f-numbers, but also the h-numbers. In other words, McMullen's Upper Bound Theorem (UBT for short) asserts that for every simplicial d-polytope P with n vertices,

$$h_i(P) \leq h_i\big(C_d(n)\big) \quad \text{for all } 0 \leq i \leq d.$$

The f-version of the UBC follows right away since the f-numbers are nonnegative linear combinations of the h-numbers. Furthermore, the f- and h-versions of the UBT can be stated as explicit bounds on the f-numbers (h-numbers, resp.) by using that

$$h_i\big(C_d(n)\big) = h_{d-i}\big(C_d(n)\big) = \binom{n-d+i-1}{i} \quad \text{for all } 0 \leq i \leq d/2.$$

Some additional remarks are in order. A simplicial $(d-1)$-sphere (or a simplicial $(d-1)$-ball) Δ is called *shellable* if its facets can be ordered $F_1, F_2, \dots, F_m$ in such a way that for every $i < m$, the simplicial complex generated by the facets $F_1, \dots, F_i$ is a simplicial $(d-1)$-ball; such an ordering of facets is called a *shelling*. While many simplicial spheres are not shellable **[35, 52]**, all spheres that arise as the boundary complexes of polytopes are. The h-numbers of a shellable complex Δ have a simple and well-known interpretation in terms of any shelling of Δ, see **[101, SECTION 8.3]**. The Dehn–Sommerville relations for all shellable spheres are a consequence of this interpretation and the following easy fact: if $F_1, F_2, \dots, F_m$ is a shelling order of facets of a simplicial sphere, then so is the reverse order $F_m, F_{m-1}, \dots, F_1$.

A far-reaching generalization of the UBT for full-dimensional subcomplexes of the boundary complex of a simplicial d-polytope was proved by Kalai **[42]**; another generalization is due to Björner **[14]**.

4. SPHERES, MANIFOLDS, AND EULERIAN COMPLEXES

For a simplicial complex Δ and its face F, the *link of F in Δ* is the following subcomplex of Δ that captures the local behavior of Δ near F:

$$\mathrm{lk}(F, \Delta) := \{G \in \Delta \mid G \cap F = \emptyset,\ G \cup F \in \Delta\}.$$

In particular, $\mathrm{lk}(\emptyset, \Delta) = \Delta$. Understanding how various algebraic, topological, and combinatorial properties of links of nonempty faces affect the properties of the entire complex is a common theme in this part of combinatorics.

Klee **[47]** introduced *Eulerian complexes* as a combinatorial analog and a vast generalization of simplicial spheres: a pure $(d-1)$-dimensional simplicial complex Δ is Eulerian if for every face F of Δ, including the empty face, $\mathrm{lk}(F, \Delta)$ has the same Euler characteristic as a $(d-|F|-1)$-dimensional sphere, $S^{d-|F|-1}$. (In particular, every ridge is in exactly two facets.) In addition to simplicial spheres, the class of Eulerian complexes includes among others all *Eulerian manifolds*, that is, all odd-dimensional simplicial manifolds, and all even-dimensional simplicial manifolds whose Euler characteristic is two.

Klee **[47]** proved that the Dehn–Sommerville relations $h_i = h_{d-i}$ for $0 \le i \le d$, hold for *all* Eulerian complexes of dimension $d-1$.[1] His proof relied on the Euler relation for the links of faces as well as on the observation that for a simplicial complex Δ and $j > i$, every $(j-1)$-face F of Δ contains exactly $\binom{j}{i}$ faces of dimension $i-1$ while every $(i-1)$-face G of Δ is contained in $f_{j-i-1}(\mathrm{lk}(G, \Delta))$ faces of dimension $j-1$. Klee then applied the Dehn–Sommerville relations along with some results from extremal combinatorics to prove the following astonishing fact: the assertion of the Upper Bound Theorem continues to hold

---

1 Klee also established a version of the Dehn–Sommerville relations for *semi-Eulerian complexes*, i.e., pure complexes all of whose vertex links are Eulerian. Very recently Sawaske and Xue **[86]** extended this result to arbitrary pure simplicial complexes by expressing $h_{d-i}(\Delta) - h_i(\Delta)$ in terms of the Euler characteristics of links of faces of Δ.

for *all Eulerian simplicial complexes* provided they have sufficiently many vertices ($d^2/2$ is enough), see [48]. In view of this result, Klee [48] proposed the following far reaching extension of Motzkin's UBC:

Conjecture 4.1. *Let Δ be an Eulerian simplicial complex of dimension $d-1$ with n vertices. Then $f_i(\Delta) \le f_i(C_d(n))$ for all $1 \le i \le d-1$.*

While in this generality the conjecture remains wide open, at present it is known to hold for all simplicial spheres[2] (Stanley [94]), all Eulerian manifolds (Novik [68, 69]), and even some pseudomanifolds with very mild singularities (Hersh and Novik [37], and Novik and Swartz [73]).

Stanley's proof of the UBT for simplicial spheres relied on the theory of Cohen–Macaulay rings. In fact, it was one of the first applications of commutative algebra to combinatorics. Here are some highlights of Stanley's proof. Let Δ be a simplicial complex on the vertex set $[n]$. Consider the polynomial ring $\mathbb{k}[X] = \mathbb{k}[x_1, \dots, x_n]$ over an infinite field $\mathbb{k}$. Let $I_\Delta = (x_{i_1} \cdots x_{i_s} \mid \{i_1 < i_2 < \cdots < i_s\} \notin \Delta) \subset \mathbb{k}[X]$ be the squarefree ideal generated by *non-faces* of Δ. The *face ring* of Δ (also known as the *Stanley–Reisner ring* of Δ) is the quotient ring $\mathbb{k}[\Delta] := \mathbb{k}[X]/I_\Delta$.

The face ring of Δ is a finitely-generated standard graded $\mathbb{k}$-algebra. We denote by $\mathbb{k}[\Delta]_j$ its jth graded component. Stanley's and Hochster's insight (independently from each other) in defining this ring [38, 94] was that algebraic properties of $\mathbb{k}[\Delta]$ reflect many combinatorial and topological properties of Δ. For instance, if Δ is $(d-1)$-dimensional, then the Hilbert series of $\mathbb{k}[\Delta]$, $\mathrm{Hilb}_{\mathbb{k}}(\mathbb{k}[\Delta], t) := \sum_{j=0}^{\infty} \dim_{\mathbb{k}} \mathbb{k}[\Delta]_j \cdot t^j$, is equal to

$$\sum_{i=0}^{d} \frac{f_{i-1}(\Delta) \cdot t^i}{(1-t)^i} = \frac{\sum_{i=0}^{d} h_i(\Delta) \cdot t^i}{(1-t)^d}.$$

Using techniques from homological algebra, Reisner [82] proved that if Δ is a simplicial $(d-1)$-sphere, then $\mathbb{k}[\Delta]$ is Cohen–Macaulay. This means that a sequence $\theta_1, \dots, \theta_d$ of *generic* linear forms in $\mathbb{k}[\Delta]$ is a *regular sequence*, i.e., θ_{s+1} is a nonzero divisor on $\mathbb{k}[\Delta]/(\theta_1, \dots, \theta_s)$ for all $0 \le s \le d-1$. Put differently, for every $0 \le s \le d-1$ and $j > 0$, the following sequence of $\mathbb{k}$-vector spaces is exact:

$$0 \to \mathbb{k}[\Delta]/(\theta_1, \dots, \theta_s)_{j-1} \xrightarrow{\cdot\theta_{s+1}} \mathbb{k}[\Delta]/(\theta_1, \dots, \theta_s)_j \to \mathbb{k}[\Delta]/(\theta_1, \dots, \theta_s, \theta_{s+1})_j \to 0. \quad (4.1)$$

The Cohen–Macaulayness of $\mathbb{k}[\Delta]$ is a key to Stanley's proof of the UBT. Indeed, standard manipulations with Hilbert series using (4.1) show that for a simplicial $(d-1)$-sphere Δ,

$$\mathrm{Hilb}_{\mathbb{k}}\big(\mathbb{k}[\Delta]/(\theta_1, \dots, \theta_d), t\big) = (1-t)^d \,\mathrm{Hilb}_{\mathbb{k}}\big(\mathbb{k}[\Delta], t\big) = \sum_{i=0}^{d} h_i(\Delta) \cdot t^i. \quad (4.2)$$

---

2 As we will see in Section 7, for $d \ge 4$, there are many more simplicial $(d-1)$-spheres than simplicial d-polytopes.

The fact that $\Bbbk[\Delta]/(\theta_1, \ldots, \theta_d)$ is generated as a $\Bbbk$-algebra by $n-d$ elements of degree one then implies that $h_i(\Delta)$ cannot exceed the number of monomials of degree i in $n-d$ variables. The number of such monomials is $\binom{n-d+i-1}{i}$, which coincides with $h_i(C_d(n))$ when $i \leq d/2$. This observation and the Dehn–Sommerville relations yield the UBT for simplicial spheres.

How does one extend Stanley's result to Eulerian manifolds? One immediate obstacle is that the face rings of simplicial manifolds are in general not Cohen–Macaulay. However, they satisfy a weaker property, known as *Buchsbaumness*, and Buchsbaum face rings are reasonably well understood **[87, 97]**, see also **[71, 72]** for more recent results.[3] Specifically, for a simplicial $(d-1)$-manifold Δ and a sequence $\theta_1, \ldots, \theta_d$ of generic linear forms in $\Bbbk[\Delta]$, Schenzel **[87]** computed the dimensions of kernels of maps $\cdot\theta_{s+1} : \Bbbk[\Delta]/(\theta_1, \ldots, \theta_s)_{j-1} \to \Bbbk[\Delta]/(\theta_1, \ldots, \theta_s)_j$. He then used this result to derive the following formula for the Hilbert function of the quotient $\Bbbk(\Delta) := \Bbbk[\Delta]/(\theta_1, \ldots, \theta_d)$ (such quotient is called an *Artinian reduction* of $\Bbbk[\Delta]$):

$$\dim_{\Bbbk} \Bbbk(\Delta)_i = h_i(\Delta) + \binom{d}{i}\big(\beta_{i-2}(\Delta) - \beta_{i-3}(\Delta) + \cdots + (-1)^i \beta_0(\Delta)\big) \quad \text{for all } 0 \leq i \leq d.$$

We denote the right-hand side of this equation by $h'_i(\Delta)$. Here $\beta_j(\Delta)$ is the dimension of the jth reduced simplicial homology of Δ computed with coefficients in $\Bbbk$. In particular, if Δ is a simplicial sphere, then $h'_i(\Delta) = h_i(\Delta)$ for all i (which recovers Stanley's formula (4.2)).

Now consider the socle $\mathcal{S}$ of $\Bbbk(\Delta)$, i.e., the collection of elements μ of $\Bbbk(\Delta)$ such that $x_\ell \cdot \mu = 0$ for all variables $x_\ell \in X$. In the spirit of Schenzel's results, Novik and Swartz **[72]** proved that the dimension of $\mathcal{S}_i$ is at least $\binom{d}{i}\beta_{i-1}(\Delta)$. (For a weaker bound that suffices to prove the UBT, see **[68]**). As $\Bbbk(\Delta)/(\mathcal{S}_i)$ is a standard graded $\Bbbk$-algebra,[4] the dimensions of its homogeneous components satisfy Macaulay's inequalities **[56]**, **[96, PAGE 56]**. In particular, one obtains an upper bound on $h'_{i+1}(\Delta)$ in terms of $h'_i - \binom{d}{i}\beta_{i-1}$. These bounds (together with the Dehn–Sommerville relations) in turn lead to the desired upper bounds on the h-numbers of an Eulerian manifold Δ.

Related to the UBT is a conjecture by Kühnel **[49, CONJECTURE B]**. It asserts that the *reduced Euler characteristic of a simplicial $2k$-manifold Δ on n vertices satisfies*

$$(-1)^k \binom{2k+1}{k}\big(\tilde{\chi}(\Delta) - 1\big) \leq \binom{n-k-2}{k+1}.$$

Moreover, equality holds if and only if Δ is $(k+1)$-neighborly. Kühnel's conjecture was proved by Novik and Swartz **[72]** using machinery similar to that outlined above. What is important to note is that while an Eulerian $2k$-manifold cannot be $(k+1)$-neighborly unless it is the boundary of a simplex, there do exist non-Eulerian simplicial $2k$-manifolds that are

---

3 For an alternative very recent treatment of Cohen-Macaulay and Buchsbaum face rings, where the use of local cohomology and other homological algebra techniques is replaced with a more topological approach, see **[5]**.

4 A much stronger result **[5, 71]** asserts that if Δ is a connected simplicial $(d-1)$-manifold orientable over $\Bbbk$, then $\Bbbk(\Delta)/\bigoplus_{i=0}^{d-1} \mathcal{S}_i$ is a Poincaré duality algebra.

$(k+1)$-neighborly. For surfaces, there are the 2-neighborly triangulations in [40, 84] (such as the six-vertex triangulation of the real projective plane). For $k > 1$, examples include the nine-vertex triangulation of the complex projective plane [50] and several 13-vertex triangulations of $S^3 \times S^3$ [55]; for additional examples see [15, 21].

5. WITT SPACES

Computations similar to those sketched near the end of the previous section can be used to show that the bounds of the UBC continue to hold for some non-Eulerian manifolds, most notably, they hold for all even-dimensional manifolds with vanishing middle homology (Novik [68]).[5] In light of this result, Kalai [43] proposed to extend the UBC to complexes triangulating Witt spaces. The notion of Witt spaces was introduced by Siegel [90]; it relies on Goresky and MacPherson's intersection homology theory [30, 31], namely the theory that was developed as a generalization of the Poincaré–Lefschetz theory to stratified singular spaces such as piecewise-linear pseudomanifolds. Here we only consider simplicial strata. A simplicial *pseudomanifold* is a pure simplicial complex Δ such that every ridge of Δ is contained in exactly two facets. An oriented simplicial pseudomanifold Δ is a *Witt space* if its intersection homology groups with respect to lower-middle perversity $\bar{m} = (0, 0, 1, 1, 2, 2, \ldots)$ have the following property: $IH_j^{\bar{m}}(\Delta'; \mathbb{Q}) = 0$ for every j and every $2j$-dimensional subcomplex $\Delta' \subset \Delta$ that is the link of a nonempty face of Δ. Kalai's conjecture posits the following:

Conjecture 5.1. *Let Δ be a simplicial complex that is a Witt space. Assume that Δ is $(d-1)$-dimensional and has n vertices. If $d-1$ is even, assume further that $IH_{(d-1)/2}^{\bar{m}}(\Delta) = 0$. Then $f_i(\Delta) \leq f_i(C_d(n))$ for all $1 \leq i \leq d-1$.*

An even stronger conjecture (in the spirit of [14, 42]) asserts that if Δ is a $(d-1)$-dimensional Witt space and $f_i(\Delta) \leq f_i(C_d(n'))$ for some i and n', then $f_j(\Delta) \leq f_j(C_d(n'))$ for all $j > i$.

At present, Conjecture 5.1 is known to hold for even-dimensional simplicial manifolds with vanishing middle homology (even on the level of the h-numbers) [68] and also for odd-dimensional pure complexes all of whose vertex links are manifolds with vanishing middle homology (but only on the level of the f-numbers) [37]. It is open in all other cases. The main difficulty seems to be our lack of understanding how various topological characteristics other than simplicial homology can be traced in the face rings.

6. THE G-CONJECTURE

This part of the story would be incomplete if we do not mention some very recent spectacular developments on the g-conjecture. The g-conjecture posits a characterization of the set of f-vectors of simplicial spheres; as such, it contains the UBC for spheres as a special

---

5 On the other hand, the existence of $(k+1)$-neighborly $2k$-manifolds that are not boundaries of simplices demonstrates that not all simplicial manifolds satisfy the bounds of the UBC.

case. (For the class of simplicial polytopes the g-conjecture was proposed by McMullen [59].) The conjecture states that an integer vector $(h_0, h_1, h_2, \dots, h_d)$ is the h-vector of a simplicial $(d-1)$-sphere if and only if (i) $h_j = h_{d-j}$ for all $0 \le j \le d$, (ii) $1 = h_0 \le h_1 \le \dots \le h_{\lfloor d/2 \rfloor}$, and (iii) the numbers $(g_j := h_j - h_{j-1})_{j=0}^{\lfloor d/2 \rfloor}$ form an M-sequence, i.e., they satisfy Macaulay's inequalities. (Here $g_0 := 1$.) The sufficiency of these conditions was established by Billera and Lee [13]. The necessity of conditions for the case of simplicial polytopes was proved by Stanley [95]; a more elementary proof was found by McMullen [60, 61], see also [25]. The necessity of conditions for simplicial spheres remained open until very recently and was considered one of the most outstanding problems in the field.

The recent striking news is that Adiprasito [3], Papadakis and Petrotou [79], and Adiprasito, Papadakis, and Petrotou [4] proved the g-conjecture for simplicial spheres. Along the way, they established surprising algebraic properties of Artinian reductions of face rings of simplicial spheres. For instance, [79] asserts that if Δ is a simplicial $(d-1)$-sphere and $\mathbb{k}$ is a field of characteristic two, then there exists a purely transcendental field extension $\mathbb{K}$ of $\mathbb{k}$ and linear forms $\theta_1, \dots, \theta_d \in \mathbb{K}[\Delta]$ such that for every nonzero homogeneous element $u \in \mathbb{K}[\Delta]/(\theta_1, \dots, \theta_d)$ of degree at most $d/2$, its square u^2 is also *nonzero*. The paper [4] provides far reaching generalizations of these algebraic results to face rings of much more general simplicial complexes such as normal pseudomanifolds. It is now more pressing than ever to compute the Hilbert functions of generic Artinian reductions of face rings and certain further quotients of these rings for the purpose of understanding complexes with singularities more complicated than those studied in [73, 74].

7. NUMBERS OF NEIGHBORLY POLYTOPES AND NEIGHBORLY SPHERES

As we saw in Section 3, the cyclic d-polytopes have the remarkable property of being $\lfloor d/2 \rfloor$-neighborly. At the same time, it follows easily from the Dehn–Sommerville relations that no simplicial $(d-1)$-sphere except for the boundary of a simplex can be $(\lfloor d/2 \rfloor + 1)$-neighborly. This naturally leads to the question of how rare, or how common, the property of being $\lfloor d/2 \rfloor$-neighborly is in the class of simplicial d-polytopes (or $(d-1)$-spheres). To make this discussion more precise, we first need to understand how many combinatorial types of simplicial d-polytopes (or $(d-1)$-spheres) with n *labeled* vertices there are. We denote these numbers by $c(d, n)$ and $s(d, n)$, respectively. (That is, we fix the vertex set to be $[n]$ and count the number of relevant simplicial complexes up to equality. In the case of *unlabeled* vertices, we count the number of complexes up to isomorphism.)

We say that a simplicial $(d-1)$-sphere is *polytopal* if it is isomorphic to the boundary complex of a d-polytope. It follows from Steinitz's theorem, see [34, CHAPTER 13], that all simplicial (and even polyhedral) 2-spheres are polytopal. Hence $c(3, n) = s(3, n)$. The asymptotic behavior of $s(3, n)$ (as well as of the number of unlabeled 2-spheres with n vertices) was worked out by Tutte [98, 99], Brown [16], and Richmond and Wormland [83].

For $d \ge 4$, the results become more surprising. While by a result of Mani [57], every simplicial $(d-1)$-sphere with $n \le d + 3$ vertices is polytopal, there are examples of nonpolytopal simplicial spheres already for $(d,n)=(4,8)$, see Grünbaum [34, SECTION 11.5] and Barnette [8]. In fact, Goodman and Pollack [29], followed by the work of Alon [7], proved that there are far fewer polytopes than what was expected at the time:

$$\big((n/d)-1\big)^{\lfloor d/2\rfloor n/2}\le c(d,n)\le n^{d(d+1)n}.$$

In other words, for a fixed d, $c(d,n)=2^{\Theta(n\log n)}$.

The proof of the inequality $c(d,n)\le n^{d(d+1)n}$ relies on a theorem of Milnor [64] that gives bounds on the sum of the Betti numbers of real algebraic varieties. Assume n is even. Alon's construction [7] providing lower bounds on $c(d,n)$ starts with the cyclic polytope $C_d(n/2)$ on the first $n/2$ vertices; he then adds the last $n/2$ labeled vertices in all possible ways by placing each of them close to a facet of $C_d(n/2)$. That for $n>2d$, $f_{d-1}(C_d(n/2))$ is at least $((n/d)-1)^{\lfloor d/2\rfloor}$ implies that $c(d,n)\ge((n/d)-1)^{\lfloor d/2\rfloor n/2}$.

In striking contrast to these results, Kalai [41] proved that there is an enormous number of simplicial spheres: for $d\ge 5$, $s(d,n)\ge 2^{\Omega(n^{\lfloor (d-1)/2\rfloor})}$. Furthermore, Pfeifle and Ziegler [81] showed that $s(4,n)\ge 2^{\Omega(n^{5/4})}$. The current record on the number of odd-dimensional simplicial spheres is due to Nevo, Santos, and Wilson [67] who verified that $s(2k,n)\ge 2^{\Omega(n^k)}$ for all $k\ge 2$. On the other hand, Stanley's UBT for spheres implies that $s(d,n)\le 2^{O(n^{\lfloor d/2\rfloor}\log n)}$ (see [41]). To summarize, the current best bounds on $s(d,n)$ are

$$2^{\Omega(n^{\lfloor d/2\rfloor})}\le s(d,n)\le 2^{O(n^{\lfloor d/2\rfloor}\log n)}\quad\text{for all } d\ge 4.$$

What proportion of simplicial d-polytopes ($(d-1)$-spheres) are $\lfloor d/2\rfloor$-neighborly? While Motzkin [65] believed that $C_d(n)$ is the only $\lfloor d/2\rfloor$-neighborly d-polytope on n vertices, Shemer [89] introduced a *sewing construction* and used it to prove that even the number of distinct unlabeled $\lfloor d/2\rfloor$-neighborly d-polytopes on n vertices is at least $n^{a_d n}$, where $\lim_{d\to\infty} a_d=1/2$. Generalizing Shemer's sewing construction, Padrol [78] greatly improved this bound: he constructed on the order of $n^{dn/2}$ labeled $\lfloor d/2\rfloor$-neighborly d-polytopes on n vertices (for even d). This lower bound on the number of $\lfloor d/2\rfloor$-neighborly d-polytopes is, of course, also a lower bound on $c(d,n)$. In fact, it is the *current best* lower bound on $c(d,n)$. This state of affairs seems to indicate that the property of being neighborly is very common among polytopes.

Along the same lines, Kalai [41] speculated that the number $\mathrm{sn}(d,n)$ of $\lfloor d/2\rfloor$-neighborly simplicial $(d-1)$-spheres with n labeled vertices is very large and posited the following conjecture:

Conjecture 7.1. *For all $d\ge 4$, $\lim_{n\to\infty}(\log \mathrm{sn}(d,n)/\log s(d,n))=1$.*

In his paper [78], Padrol also constructed a large number of $\lfloor d/2\rfloor$-neighborly $(d-1)$-spheres arising from nonrealizable oriented matroids. Yet, Padrol's bound only implied that $\mathrm{sn}(d,n)\ge 2^{\Omega(n\log n)}$. While we are still very far from being able to shed light on Conjecture 7.1, very recently Novik and Zheng [77] proved that for all $d\ge 5$,

$$\mathrm{sn}(d,n)\ge 2^{\Omega(n^{\lfloor (d-1)/2\rfloor})}. \tag{7.1}$$

Note that for $d \geq 5$, the number of combinatorial types of unlabeled $\lfloor d/2 \rfloor$-neighborly $(d-1)$-spheres on n vertices is also at least $2^{\Omega(n^{\lfloor (d-1)/2 \rfloor})}$. Indeed, dividing the lower bound by $n! = 2^{O(n \log n)}$ does not affect its asymptotic growth if $d \geq 5$.

We now sketch some of the ideas used by Kalai **[41]** to show that for odd $d \geq 5$, $s(d,n) \geq 2^{\Omega(n^{(d-1)/2})}$. We then discuss some additional ideas needed to modify Kalai's construction in order to prove (7.1) for odd d. For the rest of this section, we treat the boundary complex of $C_d(n)$ as an abstract simplicial complex. In particular, a vertex $M(t_i)$ of $C_d(n)$ is identified with $i \in [n]$, and a face $\mathrm{conv}(M(t_i) \mid i \in I)$ with $I \subset [n]$. For integers $a < b$, we write $[a,b]$ to denote the set $\{a, a+1, \ldots, b\}$. If B is a simplicial d-ball, then the *boundary of* B, ∂B, is the $(d-1)$-dimensional subcomplex of B generated by ridges of B that are contained in exactly one facet of B.

Kalai's construction is a generalization of a construction used by Billera and Lee **[13]** to prove the sufficiency of conditions of the g-conjecture. It starts with the family $\mathcal{F} = \mathcal{F}_{2k}(n)$ of all $2k$-subsets of $[n]$ (here $k \geq 2$ is a fixed integer) of the form

$$\mathcal{F} = \big\{\{i_1, i_1+1, i_2, i_2+1, \ldots, i_k, i_k+1\} \mid 1 \leq i_1 < i_1+1 < i_2 < \cdots < i_k < i_k+1 \leq n\big\}.$$

By Gale's evenness condition **[26]**, each $F \in \mathcal{F}$ is a facet of $C_{2k}(n)$. Partially order the set $\mathcal{F}$ by $\{j_1, \ldots, j_{2k}\} \leq_p \{\ell_1, \ldots, \ell_{2k}\}$ if $j_1 \leq \ell_1, \ldots, j_{2k} \leq \ell_{2k}$. For an antichain $\mathcal{A}$ in the poset $(\mathcal{F}, \leq_p)$, let $\mathcal{F}(\mathcal{A})$ be the order ideal of $\mathcal{F}$ generated by $\mathcal{A}$ and let $B(\mathcal{A})$ be the simplicial complex whose facets are the elements of $\mathcal{F}(\mathcal{A})$. For instance, if $k = 2$, $n = 8$, and $\mathcal{A} = \{\{1,2,7,8\}, \{2,3,6,7\}, \{3,4,5,6\}\}$, then $\mathcal{F}(\mathcal{A})$ consists of

$$\begin{aligned}&\{1,2,7,8\}, \{1,2,6,7\}, \{1,2,5,6\}, \{1,2,4,5\}, \{1,2,3,4\},\\ &\qquad\{2,3,6,7\}, \{2,3,5,6\}, \{2,3,4,5\}, \{3,4,5,6\},\end{aligned}$$

and $B(\mathcal{A})$ is generated by these nine facets.

Kalai **[41]** proved that for every (nonempty) antichain $\mathcal{A}$, $B(\mathcal{A})$ is a shellable $(2k-1)$-ball, that all vertices of $B(\mathcal{A})$ are on the boundary of $B(\mathcal{A})$, and that the boundaries $\partial(B(\mathcal{A}))$ and $\partial(B(\mathcal{A}'))$ of two such balls coincide if and only if $\mathcal{A} = \mathcal{A}'$. Estimating the number of antichains, he concluded that there are at least $2^{\Omega(n^{k-1})}$ such balls with exactly n vertices. Their boundaries provide us with the desired number of distinct $(2k-2)$-spheres with n (labeled) vertices. Kalai called these balls *squeezed balls* and their boundaries *squeezed spheres*.

To prove (7.1), the trick is to consider *differences* of appropriately chosen squeezed balls. To do so, for an antichain $\mathcal{A}$ in $\mathcal{F}$, define

$$\begin{aligned}\mathcal{A} - \mathbf{1} := \big\{&\{i_1 - 1, i_1, i_2 - 1, i_2, \ldots, i_{2k} - 1, i_{2k}\} \mid \{i_1, i_1+1, \ldots, i_{2k}, i_{2k}+1\}\\ &\in \mathcal{A} \text{ and } i_1 > 1\big\}.\end{aligned}$$

Let $B_{\mathcal{A}}$ be the simplicial complex whose facets are the elements of $\mathcal{F}(\mathcal{A}) \setminus \mathcal{F}(\mathcal{A} - \mathbf{1})$. In the above example $\mathcal{A} = \{\{1,2,7,8\}, \{2,3,6,7\}, \{3,4,5,6\}\}$, $\mathcal{A} - \mathbf{1} = \{\{1,2,5,6\}, \{2,3,4,5\}\}$, and the facets of $B_{\mathcal{A}}$ are $\{1,2,7,8\}$, $\{1,2,6,7\}$, $\{2,3,6,7\}$, $\{2,3,5,6\}$, and $\{3,4,5,6\}$.

One now checks, see **[77]**, that the following results, paralleling Kalai's theorem, hold: if $\mathcal{A}$ is an antichain in $\mathcal{F}$ that contains $[1,2] \cup [n-2k+3, n]$ as an element, then

$B_{\mathcal{A}}$ is a simplicial $(2k-1)$-ball that has n vertices; this ball is $(k-1)$-*neighborly* and all faces of $B_{\mathcal{A}}$ of dimension $\le k-1$ are on the boundary of $B_{\mathcal{A}}$; furthermore, the boundaries of two such balls $B_{\mathcal{A}}$ and $B_{\mathcal{A}'}$ coincide if and only if $\mathcal{A} = \mathcal{A}'$. We refer to $B_{\mathcal{A}}$ as a *relative squeezed ball* and to its boundary $\partial B_{\mathcal{A}}$ as a *relative squeezed sphere*. The bound (7.1) for odd d follows, since relative squeezed spheres are $(k-1)$-neighborly $(2k-2)$-spheres with n vertices and there are at least $2^{\Omega(n^{k-1})}$ of them.

A simplicial d-ball B all of whose faces of dimension $\le d-i-1$ lie on the boundary of B is called i-*stacked*. In particular, all $(2k-1)$-balls $B_{\mathcal{A}}$ are $(k-1)$-stacked. The notion of stackedness takes its origins in the Generalized Lower Bound Theorem [63, 66, 95].

While at present Conjecture 7.1 is likely out of reach, establishing the bound $\mathrm{sn}(d,n) \ge 2^{\Omega(n^{\lfloor d/2 \rfloor})}$ might be a more feasible goal. We would also like to mention that while Kalai's squeezed balls are shellable, and so are all squeezed spheres [51], the question of whether relative squeezed balls and relative squeezed spheres are shellable is open.

8. THE UPPER BOUND THEOREM FOR CENTRALLY SYMMETRIC SPHERES

We now shift our focus to a fascinating subclass of simplicial complexes, that of centrally symmetric complexes. Some definitions are in order. A polytope $P \subset \mathbb{R}^d$ is *centrally symmetric* (cs, for short) if $P = -P$. In the same spirit, a simplicial complex Δ is *centrally symmetric* (cs, for short) if the vertex set of Δ is endowed with a *free involution* α that induces a free involution on the set of all nonempty faces of Δ. In more detail, for every nonempty face $F \in \Delta$, the following holds:

$$\alpha(F) \in \Delta, \quad \alpha(F) \neq F, \quad \text{and} \quad \alpha\big(\alpha(F)\big) = F.$$

A complex Δ is a cs simplicial sphere if Δ is both a simplicial sphere and a cs complex. For instance, if P is a cs simplicial polytope, then the boundary complex ∂P of P with the map $\alpha(v) = -v$ is a cs simplicial sphere.

To simplify notation, for a cs simplicial complex Δ and a face $F \in \Delta$, we write $\alpha(F) = -F$ and refer to F and $-F$ as antipodal faces of Δ. In particular, if Δ is a cs complex with $2n$ vertices, we usually assume that the vertex set of Δ is $V_n = \{\pm 1, \pm 2, \dots, \pm n\}$. Any cs complex Δ with $2n$ vertices can be naturally associated with a subcomplex of $\partial \mathcal{C}_n^*$, where $\mathcal{C}_n^* = \mathrm{conv}(\pm e_1, \pm e_2, \dots, \pm e_n)$ is the n-dimensional *cross-polytope*. (Here $e_1, \dots, e_n$ are the endpoints of the standard basis of $\mathbb{R}^n$.)

Our discussion in this and the next sections is aimed at and motivated by the following questions: What restrictions does being cs impose on the f-vectors of cs simplicial spheres and cs polytopes? What are the cs analogs of the UBT? Is there a cs version of the cyclic polytope?

To start, note that if Δ is a cs complex and v is a vertex of Δ, then v and $-v$ never form an edge. Thus the notion of neighborliness requires minor adjustments: a cs simplical complex Δ is *cs-k-neighborly* if every set of k of its vertices, no two of which are antipodes, is a face of Δ. Some examples: $\partial \mathcal{C}_d^*$ is cs-d-neighborly, while (the boundary complex of) conv($\pm e_1, \pm e_2, \dots, \pm e_d, \pm \sum_{i=1}^{d} e_i$), which is a cs d-polytope with $2(d+1)$ vertices, is cs-$\lfloor d/2 \rfloor$-neighborly. This latter example is due to McMullen and Shephard [62].

What can be said about neighborliness of cs d-polytopes and cs $(d-1)$-spheres with more than $2(d+1)$ vertices? At this point some striking discrepancies between the cs and non-cs worlds start to emerge. On one hand, Grünbaum [34, SECTION 6.4] (for $d = 4$) and McMullen and Shephard [62] (for general d) proved that in contrast to the non-cs case, a cs polytope with at least $2(d+2)$ vertices cannot be cs-$(\lfloor (d+1)/3 \rfloor + 1)$-neighborly. On the other hand, Grünbaum [32, 33] constructed cs simplicial 3-spheres with 12 vertices that are cs-2-neighborly, thus leaving open the possibility that cs-$\lfloor d/2 \rfloor$-neighborly simplicial $(d-1)$-spheres with an arbitrary large number of vertices may exist.

An additional incentive for investigating if highly neighborly cs spheres exist comes from the following theorem due to Adin [2] and Stanley (unpublished): *in the class of all cs simplicial $(d-1)$-spheres with $2n$ vertices, a cs-$\lfloor d/2 \rfloor$-neighborly sphere simultaneously maximizes all the face numbers assuming such a sphere exists.* (The proof uses face rings and is similar to Stanley's proof of the UBT for spheres discussed in Section 4.)

Does such a sphere exist? In 1995, Jockusch [39] gave a positive answer for $d = 4$: he showed that for *every* value of $n \geq 4$, there is a cs simplicial 3-sphere with $2n$ vertices that is cs-2-neighborly. A few years later, for each $d \leq 7$, Lutz [54] found (by a computer search) several cs simplicial $(d-1)$-spheres with $2(d+2)$ vertices that are cs-$\lfloor d/2 \rfloor$-neighborly. Recently, building on the work of Jockusch [39], Novik and Zheng [75] provided a complete answer: for *all* values of $d \geq 4$ and $n \geq d$, there exists a cs simplicial $(d-1)$-sphere with $2n$ vertices, Δ_n^{d-1}, that is cs-$\lfloor d/2 \rfloor$-neighborly. Combined with the work of Adin and Stanley, this result completely resolved the upper bound problem for cs simplicial spheres.

The construction of Δ_n^{d-1} is quite involved and uses induction on both $d \geq 2$ and $n \geq d$. One key idea of the construction is for all $d, n \geq d$, and $i \leq \lfloor d/2 \rfloor - 1$, to define (by triple induction) an auxiliary simplicial $(d-1)$-ball, $B_n^{d-1,i} \subset \partial \mathcal{C}_n^*$, on the vertex set $\{\pm 1, \dots, \pm n\}$, that is both i-stacked and cs-i-neighborly. (Recall that i-stacked balls were defined at the end of Section 7. While $B_n^{d-1,i} \subset \partial \mathcal{C}_n^*$ is not a cs complex, it is a subcomplex of one; hence the definition of cs-i-neighborliness still makes sense.) In the case of $d = 4$, the construction reduces to Jockusch's construction. A curious property of Δ_n^{d-1} worth mentioning is that the link of $\{n-1, n\}$ in Δ_n^{2k+1} is the complex Δ_{n-2}^{2k-1}, while the link of $\{n-2, n-1, n\}$ in Δ_n^{2k+2} is Δ_{n-3}^{2k-1}. Another property worth mentioning is that in addition to being cs-k-neighborly, the sphere Δ_n^{2k-1} is also *k-stacked*, i.e., it is the boundary of a k-stacked ball.

By results of McMullen and Shephard [62], for $d \geq 4$ and $n \geq d+2$, the complex Δ_n^{d-1} is not isomorphic to the boundary complex of a *cs* polytope. This still leaves open the question of whether Δ_n^{d-1} can be realized as the boundary complex of some non-cs polytope. The answer was recently provided by Pfeifle [80] who proved that for all $d \geq 4$ and $n \geq d+1$ (including $n = d+1$), the complex Δ_n^{d-1} is *not* polytopal! This is quite a remarkable achievement because, while most of simplicial spheres are not polytopal (see Section 7), determining whether a particular simplicial sphere is polytopal or not is very hard. For his proof, Pfeifle introduced a new method for finding a nonrealizability certificate of a simplicial sphere; these certificates involve combinations of Plücker relations.

Now that the existence of cs $(d-1)$-spheres with arbitrarily many vertices that are cs-$\lfloor d/2 \rfloor$-neighborly is established, a new tantalizing question is: For a fixed $d \geq 4$, how many pairwise nonisomorphic such cs spheres with $2n$ vertices are there? In light of Section 7, it is very tempting to conjecture that there are at least $2^{\Omega(n^{\lfloor (d-1)/2 \rfloor})}$ of them. This is wide open at present. Indeed, our current knowledge on this subject is as follows **[76]**: while for $d = 4$ and 5, there are at least $\Omega(2^n)$ such nonisomorphic cs spheres, for $d = 2k > 4$ and $n \gg 0$, only two nonisomorphic constructions are available at present; they are the edge links of $\{n+1, n+2\}$ and $\{1, 2\}$ in Δ_{n+2}^{2k+1}. For $d = 2k+1$, there are three such constructions: the suspensions of the two complexes just mentioned (but with $2(n-1)$ vertices) and Δ_n^{2k}.

Another natural question is whether a cs analog of Klee's UBC holds. Specifically, is it true that in the class of all *cs* Eulerian simplicial complexes of dimension $d-1$ with $2n$ vertices, the complex Δ_n^{d-1} simultaneously maximizes all the face numbers?

It is also worth mentioning that parallel to Kühnel's conjecture (see Section 4) is Sparla's conjecture on the Euler characteristic of cs simplicial $2k$-manifolds **[92, 93]**. This conjecture is still open for manifolds with fewer than $6k+4$ vertices (see **[46, 69]** for the state-of-the-art). On a related note, there do exist non-Eulerian cs simplicial $2k$-manifolds that are cs-$(k+1)$-neighborly. A construction of such a cs $(4k+4)$-vertex triangulation of $S^k \times S^k$ for each $k \geq 1$ is given in **[46]**.

9. HOW NEIGHBORLY CAN A CS POLYTOPE BE?

We now arrive at the most mysterious part of the story: trying to understand possible neighborliness of cs polytopes, as well as trying to come up with tight upper bounds on face numbers of cs polytopes.

As was mentioned in Section 8, a cs d-polytope with at least $2(d+2)$ vertices cannot be cs-$(\lfloor (d+1)/3 \rfloor + 1)$-neighborly **[62]**. To prove this result, McMullen and Shephard developed a notion of *cs transforms* of cs polytopes, which is a cs analog of the celebrated Gale diagrams. A cs transform associates with a cs set $V = \{\pm v_1, \ldots, \pm v_m\} \subset \mathbb{R}^d$ a certain cs set $\overline{V} = \{\pm \bar{v}_1, \ldots, \pm \bar{v}_m\} \subset \mathbb{R}^{m-d}$ in such a way that, given $\overline{V}$, one can check whether it is a cs transform of the vertex set of a cs polytope, and, if this is the case, one can read the vertex sets of the faces of this polytope from $\overline{V}$.

How neighborly can a cs polytope be? We let $k(d,n)$ denote the largest integer k such that there exists a cs d-polytope with $2(d+n)$ vertices that is cs-k-neighborly. In view of their results that $k(d,1) = \lfloor d/2 \rfloor$ and $k(d,2) = \lfloor (d+1)/3 \rfloor$, McMullen and Shephard **[62]** conjectured that $k(d,n) \leq \lfloor (d+n-1)/(n+1) \rfloor$ for all $n \geq 3$. In particular, according to this conjecture, $k(d,d) = 1$, so if the conjecture holds, a cs d-polytope with $4d$ vertices cannot even be cs-2-neighborly. The conjecture was quickly refuted by Halsey **[36]** and then by Schneider **[88]**, but only for $d \gg n$. In a positive direction, Burton **[18]** proved that a cs d-polytope with a sufficiently large number of vertices ($\approx (d/2)^{d/2}$) indeed cannot even be cs-2-neighborly.

The field lay dormant for a few decades, until Donoho et al. [23, 24], see also [85], discovered some amazing connections between cs polytopes with many faces and seemingly unrelated areas of error-correcting codes and sparse signal reconstruction. In particular, Donoho [23] proved that there exists a positive constant ρ such that for large d, $k(d,d) \geq \rho d$. More specifically, he showed that the orthogonal projection of the cross-polytope $\mathcal{C}^*_{2d}$ onto a d-dimensional subspace of $\mathbb{R}^{2d}$, chosen uniformly at random, is with high probability at least cs-$\lfloor \rho d \rfloor$-neighborly.

Linial and Novik [53], following the work of Donoho, established the asymptotics of $k(d,n)$. They proved that there exist constants $C_1, C_2 > 0$ independent of d and n such that

$$\frac{C_1 d}{1+\log((d+n)/d)} \leq k(d,n) \leq \frac{C_2 d}{1+\log((d+n)/d)}. \tag{9.1}$$

The lower bound $k(d,n) \geq \frac{C_1 d}{1+\log((d+n)/d)}$ was also proved independently and at about the same time by Rudelson and Vershynin [85]. Both proofs of the lower bound relied on probabilistic arguments or, more precisely, on "high-dimensional" paradoxes such as Kašin's theorem [45] and Garnaev–Gluskin's theorem [27]. Consider the Grassmannian manifold $G_{n,d+n}$ endowed with the normed unitary invariant measure. Garnaev–Gluskin's theorem asserts that an n-dimensional subspace L of $\mathbb{R}^{d+n}$, chosen uniformly at random, is with positive probability "almost Euclidean", meaning that for all $x \in L \setminus \{0\}$, the ratio $\|x\|_2/\|x\|_1$ is bounded from above by $\tilde{C}\sqrt{\frac{1+\log((d+n)/d)}{d}}$ for some absolute constant $\tilde{C} > 0$. Via cs transforms, the existence of such a subspace L implies the existence of a cs-k-neighborly d-polytope with $2(d+n)$ vertices where k is given by the left-hand side of (9.1); see [53].

While proofs using probabilistic arguments are very beautiful, some disadvantages of such proofs are that they only show existence rather than provide explicit constructions and they only produce asymptotic bounds rather than exact values. Recently some progress has been made on understanding the maximum possible number of vertices that a cs-2-neighborly d-polytope can have. Although we still do not know the exact value, we now know it up to a factor of two: *for $d \geq 2$, there exists a cs d-polytope with $2^{d-1}+2$ vertices that is cs-2-neighborly* [70]; *on the other hand, for $d \geq 3$, no cs polytope with 2^d or more vertices can be cs-2-neighborly* [53]. Written in terms of $k(d,n)$, this result says that, for $d \geq 3$, $k(d,n) \geq 2$ for all $n \leq 2^{d-2}+1-d$ while $k(d,n) = 1$ for all $n \geq 2^{d-1}-d$.

The result that for $d \geq 3$, no cs polytope with 2^d or more vertices can be cs-2-neighborly is due to Linial and Novik [53]. The proof has two ingredients. The first is a simple observation that the vertex set $V \subset \mathbb{R}^d$ of a cs-2-neighborly polytope is *antipodal*, i.e., for every two vertices $x, y \in V$ $(x \neq y)$, there exist two distinct parallel hyperplanes H_x and H_y such that $x \in H_x$, $y \in H_y$, and all elements of V lie in the closed strip defined by H_x and H_y. The second ingredient is a celebrated theorem of Danzer and Grünbaum [22] (see also [6, CHAPTER 17]) asserting that an antipodal set in $\mathbb{R}^d$ has at most 2^d points, and it has exactly 2^d points if and only if it is the vertex set of a parallelotope (which is not cs-2-neighborly, unless $d = 2$).

The existence of a cs d-polytope with $2^{d-1}+2$ vertices that is cs-2-neighborly was established by Novik [70]. The construction of such a polytope is a modification of a recent construction due to Gerencsér and Harangi [28] of an acute set in $\mathbb{R}^d$ of size $2^{d-1}+1$. Informally, the description is as follows: start with the vertex set V of the $(d-1)$-cube $[-1,1]^{d-1}$ embedded in the coordinate hyperplane $\mathbb{R}^{d-1}\times\{0\}$. Then use the extra dimension to perturb the vertices in such a way that the resulting set V' is "almost acute," i.e., V' is cs and for every $v,u,w\in V'$ with $v\neq -w$, the angle $\angle vuw$ is acute. Adding to V' a pair of antipodes of the form $\pm(0,0,\dots,0,c)$, where $0<c\in\mathbb{R}$ is sufficiently large, creates the vertex set of a cs d-polytope that is cs-2-neighborly and has a desired number of vertices; see [70] for details.

To summarize our discussion, for $d\geq 3$, the maximum number of vertices, m_d, that a cs-2-neighborly d-polytope can have lies in the interval $[2^{d-1}+2, 2^d-2]$. The value of m_3 is $6=2^2+2$ as the only cs-2-neighborly 3-polytope is an octahedron. The value of m_4 is $10=2^3+2$: this is a consequence of Grünbaum's result that a cs 4-polytope with 12 vertices cannot be cs-2-neighborly. The exact values of m_d for $d\geq 5$ are unknown at present.

10. TOWARDS AN UPPER BOUND CONJECTURE FOR CS POLYTOPES

What is the largest number $\mathrm{fmax}(d,N;i)$ of i-faces that a cs d-polytope with $N=2n$ vertices can have? The discussion in the previous section indicates that at present we are very far from even being able to pose a plausible conjecture, even in the case of $i=1$. Instead, we can try to ask for asymptotic bounds on $\mathrm{fmax}(d,N;i)$.

For the case of $i=1$, the best to-date bounds are:

$$\frac{3}{4}\cdot\frac{N^2}{2}-O(N)\leq \mathrm{fmax}(4,N;1)\leq\frac{15}{16}\cdot\frac{N^2}{2}, \tag{10.1}$$

and for any even $d=2k>4$,

$$\left(1-\frac{1}{3}(\sqrt{3})^{-d}\right)\cdot\binom{N}{2}\leq \mathrm{fmax}(d,N;1)\leq(1-2^{-d})\cdot\frac{N^2}{2}. \tag{10.2}$$

The upper bounds are due to Barvinok and Novik [11]. Their proof involves a careful use of a volume trick similar to that utilized in the proof of the Danzer–Grünbaum theorem [22]. The lower bounds were established by Barvinok, Lee, and Novik [9]; they rely on a construction that we sketch below.

An idea for such a construction arose from trying to come up with a cs analog of the cyclic polytope. Recall that the cyclic polytope is the convex hull of n points on the dth moment curve. It is a result of Gale [26] that for $d=2k$, the convex hull of n points on the *trigonometric moment curve* $\mathcal{T}_k=(\cos t,\sin t,\cos 2t,\sin 2t,\dots,\cos kt,\sin kt)$ has the same combinatorial type as the cyclic polytope. The curve $\mathcal{T}_k$ does not suit our purpose since it is not symmetric, but this is easily rectified if one considers only odd multiples of t.

Specifically, consider the *symmetric moment curve*

$$U_k(t)=\big(\cos t,\sin t,\cos 3t,\sin 3t,\dots,\cos(2k-1)t,\sin(2k-1)t\big)\quad\text{for } t\in\mathbb{R},$$

or one of its variants such as

$$\Phi_k(t)=(\cos t,\sin t,\cos 3t,\sin 3t,\cos 3^2t,\sin 3^2t,\dots,\cos 3^{k-1}t,\sin 3^{k-1}t)\quad\text{for } t\in\mathbb{R}.$$

Since $U_k(t + 2\pi) = U_k(t)$ and similarly for Φ_k, from this point on, we consider both curves as defined on the unit circle $S^1 = \mathbb{R}/2\pi\mathbb{Z}$. Furthermore, since t and $t + \pi$ form a pair of opposite points on S^1 (for all $t \in S^1$) and since $U_k(t + \pi) = -U_k(t)$, it follows that for each choice of $0 \leq t_1 < \cdots < t_n < \pi$, the convex hull of $\{U_k(t_i), U(t_i + \pi) \mid i \in [n]\}$ is a cs polytope; a similar statement holds for Φ_k. Polytopes with vertices on $U_2 = \Phi_2$ (among certain more general 4-polytopes) were introduced and analyzed by Smilansky [91]. Polytopes with vertices on U_k, for general k, were studied in [10, 11, 100]; they are known as *bicyclic polytopes*. Polytopes with vertices on Φ_k were defined in [9]. One property of bicyclic $2k$-polytopes worth mentioning is that they are *locally k-neighborly*: if the set $\{t_{i_1}, \ldots, t_{i_k}\}$ is contained in an arc of S^1 of length $\pi/2$, then $\{U_k(t_{i_1}), \ldots, U_k(t_{i_k})\}$ is the vertex set of a face. For some applications of bicyclic polytopes to topology, see [1].

To obtain the lower bound on $\mathrm{fmax}(4, N; 1)$ promised in (10.1), take a cs subset X of S^1 consisting of four clusters of points, each of size $N/4$, with the jth cluster lying on a small arc containing $j\pi/2$. The cs 4-polytope $\mathrm{conv}(U_2(x) \mid x \in X)$ has at least $\frac{1}{2} \cdot N(\frac{3}{4}N - 1) \approx \frac{3}{4}\binom{N}{2}$ edges. Similarly, for $k > 2$, consider $A = 2(3^{k-1} - 1)$ equally spaced points $p_1, \ldots, p_A$ on S^1. Replace each p_j with a cluster of N/A points lying on a small arc containing p_j in such a way that the resulting set V is cs. The convex hull of $\Phi_k(V)$ is then a cs $2k$-polytope that verifies the lower bound of (10.2); see [9] for details.

For $i > 2$, the gap between the current best upper and lower bounds on $\mathrm{fmax}(d, N; i - 1)$ is so much worse than the gap for the number of edges that instead of stating the bounds here, we merely refer the reader to [9] for the lower bound and to [11] for the upper bound.

To conclude, we want to emphasize once again that in sharp contrast with the situation for cs spheres, at the moment we are nowhere near having a good handle on the upper bound type results for cs polytopes, and not for the lack of effort. We do not even know what is the largest number of edges that a cs 4-polytope with $N = 2n$ vertices can have. In fact, for $d \geq 6$ and $N \geq 2(d + 2)$, we do not even know if in the class of cs d-polytopes with N vertices, there is a polytope that simultaneously maximizes all the f-numbers. What we seem to be lacking is new constructions (either explicit or probabilistic) of cs polytopes, and, in particular, constructions that may improve the lower bounds given in (10.1) and (10.2): in light of the main result of [70], we believe that $\mathrm{fmax}(d, N; 1)$ is closer to the right-hand side of (10.2) than to the left one.

ACKNOWLEDGMENTS

The author is grateful to Bennet Goeckner, Gil Kalai, Steve Klee, and Hailun Zheng for their comments on the preliminary versions of this paper.

FUNDING

This work was partially supported by NSF grants DMS-1664865 and DMS-1953815, and by Robert R. & Elaine F. Phelps Professorship in Mathematics.

REFERENCES

[1] H. Adams, J. Bush, and F. Frick, Metric thickenings, Borsuk–Ulam theorems, and orbitopes. *Mathematika* **66** (2020), no. 1, 79–102.

[2] R. M. Adin, *Combinatorial structure of simplicial complexes with symmetry*. PhD thesis, Hebrew University, Jerusalem, 1991.

[3] K. Adiprasito, Combinatorial Lefschetz theorems beyond positivity. 2018, arXiv:1812.10454.

[4] K. Adiprasito, S. A. Papadakis, and V. Petrotou, Anisotropy, biased pairings, and the Lefschetz property for pseudomanifolds and cycles. 2021, arXiv:2102.03659.

[5] K. Adiprasito and G. Yashfe, The partition complex: an invitation to combinatorial commutative algebra. In *Surveys in combinatorics 2021*, pp. 1–41, London Math. Soc. Lecture Note Ser. 470, Cambridge Univ. Press, Cambridge, 2021.

[6] M. Aigner and G. M. Ziegler, *Proofs from the book*. 6th edn., Springer, Berlin, 2018.

[7] N. Alon, The number of polytopes, configurations and real matroids. *Mathematika* **33** (1986), no. 1, 62–71.

[8] D. Barnette, Diagrams and Schlegel diagrams. In *Combinatorial structures and their applications (Proc. Calgary Internat. Conf., Calgary, Alta)*, pp. 1–4, Gordon and Breach, New York, 1970.

[9] A. Barvinok, S. J. Lee, and I. Novik, Explicit constructions of centrally symmetric k-neighborly polytopes and large strictly antipodal sets. *Discrete Comput. Geom.* **49** (2013), no. 3, 429–443.

[10] A. Barvinok, S. J. Lee, and I. Novik, Neighborliness of the symmetric moment curve. *Mathematika* **59** (2013), no. 1, 223–249.

[11] A. Barvinok and I. Novik, A centrally symmetric version of the cyclic polytope. *Discrete Comput. Geom.* **39** (2008), no. 1–3, 76–99.

[12] L. J. Billera and A. Björner, Face numbers of polytopes and complexes. In *Handbook of discrete and computational geometry*, pp. 449–475, Discrete Math. Appl., 3rd edn., CRC Press, Boca Raton, FL, 2017.

[13] L. J. Billera and C. W. Lee, A proof of the sufficiency of McMullen's conditions for f-vectors of simplicial convex polytopes. *J. Combin. Theory Ser. A* **31** (1981), 237–255.

[14] A. Björner, A comparison theorem for f-vectors of simplicial polytopes. *Pure Appl. Math. Q.* **3** (2007), no. 1, Special Issue: In honor of Robert D. MacPherson. Part 3, 347–356.

[15] U. Brehm and W. Kühnel, 15-vertex triangulations of an 8-manifold. *Math. Ann.* **294** (1992), no. 1, 167–193.

[16] W. G. Brown, Enumeration of non-separable planar maps. *Canad. J. Math.* **15** (1963), 526–545.

[17] H. Bruggesser and P. Mani, Shellable decompositions of cells and spheres. *Math. Scand.* **29** (1971), 197–205.

[18] G. R. Burton, The nonneighbourliness of centrally symmetric convex polytopes having many vertices. *J. Combin. Theory Ser. A* **58** (1991), 321–322.

[19] C. Carathéodory, Über den Variabilitätsbereich der Koeffizienten von Potenzreihen, die gegebene Werte nicht annehmen. *Math. Ann.* **64** (1907), no. 1, 95–115.

[20] C. Carathéodory, Über den Variabilitatsbereich det Fourierschen Konstanten von positiven harmonischen Furktionen. *Ren. Circ. Mat. Palermo* **32** (1911), 193–217.

[21] M. Casella and W. Kühnel, A triangulated $K3$ surface with the minimum number of vertices. *Topology* **40** (2001), no. 4, 753–772.

[22] L. Danzer and B. Grünbaum, Über zwei Probleme bezüglich konvexer Körper von P. Erdős und von V. L. Klee. *Math. Z.* **79** (1962), 95–99.

[23] D. L. Donoho, High-dimensional centrally symmetric polytopes with neighborliness proportional to dimension. *Discrete Comput. Geom.* **35** (2006), no. 4, 617–652.

[24] D. L. Donoho and J. Tanner, Exponential bounds implying construction of compressed sensing matrices, error-correcting codes, and neighborly polytopes by random sampling. *IEEE Trans. Inf. Theory* **56** (2010), no. 4, 2002–2016.

[25] B. Fleming and K. Karu, Hard Lefschetz theorem for simple polytopes. *J. Algebraic Combin.* **32** (2010), no. 2, 227–239.

[26] D. Gale, Neighborly and cyclic polytopes. In *Proc. Sympos. Pure Math.*, pp. 225–232, VII, Amer. Math. Soc., Providence, RI, 1963.

[27] A. Y. Garnaev and E. D. Gluskin, The widths of a Euclidean ball. *Dokl. Akad. Nauk SSSR* **277** (1984), no. 5, 1048–1052.

[28] B. Gerencsér and V. Harangi, Acute sets of exponentially optimal size. *Discrete Comput. Geom.* **62** (2019), no. 4, 775–780.

[29] J. E. Goodman and R. Pollack, Upper bound for configurations and polytopes in $\mathbb{R}^d$. *Discrete Comput. Geom.* **1** (1986), 219–227.

[30] M. Goresky and R. MacPherson, Intersection homology theory. *Topology* **19** (1980), 135–162.

[31] M. Goresky and R. MacPherson, Intersection homology. II. *Invent. Math.* **72** (1983), no. 1, 77–129.

[32] B. Grünbaum, The importance of being straight. In *Proc. Twelfth Biennial Sem. Canad. Math. Congr. on Time Series and Stochastic Processes; Convexity and Combinatorics (Vancouver, BC, 1969)*, pp. 243–254, Canad. Math. Congr., Montreal, Que, 1970.

[33] B. Grünbaum, On combinatorial spheres. In *Combinatorial structures and their applications (Proc. Calgary Internat. Conf., Calgary, Alta., 1969)*, pp. 119–122, Gordon and Breach, New York, 1970.

[34] B. Grünbaum, *Convex polytopes*. 2nd edn., Grad. Texts in Math. 221, Springer, New York, 2003.

[35] M. Hachimori and G. M. Ziegler, Decompositons of simplicial balls and spheres with knots consisting of few edges. *Math. Z.* **235** (2000), no. 1, 159–171.

[36] E. R. Halsey, *Zonotopal complexes on the d-cube*. PhD thesis, University of Washington, 1972.

[37] P. Hersh and I. Novik, A short simplicial h-vector and the upper bound theorem. *Discrete Comput. Geom.* **28** (2002), no. 3, 283–289.

[38] M. Hochster, Cohen–Macaulay rings, combinatorics, and simplicial complexes. In *Ring theory, II (Proc. Second Conf., Univ. Oklahoma, Norman, OK, 1975)*, pp. 171–223, Lect. Notes Pure Appl. Math. 26, Dekker, New York, 1977.

[39] W. Jockusch, An infinite family of nearly neighborly centrally symmetric 3-spheres. *J. Combin. Theory Ser. A* **72** (1995), no. 2, 318–321.

[40] M. Jungerman and G. Ringel, Minimal triangulations on orientable surfaces. *Acta Math.* **145** (1980), no. 1–2, 121–154.

[41] G. Kalai, Many triangulated spheres. *Discrete Comput. Geom.* **3** (1988), no. 1, 1–14.

[42] G. Kalai, The diameter of graphs of convex polytopes and f-vector theory. In *Applied geometry and discrete mathematics*, pp. 387–411, DIMACS Ser. Discrete Math. Theoret. Comput. Sci. 4, Amer. Math. Soc., Providence, RI, 1991.

[43] G. Kalai, Algebraic shifting. In *Computational commutative algebra and combinatorics*, edited by T. Hibi, pp. 121–163, Adv. Stud. Pure Math. 33, Mathematical Society of Japan, Tokyo, 2002.

[44] G. Kalai, Polytope skeletons and paths. In *Handbook of discrete and computational geometry*, pp. 505–532, Discrete Math. Appl., 3rd edn., CRC Press, Boca Raton, FL, 2017.

[45] B. S. Kašin, The widths of certain finite-dimensional sets and classes of smooth functions. *Izv. Ross. Akad. Nauk Ser. Mat.* **41** (1977), no. 2, 334–351.

[46] S. Klee and I. Novik, Centrally symmetric manifolds with few vertices. *Adv. Math.* **229** (2012), 487–500.

[47] V. Klee, A combinatorial analogue of Poincaré's duality theorem. *Canad. J. Math.* **16** (1964), 517–531.

[48] V. Klee, On the number of vertices of a convex polytope. *Canad. J. Math.* **16** (1964), 701–720.

[49] W. Kühnel, *Tight polyhedral submanifolds and tight triangulations*. Lecture Notes in Math. 1612, Springer, Berlin, 1995.

[50] W. Kühnel and G. Lassmann, The unique 3-neighborly 4-manifold with few vertices. *J. Combin. Theory Ser. A* **35** (1983), no. 2, 173–184.

[51] C. W. Lee, Kalai's squeezed spheres are shellable. *Discrete Comput. Geom.* **24** (2000), 391–396.

[52] W. B. R. Lickorish, Unshellable triangulations of spheres. *European J. Combin.* **12** (1991), no. 6, 527–530.

[53] N. Linial and I. Novik, How neighborly can a centrally symmetric polytope be? *Discrete Comput. Geom.* **36** (2006), 273–281.

[54] F. H. Lutz, *Triangulated manifolds with few vertices and vertex-transitive group actions*. Dissertation, Technische Universität Berlin, Berlin, 1999.

[55] F. H. Lutz, Triangulated manifolds with few vertices: Combinatorial manifolds. 2005, arXiv:math/0506372.

[56] F. S. Macaulay, Some properties of enumeration in the theory of modular systems. *Proc. Lond. Math. Soc.* **26** (1927), 531–555.

[57] P. Mani, Spheres with few vertices. *J. Combin. Theory Ser. A* **13** (1972), 346–352.

[58] P. McMullen, The maximum numbers of faces of a convex polytope. *Mathematika* **17** (1970), 179–184.

[59] P. McMullen, The numbers of faces of simplicial polytopes. *Israel J. Math.* **9** (1971), 559–570.

[60] P. McMullen, On simple polytopes. *Invent. Math.* **113** (1993), no. 2, 419–444.

[61] P. McMullen, Weights on polytopes. *Discrete Comput. Geom.* **15** (1996), no. 4, 363–388.

[62] P. McMullen and G. C. Shephard, Diagrams for centrally symmetric polytopes. *Mathematika* **15** (1968), 123–138.

[63] P. McMullen and D. W. Walkup, A generalized lower-bound conjecture for simplicial polytopes. *Mathematika* **18** (1971), 264–273.

[64] J. Milnor, On the Betti numbers of real varieties. *Proc. Amer. Math. Soc.* **15** (1964), 275–280.

[65] T. S. Motzkin, Comonotone curves and polyhedra. *Bull. Amer. Math. Soc.* **63** (1957), 35.

[66] S. Murai and E. Nevo, On the generalized lower bound conjecture for polytopes and spheres. *Acta Math.* **210** (2013), no. 1, 185–202.

[67] E. Nevo, F. Santos, and S. Wilson, Many triangulated odd-dimensional spheres. *Math. Ann.* **364** (2016), no. 3–4, 737–762.

[68] I. Novik, Upper bound theorems for homology manifolds. *Israel J. Math.* **108** (1998), 45–82.

[69] I. Novik, On face numbers of manifolds with symmetry. *Adv. Math.* **192** (2005), 183–208.

[70] I. Novik, From acute sets to centrally symmetric 2-neighborly polytopes. *SIAM J. Discrete Math.* **32** (2018), no. 3, 1572–1576.

[71] I. Novik and E. Swartz, Gorenstein rings through face rings of manifolds. *Compos. Math.* **145** (2009), 993–1000.

[72] I. Novik and E. Swartz, Socles of Buchsbaum modules, posets and complexes. *Adv. Math.* **222** (2009), 2059–2084.

[73] I. Novik and E. Swartz, Face numbers of pseudomanifolds with isolated singularities. *Math. Scand.* **110** (2012), no. 2, 198–222.

[74] I. Novik and E. Swartz, g-vectors of manifolds with boundary. *Algebraic Combin.* **3** (2020), no. 4, 887–911.

[75] I. Novik and H. Zheng, Highly neighborly centrally symmetric spheres. *Adv. Math.* **370** (2020), 107238, 16 pp.

[76] I. Novik and H. Zheng, New families of highly neighborly centrally symmetric spheres. 2020, arXiv:2005.01155.

[77] I. Novik and H. Zheng, Many neighborly spheres. *Math. Ann.* (to appear).

[78] A. Padrol, Many neighborly polytopes and oriented matroids. *Discrete Comput. Geom.* **50** (2013), no. 4, 865–902.

[79] S. A. Papadakis and V. Petrotou, The characteristic 2 anisotropicity of simplicial spheres. 2020, arXiv:2012.09815.

[80] J. Pfeifle, Positive Plücker tree certificates for non-realizability. 2020, arXiv:2012.11500.

[81] J. Pfeifle and G. M. Ziegler, Many triangulated 3-spheres. *Math. Ann.* **330** (2004), no. 4, 829–837.

[82] G. A. Reisner, Cohen–Macaulay quotients of polynomial rings. *Adv. Math.* **21** (1976), no. 1, 30–49.

[83] L. B. Richmond and N. C. Wormald, The asymptotic number of convex polyhedra. *Trans. Amer. Math. Soc.* **273** (1982), no. 2, 721–735.

[84] G. Ringel, Wie man die geschlossenen nichtorientierbaren Flächen in möglichst wenig Dreiecke zerlegen kann. *Math. Ann.* **130** (1955), 317–326.

[85] M. Rudelson and R. Vershynin, Geometric approach to error correcting codes and reconstruction of signals. *Int. Math. Res. Not.* **64** (2005), 4019–4041.

[86] C. Sawaske and L. Xue, Non-Eulerian Dehn–Sommerville relations. *Mathematika* **67** (2021), no. 2, 257–287.

[87] P. Schenzel, On the number of faces of simplicial complexes and the purity of Frobenius. *Math. Z.* **178** (1981), no. 1, 125–142.

[88] R. Schneider, Neighbourliness of centrally symmetric polytopes in high dimensions. *Mathematika* **22** (1975), 176–181.

[89] I. Shemer, Neighborly polytopes. *Israel J. Math.* **43** (1982), no. 4, 291–314.

[90] P. H. Siegel, Witt spaces: a geometric cycle theory for KO-homology at odd primes. *Amer. J. Math.* **105** (1983), no. 5, 1067–1105.

[91] Z. Smilansky, Convex hulls of generalized moment curves. *Israel J. Math.* **52** (1985), no. 1–2, 115–128.

[92] E. Sparla, *Geometrische und kombinatorische Eigenschaften triangulierter Mannigfaltigkeiten*. Ber. Math.. Verlag Shaker, Aachen, 1997.

[93] E. Sparla, An upper and a lower bound theorem for combinatorial 4-manifolds. *Discrete Comput. Geom.* **19** (1998), 575–593.

[94] R. P. Stanley, The upper bound conjecture and Cohen–Macaulay rings. *Stud. Appl. Math.* **54** (1975), 135–142.

[95] R. P. Stanley, The number of faces of a simplicial convex polytope. *Adv. Math.* **35** (1980), 236–238.

[96] R. P. Stanley, *Combinatorics and commutative algebra*. Progr. Math., Birkhäuser, Boston, Inc., Boston, MA, 1996.

[97] J. Stückrad and W. Vogel, *Buchsbaum rings and applications*. Springer, Berlin, 1986.

[98] W. T. Tutte, A census of planar triangulations. *Canad. J. Math.* **14** (1962), 21–38.

[99] W. T. Tutte, On the enumeration of convex polyhedra. *J. Combin. Theory Ser. B* **28** (1980), no. 2, 105–126.

[100] C. Vinzant, Edges of the Barvinok–Novik orbitope. *Discrete Comput. Geom.* **46** (2011), no. 3, 479–487.

[101] G. M. Ziegler, *Lectures on polytopes*. Grad. Texts in Math. 152, Springer, New York, 1995.

ISABELLA NOVIK

Department of Mathematics, University of Washington, Seattle, WA 98195-4350, USA, novik@uw.edu

This Page Intentionally Left Blank

RESTRICTED PROBLEMS IN EXTREMAL COMBINATORICS

MATHIAS SCHACHT

ABSTRACT

Extremal combinatorics is a central research area in discrete mathematics. The field can be traced back to the work of Turán and it was established by Erdős through his fundamental contributions and his uncounted guiding questions. Since then it has grown into an important discipline with strong ties to other mathematical areas such as theoretical computer science, number theory, and ergodic theory.

We focus on extremal problems for *hypergraphs*, which were introduced by Turán. After solving the analogous question for graphs, Turán asked to determine the maximum cardinality of a set E of 3-element subsets of a given n-element set V such that for any 4 elements of V at least one triple is missing in E. This innocent looking problem is still open and, despite a great deal of effort over the last 80 years, our knowledge is still somewhat limited. We consider a variant of the problem by imposing additional restrictions on the distribution of the 3-element subsets in E. These additional assumptions yield a finer control over the corresponding extremal problem. In fact, this leads to many interesting and more manageable problems, some of which were already considered by Erdős and Sós in the 1980s. The additional assumptions on the distribution of the 3-element subsets are closely related to the theory of *quasirandom discrete structures*, which was pioneered by Szemerédi and became a central theme in the field. In fact, the hypergraph extensions by Gowers and by Rödl et al. of the *regularity lemma* provide essential tools for this line of research.

MATHEMATICS SUBJECT CLASSIFICATION 2020

Primary 05C35; Secondary 05C65, 05C80

KEYWORDS

Hypergraphs, extremal combinatorics, Turán's problem, quasirandomness

ICM INTERNATIONAL CONGRESS OF MATHEMATICIANS 2022 JULY 6–14

Proc. Int. Cong. Math. 2022, Vol. 6, pp. 4646–4658
DOI 10.4171/ICM2022/193

Published by EMS Press

1. INTRODUCTION

Extremal and probabilistic combinatorics is an important area of discrete mathematics with strong ties to Ramsey theory, random graph theory, number theory, theoretical computer science, and ergodic theory. This branch and those connections are central in discrete mathematics and have seen strong developments in the last few decades.

A prime example in that direction is Szemerédi's celebrated density theorem on arithmetic progressions [47]. Its connection to extremal problems for graphs and hypergraphs, provided by the *removal lemma*, was the source for some of the most important developments in the field and led to powerful techniques in extremal combinatorics, which include *Szemerédi's regularity lemma* for graphs [48], its extensions to hypergraphs due to Gowers [26] and Rödl et al. [33, 44], the systematic study of *quasirandom discrete structures* by Thomason [49, 50] and Chung, Graham, and Wilson [8], and the notion of *limits of sequences of graphs and hypergraphs* pioneered by Lovász and Szegedy [9, 30, 31]. Moreover, Szemerédi's theorem has interesting connections to *ergodic theory* (established by Furstenberg and his collaborators [20–22]), to *harmonic analysis* (see, e.g., the work of Roth [45, 46], Gowers [25], Bourgain [6], and others), and to *number theory* (see, e.g., the Green–Tao theorem [27]).

Pivotal in those works was the understanding of suitable notions of *quasirandomness* of discrete structures. A quasirandom structure resembles a truly random object by sharing significant properties with it. The systematic study of quasirandom graphs was initiated by Thomason [49, 50] and Chung, Graham, and Wilson [8]. Those authors considered sequences of deterministic (finite) graphs $G_n = (V_n, E_n)$ with the number of vertices $|V_n|$ tending to infinity with n and with *density* $|E_n|/\binom{|V_n|}{2}$ close to some constant $p \in [0, 1]$. Such a sequence of graphs is quasirandom if it shares some important properties with the binomial random graph $G(n, p)$ of the same density, i.e., G_n has some of the significant properties which hold for $G(n, p)$ with high probability. One of the key properties of $G(n, p)$ is its uniform edge distribution, and Thomason chose a quantitative version of it to define quasirandom graphs. The Chung–Graham–Wilson theorem established a deterministic equivalence between the uniform edge distribution and several other significant properties, including large spectral gap for the eigenvalues of the adjacency matrix, the number of cycles of length four appearing as a subgraph, and the expected number of copies of subgraphs of any fixed isomorphism type.

We consider extremal problems for uniform hypergraphs. The classical extremal problem for hypergraphs, already posed by Turán [51] about 80 years ago, turned out to be notoriously hard and, despite a great deal of effort, our current knowledge is still somewhat limited. We investigate a variant of the classical problem by imposing additional restrictions on the distribution of the hyperedges. Roughly speaking, we shall consider *uniformly dense* hypergraphs, i.e., hypergraphs which induce on large sets of vertices at least a given edge density. This additional assumption yields a better control over the corresponding extremal problem. This leads to many interesting and sometimes more manageable subproblems, some of which were already considered by Erdős and Sós [12, 15]. In particular, those additional assumptions on the hyperedge distribution are closely related to the theory of quasirandom hypergraphs and make these problems amenable to the regularity method for hypergraphs. Extremal problems of this type were investigated in [4,7,23,24,32,37–43].

2. EXTREMAL PROBLEMS FOR GRAPHS AND HYPERGRAPHS

Given a fixed graph F, a classical problem in extremal graph theory asks for the maximum number of edges that a (large) graph G on n vertices containing no copy of F can have. More formally, for a fixed graph F let the *extremal number* $\mathrm{ex}(n, F)$ be the number $|E|$ of edges of an F-free graph $G = (V, E)$ on $|V| = n$ vertices with the maximum number of edges. It is well known and not hard to observe that the sequence $\mathrm{ex}(n, F)/\binom{n}{2}$ is decreasing. Consequently, one may define the *Turán density*

$$\pi(F) = \lim_{n\to\infty} \frac{\mathrm{ex}(n, F)}{\binom{n}{2}},$$

which describes the maximum density of large F-free graphs. The systematic study of these extremal parameters was initiated by Turán [51], who determined $\mathrm{ex}(n, K_t)$ for complete graphs K_t. Recalling that the chromatic number $\chi(F)$ of a graph F is the minimum number of colors one can assign to the vertices of F in such a way that any two vertices connected by an edge receive distinct colors, it follows from a result of Erdős and Stone [16] that

$$\pi(F) = \frac{\chi(F) - 2}{\chi(F) - 1}, \tag{2.1}$$

while the connection with the chromatic number first appeared in the work of Erdős and Simonovits [14]. In particular, the value of $\pi(F)$ can be calculated in finite time. It also follows that the set $\Pi^{(2)} = \{\pi(F)\colon F \text{ is a graph}\}$ of all Turán densities of graphs is given by

$$\Pi^{(2)} = \left\{0, \frac{1}{2}, \frac{2}{3}, \ldots, \frac{t-2}{t-1}, \ldots\right\}.$$

Already in his original work [51], Turán asked for hypergraph extensions of these extremal problems. We mainly restrict ourselves here to 3-*uniform hypergraphs* $H = (V, E)$, where $V = V(H)$ is a finite set of *vertices* and the set of *hyperedges* $E = E(H) \subseteq V^{(3)}$, where $V^{(3)} = \{e \subseteq V\colon |e| = 3\}$ is a collection of 3-element sets of vertices. Despite considerable effort, even for 3-uniform hypergraphs F, no similar characterization (as in the graph case) is known. In fact, it is known that the corresponding set $\Pi^{(3)}$ of Turán densities for 3-uniform hypergraphs is much more complicated and, in particular as a subset of the reals, it is not well-ordered (see, e.g., [19] and [34]). Determining the value of $\pi(F)$ is a well known and hard problem even for "simple" hypergraphs like the complete 3-uniform hypergraph $K_4^{(3)}$ on four vertices and $K_4^{(3)-}$, the hypergraph with four vertices and three hyperedges. Currently the best known bounds for these Turán densities are

$$\frac{5}{9} \leq \pi(K_4^{(3)}) \leq 0.5615 \quad \text{and} \quad \frac{2}{7} \leq \pi(K_4^{(3)-}) \leq 0.2871,$$

where the lower bounds are given by what is believed to be optimal constructions due to Turán (see, e.g., [11]) and Frankl and Füredi [18]. The stated upper bounds are due to Razborov [36], Baber [1], and Baber and Talbot [2], and their proofs are based on the *flag algebra method* introduced by Razborov [35]. For a thorough discussion of Turán-type results and problems for hypergraphs we refer to the survey of Keevash [28].

3. HYPERGRAPHS UNIFORMLY DENSE ON SETS OF VERTICES

Erdős and Sós (see, e.g., [12,15]) suggested a variant, where one restricts to F-free hypergraphs H that are *uniformly dense* on large subsets of the vertices.

Definition 3.1. For reals $d \in [0,1]$ and $\eta > 0$, we say a 3-uniform hypergraph $H = (V, E)$ is *$(d, \eta, \therefore)$-dense* if all subsets $X, Y, Z \subseteq V$ induce at least

$$d|X||Y||Z| - \eta|V|^3$$

triples $(x, y, z) \in X \times Y \times Z$ such that $\{x, y, z\}$ is a hyperedge of H.

Restricting to $\therefore$-dense hypergraphs, the appropriate Turán density $\pi_{\therefore}(F)$ for a given hypergraph F can be defined as

$$\pi_{\therefore}(F) = \sup\big\{d \in [0,1]\colon \text{for every } \eta > 0 \text{ and } n \in \mathbb{N} \text{ there exists}$$
$$\text{a 3-uniform, } F\text{-free, } (d, \eta, \therefore)\text{-dense hypergraph } H \text{ with } |V(H)| \geq n\big\},$$

and we obtain from the definitions that

$$\pi(F) \geq \pi_{\therefore}(F)$$

for every 3-uniform hypergraph F.

We first note that these Turán densities are nontrivial, i.e., there exist hypergraphs F such that $\pi_{\therefore}(F) > 0$, as the following examples show, which can be traced back to the work of Erdős and Hajnal [13].

Example 3.2. Consider a random tournament T_n on the vertex set $[n] = \{1, \dots, n\}$, i.e., an orientation of all edges of the complete graph on the first n positive integers such that each of the two directions (i, j) or (j, i) of every pair of vertices $\{i, j\}$ is chosen independently with probability $1/2$. Given such a tournament T_n, we define the 3-uniform hypergraph $H(T_n)$ on the same vertex set, by including the triple $\{i, j, k\}$ in $E(H(T_n))$ if these three vertices span a cyclically oriented cycle of length three, i.e., $\{i, j, k\} \in E(H(T_n))$ if either (i, j), (j, k), and (k, i) are all in $E(T_n)$ or (i, k), (k, j), and (j, i) are all in $E(T_n)$. It is easy to check that for every $\eta > 0$ with probability tending to 1 as $n \to \infty$ the hypergraph $H(T_n)$ is $(1/4, \eta, \therefore)$-dense. Moreover, no hypergraph H obtained from a tournament in this way contains three hyperedges on four vertices, i.e., every such H is $K_4^{(3)-}$-free and this establishes $\pi_{\therefore}(K_4^{(3)-}) \geq 1/4$.

It was shown by Glebov, Král', and Volec [24] that indeed this construction is essentially optimal by providing a matching upper bound.

Theorem 3.3 (Glebov, Král', and Volec). *We have* $\pi_{\therefore}(K_4^{(3)-}) = 1/4$.

The proof in [24] is computer-assisted and based on flag-algebras. With Reiher and Rödl [41], we obtained an alternative proof, which relies on the regularity method for hypergraphs.

Note that $K_4^{(3)-}$ can be described as the hypergraph given by one vertex a having a triangle as its *link graph*, i.e., the graph consisting of the pairs of vertices that together with a form a hyperedge. From that point of view the following problem asks for a natural extension of Theorem 3.3.

Problem 3.4. *For $t \geq 3$, let S_t be the 3-uniform hypergraph on $t+1$ vertices $a, u_1, \dots, u_t$ such that $\{a, u_i, u_j\}$ is a hyperedge for all $1 \leq i < j \leq t$. Determine $\pi_{\therefore}(S_t)$ for $t \geq 4$.*

In [41] it is shown that

$$\frac{t^2 - 5t + 7}{(t-1)^2} \leq \pi_{\therefore}(S_t) \leq \left(\frac{t-2}{t-1}\right)^2,$$

which for $t = 3$ recovers Theorem 3.3 since $S_3 = K_4^{(3)-}$. For the first open case $t = 4$, we have $\frac{1}{3} \leq \pi_{\therefore}(S_4) \leq \frac{4}{9}$, and it would be interesting to close this gap and to find the extremal structures for this problem.

Another intriguing problem concerns $K_4^{(3)}$, the so-called *tetrahedron*. The following random construction of Rödl [43] shows that $\pi_{\therefore}(K_4^{(3)}) \geq 1/2$, and Erdős [12] suggested that this might be best possible.

Example 3.5. Given any map $\varphi\colon [n]^{(2)} \to \{\text{red}, \text{green}\}$, we define the 3-uniform hypergraph H_φ with vertex set $[n]$ by putting a triple $\{i, j, k\}$ with $i < j < k$ into $E(H_\varphi)$ if and only if the colors of the two pairs $\{i, j\}$ and $\{i, k\}$ differ. Irrespective of the choice of the coloring φ, the hypergraph H_φ contains no tetrahedra: for if a, b, c, and d are any four distinct vertices, say with $a = \min(a, b, c, d)$, then it is impossible for all three of the pairs $\{a, b\}$, $\{a, c\}$, and $\{a, d\}$ to have distinct colors, whence not all three of the triples $\{a, b, c\}$, $\{a, b, d\}$, and $\{a, c, d\}$ can be hyperedges of H_φ. Moreover, it was noticed in [43] that if the coloring φ is chosen uniformly at random, then for any $\eta > 0$ the hypergraph H_φ is with high probability $(1/2, \eta, \therefore)$-dense as n tends to infinity. This is easily checked using standard tail estimates for binomial distributions. In other words, this examples show that $\pi_{\therefore}(K_4^{(3)}) \geq \frac{1}{2}$ holds.

It is believed that this construction is optimal, which leads to the following beautiful problem suggested by Erdős [12].

Problem 3.6. *Show that* $\pi_{\therefore}(K_4^{(3)}) = \frac{1}{2}$.

There is some evidence in support of that conjecture. Recently, Balogh, Clemen, and Lidický [3] showed that

$$\pi_{\therefore}(K_4^{(3)}) \leq 0.529$$

and, hence, $\pi_{\therefore}(K_4^{(3)})$ is strictly smaller than the Turán density $\pi(K_4^{(3)})$. In joint work with Reiher and Rödl, we were able to resolve Problem 3.6 affirmatively for a slightly stronger notion of uniform edge distribution [38], which is also satisfied by the hypergraphs from Examples 3.2 and 3.5 (see Theorem 4.2 below). The construction in Example 3.5 can be extended for arbitrary cliques and this leads to the following general problem.

Problem 3.7. *For every fixed integer $t \geq 4$, show that $\pi_{\therefore}(K_t^{(3)}) = \frac{t-3}{t-2}$.*

This problem seems to be one of the main problems in the area. However, for $t = 6$ a second different lower bound constructions is known (see [41, CONCLUDING REMARKS]), which may indicate that the general problem might be more challenging.

Example 3.8. Similarly as in Example 3.5, we consider a random 2-coloring φ of $[n]^{(2)}$. However, this time we include all triples as hyperedges in H_φ if the three underlying pairs are not all of the same color. Again it is easy to check that for every $\eta > 0$ the hypergraph H_φ is with high probability $(3/4, \eta, \therefore)$-dense and, due to the first nontrivial instance of Ramsey's theorem, it is also $K_6^{(3)}$-free.

Very recently, Bucić, Cooper, Kráľ', Mohr, and Munhá Correia [7] could determine the $\pi_{\therefore}(C_\ell)$ for hypergraph cycles. Here a hypergraph cycle C_ℓ for $\ell \geq 4$ is defined by

$$V(C_\ell) = \mathbb{Z}/\ell\mathbb{Z} \quad \text{and} \quad E(C_\ell) = \big\{\{i, i+1, i+2\}: i \in \mathbb{Z}/\ell\mathbb{Z}\big\}.$$

Note that for $\ell = 4$ we have $C_4 = K_4^{(3)}$ and the best known lower bound $\pi_{\therefore}(C_4) \geq 1/2$ is given by Example 3.5. For $\ell = 5$, Reiher [37] gave an example which shows $\pi_{\therefore}(C_5) \geq 4/27$. On the other hand, for ℓ divisible by 3, the hypergraph cycle C_ℓ is tripartite, and it follows from the definition and the work of Erdős [10] that

$$\pi_{\therefore}(C_{3k}) \leq \pi(C_{3k}) = 0$$

for every $k \geq 2$. Bucić et al. [7] showed that the construction of Reiher is optimal and established the same bound for all $\ell \geq 5$ that are not divisible by 3.

Theorem 3.9 (Bucić et al.). *For every $\ell \geq 5$ with $\ell \not\equiv 0 \pmod 3$, we have $\pi_{\therefore}(C_\ell) = 4/27$.*

Besides determining $\pi_{\therefore}(\cdot)$ for particular hypergraphs, as in the problems and results above, it would be interesting to study the set $\Pi_{\therefore}^{(3)} = \{\pi_{\therefore}(F): F \text{ is a 3-uniform hypergraph}\}$ of all such Turán densities. In that direction as a first problem one may consider the smallest nonzero value. In [10] Erdős showed that $\pi(F) = 0$ if and only if F is tripartite and from this characterization it follows that the smallest nonzero classical Turán density is at least $2/9$. It was proved in [5, 17] that it is in fact $2/9$. For $\pi_{\therefore}(\cdot)$, we showed in [39], similarly as Erdős for $\pi(\cdot)$, a characterization of the hypergraphs F with $\pi_{\therefore}(F) = 0$.

Theorem 3.10. *For a 3-uniform hypergraph F, the following are equivalent:*

(a) $\pi_{\therefore}(F) = 0$.

(b) *There is an enumeration of the vertex set $V(F) = \{v_1, \dots, v_f\}$ and there is a three-coloring $\varphi: \partial F \to \{red, blue, green\}$ of the pairs of vertices ∂F covered by hyperedges of F such that every hyperedge $\{v_i, v_j, v_k\} \in E(F)$ with $i < j < k$*

satisfies

$$\varphi(v_i, v_j) = \textit{red}, \quad \varphi(v_i, v_k) = \textit{blue}, \quad \textit{and} \quad \varphi(v_j, v_k) = \textit{green}.$$

This characterization implies that the smallest nonzero Turán density in this context is at least $1/27$.

Corollary 3.11. *If a hypergraph F satisfies $\pi_{\therefore}(F) > 0$, then $\pi_{\therefore}(F) \geq \frac{1}{27}$.*

Proof. Given a positive integer n, consider a three-coloring $\varphi\colon [n]^{(2)} \to \{\text{red, blue, green}\}$ of the pairs of the first n positive integers. We define a hypergraph H_φ with vertex set $[n]$ by regarding a triple $\{i, j, k\}$ with $1 \leq i < j < k \leq n$ as being a hyperedge if and only if $\varphi(i, j) = \text{red}$, $\varphi(i, k) = \text{blue}$, and $\varphi(j, k) = \text{green}$. Standard probabilistic arguments show that when φ is chosen uniformly at random, then for any fixed $\eta > 0$ the probability that H_φ is $(1/27, \eta, \therefore)$-dense tends to 1 as n tends to infinity. On the other hand, as F does not satisfy condition (b) from Theorem 3.10, it is in a deterministic sense the case that F is never a subgraph of H_φ no matter how large n becomes. Thus we have indeed $\pi_{\therefore}(F) \geq \frac{1}{27}$. ■

Recently, Garbe, Král', and Lamaison [23] complemented Corollary 3.11 and established a matching upper bound for the smallest nonzero value of $\pi_{\therefore}(\cdot)$.

Theorem 3.12 (Garbe, Král', and Lamaison). *There is a hypergraph F with $\pi_{\therefore}(F) = 1/27$.*

It seems plausible that $\Pi^{(3)}_{\therefore}$ is structurally similar to $\Pi^{(2)}$. For example, the construction in Example 3.5 in some sense transfers the extremal example for triangle-free graphs into our context here. In contrast to $\Pi^{(3)}$ we put forward the following problem.

Problem 3.13. *Show that $\Pi^{(3)}_{\therefore} = \{\pi_{\therefore}(F)\colon F \text{ is a 3-uniform hypergraph}\}$ is well-ordered as a subset of the reals.*

Another intriguing open problem from [39] concerns the comparison of $\pi_{\therefore}(F)$ with $\pi(F)$.

Problem 3.14. *Is $\pi_{\therefore}(F) < \pi(F)$ for every 3-uniform hypergraph F with $\pi(F) > 0$?*

Roughly speaking, this questions has an affirmative answer, if no 3-uniform hypergraph F with positive Turán density has an extremal hypergraph H that is uniformly dense with respect to large vertex sets $U \subseteq V(H)$ (see also [15, PROBLEM 7] for a related assertion). Problem 3.14 is motivated by the fact that currently all known extremal constructions for such 3-uniform hypergraphs F are obtained from blow-ups or iterated blow-ups of smaller hypergraphs, which fail to be $(d, \eta, \therefore)$-dense for all $d > 0$ and sufficiently small $\eta > 0$, which may suggest that the answer to Problem 3.14 is affirmative.

The work in [7, 39, 41] may indicate that the regularity method for hypergraphs provides a suitable approach for the problems stated in this section. Moreover, those proofs require Ramsey-type arguments and new results from extremal graph theory which are of independent interest.

4. HYPERGRAPHS UNIFORMLY DENSE ON VERTICES AND PAIRS

In Definition 3.1 we defined uniform hyperedge distribution with respect to vertex sets, and Examples 3.2, 3.5, and 3.8 showed that this notion alone with density bounded away from 0 does not suffice to embed arbitrary 3-uniform hypergraphs. In contrast, it follows that if we define uniform hyperedge density with respect to sets of pairs, then such a notion would allow the embedding of arbitrary fixed 3-uniform hypergraphs (see, e.g., [29]) and, in fact, these considerations led to the more involved concepts in the hypergraph regularity projects of Gowers and Rödl et al.

Moreover, there seem to be at least two intermediate variants of uniformly dense hypergraphs. In connection with extremal problems, those notions were already partly investigated in [4, 38, 42] and they lead to several interesting problems. The first strengthening is the following stronger concept of uniformly dense hypergraphs, where we "replace" the two sets Y and Z from Definition 3.1 by an arbitrary set of pairs P.

Definition 4.1. For reals $d \in [0,1]$ and $\eta > 0$, we say a 3-uniform hypergraph $H = (V, E)$ is $(d, \eta, \overset{\cdot}{\cdot\!-\!\cdot})$-dense if for every subset $X \subseteq V$ of vertices and every subset of pairs of vertices $P \subseteq V \times V$ the number $e_{\overset{\cdot}{\cdot\!-\!\cdot}}(X, P)$ of pairs $(x, (y, z)) \in X \times P$ with $\{x, y, z\} \in E$ satisfies

$$e_{\overset{\cdot}{\cdot\!-\!\cdot}}(X, P) \geq d|X||P|| - \eta|V|^3.$$

Since for any hypergraph $H = (V, E)$ and sets $X, Y, Z \subseteq V$ we may apply the definition for X and $P = Y \times Z$, it follows from these definitions that $(d, \eta, \overset{\cdot}{\cdot\!-\!\cdot})$-dense hypergraphs are also $(d, \eta, \therefore)$-dense. Moreover, we can introduce the corresponding Turán density $\pi_{\overset{\cdot}{\cdot\!-\!\cdot}}(F)$ for a given hypergraph F by

$$\pi_{\overset{\cdot}{\cdot\!-\!\cdot}}(F) = \sup\{d \in [0,1]\colon \text{for every } \eta > 0 \text{ and } n \in \mathbb{N} \text{ there exists}$$
$$\text{a 3-uniform, } F\text{-free, } (d, \eta, \overset{\cdot}{\cdot\!-\!\cdot})\text{-dense hypergraph } H \text{ with } |V(H)| \geq n\}$$

and obtain from the definitions that

$$\pi(F) \geq \pi_{\therefore}(F) \geq \pi_{\overset{\cdot}{\cdot\!-\!\cdot}}(F)$$

for every 3-uniform hypergraph F. One can check that the random constructions in Examples 3.2 and 3.5 also give lower bounds for $\pi_{\overset{\cdot}{\cdot\!-\!\cdot}}(K_4^{(3)-})$ and $\pi_{\overset{\cdot}{\cdot\!-\!\cdot}}(K_4^{(3)})$, as the constructed hypergraphs in these examples are also $\overset{\cdot}{\cdot\!-\!\cdot}$-dense. In particular, we have $\pi_{\overset{\cdot}{\cdot\!-\!\cdot}}(K_4^{(3)}) \geq 1/2$ and a matching upper bound was proved in joint work with Reiher and Rödl [38], which can be viewed as some evidence towards an affirmative answer for Problem 3.6.

Theorem 4.2. *We have* $\pi_{\overset{\cdot}{\cdot\!-\!\cdot}}(K_4^{(3)}) = 1/2$.

Moreover, these considerations naturally suggest that a possible first step towards Problems 3.7, 3.4, and 3.13 is to consider these problems for $\pi_{\overset{\cdot}{\cdot\!-\!\cdot}}(\cdot)$.

Problem 4.3. (i) *Show that* $\pi_{\overset{\cdot}{\cdot\!-\!\cdot}}(K_t^{(3)}) = \frac{t-3}{t-2}$ *for every* $t > 4$.

(ii) *Determine* $\pi_{\overset{\cdot}{\cdot\!-\!\cdot}}(S_t)$ *for* $t \geq 4$.

(iii) *Show that* $\Pi^{(3)}_{\underline{\therefore}} = \{\pi_{\underline{\therefore}}(F)\colon F$ *is a* 3*-uniform hypergraph*$\}$ *is well-ordered as a subset of the reals.*

Finding the smallest nonzero value of $\pi_{\underline{\therefore}}(\cdot)$ would also be of high interest. However, the situation here is less clear and maybe as a first step it would be useful to establish a meaningful characterization of the hypergraphs F with $\pi_{\underline{\therefore}}(F) = 0$. By definition, the set of those hypergraphs must contain all hypergraphs F with $\pi_{\therefore}(F) = 0$, but finding a useful characterization appears to be an interesting problem on its own.

Problem 4.4. *Find a useful characterization of the* 3*-uniform hypergraphs* F *with* $\pi_{\underline{\therefore}}(F) = 0$.

5. HYPERGRAPHS UNIFORMLY DENSE ON PAIRS OF SETS OF PAIRS

The following further strengthening of the notion of $\underline{\therefore}$-dense hypergraphs is in some sense the strongest nontrivial uniform density condition for extremal problems in 3-uniform hypergraphs.

Definition 5.1. For reals $d \in [0, 1]$ and $\eta > 0$, we say a 3-uniform hypergraph $H = (V, E)$ is $(d, \eta, \wedge)$-dense if for any two subsets of pairs $P, Q \subseteq V \times V$ the number $e_{\wedge}(P, Q)$ of pairs of pairs $((x, y), (x, z)) \in P \times Q$ with $\{x, y, z\} \in E$ satisfies

$$e_{\wedge}(P, Q) \geq d\,|\mathcal{K}_{\wedge}(P, Q)| - \eta|V|^3,$$

where $\mathcal{K}_{\wedge}(P, Q)$ denotes the set of pairs in $P \times Q$ of the form $((x, y), (x, z))$.

The corresponding Turán density $\pi_{\wedge}(F)$ can be defined similarly as above by

$$\pi_{\wedge}(F) = \sup\big\{d \in [0, 1]\colon \text{for every } \eta > 0 \text{ and } n \in \mathbb{N} \text{ there exists}$$
$$\text{a 3-uniform, } F\text{-free, } (d, \eta, \wedge)\text{-dense hypergraph } H \text{ with } |V(H)| \geq n\big\}$$

and again the definition ensures that

$$\pi(F) \geq \pi_{\therefore}(F) \geq \pi_{\underline{\therefore}}(F) \geq \pi_{\wedge}(F).$$

With respect to cliques $K_t^{(3)}$, parameter $\pi_{\wedge}(\cdot)$ behaves differently and grows much more slowly. In **[42]**, together with Reiher and Rödl, we could show the following upper bound.

Theorem 5.2. *For every* $t \geq 2$,

$$\pi_{\wedge}(K^{(3)}_{2^t}) \leq \frac{t-2}{t-1},$$

which is tight for $t = 2, 3,$ *and* 4.

Maybe somewhat surprisingly, **[42]** establishes the precise value of $\pi_{\wedge}(K_s^{(3)})$ for $s \in \{4, 6, 7, 8, 11, 12, \ldots, 16\}$, but the cases $s = 5$, 9, and 10, were left open. Very recently, in joint work with Berger, Piga, Reiher, and Rödl **[4]**, we could resolve the case $s = 5$ and showed

$$\pi_{\wedge}(K_5^{(3)}) = \frac{1}{3}.$$

Comparing Theorem 5.2 with the known (or believed to be optimal) lower bounds for $\pi_{\underline{\therefore}}(K_{t+1}^{(3)})$, we have

$$\pi_{\wedge}(K_{2^t}^{(3)}) \le \frac{t-2}{t-1} \le \pi_{\underline{\therefore}}(K_{t+1}^{(3)}) \le \pi_{\therefore}(K_{t+1}^{(3)}).$$

So in particular, the Turán densities for $\wedge$-dense hypergraphs for $K_t^{(3)}$ grow much slower compared to $\underline{\therefore}$-dense or $\therefore$-dense hypergraphs.

Maybe the most urgent questions related to $\pi_{\wedge}(\cdot)$ are an appropriate version of Problem 4.4 and determining $\pi_{\wedge}(K_s^{(3)})$ for the missing small values of $s = 9$ and 10 mentioned above.

Problem 5.3. (i) *Find a useful characterization of the* 3*-uniform hypergraphs F with $\pi_{\wedge}(F) = 0$.*

(ii) *Determine $\pi_{\wedge}(K_s^{(3)})$ for $s = 9$ and* 10.

It seems plausible that by combining the main ideas from [42] and [4] one can derive the improved upper bound

$$\pi_{\wedge}(K_{10}^{(3)}) \le \frac{3}{5}.$$

More generally, one may show that if $\pi_{\wedge}(K_s^{(3)}) \le \alpha$, then $\pi_{\wedge}(K_{2s}^{(3)}) \le \frac{1}{2-\alpha}$, which was suggested by Reiher in [37].

Determining the value $\pi_{\wedge}(K_s^{(3)})$ for large values of s might be a challenging problem, and one may first focus on the asymptotic behavior. For every $s \ge 3$, Theorem 5.2 tells us

$$\pi_{\wedge}(K_s^{(3)}) \le 1 - \frac{1}{\log_2(s)}. \tag{5.1}$$

For a lower bound, we consider the following well-known random construction.

Example 5.4. For $r \ge 2$, we consider random hypergraphs $H_\varphi = (V, E_\varphi)$ with the edge set defined by the nonmonochromatic triangles of a random r-coloring $\varphi\colon V^{(2)} \to [r]$ for a sufficiently large vertex set V. It is easy to check that for any fixed $\eta > 0$ with high probability such hypergraphs H_φ are $(\frac{r-1}{r}, \eta, \wedge)$-dense. On the other hand, if s is at least as large as $R(3; r)$, the r-color Ramsey number for graph triangles, then every such H_φ is $K_s^{(3)}$-free.

Consequently, Example 5.4 yields

$$\pi_{\wedge}(K_s^{(3)}) \ge 1 - \frac{1}{r}, \quad \text{whenever } s \ge R(3; r)$$

and, using the simple upper bound $R(3; r) \le 3r!$, we arrive at

$$\pi_{\wedge}(K_s^{(3)}) \ge 1 - \frac{\log_2 \log_2(s)}{\log_2(s)} \tag{5.2}$$

for sufficiently large s. Comparing the bounds in (5.1) and (5.2) leads to the following problem.

Problem 5.5. *Determine the asymptotic behavior of* $1 - \pi_{\wedge}(K_s^{(3)})$.

ACKNOWLEDGMENTS

The work on extremal problems for uniformly dense hypergraphs is part of a joint research project with Christian Reiher and Vojtěch Rödl and my gratitude to both of them goes beyond our collaboration.

FUNDING

This work was partially supported by the European Research Council (PEPCo 724903).

REFERENCES

[1] R. Baber, Turán densities of hypercubes. 2012, arXiv:1201.3587.

[2] R. Baber and J. Talbot, Hypergraphs do jump. *Combin. Probab. Comput.* **20** (2011), no. 2, 161–171.

[3] J. Balogh, F. C. Clemen, and B. Lidický, Hypergraph Turán problems in ℓ_2-norm. In *Surveys in combinatorics 2022*, edited by A. Nixon and S. Prendiville, pp. 21–63, London Math. Soc. Lecture Note Ser. 481, Cambridge Univ. Press, Cambridge, 2022.

[4] S. Berger, S. Piga, Chr. Reiher, V. Rödl, and M. Schacht, Turán density of cliques of order five in 3-uniform hypergraphs with quasirandom links. 2022, arXiv:2206.07354.

[5] B. Bollobás, Three-graphs without two triples whose symmetric difference is contained in a third. *Discrete Math.* **8** (1974), 21–24.

[6] J. Bourgain, Roth's theorem on progressions revisited. *J. Anal. Math.* **104** (2008), 155–192.

[7] M. Bucić, J. W. Cooper, D. Král', S. Mohr, and D. Munhá Correia, Uniform Turán density of cycles. 2021, arXiv:2112.01385.

[8] F. R. K. Chung, R. L. Graham, and R. M. Wilson, Quasi-random graphs. *Combinatorica* **9** (1989), no. 4, 345–362.

[9] G. Elek and B. Szegedy, A measure-theoretic approach to the theory of dense hypergraphs. *Adv. Math.* **231** (2012), no. 3–4, 1731–1772.

[10] P. Erdős, On extremal problems of graphs and generalized graphs. *Israel J. Math.* **2** (1964), 183–190.

[11] P. Erdős, Paul Turán, 1910–1976: His work in graph theory. *J. Graph Theory* **1** (1977), no. 2, 97–101.

[12] P. Erdős, Problems and results on graphs and hypergraphs: similarities and differences. In *Mathematics of Ramsey theory*, edited by J. Nešetřil and V. Rödl, pp. 12–28, Algorithms Combin. 5, Springer, Berlin, 1990.

[13] P. Erdős and A. Hajnal, On Ramsey like theorems. Problems and results. In *Combinatorics*, edited by D. J. A. Walsh and D. R. Woodall, pp. 123–140, Inst. Math. Appl., Southend-on-Sea, 1972.

[14] P. Erdős and M. Simonovits, A limit theorem in graph theory. *Studia Sci. Math. Hungar.* **1** (1966), 51–57.

[15] P. Erdős and V. T. Sós, On Ramsey–Turán type theorems for hypergraphs. *Combinatorica* **2** (1982), no. 3, 289–295.

[16] P. Erdős and A. H. Stone, On the structure of linear graphs. *Bull. Amer. Math. Soc.* **52** (1946), 1087–1091.

[17] P. Frankl and Z. Füredi, A new generalization of the Erdős–Ko–Rado theorem. *Combinatorica* **3** (1983), no. 3–4, 341–349.

[18] P. Frankl and Z. Füredi, An exact result for 3-graphs. *Discrete Math.* **50** (1984), no. 2–3, 323–328.

[19] P. Frankl and V. Rödl, Hypergraphs do not jump. *Combinatorica* **4** (1984), no. 2–3, 149–159.

[20] H. Furstenberg, Ergodic behavior of diagonal measures and a theorem of Szemerédi on arithmetic progressions. *J. Anal. Math.* **31** (1977), 204–256.

[21] H. Furstenberg and Y. Katznelson, An ergodic Szemerédi theorem for commuting transformations. *J. Anal. Math.* **34** (1978), 275–291.

[22] H. Furstenberg and Y. Katznelson, A density version of the Hales–Jewett theorem. *J. Anal. Math.* **57** (1991), 64–119.

[23] F. Garbe, D. Kráľ, and A. Lamaison, Hypergraphs with minimum positive uniform Turán density. *Israel J. Math.* (2022, to appear). arXiv:2105.09883.

[24] R. Glebov, D. Kráľ, and J. Volec, A problem of Erdős and Sós on 3-graphs. *Israel J. Math.* **211** (2016), no. 1, 349–366.

[25] W. T. Gowers, A new proof of Szemerédi's theorem. *Geom. Funct. Anal.* **11** (2001), no. 3, 465–588.

[26] W. T. Gowers, Hypergraph regularity and the multidimensional Szemerédi theorem. *Ann. of Math. (2)* **166** (2007), no. 3, 897–946.

[27] B. Green and T. Tao, The primes contain arbitrarily long arithmetic progressions. *Ann. of Math. (2)* **167** (2008), no. 2, 481–547.

[28] P. Keevash, Hypergraph Turán problems. In *Surveys in combinatorics 2011*, edited by R. Chapman, pp. 83–139, London Math. Soc. Lecture Note Ser. 392, Cambridge University Press, Cambridge, 2011.

[29] Y. Kohayakawa, V. Rödl, and J. Skokan, Hypergraphs, quasi-randomness, and conditions for regularity. *J. Combin. Theory Ser. A* **97** (2002), no. 2, 307–352.

[30] L. Lovász, *Large networks and graph limits*. Amer. Math. Soc. Colloq. Publ. 60, American Mathematical Society, Providence, RI, 2012, xiv+475 pp.

[31] L. Lovász and B. Szegedy, Limits of dense graph sequences. *J. Combin. Theory Ser. B* **96** (2006), no. 6, 933–957.

[32] D. Mubayi and V. Rödl, Supersaturation for Ramsey-Turán problems. *Combinatorica* **26** (2006), no. 3, 315–332.

[33] B. Nagle, V. Rödl, and M. Schacht, The counting lemma for regular k-uniform hypergraphs. *Random Structures Algorithms* **28** (2006), no. 2, 113–179.

[34] O. Pikhurko, On possible Turán densities. *Israel J. Math.* **201** (2014), no. 1, 415–454.

[35] A. A. Razborov, Flag algebras. *J. Symbolic Logic* **72** (2007), no. 4, 1239–1282.

[36] A. A. Razborov, On 3-hypergraphs with forbidden 4-vertex configurations. *SIAM J. Discrete Math.* **24** (2010), no. 3, 946–963.

[37] Chr. Reiher, Extremal problems in uniformly dense hypergraphs. In *Selected papers of EuroComb17*, edited by M. Drmota, M. Kang, C. Krattenthaler, and J. Nešetřil, European J. Combin. 88, 2020.

[38] Chr. Reiher, V. Rödl, and M. Schacht, Embedding tetrahedra into quasirandom hypergraphs. *J. Combin. Theory Ser. B* **121** (2016), 229–247.

[39] Chr. Reiher, V. Rödl, and M. Schacht, Hypergraphs with vanishing Turán density in uniformly dense hypergraphs. *J. Lond. Math. Soc. (2)* **97** (2018), no. 1, 77–97.

[40] Chr. Reiher, V. Rödl, and M. Schacht, On a generalisation of Mantel's theorem to uniformly dense hypergraphs. *Int. Math. Res. Not. IMRN* **16** (2018), 4899–4941.

[41] Chr. Reiher, V. Rödl, and M. Schacht, On a Turán problem in weakly quasirandom 3-uniform hypergraphs. *J. Eur. Math. Soc. (JEMS)* **20** (2018), no. 5, 1139–1159.

[42] Chr. Reiher, V. Rödl, and M. Schacht, Some remarks on $\pi_{\wedge}$. In *Connections in discrete mathematics*, edited by S. Butler, J. Cooper, and G. Hurlbert, pp. 214–239, Cambridge University Press, Cambridge, 2018.

[43] V. Rödl, On universality of graphs with uniformly distributed edges. *Discrete Math.* **59** (1986), no. 1–2, 125–134.

[44] V. Rödl and J. Skokan, Regularity lemma for k-uniform hypergraphs. *Random Structures Algorithms* **25** (2004), no. 1, 1–42.

[45] K. F. Roth, Sur quelques ensembles d'entiers. *C. R. Acad. Sci. Paris* **234** (1952), 388–390.

[46] K. F. Roth, On certain sets of integers. *J. Lond. Math. Soc.* **28** (1953), 104–109.

[47] E. Szemerédi, On sets of integers containing no k elements in arithmetic progression. *Acta Arith.* **27** (1975), 199–245.

[48] E. Szemerédi, Regular partitions of graphs. In *Problèmes combinatoires et théorie des graphes*, edited by J.-C. Bermond, J.-C. Fournier, M. Las Vergnas, and D. Sotteau, pp. 399–401, Colloq. Int. Cent. Natl. Rech. Sci. 260, CNRS, Paris, 1978.

[49] A. Thomason, Pseudorandom graphs. In *Random graphs'85 (Poznań, 1985)*, edited by M. Karoński and Z. Palka, pp. 307–331, North-Holl. Math. Stud. 144, North-Holland, Amsterdam, 1987.

[50] A. Thomason, Random graphs, strongly regular graphs and pseudorandom graphs. In *Surveys in combinatorics 1987*, edited by C. Whitehead, pp. 173–195, London Math. Soc. Lecture Note Ser. 123, Cambridge University Press, Cambridge, 1987.

[51] P. Turán, Eine Extremalaufgabe aus der Graphentheorie. *Mat. Fiz. Lapok* **48** (1941), 436–452.

MATHIAS SCHACHT

Fachbereich Mathematik, Universität Hamburg, Hamburg, Germany, schacht@math.uni-hamburg.de

This Page Intentionally Left Blank

GRAPHS OF LARGE CHROMATIC NUMBER

ALEX SCOTT

ABSTRACT

The chromatic number has been a fundamental topic of study in graph theory for more than 150 years. Graph coloring has a deep combinatorial theory and, as with many NP-hard problems, is of interest in both mathematics and computer science. An important challenge is to understand graphs with very large chromatic number. The chromatic number tells us something global about the structure of a graph: if G has small chromatic number then it can be partitioned into a few very simple pieces. But what if G has large chromatic number? Is there anything that we can say about its local structure? In particular, are there particular substructures that it must contain? In this paper, we will discuss recent progress and open problems in this area.

MATHEMATICS SUBJECT CLASSIFICATION 2020

Primary 05C15; Secondary 05C69, 05C75, 05D10

KEYWORDS

Graph coloring, induced subgraphs, χ-boundedness, Erdős-Hajnal Conjecture

INTERNATIONAL CONGRESS OF MATHEMATICIANS 2022 JULY 6–14

Proc. Int. Cong. Math. 2022, Vol. 6, pp. 4660–4680
DOI 10.4171/ICM2022/149

Published by EMS Press

1. INTRODUCTION

The chromatic number has been a fundamental topic of study in graph theory for more than 150 years. For example, the famous Four Color Conjecture, which states that every graph that can be embedded in the plane has chromatic number at most 4, was first raised in the 1850s by Francis Guthrie (a student of Augustus de Morgan), and was finally proved in the 1970s by Appel and Haken [4], in one of the first computer-assisted proofs. Attempts to solve the conjecture led to Birkhoff's [9] development in 1912 of the chromatic polynomial, which counts the number of k-colorings of a graph G. The chromatic polynomial was generalized by Tutte [85] to what is now known as the Tutte polynomial, which is closely connected to the Ising model and Potts model in statistical physics (see Fortuin and Kasteleyn [43], Sokal [83]), the random cluster model in probability (see Grimmett [44]) and the Jones polynomial in knot theory (see Jones [51]).

However, many fundamental questions about graph coloring remain. A particular challenge is to understand graphs with very large chromatic number. The chromatic number says something about the global structure of a graph: if G has small chromatic number then it can be partitioned into a few very simple pieces. But what if G has large chromatic number? Is there anything that we can say about its local structure? In particular, are there particular substructures that it must contain?

We will need a few definitions. Let G be a graph with vertex set $V = V(G)$ and edge set $E = E(G)$ (all graphs in this paper are finite). A *complete graph* is a graph in which every pair of vertices is joined. A *stable set* (or *independent set*) in G is a set $S \subseteq V$ such that no two vertices of S are adjacent in G. The *clique number* $\omega(G)$ of G is the maximum number of vertices in a complete subgraph of G; and the *stability number* $\alpha(G)$ is the largest number of vertices in a stable set in G. A k*-coloring* of a graph is function from its vertices to $\{1, \dots, k\}$ so that adjacent vertices have different colors. The *chromatic number* $\chi(G)$ of G is the smallest integer k such that G has a k-coloring.

Graphs with chromatic number at most 2 are easily characterized: they are the graphs that do not contain an odd cycle. But for $k \geq 3$, there does not appear to be any simple structural characterization even of the minimal graphs with chromatic number more than k (see [10]). The algorithmic problem of k-colorability is well known to be NP-complete for $k \geq 3$, and was one of Karp's celebrated list [54] of 21 NP-complete problems; indeed, for $\varepsilon > 0$, it is NP-hard even to approximate the chromatic number within a factor $n^{1-\varepsilon}$, where n is the number of vertices. As with many NP-hard problems, graph coloring has a correspondingly deep combinatorial theory, and it has been the focus of extensive study in both mathematics and computer science, and understanding the connections between graph structure and chromatic number has been one of the fundamental goals of structural graph theory in the last 30 years.

Let us clarify the notion of *substructure*. A graph H is a *subgraph* of a graph G if $V(H) \subseteq V(G)$ and $E(H) \subseteq E(G)$. Thus H is obtained from G by deleting vertices and edges. We say that H is an *induced subgraph of* G if $V(H) \subset V(G)$, and $E(H)$ consists of the edges of G that are contained in $V(H)$ (and then H is the subgraph of G *induced by* $V(H)$). For example, every graph is a subgraph of some complete graph; but if G is a complete graph then all of its induced subgraphs are complete graphs. In this paper we will be concerned primarily with *induced* subgraphs. We say that a graph G is H*-free* if G does not contain an induced subgraph that is isomorphic to H (more informally, if G does not contain an *induced copy* of H).

So, what can we say about the induced subgraphs of a graph with large chromatic number? One possibility is that G might itself be complete, in which case it only contains complete graphs as induced subgraphs. But what if G does not contain a large complete subgraph: are there particular structures that have to appear as induced subgraphs? In this paper we will be interested in statements of the form:

> *Every graph with sufficiently large chromatic number contains either a complete subgraph on k vertices or an induced ***.*

Equivalently, we will often say:

> *If G contains neither a complete subgraph on k vertices nor an induced *** then it has bounded chromatic number.*

The question is: what can we put in place of the asterisks?

The rest of this paper is organized as follows. In Section 2 we look at whether graphs with large chromatic number need to contain large complete subgraphs (they do not). In the next few sections, we investigate the relationship between chromatic number and clique number: after discussing perfect graphs in Section 3, we introduce χ-bounded classes in Section 4 and look at the effects of forbidding a single induced subgraph. In Section 5 we look at induced cycles in graphs of large chromatic number, and then Section 6 considers more complex induced subgraphs. Section 7 discusses the Erdős-Hajnal Conjecture, and Section 8 looks at its connection with polynomially χ-bounded classes. Finally, in Section 9, we compare the effects of excluding induced subgraphs with the effects of excluding graph minors.

2. GIRTH AND CHROMATIC NUMBER

Suppose that a graph G has huge chromatic number. Are there induced subgraphs that it must contain? Perhaps the first question of this type to ask is: does every graph of large chromatic number contain a large complete subgraph? This question was answered in the negative in the 1940s by Tutte (writing as Blanche Descartes [33, 34]), who showed that there are triangle-free graphs with arbitrarily large chromatic number. Many constructions are now known. For example, there is a simple construction of Mycielski from the 1950s [66]: given a graph G with vertices $\{v_1, \dots, v_k\}$ we define a new graph $M(G)$ with vertices $\{x_1, \dots, x_k, y_1, \dots, y_k, z\}$; for each edge $v_i v_j$ of G, the graph $M(G)$ has edges $x_i x_j$, $y_i x_j$ and $x_i y_j$ (but not $y_i y_j$), and z is adjacent to all the y_i. It is straightforward to check that $\chi(M(G)) = \chi(G) + 1$, and if G is triangle-free then so is $M(G)$. Thus starting with $G_1 = K_1$

and inductively defining $G_{i+1} = M(G_i)$, we obtain a sequence of triangle-free graphs G_i with $\chi(G_i) = i$ for each i.

So graphs of large chromatic number need not have large complete subgraphs. But perhaps they must have short cycles? Or can they instead be "locally tree-like"? It turns out that the latter is the case. A *cycle of length* $k \geq 3$ is a graph with vertices $x_1, \ldots, x_k$ and edges $x_i x_{i+1}$ for each i (where indices are taken modulo k). The *girth* of a graph is the smallest k such that G contains a cycle of length k as a subgraph. In one of the earliest applications of probability in graph theory, Erdős [35] showed that there are graphs with arbitrarily large girth and chromatic number. Indeed, consider a random graph on n vertices in which every edge is present with probability $(\log n)/n$. A simple first moment argument shows that, with positive probability, only $o(n)$ vertices are contained in short cycles, while any stable set has size at most $o(n)$ (so the chromatic number is large, as a coloring is a partition into stable sets). Deleting all vertices in short cycles gives the desired graph.

Constructing *explicit* examples of graphs with large girth and chromatic number is rather harder. There is a pretty example of a graph with large chromatic number and no short *odd* cycles: the *Kneser graph* $K(n, k)$ has as its vertex set all k-sets contained in $\{1, \ldots, n\}$, with A and B adjacent if and only if they are disjoint [57]. It is easy to check that if $n = 2k + t$ then there are no odd cycles of length less than about n/t. It is rather harder to show that Kneser graphs can have large chromatic number: in fact, it turns out that for $k > n/2$, the Kneser graph $K(n, k)$ has chromatic number exactly $n - 2k + 2$. This was proved in a celebrated paper of Lovász [63], which developed the connection between chromatic number and the topology of the neighborhood complex of a graph; shortly afterwards, Bárány [6] found a second beautiful (and surprisingly simple) topological proof.

There are now a number of explicit constructions of graphs with large chromatic number and *no* short cycles. These include a construction of Lovász [62]; the Ramanujan graphs of Lubotzky, Phillips, and Sarnak [64]; a construction of Nešestřil and Rödl [67] using the "amalgamation method"; and an ingenious recent construction of Alon, Kostochka, Reiniger, West, and Zhu [2] based on careful augmentation of trees.

Even here, though, there are basic questions that remain. For example, the following question of Erdős and Hajnal [36] concerning (not necessarily induced) subgraphs has been open for 50 years.

Conjecture 2.1. *For every pair of positive integers k, t, every graph of sufficiently large chromatic number contains a subgraph with chromatic number more than t and girth more than k.*

The best current result is due to Rödl [72], who proved the conjecture for $k = 3$.

3. PERFECT GRAPHS

Recall that the *clique number* $\omega(G)$ of a graph G is the maximum number of vertices in a complete subgraph in G. Every graph has chromatic number at least as large as its clique number, as the vertices in any clique must all have different colors in a proper coloring. But when is the chromatic number larger than the clique number?

Here are two examples where this happens:

- Let C be an cycle of odd length. Then $\chi(C) = 3$ and (unless C is a triangle) $\omega(C) = 2$.
- Let $\overline{C}$ be the complement of a cycle of odd length. Then it can be checked that (unless C is a triangle) $\chi(\overline{C}) > \omega(\overline{C})$.

Let us say that an induced subgraph of a graph G is a *hole (in G)* if it is a cycle of length at least four, and an *antihole* if it is the complement of a cycle of length at least four (or, equivalently, if it corresponds to a hole in the complement of G). A hole or antihole is *odd* if it has an odd number of vertices. In the 1960s, Claude Berge [7] conjectured that the minimal graphs with chromatic number larger than clique number are precisely the odd holes and odd antiholes. This became known as the Strong Perfect Graph Conjecture, and was a central problem in structural graph theory for many years.

The conjecture was finally resolved by Chudnovsky, Robertson, Seymour, and Thomas in 2006 [19]:

Theorem 3.1. *If the chromatic number of G is larger than its clique number, then G contains an odd hole or an odd antihole.*

The proof of the Strong Perfect Graph Theorem is a tour de force of structural techniques. The details are rather complicated, but the strategy of the proof is to show that if G has no odd holes and no odd antiholes, then either it belongs to one of a small number of well-understood "basic classes" of graphs, or it has a "nice" decomposition into smaller graphs.

This type of approach is frequently used in structural graph theory and has been remarkably successful in understanding a wide variety of graph classes, but it is only effective when the classes being examined have some sort of nice structure. In the rest of this paper, we will mostly be interested in larger classes, where it is unlikely that there are nice decomposition results, and so very different techniques need to be used.

4. χ-BOUNDED CLASSES AND THE GYÁRFÁS–SUMNER CONJECTURE

The Strong Perfect Graph Theorem characterizes when the chromatic number $\chi(G)$ is larger than the clique number $\omega(G)$, but what induced subgraphs can we get when the chromatic number is *much* larger than the clique number? In the 1980s, András Gyárfás wrote an influential paper, *Problems from the world surrounding perfect graphs*, in which he initiated the systematic investigation of this question, using the language of χ-bounded classes. Gyárfás laid out a research programme for the study of χ-bounded classes and made a sequence of important conjectures, many of which have been resolved only in the last few years.

We will always be concerned with *hereditary* classes, namely those that are closed under taking induced subgraphs.

Definition 4.1. A hereditary class $\mathcal{G}$ of graphs is *χ-bounded* if there is a function $f : \mathbb{N} \to \mathbb{N}$ such that $\chi(G) \leq f(\omega(G))$ for every $G \in \mathcal{G}$ (see [46,79]).

The class of all graphs is not χ-bounded, as there are triangle-free graphs with arbitrarily large chromatic number (so we cannot even define $f(3)$). So any χ-bounded class must exclude at least one induced subgraph. In this section, we look at the question: when is it enough to forbid a single induced subgraph H?

Let us fix a graph H. If H contains a cycle C then the class of H-free graphs is not χ-bounded: we know from Section 2 that there are graphs with arbitrarily large girth and chromatic number (if their girth is more than the length of C, they do not contain a copy of H). So the interesting case is when H is acyclic, i.e., a *forest*. This is the subject of the well-known *Gyárfás–Sumner Conjecture* [45,84]:

Conjecture 4.2. *For every forest H the class of H-free graphs is χ-bounded.*

The Gyárfás–Sumner Conjecture can equivalently be stated as follows:

Conjecture 4.3. *For every forest H and every $k \geq 1$, every graph with sufficiently large chromatic number contains either a complete graph on k vertices or an induced copy of H.*

The conjecture has proved extremely resistant. It is not hard to show that it suffices to consider the case when H is a tree (for a forest F, the class of F-free graphs is χ-bounded if and only if the class of H-free graph is χ-bounded for every component H of F). But the conjecture is only known for a few quite special trees, for example:

- If H is a star, then it follows easily from Ramsey's theorem.
- If H is a path then there is a simple and elegant argument due to Gyárfás [45]. It is also known for the broom [45] and the double broom [22].
- It is true if H is a tree of radius 2: the triangle-free case was proved by Gyárfás, Szemerédi, and Tuza [48], and the general case by Kierstead and Penrice [55].
- It is known for some special trees of radius three [56,78].

In most cases, the proofs are quite intricate. However, the argument when H is a path is simple and elegant, so let us sketch it. Suppose we are looking for a path P with t vertices, and G is a graph with huge chromatic number that does not contain an induced copy of P or a complete subgraph on k vertices. By induction, we may assume that for every vertex v in G, its neighborhood $N(v)$ has small chromatic number (as it does not contain a complete subgraph on $k-1$ vertices). We may also assume that G is connected, by just considering the component with largest chromatic number. Now choose any vertex x_1. If we delete x_1 and its neighbors from G, then the remaining graph falls into components $C_1, \ldots, C_r$ for some r, and as the neighborhood of x_1 has small chromatic number and G has large chromatic number, one of the these components (say C_1) must also have large chromatic number. Since G is connected, there must be some vertex x_2 that is both adjacent to x_1 and has a neighbor in C_1. We focus on x_2 and C_1 and repeat the argument, deleting neighbors of x_2 from C_1, choosing a component C of the remainder with large chromatic number, and choosing a vertex x_3 that is adjacent to x_2 and has neighbors in C. Continuing in this way, we walk into the graph, always heading towards a region with large chromatic number, and build an induced path along the way. This type of argument crops up repeatedly, and has become known as the *Gyárfás path argument*.

Another (rather more complicated) technique is the *method of templates*. Building on the work of Gyárfás, Szemerédi, and Tuza **[48]**, this was developed by Kierstead and Penrice **[55]**. The idea is to look for complete multipartite subgraphs with (large) constant size. Thus we look for a large complete bipartite graph, or a (slightly less) large complete tripartite graph, and so on. The process terminates as we are assuming that there is no complete graph on k vertices. A *template* T consists of one of these subgraphs, say K, together with all the vertices that are moderately dense to K (in some appropriate sense). We define a sequence $T_1, T_2, \dots$ of templates by repeatedly choosing one that is maximal (by some measure), deleting it from the graph, and then looking at templates in the graph that remains. When we are finished, we are left with a graph containing no templates, and show that it has small chromatic number. The key now is to show that edges between the templates we removed are rather restricted: there is usually quite a complex argument to partition and "clean up" the templates into progressively more simply structured pieces, until all the pieces have small chromatic number (and so we are done, as we have partitioned the whole graph into a bounded number of pieces with small chromatic number).

The method of templates has been a powerful approach for handling small-radius trees, but at present there seem to be significant technical obstacles to extending it to trees of radius more than 3. It is worth noting that the base case (finding complete bipartite graphs) is not straightforward, but can usually be handled with the following result (proved by Rödl but not published):

Theorem 4.4. *For every k and t, and every tree T, every graph with sufficiently large chromatic number contains either K_k, $K_{t,t}$ or T as an induced subgraph.*

Here K_k denotes the complete graph on k vertices, and $K_{t,t}$ denotes the complete bipartite graph with t vertices in each class. Kierstead and Penrice **[55]** strengthened Theorem 4.4, showing that such graphs have bounded degeneracy.[1] See Theorem 8.5 below for a further strengthening.

Perhaps the most general result related to the Gyárfás–Sumner conjecture concerns *induced subdivisions* of forests. We say that a graph F is a *subdivision* of a graph H (or is *homeomorphic to* H) if F can be obtained from H by adding vertices along the edges, or equivalently by replacing some subset of the edges by paths. For example, every cycle is a subdivision of a triangle. The following weakening of the Gyárfás–Sumner conjecture was obtained in **[73]**.

---

1 The *degeneracy* of a graph G is the maximum integer r such that every subgraph of G has a vertex with degree at most r. The chromatic number of a graph is at most one more than its degeneracy.

Theorem 4.5. *Let H be a forest. The class of graphs that do not contain an induced copy of any subdivision of H is χ-bounded.*

A special case of this implies the Gyárfás–Sumner conjecture when H is a subdivision of a star (as any subdivision of H contains an induced copy of H).

In fact, something slightly stronger than Theorem 4.5 was shown in [73]:

Theorem 4.6. *For every forest H there is a finite list $H_1, \ldots, H_t$ of subdivisions of H such that the class of graphs that do not contain an induced copy of any H_i is χ-bounded.*

5. HOLES IN GRAPHS OF LARGE CHROMATIC NUMBER

What other structures must appear in graphs of large chromatic number? If we do not forbid a forest, then the existence of graphs with large girth and chromatic number implies that it is not enough to forbid a single induced subgraph, or indeed any finite list of induced subgraphs. Perhaps the simplest example of this is where we forbid a collection of induced holes (i.e., induced cycles of length at least four). Gyárfás made several important conjectures concerning holes, and we will focus on these in this section.

The Strong Perfect Graph Theorem tells us that the class of graphs with no odd holes and no odd antiholes is χ-bounded, and furthermore with the best possible function $f(\omega) = \omega$. Long before this theorem was proved, Gyárfás conjectured that for χ-boundedness it would suffice to exclude only odd holes.

Conjecture 5.1. *The class of graphs with no odd holes is χ-bounded.*

He also conjectured that it would be enough to exclude only long holes; and more adventurously that it would be enough to exclude *long* odd holes:

Conjecture 5.2. *For every integer t, the class of graphs with no holes of length more than t is χ-bounded.*

Conjecture 5.3. *For every integer t, the class of graphs with no odd holes of length more than t is χ-bounded.*

For some time, there was no progress on these conjectures. As noted earlier, the structural techniques used to prove the Strong Perfect Graph theorem rely on the fact that perfect graphs have nice structural features, and a minimum counterexample to the theorem can be decomposed in some nice way. The larger classes considered by Gyárfás have much wilder structure, and do not appear to be amenable to decomposition techniques. So a different approach is required.

For a long time, all three conjectures appeared intractable. However, the three conjectures have now been proved: the first was proved in a paper with Seymour [74], giving the following bound.

Theorem 5.4. *For $k \geq 1$, every graph with chromatic number at least $2^{2^{k+2}}$ contains either a complete subgraph on k vertices or an odd hole.*

The doubly exponential bound is probably far from best possible. Indeed, there is only one obstacle in the proof that causes the bound to jump from single to double exponential, and it seems likely that this could be circumvented with new ideas. One approach would be to prove the Hoàng–McDiarmid conjecture [50], which says:

Conjecture 5.5. *Let G be a graph with no odd hole and at least one edge. Then the vertices of G can be partitioned into two sets such that every maximum clique in G intersects both sets.*

Conjecture 5.5 would imply immediately that graphs without odd holes satisfy $\chi(G) \leq 2^{\omega(G)}$.

The class of graphs without odd holes has also been of significant algorithmic interest. Following the proof of the Strong Perfect Graph Theorem, Chudnovsky, Cornuéjols, Liu, Seymour, and Vušković [18] showed that there is a polynomial-time algorithm to recognize perfect graphs (i.e., graphs with no odd hole and no odd antihole). However, it was only recently shown that it was shown that there is a polynomial-time algorithm to test for the presence of an odd hole [26]; indeed, the problem had been open since the 1980s (and there was reason to expect that the problem might be hard, as Bienstock showed that testing for the presence of an odd hole containing a specific vertex is NP-complete [8]). In subsequent work, it has been shown that finding a shortest odd hole [23] and an odd hole of at least a fixed length [25] can also be solved in polynomial time.

Conjectures 5.2 and 5.3 have also now been proved. The second conjecture was proved in a paper with Chudnovsky and Seymour [21], and the third in a paper with Chudnovsky, Seymour, and Spirkl [27] (both with significantly larger bounds). But this raises a natural further question: why ask only for *odd* holes? In the light of the (then) Strong Perfect Graph Conjecture, it was very natural for Gyárfás and others to think about holes of odd or even parity. However, motivated by topological considerations, Kalai and Meshulam [52, 53] also conjectured that the class of graphs with no triangle and no hole of length divisible by 3 does not contain graphs of arbitrarily large chromatic number. This was proved in a breakthrough paper of Bonamy, Charbit, and Thomassé [12].

It turns out that much stronger results hold. The current state-of-the-art is the following, which was proved in a paper with Seymour [76].

Theorem 5.6. *For all integers $k \geq 0$ and $\ell \geq 1$, the class of graphs with no hole of length k modulo ℓ is χ-bounded.*

As an application, this resolves two further conjectures of Kalai and Meshulam, connecting the chromatic number of a graph with the homology of its independence complex.

It seems likely that even stronger results are true. Indeed, perhaps we can break away from parity conditions altogether and just use some sort of density condition:

Conjecture 5.7. *Let $A \subset \mathbb{N}$ be an infinite set with bounded gaps. Then the class of graphs that do not contain a hole of any length in A is χ-bounded.*

This has been proved in the special case of triangle-free graphs (in another paper with Seymour [75]). The proof is long and complicated, and extending it to the general case will require significant new ideas.

It would also be very interesting to answer the following question:

Conjecture 5.8. *Is there a set $A \subset \mathbb{N}$ with upper density 0 such that the class of graphs that do not contain any hole with length in A is χ-bounded?*

What can be said about the techniques? Proving these results has required a substantially different toolbox from the decomposition techniques used to study perfect graphs. The methods use a mixture of structural and extremal techniques, and can perhaps be thought of as a "rougher structural" approach.

A useful framework is provided by using the *local chromatic number*. For an integer $r \geq 0$ and a graph G, we define the *r-local chromatic number $\chi^{(r)}(G)$ of G* to be the maximum of $\chi(B)$ over all subgraphs B induced by r-balls in G (using the shortest path metric). The relationship between $\chi^{(r)}(G)$ and $\chi(G)$ is interesting: roughly speaking, it is interesting to distinguish between graphs in which some small ball has large chromatic number, and graphs where the chromatic number is not visible locally (for instance, if the graph is locally treelike) so that it is somehow "spread out" in the graph. More precisely, given any graph G with very large chromatic number, it is possible to drop to an induced subgraph G' with one of the two following properties:

- G' has large chromatic number and small r-local chromatic number;
- G' has large chromatic number, and every induced subgraph of G' with large chromatic number contains an r-ball with large chromatic number.

This framework was introduced in [73], and has been the starting point for many subsequent proofs. The "local" and "spread-out" cases have very different structural behaviors, and usually require very different methods.

6. INDUCED SUBDIVISIONS AND GEOMETRIC CONSTRUCTIONS

So far, we have discussed forests and cycles. But it is natural to ask whether we can ask for more complicated local structures. In 1997, it was conjectured in [73] that if we allow subdivisions, then any structure can be found:

Conjecture 6.1. *For every graph H, the class of graphs with no induced subdivision of H is χ-bounded.*

Equivalently, the conjecture claims that any graph with large chromatic number contains either a large clique or an induced subdivision of H. When H is a forest, this is true by Theorem 4.5 [73]; and when H is a cycle, this follows from the truth of Conjecture 5.2 [21, 46]. Motivated by Conjecture 6.1, Kühn and Osthus [58] also proved the following beautiful result, showing that if we forbid a complete *bipartite* graph then large minimum degree is already enough.

Theorem 6.2. *For every graph H and positive integer k, every graph with sufficiently large minimum degree contains either a complete graph K_k, a complete bipartite graph $K_{k,k}$ or a subdivision of H as an induced subgraph.*

As we will see below, Conjecture 6.1 ultimately turned out to be incorrect, but it remained open for more than 15 years. Part of the difficulty in finding a counterexample lies in the fact that we do not have many ways to generate structured examples of graphs with large chromatic number. For, example, random graphs provide a simple way to create graphs with large chromatic number; but typically they also have good expansion and connectivity properties, and contain subdivisions of any fixed graph H. And while there are many ways to construct examples of graphs with large chromatic number and (for example) no triangles, it is similarly hard to constrain their larger-scale structure.

One fruitful line of construction has come from considering geometric graphs. It is not enough to consider graphs embeddable on a fixed surface, as these have bounded chromatic number (this can be deduced easily from Euler's formula, which implies that graphs embeddable on a fixed surface have bounded degeneracy). But it is rather more interesting to consider *intersection graphs*: these have vertex set consisting of a family $\mathcal{C}$ of sets, with $A, B \in \mathcal{C}$ adjacent if $A \cap B$ is nonempty.

An important example is given by the intersection graph of a collection of axis-aligned boxes in $\mathbb{R}^d$. When $d = 1$, we obtain the family of *interval graphs*. These are well-known to be perfect [49]. When $d = 2$, we are considering intersections of rectangles in the plane: Asplund and Grünbaum [5] showed that these satisfy $\chi(G) = O(\omega(G)^2)$ (recently improved to $O(\omega \log \omega)$ by Chalermsook and Walczak [15]). However, in three dimensions more happens: Burling constructed triangle-free intersection graphs of boxes in three dimensions with arbitrarily large chromatic number ([14]; see also [69]). It follows that intersection graphs of families of boxes in d-dimensions are χ-bounded for $d = 1, 2$, but not for $d \geq 3$.

A larger class of two-dimensional intersection graphs is provided by the family of *string graphs*, namely intersection graphs of curves in the plane (see, for example, [65]). Many special families of string graphs have been of interest. For example, the intersection graphs of straight line segments in the plane: Erdős asked in the 1970s whether this family is χ-bounded. It was a surprise when, in 2014, Pawlik, Kozik, Krawczyk, Lasoń, Micek, Trotter, and Walczak [68] came up with a way to represent Burling's graphs in two dimensions. Their beautiful construction shows the following.

Theorem 6.3. *There are triangle-free intersection graphs of line segments in the plane that have arbitrarily large chromatic number.*

As a corollary, Conjecture 6.1 does not hold (for example, for any graph H that is obtained by subdividing every edge of a nonplanar graph). However, it remains an interesting problem to determine when the conjecture does hold. Chalopin, Esperet, Li, and Ossona de Mendez [16] analyzed the construction from [68] in detail, further limiting the graphs that could satisfy Conjecture 6.1. In the case of string graphs, the problem was completely solved in [24], which also proved the following result:

Theorem 6.4. *Every string graph with large chromatic number contains a 2-ball with large chromatic number.*

Perhaps this is a necessary feature in any family of counterexamples to Conjecture 6.1? The following resuscitation of that conjecture is proposed in [77]. Informally:

Conjecture 6.5. *For every graph H, every graph with large chromatic number contains either a 2-ball with large chromatic number or an induced subdivision of H.*

7. THE ERDŐS–HAJNAL CONJECTURE

In this section, we look at the largest complete subgraph or independent set in a graph. Frank Ramsey [70] showed in 1930 that every infinite graph contains an infinite complete subgraph or stable set. The finite version of this result is the following:

Theorem 7.1. *For every $k \geq 1$ there is an integer $R(k)$ such that every graph with at least $R(k)$ vertices contains a complete subgraph or stable set of size k.*

So "large" graphs contain "large" homogeneous structures. But how large is large? Ramsey gave an explicit bound on $R(k)$, but a nice quantitative version of Ramsey's Theorem was proved by Erdős and Szekeres [40] in the 1930s:

Theorem 7.2. *Every graph with at least $\binom{s+t-2}{s-1}$ vertices contains either a complete subgraph of size s or a stable set of size t.*

By taking $s = t$, it follows that every graph on n vertices contains a complete subgraph or stable set of size at least $c_1 \log n$. On the other hand, by considering random graphs, it is not hard to show that most graphs on n vertices do not contain a complete subgraph or stable set of size more than $c_2 \log n$.

How does the picture change if we know something about the local structure of a graph? Erdős and Hajnal speculated in the 1980s [37, 38] that H-free graphs exhibit a very different behavior:

Conjecture 7.3. *For every graph H, there is a constant $c = c(H) > 0$ such that the following holds: every H-free graph with n vertices has a complete subgraph or stable set with at least n^c vertices.*

In other words, if we exclude any induced subgraph then the largest stable set or complete subgraph that must occur jumps in size from logarithmic to polynomial. The Erdős–Hajnal Conjecture has become one of the central conjectures in graph theory.

Despite considerable work, Conjecture 7.3 is only known for a small family of graphs. There are a few small examples: complete graphs (this follows from the quantitative form of Ramsey's Theorem 7.2), the four-vertex path P_4 (P_4-free graphs are perfect), and the bull (Chudnovsky and Safra [28]). The class of graphs H for which the Erdős–Hajnal Conjecture holds also satisfies two closure properties:

- It follows immediately from the statement of the conjecture that if it holds for H then it also holds for $\overline{H}$.
- Alon, Pach, and Solymosi [3] proved that the class of graphs H for which Erdős–Hajnal holds is closed under substitution (the operation of *substituting* a graph F for a vertex x of H deletes x and replaces it with a copy of F; every new vertex is joined to every vertex that was adjacent to x).

Recently, a new graph was added to the list [29]:

Theorem 7.4. *The Erdős–Hajnal Conjecture holds when H is a cycle of length 5.*

This was of particular interest, as it had been highlighted as an important case both by Erdős and Hajnal [38] and Gyárfás [47], and was part of the original motivation for the conjecture. However, the conjecture remains open even for the five vertex path and the best bound known for general graphs is due to Erdős and Hajnal [38], who showed that the conjecture holds with $e^{c\sqrt{\log n}}$ in place of n^c.

There has been substantial recent progress in looking at analogues of the Erdős–Hajnal Conjecture with more than one excluded graph. A hereditary class $\mathcal{G}$ of graphs has the *Erdős–Hajnal property* if there is $c > 0$ such that every $G \in \mathcal{G}$ has a stable set or complete subgraph with at least $|G|^c$ vertices. Thus the Erdős–Hajnal Conjecture says that the class of H-free graphs has the Erdős–Hajnal property.

One approach to proving that graph classes satisfy the Erdős–Hajnal property has been through looking at large *bipartite* structures. Disjoint sets A, B of vertices in a graph G are *complete* if G contains all edges between A and B and *anticomplete* if G contains no edges between A and B. There is a substantial body of work on finding this type of structure in various graph classes (see [32, 39] and the sequence of papers starting with [28]). It is particularly helpful when it is possible to find *linear* complete or anticomplete pairs. A hereditary class $\mathcal{G}$ of graphs has the *strong Erdős–Hajnal property* if there is $\delta > 0$ such that every $G \in \mathcal{G}$ has disjoint sets A, B of at least δn vertices such that the pair (A, B) is either complete or anticomplete.

The strong Erdős–Hajnal property is useful for the following reason:

Lemma 7.5. *The strong Erdős–Hajnal property implies the Erdős–Hajnal property.*

So when does the strong Erdős–Hajnal property hold for the class of H-free graphs? By considering sparse random graphs (for instance, with $p \sim \log n/n$), it can be seen that H must be a forest; on the other hand, by considering complements of sparse random graphs, it follows that the complement of H must also be a forest. But if both H and its complement are forests, then H has at most four vertices, and the conjecture is already known for these cases. So it seems that the strategy gives us nothing. But here is an interesting result of Bousquet, Lagoutte, and Thomassé [13]:

Theorem 7.6. *For every positive integer t, the class of graphs G such that neither G nor its complement contains a t-vertex path as an induced subgraph satisfies the strong Erdős–Hajnal property.*

Thus the strong Erdős–Hajnal property holds if we exclude *two* graphs: one sparse (a path on t vertices) and one dense (the complement of a path on t vertices). Theorem 7.6 was extended by Choromanski, Falik, Liebenau, Patel, and Pilipczuk [17], and then further by Liebenau, Pilipczuk, Seymour, and Spirkl [60]. An optimal result was given in [28]:

Theorem 7.7. *Let T be a forest. Then the class of graphs G such that neither G nor its complement contains an induced copy of T satisfies the strong Erdős–Hajnal property.*

Since we must exclude both a forest and the complement of a forest to obtain the strong Erdős–Hajnal property, the result characterizes all hereditary classes that are defined by a finite set of excluded subgraphs and satisfy the strong Erdős–Hajnal property. (See [30] for an analogous result where we forbid all induced subdivisions of a single graph H in both G and its complement.)

We end the section by noting that there is a natural connection between the Erdős–Hajnal Conjecture and problems about χ-boundedness such as the Gyárfás–Sumner Conjecture: a graph with small chromatic number must contain large stable sets (as a coloring is a partition into stable sets); and the Erdős–Hajnal Conjecture tells us that H-free graphs have "large" cliques or stable sets. However, there is no immediate implication. For example, the class of triangle-free graphs has the Erdős–Hajnal Property, but contains graphs of arbitrarily large chromatic number. And Theorem 5.4 shows that the class of graphs with no odd holes is χ-bounded, but the bounds do not imply anything like polynomial behavior of cliques or stable sets. However, under some conditions it is possible to deduce the Erdős–Hajnal Property from χ-boundedness: we will discuss this in the next section.

8. POLYNOMIAL BOUNDS AND ESPERET'S CONJECTURE

So far, we have discussed classes in which the chromatic number is bounded as a function of the clique number, without considering what type of function provides the bound. In most of the results we have mentioned, the proofs give multiply exponential functions, either because there are repeated applications of Ramsey-type results, or because there is some blowup at the inductive step. In this section, we will be concerned with *polynomially χ-bounded* classes, namely classes $\mathcal{G}$ for which there is a *polynomial* function f such that $\chi(G) \leq f(\mathrm{cl}(G))$ for every $G \in \mathcal{G}$. Polynomially χ-bounded classes are of particularly interest because of their connection to the Erdős–Hajnal Conjecture: it follows immediately that any polynomially χ-bounded class has the Erdős–Hajnal property.

Esperet [41] made the remarkable (and provocative) conjecture that all χ-bounded classes are polynomially χ-bounded:

Conjecture 8.1. *If a hereditary class $\mathcal{G}$ is χ-bounded then it is polynomially χ-bounded.*

If Esperet's conjecture is true, then the Gyárfás–Sumner could be strengthened to the following:

Conjecture 8.2. *For every forest H, the class of H-free graphs is polynomially χ-bounded.*

Since the Gyárfás–Sumner Conjecture is only known for some small families of trees, the polynomial Gyárfás–Sumner Conjecture looks very challenging (and may well turn out to be incorrect). However, there has been some progress, and it is known for a few very small trees [81]:

Theorem 8.3. *The polynomial Gyárfás–Sumner Conjecture holds for every tree of diameter at most 3.*

Paths form a particularly interesting case. Let P_k be the path on k vertices. Graphs that are P_3- or P_4-free are well known to be perfect, so the polynomial Gyárfás–Sumner Conjecture follows immediately. However, in general, the best bounds are exponential, even when excluding paths. The current borderline case is the five vertex path, where until recently the best bound was exponential [42]. This was improved in [82]:

Theorem 8.4. *Every graph with chromatic number at least $k^{\log_2 k}$ contains either a clique on k vertices or an induced path on five vertices.*

This is just slightly superpolynomial, but it is not yet small enough to prove the Erdős–Hajnal Conjecture for P_5.

Polynomial bounds are also known when a tree and a complete bipartite graph are excluded. The following result [80] strengthens Theorem 4.4 and answers a question of Bonamy, Bousquet, Pilipczuk, Rzazewski, Thomassé, and Walczak [11]:

Theorem 8.5. *For every tree T, there is a polynomial $p(t)$ such that, for every $t \geq 1$, every graph with minimum degree at least $p(t)$ contains either an induced copy of T or a (not necessarily induced) copy of $K_{t,t}$.*

It seems likely that even more could hold. Indeed, Paul Seymour and I conjecture the following strengthening of Theorem 6.2:

Conjecture 8.6. *For every graph H there is a polynomial $p(t)$ such that, for every $t \geq 1$, every graph with minimum degree at least $p(t)$ contains either an induced subdivision of H or a (not necessarily induced) copy of $K_{t,t}$.*

9. GRAPH MINORS AND INDUCED SUBGRAPHS

Throughout this paper, we have been looking at the large-scale structural consequences of forbidding one or more induced subgraphs. In this final section, we compare this with the effects of excluding graph minors. A graph H is a *minor* of a graph G if H can be obtained from G by contracting edges and deleting edges and vertices (a *contraction* of an edge xy replaces the vertices x and y by a single vertex z adjacent to all other vertices that were previously adjacent to x or to y; for simplicity, we will ignore loops and parallel edges). A class $\mathcal{G}$ of graphs is *minor-closed* if, whenever $G \in \mathcal{G}$ then all its minors are in $\mathcal{G}$.

Minor-closed classes arise in many contexts: for example, the class of all graphs embeddable on a fixed surface is minor-closed. For the plane, Wagner [86] proved the following result (which also follows from work of Kuratowski [59]):

Theorem 9.1. *A graph is planar if and only if it does not contain a minor of K_5 or $K_{3,3}$.*

The theory of graph minors was developed in a major series of papers by Robertson and Seymour. A celebrated result in this theory is the following [71]:

Theorem 9.2. *Let $G_1, G_2, \ldots$ be an infinite sequence of graphs. Then there are $i < j$ such that G_i is a minor of G_j.*

In other words, finite graphs are well-quasiordered under the excluded minor relation. A corollary of this is a vast extension of Theorem 9.1: for any class $\mathcal{G}$ of graphs that is closed under minors, there is a finite set $\mathcal{M}$ of graphs such that a graph G is in $\mathcal{G}$ if and only if it does not contain any graph in $\mathcal{M}$ as a minor. In other words, any minor-closed class has a finite set of minimal excluded minors.

A central result in the theory of graph minors is the *Graph Minor Structure Theorem*, which states (very roughly) that for every fixed graph H, any H-minor-free graph can be obtained by gluing together (in a treelike way) a sequence of graphs that can (almost) be embedded in surfaces of bounded genus. This is not a structural description, but can be thought of as an approximate structure theorem: the class $\mathcal{G}$ of H-minor-free graphs is contained in a class $\mathcal{G}'$ in which the graphs can all be built in a certain way, and which does not contain graphs that are much more "complex" than H.

So can anything similar be said for induced subgraphs? The class of finite graphs is not well-quasiordered by the induced subgraph relation: consider, for example, the class of cycles. So no theorem directly analogous to Theorem 9.2 can hold (see, for example, [61] for further discussion). On the positive side, there are good structural descriptions of H-free graphs for some very small H, although precise structural descriptions look intractable for larger H. For arbitrary H, it is known that every H-free graph can be partitioned into a bounded number of pieces that are either dense or sparse [31]; and there is a great deal known about the structure of *typical* H-free graphs (see, for instance, [1]). But what is really missing is an analogue for induced subgraphs of the Graph Minor Structure Theorem.

At the moment, it is not yet clear what such a theorem would say: what would the "basic" graph classes be? How would they be glued together? And would the theory describe the whole graph, or just some suitably well-structured "core"? But such a theorem could draw together a large body of work, and would have widespread applications. An essential part of this theory will be understanding the relationship between chromatic number and induced subgraphs; the size of cliques and independent sets will also be crucial. The Gyárfás–Sumner and Erdős–Hajnal Conjectures are major challenges in our understanding of induced subgraphs, and resolving either of them would be a substantial milestone in the development of a more general theory.

ACKNOWLEDGMENTS

We thank Paul Seymour, for helpful comments and discussion.

FUNDING

Research supported by EPSRC grant EP/V007327/1.

REFERENCES

[1] N. Alon, J. Balogh, B. Bollobás, and R. Morris, The structure of almost all graphs in a hereditary property. *J. Combin. Theory Ser. B* **101** (2011), 85–110.

[2] N. Alon, A. Kostochka, B. Reiniger, D. West, and X. Zhu, Coloring, sparseness, and girth. *Israel J. Math.* **214** (2016), 315–331.

[3] N. Alon, J. Pach, and J. Solymosi, Ramsey-type theorems with forbidden subgraphs. *Combinatorica* **21** (2001), 155–170.

[4] K. Appel and E. Haken, Planar map is four colorable. I. Discharging. *Illinois J. Math.* **21** (1977), 429–490. Every planar map is four colorable. II. Reducibility. *Illinois J. Math.* **21** (1977), 491–567.

[5] E. Asplund and B. Grünbaum, On a coloring problem. *Math. Scand.* **8** (1960), 181–188.

[6] I. Bárány, A short proof of Kneser's conjecture. *J. Combin. Theory Ser. A* **25** (1978), 325–326.

[7] C. Berge, Färbung von Graphen, deren sämtliche bzw. deren ungerade Kreise starr sind. *Wiss. Z., Martin-Luther-Univ. Halle-Wittenb., Math.-Nat.wiss. Reihe* **10** (1961), 114.

[8] D. Bienstock, On the complexity of testing for odd holes and induced odd paths. *Discrete Math.* **90** (1991), 85–92; D. Bienstock, Corrigendum: On the complexity of testing for odd holes and induced odd paths. *Discrete Math.* **102** (1992), 109.

[9] G. Birkhoff, A determinantal formula for the number of ways of coloring a map. *Ann. of Math.* **14** (1912), 42–46.

[10] B. Bollobás, *Extremal graph theory*. Academic Press Inc., London–New York, 1978, xx+488pp.

[11] M. Bonamy, N. Bousquet, M. Pilipczuk, P. Rzazewski, S. Thomassé, and B. Walczak, Degeneracy of P_t-free and $C_{\geq t}$-free graphs with no large complete bipartite subgraphs. 2021, arXiv:2012.03686.

[12] M. Bonamy, P. Charbit, and S. Thomassé, Graphs with large chromatic number induce $3k$-cycles. 2014, arXiv:1408.2172.

[13] N. Bousquet, A. Lagoutte, and S. Thomassé, The Erdős–Hajnal conjecture for paths and antipaths. *J. Combin. Theory Ser. B* **113** (2015), 261–264.

[14] J. Burling, *On coloring problems of families of polytopes*. Ph.D. thesis, University of Colorado, 1965.

[15] P. Chalermsook and B. Walczak, Coloring and maximum weight independent set of rectangles. In *Proceedings of the 2021 ACM–SIAM Symposium on Discrete Algorithms (SODA)*, pp. 860–868, SIAM, 2021.

[16] J. Chalopin, L. Esperet, Z. Li, and P. Ossona de Mendez, Restricted frame graphs and a conjecture of Scott. *Electron. J. Combin.* **23** (2016), #P1.30.

[17] K. Choromanski, D. Falik, A. Liebenau, V. Patel, and M. Pilipczuk, Excluding hooks and their complements. *Electron. J. Combin.* **25** (2018), #P3.27.

[18] M. Chudnovsky, G. Cornuéjols, X. Liu, P. Seymour, and K. Vušković, Recognizing Berge graphs. *Combinatorica* **25** (2005), 143–186.

[19] M. Chudnovsky, N. Robertson, P. Seymour, and R. Thomas, The strong perfect graph theorem. *Ann. of Math.* **164** (2006), 51–229.

[20] M. Chudnovsky and M. Safra, The Erdős–Hajnal conjecture for bull-free graphs. *J. Combin. Theory Ser. B* **98** (2008), 1301–1310.

[21] M. Chudnovsky, A. Scott, and P. Seymour, Induced subgraphs of graphs with large chromatic number. III. Long holes. *Combinatorica* **37** (2017), 1057–72.

[22] M. Chudnovsky, A. Scott, and P. Seymour, Induced subgraphs of graphs with large chromatic number. XII. Distant stars. *J. Graph Theory* **92** (2019), 237–254.

[23] M. Chudnovsky, A. Scott, and P. Seymour, Finding a shortest odd hole. *ACM Trans. Algorithms* **17** (2021), 13, 21 pp.

[24] M. Chudnovsky, A. Scott, and P. Seymour, Induced subgraphs of graphs with large chromatic number. V. Chandeliers and strings. *J. Combin. Theory Ser. B* **150** (2021), 195–243.

[25] M. Chudnovsky, A. Scott, and P. Seymour, Detecting a long odd hole. *Combinatorica* (to appear).

[26] M. Chudnovsky, A. Scott, P. Seymour, and S. Spirkl, Detecting an odd hole. *J. ACM* **67** (2020), no. 1, 5, 12 pp.

[27] M. Chudnovsky, A. Scott, P. Seymour, and S. Spirkl, Induced subgraphs of graphs with large chromatic number. VIII. Long odd holes. *J. Combin. Theory Ser. B* **140** (2020), 84–97.

[28] M. Chudnovsky, A. Scott, P. Seymour, and S. Spirkl, Pure pairs. I. Trees and linear anticomplete pairs. *Adv. Math.* **375** (2020), 107396.

[29] M. Chudnovsky, A. Scott, P. Seymour, and S. Spirkl, Erdős–Hajnal for graphs with no 5-hole. 2021, arXiv:2102.04994.

[30] M. Chudnovsky, A. Scott, P. Seymour, and S. Spirkl, Pure pairs. II. Excluding all subdivisions of a graph. *Combinatorica* **41** (2021), 379–405.

[31] M. Chudnovsky, A. Scott, P. Seymour, and S. Spirkl, Strengthening Rödl's theorem. 2021, arXiv:2105.07370.

[32] D. Conlon, J. Fox, and B. Sudakov, Recent developments in graph Ramsey theory. In *Surveys in combinatorics 2015*, pp. 49–118, Cambridge University Press, 2015.

[33] B. Descartes, A three colour problem. *Eureka* **21** (1947). (Solution March 1948).

[34] B. Descartes, Solution to Advanced Problem no. 4526. *Amer. Math. Monthly* **61** (1954), 352.

[35] P. Erdős, Graph theory and probability. *Canad. J. Math.* **11** (1959), 34–38.

[36] P. Erdős, Problems and results in combinatorial analysis and graph theory. In *Proof techniques in graph theory*, pp. 27–35, Academic Press, 1969.

[37] P. Erdős and A. Hajnal, On spanned subgraphs of graphs. In *Contributions to graph theory and its applications (Internat. Colloq., Oberhof, 1977)*, pp. 80–96, Tech. Hochschule Ilmenau, Ilmenau, 1977 (German).

[38] P. Erdős and A. Hajnal, Ramsey-type theorems. *Discrete Appl. Math.* **25** (1989), 37–52.

[39] P. Erdős, A. Hajnal, and J. Pach, A Ramsey-type theorem for bipartite graphs. *Geombinatorics* **10** (2000), 64–68.

[40] P. Erdős and G. Szekeres, A combinatorial problem in geometry. *Compos. Math.* **2** (1935), 463–470.

[41] L. Esperet, *Graph colorings, flows and perfect matchings*. Habilitation thesis, Université Grenoble Alpes (2017), 24.

[42] L. Esperet, L. Lemoine, F. Maffray, and G. Morel, The chromatic number of $\{P_5, K_4\}$-free graphs. *Discrete Math.* **313** (2013), 743–754.

[43] C. Fortuin and P. Kasteleyn, On the random-cluster model, I: Introduction and relation to other models. *Physica* **57** (1972), 536–564.

[44] G. Grimmett, *The random-cluster model*. Grundlehren Math. Wiss. 333, Springer, New York, 2006.

[45] A. Gyárfás, On Ramsey covering-numbers. In *Infinite and finite sets (Colloq., Keszthely, 1973; dedicated to P. Erdős on his 60th birthday), Vol. II*, pp. 801–816, Colloq. Math. Soc. Janos Bolyai 10, North-Holland, Amsterdam, 1975.

[46] A. Gyárfás, Problems from the world surrounding perfect graphs. In *Proceedings of the international conference on combinatorial analysis and its applications (Pokrzywna, 1985)*, pp. 413–441, Zastos. Mat. 19, 1987.

[47] A. Gyárfás, Reflections on a problem of Erdős and Hajnal. In *The Mathematics of Paul Erdős*, edited by R. L. Graham and J. Nešetřil, pp. 93–98, Algorithms Combin. II 14, Springer, Heidelberg, 1997.

[48] A. Gyárfás, E. Szemerédi, and Zs. Tuza, Induced subtrees in graphs of large chromatic number. *Discrete Math.* **30** (1980), 235–344.

[49] G. Hajós, Über eine Art von Graphen. *Internat. Math. Nachr.* **11** (1957), 65.

[50] C. Hoáng and C. McDiarmid, On the divisibility of graphs. *Discrete Math.* **242** (2002), 145–156.

[51] V. Jones, A polynomial invariant for knots via von Neumann algebras. *Bull. Amer. Math. Soc.* **12** (1985), 103–112.

[52] G. Kalai, https://gilkalai.wordpress.com/2014/12/19/when-a-few-colors-suffice/.

[53] G. Kalai, https://gilkalai.files.wordpress.com/2010/10/es.pdf.

[54] R. Karp, Reducibility among combinatorial problems. In *Complexity of computer computations*, edited by R. E. Miller, J. W. Thatcher, and J. D. Bohlinger, pp. 85–103, Plenum, New York, 1972.

[55] H. A. Kierstead and S. G. Penrice, Radius two trees specify χ-bounded classes. *J. Graph Theory* **18** (1994), 119–129.

[56] H. A. Kierstead and Y. Zhu, Radius three trees in graphs with large chromatic number. *SIAM J. Discrete Math.* **17** (2004), 571–581.

[57] M. Kneser, Aufgabe 360. *Jahresber. Dtsch. Math.-Ver.* **58** (1955), 27.

[58] D. Kühn and D. Osthus, Induced subdivisions in $K_{s,s}$-free graphs of large average degree. *Combinatorica* **24** (2004), 287–304.

[59] K. Kuratowski, Sur le probléme des courbes gauches en topologie. *Fund. Math.* **15** (1930), 271–283.

[60] A. Liebenau, M. Pilipczuk, P. Seymour, and S. Spirkl, Caterpillars in Erdős–Hajnal. *J. Combin. Theory Ser. B* **136** (2019), 33–43.

[61] C.-H. Liu and R. Thomas, Robertson's conjecture I. Well-quasi-ordering bounded tree-width graphs by the topological minor relation. 2020, arXiv:2006.00192.

[62] L. Lovász, On chromatic number of finite set-systems. *Acta Math. Acad. Sci. Hung.* **19** (1968), 59–67.

[63] L. Lovász, Kneser's conjecture, chromatic number, and homotopy. *J. Combin. Theory Ser. A* **25** (1978), 319–324.

[64] A. Lubotzky, R. Phillips, and P. Sarnak, Ramanujan graphs. *Combinatorica* **8** (1988), 261–277.

[65] J. Matoušek, String graphs and separators. In *Geometry, structure and randomness in combinatorics*, pp. 61–97, CRM Series 18, Ed. Norm, Pisa, 2015.

[66] J. Mycielski, Sur le coloriage des graphes. *Colloq. Math.* **3** (1955), 161–162.

[67] J. Nešetřil and V. Rödl, Chromatically optimal rigid graphs. *J. Combin. Theory Ser. B* **46** (1989), 133–141.

[68] A. Pawlik, J. Kozik, T. Krawczyk, M. Lasoń, P. Micek, W. T. Trotter, and B. Walczak, Triangle-free intersection graphs of line segments with large chromatic number. *J. Combin. Theory Ser. B* **105** (2014), 6–10.

[69] P. Pournajafi and N. Trotignon, Burling graphs revisited—Part 1: New characterizations. 2021, arXiv:2104.07001; Burling graphs revisited—Part 2: Structure. 2021, arXiv:2106.16089.

[70] F. Ramsey, On a problem of formal logic. *Proc. Lond. Math. Soc.* **30** (1930), 264–286.

[71] N. Robertson and P. Seymour, Graph Minors. XX. Wagner's conjecture. *J. Combin. Theory Ser. B* **92** (2004), 325–357.

[72] V. Rödl, On the chromatic number of subgraphs of a given graph. *Proc. Amer. Math. Soc.* **64** (1977), 370–371.

[73] A. Scott, Induced trees in graphs of large chromatic number. *J. Graph Theory* **24** (1997), 297–311.

[74] A. Scott and P. Seymour, Induced subgraphs of graphs with large chromatic number. I. Odd holes. *J. Combin. Theory Ser. B* **121** (2016), 68–84.

[75] A. Scott and P. Seymour, Induced subgraphs of graphs with large chromatic number. IV. Consecutive holes. *J. Combin. Theory Ser. B* **132** (2018), 180–235.

[76] A. Scott and P. Seymour, Induced subgraphs of graphs with large chromatic number. X. Holes with specific residue. *Combinatorica* **39** (2019), 1105–1132.

[77] A. Scott and P. Seymour, Induced subgraphs of graphs with large chromatic number. VI. Banana trees. *J. Combin. Theory Ser. B* **145** (2020), 487–510.

[78] A. Scott and P. Seymour, Induced subgraphs of graphs with large chromatic number. XIII. New brooms. *European J. Combin.* **84** (2020), 103024.

[79] A. Scott and P. Seymour, A survey of χ-boundedness. *J. Graph Theory* **95** (2020), 473–504.

[80] A. Scott, P. Seymour, and S. Spirkl, Polynomial bounds for chromatic number. I. Excluding a biclique and an induced tree. 2021, arXiv:2104.07927.

[81] A. Scott, P. Seymour, and S. Spirkl, Polynomial bounds for chromatic number. III. Excluding a double star. 2021, arXiv:2108.07066.

[82] A. Scott, P. Seymour, and S. Spirkl, Polynomial bounds for chromatic number. IV. A near-polynomial bound for excluding the five-vertex path. 2021, arXiv:2110.00278.

[83] A. D. Sokal, Bounds on the complex zeros of (di)chromatic polynomials and Potts-model partition functions. *Combin. Probab. Comput.* **10** (2001), 41–77.

[84] D. P. Sumner, Subtrees of a graph and chromatic number. In *The theory and applications of graphs*, edited by G. Chartrand, pp. 557–576, John Wiley & Sons, New York, 1981.

[85] W. Tutte, A contribution to the theory of chromatic polynomials. *Canad. J. Math.* **6** (1954), 80–91.

[86] K. Wagner, Über eine Eigenschaft der ebenen Komplexe. *Math. Ann.* **114** (1937), 570–590.

ALEX SCOTT

Mathematical Institute, University of Oxford, Oxford OX2 6GG, United Kingdom, scott@maths.ox.ac.uk

This Page Intentionally Left Blank

LOCAL-VS-GLOBAL COMBINATORICS

ASAF SHAPIRA

ABSTRACT

Many of the most outstanding open problems in combinatorics relate the local and global properties of large discrete structures. The research aimed at solving these questions led to some of the most important developments in this area, as well as in related areas such as theoretical computer science, additive number theory, and harmonic analysis. In this paper we discuss some of these advances and mention several open problems.

MATHEMATICS SUBJECT CLASSIFICATION 2020

Primary 05C35; Secondary 05D99, 05C65, 05C85, 68R10, 68W20

KEYWORDS

Regularity lemma, quasi randomness, removal lemma, property testing, graph theory, hypergraphs

INTERNATIONAL CONGRESS OF MATHEMATICIANS 2022 JULY 6–14

Proc. Int. Cong. Math. 2022, Vol. 6, pp. 4682–4708
DOI 10.4171/ICM2022/36

Published by EMS Press

1. INTRODUCTION

Extremal combinatorics is one of the fastest growing areas of research within discrete mathematics. Questions in this area deal with the asymptotic relations between various parameters of large discrete structures such as graphs, hypegraphs, permutations, sets of integers, etc. This area has grown tremendously in the past few decades, both in depth and in breadth, and supplied many spectacular results that affected various other areas of mathematics, such as number theory, group theory, probability theory, information theory, harmonic analysis, and theoretical computer science. Many key insights that were developed in order to solve some of the core problems in extremal combinatorics were later exported to other areas. Perhaps the prime example is Szemerédi's theorem [94], stating that dense sets of integers contain arbitrarily long arithmetic progressions. This theorem motivated some of the most important investigations in extremal combinatorics such as the regularity method in graphs [95] and hypergraphs [54, 73, 81], the theory of quasirandom graphs [23, 98], and the theory of graph limits [68]. Szemerédi's theorem also motivated the development of tools in other areas such as ergodic theory (the multiple recurrence theorem [39, 40]), harmonic analysis (the Gowers norms [53]), number theory (the Green–Tao theorem [57]), and theoretical computer science (the PCP theorem [13, 14] and property testing [48]). See [96] for a more detailed discussion.

In this paper we describe a variety of results and open problems in extremal combinatorics relating local and global properties of graphs and hypergraphs. The first set of problems is related to one of the most influential open problems in extremal combinatorics. To state it, we need the following definitions. An r-graph $\mathcal{H} = (V, E)$ consists of a ground set V (the vertices) and a collection of subsets E (the edges) where each edge in E contains r distinct vertices of V. When $r = 2$, we will use the term *graph* and denote graphs by G. An r-graph $\mathcal{H}$ is *linear* if every pair of vertices $u, v \in V$ belongs to at most one edge of E. A (v, e)-*configuration* in $\mathcal{H}$ is a set of e edges whose union contains at most v vertices. The following conjecture was raised 50 years ago by Brown, Erdős, and Sós [21, 22]. In the next statement, and in the rest of the paper, we use standard $O/\Omega/\Theta/o$ notation.

Conjecture 1.1 (Brown–Erdős–Sós conjecture). *Fix $e \geq 3$ and suppose $\mathcal{H}$ is an n-vertex linear 3-graph without $(e+3, e)$-configurations. Then $\mathcal{H}$ has $o(n^2)$ edges.*

Note that as in a (v, e)-configuration we fix the number of edges and only bound the number of vertices, such a configuration is a locally dense subset of $\mathcal{H}$. Since a linear $\mathcal{H}$ clearly has at most n^2 edges, what the above conjecture states is that if $\mathcal{H}$ is locally sparse then it is also globally sparse. It is easy to see that if Conjecture 1.1 holds for linear 3-graphs then it holds for arbitrary 3-graphs.

The second set of problems we cover revolves around the *triangle removal lemma* of Ruzsa and Szemerédi [85], devised for the purpose of proving Conjecture 1.1 for the special case $e = 3$, which is widely considered to be one of the cornerstone results of extremal combinatorics. In what follows, a triangle in a graph $G = (V, E)$ is a triple of vertices u, v, w so that $(u, v), (u, w)(v, w) \in E$. A graph is *triangle-free* if it contains no triangle.

Theorem 1.2 (Triangle removal lemma). *For every $\varepsilon > 0$, there is* $\operatorname{Rem}(\varepsilon)$ *so that if G is an n-vertex graph with the property that one should remove at least εn^2 of its edges in order to make it triangle-free, then G contains at least $n^3/\operatorname{Rem}(\varepsilon)$ triangles.*

Note that if G has $n^3/\operatorname{Rem}(\varepsilon)$ triangles, then a random subset of (about) $\operatorname{Rem}(\varepsilon)$ vertices contains a triangle with probability at least $2/3$. In particular, this means that no matter how large n is, most subsets of vertices of size $\operatorname{Rem}(\varepsilon)$ are not triangle-free. We can thus interpret Theorem 1.2 as stating that if G is globally far from being triangle-free, then it is also locally far from being triangle-free.

The third set of problems we discuss is related to the celebrated *regularity lemma* of Szemerédi [95]. One of the first applications of this lemma was the proof Theorem 1.2 in [85]. Since then it has become one of the most important tools for solving extremal problems in graph theory (see [88]). To state it we need a few definitions. Suppose G is a graph and A, B are two disjoint subsets of V. We use $e(A, B)$ to denote the number of edges of E that connect a vertex in A with a vertex in B. We also let $d(A, B) = e(A, B)/|A||B|$ denote the *edge density* between A, B. Finally, we say that the pair (A, B) is ε-regular if $|d(A, B) - d(A', B')| \le \varepsilon$ for every pair $A' \subseteq A$, $B' \subseteq B$ satisfying $|A'| \ge \varepsilon|A|$ and $|B'| \ge \varepsilon|B|$. A partition $V_1, \dots, V_k$ of the vertices of G into k sets is called ε-regular if all but εk^2 of the pairs (V_i, V_j) are ε-regular and all the sets are of equal size n/k (or of sizes $\lfloor n/k \rfloor$ and $\lceil n/k \rceil$). The *order* of such a partition is the number of sets V_i in it (i.e., k above).

Theorem 1.3 (Szemerédi's regularity lemma). *For every $\varepsilon > 0$, there is an $M = M(\varepsilon)$ so that every graph has an ε-regular partition of order k with $1/\varepsilon \le k \le M$.*

The rest of the paper is organized as follows. In Section 2 we discuss Conjecture 1.1 and many related questions. In Section 3 we discuss Theorem 1.3 and many of its variants along with their applications. Finally, variants of Theorem 1.2, and their relations to problems in theoretical computer science, are described in Sections 3 and 4. Since it is clearly impossible to cover all themes related to the subject of this paper, or even those related to the above three topics, many important results will be left out.

2. THE BROWN–ERDŐS–SÓS CONJECTURE

In this section we describe several results and open problems related to Conjecture 1.1. We will henceforth use the acronym BESC. Let $f_r(n, v, e)$ denote the largest number of edges in a linear r-graph on n vertices that contains no (v, e)-configuration. Note that Conjecture 1.1 is the statement that $f_3(n, e+3, e) = o(n^2)$. Despite much effort by many researchers, Conjecture 1.1 is wide open, having only been settled for $e = 3$ by Ruzsa and Szemerédi [85] in what has become known as the (6, 3)-theorem. To get some perspective on the significance of this special case of Conjecture 1.1, let us just mention that besides its relation to Theorems 1.2 and 1.3 mentioned above, the (6, 3)-theorem implies Roth's theorem [82] on 3-term arithmetic progressions in dense sets of integers (see the next subsection). As another indication of the importance of this problem, we note that one of the main driving forces for proving the celebrated hypergraph removal lemma (see Section 3.3) was the hope that it would lead to a proof of Conjecture 1.1.

2.1. Approximate versions of BESC

At present we seem to be quite far from proving Conjecture 1.1. As an indication of the difficulty of Conjecture 1.1 for $e > 3$, let us mention that already the case $e = 4$ (i.e., the statement $f_3(n, 7, 4) = o(n^2)$) implies the notoriously difficult Szemerédi's theorem **[93]** for 4-term arithmetic progressions, see **[28]**. It is thus natural to look for approximate versions of this conjecture. Namely, given $e \geq 3$, find the smallest $d = d(e)$ such that $f_3(n, e + d, e) = o(n^2)$. Until very recently, the best result of this type was obtained about 15 years ago by Sárközy and Selkow **[87]**, who proved that

$$f_3\big(n, e + 2 + \lfloor \log_2 e \rfloor, e\big) = o(n^2). \tag{2.1}$$

Since the result of **[87]**, the only advance was obtained by Solymosi and Solymosi **[91]**, who improved the bound for $e = 10$ from $f_3(n, 15, 10) = o(n^2)$ (which follows from (2.1)), to $f_3(n, 14, 10) = o(n^2)$. The first improvement over (2.1) for all large enough e was obtained recently in **[26]**. Moreover, it shows that one can replace the $\lfloor \log_2 e \rfloor$ "error term" in (2.1) by a much smaller, sublogarithmic, term.

Theorem 2.1. *For every $e \geq 3$,*

$$f_3(n, e + 18 \log e / \log \log e, e) = o(n^2).$$

The main idea of **[87]** in their proof of (2.1) was the following: The triangle removal lemma actually shows that a linear 3-graph with $\Omega(n^2)$ edges has many $(6, 3)$-configurations. One then defines an auxiliary graph based on these $(6, 3)$-configurations, and uses the triangle removal lemma again in order to double a $(6, 3)$-configuration into a configuration with 7 edges, and so on. The caveat is that each time the number of edges is doubled, the difference between v and e increases by 1, resulting in the $\log_2 e$ error term. The main idea of **[91]** was to use the 3-*graph removal lemma* (see Theorem 3.8), which is the extension of Theorem 1.2 to 3-graphs, in order to perform a *single* iteration in the style of **[87]**, which instead of doubling the number of edges, (roughly) triples it. The main novelty in **[26]** is in managing to perform these multiplications an unbounded number of times. To do so, the r-*graph removal lemma* is used (for all uniformities r) in order to sequentially multiply the number of edges by $3, 4, 5, \ldots$. The main difficulty is in ensuring that each time one multiplies a configuration, the difference between v and e increases only by 1. Another challenge is in making sure the configuration has exactly e edges.

Conjecture 1.1 has a more general form (see **[87]**), stating that for every $2 \leq k < r$ and $e \geq 3$ we have $f_r(n, (r - k)e + k + 1, e) = o(n^k)$. However, as noted in **[26]**, this more general version is, in fact, equivalent to the special case stated as Conjecture 1.1 (corresponding to $k = 2$ and $r = 3$). Moreover, any approximate version of Conjecture 1.1, like that stated in Theorem 2.1, gives analogous approximate versions of the general conjecture.

2.2. Lower bounds for BESC

As we mentioned at the end of the previous subsection, the general form of the BESC states that for every $2 \le k < r$ and $e \ge 3$ we have $f_r(n, (r-k)e + k + 1, e) = o(n^k)$. It is also widely believed (see [28, 29]) that the following lower bound holds.

Conjecture 2.2. *For every $2 \le k < r$ and $e \ge 3$, we have $f_r(n, (r-k)e + k + 1, e) > n^{k-o(1)}$.*

Given the difficulty of proving an upper bound for the BESC, one might expect that Conjecture 2.2 would be relatively easy to resolve. As it turns out, this is not the case. Ruzsa and Szemerédi [85] gave an ingenious construction showing that

$$f_3(n, 6, 3) \ge \Omega\big(n \cdot r_3(n)\big) \ge n^{2-o(1)}, \tag{2.2}$$

where $r_3(n)$ is the size of the largest subset of the first n integers without a 3-term arithmetic progression, and the second inequality follows from the well-known construction of Behrend [16] showing that $r_3(n) > n^{1-o(1)}$. Observe that combined with the fact that $f_3(n, 6, 3) = o(n^2)$ mentioned above, this implies Roth's theorem [82] stating that $r_3(n) = o(n)$. This establishes Conjecture 2.2 for $e = 3$, $k = 2$, $r = 3$. Erdős, Frankl, and Rödl [29] later extended this to arbitrary $r \ge 3$ (and $e = 3$, $k = 2$ as in [85]). A result of [8] then verified Conjecture 2.2 for $e = 3$ and arbitrary $2 \le k < r$.

The key idea in the above results, which handle the case $e = 3$, is to start with a set of integers $X \subseteq [n]$, and construct a Cayley-type r-graph $\mathcal{H}$ in such a way that one can "extract" from any $((r-k)3 + k + 1, 3)$-configuration in $\mathcal{H}$ a nontrivial solution to an equation of the form $ax + by = (a+b)z$ with bounded a, b and $x, y, z \in X$. A simple generalization of [16] shows that there are sets $X \subseteq [n]$ of size $n^{1-o(1)}$ without nontrivial solutions to equations of this type, which can be used to give a bound as in (2.2). The reason why Conjecture 2.2 becomes much harder when $e > 3$ is that when handling more than 3 edges, the linear equation E we can extract from a $((r-k)e + k + 1, e)$-configuration might be one for which there is no $X \subseteq [n]$ of size $n^{1-o(1)}$ without a solution of E. For example, if the equation is $x + y = z + w$ then a set without nontrivial solutions has size $O(\sqrt{n})$.

Let us focus then on the case $e > 3$ and $k = 2$ and $r = 3$. It is easy to check that every $(7, 4)$ or $(8, 5)$-configuration contains a $(6, 3)$-configuration, hence the bounds $f_3(n, 7, 4), f_3(n, 8, 5) \ge n^{2-o(1)}$ follow from (2.2). The situation becomes much harder when $e = 6$, since there is a $(9, 6)$-configuration which does not contain a $(6, 3)$-configuration. Indeed, this is the 3×3 *grid*, denoted $\mathcal{G}_{3\times 3}$, namely the 3-graph whose vertices are the nine points in a 3×3 point array, and whose edges correspond to the 6 horizontal and vertical lines of this array. Let $\mathcal{T}$ denote the 3-graph with vertices $1, 2, 3, 4, 5, 6$ and edges $\{1, 2, 3\}, \{3, 4, 5\}, \{5, 6, 1\}$ (this is the unique linear $(6, 3)$-configuration). It is not hard to verify that every linear $(9, 6)$-configuration (in a 3-graph) either contains a copy of $\mathcal{T}$ or is isomorphic to $\mathcal{G}_{3\times 3}$. Hence, to prove that $f(n, 9, 6) \ge n^{2-o(1)}$, it would suffice to construct a linear 3-graph with $n^{2-o(1)}$ edges and no copy of either $\mathcal{G}_{3\times 3}$ or $\mathcal{T}$.

The above facts led Füredi and Ruszinkó [38] to study various extremal problems related to $\mathcal{G}_{3\times 3}$. In particular, they conjectured that there is a $\mathcal{G}_{3\times 3}$-free linear 3-graph with $(1/6 - o(1))n^2$ edges. Using a standard probabilistic alterations argument, Füredi and Ruszinkó [38] constructed such a 3-graph with $\Omega(n^{1.8})$ edges. This was slightly improved (as a special case of a more general result) to $\Omega(n^{1.8} \log^{1/5} n)$ in [88]. The following result of [44] makes a significant progress on the conjecture of Füredi and Ruszinkó [38], by improving these results to $\Omega(n^2)$. We, in fact, have the following more general statement.

Theorem 2.3. *For a prime p, and two sets $X, A \subseteq \mathbb{F}_p$, define the following 3-partite 3-graph $\mathcal{H} = \mathcal{H}(X, A)$ on vertex sets V_1, V_2, V_3 where we think of each V_i as a copy of $\mathbb{F}_p$: for every $x \in X$ and $a \in A$, place a 3-edge containing the vertices $x \in V_1$, $x + a \in V_2$ and $x + a^2 \in V_3$ (all operations over $\mathbb{F}_p$). Then, every pair of vertices of $\mathcal{H}$ belongs to at most 2 edges. Also, if there are no $x_1, x_2 \in X$ and $a \in A$ satisfying*

$$4x_1 + 4a \equiv 4x_2 + 1 \pmod{p} \tag{2.3}$$

then $\mathcal{H}$ is $\mathcal{G}_{3\times3}$-free, and if A has no solution to the equation

$$a + b^2 - b \equiv c^2 \pmod{p} \tag{2.4}$$

in distinct $a, b, c \in A$ then $\mathcal{H}$ is $\mathcal{T}$-free.

It is easy to see that the sets $X = A = \{1, \dots, \lfloor p/8 \rfloor\}$ do not contain a solution to (2.3), hence $\mathcal{H} = \mathcal{H}(X, A)$ has $\Omega(n^2)$ edges and no copy of $\mathcal{G}_{3\times3}$. Also, since each pair of vertices belongs to at most 2 edges we get that there is also a *linear* $\mathcal{H}$ with the same properties, thus establishing the improved bound for the Füredi–Ruszinkó conjecture stated before Theorem 2.3. The second assertion Theorem 2.3 thus leads to the following problem:

Problem 2.4. Is there $A \subseteq \mathbb{F}_p$ of size $p^{1-o(1)}$ without a solution of (2.4) in distinct a, b, c?

If a set A as above exists, then, to prove that $f(n, 9, 6) \geq n^{2-o(1)}$, one would just need to find $X \subseteq \mathbb{F}_p$ of size $p^{1-o(1)}$ so that A and X have no solution to (2.3). Also, in the spirit of [84], it seems interesting to further study the size of the largest subsets of $\mathbb{F}_p$ without nontrivial solutions to other polynomial equations.

Conjectures 1.1 and 2.2 state nearly matching lower and upper bounds. We conclude this subsection by recalling a problem of Erdős [28], who asked if the exact asymptotic formula $f(n, e+3, e) = \Theta(n \cdot r_e(n))$ holds, where $r_e(n)$ denotes the size of the largest subset of the first n integers without an e-term arithmetic progression. As of now, the upper bound is not known for any $e \geq 3$, while the lower bound is known only for $e = 3, 4, 5$.

2.3. The Gowers–Long conjecture

The BESC states that a 3-graph without $(e+3, e)$-configurations has $o(n^2)$ edges. By (2.2), already when $e = 3$, one cannot reduce this to $n^{2-\delta}$ for some absolute $\delta > 0$. Gowers and Long [55] conjectured that for sparser configurations, such a bound is attainable.

Conjecture 2.5. *For every $e \geq 3$, there is a $\delta = \delta(e) > 0$ so that $f_3(n, e+4, e) = O(n^{2-\delta})$.*

In the previous subsection we discussed the approximate versions of BESC obtained in [26] and [87]. The following result of [41] gives an analogous approximate version of the Gowers–Long conjecture.

Theorem 2.6. *For every $e \geq 3$, there is a $\delta = \delta(e) > 0$ so that*

$$f_3(n, e + O(\log e), e) = O(n^{2-\delta}).$$

Recall that prior to Theorem 2.1 of [26], the state-of-the-art result for the BESC was inequality (2.1) of Sárközy–Selkow [87]. The above theorem then shows that with an error term close to that of Sárközy–Selkow, one can in fact improve the $o(n^2)$ bound on the number of edges to $O(n^{2-\delta})$.

As we noted after (2.2), one of the surprising implications of the (6, 3)-theorem is Roth's theorem. A result of this nature due to Gowers and Long [55] then states that a positive answer to Conjecture 2.5 for $e = 5$ would imply that for some $c > 0$, every $S \subseteq [n]$ of size n^{1-c} contains a nontrivial solution (i.e., one where not all integers are equal) of the equation $2x_1 + 2x_3 = x_3 + 3x_4$. This is related to a famous problem of Ruzsa [84], asking if for every linear equation E of the form $\sum_i a_i x_i = 0$ with $\sum_i a_i = 0$, and such that no (nonempty) proper subset of the coefficients a_i sums to 0, there is $X \subseteq [n]$ of size $n^{1-o(1)}$ without a nontrivial solution of E. As of now, a positive answer is only known when all but one of the a_i's are positive, via a straightforward generalization of Behrend's construction [16]. It would be very interesting to show that the relation between Ruzsa's problem and Conjecture 2.5 can be extended to other equations.

2.4. A Ramsey variant of BESC

Given the difficulty of the BESC, it is natural to look at various relaxations of it. For example, instead of looking at arbitrary r-graphs, one can look at those arising from a group (see [74] and its references). We now state another natural simplification of BESC that was recently suggested by Conlon and Nenadov. We say that a linear r-graph is *complete* if every pair of vertices of V belongs to exactly one edge of E. Such an object is sometimes called an r-Steiner system (when $r = 3$ this is a *Steiner triple system*). Conlon and Nenadov then suggested the following weaker Ramsey-type version of the BESC, namely proving that the following holds for every $r \geq 3$, $e \geq 3$, $c \geq 2$, and large enough $n \geq n_0(r, e, c)$: If $\mathcal{H}$ is an n-vertex complete linear r-graph then in every c-coloring of its edges one can find e edges of the same color, which are spanned by at most $(r-2)e + 3$ vertices. Note that BESC implies the above statement, as it gives the required monochromatic configuration in the most popular color. It is easy to see that this problem has a positive answer when $c = 1$ or when $e = 3$. The problem is wide open already when $e = 4$. The following result of [90] gives a positive answer to the Conlon–Nenadov problem assuming r is large enough.

Theorem 2.7. *For every integer c, there exists an $r_0 = r_0(c)$ such that for every $r \geq r_0$ and integer $e \geq 3$ there exists $n_0 = n_0(c, r, e)$ such that every c-coloring of a complete linear r-graph on $n > n_0$ vertices contains a monochromatic $((r-2)e + 3, e)$-configuration.*

In the natural case $c = 2$, it was further shown in [90] that $r_0(2) \leq 4$. The above results were recently generalized in [64], building on some of the tools and ideas introduced in [90]. One of these tools was an auxiliary graph which helps one "grow" $((r-2)e + 3, e)$-configurations, for $e = 3, 4, \ldots$, and thus prove Theorem 2.7. Perhaps one take-home message of [98] is that even when considering the Ramsey relaxation of the BESC, and even after adding the assumption that $r \geq r_0(c)$, one still has to work quite hard in order to find the $((r-2)e+3, e)$-configurations of the BESC.

3. VARIANTS OF THE REGULARITY LEMMA AND THEIR APPLICATIONS

In this section we discuss several variants of the regularity lemma and their relation to three of the most well-studied variants of the removal lemma. We will need the following definitions. For a fixed graph property $\mathcal{P}$, the *distance* of an n-vertex graph G from satisfying $\mathcal{P}$ is the smallest number of edges modifications (i.e., addition and removal) needed to turn G into an n-vertex graph satisfying $\mathcal{P}$. We say that an n-vertex graph is ε*-far* from satisfying $\mathcal{P}$ if G's distance from $\mathcal{P}$ is at least εn^2.

3.1. Triangle removal using weak regularity lemmas

In this subsection we focus on the triangle removal lemma, and more concretely, on the quantitative bounds for the function $\text{Rem}(\varepsilon)$ introduced in Lemma 1.2. Actually, all the results will hold for the more general *graph removal lemma*, which states that for every graph H there is a function $\text{Rem}_H(\varepsilon)$ so that if G is ε-far from being H-free then G contains $n^h/\text{Rem}_H(\varepsilon)$ copies of H, where $h = |V(H)|$. In what follows, we will use $\text{twr}(x)$ to denote the tower function, namely a tower of exponents of height x. For example, $\text{twr}(3) = 2^{2^2}$.

The original proof of the triangle removal lemma in [85], and of its generalization to every fixed H [3], relied on Szemerédi's regularity lemma [95] and gave the bound $\text{Rem}_H(\varepsilon) \leq M(\text{poly}(\varepsilon))$. Let us sketch this proof when H is the triangle (the proof for general H is almost identical). Given G that is ε-far from being H-free, one first invokes the regularity lemma (with $\varepsilon/10$) in order to obtain an ε-regular partition of $V(G)$. One then removes from G edges that are either (i) inside one of the sets V_i, or (ii) connect V_i to V_j so that (V_i, V_j) is not $\varepsilon/10$-regular, or (iii) connect V_i to V_j so that $d(V_i, V_j) \leq \varepsilon/5$. Since this "cleaning process" removes less than εn^2 edges, at least one triangle remains in the new graph. By the nature of the cleaning process, there must be V_i, V_j, V_k so that this triangle has one vertex in each of these sets, and so that all three pairs of sets have density at least $\varepsilon/5$ and are $\varepsilon/10$-regular. One then invokes the so called *counting lemma* (see, e.g., Lemma 3.2 in [4]) in order to show that such a triple of sets V_i, V_j, V_k in fact contains $\text{poly}(\varepsilon)|V_i||V_j||V_k|$ many triangles. Since each of these three sets has size at least $n/M(\varepsilon/10)$, we conclude that G has at least $n^3/M^3(\text{poly}(\varepsilon))$ triangles. Unfortunately, Szemerédi's proof of his lemma gave the bound $M(\varepsilon) \leq \text{twr}(\text{poly}(1/\varepsilon))$, which combined with the preceding proof gives

$$\text{Rem}_H(\varepsilon) \leq \text{twr}\big(\text{poly}(1/\varepsilon)\big). \tag{3.1}$$

As to lower bounds, extending a construction of [85] for triangle-freeness, Alon [1] proved that for every nonbipartite H, there is $C = C(H)$ satisfying

$$\text{Rem}_H(\varepsilon) \geq (1/\varepsilon)^{C \log(1/\varepsilon)}. \tag{3.2}$$

There are two natural approaches for improving (3.1). The first would be to obtain better bounds for the regularity lemma, namely for $M(\varepsilon)$, which would immediately lead to improved bounds for $\mathrm{Rem}_H(\varepsilon)$. This, as well as numerous other applications of the regularity lemma, gave the hope that one can find a new proof of this lemma with significantly better quantitative bounds. These hopes were unfortunately shattered when Gowers [52] famously proved that $M(\varepsilon) \geq \mathrm{twr}(\mathrm{poly}(1/\varepsilon))$. The current record lower bound was obtained in [34] who showed that $M(\varepsilon) \geq \mathrm{twr}(1/\varepsilon^2)$, while a much shorter and simpler proof of Gowers's lower bound appears in [70].

Given the above, the second approach for improving (3.1) was to find a proof of the removal lemma that avoids the regularity lemma. This problem was open for many years until the breakthrough result of Fox [31], who found a new proof showing that

$$\mathrm{Rem}_H(\varepsilon) \leq \mathrm{twr}\big(O\big(\log(1/\varepsilon)\big)\big). \tag{3.3}$$

Fox's proof used an ad-hoc argument which was simplified in [25]. Given the simplicity of the proof of the removal lemma using the regularity lemma (sketched above), it is natural to ask if there is a weaker version of the regularity lemma, which is strong enough for proving the removal lemma (in a way similar to the original proof), yet weak enough so as to yield better bounds. Before describing two such proofs we should point out that the idea of devising weak regularity lemmas for specific applications was used before. Two notable such examples are the Frieze–Kannan *weak regularity lemma* [37] and the Duke–Lefman–Rödl [27] *cylinder lemma*. Below we describe two weaker versions of the regularity lemma which do produce bounds better than (3.1) by giving alternative proofs of (3.3).

Finding a regular partition of only part of the graph. Recall that when deriving the removal lemma from the regularity lemma, we only needed the 3 pairs among V_i, V_j, V_k to be ε-regular. The reason why the bound was so weak is that these sets are of size $n/\mathrm{twr}(\mathrm{poly}(1/\varepsilon))$. A special case of the cylinder lemma [27], shows that given an n-vertex graph, one can find three sets V_i, V_j, V_k so that each of the three pairs among them is ε-regular and the three sets have size $n/2^{\mathrm{poly}(1/\varepsilon)}$. Unfortunately, it does not appear that this lemma can be used to prove the triangle removal lemma since there is no structural connection between the 3 sets and G. In their work on the induced removal lemma (see Section 3.2), Alon, Fischer, Krivelevich, and Szegedy [4] proved the following related theorem.

Theorem 3.1. *For every $\varepsilon > 0$ and h, there is an $S = S(\varepsilon, h)$ so that every graph G has an equipartition $\{V_1, \ldots, V_k\}$ and a collection of subsets $W_1 \subseteq V_1, \ldots, W_k \subseteq V_k$ satisfying:*

(i) $\sum_{1 \leq i < j \leq k} |d(V_i, V_j) - d(W_i, W_j)| \leq \varepsilon k^2$;

(ii) *All pairs (W_i, W_j) are ε^h-regular;*

(iii) $|W_i| \geq |V(G)|/S$.

Furthermore, we have $\mathrm{Rem}_H(\varepsilon) \leq S(\mathrm{poly}(\varepsilon), h)$, where $h = |V(H)|$.

We note that the proof of the "furthermore" part of this theorem is almost identical to the way one derives the removal lemma from the regularity lemma, as we sketched above.

Note that item (i) gives us the required relation between G and the sets $W_1, \ldots, W_k$, which is missing when applying the cylinder lemma [27].

Alon et al. [4] obtained a wowzer-type upper bound for $S(\varepsilon, h)$, where wowzer is the iterated version of the tower function. This wowzer-type bound resulted from using the *strong regularity lemma* which was introduced in [4]. It was later proved [24, 62] that the wowzer-type bounds for the strong regularity lemma are unavoidable, thus ruling out the possibility of improving the bounds for Theorem 3.1 by improving the bounds for the strong regularity lemma. A new proof of Theorem 3.1 was obtained by Conlon and Fox [24] who used the cylinder lemma [27] in a sophisticated way in order to prove that $S(\varepsilon, h) \leq \mathrm{twr}(\mathrm{poly}(1/\varepsilon))$. Note that even this improved bound does not give an improvement over (3.1), and as we mention below in Theorem 3.6, for *general graphs* this improved bound is the best possible. Hence, it appears as if Theorem 3.1 cannot be used to improve (3.1). However, by combining the ideas of [24] with those of [71], the following result was obtained in [86].

Theorem 3.2. *If G in Theorem 3.1 has $O(\varepsilon n^2)$ edges then $S(\varepsilon, h) \leq \mathrm{twr}(O(\log(1/\varepsilon)))$.*

Using the theorem above, and a minor variant of the proof of the removal lemma from the regularity lemma sketched above, one obtains (3.3) but only for graphs with $O(\varepsilon n^2)$ edges. So to reprove (3.3) in full generality it remains to prove that if G is ε-far from being H-free, then G has a subgraph G' with $O(\varepsilon n^2)$ edges which is $\Omega(\varepsilon)$-far from being H-free. Indeed, we could then apply the statement that holds only for graphs with $O(\varepsilon n^2)$ edges, and then use the fact that every copy of H in G' is also a copy of H in G. To find such a G', we first note that if G has δn^2 edge-disjoint copies of H, then G is clearly δ-far from being H-free, and that conversely, if G is ε-far from being H-free, then it contains at least $\varepsilon n^2/h^2$ edge-disjoint copies of H (where $h = |V(H)|$). Hence, taking G' to be the union of these edge-disjoint copies of H we obtain the required subgraph of G.

Modifying the graph. Recall that when proving the removal lemma, we first obtained an ε-regular partition of the graph, then removed edges from G, and then found many triangles in the new graph G'. Since we are already finding triangles in a modified version of G, one can ask if instead of finding a regular partition of G (which might be hard by Gowers's lower bound), it is enough to find a regular partition of a modified version of G. A version of the regularity lemma called the *regular approximation lemma* achieves this task.

Theorem 3.3. *For every $\varepsilon, \delta > 0$, there is a $T = T(\varepsilon, \delta)$ so that one can add/remove from an n-vertex graph at most δn^2 edges so that the new graph G' has a partition of order at most T in which* all *pairs are ε-regular.*

The first proofs of this lemma [69, 78] supplied (at best) only wowzer-type bounds. A better tower-type bound was obtained by Conlon and Fox [24]. Interestingly, the tower dependence is only on δ and not on ε. Unfortunately, for proving the removal lemma, one has to take $\delta \approx \varepsilon$, and so we again obtain (3.1). However, the following result of [71], which is a variant tailored for sparse graphs and appropriately dubbed the *sparse regular approximation lemma*, supplies a better bound.

Theorem 3.4.

Suppose G is a graph with $O(\varepsilon n^2)$ edges. Then one can add/delete from G at most $\varepsilon n^2/100$ edges so that the resulting graph G' has an ε^3-regular partition of order at most $\mathrm{twr}(O(\log 1/\varepsilon))$.

Let us briefly describe how the above theorem can be used to prove (3.3). First, as described after Theorem 3.2, it is enough to consider only graphs with $O(\varepsilon n^2)$ edges. Given such a G, we apply Theorem 3.4 to obtain G'. Since G' was obtained using few edge modifications, it is $\varepsilon/2$-far from being triangle free. We can now repeat the same argument used to derive the removal lemma from the regularity lemma, to infer that G' contains $n^h/\mathrm{twr}(O(\log 1/\varepsilon))$ copies of H (the improved bound comes from Theorem 3.4). Since we are allowed to add edges to G when producing G', one needs to be careful here since G' might contain "ghost" triangles that do not belong to G. However, it is not hard to show that at least half of the triangles in G' also belong to G thus completing the proof.

3.2. Improved bounds for the induced removal lemma?

We now consider the so called *induced removal lemma*, which is the induced variant of the removal lemma we discussed above. It states that for every fixed graph H, there is $\mathrm{Rem}^*_H(\varepsilon)$ so that if G is ε-far from being induced H-free then G has at least $n^h/\mathrm{Rem}^*_H(\varepsilon)$ induced copies of H. The fact that for every H such a function $\mathrm{Rem}^*_H(\varepsilon)$ exists was first obtained in [4] using Theorem 3.1. More precisely, what they proved was that $\mathrm{Rem}^*_H(\varepsilon) \leq S(\mathrm{poly}(\varepsilon), h)$. As we noted in the previous subsection, a tower-type bound for Theorem 3.1 was obtained in [24] giving the improved $\mathrm{twr}(\mathrm{poly}(1/\varepsilon))$ upper bound for the induced removal lemma. Conlon and Fox later raised [25] the following natural problem, asking if one can extend (3.3) to the more difficult setting of the induced removal lemma.

Problem 3.5. Show that for every H we have $\mathrm{Rem}^*_H(\varepsilon) \leq \mathrm{twr}(O(\log(1/\varepsilon)))$.

Since we know that $\mathrm{Rem}^*_H(\varepsilon) \leq S(\mathrm{poly}(\varepsilon), h)$, it is natural to try and resolve the above problem by further reducing the upper bound for Theorem 3.1 to $\mathrm{twr}(O(\log(1/\varepsilon)))$. Recall that Theorem 3.2 shows that such a bound *is* attainable for graphs with $O(\varepsilon n^2)$ edges, implying a positive answer for Problem 3.5 for graphs with this many edges. Unfortunately, a recent result of [67] shows that such a bound is not attainable in general.

Theorem 3.6. *There is a* $\mathrm{twr}(\mathrm{poly}(1/\varepsilon))$ *lower bound for $S(\varepsilon, 10)$ in Theorem* 3.1.

Another approach for resolving Problem 3.5 is to reduce the general case of bounding $\mathrm{Rem}^*_H(\varepsilon)$ to the case where G has $O(\varepsilon n^2)$ edges, since for this special case we can resolve Problem 3.5. As we observed after Theorem 3.2, such a reduction is easy to obtain for the (noninduced) removal lemma. There is a very natural way to try and extend that argument to the setting of induced H-freeness. We say that two induced copies of H in G are *pair-disjoint* if they share at most one vertex. As in the case of H-freeness, it is clear that if G contains εn^2 pair-disjoint induced copies of H then G is ε-far from being induced H-free. Perhaps surprisingly, the converse is not true. For example, one can construct a graph that is

ε-far from being induced C_4-free, yet it contains only $O(\varepsilon^2 n^2)$ pair-disjoint induced copies of C_4. However, it is natural to ask if the following approximate result holds.

Problem 3.7. Show that if G is ε-far from being induced H-free then G contains at least $\text{poly}(\varepsilon) \cdot n^2$ pair-disjoint induced copies of H.

It is a simple corollary of the induced removal lemma itself that if G is ε-far from being induced H-free then G contains $\Omega(n^2)$ pair-disjoint copies of H, but the hidden constant has a tower-type dependence on ε. The question is if this can be made polynomial in ε. Besides being a natural problem, solving Problem 3.7 would also lead to a solution of Problem 3.5. This can be proved using the ideas of [71]. A much simpler argument was noted independently by Jacob Fox (private communication).

3.3. The hypergraph regularity lemma

The following lemma is the natural generalization of Lemma 1.2 (the triangle removal lemma) to r-graphs. Here, $K_{r+1}^{(r)}$ denotes the complete r-graph on $r+1$ vertices, namely, a set of $r+1$ vertices containing all possible $r+1$ r-edges (so $K_3^{(2)}$ is a triangle).

Theorem 3.8 (Hypergraph removal lemma). *For every $r \geq 2$ and $\varepsilon > 0$, there is an* $\text{Rem}_r(\varepsilon)$ *so that the following holds. Suppose $\mathcal{H}$ is an n-vertex r-graph with the property that one should remove at least εn^r of its edges to make it $K_{r+1}^{(r)}$-free. Then $\mathcal{H}$ contains at least $n^{r+1}/\text{Rem}_r(\varepsilon)$ copies of $K_{r+1}^{(r)}$.*

The first to conjecture the above extension of the triangle removal lemma were Erdős, Frankl, and Rödl [29] in the 1980s. One of the main motivations for obtaining Theorem 3.8 was the observation of Frankl and Rödl [36] (see also [92]) that it would give an alternative proof of Szemerédi's theorem for progressions of arbitrary length. Another motivation was the hope that it would lead to a solution of Conjecture 1.1.

As we discussed earlier, the proof of the triangle removal lemma relied on Szemerédi's regularity lemma. The quest for an r-graph regularity lemma that would allow one to prove Theorem 3.8 took about 20 years. The first milestone was the result of Frankl and Rödl [36], who obtained a regularity lemma for 3-graphs and using it proved Theorem 3.8 for $r = 3$. About 10 years later, the approach of [36] was extended to r-graphs (for arbitrary $r \geq 2$) by Rödl, Skokan, Nagle, and Schacht [73, 81]. At the same time, Gowers [54] obtained an alternative version of the regularity lemma for r-graphs. Shortly after, Tao [97] and Rödl and Schacht [77,78] obtained two additional versions of the lemma. A more detailed discussion appears in [75].

For the next discussion, we need to introduce the functions comprising the Ackermann hierarchy. Set $A_1(x) = 2^x$ and, for every $r \geq 2$, define the function $A_r(x)$ to be the result of iterating A_{r-1} on itself x times. So $A_2(x)$ is the tower function and $A_3(x)$ is the wowzer function. We refer to A_r as the rth Ackermann function.

Although the above mentioned regularity lemmas for r-graphs are quite different from each other, they all involve constants that grow as fast as the rth Ackermann function.

As a result, all proofs of the removal lemma for r-graphs give (at best) a bound of the form $\mathrm{Rem}_r(\varepsilon) \le A_r(1/\varepsilon)$. This leads to the following open problem:

Problem 3.9. Obtain primitive recursive bounds for the r-graph removal lemma. That is, show that there is a universal constant r_0 so that $\mathrm{Rem}_r(\varepsilon) \le A_{r_0}(1/\varepsilon)$ for every $r \ge 2$.

It is natural to first ask if one can resolve the above problem simply by improving the constants involved in one of the r-graph regularity lemmas mentioned above, which are known to imply the removal lemma. As mentioned above, Gowers [52] proved that for $r = 2$, one cannot obtain better than $A_2(\mathrm{poly}(1/\varepsilon))$ bounds for the graph regularity lemma. This result was extended to all $r \ge 2$ in [72].

Theorem 3.10. *For every $r \ge 2$, there is an $A_r(\log(1/\varepsilon))$ lower bound for the r-graph regularity lemma.*

Hence, if one wishes to improve the A_r-type bounds for the r-graph removal lemma, one has to develop new variants of the r-graph regularity lemma that, on the one hand, are strong enough to prove Theorem 3.8, and, on the other hand, are weak enough to yield better bounds.

The main challenge in proving Theorem 3.10 is facilitating an inductive approach (on r): one has to prove a stronger lower bound, showing that even very weak versions of the r-graph regularity lemma cannot give bounds better than $A_r(\log(1/\varepsilon))$. Among other things, one has to show that the lower bound holds even if one is allowed to change, say, a 0.01-fraction of the edges. As the reader might recall, this is exactly the type of regularity lemma mentioned in Theorem 3.4, where we stated an upper bound for such a weak version of the lemma. As part of the proof in [72], the following matching lower bound was obtained.

Theorem 3.11. *The* $\mathrm{twr}(O(\log 1/\varepsilon))$ *upper bound in Theorem* 3.4 *is tight.*

Returning to the discussion in Section 3.1, Theorem 3.11 implies that the approach described prior to Theorem 3.4 cannot improve (3.3).

4. VARIANTS OF THE REMOVAL LEMMA

We next describe several problems related to the triangle removal lemma mentioned in Section 1. Since its inception [85], the triangle removal lemma was extended in various ways. Two such generalizations, the general removal lemma and the induced removal lemma, were discussed in the previous section. These extensions culminated in the following result of [9], where a graph property is *hereditary* if it is closed under removal of vertices.

Theorem 4.1. *For every hereditary property $\mathcal{P}$, there is a function* $\mathrm{Rem}_{\mathcal{P}}(\varepsilon)$, *so that if a graph G is ε-far from satisfying $\mathcal{P}$, then a random and uniform sample of* $\mathrm{Rem}_{\mathcal{P}}(\varepsilon)$ *vertices from G spans a graph not satisfying $\mathcal{P}$ with probability at least* $2/3$.

Given a (possibly infinite) family of graphs $\mathcal{F}$, let $\mathcal{P}^*_{\mathcal{F}}$ denote the property of being induced $\mathcal{F}$-free, namely, not containing an induced copy of any $F \in \mathcal{F}$. It is clear that the family of hereditary properties coincides with the family of properties $\mathcal{P}^*_{\mathcal{F}}$, so Theorem 4.1 is the most general version of the removal lemma one can hope to prove. The fact that Theorem 4.1 is indeed a generalization of the removal lemma and the induced removal lemma follows from the reasoning in the paragraph following the statement of Theorem 1.2. The reason we change gears here is that, when $\mathcal{F}$ is infinite, stating the removal lemma for $\mathcal{P}^*_{\mathcal{F}}$ in the style of Theorem 1.2 becomes cumbersome. The same applies to Problem 4.2 below.

4.1. A theoretical computer science interlude

Although the removal lemmas we discuss below have purely combinatorial statements, part of the motivation leading to these results came from questions in theoretical computer science, more specifically from the area of *graph property testing* [48]. The interplay between this area and extremal combinatorics has been extremely fruitful, with many questions raised in one area motivating the development of new tools in the other. Examples of tools are the weak regularity lemma of Frieze and Kannan [37], the conditional regularity lemma of Alon, Fischer, and Newman [5], and the notion of partition oracles [59]. A comprehensive discussion on the combinatorial aspects can be found in Lovász's book [68] and on the more algorithmic aspects in Goldreich's book [48]. In this subsection we give a brief background on this area.

Classical models of computation ask for an algorithm that can decide if an input satisfies some property $\mathcal{P}$, for example, whether an input graph G is planar, or whether an input matrix A is invertible. It is easy to see that in this case we have to read the entire input at least once, for example, because deleting a single edge of the graph might make it planar, or because changing a single entry of A might make it invertible. Due to the need to analyze huge inputs, which might be too costly to scan even once, researchers introduced a new type of algorithms, called *property testers*, that solve only relaxed versions of the classical decision problem, but do so extremely fast. These are randomized algorithms whose goal is to distinguish (with high probability, say, $2/3$) between objects satisfying some fixed property $\mathcal{P}$ and those that are ε-far from satisfying it. The study of such problems originated in the seminal papers of Rubinfeld and Sudan [83], Blum, Luby, and Rubinfeld [18], and Goldreich, Goldwasser, and Ron [50]. Below are the precise definitions related to property testing of graphs.

We say that a graph property is *testable* if there is a function $q_{\mathcal{P}}(\varepsilon)$ so that by sampling a set of vertices S of size $q_{\mathcal{P}}(\varepsilon)$ from a graph G, one can distinguish with probability at least $2/3$ between the following two cases (i) G satisfies $\mathcal{P}$ and (ii) G is ε-far from $\mathcal{P}$. So the fact that $\mathcal{P}$ is testable means that we can distinguish a graph $G \in \mathcal{P}$ from G that is ε-far from $\mathcal{P}$ while looking at a subgraph of G of constant size! It is not hard to see that the triangle removal lemma is equivalent to the statement that triangle-freeness is a testable property, and that bounding $\mathrm{Rem}(\varepsilon)$ is equivalent to bounding the corresponding function $q(\varepsilon)$.

Some of the most important questions in property testing are those asking if general families of properties are testable. Observe that Theorem 4.1 implies that every hereditary graph property $\mathcal{P}$ is testable. Indeed, the algorithm for testing $\mathcal{P}$ simply samples a set S of $\mathrm{Rem}_{\mathcal{P}}(\varepsilon)$ vertices. If the graph spanned by S satisfies $\mathcal{P}$ then the algorithm declares that the input satisfies $\mathcal{P}$, otherwise it declares the input is ε-far from $\mathcal{P}$. Since $\mathcal{P}$ is hereditary, if G satisfies $\mathcal{P}$, the algorithm will declare this with probability 1. On the other hand, the definition of $\text{Rem}_{\mathcal{P}}(\varepsilon)$ guarantees that if G is ε-far from $\mathcal{P}$, the algorithm will declare this with probability at least $2/3$.

As we noted above, the algorithm for testing a hereditary property always answers correctly when the input belongs to $\mathcal{P}$. Such an algorithm is said to have one-sided error. It was shown in [9] that hereditary properties are (essentially) the only properties that can be tested by a one-sided error algorithm. A characterization of the properties that can be tested in the more general setting of two-sided error algorithms was obtained in [6] and [20]. These results were extended to r-graphs in [15, 61, 79].

It should be noted that while the algorithms defined above have running time that depends only on ε (and are independent of $|V(G)|$), the dependence on ε might be enormous. Indeed, as we discussed in Section 3.1, even in the special case of $\mathcal{P}$ being triangle-freeness, the running time of the testing algorithm is given by the tower-type function in (3.3). Furthermore, a result of [10] shows that there are properties $\mathcal{P}$ for which $\text{Rem}_{\mathcal{P}}(\varepsilon)$ grows faster than any recursive function.

While the results discussed above give rather satisfactory *qualitative* answers, by the previous paragraph they give very poor *quantitative* answers. Hence, once we know that a property is testable, the next natural question is whether we can obtain a "reasonable" bound for $q_{\mathcal{P}}(\varepsilon)$. As in many questions, the natural definition of reasonable is polynomial. We thus say that $\mathcal{P}$ is *easily testable* if it is testable with a polynomial sample, that is, if $q_{\mathcal{P}}(\varepsilon) = \text{poly}(1/\varepsilon)$. One of the most important open problems in this area was popularized by Goldreich [48] and by Alon and Fox [7], who asked for a characterization of the easily testable graph properties. Currently, this problem is open even when restricted to hereditary properties. Note that, by the above discussion, proving that a hereditary property $\mathcal{P}$ is easily testable is equivalent to proving that in Theorem 4.1 we have $\text{Rem}_{\mathcal{P}}(\varepsilon) = \text{poly}(1/\varepsilon)$. This leads to the following open problem.

Problem 4.2. Characterize hereditary graph properties $\mathcal{P}$ for which $\text{Rem}_{\mathcal{P}}(\varepsilon) = \text{poly}(1/\varepsilon)$.

This line of research was initiated by Alon [1], who proved that if $\mathcal{P}_H$ is the property of being H-free, then $\mathcal{P}_H$ is easily testable if and only if H is bipartite. Another notable early result was obtained by Goldreich, Goldwasser, and Ron [50] who proved that for any fixed k, the property of being k-colorable is easily testable. This was a major improvement over an earlier result of Rödl and Duke [76] who used the regularity lemma and (implicitly) gave a tower-type upper bound for testing k-colorability. In the next subsection we discuss recent progress related to Problem 4.2.

We finally mention another variant of Theorem 4.1. It is natural to ask if using a sample of vertices of constant size, one can not only detect if an input G is ε-far from $\mathcal{P}$, but further *estimate* G's distance from $\mathcal{P}$. In the literature on property testing, this is called *tolerant testing*. Such a result was obtained for monotone properties in [11], and for all testable properties, and in particular all hereditary properties, in [30]. A recent result of [60] further shows how to efficiently transform any bound for $\mathrm{Rem}_{\mathcal{P}}(\varepsilon)$ into a bound for tolerantly testing $\mathcal{P}$.

4.2. Removal lemmas with polynomial bounds

In this subsection we describe the progress towards Problem 4.2. It will be more convenient to think of a hereditary property in terms of its forbidden induced subgraphs, that is, represent it as $\mathcal{P}^*_{\mathcal{F}}$, as defined after Theorem 4.1. When $\mathcal{F}$ consists of a single graph F we will use the notation $\mathcal{P}^*_F$.

We first consider hereditary properties $\mathcal{P}^*_{\mathcal{F}}$ with finite $\mathcal{F}$. Recall that a graph F is *bipartite* if $V(F)$ can be partitioned into two sets A, B that are both independent, that is, contain no edges. A graph F is *cobipartite* if $V(F)$ can be partitioned into two complete graphs A, B. Finally, F is *split* if $V(F)$ can be partitioned into A, B, one independent and the other complete. The following result is proved in **[45]**.

Theorem 4.3. *If $\mathcal{F}$ is a finite family of graphs that contains a bipartite graph, a cobipartite graph and a split graph then $\mathcal{P}^*_{\mathcal{F}}$ is easily testable.*

As discussed in **[45]**, many known and new results can be derived from Theorem 4.3. For example, Alon and Fox **[7]**, using a somewhat involved ad hoc argument, proved that the property of being induced P_4-free (P_4 is the path on 4 vertices) is easily testable. This follows immediately from Theorem 4.3 since P_4 is bipartite, cobipartite, and split.

The next theorem from **[45]** shows that the sufficient condition in Theorem 4.3 is almost necessary.

Theorem 4.4. *Let $\mathcal{F}$ be a finite family for which $\mathcal{P}^*_{\mathcal{F}}$ is easily testable. Then $\mathcal{F}$ contains a bipartite graph and a cobipartite graph.*

As in the case of Theorem 4.3, the above theorem can also be used in order to obtain many previous results showing that certain properties are not easily testable. Having given both a necessary and a sufficient condition, it is natural to ask if one of them in fact characterizes the finite families $\mathcal{F}$ for which $\mathcal{P}^*_{\mathcal{F}}$ is easily testable. Unfortunately, it was proved in **[45]** that none of them is a characterization. Hence, even the special case of Problem 4.2, that of characterizing the finite families of graph $\mathcal{F}$ for which $\mathcal{P}^*_{\mathcal{F}}$ is easily testable, is still open.

In addition to the above results concerning finite $\mathcal{F}$, **[45]** also obtained a sufficient condition guaranteeing that $\mathcal{P}^*_{\mathcal{F}}$ is easily testable for general families $\mathcal{F}$. Instead of describing this condition, we discuss a corollary of it, which concerns the family of *semialgebraic* graph properties. A semialgebraic graph property $\mathcal{P}$ is given by an integer $k \geq 1$, a set of real $2k$-variate polynomials $f_1, \ldots, f_t \in \mathbb{R}[x_1, \ldots, x_{2k}]$ and a Boolean function $\Phi : \{\text{true}, \text{false}\}^t \to \{\text{true}, \text{false}\}$. A graph G satisfies a property $\mathcal{P}$ if one can assign a point $p_v \in \mathbb{R}^k$ to each vertex $v \in V(G)$ in such a way that a pair of distinct vertices u, v are adjacent if and only if

$$\Phi\big(f_1(p_u, p_v) \geq 0, \ldots, f_t(p_u, p_v) \geq 0\big) = \text{true}.$$

In the expression $f_i(p_u, p_v)$, we substitute p_u into the first k variables of f_i and p_v into the last k variables of f_i.

Some examples of semialgebraic graph properties are those that correspond to being an intersection graph of certain semialgebraic sets in $\mathbb{R}^k$. For example, a graph is an *interval graph* if one can assign an interval in $\mathbb{R}$ to each vertex so that u, v are adjacent iff their intervals intersect. Similarly, a graph is a *unit disc graph* if it is the intersection graph of unit discs in $\mathbb{R}^2$.

The family of semialgebraic graph properties has been extensively studied by many researchers, see, e.g., [35] and its references. Alon conjectured that every semialgebraic graph property is easily testable. This conjecture was verified in [45].

Theorem 4.5. *Every semialgebraic graph property is easily testable.*

The proofs of Theorems 4.3 and 4.5 use the *conditional regularity lemma* of Alon, Fischer, and Newman [5]. This variant of the regularity lemma states that if there is a fixed bipartite graph H so that the graph G has no induced copy of H (when considering only the edges connecting the two sided of H), then G has an ε-regular partition of size only $\text{poly}(1/\varepsilon)$. In fact, the same statement holds if G has only a few copies of H. One of the key steps in the proofs of Theorems 4.3 and 4.5 is then to show that for every relevant property $\mathcal{P}$, an appropriate H as above exists. A related strategy was taken in [32] in the setting of tournaments.

We conclude this subsection with a problem of Alon [1], who asked to characterize the graphs H for which the property of being induced H-free is easily testable. It can be easily checked that Theorems 4.3 and 4.4 can be used to answer this question for all graphs except C_4. There is an interesting reason why this case remains elusive. As we mentioned in the previous paragraph, in the proof of Theorem 4.3 we use the fact that graphs satisfying the properties in its statement are guaranteed to have ε-regular partitions of order $\text{poly}(1/\varepsilon)$. The reason why induced C_4-freeness is harder is that a graph might satisfy this property and still only have regular partitions as in Gowers's example [52] (i.e., having tower-type size). To see this, one just has to note that every split graph (defined before Theorem 4.3) is induced C_4-free, and that one can assume that Gowers's example is a bipartite graph. Hence, taking this graph, and turning one of its independent sets into a complete graph, gives the required example.

Alon and Fox [7] asked if one can improve the tower-type bounds for induced C_4-freeness, which follow from the bound on $\text{Rem}^*_H(\varepsilon)$ discussed in Section 3.2. The following result of [42] improved this to a mere exponential bound.

Theorem 4.6. $\text{Rem}_{C_4}(\varepsilon) \leq 2^{\text{poly}(1/\varepsilon)}$.

The problem of improving this bound to $\text{poly}(1/\varepsilon)$ remains open.

4.3. Removal lemmas of prescribed growth and generalized Turán problems

In the previous parts of this paper, we have mentioned that there are various types of lower and upper bounds for the function $\text{Rem}_{\mathcal{P}}(\varepsilon)$. However, in all cases we could either prove that this function is polynomial (as in the previous subsection) or we had a huge tower-type difference between the best lower and upper bounds (e.g., when $\mathcal{P}$ is triangle-freeness, see (3.1) and (3.2)). This raises the natural question of finding, for a given growth function f, a property $\mathcal{P}$ for which $\mathrm{Rem}_{\mathcal{P}}(\varepsilon) \approx f(\varepsilon)$. A further motivation for this problem comes from theoretical computer science (see Section 4.1). One of the most basic results in this area is the *time hierarchy theorem*, stating (roughly) that for every (natural) function f, there are computational tasks requiring time $f(n)$ on inputs of size n. There are other theorems of this type with respect to memory usage, random bits, etc. Goldreich [48] asked for such a hierarchy theorem for the query complexity of testing graph properties. As we discussed in Section 4.1, the query complexity of testing a hereditary $\mathcal{P}$ with one-sided error is given by $\mathrm{Rem}_{\mathcal{P}}(\varepsilon)$. Hence, the following theorem from [43] gives a hierarchy theorem for testing graph properties with one-sided error.

Theorem 4.7. *For every decreasing $f : (0,1) \to \mathbb{N}$ satisfying $f(x) \geq 1/x$, there is a hereditary graph property $\mathcal{P}$ satisfying $f(\varepsilon) \leq \mathrm{Rem}_{\mathcal{P}}(\varepsilon) \leq \varepsilon^{-14} f(\varepsilon/c)$, where c is an absolute constant.*

As an immediate application of the above theorem, we see that there is a property $\mathcal{P}$ for which $\mathrm{Rem}_{\mathcal{P}}(\varepsilon) = 2^{\Theta(1/\varepsilon)}$ or one for which $\mathrm{Rem}_{\mathcal{P}}(\varepsilon) = \mathrm{twr}(\Theta(1/\varepsilon))$. The properties used in the proof of Theorem 4.7 are quite simple. Given f, the property $\mathcal{P}$ is that of not containing a cycle whose length belongs to the set of integers $\{a_1, a_2, \ldots\}$ where $a_1 = 3$ and for every $i \geq 1$ we define $a_{i+1} = 2f(1/2(a_i + 2)^2) + 1$.

While the properties used in the proof of Theorem 4.7 are simple, the proof that they satisfy its assertion is more complicated, and relies on a theorem we describe below. Turán's Theorem [100], one of the cornerstone results in graph theory, determines the maximum number of edges in an n-vertex graph that does not contain a K_t (the complete graph on t vertices). Turán's problem is the following more general question: for a fixed graph H and an integer n, what is the maximum number of edges in an n-vertex H-free graph? This quantity is denoted by $\mathrm{ex}(n, H)$. Estimating $\mathrm{ex}(n, H)$ for various graphs H is one of the most well-studied problems in graph theory. Alon and Shikhelman [12] have recently initiated the systematic study of the following natural generalization of $\mathrm{ex}(n, H)$; for fixed graphs H and T, estimate $\mathrm{ex}(n, T, H)$, which is the maximum number of copies of T in an n-vertex graph that contains no copy of H. Note that $\mathrm{ex}(n, H) = \mathrm{ex}(n, K_2, H)$. For the sake of brevity, we refer the reader to [12] for more background and motivation. Let us just mention that this family of problems is also related to those discussed in Section 2 since it is not hard to see that if we set D to be the graph comprising of two triangles sharing an edge, then $\mathrm{ex}(n, K_3, D) = \Theta(f_3(n, 6, 3))$.

Some of the most well-studied graphs analyzed in the setting of Turán problems are cycles. In the setting of generalized Turán problems, Bollobás and Györi [19] and Györi and Li [58] obtained tight bounds for $\mathrm{ex}(n, C_3, C_5)$ and $\mathrm{ex}(n, C_3, C_{2k+1})$. The main result of [43] was a tight bound for $\mathrm{ex}(n, C_k, C_\ell)$ for all pairs k, ℓ. For odd cycles, it states the following.

Theorem 4.8. *For every $2 \leq k < \ell$, we have $\mathrm{ex}(n, C_{2k+1}, C_{2\ell+1}) = \Theta_k(\ell^{k+1} n^k)$.*

4.4. Three generalizations of induced $\mathcal{F}$-freeness

Removal for linear combinations of subgraph statistics. For a fixed integer h, let us assign a weight $w_H \in [0, 1]$ to every graph H on h vertices, and then collect all these weights into a sequence denoted $\overline{w}$. Let $d_H(G)$ denote the fraction of subsets of $V(G)$ of size h that induce a copy of H. Given h, a sequence of weights $\overline{w}$ as above, and $c \geq 0$, we say that a graph G satisfies $\mathcal{P}_{h,\overline{w},c}$ if $\sum_H w_H \cdot d_H(G) \leq c$. Note that if $c = 0$ and the only nonzero entry of $\overline{w}$ is w_H, then $\mathcal{P}_{h,\overline{w},c}$ is the property of being induced H-free. In a similar manner, for every finite family of graphs $\mathcal{F}$, we can encode the property of being induced $\mathcal{F}$-free. In particular, for every H, we can encode the property of being (not necessarily induced) H-free. Goldreich and Shinkar **[51]** conjectured that one can extend the induced removal lemma **[4]** by proving that every property $\mathcal{P}_{h,\overline{w},c}$ is testable. They in fact conjectured that these properties can be tested using a very restricted type of testing algorithm. A result of **[47]** shows that some of these properties are not testable at all.

Theorem 4.9. *There is a property $\mathcal{P}_{4,\overline{w},\frac{5}{16}}$ which is not testable with $n^{1/100}$ queries.*

To prove the above theorem, it is shown in **[47]** how to define a vector $\overline{w}$ so that the resulting property $\mathcal{P}_{4,\overline{w},\frac{5}{16}}$ encodes the property of being a quasirandom graph in the sense of Chung, Graham, and Wilson **[23]**. It remains an open problem to decide if the properties $\mathcal{P}_{h,\overline{w},c}$ can at least be tested using $o(n^2)$ edges queries.

Removal against an arbitrary distribution. Suppose G_1, G_2 are two graphs on the same vertex-set V and $\mathcal{D}$ is a distribution on V. The distance between G_1 and G_2 with respect to $\mathcal{D}$ is then defined to be $\sum_{\{x,y\} \in E(G_1) \Delta E(G_2)} \mathcal{D}(x) \cdot \mathcal{D}(y)$. We say that the pair $(G, \mathcal{D})$ is ε-far from satisfying a graph property $\mathcal{P}$ if for every $G' \in \mathcal{P}$, the distance between G and G' with respect to $\mathcal{D}$ is at least ε. Observe that the above definition generalizes the definitions we presented at the beginning of Section 3 which correspond to the uniform distribution over $V(G)$, that is, the one that assigns every vertex a weight of $1/n$. Now, for a given hereditary property $\mathcal{P}$ and $\varepsilon > 0$ we let $\text{Rem}'_{\mathcal{P}}(\varepsilon)$ be the smallest integer so that for *every* distribution $\mathcal{D}$, if G is ε-far from $\mathcal{P}$ with respect to $\mathcal{D}$, then a sample of $\text{Rem}'_{\mathcal{P}}(\varepsilon)$ vertices from $V(G)$, sampled *according to* $\mathcal{D}$, induces a graph not satisfying $\mathcal{P}$ with probability at least $2/3$. The order of quantifiers here is crucial; the definition of $\text{Rem}'_{\mathcal{P}}(\varepsilon)$ requires that it would suffice for *every* $\mathcal{D}$. It is again clear that the above definition is much stronger than the one introduced at the beginning of Section 4, since $\text{Rem}_{\mathcal{P}}(\varepsilon)$ only applies to the uniform distribution.

A priori it is not clear why a function $\text{Rem}'_{\mathcal{P}}(\varepsilon)$ as above should exist for any (interesting) hereditary property. Goldreich **[49]** proved that such a function indeed exists for several types of hereditary properties. The main motivation for his study was that similar algorithmic tasks have been studied in many other settings, where they are called *distribution-free algorithms*, see **[49]** for more background. Goldreich asked if a function $\text{Rem}'_{\mathcal{P}}(\varepsilon)$ as above exists for every hereditary $\mathcal{P}$. To answer this question, we need an important definition. We say that a graph property $\mathcal{P}$ is *extendable* if for every graph G satisfying $\mathcal{P}$, we can add to G a new vertex v and connect it to $V(G)$ in such a way that the resulting graph will also satisfy $\mathcal{P}$. The following answer to Goldreich's problem was given in **[46]**.

orem 4.10.

If $\mathcal{P}$ is hereditary, then $\text{Rem}'_{\mathcal{P}}(\varepsilon)$ *exists if and only if $\mathcal{P}$ is extendable.*

It was also proved in [46] that several natural restrictions on $\mathcal{D}$ guarantee that $\text{Rem}'_{\mathcal{P}}(\varepsilon)$ exists for every hereditary $\mathcal{P}$. For example, this is the case if we assume that $\max_{v \in V(G)} \mathcal{D}(v) = o(1)$ or if we assume that $\min_{v \in V(G)} \mathcal{D}(v) = \Omega(1/|V(G)|)$. At a high level, the proof of Theorem 4.10 in [46] follows the framework of [9], but the fine details differ substantially. The proof in [9] uses Szemerédi's regularity lemma and its variants in order to handle every hereditary property, but only with respect to the uniform distribution. In [46] a new version of the regularity lemma is introduced, which takes into account the weight function $\mathcal{D}$, yet produces bounds that are independent of $\mathcal{D}$ (for extendable properties).

Removal for ordered graphs and matrices. For a fixed $k \times k$ matrix H with $0/1$ entries, we say that an $n \times n$ matrix A is H-free if there are no $r_1 < \cdots < r_k$ and $c_1 < \cdots < c_k$ so that $A_{r_i,c_j} = H_{i,j}$ for every $i, j \in [k]$. We define A to be ε-far from being H-free if one should change at least εn^2 of its entries to make it H-free. Observe that the matrix property of being H-free depends on the ordering of rows and columns. This is in sharp contrast to the graph property of being H-free which is independent of the "names" (or the ordering) of the vertices.

Alon, Fischer, and Newman [5] asked if the graph removal lemma can be extended to the setting of matrices, that is, if every A that is ε-far from being H-free contains at least $n^{2k} / \text{Rem}_H(\varepsilon)$ copies of H. Alon, Ben-Eliezer, and Fischer [2] recently gave a positive answer to this question, showing that $\text{Rem}_H(\varepsilon)$ can be bounded by a wowzer function of ε. Using the methods discussed in Section 3 this can probably be reduced to a tower-type bound. But the following problem is still open.

Problem 4.11. Obtain $\text{poly}(1/\varepsilon)$ bounds for the matrix removal lemma.

We should point out that Alon, Fischer, and Newman [5] obtained a polynomial bound for the *unlabeled* variant of the matrix removal lemma, that is, one where the order of rows/columns of H does not matter. Equivalently, the result of [5] gives a polynomial bound for the induced removal lemma (see Section 3.2) in bipartite graphs. In this case the input G is an $n \times n$ bipartite graph, and G has an induced copy of a bipartite H on vertex sets U_1, U_2 if it has an induced copy in which $U_1 \subseteq V_1$ and $U_2 \subseteq V_2$, or vice versa. This efficient removal lemma was instrumental in the results described in Section 4.2.

4.5. Arithmetic removal lemmas for linear equations and functions

Considering the importance of the removal lemma, it is natural to ask if analogous results can be obtained in other settings. A notable example is the removal lemma for linear equations over groups obtained by Green [56]. A significantly simpler proof of Green's result was obtained by Král', Serra, and Vena [65] who derived it from the graph removal lemma. To state Green's result, we need a few natural generalizations of the notions we used in the setting of graphs. Let $S \subseteq [n]$ be a set of integers, let M be an $\ell \times t$ integer matrix, and $b \in \mathbb{N}^{\ell}$ an integer vector. We say that S is (M, b)-free if there is no $x \in S^t$ satisfying $Mx = b$, and say that S is ε-far from being (M, b)-free if we need to remove at least εn of its elements to make it (M, b)-free. Finally, we say that the pair (M, b) has the *removal property* if there is a function $\mathrm{Rem}_{M,b}(\varepsilon)$ so that if S is ε-far from being (M, b)-free, then S^t contains at least $n^{t-\ell}/\mathrm{Rem}_{M,b}(\varepsilon)$ vectors x satisfying $Mx = b$. Green's result then states that for $\ell = 1$ (i.e., for a single equation), every pair $(M, 0)$ has the removal property. He conjectured [56] that, for every M, the pair $(M, 0)$ has the removal property. Green's conjecture was verified in the following stronger form independently by [66] and [89]. Both proofs rely on Theorem 3.8.

Theorem 4.12. *Every pair (M, b) has the removal property.*

We conclude by describing an extension of Theorem 4.1 from the setting of graphs to the setting of boolean functions $f : \mathbb{F}_2^n \to \{0, 1\}$. Let $\mathcal{P}$ be a property of such functions, and say that f is ε-far from satisfying $\mathcal{P}$ if one should change the truth table of f in at least $\varepsilon 2^n$ places to make it satisfy $\mathcal{P}$. Let $\mathcal{T}$ be the property of such functions indicating that there is no pair $x, y \in \mathbb{F}_2^n$ so that $f(x) = f(y) = f(x + y) = 1$. Then Green's result [56] (mentioned above) implies that if f is ε-far from $\mathcal{T}$, then there are $2^{2n}/\mathrm{Rem}(\varepsilon)$ many pairs x, y witnessing this fact. Green's proof gave a tower-type bound for $\mathrm{Rem}(\varepsilon)$, which was improved to $\mathrm{poly}(1/\varepsilon)$ by [33], using tools related to the solution of the famous cap-set conjecture (see the discussion in [33]).

It is natural to ask for a unifying explanation for why property $\mathcal{T}$ above obeys a removal lemma. Such a systematic study was initiated by Kaufman and Sudan [63] who emphasized the role of invariance. Observe that a key feature in graph properties is that vertex names do not play a role, or more formally, they are closed under isomorphism. This motivated [17] to conjecture that a result analogous to Theorem 4.1 should hold in the setting of boolean functions. To state it we need two definitions. A property of boolean functions is *linear-invariant* if for every $f \in \mathcal{P}$ and any linear transformation $L : \mathbb{F}_2^n \to \mathbb{F}_2^n$ we have $f \circ L \in \mathcal{P}$ where $(f \circ L)(x) = f(L(x))$. We also say that a linear invariant $\mathcal{P}$ is *subspace-hereditary* if for every $f \in \mathcal{P}$ and every linear subspace U of $\mathbb{F}_2^n$ the restriction $f_{|U} \in \mathcal{P}$. We can thus think of linear-invariant subspace-hereditary properties as the analogue of hereditary graph properties. To further emphasize this analogy, it was observed in [17] that just as hereditary properties are those characterized by a (possibly infinite) family of forbidden induced subgraphs (as discussed after Theorem 4.1), then every linear-invariant subspace-hereditary property can be characterized by a (possibly infinite) family of forbidden "patterns" like the one forbidden in the above $\mathcal{T}$. The conjecture raised in [17] was that every linear-invariant subspace-hereditary property of boolean functions has a removal lemma. This conjecture was recently verified by Tidor and Zhao [99].

ACKNOWLEDGMENTS

I am grateful to Noga Alon, Lior Gishboliner, and Yevgeny Levanzov for helpful remarks on a preliminary version of this manuscript.

FUNDING

Supported in part by ISF Grant 1028/16, ERC Consolidator Grant 863438 and NSF-BSF Grant 2019.

REFERENCES

[1] N. Alon, Testing subgraphs in large graphs. *Random Structures Algorithms* **21** (2002), 359–370.

[2] N. Alon, O. Ben-Eliezer, and E. Fischer, Testing hereditary properties of ordered graphs and matrices. In *IEEE 58th Annual Symposium on Foundations of Computer Science (FOCS)*, pp. 848–858, 2017.

[3] N. Alon, R. A. Duke, H. Lefmann, V. Rödl, and R. Yuster, The algorithmic aspects of the Regularity Lemma. *J. Algorithms* **16** (1994), 80–109.

[4] N. Alon, E. Fischer, M. Krivelevich, and M. Szegedy, Efficient testing of large graphs. *Combinatorica* **20** (2000), 451–476.

[5] N. Alon, E. Fischer, and I. Newman, Testing of bipartite graph properties. *SIAM J. Comput.* **37** (2007), 959–976.

[6] N. Alon, E. Fischer, I. Newman, and A. Shapira, A combinatorial characterization of the testable graph properties: it's all about regularity. *SIAM J. Comput.* **39** (2009), 143–167.

[7] N. Alon and J. Fox, Easily testable graph properties. *Combin. Probab. Comput.* **24** (2015), 646–657.

[8] N. Alon and A. Shapira, On an extremal hypergraph problem of Brown, Erdős and Sós. *Combinatorica* **26** (2006), 627–645.

[9] N. Alon and A. Shapira, A characterization of the (natural) graph properties testable with one-sided error. *SIAM J. Comput.* **37** (2008), 1703–1727.

[10] N. Alon and A. Shapira, A separation theorem in property testing. *Combinatorica* **28** (2008), 261–281.

[11] N. Alon, A. Shapira, and B. Sudakov, Additive approximation for edge-deletion problems. *Ann. of Math.* **170** (2009), 371–411.

[12] N. Alon and C. Shikhelman, Many T copies in H-free graphs. *J. Combin. Theory Ser. B* **121** (2016), 146–172.

[13] S. Arora, C. Lund, R. Motwani, M. Sudan, and M. Szegedy, Proof verification and the hardness of approximation problems. *J. ACM* **45** (1998), 501–555.

[14] S. Arora and S. Safra, Probabilistic checking of proofs: a new characterization of NP. *J. ACM* **45** (1998), 70–122.

[15] T. Austin and T. Tao, Testability and repair of hereditary hypergraph properties. *Random Structures Algorithms* **36** (2010), 373–463.

[16] F. A. Behrend, On sets of integers which contain no three terms in arithmetic progression. *Proc. Natl. Acad. Sci. USA* **32** (1946), 331–332.

[17] A. Bhattacharyya, E. Grigorescu, and A. Shapira, A unified framework for testing linear-invariant properties. *Random Structures Algorithms* **46** (2015), 232–260.

[18] M. Blum, M. Luby, and R. Rubinfeld, Self-testing/correcting with applications to numerical problems. *J. Comput. System Sci.* **47** (1993), 549–595.

[19] B. Bollobás and E. Györi, Pentagons vs. triangles. *Discrete Math.* **308** (2008), 4332–4336.

[20] C. Borgs, J. Chayes, L. Lovász, V. T. Sós, B. Szegedy, and K. Vesztergombi, Graph limits and parameter testing. In *Proc. of STOC 2006*, pp. 261–270, 2006.

[21] W. G. Brown, P. Erdős, and V. T. Sós, On the existence of triangulated spheres in 3-graphs and related problems. *Period. Math. Hungar.* **3** (1973), 221–228.

[22] W. G. Brown, P. Erdős, and V. T. Sós, Some extremal problems on r-graphs. In *New Directions in the Theory of Graphs, Proc. 3rd Ann Arbor Conference on Graph Theory*, pp. 55–63, Academic Press, New York, 1973.

[23] F. R. K. Chung, R. L. Graham, and R. M. Wilson, Quasi-random graphs. *Combinatorica* **9** (1989), 345–362.

[24] D. Conlon and J. Fox, Bounds for graph regularity and removal lemmas. *Geom. Funct. Anal.* **22** (2012), 1191–1256.

[25] D. Conlon and J. Fox, Graph removal lemmas. In *Surveys in Combinatorics*, pp. 1–50, Cambridge University Press, 2013.

[26] D. Conlon, L. Gishboliner, Y. Levanzov, and A. Shapira, A new bound for the Brown–Erdős–Sós problem, submitted.

[27] R. Duke, H. Lefmann, and V. Rödl, A fast approximation algorithm for computing the frequencies of subgraphs in a given graph. *SIAM J. Comput.* **24** (1995), 598–620.

[28] P. Erdős, Problems and results in combinatorial number theory. In *Journées Arithmétiques de Bordeaux (Conf., Univ. Bordeaux, Bordeaux, 1974)*, pp. 295–310, Astérisque 24–25, Soc. Math. France, Paris, 1975.

[29] P. Erdős, P. Frankl, and V. Rödl, The asymptotic number of graphs not containing a fixed subgraph and a problem for hypergraphs having no exponent. *Graphs Combin.* **2** (1986), 113–121.

[30] E. Fischer and I. Newman, Testing versus estimation of graph properties. *SIAM J. Comput.* **37** (2007), 482–501.

[31] J. Fox, A new proof of the graph removal lemma. *Ann. of Math.* **174** (2011), 561–579.

[32] J. Fox, L. Gishboliner, A. Shapira, and R. Yuster, The Removal Lemma for Tournaments. *J. Combin. Theory Ser. B* **136** (2019), 110–134.

[33] J. Fox and L. M. Lovász, A tight bound for Green's arithmetic triangle removal lemma in vector spaces. *Adv. Math.* **321** (2017), 287–297.

[34] J. Fox and L. M. Lovász, A tight lower bound for Szemerédi's regularity lemma. *Combinatorica* **37** (2017), 911–951.

[35] J. Fox, J. Pach, and A. Suk, A polynomial regularity lemma for semi-algebraic hypergraphs andits applications in geometry and property testing. *SIAM J. Comput.* **45**, 2199–2223.

[36] P. Frankl and V. Rödl, Extremal problems on set systems. *Random Structures Algorithms* **20** (2002), 131–164.

[37] A. Frieze and R. Kannan, Quick approximation to matrices and applications. *Combinatorica* **19** (1999), 175–220.

[38] Z. Füredi and M. Ruszinkó, Uniform hypergraphs containing no grids. *Adv. Math.* **240** (2013), 302–324.

[39] H. Furstenberg, Ergodic behaviour of diagonal measures and a theorem of Szemerédi on arithmetic progressions. *J. Anal. Math.* **31** (1997), 204–256.

[40] H. Furstenberg and Y. Katznelson, An ergodic Szemerédi theorem for commuting transformations. *J. Anal. Math.* **34** (1978), 275–291.

[41] L. Gishboliner, Y. Levanzov, and A. Shapira, An approximate version of the Gowers–Long conjecture, submitted.

[42] L. Gishboliner and A. Shapira, Efficient removal without efficient regularity. *Combinatorica* **39** (2019), 639–658.

[43] L. Gishboliner and A. Shapira, A generalized Turán problem and its applications. *Int. Math. Res. Not.* **11** (2020), 3417–3452.

[44] L. Gishboliner and A. Shapira, Constructing dense grid-free linear 3-graphs. *Proc. Amer. Math. Soc.*, to appear.

[45] L. Gishboliner and A. Shapira, Removal lemmas with polynomial bounds. *Int. Math. Res. Not.*, to appear.

[46] L. Gishboliner and A. Shapira, Testing graphs against an unknown distribution. *Israel J. Math.*, to appear.

[47] L. Gishboliner, A. Shapira, and H. Stagni, Testing Linear Inequalities of Subgraph Statistics. *Random Structures Algorithms* **58** (2021), 468–479.

[48] O. Goldreich, *Introduction to Property Testing*. Cambridge University Press, 2017.

[49] O. Goldreich, Testing graphs in vertex-distribution-free models. In *Proc. STOC 2016*, pp. 527–534, 2016.

[50] O. Goldreich, S. Goldwasser, and D. Ron, Property testing and its connection to learning and approximation. *J. ACM* **45** (1998), 653–750.

[51] O. Goldreich and I. Shinkar, Two-sided error proximity oblivious testing. *Random Structures Algorithms* **48** (2016), 341–383.

[52] W. T. Gowers, Lower bounds of tower type for Szemerédi's uniformity lemma. *Geom. Funct. Anal.* **7** (1997), 322–337.

[53] W. T. Gowers, A new proof of Szemerédi's theorem. *Geom. Funct. Anal.* **11** (2001), 465–588.

[54] W. T. Gowers, Hypergraph regularity and the multidimensional Szemerédi theorem. *Ann. of Math.* **166** (2007), 897–946.

[55] W. T. Gowers and J. Long, The length of an s-increasing sequence of r-tuples. *Combin. Probab. Comput.*, to appear.

[56] B. Green, A Szemerédi-type regularity lemma in abelian groups, with applications. *Geom. Funct. Anal.* **15** (2005), 340–376.

[57] B. Green and T. Tao, The primes contain arbitrarily long arithmetic progressions. *Ann. of Math.* **167** (2008), 481–547.

[58] E. Györi and H. Li, The maximum number of triangles in C_{2k+1}-free graphs. *Combin. Probab. Comput.* **21** (2012), 187–191.

[59] A. Hassidim, J. Kelner, H. Nguyen, and K. Onak, Local graph partitions for approximation and testing. In *50th Annual IEEE Symposium on Foundations of Computer Science (FOCS)*, pp. 22–31, 2009.

[60] C. Hoppen, Y. Kohayakawa, R. Lang, H. Lefmann, and H. Stagni, On the query complexity of estimating the distance to hereditary graph properties. *SIAM J. Discrete Math.* **35** (2021), 1238–1251.

[61] F. Joos, J. Kim, D. Kuhn, and D. Osthus, A characterization of testable hypergraph properties. In *IEEE 58th Annual Symposium on Foundations of Computer Science (FOCS)* pp. 859–867, 2017.

[62] S. Kalyanasundaram and A. Shapira, A wowzer type lower bound for the strong regularity lemma. *Proc. Lond. Math. Soc.* **106** (2013), 621–649.

[63] T. Kaufman and M. Sudan, Algebraic property testing: the role of invariance. In *Proc. 40th Annual ACM Symposium on the Theory of Computing (STOC)*, pp. 403–412, 2008.

[64] P. Keevash and J. Long, The Brown–Erdős–Sós Conjecture for hypergraphs of large uniformity. *Proc. Amer. Math. Soc.*, to appear.

[65] D. Král', O. Serra, and L. Vena, A combinatorial proof of the removal lemma for groups. *J. Combin. Theory Ser. A* **116** (2009), 971–978.

[66] D. Král', O. Serra, and L. Vena, A removal lemma for systems of linear equations over finite fields. *Israel J. Math.* **187** (2012), 193–207.

[67] A. Lev-Ran and A. Shapira, A lower bound for the strong cylinder lemma, in preparation.

[68] L. Lovász, *Large networks and graph limits*. AMS, 2012.

[69] L. Lovász and B. Szegedy, Szemerédi's lemma for the analyst. *Geom. Funct. Anal.* **17** (2007), 252–270.

[70] G. Moshkovitz and A. Shapira, A short proof of Gowers' lower bound for the regularity lemma. *Combinatorica* **36** (2016), 187–194.

[71] G. Moshkovitz and A. Shapira, A sparse regular approximation lemma. *Trans. Amer. Math. Soc.* **371** (2019), 6779–6814.

[72] G. Moshkovitz and A. Shapira, A tight bound for hypergraph regularity. *Geom. Funct. Anal.* **29** (2019), 1531–1578.

[73] B. Nagle, V. Rödl, and M. Schacht, The counting lemma for regular k-uniform hypergraphs. *Random Structures Algorithms* **28** (2006), 113–179.

[74] R. Nenadov, B. Sudakov, and M. Tyomkyn, Proof of the Brown–Erdős–Sós conjecture in groups. *Math. Proc. Cambridge Philos. Soc.* **169** (2020), 323–333.

[75] V. Rödl, Quasi-randomness and the regularity method in hypergraphs. In *Proceedings of the International Congress of Mathematicians (ICM) 1*, pp. 571–599, 2015.

[76] V. Rödl and R. Duke, On graphs with small subgraphs of large chromatic number. *Graphs Combin.* **1** (1985), 91–96.

[77] V. Rödl and M. Schacht, Regular partitions of hypergraphs: counting lemmas. *Combin. Probab. Comput.* **16** (2007), 887–901.

[78] V. Rödl and M. Schacht, Regular partitions of hypergraphs: regularity lemmas. *Combin. Probab. Comput.* **16** (2007), 833–885.

[79] V. Rödl and M. Schacht, Generalizations of the removal lemma. *Combinatorica* **29** (2009), 467–501.

[80] V. Rödl and M. Schacht, Regularity lemmas for graphs. In *Fete of Combinatorics and Computer Science*, pp. 287–325, Bolyai Soc. Math. Stud. 20, 2010.

[81] V. Rödl and J. Skokan, Regularity lemma for k-uniform hypergraphs. *Random Structures Algorithms* **25** (2004), 1–42.

[82] K. F. Roth, On certain sets of integers (II). *J. Lond. Math. Soc.* **29** (1954), 20–26.

[83] R. Rubinfeld and M. Sudan, Robust characterizations of polynomials with applications to program testing. *SIAM J. Comput.* **25** (1996), 252–271.

[84] I. Ruzsa, Solving a linear equation in a set of integers I. *Acta Arith.* **65** (1993), 259–282.

[85] I. Ruzsa and E. Szemerédi, Triple systems with no six points carrying three triangles. In *Combinatorics (Keszthely, 1976), Volume II*, pp. 939–945, Colloq. Math. Soc. János Bolyai 18, North Holland, Amsterdam, 1978.

[86] S. Sapir and A. Shapira, The induced removal lemma in sparse graphs. *Combin. Probab. Comput.* **29** (2020), 153–162.

[87] G. N. Sárközy and S. Selkow, An extension of the Ruzsa–Szemerédi theorem. *Combinatorica* **25** (2004), 77–84.

[88] C. Shangguan and I. Tamo, Sparse hypergraphs with applications to coding theory. *SIAM J. Discrete Math.* **34** (2020), 1493–1504.

[89] A. Shapira, A proof of Green's conjecture regarding the removal properties of sets of linear equations. *J. Lond. Math. Soc.* **81** (2010), 355–373.

[90] A. Shapira and M. Tyomkyn, A Ramsey variant of the Brown–Erdős–Sós conjecture. *Bull. Lond. Math. Soc.*

[91] D. Solymosi and J. Solymosi, Small cores in 3-uniform hypergraphs. *J. Combin. Theory Ser. B* **122** (2017), 897–910.

[92] J. Solymosi, A note on a question of Erdős and Graham. *Combin. Probab. Comput.* **13** (2004), 263–267.

[93] E. Szemerédi, On sets of integers containing no four elements in arithmetic progression. *Acta Math. Acad. Sci. Hung.* **20** (1969), 89–104.

[94] E. Szemerédi, On sets of integers containing no k elements in arithmetic progression. *Acta Arith.* **27** (1975), 199–245.

[95] E. Szemerédi, Regular partitions of graphs. In *Proc. Colloque Inter. CNRS*, pp. 399–401, 1978.

[96] T. Tao, The dichotomy between structure and randomness, arithmetic progressions, and the primes. In *Proc. Intern. Congress of Math. I*, pp. 581-–608, Eur. Math. Soc., Zurich, 2006.

[97] T. Tao, A variant of the hypergraph removal lemma. *J. Combin. Theory Ser. A* **113** (2006), 1257–1280.

[98] A. Thomason, Pseudorandom graphs. In *Random graphs '85 (Poznań, 1985)*, pp. 307–331, North-Holl. Math. Stud. 144, North-Holland, Amsterdam, 1987.

[99] J. Tidor and Y. Zhao, Testing linear-invariant properties. In *IEEE 61st Annual Symposium on Foundations of Computer Science (FOCS)*, pp. 1180–1190, 2020.

[100] P. Turán, On an extremal problem in graph theory. *Mat. Fiz. Lapok* **48** (1941), 436–452.

ASAF SHAPIRA

School of Mathematical Sciences, Tel Aviv University, Tel Aviv, 6997801, Israel, asafico@tauex.tau.ac.il

This Page Intentionally Left Blank

THE POSITIVE GRASSMANNIAN, THE AMPLITUHEDRON, AND CLUSTER ALGEBRAS

LAUREN K. WILLIAMS

ABSTRACT

The *positive Grassmannian* $\mathrm{Gr}_{k,n}^{\geq 0}$ is the subset of the real Grassmannian where all Plücker coordinates are nonnegative. It has a beautiful combinatorial structure as well as connections to statistical physics, integrable systems, and scattering amplitudes. The *amplituhedron* $\mathcal{A}_{n,k,m}(Z)$ is the image of the positive Grassmannian $\mathrm{Gr}_{k,n}^{\geq 0}$ under a positive linear map $\mathbb{R}^n \to \mathbb{R}^{k+m}$. We will explain how ideas from oriented matroids, tropical geometry, and cluster algebras shed light on the structure of the positive Grassmannian and the amplituhedron.

MATHEMATICS SUBJECT CLASSIFICATION 2020

Primary 05E14; Secondary 13F60, 14M15, 52B40, 14T90, 14J81

KEYWORDS

Positive Grassmannian, amplituhedron, cluster algebras, tropical geometry, matroid, positroid

ICM INTERNATIONAL CONGRESS OF MATHEMATICIANS 2022 JULY 6–14

Proc. Int. Cong. Math. 2022, Vol. 6, pp. 4710–4737
DOI 10.4171/ICM2022/126

Published by EMS Press

1. INTRODUCTION

The *totally nonnegative Grassmannian* $\mathrm{Gr}_{k,n}^{\geq 0}$ or (informally) the *positive Grassmannian* [40, 47] can be defined as the subset of the real Grassmannian $\mathrm{Gr}_{k,n}$ where all Plücker coordinates are nonnegative. It has a beautiful decomposition into *positroid cells* [47, 49, 52], where each cell is obtained by specifying that certain Plücker coordinates are strictly positive and the rest are zero. Since the work of Lusztig [40] and Postnikov [47, 48], there has been an extensive study of the positive Grassmannian, including approaches involving cluster algebras, tropical geometry, and matroids and the moment map.

Remarkably, the positive Grassmannian has several applications in theoretical physics. For example, the stationary distribution of the *asymmetric simple exclusion process* (a model for particles hopping on a one-dimensional lattice with open boundaries), can be described in terms of cells of $\mathrm{Gr}_{k,n}^{\geq 0}$ [13]. In a different direction, each point C of the real Grassmannian gives rise to a *soliton solution of the KP equation* (modeling interaction patterns of shallow water waves), whose asymptotics are determined by the matroid of C, and which is regular for all times t if and only if C lies in the positive Grassmannian [35, 36]. In yet a third direction, the positive Grassmannian encodes most of the physical properties of *scattering amplitudes in planar* $\mathcal{N} = 4$ *super Yang–Mills theory* [3, 4, 12] (which compute probabilities that certain particles are produced in a collision involving other particles). This insight combined with an idea of Hodges [29] led Arkani-Hamed and Trnka to introduce the *amplituhedron* [7], defined as the image of the positive Grassmannian under the *amplituhedron map*. In particular, any $n \times (k+m)$ matrix Z whose maximal minors are positive induces a map $\tilde{Z}$ from $\mathrm{Gr}_{k,n}^{\geq 0}$ to the Grassmannian $\mathrm{Gr}_{k,k+m}$, whose image (of full dimension km) is the *amplituhedron* $\mathcal{A}_{n,k,m}(Z)$ [7]. When $m = 4$, the *BCFW recurrence* [11] for computing scattering amplitudes can be used to produce collections of $4k$-dimensional cells in $\mathrm{Gr}_{k,n}^{\geq 0}$ whose images conjecturally subdivide ("tile" or "triangulate") the amplituhedron.

The amplituhedron $\mathcal{A}_{n,k,m}(Z)$ generalizes the positive Grassmannian (obtained when $k + m = n$), cyclic polytopes (when $k = 1$) [7], and cyclic hyperplane arrangements (when $m = 1$) [32]. Moreover, the amplituhedron has intriguing and beautiful mathematical properties, many of them conjectural. For instance, we conjecture that for even m, the number of top-dimensional strata comprising a tiling of $\mathcal{A}_{n,k,m}(Z)$ is equal to the number of plane partitions contained in the $k \times (n-k-m) \times \frac{m}{2}$ box [33]. As another example, despite the fact that they have different dimensions and one of them is not a polytope, the hypersimplex $\Delta_{k+1,n}$ and the amplituhedron $\mathcal{A}_{n,k,2}(Z)$ are closely related: for example, *T-duality* gives a bijection between positroid tilings of $\Delta_{k+1,n}$ and positroid tilings of $\mathcal{A}_{n,k,2}(Z)$ [39, 46].

In this article we explain how ideas from the theory of matroids, tropical geometry, and cluster algebras shed light on the structure of positive Grassmannians and amplituhedra. We start in Section 2 by introducing the matroid stratification of the Grassmannian and the positroid cell decomposition of the positive Grassmannian. Given a surjective map ϕ from a cell complex X onto another topological space Y, we also introduce the notion of *ϕ-induced tiling* of Y, which we will study in the case that X is the positive Grassmannian (and call a *positroid tiling*). In Section 3 we study positroid tilings when ϕ is the moment map, which are subdivisions of the hypersimplex into positroid polytopes, and are related to the positive tropical Grassmannian. In Section 4 we introduce the amplituhedron, giving two equivalent definitions, defining natural coordinates, characterizing its points when $m = 1$ and 2, and defining its *sign stratification*, which is an analogue of the matroid stratification. In Section 5 we then study positroid tilings when ϕ is the amplituhedron map. We give a conjectural link to plane partitions, and discuss the positroid cells on which the amplituhedron map is injective. In Section 6 we explain a mysterious notion called *T-duality*, which relates positroid tiles and tilings of the hypersimplex $\Delta_{k+1,n}$ to positroid tiles and tilings for the amplituhedron $\mathcal{A}_{n,k,2}(Z)$. One manifestation of this duality is the fact that the number of realizable sign strata of $\mathcal{A}_{n,k,2}(Z)$ equals the volume of $\Delta_{k+1,n}$ (an Eulerian number). Finally, in Section 7 we present several connections between the amplituhedron and cluster algebras, proved for $m = 2$ but conjectural in general.

A great many mathematicians and physicists have made tremendous contributions to the study of the positive Grassmannian and amplituhedron; it is impossible to give a complete account here. The results described below in which I played a role are joint with various collaborators including F. Ardila, S. Karp, T. Lukowski, M. Parisi, K. Rietsch, F. Rincón, M. Sherman-Bennett, D. Speyer, K. Talaska, E. Tsukerman, and Y. Zhang.

2. THE POSITIVE GRASSMANNIAN AND MATROID STRATIFICATION

2.1. The Grassmannian and the matroid stratification

The *Grassmannian* $\mathrm{Gr}_{k,n} = \mathrm{Gr}_{k,n}(\mathbb{K})$ is the space of k-dimensional subspaces of an n-dimensional vector space $\mathbb{K}^n$. Let $[n]$ denote $\{1, \dots, n\}$, and $\binom{[n]}{k}$ denote the set of k-element subsets of $[n]$. We can represent a point $V \in \mathrm{Gr}_{k,n}$ as the row-span of a full-rank $k \times n$ matrix C; then, for $I \in \binom{[n]}{k}$, we let $p_I(V)$ be the $k \times k$ minor of C occupying the columns in I. The $p_I(V)$ are called the *Plücker coordinates* of V, and are independent of the choice of matrix representative C (up to common rescaling). The map $V \mapsto \{p_I(V)\}_{I \in \binom{[n]}{k}}$ embeds $\mathrm{Gr}_{k,n}$ into projective space. We will occasionally *identify C with its row-span.*

Definition 2.1. A *matroid* $\mathcal{M}$ is a pair $(E, \mathcal{B})$, where E is a finite set and $\mathcal{B}$ a nonempty collection of subsets of E called *bases*, such that if B_1, B_2 are distinct bases and $b_1 \in B_1 \setminus B_2$, then there exists an element $b_2 \in B_2 \setminus B_1$ such that $(B_1 \setminus \{b_1\}) \cup \{b_2\}$ is a basis.

Matroid theory originated in the 1930s as a combinatorial model that keeps track of, and abstracts, the dependence relations among a set of vectors.

Definition 2.2. Any full-rank $k \times n$ matrix C (with entries in a field $\mathbb{K}$), and consequently any point $C \in \mathrm{Gr}_{k,n}(\mathbb{K})$, gives rise to a matroid $\mathcal{M}(C) := ([n], \mathcal{B})$, where $\mathcal{B} = \{I \in \binom{[n]}{k} \mid p_I(C) \neq 0\}$. Such matroids are called *realizable* or *representable* over $\mathbb{K}$.

Example 2.3. Consider the full rank matrix

$$C = \begin{pmatrix} 1 & 0 & -1 & -2 \\ 0 & 1 & 2 & 4 \end{pmatrix} \quad \text{(or the corresponding point } C \in \mathrm{Gr}_{2,4}\text{)}.$$

Here $p_{12}(C) = 1$, $p_{13}(C) = 2$, $p_{14}(C) = 4$, $p_{23}(C) = 1$, $p_{24}(C) = 2$, and $p_{34}(C) = 0$.

The corresponding matroid is $\mathcal{M}(C) = \{[4], \mathcal{B}\}$ where $\mathcal{B} = \{12, 13, 14, 23, 24\}$.

In what follows, we will be concerned with the *real* Grassmannian $\mathrm{Gr}_{k,n} = \mathrm{Gr}_{k,n}(\mathbb{R})$. While every full rank matrix gives rise to a matroid, there are many matroids which are *not* realizable (say over $\mathbb{R}$), that is, they cannot be realized by (real) matrices. The *non-Pappus matroid* is a matroid which is not realizable over any field.

The *matroid stratification* of the Grassmannian is the decomposition of $\mathrm{Gr}_{k,n}$ into strata consisting of all points with the same matroid. While this stratification has many beautiful properties [24], we also know that by Mnëv's universality theorem [43], a matroid stratum can have topology as bad as that of any algebraic variety!

One running theme in this article will be that matroids and the matroid stratification of the Grassmannian can exhibit pathological behavior, but when one adds the adjective "positive" to the picture, this bad behavior is replaced by the nicest possible statements.

2.2. The positive Grassmannian

Definition 2.4 ([40, 47]). We say that $V \in \mathrm{Gr}_{k,n}$ is *totally nonnegative* if (up to a global change of sign) $p_I(V) \geq 0$ for all $I \in \binom{[n]}{k}$. Similarly, V is *totally positive* if $p_I(V) > 0$ for all $I \in \binom{[n]}{k}$. We let $\mathrm{Gr}_{k,n}^{\geq 0}$ and $\mathrm{Gr}_{k,n}^{>0}$ denote the set of totally nonnegative and totally positive elements of $\mathrm{Gr}_{k,n}$, respectively; $\mathrm{Gr}_{k,n}^{\geq 0}$ is called the *totally nonnegative Grassmannian*, or sometimes just the *positive Grassmannian*.

Note that the matrix C from Example 2.3 represents an element of $\mathrm{Gr}_{2,4}^{\geq 0}$.

The positive and nonnegative parts of a generalized partial flag variety G/P were first introduced by Lusztig [40], who gave a Lie-theoretic definition of $(G/P)_{>0}$ and $(G/P)_{\geq 0} := \overline{(G/P)_{>0}}$. Postnikov [47] subsequently defined $\mathrm{Gr}_{k,n}^{\geq 0}$ as in Definition 2.4. These definitions agree when $G/P = \mathrm{Gr}_{k,n}$ [51], [64, COROLLARY 1.2].

While the positive Grassmannian was introduced rather recently, the theory of totally positive matrices is much older. In fact, one can use results of Gantmakher and Krein [23] from 1950 to characterize $\mathrm{Gr}_{k,n}^{\geq 0}$ and $\mathrm{Gr}_{k,n}^{>0}$ in terms of sign variation [31].

Definition 2.5. Given $v \in \mathbb{R}^n$, let $\mathrm{var}(v)$ be the number of sign changes of v, when v is viewed as a sequence of n numbers and zeros are ignored. We also define

$$\overline{\mathrm{var}}(v) := \max\{\mathrm{var}(w) : w \in \mathbb{R}^n \text{ such that } w_i = v_i \text{ for all } i \in [n] \text{ with } v_i \neq 0\},$$

i.e., $\overline{\mathrm{var}}(v)$ is the maximum number of sign changes after we choose a sign for each $v_i = 0$.

For example, if $v := (2, 0, 2, -1) \in \mathbb{R}^4$, then $\mathrm{var}(v) = 1$ and $\overline{\mathrm{var}}(v) = 3$.

The following result is based on [23, THEOREMS V.3, V.7, V.1, V.6].

Theorem 2.6 ([31, THEOREM 1.1]). *Let $V \in \mathrm{Gr}_{k,n}$ with orthogonal complement $V^\perp \in \mathrm{Gr}_{n-k,n}$.*

(i) $V \in \mathrm{Gr}_{k,n}^{\geq 0} \iff \mathrm{var}(v) \leq k-1 \quad \forall v \in V \iff \overline{\mathrm{var}}(w) \geq k \quad \forall w \in V^\perp$.

(ii) $V \in \mathrm{Gr}_{k,n}^{>0} \iff \overline{\mathrm{var}}(v) \leq k-1 \quad \forall v \in V \setminus \{0\} \iff \mathrm{var}(w) \geq k \quad \forall w \in V^\perp \setminus \{0\}$.

2.3. The positroid cell decomposition

Despite the fact that the topology of matroid strata can be very bad, Postnikov realized that if one intersects these strata with the positive Grassmannian, one obtains a *cell decomposition* [47]. In fact, it is a regular CW decomposition [21, 49, 53, 67].

Theorem 2.7 ([47]). *For* $\mathcal{M} \subseteq \binom{[n]}{k}$, *let*

$$S_{\mathcal{M}} := \{V \in \mathrm{Gr}_{k,n}^{\geq 0} \mid p_I(V) > 0 \text{ if and only if } I \in \mathcal{M}\}.$$

Then $\mathrm{Gr}_{k,n}^{\geq 0} = \cup S_{\mathcal{M}}$ *is a cell decomposition, i.e., each* $S_{\mathcal{M}}$ *is an open ball.*

If $S_{\mathcal{M}} \neq \emptyset$, *we call* $\mathcal{M}$ *a* positroid *and* $S_{\mathcal{M}}$ *its* positroid cell.

More generally, Rietsch gave a cell decomposition of $(G/P)_{\geq 0}$ [52]. When $G/P = \mathrm{Gr}_{k,n}$, the two cell decompositions agree [64, COROLLARY 1.2].

As shown in [47] and explained below, the cells of $\mathrm{Gr}_{k,n}^{\geq 0}$ can be indexed by combinatorial objects such as *decorated permutations* π or move-equivalence classes of *plabic graphs* G, see, e.g., [19, CHAPTER 7]. We will correspondingly refer to such cells as S_π and S_G.

Definition 2.8. A *decorated permutation* on $[n]$ is a permutation $\pi \in S_n$ whose fixed points are each colored either black ("loop") or white ("coloop"). We denote a black fixed point i by $\pi(i) = \underline{i}$, and a white fixed point i by $\pi(i) = \overline{i}$. An *antiexcedance* of a decorated permutation π is an element $i \in [n]$ such that either $\pi^{-1}(i) > i$ or $\pi(i) = \overline{i}$. We say that a decorated permutation on $[n]$ is of *type* (k, n) if it has k antiexcedances.

For example, $\pi = (3, \underline{2}, 5, 1, 6, 8, \overline{7}, 4)$ has a loop in position 2 and a coloop in position 7. Its antiexcedances are 1, 4, and 7.

Definition 2.9. To a $k \times n$ matrix C with columns $(c^1, \dots, c^n)$ representing an element of $\mathrm{Gr}_{k,n}^{\geq 0}$, we associate a decorated permutation $\pi := \pi_C$ of type (k, n) as follows. We set $\pi(i) := j$ to be the label of the first column j such that $c^i \in \mathrm{span}\{c^{i+1}, c^{i+2}, \dots, c^j\}$, where the columns are listed in cyclic order (going from c^n to c^1 if $i + 1 > j$). If $c^i = \mathbf{0}$, then i is a *loop* of matroid $\mathcal{M}(C)$ and we set $\pi(i) = \underline{i}$, and if c^i is not in the span of the other column vectors, then i is a *coloop* of $\mathcal{M}(C)$ and we set $\pi(i) = \overline{i}$.

The construction above gives a well-defined map from $\mathrm{Gr}_{k,n}^{\geq 0}$ to decorated permutations of type (k, n). If C is the matrix from Example 2.3, then $\pi_C = (3, 1, 4, 2)$.

Proposition 2.10. *Let* π *be a decorated permutation of type* (k, n), *and let*

$$S_\pi = \{C \in \mathrm{Gr}_{k,n}^{\geq 0} \mid \pi_C = \pi\}.$$

Then S_π *is a positroid cell, and all positroid cells of* $\mathrm{Gr}_{k,n}^{\geq 0}$ *have the form* S_π *for some decorated permutation* π *of type* (k, n).

Definition 2.11. A *planar bicolored graph* (or *plabic graph*) is a planar graph G properly embedded into a closed disk with (uncolored) vertices lying on the boundary of the disk labeled $1, \dots, n$ in clockwise order for some positive n, such that: each boundary vertex is

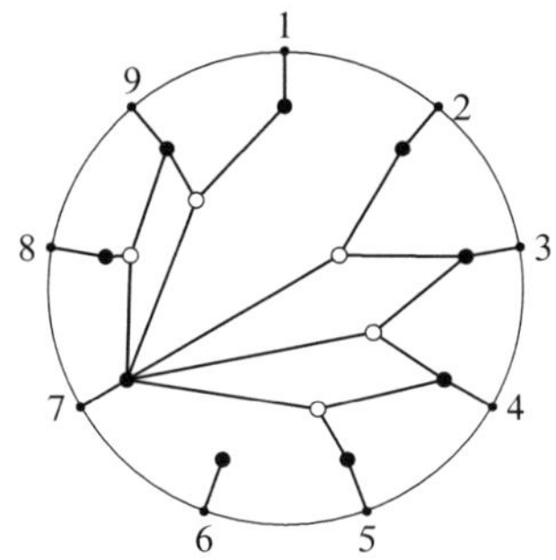

FIGURE 1
A plabic graph G with $\pi_G = (8, 5, 9, 2, 3, \underline{6}, 4, 1, 7)$. It has a black lollipop at 6.

incident to a single edge; each internal vertex is colored black or white; and each internal vertex is connected by a path to some boundary vertex. See Figure 1.

If a boundary vertex i is attached to an edge whose other endpoint is a leaf, we call this component a *lollipop*. *We will assume that G has no internal leaves except for lollipops.*

We next describe some local moves on plabic graphs, see Figure 2.

(M1) Square Move. If there is a square formed by four trivalent vertices whose colors alternate, then we can switch the colors of these four vertices.

(M2) Two adjacent internal vertices of the same color can be merged. Alternatively, we can split an internal vertex into two vertices of the same color joined by an edge.

(M3) We can remove/add degree 2 vertices, as shown.

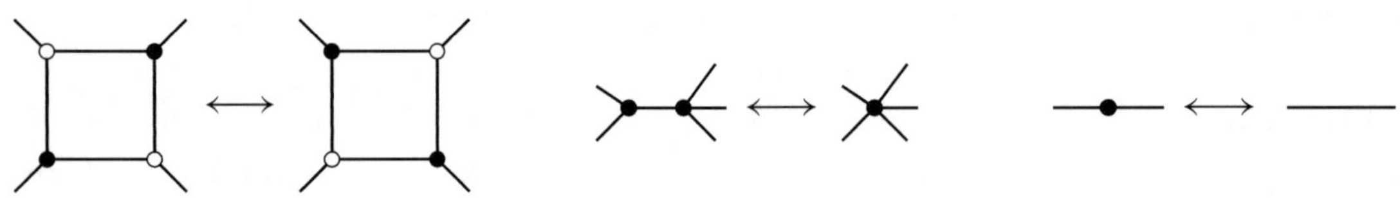

FIGURE 2
Moves (M1), (M2), (M3) on plabic graphs.

Definition 2.12. Two plabic graphs are *move-equivalent* if they can be obtained from each other by moves (M1)–(M3). A plabic graph is *reduced* if there is no graph move-equivalent to it in which two adjacent vertices u and v are connected by more than one edge.

Definition 2.13 and Proposition 2.16 give several ways to read a positroid off of a plabic graph. The positroid depends only on the move-equivalence class of the plabic graph.

Definition 2.13. Let G be a reduced plabic graph as above with boundary vertices $1, \ldots, n$. For each boundary vertex $i \in [n]$, we follow a path along the edges of G starting at i, turning (maximally) right at every internal black vertex, and (maximally) left at every internal white vertex. This path ends at some boundary vertex $\pi(i)$. By **[47, SECTION 13]**, the fact that G is reduced implies that each fixed point of π is attached to a lollipop; we color each fixed point by the color of its lollipop. In this way we obtain the *(decorated) trip permutation* $\pi_G = \pi$ of G. We say that G is of *type* (k, n), where k is the number of antiexcedances of π_G.

In Figure 1 we have $\pi_G = (8, 5, 9, 2, 3, \underline{6}, 4, 1, 7)$, which has $k = 5$ antiexcedances.

Theorem 2.14 (Fundamental theorem of reduced plabic graphs, **[47, THEOREM 13.4]**, see also **[19, THEOREM 7.4.25]**). *Let G and G' be reduced plabic graphs. Then G and G' are move-equivalent if and only if G and G' have the same decorated trip permutation.*

Definition 2.15. Let G be a bipartite plabic graph in which each boundary vertex is incident to a white vertex. An *almost perfect matching* of G is a subset M of edges such that each internal vertex is incident to exactly one edge in M (and each boundary vertex i is incident to either one or no edges in M). We let $\partial M = \{i \mid i \text{ is incident to an edge of } M\}$.

Given a plabic graph, we can use move (M3) to ensure that the resulting graph is bipartite and that each boundary vertex is incident to a white vertex. (Note that we can think of such a graph as a bipartite graph G in which all boundary vertices are colored black.)

Proposition 2.16 (**[47, PROPOSITION 11.7, LEMMA 11.10]**). *Let G be a bipartite plabic graph such that each boundary vertex is incident to a white vertex. Let*

$$\mathcal{M}(G) = \{\partial M \mid M \text{ an almost perfect matching of } G\}.$$

If $\mathcal{M}(G)$ is nonempty, then $\mathcal{M}(G)$ is the set of bases of a positroid on $[n]$. Moreover, all positroids arise from plabic graphs.

See Figure 3 for an example.

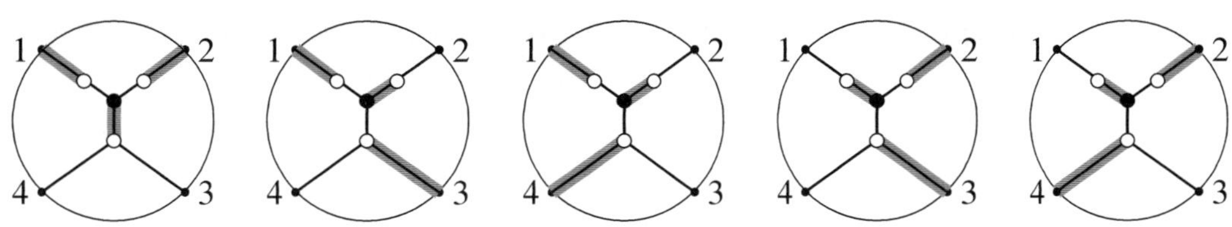

FIGURE 3
A bipartite plabic graph G_1 with $\pi_{G_1} = (3, 1, 4, 2)$ which has five almost-perfect matchings. The corresponding positroid is $([4], \mathcal{M}(G_1))$ where $\mathcal{M}(G_1) = \{12, 13, 14, 23, 24\}$.

Postnikov used plabic graphs to give parameterizations of cells of $\mathrm{Gr}_{k,n}^{\geq 0}$ **[47]**; this result can be recast in terms of *flows* **[63]** or as a variant of a theorem of Kasteleyn **[56]**.

Theorem 2.17 (**[47, 56, 63]**). *Let G be a bipartite plabic graph with n boundary vertices, all of which are colored black. Suppose G has at least one almost perfect matching M_0, and let $k = |\partial M_0|$. Let $w : \mathrm{Edges}(G) \to \mathbb{R}_{>0}$ be any weight function, and for M an almost perfect* *matching, let $w(M) := \prod_{e \in M} w_e$, where w_e denotes the weight of edge e. Then there is a $k \times n$ matrix $L = L(w)$ representing a point of $\mathrm{Gr}_{k,n}^{\geq 0}$ such that*

$$p_I(L) = \sum_{M : \partial M = I} w(M) \quad \textit{for all } I \in \binom{[n]}{k}.$$

Moreover, if we let w vary over weight functions, we obtain a positroid cell

$$S_G := \{L(w) \mid w : \mathrm{Edges}(G) \to \mathbb{R}_{>0}\}.$$

If G is a tree, we call S_G a *tree positroid cell.*

Remark 2.18. If G is a plabic graph as in Theorem 2.17 which is reduced, with decorated permutation π_G and almost perfect matchings $\mathcal{M}(G)$, we have that $S_G = S_{\mathcal{M}(G)} = S_{\pi_G}$ [47]. So we can index positroid cells by plabic graphs, bases, or decorated permutations.

2.4. ϕ-induced subdivisions and positroid tilings

Given a surjective map $\phi : X \to Y$ from a cell complex X onto a topological space Y, it is natural to try to decompose Y using images of cells under ϕ.

Definition 2.19. Let $X = \bigsqcup_\pi S_\pi$ be a cell complex and let $\phi : X \to Y$ be a continuous surjective map onto Y, a d-dimensional cell complex or subset thereof. We define a *ϕ-induced dissection of Y* to be a collection $\{\overline{\phi(S_\pi)} \mid \pi \in \mathcal{C}\}$ of images of cells of X, indexed by the set $\mathcal{C}$, such that their union $\bigcup_{\pi \in \mathcal{C}} \overline{\phi(S_\pi)}$ equals Y, the interiors are pairwise disjoint, i.e., $\phi(S_\pi) \cap \phi(S_{\pi'}) = \emptyset$ for $\pi \neq \pi' \in \mathcal{C}$, and $\dim(\phi(S_\pi)) = d$ for all $\pi \in \mathcal{C}$;

We call a dissection $\{\overline{\phi(S_\pi)} \mid \pi \in \mathcal{C}\}$ a *ϕ-induced tiling* if additionally ϕ is injective on each S_π for $\pi \in \mathcal{C}$. And we call a dissection $\{\overline{\phi(S_\pi)} \mid \pi \in \mathcal{C}\}$ a *ϕ-induced subdivision* if whenever $\overline{\phi(S_\pi)} \cap \overline{\phi(S_{\pi'})} \neq \emptyset$, this intersection equals $\overline{\phi(S_{\pi''})}$, where $S_{\pi''}$ lies in $\overline{S_\pi} \cap \overline{S_{\pi'}}$.

When $\phi : X \to Y$ is an affine projection of convex polytopes, the above notion of ϕ-induced subdivision recovers Billera–Sturmfels' notion of ϕ-induced polyhedral subdivision [9]. The subdivisions which are also ϕ-induced tilings are their *tight* ϕ-induced subdivisions. The relation to polyhedral subdivisions suggests a number of questions. What can one say about the *Baues poset* of ϕ-induced subdivisions, partially ordered by refinement? What can one say about the *flip graph*, the restriction of the Hasse diagram of $\omega(\phi : X \to Y)$ to elements of rank 0 and 1? When is it connected? Is there an analogue of *fiber polytopes* [9] (perhaps along the lines of [41]), which control the *coherent* π-induced subdivisions?

Definition 2.20. In Definition 2.19, let X be the positive Grassmannian $\mathrm{Gr}_{k,n}^{\geq 0}$ with its positroid cell decomposition. For S_π a d-dimensional positroid cell, we say that $\overline{\phi(S_\pi)}$ is a *positroid tile* (for ϕ) if ϕ is injective on S_π. We will refer to ϕ-induced dissections, tilings, and subdivisions, as *positroid dissections, tilings*, and *subdivisions*. Our notion of positroid subdivision is closely related to the *good dissections* studied in [39].

In this article we will take X to be the positive Grassmannian, and consider the case where ϕ is the *moment map* (Section 3) or the *amplituhedron map* (Section 5).

3. THE MOMENT MAP AND POSITROID TILINGS

The foundational 1987 paper of Gelfand–Goresky–MacPherson–Serganova [24] initiated the study of the Grassmannian and its matroid stratification via the moment map. Here we will consider the restriction of the moment map to the positive Grassmannian.

Given a subset $I \subset [n]$ and a point $x \in \mathbb{R}^n$, we use the notation $x_I := \sum_{i\in I} x_i$. We also let $e_I := \sum_{i\in I} e_i \in \mathbb{R}^n$, where $\{e_1, \ldots, e_n\}$ is the standard basis of $\mathbb{R}^n$.

The torus $T = (\mathbb{C}^*)^n$ acts on $\mathrm{Gr}_{k,n}$ by scaling the columns of a matrix representative C. This torus action gives rise to a *moment map* $\mu : \mathrm{Gr}_{k,n} \to \mathbb{R}^n$.

Definition 3.1. Let $C \in \mathrm{Gr}_{k,n}$. The *moment map* $\mu : \mathrm{Gr}_{k,n} \to \mathbb{R}^n$ is defined by

$$\mu(C) = \frac{\sum_{I\in\binom{[n]}{k}} |p_I(C)|^2 e_I}{\sum_{I\in\binom{[n]}{k}} |p_I(C)|^2}.$$

Let TC denote the orbit of C under the action of T, and $\overline{TC}$ its closure. It follows from [26] that the image $\mu(\overline{TC})$ is a convex polytope, whose vertices are the images of the torus-fixed points. This polytope is the *matroid polytope* $\Gamma_{\mathcal{M}(C)}$ [24], as defined below.

Definition 3.2. Given a matroid $\mathcal{M} = ([n], \mathcal{B})$, the (basis) *matroid polytope* $\Gamma_{\mathcal{M}}$ is the convex hull of the indicator vectors of the bases of $\mathcal{M}$,

$$\Gamma_{\mathcal{M}} := \mathrm{conv}\{e_B \mid B \in \mathcal{B}\} \subset \mathbb{R}^n.$$

A special case of a matroid polytope is the *hypersimplex* $\Delta_{k,n}$, the convex hull of all points e_I for $I \in \binom{[n]}{k}$. We have that $\mu(\mathrm{Gr}_{k,n}) = \Delta_{k,n}$.

3.1. Classification of positroid tiles for the moment map

Now let us consider the positive analogues of some of the above objects. If $\mathcal{M}$ is a *positroid*, the matroid polytope $\Gamma_{\mathcal{M}}$ is called a *positroid polytope*. If one restricts the moment map to $\mathrm{Gr}_{k,n}^{\geq 0}$, one can show that the moment map image $\mu(\overline{S_{\mathcal{M}}}) = \overline{\mu(S_{\mathcal{M}})}$ of $S_{\mathcal{M}}$ is precisely the corresponding positroid polytope $\Gamma_{\mathcal{M}}$ [65, PROPOSITION 7.10]. In particular, the moment map image of $\mathrm{Gr}_{k,n}^{\geq 0}$ is again the hypersimplex $\Delta_{k,n}$. If $\mathcal{M}$ is the positroid associated to a cell S_π or S_G, we also use the notation Γ_π and Γ_G to refer to $\Gamma_{\mathcal{M}}$. See Figure 4.

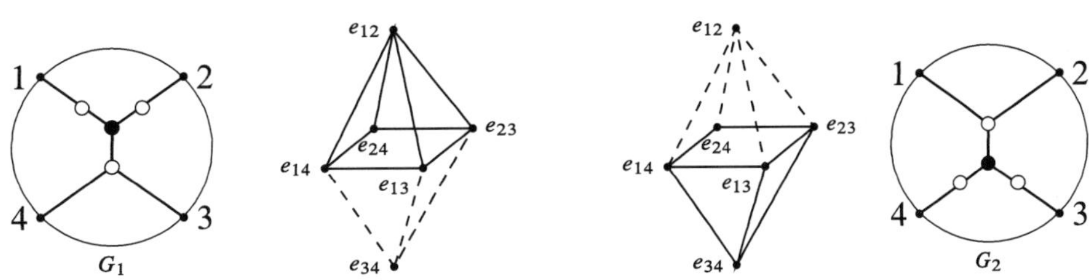

FIGURE 4
Positroid polytopes Γ_{G_1} and Γ_{G_2} associated to graphs G_1 and G_2, cf. Figure 3.

Applying Definition 2.20 to the moment map $\mu : \mathrm{Gr}_{k,n}^{\geq 0} \to \Delta_{k,n}$ onto the hypersimplex, a polytope of dimension $n-1$, we see that a positroid tile is the (closure of the) image of an $(n-1)$-dimensional positroid cell on which the moment map is injective.

Proposition 3.3 ([39, PROPOSITION 3.16], based on [1, 65]). *The positroid tiles for the moment map are exactly the positroid polytopes Γ_G associated to the tree positroid cells S_G. Two positroid tiles Γ_G and $\Gamma_{G'}$ are the same if and only if G and G' are related by move (M2).*

3.2. The positive tropical Grassmannian and positroid subdivisions

How can one produce positroid tilings and, more generally, positroid subdivisions of the hypersimplex $\Delta_{k,n}$? One way is to use the *positive tropical Grassmannian*.

The *tropical Grassmannian* $\mathrm{Trop}\,\mathrm{Gr}_{k,n}$ [27, 34, 57] is the space of *realizable tropical linear spaces*, obtained by applying the valuation map to elements of the Grassmannian $\mathrm{Gr}_{k,n}(K)$ over the field $K = \mathbb{C}\{\{t\}\}$ of *Puiseux-series*. Meanwhile the *Dressian* $\mathrm{Dr}_{k,n}$ is the space of *tropical Plücker vectors* $P = \{P_I\}_{I \in \binom{[n]}{k}}$, also known as *valuated matroids*. Thinking of each $P \in \mathrm{Dr}_{k,n}$ as a *height function* on the vertices of the hypersimplex $\Delta_{k,n}$, one can show that the Dressian parameterizes regular matroid subdivisions $\mathcal{D}_P$ of $\Delta_{k,n}$ [30, 56], which in turn are dual to *abstract tropical linear spaces* [56].

There are positive notions of both of the above spaces. The *positive tropical Grassmannian* [58] is the space of *realizable positive tropical linear spaces*, obtained by applying the valuation map to Puiseux-series valued elements of the positive Grassmannian. The *positive Dressian* is the space of *positive tropical Plücker vectors*.

Definition 3.4. We say $P = \{P_I\}_{I \in \binom{[n]}{k}} \in \mathbb{R}^{\binom{[n]}{k}}$ is a *positive tropical Plücker vector* if for each three-term Plücker relation, it lies in the positive part of the associated tropical hypersurface: for $1 \leq a < b < c < d \leq n$ and $S \in \binom{[n]}{k-2}$ disjoint from $\{a, b, c, d\}$, either

$$P_{Sac} + P_{Sbd} = P_{Sab} + P_{Scd} \leq P_{Sad} + P_{Sbc} \quad \text{or}$$
$$P_{Sac} + P_{Sbd} = P_{Sad} + P_{Sbc} \leq P_{Sab} + P_{Scd}.$$

The *positive Dressian* $\mathrm{Dr}_{k,n}^{+}$ is the set of positive tropical Plücker vectors.

In general, the Dressian $\mathrm{Dr}_{k,n}$ contains the tropical Grassmannian $\mathrm{Gr}_{k,n}$ but is much larger [28]. However, the situation for their positive parts is different, see [59] and [5].

Theorem 3.5 ([59, THEOREM 3.9]). *The positive tropical Grassmannian* $\mathrm{Trop}^{+}\,\mathrm{Gr}_{k,n}$ *equals the positive Dressian* $\mathrm{Dr}_{k,n}^{+}$. *That is, all abstract positive tropical linear spaces are realizable.*

There are two natural fan structures on the Dressian—the Plücker fan and the secondary fan—which coincide [45]. The cones of $\mathrm{Trop}^{+}\,\mathrm{Gr}_{k,n}$ control the *regular subdivisions* of $\Delta_{k,n}$ into positroid polytopes, with maximal cones giving rise to positroid tilings.

Consider a point $P = \{P_I\}_{I \in \binom{[n]}{k}} \in \mathbb{R}^{\binom{[n]}{k}}$, which we also think of as a real-valued function $\{e_I\} \mapsto P_I$ on the vertices of $\Delta_{k,n}$. We define a polyhedral subdivision $\mathcal{D}_P$ of $\Delta_{k,n}$ as follows: consider the points $(e_I, P_I) \in \Delta_{k,n} \times \mathbb{R}$ and take their convex hull. Take the lower faces (those whose outwards normal vector have last component negative) and project them down to $\Delta_{k,n}$. This gives us the *regular subdivision* $\mathcal{D}_P$ of $\Delta_{k,n}$. Note that $\mathcal{D}_P$ is a *polytopal subdivision* and that its facets (top-dimensional faces) comprise a positroid subdivision for the moment map (cf. Definition 2.20).

Theorem 3.6 ([39, THEOREM 9.12], see also [5]). *The point $P = \{P_I\}_{I \in \binom{[n]}{k}}$ is a positive tropical Plücker vector if and only if every face of $\mathcal{D}_P$ is a positroid polytope.*

Example 3.7. Consider a positive tropical Plücker vector $P = \{P_I\}_{I \in \binom{[4]}{2}} \in \mathbb{R}^{\binom{[4]}{2}}$. If we have $P_{13} + P_{24} = P_{14} + P_{23} < P_{12} + P_{34}$, then P induces the subdivision $\mathcal{D}_P$ of $\Delta_{2,4}$ into the two square pyramids shown in Figure 4. If $P_{13} + P_{24} = P_{12} + P_{34} < P_{14} + P_{23}$, we get the subdivision of $\Delta_{2,4}$ into two square pyramids separated by the square with vertices $e_{12}, e_{13}, e_{24}, e_{34}$, see Figure 7. These are both positroid tilings of $\Delta_{2,4}$.

Positroid tilings are particularly nice. The following result refines [59, THEOREM 6.6].

Theorem 3.8 ([59, THEOREM 6.6]). *Let $P \in \mathrm{Trop}^+ \mathrm{Gr}_{k,n}$ and consider the positroid subdivision $\mathcal{D}_P$. The following statements are equivalent:*

- *The facets of $\mathcal{D}_P$ comprise a positroid tiling.*
- *The facets of $\mathcal{D}_P$ comprise a finest positroid subdivision.*
- *Every face of $\mathcal{D}_P$ is the matroid polytope of a series-parallel matroid.*
- *Every octahedron in $\mathcal{D}_P$ is subdivided.*

Combining Theorem 3.8 with Speyer's f-vector theorem [55] gives the following.

Corollary 3.9. *Let $\mathcal{D}_P$ be a positroid tiling of $\Delta_{k,n}$ as in Theorem 3.8. Then this polyhedral subdivision contains precisely $\frac{(n-c-1)!}{(k-c)!(n-k-c)!(c-1)!}$ interior faces of dimension $n - c$. In particular, $\mathcal{D}_P$ consists of precisely $\binom{n-2}{k-1}$ positroid tiles.*

3.3. Realizability of positively oriented matroids

Positroid polytopes and the positive tropical Grassmannian have applications to realizability questions. One of the early goals of matroid theory was to find axioms that characterized the realizable matroids, i.e., those that arise from full rank matrices as in Definition 2.2. However, this problem (over a field of characteristic 0) is now considered to be intractible [42, 66]: in Vámos's words, "the missing axiom of matroid theory is lost forever."

There is a notion of *oriented matroids*, in which bases have signs: an oriented matroid is a matroid $([n], \mathcal{B})$ together with a *chirotope*, a function $\chi : [n]^k \to \{0, +, -\}$ obeying certain axioms which roughly encode the three-term Plücker relations [10]. If M is a full rank $k \times n$ real matrix, the function taking $(i_1, i_2, \ldots, i_k)$ to the sign of the minor using columns $(i_1, i_2, \ldots, i_k)$ is a chirotope, and the realizable oriented matroids are precisely the chirotopes occurring in this way. Thus, if M represents a point of the positive Grassmannian, then M gives a chirotope χ with $\chi(i_1, i_2, \ldots, i_k) \in \{0, +\}$ for $1 \le i_1 < i_2 < \cdots < i_k \le n$.

A *positively oriented matroid* is a chirotope χ such that $\chi(i_1, i_2, \dots, i_k) \in \{0, +\}$ for $1 \leq i_1 < i_2 < \cdots < i_k \leq n$. The following was conjectured by da Silva in 1987 [14].

Theorem 3.10 ([2,59]). *Every positively oriented matroid is realizable. In other words, every positively oriented matroid is a positroid.*

The first proof used the geometry of positroid polytopes [2]. The second used the fact that if $P \in \operatorname{Trop}\operatorname{Gr}_{k,n}$, every face of $\mathcal{D}_P$ corresponds to a realizable matroid [59].

4. THE AMPLITUHEDRON AND THE SIGN STRATIFICATION

Building on [3], Arkani-Hamed and Trnka [7] introduced the *(tree) amplituhedron*, which is the image of the positive Grassmannian under a positive linear map. Let $\operatorname{Mat}_{n,p}^{>0}$ denote the set of $n \times p$ matrices whose maximal minors are positive. Throughout this section we will make the convention that k, n, m are positive integers such that $k + m \leq n$.

Definition 4.1. Let $Z \in \operatorname{Mat}_{n,k+m}^{>0}$. The *amplituhedron map* $\tilde{Z} : \operatorname{Gr}_{k,n}^{\geq 0} \to \operatorname{Gr}_{k,k+m}$ is defined by $\tilde{Z}(C) := CZ$, where C is a $k \times n$ matrix representing an element of $\operatorname{Gr}_{k,n}^{\geq 0}$, and CZ is a $k \times (k+m)$ matrix representing an element of $\operatorname{Gr}_{k,k+m}$. The *amplituhedron* $\mathcal{A}_{n,k,m}(Z) \subset \operatorname{Gr}_{k,k+m}$ is the image $\tilde{Z}(\operatorname{Gr}_{k,n}^{\geq 0})$.

The condition $Z \in \operatorname{Mat}_{n,k+m}^{>0}$ ensures that $\operatorname{rank}(CZ) = k$ and hence $\tilde{Z}$ is well defined [7]. This condition can be relaxed: the map $\tilde{Z}$ is well defined if and only if $\operatorname{var}(v) \geq k$ for all nonzero $v \in \ker(Z)$ [31, THEOREM 4.2].

If $k + m = n$, $\mathcal{A}_{n,k,m}(Z)$ is isomorphic to $\operatorname{Gr}_{k,k+m}^{\geq 0}$. The amplituhedron $\mathcal{A}_{n,k,m}(Z)$ has full dimension km inside $\operatorname{Gr}_{k,k+m}$, but it does not lie inside $\operatorname{Gr}_{k,k+m}^{\geq 0}$ in general.

Example 4.2. If $k = 1$ and $m = 2$, $\mathcal{A}_{n,1,2}(Z)$ is a polygon in projective space $\mathbb{P}^2$. To see this, let $Z_1, \dots, Z_n$ denote the rows of $Z \in \operatorname{Mat}_{n,3}^{>0}$; we can think of each Z_i as a point in $\mathbb{P}^2$. Since $Z \in \operatorname{Mat}_{n,3}^{>0}$, the points $Z_1, \dots, Z_n$ are arranged in convex position like the vertices of a polygon, see Figure 8. Elements $C \in \operatorname{Gr}_{1,n}^{\geq 0}$ are nonzero vectors with coordinates in $\mathbb{R}_{\geq 0}$. If $C = e_i$ then $CZ = Z_i$, so all points Z_i lie in $\mathcal{A}_{n,1,2}$. As C varies over $\operatorname{Gr}_{1,n}^{\geq 0}$, the image CZ varies over all convex combinations of the Z_i's, and hence $\mathcal{A}_{n,1,2}(Z)$ is an n-gon. More generally, it follows from [61] that $\mathcal{A}_{n,1,m}(Z)$ is a *cyclic polytope* in $\mathbb{P}^m$.

Physicists are most interested in the amplituhedron $\mathcal{A}_{n,k,4}(Z)$. Since $\mathcal{A}_{n,k,m}(Z) \subset \operatorname{Gr}_{k,k+m}$, when m is small it is sometimes more convenient to take orthogonal complements and work with $\operatorname{Gr}_{m,k+m}$ instead of $\operatorname{Gr}_{k,k+m}$. This motivates the following definition of the *B-amplituhedron*, which is homeomorphic to the "A-amplituhedron."

Definition 4.3 ([32, DEFINITION 3.8]). Choose $Z \in \operatorname{Mat}_{n,k+m}^{>0}$, and let $W \in \operatorname{Gr}_{k+m,n}^{>0}$ be the column span of Z. We define the *B-amplituhedron* to be

$$\mathcal{B}_{n,k,m}(W) := \{V^{\perp} \cap W \mid V \in \operatorname{Gr}_{k,n}^{\geq 0}\} \subset \operatorname{Gr}_m(W),$$

where $\operatorname{Gr}_m(W)$ denotes the Grassmannian of m-planes in W.

The idea behind the identification $\mathcal{B}_{n,k,m}(W) \cong \mathcal{A}_{n,k,m}(Z)$ is that we obtain $\mathcal{B}_{n,k,m}(W) \subset \mathrm{Gr}_m(W) \subset \mathrm{Gr}_{m,n}$ from $\mathcal{A}_{n,k,m}(Z) \subset \mathrm{Gr}_{k,k+m}$ by taking orthogonal complements in $\mathbb{R}^{k+m}$, then applying an isomorphism between $\mathbb{R}^{k+m}$ and W so that our subspaces lie in W, not $\mathbb{R}^{k+m}$.

Proposition 4.4 ([32, LEMMA 3.10 AND PROPOSITION 3.12]). *Choose $Z \in \mathrm{Mat}_{n,k+m}^{>0}$, and let $W \in \mathrm{Gr}_{k+m,n}^{>0}$ be the column span of Z. We define a map $f_Z : \mathrm{Gr}_m(W) \to \mathrm{Gr}_{k,k+m}$ by*

$$f_Z(X) := Z(X^\perp) = \{Z(x) : x \in X^\perp\}.$$

Then f_Z restricts to a homeomorphism from $\mathcal{B}_{n,k,m}(W)$ onto $\mathcal{A}_{n,k,m}(Z)$, sending $V^\perp \cap W$ to $\tilde{Z}(V)$ for all $V \in \mathrm{Gr}_{k,n}^{\geq 0}$.

We next discuss coordinates on $\mathcal{A}_{n,k,m}(Z) \subset \mathrm{Gr}_{k,k+m}$ and $\mathcal{B}_{n,k,m}(W) \subset \mathrm{Gr}_{m,n}$.

Definition 4.5. Choose $Z \in \mathrm{Mat}_{n,k+m}^{>0}$ with rows $Z_1, \dots, Z_n \in \mathbb{R}^{k+m}$. Given a matrix Y with rows $y_1, \dots, y_k$ representing an element of $\mathrm{Gr}_{k,k+m}$, and $\{i_1, \dots, i_m\} \subset [n]$, we define the *twistor coordinate*, denoted

$$\langle Y Z_{i_1} Z_{i_2} \cdots Z_{i_m} \rangle \quad \text{or} \quad \langle y_1, \dots, y_k, Z_{i_1}, \dots, Z_{i_m} \rangle,$$

to be the determinant of the matrix with rows $y_1, \dots, y_k, Z_{i_1}, \dots, Z_{i_m}$.

If $I = \{i_1 < \cdots < i_m\}$, we also use $\langle Y Z_I \rangle$ to denote $\langle Y Z_{i_1} Z_{i_2} \cdots Z_{i_m} \rangle$.

Lemma 4.6 shows that the twistor coordinates of the amplituhedron $\mathcal{A}_{n,k,m}(Z) \subset \mathrm{Gr}_{k,k+m}$ are equal to the Plücker coordinates of the B-amplituhedron $\mathcal{B}_{n,k,m}(W) \subset \mathrm{Gr}_{m,n}$.

Lemma 4.6 ([32, (3.11)]). *If we let $Y := f_Z(X)$ in Proposition 4.4, we have*

$$p_I(X) = \langle Y Z_I \rangle \quad \text{for all } I \in \binom{[n]}{m}. \tag{4.7}$$

By Definition 4.3 and Theorem 2.6, if $X \in \mathcal{B}_{n,k,m}(W)$ and $w \in X \setminus \{0\}$ then

$$k \leq \overline{\mathrm{var}}(w) \leq k + m - 1.$$

When $m = 1$ this leads to the following sign variation description of the amplituhedron.

Theorem 4.8 ([32, COROLLARY 3.19] and Lemma 4.6). *We have*

$$\mathcal{B}_{n,k,1}(W) = \{w \in \mathbb{P}(W) \mid \overline{\mathrm{var}}(w) = k\} \subset \mathbb{P}(W) \quad \text{or, equivalently,} \tag{4.9}$$

$$\mathcal{A}_{n,k,1}(Z) = \{Y \in \mathrm{Gr}_{k,k+1} \mid \overline{\mathrm{var}}(\langle Y Z_1 \rangle, \langle Y Z_2 \rangle, \dots, \langle Y Z_n \rangle) = k\}. \tag{4.10}$$

Moreover, $\mathcal{A}_{n,k,1}(Z) \cong \mathcal{B}_{n,k,1}(W)$ can be identified with the complex of bounded faces of a cyclic hyperplane arrangement.

To make the latter identification, we choose a basis $w^{(0)}, w^{(1)}, \dots, w^{(k)}$ for $W \subset \mathbb{R}^n$ such that $\mathrm{span}(w^{(1)}, \dots, w^{(k)}) \in \mathrm{Gr}_{k,n}^{>0}$ and $w^{(0)}$ is *positively oriented* as in [32, DEFINITION 6.10]. Then we let $\mathcal{H}^W$ be the hyperplane arrangement in $\mathbb{R}^k$ with hyperplanes

$$H_i := \{x \in \mathbb{R}^k \mid w_i^{(1)} x_1 + \cdots + w_i^{(k)} x_k + w_i^{(0)} = 0\} \quad \text{for } i \in [n].$$

By [32, THEOREM 6.16], the map

$$\Psi_{\mathcal{H}^W} : \mathbb{R}^k \to \mathbb{P}(W), \quad x \mapsto \langle x_1 w^{(1)} + \cdots + x_k w^{(k)} + w^{(0)} \rangle$$

is a homeomorphism from the bounded complex $B(\mathcal{H}^W)$ of $\mathcal{H}^W$ to $\mathcal{B}_{n,k,1}(W)$. Moreover, if we partition the elements $w = (w_1, \dots, w_n) \in \mathcal{B}_{n,k,1}(W)$ based on the signs of the w_i, we obtain the bounded faces of $B(\mathcal{H}^W)$. See Figure 5.

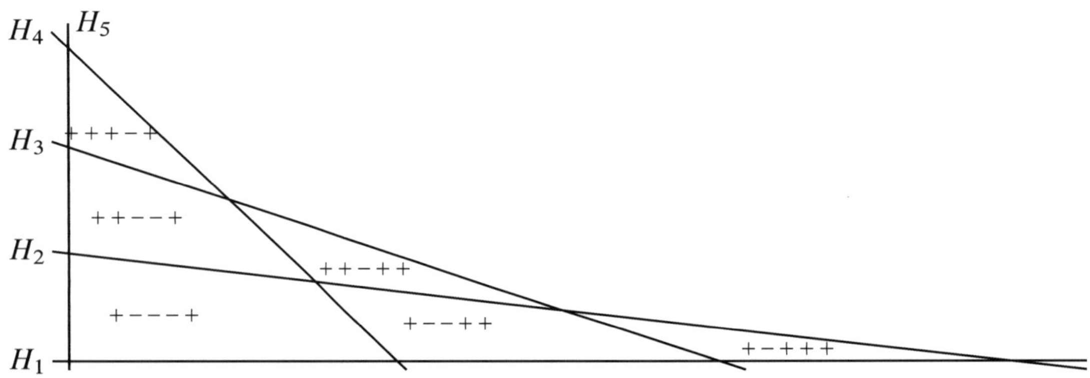

FIGURE 5
The hyperplane arrangement $\mathcal{H}^W$ from $\mathcal{B}_{5,2,1}(W) \cong B(\mathcal{H}^W)$, where $w^{(0)} = (0, -9, -6, -3, 0)$, $w^{(1)} = (0, 1, 1, 1, 1)$, and $w^{(2)} = (1, 9, 3, 1, 0)$. Its bounded faces are labeled by sign vectors. Here we have labeled only the maximal faces.

Before turning to the case $m = 2$, we need to introduce some notation.

Remark 4.11. Let $Z \in \mathrm{Mat}_{n,p}^{\geq 0}$ with $n \geq p$ have rows $Z_1, Z_2, \dots, Z_n$, and let $\hat{Z}_i$ denote $(-1)^{p-1} Z_i$. Then the matrix with rows $Z_2, \dots, Z_n, \hat{Z}_1$ also lies in $\mathrm{Mat}_{n,p}^{>0}$. Thus matrices with maximal minors nonnegative (and elements of $\mathrm{Gr}_{k,n}^{\geq 0}$) exhibit a *twisted cyclic symmetry*.

The following sign variation description of $\mathcal{A}_{n,k,2}(Z)$ was conjectured in [6].

Theorem 4.12 ([46, THEOREM 5.1]). *Fix $k < n$ and $Z \in \mathrm{Mat}_{n,k+2}^{>0}$. Let*

$$\mathcal{F}_{n,k,2}^{\circ}(Z) := \{Y \in \mathrm{Gr}_{k,k+2} \mid \langle Y Z_i Z_{i+1} \rangle > 0 \text{ for } 1 \leq i \leq n-1, \text{ and } \langle Y Z_n \hat{Z}_1 \rangle > 0, \\ \text{and } \mathrm{var}(\langle Y Z_1 Z_2 \rangle, \langle Y Z_1 Z_3 \rangle, \dots \langle Y Z_1 Z_n \rangle) = k.\}$$

Then $\mathcal{A}_{n,k,2}(Z) = \overline{\mathcal{F}_{n,k,2}^{\circ}(Z)}$.

To generalize (4.10) and Theorem 4.12 for $m > 2$, we first observe the following [6].

Lemma 4.13. *Let $Z \in \mathrm{Mat}_{n,k+m}^{>0}$ and $Y = CZ$ for $C \in \mathrm{Gr}_{k,n}^{>0}$. Let $r = \lfloor \frac{m}{2} \rfloor$.*

If m is even, $\langle Y Z_I \rangle > 0 \quad \forall I := \{i_1 < i_1 + 1 < i_2 < i_2 + 1 < \cdots < i_r < i_r + 1\} \in \binom{[n]}{m}$

and $\langle Y Z_I Z_n \hat{Z}_1 \rangle > 0 \quad \forall I := \{i_1 < i_1 + 1 < \cdots < i_{r-1} < i_{r-1} + 1\} \in \binom{[2, n-1]}{m-2}$.

If m is odd, $(-1)^k\langle YZ_I\rangle > 0 \quad \forall I := \{1 = i_0 < i_1 < i_1+1 < \cdots < i_r < i_r+1\} \in \binom{[n]}{m}$

and $\langle YZ_I\rangle > 0 \quad \forall I := \{i_1 < i_1+1 < \cdots < i_r < i_r+1 < i_{r+1} = n\} \in \binom{[n]}{m}$.

Proof. The lemma follows from the fact that $Z \in \mathrm{Mat}^{>0}_{n,k+m}$ and

$$\langle CZ, Z_{i_1}, \dots, Z_{i_m}\rangle = \sum_{J=\{j_1<\cdots<j_k\}\in\binom{[n]}{k}} p_J(C)\langle Z_{j_1}, \dots, Z_{j_k}, Z_{i_1}, \dots, Z_{i_m}\rangle, \tag{4.14}$$

by considering how many swaps are necessary to put $\{j_1, \dots, j_k, i_1, \dots, i_m\}$ in order. ■

The following conjecture is (up to taking the closure) the statement [6, (5.14)].

Conjecture 4.15 ([6, (5.14)]). *Fix $Z \in \mathrm{Mat}^{>0}_{n,k+m}$, and define*

$$\mathcal{F}^{\circ}_{n,k,m}(Z) := \big\{Y \in \mathrm{Gr}_{k,k+m} \mid \textit{the conclusion of Lemma 4.13 holds, and}$$
$$\mathrm{var}\big(\langle YZ_1 \dots Z_{m-1}Z_m\rangle, \langle YZ_1 \dots Z_{m-1}Z_{m+1}\rangle, \dots, \langle YZ_1 \dots Z_{m-1}Z_n\rangle\big) = k\big\}.$$

Then $\mathcal{A}_{n,k,m}(Z) = \overline{\mathcal{F}^{\circ}_{n,k,m}(Z)}$.

To see that $\mathcal{A}_{n,k,m}(Z) \subseteq \overline{\mathcal{F}^{\circ}_{n,k,m}(Z)}$, see [6, SECTION 5.4] or [32, COROLLARY 3.21].

Given the evident importance of signs of coordinates, and taking inspiration from the $m = 1$ example, we define the *sign stratification* for the amplituhedron; this stratification is closely related to the *oriented matroid stratification* of the Grassmannian.

Definition 4.16. Let $\sigma = (\sigma_I) \in \{0, +, -\}^{\binom{n}{m}}$ be a nonzero sign vector with coordinates indexed by subsets $I \in \binom{[n]}{m}$. We consider σ *modulo multiplication by* ± 1 (since Plücker and twistor coordinates are coordinates in projective space). Set

$$\mathcal{A}^{\sigma}_{n,k,m}(Z) := \big\{Y \in \mathcal{A}_{n,k,m}(Z) \mid \mathrm{sign}\langle YZ_I\rangle = \sigma_I \text{ for all } I\big\} \quad \text{and}$$
$$\mathcal{B}^{\sigma}_{n,k,m}(W) := \big\{X \in \mathcal{B}_{n,k,m}(W) \mid \mathrm{sign}\big(p_I(X)\big) = \sigma_I \text{ for all } I\big\}.$$

We call $\mathcal{A}^{\sigma}_{n,k,m}(Z)$ (respectively, $\mathcal{B}^{\sigma}_{n,k,m}(W)$) an *(amplituhedron) sign stratum.* Clearly,

$$\mathcal{A}_{n,k,m}(Z) = \bigsqcup_{\sigma} \mathcal{A}^{\sigma}_{n,k,m}(Z) \quad \text{and} \quad \mathcal{B}_{n,k,m}(W) = \bigsqcup_{\sigma} \mathcal{B}^{\sigma}_{n,k,m}(W).$$

If $\sigma \in \{+, -\}^{\binom{n}{m}}$, we call $\mathcal{A}^{\sigma}_{n,k,m}(Z)$ and $\mathcal{B}^{\sigma}_{n,k,m}(W)$ open *(amplituhedron) chambers.*

For arbitrary σ, $\mathcal{A}^{\sigma}_{n,k,m}(Z)$ (or $\mathcal{B}^{\sigma}_{n,k,m}(W)$) may be empty. We call $\mathcal{A}^{\sigma}_{n,k,m}$ *realizable* if there is some Z for which $\mathcal{A}^{\sigma}_{n,k,m}(Z)$ is nonempty. It is an open problem to classify the realizable chambers/strata of the amplituhedron for $m > 2$. When $m = 1$, the realizable strata are precisely the $\mathcal{A}^{\sigma}_{n,k,1}(Z)$ with $\overline{\mathrm{var}}(\sigma) = k$ (cf. (4.10) and [32]). When $m = 2$, we will show in Section 6 that the realizable chambers of $\mathcal{A}_{n,k,2}$ are counted by the *Eulerian numbers*. This is related to the fact that the volume of $\Delta_{k+1,n}$ is the Eulerian number.

5. THE AMPLITUHEDRON MAP AND POSITROID TILINGS

In this section we begin our discussion of positroid tilings of the amplituhedron, cf. Definition 2.20 with $\phi = \tilde{Z}$. These have also been called *(positroid) triangulations*.

Definition 5.1. Choose $Z \in \mathrm{Mat}_{n,k+m}^{>0}$. Given a positroid cell S_π of $\mathrm{Gr}_{k,n}^{\geq 0}$, we let $Z_\pi^\circ := \tilde{Z}(S_\pi)$ and $Z_\pi := \overline{\tilde{Z}(S_\pi)} = \tilde{Z}(\overline{S_\pi})$, and we refer to Z_π and Z_π° as *Grasstopes* and *open Grasstopes*, respectively. As in Definition 2.20, we call Z_π and Z_π° a *(positroid) tile* and an *open (positroid) tile* for $\mathcal{A}_{n,k,m}(Z)$ if $\dim(S_\pi) = km$ and $\tilde{Z}$ is injective on S_π. Since positroid cells are also indexed by plabic graphs, we will also use Z_G and Z_G° to denote the Grasstopes associated to the positroid cell S_G.

Images of positroid cells under the map $\tilde{Z}$ have been studied since [7], where the authors conjectured that the images of various *BCFW* collections of $4k$-dimensional cells in $\mathrm{Gr}_{k,n}^{\geq 0}$ give a "triangulation" (positroid tiling) of the amplituhedron $\mathcal{A}_{n,k,4}(Z)$. (For the "canonical" BCFW tiling studied in [33], this conjecture has been recently proved in a beautiful paper of Even-Zohar–Lakrec–Tessler [17].)

The positroid tiles for $\mathcal{A}_{n,k,m}(Z)$ were classified for $m = 1$ in [32, THEOREM 8.10]. For $m = 2$ and $k = 1$ (cf. Example 4.2), the amplituhedron $\mathcal{A}_{n,1,2}(Z)$ is a convex n-gon in $\mathbb{P}^2$. The positroid tiles are exactly the triangles on vertices $Z_1, \dots, Z_n$ of the polygon, and positroid tilings of $\mathcal{A}_{n,1,2}(Z)$ are just ordinary triangulations of a polygon. More generally, the positroid tiles have been classified for $m = 2$ in [46, THEOREM 4.25], as we now describe.

5.1. Classification of positroid tiles for the amplituhedron map when $m = 2$

When $m = 2$, positroid tiles are in bijection with *bicolored subdivisions* of an n-gon.

Definition 5.2. Let $\mathbf{P}_n$ be a convex n-gon with boundary vertices labeled from 1 to n in clockwise order. A *bicolored triangulation* $\mathcal{T}$ of $\mathbf{P}_n$ is a triangulation whose edges connect vertices of $\mathbf{P}_n$ and whose triangles are colored black or white. If $\mathcal{T}$ has exactly k black triangles, we say it has *type* (k, n). Two bicolored triangulations $\mathcal{T}$ and $\mathcal{T}'$ are *equivalent* if the union of the black triangles is the same for both of them. If we erase the diagonals separating pairs of like-colored triangles in $\mathcal{T}$ sharing an edge, we obtain a *bicolored subdivision* $\overline{\mathcal{T}}$ of $\mathbf{P}_n$ into black and white polygons, which represents the equivalence class of $\mathcal{T}$.

Given $\mathcal{T}$ as above, we build a labeled bipartite graph $\hat{G}(\mathcal{T})$ by placing black boundary vertices at the vertices of the n-gon, and placing a trivalent white vertex in the middle of each black triangle, connecting it to the three vertices of the triangle. See Figure 6. We can think of $\hat{G}(\mathcal{T})$ as a *plabic graph* if we enclose it in a slightly larger disk and add n edges connecting each black vertex to the boundary of the disk. See Figure 6.

Note that if $\mathcal{T}$ and $\mathcal{T}'$ are equivalent, $\hat{G}(\mathcal{T})$ and $\hat{G}(\mathcal{T}')$ are move-equivalent. Bicolored triangulations and subdivisions are special cases of *plabic tilings* [44].

The following statement refines a conjecture from [38].

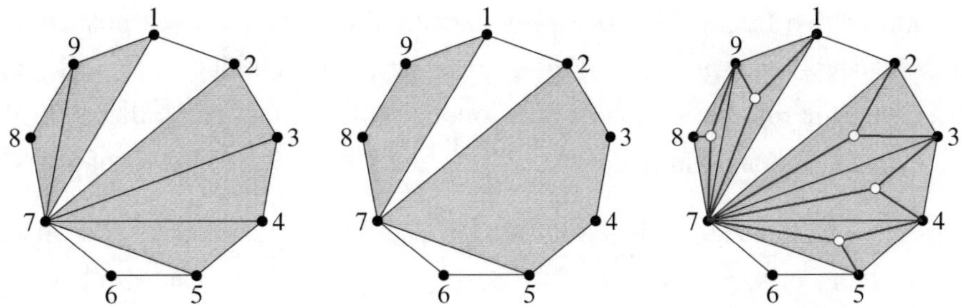

FIGURE 6
A bicolored triangulation $\mathcal{T}$; the corresponding bicolored subdivision $\overline{\mathcal{T}}$; the graph $\hat{G}(\mathcal{T})$.

Theorem 5.3 ([46, THEOREM 4.25]). *Fix $k < n$ and $Z \in \mathrm{Mat}_{n,k+2}^{>0}$. Then $\tilde{Z}$ is injective on the $2k$-dimensional cell $S_{\mathcal{M}}$ if and only if $S_{\mathcal{M}} = S_{\hat{G}(\mathcal{T})}$ for some bicolored triangulation $\mathcal{T}$ of type (k, n). That is, the positroid tiles for $\mathcal{A}_{n,k,2}$ are exactly the Grasstopes $Z_{\hat{G}(\mathcal{T})}$.*

It is an open problem to classify positroid tiles of $\mathcal{A}_{n,k,m}(Z)$ for $m > 2$.

5.2. Numerology of positroid tilings of the amplituhedron

When $m = 4$, each (conjectural) BCFW tiling of $\mathcal{A}_{n,k,4}(Z)$ has cardinality equal to the *Narayana number* $N_{n-3,k+1} = \frac{1}{n-3}\binom{n-3}{k+1}\binom{n-3}{k}$. What should be the cardinality for $m \neq 4$? Table 1 gives data about special cases studied thus far.

Special case	Cardinality of tiling of $\mathcal{A}_{n,k,m}(Z)$	Explanation
$m = 0$ or $k = 0$ or $k + m = n$	1	$\mathcal{A}$ is a point or $\mathcal{A} \cong \mathrm{Gr}_{k,n}^{\geq 0}$
$m = 1$	$\binom{n-1}{k}$	[32]
$m = 2$	$\binom{n-2}{k}$	[6, 8, 46]
$m = 4$	$\frac{1}{n-3}\binom{n-3}{k+1}\binom{n-3}{k}$	[7, 17]
$k = 1$, m even	$\binom{n-1-\frac{m}{2}}{\frac{m}{2}}$	$\mathcal{A} \cong$ cyclic polytope $C(n, m)$

TABLE 1
Cardinalities of tilings of $\mathcal{A}_{n,k,m}(Z)$.

As we will see later, the appearance of the number $\binom{n-2}{k}$ in the $m = 2$ row of the table is related to the appearance of the number $\binom{n-2}{k}$ in Corollary 3.9.

The special cases in the table led us to make the following intriguing conjecture.

Conjecture 5.4 ([33, CONJECTURE 8.1]). *If m is even, the cardinality of a positroid tiling of the amplituhedron $\mathcal{A}_{n,k,m}(Z)$ is $M(k, n-k-m, \frac{m}{2})$, where*

$$M(a,b,c) := \prod_{i=1}^{a}\prod_{j=1}^{b}\prod_{k=1}^{c}\frac{i+j+k-1}{i+j+k-2}$$

is the number of plane partitions *contained in an $a \times b \times c$ box.*

For odd m, we believe the *maximum* cardinality achieved by a positroid tiling of $\mathcal{A}_{n,k,m}(Z)$ is $M(k, n-k-m, \lceil \frac{m}{2} \rceil)$. This is consistent with results for $m = 1$ [32] and the fact that for odd m, the number of top-dimensional simplices in a triangulation of the cyclic polytope $C(n, m)$ can lie anywhere between $\binom{n-1-\frac{m+1}{2}}{\frac{m-1}{2}}$ and $\binom{n-\frac{m+1}{2}}{\frac{m+1}{2}}$ [50, COROLLARY 1.2(II)].

Remark 5.5. Clearly, $M(a, b, c)$ is symmetric in a, b and c. Conjecture 5.4 thus suggests that for even m, there is a symmetry of (the positroid tilings of) the amplituhedron $\mathcal{A}_{n,k,m}$ that allows one to exchange k, $n-k-m$, and $\frac{m}{2}$. The symmetry between k and $n-k-m$ is called *parity duality* and was subsequently verified in [22].

Remark 5.6. The numbers $M(a, b, c)$ also count collections of c noncrossing lattice paths inside an $a \times b$ rectangle; rhombic tilings of a hexagon with side lengths a, b, c, a, b, c; and perfect matchings of a certain honeycomb lattice. See [33, SECTION 8.1].

6. T-DUALITY AND POSITROID TILINGS OF $\Delta_{k+1,n}$ AND $\mathcal{A}_{n,k,2}$

In this section we will fix $1 \le k \le n-2$, and explore a mysterious duality between the hypersimplex $\Delta_{k+1,n}$—an $(n-1)$-dimensional polytope in $\mathbb{R}^n$—and the amplituhedron $\mathcal{A}_{n,k,2}(Z)$—a $2k$-dimensional (non-polytopal) subset of $\mathrm{Gr}_{k,k+2}$ [39, 46]. This duality was first discovered in [39], after we observed that for small k and n, the f-vector of the positive tropical Grassmannian $\mathrm{Trop}^+ \mathrm{Gr}_{k+1,n}$ [58] agrees with the numbers of positroid subdivisions of $\mathcal{A}_{n,k,2}(Z)$ [39, SECTION 11]. By Theorem 3.6, the cones of $\mathrm{Trop}^+ \mathrm{Gr}_{k+1,n}$ parameterize the regular positroid subdivisions of $\Delta_{k+1,n}$, so this leads to the idea that positroid subdivisions of $\Delta_{k+1,n}$ and $\mathcal{A}_{n,k,2}(Z)$ must be related [39].

Example 6.1. Continuing Example 3.7, there are two maximal cones of $\mathrm{Trop}^+ \mathrm{Gr}_{2,4}$, defined by the inequalities $P_{13} + P_{24} = P_{14} + P_{23} < P_{12} + P_{34}$ and $P_{13} + P_{24} = P_{12} + P_{34} < P_{14} + P_{23}$. These two cones give rise to the two subdivisions of $\Delta_{2,4}$ shown in Figure 7. These subdivisions are both positroid tilings for the moment map (and there are no others). Meanwhile, the amplituhedron $\mathcal{A}_{4,1,2}(Z)$ is a quadrilateral, which has precisely two positroid tilings for the amplituhedron map, as shown in Figure 8.

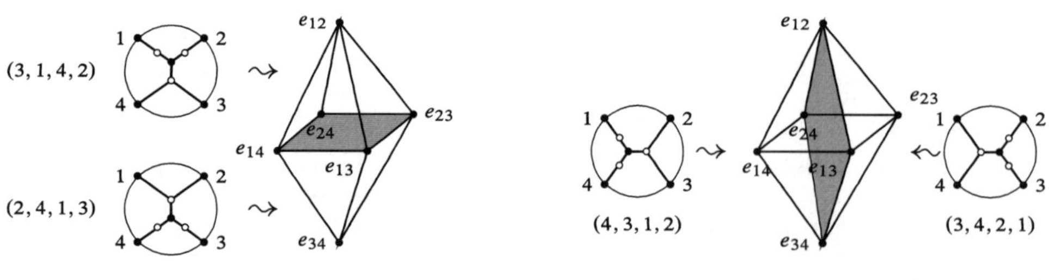

FIGURE 7
The two positroid tilings of $\Delta_{2,4}$ for the moment map. The plabic graphs specify the positroid cells whose images are the positroid tiles (positroid polytopes).

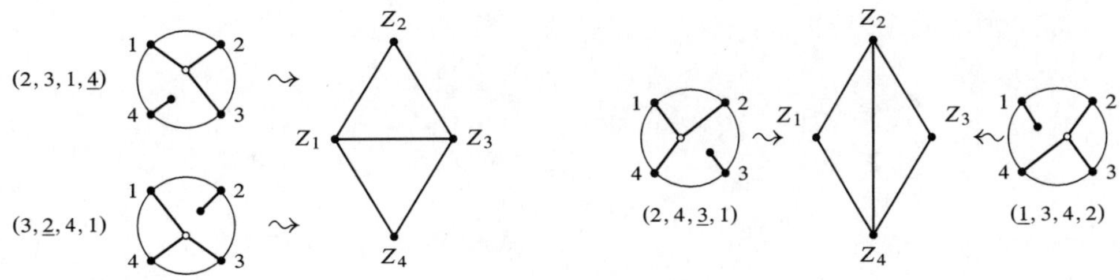

FIGURE 8

The two positroid tilings of $\mathcal{A}_{4,1,2}(Z)$ for $\tilde{Z}$. The plabic graphs specify the positroid cells whose images are the positroid tiles (Grasstopes).

Definition 6.2. Let $\pi = (a_1, a_2, \dots, a_n)$ be a loopless decorated permutation (written in one-line notation). Its *T-dual* decorated permutation is $\hat{\pi} : i \mapsto \pi(i-1)$, so that $\hat{\pi} = (a_n, a_1, a_2, \dots, a_{n-1})$. Any fixed points in $\hat{\pi}$ are declared to be loops.[1]

For example, the four permutations $(3,1,4,2)$, $(2,4,1,3)$, $(4,3,1,2)$, $(3,4,2,1)$ labeling the positroid tilings of $\Delta_{2,4}$ in Figure 7 are loopless. Their T-dual images are $(2,3,1,\underline{4})$, $(3,\underline{2},4,1)$, $(2,4,\underline{3},1)$, and $(\underline{1},3,4,2)$—precisely the permutations labeling the positroid tilings of $\mathcal{A}_{4,1,2}(Z)$ in Figure 8!

The T-duality map appears in **[33, 39]**, and is a version of an $m = 4$ map from **[3]**.

Proposition 6.3 (**[39, Lemma 5.2]** and **[46, Proposition 8.1]**). *T-duality is a bijection between loopless cells of* $\mathrm{Gr}^{\geq 0}_{k+1,n}$ *and coloopless cells of* $\mathrm{Gr}^{\geq 0}_{k,n}$. *Moreover, it is a poset isomorphism: we have* $S_\mu \subset \overline{S_\pi}$ *if and only if* $S_{\hat{\mu}} \subset \overline{S_{\hat{\pi}}}$.

One can also describe T-duality as a map on reduced plabic graphs G; we say G is *black-trivalent* (*white-trivalent*) if all of its interior black (white) vertices are trivalent.

Definition 6.4 (**[46, Definition 8.6]**). Let G be a reduced black-trivalent plabic graph. The *T-dual* of G, denoted $\hat{G}$, is the graph obtained as follows (see Figure 9).

- In each face f of G, place a black vertex $\hat{b}(f)$.
- "On top of" each black vertex b of G, place a white vertex $\hat{w}(b)$;
- For each black vertex b of G incident to face f, add edge $(\hat{w}(b), \hat{b}(f))$;
- Put $\hat{i}$ on the boundary of G between vertices $i-1$ and i and draw an edge from $\hat{i}$ to $\hat{b}(f)$, where f is the adjacent boundary face.

Note that the graphs in Figures 8 and 7 are related by T-duality.

1 Our use of the "hat" notation here is unrelated to that from Remark 4.11.

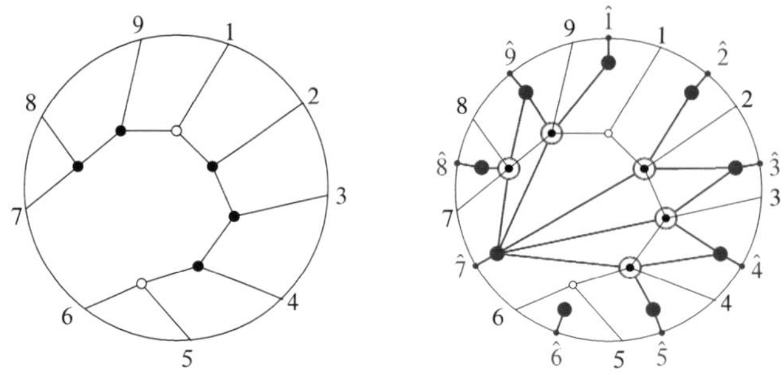

FIGURE 9

(Left) A plabic graph G with trip permutation $\pi_G = (5, 9, 2, 3, 6, 4, 1, 7, 8)$; (Right) G with the T-dual graph $\hat{G}$ superimposed. We have $\pi_{\hat{G}} = (8, 5, 9, 2, 3, \underline{6}, 4, 1, 7)$.

Proposition 6.5 ([46, PROPOSITION 8.8]). *Let G be a reduced black-trivalent plabic graph with trip permutation $\pi_G = \pi$. Then $\hat{G}$ is a reduced white-trivalent plabic graph with $\pi_{\hat{G}} = \hat{\pi}$.*

T-duality provides a link between positroid tilings of $\mathcal{A}_{n,k,2}(Z)$ and $\Delta_{k+1,n}$. The first result is that T-duality gives a bijection between positroid tiles of $\Delta_{k+1,n}$ (see Proposition 3.3) and positroid tiles of $\mathcal{A}_{n,k,2}$ (see Theorem 5.3).

Given a bicolored triangulation $\mathcal{T}$, we define $\text{area}(a \to b)$ to be the number of black triangles to the left of $a \to b$ in any triangulation of $\overline{\mathcal{T}}$ compatible with $a \to b$.

Theorem 6.6 ([46, SECTION 8]). *Given a bicolored triangulation $\mathcal{T}$ of type (k, n), we can read off T-dual graphs G and $\hat{G}$ giving positroid tiles Γ_G and $Z_{\hat{G}}$ as follows:*

- *$G := G(\mathcal{T})$ is the dual graph of $\mathcal{T}$, as shown at the left of Figure 10.*
- *$\hat{G} := \hat{G}(\mathcal{T})$ is the graph from Definition 5.2, as shown at the right of Figure 10.*

This correspondence gives a bijection between positroid tiles of $\Delta_{k+1,n}$ and $\mathcal{A}_{n,k,2}$, both of which depend only on $\overline{\mathcal{T}}$.

Moreover, if we let $h \to j$ (with $h < j$) range over arcs of $\mathcal{T}$, the inequalities

(1) $\text{area}(h \to j) + 1 \geq x_h + x_{h+1} + \cdots + x_{j-1} \geq \text{area}(h \to j)$ *for* $x \in \Gamma_{G(\mathcal{T})}$;

(2) $(-1)^{\text{area}(h \to j)} \langle Y Z_h Z_j \rangle \geq 0$ *for* $Y \in Z_{\hat{G}(\mathcal{T})}$

cut out the positroid tiles $\Gamma_{G(\mathcal{T})}$ and $Z_{\hat{G}(\mathcal{T})}$.

We now explain how Eulerian numbers enter the story.

Definition 6.7. Let $w \in S_n$. We call a letter $i \geq 2$ in w a *left descent* if $w^{-1}(i) < w^{-1}(i-1)$. And we say that $i \in [n]$ in w is a *cyclic left descent* if either $i \geq 2$ is a left descent of w or if $i = 1$ and $w^{-1}(1) < w^{-1}(n)$. Let $\text{cDes}_{\text{L}}(w)$ denote the set of cyclic left descents of w.

Let $D_{k+1,n}$ be the set of permutations $w \in S_n$ with $k+1$ cyclic descents and $w_n = n$. Note that $|D_{k+1,n}|$ equals the *Eulerian number* $E_{k,n-1} := \sum_{\ell=0}^{k+1} (-1)^{\ell} \binom{n}{\ell} (k+1-\ell)^{n-1}$.

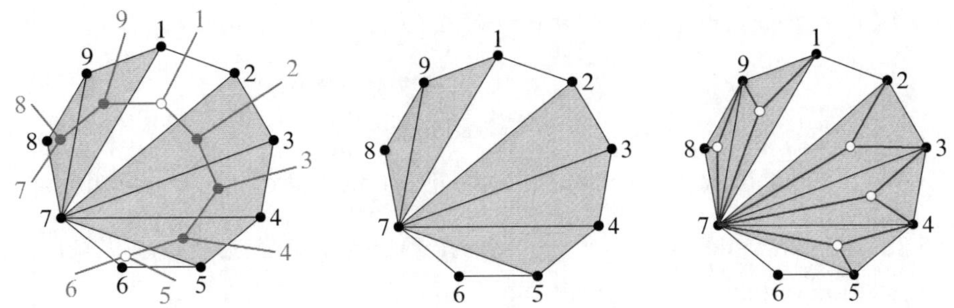

FIGURE 10

In the middle: a bicolored triangulation $\mathcal{T}$, with the dual graph $G(\mathcal{T})$ to its left, and the T-dual graph $\hat{G}(\mathcal{T})$ to its right.

Definition 6.8. For $w \in D_{k+1,n}$, let $w^{(a)}$ denote the cyclic rotation of w ending at a. Let $I_1 = I_1(w) := \mathrm{cDes_L}(w)$ and for $2 \le r \le n$, let $I_r = I_r(w) := \mathrm{cDes_L}(w^{(r-1)})$. We then define the *w-simplex* Δ_w to be the convex hull $\Delta_w := \mathrm{conv}(e_{I_1}, \dots, e_{I_n}) \subseteq \Delta_{k+1,n}$.

Example 6.9. If $w = (1, 3, 2, 4)$, then we have $I_1 = \{1, 3\}$, $I_2 = \{2, 3\}$, $I_3 = \{3, 4\}$, and $I_4 = \{2, 4\}$, so $\Delta_w = \mathrm{conv}(e_{13}, e_{23}, e_{34}, e_{24})$. See Figure 7.

Stanley gave the first triangulation of the hypersimplex [60], see also [62] and [37].

Proposition 6.10 ([60]). *We have*

$$\Delta_{k+1,n} = \bigcup_{w \in D_{k+1,n}} \Delta_w.$$

Example 6.11. For example, $\Delta_{24} = \Delta_{1324} \cup \Delta_{2134} \cup \Delta_{2314} \cup \Delta_{3124}$. This decomposition refines the two positroid tilings shown in Figure 7 (and this holds in general [37]).

Definition 6.12. Let $w \in D_{k+1,n}$ and let $I_1, \dots, I_n$ be as in Definition 6.8. We define $\hat{\Delta}^\circ_w(Z)$ to be the open amplituhedron chamber consisting of $Y \in \mathcal{A}_{n,k,2}(Z)$ such that for $a = 1, \dots, n$, the sign flips of the sequence

$$(\langle YZ_a\hat{Z}_1\rangle, \langle YZ_a\hat{Z}_2\rangle, \dots, \langle YZ_a\hat{Z}_{a-1}\rangle, \langle YZ_aZ_a\rangle, \langle YZ_aZ_{a+1}\rangle, \dots, \langle YZ_aZ_n\rangle)$$

occur precisely in positions $I_a \setminus \{a\}$, where we say a sign flip occurs in position j if $\langle YZ_aZ_j\rangle$ and $\langle YZ_aZ_{j+1}\rangle$ are nonzero and have different signs (if $j = n$ we consider $j + 1 = 1$).

We refer to $\hat{\Delta}^\circ_w(Z)$ and $\hat{\Delta}_w(Z) := \overline{\hat{\Delta}^\circ_w(Z)}$ as open and closed *w-chambers*.

Example 6.13. If $w = (1, 3, 2, 4)$, then $I_1 = \{1, 3\}$, $I_2 = \{2, 3\}$, $I_3 = \{3, 4\}$, and $I_4 = \{2, 4\}$, so $\hat{\Delta}^\circ_w(Z)$ consists of $Y \in \mathcal{A}_{n,1,2}(Z)$ such that $\langle YZ_1Z_4\rangle < 0$, $\langle YZ_2Z_4\rangle < 0$, and the other four $\langle YZ_iZ_j\rangle$ with $i < j$ are positive. In Figure 8, this corresponds to the triangle with vertices Z_3, Z_4, and the point where the two diagonals of the quadrilateral intersect.

For some choices of Z, $\hat{\Delta}^\circ_w(Z)$ can be empty. However, for each $w \in D_{k+1,n}$ one can find explicit matrices $Z \in \mathrm{Mat}^{>0}_{n,k+2}$ such that $\hat{\Delta}^\circ_w(Z)$ is nonempty [46, SECTION 11]. Moreover, the w-chambers are the only amplituhedron chambers which are realizable.

Theorem 6.14 ([46, THEOREM 10.10]). *For any $Z \in \mathrm{Mat}_{n,k+2}^{>0}$, we have*

$$\mathcal{A}_{n,k,2}(Z) = \bigcup_{w \in D_{k+1,n}} \hat{\Delta}_w(Z).$$

By Corollary 6.6, each positroid tile $Z_{\hat{G}}$ is a union of closed w-chambers.

Example 6.15. For example, $\mathcal{A}_{4,1,2}(Z) = \hat{\Delta}_{1324}(Z) \cup \hat{\Delta}_{2134}(Z) \cup \hat{\Delta}_{2314}(Z) \cup \hat{\Delta}_{3124}(Z)$. This decomposition refines the two positroid tilings shown in Figure 8.

Given that positroid tiles $\Gamma_G \subset \Delta_{k+1,n}$ are unions of w-simplices, and positroid tiles $Z_{\hat{G}} \subset \mathcal{A}_{n,k,2}(Z)$ are unions of w-chambers, the following is the key to proving that positroid tilings of $\Delta_{k+1,n}$ and $\mathcal{A}_{n,k,2}(Z)$ are in bijection.

Proposition 6.16 ([46, PROPOSITION 11.1]). *Let $Z \in \mathrm{Mat}_{n,k+2}^{>0}$. Suppose $w \in D_{k+1,n}$ and that $\hat{\Delta}_w(Z) \neq \emptyset$. For any positroid tile Γ_π, $\Delta_w \subset \Gamma_\pi$ if and only if $\hat{\Delta}_w(Z) \subset Z_{\hat{\pi}}$.*

Figures 7 and 8 illustrate the fact that the two positroid tilings of $\mathcal{A}_{4,1,2}(Z)$ are related to the two positroid tilings of $\Delta_{2,4}$ by T-duality. The following result, first conjectured in [39, CONJECTURE 6.9], generalizes this example to arbitrary k and n.

Theorem 6.17 ([46]). *A collection $\{\Gamma_\pi\}$ of positroid polytopes in $\Delta_{k+1,n}$ gives a positroid tiling of $\Delta_{k+1,n}$ if and only if for all $Z \in \mathrm{Mat}_{n,k+2}^{>0}$, the collection $\{Z_{\hat{\pi}}\}$ of Grasstopes gives a positroid tiling of $\mathcal{A}_{n,k,2}(Z)$.*

In light of the fact that $\Delta_{k+1,n}$ is an $(n-1)$-dimensional polytope, and $\mathcal{A}_{n,k,2}(Z)$ is a $2k$-dimensional nonpolytopal subset of $\mathrm{Gr}_{k,k+2}$, we find Theorem 6.17 very surprising!

We believe that more generally, Theorem 6.17 extends to give a bijection between positroid dissections (respectively, subdivisions) of $\Delta_{k+1,n}$ and positroid dissections (respectively, subdivisions) of $\mathcal{A}_{n,k,2}(Z)$, see [39, CONJECTURES 6.9 AND 8.8].

Given that $\mathrm{Trop}^+ \mathrm{Gr}_{k+1,n}$ controls the regular positroid subdivisions of $\Delta_{k+1,n}$, which are (conjecturally) in bijection with positroid subdivisions of $\mathcal{A}_{n,k,2}(Z)$, it is natural to ask: can we make a direct connection between $\mathrm{Trop}^+ \mathrm{Gr}_{k+1,n}$ and $\mathcal{A}_{n,k,2}(Z)$? Is there a way to think of points of $\mathrm{Trop}^+ \mathrm{Gr}_{k+1,n}$ as giving "height functions" for $\mathcal{A}_{n,k,2}(Z)$?

7. THE AMPLITUHEDRON AND CLUSTER ALGEBRAS

Cluster algebras are a remarkable class of commutative rings introduced by Fomin and Zelevinsky [18,20], see also [19]. Many homogeneous coordinate rings of "nice" algebraic varieties have a cluster algebra structure, including the Grassmannian [54]. Starting in 2013, various authors connected scattering amplitudes of planar $\mathcal{N} = 4$ super-Yang–Mills theory to cluster algebras [16, 25, 38, 46]. In this section we explain several connections between the amplituhedron $\mathcal{A}_{n,k,m}(Z) \subset \mathrm{Gr}_{k,k+m}$ and the cluster algebra structure on $\mathrm{Gr}_{m,n}$.

7.1. Cluster adjacency for facets of positroid tiles

Facets of positroid tiles in $\mathcal{A}_{n,k,m}(Z)$ are related to the cluster structure on $\mathrm{Gr}_{m,n}$.

Definition 7.1. Let Z_π be a Grasstope of $\mathcal{A}_{n,k,m}(Z)$. We say that $Z_{\pi'}$ is a *facet* of Z_π if it is maximal by inclusion among the Grasstopes satisfying the following three properties: $Z_{\pi'} \subset \partial Z_\pi$; the cell $S_{\pi'}$ is contained in $\overline{S_\pi}$; and $Z_{\pi'}$ has codimension 1 in Z_π.

Recall from [19, 20] that the cluster variables for $\mathrm{Gr}_{2,n}$ are the Plücker coordinates p_{ij}, and a collection of Plücker coordinates is *compatible* if the corresponding diagonals in an n-gon are noncrossing. When $m = 2$, we have the following theorem (whose first paragraph is the *cluster adjacency conjecture* from [38]).

Theorem 7.2 ([46, THEOREM 9.12]). *Let $Z_{\hat{G}(\mathcal{T})}$ be a positroid tile of $\mathcal{A}_{n,k,2}(Z)$. Each facet lies on a hypersurface $\langle YZ_iZ_j\rangle = 0$, and the collection of Plücker coordinates $\{p_{ij}\}_{\hat{G}(\mathcal{T})}$ corresponding to facets is a collection of compatible cluster variables for* $\mathrm{Gr}_{2,n}$.

If p_{hl} is compatible with $\{p_{ij}\}_{\hat{G}(\mathcal{T})}$, then $\langle YZ_hZ_l\rangle$ has a fixed sign on $Z^\circ_{\hat{G}(\mathcal{T})}$.

For $m > 2$ the Grassmannian $\mathrm{Gr}_{m,n}$ has infinitely many cluster variables, each of which can be written as a polynomial $Q(p_I)$ in the $\binom{n}{m}$ Plücker coordinates. Meanwhile, each facet of a positroid tile of the amplituhedon $\mathcal{A}_{n,k,m}(Z)$ lies on a hypersurface defined by the vanishing of some polynomial $Q(\langle YZ_I\rangle)$ in the $\binom{n}{m}$ twistor coordinates $\langle YZ_I\rangle_{I\in\binom{[n]}{m}}$.

Conjecture 7.3 ([46, CONJECTURE 6.2]). *Let Z_π be a positroid tile of $\mathcal{A}_{n,k,m}(Z)$ and let*

$$\mathrm{Facet}(Z_\pi) := \big\{Q(p_I) \mid \text{a facet of } Z_\pi \text{ lies on the hypersurface } Q(\langle YZ_I\rangle) = 0\big\},$$

where Q is a polynomial in the $\binom{n}{m}$ Plücker coordinates. Then each $Q \in \mathrm{Facet}(Z_\pi)$ is a cluster variable for $\mathrm{Gr}_{m,n}$, and $\mathrm{Facet}(Z_\pi)$ consists of compatible cluster variables. Moreover, if $\tilde{Q}$ is a cluster variable compatible with $\mathrm{Facet}(Z_\pi)$, the polynomial $\tilde{Q}(\langle YZ_I\rangle)$ in twistor coordinates has a fixed sign on Z°_π.

7.2. Positroid tiles as totally positive parts of cluster varieties

Following [46, SECTION 6.2], we now build a cluster variety $\mathcal{V}_{\overline{\mathcal{T}}}$ in $\mathrm{Gr}_{k,k+2}(\mathbb{C})$ for each positroid tile $Z_{\hat{G}(\overline{\mathcal{T}})}$ of $\mathcal{A}_{n,k,2}(Z)$. The positroid tile $Z^\circ_{\hat{G}(\overline{\mathcal{T}})}$ is exactly the *totally positive part* of $\mathcal{V}_{\overline{\mathcal{T}}}$ (in the sense of [18]).

Fix a bicolored subdivision $\overline{\mathcal{T}}$ of type (k,n), with black polygons $P_1,\dots,P_r$. Let $\mathcal{S}(\overline{\mathcal{T}})$ denote the set of all bicolored triangulations represented by $\overline{\mathcal{T}}$. For each black polygon P_i, fix a *distinguished boundary arc* $h_i \to j_i$ with $h_i < j_i$ in the boundary of P_i. We will build $\mathcal{V}_{\overline{\mathcal{T}}}$ by defining seeds in the field of rational functions on $\mathrm{Gr}_{k,k+2}(\mathbb{C})$.

Definition 7.4 (Cluster variables). Let $a \to b$ with $a < b$ be an arc which is contained in a black polygon P_i and is not the distinguished boundary arc $h_i \to j_i$. We define

$$x_{ab} := \frac{(-1)^{\mathrm{area}(a\to b)}\langle YZ_aZ_b\rangle}{(-1)^{\mathrm{area}(h_i\to j_i)}\langle YZ_{h_i}Z_{j_i}\rangle},$$

which is a rational function on $\mathrm{Gr}_{k,k+2}(\mathbb{C})$.

Definition 7.5 (Seeds). Let $\mathcal{T} \in \mathcal{S}(\overline{\mathcal{T}})$. We define the quiver $Q_{\mathcal{T}}$ as follows:

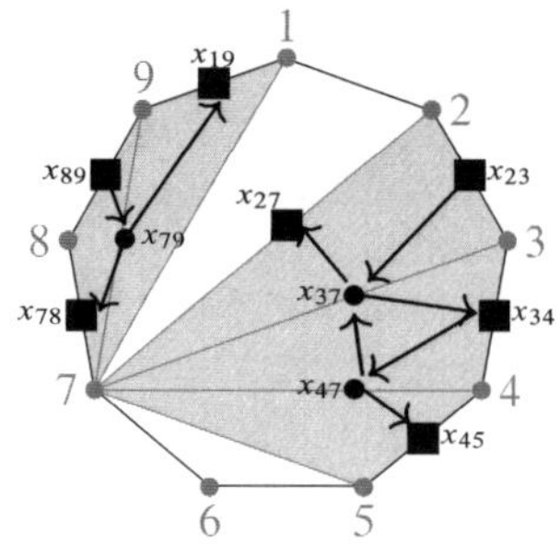

FIGURE 11
In gray, a bicolored triangulation $\mathcal{T}$. In black, the seed $\Sigma_{\mathcal{T}}$. The distinguished boundary arcs are $1 \to 7$ and $5 \to 7$.

- Place a frozen vertex on each non-distinguished boundary arc of $P_1, \dots, P_r$ and a mutable vertex on every other arc (a "black arc") bounding a triangle of $\mathcal{T}$.
- If arcs $a \to b$, $b \to c$, $c \to a$ form a triangle, we put arrows in Q between the corresponding vertices, going clockwise around the triangle.

We label the vertex of $Q_{\mathcal{T}}$ on arc $a \to b$ of $\mathcal{T}$ with the function x_{ab}. The collection of vertex labels is the *(extended) cluster* $\mathbf{x}_{\mathcal{T}}$. The pair $(Q_{\mathcal{T}}, \mathbf{x}_{\mathcal{T}})$ is the *seed* $\Sigma_{\mathcal{T}}$.

See Figure 11 for an example. Note that the cluster $\mathbf{x}_{\mathcal{T}}$ has size $2k$.

Theorem 7.6 ([46, SECTION 6.2]). *Let $\mathcal{T} \in \mathcal{S}(\overline{\mathcal{T}})$. Then*

$$\mathcal{V}_{\mathcal{T}} := \Big\{ Y \in \mathrm{Gr}_{k,k+2}(\mathbb{C}) : \prod_{a \to b \text{ black arc of } \mathcal{T}} \langle Y Z_a Z_b \rangle \neq 0 \Big\}$$

is birational to an algebraic torus of dimension $2k$, and its field of rational functions is the field $\mathbb{C}(\mathbf{x}_{\mathcal{T}})$ of rational functions in the cluster $\mathbf{x}_{\mathcal{T}}$.

The set $\mathcal{V}_{\overline{\mathcal{T}}} := \bigcup_{\mathcal{T} \in \mathcal{S}(\overline{\mathcal{T}})} \mathcal{V}_{\mathcal{T}}$ is a cluster variety in $\mathrm{Gr}_{k,k+2}(\mathbb{C})$. In particular, if $\mathcal{T}$ and $\mathcal{T}'$ are related by flipping arc $a \to b$, seeds $\Sigma_{\mathcal{T}}$ and $\Sigma_{\mathcal{T}'}$ are related by mutation at x_{ab}.

The positive part $$\mathcal{V}_{\overline{\mathcal{T}}}^{>0} := \big\{ Y \in \mathcal{V}_{\overline{\mathcal{T}}} : x_{ab}(Y) > 0 \text{ for all cluster variables } x_{ab} \big\}$$

of the cluster variety $\mathcal{V}_{\overline{\mathcal{T}}}$ is equal to the positroid tile $Z^{\circ}_{\hat{G}(\overline{\mathcal{T}})}$.

We conjecture that for even m, each positroid tile of $\mathcal{A}_{n,k,m}(Z)$ can be realized as the totally positive part of a cluster variety in $\mathrm{Gr}_{k,k+m}(\mathbb{C})$.

8. FUTURE DIRECTIONS

There are other geometric objects related to the amplituhedron, including the *loop amplituhedron* [7], and the *momentum amplituhedron* (defined for $m = 4$ in [15] and for even m in [39]). It would be interesting to explore these objects systematically as above.

ACKNOWLEDGMENTS

Main references for this article include [32, 33, 39, 46, 59]. It is a pleasure to thank my collaborators, especially S. Karp, T. Lukowski, M. Parisi, M. Sherman-Bennett, D. Speyer, and Y. Zhang. I have also benefited enormously from discussions and collaborations with many others, including F. Ardila, N. Arkani-Hamed, S. Fomin, A. Postnikov, K. Rietsch, F. Rincón, and B. Sturmfels. Finally, I thank J. Boretsky, A. Burcroff, C. Defant, S. Fomin, P. Galashin, J. Gao, S. Karp, B. Keller, M. Parisi, M. Sherman-Bennett, and R. Vlad for helpful comments on this manuscript, and the Radcliffe Institute for (virtual) hospitality.

FUNDING

This work was partially supported by NSF grants DMS-1854316 and DMS-1854512.

REFERENCES

[1] F. Ardila, F. Rincón, and L. Williams, Positroids and non-crossing partitions. *Trans. Amer. Math. Soc.* **368** (2016), no. 1, 337–363.

[2] F. Ardila, F. Rincón, and L. Williams, Positively oriented matroids are realizable. *J. Eur. Math. Soc. (JEMS)* **19** (2017), no. 3, 815–833.

[3] N. Arkani-Hamed, J. Bourjaily, F. Cachazo, A. Goncharov, A. Postnikov, and J. Trnka, *Grassmannian geometry of scattering amplitudes*. Cambridge University Press, Cambridge, 2016.

[4] N. Arkani-Hamed, F. Cachazo, C. Cheung, and J. Kaplan, A duality for the S matrix. *J. High Energy Phys.* **2010** (2010), 20.

[5] N. Arkani-Hamed, T. Lam, and M. Spradlin, Positive configuration space. 2021, DOI 10.1007/s00220-021-04041-x.

[6] N. Arkani-Hamed, H. Thomas, and J. Trnka, Unwinding the amplituhedron in binary. *J. High Energy Phys.* **2018** (2018), 16.

[7] N. Arkani-Hamed and J. Trnka, The amplituhedron. *J. High Energy Phys.* **10** (2014), 33.

[8] H. Bao and X. He, The $m = 2$ amplituhedron. 2019, arXiv:1909.06015.

[9] L. J. Billera and B. Sturmfels, Fiber polytopes. *Ann. of Math. (2)* **135** (1992), no. 3, 527–549.

[10] A. Björner, M. Las Vergnas, B. Sturmfels, N. White, and G. M. Ziegler, *Oriented matroids. Second edn*. Encyclopedia Math. Appl. 46, Cambridge University Press, Cambridge, 1999.

[11] R. Britto, F. Cachazo, B. Feng, and E. Witten, Direct proof of the tree-level scattering amplitude recursion relation in Yang-Mills theory. *Phys. Rev. Lett.* **94** (2005), no. 18, 181602.

[12] M. Bullimore, L. Mason, and D. Skinner, Twistor-strings, Grassmannians and leading singularities. *J. High Energy Phys.* **2010** (2010), 70.

[13] S. Corteel and L. K. Williams, Tableaux combinatorics for the asymmetric exclusion process. *Adv. in Appl. Math.* **39** (2007), no. 3, 293–310.

[14] I. P. da Silva, *Quelques propriétés des matroides orientés*. Ph.D. Dissertation, Université Paris VI, 1987.

[15] D. Damgaard, L. Ferro, T. Łukowski, and M. Parisi, The momentum amplituhedron. *J. High Energy Phys.* **2019** (2019), 42.

[16] J. Drummond, J. Foster, and Ö. Gürdoğan, Cluster adjacency properties of scattering amplitudes in $N = 4$ supersymmetric Yang–Mills theory. *Phys. Rev. Lett.* **120** (2018), no. 16, 161601.

[17] C. Even-Zohar, T. Kalrec, R. Tessler, The Amplituhedron BCFW triangulation. 2021, arXiv:2112.02703.

[18] S. Fomin, Total positivity and cluster algebras. In *Proceedings of the international congress of mathematicians. Volume II*, pp. 125–145, Hindustan Book Agency, New Delhi, 2010.

[19] S. Fomin, L. Williams, and A. Zelevinsky, Introduction to cluster algebras. Chapters 1–3, 4–5, 6, 7. 2021, arXiv:1608.05735, arXiv:1707.07190, arXiv:2008.09189, arXiv:2106.02160.

[20] S. Fomin and A. Zelevinsky, Cluster algebras I. Foundations. *J. Amer. Math. Soc.* **15** (2002), no. 2, 497–529.

[21] P. Galashin, S. N. Karp, and T. Lam, Regularity theorem for totally nonnegative flag varieties. 2019, arXiv:1904.00527.

[22] P. Galashin and T. Lam, Parity duality for the amplituhedron. *Compos. Math.* **156** (2020), no. 11, 2207–2262.

[23] F. P. Gantmacher and M. G. Krein, *Oscillation matrices and kernels and small vibrations of mechanical systems. Revised edn.* AMS Chelsea Publishing, Providence, RI, 2002.

[24] I. M. Gelfand, R. M. Goresky, R. D. MacPherson, and V. V. Serganova, Combinatorial geometries, convex polyhedra, and Schubert cells. *Adv. Math.* **63** (1987), no. 3, 301–316.

[25] J. Golden, A. B. Goncharov, M. Spradlin, C. Vergu, and A. Volovich, Motivic amplitudes and cluster coordinates. *J. High Energy Phys.* **2014** (2014), 91.

[26] V. Guillemin and S. Sternberg, Convexity properties of the moment mapping. *Invent. Math.* **67** (1982), no. 3, 491–513.

[27] P. Hacking, S. Keel, and J. Tevelev, Compactification of the moduli space of hyperplane arrangements. *J. Algebraic Geom.* **15** (2006), no. 4, 657–680.

[28] S. Herrmann, A. N. Jensen, M. Joswig, and B. Sturmfels, How to draw tropical planes. *Electron. J. Combin.* **16** (2008).

[29] A. Hodges, Eliminating spurious poles from gauge-theoretic amplitudes. *J. High Energy Phys.* **2013** (2013), 135.

[30] M. M. Kapranov, Chow quotients of Grassmannians. I. In *I. M. Gelfand seminar*, pp. 29–110, Adv. Sov. Math. 16, Amer. Math. Soc., Providence, RI, 1993.

[31] S. N. Karp, Sign variation, the Grassmannian, and total positivity. *J. Combin. Theory Ser. A* **145** (2017), 308–339.

[32] S. N. Karp and L. K. Williams, The $m = 1$ amplituhedron and cyclic hyperplane arrangements. *Int. Math. Res. Not. IMRN* **5** (2019), 1401–1462.

[33] S. N. Karp, L. K. Williams, and Y. X. Zhang, Decompositions of amplituhedra. *Ann. Inst. Henri Poincaré D* **7** (2020), no. 3, 303–363.

[34] S. Keel and J. Tevelev, Geometry of Chow quotients of Grassmannians. *Duke Math. J.* **134** (2006), no. 2, 259–311.

[35] Y. Kodama and L. Williams, The Deodhar decomposition of the Grassmannian and the regularity of KP solitons. *Adv. Math.* **244** (2013), 979–1032.

[36] Y. Kodama and L. Williams, KP solitons and total positivity for the Grassmannian. *Invent. Math.* **198** (2014), no. 3, 637–699.

[37] T. Lam and A. Postnikov, Alcoved polytopes. I. *Discrete Comput. Geom.* **38** (2007), no. 3, 453–478.

[38] T. Łukowski, M. Parisi, M. Spradlin, and A. Volovich, Cluster adjacency for $m = 2$ Yangian invariants. *J. High Energy Phys.* **2019** (2019), 158.

[39] T. Łukowski, M. Parisi, and L. K. Williams, The positive tropical Grassmannian, the hypersimplex, and the $m = 2$ amplituhedron. 2020, arXiv:2002.06164.

[40] G. Lusztig, Total positivity in reductive groups. In *Lie theory and geometry*, pp. 531–568, Progr. Math. 123, Birkhäuser Boston, Boston, MA, 1994.

[41] L. Mathis and C. Meroni, Fiber convex bodies. 2021, arXiv:2002.06164.

[42] D. Mayhew, M. Newman, and G. Whittle, Yes, the 'missing axiom' of matroid theory is lost forever. *Trans. Amer. Math. Soc.* **370** (2018), no. 8, 5907–5929.

[43] N. E. Mnëv, The universality theorems on the classification problem of configuration varieties and convex polytopes varieties. In *Topology and geometry—Rohlin seminar*, pp. 527–543, Lecture Notes in Math. 1346, Springer, Berlin, 1988.

[44] S. Oh, A. Postnikov, and D. E. Speyer, Weak separation and plabic graphs. *Proc. Lond. Math. Soc. (3)* **110** (2015), no. 3, 721–754.

[45] J. A. Olarte, M. Panizzut, and B. Schröter, On local Dressians of matroids. In *Algebraic and geometric combinatorics on lattice polytopes*, pp. 309–329, World Sci. Publ., Hackensack, NJ, 2019.

[46] M. Parisi, M. Sherman-Bennett, and L. K. Williams, The $m = 2$ amplituhedron and the hypersimplex: signs, clusters, triangulations, Eulerian numbers. 2021, arXiv:2104.08254.

[47] A. Postnikov, Total positivity, Grassmannians, and networks. 2006, arXiv:math/0609764.

[48] A. Postnikov, Positive Grassmannian and polyhedral subdivisions. In *Proceedings of the international congress of mathematicians—Rio de Janeiro 2018. Vol. IV. Invited lectures*, pp. 3181–3211, World Sci. Publ., Hackensack, NJ, 2018.

[49] A. Postnikov, D. Speyer, and L. Williams, Matching polytopes, toric geometry, and the totally non-negative Grassmannian. *J. Algebraic Combin.* **30** (2009), no. 2, 173–191.

[50] J. Rambau, Triangulations of cyclic polytopes and higher Bruhat orders. *Mathematika* **44** (1997), no. 1, 162–194.

[51] K. Rietsch, Private communication, 2009.

[52] K. C. Rietsch, *Total positivity and real flag varieties*. Ph.D. thesis, Massachusetts Institute of Technology, 1998.

[53] K. Rietsch and L. Williams, Discrete Morse theory for totally non-negative flag varieties. *Adv. Math.* **223** (2010), no. 6, 1855–1884.

[54] J. S. Scott, Grassmannians and cluster algebras. *Proc. Lond. Math. Soc. (3)* **92** (2006), no. 2, 345–380.

[55] D. E. Speyer, A matroid invariant via the K-theory of the Grassmannian. *Adv. Math.* **221** (2009), no. 3, 882–913.

[56] D. E. Speyer, Variations on a theme of Kasteleyn, with application to the totally nonnegative Grassmannian. *Electron. J. Combin.* **23** (2016), no. 2, 2.24.

[57] D. Speyer and B. Sturmfels, The tropical Grassmannian. *Adv. Geom.* **4** (2004), no. 3, 389–411.

[58] D. Speyer and L. Williams, The tropical totally positive Grassmannian. *J. Algebraic Combin.* **22** (2005), no. 2, 189–210.

[59] D. Speyer and L. K. Williams, The positive Dressian equals the positive tropical Grassmannian. *Trans. Amer. Math. Soc. Ser. B* **8** (2021), 330–353.

[60] R. Stanley, Eulerian partitions of a unit hypercube. In *Higher combinatorics: Proceedings of the NATO Advanced Study Institute held in Berlin, September 1–10, 1976*, edited by M. Aigner, p. 49, D. Reidel Publishing Co., Dordrecht–Boston, Mass, 1977.

[61] B. Sturmfels, Totally positive matrices and cyclic polytopes. In *Proceedings of the Victoria conference on combinatorial matrix analysis (Victoria, BC, 1987)*, *Lin. Alg. Appl.* **107** (1988), pp. 275–281.

[62] B. Sturmfels, *Gröbner bases and convex polytopes*. Univ. Lecture Ser. 8, American Mathematical Society, Providence, RI, 1996.

[63] K. Talaska, A formula for Plücker coordinates associated with a planar network. *Int. Math. Res. Not. IMRN* **2008** (2008).

[64] K. Talaska and L. Williams, Network parametrizations for the Grassmannian. *Algebra Number Theory* **7** (2013), no. 9, 2275–2311.

[65] E. Tsukerman and L. Williams, Bruhat interval polytopes. *Adv. Math.* **285** (2015), 766–810.

[66] P. Vámos, The missing axiom of matroid theory is lost forever. *J. Lond. Math. Soc.* **18** (1978), 403–408.

[67] L. K. Williams, Shelling totally nonnegative flag varieties. *J. Reine Angew. Math.* **609** (2007), 1–21.

LAUREN K. WILLIAMS

Department of Mathematics, 1 Oxford Street, Cambridge, MA 02138, USA,
williams@math.harvard.edu

This Page Intentionally Left Blank

14. MATHEMATICS OF COMPUTER SCIENCE

SPECIAL LECTURES

DIFFERENTIAL PRIVACY: GETTING MORE FOR LESS

CYNTHIA DWORK

ABSTRACT

The key to the success of differential privacy, now the gold standard for privacy-preserving data analysis, is the ability to quantify and reason about cumulative privacy loss over many differentially private interactions. When upper bounds on privacy loss are loose, the deployment of the algorithms is by definition conservative. Under high levels of composition, much potential utility is lost. We survey two general approaches to getting more utility: privacy amplification methods, which are algorithmic, and definitional methods, which admit a wider class of algorithms and lead to tighter analyses of existing algorithms.

MATHEMATICS SUBJECT CLASSIFICATION 2020

Primary 68P27; Secondary 68W20, 68Q32, 68T05

KEYWORDS

Differential privacy, private data analysis, privacy-preserving data analysis, machine learning, private machine learning

INTERNATIONAL CONGRESS OF MATHEMATICIANS 2022 JULY 6–14

Proc. Int. Cong. Math. 2022, Vol. 6, pp. 4740–4761
DOI 10.4171/ICM2022/196

Published by EMS Press

1. INTRODUCTION

The Fundamental Law of Information Recovery states, informally, that "overly accurate" estimates of "too many" statistics completely destroy privacy ([11] *et sequelae*; see [21] for a survey). Differential privacy is a mathematically rigorous definition of privacy tailored to analysis of large datasets and equipped with a formal measure of privacy loss [12,15–18]. Differentially private algorithms take as input a parameter, typically called ε, that caps the permitted privacy loss in any execution of the algorithm and offers a concrete privacy/utility tradeoff. The key to differential privacy's success is the ability to reason about cumulative privacy loss as the data are analyzed and reanalyzed, that is, we can understand its behavior under *composition*. This permits modular construction of differentially private algorithms from simple differentially private building blocks; in other words, differential privacy is *programmable*. The art of differentially private algorithm design is to obtain as much utility as possible, while minimizing cumulative privacy loss.

A statistic is a quantity computed from a sample. Statistics "feel" private for the same reasoning that statistics works as a discipline, meaning that we expect we will obtain approximately the same outcome independent of the actual sample chosen, provided that proper sampling procedures are followed and the sample is sufficiently large. In this sense, the statistic is not "about" the members of the sample, but instead it is a quantity that describes the population as a whole. Relatedly, statistics feel private because of the privacy of the sample: "I could have opted out," "no one knows that I am in the sample." Differential privacy adheres to this intuition, maintaining the "I could have opted out" semantics even when the computations are carried out on the entire population, as in a census.

Roughly speaking, differential privacy ensures that the outcome of any analysis on a dataset x is distributed very similarly to the outcome on any neighboring dataset y that differs from x in just one row. That is, differentially private algorithms are randomized, giving rise to a probability distribution $\mathcal{A}(x)$ when run on dataset x; the definition requires that the *max divergence* between these two distributions $\mathcal{A}(x)$, $\mathcal{A}(y)$ (the maximum log odds ratio for any event $S \in \text{Range}(\mathcal{A})$, also known as the maximum *privacy loss*; Definition 2.2 below) is bounded by a privacy parameter ε. This absolute guarantee on the maximum privacy loss is now known as *pure* differential privacy.

This absolute bound on privacy loss is a conservative estimate of the privacy offered by any given execution of the algorithm: for $\varepsilon \leq 1$, the expected privacy loss of an arbitrary ε-differentially private algorithm is bounded by ε^2 [19,27]. Under very high levels of composition, which is the norm in machine learning and in continual, industrial-scale tracking, as is common with cell phone location and usage data, any looseness in the bounds is severely compounded. Thus, analytical techniques that more tightly capture the behavior of the privacy loss random variable quickly lead to improved utility through a more informed selection of the parameters.

Beginning with [13], a large body of work examines what can be achieved by providing high probability, rather than absolute, bounds on the privacy loss. Relaxing the protection goal – that is, relaxing the definition of privacy-preserving analysis – provides opportunities for more refined analyses, and the choice of relaxation may rightfully be influenced by the analytical tools enabled by the definition. In many "workhorse" cases we are interested in better understanding of very large numbers of invocations of the *same* algorithm, applied to the *same* data. Statistical queries ("What fraction of the people in the dataset satisfy property P?"), and gradient descent, central to data analytics and modern machine learning, respectively, are exemplars of this phenomenon [1–3, 6, 10, 15, 24–26, 34, 37].

This informal note reviews definitions and sketches some exciting recent directions in *privacy amplification*. In Section 2 we establish the notation used throughout, and motivate the definition of pure differential privacy. Section 3 describes some basic differentially private primitives. Section 4 discusses four relaxations and provides some comparisons between them. In Section 5 we provide intuition for three amplification techniques. Section 6 describes some applications.

2. PURE DIFFERENTIAL PRIVACY

Let $\mathcal{U}$ denote a unverse of data records, where each $u \in \mathcal{U}$ corresponds to a possible value of the data of an individual. *Datasets* are multisets of draws from $\mathcal{U}$, and *adjacent* datasets differ in the data of just one individual. There are two notions of adjacency. In *replace-one* adjacency the two sets have the same cardinality and agree on all but (possibly) one element; in *add/remove* adjacency one set is contained in the other, and the larger has the data of just one additional individual. The distinction rarely matters, beyond a factor of two in the privacy loss bounds.

(Useful) differentially private algorithms are necessarily randomized [15]. For an algorithm $\mathcal{A}$ and dataset x, we let $\mathcal{A}(x)$ denote the probability distribution on outcomes of the randomized $\mathcal{A}$ operating on x. All probabilities and expectations are taken over the coin flips of the algorithms.

Definition 2.1 (Differential privacy [15]). For $\varepsilon \geq 0$, Algorithm $\mathcal{A}$ is ε-differentially private if, for all adjacent datasets $x, y \in \mathcal{U}^*$ and for any event C in the range of $\mathcal{A}$.

$$\Pr\big[\mathcal{A}(x) \in C\big] \leq e^{\varepsilon}\Pr\big[\mathcal{A}(y) \in C\big] \tag{2.1}$$

where the probabilities are taken over the randomness of $\mathcal{A}$.

Note that the definition is symmetric in x and y and is therefore equivalent to the condition $|\log \frac{\Pr[\mathcal{A}(x)\in C]}{\Pr[\mathcal{A}(y)\in C]}| \leq \varepsilon$. Note also that differential privacy is a worst-case notion: the probabilities are over the randomness of the algorithms, not the choice of the datasets.

Composition refers to running multiple differentially private algorithms on the same dataset, and publishing the outputs at each step.[1] It is easily verified that for any algorithms $\mathcal{A}$ and $\mathcal{A}'$ that are ε and ε' differentially private, respectively, the composition that on input x first runs $\mathcal{A}(x)$ and then runs $\mathcal{A}'(x)$, publishing both outputs, is $(\varepsilon + \varepsilon')$-differentially

1 Composition can be defined more broadly, for example, to cover analyses carried out independently on overlapping datasets; see [20].

private [15]. Moreover, this is true even if $\mathcal{A}'$ is chosen adaptively, after having seen the output of $\mathcal{A}(x)$. Thus, differential privacy is closed under composition: composition does not destroy privacy, but it does, eventually, erode it.

Definition 2.2 (Privacy loss random variable). Let $\mathcal{A}$ be an algorithm and let x, y be adjacent datasets. For all $\xi \in \text{Range}(\mathcal{A})$, the *privacy loss of an outcome* ξ *with respect to* y *when running on* x, denoted $L_{x,y,\xi}$ is the ratio

$$L_{x,y,\xi} = \log \frac{\Pr[\mathcal{A}(x) = \xi]}{\Pr[\mathcal{A}(y) = \xi]}. \tag{2.2}$$

For continuous output spaces, the probabilities above are replaced by the probability density functions.

This definition is not symmetric in x and y, and any event that can occur with nonzero probability when running $\mathcal{A}(x)$ but cannot appear when running $\mathcal{A}(y)$ has infinite privacy loss: if an adversary observes such an event, then it knows, for certain, that the input dataset is not y.

Fix any adjacent x and y, and consider an execution of $\mathcal{A}(x)$ resulting in an output ξ. The loss $L_{x,y,\xi}$ might be positive – which is the case when ξ is more likely under $\mathcal{A}(x)$ than under $\mathcal{A}(y)$ – or it may be negative. The fact that privacy loss can be negative leads to cancellation when we run multiple algorithms: the cumulative privacy loss random variable exhibits a martingale-like behavior, and is tightly concentrated around its expectation. This phenomenon is captured by the *Advanced Composition Theorem* [20], stated in Section 4.

Differential privacy enjoys several other properties, in particular (1) it is "future-proof," meaning it is closed under postprocessing; no amount of computation after the fact, and no auxiliary information obtained from other sources, can increase the realized privacy loss; (2) bounds on privacy loss for *groups* degrades gracefully with the size of the group, and an ε-differentially private algorithm is automatically $k\varepsilon$-differentially private for groups of size k.

3. TWO PRIMITIVES

We briefly describe two primitives, or building blocks, that yield pure differentially private algorithms, which we will need for this note.

Definition 3.1 (Counting queries, statistical queries). A counting query is constructed from a predicate, that is, a mapping $q : \mathcal{U} \to \{0, 1\}$. When applied to a dataset, the counting query is asking how many individuals in the dataset are mapped to 1 ("satisfy q"). The associated statistical query is the fraction of members of the dataset that are mapped to 1.

Randomized Response is a generalization of a technique introduced by Warner to conduct surveys about embarrassing or illegal behavior; it provides plausible deniability, allowing individuals to self-report in a randomized way that is biased towards the truth [36]. In this mechanism, the curator/researcher/analyst does not see the private data: individuals privatize their information before releasing it to a not-necessarily trusted analyst. Randomized response forms the lion's share of differential privacy in industrial use, where it is viewed as "shifting the trust boundary to the client," absolving the server of risk (and, it is perhaps hoped, liability) of privacy violation even if the server is compromised.

In its simplest form, for $u \in \mathcal{U} = \{0, 1\}$, Randomized Response is the following algorithm [36]:

$\mathcal{RR}(\mathbf{u} \in \{\mathbf{0}, \mathbf{1}\}, \ \mathbf{p} \in [\mathbf{0}, \mathbf{1}])$
$b \leftarrow \text{Ber}(p)$
If $b = 1$ **then** output a random draw from $\text{Ber}(1/2)$;
Else output u.

Here, the notation $\text{Ber}(p)$ denotes the Bernoulli distribution with parameter p. This algorithm, which ensures ε-differential privacy whenever $p \geq 2/(1 + e^{\varepsilon})$ [15], operates on a dataset of size $n = 1$. Given a collection $v = \{v_1, \dots, v_n\}$, where v_i is obtained by running $\mathcal{RR}(u_i, p)$ (either because individual i ran this algorithm on its own data before sending the result to the server, or because the server has all the data $\{u_1, \dots, u_n\}$ and has calculated the v_i's as intermediate results), the fraction of the number of u_i that satisfy property q can be estimated by "reverse-engineering" the reported statistics as follows. Let $T = \sum_i v_i$. Approximately $pn/2$ of the observed ones in v come from random draws from $\text{Ber}(1/2)$, and so (most of) the remaining 1's come from the truthful responses. Thus, the fraction of positive x_i's is approximately $\frac{T-(pn/2)}{(1-p)n}$. The expected error is $\Theta(\frac{1}{\varepsilon\sqrt{n}})$; when rescaled for a counting query, we get $\Theta(\frac{\sqrt{n}}{\varepsilon})$.

Definition 3.2 (L_1- and L_2-sensitivity). Let $f : \mathcal{U}^* \to \mathbb{R}^d$ be an arbitrary function. The L_1-sensitivity of f is the maximum, over all adjacent datasets x, y, of $\|f(x) - f(y)\|_1$. The L_2-sensitivity of f is the maximum, over all adjacent datasets x, y, of $\|f(x) - f(y)\|_2$.

Sensitivity is a property of the function, and does not depend on the specific dataset to which the function is to be applied.

We will also make use of the 0-centered Laplace distribution, $\text{Lap}(b)$, which has density function $\mu(x) = \frac{1}{2b}e^{-|x|/b}$.

Theorem 3.1 (Laplace mechanism [15]). *Let $f : \mathcal{U}^* \to \mathbb{R}^d$ be an arbitrary function, and let Δ_1 denote its L_1-sensitivity. Then the mechanism that, on input dataset x and privacy parameter ε, computes $f(x)$ and adds an independent draw from* $\text{Lap}(\Delta_1/\varepsilon)$ *to each coordinate, satisfies ε-differential privacy. That is,*

$$\mathcal{A}(\varepsilon, x) = f(x) + \left(\text{Lap}\left(\frac{\Delta_1}{\varepsilon}\right)\right)^d \tag{3.1}$$

is ε-differentially private.

The L_1-sensitivity of a counting query is 1; the expected error for this mechanism is $O(\frac{1}{\varepsilon})$, a substantial improvement over randomized response. Indeed, for counting queries, the error introduced for privacy by the Laplace mechanism is less than the sampling error. In this sense, privacy is "for free."

A natural question is whether addition of Gaussian noise $\mathcal{N}(0, \sigma^2 \mathbb{I}_d)$ also yields differential privacy. It does not, even in one dimension. Let the probability densities of the distributions $\mathcal{N}(0, \sigma^2)$ and $\mathcal{N}(1, \sigma^2)$ at any $t \in \mathbb{R}$ be denoted $\mu_0(t)$ and $\mu_1(t)$, respectively. Then there is no fixed bound on the ratio $\frac{\mu_1(t)}{\mu_0(t)}$ as $|t|$ grows. On the other hand, by choosing σ to be sufficiently large as a function of the sensitivity and ε, we can control the likelihood of *failing* to satisfy privacy loss bounded by ε. This motivates the first of the relaxations of pure differential privacy discussed in the next section.

4. RELAXATIONS OF PURE DIFFERENTIAL PRIVACY

In this section we will discuss several relaxations of differential privacy. The differences between the various notions come down to how they treat very low probability events. A detailed discussion appears in [7].

4.1. Approximate differential privacy

Definition 4.1 (Approximate (or (ε, δ)) differential privacy [13]). For $\varepsilon, \delta \geq 0$, Algorithm $\mathcal{A}$ is (ε, δ)-differentially private if, for all adjacent datasets $x, y \in \mathcal{U}^*$ and for any event C in the range of $\mathcal{A}$,

$$\Pr\big[\mathcal{A}(x) \in C\big] \leq e^{\varepsilon} \Pr\big[\mathcal{A}(y) \in C\big] + \delta, \tag{4.1}$$

where the probabilities are taken over the randomness of $\mathcal{A}$.

When $\delta = 0$, this recovers the definition of pure differential privacy. When $\delta > 0$, even if it is negligibly small, this relaxation provides what amounts to a switch of quantification order: pure differential privacy ensures that on every execution of $\mathcal{A}(x)$ the observed outcome will be essentially equally likely on *all* adjacent datasets y simultaneously. In approximate differential privacy, for any specific y adjacent to x, it is extremely unlikely *ex ante* that the observed value $\mathcal{A}(x)$ will be one that is much more or much less likely when the dataset is x rather than when the dataset is y, but given an output $\xi \leftarrow \mathcal{A}(x)$ it might be possible to find *some* y such that ξ is, say, much more likely to be produced on x than it is on y. For this y, $L_{x,y,\xi}$ would be very large.

Although (ε, δ)-differential privacy satisfies a simple composition theorem analogous to simple composition for pure differential privacy – the epsilons and the deltas add up – it, like pure differential privacy, enjoys the benefits of the Advanced Composition Theorem of Dwork, Rothblum, and Vadhan [20], stated next.

Theorem 4.1 (Advanced composition [20]). *For all $\varepsilon, \delta, \delta' \geq 0$, the class of (ε, δ) differentially private mechanisms satisfies $(\varepsilon', k\delta + \delta')$-differential privacy under adaptive k-fold composition for*

$$\varepsilon' = \sqrt{2k \ln(1/\delta')} \cdot \varepsilon + k\varepsilon(e^{\varepsilon} - 1). \tag{4.2}$$

In Advanced Composition, the dependence on the degree k of composition is of order $\sqrt{k}$, rather than linear in k. Observe that the theorem yields a host of bounds; for each value of $\delta' > 0$, one obtains a corresponding value of ε', and *vice versa*.

Remark 4.2. The coefficient $\varepsilon(e^{\varepsilon} - 1)$ can be improved by a factor of 2 [19,27]; tight bounds are obtained in [27] for the homogeneous case (same epsilons and deltas). The optimal composition bound in the nonhomogeneous case is very hard (#-P hard) to compute [32].

While the Gaussian mechanism, described next, cannot offer pure differential privacy, it yields approximate differential privacy.

Theorem 4.3 (Gaussian mechanism [13]). *Let $f : \mathcal{U}^* \to \mathbb{R}^d$ be an arbitrary function, and let Δ_2 denote its L_2-sensitivity. Then the mechanism that, on input dataset x and privacy parameter ε, computes $f(x)$ and adds an independent draw from $\mathcal{N}(0, c^2(\Delta_2/\varepsilon)^2\mathbb{I}_d)$ to each coordinate, satisfies (ε, δ)-differential privacy whenever $c^2 \geq 2\ln(2/\delta)$. That is, for any such c,*

$$\mathcal{A}(\varepsilon, x) = f(x) + \left(\mathcal{N}\left(0, c^2(\Delta_2/\varepsilon)^2\mathbb{I}_d\right)\right)^d \tag{4.3}$$

is (ε, δ)-differentially private.[2]

Approximate differential privacy can also provide an appealing bridge to *robust statistics* through the so-called *Propose-Test-Release* paradigm [14]. Focusing here on the one-dimensional case, in this framework, one runs a differentially private algorithm to test whether the dataset x is far, in Hamming distance, from all datasets y for which $|f(x) - f(y)|$ is larger than some fixed Δ. If so, the algorithm releases $f(x) + \text{Lap}(\frac{\Delta}{\varepsilon})$, and otherwise it releases a special output $\perp$.[3] Such an algorithm has a risk of a false positive, which would lead to an inadequate parameter for the Laplace draw, giving rise to the additive δ term. The Propose-Test-Release paradigm is useful when we expect the statistic of interest to be insensitive to small changes on datasets seen in practice, despite the worst-case sensitivity being high.

4.2. (t)Concentrated and Rényi differential privacy

Consider a very undesirable randomized algorithm that draws $b \sim \text{Ber}(10^{-6})$ and proceeds to either release the entire dataset, if $b = 1$; or else outputs the empty string, if $b = 0$. This algorithm is $(0, 10^{-6})$-differentially private, but this does not sound like a good idea. For these and other reasons, we think in terms of choosing δ to be cryptographically small. The variants considered in this section are *strengthenings* of approximate differential privacy that preclude such "death and destruction" behavior.

An investigation of the Gaussian mechanism reveals that it never results in catastrophic – that is, infinite – privacy loss. In fact, the distribution of the privacy loss random

2 This bound on c is not tight.

3 The Hamming distance between two datasets is the number of elements on which they differ, so Hamming distance to a set is a sensitivity-1 query.

variable for the Gaussian mechanism is itself a Gaussian! Roughly speaking, the probability of privacy loss exceeding its expectation by $k\varepsilon$ falls exponentially in $k^2/2$. Dwork and Rothblum proposed this as a new relaxation of pure differential privacy, which they called *Concentrated Differential Privacy* (CDP). Concentrated differential privacy requires that the privacy loss random variable be sub-Gaussian [19]. The compelling motivation is that the Gaussian mechanism with a scale σ^2 that is independent of δ "cuts corners" in a way that has no privacy cost under high levels of composition.

To gain a little insight, suppose we will have T applications of the Gaussian mechanism, and we want an overall guarantee of (ε, δ)-differential privacy. Then, using the advanced composition theorem, one can choose $\varepsilon_0 \approx \varepsilon/\sqrt{T \ln(1/\delta)}$ and $\delta_0 = \delta/T$ for the base mechanism. Speaking intuitively, this ensures that each invocation of the Gaussian mechanism is likely to have privacy loss whose absolute value is bounded by ε_0. So long as these bounds hold simultaneously, we can apply the Azuma–Hoeffding bound as in the proof of the Advanced Composition Theorem to bound the cumulative privacy loss. However, even if we allow some small violations of these individual ε_0 bounds, the cumulative loss will still (likely) exhibit sufficient cancellation, when T is large, and this is what is exploited in concentrated differential privacy. In practical terms, it gets the $\sqrt{\log(1/\delta)}$ term out of σ in the Gaussian mechanism, greatly improving accuracy when δ is small.

Bun and Steinke [8] continued this line of investigation, proposing a relaxation of Concentrated Differential Privacy with similar intuition and closure under postprocessing (unlike [19]). Their definition, based on *Rényi divergence*, defined next, is the variant of differential privacy deployed by US Census Bureau for the 2020 Decennial Census.

Definition 4.2 (Rényi divergence between distributions). The Rényi divergence of order $\alpha \in (1, \infty)$ between distributions P and Q over a sample space Ω (with $P \ll Q$)[4] is defined to be

$$D_\alpha(P \| Q) = \frac{1}{\alpha - 1} \ln \int \left(\frac{P(z)}{Q(z)}\right)^\alpha Q(z)\, dz. \tag{4.4}$$

We follow the convention that $0/0 = 1$. Also, if $P \not\ll Q$, we define the divergence to be infinite. Rényi divergence of order $\alpha = 1$ and $\alpha = \infty$ is defined by continuity.

Definition 4.3 (Zero-concentrated differential privacy (zCDP) [8]). A randomized algorithm $\mathcal{A} : \mathcal{U}^n \to \mathcal{Y}$ satisfies (ξ, ρ)-zero-concentrated differential privacy if for all adjacent $x, x' \in \mathcal{U}^n$ and all $\alpha \in (1, \infty)$,

$$D_\alpha\big(\mathcal{A}(x) \| \mathcal{A}(x')\big) \le \xi + \rho\alpha. \tag{4.5}$$

It is common to take $\xi = 0$, which results in a single-parameter formulation, ρ-zCDP, that is easy to work with.

The next notion is a further relaxation that constrains only some fixed lower-order set of divergences instead of all $\alpha \in (1, \infty)$.

---

4 $P(S) = 0$ whenever $Q(S) = 0$ for all measurable sets S; that is, P is absolutely continuous with respect to Q.

Definition 4.4 (Rényi differential privacy [30]). A mechanism $\mathcal{A}$ satisfies (α, ε)-*Rényi differential privacy* (RDP) if for all adjacent datasets x, x', $D_\alpha(\mathcal{A}(x) \| \mathcal{A}(x')) \leq \varepsilon$.

For $\alpha' \in [1, \alpha)$, one has $D_{\alpha'}(P \| Q) < D_\alpha(P \| Q)$ [30], thus if $\mathcal{A}$ satisfies (α, ε)-RDP then it satisfies (α', ε) for all $\alpha' \in (1, \alpha]$.

We introduce one final variation, truncated concentrated differential privacy (tCDP), which lies between zCDP and RDP [7].

Definition 4.5 (Truncated concentrated differential privacy [7]). Let $\rho > 0$ and $\omega > 1$. A randomized algorithm $\mathcal{A} : \mathcal{U}^n \to \mathcal{Y}$ satisfies ω-truncated ρ-concentrated differential privacy (or (ρ, ω)-tCDP) if, for all adjacent datasets $x, x' \in \mathcal{U}^n$, $\forall \alpha \in (1, \omega)$,

$$D_\alpha\big(\mathcal{A}(x) \| \mathcal{A}(x')\big) \leq \rho\alpha. \tag{4.6}$$

Setting $\omega = \infty$ exactly recovers the definition of ρ-zCDP.

4.3. Some relationships among the relaxations

Following the discussion in [7], for adjacent datasets x, x', we examine the random variable $Z = f(\mathcal{A}(x))$, where $f(\zeta) = \ln(\Pr[\mathcal{A}(x) = \zeta]/\Pr[\mathcal{A}(x') = \zeta])$; this is simply the privacy loss random variable $L_{x,x',\zeta}$.

- Pure ε-differential privacy requires $Z \leq \varepsilon$.
- ρ-zCDP requires that Z is sub-Gaussian: the tail behavior of Z should be like that of $\mathcal{N}(\rho, 2\rho)$, with $\Pr[Z > t + \rho] \leq e^{-t^2/(4\rho)}$ for all $t \geq 0$.
- (ρ, ω)-tCDP also requires Z to be sub-Gaussian near the origin, but only subexponential in its tails. That is, $\Pr[Z > t + \rho] \leq e^{-t^2/(4\rho)}$ for all $t \in [0, 2\rho(\omega - 1)]$, and for $t > 2\rho(\omega - 1)$, we have $\Pr[Z > t + \rho] \leq e^{(\omega-1)^2\rho} \cdot e^{-(\omega-1)t}$.
- (ω, ε)-Rényi differential privacy requires $\Pr[Z > t + \varepsilon] \leq e^{-(\omega-1)t}$.
- Up to constant factors (on δ), (ε, δ)-differential privacy requires $\Pr[Z > \varepsilon] \leq \delta$.

See [30] for further discussion.

All the variants introduced in this subsection enjoy pleasant composition bounds (recall that (ε, δ)-differential privacy is governed by the Advanced Composition Theorem (Theorem 4.1)):

Theorem 4.4 (Composition bounds [7, 8, 30]).

(1) *Let $\mathcal{A} : \mathcal{U}^n \to \mathcal{Y}$ and $\mathcal{A}' : \mathcal{U}^n \to \mathcal{Y}$ satisfy (ξ, ρ)-zCDP and (ξ', ρ')-zCDP, respectively. Then their composition satisfies $(\xi + \xi', \rho + \rho')$-zCDP* [8].

(2) *Let $\mathcal{A} : \mathcal{U}^n \to \mathcal{Y}$ and $\mathcal{A}' : \mathcal{U}^n \to \mathcal{Y}$ satisfy (ρ, ω)-tCDP and (ρ', ω')-tCDP, respectively. Then their composition satisfies $(\rho + \rho', \min\{\omega, \omega'\})$-tCDP* [7].

(3) *Let $\mathcal{A} : \mathcal{U}^n \to \mathcal{Y}$ and $\mathcal{A}' : \mathcal{U}^n \to \mathcal{Y}$ satisfy (α, ε)-RDP and (α, ε')-RDP, respectively. Then their composition satisfies $(\alpha, \varepsilon + \varepsilon')$-RDP* [30].

Moreover, by analogues of the data processing inequalities [35], zCDP, tCDP, and RDP are closed under postprocessing.

Remark 4.5 (Group privacy for the relaxations). As noted above, pure differential privacy yields privacy for groups of size k with a factor k increase in the privacy loss bound. For (ε, δ)-differential privacy, the second term deteriorates markedly, and we obtain $(k\varepsilon, ke^{(k-1)}\delta)$-differential privacy. Also ρ-zCDP yields ρk^2-zCDP for groups of size k [8]; (ρ, ω)-tCDP yields $(k^2\rho, \omega/k)$-tCDP for groups of size $k \leq \omega$, and this is tight. tCDP also provides group privacy that degrades gracefully for larger groups, but the degradation is worse than for smaller groups [7]. The situation for RDP is more complex; see [30] and more recent results in [31].

Remark 4.6 (Canonical noise distributions). Different variants of differential privacy have different "canonical" noise distribution for low-sensitivity, real-valued queries. In the case of pure differential privacy, this is the Laplace distribution; for zCDP, it is the Gaussian. Bun *et al.* suggest that for (ε, δ)-differential privacy this could be the Laplace distribution with standard deviation Δ/ε, but with its support truncated to the interval $[\pm O(\Delta \log(1/\delta)/\varepsilon)]$ [7] (although, a truncated Gaussian also works so perhaps "canonical" is wide of the mark for this case). For tCDP, Bun *et al.* suggest the sinh-*normal distribution* for parameters $\sigma, A > 0$,

$$Z \leftarrow A \cdot \operatorname{arcsinh}\left(\frac{\sigma}{A} \cdot \mathcal{N}(0,1)\right). \tag{4.7}$$

This is just the Gaussian $\mathcal{N}(0, \sigma^2)$ with exponentially faster tail decay. The tails of this distribution decay doubly exponentially, rather than just in a sub-Gaussian manner, and the value of A determines where the transition from linear to logarithmic occurs.

With this sinh-normal noise distribution it is sometimes possible to obtain significantly more accurate results. We give one example here.

In a *histogram* query, the universe $\mathcal{U}$ is partitioned into some number k of disjoint cells, and the query is asking, for a dataset x, how many elements of x lie in each of the cells. Although the number of cells is very large, the sensitivity of the query is only 1, as adding or removing a single individual can change the count of at most one cell, and that change is bounded by 1. Histogram queries are the workhorse of official statistics, and the question of how accurately one can privately answer histogram queries is well studied. For pure differential privacy, we can, with high probability, achieve error $\Theta(\frac{\log k}{\varepsilon})$ by adding independent Laplace draws to the count for each cell. For (ε, δ)-differential privacy, this can be improved to $\Theta(\frac{\log(1/\delta)}{\varepsilon})$ (truncated Laplace noise). For z-CDP, we get $\Theta(\sqrt{\frac{\log k}{\rho}})$ (Gaussian noise). For tCDP, using noise from the sinh-normal distribution, we obtain error $O(\omega \log \log k)$.

5. PRIVACY AMPLIFICATON TECHNIQUES

5.1. Amplification by subsampling

Consider a statistical query specified by a predicate q. In *subsampling*, one first chooses a random subset $S \subseteq x$ of the dataset, for example, by selecting each element for inclusion with a fixed probability p, and then outputs an ε-differentially private estimate of the statistical query performed on S. What can we say about the privacy of this algorithm? Consider a pair of adjacent datasets x, y, and let u be an element in x but not in y. The probability that u is selected to S is only p. If it is not included then, speaking intuitively, it will incur no privacy loss; if it is included, then its privacy loss is bounded by ε. This informal argument suggests a privacy loss of $p\varepsilon < \varepsilon$, which is *almost* the right answer (it is off by roughly a factor of at most 2). Moreover, there is nothing special about statistical queries. There is one caveat, however, and this harkens back to our earlier discussion of why statistics *feel* private: privacy amplification requires secrecy of the subsample.

Privacy via subsampling was formalized by Beimel, Brenner, Kasiviswanathan, and Nissim [4], who describe it as implicit in [28].

Theorem 5.1 (Privacy via subsampling [4, 28]). *Let $\mathcal{A}$ be an ε^*-differentially private algorithm. Construct an algorithm $\mathcal{B}$ that, on input a dataset $x = \{x_1, \ldots, x_n\}$, creates a new dataset y by including each x_i independently with probability*

$$\frac{e^{\varepsilon} - 1}{e^{\varepsilon^*} - e^{\varepsilon - \varepsilon^*} - 1} \tag{5.1}$$

and then runs $\mathcal{A}(y)$. Then $\mathcal{B}$ is ε-differentially private.

Remark 5.2. Similar results to those obtained in Theorem 5.1 hold for (ε, δ)-DP, in part because of the δ "escape hatch." However, amplification by subsampling is not so easy for concentrated or Rényi differential privacy. This was the motivation for tCDP; see [7] for precise bounds. The special case of the subsampled Gaussian mechanism is treated in [1, 31].

5.2. Amplification by shuffling

Randomized Response (Section 3) is used in the distributed setting (your cell phone), where clients apply privacy-preserving randomization before sending the (randomized) information to the server. No attempt is made to disassociate a response from the client (you) that sent it. In the *shuffling model*, the assorted responses are assumed to be randomly permuted (somehow, by someone), severing individuals from their randomized information. These responses can then be analyzed without further application of differential privacy, as the analysis is simply a form of post-processing. The *shuffle model* emerged from a series of works, inspired by the very thoughtful PROCHLO paper of Bittau et al. [5] exploring the practical considerations of privacy-preserving computations in internet-scale systems, and formalized by Erlingsson, Feldman, Mironov, Talwar, and Thakurta [22] and, slightly differently, by Cheu, Smith, Ullman, Zeber, and Zhiyaev [9]. In the next section we will describe a simple algorithm for statistical queries due to Cheu et al. [9] in the shuffle model, and we focus here on the definition used in that work.

There are n users, each with a datum $x_i \in \mathcal{U}$. The term *dataset* now refers to the union of the x_i, $i \in [n]$, although in this model the dataset is held in distributed form, with user i holding data u_i. Protocols in this model consist of three parts:

(1) A *randomizer* $\mathcal{R} : \mathcal{U} \to \mathcal{Y}^m$ maps individual data points to an m-tuple of values in an arbitrary range $\mathcal{Y}$;

(2) A *shuffler* $\mathcal{S} : \mathcal{Y}^* \to \mathcal{Y}^*$ applies a random permutation to the set of all messages in its first argument. In our context, if each of n individuals applies a randomizer that yields m elements of $\mathcal{Y}$, the shuffler will apply a random permutation to the set of nm messages. This breaks up, into m unlinked messages, the m-tuple produced by any given individual, in addition to severing the individual-to-message(s) connection;

(3) An analyzer $\mathcal{A} : \mathcal{Y}^* \to \mathcal{Z}$ attempts to estimate some function $f(x_1, \dots, x_n)$ from these messages.

The algorithmic task is to define $\mathcal{R}$ and $\mathcal{A}$ so that the former ensures differential privacy, while in conjunction with the latter it permits accurate estimation.

Definition 5.1 (Differential privacy in the shuffled model [9]). A protocol $P = (\mathcal{R}, \mathcal{S}, \mathcal{A})$ is (ε, δ)-differentially private if the algorithm $\mathcal{S}(\mathcal{R}(x_1), \dots, \mathcal{R}(x_n))$ is (ε, δ)-differentially private.

In a slightly different formulation, Erlingsson et al. showed that shuffling improves ordinary randomized response by a factor of $\Omega(\sqrt{\frac{\log(1/\delta)}{n}})$ [22]; simplifications and the optimal dependence on ε appear in [23]. In Section 6 we will see a special case of this improvement, due to [9]. This is an enormous gain for internet-scale n. Deploying shufflers would go a long way toward remedying the cumulative privacy erosion of continual facially differentially private (but with very large ϵ) monitoring, e.g., of phones, browser activity, and activities within app usage.

5.3. Amplification by secrecy of the journey

The intuition behind this approach, which in the literature is referred to as *privacy amplification by iteration* [24], is that privacy is enhanced when the intermediate steps of the algorithm are kept secret. The motivating scenario is private gradient descent. Standard privacy analyses call for each round to satisfy some privacy guarantee, and then to apply composition to the sequence of all rounds. This will work even though it permits the intermediate results to be public, but this is also potentially wasteful: the analyst only needs the result of the final iteration.

The power of keeping intermediate state private was exploited in privacy via subsampling, which relies on the privacy of the subsample, and in the subsample-and-aggregate framework of Nissim, Raskhodnikova, and Smith [33], which provides a method of achieving differential privacy even if the statistic to be computed has large, or difficult to analyze, sensitivity. In subsample-and-aggregate, the algorithm partitions the dataset into disjoint subsamples, computes the statistic on each subsample without privacy, and then applies a privacy-preserving aggregation mechanism to combine the results. The intuition is that each data point is contained in only one cell of the partitioning, and therefore can affect only one input to the aggregator.

6. APPLICATIONS

In the standard setup for gradient descent in machine learning, we have a dataset $x = \{x_1, \dots, x_n\}$, where each $x_i \in \mathbb{R}^p$, a convex body $\mathcal{K} \in \mathbb{R}^d$, a starting point $\omega \in \mathcal{K}$, a loss function $L : \mathbb{R}^d \times (\mathbb{R}^p)^n \to \mathbb{R}$. Typically, $L(\omega, x) = \sum_i \ell(\omega, x_i)$ for a convex, Lipschitz, loss $\ell : \mathbb{R}^d \times \mathbb{R}^p \to \mathbb{R}$.

When the $x_i \sim \mathcal{D}$, for a distribution $\mathcal{D}$ on the underlying population, the goal is often convex risk minimization: to find an approximate minimizer of $L(\xi, \mathcal{D})$. The empirical risk is the difference between the population minimizer ξ^* and the empirical minimizer $\hat{\xi}$, and differential privacy adds to the nonprivate empirical risk, since release of $\hat{\xi}$ would be disclosive.

Fix a number of iterations T, and *step sizes* η_t indexed by the iteration number $t \in [T]$. For any $z \in \mathbb{R}^d$, let $\Pi_{\mathcal{K}}(z)$ denote the projection of z onto the convex body $\mathcal{K}$. After setting ω_0 to the starting point ω, each iteration of projected gradient descent has the form

$$\omega_{t+1} \leftarrow \Pi_{\mathcal{K}}\big(\omega_t - \eta_t \cdot \nabla_\omega L(\omega_t, x)\big) = \Pi_{\mathcal{K}}\Big(\omega_t - \eta_t \cdot \sum_{i \in [n]} \nabla_\omega \ell(\omega_t, x_i)\Big) \tag{6.1}$$

This is easily made differentially private by introducing appropriately scaled Gaussian noise before multiplication by the stepsize and projection,

$$\omega_{t+1} \leftarrow \Pi_{\mathcal{K}}\Big(\omega_t - \eta_t \cdot \Big[\sum_{i \in [n]} \nabla_\omega \ell(\omega_t, x_i) + \mathcal{N}\big(0, \sigma^2 \mathbb{I}_d\big)\Big]\Big). \tag{6.2}$$

Ensuring that σ is sufficiently large to provide privacy for the gradient computation

$$\sum_{i \in [n]} \nabla_\omega \ell(\omega_t, x_i) \tag{6.3}$$

suffices, as multiplication by η_t and projection onto $\mathcal{K}$ are postprocessing steps (provided the value of η_t is independent of x, which is typical). Note that this algorithm does not require that the loss function ℓ be differentiable, and may be run with any subgradient of ℓ.

There are many variations of the basic approach (for example, full-batch, mini-batch, stochastic), as well as of the problem statement, e.g., assumptions on the functions $f(\cdot, x_i)$, and we will not compare the bounds obtained by the algorithms we discuss, instead focusing on the privacy arguments.

6.1. Privacy via subsampling in gradient descent

For a large dataset, the computational cost of each iteration of full-batch gradient descent (Equation (6.2)) would be prohibitive. Bassily, Smith, and Thakurta [3], in an elegant feat of differential privacy sleight of hand, proposed instead a variant in which at each iteration an element z is chosen uniformly from the dataset $x = \{x_1, \dots, x_n\}$, and only z will be used in the gradient computation at that iteration:

$$z \sim \{x_1, \dots, x_n\}, \tag{6.4}$$

$$\omega_{t+1} \leftarrow \Pi_{\mathcal{K}}\big(\omega_t - \eta_t \cdot \big[n\nabla_\omega \ell(\omega_t, z) + \mathcal{N}\big(0, \sigma^2 \mathbb{I}_d\big)\big]\big). \tag{6.5}$$

Observe that the scale and expectation of the simple computation $n\nabla_\omega \ell(\omega_t, z)$ and the expensive computation $\sum_{i\in[n]} \nabla_\omega \ell(\omega_t, x_i)$ are respectively identical.

The computational advantage is enormous: savings of a factor of n! On the other hand, the sensitivity of the computation has increased by a factor of n. This is because the contribution, for the selected x_i, to the gradient is n times what it was in the original computation. This is counteracted by privacy amplification via subsampling results for (ε, δ)-differential privacy. On a dataset of size n, the subsampled dataset of size 1 corresponds to a selection probability of $q = 1/n$.

Using this and advanced composition, Bassily et al. show that (ε, δ)-differential privacy can be achieved when

$$\sigma^2 = O\left(\frac{n^2 \log(n/\delta) \log(1/\delta)}{\varepsilon^2}\right) \tag{6.6}$$

even when running for n^2 rounds,[5] where the constants include the diameter of $\mathcal{K}$ and the Lipshitz bounds on f.

6.2. Privacy amplification via shuffling

Recall the randomized response primitive for Boolean inputs: each individual chooses $b \sim \mathrm{Ber}(p)$ and, if $b = 1$, answers with a random draw from $\mathrm{Ber}(1/2)$ and otherwise answers truthfully. As noted earlier, $p \geq 2/(1 + e^\varepsilon)$ suffices to ensure ε-differential privacy. Recall further that in the shuffling model individuals randomize their responses and then these responses are randomly shuffled into a pool of messages.

Consider the randomizer defined by running the randomized response with (tiny) randomization parameter $p = \frac{\kappa \ln(1/\delta)}{n\varepsilon^2}$, for some constant κ. With such a small value of p, most participants will report truthfully, but a handful will respond randomly. The analyzer simply sums the randomized values to obtain an approximate total count of ones.

Cheu et al. proposed and analyzed this algorithm with the following intuition [9]. We can think of the initial Bernoulli draw as partitioning the participants into a small set H of *noise makers*, and its complement, whose members respond truthfully. The noise makers create their noise in the second Bernoulli draw; the sum of their $\mathrm{Ber}(1/2)$ draws follows a binomial distribution $B(|H|, 1/2)$. Concentration bounds for the first Bernoulli control the

---

5 Bassily et al. note, "Even nonprivate first-order algorithms – i.e., those based on gradient measurements – must learn information about the gradient at $\Omega(n^2)$ points to get risk bounds that are independent of n (this follows from "oracle complexity" bounds showing that $1/\sqrt{T}$ convergence rate is optimal...)".

size of H. The noise is then added to the sum of the truthful responses from those not in H. For sufficiently large $|H|$, this yields (ε, δ)-differential privacy.

The algorithm has expected error of order $O(\frac{1}{\varepsilon}\sqrt{\ln \frac{1}{\delta}})$ for counting queries (cf. $\Theta(1/\varepsilon)$ for the Laplace mechanism and $\Theta(\sqrt{n}/\varepsilon)$ for randomized response). Moreover, it is *succinct*: each participant sends only a single bit to the shuffler.

6.3. Privacy of the journey

While composition theorems permit modular "analysis by parts" of complex differentially private algorithms, there is one sense in which they may be overly conservative: they provide guarantees for cumulative privacy loss at every step of the computation, even when the intermediate results are not released. Gradient descent is a case in point: why use privacy parameters that permit every intermediate ω_t to be released, when the data analyst only cares about the final value ω_T? In other words, the destination, and not the journey, is the sole object of interest in gradient descent. Can we exploit this? For the case of gradient descent, the answer is positive, and two lovely lines of work reach the same conclusions via very different proof techniques.

Consider a variant of noisy stochastic gradient descent in which dataset elements are processed sequentially: x_1, then x_2, and so on, for $T = n$ rounds:

$$\omega_{t+1} \leftarrow \Pi_{\mathcal{K}}\big(\omega_t - \eta_t \cdot \big[n\nabla_\omega \ell(\omega_t, x_{t+1}) + \mathcal{N}\big(0, \sigma^2 \mathbb{I}_d\big)\big]\big). \tag{6.7}$$

Consider the element x_1, and observe that its impact on the evolving computation does not end with the first iteration, even though that is the only step in which it appears as an argument to ℓ. This is because x_1 has an impact on the choice of ω_1, which in turn affects ω_2, and so on. Nonetheless, from the perspective of x_1, everything after the first iteration is just postprocessing, so adding noise scaled to the sensitivity of the gradient descent step during the first iteration suffices to protect privacy.

However, keeping the journey secret leads to the following speculation. Suppose we add only half the requisite noise, $\mathcal{N}(0, (\sigma/2)^2\mathbb{I}_d)$ in the first iteration, and another half after the second iteration, when the algorithm operates on x_2. After the second step, we will have added *two* Gaussian samples: the first sample during the first iteration, and the second during the second iteration. This is equivalent to a draw from $\mathcal{N}(0, \sigma^2\mathbb{I}_d)$. So perhaps x_1 is completely protected.

This intuition was made rigorous by Feldman, Mironov, Talwar, and Thakurta, who introduced a powerful notion that interpolates between a metric distance on the output space $\mathcal{K} \subset \mathbb{R}^d$ and the information-theoretic Rényi divergence D_α on distributions of outputs on neighboring datasets, together with two key lemmata that manipulate this quantity [24].

Definition 6.1 (Contractive mapping). A function $\psi : \mathbb{R}^d \to \mathbb{R}^d$ is *contractive* if it is 1-Lipschitz.

In this note, the contractive functions of interest are those computed at each iteration of the noisy gradient descent algorithm. Later, we will make use of the following facts:

Proposition 6.1 ([24]). *For suitable learning rates, the steps at the heart of projected gradient descent are contractions:*

(1) *For convex $\mathcal{K} \in \mathbb{R}^d$, the projection $\Pi_{\mathcal{K}}(x) = \arg\min_{y\in\mathcal{K}} \|x - y\|_2$ is a contraction.*

(2) *Let $f : \mathbb{R}^d \to \mathbb{R}$ be convex and β-smooth. For $\eta \le 2/\beta$, the function $\psi(w) = w - \eta\nabla_w f(w)$ is a contraction.*

Given a starting point $\mathcal{X}_0$, a sequence of contractive maps $\Psi_1, \dots, \Psi_T$, and sequence of noise distributions $\{\zeta_t\}$, we define a *Contractive Noisy Iteration* (CNI) as

$$\mathcal{X}_{t+1} = \Psi_{t+1}(X_t) + Z_{t+1}, \tag{6.8}$$

where Z_{t+1} is drawn independently from ζ_t, and denote the output of this process as

$$\mathrm{CNI}_T\big(\mathcal{X}_0, \{\Psi_t\}, \{\zeta_t\}\big) \tag{6.9}$$

Let us consider what happens when each Ψ_t is the identity map and each $\zeta_t = \mathcal{N}(0, \sigma^2\mathbb{I}_d)$. In this case, $\mathcal{X}_{t+1} = \Psi_{t+1}(\mathcal{X}_t) + \mathcal{N}(0, \sigma^2\mathbb{I}_d) = \mathcal{X}_t + \mathcal{N}(0, \sigma^2\mathbb{I}_d)$, whence by induction and the fact that the sum of T mean-centered Gaussians of noise scale σ has distribution $\mathcal{N}(0, T\sigma^2\mathbb{I}_d)$, we obtain $X_T = X_0 + \mathcal{N}(0, T\sigma^2\mathbb{I}_d)$.

With this in mind we note that if $\|\mathcal{X}_0 - \mathcal{X}_0'\|_2 \le 1$, then, letting

$$X_T = \mathrm{CNI}_T\big(\mathcal{X}_0, \{I\}, \mathcal{N}(0, \sigma^2\mathbb{I}_d)\big), \tag{6.10}$$

$$X_T' = \mathrm{CNI}_T\big(\mathcal{X}_0', \{I\}, \mathcal{N}(0, \sigma^2\mathbb{I}_d)\big), \tag{6.11}$$

we have

$$D_\alpha(\mathbb{P}_{\mathcal{X}_T} \| \mathbb{P}_{\mathcal{X}_T'}) \le \frac{\alpha}{2T\sigma^2}.$$

Feldman et al. show that the identity case is the worst case. Although they consider arbitrary noise distributions ζ, we will confine our attention to Gaussian noise. They make heavy use of the following fact.

Fact 6.1. *For all $x \in \mathbb{R}^d$,*

$$D_\alpha\big(\mathcal{N}(0, \sigma^2\mathbb{I}_d) \| \mathcal{N}(x, \sigma^2\mathbb{I}_d)\big) = \frac{\alpha\|x\|^2}{2\sigma^2}. \tag{6.12}$$

It is helpful to keep the application in mind: we have two adjacent datasets $x, x' \in (\mathbb{R}^p)^n$. We imagine running noisy gradient descent on them in parallel, starting from a common point $\omega_0 = \omega_0' \in \mathbb{R}^d$. We are interested in the probability distributions on ω_t and ω_t' for $t \in [T]$. Due to the addition of noise, the values of ω_t and ω_t' are random variables, denoted $\mathcal{X}_t$ and $\mathcal{X}_t'$, respectively. We now wish to bound the α-Rényi divergence of the distributions of these two random variables.

Definition 6.2. For distributions P, Q over $\mathbb{R}^d$, the ∞-Wasserstein distance $\mathcal{W}_\infty(P, Q)$ is the smallest real number given by

$$\mathcal{W}_\infty(P, Q) = \inf_{\gamma\in\Gamma(P,Q)} \operatorname*{ess\,sup}_{(x,y)\sim\gamma} \|x - y\|_2, \tag{6.13}$$

where $(x, y) \sim \gamma$ means that the essential supremum is taken relative to measure γ; here Γ is the collection of couplings of P and Q.

The next quantity, *shifted Rényi divergence*, defined in [24], is a hybrid distance notion that interpolates between metric distances between points in $\mathbb{R}^d$ and distributional divergences.

Definition 6.3 (Shifted Rényi divergence [24]). Let P, Q be distributions defined on a Banach space $(\mathcal{Z}, \|\cdot\|)$. For parameters $z \geq 0$ and $\alpha \geq 1$, the *z-shifted Rényi divergence* between P and Q is defined as

$$D_\alpha^{(z)}(P\|Q) = \inf_{P' : \mathcal{W}_\infty(P,P') \leq z} D_\alpha(P'\|Q). \tag{6.14}$$

To understand this, consider the Wasserstein ball of radius z around P. Then P' minimizes, among all distributions in this ball, the α-Rényi divergence to Q. Note that the larger the ball, the smaller the shifted divergence, since increasing the radius only adds to the collection of candidates from which to choose P'. Moreover, when the radius is so large that the ball includes Q, the shifted divergence is zero, since the divergence will be minimized at $P' = Q$.

Lemma 6.1 ([24]). *For all $s > 0$ simultaneously,*

$$D_\alpha^{(z-s)}\big(P + \mathcal{N}(0, \sigma^2 \mathbb{I}_d)\|Q + \mathcal{N}(0, \sigma^2 \mathbb{I}_d)\big) \leq D_\alpha^{(z)}(P\|Q) + \frac{\alpha s^2}{2\sigma^2}. \tag{6.15}$$

In other words, letting $\tilde{P} = P + \mathcal{N}(0, \sigma^2 \mathbb{I}_d)$ and $\tilde{Q} = Q + \mathcal{N}(0, \sigma^2 \mathbb{I}_d)$, we reduce the shift amount (the superscript (z) is decreased to $(z - s)$), which corresponds to a stronger requirement (smaller Wasserstein ball), paying a divergence price of $\frac{\alpha s^2}{2\sigma^2}$. Figuratively, we are drawing a ball of smaller radius $(z - s < z)$ around a distribution $\tilde{P}$ that is close to P, and finding the distribution $\tilde{P}'$ within that ball that is closest to $\tilde{Q} = Q + \mathcal{N}(0, \sigma^2 \mathbb{I}_d)$. The noise distribution is fixed; the flexibility over the choice of s should be thought of as providing an opportunity for creative divergence accounting.

Notation. In the sequel, we sometimes abuse notation by writing $\mathcal{X}$ instead of $\mathbb{P}_{\mathcal{X}}$, identifying the random variable with its distribution.

We next observe that contraction reduces shifted divergence.

Lemma 6.2 ([24]). *Let Ψ, Ψ' be contractive maps on $(\mathcal{Z}, \|\cdot\|)$. If $\sup_x \|\Psi(x) - \Psi'(x)\| \leq s$ then for random variables $\mathcal{X}$, and $\mathcal{X}'$ over $\mathcal{Z}$,*

$$D_\alpha^{(z+s)}\big(\Psi(\mathcal{X})\|\Psi'(\mathcal{X}')\big) \leq D_\alpha^{(z)}(\mathcal{X}\|\mathcal{X}'). \tag{6.16}$$

Theorem 6.3 ([24], as stated informally in [2, PROPOSITION 2.17]). *Let $\mathcal{X}_T$ and $\mathcal{X}'_T$ denote the outputs of CNI($\mathcal{X}_0, \{\phi_t\}, \{\xi_t\}$) and CNI($\mathcal{X}_0, \{\phi'_t\}, \{\xi_t\}$) where $\xi_t = \mathcal{N}(0, \sigma_t^2 \mathbb{I}_d)$. Denote $s_t = \sup_x \|\phi_t(x) - \phi'_t(x)\|$, and consider any sequence $a_1, \ldots, a_T$ such that $z_t = \sum_{i=1}^{t}(s_i - a_i)$ is nonnegative for all t and satisfies $z_T = 0$. Then*

$$D_\alpha(\mathbb{P}_{\mathcal{X}_T}\|\mathbb{P}_{\mathcal{X}'_T}) \leq \frac{\alpha}{2}\sum_{t=1}^{T}\frac{a_t^2}{\sigma_t^2}. \tag{6.17}$$

We can now sketch the analysis in [24] of projected stochastic gradient descent for a fixed choice σ of noise scale. Our starting assumption $\mathcal{X}_0 = \mathcal{X}'_0$ ensures that $\mathcal{W}_\infty(\mathcal{X}_0, \mathcal{X}'_0) \leq 1$ which in turn is equivalent to $D^{(1)}_\alpha(\mathcal{X}_0 \| \mathcal{X}'_0) = 0$. Our desired conclusion is a bound on $D_\alpha(\mathcal{X}_T \| \mathcal{X}'_T) = D^{(0)}_\alpha(\mathcal{X}_T \| \mathcal{X}'_T)$. Adding Gaussian noise allows us to reduce the shift amount (Lemma 6.1), while recording a divergence cost: the greater the shift reduction, the higher the privacy cost. Taking gradient descent steps moves us towards our computational goal but increases the shift amount (Lemma 6.2).

Projected noisy stochastic gradient descent

Input: Dataset $x = \{x_1, \dots, x_n\}$; $f : \mathcal{K} \times \mathcal{U} \to \mathbb{R}$ a convex function in the first parameter; learning rate η; starting point $\omega_0 \in \mathcal{K}$; noise parameter σ.

For $t \in \{0, \dots, n-1\}$:

$v_{t+1} \leftarrow \omega_t - \eta(\nabla_\omega f(\omega_t, x_{t+1}) + Z)$, where $Z \sim \mathcal{N}(0, \sigma^2 \mathbb{I}_d)$,

$\omega_{t+1} \leftarrow \Pi_{\mathcal{K}}(v_{t+1})$

End For

Return the final iterate ω_n.

Assuming f is convex and β-smooth in its first parameter, the gradient step is contractive whenever $\eta \leq 2/\beta$.

Let $x, x' \in \mathcal{U}^n$ be adjacent, and let t be the unique index in which they differ. For dataset x, we can define the contractive noisy iteration by the initial point ω_0, the sequence of functions $g_i(\omega) = \Pi_{\mathcal{K}}(\omega) - \eta \nabla f(\Pi_{\mathcal{K}}(\omega), x_i)$ and sequence of noise distributions $\zeta_i \sim \mathcal{N}(0, (\eta\sigma)^2 \mathbb{I}_d)$. The CNI is defined analogously for dataset x', but with $g'_t(\omega) = \Pi_{\mathcal{K}}(\omega) - \eta \nabla f(\Pi_{\mathcal{K}}(\omega), x'_t)$. By assumption, $f(\omega, z)$ is L-Lipschitz for every $\omega \in \mathcal{K}$ and $z \in \mathcal{U}$, and therefore

$$\sup_\omega \left\| g_t(\omega) - g'_t(\omega) \right\|_2 \leq 2\eta L. \tag{6.18}$$

We choose $a_1, \dots, a_{t-1} = 0$, that is, paying no divergence costs for the first $t-1$ noise additions and obtaining no shift reductions, and $a_t, \dots, a_n = \frac{2\eta L}{n-t+1}$ for the remaining steps, and noting that the contractive map in the tth iteration increases the shift by $s = 2\eta L$, and there are no further increases because the datasets agree on the remaining steps. A simple induction shows that at every step the shift parameter is nonnegative, while the shift parameter at the end of step n is $z_n = 0$, yielding a bound on divergence at the final step of

$$D_\alpha(\mathcal{X}_n \| \mathcal{X}'_n) \leq \frac{\alpha}{2\eta^2\sigma^2} \sum_{i=1}^{n} a_i^2 \leq \frac{2\alpha L^2}{\sigma^2 \cdot (n-t+1)}. \tag{6.19}$$

This yields $(\alpha, \frac{\alpha 2L^2}{\sigma^2(n+1-t)})$-Rényi differential privacy for the tth element. Observe that this bound echoes our earlier discussion of the privacy protection for elements processed early, that is, x_i for small i, and elements processed later. The smaller t in this bound, the larger the denominator, yielding smaller divergence, which captures the privacy loss.

Feldman et al. consider various methods of employing this basic mechanism, or some simple variants, to remedy the reduced protections for x_t when t is large. For example, suppose (for some reason) we have access to a modest sample $\{y_1, \ldots, y_m\}$ of *nonprivate* data drawn from the same distribution as the members of the dataset. Then one could run the algorithm on the augmented dataset $\{x_1, \ldots, x_n, y_1, \ldots, y_m\}$, keeping the iterates secret and only making public the final $(n+m)$th iterate.

6.3.1. Very large numbers of iterations

Suppose we wish to run a CNI process for more than $T = n$ rounds. Theorem 6.3 above (see [24]) says that $D_\alpha(\mathbb{P}_{X_T} \| \mathbb{P}_{X'_T}) \le \frac{\alpha}{2} \sum_{t=1}^{T} \frac{a_t^2}{\sigma_t^2}$, which goes to infinity as T grows.

In very recent papers, two lines of work show this dependence on T can be avoided. Breakthrough results of Chourasia, Ye, and Shokri (for the nonstochastic case) [10], followed by Ye and Shokri [37] and Ryffel, Bach, and Pointcheval [34], use a diffusion argument to prove that Projected Noisy Stochastic Gradient Descent has a privacy loss that converges as $T \to \infty$, provided the smooth loss functions are also strongly convex. The intuition is that Projected Noisy-SGD is a discretization of a continuous-time algorithm with bounded privacy loss; in particular, it can be viewed as the Stochastic Gradient Langevin Dynamics algorithm, which is a discretization of a continuous-time Markov process whose stationary distribution is equivalent to the differentially private *exponential mechanism* [29].

Using different techniques, Altschuler and Talwar [2] combine the privacy amplification via iteration techniques discussed above with privacy amplification via subsampling for the Gaussian mechanism to also obtain finite privacy loss as T goes to infinity; moreover, they are able to remove the strong convexity assumption.

Theorem 6.4 (Informal statement from [2]). *Let $x = \{x_1, \ldots, x_n\} \in \mathcal{U}^n$, whether each x_i defines a convex L-Lipschitz, and M-smooth loss function $f(\cdot, x_i)$ on a convex region $\mathcal{K} \subset \mathbb{R}^d$ of diameter D. For a large range of parameters, Projected Noisy-SGD, when run for T iterations, satisfies (α, ε)-Rényi differential privacy for*

$$\varepsilon \le \frac{\alpha L^2}{n^2 \sigma^2} \min\left\{T, \frac{Dn}{L\eta}\right\}, \tag{6.20}$$

and this bound is tight up to a constant factor.

The proof of privacy exploits the diameter on the constraint set, as follows. Noisy-SGD updates on adjacent datasets will eventually diverge to maximally distant points. At that time, which can be shown to be of order $\frac{Dn}{L\eta}$, their *shifted* divergence will be zero! Thus, the proof of privacy only needs to be concerned with the final $T - \bar{T}$ iterations.

ACKNOWLEDGMENTS

We thank Kunal Talwar, Franklyn Wang, and the members of the reading group: Silvia Casacuberta-Puig, Zhun Deng, Elbert Du, Conlan Olson, Manish Raghavan, Pranay Tankala, and Linjun Zhang, for helpful discussions, and David Fuchs and Daniel Lee for helpful comments on earlier versions of this manuscript.

FUNDING

We gratefully acknowledge the support of the Sloan Foundation grant, Towards Practicing Privacy.

REFERENCES

[1] M. Abadi, A. Chu, I. Goodfellow, H. B. McMahan, I. Mironov, K. Talwar, and L. Zhang, Deep learning with differential privacy. In *Proceedings of the 2016 ACM SIGSAC conference on computer and communications security*, pp. 308–318, ACM, 2016.

[2] J. M. Altschuler and K. Talwar, Privacy of noisy stochastic gradient descent: more iterations without more privacy loss. 2022, arXiv:2205.13710.

[3] R. Bassily, A. Smith, and A. Thakurta, Private empirical risk minimization: Efficient algorithms and tight error bounds. In *2014 IEEE 55th annual symposium on foundations of computer science*, pp. 464–473, IEEE, 2014.

[4] A. Beimel, H. Brenner, S. P. Kasiviswanathan, and K. Nissim, Bounds on the sample complexity for private learning and private data release. *Mach. Learn.* **94** (2014), no. 3, 401–437.

[5] A. Bittau, Ú. Erlingsson, P. Maniatis, I. Mironov, A. Raghunathan, D. Lie, M. Rudominer, U. Kode, J. Tinnes, and B. Seefeld, Prochlo: strong privacy for analytics in the crowd. In *Proceedings of the 26th ACM symposium on operating systems principles*, pp. 441–459, ACM, 2017.

[6] A. Blum, K. Ligett, and A. Roth, A learning theory approach to non-interactive database privacy. In *Proceedings of the 40th ACM SIGACT symposium on theory of computing*, ACM, 2008.

[7] M. Bun, C. Dwork, G. N. Rothblum, and T. Steinke, Composable and versatile privacy via truncated CDP. In *Proceedings of the 50th annual ACM SIGACT symposium on theory of computing*, pp. 74–86, ACM, 2018.

[8] M. Bun and T. Steinke, Concentrated differential privacy: Simplifications, extensions, and lower bounds. In *Theory of cryptography conference*, pp. 635–658, Springer, 2016.

[9] A. Cheu, A. Smith, J. Ullman, D. Zeber, and M. Zhilyaev, Distributed differential privacy via shuffling. In *Annual international conference on the theory and applications of cryptographic techniques*, pp. 375–403, Springer, 2019.

[10] R. Chourasia, J. Ye, and R. Shokri, Differential privacy dynamics of Langevin diffusion and noisy gradient descent. *Adv. Neural Inf. Process. Syst.* **34** (2021), 14771–14781.

[11] I. Dinur and K. Nissim, Revealing information while preserving privacy. In *Proceedings of the 22nd ACM SIGMOD-SIGACT-SIGART symposium on principles of database systems*, pp. 202–210, ACM, 2003.

[12] C. Dwork, Differential privacy. In *Proceedings of the international colloquium on automata, languages and programming (ICALP)*, pp. 1–12, Springer, 2006.

[13] C. Dwork, K. Kenthapadi, F. McSherry, I. Mironov, and M. Naor, Our data, ourselves: privacy via distributed noise generation. In *Advances in cryptology – EUROCRYPT 2006, 25th annual international conference on the theory and applications of cryptographic techniques, St. Petersburg, Russia, May 28 – June 1, 2006, proceedings*, pp. 486–503, Springer, 2006.

[14] C. Dwork and J. Lei, Differential privacy and robust statistics. In *Proceedings of the 41st ACM SIGACT symposium on theory of computing*, ACM, 2009.

[15] C. Dwork, F. McSherry, K. Nissim, and A. Smith, Calibrating noise to sensitivity in private data analysis. In *Theory of cryptography conference*, pp. 265–284, Springer, 2006.

[16] C. Dwork, F. McSherry, K. Nissim, and A. Smith, Calibrating noise to sensitivity in private data analysis. *J. Priv. Confid.* **7** (2016), no. 3, 17–51.

[17] C. Dwork and M. Naor, On the difficulties of disclosure prevention in statistical databases or the case for differential privacy. *J. Priv. Confid.* **2** (2010), no. 1.

[18] C. Dwork and A. Roth, The algorithmic foundations of differential privacy. *Found. Trends Theor. Comput. Sci.* **9** (2014), no. 3–4, 211–407.

[19] C. Dwork and G. N. Rothblum, Concentrated differential privacy. 2016, arXiv:1603.01887.

[20] C. Dwork, G. N. Rothblum, and S. P. Vadhan, Boosting and differential privacy. In *Proceedings of the 51st IEEE symposium on foundations of computer science*, pp. 51–60, IEEE, 2010.

[21] C. Dwork, A. Smith, T. Steinke, and J. Ullman, Exposed! A survey of attacks on private data. *Annu. Rev. Stat. Appl.* **4** (2017), no. 1, 61–84.

[22] Ú. Erlingsson, V. Feldman, I. Mironov, A. Raghunathan, K. Talwar, and A. Thakurta, Amplification by shuffling: From local to central differential privacy via anonymity. In *Proceedings of the thirtieth annual ACM–SIAM symposium on discrete algorithms*, pp. 2468–2479, SIAM, 2019.

[23] V. Feldman, A. McMillan, and K. Talwar, Hiding among the clones: A simple and nearly optimal analysis of privacy amplification by shuffling. In *2021 IEEE 62nd annual symposium on foundations of computer science (FOCS)*, pp. 954–964, IEEE, 2022.

[24] V. Feldman, I. Mironov, K. Talwar, and A. Thakurta, Privacy amplification by iteration. In *2018 IEEE 59th annual symposium on foundations of computer science (FOCS)*, pp. 521–532, IEEE Computer Society, 2018.

[25] M. Hardt and G. N. Rothblum, A multiplicative weights mechanism for interactive privacy-preserving data analysis. In *Proceedings of the 51st annual IEEE symposium on foundations of computing*, IEEE, 2010.

[26] M. Hardt, G. N. Rothblum, and R. A. Servedio, Private data release via learning thresholds. In *Proceedings of the twenty-third annual ACM–SIAM symposium on discrete algorithms*, pp. 168–187, SIAM, 2012.

[27] P. Kairouz, S. Oh, and P. Viswanath, The composition theorem for differential privacy. *International Conference on Machine Learning*, pp. 1376–1385, PMLR, 2015.

[28] S. P. Kasiviswanathan, H. K. Lee, K. Nissim, S. Raskhodnikova, and A. Smith, What can we learn privately? *SIAM J. Comput.* **40** (2011), no. 3, 793–826.

[29] F. McSherry and K. Talwar, Mechanism design via differential privacy. In *48th annual IEEE symposium on foundations of computer science (FOCS'07)*, pp. 94–103, IEEE, 2007.

[30] I. Mironov, Rényi differential privacy. In *2017 IEEE 30th computer security foundations symposium (CSF)*, pp. 263–275, IEEE, 2017.

[31] I. Mironov, K. Talwar, and L. Zhang, Rényi differential privacy of the sampled gaussian mechanism. 2019, arXiv:1908.10530.

[32] J. Murtagh and S. Vadhan, The complexity of computing the optimal composition of differential privacy. In *Theory of cryptography conference*, pp. 157–175, Springer, 2016.

[33] K. Nissim, S. Raskhodnikova, and A. Smith, Smooth sensitivity and sampling in private data analysis. In *Proceedings of the thirty-ninth annual ACM SIGACT symposium on theory of computing*, pp. 75–84, ACM 2007.

[34] T. Ryffel, F. Bach, and D. Pointcheval, Differential privacy guarantees for stochastic gradient Langevin dynamics. 2022, arXiv:2201.11980.

[35] T. van Erven and P. Harremos, Rényi divergence and Kullback–Leibler divergence. *IEEE Trans. Inf. Theory* **60** (2014), no. 7, 3797–3820.

[36] S. L. Warner, Randomized response: a survey technique for eliminating evasive answer bias. *J. Amer. Statist. Assoc.* **60** (1965), no. 309, 63–69.

[37] J. Ye and R. Shokri, Differentially private learning needs hidden state (or much faster convergence). 2022, arXiv:2203.05363.

CYNTHIA DWORK

Department of Computer Science, Harvard John A. Paulson School of Engineering and Applied Sciences, Harvard University, 150 Western Avenue, Allston, MA 02134, USA, dwork@seas.harvard.edu

INDISTINGUISHABILITY OBFUSCATION

AAYUSH JAIN, HUIJIA LIN, AND AMIT SAHAI

ABSTRACT

At least since the initial public proposal of public-key cryptography based on computational hardness conjectures (Diffie and Hellman, 1976), cryptographers have contemplated the possibility of a "one-way compiler" that translates computer programs into "incomprehensible" but equivalent forms. And yet, the search for such a "one-way compiler" remained elusive for decades. We examine a formalization of this concept with the notion of indistinguishability obfuscation ($i\mathcal{O}$). Roughly speaking, $i\mathcal{O}$ requires that the compiled versions of any two equivalent programs (with the same size and running time) be indistinguishable to any efficient adversary. Finally, we show how to construct $i\mathcal{O}$ in such a way that we can prove the security of our $i\mathcal{O}$ scheme based on well-studied computational hardness conjectures in cryptography.

MATHEMATICS SUBJECT CLASSIFICATION 2020

Primary 94A60; Secondary 68Q17, 68Q17, 11H99, 11T71

KEYWORDS

Indistinguishability obfuscation, well-studied assumptions, learning parity with noise, pairing groups, pseudorandom generators, learning with errors

ICM INTERNATIONAL CONGRESS OF MATHEMATICIANS 2022 JULY 6–14

Proc. Int. Cong. Math. 2022, Vol. 6, pp. 4762–4798
DOI 10.4171/ICM2022/86

Published by EMS Press

1. INTRODUCTION

Consider the polynomial $f_1(x, y) \in \mathbb{Z}[x, y]$ that is computed as follows:

$$f_1(x, y) = (x + y)^{16} - (x - y)^{16}.$$

Alternatively, contemplate the polynomial $f_2(x, y) \in \mathbb{Z}[x, y]$ that is computed via:

$$\begin{aligned} f_2(x, y) = {} & 32x^{15}y + 1120x^{13}y^3 + 8736x^{11}y^5 + 22880x^9y^7 \\ & + 22880x^7y^9 + 8736x^5y^{11} + 1120x^3y^{13} + 32xy^{15}. \end{aligned}$$

A calculation shows that f_1 and f_2 are, in fact, the same polynomial, computed in two different ways. Indeed, the expressions f_1 and f_2 above are special cases of *arithmetic circuits*, which precisely represent "ways to compute a polynomial."

What if we wanted to hide all implementation choices made when creating such an arithmetic circuit for a particular polynomial? An easy way to do that would be to first convert our polynomial into a canonical form, and then implement the canonical form as an arithmetic circuit. Indeed, the description of f_2 above can be seen as a canonical representation of the polynomial as a sum of monomials with regard to a natural monomial ordering. However, as this example illustrates, canonical forms can be substantially more complex than other implementations of the same polynomial. For polynomials in n variables, the loss in efficiency can be exponential in n. This would often make computing the canonical form—or indeed, even writing it down—infeasible.

A pseudocanonical form. Given that computing canonical forms can be infeasible, what is there to do? Here, following [22], we draw an analogy to the notion of pseudorandomness. When truly random values are not available, we can instead aim to produce values that "look random" by means of a pseudorandom generator. That is, we require that no efficient algorithm can distinguish between truly random values and the output of our pseudorandom generator.

Now, for two arithmetic circuits g_1 and g_2 that compute the same underlying polynomial, a true canonical form $\mathsf{Canonical}(g_1)$ would be identical to the canonical form of $\mathsf{Canonical}(g_2)$. Instead, we would ask that a pseudocanonical form $\mathsf{PseudoCanonical}(g_1)$ would simply be indistinguishable from the pseudocanonical form $\mathsf{PseudoCanonical}(g_2)$, to all efficient algorithms that were given g_1 and g_2 as well. Observe that unless there are actual efficiently computable canonical forms for all arithmetic circuits—which we do not believe to be true—it must be that such a PseudoCanonical operator is randomized, and outputs a *probability distribution* over arithmetic circuits computing the same polynomial.

The computing lens. Let us now step back, and view the problem stated above through the lens of computing. The classic theory of computation (see, e.g., [46]) tells us that general computer programs can be converted into equivalent polynomials (albeit over finite fields, which we will focus on implicitly in the sequel). So the pseudocanonicalization question posed above is equivalent to the pseudocanonicalization question for general computer programs. Indeed, the question of hiding implementation details within a computer program has a long history, dating at least as far back as the groundbreaking 1976 work of [50] introducing the concept of public-key cryptography. Historically, this problem has been called "program obfuscation," albeit it was typically discussed in an ill-defined form. Discussed in these vague terms, it was folklore that truly secure program obfuscation would have revolutionary applications to computing, especially for securing intellectual property. The work of [22] gave a formal treatment of this problem, and proved the impossibility of strong forms of general-purpose program obfuscation. This work also formalized the pseudo-canonicalization problem discussed above via the notion of *indistinguishability obfuscation* ($i\mathcal{O}$). Writing now in the language of Boolean circuits, we define the problem as follows:

Definition 1.1 (Indistinguishability obfuscator (iO) for circuits [22]). A probabilistic polynomial-time algorithm $i\mathcal{O}$ is called a secure indistinguishability obfuscator for polynomial-sized circuits if the following holds:

- (*Completeness*) For every $\lambda \in \mathbb{N}$, every circuit C with input length n, and every input $x \in \{0,1\}^n$, we have that
$$\Pr\big[\tilde{C}(x) = C(x) : \tilde{C} \leftarrow i\mathcal{O}(1^\lambda, C)\big] = 1.$$
- (*Indistinguishability*) For every two ensembles $\{C_{0,\lambda}\}_{\lambda\in\mathbb{Z}^+}$ and $\{C_{1,\lambda}\}_{\lambda\in\mathbb{Z}^+}$ of polynomial-sized circuits that have the same size, input length, and output length, and are functionally equivalent, that is, $\forall\lambda \in \mathbb{Z}^+$, $C_{0,\lambda}(x) = C_{1,\lambda}(x)$ for every input x, the distributions $i\mathcal{O}(1^\lambda, C_{0,\lambda})$ and $i\mathcal{O}(1^\lambda, C_{1,\lambda})$ are computationally indistinguishable, that is, for every efficient polynomial-time algorithm D and for every constant $c > 0$, there exists a constant $\lambda_0 \in \mathbb{Z}^+$ such that, for all $\lambda > \lambda_0$, we have
$$\big|\Pr\big[D(i\mathcal{O}(1^\lambda, C_{0,\lambda}) = 1\big] - \Pr\big[D(i\mathcal{O}(1^\lambda, C_{1,\lambda}) = 1\big]\big| \le \frac{1}{\lambda^c}.$$

As we discuss below in Section 1.2, indeed $i\mathcal{O}$ as a formalization of pseudo-canonicalization lived up to the folklore promise of software obfuscation: there was, and still is, a large research community studying novel applications of $i\mathcal{O}$.

In contrast, demonstrating the feasibility of constructing $i\mathcal{O}$ proved far more challenging. Often one expects that theory will lag behind practice, and given the folklore promise of software obfuscation, one might expect that over the years perhaps clever programmers had come up with heuristic approaches to software obfuscation that resisted attack. The reality is the opposite. Indeed, in 2021 the third annual White Box Cryptography contest was held to evaluate heuristic methods for software obfuscation, and every one of the 97 submitted obfuscations was broken before the contest ended [44].

A large body of theoretical work, starting with the pioneering work of [55], has attempted to construct $i\mathcal{O}$ using mathematical tools. However, prior to the result [68] by the authors of this article, all previous mathematical approaches to constructing $i\mathcal{O}$ relied on new, unproven mathematical assumptions, many of which turned out to be false. We survey this work in Section 1.3 below.

We would like to build $i\mathcal{O}$ whose security rests upon cryptographic hardness assumptions that have stood the test of the time, have a long history of study, and are widely believed to be true. The main result of our works [68,69] is the construction of an $i\mathcal{O}$ scheme from three well-studied assumptions. We discuss this in more detail next.

Informal Theorem 1.1 ([68,69]). Under the following assumptions[1]:

- *the Learning Parity with Noise (*LPN*) assumption over general prime fields $\mathbb{Z}_p$ with polynomially many* LPN *samples and error rate $1/k^{\delta}$, where k is the dimension of the* LPN *secret, and $\delta > 0$ is any constant;*
- *the existence of a Boolean Pseudorandom Generator (*PRG*) in* NC^0 *with stretch $n^{1+\tau}$, where n is the length of the* PRG *seed, and $\tau > 0$ is any constant;*
- *the Decision Linear (*DLIN*) assumption on symmetric bilinear groups of prime order,*

indistinguishability obfuscation for all polynomial-size circuits exists.

The three assumptions above (discussed further below in Section 1.1) are based on computational problems with a long history of study, rooted in complexity, coding, and number theory. Further, they were introduced for building basic cryptographic primitives (such as public key encryption), and have been used for realizing a variety of cryptographic goals that have nothing to do with $i\mathcal{O}$.

1.1. Assumptions in more detail

We now describe each of the assumptions we need in more detail and briefly survey their history.

The DLIN assumption. The Decisional Linear assumption (DLIN) is stated as follows: For an appropriate λ-bit prime p, two groups $\mathbb{G}$ and $\mathbb{G}_T$ of order p are chosen such that there exists an efficiently computable nontrivial symmetric bilinear map $e : \mathbb{G} \times \mathbb{G} \to \mathbb{G}_T$. A canonical generator g for $\mathbb{G}$ is also computed. Following the tradition of cryptography, we describe the groups above using multiplicative notation, even though they are cyclic. The DLIN assumption requires that the following computational indistinguishability holds:

$$\{(g^x, g^y, g^{xr}, g^{ys}, g^{r+s}) \mid x, y, r, s \leftarrow \mathbb{Z}_p\}$$
$$\approx_c \{(g^x, g^y, g^{xr}, g^{ys}, g^z) \mid x, y, r, s, z \leftarrow \mathbb{Z}_p\}.$$

This assumption was first introduced in the 2004 work of Boneh, Boyen, and Shacham [31], and instantiated using appropriate elliptic curves. Since then DLIN and assumptions implied by DLIN have seen extensive use in a wide variety of applications throughout cryptography, such as Identity-Based Encryption, Attribute-Based Encryption, Functional Encryption for degree 2 polynomials, Noninteractive Zero Knowledge, etc. (see, e.g., [25,38, 62,89]).

1 For technical reasons, we need to hardness of these assumptions to be such that no polynomial-time adversaries have beyond subexponentially small advantage in breaking the hardness of the underlying problems.

The existence of PRGs in NC^0. The assumption of the existence of a Boolean Pseudorandom Generator PRG in NC^0 states that there exists a Boolean function $\mathsf{G} : \{0,1\}^n \to \{0,1\}^m$ where $m = n^{1+\tau}$ for some constant $\tau > 0$, and where each output bit computed by G depends on a constant number of input bits, such that the following computational indistinguishability holds:

$$\{\mathsf{G}(\sigma) \mid \sigma \leftarrow \{0,1\}^n\} \approx_c \{y \mid y \leftarrow \{0,1\}^m\}.$$

Pseudorandom generators are a fundamental primitive in their own right, and have vast applications throughout cryptography. PRGs in NC^0 are tightly connected to the fundamental topic of Constraint Satisfaction Problems (CSPs) in complexity theory, and were first proposed for cryptographic use by Goldreich [49,61,65] 20 years ago. The complexity theory and cryptography communities have jointly developed a rich body of literature on the cryptanalysis and theory of constant-locality Boolean PRGs [10,12,13,16,17,30,45,48,49,61,73,86,87].

LPN over large fields. The Learning Parity with Noise LPN assumption over finite fields $\mathbb{Z}_p$ is a decoding problem. The standard LPN assumption with respect to subexponential-size modulus p, dimension ℓ, sample complexity n, and a noise rate $r = 1/\ell^\delta$ for some $\delta \in (0,1)$ states that the following computational indistinguishability holds:

$$\{\boldsymbol{A}, \boldsymbol{s} \cdot \boldsymbol{A} + \boldsymbol{e} \bmod p \mid \boldsymbol{A} \leftarrow \mathbb{Z}_p^{\ell \times n}, \boldsymbol{s} \leftarrow \mathbb{Z}_p^{1 \times \ell}, \boldsymbol{e} \leftarrow \mathcal{D}_r^{1 \times n}\}$$
$$\approx_c \{\boldsymbol{A}, \boldsymbol{u} \mid \boldsymbol{A} \leftarrow \mathbb{Z}_p^{\ell \times n}, \boldsymbol{u} \leftarrow \mathbb{Z}_p^{1 \times n}\}.$$

Above $e \leftarrow \mathcal{D}_r$ is a generalized Bernoulli distribution, i.e., e is sampled randomly from $\mathbb{Z}_p$ with probability $1/\ell^\delta$ and set to be 0 with probability $1 - 1/\ell^\delta$. We consider polynomial sample complexity $n(\ell)$, and the modulus p is an arbitrary subexponential function in ℓ.

The origins of the LPN assumption date all the way back to the 1950s: the works of Gilbert [60] and Varshamov [95] showed that random linear codes possessed remarkably strong minimum distance properties. However, since then, very little progress has been made in efficiently decoding random linear codes under random errors. The LPN over fields assumption above formalizes this, and was introduced over $\mathbb{Z}_2$ for cryptographic uses in 1994 [29], and formally defined for general finite fields and parameters in 2009 [66], under the name "Assumption 2."

While in [66] the assumption was used when the error rate was assumed to be a constant, in fact, polynomially low error (in fact, $\delta = 1/2$) has an even longer history in the LPN literature: it was used by Alekhnovitch in 2003 [4] to construct public-key encryption with the field $\mathbb{Z}_2$, and used to build public-key encryption over $\mathbb{Z}_p$ in 2015 [11]. The exact parameter settings that we describe above, with both general fields and inverse polynomial error rate corresponding to an arbitrarily small constant $\delta > 0$, were explicitly posed by [35], in the context of building efficient secure two-party and multiparty protocols for arithmetic computations.

Recently, the LPN assumption has led to a wide variety of applications (see, for example, [11, 14, 35, 37, 52, 59, 66]). A comprehensive review of known attacks on LPN over large fields, for the parameter settings we are interested in, was given in [35, 36]. For our parameter setting, the running time of all known attacks is $\Omega(2^{\ell^{1-\delta}})$, for any choice of the constant $\delta \in (0,1)$ and for any polynomial number of samples $n(\ell)$.

On search vs. decision versions of our assumptions. Except for the DLIN assumption, the other two assumptions that we make can be based on search assumptions.

The LPN over $\mathbb{Z}_p$ assumption we require is implied by the subexponential hardness of its corresponding search versions [29,82,83,91]. As summarized in [94], there is a search-to-decision reduction[2] whose sample complexity is $m = \mathrm{poly}(\dim(s), m', 1/\varepsilon)$ (namely, polynomial in the dimension $\dim(s)$ of the secret, sample complexity m' of the decision version, and the inverse of the distinguishing gap ε) and runtime $\mathrm{poly}(\dim(s), p, m)$. In this work, we need the pseudorandomness of (polynomially many) LPN samples to hold against polynomial-time adversaries, with a *subexponential* distinguishing gap. We can further set the modulus p to an arbitrarily small subexponential function[3] in $\dim(s)$. Decisional LPN with such parameters are implied by the following subexponential search LPN assumption: There is a constant $\gamma > 0$ such that no subexponential-time $2^{\dim(s)^\gamma}$ adversary, given a subexponential $2^{\dim(s)^\gamma}$ number of samples, can recover s with noticeable probability.

The works of [10,16] showed that the one-wayness of *random local functions* implies the existence of PRGs in NC^0. More precisely, for a length parameter $m = m(n)$, a locality parameter $d = O(1)$, and a d-ary predicate $Q : \{0,1\}^d \to \{0,1\}$, a distribution $\mathcal{F}_{Q,m}$ samples a d-local function $f_{G,Q} : \{0,1\}^d \to \{0,1\}$ by choosing a random d-uniform hypergraph G with n nodes and m hyperedges, where each hyperedge is chosen uniformly and independently at random. The ith output bit of $f_{G,Q}$ is computed by evaluating Q on the d input bits indexed by nodes in the ith hyperedge. The one-wayness of $\mathcal{F}_{Q,m}$ for proper choices of Q, m has been conjectured and studied in [12,45,61,86]. The works of [10,16] showed how to construct a family of PRG in NC^0 with polynomial stretch based on the one-wayness of $\mathcal{F}_{Q,m}$ for any Q that is sensitive (i.e., some input bit i of Q has full influence) and any $m = n^{1+\delta}$ with $\delta > 0$. The constructed PRGs have negligible distinguishing advantage and the reduction incurs a multiplicative polynomial security loss. Therefore, the subexponential pseudorandomness of PRG in NC^0 that we need is implied by the existence of $\mathcal{F}_{Q,m}$ that is hard to invert with noticeable probability by adversaries of some subexponential size.

1.2. Applications of $i\mathcal{O}$

The notion of $i\mathcal{O}$ occupies an intriguing and influential position in complexity theory and cryptography. Interestingly, if $\mathsf{NP} \subseteq \mathsf{BPP}$, then $i\mathcal{O}$ exists for the class of all polynomial-size circuits because if $\mathsf{NP} \subseteq \mathsf{BPP}$, then it is possible to efficiently compute a canonical form for any function computable by a polynomial-size circuit. On the other hand, if $\mathsf{NP} \not\subseteq \mathsf{io\text{-}BPP}$, then in fact the existence of $i\mathcal{O}$ for polynomial-size circuits implies that one-way functions exist [71]. A large body of work has shown that $i\mathcal{O}$ plus one-way func-

---

2 Importantly, this reduction is oblivious to the distribution of the errors and hence applies to both LWE and LPN.

3 In the construction, we set $p = \Theta(2^\lambda)$ and $\dim(s)$ to a large enough polynomial in λ.

tions imply a vast array of cryptographic objects, so much so that iO has been conjectured to be a "central hub" [71, 92] for cryptography.

An impressive list of fascinating new cryptographic objects are only known under $i\mathcal{O}$ or related objects such as functional encryption and witness encryption. Hence, our construction of $i\mathcal{O}$ from well-founded assumptions immediately implies these objects from the same assumptions. Below, we highlight a small subset of these implications as corollaries. In all the applications, by λ we denote the security parameter.

Corollary 1.1 (Informal). *Assuming the* subexponential *hardness of the three assumptions in Theorem* 1.1, *we have:*

- *Multiparty noninteractive key exchange in the plain model (without trusted setup), e.g.,* [33, 70]*;*
- *Selectively sound and perfectly zero-knowledge Succinct Noninteractive ARGument (ZK-SNARG) for any NP language with statements up to a bounded polynomial size in the CRS model, where the CRS size is* $\mathsf{poly}(\lambda)(n+m)$*,* n, m *are upper bounds on the lengths of the statements and witnesses, and the proof size is* $\mathsf{poly}(\lambda)$ [92]*;*
- *(Symmetric or asymmetric) multilinear maps with bounded polynomial multilinear degrees, following* [3, 53]*, and a self-bilinear map over composite and unknown order group, assuming additionally the polynomial hardness of factoring* [97]*;*
- *Witness Encryption (WE) for any NP language, following as a special case of* $i\mathcal{O}$ *for polynomial size circuits;*
- *Secret sharing for any monotone function in NP* [72]*;*
- *Fully homomorphic encryption scheme for unbounded-depth polynomial size circuits (without relying on circular security), assuming slightly superpolynomial hardness of the assumptions above* [41]*;*
- *Hardness of finding Nash equilibrium (more generally, for the class PPAD)* [27].

1.3. Prior work on the feasibility of $i\mathcal{O}$

There is a rich landscape of research on conjectured constructions of $i\mathcal{O}$. Despite being posed as a question at least 20 years ago [22, 50], the first candidate mathematical construction came only in 2013, through the work of [55]. This construction relied on a newly constructed primitive called multilinear maps [54], which is a generalization of a bilinear maps where one could do high degree computations in the exponents. Soon after, several different candidates for multilinear maps were proposed [47, 57] and many other constructions of $i\mathcal{O}$ were proposed. This propelled a huge body of constructions of $i\mathcal{O}$ relying on multilinear maps and related ideas (e.g., [18, 20, 39, 43, 47, 51, 54, 55, 57, 84, 85, 90].) Unfortunately, all these works suffered from one of the three main problems:

- Most constructions were heuristic in the sense that they were just conjectured to be secure. There was no simple assumption on the multilinear maps on which you could base security.

- Sometimes security was based on some new assumption, but it was a new assumption proposed solely for proving that the construction was secure. Such assumptions lacked a long history of study.

- Most of the time, in the above both cases there were actually cycles of attacks and fixes on the constructions and/or underlying assumptions (e.g., [19,21,42,43,63,81,84,85]) which reduced our confidence further.

With this, the focus shifted to trying to minimize the degree of the multilinear map needed, with the goal of eventually reaching degree 2. In a beautiful line of work [9,74,75,79,80], it was shown that $i\mathcal{O}$ can be constructed just from succinct assumptions on degree-3 multilinear maps. Unfortunately, the candidates for degree-3 multilinear maps were the same as the candidates for high-degree multilinear maps and suffered from the same class of attacks as before.

Soon after, a line of work [1,2,6,8,56,67,76] constructed $i\mathcal{O}$ relying on bilinear maps, along with new kinds of pseudorandom generators. These assumptions were much simpler to state than before. Even though earlier proposals for some of those pseudorandom generators were attacked [19,21,81], exploring the limits of those attacks helped us design $i\mathcal{O}$ based on new but simple-to-state assumptions [6,56,67] that resisted all known attacks. However, these assumptions were newly stated and did not have a long history of study.

Therefore, building upon [6,8,56,67,76], these works culminated finally in our recent works [68,69], which managed to construct $i\mathcal{O}$ from the three assumptions in Theorem 1.1. This eliminated the need for making any new unstudied hardness assumptions. We now discuss some of the main open problems in the space of $i\mathcal{O}$ constructions.

1.4. Open problems

Our work places $i\mathcal{O}$ on firm foundations with respect to the assumptions it is based on, thereby answering the main feasibility question for the primitive (until we resolve the P vs. NP question). However, there are many important open questions that remain to be answered:

- *Concrete efficiency.* Our work first builds the notion of functional encryption and then boosts this object to $i\mathcal{O}$ via a complex transformation [5,28]. As a result, the final construction is quite complex. A highly important question that remains open is the following one: Is it possible to construct $i\mathcal{O}$ either by fine-tuning our approach, or otherwise (as in [23,58]) in a way that the resulting scheme yields concrete implementable efficiency? For this question, as a first step, it is even interesting if the construction rests upon new assumptions as long as the assumptions are rigorously cryptanalyzed.

- *Postquantum* $i\mathcal{O}$. Our work relies on bilinear maps (in a somewhat crucial way). As a result of that, our construction is broken in polynomial time using a quantum computer. Therefore, an important and a natural question to ask here is if we can build $i\mathcal{O}$ on any combination of well-studied postquantum assumptions such as LWE, LPN, or PRG in NC^0. This is indeed an active area of research.
- $i\mathcal{O}$ *for quantum circuits.* All known constructions of $i\mathcal{O}$ support only classical circuits. If quantum computers come one day, an interesting question is to construct an $i\mathcal{O}$ scheme that can be used to actually obfuscate quantum circuits. There are some results in restricted models [24, 40], but none of the known constructions work to obfuscate general quantum programs.
- *Understanding assumptions better.* We are still in the early stages of understanding the feasibility of $i\mathcal{O}$. An immediate question that arises out of work is to identify essential and nonessential assumptions out of the three assumptions, and if any of the assumptions can be replaced by another. Identifying if there is any other substantially different approach that also yields $i\mathcal{O}$ from well-studied assumptions will also shed light on this question.

2. TECHNICAL OVERVIEW: HOW TO CONSTRUCT $i\mathcal{O}$?

Below, we describe a very high-level overview of the main technical ideas implying $i\mathcal{O}$. For simplicity of exposition, we choose the simplest path to $i\mathcal{O}$ that we are aware of. This overview is based on a combination of ideas from [68] and [69]. However, for simplicity, the route discussed below would require one additional assumption to the three stated above (See Theorem 1.1)—namely, the Learning With Errors (LWE) [91] assumption. However, we do not actually discuss the exact technical reasons for needing LWE, as this assumption is actually unnecessary [69].

2.1. Preliminaries

Let us start with introducing some basic notation. Let $\mathsf{size}(X)$ indicate the length of the binary description of an object X (e.g., a string, a circuit, or truth table). Throughout, we consider Boolean functions or circuits or algorithms mapping n-bit binary strings to m-bit binary strings, for some $n, m \in \mathbb{Z}^+$. Let $\mathsf{time}(A, x)$ denote the running time of an algorithm (or circuit) A on an input x (in the case of a circuit C, $\mathsf{time}(C, x)$ is the same as $\mathsf{size}(C)$). We say that an algorithm or circuit is efficient if its running time is bounded by a (fixed) polynomial in the length of the input, that is, $\mathsf{time}(C, x) = \mathsf{size}(x)^c$ for some positive integer $c \in \mathbb{Z}^+$. When we only care about the existence of a constant and the concrete value is not important, we write $O(1)$ in place of that constant, e.g., $\mathsf{time}(C, x) = \mathsf{size}(x)^{O(1)}$ (following the big-O notation in complexity theory).

Our goal is designing an efficient *randomized* algorithm, called the obfuscator $\mathcal{O}$, that, given a Boolean circuit $C : \{0, 1\}^n \to \{0, 1\}$ with size $s \leq n^{O(1)}$, referred to as the original circuit, outputs another Boolean circuit $\hat{C} : \{0,1\}^n \to \{0,1\}$, called the obfuscated circuit, such that the following three properties hold:

- (*Correctness*) The obfuscated circuit $\hat{C}$ is *functionally equivalent* to the original circuit C, denoted as $\hat{C} \equiv C$, meaning that for every $x \in \{0,1\}^n$, $\hat{C}(x) = C(x)$. Correctness must hold no matter what random coins the obfuscator $\mathcal{O}$ uses.
- (*Efficiency*) The obfuscator is efficient, meaning $\mathcal{O}$ runs in time polynomial in the size of the original circuit, namely, $\mathsf{time}(\mathcal{O}, C) = \mathsf{size}(C)^{O(1)}$.
- (*Security*) The obfuscated circuit $\hat{C}$ hides the implementation details in the original circuit C. This is formalized as follows: for every two *equally-sized* and *functionally-equivalent* circuits C_0 and C_1 (i.e., $\mathsf{size}(C_0) = \mathsf{size}(C_1)$ and $C_0 \equiv C_1$), the obfuscated circuits $\hat{C}_0$ and $\hat{C}_1$ are *computationally hard to distinguish*.

In the above by *distinguish* we mean having an algorithm D acting as a distinguisher and which, given an obfuscated circuit $\hat{C}$ generated from C_0 or C_1 chosen at random with equal probability, tries to determine which of C_0 and C_1 is the original circuit. By *computationally hard* we mean that no *efficient* distinguisher D can do much better than random guessing, that is, the probability of guessing correctly is bounded by $\frac{1}{2} + \varepsilon$ for some very small ε. And we say that $\hat{C}_0$ and $\hat{C}_1$ are computationally indistinguishable, which intuitively implies that they hide all implementation differences between C_0 and C_1 to computationally limited adversaries. However, computationally *unlimited* adversaries may well be able to distinguish them. We focus on computational security, since if $i\mathcal{O}$ with security against computationally unlimited adversaries were to exist, this would imply a collapse of the polynomial hierarchy [34] in complexity theory, a collapse which is widely conjectured to be false.

Next, we give an informal overview of how to construct $i\mathcal{O}$ from well-studied assumptions. In Section 2.2, we describe first how to reduce the task of constructing $i\mathcal{O}$ that compiles general Boolean circuits to a much simpler task—building $xi\mathcal{O}$ (introduced shortly) for specific simple circuits. In Section 2.3, we illustrate how this simplified task connects with *bilinear pairing groups*. This overview paints the overall blueprint. In the next section, Section 3.2, we will zoom into the key ideas that bridge the simpler task with bilinear pairing groups. These ideas are the last jigsaw pieces that complete the construction of $i\mathcal{O}$, which appeared in our latest works [68, 69].

2.2. Simplifying the task of $i\mathcal{O}$

Perhaps the simplest starting point is the following: If there is no restriction on the time the obfuscator $\mathcal{O}$ can take, then there is an extremely intuitive obfuscator: the obfuscated circuit is the *truth table* of the original circuit. The truth table TT_C of a Boolean circuit C is an array indexed by inputs, where $\mathsf{TT}_C[x] = C(x)$. It can be computed in time $2^n \cdot s$ if the input length is n and circuit size is s. Perfect security comes from the fact that, for any two functionally equivalent circuits $C_0 \equiv C_1$, their truth tables are identical $\mathsf{TT}_{C_0} = \mathsf{TT}_{C_1}$ and hence impossible to distinguish.

To put simply, a truth table is a *canonical form* of all circuits producing it. While outputting the truth table satisfies the correctness and security requirement of $i\mathcal{O}$, the obvious flaw with this is that the obfuscator is far from efficient: The running time of an $i\mathcal{O}$ scheme should be $s^{O(1)}$, rather than $2^n \cdot s$. The input length n of a Boolean circuit can be close to its size s, and hence the time to compute a truth table is exponentially large! This inefficiency is likely inherent, as efficient methods of finding canonical forms of circuits implies the collapse of the polynomial hierarchy, which is widely conjectured to be false.

Therefore, to improve efficiency, we seek canonical forms that fool computationally limited adversaries. Naturally, we start with a more humble goal:

Can we improve efficiency slightly to, say $2^{n(1-\varepsilon)} \cdot s^{O(1)}$ for some small $\varepsilon > 0$? What does $i\mathcal{O}$ with such nontrivial efficiency imply?

Simplification 1: obfuscation with nontrivial exponential efficiency. $i\mathcal{O}$ with nontrivial efficiency was studied in **[26, 77, 78]**. Surprisingly, their authors showed that very modest improvement on efficiency—captured in the notion of exponential-efficiency $i\mathcal{O}$, or $xi\mathcal{O}$ for short—is actually enough to construct completely efficient (polynomial time) $i\mathcal{O}$:

- $xi\mathcal{O}$ is an obfuscator $\mathcal{O}$ whose running time is still "trivial" $(2^n \cdot s)^{O(1)}$, but outputs an obfuscated circuit $\hat{C}$ of "nontrivial" size $2^{n(1-\varepsilon)} \cdot s^{O(1)}$ for some $\varepsilon > 0$.[4]

We can think of $xi\mathcal{O}$ (as well as $i\mathcal{O}$) as a special kind of encryption method, where the obfuscated circuit $\hat{C}$ is a "ciphertext" of the original circuit C, also denoted as $\mathsf{spCT}(C) = \hat{C}$, such that

- the ciphertext $\mathsf{spCT}(C)$ hides all information about C, except that it lets anyone with access to it learn the truth table of C. This is unlike normal encryption that reveals no information of the encrypted message.
- The size of the ciphertext $\mathsf{spCT}(C)$ is $2^{(1-\varepsilon)\cdot n}$ for some $0 < \varepsilon < 1$. So it can be viewed as a (slightly) compressed version of the truth table (that reveals no other information of C to computationally limited adversaries).

Such a special encryption scheme is known as *functional encryption* **[32, 88, 93]**, which controls precisely which information of the encrypted message is revealed, and hides all other information. This notion is tightly connected with $xi\mathcal{O}$ and $i\mathcal{O}$, and, in fact, the implication of $xi\mathcal{O}$ to $i\mathcal{O}$ goes via the notion of functional encryption **[5, 7, 28]**.

When viewing $\mathsf{spCT}(C)$ as a compressed version of the truth table. It becomes clear why even slight compression is powerful enough to imply $i\mathcal{O}$: The idea is keeping compressing iteratively until the size of the special ciphertext becomes polynomial. The works

4 We note that $xi\mathcal{O}$ should be distinguished from the Minimal Circuit Size Problem (MCSP) in complexity theory, which asks to compute the circuit complexity of a function, given as a truth table TT. In contrast, the obfuscator is given a small circuit C as input.

of [5,28,77,78] turn this high-level idea into an actual proof that $xi\mathcal{O}$ implies $i\mathcal{O}$[5] and allows us to focus on constructing $xi\mathcal{O}$, or equivalently, the special encryption described above.

Simplification 2: it suffices to obfuscate simple circuits. Unfortunately, despite the efficiency relaxation, it is still unclear how to obfuscate general Boolean circuits, which can be complex. Naturally, we ask:

> *Can we obfuscate simple subclasses of circuits?*
> *What does $xi\mathcal{O}$ for simple circuits imply?*

It turns out that it suffices to focus on an extremely simple class of circuits $\mathsf{C} = \mathsf{NC}^0$, where NC^0 is the set of all circuits with constant *output locality*, meaning every output bit depends on a constant number of input bits. To do so, we will rely on two cryptographic tools, *randomized encodings* and *Pseudorandom Generators (PRGs)*.

Randomized encoding in NC^0. A randomized encoding (RE) scheme consists of two efficient algorithms (RE, Decode). It gives a way to represent a complex circuit $C(\cdot)$ by a much simpler *randomized* circuit $\mathsf{RE}_C(\cdot;\cdot) := \mathsf{RE}(C,\cdot;\cdot)$ such that

- (*Correctness*) For every input x, the output π_x of $\mathsf{RE}(C, x; r)$ produced using uniformly random coins r encodes the correct output; in other words, there exists an efficient decoding algorithm Decode such that $\mathsf{Decode}(\pi) = C(x)$.
- (*Security*) π_x reveals no information of C beyond the output of x to efficient adversaries.
- (*Simplicity*) RE is a simple circuit by some measure of simplicity. Classic works [15,64,98] showed that RE can be simply an NC^0 circuit, when assuming the existence of PRGs in NC^0 like we are.

The correctness and security of the randomized encoding suggests that, instead of directly obfuscating a general circuit C, we can alternatively obfuscate a circuit D that on input x outputs an encoding π_x, which reveals only $C(x)$. The potential benefit is that D depending on RE and C should be simply an NC^0 circuit. Hence, it would suffice to construct $xi\mathcal{O}$ for simple NC^0 circuits!

To make the above idea go through, there are, however, a few wrinkles to be ironed out. The issue is that the security of randomized encoding only holds if an encoding π_x is generated using fresh random coins. There are concrete attacks that learn information of C beyond the outputs, when π_x is generated using nonuniform random coins, or when two encoding π_x and $\pi_{x'}$ are generated using correlated random coins. If the truth table of the

---

5 This implication, however, comes with a quantitative weakening in the security. To obtain $i\mathcal{O}$ that is secure against adversaries that run in time polynomial in the input length, the original $xi\mathcal{O}$ needs to be secure against adversaries that run in time subexponential 2^{n^ε} for some $\varepsilon \in (0, 1)$ in its input length n.

circuit D contains an encoding π_x for every input x (i.e., $\mathsf{TT}_D[x] = \pi_x$) the random coins for generating these encoding must be embedded in D, that is,

$$D = D_{C,\mathsf{RE},r_1,r_2,\ldots,r_{2^n}} \quad \text{such that} D(x) = \mathsf{RE}(C, x; r_x).$$

Such a circuit D has size at least 2^n. In particular, we cannot hope to "compress" the random coins $r_1, r_2, \ldots, r_{2^n}$ into $2^{n(1-\varepsilon)}$ space, which is the target size of the obfuscated circuit. To resolve this problem, we will use a Pseudorandom Generator.

Pseudorandom generator in NC^0. A pseudorandom generator is a Boolean function $\mathsf{PRG} : \{0,1\}^n \to \{0,1\}^m$ that takes as input a uniformly random string $sd \in \{0,1\}^n$, called a *seed*, and produces a polynomially longer output $r \in \{0,1\}^m$, where $m = n^{1+\rho}$ for some constant $\rho > 0$, such that r is indistinguishable from a uniformly random m-bit string to computationally limited adversaries. Pseudorandom generators are among the most basic cryptographic primitives and have been extensively studied. Among these studies is a beautiful line of works, initiated by [61], investigating pseudorandom generators in NC^0, for which there are several candidates, including those proposed in [17, 86, 87].

Equipped with a pseudorandom generator in NC^0, we can now replace uniformly random coins $r_1, r_2, \ldots, r_{2^n}$ with pseudorandom coins expanded from a much shorter seed sd of length roughly $2^{n/(1+\rho)} = 2^{n(1-\varepsilon')}$ for some $\varepsilon' \in (0,1)$.[6] This gives a variant of the circuit D above:

$$D' = D'_{C,\mathsf{RE},\mathsf{PRG},sd} \quad \text{such that } D'(x) = \mathsf{RE}\big(C, x; \big(r_x = \mathsf{PRG}_x(sd)\big)\big),$$

where $\mathsf{PRG}(sd)$ expands to output $r_1, \ldots, r_{2^n}$ and $r_x = \mathsf{PRG}_x(sd)$ is the xth chunk of the output. Thanks to the fact that both PRG and RE are in NC^0, so is D'.

Moving forward, it suffices to devise a way to encrypt $\mathsf{spCT}(C, sd)$, from which we can expand out the truth table of D', while hiding all other information of C and sd:

$$C, sd \xrightarrow{G_{\mathsf{RE},\mathsf{PRG}}} \mathsf{TT}_{D'}, \quad \mathsf{spCT}(C, sd) \xrightarrow{\text{Expand}} \mathsf{TT}_{D'}.$$

Note that the special encryption only needs to hide (C, sd) instead of the entire description of circuit D' since RE and PRG are public algorithms.

2.3. Special encryption for NC^0 mappings

In our works [68, 69], we constructed the needed special encryption for general NC^0 mappings $G : \{0,1\}^l \to \{0,1\}^m$, where ciphertext $\mathsf{spCT}(X)$ reveals the output of G on X while hiding all other information of X, such that $\mathsf{size}(\mathsf{spCT}(X)) \sim \mathsf{size}(X) + m^{1-\delta}$ for some $\delta > 0$. In this overview, we describe half of the ideas behind our construction, which connects the special encryption with bilinear pairing groups and a new object called structured-seed pseudorandom generator. The other half of ideas is explained in Section 3.2, captured by the construction of structured-seed pseudorandom generator.

6 More precisely, the length of sd is $(2^n s^{O(1)})^{1/(1+\rho)}$ since each r_x is an $s^{O(1)}$-bit string instead of a single bit. However, the dominant term is $2^{n/(1+\rho)}$ as s is only polynomial in n.

Connection with bilinear pairing groups. As a starting point, suppose that the function G is really simple—simple enough so that it can be computed by a degree-2 polynomial mapping $Q : \mathbb{Z}_p^l \to \mathbb{Z}_p^m$ over the finite field $\mathbb{Z}_p$ for some prime modulus p. Namely,

$$\forall X \in \{0,1\}^l, \quad G(X) = Q(X),$$
$$\text{where } Q_i(X) = \Big(\sum_{j,k} \alpha_{j,k} X_j \cdot X_k + \sum_k \beta_k X_k + \gamma\Big) \bmod p.$$

Then, the special encryption can be implemented using bilinear pairing groups as shown in [9,75].

Bilinear pairing groups. At a high-level, pairing groups allow for computing quadratic polynomials over secret encoded values and reveal only whether the output is zero or not. More specifically, they consist of cyclic groups $\mathbb{G}_1$, $\mathbb{G}_2$, and $\mathbb{G}_T$ with generators g_1, g_2, and g_T, respectively, and all of order p; $\mathbb{G}_1$ and $\mathbb{G}_2$ are referred to as the source groups and $\mathbb{G}_T$ as the target group. (In some instantiations, the two source groups are the same group, which is called symmetric pairing groups.) They support the following operations:

- (*Encode*) For every group $\mathbb{G}_i$, one can compute g_i^x for $x \in \mathbb{Z}_p$. The group element g_i^x is viewed as an *encoding* of x in the group $\mathbb{G}_i$.
- (*Group Operation*) In each group $\mathbb{G}_i$, one can perform the group operation to get $g_i^{x_1} \circ g_i^{x_2} = g_i^{x_1+x_2}$, corresponding to "homomorphic" addition modulo p in the exponent. Following the tradition of cryptography, we write the group operation multiplicatively.
- (*Bilinear Pairing*) Given two source group elements g^{x_1} and $g_2^{x_2}$, one can efficiently compute a target group element $g_T^{x_1 \cdot x_2}$, using the so-called pairing operation $e(g_1^{x_1}, g_2^{x_2}) = g_T^{x_1 \cdot x_2}$. This corresponds to "homomorphic" multiplication modulo p in the exponent. However, after multiplication, we obtain an element in the target group which cannot be paired anymore.
- (*Zero Testing*) Given a group element g_i^x, one can always tell if the encoded value is $x = 0$, by comparing the group element with the identity in G_i. Similarly, one can do equality testing to see if $g_i^x = g_i^c$ for any c.

Combining above abilities gives a "rudimentary" special encryption scheme that supports evaluation of degree-2 polynomials. A ciphertext of $X \in \mathbb{Z}_p^l$ includes encodings of every element X_l in both source groups, $((g_1^{X_1}, \dots g_1^{X_l}), (g_2^{X_1}, \dots g_2^{X_l}))$. Given these, one can "homomorphically" compute a quadratic mapping Q to obtain an encoding of the output $y = Q(X)$ in the target group $(g_T^{y_1}, \dots, g_T^{y_m})$ (without knowing the encoded input X at all); finally, if the output y happens to be a *bit string*, one can learn y in the clear via zero-testing. In summary,

$$((g_1^{X_1}, \dots g_1^{X_l}), (g_2^{X_1}, \dots g_2^{X_l})) \xrightarrow{\text{Expand}} Q(x), \quad \text{if } Q(x) \in \{0,1\}^m.$$

This fulfills the correctness of the special encryption. What about security? The ciphertext must hide all information about X beyond what is revealed by $Q(x)$, for which we resort to the security of pairing groups. For simplicity of this overview, let us gain some security intuition by assuming the strongest hardness assumptions pertaining to the pairing groups, known as the generic group model. Think of encoding as black boxes and the only way of extracting information (of the encoded values) is by performing (a combination of) the above four operations. In this model, given g^x for a secret and random $x \in \mathbb{Z}_p$, no efficient adversary can learn x. Extending further, if X were random (in $\mathbb{Z}_p$) subject to $Q(X) = y$, then the encoding would reveal only y and hide all other information of X. The only issue is that X in our example is not random—it is binary. Nevertheless, the works of [9,75] designed clever ways of *randomizing* an arbitrary input X, to a random input $\bar{X}$ subject to the only condition that an appropriate quadratic mapping $\bar{Q}$ reveals the right output $\bar{Q}(\bar{X}) = Q(X)$. Therefore, security is attained. We refer the reader to [9,75] for details about how such randomization works; in fact, it is possible to rely on much weaker assumption than the generic group model [75,96].

Challenges beyond degree 2. Unfortunately, the mapping $G_{\mathsf{RE,PRG}}$ we care about can only be computed by polynomials of degree much larger than 2. It is known that a Boolean function with output locality l can be computed by a multilinear degree l polynomial over $\mathbb{Z}_p$. However, the locality of Boolean PRG we need is at least 5 [86] and known randomized encodings have locality at least 4 [15].

Key idea: the preprocessing model. To overcome this challenge, our first idea is using preprocessing of inputs to help reduce the degree of computation. Instead of directly computing $G(X)$ in one shot, we separate it into two steps: First, the input is preprocessed $X' = \mathsf{pre}(X; r)$ in a randomized way using fresh random coins r, then the output is computed from the preprocessed input $y = Q(X')$. The idea is that the preprocessing can perform complex transformations on the input in order to help the computation later. The only constraint is that the preprocessing should not increase the size of its input too much, that is, $\mathsf{size}(X') \sim \mathsf{size}(X) + m^{1-\delta}$, for some $\delta > 0$. As such, it suffices to encrypt the preprocessed input $\mathsf{spCT}(X')$, from which one can recover the desired output $y = G(X)$, by evaluating a (hopefully) simpler function Q. Unfortunately, because of the restriction on the size of X', it is unclear how preprocessing alone can help.

Our second idea is to further relax the preprocessing model to allow the preprocessed input to contain a public part and a secret part $X' = (P, S)$. Importantly, the public part P should reveal no information about X to computationally limited adversaries. (In contrast, P and S together reveal X completely.) Moreover, we allow the second stage computation $Q(P, S)$ to have arbitrary constant degree in P and only restrict its degree on S to 2, that is,

$$Q_i(P, S) = \Big(\sum_{j,k} \alpha_{j,k}(P) \cdot S_j \cdot S_k + \sum_k \beta_k(P) \cdot S_k + \gamma(P)\Big) \bmod p,$$

where $\alpha_{j,k}, \beta_k, \gamma$ are constant-degree polynomials.

It turns out that the techniques alluded to above for special encryption for degree 2 computations can be extended (see [6,56,67,96]), so that given ciphertext $\mathsf{spCT}(S)$ and P in the clear, one can homomorphically compute Q and thereby learn the output $G(X)$.

In [68,69], we show how to compute any NC^0 Boolean function G in such a preprocessing model, assuming the Learning Parity with Noises assumption over general fields, which completes the puzzle of $i\mathcal{O}$. When applied to specific cryptographic tools, our techniques give interesting new objects. For instance, it converts any PRG in NC^0 into what we call structured-seed PRG. Given a preprocessed seed $(P, S) = \mathsf{PRG}(sd; r')$, the structured-seed PRG expands out a polynomially longer output $r = \mathsf{sPRG}(P, S)$, where the computation has only degree-2 in the private seed S, and the output r is pseudorandom given the public seed P. In the next section, we describe how to do preprocessing in the context of constructing structured-seed PRG. The same ideas can be extended to handle general NC^0 functions.

3. STRUCTURED SEED PRG

In this section we define and construct our main object, namely a structured seed PRG (sPRG). But before we do that, we introduce a few preliminaries. For any distribution $\mathcal{X}$, we denote by $x \leftarrow \mathcal{X}$ the process of sampling a value x from the distribution $\mathcal{X}$. Similarly, for a set X, we denote by $x \leftarrow X$ the process of sampling x from the uniform distribution over X. For an integer $n \in \mathbb{N}$, we denote by $[n]$ the set $\{1, \ldots, n\}$. A function $\mathsf{negl} : \mathbb{N} \to \mathbb{R}$ is said to be a negligible function if for every constant $c > 0$ there exists an integer N_c such that $\mathsf{negl}(\lambda) < \lambda^{-c}$ for all $\lambda > N_c$.

Throughout, when we refer to polynomials in the security parameter, we mean constant degree polynomials that take positive value on nonnegative inputs. We denote by $\mathsf{poly}(\lambda)$ an arbitrary polynomial in λ satisfying the above requirements of nonnegativity. We denote vectors by bold letters such as $\boldsymbol{b}$ and $\boldsymbol{u}$. Matrices will be denoted by capitalized bold letters for such as $\boldsymbol{A}$ and $\boldsymbol{M}$. For any $k \in \mathbb{N}$, we denote by the tensor product $\boldsymbol{v}^{\otimes k} = \underbrace{\boldsymbol{v} \otimes \cdots \otimes \boldsymbol{v}}_{k}$ to be the standard tensor product, but converted back into a vector. This vector contains all the monomials in the variables inside $\boldsymbol{v}$ of degree exactly k.

For any two polynomials $a(\lambda, n), b(\lambda, n) : \mathbb{N} \times \mathbb{N} \to \mathbb{R}^{\geq 0}$, we say that a is polynomially smaller than b, denoted as $a \ll b$, if there exist an $\varepsilon \in (0, 1)$ and a constant $c > 0$ such that $a < b^{1-\varepsilon} \cdot \lambda^c$ for all large enough $n, \lambda \in \mathbb{N}$. The intuition behind this definition is to think of n as being a sufficiently large polynomial in λ.

Multilinear representation of polynomials and representation over $\mathbb{Z}_p$. A straightforward fact from analysis of Boolean functions is that every NC^0 function $F : \{0,1\}^n \to \{0,1\}$ can be represented by a unique constant degree multilinear polynomial $f \in \mathbb{Z}[x_1, \ldots, x_n]$ that agrees with F over $\{0,1\}^n$. At times, we will also interpret $f(\boldsymbol{x})$ as a polynomial over $\mathbb{Z}_p$ for some prime p. This is done by actually reducing coefficients of f modulo p, and then evaluating the same multilinear polynomial over $\mathbb{Z}_p$. Observe that for every $\boldsymbol{x} \in \{0,1\}^n$, $f(\boldsymbol{x}) = f(\boldsymbol{x}) \bmod p$, as the for every $\boldsymbol{x} \in \{0,1\}^n$, $f(\boldsymbol{x}) \in \{0,1\}$. Furthermore, given any

NC^0 function F, finding these representations over $\mathbb{Z}$, as well as $\mathbb{Z}_p$, takes polynomial time. We now describe the notion of computational indistinguishability.

Definition 3.1 (ε-indistinguishability). We say that two ensembles $\mathcal{X} = \{\mathcal{X}_\lambda\}_{\lambda\in\mathbb{N}}$ and $\mathcal{Y} = \{\mathcal{Y}_\lambda\}_{\lambda\in\mathbb{N}}$ are ε-indistinguishable where $\varepsilon : \mathbb{N} \to [0, 1]$ if for every nonnegative polynomial $\mathsf{poly}(\cdot)$ and any adversary $\mathcal{A}$ running in time bounded by $\mathsf{poly}(\lambda)$ it holds that, for every sufficiently large $\lambda \in \mathbb{N}$,

$$\Big| \Pr_{x\leftarrow\mathcal{X}_\lambda} \big[\mathcal{A}(1^\lambda, x) = 1\big] - \Pr_{y\leftarrow\mathcal{Y}_\lambda} \big[\mathcal{A}(1^\lambda, y) = 1\big]\Big| \le \varepsilon(\lambda).$$

We say that two ensembles are indistinguishable if they are ε-indistinguishable for some ε that is a negligible function, and subexponentially indistinguishable if they are ε-indistinguishable for $\varepsilon(\lambda) = 2^{-\lambda^c}$ for some positive constant c.

We now formally define our LPN assumption **[11,29,35,66]**.

Definition 3.2 (δ-LPN assumption, **[11, 29, 35, 66]**). Let $\delta \in (0, 1)$. We say that the δ-LPN assumption is true if the following holds: For any constant $\eta_p > 0$, any function $p : \mathbb{N} \to \mathbb{N}$ such that, for every $\ell \in \mathbb{N}$, $p(\ell)$ is a prime of ℓ^{η_p} bits, any constant $\eta_n > 0$, we set $p = p(\ell)$, $n = n(\ell) = \ell^{\eta_n}$, and $r = r(\ell) = \ell^{-\delta}$, and we require that the following two distributions are computationally indistinguishable:

$$\big\{(\boldsymbol{A}, \boldsymbol{b} = \boldsymbol{s}\cdot\boldsymbol{A} + \boldsymbol{e}) \mid \boldsymbol{A} \leftarrow \mathbb{Z}_p^{\ell\times n}, \boldsymbol{s} \leftarrow \mathbb{Z}_p^{1\times\ell}, \boldsymbol{e} \leftarrow \mathcal{D}_r^{1\times n}(p)\big\}_{\ell\in\mathbb{N}},$$
$$\big\{(\boldsymbol{A}, \boldsymbol{u}) \mid \boldsymbol{A} \leftarrow \mathbb{Z}_p^{\ell\times n}, \boldsymbol{u} \leftarrow \mathbb{Z}_p^{1\times n}\big\}_{\ell\in\mathbb{N}}.$$

In addition, we say that subexponential δ-LPN holds if the two distributions above are subexponentially indistinguishable.

We now define the notion of an sPRG.

3.1. Definition of structured-seed PRG

Definition 3.3 (Syntax of structured-seed pseudorandom generators (sPRG)). Let $\tau > 1$. A structured-seed Boolean PRG (sPRG) with polynomial stretch τ is defined by the following PPT algorithms:

- $\mathsf{IdSamp}(1^n, p)$ takes as input the input length parameter n and a prime p. It samples a function index I.
- $\mathsf{SdSamp}(I)$ jointly samples two strings, a public seed and a private seed, $sd = (P, S)$, which are both vectors of dimension $\ell_{sd} = O(n)$ over $\mathbb{Z}_p$.
- $\mathsf{Eval}(I, sd)$ computes a string in $\{0, 1\}^m$. Here $m = n^\tau$.

Looking ahead, the prime p, that we choose, is set to the order of the bilinear group which has a bit length of $n^{\Theta(1)}$.

Definition 3.4 (Security of sPRG). A structured-seed Boolean PRG, sPRG, satisfies the security requirement if, for any constant $\rho > 0$, any function $p : \mathbb{N} \to \mathbb{Z}$ that takes as input a number $k \in \mathbb{N}$ and outputs a k^ρ bit prime $p(k)$, any $n \in \mathbb{N}$, with probability $1 - o(1)$ over $\mathsf{IdSamp}(1^n, p = p(n)) \to I$, it holds that the following distributions are $\varepsilon(n)$ indistinguishable:

$$\{I, P, \mathsf{Eval}(I, P, S) \mid I \leftarrow \mathsf{IdSamp}(1^n, p), sd \leftarrow \mathsf{SdSamp}(I)\},$$
$$\{I, P, \boldsymbol{r} \mid I \leftarrow \mathsf{IdSamp}(1^n, p), sd \leftarrow \mathsf{SdSamp}(I), \boldsymbol{r} \leftarrow \{0,1\}^{m(n)}\},$$

where $\varepsilon(n)$ is a negligible function. Further, we say that sPRG is subexponentially secure if $\varepsilon(n) = 2^{-n^{\Omega(1)}}$.

Definition 3.5 (Complexity and degree of sPRG). Let $\rho > 0$, $d_1, d_2 \in \mathbb{N}$ be any constants, and $p : \mathbb{N} \to \mathbb{Z}$ be any function that maps an integer k into a k^ρ bit prime $p(k)$. An sPRG has degree d_1 in public seed P and degree d_2 in S over $\mathbb{Z}_p$, denoted as $\mathsf{sPRG} \in (\deg d_1, \deg d_2)$, if for every I in the support of $\mathsf{IdSamp}(1^n, p = p(n))$, there exist efficiently generatable polynomials $g_{I,1}, \ldots, g_{I,m}$ over $\mathbb{Z}_p$ such that:

- $\mathsf{Eval}(I, (P, S)) = (g_{I,1}(P, S), \ldots, g_{I,m}(P, S))$, and
- the maximum degree of each $g_{I,j}$ over P is d_1, while the maximum degree of $g_{I,j}$ over S is d_2.

We remark that the above definition generalizes the standard notion of families of PRGs in two aspects: (1) the seed consists of a public and a private parts, jointly sampled and arbitrarily correlated, and (2) the seed may not be uniform. Therefore, we obtain the standard notion as a special case.

Definition 3.6 (Pseudorandom generators, degree, and locality). A (uniform-seed) Boolean PRG (PRG) is an sPRG with a seed sampling algorithm $\mathsf{SdSamp}(I)$ that outputs a public seed P that is an empty string and a uniformly random private seed $S \leftarrow \{0,1\}^n$. Let $d, c \in \mathbb{N}$. The PRG has multilinear degree d if, for every $n \in \mathbb{N}$ and I in the support of $\mathsf{IdSamp}(1^n)$, we have that $\mathsf{Eval}(I, sd)$ can be written as an $m(n)$-tuple of degree-d polynomials over $\mathbb{Z}$ in S. It has constant locality c if, for every $n \in \mathbb{N}$ and I in the support of $\mathsf{IdSamp}(1^n)$, every output bit of $\mathsf{Eval}(I, sd)$ depends on at most c bits of S.

In what follows next we will construct an sPRG from the LPN assumption and the existence of PRG in NC^0. For the ease of exposition, we will actually use Goldreich's PRG candidate [61] which is the most well-known conjectured PRG in NC^0.

Definition 3.7 (Goldreich's PRG). A Goldreich PRG of locality c mapping n bits to m bits is described using a predicate $f : \{0,1\}^c \to \{0,1\}$ and a hypergraph $H = \{Q_1, \ldots, Q_m\}$ where each Q_i is a randomly chosen ordered subset of $[n]$ of size c. The index I consists of f and H. Further, on input $\boldsymbol{x} \in \{0,1\}^n$, $\mathsf{PRG.Eval}(I, \boldsymbol{x}) = \boldsymbol{y}$, where $\boldsymbol{y} = (y_1, \ldots, y_m)$. Here each $y_i = f(x_{i_1}, \ldots x_{i_c})$ where $Q_i = (i_1, \ldots, i_c)$.

Remark 3.1. In a Goldreich's PRG, when the hypergraph is randomly chosen, with probability $\frac{1}{n^{O(1)}}$ it fails to be an expander. With probability $1 - o(1)$, the hypergraph has appropriate expansion properties. Under such conditions, the security of Goldreich PRGs has been very well studied [10, 12, 13, 16, 17, 30, 45, 48, 49, 61, 73, 86, 87] and is widely believed to hold. This is the precise reason in the security definition, we require pseudorandomness to hold with probability $1 - o(1)$ over the choice of I.

3.2. Construction of structured-seed PRG

Now we show how to construct our sPRG. We prove:

Theorem 3.1. *Let $d \in \mathbb{N}$, $\delta \in (0, 1)$, and $\tau > 1$ be constants. Then, assuming the following:*

- *the security of locality d Goldreich's* PRG *with stretch τ, and*
- *the δ-*LPN*-assumption,*

there exists an sPRG *with polynomial stretch in* $(\deg d, \deg 2)$. *Additionally, if both assumptions are subexponentially secure, then so is the* sPRG.

We first give an overview and then dive into the construction.

Technical overview. We start with a Goldreich PRG PRG = (IdSamp, Eval) with stretch τ.

Such a PRG is associated with a d-local predicate $f : \{0,1\}^d \to \{0,1\}$. Recall now how the index sampling IdSamp works. The IdSamp algorithm on input n outputs a random hypergraph H.

We start by observing that on any input $\boldsymbol{\sigma} \in \{0,1\}^n$, $\boldsymbol{y} = \mathsf{PRG.Eval}(I, \boldsymbol{\sigma})$ can be computed by degree d multilinear polynomials over $\mathbb{Z}$ as f is a d local predicate. Our high-level strategy is to somehow "preprocess" $\boldsymbol{\sigma}$ into two vectors (P, S) of small dimension (preferably $O(n)$, but anything sublinear in m works) such that $\boldsymbol{y} \in \{0,1\}^m$ can be computed in $(\deg d, \deg 2)$. Thereby this will have an effect of transferring complexity of computation to the public input. To achieve this, we preprocess $\boldsymbol{\sigma}$ into appropriate public and private seeds (P, S) and leverage the LPN assumption over $\mathbb{Z}_p$ (which is the input to sPRG.IdSamp) to show that the seed is hidden.

Our first idea towards this is that we can "encrypt" the seed $\boldsymbol{\sigma}$ using LPN samples over $\mathbb{Z}_p$ as follows:

$$\begin{aligned} \text{Sample:} \quad & \boldsymbol{A} \leftarrow \mathbb{Z}_p^{\ell \times n}, \boldsymbol{s} \leftarrow \mathbb{Z}_p^{1 \times \ell}, \boldsymbol{e} \leftarrow \mathcal{D}_r^{1 \times n}(p), \\ \text{Add to the function index } I': \quad & \boldsymbol{A}, \\ \text{Add to public seed } P: \quad & \boldsymbol{b} = \boldsymbol{s}\boldsymbol{A} + \boldsymbol{e} + \boldsymbol{\sigma}, \end{aligned}$$

where ℓ is set to be the dimension for the LPN samples. It is set so that $\ell^{\lceil \frac{d}{2} \rceil} = n$; $\mathcal{D}_r(p)$ is a distribution that samples randomly from p with probability r, and 0 otherwise. Finally, $r = \ell^{-\delta}$.

It follows directly from LPN assumption that $(\boldsymbol{A}, \boldsymbol{b})$ is pseudorandom and hides $\boldsymbol{\sigma}$. Furthermore, due to the sparsity of LPN noises, the vector $\boldsymbol{\sigma} + \boldsymbol{e}$ differs from $\boldsymbol{\sigma}$ only at an $\ell^{-\delta}$ fraction of components—thus it is a sparsely erroneous version of the seed. For $i \in [m]$, let $f_i(\boldsymbol{\sigma})$ be the locality d, degree d polynomial over $\mathbb{Z}$ such that $y_i = f_i(\boldsymbol{\sigma})$. Then, for

$i \in [m]$, consider the following polynomial:

$$h_i(\boldsymbol{b}, \overline{\boldsymbol{s}}^{\otimes \lceil \frac{d}{2} \rceil}) = f_i(\boldsymbol{b} - \boldsymbol{s}\boldsymbol{A}), \quad \overline{\boldsymbol{s}} = \boldsymbol{s}||1.$$

Above we first interpret f_i over $\mathbb{Z}$ into the field $\mathbb{Z}_p$ by simply reducing coefficients mod p, and then compute as given. Observe that f_i is a degree d polynomial in $\boldsymbol{b}$ and $\boldsymbol{s}$, therefore its degree over $\boldsymbol{b}$ is d and over $\overline{\boldsymbol{s}}^{\otimes \lceil \frac{d}{2} \rceil}$ is two. Thus h_i is a $(\deg d, \deg 2)$ polynomial. Observe that $h_i(\boldsymbol{b}, \overline{\boldsymbol{s}}^{\otimes \lceil \frac{d}{2} \rceil}) = f_i(\boldsymbol{\sigma} + \boldsymbol{e})$. The main point of this is that if we set the polynomial map $G^{(1)} = (G_1^{(1)}, \dots, G_m^{(1)})$ by letting each $G_i^{(1)} = h_i$, and set the private seed $S = \overline{\boldsymbol{s}}^{\otimes \lceil \frac{d}{2} \rceil}$, then

$$G^{(1)}(P, S) = \mathsf{PRG.Eval}_I(\boldsymbol{\sigma} + \boldsymbol{e}).$$

The reason $G^{(1)}$ is interesting is because $\boldsymbol{e}$ is sparse. With probability $1 - 2^{-n^{\Omega(1)}}$, it is nonzero at $O(n\ell^{-\delta})$ locations. As a consequence, for any given $i \in [m]$, $f_i(\boldsymbol{\sigma}) = f_i(\boldsymbol{\sigma} + \boldsymbol{e})$ with all but $O(\ell^{-\delta})$ probability as f_i is a d local function depending on d randomly chosen inputs. Since in the hypergraph H of the Goldreich PRG, each set Q_i is chosen independently, every output is independently error prone with probability $O(\ell^{-\delta})$. Because of this, due to Chernoff style concentration bounds, out of m outputs, with probability $1 - 2^{-n^{\Omega(1)}}$, all but $T = O(m\ell^{-\delta})$ outputs are error prone.

This gives us as a nice candidate for sPRG that satisfies almost all properties! The dimension of S and P is $O(n)$ which is sublinear in m, and it can be computed by a degree $(\deg d, \deg 2)$ polynomial $G^{(1)}$. We would be done if we could somehow force the output to be correct on all the m coordinates. For the rest of the overview, we refer to the indices $i \in [m]$ such that $f_i(\boldsymbol{\sigma}) \neq f_i(\boldsymbol{\sigma} + \boldsymbol{e})$ as bad indices/outputs.

To correct errors, we further modify the polynomial and include more preprocessed information in the private seeds. We describe a sequence of ideas that lead to the final correction method, starting with two wrong ideas that illustrate the difficulties we will overcome:

- The first wrong idea is correcting by adding the difference $\mathsf{Corr} = \boldsymbol{y} - \boldsymbol{y}'$ between the correct and erroneous outputs, $\boldsymbol{y} = \mathsf{Eval}_I(\boldsymbol{\sigma})$ and $\boldsymbol{y}' = \mathsf{Eval}_I(\boldsymbol{\sigma} + \boldsymbol{e})$; we refer to Corr as the *correction vector*. To obtain the correct output, evaluation can compute the polynomial map $G^{(1)}(\boldsymbol{b}, (\overline{\boldsymbol{s}}^{\otimes \lceil \frac{d}{2} \rceil})) + \mathsf{Corr}$. The problem is that Corr must be included in the seed, but it is as long as the output and would destroy expansion. Thus, we have to make use of the fact that Corr is sparse.

- To fix expansion, the second wrong idea is adding correction only for bad outputs, so that the seed only stores nonzero entries in Corr. Recall that Corr is sparse with at most T nonzero elements. More precisely, the jth output can be computed as $G_j^{(1)}(\boldsymbol{b}, (\overline{\boldsymbol{s}}^{\otimes \lceil \frac{d}{2} \rceil})) + \mathsf{Corr}_j$ if output j is bad and without adding Corr_j otherwise. This fixes expansion, but now the evaluation polynomial depends on the location of bad outputs. If these locations are included in public information, this would leak information of the location of LPN noises, and jeopardize security. If, on the other hand, the locations are included in the private seed, then it is unclear how to maintain the requirement that the polynomial map computes only a degree-two polynomial in the private seed.

These two wrong ideas illustrate the tension between the expansion and security of our sPRG. Our construction takes care of both, by *compressing* the correction vector Corr to be polynomially shorter than the output and stored in the seed, and *expanding* it back during evaluation in a way that is oblivious of the location of bad output bits. This is possible thanks to the sparsity of the correction vector and the allowed degree-two computation on the private seed. We first illustrate our idea with the help of a simple case.

Simple case 1: much fewer than $\sqrt{m}$ bad outputs. Suppose hypothetically that the number of bad outputs is bounded by z which is much smaller than $\sqrt{m}$. Thus, if we convert Corr into a $\sqrt{m} \times \sqrt{m}$ matrix,[7] it has low rank z. We can then factorize Corr into two matrixes $\mathbf{U}$ and $\mathbf{V}$ of dimensions $\sqrt{m} \times z$ and $z \times \sqrt{m}$, respectively, such that $\mathsf{Corr} = \mathbf{UV}$, and compute the correct output as follows:

$$\forall j \in [m], \quad G_j^{(2)}\big(\boldsymbol{b}, \big(\overline{\boldsymbol{s}}^{\otimes \lceil \frac{d}{2} \rceil}, \mathbf{U}, \mathbf{V}\big)\big) = G_j^{(1)}\big(\boldsymbol{b}, \big(\overline{\boldsymbol{s}}^{\otimes \lceil \frac{d}{2} \rceil}\big)\big) + (\mathbf{UV})_{k_j, l_j},$$

where (k_j, l_j) is the corresponding index of the output bit j in the $\sqrt{m} \times \sqrt{m}$ matrix. When $z \ll \sqrt{m}$, the matrices $\mathbf{U}, \mathbf{V}$ have $2z\sqrt{m}$ field elements, which is polynomially smaller than $m = n^{\tau}$. As such, $G^{(2)}$ is expanding. Moreover, observe that $G^{(2)}$ has only degree 2 in the private seed and is completely oblivious of where the bad outputs are.

While the idea above works for fewer than $\sqrt{m}$ bad outputs, it does not work for the case we are dealing with. We have $T = \Theta(m\ell^{-\delta})$ bad outputs. Nevertheless, we show that a similar idea works for this case.

T bad outputs. The above method, however, cannot handle more than $\sqrt{m}$ bad outputs, whereas the actual number of bad outputs can be up to $T = \Omega(m/\ell^{\delta})$, much larger than $\sqrt{m}$ since δ is an arbitrarily small constant. Consider another hypothetical case where the bad outputs are evenly spread in the following sense: suppose that if we divide the matrix Corr into m/ℓ^{δ} blocks, each of dimension $\ell^{\delta/2} \times \ell^{\delta/2}$, there are at most ℓ^{ρ} bad outputs in each block where $\rho > 0$ is a really small constant (say $\delta/10$). In this case, we can "compress" each block of Corr separately using the idea from case 1. More specifically, for every block $i \in [m/\ell^{\delta}]$, we factor it into $\mathbf{U}_i \mathbf{V}_i$, with dimensions $\ell^{\delta/2} \times \ell^{\rho}$ and $\ell^{\rho} \times \ell^{\delta/2}$, respectively, and correct bad outputs as follows:

$$\forall j \in [m], \quad G_j^{(2)}\big(\boldsymbol{b}, \big(\overline{\boldsymbol{s}}^{\otimes \lceil \frac{d}{2} \rceil}, (\mathbf{U}_i, \mathbf{V}_i)_{i \in [\frac{m}{\ell^{\delta}}]}\big)\big) = G_j^{(1)}\big(\boldsymbol{b}, \big(\overline{\boldsymbol{s}}^{\otimes \lceil \frac{d}{2} \rceil}\big)\big) + (\mathbf{U}_{i_j} \mathbf{V}_{i_j})_{k_j, l_j},$$

where i_j is the block that output j belongs to, and $(k_j, l_j) \in [\ell^{\delta/2}] \times [\ell^{\delta/2}]$ is its index within this block. We observe that $G^{(2)}$ is expanding, since each matrix $\mathbf{U}_i$ or $\mathbf{V}_i$ has $\ell^{\delta/2+\rho}$ field elements, and the total number of elements is $\ell^{\delta/2+\rho} \cdot \frac{m}{\ell^{\delta}}$, which is polynomially smaller than m as long as δ is positive and m is polynomially related to ℓ. Moreover, $G^{(2)}$ is oblivious of the location of bad outputs just as in case 1.

This completely solves our problem except that we need to ensure that the bad outputs are well spread out in the manner described above. Our main observation here is that this

---

7 Any injective mapping from a vector to a matrix that is efficient to compute and invert will do.

is ensured due to the fact that in a Goldreich's PRG candidate the input dependence hypergraph $Q_1, \ldots, Q_m$ is randomly chosen. Therefore, once we fix the location of the nonzero errors locations inside $\boldsymbol{e}$ (where with high probability $O(n\ell^{-\delta})$ locations are nonzero), in every block of ℓ^δ output bits, each entry j is independently nonzero with probability $O(\ell^{-\delta})$. Thus, in expectation each block has a constant number of bad output bits. More so, due to the Chernoff bound, it can be seen that with probability $1 - 2^{-\ell^{\Omega(1)}}$, each has at most ℓ^ρ nonzero elements. Thus, our construction can be summarized as follows:

Step 1: Assign outputs. We partition the outputs into B buckets, via a mapping $\phi_{\mathsf{bkt}} : [m] \to [B]$. The number of buckets is set to $B = m/\ell^\delta$ and the number of elements in each bucket is set to be ℓ^δ so that they exactly form partition of m. The mapping ϕ_{bkt} simply divides m by B, and outputs the remainder. Since the error $\boldsymbol{e}$ is chosen to be from the LPN error distribution and the hypergraph H of the PRG is randomly chosen, by a Chernoff-style argument, we can show that in each bucket out of ℓ^δ output bits, at most t of them are bad, except with probability $1 - 2^{-t^{\Omega(1)}}$. We will set $t = \ell^\rho$ for a tiny constant $\rho > 0$.

Step 2: Compress the buckets. Next, we organize each bucket i into a matrix $\mathbf{M}_i$ of dimension $\ell^{\delta/2} \times \ell^{\delta/2}$ and then compute its factorization $\mathbf{M}_i = \mathbf{U}_i \mathbf{V}_i$, where $\mathbf{U}_i, \mathbf{V}_i$ are matrices of dimensions $\ell^{\delta/2} \times t$ and $t \times \ell^{\delta/2}$, respectively. To form matrix $\mathbf{M}_i$, we use another mapping $\phi_{\mathsf{ind}} : [m] \to [\ell^{\delta/2}] \times [\ell^{\delta/2}]$ to assign each output bit j to an index (k_j, l_j) in the matrix of the bucket i_j it is assigned to. This assignment must guarantee that no two output bits in the same bucket (assigned according to ϕ_{bkt}) have the same index. One such way to compute $\phi_{\mathsf{ind}}(j)$ is to divide $j \in [m]$ by B. The remainder is set as $\phi_{\mathsf{bkt}}(j)$, and the quotient is divided further by $\ell^{\delta/2}$. The quotient and the remainder from this division are set as the resulting indices (k_j, l_j). Once we have this, $(\mathbf{M}_i)_{k,l}$ is set to Corr_j if there is j such that $\phi_{\mathsf{bkt}}(j) = i$ and $\phi_{\mathsf{ind}}(j) = (k, l)$. Since every matrix $\mathbf{M}_i$ has at most t nonzero entries, we can factor them and compute the correct output as:

$$\forall j \in [m], \quad G_j^{(2)}(\boldsymbol{b}, \underbrace{(\overline{s}^{\otimes \lceil \frac{d}{2} \rceil}, (\mathbf{U}_i, \mathbf{V}_i)_{i \in [B]})}_{S})$$
$$= G_j^{(1)}(\boldsymbol{b}, (\overline{s}^{\otimes \lceil \frac{d}{2} \rceil})) + (\mathbf{U}_{\phi_{\mathsf{bkt}}(j)} \cdot \mathbf{V}_{\phi_{\mathsf{bkt}}(j)})_{\phi_{\mathsf{ind}}(j)},$$

$G^{(2)}$ is expanding because the number of field elements in $\mathbf{U}_i$'s and $\mathbf{V}_i$'s are much smaller than m, namely $2t\ell^{\delta/2}B = O(m\ell^{-\delta/2+\rho}) \ll m$. We set $I' = (I, \boldsymbol{A}, \phi_{\mathsf{bkt}}, \phi_{\mathsf{ind}})$.

Step 3: Zeroize if uneven buckets. Finally, to deal with the low probability event that some bucket contains more than t bad outputs, we introduce a new variable called a flag. If this occurs, our sPRG sets P and S as all-zero vectors. In this case the evaluation always outputs 0. This gives us our candidate sPRG. For security, observe that the polynomial map $G^{(2)}$ is independent of the location of LPN noises. With probability $1 - 2^{-n^{\Omega(1)}}$, the evaluation results in output $\boldsymbol{y}$. Therefore, by the LPN over $\mathbb{Z}_p$ assumption, the seed $\boldsymbol{\sigma}$ of PRG is hidden and the security of PRG ensures that the output is pseudorandom when it is not all zero (which occurs with a subexponentially small probability). We now proceed to the formal construction and proof.

Construction. We now formally describe our scheme. Assume the premise of the theorem. Let (IdSamp, Eval) be the function index sampling algorithm and evaluation algorithm for the Goldreich PRG. Recall that its seed consists of only a private seed sampled uniformly at random.

We first introduce and recall some notation. The construction is parameterized by

- the input length n and output length $m = n^{\tau}$ of the Goldreich PRG (PRG),
- the stretch $\tau > 1$ and degree/locality d of the Goldreich PRG,
- the LPN secret dimension $\ell = n^{1/\lceil d/2 \rceil}$ and the error probability $r = \ell^{-\delta}$,
- a slack parameter $t = \ell^{\rho}$ used for bounding the number of bad outputs in each bucket,
- a parameter $B = m/\ell^{\delta}$ that indicates the number of buckets used, and
- a parameter $c = \ell^{\delta}$ that indicates the capacity of each bucket; it is set so that $c \cdot B = m$,
- a parameter η, which is the dimension of each bucket; we set $\eta = \sqrt{c}$,
- assignment function $\phi_{\mathsf{bkt}} : [m] \to [B]$ that is computed by dividing input j by B and returning its remainder,
- assignment function $\phi_{\mathsf{ind}} : [m] \to [\eta]$ that is computed by dividing input $j \in [m]$ by B and dividing further the quotient with η, and returning the quotient and the remainder of this division.

$I' \leftarrow \mathsf{IdSamp}'(1^{n'}, p)$: Generate the public index as follows:

- Sample $I \leftarrow \mathsf{PRG.IdSamp}(1^n)$ and $\boldsymbol{A} \leftarrow \mathbb{Z}_p^{\ell \times n}$.
- Output $I' = (I, \boldsymbol{\phi} = (\phi_{\mathsf{bkt}}, \phi_{\mathsf{ind}}), \boldsymbol{A})$.

(Note that the PRG seed length n below is an efficiently computable polynomial in n', and can be inferred from the next seed sampling algorithm. See Claim 3.1 for the exact relationship between n and n'.)

$sd \leftarrow \mathsf{SdSamp}'(I')$: Generate the seed as follows:

- Sample a PRG seed $\boldsymbol{\sigma} \leftarrow \{0, 1\}^n$.

- Prepare samples of LPN over $\mathbb{Z}_p$: Sample $\boldsymbol{s} \leftarrow \mathbb{Z}_p^{1\times\ell}$, $\boldsymbol{e} \leftarrow \mathcal{D}_r^{1\times n}(p)$, and set

$$\boldsymbol{b} = \boldsymbol{s}\boldsymbol{A} + \boldsymbol{\sigma} + \boldsymbol{e}.$$

- Find indices $i \in [n]$ of seed bits where $\boldsymbol{\sigma} + \boldsymbol{e}$ and $\boldsymbol{\sigma}$ differ, which are exactly these indices where $\boldsymbol{e}$ is not 0, and define

$$\mathsf{ERR} = \{i \mid \sigma_i + e_i \neq \sigma_i\} = \{i \mid e_i \neq 0\}.$$

  We say a seed index i is *erroneous* if $i \in \mathsf{ERR}$. Since LPN noise is sparse, errors are sparse.

- Find indices $j \in [m]$ of outputs that depend on one or more erroneous seed indices. Let Vars_j denote the indices of seed bits that the jth output of Eval_I depends on. Define

$$\mathsf{BAD} = \left\{j \mid |\mathsf{Vars}_j \cap \mathsf{ERR}| \geq 1\right\}.$$

  We say an output index j is bad if $j \in \mathsf{BAD}$, and good otherwise.

- Set $\mathsf{flag} = 0$ if there is some bucket containing too many bad outputs: $\exists i \in [B]$, $|\phi_{\mathsf{bkt}}^{-1}(i) \cap \mathsf{BAD}| > t$. Otherwise, set $\mathsf{flag} = 1$.

- Compute the outputs of PRG on inputting the correct seed and the erroneous seed, $\boldsymbol{y} = \mathsf{PRG.Eval}_I(\boldsymbol{\sigma})$ and $\boldsymbol{y}' = \mathsf{PRG.Eval}_{\boldsymbol{I}}(\boldsymbol{\sigma} + \boldsymbol{e})$. Set the correction vector $\mathsf{Corr} = \boldsymbol{y} - \boldsymbol{y}'$.

- Construct matrices $\mathbf{M}_1, \ldots, \mathbf{M}_B$ by setting

$$\forall j \in [m], \quad (\mathbf{M}_{\phi_{\mathsf{bkt}}(j)})_{\phi_{\mathsf{ind}}(j)} = \mathsf{Corr}_j.$$

  Every other entry is set to 0.

- "Compress" matrices $\mathbf{M}_1, \ldots, \mathbf{M}_B$ as follows:

  - If $\mathsf{flag} = 1$, for every $i \in [B]$, compute factorization

$$\mathbf{M}_i = \mathbf{U}_i \mathbf{V}_i, \quad \mathbf{U}_i \in \mathbb{Z}_p^{\eta\times t}, \mathbf{V}_i \in \mathbb{Z}_p^{t\times\eta}.$$

    This factorization exists because, when $\mathsf{flag} = 1$, each bucket has at most t nonzero entries, and therefore its rank is less than or equal to t.

  - If $\mathsf{flag} = 0$, for every $i \in [B]$, set $\mathbf{U}_i$ and $\mathbf{V}_i$ to be 0 matrices.

- Set the public seed to

$$P = (\boldsymbol{b} \cdot \mathsf{flag}).$$

  This means that, if $\mathsf{flag} = 0$, P is the all-zero vector in $\mathbb{Z}_p^n$.

- Prepare the private seed S as follows. Let $\overline{s} = s||1$ and set

$$S = \left(\mathsf{flag} \cdot \overline{s}^{\otimes \lceil \frac{d}{2} \rceil}, \{\mathbf{U}_i, \mathbf{V}_i\}_{i \in [B]}\right). \tag{3.1}$$

This means that, if $\mathsf{flag} = 0$, S is the all-zero vector over $\mathbb{Z}_p$. Output $sd = (P, S)$ as $\mathbb{Z}_p$ elements.

$\boldsymbol{y} \to \mathsf{Eval}'(I', sd)$: Compute $\boldsymbol{y} \leftarrow \mathsf{Eval}(I, \boldsymbol{\sigma})$ and output $\boldsymbol{z} = \mathsf{flag} \cdot \boldsymbol{y}$. Looking ahead, $\mathsf{flag} = 1$ will happen with all but subexponentially small probability. This computation is done via a polynomial map $G^{(2)}$:

- Every output bit of Eval is a linear combination of degree d monomials (without loss of generality, assume that all monomials have exactly degree d which can be done by including 1 in the seed $\boldsymbol{\sigma}$).

  Notation. Let us introduce some notation for monomials. A monomial h on a vector $\boldsymbol{a}$ is represented by the set of indices $h = \{i_1, i_2, \ldots, i_k\}$ of variables used in it; h evaluated on $\boldsymbol{a}$ is $\prod_{i \in h} a_i$ if $h \neq \emptyset$ and 1 otherwise. We will use the notation $a_h = \prod_{i \in h} a_i$. We abuse notation to also use a polynomial g to denote the set of monomials involved in its computation; hence $h \in g$ says monomial h has a nonzero coefficient in g.

  With the above notation, we can write Eval as

$$\forall j \in [m], \quad y_j = \mathsf{Eval}_j(\boldsymbol{\sigma}) = L_j((\sigma_h)_{h \in \mathsf{Eval}_j}), \quad \text{for a linear } L_j.$$

- $(\boldsymbol{A}, \boldsymbol{b} = \boldsymbol{s}\boldsymbol{A} + \boldsymbol{x})$ in the public seed encodes $\boldsymbol{x} = \boldsymbol{\sigma} + \boldsymbol{e}$. Therefore, we can compute every monomial x_v as follows:

$$x_i = \langle \boldsymbol{c}_i, \overline{\boldsymbol{s}} \rangle, \quad \boldsymbol{c}_i = -\boldsymbol{a}_i^{\mathrm{T}} || b_i, \boldsymbol{a}_i \text{ is the } i\text{th column of } \boldsymbol{A},$$
$$x_v = \langle \otimes_{i \in v} \boldsymbol{c}_i, \otimes_{i \in v} \overline{\boldsymbol{s}} \rangle.$$

  (Recall that $\otimes_{i \in v} \boldsymbol{z}_i = \boldsymbol{z}_{i_1} \otimes \cdots \otimes \boldsymbol{z}_{i_k}$ if $v = \{i_1, \ldots, i_k\}$ and is not empty; otherwise, it equals 1.) Combining with the previous step, we obtain a polynomial $G^{(1)}(\boldsymbol{b}, S)$ that computes $\mathsf{Eval}(\boldsymbol{\sigma} + \boldsymbol{e})$:

$$G_j^{(1)}(\boldsymbol{b}, S) := L_j\left(\left(\langle \otimes_{i \in v} \boldsymbol{c}_i, \otimes_{i \in v} \overline{\boldsymbol{s}} \rangle\right)_{v \in \mathsf{Eval}_j}\right). \tag{3.2}$$

  Note that $G^{(1)}$ implicitly depends on $\boldsymbol{A}$ contained in I'. Since all relevant monomials v have degree d, we have that $G^{(1)}$ has degree at most d in P, and degree 2 in S. The latter follows from the fact that S contains $\overline{\boldsymbol{s}}^{\otimes \lceil \frac{d}{2} \rceil}$, and hence $S \otimes S$ contains all monomials in $\boldsymbol{s}$ of total degrees d.

  Since only bad outputs depend on erroneous seed bits such that $\sigma_i + e_i \neq \sigma_i$, we have that the output of $G^{(1)}$ agrees with the correct output $\boldsymbol{y} = \mathsf{Eval}(\boldsymbol{\sigma})$ on all good output bits:

$$\forall j \notin \mathsf{BAD}, \quad \mathsf{Eval}_j(\boldsymbol{\sigma}) = G_j^{(1)}(\boldsymbol{b}, S).$$

- To further correct bad output bits, we add to $G^{(1)}$ all the expanded correction vectors as follows:

$$\begin{aligned} G_j^{(2)}(P,S) &:= G_j^{(1)}(\boldsymbol{b},S) + \big(\mathbf{U}_{\phi_{\mathsf{bkt}}(j)}\mathbf{V}_{\phi_{\mathsf{bkt}}}(j)\big)_{\phi_{\mathsf{ind}}(j)} \\ &= G_j^{(1)}(\boldsymbol{b},S) + (\mathbf{M}_{\phi_{\mathsf{bkt}}(j)})_{\phi_{\mathsf{ind}}(j)}. \end{aligned}$$

  We have that $G^{(2)}$ agrees with the correct output $\boldsymbol{y} = \mathsf{Eval}(\boldsymbol{\sigma})$ if $\mathsf{flag} = 1$. This is because, under the condition for $\mathsf{flag} = 1$, every entry j in the correction vector Corr_j is placed at entry $(\mathbf{M}_{\phi_{\mathsf{bkt}}(j)})_{\phi_{\mathsf{ind}}(j)}$. Adding it back as above produces the correct output. Observe that the function is quadratic in S and degree d in the public component of the seed P.

- When $\mathsf{flag} = 0$, however, sPRG needs to output all zero. This happens because P and S are both all-zero vectors and $G^{(2)}$ does not use a constant term. At last, $G^{(2)}$ has degree d in the public seed, and only degree 2 in the private seed, as desired.

Analysis of stretch. We derive a set of constraints, under which sPRG has polynomial stretch. Recall that PRG output length is $m = n^{\tau}$, degree d, LPN secret dimension $\ell = n^{1/\lceil d/2\rceil}$, modulus p is a prime, and the slack parameter $t = \ell^{\rho}$.

Claim 3.1. *For the parameters as set in the Construction,* sPRG *has stretch of τ' for some constant $\tau' > 1$.*

Proof. Let us start by analyzing the dimension of the public and private seeds:

- The public seed contains $P = (\boldsymbol{b} \cdot \mathsf{flag})$ and has $O(n)$ field elements.
- The private seed S contains S_1, S_2 as follows:

$$S_1 = \mathsf{flag} \cdot \big(\overline{\boldsymbol{s}}^{\otimes\lceil \frac{d}{2}\rceil}\big), \quad S_2 = \{\mathbf{U}_i, \mathbf{V}_i\}_{i\in[B]}.$$

  The dimension of S_1 is $O(n)$ as $\ell = n^{\frac{1}{\lceil \frac{d}{2}\rceil}}$, and S_2 consists of $O(B\cdot\eta\cdot t)$ field elements. This consist of $O(m\ell^{-(\delta/2-\rho)})$ field elements. Because $m = n^{\tau}$ and $\ell = n^{\frac{1}{\lceil \frac{d}{2}\rceil}}$, we have:

$$\begin{aligned} \mathsf{dim}(S_1) &= O(n), \\ \mathsf{dim}(S_2) &= O\big(n^{\tau - \frac{(\delta/2-\rho)}{\lceil \frac{d}{2}\rceil}}\big), \\ \mathsf{dim}(P,S) &= O\big(n + n^{\tau - \frac{2\delta}{5\cdot\lceil \frac{d}{2}\rceil}}\big). \end{aligned}$$

The last equality uses $\rho = \delta/10$. We set $n' = n + n^{\tau - \frac{2\delta}{5\lceil \frac{d}{2}\rceil}}$, and therefore $m = n'^{\tau'}$ for some $\tau' > 1$. This concludes the proof. ∎

Proof of pseudorandomness. We prove the following proposition which implies that sPRG is secure.

Proposition 3.1. *Let $\delta > 0$, $\tau > 1$, $d \in \mathbb{N}$, and $\beta \geq 0$ be constants. Assume the following assumptions hold:*

- δ-LPN, *and*
- PRG *is a secure Goldreich PRG with stretch τ and locality d.*

Then, for any prime generating function $p : \mathbb{N} \to \mathbb{N}$ that takes as input an integer k and outputs a prime p of bit length k^β, we have the following. Let $n \in \mathbb{N}$ and $p = p(n)$, then it holds that, with probability $1 - o(1)$ over $I \leftarrow \mathsf{IdSamp}(1^n)$,

$$\{(I', P, \boldsymbol{z}) : \boldsymbol{A} \leftarrow \mathbb{Z}_p^{\ell \times n}, I' = (I, \boldsymbol{\phi}, \boldsymbol{A}, p), (P, S) \leftarrow \mathsf{SdSamp}'(I'), \boldsymbol{z} \leftarrow \mathsf{Eval}'(I, sd)\},$$
$$\{(I', P, \boldsymbol{r}) : \boldsymbol{A} \leftarrow \mathbb{Z}_p^{\ell \times n}, I' = (I, \boldsymbol{\phi}, \boldsymbol{A}, p), (P, S) \leftarrow \mathsf{SdSamp}'(I'), \boldsymbol{r} \leftarrow \{0,1\}^m\},$$

are computationally indistinguishable. Further assuming that the assumptions are subexponentially secure, these distributions are subexponentially indistinguishable.

We first recall the structure of P and the evaluation $\boldsymbol{z}$: P consists of $\mathsf{flag} \cdot P$, where $\boldsymbol{b} = \boldsymbol{s}\boldsymbol{A} + \boldsymbol{e} + \boldsymbol{\sigma}$ and $\boldsymbol{\sigma}$ is a randomly generated PRG seed. As shown in the correctness proof, $\boldsymbol{z} = \mathsf{flag} \cdot \boldsymbol{y}$, where $\boldsymbol{y} = \mathsf{Eval}(I, \boldsymbol{\sigma})$. If flag is always equal to 1, the proof becomes trivial: P is pseudorandom due to LPN assumption and therefore it computationally hides $\boldsymbol{\sigma}$. Secondly, once $\boldsymbol{\sigma}$ is hidden with probability $1 - o(1)$ over choice of I, $\boldsymbol{y} = \mathsf{Eval}(I, \boldsymbol{\sigma})$ is computationally indistinguishable to a random string $\boldsymbol{r}$. We observe that $\mathsf{flag} = 1$, with all but $2^{-n^{\Omega(1)}}$ probability, by the random choice of hypergraph underlying the PRG. We thus have (proof omitted, see [68] for details):

Lemma 3.2. *In the* sPRG *construction,* $\Pr[\mathsf{flag} = 1] = 1 - 2^{-n^{\Omega(1)}}$.

We now list a few hybrid experiments, $\mathsf{H}_0, \mathsf{H}_1, \mathsf{H}_2$, and H_3, where the first corresponds to the first distribution in the proposition, and the last corresponds to the second distribution in the proposition. We abuse notation to also use H_i to denote the output distribution of the hybrid.

Hybrid H_0 samples $(I', P, \boldsymbol{y})$ honestly as in the first distribution, that is,

$$\begin{aligned}
\text{Sample:}\quad & \boldsymbol{A} \leftarrow \mathbb{Z}_p^{\ell \times n}, \boldsymbol{s} \leftarrow \mathbb{Z}_p^{1 \times \ell}, \boldsymbol{e} \leftarrow \mathcal{D}_r^{1 \times n}(p), \boldsymbol{\sigma} \leftarrow \{0,1\}^n \\
& \boldsymbol{b} = \boldsymbol{s}\boldsymbol{A} + \boldsymbol{e} + \boldsymbol{\sigma}, I \leftarrow \mathsf{IdSamp}(1^n), \boldsymbol{y} = \mathsf{Eval}_I(\boldsymbol{\sigma}) \\
\text{Output:}\quad & I, \boldsymbol{\phi}, \boldsymbol{A}, P = \mathsf{flag} \cdot \boldsymbol{b}, \mathsf{flag} \cdot \boldsymbol{y} \\
& \text{where } \mathsf{flag} = 1 \text{ iff:} \\
& \forall i \in [B], \left|\phi_{\mathsf{bkt}}^{-1}(i) \cap \mathsf{BAD}\right| \leq \ell^\rho
\end{aligned}$$

Hybrid H_1 computes the distribution as before, except that we set $\mathsf{flag} = 1$:

$$\begin{aligned}\text{Sample:}\quad & \boldsymbol{A} \leftarrow \mathbb{Z}_p^{\ell\times n}, \boldsymbol{s} \leftarrow \mathbb{Z}_p^{1\times\ell}, \boldsymbol{e} \leftarrow \mathcal{D}_r^{1\times n}(p), \boldsymbol{\sigma} \leftarrow \{0,1\}^n \\ & \boldsymbol{b} = \boldsymbol{s}\boldsymbol{A} + \boldsymbol{e} + \boldsymbol{\sigma}, I \leftarrow \mathsf{IdSamp}(1^n), \boldsymbol{y} = \mathsf{Eval}_I(\boldsymbol{\sigma}) \\ \text{Output:}\quad & I, \boldsymbol{\phi}, \boldsymbol{A}, P = \boldsymbol{b}, \boldsymbol{y}\end{aligned}$$

Note that Hybrid H_0 and Hybrid H_1 are statistically indistinguishable with statistical distance $2^{-n^{\Omega(1)}}$ due to Lemma 3.2.

Hybrid H_2 computes $\boldsymbol{b}$ by sampling it as a random vector over $\mathbb{Z}_p$:

$$\begin{aligned}\text{Sample:}\quad & \boldsymbol{A} \leftarrow \mathbb{Z}_p^{\ell\times n}, \boldsymbol{\sigma} \leftarrow \{0,1\}^n \\ & \boldsymbol{b} \leftarrow \mathbb{Z}_p^n, I \leftarrow \mathsf{IdSamp}(1^n), \boldsymbol{y} = \mathsf{Eval}_I(\boldsymbol{\sigma}) \\ \text{Output:}\quad & I, \boldsymbol{\phi}, \boldsymbol{A}, P = \boldsymbol{b}, \boldsymbol{y}\end{aligned}$$

Note that Hybrid H_1 and Hybrid H_2 are computationally indistinguishable due to the security of the LPN assumption. The only difference between the hybrids is how $\boldsymbol{b}$ is generated. In Hybrid H_1, $\boldsymbol{b}$ is generated by sampling $\boldsymbol{s}$ and computing $\boldsymbol{b} = \boldsymbol{s}\boldsymbol{A} + \boldsymbol{e} + \boldsymbol{\sigma}$ where $\boldsymbol{e}$ is generated using LPN error distribution, and in Hybrid H_2 it is generated by sampling $\boldsymbol{b}$ by first sampling a uniform vector $\boldsymbol{u}$ and then adding $\boldsymbol{\sigma}$ (which is equivalent to just sampling $\boldsymbol{b}$ uniformly). Note that $\boldsymbol{e}$ and $\boldsymbol{s}$ appear nowhere else in the hybrids. Thus, relying on a straightforward reduction, one can reduce indistinguishability of these hybrids to the security of LPN. Further, if LPN is subexponentially secure, then these hybrids are subexponentially indistinguishable.

Hybrid H_3 simply replaces $\boldsymbol{y}$ by sampling it as a random vector in $\{0,1\}^m$:

$$\begin{aligned}\text{Sample:}\quad & \boldsymbol{A} \leftarrow \mathbb{Z}_p^{\ell\times n} \\ & \boldsymbol{b} \leftarrow \mathbb{Z}_p^n, I \leftarrow \mathsf{IdSamp}(1^n), \boldsymbol{y} \leftarrow \{0,1\}^m \\ \text{Output:}\quad & I, \boldsymbol{\phi}, \boldsymbol{A}, P = \boldsymbol{b}, \boldsymbol{y}\end{aligned}$$

We now show the following claim:

Claim 3.2. *Assuming* PRG *is a secure Goldreich's PRG, then, with probability* $1 - o(1)$ *over* I, *for any probabilistic polynomial time adversary* $\mathcal{A}$,

$$\left|\Pr\left[\mathcal{A}(\mathsf{H}_2) = 1\right] - \Pr\left[\mathcal{A}(\mathsf{H}_3) = 1\right]\right| \le \varepsilon_{\mathsf{PRG}}(n),$$

where $\varepsilon_{\mathsf{PRG}}$ *is the distinguishing advantage of the* PRG.

We prove it by contradiction. Assume that for $1 - \Omega(1)$ probability over the choice of I,

$$\left|\Pr\left[\mathcal{A}(\mathsf{H}_2) = 1\right] - \Pr\left[\mathcal{A}(\mathsf{H}_3) = 1\right]\right| > \varepsilon_{\mathsf{PRG}}(n).$$

We will show that if this happens then there exists a polynomial time distinguisher D, for which with probability $1 - \Omega(1)$ over $I \leftarrow \mathsf{IdSamp}(1^n)$,

$$\begin{aligned} \big| \Pr\big[D\big(I, \mathsf{Eval}(I, \boldsymbol{\sigma} \leftarrow \{0,1\}^n)\big) = 1\big] \\ - \Pr\big[D\big(I, \mathsf{Eval}\big(I, \boldsymbol{r} \leftarrow \{0,1\}^m\big) = 1\big]\big| > \varepsilon_{\mathsf{PRG}}(n), \end{aligned}$$

thereby breaking the PRG security. We show this by building D as a reduction relying on $\mathcal{A}$. The reduction gets as input an index I, $\boldsymbol{y}$ from the PRG challenger, and samples $\boldsymbol{A} \leftarrow \mathbb{Z}_p^{n \times \ell}$ and $\boldsymbol{b} \leftarrow \mathbb{Z}_p^n$. It sends to $\mathcal{A}$ the input $(I, \boldsymbol{\phi}, \boldsymbol{A}, P = \boldsymbol{b}, \boldsymbol{y})$ and outputs whatever it outputs. Note that the view of D is identical to the adversary for the PRG game. For $\mathcal{A}$, if $\boldsymbol{y}$ is generated using PRG.Eval, then its view is identical to Hybrid H_2, otherwise it is identical to Hybrid H_3. Therefore, if $\mathcal{A}$ manages to distinguish between the hybrids with probability more than $\varepsilon_{\mathsf{PRG}}$ over $1 - o(1)$ choice of I, then D will also be able to win the PRG game security with probability more than $\varepsilon_{\mathsf{PRG}}$. Thus, the claim follows.

FUNDING

Aayush Jain was partially supported by grants listed under Amit Sahai, and a Google PhD fellowship. Huijia Lin was supported by NSF grants CNS-1528178, CNS-1929901, CNS-1936825 (CAREER), CNS-2026774, a Hellman Fellowship, a JP Morgan AI research Award, DARPA and ARO Contract No. W911NF-15-C-0236, and a subcontract No. 2017-002 through Galois. Amit Sahai was supported in part from DARPA SAFEWARE and SIEVE awards, NTT Research, NSF Frontier Award 1413955, and NSF grant 1619348, BSF grant 2012378, a Xerox Faculty Research Award, a Google Faculty Research Award, and an Okawa Foundation Research Grant. This material is based upon work supported by the Defense Advanced Research Projects Agency through Award HR00112020024 and the ARL under Contract W911NF-15-C- 0205. The views expressed are those of the authors and do not reflect the official policy or position of the Department of Defense, DARPA, ARO, Simons, Intel, Okawa Foundation, ODNI, IARPA, DIMACS, BSF, Xerox, the National Science Foundation, NTT Research, Google, or the US Government.

REFERENCES

[1] S. Agrawal, Indistinguishability obfuscation without multilinear maps: New methods for bootstrapping and instantiation. In *Eurocrypt 2019, part I*, edited by Y. Ishai and V. Rijmen, pp. 191–225, Lecture Notes in Comput. Sci. 11476, Springer, Heidelberg, 2019.

[2] S. Agrawal and A. Pellet-Mary, Indistinguishability obfuscation without maps: Attacks and fixes for noisy linear FE. In *Eurocrypt 2020, part I*, edited by V. Rijmen and Y. Ishai, pp. 110–140, Lecture Notes in Comput. Sci., Springer, Heidelberg, 2020.

[3] M. R. Albrecht, P. Farshim, D. Hofheinz, E. Larraia, and K. G. Paterson, Multilinear maps from obfuscation. In *TCC 2016-a, part I*, edited by E. Kushilevitz and T. Malkin, pp. 446–473, Lecture Notes in Comput. Sci. 9562, Springer, Heidelberg, 2016.

[4] M. Alekhnovich, More on average case vs approximation complexity. In *44th FOCS*, pp. 298–307, IEEE Computer Society Press, 2003.

[5] P. Ananth and A. Jain, Indistinguishability obfuscation from compact functional encryption. In *Crypto 2015, part I*, edited by R. Gennaro and M. J. B. Robshaw, pp. 308–326, Lecture Notes in Comput. Sci. 9215, Springer, Heidelberg, 2015.

[6] P. Ananth, A. Jain, H. Lin, C. Matt, and A. Sahai, Indistinguishability obfuscation without multilinear maps: New paradigms via low degree weak pseudorandomness and security amplification. In *Crypto 2019, part III*, edited by A. Boldyreva and D. Micciancio, pp. 284–332, Lecture Notes in Comput. Sci. 11694, Springer, Heidelberg, 2019.

[7] P. Ananth, A. Jain, and A. Sahai, Indistinguishability obfuscation from functional encryption for simple functions. *IACR Cryptol. ePrint Arch.* **730** (2015), 2015.

[8] P. Ananth, A. Jain, and A. Sahai, Indistinguishability obfuscation without multilinear maps: iO from LWE, bilinear maps, and weak pseudorandomness. *IACR Cryptol. ePrint Arch.* **2018** (2018), 615.

[9] P. Ananth and A. Sahai, Projective arithmetic functional encryption and indistinguishability obfuscation from degree-5 multilinear maps. In *Eurocrypt 2017, part I*, edited by J. Coron and J. B. Nielsen, pp. 152–181, Lecture Notes in Comput. Sci. 10210, Springer, Heidelberg, 2017.

[10] B. Applebaum, Pseudorandom generators with long stretch and low locality from random local one-way functions. *SIAM J. Comput.* **42** (2013), no. 5, 2008–2037.

[11] B. Applebaum, J. Avron, and C. Brzuska, Arithmetic cryptography: extended abstract. In *ITCS 2015*, edited by T. Roughgarden, pp. 143–151, ACM Press, 2015.

[12] B. Applebaum, B. Barak, and A. Wigderson, Public-key cryptography from different assumptions. In *42nd ACM STOC*, edited by L. J. Schulman, pp. 171–180, ACM Press, 2010.

[13] B. Applebaum, A. Bogdanov, and A. Rosen, A dichotomy for local small-bias generators. In *TCC 2012*, edited by R. Cramer, pp. 600–617, Lecture Notes in Comput. Sci. 7194, Springer, Heidelberg, 2012.

[14] B. Applebaum, I. Damgård, Y. Ishai, M. Nielsen, and L. Zichron, Secure arithmetic computation with constant computational overhead. In *Crypto 2017, part I*, edited by J. Katz and H. Shacham, pp. 223–254, Lecture Notes in Comput. Sci. 10401, Springer, Heidelberg, 2017.

[15] B. Applebaum, Y. Ishai, and E. Kushilevitz, Cryptography in NC^0. In *FOCS*, pp. 166–175, IEEE Computer Society, 2004.

[16] B. Applebaum and E. Kachlon, Sampling graphs without forbidden subgraphs and unbalanced expanders with negligible error. In *60th FOCS*, edited by D. Zuckerman, pp. 171–179, IEEE Computer Society Press, 2019.

[17] B. Applebaum and S. Lovett, Algebraic attacks against random local functions and their countermeasures. In *48th ACM STOC*, edited by D. Wichs and Y. Mansour, pp. 1087–1100, ACM Press, 2016.

[18] S. Badrinarayanan, E. Miles, A. Sahai, and M. Zhandry, Post-zeroizing obfuscation: new mathematical tools, and the case of evasive circuits. In *Advances in Cryptology – EUROCRYPT 2016 – 35th Annual International Conference on the Theory and Applications of Cryptographic Techniques, Vienna, Austria, May 8–12, 2016, Proceedings, Part II*, pp. 764–791, Lecture Notes in Comput. Sci. 9666, Springer, 2016.

[19] B. Barak, Z. Brakerski, I. Komargodski, and P. K. Kothari, Limits on low-degree pseudorandom generators (or: Sum-of-squares meets program obfuscation). In *Eurocrypt 2018, part II*, edited by J. B. Nielsen and V. Rijmen, pp. 649–679, Lecture Notes in Comput. Sci. 10821, Springer, Heidelberg, 2018.

[20] B. Barak, S. Garg, Y. T. Kalai, O. Paneth, and A. Sahai, Protecting obfuscation against algebraic attacks. In *Eurocrypt 2014*, edited by P. Q. Nguyen and E. Oswald, pp. 221–238, Lecture Notes in Comput. Sci. 8441, Springer, Heidelberg, 2014.

[21] B. Barak, S. B. Hopkins, A. Jain, P. Kothari, and A. Sahai, Sum-of-squares meets program obfuscation, revisited. In *Eurocrypt 2019, part I*, edited by Y. Ishai and V. Rijmen, pp. 226–250, Lecture Notes in Comput. Sci. 11476, Springer, Heidelberg, 2019.

[22] B. Barak, O. Goldreich, R. Impagliazzo, S. Rudich, A. Sahai, S. P. Vadhan, and K. Yang, On the (im)possibility of obfuscating programs. In *Crypto 2001*, edited by J. Kilian, pp. 1–18, Lecture Notes in Comput. Sci. 2139, Springer, Heidelberg, 2001.

[23] J. Bartusek, Y. Ishai, A. Jain, F. Ma, A. Sahai, and M. Zhandry, Affine determinant programs: a framework for obfuscation and witness encryption. In *ITCS 2020*, edited by T. Vidick, LIPIcs. Leibniz Int. Proc. Inform. 151, pp. 82:1–82:39, 2020.

[24] J. Bartusek and G. Malavolta, Candidate obfuscation of null quantum circuits and witness encryption for QMA. *IACR Cryptol. ePrint Arch.* **2021** (2021), 421.

[25] A. Bishop, A. Jain, and L. Kowalczyk, Function-hiding inner product encryption. In *Asiacrypt 2015, part I*, edited by T. Iwata and J. H. Cheon, pp. 470–491, Lecture Notes in Comput. Sci. 9452, Springer, Heidelberg, 2015.

[26] N. Bitansky, R. Nishimaki, A. Passelègue, and D. Wichs, From cryptomania to obfustopia through secret-key functional encryption. In *TCC 2016-b, part II*, edited by M. Hirt and A. D. Smith, pp. 391–418, Lecture Notes in Comput. Sci. 9986, Springer, Heidelberg, 2016.

[27] N. Bitansky, O. Paneth, and A. Rosen, On the cryptographic hardness of finding a Nash equilibrium. In *56th FOCS*, edited by V. Guruswami, pp. 1480–1498, IEEE Computer Society Press, 2015.

[28] N. Bitansky and V. Vaikuntanathan, Indistinguishability obfuscation from functional encryption. In *56th FOCS*, edited by V. Guruswami, pp. 171–190, IEEE Computer Society Press, 2015.

[29] A. Blum, M. L. Furst, M. J. Kearns, and R. J. Lipton, Cryptographic primitives based on hard learning problems. In *Crypto'93*, edited by D. R. Stinson, pp. 278–291, Lecture Notes in Comput. Sci. 773, Springer, Heidelberg, 1994.

[30] A. Bogdanov and Y. Qiao, On the security of Goldreich's one-way function. *Comput. Complexity* **21** (2012), no. 1, 83–127.

[31] D. Boneh, X. Boyen, and H. Shacham, Short group signatures. In *Crypto 2004*, edited by M. Franklin, pp. 41–55, Lecture Notes in Comput. Sci. 3152, Springer, Heidelberg, 2004.

[32] D. Boneh, A. Sahai, and B. Waters, Functional encryption: definitions and challenges. In *TCC 2011*, edited by Y. Ishai, pp. 253–273, Lecture Notes in Comput. Sci. 6597, Springer, Heidelberg, 2011.

[33] D. Boneh and M. Zhandry, Multiparty key exchange, efficient traitor tracing, and more from indistinguishability obfuscation. In *Crypto 2014, part I*, edited by J. A. Garay and R. Gennaro, pp. 480–499, Lecture Notes in Comput. Sci. 8616, Springer, Heidelberg, 2014.

[34] R. B. Boppana, J. Håstad, and S. Zachos, Does co-NP have short interactive proofs? *Inform. Process. Lett.* **25** (1987), no. 2, 127–132.

[35] E. Boyle, G. Couteau, N. Gilboa, and Y. Ishai, Compressing vector OLE. In *ACM CCS 2018*, edited by D. Lie, M. Mannan, M. Backes, and X. Wang, pp. 896–912, ACM Press, 2018.

[36] E. Boyle, G. Couteau, N. Gilboa, Y. Ishai, L. Kohl, and P. Scholl, Correlated pseudorandom functions from variable-density LPN In *61st IEEE Annual Symposium on Foundations of Computer Science, FOCS 2020, Durham, NC, USA, November 16–19, 2020*, pp. 1069–1080, IEEE, 2020.

[37] E. Boyle, G. Couteau, N. Gilboa, Y. Ishai, L. Kohl, P. Rindal, and P. Scholl, Efficient two-round OT extension and silent non-interactive secure computation. In *ACM CCS 2019*, edited by L. Cavallaro, J. Kinder, X. Wang, and J. Katz, pp. 291–308, ACM Press, 2019.

[38] Z. Brakerski, Y. T. Kalai, J. Katz, and V. Vaikuntanathan, Overcoming the hole in the bucket: public-key cryptography resilient to continual memory leakage. In *51st FOCS*, pp. 501–510, IEEE Computer Society Press, 2010.

[39] Z. Brakerski and G. N. Rothblum, Virtual black-box obfuscation for all circuits via generic graded encoding. In *Theory of Cryptography – 11th Theory of Cryptography Conference, TCC 2014, San Diego, CA, USA, February 24–26, 2014. Proceedings*, pp. 1–25, Lecture Notes in Comput. Sci. 8349, Springer, 2014.

[40] A. Broadbent and R. A. Kazmi, Indistinguishability obfuscation for quantum circuits of low t-count. *IACR Cryptol. ePrint Arch.* **2020** (2020), 639.

[41] R. Canetti, H. Lin, S. Tessaro, and V. Vaikuntanathan, Obfuscation of probabilistic circuits and applications. In *TCC 2015, part II*, edited by Y. Dodis and J. B. Nielsen, pp. 468–497, Lecture Notes in Comput. Sci. 9015, Springer, Heidelberg, 2015.

[42] J. H. Cheon, K. Han, C. Lee, H. Ryu, and D. Stehlé, Cryptanalysis of the multilinear map over the integers. In *Advances in Cryptology – EUROCRYPT 2015 – 34th Annual International Conference on the Theory and Applications of Cryptographic Techniques, Sofia, Bulgaria, April 26–30, 2015, Proceedings, Part I*, pp. 3–12, Lecture Notes in Comput. Sci. 9056, Springer, 2015.

[43] J. H. Cheon, C. Lee, and H. Ryu, Cryptanalysis of the new clt multilinear maps. Cryptology ePrint Archive, Report 2015/934, 2015. http://eprint.iacr.org/.

[44] WhibOx Contest 2021—CHES 2021 Challenge. 2021, https://whibox.io/contests/2021/.

[45] J. Cook, O. Etesami, R. Miller, and L. Trevisan, Goldreich's one-way function candidate and myopic backtracking algorithms. In *TCC 2009*, edited by O. Reingold, pp. 521–538, Lecture Notes in Comput. Sci. 5444, Springer, Heidelberg, 2009.

[46] S. A. Cook, The complexity of theorem-proving procedures. In *Proceedings of the 3rd annual ACM symposium on theory of computing, May 3–5, 1971, Shaker Heights, Ohio, USA*, edited by M. A. Harrison, R. B. Banerji, and J. D. Ullman, pp. 151–158, ACM Press, 1971.

[47] J.-S. Coron, T. Lepoint, and M. Tibouchi, Practical multilinear maps over the integers. In *Crypto 2013, part I*, edited by R. Canetti and J. A. Garay, pp. 476–493, Lecture Notes in Comput. Sci. 8042, Springer, Heidelberg, 2013.

[48] G. Couteau, A. Dupin, P. Méaux, M. Rossi, and Y. Rotella, On the concrete security of Goldreich's pseudorandom generator. In *Asiacrypt 2018, part II*, edited by T. Peyrin and S. Galbraith, pp. 96–124, Lecture Notes in Comput. Sci. 11273, Springer, Heidelberg, 2018.

[49] M. Cryan and P. B. Miltersen, On pseudorandom generators in NC. In *MFCS 2001*, pp. 272–284, Lecture Notes in Comput. Sci. 2136, Springer, 2001.

[50] W. Diffie and M. E. Hellman, New directions in cryptography. *IEEE Trans. Inf. Theory* **22** (1976), no. 6, 644–654.

[51] N. Döttling, S. Garg, D. Gupta, P. Miao, and P. Mukherjee, Obfuscation from low noise multilinear maps. *IACR Cryptol. ePrint Arch.* **2016** (2016), 599.

[52] N. Döttling, S. Ghosh, J. B. Nielsen, T. Nilges, and R. Trifiletti, TinyOLE: efficient actively secure two-party computation from oblivious linear function evaluation. In *ACM CCS 2017*, edited by B. M. Thuraisingham, D. Evans, T. Malkin, and D. Xu, pp. 2263–2276, ACM Press, 2017.

[53] P. Farshim, J. Hesse, D. Hofheinz, and E. Larraia, Graded encoding schemes from obfuscation. In *PKC 2018, part II*, edited by M. Abdalla and R. Dahab, pp. 371–400, Lecture Notes in Comput. Sci. 10770, Springer, Heidelberg, 2018.

[54] S. Garg, C. Gentry, and S. Halevi, Candidate multilinear maps from ideal lattices. In *Eurocrypt 2013*, edited by T. Johansson and P. Q. Nguyen, pp. 1–17, Lecture Notes in Comput. Sci. 7881, Springer, Heidelberg, 2013.

[55] S. Garg, C. Gentry, S. Halevi, M. Raykova, A. Sahai, and B. Waters, Candidate indistinguishability obfuscation and functional encryption for all circuits. In *54th FOCS*, pp. 40–49, IEEE Computer Society Press, 2013.

[56] R. Gay, A. Jain, H. Lin, and A. Sahai, Indistinguishability obfuscation from simple-to-state hard problems: New assumptions, new techniques, and simplification. In *EUROCRYPT 2021*, pp. 97–126, Lecture Notes in Comput. Sci. 12698, Springer, 2021.

[57] C. Gentry, S. Gorbunov, and S. Halevi, Graph-induced multilinear maps from lattices. In *TCC 2015, part II*, edited by Y. Dodis and J. B. Nielsen, pp. 498–527, Lecture Notes in Comput. Sci. 9015, Springer, Heidelberg, 2015.

[58] C. Gentry, C. S. Jutla, and D. Kane, Obfuscation using tensor products. *Electron. Colloq. Comput. Complex.* **25** (2018), 149.

[59] S. Ghosh, J. B. Nielsen, and T. Nilges, Maliciously secure oblivious linear function evaluation with constant overhead. In *Asiacrypt 2017, part I*, edited by T. Takagi and T. Peyrin, pp. 629–659, Lecture Notes in Comput. Sci. 10624, Springer, Heidelberg, 2017.

[60] E. N. Gilbert, A comparison of signalling alphabets. *Bell Syst. Tech. J.* **31** (1952), no. 3, 504–522.

[61] O. Goldreich, Candidate one-way functions based on expander graphs. *Electron. Colloq. Comput. Complex.* **7** (2000), no. 90.

[62] J. Groth and A. Sahai, Efficient non-interactive proof systems for bilinear groups. In *Eurocrypt 2008*, edited by N. P. Smart, pp. 415–432, Lecture Notes in Comput. Sci. 4965, Springer, Heidelberg, 2008.

[63] Y. Hu and H. Jia, Cryptanalysis of GGH map. *IACR Cryptol. ePrint Arch.* **2015** (2015), 301.

[64] Y. Ishai and E. Kushilevitz, Perfect constant-round secure computation via perfect randomizing polynomials. In *ICALP*, pp. 244–256, Lecture Notes in Comput. Sci. 2380, Springer, 2002.

[65] Y. Ishai, E. Kushilevitz, R. Ostrovsky, and A. Sahai, Cryptography with constant computational overhead. In *40th ACM STOC*, edited by R. E. Ladner and C. Dwork, pp. 433–442, ACM Press, 2008.

[66] Y. Ishai, M. Prabhakaran, and A. Sahai, Founding cryptography on oblivious transfer—efficiently. In *Crypto 2008*, edited by D. Wagner, pp. 572–591, Lecture Notes in Comput. Sci. 5157, Springer, Heidelberg, 2008.

[67] A. Jain, H. Lin, C. Matt, and A. Sahai, How to leverage hardness of constant-degree expanding polynomials over $\mathbb{R}$ to build $i\mathcal{O}$. In *Eurocrypt 2019, part I*, edited by Y. Ishai and V. Rijmen, pp. 251–281, Lecture Notes in Comput. Sci. 11476, Springer, Heidelberg, 2019.

[68] A. Jain, H. Lin, and A. Sahai, Indistinguishability obfuscation from well-founded assumptions. In *STOC'21: 53rd annual ACM SIGACT symposium on theory of computing, virtual event, Italy, June 21–25, 2021*, edited by S. Khuller and V. V. Williams, pp. 60–73, ACM Press, 2021.

[69] A. Jain, H. Lin, and A. Sahai, Indistinguishability obfuscation without lattices. Unpublished manuscript, 2021.

[70] D. Khurana, V. Rao, and A. Sahai, Multi-party key exchange for unbounded parties from indistinguishability obfuscation. In *Asiacrypt 2015, part I*, edited by T. Iwata and J. H. Cheon, pp. 52–75, Lecture Notes in Comput. Sci. 9452, Springer, Heidelberg, 2015.

[71] I. Komargodski, T. Moran, M. Naor, R. Pass, A. Rosen, and E. Yogev, One-way functions and (im)perfect obfuscation. In *55th FOCS*, pp. 374–383, IEEE Computer Society Press, 2014.

[72] I. Komargodski, M. Naor, and E. Yogev, Secret-sharing for NP. In *Asiacrypt 2014, part II*, edited by P. Sarkar and T. Iwata, pp. 254–273, Lecture Notes in Comput. Sci. 8874, Springer, Heidelberg, 2014.

[73] P. K. Kothari, R. Mori, R. O'Donnell, and D. Witmer, Sum of squares lower bounds for refuting any CSP. In *49th ACM STOC*, edited by H. Hatami, P. McKenzie, and V. King, pp. 132–145, ACM Press, 2017.

[74] H. Lin, Indistinguishability obfuscation from constant-degree graded encoding schemes. In *Eurocrypt 2016, part I*, edited by M. Fischlin and J.-S. Coron, pp. 28–57, Lecture Notes in Comput. Sci. 9665, Springer, Heidelberg, 2016.

[75] H. Lin, Indistinguishability obfuscation from SXDH on 5-linear maps and locality-5 PRGs. In *Crypto 2017, part I*, edited by J. Katz and H. Shacham, pp. 599–629, Lecture Notes in Comput. Sci. 10401, Springer, Heidelberg, 2017.

[76] H. Lin and C. Matt, Pseudo flawed-smudging generators and their application to indistinguishability obfuscation. *IACR Cryptol. ePrint Arch.* **2018** (2018), 646.

[77] H. Lin, R. Pass, K. Seth, and S. Telang, Indistinguishability obfuscation with non-trivial efficiency. In *PKC 2016, part II*, edited by C.-M. Cheng, K.-M. Chung, G. Persiano, and B.-Y. Yang, pp. 447–462, Lecture Notes in Comput. Sci. 9615, Springer, Heidelberg, 2016.

[78] H. Lin, R. Pass, K. Seth, and S. Telang, Output-compressing randomized encodings and applications. In *TCC 2016-a, part I*, edited by E. Kushilevitz and T. Malkin, pp. 96–124, Lecture Notes in Comput. Sci. 9562, Springer, Heidelberg, 2016.

[79] H. Lin and S. Tessaro, Indistinguishability obfuscation from trilinear maps and block-wise local PRGs. In *Crypto 2017, part I*, edited by J. Katz and H. Shacham, pp. 630–660, Lecture Notes in Comput. Sci. 10401, Springer, Heidelberg, 2017.

[80] H. Lin and V. Vaikuntanathan, Indistinguishability obfuscation from DDH-like assumptions on constant-degree graded encodings. In *57th FOCS*, edited by I. Dinur, pp. 11–20, IEEE Computer Society Press, 2016.

[81] A. Lombardi and V. Vaikuntanathan, Limits on the locality of pseudorandom generators and applications to indistinguishability obfuscation. In *TCC 2017, part I*, edited by Y. Kalai and L. Reyzin, pp. 119–137, Lecture Notes in Comput. Sci. 10677, Springer, Heidelberg, 2017.

[82] D. Micciancio and P. Mol, Pseudorandom knapsacks and the sample complexity of LWE search-to-decision reductions. In *Crypto 2011*, edited by P. Rogaway, pp. 465–484, Lecture Notes in Comput. Sci. 6841, Springer, Heidelberg, 2011.

[83] D. Micciancio and C. Peikert, Hardness of SIS and LWE with small parameters. In *Crypto 2013, part I*, edited by R. Canetti and J. A. Garay, pp. 21–39, Lecture Notes in Comput. Sci. 8042, Springer, Heidelberg, 2013.

[84] E. Miles, A. Sahai, and M. Zhandry, Annihilation attacks for multilinear maps: Cryptanalysis of indistinguishability obfuscation over GGH13. In *Advances in cryptology—CRYPTO*, 2016.

[85] B. Minaud and P.-A. Fouque, Cryptanalysis of the new multilinear map over the integers. Cryptology ePrint Archive, Report 2015/941, 2015. http://eprint.iacr.org/.

[86] E. Mossel, A. Shpilka, and L. Trevisan, On e-biased generators in NC^0. In *44th FOCS*, pp. 136–145, IEEE Computer Society Press, 2003.

[87] R. O'Donnell and D. Witmer, Goldreich's PRG: evidence for near-optimal polynomial stretch. In *IEEE 29th conference on computational complexity, CCC 2014, Vancouver, BC, Canada, June 11–13, 2014*, pp. 1–12, IEEE Computer Society, 2014.

[88] A. O'Neill, Definitional issues in functional encryption. *IACR Cryptol. ePrint Arch.* **2010** (2010), 556.

[89] T. Okamoto and K. Takashima, Fully secure functional encryption with general relations from the decisional linear assumption. In *Crypto 2010*, edited by T. Rabin, pp. 191–208, Lecture Notes in Comput. Sci. 6223, Springer, Heidelberg, 2010.

[90] R. Pass, K. Seth, and S. Telang, Indistinguishability obfuscation from semantically-secure multilinear encodings. In *Advances in Cryptology – CRYPTO 2014 – 34th Annual Cryptology Conference, Santa Barbara, CA, USA, August 17–21, 2014, Proceedings, Part I*, pp. 500–517, Lecture Notes in Comput. Sci. 8616, Springer, 2014.

[91] O. Regev, On lattices, learning with errors, random linear codes, and cryptography. In *Proceedings of the 37th Annual ACM Symposium on Theory of Computing, Baltimore, MD, USA, May 22–24, 2005*, pp. 84–93, ACM, 2005.

[92] A. Sahai and B. Waters, How to use indistinguishability obfuscation: deniable encryption, and more. In *46th ACM STOC*, edited by D. B. Shmoys, pp. 475–484, ACM Press, 2014.

[93] A. Sahai and B. R. Waters, Fuzzy identity-based encryption. In *Eurocrypt 2005*, edited by R. Cramer, pp. 457–473, Lecture Notes in Comput. Sci. 3494, Springer, Heidelberg, 2005.

[94] V. Vaikunthananthan, Cs 294 lecture 4: worst-case to average-case reductions for LWE. Class at UC Berkeley, 2020, http://people.csail.mit.edu/vinodv/CS294/lecture4.pdf.

[95] R. Varshamov, Estimate of the number of signals in error correcting codes. *Dokl. Akad. Nauk SSSR* **117** (1957), 739–741.

[96] H. Wee, Functional encryption for quadratic functions from k-LIN, revisited. In *Theory of Cryptography – 18th International Conference, TCC 2020, Durham, NC, USA, November 16–19, 2020, Proceedings, Part I*, pp. 210–228, Lecture Notes in Comput. Sci. 12550, Springer, 2020.

[97] T. Yamakawa, S. Yamada, G. Hanaoka, and N. Kunihiro, Self-bilinear map on unknown order groups from indistinguishability obfuscation and its applications. In *Crypto 2014, part II*, edited by J. A. Garay and R. Gennaro, pp. 90–107, Lecture Notes in Comput. Sci. 8617, Springer, Heidelberg, 2014.

[98] A. C.-C. Yao, How to generate and exchange secrets (extended abstract). In *27th Annual Symposium on Foundations of Computer Science, Toronto, Canada, 27–29 October 1986*, pp. 162–167, IEEE Computer Society, 1986.

AAYUSH JAIN

UCLA, Los Angeles, CA, USA; and NTT Research, Sunnyvale, CA, USA, aayushjain1728@gmail.com

HUIJIA LIN

University of Washington, Seattle, WA, USA, rachel@cs.washington.edu

AMIT SAHAI

UCLA, Los Angeles, CA, USA, sahai@cs.ucla.edu

This Page Intentionally Left Blank

SIMULATION-BASED SEARCH

DAVID SILVER AND ANDRE BARRETO

ABSTRACT

Planning is one of the oldest and most important problems in artificial intelligence. Simulation-based search algorithms, such as AlphaZero, have achieved superhuman performance in chess and Go and are used widely in real-world applications of planning. In this paper we provide a unified framework for simulation-based search. Algorithms in this framework interleave operators for policy evaluation (better estimating the value function of the current policy) and policy improvement (using the value function to form a better policy). These operators are applied to states and actions that are sampled in sequential trajectories, and that may branch recursively into other sampled trajectories. The value function and policy may also be represented by a function approximator. Our framework includes a broad family of search algorithms that includes Monte-Carlo tree search, sparse sampling, nested Monte-Carlo search, classification-based policy iteration, and AlphaZero.

MATHEMATICS SUBJECT CLASSIFICATION 2020

Primary 68T05; Secondary 68T20, 68T37, 90C39, 90C40

KEYWORDS

Planning, reinforcement learning, Markov decision processes, Monte-Carlo tree search, Monte-Carlo simulation

ICM INTERNATIONAL CONGRESS OF MATHEMATICIANS 2022 JULY 6–14

Proc. Int. Cong. Math. 2022, Vol. 6, pp. 4800–4819
DOI 10.4171/ICM2022/180

1. INTRODUCTION

One of the oldest and most important problems in artificial intelligence is to select an action by looking ahead. Such planning methods are ubiquitous across several decades of artificial intelligence research, and have achieved superhuman level performance in chess [9] and Go [34], as well as contributing to real-world applications such as process control [15], robotics [24], and logistics [26].

Planning algorithms may be described by successive applications of operators that propagate information from subsequent states back to a previous state. The nature of planning is determined both by the nature of those operators, and by the order in which they are applied. A large family of planning algorithms may be understood as instances of *generalized policy iteration* [39]. In these algorithms, operators alternate between policy evaluation (better estimating the value function of the current policy) and policy improvement (using the value function to form a better policy). If operators of both types are applied repeatedly to all states, this procedure will result in the optimal value function and an optimal policy for any Markov decision process.

The precise order in which operators are applied, known as the *search control* method, has a significant impact on the efficiency of the algorithm. While many approaches to search control exist, we focus on simulation-based search. In this approach, actions and state transitions are sampled in sequential trajectories; this allows simulation-based search algorithms to look many steps ahead. Simulation-based search algorithms such as Monte-Carlo tree search [12] have been successful in large and complex planning problems such as the game of Go. We also consider recursive simulation algorithms that branch into many child simulations before backtracking to the parent. This ensures that the dependencies of an operator are accurately computed, by recursive simulation, before that operator is applied.

Many complex problems are intractable to exact solution. The state space may be too large to explicitly represent all states. In this case, the value function or policy may be represented by a function approximator such as a neural network. We show that some of today's most powerful planning algorithms, such as AlphaZero [36], can be understood as instances of generalized policy iteration using recursive simulation and function approximation.

The contribution of this paper is a unified understanding of simulation-based search algorithms. Algorithms in this framework are elucidated by three complementary mechanisms. First, equations are provided, based on the application of operators to the joint space of value functions and policies. Second, diagrams are provided, akin to backup diagrams [39], that show the states and actions used by an operator. Third, pseudocode is provided for several key algorithms.

2. RELATED WORK

Extensive literature exists on policy iteration and value iteration methods, e.g., [6,27]. However, there is little discussion of search control in this literature. The relationship of policy and value iteration to AlphaZero is discussed in [5], including an elegant exposition of their relationship to Newton's method. However, search control is not discussed in depth.

Operators that act upon a value function, such as the Bellman operator, are thoroughly analyzed within the dynamic programming literature [6, 27]. Several operators that act upon a policy are introduced in [17]. Generalized operators that unify the treatment of maximizing and minimaximizing operators, among others, are discussed in [23]. The treatment of generalized policy iteration using operators that act jointly upon a value function and policy may be novel to this paper.

Tree-based search algorithms are extensively analyzed in the search literature, e.g., [10]. Simulation-based search algorithms are discussed in [7] and their relationship to reinforcement learning is discussed in [31,32]. Several search algorithms have combined elements of both depth-first search and simulation, e.g., [11,29,32,36] but there has been little prior discussion of the common principles underlying these algorithms.

3. OPERATORS

We consider a discounted Markov decision process (MDP) with a finite state space $\mathcal{S}$ and a finite action space $\mathcal{A}$ [27]. The MDP has reward and transition dynamics $R, S' \sim \varepsilon(s,a)$, where ε is a joint probability distribution over reward $R \in \mathbb{R}$ and next state $S' \in \mathcal{S}$ conditioned on current state $s \in \mathcal{S}$ and action $a \in \mathcal{A}$. The discount factor of the MDP is $\gamma \in [0, 1)$. A policy $\pi : \mathcal{S} \to \Delta^{|\mathcal{A}|-1}$ specifies the probability of selecting each action $a \in \mathcal{S}$ in every state $s \in \mathcal{S}$; here $\Delta^{|\mathcal{A}|-1}$ is the probability simplex of dimension $|\mathcal{A}| - 1$. We will use Π to denote a complete metric space composed of all possible policies.

Let $\mathbb{Q}$ be a complete metric space whose elements are action–value functions $q : \mathcal{S} \times \mathcal{A} \to \mathbb{R}$ that map a state $s \in \mathcal{S}$ and an action $a \in \mathcal{A}$ onto a scalar value. We will henceforth simply refer to action-value functions as value functions. The true value function[1] of a policy π, denoted by $q_\pi \in \mathbb{Q}$, is defined as $q_\pi(s,a) = \mathbb{E}_{\pi,\varepsilon}[\sum_{k=0}^{\infty} \gamma^k R_{t+k+1} | S_t = s, A_t = a]$, where $\mathbb{E}_{\pi,\varepsilon}[\cdot]$ denotes expectation over the Markov process $S_t, A_t, R_{t+1}, S_{t+1}, A_{t+1}, R_{t+2}, \ldots$ where $R_{t+1}, S_{t+1} \sim \varepsilon(S_t, A_t)$ and $A_t \sim \pi(S_t)$ for $t = 1, 2, \ldots$ The optimal value function is defined as $q^*(s,a) = \max_{\pi \in \Pi} q_\pi(s,a)$ for all $(s,a) \in \mathcal{S} \times \mathcal{A}$; it is well known that such a function always exists and is unique [27]. An optimal policy π^* is any policy that achieves the maximum value for all states and actions, that is, any policy whose value function is q^*.

We consider *operators* $\boldsymbol{o} : \mathbb{Q} \times \Pi \to \mathbb{Q} \times \Pi$ that transform functions and policies into other functions and policies. When the same operator is applied multiple times, we use the shorthand $\boldsymbol{o}^n(q, \pi) = (\boldsymbol{o}\boldsymbol{o} \ldots \boldsymbol{o})(q, \pi)$. When composing different operators, we use $(\prod_{i=1}^{n} \boldsymbol{o}_n)(q, \pi)$ to denote the sequence of operators $\boldsymbol{o}_1 \boldsymbol{o}_2 \ldots \boldsymbol{o}_n$ applied to (q, π) from right to left, $(\prod_{i=1}^{n} \boldsymbol{o}_n)(q, \pi) = (\boldsymbol{o}_1 \boldsymbol{o}_2 \ldots \boldsymbol{o}_n)(q, \pi) = \boldsymbol{o}_1(\ldots(\boldsymbol{o}_n(q, \pi)))$.

1 We will refer to any $q \in \mathbb{Q}$ as a value function, and use the term "true" value function to denote the specific value function corresponding to an expected return.

We begin with two primitive operators: *evaluation* operators that update the function q, but leave the policy π unchanged, and *improvement* operators that update π but leave q unchanged. An evaluation operator $\boldsymbol{e}(q, \pi)$ has the property that it moves q closer to the value function of π, such that a sequence of n applications of that operator converges to the true value function of π as n grows, $\lim_{n\to\infty} \boldsymbol{e}^n(q, \pi) = (q_\pi, \pi)$. A canonical example of an evaluation operator would be the one-step Bellman expectation operator, $\boldsymbol{e}_B(q, \pi) = (Bq, \pi)$ where $(Bq)(s, a) = \mathbb{E}_{R,S'\sim\mathcal{E}(s,a),A'\sim\pi(S')}[R + \gamma q(S', A')]$.

An improvement operator $\boldsymbol{i}$ is an operator that, when applied to a policy π and its value function q_π, produces a new policy π' whose value function is at least as large, for all state-action pairs, as that of π: $\boldsymbol{i}(q_\pi, \pi) = (q_\pi, \pi')$ such that $q_{\pi'} \geq q_\pi$ with equality if and only if $q_\pi = q^*$. A canonical example of an improvement operator would be the greedy operator, $\boldsymbol{i}^*(q_\pi, \pi)$, which produces a new policy defined as $\pi'(s) \in \operatorname{argmax}_{a\in\mathcal{A}} q_\pi(s, a)$ for all $s \in \mathcal{S}$; however, many other improvement operators are possible [17].

By alternating evaluation and improvement operators, one can compute an optimal policy for an MDP. For example, the well known value iteration algorithm can be understood as successive applications of the operators $\boldsymbol{e}_B$ and $\boldsymbol{i}^*$; starting from any function q, the sequence $(\boldsymbol{e}_B\boldsymbol{i}^*)^n(q, \cdot)$ approaches (q^*, π^*) as $n \to \infty$. If instead we consider the sequence $(\boldsymbol{i}^*\boldsymbol{e}_B^\infty)^n(\cdot, \pi)$, we obtain the well known policy iteration algorithm, which converges to (q^*, π^*) in a finite number of iterations.

3.1. State and state–action operators

Interesting problems typically contain many state variables (that is, the state space $\mathcal{S}$ is high-dimensional). The size of $\mathcal{S}$ grows exponentially with the number of variables, an issue known as the curse of dimensionality. Consequently, it may be infeasible to update all state-action pairs.

However, it is not necessary to apply operators to the entire state space at once. We will use $\boldsymbol{o}[s]$ or $\boldsymbol{o}[s, a](q, \pi)$ to denote a *state* or *state–action operator* that is specific to that state or state–action. An evaluation operator $(q', \cdot) = \boldsymbol{e}(q, \cdot)$ has a corresponding state–action evaluation operator $(q'', \cdot) = \boldsymbol{e}[s, a](q, \cdot)$ where the resulting value function q'' is only modified at a single state–action (s, a) and not modified for other states and actions,

$$q''(\bar{s}, \bar{a}) = \begin{cases} q'(s, a) & \text{if } \bar{s} = s \text{ and } \bar{a} = a, \\ q(s, a) & \text{otherwise.} \end{cases}$$

Similarly, an improvement operator $\boldsymbol{i}[s]$ improves the policy only at that single state s, and does not modify the policy for other states.

We will also encounter other operators $\boldsymbol{x}[s]$ or $\boldsymbol{x}[s, a] = \boldsymbol{o}[s_1, a_1] \dots \boldsymbol{o}[s_n, a_n]$ that are indexed by a state s, state–action (s, a), or other variables. These operators may be composed internally of multiple state–action operators and hence may modify the value function and policy at multiple states.

State–action operators provide great flexibility on how pairs (q, π) are updated. Let $\boldsymbol{i}$ be a generic improvement operator and let $\boldsymbol{e}$ be a generic evaluation operator. Given any sequence of states and actions $(s_1, a_1), (s_2, a_2), \dots$ that includes all state–action pairs infinitely many times, the application of $\prod_{i=1}^{n}(\boldsymbol{ie})[s_i, a_i](q, \pi)$ will approach (q^*, π^*) as $n \to \infty$. The order in which state–action pairs show up in the sequence can have a significant impact on the convergence rate. This flexible way of updating (q, π) is called *generalized policy iteration* [39].

3.2. Sample-based evaluation

The curse of dimensionality also means that computing an expectation over all successor states, such as the one appearing in the evaluation operator $\boldsymbol{e}_B$, may be infeasible. *Sample-based* evaluation operators address this issue by estimating the true value function using samples from the distribution underlying the expectation. This is achieved by decomposing evaluation into two steps: constructing a return from a sampled trajectory, and updating the value function towards the return. These two steps may be represented by a *return operator* and a *value update operator,* respectively. These operators are applied to triples (q, π, g) rather than pairs (q, π). The additional argument to the operator is a scalar, $g \in \mathbb{R}$, that is used to maintain a sample of a return, such as the total discounted return $R + \gamma R' + \gamma^2 R'' + \cdots$, that follows a state–action pair (s, a) when executing policy π.

A return operator $\boldsymbol{r}$ only modifies g, leaving q and π unaltered. The *Monte-Carlo return operator* is defined as $\boldsymbol{r}_1[r, s'](q, \pi, g) = (q, \pi, r + \gamma g)$, where r is the reward and s' is a state. When applied repeatedly to the transitions of a trajectory $S_t, A_t, R_{t+1}, \ldots, S_T$ it produces the total discounted return, $\boldsymbol{r}_1[R_{t+1}, S_{t+1}] \ldots \boldsymbol{r}_1[R_T, S_T](q, \pi, 0) = (q, \pi, \sum_{j=1}^{T-t} \gamma^{j-1} R_{t+j})$, recalling that a sequence of operators is applied from right to left, corresponding to the application of operator $\boldsymbol{r}_1$ over the sequence in reverse order.[2] More generally, we define the λ*-return operator* $\boldsymbol{r}_\lambda$ as

$$\boldsymbol{r}_\lambda[r, s'](q, \pi, g) = \big(q, \pi, r + \gamma\big(\lambda g + (1 - \lambda)v(s')\big)\big) \quad (1)$$

where $v(s) = \sum_{a \in \mathcal{A}} \pi(a|s) q(s, a)$ is the value of state s.

When applied to a sampled trajectory $S_1, A_1, R_1, \ldots, S_T$ the λ-return operator computes a geometrically weighted mixture, $(1 - \lambda) \sum_{l=1}^{T-t-1} \lambda^{l-1} G_{t:t+l} + \lambda^{T-l-1} G_{t:T}$, of l-step returns, $G_{t:t+l} = \sum_{j=t}^{l-1} \gamma^{j-t} R_{j+1} + \gamma^{l-1} v(S_{t+l})$ [38], which as a special case includes the Monte-Carlo return operator $\boldsymbol{r}_1$ when $\lambda = 1$.

Equipped with the concept of return operators, we can now introduce value update operators. These operators update the value function to approximate the return g. A canonical example of a value update operator adjusts the value function q by a fixed[3] step-size $\alpha \in (0, 1]$ in the direction of the return g, $\boldsymbol{e}_q[s, a](q, \pi, g) = (q', \pi, g)$ where $q'(s, a) = q(s, a) + \alpha(g - q(s, a))$. By composing the update with the λ-return operator, $\boldsymbol{e}_q[s, a]\boldsymbol{r}_\lambda[r, s'](q, \pi, g)$, we recover the well known temporal difference update TD(λ).

2 One could also conceive other operators, such as eligibility traces [39], that are applied to the sequence in forward order.

3 In practical algorithms, step-sizes may vary or adapt over time.

We can make any improvement operator $\boldsymbol{i}$ "compatible" with sample-based evaluation operators by simply defining $\boldsymbol{i}'(q, \pi, g) = (q, \pi', g)$, where $(\cdot, \pi') = \boldsymbol{i}(\cdot, \pi)$. In what follows we will focus exclusively on operators that operate over triples (q, π, g). We will henceforth use $\boldsymbol{i}$ as a generic improvement operator defined over (q, π, g); this may be instantiated by any improvement operator, such as the greedy improvement operator $\boldsymbol{i}^*$, or a policy gradient operator [40]. We will illustrate algorithms using value update operator $\boldsymbol{e}_q$ and a λ-return operator $\boldsymbol{r}_\lambda$, although these could in practice be replaced by other value update and return operators.

4. SEARCH CONTROL

We now turn our attention to the order in which states and actions are visited, so that operators may be applied in an efficient sequence.

Because the state space is typically very large, it is often infeasible to compute the optimal value function and optimal policy for all states. Instead, planning methods often focus upon the computation of the optimal value $q^*(s, \cdot)$ and an optimal policy $\pi^*(\cdot|s)$ for a specific state s. To solve this problem, it is sufficient to solve a local MDP $\mathcal{M}[s]$ consisting of a subset of states in the original MDP M that are reachable from state s with nonzero probability. The local MDP otherwise has the same action space, reward, and transition dynamics as the original MDP M.

Ideally, we would like to have algorithms that are guaranteed to solve the local planning problem. We will define a *sound* planning operator $\boldsymbol{x}[s]$ to be one that converges with repeated application to the optimal value function and an optimal policy (q^*, π^*) for the local MDP $\mathcal{M}[s]$,

$$\lim_{n\to\infty} \boldsymbol{x}[s]^n(q, \pi, \cdot) = (q^*, \pi^*, \cdot), \quad \text{for any } q \in \mathbb{Q}, \pi \in \mathbb{\Pi}. \tag{2}$$

Since generalized policy iteration finds the solution to general MDPs, it is also a sound algorithm for local planning, so long as all reachable states and actions are visited infinitely often. Note that the set of reachable states may be dramatically smaller than in the complete problem. For example, only a tiny fraction of possible positions in chess are reachable from a given endgame position; solving that endgame may be considerably simpler than solving the entire game. However, the number of reachable states may nevertheless still be large, so the order in which those states are visited remains of crucial importance.

4.1. Backup diagrams for search control

A *search control* strategy determines the order in which operators are applied to states and state–actions. It may be illustrated by a *backup diagram* [39] that shows the states and actions used by a search operator $\boldsymbol{x}[s]$. In these backups diagrams, large white circles represent states and small black circles represent state–actions. Arrows indicate transitions from state to state–action and from state–action to state. If arrows are labeled by the action space $\mathcal{A}$ or by the state space $\mathcal{S}$ then this denotes corresponding transitions for all actions $a \in \mathcal{A}$ or to all states that may follow a state–action pair. If an arrow is labeled by an environment ε or policy π this denotes that the successor state or action is sampled from the corresponding environment or policy (these labels may be omitted where clear from context). At most two transitions will be shown from each state or state–action. The leftmost circle indicates the root state s to which the search operator $\boldsymbol{x}[s]$ is applied (or sometimes the root state-action s, a to which the search operator $\boldsymbol{x}[s, a]$ is applied). Some search operators $\boldsymbol{x}_k[s]$ may be indexed by a level k, and are defined recursively in terms of lower-level operators $\boldsymbol{x}_{k-1}[s']$; in this case the states in the backup diagram associated with s and s' may be labeled by the corresponding operators. For example, a depth-first search operator could be represented by the following backup diagram:

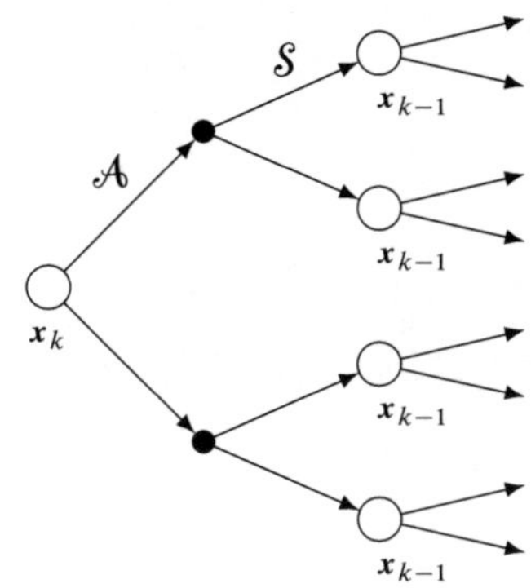

4.2. Simulation

Simulation is a search control strategy in which trajectories are generated by sampling actions from the policy and sampling next states from the environment, as represented by the following search control diagram:

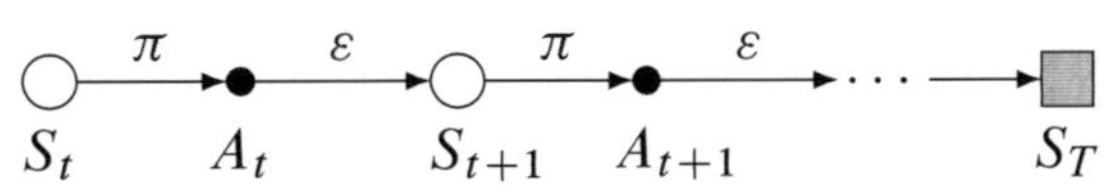

(3)

Simulation ensures that the most likely future outcomes under the current policy are explored most frequently, and may provide an effective mechanism for estimating future value, even when the state space is prohibitively large for full-width tree search. A *simulation-based search* applies evaluation and improvement operators to the sequence of states and actions encountered during simulation.

To simplify the definition of the operators, we will assume that all policies eventually reach a terminal state. The operator $\boldsymbol{x}[s]$ defined below applies evaluation and improvement operators to the sequence following state s until termination,

$$\boldsymbol{x}[s](q, \pi, g) = \begin{cases} (q, \pi, 0) & \text{if } s \text{ is terminal,} \\ \boldsymbol{i}[s]\, \boldsymbol{e}_q[s, A]\, \boldsymbol{r}_\lambda[R, S']\, \boldsymbol{x}[S'](q, \pi, g) & \text{otherwise,} \end{cases}$$

$$\text{with } A \sim \pi(s) \text{ and } R, S' \sim \varepsilon(s, A). \tag{4}$$

All the operators introduced from this point on will be based on the assumption that every policy eventually terminates. When this is not the case, one can easily modify the simulation-based operators to ensure that they terminate after T steps.

Note that, unlike the evaluation, improvement and return operators previously defined, the simulation operator $\boldsymbol{x}$ potentially modifies all of its arguments (q, π, g). This operator is typically iterated over n simulations, $\boldsymbol{x}^n[s]$, to compute the value and policy at a root state $s \in \mathcal{S}$. Pseudocode for simulation-based search with sample operators is given in Algorithm 1. The function called IMPROVE() may invoke any suitable improvement operator $\boldsymbol{i}$.

Algorithm 1 Simulation-Based Search

```
procedure SIM(ε, q, π, s)
    if s is terminal then return 0
    A ~ π(s)
    R, S' ← ε(s, A)
    v(S') ← Σ_{a∈A} π(a|S')q(S', a)
    G ← R + γ(λ SIM(ε, q, π, S') + (1 − λ)v(S'))
    q(s, A) ← q(s, A) + α(G − q(s, A))
    π(s) ← IMPROVE(π(s), q(s, ·), G)
    return G
end procedure
```

Algorithm 2 Recursive Simulation-Based Search

```
procedure RSIM(ε, q, π, s, k)
    if s is terminal or k = 0 then return 0
    for i = 1 to n do
        RSIM(ε, q, π, s, k − 1)
    end for
    A ~ π(s)                                        ▷ π may have changed
    R, S' ← ε(s, A)
    v(S') ← Σ_{a∈A} π(a|S')q(S', a)
    G ← R + γ(λ RSIM(ε, q, π, S', k) + (1 − λ)v(S'))
    q(s, A) ← q(s, A) + α(G − q(s, A))
    π(s) ← IMPROVE(π(s), q(s, ·), G)
    return G
end procedure
```

One may also compute the return from a simulation without any value update or policy improvement. These simple simulations are known as *rollouts*,

$$\boldsymbol{z}[s,a](q,\pi,g) = \begin{cases} (q,\pi,0) & \text{if } s \text{ is terminal,} \\ \boldsymbol{r}_\lambda[R,S']\,\boldsymbol{z}[S',A'](q,\pi,g) & \text{otherwise,} \end{cases}$$
$$\text{with } R, S' \sim \varepsilon(s,a) \text{ and } A' \sim \pi(S'). \tag{5}$$

Exploration with soft improvement operators. For a simulation-based search to be sound, the simulation policy π must continue to visit all states and actions infinitely often as we apply the operator multiple times. That is, if $(\cdot, \pi', \cdot) = \boldsymbol{x}^n(\cdot, \pi, \cdot)$, we want π' to select all actions with nonzero probability. This does not necessarily follow for all $\boldsymbol{x}$. For example, if we plug in the greedy improvement operator $\boldsymbol{i}^*$ in (5), the resulting π' is a deterministic policy—that is, $\pi(a|s) = 1$ for a specific $a \in \mathcal{A}$. Even when π' is not deterministic, one may want to sample actions from a distribution that ensures an appropriate level of *exploration* [39].

Exploration may be accomplished by using a *soft improvement operator* that yields a policy that selects all actions with nonzero probability. If the soft improvement operator approaches the greedy operator as the number of applications tends to infinity, $\lim_{n\to\infty} \boldsymbol{i}^n = \boldsymbol{i}^*$, then under mild conditions convergence is assured (c.f. (2)). This type of exploration is referred to as "greedy in the limit of infinite exploration" (GLIE) [37].

As an example, the ϵ-greedy operator $\boldsymbol{i} = \boldsymbol{\epsilon}\boldsymbol{i}^*$ is a soft improvement operator that introduces randomness via a noisy operator $\boldsymbol{\epsilon}(q, \pi, g) = (q, \epsilon\pi_{\text{rand}} + (1-\epsilon)\pi, g)$, where π_{rand} is any policy that selects all actions with nonzero probability. A common choice is to have π_{rand} select actions uniformly at random, $\pi_{\text{rand}}(\cdot|s) = 1/|\mathcal{A}|$. Note that, if we think of ϵ as a parameter of $\boldsymbol{\epsilon}(q, \pi, g)$ that is decreased at each application of $\boldsymbol{\epsilon}$, we obtain a GLIE operator. Alternatively, an upper-confidence rule $\boldsymbol{i}^{\text{UCB}}$ may be used to encourage exploration of uncertain values, by acting greedily with respect to an upper confidence bound on the value function, $\text{argmax}_{a\in\mathcal{A}}\, q(s,a) + u(s,a)$, where $u(s,a)$ represents value uncertainty [3].

In what follows we will define some operators using a generic improvement operator $\boldsymbol{i}$; unless noted otherwise, the reader should think of $\boldsymbol{i}$ as some instantiation of a GLIE soft improvement operator.

Example 4.1: All-Action Monte-Carlo Search

Historically, the earliest forms of simulation-based search [8,41], used for example to achieve superhuman performance in Scrabble [38], were based upon rollouts $\boldsymbol{z}[s,a]$ that start immediately after a single action a from state s. The main idea is to estimate the action–value of every action $a \in \mathcal{A}$ from the root state by the outcome of simulations starting from that action. Finally, the action with maximum value is executed. We refer to this search algorithm as *all-action Monte-Carlo search*, which can be described by the following search control diagram,

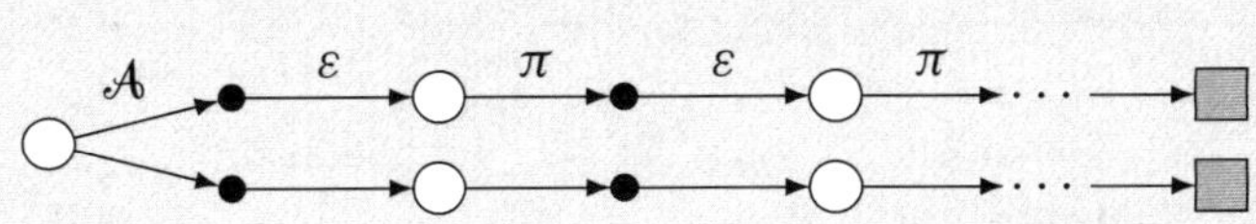

D. SILVER AND A. BARRETO

In this case a single improvement operator $\boldsymbol{i}^*$ is applied to the root state based on the values[a] estimated through the Monte-Carlo simulations:

$$\boldsymbol{x}[s](q, \pi, g) = \boldsymbol{i}^*[s] \prod_{a \in \mathcal{A}} \boldsymbol{e}_q[s, a]\boldsymbol{z}[s, a](q, \pi, g).$$

A In practice, Monte-Carlo algorithms often update the value function using a step-size $\alpha = 1/visits(s, a)$. In this case the update $\boldsymbol{e}_q$ incrementally updates the value $q(s, a)$ to the mean return following state-action s, a.

Example 4.2: Monte-Carlo Tree Search

A simulation-based search algorithm using sample operators at each state and action is known as *Monte-Carlo tree search* (MCTS) [12], as used in the first master-level 9×9 [13, 16] and 19×19 [34] Go programs. Each simulation of MCTS consists of two stages: a first stage in which sample-based evaluation and improvement operators are applied, and a rollout that samples the remainder of the trajectory. The first stage of simulation typically finishes upon reaching a previously unvisited state, while the rollout finishes upon reaching a terminal state, as shown below,

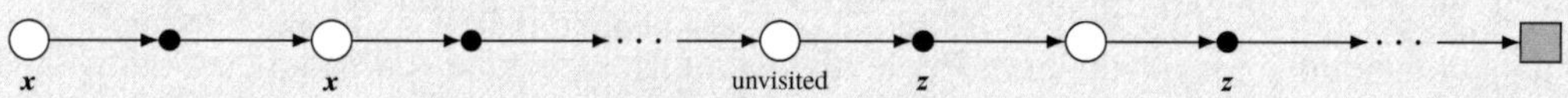

$$\boldsymbol{x}[s](q, \pi, g) = \begin{cases} \boldsymbol{i}[s]\, \boldsymbol{e}_q[s, A]\, \boldsymbol{z}[s, A](q, \pi, g) & \text{if } s \text{ is unvisited,} \\ \boldsymbol{i}[s]\, \boldsymbol{e}_q[s, A]\, \boldsymbol{r}_\lambda[R, S']\, \boldsymbol{x}[S'](q, \pi, g) & \text{otherwise,} \end{cases}$$

$$\text{with } A \sim \pi(s) \text{ and } R, S' \sim \varepsilon(s, A). \tag{6}$$

This leads to a gradual expansion of the frontier of visited states as each subsequent simulation goes one step further. When the operator $\boldsymbol{i}$ used in (6) is a GLIE soft improvement operator, MCTS is sound under mild assumptions (cf. Eq. (2)). An upper confidence bound around an estimate of q_π is often used to guide exploration [21]. As in value iteration, policies are not explicitly represented.

MCTS typically uses Monte-Carlo returns, $\boldsymbol{z}$, with $\lambda = 1$. However, a variant of MCTS that uses instead TD(λ) returns is sometimes known as *temporal-difference search* [33]

4.3. Recursive simulation

Classical search methods traverse a search tree by branching from a parent state to a child state, performing a search from the child, and then backtracking to the parent. Branching and backtracking in this manner may be advantageous because it ensures that child values are accurate before applying any operation to the parent. If the value and policy of each child are optimal, only a single operation needs to be applied to the parent to ensure optimality.

By contrast, simulation breaks the curse of dimensionality by sampling trajectories, allowing search to look deeply ahead even in large state spaces. However, it may need to visit each state multiple times.

We propose here a marriage of these two search control strategies by allowing parent simulations to branch recursively into child simulations. This produces a *recursive simulation-based search*. Each level k simulation samples a sequence of states and actions; multiple level $k-1$ simulations are invoked from each state (or state–action) of the sequence before sampling the next action. This is illustrated in the search diagram below, where each arrow labeled $\boldsymbol{x}_{k-1}$ represents a lower level simulation starting from that state, corresponding to the recursive application of the same search diagram with $k-1$,

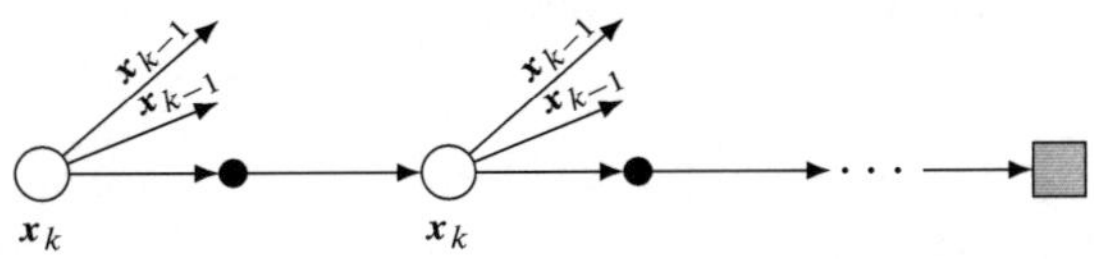

(7)

Applying improvement and evaluation operators to the states and actions of the search results in the following search algorithm:

$$\boldsymbol{x}_k[s](q,\pi,g) = \begin{cases} (q,\pi,0) & \text{if } s \text{ is terminal or } k=0, \\ \boldsymbol{i}[s]\,\boldsymbol{e}_q[s,A]\,\boldsymbol{r}_\lambda[R,S']\,\boldsymbol{x}_k[S']\,\underbrace{\boldsymbol{x}^n_{k-1}[s](q,\pi,g)}_{(\cdot,\pi',\cdot)} & \text{otherwise,} \end{cases} \tag{8}$$

where $(\cdot,\pi',\cdot) = \boldsymbol{x}^n_{k-1}[s](q,\pi,g)$, $A \sim \pi'(s)$ and $R, S' \sim \varepsilon(s,A)$.

One can think of the operator above as a simulation, akin to Eq. (5), in which the action A to be taken is sampled from a policy π' resulting from the application of the operator itself. The overall algorithm, shown in Algorithm 2, is similar to Algorithm 1 with the addition of a recursive call that corresponds to the operator $\boldsymbol{x}^n_{k-1}$ in Eq. (8) (see for example *nested Monte-Carlo tree search* **[4]**).

Example 4.3: Nested Monte-Carlo Search

Recursive simulation may be used with an all-action Monte-Carlo search. This gives rise to the more powerful algorithm of *nested Monte-Carlo search* **[11, 42]**. In this algorithm, rollouts are nested within rollouts. At level $k+1$, all possible actions are enumerated. Each action $a \in \mathcal{A}$ is evaluated by the average outcome (i.e., Monte-Carlo evaluation $\boldsymbol{e}_q\boldsymbol{z}$) of level k simulations that start from action a. The policy at the root state of the simulation selects the action with maximum value (i.e., greedy improvement $\boldsymbol{i}^*$). An instance of this algorithm is illustrated by the following search control diagram and equations:

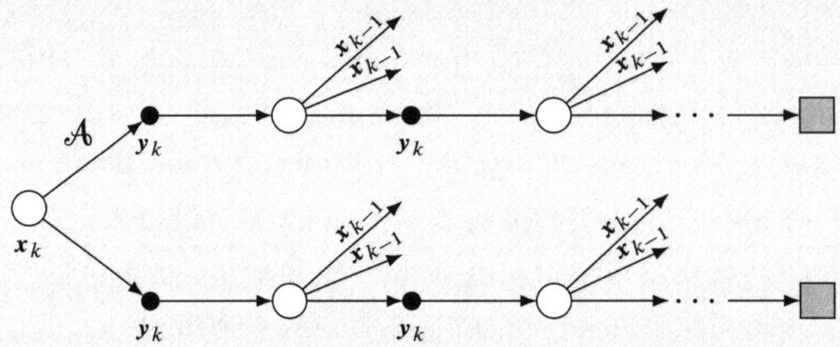

$$\boldsymbol{x}_k[s](q,\pi,g) = \boldsymbol{i}^*[s] \prod_{a \in \mathcal{A}} \boldsymbol{e}_q[s,a]\boldsymbol{y}_k[s,a](q,\pi,g),$$

$$\boldsymbol{y}_k[s,a](q,\pi,g) = \begin{cases} (q,\pi,0) \text{ if } s \text{ is terminal or } k=0, \\ \boldsymbol{r}_\lambda[R,S']\boldsymbol{y}_k[S',A']\boldsymbol{x}^n_{k-1}[S'](q,\pi,g) \text{ otherwise,} \end{cases}$$

$$\text{where } R, S' \sim \varepsilon(s,a), (\cdot,\pi',\cdot) = \boldsymbol{x}^n_{k-1}\big[S'\big](q,\pi,g), \text{ and } A' \sim \pi'(S').$$

Nested Monte-Carlo search achieved superhuman performance in Morpion solitaire [11] and a variant of Klondike solitaire [42]. Each additional level of recursion $\boldsymbol{x}_1[s], \ldots, \boldsymbol{x}_4[s]$ produced stronger results (even when taking account of the additional computational cost).

If simulations are truncated after one time-step, $\boldsymbol{y}_k[s, a](q, \pi, g) = \boldsymbol{r}_\lambda[R, S']\boldsymbol{x}^n_{k-1}[S'](q, \pi, v(S'))$, then we recover a *sparse sampling* tree search algorithm [20]

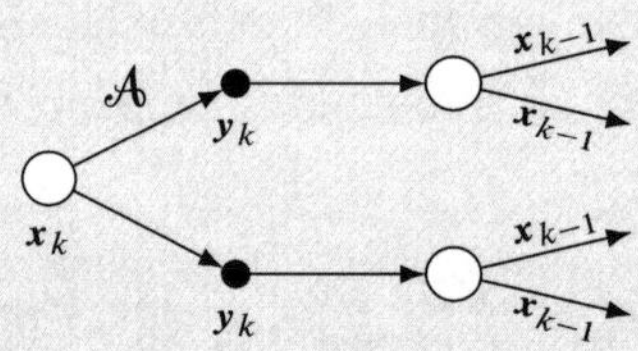

5. APPROXIMATION

Many problems are so large that they are intractable to exact methods—even when using simulation. To address this issue, we consider approximate methods that use a function approximator (such as a neural network) q_θ with parameters θ to represent a value function, or a function approximator π_η with parameters η to represent a policy. We will consider operators that aim at finding the closest representable value parameters $\theta^* = \text{argmin}_\theta\, \ell_q(q, q_\theta)$ to target value function q according to some loss ℓ_q, such as squared error, or the closest representable policy parameters $\eta^* = \text{argmin}_\eta\, \ell_\pi(\pi, \pi_\eta)$ to a target policy π according to some loss ℓ_π, such as the KL divergence.

We formalize these operators as modified versions of their counterparts which project their operands onto the space of representable value functions or policies. They do so by directly manipulating the parameters θ or η. In practice most approximation methods incrementally optimize parameters by gradient descent. We define a gradient-based evaluation operator that acts upon an approximate value function as

$$\begin{aligned} \boldsymbol{e}_{\theta}(q_{\theta}, \pi, g) &= (q_{\theta'}, \pi, g), \\ \text{where } \theta' &= \theta - \alpha \frac{\partial}{\partial \theta} \ell_q(g, q_{\theta}). \end{aligned} \tag{9}$$

Gradient descent on policy parameters may be formalized analogously by defining a generic, gradient-based improvement operator,

$$\begin{aligned} &\boldsymbol{i}_{\eta}(q, \pi_{\eta}, g) = (q, \pi_{\eta'}, g), \\ &\quad \text{where } \eta' = \eta - \alpha \frac{\partial}{\partial \eta} \ell_{\pi}(\pi', \pi_{\eta}) \quad \text{and} \quad (\cdot, \pi', \cdot) = \boldsymbol{i}(\cdot, \pi_{\eta}, \cdot). \end{aligned} \tag{10}$$

In the above equation, π' is the policy resulting from applying the generic improvement operator $\boldsymbol{i}$ to policy π_{η}. For example, a gradient-based greedy improvement operator may be instantiated, $\boldsymbol{i}^*_{\eta}(q, \pi_{\eta}, g) = (q, \pi_{\eta'}, g)$ using a corresponding greedy improvement operator $(\cdot, \pi', \cdot) = \boldsymbol{i}^*(\cdot, \pi_{\eta}, \cdot)$ to provide the target policy π' for the loss function ℓ_{π}.

Function approximation may be combined with simulation-based search by simply replacing the regular improvement and evaluation operators with their counterparts defined above.

The combination of approximation with recursive simulation-based search (see Section 4.3) yields well-known algorithms that have been successfully applied to large and complex problems, as discussed in the examples below. Algorithm 3 illustrates a canonical algorithm using function approximation.

Algorithm 3 Approximate Recursive Simulation-Based Search

procedure ARSIM($\varepsilon, \theta, \eta, s, k$)
  if s is terminal **or** $k = 0$ **then return** 0
  for $i = 1$ **to** n **do**
    ARSIM($\varepsilon, \theta, \eta, s, k-1$)
  end for
  $A \sim \pi_{\eta}(s)$ ▷ η may have changed
  $R, S' \leftarrow \varepsilon(s, A)$
  $v(S') \leftarrow \sum_{a \in \mathcal{A}} \pi(a|S') q(S', a)$
  $G \leftarrow R + \gamma(\lambda \,\text{ARSIM}(\varepsilon, \theta, \eta, S', k) + (1-\lambda) v(S'))$
  $\theta \leftarrow \theta - \alpha \frac{\partial}{\partial \theta} \ell_q(G, q_{\theta}(s, A))$
  $\pi'(s) \leftarrow \text{IMPROVE}(\pi_{\eta}, q_{\theta}(s, \cdot), G)$
  $\eta \leftarrow \eta - \alpha \frac{\partial}{\partial \eta} \ell_{\pi}(\pi'(s), \pi_{\eta}(s))$
  return G
end procedure

Algorithm 4 AlphaZero

procedure AlphaZero($\varepsilon, \theta, \eta, s, k$)
  if s is terminal **or** $k = 0$ **then return** 0
  $\pi' \leftarrow \pi_\eta$; $q' \leftarrow q_\theta$
  for $i = 1$ **to** n **do**
    $\text{MCTS}(\varepsilon, q', \pi', s, k-1)^\dagger$
  end for
  $A \sim \pi'(s)$
  $R, S' \leftarrow \varepsilon(s, A)$
  $v(S') \leftarrow \sum_{a \in \mathcal{A}} \pi(a|S')q(S', a)$
  $G \leftarrow R + \gamma(\lambda\,\text{AlphaZero}(\varepsilon, \theta, \eta, S', k) + (1-\lambda)v(S'))$
  $\theta \leftarrow \theta - \alpha \frac{\partial}{\partial \theta} \ell_q(G, q_\theta(s, A))$
  $\eta \leftarrow \eta - \alpha \frac{\partial}{\partial \eta} \ell_\pi(\pi'(s), \pi_\eta(s))$
  return G
end procedure

† MCTS is an instantiation of Sim (Algorithm 1).

Example 5.1: Dyna-k

The Dyna-k algorithm [32] is an example of recursive simulation-based search with value function approximation. At every level k it uses an approximate evaluation operator $\boldsymbol{e}_\theta$ that minimizes squared error with respect to a sampled return. This return corresponds to the outcome of a level k simulation in which actions are sampled from the improved policy resulting from multiple lower level $k-1$ simulations. The search control diagram follows the same pattern as Eq. (7).

$$\boldsymbol{x}_k[s](\pi, q_\theta, g) = \begin{cases} (q, \pi, 0) & \text{if } s \text{ is terminal or } k = 0, \\ \boldsymbol{i}^*[s]\boldsymbol{e}_\theta[s, A]\boldsymbol{r}_\lambda[R, S']\boldsymbol{x}_k[S']\boldsymbol{x}^n_{k-1}[s](\pi, q_{\theta'}, g), \end{cases}$$

with $\theta' = \theta, (\cdot, \pi', \cdot) = \boldsymbol{x}^n_{k-1}[s](\pi, q'_\theta, g), A \sim \pi'(s)$ and $R, S' \sim \varepsilon(s, A)$.

Dyna-k may also be adapted to apply function approximation to the policy (see Algorithm 3). For example, *policy gradient search* [1] utilizes a gradient-based improvement operator $\boldsymbol{i}_\eta$ based upon policy gradient algorithms [40].

In practice, the value function (or policy) is represented separately at different levels of the recursion by distinct parameters. Level $k-1$ parameters are copied from level k parameters, and are then updated based upon the level $k-1$ returns. This allows parameters to specialize to a local region of the search tree. Using multiple representations boosted performance in 9×9 Go, compared to a single representation [32].

Example 5.2: Classification-Based Policy Iteration

Classification-based policy iteration (CBPI) combines all-action Monte-Carlo search with policy approximation [22]. The rollouts of the Monte-Carlo search sample actions according to π_η,

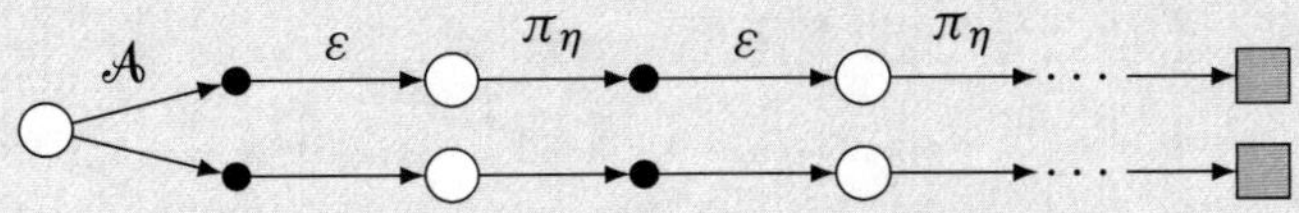

$$y[s](q, \pi_\eta, g) = i^*_\eta \prod_{a \in \mathcal{A}} e_q[s, a] z[s, a](q, \pi_\eta, g),$$

The main search consists of a second level of simulations. At every step of the main search, an all-action Monte-Carlo search is called recursively from state s, to compute an improved policy. Policy parameters η are updated by an approximate improvement operator i^*_η that minimizes a classification loss, $\ell_\pi(\pi', \pi_\eta)$, with respect to a greedy improvement operator $(\cdot, \pi', \cdot) = i^*(q, \pi, g)$,

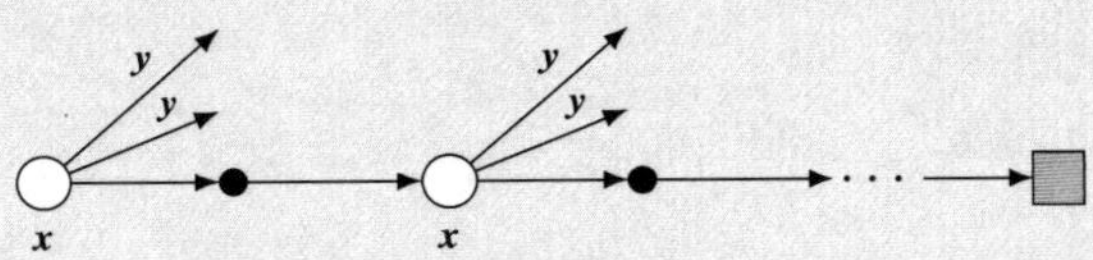

$$x[s](q, \pi_\eta, 0) = \begin{cases} (q, \pi_\eta, 0) & \text{if } s \text{ is terminal or } k = 0, \\ x[S'] y^n[s](q, \pi_\eta, g), & \end{cases}$$

where $(\cdot, \pi_{\eta'}, \cdot) = y^n[s](q, \pi_\eta, g)$, $A \sim \pi_{\eta'}(s)$ and $R, S' \sim \varepsilon(s, A)$.
Expert iteration applies CBPI to a Monte-Carlo tree search; it achieved state-of-the-art performance in Hex [2]. In practice, both expert iteration and CBPI are often combined with value function approximation [28].

Example 5.3: AlphaZero

The AlphaZero algorithm [35, 36] achieved superhuman performance across chess, Go and shogi. It is a two-level recursive simulation-based search that utilizes both value function and policy approximation at the second level.
At the first level, AlphaZero uses a Monte-Carlo tree search. Simulations finish upon reaching an unvisited state, without any rollout, at which point the return is initialized to the value function of this state.[a] To ensure adequate exploration, an improvement operator i^{UCB} selects the action that maximizes an upper confidence bound $u(s, a) \propto$

$\pi_\eta(a|s)/(visits(s,a)+1)$ that is informed by the policy π_η [34],[b]

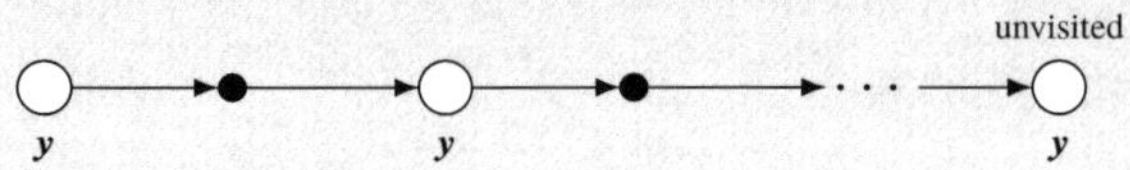

$$\boldsymbol{y}[s](q,\pi,g) = \begin{cases} \boldsymbol{i}^{\text{UCB}}[s]\,\boldsymbol{e}_q[s,A](q,\pi,v(s)) & \text{if } s \text{ is unvisited,} \\ \boldsymbol{i}^{\text{UCB}}[s]\,\boldsymbol{e}_q[s,A]\,\boldsymbol{r}_\lambda[R,S']\,\boldsymbol{y}[S'](q,\pi,g) & \text{otherwise,} \end{cases}$$

where $A \sim \pi(s)$ and $R, S' \sim \varepsilon(s,A)$.

At the second level, AlphaZero represents its value function and policy by neural networks with parameters θ and η, respectively.[c] An approximate improvement operator $\boldsymbol{i}^\eta$ is applied to the Monte-Carlo tree search operator defined above, $\boldsymbol{y}[s]$. The operator $\boldsymbol{i}^\eta$ is based on a classification loss, similar to the one used in Example 5.2, that "projects" a policy π onto the space spanned by η. An approximate evaluation operator $\boldsymbol{e}_\theta$ is applied to the Monte-Carlo return corresponding to the outcome of the high-level simulation, in a similar manner to Algorithm 3,

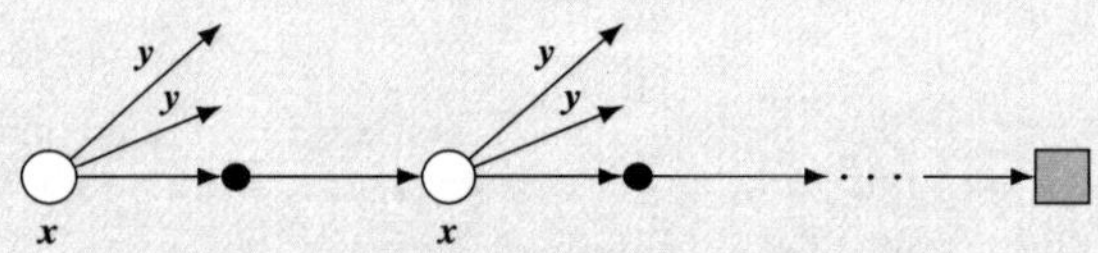

$$\boldsymbol{x}[s](q_\theta,\pi_\eta,g) = \begin{cases} (q_\theta,\pi_\eta,0) \quad \text{if } s \text{ is terminal or } k=0, \\ \boldsymbol{e}_\theta[s,A]\boldsymbol{r}_\lambda[R,S']\boldsymbol{x}[S']\underbrace{\boldsymbol{i}_\eta[s]\boldsymbol{y}^n[s](q_\theta,\pi_\eta,g)}_{(\cdot,\pi_{\eta'},\cdot)}, \end{cases}$$

$$\text{where } (\cdot,\pi_{\eta'},\cdot) = \boldsymbol{i}_\eta[s]\boldsymbol{y}^n[s](q_\theta,\pi_\eta,g),\ A \sim \pi_{\eta'}(s),\ S' \sim \varepsilon(s,A).$$

In reality, the operator $\boldsymbol{y}$ is called with copies of q_θ and π_η,[d] and the resulting policy is used as a target[e] for the gradient-based policy improvement π_η (see Algorithm 4).

A AlphaZero uses a state value function; for consistency, it is illustrated here using an action value function.

B Other policy improvement operators may also be used [14, 18].

C In practice both neural networks share the same parameters.

D The policy copy passed into the operator $\boldsymbol{y}$ may be modified by noise, $(\cdot,\pi_{\eta'},\cdot) = \boldsymbol{i}_\eta[s]\boldsymbol{y}^n[s]\epsilon(q_\theta,\pi_\eta,g)$; this ensures adequate exploration, even when using a small number of simulations [14].

E The target can also incorporate an improvement step based upon the outer return [19].

6. DISCUSSION

We have developed a framework for understanding simulation-based search algorithms in terms of their search control methods. We have seen that many algorithms can be described as instances of generalized policy iteration, interleaving evaluation and improvement operators to ensure convergence towards an optimal value function and policy. In large problems, approximate evaluation and improvement operators may also be introduced, so as to search for an approximately optimal value function and policy.

The formalism we proposed allowed us to describe many existing search algorithms – from a basic Monte-Carlo search to more sophisticated search algorithms such as AlphaZero – that nest multiple levels of simulation. However, many other strategies exist for search control that cannot be described in these simple terms. For example, many planning algorithms utilize *prioritization* to sort states according to an appropriate criterion [25]; the highest priority state is visited next.

We have presented several specific examples of simulation-based search algorithms that utilize simulation and recursion, including many of the most successful methods used in games such as chess and Go – the canonical challenges for planning. However, the framework presented in this paper also suggests a much broader space of search algorithms that combine elements of existing algorithms. For example, could AlphaZero [36] be improved by introducing deeper levels of recursion, as in nested Monte-Carlo search [11]? Or by utilizing function approximation inside its lower level Monte-Carlo tree search, as in Dyna-2 [32]? Could other evaluation and improvement operators be more effective [14, 18, 19]?

Understanding the underlying principles may also enable existing heuristic search algorithms to be replaced with sound algorithms that converge to the optimal solution under a broader range of conditions. For example, the heuristic improvement operator in AlphaZero may be replaced with a principled policy improvement operator [14]. Finally, we hope that a greater understanding of these principles may result in the development of new search algorithms that go beyond our current frontiers.

REFERENCES

[1] T. Anthony, R. Nishihara, P. Moritz, T. Salimans, and J. Schulman, Policy gradient search: Online planning and expert iteration without search trees. *CoRR* (2019), arXiv:1904.03646.

[2] T. Anthony, Z. Tian, and D. Barber, Thinking fast and slow with deep learning and tree search. In *Advances in Neural Information Processing Systems 30: Annual Conference on Neural Information Processing Systems 2017, December 4–9, 2017, Long Beach, CA, USA*, edited by U. Luxburg, S. Bengio, H. M. Wallach, R. Fergus, S. V. N. Vishwanathan, and R. Garnett, pp. 5360–5370, Curran Associates Inc., 2017.

[3] P. Auer, N. Cesa-Bianchi, and P. Fischer, Finite-time analysis of the multi-armed bandit problem. *Mach. Learn.* **47** (2002), no. 2–3, 235–256.

[4] H. Baier and M. H. M. Winands, Nested Monte-Carlo tree search for online planning in large MDPs. In *20th European Conference on Artificial Intelligence* 242, edited by L. D. Raedt, C. Bessiere, D. Dubois, P. Doherty, P. Frasconi, F. Heintz, and P. J. F. Lucas, pp. 109–114, IOS Press, 2012.

[5] D. Bertsekas, Lessons from AlphaZero for optimal, model predictive, and adaptive control. 2021, arXiv:2108.10315.

[6] D. P. Bertsekas, *Dynamic Programming and Optimal Control, volume I.* 3rd edn., Athena Scientific, Belmont, MA, USA, 2005.

[7] C. Browne, E. J. Powley, D. Whitehouse, S. M. Lucas, P. I. Cowling, P. Rohlfshagen, S. Tavener, D. P. Liebana, S. Samothrakis, and S. Colton, A survey of Monte Carlo tree search methods. *IEEE Trans. Computat. Intell. AI Games* **4** (2012), no. 1, 1–43.

[8] B. Bruegmann, Monte-Carlo Go, 1993, http://www.cgl.ucsf.edu/go/Programs/Gobble.html.

[9] M. Campbell, A. Hoane, and F. Hsu, Deep Blue. *Artificial Intelligence* **134** (2002), 57–83.

[10] M. Campbell and T. A. Marsland, A comparison of minimax tree search algorithms. *Artificial Intelligence* **20** (1983), 347–367.

[11] T. Cazenave, Nested Monte-Carlo search. In *21st International Joint Conference on Artificial Intelligence*, edited by C. Boutilier, pp. 456–461, International Joint Conferences on Artificial Intelligence, 2009.

[12] R. Coulom, Efficient selectivity and backup operators in Monte-Carlo tree search. In *5th International Conference on Computer and Games*, pp. 72–83, Springer-Verlag, 2006.

[13] R. Coulom, Computing Elo ratings of move patterns in the game of Go. In *Computer Games Workshop*, pp. 198–208, Universiteit Maastricht, 2007.

[14] I. Danihelka, A. Guez, J. Schrittwieser, and D. Silver, Policy improvement by planning with Gumbel. In *International Conference on Learning Representations*, 2022.

[15] C. E. Garcia, D. M. Prett, and M. Morari, Model predictive control: Theory and practice—A survey. *Automatica* **25** (1989), no. 3, 335–348.

[16] S. Gelly, Y. Wang, R. Munos, and O. Teytaud, Modification of UCT with patterns in Monte-Carlo Go. Technical Report 6062, INRIA, 2006.

[17] D. Ghosh, M. C. Machado, and N. L. Roux, An operator view of policy gradient methods. In *Advances in Neural Information Processing Systems 33*, edited by H. Larochelle, M. Ranzato, R. Hadsell, M. Balcan, and H. Lin, pp. 3397–3406, Curran Associates Inc., 2020.

[18] J. Grill, F. Altché, Y. Tang, T. Hubert, M. Valko, I. Antonoglou, and R. Munos, Monte-Carlo tree search as regularized policy optimization. In *37th International Conference on Machine Learning 119*, pp. 3769–3778, Association of Computing Machinery, 2020.

[19] M. Hessel, I. Danihelka, F. Viola, A. Guez, S. Schmitt, L. Sifre, T. Weber, D. Silver, and H. van Hasselt, Muesli: Combining improvements in policy optimization. In *38th International Conference on Machine Learning 139*, edited by M. Meila and T. Zhang, pp. 4214–4226, Association of Computing Machinery, 2021.

[20] M. J. Kearns, Y. Mansour, and A. Y. Ng, A sparse sampling algorithm for near-optimal planning in large Markov decision processes. *Mach. Learn.* **49** (2002), no. 2–3, 193–208.

[21] L. Kocsis and C. Szepesvari, Bandit based Monte-Carlo planning. In *15th European Conference on Machine Learning*, pp. 282–293, Springer-Verlag, 2006.

[22] M. G. Lagoudakis and R. Parr, Reinforcement learning as classification: Leveraging modern classifiers. In *20th International Conference on Machine Learning*, edited by T. Fawcett and N. Mishra, pp. 424–431, AAAI Press, 2003.

[23] M. L. Littman and C. Szepesvári, A generalized reinforcement-learning model: Convergence and applications. In *13th International Conference on Machine Learning*, edited by L. Saitta, pp. 310–318, Morgan Kaufmann, 1996.

[24] T. Lozano-Pérez, Spatial planning: A configuration space approach. In *Autonomous Robot Vehicles*, edited by I. J. Cox and G. T. Wilfong, pp. 259–271, Springer, 1990.

[25] A. W. Moore and C. G. Atkeson, Prioritized sweeping: Reinforcement learning with less data and less time. *Mach. Learn.* **13** (1993), 103–130.

[26] W. B. Powell, *Approximate dynamic programming: solving the curses of dimensionality*. 2nd edn., Wiley Ser. Probab. Stat., Wiley, Hoboken, NJ, USA, 2011.

[27] M. L. Puterman, *Markov decision processes: discrete stochastic dynamic programming*. Wiley Ser. Probab. Stat., Wiley, 1994.

[28] B. Scherrer, M. Ghavamzadeh, V. Gabillon, B. Lesner, and M. Geist, Approximate modified policy iteration and its application to the game of tetris. *J. Mach. Learn. Res.* **16** (2015), no. 49, 1629–1676.

[29] J. Schrittwieser, I. Antonoglou, T. Hubert, K. Simonyan, L. Sifre, S. Schmitt, A. Guez, E. Lockhart, D. Hassabis, T. Graepel, et al., Mastering Atari, Go, chess and shogi by planning with a learned model. *Nature* **588** (2020), no. 7839, 604–609.

[30] B. Sheppard, World-championship-caliber Scrabble. *Artificial Intelligence* **134** (2002), no. 1–2, 241–275.

[31] D. Silver, *Reinforcement Learning and Simulation-Based Search in Computer Go*. PhD thesis, University of Alberta, 2009.

[32] D. Silver, R. S. Sutton, and M. Müller, Sample-based learning and search with permanent and transient memories. In *25th International Conference on Machine Learning*, pp. 968–975, Association of Computing Machinery, 2008.

[33] D. Silver, R. S. Sutton, and M. Müller, Temporal-difference search in computer Go. *Mach. Learn.* **87** (2012), no. 2, 183–219.

[34] D. Silver, A. Huang, C. J. Maddison, A. Guez, L. Sifre, G. van den Driessche, J. Schrittwieser, I. Antonoglou, V. Panneershelvam, M. Lanctot, S. Dieleman, D. Grewe, J. Nham, N. Kalchbrenner, I. Sutskever, T. P. Lillicrap, M. Leach, K. Kavukcuoglu, T. Graepel, and D. Hassabis, Mastering the game of Go with deep neural networks and tree search. *Nature* **529** (2016), no. 7587, 484–489.

[35] D. Silver, J. Schrittwieser, K. Simonyan, I. Antonoglou, A. Huang, A. Guez, T. Hubert, L. Baker, M. Lai, A. Bolton, et al., Mastering the game of Go without human knowledge. *Nature* **550** (2017), no. 7676, 354–359.

[36] D. Silver, T. Hubert, J. Schrittwieser, I. Antonoglou, M. Lai, A. Guez, M. Lanctot, L. Sifre, D. Kumaran, T. Graepel, et al., A general reinforcement learning algorithm that masters chess, shogi, and Go through self-play. *Science* **362** (2018), no. 6419, 1140–1144.

[37] S. P. Singh, T. S. Jaakkola, M. L. Littman, and C. Szepesvári, Convergence results for single-step on-policy reinforcement-learning algorithms. *Mach. Learn.* **38** (2000), no. 3, 287–308.

[38] R. Sutton, Learning to predict by the method of temporal differences. *Mach. Learn.* **3** (1988), no. 9, 9–44.

[39] R. Sutton and A. Barto, *Reinforcement learning: an introduction*. MIT Press, 1998.

[40] R. S. Sutton, D. A. McAllester, S. P. Singh, and Y. Mansour, Policy gradient methods for reinforcement learning with function approximation. In *Advances in Neural Information Processing Systems 12*, edited by S. A. Solla, T. K. Leen, and K. Müller, pp. 1057–1063, The MIT Press, 1999.

[41] G. Tesauro and G. Galperin, On-line policy improvement using Monte-Carlo search. In *Advances in Neural Information Processing 9*, pp. 1068–1074, Curran Associates Inc., 1996.

[42] X. Yan, P. Diaconis, P. Rusmevichientong, and B. Roy, Solitaire: Man versus machine. In *Advances in Neural Information Processing Systems 17*, edited by L. Saul, Y. Weiss, and L. Bottou, MIT Press, 2004.

DAVID SILVER

DeepMind, London, UK, and University College London, London, UK, davidsilver@google.com

ANDRE BARRETO

DeepMind, London, UK, andrebarreto@google.com

BEYOND LINEAR ALGEBRA

BERND STURMFELS

ABSTRACT

Our title challenges the reader to venture beyond linear algebra in designing models and in thinking about numerical algorithms for identifying solutions. This article accompanies the author's lecture at the International Congress of Mathematicians 2022. It covers recent advances in the study of critical point equations in optimization and statistics, and it explores the role of nonlinear algebra for linear PDEs with constant coefficients.

MATHEMATICS SUBJECT CLASSIFICATION 2020

Primary 14Q30; Secondary 13P25, 35A25, 62R01

KEYWORDS

Algebraic variety, polar degrees, maximum likelihood, partial differential equation

ICM INTERNATIONAL CONGRESS OF MATHEMATICIANS 2022 JULY 6–14

Proc. Int. Cong. Math. 2022, Vol. 6, pp. 4820–4840
DOI 10.4171/ICM2022/64

Published by EMS Press

1. INTRODUCTION

Linear algebra is ubiquitous in the mathematical universe. It plays a foundational role for many models in the sciences and engineering, and its numerical methods are a driving force behind today's technologies. The power of linear algebra stems from our ability, honed through the practice of calculus, to approximate nonlinear shapes by linear spaces.

Yet, the world is nonlinear. Nonlinear equations are a natural ingredient in mathematical models for the real world. In our view, the true nonlinear nature of a phenomenon should be respected as long as possible. We argue against the common practice of passing to a linear approximation immediately. Of course, in the final step of implementing scalable algorithms, one will always employ the powerful tools of numerical linear algebra. However, in the early phase of exploring and designing a model, there is significant benefit in going beyond linear algebra. Mathematical fields such as algebraic geometry, algebraic topology, combinatorics, commutative algebra, or representation theory furnish practical tools.

The growing awareness of theoretical mathematics in applications has led to a new field called *Nonlinear Algebra*. The textbook [32] offers foundations for interested students. The aim of this lecture is to introduce research trends and discuss a few recent results. At the core of many problems lies the study of subsets of $\mathbb{R}^n$ that are defined by polynomials:

$$\begin{aligned}\{x \in \mathbb{R}^n : f_1(x) = \cdots = f_k(x) = 0, g_1(x) \geq 0, \ldots, g_l(x) \geq 0, \\ h_1(x) > 0, \ldots, h_m(x) > 0\}. \end{aligned} \tag{1.1}$$

The set (1.1) is a *basic semialgebraic set*. The Positivstellensatz [32, THEOREM 6.14] gives a criterion for deciding whether this set is empty. This seemingly theoretical criterion has become a practical numerical method, thanks to sums of squares [32, §12.3] and semidefinite programming [7]. In addition to this, there are symbolic algorithms for real algebraic geometry (cf. [5]). So, the user has a wide range of choices for working with semialgebraic sets.

In this article we disregard the inequalities in (1.1) and retain the equations only:

$$X = \{x \in \mathbb{R}^n : f_1(x) = \cdots = f_k(x) = 0\}. \tag{1.2}$$

This is a *real algebraic variety*. We wish to answer questions about X by reliable numerical computations, in particular using tools such as `Bertini` [6] or `HomotopyContinuation.jl` [10]. We focus on questions that are addressed by solving auxiliary polynomial systems with finitely many solutions, where the number of complex solutions can be determined a priori.

In Section 2 that number is the Euclidean distance degree (ED degree) of X. This governs the following question: given $u \in \mathbb{R}^n \backslash X$, which point in X is nearest to u in Euclidean distance? We derive the critical equations of this optimization problem (2.1), and we consider all solutions to these equations, both real and complex. These include all local minima and local maxima. Theorem 2.5 expresses the ED degree in terms of the polar degrees of X. Knowing these invariants allows us to find all critical points numerically, along with a proof of correctness [9]. We ask our nearest point question also for other norms, notably those given by a polytope. The polar degrees appear again, in Proposition 2.9.

Section 3 concerns algebraic varieties X that serve as models in statistics. Their points represent probability distributions. We focus on models for Gaussian distributions and discrete distributions. In these two scenarios, the ambient space $\mathbb{R}^n$ in (1.2) is replaced by the positive-definite cone PD_n and by the probability simplex Δ_n. Given any data set, we ask whether X is an appropriate model. To this end, maximum likelihood estimation (MLE) is used. This optimization problem is stated in (3.2) and (3.7). We employ nonlinear algebra **[32]** in addressing it. The number of complex critical points is the maximum likelihood degree (ML degree) of the model X. Theorem 3.7 relates this to the Euler characteristic of the underlying very affine variety. We apply this theory to a class of models arising in particle physics, namely the configuration space of m labeled points in general position in $\mathbb{P}^{k-1}$. Known ML degrees for these models are given in Theorem 3.14.

In Section 4, we turn to an analytic interpretation of the polynomial system in (1.2). The unknowns $x_1, \dots, x_n$ are replaced by differential operators $\frac{\partial}{\partial z_1}, \dots, \frac{\partial}{\partial z_n}$. The polynomials $f_1, \dots, f_k$ are viewed as linear partial differential equations (PDEs) with constant coefficients. The variety X is replaced by the space of functions $\phi(z_1, \dots, z_n)$ that are solutions to the PDE. That space is typically infinite-dimensional. Our task is to compute it. Algorithms are based on differential primary decompositions **[2, 17, 18]**. We also study linear PDEs for vector-valued functions. These are expressed by modules over a polynomial ring.

This article accompanies a lecture to be given in July 2022 at the International Congress of Mathematicians in St. Petersburg. It encourages mathematical scientists to employ polynomials in designing models and in thinking about numerical algorithms. Sections 2 and 3 are concerned with critical point equations in optimization and statistics. Section 4 offers a glimpse on how nonlinear algebra interfaces with the study of linear PDEs.

2. NEAREST POINTS ON ALGEBRAIC VARIETIES

We consider a model X that is given as the zero set in $\mathbb{R}^n$ of a collection $\{f_1, \dots, f_k\}$ of nonlinear polynomials in n unknowns $x_1, \dots, x_n$. Thus, X is a real algebraic variety. We assume that X is irreducible, that $I_X = \langle f_1, \dots, f_k \rangle$ is its prime ideal, and that the set of nonsingular real points is Zariski dense in X. The $k \times n$ Jacobian matrix $\mathcal{J} = (\partial f_i / \partial x_j)$ has rank at most c at any point $x \in X$, where $c = \mathrm{codim}(X)$, and x is *nonsingular* on X if the rank is exactly c. Explanations of these hypotheses are found in Chapter 2 of the textbook **[32]**.

The following optimization problem arises in many applications. Given a data point $u \in \mathbb{R}^n \backslash X$, compute the distance to the model X. Thus, we seek a point x^* in X that is closest to u. The answer depends on the chosen metric. One might choose the Euclidean distance, a p-norm **[29]**, or polyhedral norms, such as those arising in optimal transport **[15]**. In all of these cases, the solution x^* can be found by solving a system of polynomial equations.

We begin by discussing the *Euclidean distance (ED) problem*, which is as follows:

$$\text{minimize } \sum_{i=1}^{n} (x_i - u_i)^2 \text{ subject to } x \in X. \tag{2.1}$$

We now derive the critical equations for (2.1). The *augmented Jacobian matrix* $\mathcal{AJ}$ is the $(k+1) \times n$ matrix obtained by placing the row $(x_1 - u_1, \ldots, x_n - u_n)$ atop the Jacobian matrix $\mathcal{J}$. We form the ideal generated by its $(c+1) \times (c+1)$ minors, we add the ideal of the model I_X, and we then saturate [19, (2.1)] that sum by the ideal of $c \times c$ minors of $\mathcal{J}$. The result is the *critical ideal* $\mathcal{C}_{X,u}$ of the model X with respect to the data u. The variety of $\mathcal{C}_{X,u}$ is the set of critical points of (2.1). For random data u, this variety is finite and it contains the optimal solution x^*, provided the latter is attained at a nonsingular point of X.

The algebro-geometric approach to the ED problem was pioneered in a project with Draisma, Horobeţ, Ottaviani, and Thomas [19]. That article introduced the *ED degree* of X. This is the cardinality of the complex algebraic variety in $\mathbb{C}^n$ defined by the critical ideal $\mathcal{C}_{X,u}$. The ED degree of a model X measures the difficulty of solving the ED problem for X.

Example 2.1 (Space curves). Fix $n = 3$ and let X be the curve in $\mathbb{R}^3$ defined by two general polynomials f_1 and f_2 of degrees d_1 and d_2 in x_1, x_2, x_3. The augmented Jacobian matrix is

$$\mathcal{AJ} = \begin{pmatrix} x_1 - u_1 & x_2 - u_2 & x_3 - u_3 \\ \partial f_1/\partial x_1 & \partial f_1/\partial x_2 & \partial f_1/\partial x_3 \\ \partial f_2/\partial x_1 & \partial f_2/\partial x_2 & \partial f_2/\partial x_3 \end{pmatrix}. \tag{2.2}$$

For random data $u \in \mathbb{R}^3$, the ideal $\mathcal{C}_{X,u} = \langle f_1, f_2, \det(\mathcal{AJ})\rangle$ has $d_1 d_2 (d_1 + d_2 - 1)$ zeros in $\mathbb{C}^3$, by Bézout [32, THEOREM 2.16]. Hence the ED degree of X equals $d_1 d_2 (d_1 + d_2 - 1)$. This can also be seen using the general formula from algebraic geometry in [19, COROLLARY 5.9]. If X is a general smooth curve of degree d and genus g, then $\mathrm{EDdegree}(X) = 3d + 2g - 2$. The above curve in 3-space has degree $d = d_1 d_2$ and genus $g = d_1^2 d_2/2 + d_1 d_2^2/2 - 2d_1 d_2 + 1$.

Here is a general upper bound on the ED degree in terms of the given polynomials.

Proposition 2.2. *Let X be a variety of codimension c in $\mathbb{R}^n$ whose ideal I_X is generated by polynomials $f_1, f_2, \ldots, f_c, \ldots, f_k$ of degrees $d_1 \geq d_2 \geq \cdots \geq d_c \geq \cdots \geq d_k$. Then*

$$\mathrm{EDdegree}(X) \leq d_1 d_2 \cdots d_c \cdot \sum_{i_1 + i_2 + \cdots + i_c \leq n-c} (d_1 - 1)^{i_1} (d_2 - 1)^{i_2} \cdots (d_c - 1)^{i_c}. \tag{2.3}$$

Equality holds when X is a generic complete intersection of codimension c (hence $c = k$).

This appears in [19, PROPOSITION 2.6]. We can derive it as follows. Bézout's Theorem ensures that the degree of the variety X is at most $d_1 d_2 \cdots d_c$. The entries in the ith row of the matrix $\mathcal{AJ}$ are polynomials of degrees $d_i - 1$. The degree of the variety of $(c+1) \times (c+1)$ minors of $\mathcal{AJ}$ is at most the sum in (2.3). The intersection of that variety with X is our set of critical points, and the cardinality of that set is bounded by the product of the two degrees. Generically, that intersection is a complete intersection and inequality (2.3) is attained.

Formulas or a priori bounds for the ED degree are important when studying exact solutions to the optimization problem (2.1). The paradigm is to compute all complex critical points, by either symbolic or numerical methods, and to then extract one's favorite real solutions among these. This leads, for instance, to all local minima in (2.1). The ED degree is an upper bound on the number of real critical points. This bound is generally not tight.

Example 2.3. Consider the case $n = 2$, $c = 1$, $d_1 = 4$ in Proposition 2.2, where X is a quartic curve in the plane $\mathbb{R}^2$. The number of complex critical points is $\text{EDdegree}(X) = 16$. But, they cannot be all real. For an illustration, consider the *Trott curve* $X = V(f)$, defined by

$$f = 144(x_1^4 + x_2^4) - 225(x_1^2 + x_2^2) + 350x_1^2x_2^2 + 81.$$

For general data $u = (u_1, u_2)$ in $\mathbb{R}^2$, we find 16 complex solutions to the critical equations $f = \frac{\partial f}{\partial x_2}(x_1 - u_1) - \frac{\partial f}{\partial x_1}(x_2 - u_2) = 0$. For u near the origin, eight of them are real. For $u = (\frac{7}{8}, \frac{1}{100})$, which is inside the rightmost oval, there are 10 real critical points. The two scenarios are shown in Figure 1. Local minima are green, while local maxima are purple. For $u = (2, \frac{1}{100})$, to the right of the rightmost oval, the number of real critical points is 12.

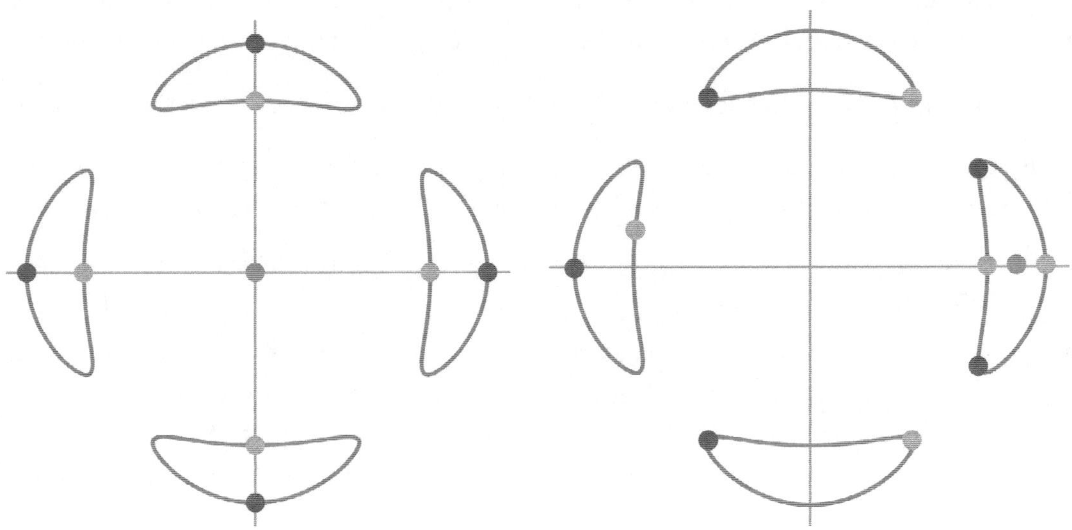

FIGURE 1
ED problems on the Trott curve: configurations of eight (left) or ten (right) critical points.

In general, our task is to compute the zeros of the critical ideal $C_{X,u}$. Algorithms for this computation can be either symbolic or numerical. Symbolic methods usually rest on the construction of a Gröbner basis, to be followed by a floating-point computation to extract the solutions. In recent years, numerical methods have become increasingly popular. These are based on homotopy continuation. Two notable packages are `Bertini` **[6]** and `HomotopyContinuation.jl` **[10]**. The ED degree is important here because it indicates how many paths need to be tracked to solve (2.1). We next illustrate current capabilities.

Example 2.4. Suppose X is defined by $c = k = 3$ random polynomials in $n = 7$ variables, for a range of degrees d_1, d_2, d_3. The table below lists the ED degree in each case, and the times used by `HomotopyContinuation.jl` to compute and certify all critical points in $\mathbb{C}^7$.

$d_1\, d_2\, d_3$	3 2 2	3 3 2	3 3 3	4 2 2	4 3 2	4 3 3	4 4 2	4 4 3
EDdegree	1188	3618	9477	4176	10152	23220	23392	49872
Solving (s)	3.849	21.06	61.51	31.51	103.5	280.0	351.5	859.3
Certifying (s)	0.390	1.549	4.653	2.762	7.591	17.16	21.65	50.07

Here we represent $C_{X,u}$ by a system of 10 equations in 10 variables. In addition to the three equations $f_1 = f_2 = f_3 = 0$ in $x_1, \ldots, x_7$, we take the seven equations $(1, y_1, y_2, y_3) \cdot \mathcal{A}\mathcal{J} = 0$. Here y_1, y_2, y_3 are new variables. These ensure that the 4×7 matrix $\mathcal{A}\mathcal{J}$ has rank ≤ 3. In all cases the timings include the certification step [9] that proves correctness and completeness. These computations were performed using `HomotopyContinuation.jl` v2.5.6 on a 16 GB MacBook Pro with an Intel Core i7 processor working at 2.6 GHz. They suggest that our critical equations can be solved fast and reliably, with proof of correctness, when the ED degree is less than 50000. For even larger numbers of solutions, success with numerical path tracking will depend on the specific structure of the problem. If the discriminant is well-behaved, then larger ED degrees are feasible. An example of this appears in [34, TABLE 1].

We next present a general formula for ED degrees in terms of projective geometry.

Theorem 2.5. *If X meets both the hyperplane at infinity and the isotropic quadric transversally, then* EDdegree(X) *equals the sum of the polar degrees of the projective closure of X.*

The *projective closure* of $X \subset \mathbb{R}^n$ is its Zariski closure in the complex projective space $\mathbb{P}^n$, which we will also denote by X. Theorem 2.5 appears in [19, PROPOSITION 6.10]. The hypothesis is stated precisely in [19, EQUATION (6.4)]. It holds for all X after a general linear change of coordinates. We now explain what the polar degrees of a variety $X \subset \mathbb{P}^n$ are. Points h in the dual projective space $(\mathbb{P}^n)^\vee$ represent hyperplanes $\{x \in \mathbb{P}^n : h_0x_0 + \cdots + h_nx_n = 0\}$. We are interested in all pairs (x, h) in $\mathbb{P}^n \times (\mathbb{P}^n)^\vee$ such that x is a nonsingular point of X and h is tangent to X at x. The Zariski closure of this set is the *conormal variety* $N_X \subset \mathbb{P}^n \times (\mathbb{P}^n)^\vee$.

It is known that N_X has dimension $n - 1$, and if X is irreducible then so is N_X. The image of N_X under projection onto the second factor is the dual variety $X^\vee$. The role of $x \in \mathbb{P}^n$ and $h \in (\mathbb{P}^n)^\vee$ can be swapped. The following biduality relations [22, §I.1.3] hold:

$$N_X = N_{X^\vee} \quad \text{and} \quad (X^\vee)^\vee = X.$$

The class of N_X in the cohomology ring $H^*(\mathbb{P}^n \times (\mathbb{P}^n)^\vee, \mathbb{Z}) = \mathbb{Z}[s,t]/\langle s^{n+1}, t^{n+1}\rangle$ has the form

$$[N_X] = \delta_1(X)s^n t + \delta_2(X)s^{n-1}t^2 + \delta_3(X)s^{n-2}t^3 + \cdots + \delta_n(X)st^n.$$

The coefficients $\delta_i(X)$ of this binary form are nonnegative integers, known as *polar degrees*.

Remark 2.6. The polar degrees satisfy $\delta_i(X) = \#(N_X \cap (L \times L'))$, where $L \subset \mathbb{P}^n$ and $L' \subset (\mathbb{P}^n)^\vee$ are general linear subspaces of dimensions $n + 1 - i$ and i, respectively. This geometric interpretation implies that $\delta_i(X) = 0$ for $i < \mathrm{codim}(X^\vee)$ and for $i > \dim(X) + 1$.

Example 2.7. Let X be a general surface of degree d in $\mathbb{P}^3$. Its dual $X^\vee$ is a surface of degree $d(d-1)^2$ in $(\mathbb{P}^3)^\vee$. The conormal variety N_X is a surface in $\mathbb{P}^3 \times (\mathbb{P}^3)^\vee$, with class

$$[N_X] = d(d-1)^2 s^3 t + d(d-1)s^2 t^2 + d\, st^3.$$

The sum of the three polar degrees equals $\mathrm{EDdegree}(X) = d^3 - d^2 + d$; see Proposition 2.2.

Theorem 2.5 allows us to compute the ED degree for many interesting varieties, e.g., using Chern classes [19, THEOREM 5.8]. This is relevant for applications in machine learning [11] which rest on low-rank approximation of matrices and tensors with special structure [33].

The discussion so far was restricted to the Euclidean norm. But, we can measure distances in $\mathbb{R}^n$ with any other norm $\|\cdot\|$. Our optimization problem (2.1) extends naturally:

$$\text{minimize } \|x - u\| \text{ subject to } x \in X. \tag{2.4}$$

The unit ball $B = \{x \in \mathbb{R}^n : \|x\| \leq 1\}$ is a centrally-symmetric convex body. Conversely, every centrally-symmetric convex body B defines a norm, and we can paraphrase (2.4) as follows:

$$\text{minimize } \lambda \text{ subject to } \lambda \geq 0 \text{ and } (u + \lambda B) \cap X \neq \emptyset. \tag{2.5}$$

If the boundary of B is smooth and algebraic then we express the critical equations as a polynomial system. This is derived as before, but we now replace the first row of the augmented Jacobian matrix $\mathcal{AJ}$ with the gradient of the map $\mathbb{R}^n \to \mathbb{R}$, $x \mapsto \|x - u\|$.

Another case of interest arises when $\|\cdot\|$ is a *polyhedral norm*. This means that B is a centrally-symmetric polytope. Familiar examples of polyhedral norms are $\|\cdot\|_\infty$ and $\|\cdot\|_1$, where B is the cube and the crosspolytope, respectively. In optimal transport theory, one uses a Wasserstein norm [15] whose unit ball B is the polar dual of a Lipschitz polytope.

To derive the critical equations, a combinatorial stratification of the problem is used, given by the face poset of the polytope B. Suppose that X is in general position. Then $(u + \lambda^* B) \cap X = \{x^*\}$ is a singleton for the optimal value λ^* in (2.5). The point $\frac{1}{\lambda^*}(x^* - u)$ lies in the relative interior of a unique face F of the unit ball B. Let L_F denote the linear span of F in $\mathbb{R}^n$. We have $\dim(L_F) = \dim(F) + 1$. Let ℓ be any linear functional on $\mathbb{R}^n$ that attains its minimum over the polytope B at the face F. We view ℓ as a point in $(\mathbb{P}^n)^\vee$.

Lemma 2.8. *The optimal point x^* in* (2.4) *is the unique solution to the optimization problem*

$$\textit{minimize } \ell(x) \textit{ subject to } x \in (u + L_F) \cap X. \tag{2.6}$$

Proof. The general position hypothesis ensures that $u + L_F$ intersects X transversally, and x^* is a smooth point of that intersection. Moreover, x^* is a minimum of the restriction of ℓ to the variety $(u + L_F) \cap X$. By our hypothesis, this linear function is generic relative to the variety, so the number of critical points is finite and the function values are distinct. ■

Problem (2.6) amounts to linear programming over a real variety. We now determine the algebraic degree of this optimization task when F is a face of codimension i.

Proposition 2.9. *Let L be a general affine-linear space of codimension $i-1$ in $\mathbb{R}^n$ and ℓ a general linear form. The number of critical points of ℓ on $L \cap X$ is the polar degree $\delta_i(X)$.*

Proof. This result is [15, THEOREM 5.1]. The number of critical points of a linear form is the degree of the dual variety $(L \cap X)^\vee$. That degree coincides with the polar degree $\delta_i(X)$. ∎

Example 2.10. Consider (2.4) and (2.5) where X is a general surface of degree d in $\mathbb{R}^3$. The optimal face F of the unit ball B depends on the location of the data point u. This is shown for $d=2$ and $\|\cdot\|_\infty$ in Figure 2. The algebraic degree of the solution x^* equals $\delta_3(X)=d$ if $\dim(F)=0$, it is $\delta_2(X)=d(d-1)$ if $\dim(F)=1$, and it is $\delta_1(X)=d(d-1)^2$ if $\dim(F)=2$.

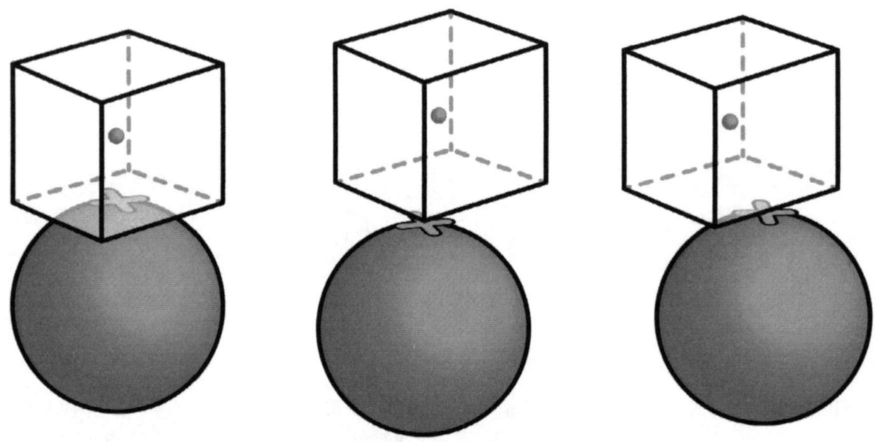

FIGURE 2
The cube is the $\|\cdot\|_\infty$ ball $\lambda^* B$ around the green point u. The variety X is the sphere. The contact point x^* is marked with a cross. The optimal face F is a facet, a vertex, or an edge.

We conclude that the conormal variety N_X and its cohomology class $[N_X]$ are key players when it comes to reliably solving the distance minimization problem for a variety X. The polar degrees $\delta_i(X)$ reveal precisely how many paths need to be tracked by numerical solvers like [6, 10] in order to find and certify [9] the optimal solution x^* to (2.1) or (2.4).

3. LIKELIHOOD GEOMETRY

The previous section was concerned with minimizing the distance from a given data point u to a model X that is described by polynomial equations. In what follows, we consider the analogous problem in the setting of algebraic statistics [36], where the model X represents a family of probability distributions. Distance to u is replaced by the log-likelihood function.

The two scenarios of most interest for statisticians are Gaussian models and discrete models. We shall discuss them both, beginning with the Gaussian case. Let PD_n denote the open convex cone of positive-definite symmetric $n \times n$ matrices. Given a mean vector $\mu \in \mathbb{R}^n$ and a covariance matrix $\Sigma \in \mathrm{PD}_n$, the associated *Gaussian distribution* on $\mathbb{R}^n$ has the density

$$f_{\mu,\Sigma}(x) := \frac{1}{\sqrt{(2\pi)^n \det \Sigma}} \cdot \exp\left(-\frac{1}{2}(x-\mu)^T \Sigma^{-1}(x-\mu)\right).$$

We fix a model $Y \subset \mathbb{R}^n \times \mathrm{PD}_n$ that is defined by polynomial equations in (μ, Σ). Suppose we are given N samples $U^{(1)}, \dots, U^{(N)}$ in $\mathbb{R}^n$. These are summarized in the *sample mean* $\bar{U} = \frac{1}{N}\sum_{i=1}^N U^{(i)}$ and in the *sample covariance matrix* $S = \frac{1}{N}\sum_{i=1}^N (U^{(i)} - \bar{U})(U^{(i)} - \bar{U})^T$. Given these data, the log-likelihood is the following function in the unknowns (μ, Σ):

$$\ell(\mu, \Sigma) = -\frac{N}{2} \cdot \left[\log\det\Sigma + \mathrm{trace}(S\Sigma^{-1}) + (\bar{U}-\mu)^T \Sigma^{-1}(\bar{U}-\mu)\right]. \tag{3.1}$$

The task of likelihood inference is to minimize this function subject to $(\mu, \Sigma) \in Y$.

There are two extreme cases. First, consider a model where Σ is fixed to be the identity matrix Id_n. Then $Y = X \times \{\mathrm{Id}_n\}$ and we are supposed to minimize the Euclidean distance from the sample mean $\bar{U}$ to the variety X in $\mathbb{R}^n$. This is precisely our problem (2.1).

We instead focus on the second case, the family of *centered Gaussians*, where μ is fixed at zero. The model has the form $\{0\} \times X$, where X is a variety in the space $\mathrm{Sym}_2(\mathbb{R}^n)$ of symmetric $n \times n$ matrices. Following [36, PROPOSITION 7.1.10], our task is now as follows:

$$\text{minimize the function } \Sigma \mapsto \log\det\Sigma + \mathrm{trace}(S\Sigma^{-1}) \quad \text{subject to } \Sigma \in X. \tag{3.2}$$

Using the concentration matrix $K = \Sigma^{-1}$, we can write this equivalently as follows:

$$\text{maximize the function } \Sigma \mapsto \log\det K - \mathrm{trace}(SK) \quad \text{subject to } K \in X^{-1}. \tag{3.3}$$

Here the variety X^{-1} is the Zariski closure of the set of inverses of all matrices in X.

The critical equations of the optimization problem (3.3) can be written as polynomials, since the partial derivatives of the logarithm are rational functions. These equations have finitely many complex solutions. Their number is the *ML degree* of the model X^{-1}.

Let $\mathcal{L} \subset \mathrm{Sym}_2(\mathbb{R}^n)$ be a linear space of symmetric matrices (LSSM), whose general element is assumed to be invertible. We are interested in the models $X^{-1} = \mathcal{L}$ and $X = \mathcal{L}$. It is convenient to use primal–dual coordinates (Σ, K) to write the respective critical equations.

Proposition 3.1. *Fix an LSSM $\mathcal{L}$ and its orthogonal complement $\mathcal{L}^\perp$ for the inner product $\langle X, Y\rangle = \mathrm{trace}(XY)$. The critical equations for the* linear concentration model $X^{-1} = \mathcal{L}$ *are*

$$K \in \mathcal{L}, \quad K\Sigma = \mathrm{Id}_n, \quad \textit{and} \quad \Sigma - S \in \mathcal{L}^\perp. \tag{3.4}$$

The critical equations for the linear covariance model $X = \mathcal{L}$ *are*

$$\Sigma \in \mathcal{L}, \quad K\Sigma = \mathrm{Id}_n, \quad \textit{and} \quad KSK - K \in \mathcal{L}^\perp. \tag{3.5}$$

Proof. This is well known in statistics. For proofs see [35, PROPOSITIONS 3.1 AND 3.3]. ■

The system (3.4) is linear in K, but the last group of equations in (3.5) is quadratic in K. The numbers of complex solutions are the *ML degree* of $\mathcal{L}$ and the *reciprocal ML degree* of $\mathcal{L}$. The former is smaller than the latter, and (3.4) is easier to solve than (3.5).

Example 3.2. Let $n = 4$ and let $\mathcal{L}$ be a generic LSSM of dimension k. Our degrees are as follows:

$k = \dim(\mathcal{L})$	2	3	4	5	6	7	8	9
ML degree	3	9	17	21	21	17	9	3
reciprocal ML degree	5	19	45	71	81	63	29	7

These numbers and many more appear in [35, TABLE 1].

ML degrees and the reciprocal ML degrees have been studied intensively in the recent literature, both for generic and special spaces $\mathcal{L}$. See [3,8,21] and the references therein. We now present an important result due to Manivel, Michałek, Monin, Seynnaeve, Vodička, and Wiśniewski. Theorem 3.3 paraphrases highlights from their articles [30, 31].

Theorem 3.3. *The ML degree of a generic linear subspace $\mathcal{L}$ of dimension k in* $\mathrm{Sym}_2(\mathbb{R}^n)$ *is the number of quadrics in $\mathbb{P}^{n-1}$ that pass through $\binom{n+1}{2} - k$ general points and are tangent to $k-1$ general hyperplanes. For fixed k, this number is a polynomial in n of degree $k-1$.*

Proof. The first statement is [31, COROLLARY 2.6 (4)], here interpreted classically in terms of Schubert calculus. For a detailed discussion, see the introduction of [30]. The second statement appears in [30, THEOREM 1.3 AND COROLLARY 4.13]. It proves a conjecture of Sturmfels and Uhler. ■

Example 3.4 ($n = 4$). Fix $10 - k$ points and $k - 1$ planes in $\mathbb{P}^3$. We seek quadratic surfaces containing the points and tangent to the planes. This imposes 9 constraints on $\mathbb{P}(\mathrm{Sym}_2(\mathbb{C}^4)) \simeq \mathbb{P}^9$. Passing through a point is a linear equation. Being tangent to a plane is a cubic equation. Bézout's Theorem suggests that there could be 3^{k-1} solutions. This is correct for $k \leq 3$ but it overcounts for $k \geq 4$. Indeed, in Example 3.2 we see $17, 21, 21, \ldots$ instead of $27, 81, 243, \ldots$

The intersection theory in [30, 31] leads to formulas for the ML degrees of linear Gaussian models. From this we obtain provably correct numerical methods for maximum likelihood estimation. Namely, after computing critical points as in [35], we can certify them as in [9]. Since the ML degree is known, one can be sure that all solutions have been found.

We now shift gears and turn our attention to discrete statistical models. We take the state space to be $\{0, 1, \ldots, n\}$. The role of the cone PD_n is played by the probability simplex

$$\Delta_n = \{p = (p_0, p_1, \ldots, p_n) \in \mathbb{R}^{n+1} : p_0 + p_1 + \cdots + p_n = 1 \text{ and } p_0, p_1, \ldots, p_n > 0\}. \tag{3.6}$$

Our model is a subset X of Δ_n defined by polynomial equations. As before, for venturing beyond linear algebra, we identify X with its Zariski closure in complex projective space $\mathbb{P}^n$.

We shall present the algebraic approach to maximum likelihood estimation (MLE). See [14, 20, 25, 27, 28, 36] and references therein. Suppose we are given N i.i.d. samples. These are summarized in the data vector $u = (u_0, u_1, \ldots, u_n)$ where u_i is the number of times state i was observed. Note that $N = u_0 + \cdots + u_n$. The associated log-likelihood function equals

$$\ell_u : \Delta_n \to \mathbb{R}, \quad p \mapsto u_0 \cdot \log(p_0) + u_1 \cdot \log(p_1) + \cdots + u_n \cdot \log(p_n).$$

Performing MLE for the model X means solving the following optimization problem:

$$\text{maximize } \ell_u(p) \text{ subject to } p \in X. \tag{3.7}$$

The *ML degree* of X is the number of complex critical points of (3.7) for generic data u. The optimal solution is denoted $\hat{p}$ and called the *maximum likelihood estimate* for the data u.

The critical equations for (3.7) are similar to those of (2.1). Let $I_X = \langle f_1, \ldots, f_k \rangle + \langle p_0 + p_1 + \cdots + p_n - 1 \rangle$ be the defining ideal of the model. Let $\mathcal{J} = (\partial f_i / \partial p_j)$ denote the Jacobian matrix of size $(k+1) \times (n+1)$, and set $c = \operatorname{codim}(X)$. The augmented Jacobian $\mathcal{A}\mathcal{J}$ is obtained by prepending one more row, namely the gradient of the objective function

$$\nabla \ell_u = (u_0/p_0, u_1/p_1, \ldots, u_n/p_n).$$

To obtain the critical equations, enlarge I_X by the $c \times c$ minors of the $(k+2) \times (n+1)$ matrix $\mathcal{A}\mathcal{J}$, then clear denominators, and finally remove extraneous components by saturation.

Example 3.5 (Space curves). Let $n = 3$ and X the curve in Δ_3 defined by two general polynomials f_1 and f_2 of degrees d_1 and d_2 in p_0, p_1, p_2, p_3. The augmented Jacobian matrix is

$$\mathcal{A}\mathcal{J} = \begin{pmatrix} u_0/p_0 & u_1/p_1 & u_2/p_2 & u_3/p_3 \\ 1 & 1 & 1 & 1 \\ \partial f_1/\partial p_0 & \partial f_1/\partial p_1 & \partial f_1/\partial p_2 & \partial f_1/\partial p_3 \\ \partial f_2/\partial p_0 & \partial f_2/\partial p_1 & \partial f_2/\partial p_2 & \partial f_2/\partial p_3 \end{pmatrix}. \tag{3.8}$$

Clearing denominators amounts to multiplying the ith column by p_i, so the determinant contributes a polynomial of degree $d_1 + d_2 + 1$ to the critical equations. Since the generators of I_X have degrees $d_1, d_2, 1$, we conclude that the ML degree of X equals $d_1 d_2 (d_1 + d_2 + 1)$.

The following MLE analogue to Proposition 2.2 is established in [25, THEOREM 5].

Proposition 3.6. *Let X be a model of codimension c in Δ_n whose ideal I_X is generated by polynomials $f_1, f_2, \ldots, f_c, \ldots, f_k$ of degrees $d_1 \geq d_2 \geq \cdots \geq d_c \geq \cdots \geq d_k$. Then*

$$\text{MLdegree}(X) \leq d_1 d_2 \cdots d_c \cdot \sum_{i_1 + i_2 + \cdots + i_c \leq n-c} d_1^{i_1} d_2^{i_2} \cdots d_c^{i_c}. \tag{3.9}$$

Equality holds when X is a generic complete intersection of codimension c (hence $c = k$).

We next present the MLE analogue to Theorem 2.5. The role of the polar degrees is now played by the Euler characteristic. Consider X in the complex projective space $\mathbb{P}^n$, and let X^o be the open subset of X that is obtained by removing $\{p_0 p_1 \cdots p_n(\sum_{i=0}^n p_i) = 0\}$. We recall from **[26,27]** that a *very affine variety* is a closed subvariety of an algebraic torus $(\mathbb{C}^*)^r$.

Theorem 3.7. *Suppose that the very affine variety X^o is nonsingular. The ML degree of the model X equals the signed Euler characteristic $(-1)^{\dim(X)} \cdot \chi(X^o)$ of the manifold X^o.*

Proof and discussion. This was proved with a further smoothness assumption in **[14, THEOREM 19]**, and in full generality in **[26, THEOREM 1]**. If X^o is singular then the Euler characteristic can be replaced by the Chern–Schwartz–MacPherson class, as shown in **[26, THEOREM'2]**. ■

Of special interest is the case when the ML degree is equal to one. This means that the estimate $\hat{p}$ is a rational function of the data u. Here are two examples where this happens.

Example 3.8 ($n = 3$). The independence model for two binary random variables is a quadratic surface X in the tetrahedron Δ_3. This model is described by the constraints

$$\det \begin{bmatrix} p_0 & p_1 \\ p_2 & p_3 \end{bmatrix} = 0 \quad \text{and} \quad p_0 + p_1 + p_2 + p_3 = 1 \quad \text{and} \quad p_0, p_1, p_2, p_3 > 0.$$

Consider data $u = \left[\begin{smallmatrix} u_0 & u_1 \\ u_2 & u_3 \end{smallmatrix}\right]$ of *sample size* $|u| = u_0 + u_1 + u_2 + u_3$. The ML degree of the surface X equals one because the MLE $\hat{p}$ is a rational function of the data, namely

$$\begin{aligned} \hat{p}_0 &= |u|^{-2}(u_0+u_1)(u_0+u_2), & \hat{p}_1 &= |u|^{-2}(u_0+u_1)(u_1+u_3), \\ \hat{p}_2 &= |u|^{-2}(u_2+u_3)(u_0+u_2), & \hat{p}_3 &= |u|^{-2}(u_2+u_3)(u_1+u_3). \end{aligned} \tag{3.10}$$

In words, we multiply the row sums with the column sums in the empirical distribution $\frac{1}{|u|}u$.

Example 3.9 ($n = 2$). Given a biased coin, we perform the following experiment: *Flip a biased coin. If it shows heads, flip it again.* The outcome is the number of heads: 0, 1, or 2. This simple model is visualized in Figure 3.

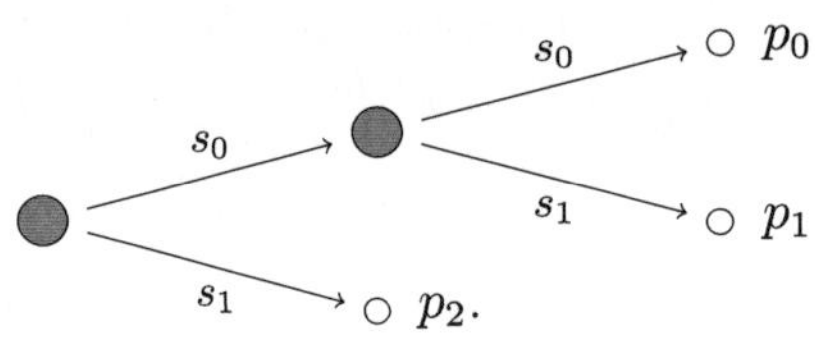

FIGURE 3
Probability tree that describes the coin toss model in Example 3.9.

If s is the bias of the cone, then the model is the parametric curve X given by

$$(0, 1) \to X \subset \Delta_2, \quad s \mapsto \left(s^2, s(1 - s), 1 - s\right).$$

This model is the conic $X = V(p_0 p_2 - (p_0 + p_1)p_1) \subset \mathbb{P}^2$. The MLE is given by the formula

$$(\hat{p}_0, \hat{p}_1, \hat{p}_2) = \left(\frac{(2u_0 + u_1)^2}{(2u_0+2u_1+u_2)^2}, \frac{(2u_0+u_1)(u_1+u_2)}{(2u_0 + 2u_1 + u_2)^2}, \frac{u_1 + u_2}{2u_0+2u_1+u_2} \right). \tag{3.11}$$

Since the coordinates of $\hat{p}$ are rational functions, the ML degree of X is equal to one.

The following theorem explains what we saw in equations (3.10) and (3.11):

Theorem 3.10. *If $X \subset \Delta_n$ is a model of ML degree one, so $\hat{p}$ is a rational function of u, then each coordinate $\hat{p}_i$ is an alternating product of linear forms with positive coefficients.*

Proof and discussion. This was shown for very affine varieties in [27]. It was adapted to statistical models in [20]. These articles offer precise statements via Horn uniformization for A-discriminants [22], i.e., hypersurfaces dual to toric varieties. See also [28, COROLLARY 3.12]. ■

This section concludes with a connection to scattering amplitudes in particle physics that was discovered recently in [34]. We consider the *CEGM model*, due to Cachazo and his collaborators [12, 13]. The role of the data vector u is played by the Mandelstam invariants. This theory rests on the space X^o of m labeled points in general position in $\mathbb{P}^{k-1}$, up to projective transformations. Consider the action of the torus $(\mathbb{C}^*)^m$ on the Grassmannian $\mathrm{Gr}(k,m) \subset \mathbb{P}^{\binom{m}{k}-1}$. Let $\mathrm{Gr}(k,m)^o$ be the open Grassmannian where all Plücker coordinates are nonzero. The CEGM model is the $(k-1)(m-k-1)$-dimensional manifold

$$X^o = \mathrm{Gr}(k,m)^o/(\mathbb{C}^*)^m. \tag{3.12}$$

Proposition 3.11. *The variety X^o is very affine, with coordinates given by the $k\times k$ minors of*

$$M_{k,m} = \begin{bmatrix} 0 & 0 & 0 & \cdots & 0 & (-1)^k & 1 & 1 & 1 & \cdots & 1 \\ 0 & 0 & 0 & \cdots & (-1)^{k-1} & 0 & 1 & x_{1,1} & x_{1,2} & \cdots & x_{1,m-k-1} \\ \vdots & \vdots & \vdots & ⋰ & \vdots & \vdots & \vdots & \vdots & \vdots & \ddots & \vdots \\ 0 & 0 & -1 & \cdots & 0 & 0 & 1 & x_{k-3,1} & x_{k-3,2} & \cdots & x_{k-3,m-k-1} \\ 0 & 1 & 0 & \cdots & 0 & 0 & 1 & x_{k-2,1} & x_{k-2,2} & \cdots & x_{k-2,m-k-1} \\ -1 & 0 & 0 & \cdots & 0 & 0 & 1 & x_{k-1,1} & x_{k-1,2} & \cdots & x_{k-1,m-k-1} \end{bmatrix}. \tag{3.13}$$

To be precise, the coordinates on $X^o \subset (\mathbb{C}^*)^{\binom{m}{k}}$ are the nonconstant minors $p_{i_1 i_2 \cdots i_k}$.

Following [1, EQUATION (4)], the antidiagonal matrix in the left $k \times k$ block of $M_{k,m}$ is chosen so that each unknown $x_{i,j}$ is precisely equal to $p_{i_1 i_2 \cdots i_k}$ for some $i_1 < i_2 < \cdots < i_k$. The *scattering potential* for the CEGM model is the following multivalued function on X^o:

$$\ell_u = \sum_{i_1, i_2, \ldots, i_k} u_{i_1 i_2 \cdots i_k} \cdot \log(p_{i_1 i_2 \cdots i_k}). \tag{3.14}$$

The critical point equations, known as *scattering equations* [1, EQUATION (7)], are given by

$$\frac{\partial \ell_u}{\partial x_{i,j}} = 0 \quad \text{for } 1 \leq i \leq k-1 \text{ and } 1 \leq j \leq m-k-1. \tag{3.15}$$

These are equations of rational functions. Solving these equations is the agenda in [12, 13, 34].

Corollary 3.12. *The number of complex solutions to (3.15) is the ML degree of the CEGM model X^o. This number equals the signed Euler characteristic $(-1)^{(k-1)(m-k-1)} \cdot \chi(X^o)$.*

Example 3.13 ($k = 2, m = 6$). The very affine threefold X^o is embedded in $(\mathbb{C}^*)^9$ via

$$p_{24} = x_1, \quad p_{25} = x_2, \quad p_{26} = x_3, \quad p_{34} = x_1 - 1, \quad p_{35} = x_2 - 1,$$
$$p_{36} = x_3 - 1, \quad p_{45} = x_2 - x_1, \quad p_{46} = x_3 - x_1, \quad p_{56} = x_3 - x_2.$$

These nine coordinates on $X^o \subset (\mathbb{C}^*)^9$ are the nonconstant 2×2 minors of our matrix

$$M_{2,6} = \begin{bmatrix} 0 & 1 & 1 & 1 & 1 & 1 \\ -1 & 0 & 1 & x_1 & x_2 & x_3 \end{bmatrix}.$$

The scattering potential is the analogue to the log-likelihood function in statistics:

$$\ell_u = u_{24}\log(p_{24}) + u_{25}\log(p_{25}) + \cdots + u_{56}\log(p_{56}).$$

This function has six critical points in X^o. Hence $\text{MLdegree}(X^o) = -\chi(X^o) = 6$.

We now examine the number of critical points of the scattering potential (3.14).

Theorem 3.14. *The known values of the ML degree for the CEGM model* (3.12) *are as follows. For* $k = 2$, *the ML degree equals* $(m-3)!$ *for all* $m \geq 4$. *For* $k = 3$, *it equals* $2, 26, 1272, 188112, 74570400$ *for* $m = 5, 6, 7, 8, 9$, *respectively, and for* $k = 4$, $m = 8$ *it equals* 5211816.

Proof. We refer to [1, EXAMPLE 2.2], [1, THEOREM 5.1] and [1, THEOREM 6.1] for $k = 2, 3, 4$. ■

Knowing these ML degrees helps in solving the scattering equations reliably. We demonstrated in [1,34] how this can be done in practice with `HomotopyContinuation.jl` [9,10]. For instance, we see in [34, TABLE 1] that the $10! = 3628800$ solutions for $k = 2, m = 13$ are found in under one hour. See [1, SECTION 6] for the solution in the challenging case $k = 4$, $m = 8$.

4. NONLINEAR ALGEBRA MEETS LINEAR PDES

In his 1938 article on the foundations of algebraic geometry, Wolfgang Gröbner introduced differential operators to characterize membership in a polynomial ideal. He solved this for zero-dimensional ideals using Macaulay's inverse systems [24]. Gröbner wanted this for all ideals, ideally with algorithmic methods. This was finally achieved in the article [18].

Analysts made substantial contributions to this subject. In the 1960s, Leon Ehrenpreis and Victor Palamodov studied solutions to linear partial differential equations (PDEs) with constant coefficients. A main step was the characterization of membership in a primary ideal by Noetherian operators. This led to their celebrated *Fundamental Principle*. That result is presented in Theorem 4.4. For background reading, see [2,17,18] and their references.

Example 4.1 ($n = 3$). We give an illustration by exploring a progression of four questions.
Question 1: What are the solutions to the system of equations $x_1^2 = x_2^2 = x_1x_3 - x_2x_3^2 = 0$?
Question 2: Determine all functions $\phi(z_1, z_2, z_3)$ that satisfy the following three linear PDEs:

$$\frac{\partial^2\phi}{\partial z_1^2} = \frac{\partial^2\phi}{\partial z_2^2} = \frac{\partial^2\phi}{\partial z_1 \partial z_3} - \frac{\partial^3\phi}{\partial z_2 \partial z_3^2} = 0.$$

Question 3: Which polynomials lie in the ideal

$$I = \langle x_1^2, x_2^2, x_1 - x_2x_3 \rangle \cap \langle x_1^2, x_2^2, x_3 \rangle? \tag{4.1}$$

Question 4: Describe the geometry of the subscheme $V(I)$ of affine 3-space given by (4.1).

Here are our answers to these four questions. Notice how they are intertwined:

Answer 1: Assuming that $x_i^2 = 0$ implies $x_i = 0$, the equations are equivalent to $x_1 = x_2 = 0$. Their solution set is a line through the origin in 3-space, namely the x_3-axis.

Answer 2: The solutions to these PDEs are precisely the functions $\phi(z)$ that have the form

$$\phi(z_1, z_2, z_3) = \xi(z_3) + \big(z_2\psi(z_3) + z_1\psi'(z_3)\big) + \alpha z_1 z_2 + \beta z_1, \tag{4.2}$$

where α, β are constants, and ξ and ψ are differentiable functions in one variable.

Answer 3: A polynomial f is in the ideal I if and only if the following four conditions hold: Both f and $\frac{\partial f}{\partial x_2} + x_3\frac{\partial f}{\partial x_1}$ vanish on the x_3-axis, and both $\frac{\partial^2 f}{\partial x_1 x_2}$ and $\frac{\partial f}{\partial x_1}$ vanish at the origin.

Answer 4: This scheme is a double x_3-axis together with an embedded point of length two at the origin. Hence I has arithmetic multiplicity four: two for the line and two for the point.

Answer 4 reveals the multiplicity structure on the naive solution set in Answer 1. This is characterized by four features, one for each differential condition in Answer 3. These are in natural bijection with the four summands of the general solution (4.2) in Answer 2.

We now turn to ideals I in the polynomial ring $\mathbb{C}[x] = \mathbb{C}[x_1, \ldots, x_n]$. We identify the n variables with differential operators $x_i = \partial_{z_i}$ that act on functions $\phi(z) = \phi(z_1, \ldots, z_n)$. In this manner, each I is a system of linear homogeneous PDEs with constant coefficients. This role of polynomials is the topic of Section 3.3 in the textbook [32]. The story begins in [32, LEMMA 3.25] with the following encoding of the variety $V(I)$ in the solutions to the PDE.

Lemma 4.2. *A point $a \in \mathbb{C}^n$ lies in the variety $V(I)$ if and only if the exponential function $\exp(a \cdot z) = \exp(a_1z_1 + \cdots + a_nz_n)$ is a solution to the system of linear PDE given by I.*

Since our PDEs are linear, their solution sets are linear spaces. Arbitrary $\mathbb{C}$-linear combinations of solutions are again solutions. The following proposition makes this precise.

Proposition 4.3. *Given any measure μ on the variety $V(I)$, here is a solution to our PDEs:*

$$\phi(z) = \int_{V(I)} \exp(a \cdot z)\, d\mu(a). \tag{4.3}$$

If I is a prime ideal then every solution to the PDEs admits such an integral representation.

The first part of Proposition 4.3 is straightforward. Recall that an ideal Q is *primary* if it has only one associated prime P. The second part is a special case of the following result.

Theorem 4.4 (Ehrenpreis–Palamodov). *Fix a prime ideal P in $\mathbb{C}[x]$. For any P-primary ideal Q in $\mathbb{C}[x]$, there exist polynomials $B_1, \ldots, B_m$ in $2n$ unknowns such that the function*

$$\phi(z) = \sum_{i=1}^{m} \int_{V(P)} B_i(x, z) \exp(x \cdot z)\, d\mu_i(x) \tag{4.4}$$

is a solution to the PDEs given by Q, for any measures $\mu_1, \ldots, \mu_m$ on the variety $V(P)$. Conversely, every solution $\phi(z)$ of the PDEs given by Q admits such an integral representation.

Proof. See **[17, THEOREM 3.3]** and the pointers to the analysis literature given there. ■

The polynomials $B_1(x,z), \ldots, B_m(x,z)$ are known as *Noetherian multipliers*. They depend only on the primary ideal Q, and not on the function $\phi(z)$. They encode the scheme structure imposed by Q on the irreducible variety $V(P)$. The Noetherian multipliers furnish a finite representation of a vector space that is usually infinite-dimensional, namely the space of all solutions to the PDE, within a suitable class of scalar-valued functions on n-space.

Example 4.5 ($n = 3$). Let $Q = \langle x_1^2, x_2^2, x_1 - x_2 x_3 \rangle$ be the first primary ideal in (4.1). Here $m = 2$, $B_1 = 1$, and $B_2 = x_3 z_1 + z_2$. Solutions to Q are given by the two summands in (4.4):

$$\phi_1(z) = \int 1 \cdot \exp(0z_1 + 0z_2 + x_3 z_3)\, d\mu_1(x) = \xi(z_3)$$

and

$$\begin{aligned}
\phi_2(z) &= \int (z_2 + z_1 x_3) \cdot \exp(0z_1 + 0z_2 + x_3 z_3)\, d\mu_2(x) \\
&= z_2 \int \exp(0z_1 + 0z_2 + x_3 z_3) d\mu_2(x) + z_1 \int x_3 \exp(0z_1 + 0z_2 + x_3 z_3) d\mu_2(x) \\
&= z_2 \psi(z_3) + z_1 \psi'(z_3).
\end{aligned}$$

We conclude that our solution $\phi_1(z) + \phi_2(z)$ agrees with the first two summands in (4.2).

Switching the roles of x and z, we now set $z_1 = \partial_{x_1}, \ldots, z_n = \partial_{x_n}$ in the Noetherian multipliers. Here it is important that the x-variables occur to the left of the z-variables in the monomial expansion of each $B_i(x,z)$. This results in the *Noetherian operators* $B_i(x, \partial_x)$. These operators are elements in the Weyl algebra and they act on polynomials in $\mathbb{C}[x]$. We use $\bullet$ to denote the action of differential operators on polynomials and other functions.

Proposition 4.6. *The Noetherian operators determine membership in the primary ideal Q. Namely, a polynomial $f(x)$ lies in Q if and only if $B_i(x, \partial_x) \bullet f(x)$ lies in P for $i = 1, \ldots, m$.*

Proof. This is the content of **[2, PROPOSITION 4.8]**. See also **[17, THEOREMS 3.2 AND 3.3]**. ■

Example 4.7. From B_1 and B_2 in Example 4.5, we obtain the Noetherian operators 1 and $x_3 \partial_{x_1} + \partial_{x_2}$. A polynomial f lies in Q if and only if f and $(x_3 \partial_{x_1} + \partial_{x_2}) \bullet f$ are in $P = \langle x_1, x_2 \rangle$.

We have seen that Noetherian multipliers and Noetherian operators are two sides of the same coin. While the latter characterize the membership in a primary ideal, as envisioned by Gröbner **[24]**, the former furnish the general solution to the associated PDEs. A next step is the extension from primary to arbitrary ideals in the polynomial ring $R = \mathbb{C}[x]$. To be more general, we consider an arbitrary submodule M of the free module R^k. Such a submodule represents a system of linear PDEs as before, but for vector-valued functions $\phi : \mathbb{C}^n \to \mathbb{C}^k$.

For a vector $m \in R^k$, the quotient $(M : m)$ is the ideal $\{f \in R : fm \in M\}$. A prime ideal $P_i \subseteq R$ is *associated to* the module M if $(M : m) = P_i$ for some $m \in R^k$. The list of all associated primes of M is finite, say $P_1, \dots, P_s$. If $s = 1$ then M is P_1-primary. A *primary decomposition* of M is a list of primary submodules $M_1, \dots, M_s \subseteq R^k$ where M_i is P_i-primary and $M = M_1 \cap M_2 \cap \cdots \cap M_s$. The contribution of the primary module M_i to M is quantified by a positive integer m_i, called the arithmetic length of M along P_i. To define this, we consider the localization $(R_{P_i})^k / M_{P_i}$. This is a module over the local ring R_{P_i}. The *arithmetic length* is the length of the largest submodule of finite length in $(R_{P_i})^k / M_{P_i}$. The sum $m_1 + \cdots + m_s$ is denoted $\operatorname{amult}(M)$ and called the *arithmetic multiplicity* of M.

Example 4.8 ($n = 3$, $k = 1$). The ideal I in (4.1) has arithmetic multiplicity 4. The arithmetic length is $m_1 = m_2 = 2$ along each of the associated primes $P_1 = \langle x_1, x_2 \rangle$ and $P_2 = \langle x_1, x_2, x_3 \rangle$.

We now present an extension of Theorem 4.4 to PDEs for vector-valued functions. Let $V_i = V(P_i) \subset \mathbb{C}^n$ be the irreducible variety defined by the ith associated prime P_i of M.

Theorem 4.9 (Ehrenpreis–Palamodov for modules). *For any submodule $M \subset R^k$, there exist* $\operatorname{amult}(M) = \sum_{i=1}^{s} m_i$ *Noetherian multipliers: these are vectors $B_{ij} \in \mathbb{C}[x, z]^k$ such that*

$$\phi(z) = \sum_{i=1}^{s} \sum_{j=1}^{m_i} \int_{V_i} B_{ij}(x, z) \exp(x \cdot z) d\mu_{ij}(x) \tag{4.5}$$

is a solution to the PDE given by M. Here μ_{ij} are measures that are supported on the variety V_i. Conversely, every solution to that PDE admits such an integral representation.

Proof. This statement appears in **[2, THEOREM 2.2]**. Differential primary decomposition **[18, THEOREM 4.6 (I)]** shows that the number of inner summands equals the arithmetic length m_i. ■

As before, we can pass from Noetherian multipliers $B_{ij}(x, z)$ to Noetherian operators $B_{ij}(x, \partial_x)$ and obtain a differential primary decomposition of M; see **[18]** and **[2, §4]**. We write $\bullet$ for the application of a vector of differential operators to a vector of functions. This is done coordinatewise and followed by summing the coordinates. The result is a function.

Corollary 4.10. *The Noetherian operators determine membership in the module M. Namely, a vector $m \in R^k$ lies in M if and only if $B_{ij}(x, \partial_x) \bullet m(x)$ vanishes on V_i for all i, j.*

The package `NoetherianOperators` **[16]** in the software `Macaulay2` **[23]** is a convenient tool for solving the PDE given by a submodule `M` of R^k. Typing `amult(M)` gives the arithmetic multiplicity of `M`. The command `solvePDE(M)` lists all associated primes P_i along with their Noetherian multipliers $B_{ij}(x, z)$. These features are described in **[2, §5]**.

What is intended with the command `solvePDE` vastly generalizes the problem of solving systems of polynomial equations, which is central to nonlinear algebra. That point is argued in **[32, CHAPTER 3]**, which culminates with writing polynomials as PDEs. First steps towards a numerical version of `solvePDE` are discussed in **[2, §7.5]** and **[16]**. It is instructive to revisit [32, THEOREM 3.27] through the lens of Theorem 4.9. The solution space of an ideal I is finite-dimensional if and only if each V_i is a point. If, furthermore, $s = 1$ and $V_1 = \{0\}$, then the Noetherian multipliers $B_1(z), \ldots, B_{m_1}(z)$ form a basis for the solution space of I.

If we pass from ideals to modules then even the case $s = 1$, $V_1 = \mathbb{C}^n$ is quite rich and interesting, especially in connection with the theory of wave cones [4]. We close with a nontrivial example which shows what wave solutions are and how they can be constructed.

Example 4.11 ($n = 4, k = 7$). Let $R = \mathbb{C}[x]$ and let $M \subset R^7$ be the module generated by $(x_1, x_2, x_3, x_4, 0, 0, 0)$, $(0, x_1, x_2, x_3, x_4, 0, 0)$, $(0, 0, x_1, x_2, x_3, x_4, 0)$, and $(0, 0, 0, x_1, x_2, x_3, x_4)$. This module is primary with $V_1 = \mathbb{C}^4$ and $\text{amult}(M) = 3$. It represents a first-order PDE for unknown functions $\phi : \mathbb{R}^4 \to \mathbb{R}^7$. To explore solutions of M, we apply the `Macaulay2` command `solvePDE`. The code outputs three Noetherian multipliers, namely the rows of

$$\begin{bmatrix} x_2^4 - 3x_1x_2^2x_3 + x_1^2x_3^2 + 2x_1^2x_2x_4 & 2x_1^2x_2x_3 - x_1x_2^3 - x_1^3x_4 & x_1^2x_2^2 - x_1^3x_3 & -x_1^3x_2 & x_1^4 & 0 & 0 \\ x_2^3x_3 - 2x_1x_2x_3^2 - x_1x_2^2x_4 + 2x_1^2x_3x_4 & x_1^2x_3^2 - x_1x_2^2x_3 + x_1^2x_2x_4 & x_1^2x_2x_3 - x_1^3x_4 & -x_1^3x_3 & 0 & x_1^4 & 0 \\ x_2^3x_4 - 2x_1x_2x_3x_4 + x_1^2x_4^2 & -x_1x_2^2x_4 + x_1^2x_3x_4 & x_1^2x_2x_4 & -x_1^3x_4 & 0 & 0 & x_1^4 \end{bmatrix}.$$

These rows are syzygies of M. They span all syzygies as a vector space over the function field $\mathbb{R}(x)$. Solutions ϕ to the PDE can be constructed from any syzygy by applying that differential operator to any function $f(z_1, z_2, z_3, z_4)$. For instance, writing subscripts for differentiation, the first row of the matrix above gives the following solution to our PDE M:

$$\phi = (f_{2222} - 3f_{1223} + f_{1133} + 2f_{1124}, 2f_{1123} - f_{1222} - f_{1114}, f_{1122} - f_{1113}, \\ - f_{1112}, f_{1111}, 0, 0).$$

Next, we show how nonlinear algebra makes waves. Consider the Hankel matrix

$$H(u) = \begin{bmatrix} u_1 & u_2 & u_3 & u_4 \\ u_2 & u_3 & u_4 & u_5 \\ u_3 & u_4 & u_5 & u_6 \\ u_4 & u_5 & u_6 & u_7 \end{bmatrix}.$$

We identify the four entries of $x \cdot H(u)$ with the generators of M. The wave cones of [4] are the determinantal varieties $\{u \in \mathbb{P}^6 : \text{rank}(H(u)) \leq r\}$. For $r = 1$, this is the rational normal curve in $\mathbb{P}^6$. For $r = 2$, it is the secant variety to the curve, of dimension 3. For $r = 3$, it is the variety of secant planes. The latter is the quartic hypersurface $\{u \in \mathbb{P}^6 : \det(H(u)) = 0\}$. The span of our three Noetherian multipliers furnishes a parametrization of that hypersurface.

Any $u \in \mathbb{P}^6$ with $H(u)$ of low rank yields wave solutions to M. For an illustration, let

$$u = (1, 2, 4, 8, 16, 32, 64).$$

Here $H(u)$ has rank 1. Its kernel is spanned by $2e_1 - e_2, 2e_2 - e_3, 2e_3 - e_4$. For any scalar function ψ in three variables, we obtain a function that satisfies the PDE given by M, namely

$$\phi(z) = \psi(2z_1 - z_2, 2z_2 - z_3, 2z_3 - z_4) \cdot u.$$

This vector is an example of a wave solution. If we take ψ to be the Dirac distribution at the origin in $\mathbb{R}^3$ then ϕ is a distributional solution that is supported on a line in $\mathbb{R}^4$. Characterizing such low-dimensional supports of solutions is the objective of the article [4].

ACKNOWLEDGMENTS

Many thanks to Simon Telen for the computation in Example 2.4. Helpful comments on draft versions of this paper were provided by Yulia Alexandr, Claudia Fevola, Marc Härkönen, Yelena Mandelshtam, Chiara Meroni, and Charles Wang.

REFERENCES

[1] D. Agostini, et al., Likelihood degenerations. 2021, arXiv:2107.10518.

[2] R. Ait El Manssour, M. Härkönen, and B. Sturmfels, Linear PDE with constant coefficients. *Glasg. Math. J.* (2022), published online.

[3] C. Améndola, et al., The maximum likelihood degree of linear spaces of symmetric matrices. *Matematiche* **76** (2021) 535–583.

[4] A. Arroyo-Rabasa, G. De Philippis, J. Hirsch, and F. Rindler, Dimensional estimates and rectifiability for measures satisfying linear PDE constraints. *Geom. Funct. Anal.* **29** (2019), 639–658.

[5] P. Aubry, F. Rouillier, and M. Safey El Din, Real solving for positive dimensional systems. *J. Symbolic Comput.* **34** (2002), 543–560.

[6] D. Bates, J. Hauenstein, A. Sommese, and C. Wampler, *Numerically solving polynomial systems with Bertini*. Software Environ. Tools 25, SIAM, Philadelphia, 2013.

[7] G. Blekherman, P. Parrilo, and R. Thomas, *Semidefinite optimization and convex algebraic geometry*. MOS-SIAM Ser. Optim. 13, SIAM, Philadelphia, 2013.

[8] T. Boege, et al., Reciprocal maximum likelihood degrees of Brownian motion tree models. *Matematiche* **76** (2021) 383–398.

[9] P. Breiding, K. Rose, and S. Timme, Certifying zeros of polynomial systems using interval arithmetic. 2020, arXiv:2011.05000.

[10] P. Breiding and S. Timme, HomotopyContinuation.jl: A package for homotopy continuation in Julia. In *Math. Software – ICMS 2018*, pp. 458–465, Springer, 2018.

[11] J. Bruna, K. Kohn, and M. Trager, Pure and spurious critical points: a geometric study of linear networks. In *Internat. Conf. on Learning Representations*, 2020.

[12] F. Cachazo, N. Early, A. Guevara, and S. Mizera, Scattering equations: from projective spaces to tropical Grassmannians. *J. High Energy Phys.* **2019** (2019), no. 6, 039.

[13] F. Cachazo, B. Umbert, and Y. Zhang, Singular solutions in soft limits. *J. High Energy Phys.* **2020** (2020), no. 5, 148.

[14] F. Catanese, S. Hoşten, A. Khetan, and B. Sturmfels, The maximum likelihood degree. *Amer. J. Math.* **128** (2006), 671–697.

[15] T. Çelik, et al., Wasserstein distance to independence models. *J. Symbolic Comput.* **104** (2021), 855–873.

[16] J. Chen, et al., Noetherian operators in Macaulay2. 2021, arXiv:2101.01002.

[17] Y. Cid-Ruiz, R. Homs, and B. Sturmfels, Primary ideals and their differential equations. *Found. Comput. Math.* **21** (2022) 1363–1399.

[18] Y. Cid-Ruiz and B. Sturmfels, Primary decomposition with differential operators. 2021, arXiv:2101.03643.

[19] J. Draisma, et al., The Euclidean distance degree of an algebraic variety. *Found. Comput. Math.* **16** (2016), 99–149.

[20] E. Duarte, O. Marigliano, and B. Sturmfels, Discrete statistical models with rational maximum likelihood estimator. *Bernoulli* **27** (2021), 135–154.

[21] C. Eur, T. Fife, J. Samper, and T. Seynnaeve, Reciprocal maximum likelihood degrees of diagonal linear concentration models. *Matematiche* **76** (2021) 447–459.

[22] I. Gel'fand, M. Kapranov, and A. Zelevinsky, *Discriminants, resultants and multidimensional determinants*. Birkhäuser, Boston, 1994.

[23] D. Grayson and M. Stillman, Macaulay2, a software system for research in algebraic geometry, available at http://www.math.uiuc.edu/Macaulay2/.

[24] W. Gröbner, On the Macaulay inverse system and its importance for the theory of linear differential equations with constant coefficients. *ACM Commun. Comput. Algebra* **44** (2010), 20–23. [*Abh. Math. Semin. Univ. Hambg.* **12** (1937), 127–132].

[25] S. Hoşten, A. Khetan, and B. Sturmfels, Solving the likelihood equations. *Found. Comput. Math.* **5** (2005), 389–407.

[26] J. Huh, The maximum likelihood degree of a very affine variety. *Compos. Math.* **149** (2013), 1245–1266.

[27] J. Huh, Varieties with maximum likelihood degree one. *J. Algebr. Stat.* **5** (2014), 1–17.

[28] J. Huh and B. Sturmfels, Likelihood geometry, In *Combinatorial algebraic geometry*, pp. 63–117, Lecture Notes in Math. 2108, Springer, 2014.

[29] K. Kubjas, O. Kuznetsova, and L. Sodomaco, Algebraic degree of optimization over a variety with an application to p-norm distance degree. 2021, arXiv:2105.07785.

[30] L. Manivel, et al., Complete quadrics: Schubert calculus for Gaussian models and semidefinite programming. 2020, arXiv:2011.08791.

[31] M. Michałek, L. Monin, and J. Wiśniewski, Maximum likelihood degree, complete quadrics, and $\mathbb{C}^*$-action. *SIAM J. Appl. Algebra Geom.* **5** (2021), 60–85.

[32] M. Michałek and B. Sturmfels, *Invitation to nonlinear algebra*. Grad. Stud. Math. 211, American Mathematical Society, Providence, 2021.

[33] G. Ottaviani, P-J. Spaenlehauer, and B. Sturmfels, Exact solutions in structured low-rank approximation. *SIAM J. Matrix Anal. Appl.* **35** (2014), 1521–1542.

[34] B. Sturmfels and S. Telen, Likelihood equations and scattering amplitudes. *Algebr. Stat.* **12** (2021), no. 2, 167–186. 2020, arXiv:2012.05041.

[35] B. Sturmfels, S. Timme, and P. Zwiernik, Estimating linear covariance models with numerical nonlinear algebra. *Algebr. Stat.* **1** (2020), 31–52.

[36] S. Sullivant, *Algebraic statistics*. Grad. Stud. Math. 194, American Mathematical Society, Providence, 2018.

BERND STURMFELS

Max-Planck Institute for Mathematics in the Sciences, Inselstrasse 22, 04103 Leipzig, Germany, bernd@mis.mpg.de, and University of California, Berkeley CA 94720, USA, bernd@berkeley.edu

14. MATHEMATICS OF COMPUTER SCIENCE

NOWHERE TO GO BUT HIGH: A PERSPECTIVE ON HIGH-DIMENSIONAL EXPANDERS

ROY GOTLIB AND TALI KAUFMAN

This exposition is dedicated to the memory of my father, Eliezer Kaufman, and to the memory of my mother, Sarah Kaufman, who was always trying to understand the reason why things behave in a certain way.

ABSTRACT

"Nowhere to go but in" is a well-known statement of Osho. Osho meant to say that the answers to all our questions should be obtained by looking into ourselves. In a paraphrase to Osho's statement we say "Nowhere to go but high." This is meant to demonstrate that for various seemingly unrelated topics and questions, the only way to get significant progress is via the prism of a new philosophy (new field) called high-dimensional expansion. In this note we give an introduction to the high-dimensional expansion philosophy, and how it has been useful recently in obtaining progress in various questions in seemingly unrelated fields.

MATHEMATICS SUBJECT CLASSIFICATION 2020

Primary 05E45; Secondary 68R05, 05C48, 68Q87

KEYWORDS

High dimensional expanders, local to global, random walks, markov chains, quantum codes, locally testable codes, matroids.

ICM INTERNATIONAL CONGRESS OF MATHEMATICIANS 2022 JULY 6–14

Proc. Int. Cong. Math. 2022, Vol. 6, pp. 4842–4871
DOI 10.4171/ICM2022/185

Published by EMS Press

1. INTRODUCTION

What is common to the following very diverse and important themes: quantum codes, counting bases of matroids, locally testable classical codes, fast mixing of Markov chains, and Gromov's topological overlapping question? Even if you have not heard of some/any of these mentioned topics, it is clear that they emerge from completely different branches of mathematics and computer science, and hence do not seem particularly related.

The purpose of this note is to highlight the idea that all the topics mentioned above, despite seeming unrelated up until recently, are in fact strongly related via a new perspective that is obtained by introducing a new object called a *high-dimensional expander*. This object created inherent, deep relations between topics that previously seemed unrelated. This new perspective, as well as new connections between different problems via the idea of high-dimensional expanders, had recently led to various important advances on the above topics and beyond.

Our goal, in this note is *not* to provide a survey on high-dimensional expanders, but rather to give *our own* perspective on this newly emerged object, and its connections/strong implications to the above topics, namely to highlight recent advances on the said topics using the new perspective of high-dimensional expanders.

We intend to highlight how this newly emerged object (or maybe newly emerged philosophy) is, in fact, tightly related to the above notions and beyond. High-dimensional expansion has different angles, enabling one to relate these diverse questions/phenomena.

Our aim here is to present what high-dimensional expanders are and how these newly defined objects give a *unified* perspective of the topics mentioned above that, prior to the introduction of high-dimensional expanders, seemed unrelated.

High-dimensional expanders, as we will see, are a generalization of graph expanders to higher dimensions. But, as we will see, the importance of high-dimensional expanders does *not* stem from the fact that they generalize expander graphs to higher dimensions, but rather from the fact that when objects exhibit expansion in higher dimensions they also have strong *local-to-global* properties/nature.

This local-to-global behavior that high-dimensional expanders exhibit is unique to the high-dimensional case in the sense that it is not present in one-dimensional expanders. Indeed, this local-to-global philosophy is what makes them so powerful and so connected to the various topics mentioned above.

Our focus here is not on rigorous proofs, but rather on presenting different angles of the high-dimensional expansion philosophy, with their recent implications for various unrelated fields.

Structure of this note. The structure of this note will be as follows. We will start by introducing basic facts about expander graphs, with specific focus on weighted graphs which are essential to the theory of high-dimensional expanders (see Section 2). Then, in Section 3, we will move to introduce the object which is the focus of this note, namely a high-dimensional expander. We will highlight the fact that there is a notion of high-dimensional topological expansion, which generalizes the Cheeger constant for graphs, and high-dimensional spectral expansion that generalizes the spectral expansion of graphs. In graphs the Cheeger inequality says that the topological expansion is, in a sense, equivalent to spectral expansion. In higher dimensions, however, Cheeger inequality does *not* hold (the spectral definition does not imply the topological definition, and vise versa). This demonstrates some of the deepness of the high-dimensional expansion phenomenon.

We then turn to discuss high-dimensional random walks (see Section 4). We define what high-dimensional random walks are and show that if all links (i.e., local neighborhoods) of the high-dimensional expander are expanding enough then high-dimensional random walks mix rapidly.

We follow that discussion by a discussion of the local-to-global aspects of high dimensional expanders—aspects which are nonexistent in graph expansion. The local-to-global implication will hold (via different proofs) both for spectral and topological expansion. We will show that a high-dimensional simplicial complex whose every local link (i.e., local neighborhoods) are spectrally/topologically expanding must be a global spectral/topological high-dimensional expander (see Section 5 for local-to-global expansion in the spectral sense, and see Section 6 for local-to-global expansion in the topological sense).

Using the local-to-global premise and the fact that we have fast mixing of high-dimensional random walks if all the links are sufficiently expanding, we will deduce fast mixing of high-dimensional random walks from local expansion in very local neighborhoods.

Then, in Section 7, we move to show that high-dimensional expansion is a form of local testability of codes and use that towards results on local testability of codes.

We then, in Section 8, demonstrate that local testability of classical codes and quantum LDPC codes are both *born together* from a high-dimensional expander. Thus, we show that classical locally testable codes and quantum LDPC codes are connected together via the high-dimensional expansion perspective. Prior to the introduction of high-dimensional expanders, these two objects were not known to be related. We will survey some of the major recent developments in these fields that emerge from this recent viewpoint that connects them both to high-dimensional expanders.

We then turn to discuss the Gromov topological overlapping problem, its discovered relation to classical local testability, and its solution via high-dimensional expanders and their connection to locally testable codes (see Section 9).

We will then (see Section 10) introduce the Mihail–Vazirani Conjecture about counting bases of matroids that was recently resolved via high-dimensional expansion. We will introduce the conjecture and will show a *rigorous* proof of its resolution via local-to-global theorems on high-dimensional expanders and random walks.

In the end (see Section 11), we mention some topics that are not covered by our note.

2. SOME CLASSICAL FACTS ABOUT EXPANDER GRAPHS

Before we discuss high-dimensional expanders, we need a solid foundation in the original theory of expander graphs. Broadly speaking, there are two types of expanders, combinatorial expanders and spectral expanders. The former generally refers to graphs $G = (V, E)$ where all subsets $S \subset V$ *expand outward* in some sense (this includes definitions such as edge expansion, vertex expansion, or unique neighbor expansion). The latter is somewhat more of a technical condition: it requires that all "nontrivial" eigenvalues of G's adjacency matrix be of bounded size. While a priori it is not obvious that these two notions of expansion are connected, it is well known that they are (at least morally) equivalent (see, e.g., discussion of Cheeger's inequality and the expander-mixing lemma in [23]).

2.1. Weighted graphs

In order to introduce expander graphs in a way that is consistent with their higher dimensional counterparts, we begin by introducing weighted graphs. We assume that there are no isolated vertices and define weighted graphs as graphs that are equipped with a weight function for the edges that satisfies the following condition:

Definition 2.1 (Weight function for the edges). Let $G = (V, E)$. A function $\mathrm{w}_E : E \to (0, 1]$ is a weight function for the edges if $\sum_{e \in E} \mathrm{w}_E(e) = 1$.

The weight function for the edges of the graph induces a weight function over the vertices of the graph in the following way:

Definition 2.2 (Weight function for the vertices). Let $G = (V, E)$ and let w_E be a weight function for the edges of the graph. Then the following $\mathrm{w}_V : V \to (0, 1]$ is the induced weight function on the vertices of the graph:

$$\mathrm{w}_V(v) = \sum_{\substack{e \in E \\ v \in e}} \frac{1}{2} \mathrm{w}_E(e).$$

We note that this weight function over the vertices also sums up to 1 (and therefore both weight functions define a probability distribution over the edges/vertices). We can now use these two functions to define the weight function for the graph:

Definition 2.3 (Weight function for a graph). Let $G = (V, E)$ be a graph and let w_E be a weight function over the edges of the graph. Define the weight function over the graph $\mathrm{w} : \{\emptyset\} \cup V \cup E \to \mathbb{R}$ to be the function that satisfies $\mathrm{w}(\emptyset) = 1$, $\mathrm{w}|_V = \mathrm{w}_V$ and $\mathrm{w}|_E = \mathrm{w}_E$.

2.2. Spectral expansion of graphs

Let us start by describing the spectral notion of expansion. We will start by going over some basic spectral properties of weighted graphs. To start, we need to define the main object of interest, the adjacency matrix.

Definition 2.4 (Adjacency matrix). The adjacency matrix of a weighted graph $G = (V, E)$ with a weight function w is

$$A_{v,w} = \frac{\mathrm{w}(\{v, w\})}{2\,\mathrm{w}(v)}.$$

One natural (and useful) way to think about the adjacency matrix is as the transition matrix[1] of the random walk underlying G which moves from a vertex v to vertex w with probability corresponding to the edge weights incident to v. Indeed, much of this survey will focus on standard connections between properties of this walk and the eigenvalues of A, and how they generalize to higher dimensions. First, however, we need a few basic spectral facts about the adjacency matrix A.

Proposition 2.5. *Let $G = (V, E)$ be a weighted graph, and A_G its corresponding adjacency matrix. Then A_G has a spectral decomposition (i.e., an orthonormal basis of eigenvectors).*

Proof. This follows from the Cauchy's celebrated spectral theorem that any self-adjoint operator can be diagonalized. The trick is then to find an inner-product space over which A_G is self-adjoint. One can show that this is the case for the standard inner product normalized by the distribution w induces over vertices. For $f, g : V \to \mathbb{R}$, let

$$\langle f, g \rangle = \sum_{v \in V} \mathrm{w}(v) f(v) g(v).$$

A simple computation verifies that $\langle Af, g \rangle = \langle f, Ag \rangle$. ■

Now that we know A_G has a well-defined spectrum, we can examine properties of its eigenvalues. For a graph G on n vertices, denote the eigenvalues of A_G in decreasing order by $\lambda_1 \geq \cdots \geq \lambda_n$. For our purposes, we are most interested in two basic properties of these eigenvalues.

Claim 2.6. *Let $G = (V, E)$ be a weighted graph, and A_G its corresponding adjacency matrix with eigenvalues $\lambda_1 \geq \cdots \geq \lambda_n$. Then* (i) *the all 1's vector is an eigenvector satisfying $A_G \mathbb{1} = \mathbb{1}$ and* (ii) *it is also the maximal eigenvalue (in absolute value), $1 = \lambda_1 \geq \lambda_2 \geq \cdots \geq \lambda_n \geq -1$.*

It is well known that a graph G is *connected* if and only if $\lambda_2 < 1$. Spectral expansion studies a strengthening of this condition, when the second eigenvalue is *bounded away from* 1 by some constant.

Definition 2.7 (Weighted spectral expansion). We say a weighted graph $G = (V, E)$ is a (one-sided) λ-spectral expander if $\lambda_2 \leq \lambda$. The quantity $1 - \lambda_2$ is often called the *spectral gap*, and is an equivalent way to express one-sided spectral expansion.

At the beginning of this section, we mentioned that spectral expansion is intimately tied to combinatorial expansion. Intuitively, this follows from the fact that the latter can be viewed as a sort of robust connectivity. Thought of in this manner, it is no surprise that

---

1 Technically, the transpose, as we will discuss later.

since $\lambda_2 < 1$ enforces G to be connected, smaller λ_2 corresponds to stronger notions of connectivity akin to combinatorial expansion.

2.3. Topological expansion of graphs

Informally, a finite graph is called an expander if relatively many edges cross between every set of vertices and its complement.

If G is a weighted graph with w as its weight function (one can take $w_E = \frac{1}{|E|}$ to ignore the weights), then one can define the following norm over sets of vertices/edges (or the indicator of such sets):

$$\|S\|_w = \sum_{s \in S} w(s), \quad \|\mathbb{1}_S\|_w = \|S\|_w.$$

In this case, the topological expansion of (G, w) is quantified by its *Cheeger constant*, defined as

$$h(G, w) = \min_{\emptyset \neq S \subsetneq V} \frac{\|\delta \mathbb{1}_S\|_w}{\min\{\|\mathbb{1}_S\|_w, \|\mathbb{1}_{V \setminus S}\|_w\}},$$

where $\delta \mathbb{1}_S$ is a function that accepts an edge $\{u, v\}$ and returns 1 iff $\mathbb{1}_S(u) + \mathbb{1}_S(v) = 1$ (the sum is performed modulo 2). Note that the numerator of the Cheeger constant is the norm of the edges that connect S and $V \setminus S$. One says that (G, w) is an ε-expander if $h(G) \geq \varepsilon$.

We know that if a weighted graph (G, w) is a λ-spectral expander, then (G, w) is has a good Cheeger constant. We call such inequalities Cheeger inequalities and in the following establish a weighted version of the Cheeger inequality. Recall that given a weighted graph (G, w) and $A \subseteq V$, we write $\mathbb{1}_A$ for the indicator function of A.

Theorem 2.1 (Weighted Cheeger inequality). *Let (G, w) be a weighted graph which is also a λ-spectral-expander and let $A \subseteq V$. Then (see* [31, THEOREM 4.4]*)*

$$\|\delta \mathbb{1}_A\|_{\mathrm{w}} \geq 2(1 - \lambda)\|\mathbb{1}_A\|_{\mathrm{w}}\big(1 - \|\mathbb{1}_A\|_{\mathrm{w}}\big).$$

In addition (see [16, THEOREM 2.1]*), if $h(G, \mathrm{w}) \geq \epsilon$ then (G, w) is a $\sqrt{1 - \frac{\epsilon^2}{4}}$-spectral expander.*

Importantly, we have no analogue for the Cheeger inequality in higher dimensions; i.e., high-dimensional spectral expansion does not imply high-dimensional topological expansion, and the implication in the other direction also does not hold.

3. HIGH-DIMENSIONAL EXPANDERS: THE OBJECT OF STUDY

3.1. Simplicial complexes and links

In order to generalize an expander graph to higher dimensions, we first have to define an object that has higher dimension. Our object of choice is a *pure simplicial complex*. These are objects generalize graphs in two important ways: firstly, much like graphs only contain an edge if both its vertices are in the graph. If a simplicial complex contains a higher-dimensional edge then it contains all of the lower dimensional edges that are contained in it. Secondly, the complex does not include a high-dimensional isolated vertex, i.e., every edge is contained in a maximal edge.

Definition 3.1 (Pure simplicial complex). A collection of sets X is a pure simplicial complex if it satisfies the following:

- *X is a simplicial complex, i.e.,* if $\sigma \in X$ and $\tau \subset \sigma$ then $\tau \in X$.
- *X is pure, i.e.,* if $\sigma, \tau \in X$ are maximal sets (a set is *maximal* if it is not contained in any strictly larger set of the complex) in X then $|\sigma| = |\tau|$.

We call subsets in X the *faces* of X.

We define the dimension of a face as follows:

Definition 3.2 (Dimensions). Given a face $\sigma \in X$, we define the dimension of σ to be $\dim(\sigma) = |\sigma| - 1$ and the dimension of the complex to be the dimension of the maximal face in X. We also define the set of faces of a certain dimension in the following way:

$$X(i) = \{\sigma \in X : \dim(\sigma) = i\}.$$

We should also stress that $X(-1) = \{\emptyset\}$.

Definition 3.3 (Degree). The degree of a simplicial complex is the maximal number of faces on a single vertex. A family of simplicial complexes with growing number of vertices is said to have *bounded degree* if their degree is independent on the number of vertices, and remains fixed as the number of vertices in the family grows.

We note that one can think of pure d-dimensional complex as a d-uniform hypergraph with closure property.

We often refer to local neighborhoods of a simplicial complex. These are called links. Links play a major role in studying high-dimensional expanders, as much of the study is done via the local-to-global paradigm, where we study a complex by its links. Links are defined as follows.

Definition 3.4 (Links). Let X be a d-dimensional pure simplicial complex. For every i and $\tau \in X(i)$, the *link* of τ is the restriction of the complex to faces containing τ, that is,

$$X_\tau = \{\sigma \setminus \tau : \sigma \in X \text{ and } \tau \subseteq \sigma\}.$$

Put into words, X_τ is the complex that arises by selecting all d-faces that contain τ, then removing τ itself. Finally, it is often convenient to refer to the set of links for all $\tau \in X(i)$, which we will refer to as the *i-links*.

Links provide a natural method for decomposing global functions on simplicial complexes into local parts. For instance, it is not hard to see that given a function on k-faces $f : X(k) \to \mathbb{R}$, its expectation over the complex is equal to the average of its expectation over links:[2]

$$\mathop{\mathbb{E}}_{X(k)}[f] = \mathop{\mathbb{E}}_{\tau \in X(i)}\Big[\mathop{\mathbb{E}}_{\sigma \in X_\tau(k-i)}\big[f(\tau \cup \sigma)\big]\Big].$$

---

2 Formally, these expectations are defined over *weighted* complexes where each level (and link) is endowed with a distribution. We cover this in the following section.

3.2. Weighted simplicial complexes and weighted links

When defining high-dimensional expanders, we would need to work with *weighted* simplicial complexes.

Definition 3.5 (Weighted simplicial complex). A weighted pure d-dimensional simplicial complex (X, Π) is a pure d-dimensional simplicial complex X endowed with a distribution Π on faces of the maximal dimension. The weight of a k-dimensional face in the simplicial complex is then defined in the following way:

$$\mathrm{w}(\tau) = \begin{cases} \Pi(\tau) & \text{if } \tau \in X(d), \\ \frac{1}{|\tau|+1} \sum_{\substack{\sigma \in X(k+1) \\ \tau \subseteq \sigma}} \mathrm{w}(\sigma) & \text{otherwise.} \end{cases}$$

When Π is not specified, it is assumed to be uniform.

The *weighted* links of a weighted complex are, naturally, themselves weighted complexes with distributions inherited from the global distribution Π.

Definition 3.6 (Weighted links). Let (X, Π) be a d-dimensional weighted simplicial complex. For all $0 \leq i \leq d$ and $\tau \in X(i)$, the weighted link (X_τ, Π_τ) is given by:

(1) $X_\tau = \{\sigma \setminus \tau : \sigma \in X \text{ and } \tau \subseteq \sigma\}$,

(2) $\mathrm{w}_\tau(\sigma) = \Pr_{\sigma' \sim \Pi}[\sigma' = \sigma \mid \tau \subseteq \sigma'] = \frac{\mathrm{w}(\sigma \cup \tau)}{\binom{|\sigma|+|\tau|}{|\tau|} \mathrm{w}(\tau)}$.

In other words, the distribution over X_τ is simply given by normalizing Π over the top level faces of X_τ. Finally, note that we usually drop the distribution Π_τ when clear from context.

Much like in the one-dimensional case, we will be interested in defining a norm on sets of faces of some dimension. And, again, much like in the graph case, we will do so in the following way:

Definition 3.7 (Norm). For any weighted pure d-dimensional simplicial complex and every dimension i, define the following norm:

$$\forall S \subseteq X(i), \quad \|S\|_w = \sum_{s \in S} w(s), \qquad \|\mathbb{1}_S\|_w = \|S\|_w.$$

3.3. Spectral definition of high-dimensional expansion

We are now ready to give definitions of high-dimensional expanders. We will give spectral and topological definitions of high-dimensional expanders. The first definition we are going to consider is a spectral definition of high-dimensional expanders that are called *local spectral expanders*. These are simplicial complexes whose links are excellent expanders in the sense that their underlying graph is an expander. Formally, consider the following two definitions:

Definition 3.8 (Skeleton). Let X be a simplicial complex and define the i skeleton of X to be the following simplicial complex:

$$X^{(i)} := \{\sigma \in X : \dim \sigma \leq i\}.$$

Using this definition, we consider the underlying graph of the links as their 1-skeleton and arrive at the following definition that was introduced by [12, 25, 39].

Definition 3.9 (Local spectral expander [12,25,39]). A d-dimensional complex X is a λ-local spectral expander if for every $i \leq d-2$ and $\tau \in X(i)$ it holds that:

- $X_\tau^{(1)}$ is connected,
- $\lambda_2(X_\tau^{(1)}) \leq \lambda$.

A simplicial complex X will be called a *strong* local spectral expander if it is a λ-local spectral expander with $\lambda < \frac{1}{d}$. Otherwise, it is called a *weak* local spectral expander.

Note that in this definition we only regarded some of the links. This is because the rest of the links are either a set of unconnected vertices or a complex that contains only the empty face.

3.4. Topological definition of high-dimensional expansion

Another generalization of expander graphs to higher dimensions generalizes them in a topological sense. Consider the Cheeger constant definition of expansion in the graph case. The Cheeger constant is the proportion between the weight of edges that "go out" of the set and the weight of the set itself (for sufficiently small sets). In case of higher dimension, it will be useful to think of the indicator function of a set of sets. We therefore define:

Definition 3.10 (Cochains). For a weighted simplicial complex (X, w), define the set of cochains of X over an abelian group G to be

$$C^i(X; G) := G^{X(i)}.$$

For $G = \mathbb{F}_2$, this definition coincides with an indicator function for a set of sets of size $i + 1$. For the vast majority of this note, the cochains will be, indeed, defined over $\mathbb{F}_2$. We will therefore state explicitly when we are using a different underlying group. Moreover, the rest of the definitions in this section are out of the scope of this note for cochains that are defined over groups that are not $\mathbb{F}_2$. Therefore we will only present the following definitions over $\mathbb{F}_2$.

It is then natural to ask how to define, for example, a triangle leaving a set of edges (similar to an edge leaving a set of vertices). Consider the following generalization of δ:

Definition 3.11 (Coboundary operator). The *coboundary operator*

$$\delta_i : C^i(X; \mathbb{F}_2) \to C^{i+1}(X; \mathbb{F}_2)$$

is defined by

$$\delta_i F(\sigma) = \sum_{\tau \in \binom{\sigma}{i+1}} F(\tau). \tag{1}$$

In most cases, the dimension will be clear from context and therefore omitted.

Therefore we say that a triangle is leaving a set of edges if an odd number of its edges is in the set. In addition, a standard computation shows that $\delta_{i+1} \circ \delta_i = 0$. We can therefore define the spaces of *i-coboundaries* and the space of *i-cocycles* as

$$B^i = B^i(X) = B^i(X; \mathbb{F}_2) = \operatorname{Im}(\delta_{i-1}) \quad \text{and} \quad Z^i = Z^i(X) = Z^i(X; \mathbb{F}_2) = \ker(\delta_i), \tag{2}$$

respectively, where $\delta_{-2} = 0$ by convention. We have $B^i \subseteq Z^i \subseteq C^i$ because $\delta_i \circ \delta_{i-1} = 0$, and the quotient space $H^i(X; \mathbb{F}_2) = Z^i / B^i$ is the *ith cohomology* space. The space dual to $H^i(X; \mathbb{F}_2)$ is the *ith homology* space denoted as $H_i(X; \mathbb{F}_2)$. We say that X is i-dimensional $\mathbb{F}_2$*-connected* if $H^i(X; \mathbb{F}_2) = 0$.

We would now like to move on to describe the high-dimensional generalization of the Cheeger constant. Before we do that, however, we have to thoroughly inspect the denominator of the Cheeger constant. Note that in any graph there are sets that are guaranteed not to expand. These sets are the empty set and the whole graph. Therefore, in the Cheeger constant we are not looking for the absolute expansion of a set, but rather we relate the expansion of the set to how different it is from one of these trivially nonexpanding sets. In the high-dimensional case we very much do the same. Here, however, there will be more sets that are trivially nonexpanding. Specifically, these sets are coming from the dimension below. Formally, a set of the form δF where F is an $(i-1)$-dimensional set is trivially nonexpanding in the ith dimension. These sets are called the coboundaries of X and are denoted as $B^i(X)$. We therefore define coboundary expansion, the higher dimensional analogue of the edge expansion in the following way. The definition is originated in the work of Linial and Meshulam and the work of Gromov [21, 36].

Definition 3.12 (Coboundary expansion [21, 36]). Let (X, w) be a pure, d-dimensional weighted simplicial complex. Define the following generalization of the Cheeger constant:

$$h^i(X, w) = \min_{F \in C^i(X) \setminus B^i(X)} \left\{ \frac{\|\delta F\|_w}{\min_{G \in B^i(X)} \{\|F + G\|_w\}} \right\}.$$

We say that a simplicial complex is an ε-coboundary expander if $h^i(X, w) \geq \epsilon$ for every dimension.

Note here that $h^0(X, w)$ is the Cheeger constant of the graph corresponding to the one skeleton of the complex. Note further that $h^i(X, w) > 0$ iff $H^i(X; \mathbb{F}_2) = 0$.

To date, no bounded-degree coboundary expanders are known. However, in most cases a relaxation of this condition suffices, namely cosystolic expansion. These cosystolic expanders are the high-dimensional analogue of a graph with several large connected components that are all expanders. The high-dimensional analogue very much follows suite: the high-dimensional analogue of connected component is a generalization of the fact that a connected component is a set of vertices that has no outgoing edges which are, in fact, the cocycles. Therefore a complex with large connected components that are all expanders has been defined by [12, 25] as follows.

Definition 3.13 (Cosystolic expansion [12, 25]). Let (X, w) be a pure, d-dimensional simplicial complex. Consider the following two definitions:

(1) *The expansion of the connected components*

$$\tilde{h}^i(X, w) = \min_{F \in C^i(X) \setminus Z^i(X)} \left\{ \frac{\|\delta F\|_w}{\min_{G \in Z^i(X)} \{\|F + G\|_w\}} \right\}.$$

(2) *The minimal connected component*

$$\operatorname{cosyst}^i(X, w) = \min_{F \in Z^i(X) \setminus B^i(X)} \{\|F\|_w\}.$$

We say that a weighted simplicial complex is an (ϵ, μ)-cosystolic expander if for every dimension it holds that both all the connected components are expanding, i.e., $\tilde{h}^i(X, w) \geq \epsilon$, and all of the connected components are large $\operatorname{cosyst}^i(X, w) \geq \mu$.

Now that we have generalized expansion to higher dimensions, let us discuss some of the properties of high-dimensional expanders. We will start with considering one of the most important properties of high-dimensional expanders, namely the fact that high-dimensional random walks converge rapidly to their stationary distribution.

4. RANDOM WALKS ON HIGH-DIMENSIONAL EXPANDERS

High-dimensional random walks are one of the major tools in the study of high-dimensional expanders. They were introduced by Kaufman and Mass [27] and were further developed in [1,10,30,33]. They have various implications; however, in this note we are going to mention only a few. One of the most important properties of local spectral expanders is the fast convergence of random walks to their stationary distribution.

The proof of fast mixing of random walks on high-dimensional expanders is in the spirit of the local-to-global method developed by Garland [18] who has shown that a simplicial complex whose all links spectrally expand has vanishing cohomology over $\mathbb{R}$. One can think about Garland's result as studying expansion over $\mathbb{R}$ using local-to-global arguments. Essential to Garland's method is the fact that over $\mathbb{R}$ one can use self-adjoint operators and inner products which do not exist in all spaces, for example, over $\mathbb{F}_2$.

We use a method similar in spirit Garland to prove fast mixing of high-dimensional random walks.

Unlike the graph case, in which there is one canonical random walk, in a high-dimensional expander many random walks are considered. We will be interested in the *up–down walk* and the *down–up walk* of every dimension. The k-dimensional up–down walk transitions between faces of dimension k. Every step of the walk comprises two substeps, the up step and the down step. If at the beginning of the step the walk is at σ then in the up step, a $(k+1)$-dimensional face τ is chosen that contains σ with distribution proportional to its weight. Then in the down step, a k-dimensional face that is contained in τ is selected with equal probability. The down–up walk can similarly be defined as taking the step down first and then taking the step up. Formally, these are defined as follows:

Definition 4.1 (Down and up walks). Define the up walk as $U^k : C^k(X; \mathbb{R}) \to C^k(X; \mathbb{R})$, where

$$U^k F(\tau) = \mathbb{E}_{\sigma \in X(k)} [F(\sigma) \mid \sigma \subseteq \tau].$$

And the down walk $D^k : C^k(X;\mathbb{R}) \to C^{k-1}(X;\mathbb{R})$ is defined as

$$D^k F(\tau) = \mathbb{E}_{\sigma \in X(k)}\big[F(\sigma) \mid \tau \subseteq \sigma\big].$$

And the corresponding random walks were defined as:

Definition 4.2 (Up–down and down–up random walks). Define the up–down random walk and the down–up random walks as

$$\Delta_k^+ = D^{k+1}U^k \quad \text{and} \quad \Delta_k^- = U^{k-1}D^k,$$

respectively.

As previously mentioned, the convergence of these random walks is a key property of high-dimensional expanders. Since their introduction by Kaufman and Mass in [27], high-dimensional random walks have proven themselves to be the backbone of many applications of high-dimensional expander in computer science. Examples include resolution [3] of the Mihail–Vazirani conjecture [38] (which we will rigorously present in Section 10), derandomization of direct product testing [7, 10, 28], and more.

We say that these random walks converge rapidly due to:

Theorem 4.3 (Random walks converge rapidly on good enough local spectral expanders [30]). *If X is a γ-local spectral expander then*

$$\lambda_2(\Delta_{k-1}^+) = \lambda_2(\Delta_k^-) \le 1 - \frac{1}{k+1} + \frac{k}{2}\gamma.$$

Note that Theorem 4.3 yields nontrivial results for every dimension only when $\gamma \in O(\frac{1}{d^2})$, which suffices for many applications. There are, however, cases in which we might be interested in convergence of random walks and will not have such strong expansion assumptions (for example, when trying to sample an independent set with the hardcore distribution [2]). In [1] Alev and Lau relaxed this requirement and proved the following theorem:

Theorem 4.4 (Random walks converge rapidly even for weak local spectral expanders [1]). *Let X be a pure d-dimensional simplicial complex and let*

$$\gamma_i = \max\big\{\lambda_2(X_\tau^{(1)}) : \tau \in X(i)\big\}.$$

Then

$$\lambda_2(\Delta_k^-) \le 1 - \frac{\prod_{i=-1}^{k-2}(1-\gamma_i)}{k+1},$$

which is meaningful whenever all the 1-skeletons of the links are connected.

It is important to note that in both Theorems 4.3 and 4.4 the key observation is that the k-dimensional random walks can be viewed through the random walks over the links of the complex. For example, consider a step in the up–down random walk. This step corresponds to first picking a link to walk over (specifically, the link of a $(k-1)$-dimensional face). In that link, the original face is a vertex. Then the walk simply picks one of its neighbors and walks there.[3]

Recent works on high-dimensional random walks focuss on a stronger property, namely they are interested in optimal mixing time and have studied conditions under which such an optimal mixing time exists (see, e.g., [6]).

5. LOCAL-TO-GLOBAL SPECTRAL EXPANSION OF HIGH-DIMENSIONAL EXPANDERS

In this section we present a *local-to-global* property of spectral high-dimensional expanders. Specifically, we will show that if all the links are connected, global expansion can be derived from local expansion. The philosophy implemented in the proof is similar in spirit to that of Garland, where one uses self-adjoint operators and inner products defined over $\mathbb{R}$ to study expansion over $\mathbb{R}$.

However, the local-to-global descent of spectral gaps is *inherently* different from the random walks' result in the following way. The fast mixing of random walks' result assumes spectral expansion in all links including the link of the empty set (which is a global expansion condition on the whole complex!) to conclude fast mixing of random walks. Hence that result is not based only on local assumptions. In contrast, the following result about descent of spectral gaps from links to the entire complex only assumes spectral expansion in *local* links to deduce global expansion! So the following result is a *genuine* local-to-global result, while the random walk result also assumed some global property.

We emphasize, however, that the ability to get the descent of spectral expansion from the links to the global object requires the complex to be a *strong local spectral expander*. By this we mean that the complex is a γ-local spectral expander with $\gamma < \frac{1}{d}$, where d is the dimension of the complex. Namely, the descent of spectral gaps that we are going to present, and is known under the name of the "Trickling Down Theorem" is only possible under a strong local spectral expansion guarantee. This is in contrast with the previously discussed fast mixing of random walks in local spectral expanders, which we have shown to hold for any local spectral expander, not necessarily strong (see, e.g., [19] for an example of a local spectral expander that is not strong). The point we are making is that strong local spectral expansion is inherently different than nonstrong local spectral expansion, as both allow for fast mixing of random walks, but only the strong one exhibits trickling down effects. We further note that all currently known combinatorial constructions of high-dimensional expanders yield only weak spectral expanders. Constructions of strong spectral expanders are only known via algebraic means.

Theorem 5.1 (Trickling Down Theorem [39]). *Let X be a pure n-dimensional simplicial complex with connected links. If, for every vertex v, it holds that X_v is a λ-local spectral expander then X is a $\frac{\lambda}{1-\lambda}$-local spectral expander.*

---

3 Note that this yields a walk whose lazy component is larger and therefore it stays in the original face with higher probability.

We will prove the theorem by looking at the Laplacian of the adjacency matrix. We will denote the adjacency matrix of the complex by A and introduce the following:

Definition 5.2. Let $\mathscr{L}_0^+$ be the Laplacian of an up–down walk on the complex X, i.e., $\mathscr{L}_0^+ = I - A$. In addition, denote the Laplacian of the link of σ by $\mathscr{L}_{\sigma,0}^+$.

We note that $\mathscr{L}_0^+$ is self-adjoint and present the following:

Definition 5.3 (Restriction). Let σ be a face and $F \in C^k(X;\mathbb{R})$ be a cochain. Define the restriction of F to σ to be $F^\sigma \in C^k(X_\sigma;\mathbb{R})$ such that $F^\sigma(\tau) = F(\tau)$.

Lemma 5.4. *Let $F, G \in C^k(X;\mathbb{R})$ and let $0 \le l \le n-k-1$. Then*

$$\langle F, G\rangle = \mathbb{E}_{\tau \in X(l)}\big[\langle F^\tau, G^\tau\rangle\big].$$

In addition, if $F, G \in C^0(X;\mathbb{R})$ then

$$\langle \mathscr{L}_0^+ F, G\rangle = \mathbb{E}_{\sigma \in X(l)}\big[\langle \mathscr{L}_{\sigma,0}^+ F^\sigma, G^\sigma\rangle_\sigma\big].$$

This lemma is the key of the local-to-global argument. We are interested in studying the behavior of some process over the complex; in this example, the up–down random walk. In order to do so, we look at the process from a local point of view, i.e., the links of the complex. We then use properties we already know about the links in order to derive that the entire complex satisfies the property as well. We are interested in the smallest nonzero eigenvalue of $\mathscr{L}_0^+$. It would therefore be useful to understand the eigenspace whose eigenvalue is exactly 0. We would therefore consider the following projection into the eigenspace of 0:

Definition 5.5. We let $\mathscr{L}_0^-$ be the projection to the space of constant functions, formally

$$\forall F \in C^0(X;\mathbb{R}) : \mathscr{L}_0^- F(v) = \langle F, \mathbb{1}\rangle \mathbb{1} = \mathbb{E}_{u \in X(0)}\big[F(u)\big].$$

And, as with $\mathscr{L}_0^-$, we also define local versions of this operator as

$$\mathscr{L}_{u,0}^- F(v) = \langle F, \mathbb{1}_u\rangle_u \mathbb{1}_u = \mathbb{E}_{u \in X_u(0)}\big[F^u(u)\big].$$

Lemma 5.6. *For every cochain $F \in C^0(X;\mathbb{R})$, it holds that*

$$\mathscr{L}_0^+ F(v) = F(v) - \mathscr{L}_{v,0}^- F^v.$$

Proof. Notice that

$$\begin{aligned}
\mathscr{L}_0^+ F(v) &= (I-A)F(v) = F(v) - AF(v) = F(v) - \sum_{u \in X(0)} [A]_{v,u}\sigma(u) \\
&= F(v) - \sum_{\substack{u \in X(0) \\ \sigma\tau \in X(1)}} [A]_{v,u}\sigma(u) \\
&= F(v) - \sum_{u \in X_v(0)} \mathrm{w}_v(u)\sigma^v(u) = F(v) - \mathscr{L}_{v,0}^- F^v.
\end{aligned}$$

■

We are now ready to prove the trickling down theorem. We will, however, present the proof for the Laplacian instead of the actual walk operator. Theorem 5.1 can be deduced using the standard connection between the eigenvalue of an operator and its Laplacian.

Theorem 5.7. *Let X be a simplicial complex whose 1-skeleton is connected and in which, for every $v \in X(0)$, it holds that $\lambda_2(\mathscr{L}_{v,0}) \geq \lambda$. Then $\lambda_2(\mathscr{L}_0) \geq 2 - \frac{1}{\lambda}$.*

Proof. Let μ be a nontrivial eigenvalue of $\mathscr{L}_0$ with the eigenfunction F, i.e.,

$$\mathscr{L}_0 F = \mu F.$$

Note that due to Corollary 5.4, it holds that

$$\mu\|F\|^2 = \mu\langle F, F\rangle = \langle \mu F, F\rangle = \langle \mathscr{L}_0^+ F, F\rangle = \mathbb{E}_{v\in X(0)}\big[\langle \mathscr{L}_{0,v}^+ F^v, G^v\rangle_v\big]. \tag{3}$$

For every $v \in X(0)$, let $F^{v\|} = \langle F^v, \mathbb{1}^v\rangle_v \mathbb{1}^v$ be the projection of F^v to a constant on X_v and $F^{v\perp} = F^v - F^{v\|}$, and note that it is orthogonal to $F^{v\|}$. Note that for every v it holds that $F^{v\|}$ is constant over X_v and therefore

$$\mathscr{L}_{v,0}^+ F^{v\|} = (I - A)F^{v\|} = F^{v\|} - AF^{v\|} = F^{v\|} - F^{v\|} = 0,$$

where $AF^{v\|} = F^{v\|}$ due to A being an averaging operator and $F^{v\|}$ being constant. We denote by $\{G_i\}_{i\in I}$ the eigenfunction basis of $\mathscr{L}_v^+$ when excluding the constant functions over X_v (i.e., the eigenfunctions whose eigenvalue is 0). We then use the previous fact to conclude that

$$\begin{aligned}
\langle \mathscr{L}_{v,0}^+ F^v, F^v\rangle &= \langle \mathscr{L}_{v,0}^+(F^{v\perp} + F^{v\|}), F^v\rangle = \langle \mathscr{L}_{v,0}^+ F^{v\perp}, F^v\rangle = \langle F^{v\perp}, \mathscr{L}_{v,0}^+ F^v\rangle \\
&= \langle F^{v\perp}, \mathscr{L}_{v,0}^+ F^{v\perp}\rangle = \langle \mathscr{L}_{v,0}^+ F^{v\perp}, F^{v\perp}\rangle \Big\langle \mathscr{L}_v^+ \sum_{i\in I}\alpha_i G_i, F^{v\perp}\Big\rangle \\
&= \sum_{i\in I}\alpha_i\langle \mathscr{L}_v^+ G_i, F^{v\perp}\rangle = \sum_{i\in I}\alpha_i\langle \lambda_i G_i, F^{v\perp}\rangle = \sum_{i\in I}\alpha_i\lambda_i\langle G_i, F^{v\perp}\rangle \\
&\geq \lambda\sum_{i\in I}\alpha_i\langle G_i, F^{v\perp}\rangle = \lambda\Big\langle \sum_{i\in I}\alpha_i G_i, F^{v\perp}\Big\rangle = \lambda\langle F^{v\perp}, F^{v\perp}\rangle = \lambda\|F^{v\perp}\|^2.
\end{aligned}$$

We combine this with (3) to conclude that

$$\mu\|F\|^2 = \mathbb{E}_{v\in X(0)}\big[\langle \mathscr{L}_{0,v}^+ F^v, G^v\rangle_v\big] \geq \lambda\mathbb{E}_{v\in X(0)}\big[\|F^{v\perp}\|^2\big].$$

We will move on to calculate $\mathbb{E}_{v\in X(0)}[\|F^{v\perp}\|]$. Note that

$$\begin{aligned}
\|F^v\|^2 &= \langle F^v, F^v\rangle = \langle F^{v\|} + F^{v\perp}, F^{v\|} + F^{v\perp}\rangle \\
&= \langle F^{v\|}, F^{v\|}\rangle + \langle F^{v\perp}, F^{v\perp}\rangle = \|F^{v\|}\|^2 + \|F^{v\perp}\|^2.
\end{aligned} \tag{4}$$

In addition, due to Lemma 5.4,

$$\|F\|^2 = \mathbb{E}_{v\in X(0)}\big[\|F^v\|^2\big]. \tag{5}$$

We note that

$$\begin{aligned}
\mathscr{L}_v^- F^v(u) &= \mathbb{E}_{u'\in X(0)}\big[F^v(u')\big]\mathbb{1}^v(u) = \Big(\sum_{u'\in X(0)} \mathrm{w}_v(u')F^v(u')\mathbb{1}^v(u')\Big)\mathbb{1}^v(u) \\
&= \langle F^v, \mathbb{1}^v\rangle\mathbb{1}^v.
\end{aligned}$$

And conclude that

$$\mu F(v) = F(v) - F^{v\|} \Rightarrow F^{v\|} = (1-\mu)F(v).$$

We can now calculate $\mathbb{E}_{v\in X(0)}[\|F^{v\|}\|^2]$ as

$$\mathbb{E}_{v\in X(0)}\left[\left\|F^{v\|}\right\|^2\right] = (1-\mu)^2\mathbb{E}_{v\in X(0)}\left[\|F\|^2\right] = (1-\mu)^2\|F\|^2. \tag{6}$$

Combining (4)–(6), we get that

$$\begin{aligned}\mathbb{E}_{v\in X(0)}\left[\left\|F^{v\perp}\right\|\right] &= \mathbb{E}_{v\in X(0)}\left[\left\|F^{v}\right\|^2\right] - \mathbb{E}_{v\in X(0)}\left[\left\|F^{v\|}\right\|^2\right] = \|F\|^2 - (1-\mu)^2\|F\|^2 \\ &= \mu(2-\mu)\|F\|^2\end{aligned}$$

Using this, we conclude that

$$\mu\|F\|^2 \geq \lambda\mu(2-\mu)\|F\|^2.$$

And thus,

$$1 \geq \lambda(2-\mu) \Rightarrow \frac{1}{\lambda} \geq 2-\mu \Rightarrow \mu \geq 2 - \frac{1}{\lambda},$$

as claimed. ∎

Applying the trickling theorem repeatedly yields the following local-to-global result:

Corollary 5.8. *Let X be a d-dimensional pure simplicial complex. If there exists $\lambda \in (0,1]$ such that:*

- *For every $\tau \in X$ such that* $\dim(\tau) \leq d-2$, *it holds that $X_\tau^{(1)}$ is connected.*
- *For every $\tau \in X$ such that* $\dim(\tau) = d-2$, *it holds that $\lambda_2(X_\tau^{(1)}) \leq \frac{\lambda}{1+(d-1)\lambda}$.*

Then X is a λ-local spectral expander.

6. LOCAL-TO-GLOBAL TOPOLOGICAL EXPANSION OF HIGH-DIMENSIONAL EXPANDERS

In the previous section, we have seen local-to-global spectral expansion over $\mathbb{R}$. As we have explained, that result is based on Garland's philosophy using the fact that over the reals we have inner products and self-adjoint operators, which are useful in deriving the local-to-global theorem.

In order to prove local-to-global expansion in the topological sense, we need to show that a local-to-global statement occurs over finite fields (in particular, over $\mathbb{F}_2$) in the case of high-dimensional expanders. Thus, we have to deviate dramatically from the Garland paradigm that uses self-adjoint operators and inner products as they do not exist over $\mathbb{F}_2$.

The local-to-global method we develop here (that is used to prove the local-to-global expansion over $\mathbb{F}_2$) uses the following idea. It shows how to derive expansion of small sets on a complex with topologically expanding links using a newly introduced notion of *local minimality*. It then shows that expansion of large sets can be inferred from expansion of small sets in the case of a complex with high enough dimensions.

The following theorem is a central local-to-global theorem in topological high dimensional expansion. It essentially says that global topological expansion denoted as cosystolic expansion can be obtained from local topological expansion known as coboundary expansion.

We see here again the philosophy that in high-dimensional expansion we have a local-to-global deduction with some loss, i.e., from local coboundary expansion, we deduce a global cosystolic expansion, which is a weaker topological notion of expansion. Recall that, in the local-to-global spectral expansion, we have a larger loss in the spectral expansion the more we move down in the trickling procedure.

The local-to-global topological expansion theorem first appeared in [25], albeit only for dimension 2. It was then extended to any dimension in [12], and recently was extended further by [29] to show that cosystolic expansion could be defined with regard to not only binary cochains, but also to cochains that get values in any group. The work of [29] shows that, even under this more generalized setting, global cosystolic expansion could be deduced from coboundary expansion in links.

Theorem 6.1 (Local-to-global cosystolic expansion [12, 25, 29]). *For any $d, q \in N$ and $0 < \beta < 1$, there exist $0 < \lambda, \eta < 1$ such that the following holds: Let X be a d-dimensional q-bounded degree simplicial complex satisfying the following local conditions:*

- *Spectral expansion in links, i.e., X is one sided-λ-local spectral expander;*
- *Topological expansion in links, i.e., X's links are β-coboundary expanders.*

Then the $(d-1)$-skeleton of X is an (ϵ, μ)-global cosystolic expander, where

$$\epsilon = \min\left\{\eta^{2^d-1}, \frac{1}{qd^{\frac{d}{2}}}\right\} \quad \text{and} \quad \mu = \eta^{2^d-1}.$$

As previously mentioned, an important concept that was introduced in [25] and is a major tool in the analysis of the above theorem is a notion called *local minimality*.

Definition 6.2 (Local minimality). Given a weighted simplicial complex (X, w), a k-cochain $f \in C^k(X)$ is (globally) minimal if $||f|| = \min_{b \in B^k(X)}\{||f + b||\}$. Cochain f is called *locally minimal* if f_σ is minimal in X_σ for every $\sigma \in X(i)$, $0 \le i < k$.

Note that minimality implies locally minimality, but not vice versa. The main idea of [25] and its followup works was to use the notion of local minimality for showing that small sets topologically expand. Namely, for showing that a simplicial complex with local links that are both spectrally and topologically expanding is, in fact, a *small set coboundary expander*, which is defined as follows.

Definition 6.3 (Small set coboundary expander, i.e., a complex in which small sets topologically expand). A d-dimensional weighted simplicial complex (X, w) is called (ϵ, μ)-*small set coboundary expander* for some constants $0 < \epsilon, \mu \le 1$, if for every k-cochain $f \in C^k(X)$, $k < d$, with $||f|| < \mu$, it holds that $||\delta f|| > \epsilon||f||$.

Thus, [12,25,29] have used the notion of local minimality to show that a complex with expanding links is a small-set coboundary expander (i.e., it topologically expands small sets). Specifically, the following was shown:

Theorem 6.4 (Small set coboundary expansion from expanding links). *A q-bounded degree d-dimensional complex, whose links are sufficiently strong λ-local spectral expanders and β-coboundary expanders, is a d-dimensional (ϵ, μ)-small set coboundary expander where (ϵ, μ) depends on λ, β, and q.*

The next major idea of [12, 25, 29] is that there is a way to deduce expansion of all sets (in particular, large sets) from expansion of small sets, assuming the complex is of high enough dimension. Specifically, the following was shown:

Theorem 6.5 (Cosystolic expansion from small set coboundary expansion). *If (X, w) is a d-dimensional weighted simplicial complex of bounded degree that is a (ϵ, μ)-small set coboundary expander then its $(d-1)$-skeleton is a (ϵ, μ)-cosystolic expander.*

Using all the above and the known existence of bounded degree complexes with expanding links, the following was deduced:

Corollary 6.6 (There are bounded degree cosystolic expanders of every dimension [12, 25]). *For every $d \geq 1$, the d-skeleton of the $(d+1)$-dimensional Ramanujan complex* [37] *is a bounded degree cosystolic expander of dimension d.*

This local-to-global paradigm that we have introduced was proven to be very useful recently in order to deduce cosystolic expansion in covers of high-dimensional expanders [14, 20] as the links' structure of the cover of a simplicial complex is the same as in the base complex.

As we next discuss, the topological high-dimensional expansion is tightly related to local testability of codes. The local-to-global proof of the cosystolic expansion that we have discussed here can be seen, via this prism, as a method to get global local-testability of codes from local local-testability. This philosophy was then implemented in various subsequent works.

7. HIGH-DIMENSIONAL EXPANSION AND LOCAL TESTABILITY OF CODES

One of the major motivations for studying high-dimensional expanders within theoretical computer science (TCS) was the discovery of their strong relation to a central notion within TCS known as a locally testable code. Locally testable codes were extensively studied, however, their study was mostly ad hoc and there was no known mathematical phenomenon that implies local testability of codes.

The discovery that topological high-dimensional expansion is equivalent to local testability of some particular codes suggested that local testability of codes is implied by high-dimensional expansion. In order to indeed confirm this ideology, there was a need to show that (1) known locally testable codes can be explained via the high-dimensional expansion prism, and (2) a method to get new LTCs from topological high-dimensional expanders. Both of these goals were materialized using the local-to-global effect existing in high-dimensional expanders, as we discuss in the following. However, it turned out that the LTCs emerging from high-dimensional expanders are not of high enough rate.

Recent works have shown how to overcome the rate issue existing in LTCs emerging from high-dimensional expanders, and get high rate LTCs from a product of two (one-dimensional) expander graphs. These works have adjusted the local-to-global machinery developed in the realm of high-dimensional expanders, to work for products of expanders, thus overcoming the rate barrier existing in using genuine high-dimensional expanders to get LTCs. We, nevertheless, conjecture that high rate LTCs with stronger guaranties should emerge from genuine high-dimensional expanders.

We start by defining locally testable codes.

Definition 7.1 (Locally testable code). A locally testable code (LTC) is an error correcting code admitting a randomized algorithm—called a *tester*—which, given access to a word, can decide with high probability whether it is close to a codeword or not by querying just a few (i.e., $O(1)$) of its letters. A tester is called an ϵ-tester for some $\epsilon > 0$ if it accepts all codewords and the probability of rejecting a word outside the code is ϵ-proportional to its Hamming distance from the code. An LTC with an ϵ-tester is called an ϵ-locally testable code.

Topological high-dimensional expansion is a form of local testability of codes. The first discovery of the connection between high-dimensional expanders and locally testable codes was made by [26] where the authors have shown that a coboundary expansion is, in fact, equivalent to the local testability of the coboundary code.

Theorem 7.2 (Coboundary expansion is equivalent to local testability of the coboundary code [26]). *A d-dimensional complex X is an ϵ-coboundary expander iff the linear code $B^i(X)$, $i < d$ is ϵ-locally testable by the $(i+1)$-cocycle test. This test, given access to $f \in C^i(X)$, chooses a face $\sigma \in X(i+1)$ uniformly at random and accepts f if $\delta_i f(\sigma) = \sum_{\tau \in \binom{\sigma}{i+1}} f(\tau)$.*

Global local-testability from local local-testability via high-dimensional expanders. The works [12,25,29] have, in fact, shown that a global code defined over a high-dimensional expander, by requiring its projection to every link to belong to a small (local) locally-testable code, is a global locally-testable code. Namely, the local testability of the global code stems from the local testability of the small codes that compose it. In these works the global locally-testable code is $H^i(X)$ where X is a cosystolic expander, and the local locally-testable codes are $B^i(X_\sigma)$ for $\sigma \in X$.

Theorem 7.3 (Cosystolic expansion implies locally testable codes with linear distance [12,25,29]). *Let X be a d-dimensional complex which is (ϵ, μ)-cosystolic expander. Then for every $i \le d-2$, $H^i(X)$ is a ϵ-locally testable code whose normalized distance is at least μ. For $i = d-1$, the code $H^i(X)$ is of distance at least μ.*

Using the latter theorem, one can get locally testable codes associated, for example, with $H^i(X)$ of an $(i+2)$-dimensional cosystolic expander X, $i \geq 0$. However, the rate of such codes tends to be small.

Thus, the lesson is that cosystolic expanders imply locally testable codes via the *local local-testability implies global local-testability* paradigm, but such codes tend to be of small rate. Nevertheless, the philosophy of global local-testability from local one is very strong and it has recent important implications to locally testable codes as we now discuss.

Explaining and improving testability of known LTCs via high-dimensional expanders. This idea of global local-testability from local ones was used recently by [31], see also [8], to reprove the local testability of single orbit affine invariant codes [32] via the high-dimensional expansion paradigm. The new analysis has also provided tighter bounds than were previously obtained.

High rate LTCs. Another important implication of the local-to-global testability occurs in the recent works [9, 40] that have constructed high rate (good) locally testable codes from twisted products of expander graphs. Using products of expander graphs, they were able to adopt the philosophy of local-to-global testability to construct new LTCs of high rate. The importance of turning to products of expander graphs was to overcome the small rate barrier that existed in attempts to implement the local-to-global testability from two-dimensional genuine high-dimensional expanders.

Given the recent discovery of good LTCs constructed from two-dimensional objects obtained from a product of two graphs, it is natural to ask whether good LTCs could be constructed from genuine high-dimensional expanders, and what advantages such constructions might have over current ones.

On 2-LTCs from genuine high-dimensional expanders. A recent work [14] suggests a framework to get good 2-queries LTCs from genuine high-dimensional expanders by introducing expanding high-dimensional sheaves. The currently known good LTCs are of very high locality. However, for hardness of approximation, it is desirable to get 2-queries LTCs. We conjecture that LTCs emerging from genuine high-dimensional expanders should be much stronger in various respects than those constructed from products of one-dimensional graphs; however, these seem much harder to construct.

8. HIGH-DIMENSIONAL EXPANSION AND QUANTUM LDPC CODES

Good classical LDPC codes were known since the works of Gallager in the 1960s [17]. However, their quantum analogues seemed elusive until recently, the problem being that random LDPC classical codes can be shown to be good with high probability, while a quantum LDPC code is composed of two codual classical LDPC codes (see the following definition). Such a pair of codual classical LDPC codes that are both good cannot be obtained by random means, hence there was no natural source for obtaining such good codes.

Definition 8.1 (Quantum LDPC code (CSS code) [5]). A *quantum LDPC CSS code* is a quintet $C = (C_X, C_Z, \mathbb{F}^n, \Phi_X, \Phi_Z)$ such that the X-code C_X and the Z-code C_Z are subspaces of $\mathbb{F}^n$, Φ_X is a set of vectors generating $C_X^\perp$, Φ_Z is a set of vectors generating $C_Z^\perp$, and $C_X^\perp \subseteq C_Z$ (equivalently, $C_Z^\perp \subseteq C_X$). The code is LDPC if each vector in Φ_X, Φ_Z has constant (independent of n) support. The rate of C is $\dim C_X - \dim C_Z^\perp = \dim C_Z - \dim C_X^\perp$ and its distance is $\min\{d_X, d_Z\}$, where $d_X = \min\{|w|_{\text{Ham}} \mid w \in C_X - C_Z^\perp\}$ and $d_Z = \min\{\|w\|_{\text{Ham}} \mid w \in C_Z - C_X^\perp\}$; we call d_X and d_Z the X- and Z-distance, respectively. The relative distance and rate of C are its distance and rate divided by n, respectively.

Up until recently, most quantum LDPC codes were obtained from smooth topological objects with very simple local structure (surfaces, manifolds), see, e.g., the toric codes [35]. This is since in such cases there is a natural way to get a pair of such codes that the dual code is isomorphic to the primal code (thus, there is a need only to design one code). The best distance achieved by such codes has exceeded slightly $\sqrt{n}$ by the notable work of Freedman et al. [15] from about 20 years ago. Going beyond the $\sqrt{n}$ distance barrier of Freedman et al. with any rate seemed beyond reach for many years. Very recently, [13, 34] managed to get quantum LDPC codes improving the distance record of Freedman et al. to $\sqrt{n}\log^k n$ for any k, using a *novel approach based on high-dimensional expanders*. Furthermore, the codes of [13] have fast decoding algorithms.

Theorem 8.2 (Decodeable quantum LDPC codes beyond the $\sqrt{n}$ distance barrier [13]). *There exist quantum LDPC codes of distance $O(\sqrt{n}\log n)$ and rate $O(\sqrt{n}/\log n)$ that are efficiently decodeable.*

In fact, the work of [13] has used the following realization:

Quantum LDPC codes, as well as classical LTCs, are born *together* from a single object—a topological high-dimensional expander.

Theorem 8.3 (Cosystolic expanders imply simultaneously quantum codes with linear X-distance and a locally testable code with a linear distance [12, 25, 29]). *Two-dimensional cosystolic expander X implies a quantum LDPC code whose X-code corresponds to $H^1(X)$, its Z-code corresponds to $H_1(X)$; thus the X-code has linear distance. The rate of the code is* $\dim H^i(X)$. *Furthermore, $H^0(X)$ is a locally testable code with linear distance, whose rate is* $\dim H^0(X)$.

Thus, a two-dimensional cosystolic expander gives both a locally testable code corresponding to $H^0(X)$ and a quantum LDPC code corresponding to $H^1(X)$ and $H_1(X)$.

Applying this wisdom to cosystolic expanders arising from Ramanujan complexes [37] implies a quantum code with linear X-distance, logarithmic Z-distance, and nonzero rate. One can even use higher dimensional cosystolic expanders to get that the X-code both has good distance and is a locally testable code!

The idea of [13], following Hastings, was to apply the following balancing procedure that balances the X- and Z-distances to get a quantum code whose distance is a geometric average of the two.

Theorem 8.4 (Weight balancing procedure for quantum LDPC codes [13]). *There exists a way to multiply a two-dimensional cosystolic expander X with a one-dimensional expander graph to get a quantum code whose distance is the geometric average of the distances of* $H^1(X)$ *and* $H_1(X)$ *and whose rate is roughly* $\sqrt{n}$.

Following the works [13, 34], there was a flurry of improvements starting with [22] (see also [4]) that have replaced the product of a two-dimensional simplicial complex with a graph with a twisted product of two graphs. The resulted object has been a two-dimensional expanding cubical complex. These advances culminated in the work of [40] that found a way to use methods inspired by the proof of local-to-global cosystolic expansion of [25] and, in particular the local minimality concept, in order to prove linear distance for these codes, thus obtaining good quantum LDPC codes. Recall that [25] managed to prove linear distance only for the X-code. By turning to cubical complexes, [40] managed to derive the linear distance for both the X-code and Z-code. The dimension of the code arose simply from counting degrees of freedom.

Importantly, the new good LDPC quantum codes are not known to be efficiently decodeable. Another major question is whether there are good quantum LTCs. We conjecture that high-dimensional expansion will be a key towards progress on both of these questions.

9. GROMOV TOPOLOGICAL OVERLAPPING PROBLEM VIA TOPOLOGICAL HIGH-DIMENSIONAL EXPANDERS

More than a decade ago, Misha Gromov, one of the greatest mathematicians of our era, defined the notion of *topological overlapping property* of a simplicial complex and posed the following question: Are there bounded-degree simplicial complexes with the topological overlapping property? In the following we discuss how the topological high-dimensional expansion perspective, and in particular its connection to locally testable codes, has been recently used to provide a positive answer to Gromov's question by combining the works [11, 12, 25].

We start by introducing the topological overlapping property.

Definition 9.1 (Topological overlapping property [21]). A d-dimensional complex X is said to have the *topological overlapping property* if for any continuous map $F : X \to R^d$, there exists a point $p \in R^d$ such that $F^{-1}(p)$ is covered by $\epsilon > 0$ fraction of the d-faces of X.

In order to study the question on the existence of bounded-degree complexes with topological overlapping property, Gromov has introduced the notion of coboundary expansion of simplicial complexes (that was introduced independently by Linial and Mehsulam [36]). Gromov has shown that coboundary expansion implies the topological overlapping property. By doing that, he, in fact, showed that there are unbounded-degree complexes with the topological overlapping property. However, the question on the existence of bounded-degree complexes with this property remained unsolved at that point.

By the connection Gromov made between coboundary expansion (which is a notion of local testability of a code as we have explained!) and the topological overlapping question, it became evident that, in order to positively answer Gromov's question, one should find bounded-degree coboundary expanders.

The natural candidates for such bounded-degree coboundary expanders were the famous bounded-degree Ramanujan complexes [37]. Alas, these complexes are known not to be coboundary expanders. However, the local-to-global cosystolic result of [12, 25] could, in fact, be applied to these complexes to show that they possess the weaker property of cosystolic expanders, thus they come close to satisfying the condition implying the topological overlapping property.

Thus, in order to give a positive resolution to Gromov's question, one had to prove that small set coboundary expansion (which implies cosystolic expansion) is, in fact, sufficient for the topological overlapping property, and this was indeed proved in [11].

Thus, combining the works [12, 25] and [11] led to a positive resolution of Gromov's question on the existence of bounded-degree complexes with the topological overlapping property.

Another interesting aspect of this question is the following. Since we know that cosystolic expansion is a form of local testability of codes, we, in fact, see an evident connection between locally testable codes and the topological overlapping property. This connection might be proven useful in future application within theoretical computer science.

10. SAMPLING BASES OF A MATROID USING LOCAL SPECTRAL EXPANDERS

In this section we will present Anari, Liu, Gharan, and Vinzant's [3] resolution of the Mihail–Vazirani conjecture [38] by rigorously showing how rapid convergence of the down–up walk over local spectral expanders [30] was used in their solution. Before we can present the conjecture, consider the following definition:

Definition 10.1 (Matroid). A matroid $M = (X, \mathcal{I})$ is a combinatorial structure consisting of a ground set X of elements and a nonempty collection $\mathcal{I}$ of independent subsets of X satisfying:

(1) *(Heredity property)* If $T \in \mathcal{I}$ and $S \subseteq T$ then $S \in \mathcal{I}$.

(2) *(Exchange axiom)* If $T_1, T_2 \in \mathcal{I}$ and $|T_2| < |T_1|$ then there exists $i \in T_1 \setminus T_2$ such that $T_2 \cup \{i\} \in \mathcal{I}$.

We call maximal independent sets in M the *bases* of M.

Matroids are natural combinatorial objects and can be thought of as generalizations of a basis of some vector field or subforests of a graph. It is natural to try to sample a basis of a matroid (i.e., sample a basis of some vector field or a spanning tree of a graph). One natural process to try and sample a random basis of a matroid is using the "basis exchange random walk," which is a random walk over the bases of a matroid, defined as follows:

Definition 10.2 (The basis exchange random walk). The basis exchange walk is a walk between bases of a matroid whose step is defined by Algorithm 1:

Algorithm 1: Step in the basis exchange walk

Pick a member of the basis chosen uniformly at random and then remove it creating the independent set I.
Pick a new basis that contains I.
return *the new basis.*

Mihail and Vazirani conjectured the following:

Conjecture 10.3 (Mihail and Vazirani Conjecture [38]). *The basis exchange walk converges rapidly.*

We note that, due to the heredity property, one can think of a simplicial complex that comprises the independent sets of a matroid. Moreover, it is easy to see that the matroid exchange walk is the d-dimensional down–up walk! We will therefore be interested in the expansion properties of that simplicial complex, since if that complex is indeed a local spectral expander, then the base exchange walk converges rapidly and we could sample a basis of the matroid by picking some constant basis to start from (which corresponds to a face of maximal dimension) and then perform a few steps in the down–up walk. Then we can conclude that we have arrived at a random basis due to the convergence of that walk. We will show that matroids are the best possible local spectral expanders, 0-local spectral expanders. Note that the simplicial complex that is constructed by the independent sets of the matroid satisfies the following property:

Definition 10.4 (Exchange property). We say that a simplicial complex satisfies the exchange property if the following holds: For every two faces $\sigma, \tau \in X$ such that $|\sigma| < |\tau|$, there exists $v \in \tau \setminus \sigma$ such that $\sigma \cup \{v\} \in X$.

It is easy to see that if a simplicial complex satisfies the exchange property then so do all its links and skeletons. We will therefore try to use the trickling down theorem in order to prove that matroids are indeed local spectral expanders. We will start with proving that all the links of codimension 2 are 0-local spectral expanders:

Lemma 10.5. *If a graph G satisfies the exchange property and $E \neq \emptyset$ then there exists a partition of V into V_1, V_2, V_3 such that:*

- *V_1 and V_2 are not empty and independent.*
- *For all $i \neq j$, as well as for every $u \in V_i$ and $v \in V_j$, it holds that $\{u, v\} \in E$.*

Proof. Let $\{u, v\} \in E$ and consider the following sets:

$$V_u = \{w : \{v, w\} \in E, \{u, w\} \notin E\}, \quad V_v = \{w : \{u, w\} \in E, \{v, w\} \notin E\},$$
$$V_{\{u,v\}} = \{w : \{v, w\} \in E, \{u, w\} \in E\}.$$

We will show that V_u is independent (and note that V_v is also independent using the same considerations). Indeed, assume that V_u is not independent. Then there exist $w, w' \in V_u$ such that $\{w, w'\} \in E$. Therefore, due to the exchange property, either $\{v, w\} \in E$ or $\{v, w'\} \in E$, which contradicts our choice of w or w'.

We will now show that any vertex in V_u is connected to every vertex in V_v. Let $u' \in V_u$ and $v' \in V_v$. Then $u' \in V_u$, therefore $\{u', u\} \in E$. Applying the exchange property to $\{u', u\}$ and w' yields that either $\{v', v\} \in E$ or $\{u', v'\} \in E$. Therefore, due to the definition of V_u, it holds that $\{v', v\} \notin E$ and thus $\{u', v'\} \in E$.

We follow by showing that any vertex is V_u is connected to any vertex in $V_{\{u,v\}}$ (the case of V_v is analogous). Let $u' \in V_u$ and $w \in V_{\{u,v\}}$. Then $w \in V_{\{u,v\}}$ therefore $\{w, v\} \in E$. Applying the exchange property to $\{w, v\}$ and u' yields that either $\{w, u'\} \in E$ or $\{v, u'\} \in E$. Therefore, due to the definition of V_u, it holds that $\{u', v\} \notin E$ and thus $\{w, u'\} \in E$.

We finish the proof by setting:

$$V_1 = V_u \cup \{v\}, \quad V_2 = V_v \cup \{u\}, \quad V_3 = V_{\{u,v\}}$$

and noting that the properties listed in the lemma hold for these sets. ■

We will use this lemma to prove the following:

Lemma 10.6 (Graphs with the exchange property are 0-spectral expanders). *Let G be a pure graph that satisfies the exchange property then G is a 0-spectral expander.*

Proof. We will show that G is a complete partite graph and therefore a 0-spectral expander. Using Cauchy's interlacing theorem on the nonnormalized adjacency matrix and its complement yields that the nonnormalized matrix's second eigenvalue is bounded from above by 0. The normalization can then be performed by noting that it is equivalent to multiplication from the right by a positive semidefinite matrix. Then one can use Cauchy's interlacing theorem again in order to show that the resulting matrix has at most one positive eigenvalue.

We will do so by constructing the described partition using recursive application of Lemma 10.5. Start by setting $\tilde{V} = V$ and set the partition to be $\mathcal{V} = \emptyset$. While there are edges in the subgraph induced by $\tilde{V}$, do the following: apply Lemma 10.5 to $\tilde{V}$, add V_1 and V_2 to $\mathcal{V}$, and set $\tilde{V} = V_3$. Note that every set in $\mathcal{V}$ is independent due to Lemma 10.5. Also note that if $U_1, U_2 \in \mathcal{V}$ are two different sets then every vertex in U_1 is connected to every vertex in U_2. Therefore G is a complete partite graph. ■

Now all we have to check is that all of the links' 1-skeletons are connected, as we will in the following lemma:

Lemma 10.7 (Connectivity of graphs that satisfy the exchange property). *Let X be a pure simplicial complex that satisfies the exchange property. Then $X^{(1)}$ is connected.*

Proof. Let $v, u \in X^{(1)}(0)$. Then $X^{(1)}$ is pure, therefore there exists $\tilde{v}$ such that $\{v, \tilde{v}\} \in E$. Furthermore, $X^{(1)}$ is a skeleton of a complex that satisfies the exchange property and therefore it satisfies the exchange property as well, meaning that either $\{v, u\} \in E$ or $\{\tilde{v}, u\} \in E$. Therefore v and u are either connected directly or through $\tilde{v}$ and thus $X^{(1)}$ is connected. ■

Therefore we can conclude that:

Theorem 10.8. *If $M = (X, \mathcal{I})$ is a matroid then $\mathcal{I}$ is a 0-local spectral expander.*

Proof. Combining Lemmas 10.6, 10.7, and the tricking down theorem proves this theorem. ■

Therefore high-dimensional expanders can be used to resolve the Mihail–Vazirani conjecture simply by noting the following: The basis exchange walk is the down–up walk on the maximal dimension of a simplicial complex that exhibits the exchange property. Any simplicial complex that exhibits the exchange property is a 0-local spectral expander and, therefore, as we presented in Theorem 4.3, the second largest eigenvalue of the down–up walk is smaller than $1 - \frac{1}{k+1}$, which is bounded away from 1. Therefore the basis exchange walk converges rapidly.

We end this chapter by noting that there are strong connections between counting and sampling for self-reducible problems [24] and therefore being able to sample a random basis for the matroid also proves the existence of a randomized algorithm that estimates the number of bases the matroid has.

11. ADDITIONAL TOPICS THAT ARE NOT COVERED IN THIS NOTE

Before ending this note we mention some topics related to high-dimensional expansion that were *not* covered in this note. These topics include a discussion of explicit constructions of high-dimensional expanders, in particular bounded-degree constructions. Here we note that currently there are no combinatorial constructions of strong local spectral expanders, only algebraic ones. Other topics include high-dimensional expanders beyond simplicial complexes, for example, the Grassmanian complex and its properties; concentration of measure via high-dimensional expanders, Fourier analysis and hypercontractivity on high-dimensional expanders; agreement expanders and low-degree testing via high-dimensional expanders, superfast mixing of Markov chains and Glauber dynamics via strong high-dimensional expanders; unique games and high-dimensional expanders.

FUNDING

Research supported by ERC and BSF.

REFERENCES

[1] V. L. Alev and L. C. Lau, Improved analysis of higher order random walks and applications. In *Proccedings of the 52nd Annual ACM SIGACT Symposium on Theory of Computing, STOC 2020, Chicago, IL, USA, June 22–26, 2020*, pp. 1198–1211, ACM, 2020.

[2] N. Anari, K. Liu, and S. O. Gharan, Spectral independence in high-dimensional expanders and applications to the hardcore model. In *61st IEEE annual symposium on foundations of computer science, FOCS 2020, Durham, NC, USA, November 16–19, 2020*, edited by S. Irani, pp. 1319–1330, IEEE, 2020.

[3] N. Anari, K. Liu, S. O. Gharan, and C. Vinzant, Log-concave polynomials II: high-dimensional walks and an FPRAS for counting bases of a matroid. In *Proceedings of the 51st Annual ACM SIGACT Symposium on Theory of Computing, STOC 2019, Phoenix, AZ, USA, June 23–26, 2019*, pp. 1–12, ACM, 2019.

[4] N. P. Breuckmann and J. N. Eberhardt, Balanced product quantum codes. *IEEE Trans. Inf. Theory* **67** (2021), 6653–6674.

[5] A. R. Calderbank and P. W. Shor, Good quantum error-correcting codes exist. *Phys. Rev. A* **54** (1996), no. 2, 1098.

[6] M. Cryan, H. Guo, and G. Mousa, Modified log-Sobolev inequalities for strongly log-concave distributions. In *2019 IEEE 60th annual symposium on foundations of computer science (FOCS)*, pp. 1358–1370, IEEE, 2019.

[7] Y. Dikstein and I. Dinur, Agreement testing theorems on layered set systems. In *2019 IEEE 60th annual symposium on foundations of computer science (FOCS)*, pp. 1495–1524, IEEE, 2019.

[8] Y. Dikstein, I. Dinur, P. Harsha, and N. Ron-Zewi, Locally testable codes via high-dimensional expanders. 2020, arXiv:2005.01045.

[9] I. Dinur, S. Evra, R. Livne, A. Lubotzky, and S. Mozes, Locally testable codes with constant rate, distance, and locality. 2021, arXiv:2111.04808.

[10] I. Dinur and T. Kaufman, High dimensional expanders imply agreement expanders. In *2017 IEEE 58th annual symposium on foundations of computer science (FOCS)*, pp. 974–985, IEEE, 2017.

[11] D. Dotterrer, T. Kaufman, and U. Wagner, On expansion and topological overlap. In *32nd international symposium on computational geometry, SOCG 2016, June 14–18, 2016, Boston, MA, USA*, edited by S. P. Fekete and A. Lubiw, pp. 35:1–35:10, LIPIcs 51, Schloss Dagstuhl - Leibniz-Zentrum für Informatik, 2016.

[12] S. Evra and T. Kaufman, Bounded degree cosystolic expanders of every dimension. In *Proceedings of the 48th annual ACM SIGACT symposium on theory of computing, STOC 2016, Cambridge, MA, USA, June 18–21, 2016*, pp. 36–48, ACM, 2016.

[13] S. Evra, T. Kaufman, and G. Zémor, Decodable quantum LDPC codes beyond the square root distance barrier using high dimensional expanders. In *61st IEEE Annual Symposium on Foundations of Computer Science, FOCS 2020, Durham, NC, USA, November 16-19, 2020*, pp. 218–227, IEEE, 2020.

[14] U. First and T. Kaufman, (on) good 2-query locally testable codes from sheaves on high dimensional expanders, 2022.

[15] M. H. Freedman, D. A. Meyer, and F. Luo, $\mathbb{Z}_2$-systolic freedom and quantum codes. In *Mathematics of quantum computation*, pp. 287–320, Chapman & Hall/CRC, 2002.

[16] S. Friedland and R. Nabben, On Cheeger-type inequalities for weighted graphs. *J. Graph Theory* **41** (2002), no. 1, 1–17.

[17] R. G. Gallager, *Low density parity check codes*. MIT Press, Cambridge, MA, 1963.

[18] H. Garland, p-adic curvature and the cohomology of discrete subgroups of p-adic groups. *Ann. of Math.* (1973), 375–423.

[19] L. Golowich, Improved product-based high-dimensional expanders. 2021, arXiv:2105.09358.

[20] R. Gotlib and T. Kaufman, Co-boundary expansion as a method for dependency resolution with applications. 2022.

[21] M. Gromov, Singularities, expanders and topology of maps. Part 2: From combinatorics to topology via algebraic isoperimetry. *Geom. Funct. Anal.* **20** (2010), no. 2, 416–526.

[22] M. B. Hastings, J. Haah, and R. O'Donnell, Fiber bundle codes: Breaking the $N^{1/2}\operatorname{polylog}(N)$ barrier for quantum LDPC codes 2020, arXiv:2009.03921. To appear in *FOCS 2021*.

[23] S. Hoory, N. Linial, and A. Wigderson, Expander graphs and their applications. *Bull. Amer. Math. Soc.* **43** (2006), no. 04, 439–562.

[24] M. R. Jerrum, L. G. Valiant, and V. V. Vazirani, Random generation of combinatorial structures from a uniform distribution. *Theoret. Comput. Sci.* **43** (1986), 169–188.

[25] T. Kaufman, D. Kazhdan, and A. Lubotzky, Isoperimetric inequalities for Ramanujan complexes and topological expanders. *Geom. Funct. Anal.* **26** (2016), no. 1, 250–287.

[26] T. Kaufman and A. Lubotzky, High dimensional expanders and property testing. In *Innovations in Theoretical Computer Science, ITCS'14, Princeton, NJ, USA, January 12-14, 2014*, pp. 501–506, ACM, 2014.

[27] T. Kaufman and D. Mass, High dimensional random walks and colorful expansion. In *8th Innovations in Theoretical Computer Science Conference, ITCS 2017, January 9-11, 2017, Berkeley, CA, USA*, pp. 4:1–4:27, LIPIcs 67, Schloss Dagstuhl - Leibniz-Zentrum für Informatik, 2017.

[28] T. Kaufman and D. Mass, Local-to-global agreement expansion via the variance method. In *11th innovations in theoretical computer science conference, ITCS 2020, January 12–14, 2020, Seattle, Washington, USA*, edited by T. Vidick, pp. 74:1–74:14, LIPIcs 151, Schloss Dagstuhl - Leibniz-Zentrum für Informatik, 2020.

[29] T. Kaufman and D. Mass, Unique-neighbor-like expansion and group-independent cosystolic expansion. In *32nd international symposium on algorithms and computation, ISAAC 2021, December 6–8, 2021, Fukuoka, Japan*, pp. 56:1–56:17, LIPIcs 212, Schloss Dagstuhl - Leibniz-Zentrum für Informatik, 2021.

[30] T. Kaufman and I. Oppenheim, High order random walks: Beyond spectral gap. In *Approximation, randomization, and combinatorial optimization. algorithms and techniques (APPROX/RANDOM 2018), Schloss Dagstuhl-Leibniz-Zentrum für Informatik*, pp. 47:1–47:17, Schloss Dagstuhl - Leibniz-Zentrum für Informatik, 2018.

[31] T. Kaufman and I. Oppenheim, High dimensional expansion implies amplified local testability. 2021, arXiv:2107.10488.

[32] T. Kaufman and M. Sudan, Algebraic property testing: the role of invariance. In *Proceedings of the 40th annual ACM symposium on theory of computing, Victoria, British Columbia, Canada, May 17–20, 2008*, edited by C. Dwork, pp. 403–412, ACM, 2008.

[33] T. Kaufman and R. J. Tessler, Local to global high dimensional expansion and Garland's method for general posets. 2021, arXiv:2101.12621.

[34] T. Kaufman and R. J. Tessler, New cosystolic expanders from tensors imply explicit quantum LDPC codes with $\Omega(\sqrt{n}\log^k n)$ distance. In *STOC'21: 53rd annual ACM SIGACT symposium on theory of computing, virtual event, Italy, June 21–25, 2021*, pp. 1317–1329, ACM, 2021.

[35] A. Y. Kitaev, Fault-tolerant quantum computation by anyons. *Ann. Phys.* **303** (2003), no. 2, 2–30.

[36] N. Linial and R. Meshulam, Homological connectivity of random 2-complexes. *Combinatorica* **26** (2006), no. 4, 475–487.

[37] A. Lubotzky, B. Samuels, and U. Vishne, Explicit constructions of Ramanujan complexes of type a_d. *European J. Combin.* **26** (2005), no. 6, 965–993.

[38] M. Mihail and U. Vazirani, On the expansion of 0/1 polytopes. *J. Combin. Theory Ser. B* (1989).

[39] I. Oppenheim, Local spectral expansion approach to high dimensional expanders. Part I: Descent of spectral gaps. *Discrete Comput. Geom.* **59** (2018), no. 2, 293–330.

[40] P. Panteleev and G. Kalachev, Asymptotically good quantum and locally testable classical LDPC codes. 2021, arXiv:2111.03654.

ROY GOTLIB
Department of Computer Science, Bar-Ilan University, Ramat-Gan, 5290002, Israel, roy.gotlib@gmail.com

TALI KAUFMAN
Department of Computer Science, Bar-Ilan University, Ramat-Gan, 5290002, Israel, kaufmant@mit.edu

FORTY YEARS OF FREQUENT ITEMS

JELANI NELSON

ABSTRACT

We survey the last 40 years of algorithm development for finding frequent items in data streams, a line of work which surprisingly wound up developing new tools in information theory, pseudorandomness, chaining methods for bounding suprema of stochastic processes, and spectral graph theory.

MATHEMATICS SUBJECT CLASSIFICATION 2020

Primary 68Q99; Secondary 68W20

KEYWORDS

Streaming, sketching, heavy hitters, differential privacy

INTERNATIONAL CONGRESS OF MATHEMATICIANS 2022 JULY 6–14

Proc. Int. Cong. Math. 2022, Vol. 6, pp. 4872–4896
DOI 10.4171/ICM2022/154

Published by EMS Press

In many big data applications, data continuously arrives in a streaming fashion, for example, the constant stream of queries to a search engine, purchases from online vendors, or posts to social media. Such applications gave rise to the popularity of so-called *streaming algorithms*, which process such data on the fly as it arrives to later answer some queries of interest, with such algorithms often times using memory *sublinear* in the data seen so that most data is forgotten at query time. Aside from minimizing memory consumption, such sublinear memory algorithms also have the advantage of potentially being faster, since the working memory of the algorithm can fit in the faster CPU cache instead of RAM or disk, or minimize communication in distributed environments where the stream of data is sharded to many different servers for processing, which must then communicate intermediate results (or their memory footprints) to then reconstruct query answers to the aggregate data.

One of the oldest and most well-studied problems in the streaming literature is that of finding frequent items in data streams. This line of work began with an algorithm of Boyer and Moore, originally published as a technical report in 1981 and later republished a decade later [9], and still continues into the present. The results along the way have led to the development of new tools in information theory, pseudorandomness, chaining methods for bounding suprema of stochastic processes, and spectral graph theory. In this survey, we discuss some of the progress on this problem over the last 40 years.

1. THE EARLY WORK

We henceforth assume that all items in the data stream, which is finite, come from some finite universe $\mathcal{U} = [n] := \{1, \dots, n\}$. We also define the frequency histogram $f \in \mathbb{R}^n$, where f_i denotes the number of times item i was seen in the stream. Thus, seeing i in the stream corresponds to the update "$f_i \leftarrow f_i + 1$," where f is initially the zero vector. The frequent items problem then asks us to report the i such that f_i is "large" at the end of the stream.

As mentioned, the first algorithm for finding frequent items in data streams was the so-called MJRTY algorithm, discovered by Boyer and Moore in 1981 [9]. The largeness criteria used by their algorithm is that i is frequent iff $f_i > \ell/2$, where ℓ is the stream's length. That is, the MJRTY algorithm should report any i which appears a strict majority of the time (and if no such item exists, the algorithm's output is allowed to be arbitrary). Clearly, at most one majority item can possibly exist. The algorithm uses $O(1)$ memory[1] and is simple to both describe and analyze. The algorithm at any point in time stores only two things in memory: a candidate $i \in [n] \cup \{\perp\}$ for the majority element, and a counter value C which is initialized to 0. For each item j seen in the stream, if $i = j$ then C is

1 Here we measure memory in *machine words*, where a word is a unit of memory large enough to hold the name of an item ($\lceil \log_2 n \rceil$ bits) as well as the largest frequency of any item ($\lceil \log_2 \|f\|_\infty \rceil$ bits). Whenever measuring memory in bits instead, we write so explicitly.

incremented, else if $i \neq j$ then C is decremented. If C becomes nonpositive, then i is set to j and C is set to 1. Pseudocode is given in Figure 1.

```
MJRTY:
initialize().
    1.  i ← ⊥
    2.  C ← 0

update(j). // process item j in stream
    1.  if i = j:
            C ← C + 1
    2.  else:
            C ← max{0, C − 1}
    3.      if C = 0:
                i ← j
                C ← 1

query().
    1.  return i
```

```
Frequent:
initialize(k).
    1.  i_1, ..., i_{k−1} ← ⊥
    2.  C_1, ..., C_{k−1} ← 0

update(j). // process item j in stream
    1.  if ∃r such that i_r = j:
            C_r ← C_r + 1
    2.  else:
            for r = 1, ..., k − 1:
                C_r ← max{0, C_r − 1}
    3.      if ∃ r such that C_r = 0:
                pick such r arbitrarily
                i_r ← j
                C_r ← 1

query().
    1.  return {i_1, ..., i_{k−1}}\{⊥}
```

FIGURE 1
Pseudocode for MJRTY and Frequent.

Theorem 1.1. *If there exists some $i^* \in [n]$ such that $f_{i^*} > \ell/2$, then at the time of query we must have $i = i^*$.*

Not long after the development of the MJRTY algorithm, Misra and Gries developed the generalized Frequent algorithm [32], which outputs a list $L \subset [n]$ such that (1) $|L| < k$, and (2) if i is k-frequent then $i \in L$; see the pseudocode in Figure 1. Here k is a parameter that is given at the time of initialization, and we say an item is k-frequent if $f_i > \ell/k$. That is, whereas MJRTY must report any item that appears strictly more than half the time in the stream, Frequent must report any item that appears strictly more than a $1/k$ fraction of the time. The MJRTY problem thus solves the special case $k = 2$. Rather than store a single candidate frequent item i, Frequent stores $k-1$ candidate items $i_1, \dots, i_{k-1}$ together with counters $C_1, \dots, C_{k-1}$ (observe that the number of k-frequent items is at most $k-1$). When a stream item j matches one of these candidate items, its corresponding counter is incremented. Otherwise, *all* counters are decremented and some arbitrary candidate with counter zero (if one exists) is replaced with the new item j and its counter is reset to 1. From the description and pseudocode, the memory consumption of Frequent is $O(k)$, which is clearly optimal since $\Omega(k)$ memory is required since there can be up to $k-1$ frequent items and just writing down their names would take $\Omega(k)$ memory.

An alternative (randomized) algorithm is to, of course, sample: if an item appears often, then we expect it to also appear often in a substream obtained by sampling m uniformly random updates without replacement. Such a sample can be maintained in $O(m)$ memory on the fly as the stream is being updated using a fairly simple technique known as *reservoir sampling* [37], which we do not discuss in detail here. Unfortunately, it is not too hard to show that to identify all frequent items in the sense of [32], one must take $m = \Omega(k^2 \log k)$, which yields significantly worse memory consumption than the $O(k)$ memory of the Frequent algorithm, while also being randomized with a chance of error rather than providing the deterministic guarantees enjoyed by Frequent.

2. PROBLEM REFORMULATIONS AND MORE GENERAL ALGORITHMS

Rather than finding k-frequent items as defined in Section 1, one may aspire to identify a more natural set of items: the "top k" items by frequency, i.e., the k indices i with the largest f_i values (breaking ties arbitrarily). Unfortunately, we will see in Section 3 that such a task is impossible using memory sublinear in n. Instead, we try to approximate the top k set as follows. We say item i is (k, p)-*tail frequent* if

$$f_i^p > \frac{1}{k} \sum_{j=k+1}^{n} (f_j^*)^p := \frac{1}{k} \| f_{\mathrm{tail}(k)} \|_p^p .$$

Here x^* denotes the decreasing rearrangement of a vector x, i.e., x with its entries permuted so that $|x_1^*| \geq |x_2^*| \geq \cdots \geq |x_n^*|$; $x_{\mathrm{tail}(k)}$ denotes x but with its k largest entries (in magnitude) zeroed out; ties are broken arbitrarily. Similarly, we define $x_{\mathrm{head}(k)} := x - x_{\mathrm{tail}(k)}$. One sees that the number of items i which can be (k, p)-tail frequent is less than $2k$: every index in the head could be frequent, and the number of tail indices which are frequent must be strictly less than k.

Definition 2.1. For integer $k \geq 2$ and real $p \geq 1$, a (randomized) streaming algorithm is said to solve the (k, p)-*tail frequent problem* with failure probability $\delta \in (0, 1)$ if at query time it outputs a list $L \subset [n]$ such that (1) $|L| = O(k)$, and (2) with probability at least $1 - \delta$, L contains every (k, p)-tail frequent item.

When k is understood by context or not particularly relevant to the point of discussion, we sometimes refer to the (k, p)-tail frequent problem as the ℓ_p *tail heavy hitters problem*, or even more simply, the ℓ_p *heavy hitters problem*. When discussing the nontail version, we say the ℓ_p *nontail heavy hitters problem*.

It turns out the Frequent algorithm of Section 1 in fact solves the $(k, 1)$-tail frequent problem [7], i.e., it finds items that are not just more than a $1/k$ fraction of $\|f\|_1$, but even of $\|f_{\mathrm{tail}(k)}\|_1$ (with failure probability 0 in fact, as it is a deterministic algorithm). An interesting fact about the formulation of the frequent items problem in Definition 2.1 is that the notion provides a hierarchy of approximation to the actual top k problem, up to potentially changing k by a small constant factor.

Lemma 2.2. *For $p > q \geq 1$ and a frequency vector $f \in \mathbb{R}^n$, if i is (k, q)-tail frequent for f then it is also $(2k, p)$-tail frequent for f.*

Proof. Item i being (k,q)-tail frequent is equivalent to $f_i > \frac{1}{k^{1/q}}\|f_{\mathrm{tail}(k)}\|_q$. We must thus show that

$$\frac{1}{k^{1/q}}\|f_{\mathrm{tail}(k)}\|_q \geq \frac{1}{(2k)^{1/p}}\|f_{\mathrm{tail}(2k)}\|_p.$$

Define $B_j := \{jk+1,\dots,j(k+1)\}$ for $j = 0,1,\dots,n/k-1$ (we can assume that n is divisible by k without loss of generality by padding it with 0 entries, which does not affect the subsequent argument). Then

$$\begin{aligned}
\frac{1}{(2k)^{1/p}}\|f_{\mathrm{tail}(2k)}\|_p &= \left(\frac{1}{2k}\sum_{j=2}^{n/k-1}\|f^*_{B_j}\|_p^p\right)^{1/p} \\
&\leq \left(\frac{1}{2k}\sum_{j=2}^{n/k-1} k\cdot\|f_{B_j^*}\|_\infty^p\right)^{1/p} \\
&\leq \left(\frac{1}{2k}\sum_{j=2}^{n/k-1} k\cdot\left(\frac{\|f^*_{B_{j-1}}\|_q^q}{k}\right)^{p/q}\right)^{1/p} \\
&= \frac{1}{2k^{1/q}}\left(\sum_{j=1}^{n/k-2}\left(\|f^*_{B_{j-1}}\|_q^q\right)^{p/q}\right)^{1/p} \\
&\leq \frac{1}{2k^{1/q}}\left(\sum_{j=1}^{n/k-2}\|f^*_{B_{j-1}}\|_q^q\right)^{1/q} \qquad (\|z\|_p \leq \|z\|_q \text{ since } p > q) \\
&\leq \frac{1}{2k^{1/q}}\|f_{\mathrm{tail}(k)}\|_q \\
&\leq \frac{1}{k^{1/q}}\|f_{\mathrm{tail}(k)}\|_q.
\end{aligned}$$

■

Thus, up to changing k by a factor of at most 2, an algorithm which solves the p-version of the problem is strictly stronger than one for the q-version for $p \geq q$, in the sense that it is guaranteed to find at least as many items in its output list L. It is also not hard to show via Hölder's inequality that for any fixed x, the value p can be taken large enough (but finite) so that L is guaranteed to contain every element in the top k, regardless of how small its actual frequency is. Thus for some large but finite p the problem is essentially the top k problem, and we can gradually relax the problem (i.e., make it easier) by making p smaller and smaller. What then is the largest value of p for which the problem is algorithmically tractable in small memory? As we will see in Section 3, $p > 2$ requires $n^{\Omega(1)}$ memory. Meanwhile, Charikar, Chen, and Farach-Colton developed the CountSketch algorithm for the case $p = 2$ [14], using space $O(k \log n)$, which we now discuss in Section 2.1.

2.1. ℓ_2 heavy hitters and the turnstile model

We show a diagram representing the CountSketch data structure in Figure 2. Memory stores BR counters $C_{r,b}$ for $r \in [R]$, $b \in [B]$, all initialized to zero. We also pick R random functions $h_1,\dots,h_R : [n] \to [B]$ and another R random functions $\sigma_1,\dots,\sigma_R : [n] \to \{-1,1\}$. The functions are drawn independently, and each such function is drawn uniformly at random from the set of all functions mapping $[n]$ to its respective range. Note that just storing these

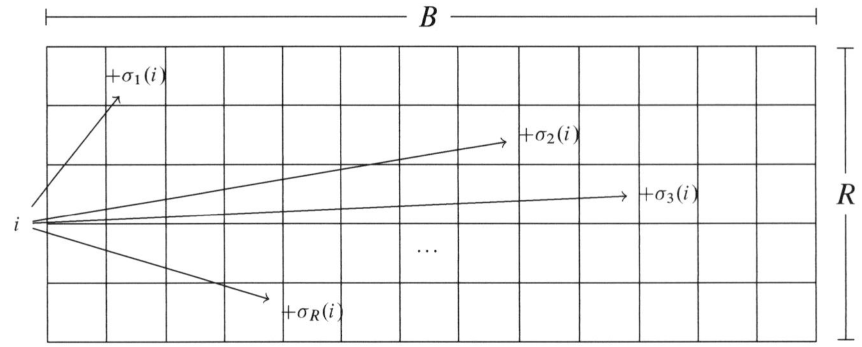

FIGURE 2
Diagram showing an update to CountSketch, when seeing item i in the stream.

functions would require an exorbitant amount of memory (more memory than simply storing the frequency histogram f explicitly in memory); we ignore the cost of storing these functions for now and address this issue later in Section 2.2. When seeing item i in the stream, for $r = 1, 2, \ldots, R$ we perform the update $C_{r,h_r(i)} \leftarrow C_{r,h_r(i)} + \sigma_r(i)$. Thus at the end of the stream, each $C_{r,j}$ will equal $\sum_{i:h_r(i)=j} \sigma_r(i) f_i$. At query time, any f_i can then be estimated as $\tilde{f}_i := \text{median}\{\sigma_r(i) C_{r,h_r(i)}\}_{r=1}^R$.

Although we are ultimately interested in answering queries for the list of frequent items, we state two different types of queries the CountSketch can answer:

- `point_query(i)` : return a value $\tilde{f}_i$ in $[f_i - \frac{1}{\sqrt{k}} \| f_{\text{tail}(k)} \|_2, f_i + \frac{1}{\sqrt{k}} \| f_{\text{tail}(k)} \|_2]$.
- `frequent()` : return a list $L \subset [n]$ such that (1) $|L| = O(k)$, and (2) if i is $(k, 2)$-tail frequent then $i \in L$.

The CountSketch has randomized correctness guarantees: for any query, there is some (tunably small) probability that its output is not correct. As mentioned above, to answer `point_query(i)` we return $\tilde{f}_i := \text{median}\{\sigma_r(i) C_{r,h_r(i)}\}_{r=1}^R$. To answer `frequent()`, we return the $2k$ coordinates i with the largest $|\tilde{f}_i|$ values. Below we show that this algorithm is correct with large probability.

Lemma 2.3. *For $B \geq 6k$, for any $1 \leq i \leq n$,*

$$\mathbb{P}\left(|\tilde{f}_i - f_i| > \sqrt{\frac{6}{B}} \| f_{tail(k)} \|_2 \right) \leq \exp(-R/16).$$

Proof. Write $\tilde{f}_{r,i} = \sigma_r(i) C_{r,h_r(i)}$. Let $H \subset [n]$ denote the locations of the largest k entries of f in magnitude so that $f_{\text{head}(k)} = f_H$. Let $\mathcal{E}_r$ be the event that $h_r(i) \notin h_r(H \backslash \{i\})$. Consider also the random variable $Z_r := \sum_{\substack{j \neq i \\ j \notin H}} \mathbb{1}\{h_r(j) = h_r(i)\} \sigma_r(j) f_j$ and let $\mathcal{E}'_r$ denote the event that $|Z_r| \leq \sqrt{6/B} \| f_{\text{tail}(k)} \|_2$.

Then by Markov's inequality,

$$\mathbb{P}(\neg \mathcal{E}_r) = \mathbb{P}\left(\left| (H \backslash \{i\}) \cap h_r^{-1}(i) \right| \geq 1 \right) \leq \frac{k}{B} \leq \frac{1}{6}.$$

Also, $\mathbb{E}\, Z_r^2 \le \|f_{\mathrm{tail}(k)}\|_2^2/B$, and thus

$$\mathbb{P}(\neg\mathcal{E}_r') = \mathbb{P}\left(|Z_r| > \sqrt{\frac{6}{B}}\|f_{\mathrm{tail}(k)}\|_2\right) = \mathbb{P}\left(Z_r^2 > \frac{6}{B}\|f_{\mathrm{tail}(k)}\|_2^2\right) < \frac{1}{6},$$

also by Markov's inequality. Thus by a union bound $\mathbb{P}(\mathcal{E}_r \wedge \mathcal{E}_r') > 2/3$. Note that when $\mathcal{E}_r \wedge \mathcal{E}_r'$ occurs, we necessarily have $|\tilde{f}_{r,i} - f_i| \le \sqrt{\frac{6}{B}}\|f_{\mathrm{tail(k)}}\|_2$. We just showed that in expectation this fails to occur for fewer than $R/3$ values of r. Thus by the Chernoff–Hoeffding bound,

$$\mathbb{P}\left(\left|\left\{r : |\tilde{f}_{r,i} - f_i| > \sqrt{\frac{6}{B}}\|f_{\mathrm{tail(k)}}\|_2\right\}\right| \ge R/2\right) \le \exp\left(-R\frac{(1/6)^2}{2(1/3)(2/3)}\right) = \exp(-R/16),$$

which implies the claim since $\tilde{f}_i$ is the median of the $\tilde{f}_{r,i}$ values over all $r \in [R]$. ■

Lemma 2.3 implies the following corollary by setting $B = 6k$, $R = \lceil 16\ln(1/\delta)\rceil$. The query time follows since the median of T numbers can be found in linear time $O(T)$ [8].

Corollary 2.4. *For any $\delta \in (0,1)$ and $k \ge 1$, there is an algorithm for answering a single call to* `point_query` *with $(k,2)$-tail error and failure probability δ using memory $O(k\log(1/\delta))$ with update time $\Theta(\log(1/\delta))$ and query time $\Theta(\log(1/\delta))$.*

A simple algorithmic reduction, which we now describe, shows how to obtain an algorithm to answer `frequent` queries in a black box way given an algorithm that solves `point_query`.

Theorem 2.5. *The CountSketch data structure with parameters $B = 54k$ and $R = \lceil 16\ln(n/\delta)\rceil$ provides a solution to the ℓ_2 heavy hitters problem with failure probability δ. The memory usage is $O(k\log(n/\delta))$, the update time is $O(\log(n/\delta))$, and the query time is $O(n\log(n/\delta))$. The output list L has size at most $18k$.*

Proof. We use the CountSketch to answer `point_query(i)` for every $i \in [n]$ to obtain $\tilde{f} = (\tilde{f}_i)_{i=1}^n$. We then define L to be the largest $18k$ entries of $\tilde{f}$ in magnitude (ties broken arbitrarily). We now analyze correctness. We show that correctness is guaranteed when we condition on the event $\|\tilde{f} - f\|_\infty \le \frac{1}{3\sqrt{k}}\|f_{\mathrm{tail}(k)}\|_2$, which happens with probability at least $1-\delta$ by Lemma 2.3 and a union bound over all $i \in [n]$. Now conditioned on this event, consider some $(k,2)$-tail frequent item i; we must show that i is in L. Note that if i' is not even $(9k,2)$-tail frequent, then necessarily $\tilde{f}_{i'} < \tilde{f}_i$. This is because $\tilde{f}_j = f_j \pm \|\tilde{f} - f\|_\infty = f_j \pm \frac{1}{3\sqrt{k}}\|f_{\mathrm{tail}(k)}\|_2$ for any j. Thus, the only items that could appear more frequent than an actual frequent item are the $(9k,2)$-tail frequent items, but since there are fewer than $18k$ such items the claim is proven. ■

Not only does the CountSketch solve the ℓ_2 tail heavy hitters problem, but it does so in a more general streaming model known as the *turnstile model*. In this model, each stream update is an (i,Δ) pair for $i \in [n]$ and $\Delta \in \mathbb{R}$ (Δ may even be negative). Such an update causes the change $f_i \leftarrow f_i + \Delta$. The previous model discussed implicitly took $\Delta = 1$ always. The definition of a $(k,2)$-tail frequent item is then similar as before except that we take absolute values: i is such a frequent item if $|f_i| > \frac{1}{\sqrt{k}}\|f_{\mathrm{tail}(k)}\|_2$.

2.2. A digression on pseudorandomness

As mentioned in Section 2.1, the CountSketch makes use of independently chosen, uniformly random functions $h_1, \dots, h_R : [n] \to [B]$ and $\sigma_1, \dots, \sigma_R : [n] \to \{-1, 1\}$. Naively storing such functions would require $\Omega(nR)$ memory, whereas the frequent items problem already admits a trivial $O(n)$ memory solution by simply storing the frequency vector f in memory explicitly. We remedy this issue by not storing perfectly random functions, but rather functions that are *pseudorandom*.

Definition 2.6. Given integer $n \geq 1$ and a finite range M, a *hash family* is simply a collection $\mathcal{H}$ of functions mapping $[n]$ to M. For integer $k \geq 1$, we say a hash family is k*-wise independent* if for all distinct $x_1, \dots, x_k \in [n]$ and all (possibly not distinct) $y_1, \dots, y_k \in M$,

$$\mathop{\mathbb{P}}_{h \in \mathcal{H}} \left(\bigwedge_{t=1}^{k} h(x_t) = y_t \right) = \frac{1}{|M|^k},$$

where h is chosen uniformly at random from $\mathcal{H}$. That is, the distribution of $(h(x_t))_{t=1}^k$ is uniform for any choice of k distinct values $x_t \in [n]$. Similarly $\mathcal{H}$ is ϵ*-almost* k*-wise independent* if the distribution of $(h(x_t))_{t=1}^k$ is ϵ-close to uniform in total variation distance for any choice of k distinct x_t.

The benefit of Definition 2.6 is that a uniformly random function from a hash family $\mathcal{H}$ can be specified using only $\lceil \log_2 |\mathcal{H}| \rceil$ bits. Thus whereas the σ_r from Section 2.1 are drawn uniformly from the set $\mathcal{H}_{[n],\{-1,1\}}$ of *all* functions from $[n]$ to $\{-1, 1\}$, requiring $\log_2 \lceil H_{n,\{-1,1\}} \rceil = n$ bits each (and even worse for the h_r), we could hope that (1) picking these hash functions from k-wise independent hash families instead still guarantees correctness of CountSketch, and (2) for small k there are k-wise independent hash families that are significantly smaller than the set of all functions mapping $[n]$ to some range. Item (2) is indeed true: if n and m are powers of 2, for example, Carter and Wegman showed that k-wise independent hash families exist mapping $\{0, 1, \dots, n-1\}$ to $\{0, 1, \dots, m-1\}$ of size only $N := \max\{n, m\}^{O(k)}$ [38], thus requiring only $O(k \log N)$ bits to specify a random function from the family. For example, one can consider the family

$$\mathcal{H}_{k,\text{poly}} = \left\{ h(x) = \left(\sum_{i=0}^{k-1} a_i x^i \right) \bmod m : a_0, \dots, a_{k-1} \in \mathbb{F}_N \right\},$$

where the arithmetic in computing $h(x)$ is done over $\mathbb{F}_N$, and the "mod m" simply identifies $\mathbb{F}_N$ with $\{0, 1\}^{\log_2 N}$ then projects to the least significant $\log_2 m$ bits. That is, $\mathcal{H}_{k,\text{poly}}$ is the set of less than degree k polynomials over $\mathbb{F}_N$ (with a mod operation after evaluation).

It can be shown that the analysis of the CountSketch in Section 2.1 only requires the h_r, σ_r to be drawn from 2-wise independent hash families, thus requiring only $O(R \log n)$ bits (i.e., $O(R)$ machine words) of memory to store all hash functions combined. Essentially, this is because the analysis of the data structure only depends on first and second moment calculations of linear forms, which are fully determined by 2-wise independence of the hash functions (note we can round B up to the nearest power of 2).

Note: The construction of $\mathcal{H}_{k,\text{poly}}$ does not actually require that n, m be powers of 2, but rather can be any prime powers. In practice, evaluation of the hash function is fastest when they are primes (and not prime powers) to allow for faster arithmetic over the finite field. The domain size n can simply be rounded up to the nearest prime. If the range size m is not a prime power, often in the analysis one can make do with ϵ-almost k-wise independence (as defined above) instead of exact k-wise independence; to achieve this, one can simply pick a prime $p > mk/\epsilon$ and pick polynomials over $\mathbb{F}_{\max\{n,p\}}$ then output the evaluation of any polynomial mod m. The number of bits to specify h is then $O(k \log(nm/\epsilon))$.

3. IMPOSSIBILITY RESULTS

So far we have discussed how to obtain and analyze algorithms for the frequent items problems, which provides an upper bound on the minimum memory required to solve the problem. In this section we focus on *lower bounds*, i.e., proving that any correct algorithm requires at least some amount of memory.

3.1. ℓ_p heavy hitters for $p > 2$

Whereas the dependence on n in the memory of CountSketch is logarithmic, it turns out that any solution to ℓ_p heavy hitters requires memory that grows polynomially with n when $p > 2$ [3]. The source of this lower bound is via reduction from a problem in *communication complexity* [29]. In the simplest communication complexity setting, there are two parties Alice and Bob. Alice receives an input $x \in \mathcal{X}$, and Bob receives $y \in \mathcal{Y}$, and they also both know some function $f : \mathcal{X} \times \mathcal{Y} \to \mathcal{Z}$. They would like to communicate back and forth (if Alice speaks first then she sends a message to Bob, who in response sends a message to Alice, who then sends a message to Bob, etc.) until some player is certain of the answer and outputs $f(x, y)$. A trivial solution is for Alice to simply send her input to Bob explicitly, taking $\lceil \log_2 |\mathcal{X}| \rceil$ bits of communication; similarly, Bob can send his input to Alice using $\lceil \log_2 |\mathcal{Y}| \rceil$ bits. The question is whether it is possible to devise a communication protocol whose total communication, that is, the sum of the lengths of all messages sent, is smaller.

In the proof of the memory lower bound for the ℓ_p heavy hitters problem [3], we imagine there are not just two parties Alice and Bob, but rather $t \geq 2$ parties $P_1, P_2, \ldots, P_t$. Each P_i receives an input from the same domain $\mathcal{X}_i$, which is the power set of $[n]$; that is, the input to P_i is some subset $S_i \subseteq [n]$. The model of communication considered is that P_i sends a message to P_{i+1}, in sequential order starting from $i = 1$, and P_t must then output its guess of the function evaluation; this model is referred to as *one-way communication*, since the players only speak once each, to the next player in turn, and there is no back-and-forth conversation. The relevant function considered to show hardness of the frequent items problem is the following partial function known as *set disjointness*:

$$\textsc{Disj}_{n,t}(S_1, S_2, \ldots, S_t) := \begin{cases} 1, & \forall i \neq j, S_i \cap S_j = \emptyset, \\ 0, & \exists x \in [n] : \forall i \neq j, S_i \cap S_j = \{x\}. \end{cases}$$

Note $\textsc{Disj}_{n,t}$ is partial since it is not defined when the pairwise intersections are not all equal (or contain more than one item). Furthermore, we will be concerned with the *randomized complexity* of the problem, in which we imagine all players share knowledge of an infinite sequence of uniform random bits, for free without any communication. P_t then need only be correct with some failure probability of at most δ, where $\delta \in (0, 1)$ is a parameter known to all parties at the beginning of the communication game. We refer to the minimum number of total bits (sum of message lengths sent by all players) required for a one-way randomized communication protocol to solve any input to $\textsc{Disj}_{n,t}$ (on the subdomain where it is defined) with failure probability at most $\delta \in (0, 1)$ as $\mathbf{R}_{\delta}^{\to,\text{pub}}(\textsc{Disj}_{n,t})$ ("pub" signifies that the randomness is public, and "$\to$" signifies that we only consider one-way communication protocols). The following theorem is due to [3,13] (see also [24,25], with a more recent lower bound in [27] for a slightly modified problem but which implies new lower bounds for heavy hitters and other problems).

Theorem 3.1. *There exists universal $\delta_0 > 0$ such that $\forall 1 \le t \le n$, $\mathbf{R}_{\delta_0}^{\to,pub}(\textsc{Disj}_{n,t}) = \Omega(n/t)$.*

Corollary 3.2. *There exists universal constant $\delta_0 > 0$ such that for $p > 2$, any randomized streaming algorithm solving the $(2, p)$-tail heavy hitters problem with failure probability at most δ_0 must use at least $\Omega(n^{1-2/p})$ bits of memory.*

Proof. Suppose there exists an algorithm $\mathcal{A}$ using at most s bits of memory which solves the $(2, p)$-tail heavy hitters problem with failure probability at most δ_0, which is the same δ_0 from the statement of Theorem 3.1. We use such $\mathcal{A}$ to define an efficient communication protocol for $\textsc{Disj}_{n,t}$ as follows for $t = \lceil (2n^{1/p}) \rceil + 1$. P_1 initializes the algorithm, then feeds $\mathcal{A}$ the stream consisting of all elements of S_1. They then take the memory state of $\mathcal{A}$, which is simply an element of $\{0, 1\}^s$, and send this state to P_2. P_2 can then continue running $\mathcal{A}$ from where it left off, and feed it as input a stream consisting of all elements of S_2, etc., for each of the first $t - 1$ players. After feeding $S_1, \dots, S_{t-1}$ to $\mathcal{A}$, the $(t - 1)$st player then queries $\mathcal{A}$ to obtain a list L of size $O(1)$ which contains all the $(2, p)$-tail frequent items. They then send $L \cap S_{t-1}$ to P_t, using $O(|L| \cdot \log n) = O(\log n)$ bits, who then outputs 1 iff $L \cap S_t = \emptyset$.

Note if there exists an x in the intersection of all S_i, then after processing $S_1, \dots, S_{t-1}$, x is $(2, p)$-tail frequent (in fact is it even non-tail frequent) since $f_x^p \ge 2n$ yet $\| f_{\text{tail}(k)} \|_p^p < n$, and thus x will be included in L with probability at least $1 - \delta_0$. Meanwhile, if all S_i have pairwise empty intersection, then the protocol will output 1 with probability 1. Thus we have given a correct protocol for $\textsc{Disj}_{n,t}$ where each player communicates at most $\max\{s, O(\log n)\}$ bits. Note that Theorem 3.1 measures total communication, and thus by an averaging argument, there must be some particular player who sends $\Omega(n/t^2)$ bits. Thus $\max\{s, O(\log n)\} = \Omega(n/t^2)$, which implies the desired lower bound on s. ■

4. STATE-OF-THE-ART ALGORITHMS

4.1. Insertion-only streams: the BPTree

Recall that in insertion-only streams, the frequency vector $f \in \mathbb{R}^n$ is updated at each time step by incrementing a particular coordinate, "$f_i \leftarrow f_i + 1$." The CountSketch provides a solution to ℓ_2 heavy hitters under such updates with high probability using memory $O(k \log n)$. In this section we outline the state-of-the-art algorithm, BPTree [10] (following the CountSieve data structure of earlier work [11]), which solves the same problem using $O(k \log k)$ memory. It is an open problem as to whether $O(k)$ memory is achievable. As this article is meant to be a survey of many results, we provide only informal arguments and not rigorous proofs. We also specifically focus on the case of failure probability $\delta < 1/3$, say (for general δ, the BPTree uses space $O(k \log(k/\delta))$).

The main approach of the BPTree (and of the earlier CountSieve) is to reduce from the problem of finding $O(k)$ frequent items to that of finding a single item that is *superheavy*.

Definition 4.1. Given a frequency vector $f \in \mathbb{R}^n$ and $C > 1$, $i \in [n]$ is C*-superheavy* if

$$f_i^2 > C \cdot \sum_{i' \neq i} f_{i'}^2.$$

One should have in mind C being a large constant, e.g., 10^5. Then, a superheavy item is one that not only contributes a noticeable fraction of the frequency vector's energy (in an ℓ_2 sense), but rather contributes almost everything.

The reduction. We show a reduction that if we have a space-S randomized algorithm $\mathcal{A}$ that identifies a C-superheavy item if one exists with probability at least $9/10$ (where C is e.g., 10^5), then we can use $\mathcal{A}$ in a black box manner to solve ℓ_2 heavy hitters with failure probability $1/3$ using space $O(S \cdot k \log k)$ (we then ultimately design such $\mathcal{A}$ with $S = O(1)$).

Now we sketch the details of the reduction. Imagine instantiating B independent copies $\mathcal{A}_1, \ldots, \mathcal{A}_B$ of algorithm $\mathcal{A}$ for B equal to some large constant times k (the constant depends on C). We also pick a hash function $h : [n] \to [B]$ at random from a 2-wise independent hash family as described in Section 2.2. An update to i is then fed to the algorithm $\mathcal{A}_{h(i)}$. It is not hard to show that if i is a $(k, 2)$-tail frequent item, then with probability at least $9/10$, over the randomness of h, i will be C-superheavy in the projected frequency vector $f_{h^{-1}(h(i))}$, and thus $\mathcal{A}_{h(i)}$ will report i with probability at least $(9/10)^2 > 4/5$. Thus in expectation, we recover 80% of the $(k, 2)$-tail frequent items. To recover them all with high probability, we repeat this basic scheme $\Theta(\log k)$ times, so that overall the total number of instantiations of $\mathcal{A}$ is $B \log k = \Theta(k \log k)$. We then return

$$L = \{ i \in [n] : \text{at least } k/2 \text{ of the } Bk \text{ instantiations of } \mathcal{A} \text{ output } i \}.$$

Then $|L| = O(k)$ simply by counting, and a Chernoff–Hoeffding bound implies that any frequent item is contained in L with probability at least $1 - 1/\operatorname{poly}(k)$; thus all are contained with probability $1 - 1/\operatorname{poly}(k)$ by a union bound.

Finding a superheavy item. The remaining task is then to identify a superheavy item in a stream with large constant probability; if one does not exist, the algorithm is allowed to behave arbitrarily. Suppose the superheavy item is $i^* \in [n]$, which we wish to learn.

Before we describe the algorithm, we first need the concept of a tracker.

Definition 4.2. Let $F : \mathbb{N}^n \to \mathbb{R}_{\geq 0}$ be a function mapping frequency vectors to nonnegative reals. Let $f^{(0)}, f^{(1)}, \ldots, f^{(\ell)} \in \mathbb{N}^n$ be the evolution of a frequency vector throughout a stream, where $f^{(t)}$ is the frequency vector after seeing the first t stream updates (where $f^{(0)} = 0$). An algorithm $\mathcal{A}$ is a *weak tracker* for F with failure probability δ and error ε if after every time step t it outputs some $\tilde{F}_t \in \mathbb{R}$ such that

$$\mathbb{P}\Big(\exists t \in [\ell],\ \big|F(f^{(t)}) - \tilde{F}_t\big| > \varepsilon \sup_{q \in [\ell]} F(f^{(q)})\Big) \leq \delta,$$

where the probability is taken over the internal randomness used by $\mathcal{A}$.

We omit the proof of the following theorem from [10]; it primarily follows from the chaining arguments of [28], but slightly modified to take bounded independence into account.

Theorem 4.3. *Let $F_2(f) = \|f\|_2^2$, and consider the algorithm $\mathcal{B}$ which stores $\Pi \in \{-1, 1\}^{m \times n}$ with Rademacher entries drawn from an 8-wise independent family for $m \geq c/\varepsilon^2$ for some sufficiently large constant c, and which provides estimates $\tilde{F}_t = \|\Pi f^{(t)}\|_2^2/m$. Then $\mathcal{B}$ is a weak tracker for F_2 with failure probability $1/10$ and error ε. Its memory usage is $O(1/\varepsilon^2)$.*

Henceforth, for ease of exposition we assume we exactly know $Q^2 := \|f^{(\ell)}\|_2^2$ before the stream even starts, where ℓ is the stream's length (the subsequent arguments can all be slightly modified if we only know Q up to a constant factor). In reality, we do not actually know Q (we do not know the future!), but this issue is circumvented in the following way. We run a weak tracker $\mathcal{B}$ for F_2 as in Theorem 4.3 with error $\varepsilon = 1/3$. We make 10 guesses in parallel that Q^2 is approximately 2^j for $j = b + 0, b + 1, \ldots, b + 9$, say, for $b = 0$. For each of these guesses independently, we run an algorithm $\mathcal{A}$ for finding the superheavy item using $S = O(1)$ memory which assumes that the corresponding guess for Q^2 was correct (up to a factor of two). Conditioned on the weak tracker succeeding, when it first reports $\tilde{F}_t \geq 2^1$, then we are certain that $Q^2 > 1$ (which is 2^0). More generally, when it first reports $\tilde{F}_t \geq 2^{b+1}$, then we are certain $Q^2 > 2^b$. In this case, we can terminate the copy of $\mathcal{A}$ which assumed $Q^2 \approx 2^b$, increment b, then start a new copy of $\mathcal{A}$ (recycling the memory from the terminated algorithm) that assumes $Q^2 \approx 2^{b+9}$ for the new value of b. The key observations are that (1) since we only operate on 10 guesses in parallel at any given time, our overall memory usage is still $O(S)$, and (2) when we instantiate a new copy of $\mathcal{A}$, if that new copy corresponds to the (approximately) correct guess of Q, then it is not hard to show that the prefix of the stream it missed processing only contained a very small constant fraction of the number of occurrences of the superheavy item and that the item must thus still be superheavy (with a slight adjustment to the constant C) in the remaining suffix of the stream.

The idea behind the algorithm $\mathcal{A}$ assuming we know Q exactly is then as follows. Knowing $Q = \|f^{(\ell)}\|_2$ exactly is equivalent to approximately knowing $Q' = (f^{(\ell)})_{i^*}$ since $(f^{(\ell)})_{i^*} \approx \|f^{(\ell)}\|_2$ due to superheaviness. We write i^* expanded in base-2 as $i^*_T i^*_{T-1} \cdots i^*_0$ for $T = \lfloor \log_2 n \rfloor$ and aim to learn these bits one at a time, starting from i_0 then moving from the least to most significant bit. The strategy to learn i^*_0 is as follows: first, we initialize two counters B_0, B_1, each to zero. We also pick a random function $\sigma : [n] \to \{-1, 1\}$ from a 4-wise independent family as described in Section 2.2. When we see i in the stream, we simply add $\sigma(i)$ to B_{i_0} (where i_j here denotes the jth least significant bit in the base-2 representation of i). We wait until the first time t that $|B^{(t)}_r| \geq Q/10$ for some $r \in \{0, 1\}$, and at that moment we declare that we have learned $i^*_0 = r$. We must iterate in some fashion to learn the remaining bits, but before we describe that, let us first get a sense for why this approach is reasonable. Consider the values of the two counters at some time t:

$$B^{(t)}_{i^*_0} = \sigma(i^*) f^{(t)}_{i^*} + \underbrace{\sum_{\substack{i \neq i^* \\ i_0 = i^*_0}} \sigma(i) f^{(t)}_i}_{\alpha}, \quad B_{1-i^*_0} = \underbrace{\sum_{i_0 = 1 - i^*_0} \sigma(i) f^{(t)}_i}_{\beta}.$$

The variances of α, β are each at most $\|f^{(t)}_{[n]\setminus\{i^*\}}\|_2^2$, which is far less than $(f^{(\ell)}_{i^*})^2 \approx Q^2$ by superheaviness. Thus we expect $|\alpha|, |\beta| \ll |f^{(\ell)}_{i^*}|$ at any fixed point in time with large probability, by the second moment method. But not only do we expect this inequality to hold at any fixed point in time, but with large probability it turns out to hold at *all* points in time, simultaneously. This fact follows by the following lemma, which can be proven by applying a Dudley-type chaining argument using limited independence (see Section 4.1.1).

Lemma 4.4. *Let $0 = y^{(0)}, y^{(1)}, \dots, y^{(T)} \in \mathbb{R}^n$ be the evolution of a frequency vector in an insertion-only stream. Let $\sigma \in \{-1, 1\}^n$ be drawn from a 4-wise independent family. Then*

$$\mathop{\mathbb{E}}_{\sigma} \sup_{0 \leq t \leq T} |\langle \sigma, y^{(t)} \rangle| = O(\|y^{(T)}\|_2).$$

Remark 4.5. Consider a random walk on the integers, starting at 0, where at every time step one decrements with probability $1/2$ and increments with probability $1/2$. Let the position of this random walk at time t be $x(t)$. Then one can model the evolution of this position in the following way: consider the stream $1, 2, 3, \dots, T$. This stream yields the sequence of frequency vectors

$$y^{(t)} = (\underbrace{1, 1, \dots, 1}_{t \text{ entries}}, 0, \dots, 0)^\top, \tag{4.1}$$

and then $x(t) = \langle \sigma, y^{(t)} \rangle$ for $\sigma \in \{-1, 1\}^T$ uniformly at random. Lemma 4.4 then implies $\mathbb{E} \sup_{1 \leq t \leq T} |x(t)| = O(\sqrt{T})$, which follows from the Lévy–Ottaviani maximal inequality (see, e.g., [30, PROPOSITION 1.1.1]). Lemma 4.4 generalizes this maximal inequality in two ways: (1) the entries of σ do not need to be independent, but rather only 4-wise independent, and (2) the lemma generalizes to the arbitrary evolution of the $y^{(t)}$ vectors where each $y^{(t+1)} - y^{(t)}$ can be any standard basis vector.

With Lemma 4.4 in hand, by Markov's inequality $|\alpha|, |\beta| \ll Q/10$ at all points in time with large constant probability. Thus, with large constant probability, (1) we will never have $|\beta| \geq Q/10$ and thus never declare some $r \neq i_0^*$, and (2) at the moment in time that we have seen the $\lceil Q/9 \rceil$th occurrence of i^* in the stream, $|\alpha|$ will be sufficiently small in magnitude that $|B_r^{(t)}| = f_{i^*}^{(t)} \pm |\alpha|$ will be at least $Q/10$. How then though do we learn *all* the bits of i^*? One we learn i_0^*, we could reset B_0, B_1 to 0 again and restart a similar process to learn i_1^*, but the argument given so far requires us to see potentially $\lceil Q/9 \rceil$ occurrences of i^* to learn a single bit of its binary representation. Thus we could only learn at most 9 of its bits this way, whereas we need to learn $\log_2 n$ bits! The idea to overcome this is to first pick a random permutation π on $[n]$ (it is possible to do this pseudorandomly as well so that π can be represented using only $O(1)$ words of memory; we omit the details). Then for every update we see in the stream, we feed $\pi(i)$ to $\mathcal{A}$ instead of i (then at the end of the entire algorithm, we apply π^{-1} to the superheavy index founded to recover i^*). This permutation has the effect that the ℓ_2^2 energy of the vector is randomly spread (in expectation). Then, after we have learned $r_{j-1}r_{j-2}\cdots r_0 = \pi(i^*)_{j-1}\pi(i^*)_{j-2}\cdots\pi(i^*)_0$ and are trying to learn $\pi(i^*)_j$, for every index i in the stream we simply ignore i (and do not feed into to $\mathcal{A}$) unless i is consistent with the bits we have learned so far, i.e., $\pi(i)_{j-1}\pi(i)_{j-2}\cdots\pi(i)_0 = r_{j-1}r_{j-2}\cdots r_0$. Intuitively this makes sense: if these bits do not match that of i^* then surely i cannot be i^*, so feeding it into $\mathcal{A}$ can only contribute to the noise α, β. Since the coordinates are randomly permuted and the energy from the nonsuperheavy item is randomly spread, by dropping a $1/2^j$ fraction of coordinates (other than i^*), the effect is that i^* only becomes *heavier* in the projected frequency vector that remains, at a geometric rate as j increases. This means that we no longer need to see $\approx Q/9$ occurrences of i^* to learn the next bit, but rather can get away with seeing a geometrically smaller number of occurrences! If we iterate in this way, then eventually there will be a unique consistent i in the remaining part of the stream (possibly because we learned all the bits of $\pi(i^*)$), and this i must be i^*. This concludes the description of $\mathcal{A}$.

4.1.1. A brief introduction to chaining arguments

We here sketch the proof of Theorem 4.4. We will be considering *Rademacher processes* defined as follows. Given a collection of vectors X, we can define a collection of random variables $(Z_x)_{x\in X}$ by $Z_x := \langle \sigma, x\rangle$, where $\sigma \in \{-1, 1\}^n$ is a vector of independent Rademachers. We now study methods for upper bounding $W(X) := \mathbb{E}\sup_{x\in X}|Z_x|$; in what remains, we assume X is a subset of the unit sphere S^{n-1}.

When reading the subsequent bounds, a good example to keep in mind is the special case of Lemma 4.4 related to a random walk on the integers of length n. That is, we define $y_t = (\sum_{i=1}^t e_i)/\sqrt{n}$ (similarly as in (4.1) but normalized to lie in the unit Euclidean ball) and $Y = \{y_t\}_{t=1}^n$. Here e_i denotes the ith standard basis vector. Then we know $W(Y) = O(1)$ by Levy's maximal inequality. In the case of independent Rademachers this can be proven simply via a simple reflection argument, but since our aim is to prove this bound even when the Rademachers are only 4-wise independent, we develop another approach.

Union bound. The first bound is via a union bound over all $x \in X$:

$$\begin{aligned}
W(X) &= \int_0^\infty \mathbb{P}\Big(\sup_x |Z_x| > u\Big) du \\
&= \int_0^{u^*} \mathbb{P}\Big(\sup_x |Z_x| > u\Big) du + \int_{u^*}^\infty \mathbb{P}\Big(\sup_x |Z_x| > u\Big) du \\
&\le u^* + \sum_{x \in X} \int_{u^*}^\infty \mathbb{P}\big(|Z_x| > u\big) du \quad \text{(union bound)} \\
&\lesssim u^* + |X| e^{-(u^*)^2/2} \\
&\lesssim \sqrt{\log |X|} \quad \big(\text{choose } u^* = \Theta\big(\sqrt{\log |X|}\big)\big).
\end{aligned} \tag{4.2}$$

Thus in the case of Y we obtain the bound $W(Y) = O(\sqrt{\log n})$, which is not sharp.

ϵ-net. Let $\mathcal{N}(X, \ell_2, \epsilon)$ denote the minimum number of ℓ_2 balls of radius ϵ required to cover X (the *covering number*), and let X' be the set of centers in such a minimum covering. Any such covering is called an *ϵ-net*, and X' is thus an ϵ-net of optimum (i.e., minimum) size. Then for any $x \in X$ let $x' \in X'$ be defined as the closest point in X' to x in ℓ_2 distance. Then

$$\begin{aligned}
\mathbb{E} \sup_{x \in X} \big|\langle \sigma, x\rangle\big| &= \mathbb{E} \sup_{x \in X} \big|\langle \sigma, x' + (x - x')\rangle\big| \\
&\le W(X') + \mathbb{E} \sup_{x \in X} \big|\langle \sigma, x - x'\rangle\big| \\
&\lesssim \log^{1/2} \mathcal{N}(X, \ell_2, \epsilon) + \epsilon \cdot \|\sigma\|_2 \quad \text{((4.2) and Cauchy–Schwarz)} \\
&\lesssim \log^{1/2} \mathcal{N}(X, \ell_2, \epsilon) + \epsilon\sqrt{n}.
\end{aligned}$$

Note one can take $\epsilon = 0$ and recover (4.2). In the case of Y, an optimal ϵ-net is $Y' = \{y_{\lfloor k\epsilon^2 n \rfloor}\}_{k=1}^{\lfloor 1/\epsilon^2 \rfloor}$, and so $\log^{1/2} \mathcal{N}(Y, \ell_2, \epsilon) = \Theta(\sqrt{\log(1/\epsilon)})$. The bound is asymptotically optimized by taking $\epsilon = \Theta(1/\sqrt{n})$, which yields the same suboptimal bound $W(X) = O(\sqrt{\log n})$ as above.

Dudley's inequality. Dudley iterates the ϵ-net approach by taking a sequence of ϵ-nets $X^0, X^2, X^3, \ldots$ where X^j is a 2^{-j}-net. Letting $x(j)$ denote the closest point in X^j to x, one can write $x = x(0) + \sum_{j=1}^\infty (x(j) - x(j-1))$. Then, taking $X^0 = \{0\}$,

$$\begin{aligned}
\mathbb{E} \sup_{x \in X} \big|\langle \sigma, x\rangle\big| &= \mathbb{E} \sup_{x \in X} \Big|\langle \sigma, x(0)\rangle + \sum_{j=1}^\infty \langle \sigma, x(j) - x(j-1)\rangle\Big| \\
&\le \sum_{j=1}^\infty \mathbb{E} \sup_{x \in X} \big|\langle \sigma, x(j) - x(j-1)\rangle\big| \\
&\lesssim \sum_{j=1}^\infty \frac{1}{2^j} \cdot \log^{1/2}\big(\mathcal{N}\big(X, \ell_2, 2^{-j}\big) \cdot \mathcal{N}\big(X, \ell_2, 2^{-(j-1)}\big)\big) \\
&\lesssim \sum_{j=1}^\infty \frac{1}{2^j} \cdot \log^{1/2} \mathcal{N}\big(X, \ell_2, 2^{-j}\big).
\end{aligned}$$

In the case of Y, the above sum is $\sum_j \sqrt{j}/2^j = O(1)$, which is correct. How can we deal with the issue though that in our case σ only has 4-wise independent entries? The key observation is that the appearance of the "$\log^{1/2}$" function in Dudley's inequality arises as the inverse of the gaussian tail of $\langle \sigma, x \rangle$ (Khintchine's inequality). If σ only has $2k$-wise independent entries, then we still have *some* tail bound from applying Markov inequality on the $(2k)$th moment (it is important that we only look at even moments, since then $|\langle \sigma, x \rangle|^{2k} = \langle \sigma, x \rangle^{2k}$, whose expectation is determined by bounded independence via expansion into a sum of monomials). This tail leads to a converging sum for $k = 2$, and hence 4-wise independence suffices. Interestingly, it was shown that 4-wise independence is necessary; Narayanan constructed a distribution over 3-wise independent Rademachers for which the conclusion of Levy's maximal inequality fails to hold [33].

Remark 4.6. Dudley's inequality is not sharp, as can be seen, for example, by taking $X = \ell_1^n$. Then $W(X) = 1$, but a calculation reveals Dudley's bound yields only $W(X) = O(\log^{3/2} n)$. A sharp approach that is correct for any X when σ is replaced by a gaussian vector is given by Fernique [22], with an asymptotically matching lower bound by Talagrand [35]. In the Rademacher case as discussed here, an upper bound was observed by Talagrand with a conjectured matching lower bound (the so-called "Bernoulli Conjecture"); that lower bound was eventually proven by Bednorz and Latała [6].

4.2. General turnstile streams: the ExpanderSketch

While the BPTree of Section 4.1 achieves an improved memory bound of $O(k \log k)$ to solve the ℓ_2 heavy hitters problem, it only works in the insertion-only model. Recall the more general *turnstile model* is one in which each update in the stream consists of a pair (i, Δ), which triggers the change $f_i \leftarrow f_i + \Delta$ (where $\Delta \in \mathbb{R}$ may even be negative). Unfortunately, a lower bound of $\Omega(k \log n)$ memory is known to hold in the turnstile model, even for the ℓ_1 heavy hitters problem [26], showing that the memory usage of CountSketch is asymptotically optimal in this more general model.

What then is there left to study in the general turnstile model? Memory turns out to not be the only resource we care about, but rather we should judge the quality of algorithms based on at least four measures of efficiency:

1. **Memory:** as already discussed.
2. **Update time:** How much time does it take the algorithm to process a new update in the stream?
3. **Query time:** At the end of the stream, when queried how long does it take the algorithm to produce the list L of frequent items?
4. **Failure probability:** Fixing the above three quantities, the lower the failure probability, the better.

Using $O(k \log n)$ memory, examining the proof of Theorem 2.5 reveals the CountSketch has update time $\Theta(\log n)$, failure probability $O(1/n^c)$ for arbitrarily large constant c (by increasing the constant in the big-Oh of the memory bound), and query time $\Theta(n \log n)$. It is this query time that we wish to improve: the output L is of size at most k, yet it takes time more than linear in the universe size n to find the items in this list! Can we obtain an algorithm that has the same asymptotic memory, update time, and failure probability as the CountSketch but with much better query time? The answer is "yes," and this is achieved by the ExpanderSketch [31], following prior work which had shown to improve the query time but at the expense of increased memory and update time [16, 17].

Theorem 4.7. *There is an algorithm for the ℓ_2 heavy hitters problem, the ExpanderSketch, which uses $O(k \log n)$ memory, and which has update time $O(\log n)$, query time $O(k \cdot \mathrm{poly}(\log n))$, and failure probability $1/n^c$ for a constant $c > 0$ that can be made arbitrarily large.*

We do not prove the theorem here but rather just give an overview of the main ideas. The main idea is to reduce to the case of small n, so that the CountSketch can then be used after the reduction. The idea behind the algorithm can be broken down into two steps. We describe Step 2 only in the case of the easier ℓ_1 heavy hitters problem, and in the so-called *strict turnstile model*, where we are promised that $f_i \geq 0$ for all i at query time. The ideas can be extended to the general turnstile model, and for the ℓ_2 version of the problem, but the details are a bit more technical and so we do not discuss them here; the simplified setting we discuss here is sufficient to highlight most of the main ideas.

Step 1. We first reduce to the case of small k: more specifically, to the case $k = O(\log n)$. This is accomplished by defining $B := \lceil k/\log n \rceil$ and picking a hash function $h : [n] \to [B]$ at random from a $\Theta(\log n)$-wise independent family. A simple argument based on Bernstein's inequality implies that any k-frequent item in the original stream will be $O(\log n)$-frequent in the projected vector $f_{h^{-1}(h(i))}$. Thus, by running a frequent items data structure $\mathcal{A}_j$ for each $j \in [B]$, we can recover the full list L as the union of the L_j returned by each $\mathcal{A}_j$.

Step 2. Due to Step 1, we can now assume that $k = O(\log n)$, and we show here how to implement each $\mathcal{A}_j$. As mentioned above, we focus only on the strict turnstile model, and for the ℓ_1 heavy hitters problem. As mentioned, the main idea is to reduce the universe size n, which we accomplish as follows. For each update (i, Δ) in the stream, we view i in base-b for $b = \mathrm{poly}(\log n)$. In this base, i has $t = O(\log_b n) = O(\log n/\log\log n)$ digits $i_{t-1}i_{t-2}\cdots i_0$. We instantiate t independent CountSketch data structures $\mathsf{CS}_0, \ldots, \mathsf{CS}_{t-1}$.[2] We would like to then feed the update (i_j, Δ) to CS_j for each $0 \leq j < t$. The reasoning is that if i is k-frequent, then i_j will be k-frequent from the viewpoint of CS_j for each j. This is because all the mass from i contributes to the frequency of i_j, plus other indices in $[n]$ with the same base-b digit in the jth position can only contribute more (this is where the strict turnstile assumption comes in, since otherwise other such indices might have frequencies with opposing sign and cause cancellation). Thus, we would like to query each CM_j to obtain i_j as frequent, then simply concatenate these digits. Note that since each CS_j only operates over a frequency histogram of dimension $b = \mathrm{poly}(\log n)$, its query time is a fast $O(b \log b) = \mathrm{poly}(\log n)$.

2 Though CountSketch solves the ℓ_2 version of the problem, we know by Lemma 2.2 that it must also solve the ℓ_p version for any $p \leq 2$ (specifically, it solves the ℓ_1 version)

There are two main issues with the above scheme. The first, and easiest to fix, is the following: recall that the CS_j are randomized data structures, and so CS_j may fail to report i_j with probability as large as $1/b^c$. It is possible to show that with probability $1/n^c$, at most 1% of the CS_j data structures fail, which means we may miss 1% of the digits of a heavy hitter i. This is easily fixed though using *error-correcting codes*. For our purposes, an error-correcting code is simply a collection $\mathcal{C}$ of at least n vectors in $[b]^{O(t)}$ such that the pairwise Hamming distances between vectors in $\mathcal{C}$ is large. In that way, given some $x \in C$ then corrupting 1% of its entries in an arbitrary way, there is a unique way to "decode" that corrupted vector to recover x. Since $|\mathcal{C}| \geq n$, there is an injection (which we call the "encoding") $\mathsf{Enc} : [n] \to C$. Then when we process update (i, Δ) in the stream, we first compute $i' = \mathsf{Enc}(i)$ and run the above scheme on i', which is represented by $t' = O(t)$ digits in base-b. With high probability we will recover 99% of these digits from the CS_j, which we can then error-correct uniquely to recover i' fully, at which point we can invert the injection Enc to recover i. Codes which let us recover from such errors with linear time encoding, error-correction, and decoding exist **[34]**, which can be used here.

The more serious issue though is that there may not be just one heavy hitter, but up to $k = O(\log n)$ of them; that is, CS_j will not only output a single i_j, but rather a list L_j of $O(\log n)$ elements in $\{0, \ldots, b-1\}$. The question is then: how do we now disentangle these lists to know which digits in different lists L_j correspond to the same $i \in [n]$? Note the number of possible combinations is $\prod_j |L_j|$, which can be as big as $k^{t'} = \mathrm{poly}(n)$. We now discuss the approach to overcoming this issue.

First, consider the following simple idea which does not quite work: pick hash functions $h_1, \ldots, h_{t'} : [n] \to [r]$ independently from a 2-wise independent hash family for $r = \log^c n$, for some large constant $c > 0$. Then when processing the update (i, Δ), rather than feeding $(\mathsf{Enc}(i)_j, \Delta)$ to CS_j, we instead feed the update $(h_j(i) \circ h_{j+1}(i) \circ \mathsf{Enc}(i)_j, \Delta)$, where $\circ$ denotes concatenation of objects. Note that from the perspective of CS_j it is receiving updates that index into a vector of length $2^{b+2r} = \mathrm{poly}(\log n)$, so its query time is still fast. The main intuition is that since the range of each h_j is $r \gg k^2$, with good probability (1) the set of frequent items are mapped injectively by h_j and thus have a unique "name" $h_j(i)$ in block j. Furthermore, since $b \gg k$, one can show that (2) the total amount of infrequent item mass that collides with i under h_j is small with large probability. We also still have that (3) CM_j succeeds with large probability. We say that any block $j \in [t']$ satisfying (1) through (3) is a "good" block.

Suppose all blocks are good. Then for each j, we first perform a filtering step on L_j: if two different returned elements have the same $h_j(i)$ values, we remove the one with the smaller estimated frequency. After this filtering, if (1)–(3) hold then L_j contains every frequent item and no items whose $h_j(i)$ values collide with any frequent item. We can then create a graph G on the vertex set $[t'] \times [2^r]$. Recall that an element of L_j will be a concatenation of three strings α, β, γ (ideally, α is a name $h_j(i)$, $\beta = h_{j+1}(i)$, and $\gamma = \mathsf{Enc}(i)_j$); each such element adds an edge in G from vertex (j, α) to $(j+1, \beta)$. Then, in the ideal situation that all blocks are good, G will contain a collection of at most k disjoint paths: one for each frequent item. We can thus recover all the frequent items by finding the connected components of G to recover these paths, concatenating the γ values along each component path to obtain a codeword, then decoding the codeword to recover the corresponding frequent item name in $[n]$.

Of course, life is not so simple: if we want success probability $1 - 1/\operatorname{poly}(n)$, then we can only condition on 99% of the blocks $j \in [t']$ being good, not them *all* being good. In such a case, however, we lose each frequent item corresponding to a path connected component of length t'. Specifically, every roughly 100 vertices along the path on average, we expect to hit a bad block j, which might cause us to miss seeing the edge from block j to $j + 1$. Bad blocks might also introduce spurious edges between path fragments corresponding to different heavy hitters. Thus, it may be not be possible to extract the vertex-disjoint paths, one corresponding to each heavy hitter, from the graph G we end up seeing after these corruptions. The main issue is that the errors introduced into G hide the underlying connected components (paths) that we were hoping to find. This final obstacle is overcome by borrowing an idea from **[23]**, but with a more sophisticated disentangling algorithm. Specifically, rather than represent each heavy hitter by a path, we represent it by a base graph H which is *robust*, in that if one deletes a small fraction of edges within H and also attaches a small number of edges to H from outside parts of G, it is still possible to identify most of H inside G. Intuitively such an H should be tightly connected internally, and in fact a clique would serve this purpose. We will want an H with constant degree though, so rather we use a *constant degree expander*. Specifically, say H has vertex set $[t']$ and is regular with degree D, and let each vertices neighbors be ordered arbitrarily. Let $\Gamma(j)_r$ be the rth neighbor of $j \in [t']$ according to H. Then now when receiving an update (i, Δ) in the stream, we feed $(h_j(i) \circ h_{\Gamma(j)_1}(i) \circ \cdots \circ h_{\Gamma(j)_D}(i) \circ \mathsf{Enc}(i)_j, \Delta)$ to CS_j for each $j \in [t']$. Thus when we recover each L_j, we hope to not only recover the random name of a heavy hitter i in block j, but also its random name in every block adjacent to j according to the expander H. In this way, we ideally recover each heavy hitter as an expander connected component in H. However, due to bad blocks each such component may be slightly corrupted as mentioned above, with some internal edges missing, and some spurious edges leading outside the component. This is overcome by developing a spectral-based graph clustering algorithm, based upon the graph version of Cheeger's inequality **[2,18]**, to recover most of the original components; we omit the details.

5. FREQUENT ITEMS WITH PRIVACY CONSTRAINTS

A new model that is increasingly gaining relevance comes from the following example. Suppose a company makes mobile devices and wishes to train better spellcheckers and autocomplete features for its messaging software. To this end, it wishes to train machine learning models based on words that its customers are texting to their contacts. The device manufacturer could accomplish this by monitoring all its customers' activity, embedding code in its messaging software which reports all text messages back to the company. Such behavior is of course problematic, as it violates most users' expectation of privacy and could even be illegal in some countries.

One way around this issue is to use *differential privacy* **[19]**, and specifically the so-called *local model* model for differential privacy (see also **[40]** for so-called federated analytics approaches). Given some database $D = \{x_1, \ldots, x_n\}$ and a randomized algorithm $\mathcal{M}(D)$ for releasing information, we say the algorithm is *ε-differentially private* if for all possible outputs M and for all "adjacent" D, D',

$$\mathbb{P}\big(\mathcal{M}(D) = M\big) \le e^{\varepsilon} \cdot \mathbb{P}\big(\mathcal{M}(D') = M\big),$$

where two databases are said to be adjacent if D, D' differ on exactly one data item x (either x is in one but not the other, or the metadata associated with x is altered between the two). An example to keep in mind is a hospital storing a database of patient records, with public health analysts wishing to query that data for their own research. Then the hospital is (possibly even legally) bound to maintain privacy of patients, which is at odds with the public health utility from obtaining that information. On the one hand, the hospital could release exact answers to all queries or even release the entire database (i.e., $\mathcal{M}(D) = D$), which would allow the analysts to determine the answer to any query they would like and thus provide them with optimal utility but at the expense of no privacy; on the other hand, the hospital could release $\mathcal{M}(D) = \bot$ (or a random string independent of D) which provides perfect privacy ($\varepsilon = 0$) but zero utility.

In the local model that is relevant for the original example with mobile devices, there is not one central server that owns the entire database, but rather the data is distributed across all devices (each device knows the words it communicated). Thus each individual device i will run its own algorithm $\mathcal{M}_i$ to decide a randomized message to send a central server (being run by the device manufacturer). Unlike the example of the hospital, in this scenario the central server is untrusted. As one might imagine for the case of training spellcheck or autocomplete software, it would be useful to know popular words, i.e., *frequent* words, that are being typed on the devices; indeed, a patent even exists on precisely such an approach **[36]**. One can then devise differentially private algorithms that allow efficient procedures for solving `point_query` and `frequent` in this model. One wishes for solutions which (1) trade off utility and privacy as efficiently as possible, (2) require low communication per device, and (3) require low processing time from the central server to answer queries given all the randomized messages it received. Finding solutions to these problems has been a very active area of research in the last several years **[1,4,5,12,15,20,21,39]**. We do not attempt to describe the latest and most efficient solutions in-depth, but rather we describe just two simple solutions for `point_query` to give a reader of the flavor of how such private algorithms look.

Below, we again assume the universe is $[n]$, and d denotes the number of devices.

Randomized response. The idea here is simple: if a device holds $x \in [n]$, then they send x to the server with some probability p, and otherwise they send a uniformly random $x \in [n]\backslash\{x\}$. Picking $p = \frac{e^{\varepsilon}}{e^{\varepsilon}+n-1}$ ensures that ε-differential privacy is satisfied.

To then produce an unbiased estimator $\tilde{f}_x$ of f_x to answer `point_query(x)`, the central server uses an estimator of the form

$$\tilde{f}_x = \sum_{i=1}^{d} \left(\alpha \cdot \mathbb{1}\{m_i = x\} + \beta\right), \tag{5.1}$$

for some $\alpha, \beta \in \mathbb{R}$, where m_i is the message sent by device i. Also, $\mathbb{1}\{\mathcal{E}\}$ is the indicator random variable for event $\mathcal{E}$. For the above estimator to be unbiased, we must have that each summand has expectation 1 when $x_i = x$ and has expectation 0 when $x_i \neq x$. Taking expectations, we thus obtain the following two linear constraints:

$$\alpha \cdot \frac{e^{\varepsilon}}{e^{\varepsilon} + U - 1} + \beta = 1 \text{ and}$$
$$\alpha \cdot \frac{1}{e^{\varepsilon} + U - 1} + \beta = 0.$$

Solving this system of two linear equations with two unknowns gives

$$\alpha = \frac{e^{\varepsilon} + n - 1}{e^{\varepsilon} - 1}, \quad \beta = -\frac{1}{e^{\varepsilon} - 1}.$$

One can also compute the variance (which is our proxy for "utility") and find

$$\mathrm{Var}[\tilde{f}_x] = \frac{e^{\varepsilon} + n - 2}{(e^{\varepsilon} - 1)^2} d + \frac{n - 2}{e^{\varepsilon} - 1} f_x.$$

The message length from each device in this protocol is $b = \lceil \log_2 n \rceil$. The query time for the server to obtain $\tilde{f}_x$ for all x is $\Theta(d + n)$. Note the dependence of the variance on n is linear, which can be quite large.

RAPPOR. We describe a simplified version of the RAPPOR scheme [20]. There are two versions depending on the specific privacy guarantees desired. In so-called *deletion privacy*, a device should be able to opt out from sending its data without the server knowing it opted out, in which case it will send a message based on some dummy input "x^*." In *replacement privacy*, we want privacy in the sense that what the server receives should be nearly indistinguishable if that device had been replaced by some other device holding some other data x. We focus only on deletion privacy, as the scheme is slightly simpler to present in this way, and for that we use the *symmetric* version of RAPPOR. In this scheme, a device holding x maps it to the standard basis vector $e_x \in \mathbb{R}^n$, also known as the *one-hot encoding* of x. The device then flips each bit of e_x independently with probability p to form its message M, which it then sends to the server (in asymmetric RAPPOR, the probabilities of flipping 0 to 1 versus 1 to 0 are different). One would expect the variance of a resulting estimator to monotonically increase as p increases from 0 to $1/2$, so the goal is to make p as small as possible while preserving privacy. To ensure privacy, we must ensure that the message that is sent has roughly the same probability for any device even if it chooses to "delete" its input and replace it with some canonical dummy input x^*.

A device that opts out of sharing its data at all will pretend that it holds $x^* = \perp$ (resulting in the vector $e_{x^*} = \vec{0}$). Consider $x \in [n]$; we must ensure that for any message M

$$e^{-\varepsilon} \cdot \mathbb{P}(M|x^*) \leq \mathbb{P}(M|x) \leq e^{\varepsilon} \cdot \mathbb{P}(M|x^*).$$

For these two inputs, the message M is either obtained by flipping bits independently in e_x or in e_{x^*}. The only index for which the resulting bit in M has differing probabilities is the index x, differing by a factor of $(1-p)/p = 1/p - 1$. Since this quantity must be at least e^ε, the smallest we can set p is $p = 1/(e^\varepsilon + 1)$.

We must now determine an unbiased estimator for the server to estimate f_x in answering `point_query(x)`. If device i sends message $m_i \in \{0,1\}^n$, we will use an estimator of the form

$$\tilde{f}_x = \sum_{i=1}^{d} \left(\alpha \cdot \mathbb{1}\{(m_i)_x = 1\} + \beta\right). \tag{5.2}$$

We focus on each summand and again after taking expectations have two linear constraints that arise from the cases $x_i = x$ and $x_i \neq x$. These constraints are

$$\alpha(1-p) + \beta = 1 \text{ and}$$
$$\alpha p + \beta = 0.$$

Some calculation then yields the solution

$$\alpha = \frac{1}{1-2p}, \quad \beta = -\frac{p}{1-2p}.$$

To compute $\mathrm{Var}[\tilde{f}_x]$, we again have independence of summands, where summands with $x_i = x$ each contribute some value A, and those with $x_i \neq x$ contribute B. The total variance is then $A \cdot f_x + B \cdot (d - f_x)$. Some computation then yields

$$\mathrm{Var}[\tilde{f}_x] = \frac{p(1-p)}{(1-2p)^2} d + \frac{p^2}{(1-2p)^2} f_x.$$

Unlike the case of Randomized Response, we do not have such a large dependence on n in the variance, and thus the utility is far superior especially for large n. The message length in this protocol is $b = n$, and the query time for the server to obtain $\tilde{f}_x$ for all x is $\Theta(dn)$; both are much worse than Randomized Response. A recent scheme of Feldman and Talwar provides a slight variant of RAPPOR which significantly reduces the message length to $O(\log n)$ bits [21].

ACKNOWLEDGMENTS

We thank Ravi Kumar, Pasin Manurangsi, Huy Lê Nguyễn, and Uri Stemmer for providing helpful feedback on a previous draft of this article.

FUNDING

The author is partially supported by NSF award CCF-1951384, ONR grant N00014-18-1-2562, ONR DORECG award N00014-17-1-2127, and a Google Faculty Research Award.

REFERENCES

[1] J. Acharya, Z. Sun, and H. Zhang, Hadamard response: estimating distributions privately, efficiently, and with little communication. In *Proceedings of the 22nd International Conference on Artificial Intelligence and Statistics (AISTATS)*, pp. 1120–1129, 2019.

[2] N. Alon and V. D. Milman, λ_1, isoperimetric inequalities for graphs, and superconcentrators. *J. Combin. Theory Ser. B* **38** (1985), no. 1, 73–88.

[3] Z. Bar-Yossef, T. S. Jayram, R. Kumar, and D. Sivakumar, An information statistics approach to data stream and communication complexity. *J. Comput. System Sci.* **68** (2004), no. 4, 702–732.

[4] R. Bassily, K. Nissim, U. Stemmer, and A. Thakurta, Practical locally private heavy hitters. *J. Mach. Learn. Res.* **21** (2020), 16:1–16:42.

[5] R. Bassily and A. D. Smith, Local, private, efficient protocols for succinct histograms. In *Proceedings of the 47th Annual ACM on Symposium on Theory of Computing (STOC)*, pp. 127–135, ACM, 2015.

[6] W. Bednorz and R. Latała, On the boundedness of Bernoulli processes. *Ann. of Math.* **3** (2014), no. 180, 1167–1203.

[7] R. Berinde, P. Indyk, G. Cormode, and M. J. Strauss, Space-optimal heavy hitters with strong error bounds. *ACM Trans. Database Syst.* **35** (2010), no. 4, 26:1–26:28.

[8] M. Blum, R. W. Floyd, V. R. Pratt, R. L. Rivest, and R. E. Tarjan, Linear time bounds for median computations. In *Proceedings of the 4th Annual ACM Symposium on Theory of Computing (STOC)*, pp. 119–124, ACM, 1972.

[9] R. S. Boyer and J. S. Moore, MJRTY: A fast majority vote algorithm. In *Automated reasoning: Essays in honor of Woody Bledsoe*, pp. 105–118, Springer, 1991.

[10] V. Braverman, S. R. Chestnut, N. Ivkin, J. Nelson, Z. Wang, and D. P. Woodruff, Bptree: An ℓ_2 heavy hitters algorithm using constant memory. In *Proceedings of the 36th ACM SIGMOD–SIGACT–SIGAI Symposium on Principles of Database Systems (PODS)*, pp. 361–376, ACM, 2017.

[11] V. Braverman, S. R. Chestnut, N. Ivkin, and D. P. Woodruff, Beating CountSketch for heavy hitters in insertion streams. In *Proceedings of the 48th Annual ACM SIGACT Symposium on Theory of Computing (STOC)*, pp. 740–753, ACM, 2016.

[12] M. Bun, J. Nelson, and U. Stemmer, Heavy hitters and the structure of local privacy. *ACM Trans. Algorithms* **15** (2019), no. 4, 51:1–51:40.

[13] A. Chakrabarti, S. Khot, and X. Sun, Near-optimal lower bounds on the multiparty communication complexity of set disjointness. In *Proceedings of the 18th Annual IEEE Conference on Computational Complexity (CCC)*, pp. 107–117, IEEE Computer Society, 2003.

[14] M. Charikar, K. C. Chen, and M. Farach-Colton, Finding frequent items in data streams. *Theoret. Comput. Sci.* **312** (2004), no. 1, 3–15.

[15] W. Chen, P. Kairouz, and A. Özgür, Breaking the communication-privacy-accuracy trilemma. In *Proceedings of the 33rd Annual Conference on Advances in Neural Information Processing Systems (NeurIPS)*, Neural Information Processing Systems Foundation, Inc., 2020.

[16] G. Cormode and M. Hadjieleftheriou, Finding frequent items in data streams. *Proc. VLDB Endow.* **1** (2008), no. 2, 1530–1541.

[17] G. Cormode and S. Muthukrishnan, An improved data stream summary: the count-min sketch and its applications. *J. Algorithms* **55** (2005), no. 1, 58–75.

[18] J. Dodziuk, Difference equations, isoperimetric inequality and transience of certain random walks. *Trans. Amer. Math. Soc.* **284** (1985), 787–794.

[19] C. Dwork, F. McSherry, K. Nissim, and A. D. Smith, Calibrating noise to sensitivity in private data analysis. In *Proceedings of the 3rd Theory of Cryptography Conference (TCC)*, pp. 265–284, Springer, 2006.

[20] Ú. Erlingsson, V. Pihur, and A. Korolova, RAPPOR: randomized aggregatable privacy-preserving ordinal response. In *Proceedings of the 2014 ACM SIGSAC Conference on Computer and Communications Security (CCS)*, pp. 1054–1067, ACM, 2014.

[21] V. Feldman and K. Talwar, Lossless compression of efficient private local randomizers. In *Proceedings of the 38th International Conference on Machine Learning (ICML)*, pp. 3208–3219, Proceedings of Machine Learning Research, 2021.

[22] X. Fernique, Regularité des trajectoires des fonctions aléatoires gaussiennes. *Lecture Notes in Math.* **480** (1975), 1–96.

[23] A. C. Gilbert, Y. Li, E. Porat, and M. J. Strauss, For-all sparse recovery in near-optimal time. In *Proceeedings of the 41st International Colloquium on Automata, Languages, and Programming (ICALP)*, pp. 538–550, Springer, 2014.

[24] A. Gronemeier, Asymptotically optimal lower bounds on the NIH-multi-party information complexity of the and-function and disjointness. In *Proceedings of the 26th International Symposium on Theoretical Aspects of Computer Science*, pp. 505–516, Leibniz-Zentrum für Informatik, 2009.

[25] T. S. Jayram, Hellinger strikes back: A note on the multi-party information complexity of AND. In *Proceedings of the 13th International Workshop on Randomization and Approximation Techniques in Computer Science (RANDOM)*, pp. 562–573, Springer, 2009.

[26] H. Jowhari, M. Saglam, and G. Tardos, Tight bounds for L_p samplers, finding duplicates in streams, and related problems. In *Proceedings of the 30th ACM SIGMOD–SIGACT–SIGART Symposium on Principles of Database Systems (PODS)*, pp. 49–58, ACM, 2011.

[27] A. Kamath, E. Price, and D. P. Woodruff, A simple proof of a new set disjointness with applications to data streams. In *Proceedings of the 36th Computational Complexity Conference (CCC)*, pp. 37:1–37:24, Leibniz-Zentrum für Informatik, 2021.

[28] F. Krahmer, S. Mendelson, and H. Rauhut, Suprema of chaos processes and the restricted isometry property. *Comm. Pure Appl. Math.* **67** (2014), no. 11, 1877–1904.

[29] E. Kushilevitz and N. Nisan, *Communication complexity*. Cambridge University Press, 1997.

[30] S. Kwapień and W. A. Woyczyński, *Random series and stochastic integrals: single and multiple*. Birkhäuser, 1992.

[31] K. G. Larsen, J. Nelson, H. L. Nguyen, and M. Thorup, Heavy hitters via cluster-preserving clustering. In *Proceedings of the 57th IEEE Annual Symposium on Foundations of Computer Science (FOCS)*, pp. 61–70, IEEE Computer Society, 2016.

[32] J. Misra and D. Gries, Finding repeated elements. *Sci. Comput. Program.* **2** (1982), no. 2, 143–152.

[33] S. Narayanan, 3-wise independent random walks can be slightly unbounded. *Random Structures Algorithms* (to appear), (2021).

[34] D. A. Spielman, Linear-time encodable and decodable error-correcting codes. *IEEE Trans. Inf. Theory* **42** (1996), no. 6, 1723–1731.

[35] M. Talagrand, Regularity of gaussian processes. *Acta Math.* **159** (1987), 99–149.

[36] A. Thakurta, A. Vyrros, U. Vaishampayan, G. Kapoor, J. Freudiger, V. Sridhar, and D. Davidson, *Learning new words*. 2017, URL https://www.google.com/patents/US9594741, US Patent 9,594,741

[37] J. S. Vitter, Random sampling with a reservoir. *ACM Trans. Math. Software* **11** (1985), no. 1, 37–57.

[38] M. N. Wegman and L. Carter, New classes and applications of hash functions. In *Proceedings of the 20th Annual Symposium on Foundations of Computer Science (FOCS)*, pp. 175–182, IEEE Computer Society, 1979.

[39] H. Wu and A. Wirth, Locally differentially private frequency estimation. 2021, CoRR, arXiv:2106.07815.

[40] W. Zhu, P. Kairouz, B. McMahan, H. Sun, and W. Li, Federated heavy hitters discovery with differential privacy. In *Proceedings of the 23rd International Conference on Artificial Intelligence and Statistics (AISTATS)*, pp. 3837–3847, Proceedings of Machine Learning Research, 2020.

JELANI NELSON

Soda Hall 633, Berkeley, CA, USA 94720-1776, minilek@berkeley.edu

This Page Intentionally Left Blank

SOME QUESTIONS RELATED TO THE REVERSE MINKOWSKI THEOREM

ODED REGEV

ABSTRACT

In this review we give an overview of some recent "reverse Minkowski" results on the geometry of lattices. Such results provide upper bounds on the number of short vectors a lattice can have, assuming that it does not have any sublattice of low determinant. We also briefly describe the proof ideas, and mention some open questions.

MATHEMATICS SUBJECT CLASSIFICATION 2020

Primary 11H06; Secondary 52C07

KEYWORDS

Reverse Minkowski, geometry of lattices, stable lattices, Epstein zeta function

INTERNATIONAL CONGRESS OF MATHEMATICIANS 2022 JULY 6–14

Proc. Int. Cong. Math. 2022, Vol. 6, pp. 4898–4912
DOI 10.4171/ICM2022/134

1. INTRODUCTION

Sphere packing is a classical question in mathematics, asking for the densest way to pack equal disjoint spheres in the n-dimensional Euclidean space, where density is defined as the fraction of space covered by the spheres. In two dimensions the optimal packing is given by the familiar hexagonal packing. The Kepler conjecture, stating that the face-centered cubic arrangement is densest in three dimensions was proven by Hales [18]. Recently the question was also resolved in dimension 8 by Viazovska [40] and in dimension 24 by Cohn et al. [6].

In all the above cases, the optimal packing is achieved by a lattice packing, i.e., spheres whose centers form a lattice (e.g., in dimension 24 it is the Leech lattice). For this and other reasons, much focus has been on understanding the geometry of lattices, especially in high dimensions. A *lattice* is a discrete subgroup of $\mathbb{R}^n$. Equivalently, it is the set of all *integer* linear combinations of some linearly independent vectors in $\mathbb{R}^n$. We typically consider full-rank lattices, which are generated by a basis of $\mathbb{R}^n$. The *determinant* or *covolume* of a full-rank lattice is the reciprocal of the number of lattice points per unit volume, in the asymptotic large volume limit,

$$\det(\mathcal{L}) = \lim_{r\to\infty} \frac{\mathrm{vol}(B_2(r))}{\mathcal{L} \cap B_2(r)},$$

where $B_2(r)$ denotes the Euclidean ball of radius r. If A is an $n \times n$ matrix whose columns form a basis of $\mathcal{L}$, then $\det(\mathcal{L}) = |\det(A)|$, hence the name. More generally, if A is an $n \times m$ matrix whose columns form a basis of the (possibly nonfull rank) lattice $\mathcal{L}$, $\det(\mathcal{L}) = \det(A^T A)^{1/2}$.

The lattice sphere packing question asks for the densest *lattice* packing in a given dimension. Notice that the largest sphere that can be packed with a given lattice has radius precisely half the length of the shortest nonzero vector in the lattice (typically denoted by λ_1). We can therefore phrase the lattice sphere packing question as follows: among all lattices $\mathcal{L} \subseteq \mathbb{R}^n$ containing asymptotically one lattice point per unit volume (i.e., with $\det(\mathcal{L}) = 1$), how large can the length of their shortest nonzero vector $\lambda_1(\mathcal{L})$ be? Minkowski's celebrated first theorem [30] bounds this length from above by the radius of a Euclidean ball of volume 2^n, which is roughly $\sqrt{2n/(\pi e)}$. (This follows immediately from the fact that a ball of volume greater than 1 cannot pack space with a lattice of determinant 1). More generally, a theorem of Blichfeldt and van der Corput [39] says that for any integer $k \geq 1$, an n-dimensional lattice with determinant 1 must contain at least $2k$ nonzero points inside the closed Euclidean ball around the origin of volume $k2^n$. For example, the ball of radius $\sqrt{n}$, whose volume is roughly $(2\pi e)^{n/2} > 4^n$, must contain at least 2^n lattice points.

The lattice packing question is part of a broader set of questions, all asking for lattices that in some sense have few short vectors assuming some fixed determinant, say 1. For instance, one can ask for the minimum of Epstein's zeta function or of the theta function [35] (see below for the definitions). Another classical related question asks for the minimum covering radius (defined as the maximum distance of a point in $\mathbb{R}^n$ from the lattice) [7].

Reverse Minkowski questions. Here we are interested in "reverse Minkowski" questions that are in some sense dual to the above questions. Specifically, instead of *minimizing* the number of short lattice vectors, we would like to *maximize* it. At first glance, this seems nonsensical: the number of short vectors in a lattice can be arbitrarily large, even if we restrict to determinant 1 lattices. For instance, consider the two-dimensional lattice generated by the vectors $(\varepsilon, 0)$ and $(0, 1/\varepsilon)$ where $\varepsilon > 0$ is arbitrarily small. Its determinant is 1, yet it has at least $1/\varepsilon$ vectors of norm at most 1.

Clearly, assuming that $\det \mathcal{L} = 1$ is not enough. We therefore impose the additional constraint that *all sublattices* of $\mathcal{L}$ have determinant at least 1. Here, by a sublattice of $\mathcal{L}$, we mean the intersection of $\mathcal{L}$ with a *lattice subspace*, i.e., a subspace spanned by lattice vectors. (Alternatively, one could define a sublattice as any discrete subgroup of $\mathcal{L}$; in all the results below, restricting to intersections with subspaces is without loss of generality.) For instance, in the example above, the 1-dimensional sublattice generated by the vector $(\varepsilon, 0)$ has determinant ε. The set of all determinant-1 lattices whose sublattices all have determinant at least 1 is known as the set of *stable* lattices and arises in a number of contexts **[16, 19, 38]**. It will play an important role below.

With this terminology in place, we can phrase the reverse Minkowski question as asking to bound from above the number of short vectors that a stable lattice can have. A precise form of this question was originally conjectured by Dadush, who was motivated by algorithmic problems related to integer programming **[22]**. Together with the present author, he went on to analyze variants of the conjecture and identified applications of the conjecture in computational complexity, cryptography, and mixing of Brownian motion **[10]**. Another application to additive combinatorics was shown in **[25]**. Dadush's conjecture was proven in **[33]**. In this review we give an overview of some of the known reverse Minkowski-style results, including a high level overview of the proof. We also present several open questions. For more details, the reader is referred to the original papers, especially **[10, 33]**. See also Bost's lecture notes **[2]** for a broader perspective.

2. REVERSE MINKOWSKI THEOREM FOR THE GAUSSIAN MASS

The main result shown in **[33]** is a reverse Minkowski theorem for the Gaussian mass, answering Dadush's original question. Here and in the rest of this review, constants are mostly arbitrary and no attempt was made to optimize them.

Theorem 2.1 (Reverse Minkowski theorem for the Gaussian mass). *For any stable lattice $\mathcal{L} \subset \mathbb{R}^n$,*

$$\rho_{1/t}(\mathcal{L}) \le \frac{3}{2}, \tag{2.1}$$

where $t := 10(\log n + 2)$.

Here, for a lattice $\mathcal{L} \subset \mathbb{R}^n$ and $s > 0$,

$$\rho_s(\mathcal{L}) := \sum_{\mathbf{y} \in \mathcal{L}} e^{-\pi \|\mathbf{y}\|^2 / s^2} \tag{2.2}$$

is the *Gaussian mass* of the lattice with *parameter* s. It is related to the theta function by $\Theta_{\mathcal{L}}(iy) := \rho_{1/\sqrt{y}}(\mathcal{L})$. An upper bound on ρ implies an upper bound on the number of short vectors in a lattice. Specifically, Theorem 2.1 immediately implies that $|\mathcal{L} \cap B_2(r)| \leq 3e^{\pi t^2 r^2}/2$ for any radius $r > 0$. In contrast to Minkowski-style theorems which provide a *lower bound* on the number of short lattice vectors, this theorem provides an *upper bound*, justifying the name "reverse Minkowski."

It is natural to ask how tight Theorem 2.1 is. Consider the lattice $\mathbb{Z}^n$, and notice that it is stable (because $\det(A^T A)$ is integer for any $A \in \mathbb{Z}^{n \times m}$ and the square root of a positive integer number is at least 1). A short calculation shows that equation (2.1) holds for t as small as $\sqrt{\log(n)/\pi} + o(1)$, but not any smaller. It is therefore possible that Theorem 2.1 holds for $t = \sqrt{\log(n)/\pi} + o(1)$. In fact, one might even conjecture the following much stronger statement, roughly saying that "$\mathbb{Z}^n$ has the most short vectors."

Question 2.2. Is it true that for all $s > 0$ and stable $\mathcal{L} \subseteq \mathbb{R}^n$, $\rho_s(\mathcal{L}) \leq \rho_s(\mathbb{Z}^n)$?

We remark that replacing $\rho_s(\mathcal{L})$ with the point counting function $|\mathcal{L} \cap B_2(s)|$ (i.e., the number of points of norm at most s) leads to a false statement, e.g., the hexagonal lattice (scaled to have determinant 1) has 7 points of norm at most $\sqrt{2}/\sqrt[4]{3}$, whereas $\mathbb{Z}^2$ has only 5. However, the question is still open for the Gaussian mass ρ_s, which is a smooth version of the point counting function.

Encouragingly, a positive answer is known for very low or high values of s, specifically, for $s \leq \sqrt{2\pi/(n+2)}$ or $s \geq \sqrt{(n+2)/(2\pi)}$ [33]. More evidence in favor of a positive answer comes from the case of the zeta function; see Theorem 3.1 below. Another piece of evidence is that better bounds are known for an important subset of stable lattices known as *unimodular* lattices. A lattice $\mathcal{L}$ is said to be unimodular if (1) $\langle \mathbf{x}, \mathbf{y} \rangle \in \mathbb{Z}$ for all $\mathbf{x}, \mathbf{y} \in \mathcal{L}$ (a property known as *integrality*); and (2) it has determinant 1. Equivalently, a lattice is unimodular if it is self-dual. Then it was shown in [34] that for all unimodular lattices (in fact, for all integral lattices), the inequality in equation (2.1) holds with $t = \sqrt{2(1 + o(1)) \log(n)/\pi}$ for some universal constant $c > 0$. Up to a constant of $\sqrt{2}$, this matches the behavior of $\mathbb{Z}^n$.

While Theorem 2.1 is stated only for stable lattices, it is possible to extend it to the set of all lattices $\mathcal{L} \subset \mathbb{R}^n$ such that $\det(\mathcal{L}') \geq 1$ for all sublattices $\mathcal{L}' \subseteq \mathcal{L}$ (i.e., we can drop the assumption that $\det(\mathcal{L}) = 1$). This is done using the so-called canonical decomposition (see [33] for details). In other words, it holds that for any $\mathcal{L}$ whose sublattices all have determinant at least 1,

$$\eta^*(\mathcal{L}) \leq 10(\log n + 2),$$

where

$$\eta^*(\mathcal{L}) := \inf\{t > 0 : \rho_{1/t}(\mathcal{L}) \leq 3/2\}$$

(known as the smoothing parameter of the dual lattice [27]). Since $\rho_{1/t}(\mathcal{L})$ is monotonically decreasing with t, goes to infinity as t goes to 0, and to 1 as t goes to infinity, the infimum is positive and achieved.

Interestingly, having a small $\eta^*(\mathcal{L})$ as above approximately characterizes the set of lattices having no sublattice with determinant less than 1. Indeed, if $\mathcal{L}'$ is a sublattice of determinant smaller than 1 of some dimension d, then by the Blichfeldt–van der Corput theorem (see Introduction), there are at least 2^d vectors of norm at most $\sqrt{d}$ in $\mathcal{L}'$. A straightforward calculation then shows that $\rho_3(\mathcal{L}) \geq \rho_3(\mathcal{L}') > 3/2$ or, equivalently, $\eta^*(\mathcal{L}) > 1/3$. (Alternatively, this fact can be shown using the Poisson summation formula.)

There is nothing special about the bound 1 on the determinants, and we can easily extend the above discussion to other values, leading to a two-sided inequality relating η^* to sublattice determinants. Namely, we claim that *for all* lattices $\mathcal{L}$,

$$\frac{1}{3} \cdot \eta_{\det}(\mathcal{L}) \leq \eta^*(\mathcal{L}) \leq 10(\log n + 2) \cdot \eta_{\det}(\mathcal{L}), \tag{2.3}$$

where

$$\eta_{\det}(\mathcal{L}) := \max_{\mathcal{L}' \subseteq \mathcal{L}} \det(\mathcal{L}')^{-1/\operatorname{rank}(\mathcal{L}')}.$$

To prove equation (2.3), note that both $\eta_{\det}(\mathcal{L})$ and $\eta^*(\mathcal{L})$ behave identically under scaling of $\mathcal{L}$ (homogeneous of degree -1), so we can assume without loss of generality that $\eta_{\det}(\mathcal{L}) = 1$, in which case equation (2.3) is precisely the statement we proved above.

3. REVERSE MINKOWSKI THEOREM FOR THE ZETA FUNCTION

Theorem 2.1 establishes a bound on the Gaussian mass of any stable lattice. It would be interesting to explore other functions in addition to the Gaussian mass. One particularly appealing choice is the Epstein zeta function.

Definition 1. For a lattice $\mathcal{L} \subset \mathbb{R}^n$ and $s > n/2$, we define the *Epstein zeta function* as the function

$$\zeta(\mathcal{L}, s) := \sum_{\mathbf{y} \in \mathcal{L} \setminus \{\mathbf{0}\}} \|\mathbf{y}\|^{-2s}.$$

Similarly to the Gaussian mass, the Epstein zeta function is a sum over lattice points of some function depending on the norm of the vector. The function decays quite rapidly, and therefore is heavily influenced by short vectors. The requirement $s > n/2$ is needed for the sum to converge. The Epstein zeta function has an analytic continuation to the complex plane except for a simple pole at $s = n/2$ [13], but in this review we only focus on $s > n/2$.

Using a proof similar to that of Theorem 2.1 (and in fact simpler, as we explain below), Eisenberg et al. recently showed the following.

Theorem 3.1 ([12]). *For any stable lattice $\mathcal{L} \subset \mathbb{R}^n$ and $s > n/2$,*

$$\zeta(\mathcal{L}, s) \leq \zeta(\mathbb{Z}^n, s),$$

with equality if and only if $\mathcal{L}$ is an orthogonal rotation of $\mathbb{Z}^n$.

Notice that unlike the case of Gaussian mass (Theorem 2.1), here we have a tight statement, showing that "$\mathbb{Z}^n$ has the most short vectors" (as quantified by the zeta function), hence answering Question 2.2 for the Epstein zeta function. We remark that a positive answer to Question 2.2 immediately implies Theorem 3.1 (so in a sense, the former is a harder question). This follows by writing the zeta function as a positive combination of Gaussian functions,

$$\zeta(\mathcal{L}, s) = \frac{2\pi^s}{\Gamma(s)} \int_0^\infty t^{-(2s+1)} (\rho_t(\mathcal{L}) - 1) \mathrm{d}t.$$

4. PROOF OVERVIEW

In this section we give a high-level overview of the proofs of Theorems 2.1 and 3.1, starting with Theorem 3.1. Both proofs follow an approach suggested by Shapira and Weiss [36].

Recall that a lattice $\mathcal{L}$ is *stable* if $\det(\mathcal{L}) = 1$ and $\det(\mathcal{L}') \geq 1$ for all sublattices $\mathcal{L}' \subseteq \mathcal{L}$. The set of stable lattices is a compact subset of the set of determinant-one lattices (under the quotient topology of $\mathrm{SL}_n(\mathbb{R})/\mathrm{SL}_n(\mathbb{Z})$). Therefore, the Epstein zeta function, being a continuous function, must attain a global maximum over the set of stable lattices. It can be shown that the Laplacian of the Epstein zeta function is positive everywhere (in fact, Epstein zeta is an eigenfunction of the Laplacian operator). It immediately follows that a global maximum $\mathcal{L}$ cannot be in the interior of the set of stable lattices. Therefore, $\mathcal{L}$ must be on the boundary of the set, which by definition, implies that there is a sublattice $\mathcal{L}_1$ of $\mathcal{L}$ such that $\det \mathcal{L}_1 = 1$ (see Figure 1). It follows immediately from the definition that $\mathcal{L}_1$ is stable. Let $\mathcal{L}_2 := \mathcal{L}/\mathcal{L}_1$ be the projection orthogonally to $\mathcal{L}_1$. It can also be shown that $\mathcal{L}_2$ is stable. Crucially, using the Poisson summation formula together with the fact that for any $q > 0$, the Fourier transform of the function $\mathbf{y} \mapsto (\|\mathbf{y}\|^2 + q)^{-s}$ is positive everywhere, it can be shown that $\zeta(\mathcal{L}, s) \leq \zeta(\mathcal{L}_1 \oplus \mathcal{L}_2, s)$. Informally speaking, "aligning the cosets" of $\mathcal{L}_1$ (see Figure 1) cannot decrease the zeta function. At this point we reduced the dimensionality of the problem, and we can continue iteratively in a similar way to split $\mathcal{L}_1$ and $\mathcal{L}_2$ into lower-dimensional stable lattices. Eventually, we arrive at the conclusion that the lattice maximizing $\zeta(\mathcal{L}, s)$ must be a direct sum of n 1-dimensional stable lattices, which is equivalent to saying that $\mathcal{L}$ is an orthogonal rotation of $\mathbb{Z}^n$. Notice that in order to continue iteratively, we need to show that the function $\mathcal{L}_1 \mapsto \zeta(\mathcal{L}_1 \oplus \mathcal{L}_2, s)$ also has a positive Laplacian (i.e., when we think of $\mathcal{L}_2$ as fixed and only vary $\mathcal{L}_1$); this turns out to indeed be the case [12]. This completes the description of the proof of Theorem 3.1.

We would like to use the same proof strategy to prove Theorem 2.1. Much of the above proof works if we replace the zeta function with the Gaussian mass. In particular, "aligning the cosets" cannot decrease the Gaussian mass, i.e., $\rho_s(\mathcal{L}) \leq \rho_s(\mathcal{L}_1 \oplus \mathcal{L}_2)$, which is proven in essentially the same way (Poisson summation formula combined with the positivity of the Fourier transform of ρ_s). Moreover, continuing iteratively is even a bit cleaner in this case since $\rho_s(\mathcal{L}_1 \oplus \mathcal{L}_2) = \rho_s(\mathcal{L}_1) \cdot \rho_s(\mathcal{L}_2)$ so once we are at the boundary, the problem truly reduces to a lower-dimensional problem. However, one serious issue is that the Gaussian mass function ρ is known to have local maxima for some parameters $s > 0$ [21]. We can therefore no longer argue as before that any global maximum must necessarily be on the boundary. (We note that for very small values of s, the Laplacian of ρ_s can be shown to

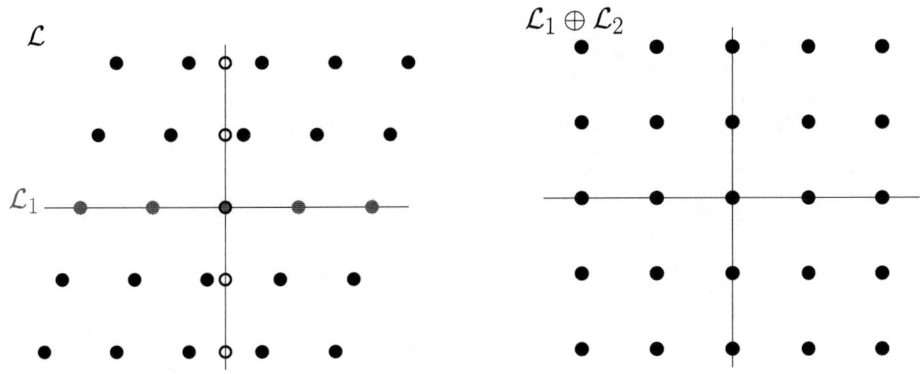

FIGURE 1
(Left) A two-dimensional stable lattice $\mathcal{L}$ (solid disks) is on the boundary of the set of stable lattices and therefore has a sublattice $\mathcal{L}_1$ of determinant 1 (red disks). Projecting $\mathcal{L}$ orthogonally to $\mathcal{L}_1$ (i.e., on the y axis) gives the lattice $\mathcal{L}_2 = \mathcal{L}/\mathcal{L}_1$, which is also stable (hollow circles). (Right) The lattice $\mathcal{L}_1 \oplus \mathcal{L}_2$.

be positive, and therefore no local maxima can exist, leading to the proof of the statement below Question 2.2; see [33]).

As a possible way around this issue, we can try to *bound ρ at local maxima*. In other words, we would like to use some property of local maxima (e.g., zero gradient) to argue that ρ cannot be too large there. We can then argue that the maximum must either be a local maximum in the interior (and then bound it as suggested here) or it must be on the boundary (in which case we can continue iteratively as before). While this approach is promising, we unfortunately do not know how to bound ρ at local maxima.

The actual proof of Theorem 2.1 follows the exact strategy described above, however, instead of working with ρ directly, it uses as a proxy another function which can be used to bound ρ from above. Namely, define the Gaussian measure of the Voronoi cell of the lattice as

$$\gamma_s\big(\mathcal{V}(\mathcal{L})\big) := \int_{\mathcal{V}(\mathcal{L})/s} e^{-\pi\|\mathbf{x}\|^2} d\mathbf{x},$$

where the Voronoi cell is the set of all points that are closer to the origin than to any other lattice vector,

$$\mathcal{V}(\mathcal{L}) := \big\{\mathbf{x} \in \mathbb{R}^n : \forall \mathbf{y} \in \mathcal{L},\ \|\mathbf{x}\| \le \|\mathbf{y} - \mathbf{x}\|\big\}.$$

It is known that $\rho_s(\mathcal{L}) \le 1/\gamma_s(\mathcal{V}(\mathcal{L}))$ [5]. Therefore, in order to prove an upper bound on $\rho_s(\mathcal{L})$, it suffices to prove a lower bound on $\gamma_s(\mathcal{V}(\mathcal{L}))$. This is achieved in [33] following the strategy suggested above. In particular, "aligning the cosets" cannot increase the Gaussian measure of the Voronoi cell, i.e., $\gamma_s(\mathcal{V}(\mathcal{L})) \ge \gamma_s(\mathcal{V}(\mathcal{L}_1 \oplus \mathcal{L}_2))$. This follows from the fact that $\mathcal{V}(\mathcal{L})$ and $\mathcal{V}(\mathcal{L}_1 \oplus \mathcal{L}_2) = \mathcal{V}(\mathcal{L}_1) \times \mathcal{V}(\mathcal{L}_2)$ are both fundamental bodies for the lattice $\mathcal{L}$, i.e., they both contain exactly one point in each coset of $\mathcal{L}$, but by definition, $\mathcal{V}(\mathcal{L})$ contains the *shortest* point in each coset, leading to the desired inequality (Figure 2). Moreover, continuing iteratively is again straightforward, since $\gamma_s(\mathcal{V}(\mathcal{L}_1 \oplus \mathcal{L}_2)) = \gamma_s(\mathcal{V}(\mathcal{L}_1) \times \mathcal{V}(\mathcal{L}_2)) = \gamma_s(\mathcal{V}(\mathcal{L}_1)) \cdot \gamma_s(\mathcal{V}(\mathcal{L}_2))$ so once we are at the boundary, the problem truly reduces

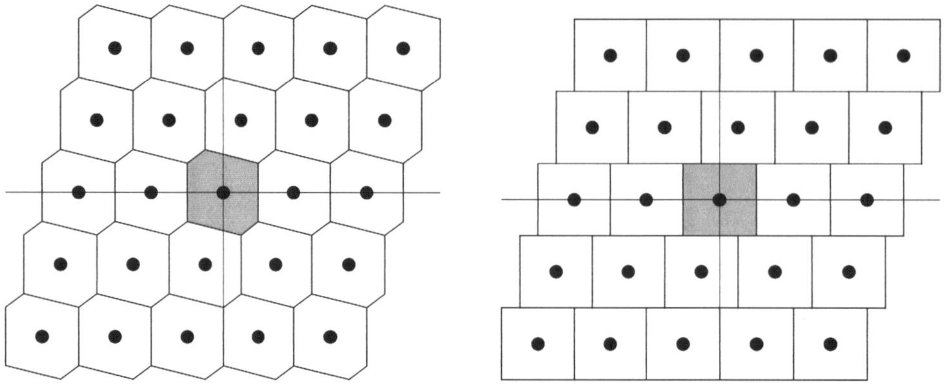

FIGURE 2
Both $\mathcal{V}(\mathcal{L})$ (left, gray) and $\mathcal{V}(\mathcal{L}_1 \oplus \mathcal{L}_2) = \mathcal{V}(\mathcal{L}_1) \times \mathcal{V}(\mathcal{L}_2)$ (right, gray) are fundamental bodies for the lattice $\mathcal{L}$.

to a lower-dimensional problem. The main technical effort is bounding the value of $\gamma_s(\mathcal{V}(\mathcal{L}))$ at local minima $\mathcal{L}$. (We do not know whether these local minima actually exist.) By considering the gradient of $\gamma_s(\mathcal{V}(\mathcal{L}))$ and using results from convex geometry [1, 8], we show that for such an $\mathcal{L}$, the convex body $\mathcal{V}(\mathcal{L})$ must be such that for all volume-preserving (determinant 1) linear transformations A, $\gamma_s(A\mathcal{V}(\mathcal{L})) \leq \gamma_s(\mathcal{V}(\mathcal{L}))$. In other words, it is in a position that maximizes its Gaussian measure. We complete the proof by using the $\ell\ell^*$ theorem [14, 24, 31], which implies that any convex body K satisfying the above must have $\gamma_s(K) \geq 2/3$ where $s = 1/(10(\log n + 2))$, as in Theorem 2.1.

5. IMPLICATIONS TO THE GEOMETRY OF VORONOI CELLS AND CONVEX BODIES

In [10], Dadush observed that Theorem 2.1 implies a certain statement about the geometry of Voronoi cells of lattices. Roughly speaking, reverse Minkowski shows that if a Voronoi cell is small (as measured by a certain Gaussian norm expectation) then there is an "explanation" of that in terms of a projection of low volume. Following [10], here we ask whether that statement might also hold for all symmetric convex bodies. (A convex body K is symmetric if $K = -K$.) We also observe that a somewhat weaker statement (where the explanation is in the form of a *slice* of low volume) is known to hold by a theorem of Milman and Pisier [29].

Definition 2 (K-norm). Let $K \subseteq \mathbb{R}^n$ be a centrally symmetric convex body. We define $\|\mathbf{x}\|_K = \min\{s \geq 0 : \mathbf{x} \in sK\}$ to be the norm on $\mathbb{R}^n$ induced by K.

Consider the quantity $\mathbb{E}[\|X\|_K]$ where $X \sim N(0, I_n)$ is a standard Gaussian vector. We can think of this quantity as measuring the size of a convex body by considering a ray starting from the origin and going in a random direction until it hits the boundary of K; we then take the expectation of the reciprocal of the (Euclidean) length of the ray. Notice that the higher this expectation, the smaller the body. What are the bodies for which this quantity is at least 1? Following [32], we now observe that any body of volume at most 1 satisfies this. By $x \gtrsim y$ we mean that $x \geq cy$ for some universal constant $c > 0$.

Lemma 5.1. *Let $K \subseteq \mathbb{R}^n$ be a symmetric convex body of volume at most* 1 *and let $\|\cdot\|_K$ be the induced norm. Then for $X \sim N(0, I_n)$,*

$$\mathbb{E}\big[\|X\|_K\big] \gtrsim 1.$$

Proof. By integrating in polar coordinates and using Jensen's inequality,

$$\begin{aligned}\mathbb{E}\big[\|X\|_K\big] &= \mathbb{E}[\|X\|_2] \int_{S^{n-1}} \|\theta\|_K \mathrm{d}\theta \\ &\geq \mathbb{E}\big[\|X\|_2\big]\bigg(\int_{S^{n-1}} \|\theta\|_K^{-n} \mathrm{d}\theta\bigg)^{-1/n} \qquad \text{(by Jensen)} \\ &= \mathbb{E}[\|X\|_2]\bigg(\frac{\mathrm{vol}_n(K)}{\mathrm{vol}_n(B_2)}\bigg)^{-1/n} \gtrsim \frac{1}{\mathrm{vol}_n(K)^{1/n}}.\end{aligned}$$ ■

In fact, more is true: instead of asking for volume at most 1, it is enough to ask for a slice of volume at most 1.

Corollary 5.2. *Let $K \subseteq \mathbb{R}^n$ be a symmetric convex body and let $\|\cdot\|_K$ be the induced norm. Then for $X \sim N(0, I_n)$, the following holds:*

$$\mathbb{E}\big[\|X\|_K\big] \gtrsim \max_{\substack{W \subseteq \mathbb{R}^n \\ d = \dim(W) \in [n]}} \frac{1}{\mathrm{vol}_d(K \cap W)^{1/d}} \tag{5.1}$$

$$\geq \max_{\substack{W \subseteq \mathbb{R}^n \\ d = \dim(W) \in [n]}} \frac{1}{\mathrm{vol}_d(\pi_W(K))^{1/d}}. \tag{5.2}$$

Proof. For the first inequality, note that

$$\mathbb{E}\big[\|X\|_K\big] = \mathbb{E}\big[\|\pi_W(X) + \pi_{W^\perp}(X)\|_K\big] \geq \mathbb{E}\big[\|\pi_W(X)\|_{K \cap W}\big],$$

by Jensen's inequality, since $\pi_W(X)$ and $\pi_{W^\perp}(X)$ are independent and $\mathbb{E}[\pi_{W^\perp}(X)] = 0$. We recover the desired lower bound by applying Lemma 5.1 to $\mathbb{E}[\|\pi_W(X)\|_{K \cap W}]$ (where we identify W with $\mathbb{R}^d$ for $d = \dim(W)$, and so $K \cap W$ is a convex body in $\mathbb{R}^d$ and $\pi_W(X)$ is distributed as $N(0, I_d)$). The second inequality is immediate from the fact that a slice of a convex body is contained in the corresponding projection. ■

It is a remarkable and nontrivial fact that follows from a theorem of Milman and Pisier [29] that, up to logarithmic terms, the right-hand side of equation (5.1) is also an *upper bound* on the expectation,

$$1 \lesssim \mathbb{E}\big[\|X\|_K\big] \min_{\substack{W \subseteq \mathbb{R}^n \\ d = \dim(W) \in [n]}} \mathrm{vol}_d(K \cap W)^{1/d} \lesssim \mathrm{poly}\log n.$$

In fact, up to a constant, the upper bound can be taken to be $\log^2(n+1)$. Here we ask whether equation (5.2) is also an upper bound on the expectation, i.e., whether for all symmetric convex bodies K it holds that

$$1 \lesssim \mathbb{E}[\|X\|_K] \min_{\substack{W \subseteq \mathbb{R}^n \\ d=\dim(W)\in[n]}} \operatorname{vol}_d\big(\pi_W(K)\big)^{1/d} \lesssim \operatorname{poly}\log n. \tag{5.3}$$

See [15, 28] for some related work.

While we do not know if equation (5.3) holds for any symmetric convex body, we now observe that it does hold for Voronoi cells of lattices. We start with an approximation of $\eta^*(\mathcal{L})$ in terms of the Voronoi cell.

Theorem 5.3 ([9]). *Let $\mathcal{L} \subseteq \mathbb{R}^n$ be an n-dimensional lattice, and let $\mathcal{V} = \mathcal{V}(\mathcal{L})$. Then for $X \sim N(0, I_n)$, we have that*

$$\mathbb{E}\big[\|X\|_{\mathcal{V}}\big] \approx \eta^*(\mathcal{L}).$$

The notation $x \approx y$ means $cy \le x \le Cy$ for some universal constants $c, C > 0$. Therefore, equation (2.3) says that for any lattice $\mathcal{L} \subset \mathbb{R}^n$ with Voronoi cell $\mathcal{V}$,

$$1 \lesssim \mathbb{E}[\|X\|_{\mathcal{V}}] \min_{\substack{W \text{ lattice subspace of } \mathcal{L} \\ d=\dim(W)\in[n]}} \big(\det(\mathcal{L}\cap W)\big)^{1/d} \lesssim 1 + \log n. \tag{5.4}$$

To complete the proof, observe that for any lattice $\mathcal{L}$ with Voronoi cell $\mathcal{V}$, and any lattice subspace W of $\mathcal{L}$,

$$\det(\mathcal{L}\cap W) = \operatorname{vol}_d\big(\mathcal{V}(\mathcal{L}\cap W)\big) \ge \operatorname{vol}_d\big(\pi_W(\mathcal{V})\big),$$

where the inequality follows from the fact

$$\begin{aligned}\mathcal{V} &= \Big\{\mathbf{x} \in \mathbb{R}^n : \langle \mathbf{x}, \mathbf{y}\rangle \le \frac{1}{2}\|\mathbf{y}\|_2^2,\ \forall \mathbf{y} \in \mathcal{L} \setminus \{\mathbf{0}\}\Big\} \\ &\subseteq \Big\{\mathbf{x} \in \mathbb{R}^n : \langle \mathbf{x}, \mathbf{y}\rangle \le \frac{1}{2}\|\mathbf{y}\|_2^2,\ \forall \mathbf{y} \in \mathcal{L}\cap W \setminus \{\mathbf{0}\}\Big\},\end{aligned}$$

and the orthogonal projection of the latter set on W is precisely $\mathcal{V}(\mathcal{L}\cap W)$.

6. COVERING RADIUS

In Section 4 we described a general strategy to bound functions on the set of stable lattices, by (1) bounding their values at local extrema and (2) analyzing lattices on the boundary by induction on dimension. We applied this strategy to two functions: the zeta function (where local maxima do not exist, making (1) trivial) and the Gaussian measure of the Voronoi cell (used as a proxy for the Gaussian mass ρ of the lattice, which we do not know how to analyze directly). It is natural to ask if there are other functions to which we can apply this strategy. Here we show one more example related to the covering radius.

The *covering radius* $\mu(\mathcal{L})$ of a lattice $\mathcal{L} \subset \mathbb{R}^n$ is the maximal distance from any point in $\mathbb{R}^n$ to the lattice or, equivalently, the minimum radius r such that $\mathcal{L} + B_2(r) = \mathbb{R}^n$. Yet another equivalent definition is $\max_{\mathbf{x}\in\mathcal{V}(\mathcal{L})} \|\mathbf{x}\|_2$. Notice that $\mu(\mathbb{Z}^n) = \sqrt{n}/2$ and analogously to Question 2.2, one can ask whether this is the maximum possible μ for a stable lattice.

Question 6.1. Is it true that for all stable $\mathcal{L} \subseteq \mathbb{R}^n$, $\mu(\mathcal{L}) \leq \mu(\mathbb{Z}^n)$?

Part of the interest in this question comes from a possible connection to integer programming, as observed by Kannan and Lovász [22] (see also [10]), as well as the connection to Minkowski's conjecture (see below) [36].

We do not know the answer to this question. It is possible, however, to derive a slightly weaker inequality directly from the statement of Theorem 2.1 using known inequalities between lattice parameters. Namely, for all stable $\mathcal{L} \subseteq \mathbb{R}^n$ it holds that [33]

$$\mu(\mathcal{L}) \leq 4\sqrt{n}(\log n + 10). \tag{6.1}$$

Below we will follow a different route, applying the strategy of Section 4 directly. Assuming a certain geometric conjecture holds, we would be able to improve on equation (6.1) and even answer Question 6.1 in the affirmative.

As was the case for the Gaussian mass ρ, we do not know how to work directly with μ, the main difficulty being bounding its value at local maxima (which were characterized in [11]). Instead, we work with the lattice parameter

$$\overline{\mu}(\mathcal{L}) := \sqrt{\frac{1}{\det(\mathcal{L})} \int_{\mathcal{V}(\mathcal{L})} \|\mathbf{x}\|^2 d\mathbf{x}}.$$

While μ considers the point farthest away from $\mathcal{L}$, $\overline{\mu}$ looks at the (L_2) average distance of a random point in space from $\mathcal{L}$. (See also [7, 17, 20, 26, 41] for more about $\overline{\mu}$.) Magazinov showed that these two parameters are quite close to each other [26].

Theorem 6.2 ([26]). *For any lattice $\mathcal{L} \subset \mathbb{R}^n$,*

$$\overline{\mu}(\mathcal{L}) \leq \mu(\mathcal{L}) \leq \sqrt{3}\overline{\mu}(\mathcal{L}).$$

The lower bound is immediate from the definition, and the upper bound is tight for the lattice $\mathbb{Z}^n$. Because of the latter, it is plausible that one could resolve Question 6.1 entirely by considering $\overline{\mu}$ (see more below). We also remark that the natural extension of the upper bound in Theorem 6.2 to all convex bodies (and not just Voronoi cells) is totally false, as can be seen by taking the ℓ_1 ball (where the maximum ℓ_2 norm of a vector is 1, yet the typical norm is only $C/\sqrt{n}$).

Our goal is therefore to bound $\overline{\mu}(\mathcal{L})$ from above for stable lattices $\mathcal{L}$. As before, "aligning the cosets" cannot decrease $\overline{\mu}$, i.e., $\overline{\mu}(\mathcal{L}) \leq \overline{\mu}(\mathcal{L}_1 \oplus \mathcal{L}_2)$. The proof is also essentially the same, namely that $\mathcal{V}(\mathcal{L})$ and $\mathcal{V}(\mathcal{L}_1 \oplus \mathcal{L}_2) = \mathcal{V}(\mathcal{L}_1) \times \mathcal{V}(\mathcal{L}_2)$ are both fundamental bodies for the lattice $\mathcal{L}$, i.e., they both contain exactly one point in each coset of $\mathcal{L}$, but by definition, $\mathcal{V}(\mathcal{L})$ contains the *shortest* point in each coset, leading to the desired inequality. Moreover, continuing iteratively is again straightforward, since $\overline{\mu}(\mathcal{L}_1 \oplus \mathcal{L}_2)^2 = \overline{\mu}(\mathcal{L}_1)^2 + \overline{\mu}(\mathcal{L}_2)^2$ so once we are at the boundary, the problem truly reduces to a lower-dimensional problem. As before, the key step in the proof is bounding local maxima $\mathcal{L}$ of $\overline{\mu}$. Using a similar proof to the one in the case of the Gaussian measure of the Voronoi cell, it can be shown that such $\mathcal{L}$ must be such that their Voronoi cell $\mathcal{V}(\mathcal{L})$ is *isotropic*. Recall that a symmetric convex body $K \subset \mathbb{R}^n$ is said to be *isotropic* if its covariance matrix is a multiple of identity, i.e., $\int_K \mathbf{x}\mathbf{x}^T d\mathbf{x} = \alpha \cdot I_n$ for some scalar $\alpha > 0$. Intuitively, this says that the

Voronoi cell is not elongated in any one direction (e.g., a square is isotropic but a rectangle is not), which one might expect should imply that $\overline{\mu}$ is small. To make this precise, define the (symmetric) isotropic constant L_n as

$$L_n^2 := \max_{d \le n} \frac{1}{d} \cdot \sup_K \int_K \|\mathbf{x}\|^2 \mathrm{d}\mathbf{x},$$

where the supremum is taken over all isotropic symmetric convex bodies $K \subset \mathbb{R}^d$ of volume one. Therefore, by following the proof strategy from Section 4 we obtain the following.

Theorem 6.3 ([33]). *For any stable lattice $\mathcal{L} \subset \mathbb{R}^n$,*

$$\mu(\mathcal{L}) \le \sqrt{3}\overline{\mu}(\mathcal{L}) \le \sqrt{3n} L_n.$$

It is known that $1/(2\sqrt{3}) \le L_n \le n^{o(1)}$ [3,4,23], the lower bound being due to the hypercube $[-1/2, 1/2]^n$. This already gives a reasonably tight upper bound on $\mu(\mathcal{L})$ for stable lattices. But perhaps L_n is even smaller? The so-called slicing conjecture implies that L_n is bounded by a universal constant. In fact, as far as we know, it is entirely possible that $L_n = 1/(2\sqrt{3})$, i.e., that the hypercube $[-1/2, 1/2]^n$ is the worst symmetric body for the slicing conjecture. If this is true, then we get that for any stable lattice $\mathcal{L} \subset \mathbb{R}^n$, $\mu(\mathcal{L}) \le \sqrt{n}/2$, which is tight for $\mathbb{Z}^n$. That is, a positive answer to Question 6.1.

Apart from being an interesting statement in its own right, it was shown by Shapira and Weiss [36] that a positive answer to Question 6.1 implies the so-called Minkowski conjecture. The conjecture asserts that for every lattice $\mathcal{L} \subset \mathbb{R}^n$ (not necessarily stable) with $\det(\mathcal{L}) = 1$ and vector $\mathbf{t} = (t_1, \dots, t_n) \in \mathbb{R}^n$,

$$\inf_{\mathbf{y} \in \mathcal{L}} \prod_i |y_i - t_i| \le 2^{-n}. \tag{6.2}$$

In order to derive the Minkowski conjecture, use the nontrivial fact that any lattice with determinant 1 can be made stable by multiplying it by a diagonal operator of determinant 1 [36,37]. Since the left-hand side of equation (6.2) is invariant under multiplication by such operators, it follows that it suffices to prove the inequality for stable lattices. But a positive answer to Question 6.1 implies that for any $\mathbf{t}$, there exists a $\mathbf{y} \in \mathcal{L}$ such that $\|\mathbf{t} - \mathbf{y}\|_2 \le \sqrt{n}/2$. The AM–GM inequality now implies that $\prod_i |y_i - t_i| \le 2^{-n}$, as desired.

ACKNOWLEDGMENTS

I am grateful to Daniel Dadush, who conjectured the existence of a reverse Minkowski theorem, and with whom much of the initial work was done [10]. I also thank Yael Eisenberg, Dan Simon, and Noah Stephens-Davidowitz for their comments on an earlier draft of this review.

FUNDING

Supported by the Simons Collaboration on Algorithms and Geometry, by a Simons Investigator Award from the Simons Foundation, and by the National Science Foundation (NSF) under Grant No. CCF-1320188. Any opinions, findings, and conclusions or recommendations expressed in this material are those of the authors and do not necessarily reflect the views of the NSF.

REFERENCES

[1] S. G. Bobkov, On Milman's ellipsoids and M-position of convex bodies. In *Concentration, functional inequalities and isoperimetry*, pp. 23–33, Contemp. Math. 545, Amer. Math. Soc., Providence, RI, 2011.

[2] J.-B. Bost, Réseaux euclidiens, séries thêta et pentes, Exp. Bourbaki **1151** (2020), 1–59, 422.

[3] J. Bourgain, On the distribution of polynomials on high-dimensional convex sets. In *Geometric aspects of functional analysis (1989–90)*, pp. 127–137, Lecture Notes in Math. 1469, Springer, Berlin, 1991.

[4] Y. Chen, An almost constant lower bound of the isoperimetric coefficient in the KLS conjecture. *Geom. Funct. Anal.* **31** (2021), no. 1, 34–61.

[5] K.-M. Chung, D. Dadush, F.-H. Liu, and C. Peikert, On the lattice smoothing parameter problem. In *2013 IEEE Conference on Computational Complexity—CCC 2013*, pp. 230–241, IEEE Computer Soc., Los Alamitos, CA, 2013.

[6] H. Cohn, A. Kumar, S. D. Miller, D. Radchenko, and M. Viazovska, The sphere packing problem in dimension 24. *Ann. of Math. (2)* **185** (2017), no. 3, 1017–1033.

[7] J. Conway and N. J. A. Sloane, *Sphere packings, lattices and groups*. Springer, New York, 1998.

[8] D. Cordero-Erausquin, M. Fradelizi, and B. Maurey, The (B) conjecture for the Gaussian measure of dilates of symmetric convex sets and related problems. *J. Funct. Anal.* **214** (2004), no. 2, 410–427.

[9] D. Dadush, *Integer programming, lattice algorithms, and deterministic volume estimation*. Ph.D. thesis, Georgia Institute of Technology, 2012.

[10] D. Dadush and O. Regev, Towards strong reverse Minkowski-type inequalities for lattices. In *57th Annual IEEE Symposium on Foundations of Computer Science—FOCS 2016*, pp. 447–456, IEEE Computer Soc., Los Alamitos, CA, 2016.

[11] M. Dutour Sikirić, A. Schürmann, and F. Vallentin, Inhomogeneous extreme forms. *Ann. Inst. Fourier (Grenoble)* **62** (2012), no. 6, 2227–2255.

[12] Y. Eisenberg, O. Regev, and N. Stephens-Davidowitz, Reverse Minkowski for the Epstein zeta function (in preparation).

[13] P. Epstein, Zur Theorie allgemeiner Zetafunctionen. *Math. Ann.* **56** (1903), no. 4, 615–644.

[14] T. Figiel and N. Tomczak-Jaegermann, Projections onto Hilbertian subspaces of Banach spaces. *Israel J. Math.* **33** (1979), no. 2, 155–171.

[15] A. Giannopoulos and E. Milman, M-estimates for isotropic convex bodies and their L_q-centroid bodies. In *Geometric aspects of functional analysis*, pp. 159–182, Lecture Notes in Math. 2116, Springer, Cham, 2014.

[16] D. R. Grayson, Reduction theory using semistability. *Comment. Math. Helv.* **59** (1984), no. 4, 600–634.

[17] V. Guruswami, D. Micciancio, and O. Regev, The complexity of the Covering Radius Problem. *Comput. Complexity* **14** (2005), no. 2, 90–121.

[18] T. C. Hales, A proof of the Kepler conjecture. *Ann. of Math. (2)* **162** (2005), no. 3, 1065–1185.

[19] G. Harder and M. S. Narasimhan, On the cohomology groups of moduli spaces of vector bundles on curves. *Math. Ann.* **212** (1975), no. 3, 215–248.

[20] I. Haviv, V. Lyubashevsky, and O. Regev, A note on the distribution of the distance from a lattice. *Discrete Comput. Geom.* **41** (2009), no. 1, 162–176.

[21] A. Heimendahl, A. Marafioti, A. Thiemeyer, F. Vallentin, and M. C. Zimmermann, Critical even unimodular lattices in the Gaussian core model. 2021, arXiv:2105.07868.

[22] R. Kannan and L. Lovász, Covering minima and lattice-point-free convex bodies. *Ann. of Math. (2)* **128** (1988), no. 3, 577–602.

[23] B. Klartag, On convex perturbations with a bounded isotropic constant. *Geom. Funct. Anal.* **16** (2006), no. 6, 1274–1290.

[24] D. R. Lewis, Ellipsoids defined by Banach ideal norms. *Mathematika* **26** (1979), no. 1, 18–29.

[25] S. Lovett and O. Regev, A counterexample to a strong variant of the polynomial Freiman–Ruzsa conjecture in Euclidean space. *Discrete Anal.* **8** (2017), 6.

[26] A. Magazinov, A proof of a conjecture by Haviv, Lyubashevsky and Regev on the second moment of a lattice Voronoi cell. *Adv. Geom.* **20** (2020), no. 1, 117–120.

[27] D. Micciancio and O. Regev, Worst-case to average-case reductions based on Gaussian measures. *SIAM J. Comput.* **37** (2007), no. 1, 267–302 (electronic).

[28] E. Milman, On the mean-width of isotropic convex bodies and their associated L_p-centroid bodies. *Int. Math. Res. Not. IMRN* **11** (2015), 3408–3423.

[29] V. D. Milman and G. Pisier, Gaussian processes and mixed volumes. *Ann. Probab.* **15** (1987), no. 1, 292–304.

[30] H. Minkowski, *Geometrie der Zahlen*. B.G. Teubner, 1910.

[31] G. Pisier, Holomorphic semigroups and the geometry of Banach spaces. *Ann. of Math. (2)* **115** (1982), no. 2, 375–392.

[32] G. Pisier, *The volume of convex bodies and Banach space geometry*. Cambridge Tracts in Math. 94, Cambridge University Press, Cambridge, 1989.

[33] O. Regev and N. Stephens-Davidowitz, A reverse Minkowski theorem. In *STOC'17—Proceedings of the 49th Annual ACM SIGACT Symposium on Theory of Computing*, pp. 941–953, ACM, New York, 2017.

[34] O. Regev and N. Stephens-Davidowitz, A reverse Minkowski theorem for integral lattices (in preparation).

[35] P. Sarnak and A. Strömbergsson, Minima of Epstein's zeta function and heights of flat tori. *Invent. Math.* **165** (2006), no. 1, 115–151.

[36] U. Shapira and B. Weiss, Stable lattices and the diagonal group. *J. Eur. Math. Soc. (JEMS)* **18** (2016), no. 8, 1753–1767.

[37] O. N. Solan, Intersections of diagonal orbits. 2016, arXiv:1612.08765.

[38] U. Stuhler, Eine Bemerkung zur Reduktionstheorie quadratischer Formen. *Arch. Math. (Basel)* **27** (1976), no. 6, 604–610.

[39] J. van der Corput, Verallgemeinerung einer Mordellschen Beweismethode in der Geometrie der Zahlen, Zweite Mitteilung. *Acta Arith.* **2** (1936), no. 1, 145–146.

[40] M. S. Viazovska, The sphere packing problem in dimension 8. *Ann. of Math. (2)* **185** (2017), no. 3, 991–1015.

[41] R. Zamir and M. Feder, On lattice quantization noise. *IEEE Trans. Inf. Theory* **42** (1996), no. 4, 1152–1159.

ODED REGEV

Courant Institute of Mathematical Sciences, 251 Mercer St., New York, NY 10012, USA, regev@cims.nyu.edu

This Page Intentionally Left Blank

MATHEMATICS OF COMPUTATION THROUGH THE LENS OF LINEAR EQUATIONS AND LATTICES

MULI (SHMUEL) SAFRA

ABSTRACT

Mathematics of computation and, in particular, *computational complexity theory*, is a fundamental research area in the intersection of computer science and mathematics.
The area revolves around classifying computational problems as feasible or alternatively as infeasible, typically in the worst-case regime.
In some related areas—and even more prominently in practice—the notion of average-case complexity is ubiquitous.
Cryptography is a prime example where proving security of protocols/primitives often necessitates average-case type hardness assumptions.
We take the choice herein to analyze these notions through the lens of *linear algebra*.
This perspective allows us to smoothly present important future research directions, as well as propose conjectures that lay a road-map for future progress.
The goal of this survey is to make research at the core of computation more accessible.
More importantly, it gives us an opportunity to naturally state open questions regarding *lattices*; a solution to which would transform our perception of computation, not only scientifically, but also practically.

MATHEMATICS SUBJECT CLASSIFICATION 2020

Primary 03D15; Secondary 11H06, 11P21, 52C07, 68W25, 11D04

KEYWORDS

Computational-complexity, PCP, lattices, linear-algebra, hardness-of-approximation, cryptography

ICM INTERNATIONAL CONGRESS OF MATHEMATICIANS 2022 JULY 6–14

Proc. Int. Cong. Math. 2022, Vol. 6, pp. 4914–4969
DOI 10.4171/ICM2022/178

1. SYNOPSIS

My presentation will cover central issues regarding the mathematics of computation, in particular the *computational complexity* of algorithmic problems, through the lens of *linear equations*. The aim is to make research at the core of computation more accessible. More importantly, though, I will take the opportunity to present open questions regarding *lattices*. These are research questions whose solutions may transform our perception of computation, not only scientifically, but also practically.

The presentation considers a series of computational problems through the lens of linear algebra, and, more specifically, as problems regarding linear equations. This formalism is flexible enough to provide a uniform perspective across well-known combinatorial optimization problems (Max-Cut is just one example); to clearly present the Unique-Games problem and associated conjecture; and to investigate fundamental advanced problems such as the Shortest-Vector Problem and the Closest-Vector Problem, over codes and, more generally, lattices. Such an approach will help clarify the seemingly subtle key differences between the feasible and infeasible.

We start with the basic framework of a system of linear equations, and the natural computational problem of solving it or at least approximating the solution. The presentation then proceeds through gradual generalizations:

(1) Consider the standard notion where one searches for an assignment to the variables that satisfies as many of the equations as possible, or at least approximating the optimal solution.

(2) Next, refine this notion and consider more general norms by which to measure the quality of an assignment. In the current context, one may think of the linear system as having the form $Mx = t$, and the distance measure being ℓ^0-norm of the difference ($\|Mx - t\|_0 \stackrel{\text{def}}{=} \Pr_i[(Mx)_i \neq t_i]$). Then, extend that notion to a general ℓ^p-norm of the difference between the solution and the target values $\|Mx - t\|_p$.

(3) Consider alternative ways to measure the quality of a solution; a natural one is to assume that the assignment must satisfy all equations, albeit the shorter the assignment, the better.

(4) Finally, further refine the setting by restricting the domain of the assignment. It is natural to define the problem over some field $\mathbb{F}$, and to restrict the domain of the variables to a subset $\mathcal{F} \subseteq \mathbb{F}$.

2. HISTORY

The problem of solving a system of linear equations has been studied extensively since the early history of mathematics—starting with al-Khwarizmi and his invention of linear algebra. The computational, algorithmic question whether a complete solution exists has long been known to be efficiently solvable via Gaussian elimination. Nevertheless, when changing the definition of the problem slightly, in order to obtain a partial solution of as many of the equations as possible, it turns out that coming up with an optimal solution is hard. In particular, an efficient algorithm that simultaneously solves almost all equations (assuming such a solution exists) would imply that P = NP, resolving the most fundamental open problem in computer science (and one of the most basic open questions of mathematics).

This problem was in fact raised by Hilbert and Ackerman **[44]** in their Entscheidungsproblem, calling for a systematic algorithm that determines the universality of any given mathematical statement. Much of the basic research into the emerging field of computer science during the 1970s and 1980s was devoted to this conundrum **[23, 45, 62]**. Consequently, many classical optimization problems were shown to be NP-hard, implying that an efficient algorithm for them would require P = NP.

Let us briefly define these classes of computational (decision) problems: P is the class of all those problems that can be computed efficiently (in polynomial time), while NP is a wider class of such problems and, for our discussion, it suffices to say that it contains many problems that are not known to be in P. An efficient solution for a problem which has been established to be NP-hard would imply that P = NP. The class coNP is the class of (decision) problems whose complement is in NP. The intersection of NP and coNP (problems in NP whose complement is in NP, too) plays an important (if yet not so clear) role in complexity theory, and even more so in cryptography.

An important caveat here is that assuming P $\neq$ NP is the preferable option for humanity: First, as it relates to the raison d'être of mathematicians, since, if P = NP, there would be no fundamental distinction between the existence of a short mathematical proof and finding it. Second, there are many useful implications to a problem (presumably) being hard, most notably in cryptography, where there is no system or secret safe from the powerful NP class.

The 1980s, through the 1990s and beyond, saw the development of novel techniques designed to facilitate efficient approximation algorithms for such problems. Since coming up with an optimal solution turns out to be unreachable, one naturally attempts to come up with the next best option, that is, a solution that comes close to optimal, namely, within some known *ratio of approximation* from optimal.

Approximation problems could also turn out to be NP-hard. This finding represents the inception of a radically different approach, which requires fundamental novel mathematics, and heralded the dawn of the PCP-era[1] **[11, 12, 36]**. Indeed, proving the hardness of approximation is vastly more complex and requires utilizing multiple connections to other mathematical fields, including probability, combinatorics, geometry, and analysis. This led to an avalanche of hardness of approximation theorems which established that for many classical optimization problems, coming up with a solution even slightly better than the best-

1 PCP is a strong characterization of the NP class via gap problems (which replace decision problems), thereby allowing proofs that approximation problems are NP-hard.

known is NP-hard. The upshot is that even approximating such a partial solution efficiently would imply the same dreaded consequence that $P = NP$.

Improving the PCP characterization of NP and proving hardness of approximation problems is a central research topic that has had far reaching implications and is still in flux [5, 7, 14, 17, 19, 26–30, 30, 31, 42, 43, 46, 48, 50, 51, 51–53, 65, 69, 73, 76]. It also has implications that extend beyond complexity, to cryptography, algorithms, and other areas.

Fortunately, most of the fundamental achievements of the PCP-era—where approximation problems are being shown to be hard via the PCP characterization of NP—can be presented clearly through the lens of linear equations. The same is true for the most fundamental open problems of the field of hardness of approximation.

As an immediate consequence, much effort that would have otherwise been spent on trying to come up with approximation algorithms is avoided, as current research indicates that they are unlikely to be efficiently solvable.

The related, but parallel, field of cryptography thrives through an endless quest for reasonable assumptions regarding the hardness of computational problems. This is essential when constructing protocols and primitives with highly intriguing properties that have both theoretical and practical uses, especially with respect to their security. In short, the idea is to hide some secret backdoor information by encoding it into some hard problem (for instance, encoding a bit string as a solution to a system of linear equations), and then to use this encoding to transfer data securely. This begs the question of whether cryptography can be based on the gold standard of $P \neq NP$.

In this context, it is worth noting that the rich theory of NP-hardness addresses only the worst-case complexity of a problem. Accordingly, an efficient algorithm that is correct on most instances of the problem (that is, on most systems of linear equations) is insufficient. Therefore, an efficient average-case algorithm for an NP-hard problem does not imply $P = NP$. This is apparently the reason we do not know how to base cryptography on the weakest possible assumption that $P \neq NP$. A cryptographic protocol chooses keys at random and, if a problem is not known to be hard in the average-case, breaking it would not imply a general algorithm according to the worst-case regime.

Thus, another question naturally arises: Is there a class of computational problems, whose hardness can be established mathematically, similarly to NP-hardness, while maintaining enough flexibility to allow for cryptography? The main topic of interest herein—lattices and lattice-based computational problems—is the prime source of such problems, bringing on a tremendous surge of creative ideas.

2.1. Lattices take center stage

Broadly, a lattice is defined as a discrete subgroup of a topological group, such that the quotient of the group by the lattice satisfies a certain finiteness property. Our interest is in a special case, which we formally define below. The mathematical background on lattices appears in Appendix B.

Historically, lattices have been investigated since the late 18th century by mathematicians such as Lagrange, Gauss, and later, Minkowski (mostly for number-theoretic applications) [68]. Interestingly, some of their theorems (especially Minkowski's) are quite relevant to various parts of our discussion and will thus be presented below in the mathematical foundation section.

In the late 20th century, lattices became a topic of active research with regards to computational aspects of various problems and, more specifically, to the complexity of solving related computational problems. Research in the area has thrived in two directions:

- the quest for algorithms to solve computational problems over lattices as efficiently as possible;
- suggestions of computational problems over lattices as candidates for problems that (presumably) cannot be solved efficiently.

The second point has proved to be more useful than the first, most notably in cryptography, which is always on the lookout for credible assumptions regarding the hardness of computational problems.

In 1982, Lenstra, Lenstra, and Lovasz [LLL or L^3] discovered an algorithm that approximates the Shortest-Vector Problem (henceforth SVP) in a lattice with a weak ratio of approximation (exponential in the dimension). SVP is one of the most fundamental computational problems regarding lattices and it will be covered thoroughly below. The LLL algorithm, which is very useful despite the weak ratio of approximation, has since been utilized for a wide variety of applications. In particular, it may be the first choice as a cryptanalysis tool in an attempt to break a cryptographic system.

In the 1990s, this area was revitalized by a groundbreaking theorem of Miklós Ajtai [4], which established a reduction from worst-case to average-case for a computational lattice problem similar to SVP. He thus turned the spotlight on lattices and, in particular, demonstrated the very interesting advantage that this type of computational problem has when it comes to being presumably hard to compute. To explain this advantage, we need to explain some basic notions in cryptography; we give a brief summary here and more details in the next subsection.

Any modern cryptographic protocol is based on a hardness assumption, that is, a computational problem that is presumed to be hard. The security of the protocol is then established by showing that breaking the protocol entails an efficient solution for the problem in question. But such problems, as we shall explain below, must always be in NP, so if $P = NP$ there would be no cryptography. Hence, any proof of security for a reasonable cryptographic protocol must assume, at the very least, that $P \neq NP$.

For cryptographic applications, however, one needs to rely on a problem that is hard on average, that is, one for which solving a random instance from a prescribed distribution is presumed to be infeasible. The assumption of $P \neq NP$ is not sufficient to guarantee this: most known hardness results for computational problems only guarantee that the worst case of the problem is difficult to solve, and in fact it is unlikely that we can find a problem that is NP-hard on average, for any probability distribution which can be algorithmically sampled. Remarkably, Ajtai showed that for certain natural lattice problems, the worst case of the problem reduces to the average case. When one has such a worst-to-average-case reduction, one can more comfortably rely on the hardness assumption for the worst-case, and, nevertheless, be able to trust and utilize the hardness of the problem on average.

Consequently, after a remarkable research journey that took more than a decade, Oded Regev [77] eventually introduced the Learning-with-Errors (LWE) problem, which is an average-case problem, and proceeded to establish that it is hard as a computational lattice problem, which can be presumed to be hard in the worst-case. Almost all current cryptography has since been based on the hardness of the LWE problem.

In the next subsection we expand on the preceding paragraphs.

2.2. Hardness assumptions

Let us elaborate on why there can be no modern cryptography without a computational hardness assumption: a protocol suggested for a cryptographic application must have a secret that only authorized entities possess, which facilitates the authorized entities sharing information inaccessible to unauthorized entities (due to their ignorance regarding the secret). But what is the mechanism that keeps the secret secured? Consider the algorithmic problem of finding the secret, assuming access to the data exchanged in the protocol. Modern cryptography, which relies on mathematical proofs of security, assumes full disclosure, that is, that all exchanges are public, and the security of the protocol is established despite adversaries having access to everything but the secret. One must therefore guarantee that it takes way too long to expose the secret, even using strong computers. But a nondeterministic adversary can "guess" the secret and verify the correctness of the guess in similar time (by attempting to decrypt the communication and checking whether the result makes sense). Hence, if $P = NP$, for any cryptographic protocol which runs in polynomial time, there exists a polynomial-time algorithm which breaks it.

Since we do not yet know that $P \neq NP$, to prove the security of a cryptosystem, we need to *assume* that a certain computational problem is hard, and then show that breaking the cryptosystem is at least as hard as solving the problem. In this case, an algorithm which breaks the cryptographic protocol (that is, exposes the secret, or just learns the secret data) efficiently can also solve the original problem efficiently, which would contradict the hardness assumption. By the above argument, this computational complexity assumption will necessarily be at least as strong as $P \neq NP$.

Many problems are by now known to be NP-hard. Thus, assuming they are hard is equivalent to assuming $P \neq NP$. Many problems—via PCP—are also known to be NP-hard to approximate. Nevertheless, when focusing on hardness assumptions for cryptographic application, one needs to choose a random secret from a distribution over all possible secrets, and, consequently, the assumption should be related to how hard the problem is to solve on average. This is in sharp contrast to the computational problems whose established NP-hardness results apply only to the worst-case of the problem (that is, how hard it is to solve the hardest instances of a given size). In fact, it is quite unlikely that we can have a problem that is

NP-hard to solve on average (there are reasons to assume that one cannot efficiently identify a subset of instances over which the problem is hard). Consequently, one has to establish security via an assumption stronger than $P \neq NP$ and, furthermore, to come up with an assumption regarding the hardness of the problem over a distribution over the inputs.

Typically, the weakest hardness assumption that enables security is that $P \neq NP \cap coNP$—this may be because one must presume hardness on average, and it is plausible that NP-hard problems are not hard on average for any reasonable, computable distribution. In contrast, some problems in $NP \cap coNP$ are known to be as hard on average as they are hard in the worst case—approximating lattice problems to within a weak enough ratio is one great source for such problems.

Ajtai's brilliant, very insightful theorem was a reduction from approximating a lattice problem (a certain variant of SVP) in the worst case to approximating it in the average case. Consequently, it is enough to assume that this problem is hard in the worst case in order to be able to rely on its hardness on average. The assumption that the problem is hard in the worst case is stronger than $P \neq NP$, but, nevertheless, has proved a rather safe hardness assumption: four decades since its introduction, there is not even a suggestion for an algorithm that could substantially improve on LLL. Moreover, the ratio of approximation that is presumed hard is much stronger (smaller) than that of LLL.

Regev's expansive and wide-ranging perspective on complexity and lattices led him to introduce the Learning-with-Errors (LWE) problem, which assumes a secret vector S that one is trying to discover (i.e., learn). All one can do is press a button, which publicly reveals a random vector x. The rules of the game could be that, in response, the secret holder must announce the inner product of S with x. This, however, would be too easy, as Gaussian elimination can be utilized to reveal the secret S within a small number of such rounds. In LWE, when the button is pushed, x is revealed as before, but what is made public is the inner-product $\langle S, x\rangle$ perturbed by some (small) error. Now, Gauss' algorithm fails, and no other efficient algorithm is known for the problem. Regev has shown a reduction from a lattice problem (a certain variant of SVP) to LWE, thereby establishing a hardness result that is very useful when proving the security of a cryptographic protocol whose hardness relies on the hardness of that lattice problem.

Assuming LWE is hard leads naturally to cryptographic applications. Indeed, the LWE problem yields a natural secret key, S, and a natural public key, the x's and the noisy inner products (we refer to [77] for a complete description of the cryptosystem).

Regev's reduction from a lattice problem to LWE is not only quite complicated but also a quantum reduction, i.e., it implies that LWE is as hard to solve as a shortest-vector-type problem for a quantum computer, but not necessarily for a classical computer. Simplifying the reduction and hopefully avoiding a quantum reduction (that is, finding a classical reduction) is one great research direction to pursue which could greatly improve our understanding of the issues involved and thereby lead to much sought-after progress.

3. FOUNDATION: A SYSTEM OF LINEAR EQUATIONS

Finding a solution for a system of linear equations is one of the most fundamental computational problems. Let us start with definitions:

Definition 3.1 (Linear equations). A system of linear equations consists of a field $\mathbb{F}$, a matrix $M \in \mathbb{F}^{m \times n}$, and a target vector $t \in \mathbb{F}^m$.

The most natural decision problem within this framework is whether a given linear system has a solution that satisfies all the equations. Of course, this is an easy problem, which can be efficiently solved via Gaussian elimination.

Consequently, given a system of linear equations, the next immediate goal is to find an assignment $x \in \mathbb{F}^n$ that satisfies as many of the equations of the system as possible. For instance, given $\Psi = [M, t, \mathbb{F}]$ a system of linear equations, let us denote by $\mathsf{val}[\Psi]$ the minimum, over all assignments $x \in \mathbb{F}^n$, of the fraction of equations that are unsatisfied,

$$\mathsf{val}[\Psi] \stackrel{\text{def}}{=} \frac{1}{m} \min_{x \in \mathbb{F}^n} \left| \left\{ i \mid [Mx]_i \neq t_i \right\} \right|.$$

One should remark here that when measuring the computational complexity of a problem, we look at the time it takes to solve it as a function of the length of the input. As the input size grows to infinity, the number of equations could also grow rapidly. It then makes sense to measure the *fraction* of equations unsatisfied. An algorithm that finds an assignment that satisfies 0.99 of the equations (leaves 0.01 unsatisfied), assuming such an assignment exists, solves it for unbounded input size.

The rest of this section surveys numerous variants of algorithmic problems over a system of linear equations, representing the most fundamental open questions in the field of the mathematics of computation. These are variants of the *Closest-Vector Problem (CVP)*—where one is given a target vector t and the algorithmic problems call for a solution coming as close as possible to satisfying the equations. Another important variant is the *Shortest-Vector Problem (SVP)*—where one assumes the target vector $t = \vec{0}$ is all 0, albeit, disallows the all 0 assignment.

The rest of this section surveys numerous variants of algorithmic problems over a system of linear equations. Here is a table of progression, with informal description—the box denotes a computational problem:

- 3.1 **Error correcting codes (ECC)**—the standard setting for computational problems regarding a system of linear equations:
  - **CVP-code search version;** $\boxed{\textbf{Compute}}$ 1—decoding a received word by searching for the closest vector.
  - **CVP-code decision;** $\boxed{\textbf{Compute}}$ 2—deciding if the received word is acceptable.
  - **SVP-code decision;** $\boxed{\textbf{Compute}}$ 3—finding the *distance* of a linear ECC.

- **Max-Cut;** Compute 4—partitioning the vertices of a graph into two sets, maximizing crossing edges.
- 3.2 **Complexity of approximation** considers approximation versions:
  - **CVP gap**$[\varepsilon, 1-\varepsilon]$ **over linear-ECC;** Compute 5—distinguishing whether the received word is acceptable or the received word is very far from acceptable.
  - **Unique games (UG)** $\mathtt{PCP}_{1\leftrightarrow 1}[\varepsilon, 1-\varepsilon]$ Compute 6—structurally restricted gap-CVP-code. It is an open problem.
  - **2-to-1 games** $\mathtt{PCP}_{2\to 1}[\varepsilon, 1-\varepsilon]$**;** Compute 7—almost linear equations (see below for precise definition). It is NP-hard.
  - **Unique games** $\mathtt{PCP}_{1\leftrightarrow 1}[\frac{1}{2}, 1-\varepsilon]$**;** Compute 8—a little harder than UG above. It is NP-hard.
- 3.3 **General norm**—one can extend all computational problems mentioned so far to general metrics.
- 3.4 **Alternative measures**
  - **Shortest integer solution;** Compute 9—finding the shortest all-satisfying assignment; or at least approximating it.
- 3.5 **Average-case complexity**
  - **Shortest integer solution average case;** Compute 10—finding the shortest all-satisfying assignment over a random system of linear equations; or at least approximating it.
  - **Learning-with-Errors (LWE);** Compute 11—assuming a secret target vector and, given a random system of linear equations plus a random, noisy version of the secret, finding the satisfying assignment.
- 3.6 **Lattices**—general sparse subdomain (discrete subgroup). For example, including full-rank matrices.
  - **CVP-lattice-search version;** Compute 12—finding the closest vector in a lattice.

3.1. Error correcting codes (ECC)

In brief, a *linear error correcting code (linear-ECC)* takes a data vector $x \in \mathbb{F}^n$ and transmits instead the vector $Mx \in \mathbb{F}^m$, where $m \gg n$. The transfer rate of information decreases, as one sends a string of length m instead of length n. Nevertheless, if vectors in the subspace $\{Mx\}$ are distinct enough, assuming not too many errors occur in transmission, the received vector $t \in \mathbb{F}^m$ is not too different from Mx and could therefore be decoded, and the original vector x extracted. The matrix M is referred to as the *generating matrix* of the linear ECC, and the question becomes whether there is a legal codeword within a certain radius of t, where the radius corresponds to the number of errors that can be handled by the code.

In other words, the decoding problem calls for an almost complete solution for a system of linear equations. Unfortunately, once one allows some equations to not be satisfied, the computational complexity of the problem becomes strikingly different. In terms of linear equations, it goes as follows:

Compute **1** (CVP-code-search). The *Closest-Vector-Search Problem over linear-ECCs* is

- **Input:** a system of linear equations $[M, t, \mathbb{F}]$ and distance $\varepsilon > 0$.
- **Goal:** find a vector generated by M within distance ε from t, $\|Mx - t\|_0 \leq \varepsilon$, assuming one exists. (Alternatively, find the one closest to t).

For a broader perspective, consider the ℓ^0 (Hamming) metric over all vectors $\mathbb{F}^m$ and, within it, the set of vectors that are spanned by M, that is, all vectors of the form Mx for $x \in \mathbb{F}^n$. Since $m \gg n$, the set of such vectors is sparse within the metric and possibly spread thin so that the distance between any pair of vectors is high; hence, finding a vector closest to a general target vector in $\mathbb{F}^m$ is nontrivial.

Our focus throughout is on this type of objects, namely, a discrete sparse subgroup which is sparse within a geometric space over some field. The field could be continuous, for example, $\mathbb{R}$; nevertheless, by default, the set of vectors spanned is discrete, as will become clear below.

The reason this computational problem is classified as concerning ECCs is that the sparse subset is the set of all legal codewords, Mx for $x \in \mathbb{F}^n$. Accordingly, the target can be thought of as the received word and the algorithm's goal is to decode it, namely, find the original transmitted codeword.

When dealing with the computational complexity of a problem, it makes more sense to consider the decision version, which then takes the following form:

Compute **2** (CVP-code). The *Closest-Vector decision problem over linear ECCs* is

- **Input:** a system of linear equations $[M, t, \mathbb{F}]$ and distance $\varepsilon > 0$.
- **Goal:** decide if there is a vector generated by M within distance ε from t—is $\mathsf{val}[M, t, \mathbb{F}] \leq \varepsilon$?

In words, given a system of linear equations, can one satisfy almost all of them?

The CVP code (**Compute** 2) problem turns out to be NP-hard for any not too large $\varepsilon > 0$. Consequently, the above search version (**Compute** 1) of finding the legal codeword of a linear-ECC closest to a given vector t (for a general generating matrix M) is also NP-hard.

A similar, yet quite distinctive, highly important computational problem regarding codes is the Shortest-Vector problem:

Compute **3** (SVP-code). The *Shortest-Vector decision problem over linear ECC* is

- **Input:** a system of linear equations $[M, \mathbb{F}]$ and distance $\varepsilon > 0$.
- **Goal:** decide if there is a *nonzero* vector generated by M within distance ε from $\vec{0}$.

In the linear ECC framework, the smallest ε for which such an x exists is the *distance* of the code.

Max-Cut. Observe that the classical Max-Cut problem is a special case of the CVP-code problem. Recall that an instance of Max-Cut consists of a graph $G = (V, E)$, and the goal is to find a bipartition of the vertices, that is, $V = L \cup R$, such that as many as possible edges go across the cut. Here is a way to phrase the Max-Cut problem as an instance of linear equations: for each vertex $v \in V$, introduce a variable x_v, set the field to be $\mathbb{F}_2$. As for the matrix M, we take it from $\mathbb{F}_2^{m \times n}$ where $m = |E|, n = |V|$, and associate each row with an edge and each column with a vertex. We set $M[e, v] = 1$ if the vertex v is an endpoint of the edge e, otherwise set $M[e, v] = 0$. Observe that M's columns are now linearly independent. Finally, let the target vector be $t = \vec{1} \in \mathbb{F}^m$. It is easy to see that for an assignment $x \in \mathbb{F}^n$, the Hamming weight of $Mx - t$ represents the number of edges missing from the cut defined by the bipartition $L = \{v \mid x_v = 0\}$, $R = \{v \mid x_v = 1\}$. Therefore, minimizing this Hamming weight corresponds exactly to maximizing the size of the cut.

Compute **4** (Max-Cut). The *Max-Cut problem* is

- **Input:** a system of linear equations $[M, \vec{1}, \mathbb{F}_2]$ and distance $\varepsilon > 0$; all rows have exactly two 1 entries while all others are 0.
- **Goal:** is there a vector generated by M within distance ε from $t = \vec{1}$? (Alternatively, find the one closest to $\vec{1}$).

Claim A.1 [folklore] below establishes that Max-Cut is NP-hard. Since this is a special case of CVP, the same is true for CVP.

Interestingly, a similar reduction applies when the field is $\mathbb{R}$ and the domain for the solution is $\mathbb{Z}$.

3.2. Complexity of approximation

An approximation algorithm abandons the option of finding an assignment that leaves the smallest possible number of equations unsatisfied. It settles instead for an assignment that leaves a larger fraction of the equations unsatisfied; still, the number of unsatisfied equations is within the ratio of approximation times the optimal.

As it turns out, however, it is hard to distinguish even between the two extreme cases: the case where almost all equations can be simultaneously satisfied versus the case where almost none can be satisfied.

To prove an approximation problem hard, one has to define a problem that is close in structure to a decision problem, nevertheless, whose hardness implies hardness of approximation. Such problems form a *gap problem* where the algorithm is required to distinguish between two extreme cases for the value of the system, but is free to return an arbitrary outcome for the in-between values.

Compute **5** (Gap CVP-code). The *Closest-Vector gap[$\varepsilon, 1-\varepsilon$] over linear-ECCs* problem is

- **Input:** a system of linear equations $[M, t, \mathbb{F}]$.
- **Goal:** distinguish between
  - **Accept:** $\mathsf{val}[M, t, \mathbb{F}] \leq \varepsilon$,
  - **Reject:** $\mathsf{val}[M, t, \mathbb{F}] > 1-\varepsilon$.

Note that in a *gap problem* the algorithm is free to accept or reject all inputs that fall within the gap; in the gap-CVP case, these are equation systems $\Psi = [M, t, \mathbb{F}]$ whose value $\mathsf{val}[\Psi]$ is between the two thresholds, $\varepsilon < \mathsf{val}[\Psi] \leq 1-\varepsilon$.

Claim 3.1. *For any $\varepsilon > 0$, there is a large enough field $\mathbb{F}$ so that CVP-code* gap[$\varepsilon, 1-\varepsilon$] *is NP-hard.*

A proof can be found at Claim A.2 below.

We have just established that the gap version of the CVP problem is hard. Numerous other versions of these problems are known to be NP-hard. In fact, for most classical approximation problems, any improved ratio of approximation, beyond the best-known efficiently achievable, is NP-hard—as a consequence of the PCP theorem. There are only a handful of those approximation problems that do not have such a result established, and most of them do have some hardness established—albeit not NP-hardness. Their hardness is relative to a problem whose complexity is not yet established, namely UG—where the hardness of UG (see below) is possibly the most fundamental open problem of the field.

Unique-games (UG). When the equations are restricted to include only two variables—in particular, where every row of the matrix M is zero except for one 1 entry and one -1 entry—it becomes the infamous Unique-Games problem, which is not known to be NP-hard despite strenuous efforts to prove such a statement:

Compute **6** (UG). The *Unique-Games [UG]* $\mathtt{PCP}_{1\leftrightarrow 1}[\varepsilon, 1-\varepsilon]$ problem is

- **Input:** a system of linear equations $[M, t, \mathbb{F}]$ so that every equation is of the form
$$x_i - x_j = t_{ij}.$$
- **Goal:** distinguish between
  - **Accept:** $\mathsf{val}[M, t, \mathbb{F}] \leq \varepsilon$,
  - **Reject:** $\mathsf{val}[M, t, \mathbb{F}] > 1-\varepsilon$.

The syntactic restriction, of every equation consisting of the difference between 2 variables required to be a prescribed value, makes it a special case of CVP and thereby potentially easier than the general form. This opened the door for considerable attacks via sophisticated algorithms, which have succeeded in reducing the time it takes to solve UG to less than a trivial exponential of the size of the input. The infamous *UG conjecture* [47] states that the problem is (NP-)hard to compute.

A grueling effort of more than a decade of work culminated in 2018 in considerable progress on this nearly two-decade-old open problem; the progress concerns the resolution of the related 2-to-1-games conjecture (a problem similar to UG, but where each equation has two optional solutions):

Compute **7** (2-to-1 is hard). The *2-to-1-Games* $\mathtt{PCP}_{2\to 1}[\varepsilon, 1-\varepsilon]$ problem is

- **Input:** a set of linear sums $[M, \mathbb{F}]$ where all constraints are of the form

$$x_i - x_j \in \{t_{ij}, t'_{ij}\}.$$

- **Goal:** distinguish between
  - **Accept:** $\mathsf{val}[M, \mathbb{F}] \le \varepsilon$,
  - **Reject:** $\mathsf{val}[M, \mathbb{F}] > 1 - \varepsilon$.

Theorem 3.1 ([53]). *The 2-to-1-Games* $\mathtt{PCP}_{2\to 1}[\varepsilon, 1-\varepsilon]$ *is NP-hard.*

As a side note, the (highly complex) proof of Theorem 3.1 starts with hardness of, again, *linear equations* over $\mathbb{F}^2$ [43] and extend it considerably. This has discouraged research attempting to show an efficient algorithm for UG. A fundamental reason for this apparent disinclination is the observation that Theorem 3.1 implies that distinguishing between the value of a UG system of linear equations being $\le \frac{1}{2}$ versus it being $> 1 - \varepsilon$ is NP-hard:

Compute **8** ($\frac{1}{2}$UG). The *Unique-Games* $\mathtt{PCP}_{1\leftrightarrow 1}[\frac{1}{2}, 1-\varepsilon]$ problem is

- **Input:** a system of linear equations $[M, t, \mathbb{F}]$ so that every equation is of the form

$$x_i - x_j = t_{ij}.$$

- **Goal:** distinguish between
  - **Accept:** $\mathsf{val}[M, t, \mathbb{F}] \le \frac{1}{2}$,
  - **Reject:** $\mathsf{val}[M, t, \mathbb{F}] > 1 - \varepsilon$.

Corollary 3.2. $\mathtt{PCP}_{1\leftrightarrow 1}[\frac{1}{2}, 1-\varepsilon]$ *is NP-hard.*

Proof. First, turn each constraint of $\mathtt{PCP}_{2\to 1}$ into two equations, for each of the possible target values. This turns a system of value $1 - \varepsilon$ into a $\mathtt{PCP}_{1\leftrightarrow 1}$ system of linear equations of value $\frac{1-\varepsilon}{2}$. Then, add a small fraction of obviously satisfied linear equations to take care of the ε. ■

All the above-mentioned sophisticated algorithms proposed so far for UG, however, produce an assignment satisfying roughly the same fraction of the equations, assuming the value of the system is $\frac{1}{2}$ in the "accept" case or the case where the value is $1-\varepsilon$. Therefore, these algorithms are ill-equipped to resolve the UG-conjecture (unless $\mathrm{P}=\mathrm{NP}$) and that direction seems to have been completely abandoned.

3.3. General norm

The next paradigm shift involves altering the measure of the distance of a solution from a perfect solution, by generalizing the ℓ^0 distance (= Hamming distance, that is, the fraction of equations left unsolved) to any norm, for example, the Euclidean ℓ^2-norm or any ℓ^p-norm. This seemingly simple technical change has a considerable impact on the computational complexity of problems, as well as on the mathematics involved.

Cables and wires. Consider the ECC paradigm we have used so far; one can think of it as a way to correct messages transmitted over a cable, which comprises numerous wires. The simplistic approach we have considered so far is that each wire can carry a binary value, namely, can be in either of two options for each clock tic. A more realistic approach is where the wires can carry some specific voltage, depending on the value transmitted, however, there is no reason to limit the number of possible voltages to two. One can have numerous plausible values, each indicating a value of a predetermined range. In that case, the transmitted voltage might come out different at the other end of the wire, nevertheless, it is more likely to change a little than to change by much. Now, recovering the transmitted message, one better utilize this to facilitate a more efficient transmission. The error is thus measured by the distance between the set of voltages transmitted, on each wire, and those received, according to some natural distance (norm). The goal is to figure out the data most likely to have been transmitted, *even if all values have changed (not by much) in transmission.*

In other words, one may take a more general perspective, nonetheless, *over the same object*; instead of looking for the assignment with the smallest number of equations left unsatisfied, which amounts to looking for a vector x so that Mx is within a small radius ℓ^0-*ball* around t. The question becomes how small a radius ball contains a vector Mx, however, *according to a general norm*

$$\mathsf{val}[M,t,\mathbb{F}] \stackrel{\text{def}}{=} \min_{x\in\mathbb{F}^n} \|Mx-t\|,$$

where the norm $\|Mx-t\|$ could be any prescribed norm that the definition of the problem calls for. In other words, take the difference between the prescribed target values t and the actual values the assignment x gives to each sum, and measure them according to some norm such as the Euclidean (ℓ^2) norm or the taxi-cab (ℓ^1) norm, that is, instead of counting how many equations are unsatisfied, measure how close the assignment is to satisfying all the equations.

As a broader perspective, considered as a discrete subgroup of a metric space, the only thing that has changed here is that the metric is now allowed to be any general metric. The set of vectors would still, by default, be a sparse set within the metric.

Consequently, the solution, as well as the computational complexity of the problem, may be different depending on the chosen norm.

Norm over fields. Note that one must assume a well-defined norm over the elements of the field $\mathbb{F}$, and, consequently, over vectors of elements of the field, otherwise the definition would be meaningless. For example, for your typical fields, say $\mathbb{R}$ ($|\alpha|$) or $\mathbb{Z}_q$ ($\min\{|\alpha|, |q - \alpha|\}$ the ℓ^1-norm is naturally defined.

From the ECC perspective, the questions corresponding to classical linear equations ask for a legal codeword closest to a given vector t *according to the Hamming distance*: In the general-norm framework, one changes the distance applied in the last formulation, but otherwise considers the same question: pick your favorite norm ℓ^p, "Is there a legal codeword close enough to the word t?", which results in the following computational question:

$$\exists x : \|Mx - t\|_p \leq \varepsilon?$$

In other words, one may consider the CVP-code-search computational problem from above (**Compute** 1) with the more general definition of val, which, by default, is interpreted over the ℓ^2-norm (even though any other norm could work, and, in fact, ℓ^1 may be more natural in such settings). Accordingly, one could define the more general versions for all the above problems (versions of CVP and SVP).

3.4. Alternative measures

Another alternative type of problems concerning the same object (a system of linear equations) calls for a complete solution for all the equations, albeit optimizing a different parameter. Let us rank distinct all-satisfying assignments according to their norm as vectors $x \in \mathbb{F}^n$, each value being an element of the field. Once the norm of elements of the field has been defined, every assignment has a well-defined norm. We may now consider an algorithmic problem that calls for an all-satisfying assignment while minimizing the norm of the assignment.

The Shortest-Integer-Solution (SIS) problem. The resulting problem—which turns out to be quite important due to the implications discussed below—is the *Shortest-Integer-Solution (SIS) problem* which is another variation on finding a solution for a system of linear equations.

For SIS, the equations must all be solved, and the parameter that measures the quality of the solution is the norm of the assignment, so that the closer those numbers are to 0, the better the solution.

This computational problem is central to worst-to-average-case reductions, and thereby natural, when it comes to hardness assumptions, which facilitates cryptographic security. Indeed, assuming SIS hardness entails the security of some *One-Way-Functions (OWF)* and *Collision-Resistant Hash-functions (CRH)* as presented below.

Let us start with a basic definition. For a matrix over a field $\mathbb{F}$, the shortest nonzero vector in the kernel of M is naturally defined as

$$\mathcal{S}[M, \mathbb{F}] \stackrel{\text{def}}{=} \min_{\substack{\vec{0} \neq x \in \mathbb{F}^n \\ Mx = \vec{0}}} \|x\|.$$

The algorithmic approximation problem searches for a short vector in the kernel of M:

Compute 9 (SIS-worst-case). The *Shortest-Integer-Solution (SIS)* SIS$[M, \mathbb{F}, \kappa]$ is

- **Input:** a matrix $M \in \mathbb{F}^{m \times n}$.
- **Goal:** find a nonzero assignment $\vec{0} \neq x \in \mathbb{F}^n$ such that $Mx = \vec{0}$ and whose norm is small, namely $0 < \|x\| \leq \kappa$.

For the last part of the definition, relating to the norm of the assignment, one may want to consider the field $\mathbb{Z}_q$, for a prime q, and a norm ℓ_p, in which case the problem is well-defined.

Therefore, the algorithmic problem is to find a relatively short solution, under the assumption that a short solution exists. As will become clear below, coming up with such an algorithm entails a solution to other classical lattice problems and, moreover, to the worst-case versions of those problems, making it natural to assume that SIS is hard.

3.5. Average-case complexity

Average-case complexity is fundamental in terms of complexity theory: we wish we had the proper tools to analyze problems, not in the worst case, but on average, over some prescribed distribution. By default, the complexity of a computational problem (for instance, the time complexity, that is, the number of steps it takes for the algorithm to solve the problem) is regarded as the worst-case complexity: given an algorithm that solves the problem, one measures, for any input length n, how long it takes for the algorithm to solve *any* input of length n. That function, from n to the upper bound on the number of steps, over all inputs of length n, is the time complexity of the algorithm. Average-case complexity, in contrast, calculates the complexity according to the number of steps that would suffice to solve inputs of some length n with high probability, or the average time it would take over inputs (according to some prescribed distribution). Average-case complexity makes more sense in general; unfortunately, our tools are insufficient for a coherent theory of the relationships between classes of average-case problems.

Formally, there is a complexity function $t(n)$ and a small parameter ε so that, for distribution Λ_n over all inputs of length n,

$$\mathop{\mathbb{E}}_{x \sim \Lambda_n} \big[\text{the time it takes for } A(x) \text{ to solve the problem}\big] \leq t(n).$$

Note that the complexity of a problem could be drastically different in the average-case compared to the worst-case. Take, for example, the Factoring problem: given an n-digit number, find out its prime-number factorization. This problem is easy to compute on average over the uniform distribution, as most numbers have small factors; in contrast, numerous security assertions rely on the problem to be hard in the worst-case, and, in fact, even on average, on some particular distributions. Nevertheless, no distribution is known over which Factoring is hard.[2] Another striking example is SAT. It is one of the basic problems shown hard to compute in the worst-case. On average, some distributions are known to follow a sharp-threshold, namely, where increasing the parameter causes the resulting formula to be satisfiable until some critical parameter when it suddenly becomes nonsatisfiable.

3.5.1. SIS

The first problem considered herein under this regime is SIS; unlike the above definition, let us consider the average-case version of the problem where instances are drawn from a distribution, and the computational complexity relates to how long it takes to solve the problem on average or with high probability:

Compute **10** (SIS-average-case). The *Shortest-Integer-Solution (SIS)* in the average-case regime is a computational problem with parameters $\mathrm{SIS}[n, m, \mathbb{F}, \kappa]$ as follows:

- **Input:** a uniformly random matrix $M \in_R \mathbb{F}^{m \times n}$.
- **Goal:** find a nonzero assignment $\vec{0} \neq x \in \mathbb{F}^n$ such that $Mx = \vec{0}$ and whose norm is small, namely $0 < \|x\| \leq \kappa$.

3.5.2. Learning-with-Errors

This is another average-case problem that is even more important than SIS. In fact, most of the recent—quite fantastic—cryptographic primitives have their proof of security that relies on the (presumed) hardness of this problem.

The Learning-with-Errors (LWE) problem was introduced by Regev in 2005 [77] and is another variant of the system of linear equations problem, in fact, it is in some sense a special case of the CVP-code-search from above.

This problem introduces a twist, as each instance is made up of two parts where one is a randomly chosen secret target vector, while the other is a system of linear equations chosen according to the distribution. The former remains a secret while the latter is made public. The goal of the algorithm is to figure out the secret.

In particular, following the linear equation framework, the secret is a vector $x \in_R \mathbb{F}^n$. Next, a random set of homogeneous linear sums is chosen as well—the matrix M—independently of x, which is made public. Lastly, the evaluation of these sums over the secret vector—only perturbed by some small noise—becomes public, too.

Routinely, one assumes a set of n variables $\vec{x}$, over a field $\mathbb{Z}_q$ for a prime number $q = q(n)$ so that $q \gg n$. In addition, let us assume a (noise) distribution ν over $[-\kappa, \kappa]$ for some $\kappa \ll q$.

---

2 Interestingly, Factoring, in its decision version, is known to be both in NP and in coNP. Hence, it is unlikely to be NP-hard as this would imply NP=coNP.

Compute **11** (LWE, search). The $\mathrm{LWE}_{\nu}[n, m, \mathbb{Z}_q, \kappa]$ is an average-case problem where

- **Secret:** a chosen secret random vector $x \in_R \mathbb{Z}_q^n$.
- **Input:** a random set of linear sums, namely, a matrix $M \in_R \mathbb{Z}_q^{m \times n}$; further, a vector of *noisy evaluations* $t = Mx + e$ where $e \sim \nu^m$ (that is, one applies the noise ν independently for each linear sum).
- **Goal:** find the secret vector x (given M and t).

The computational goal is to find the secret vector, however, for the approximate version, it suffices to find a vector x' so that $\|Mx' - t\| \leq \kappa$.

An algorithm for CVP-code clearly solves LWE—given M and t as input. This problem is a substantial special case for two reasons: first, it is an average-case problem, and, second, the target vector is not arbitrary, but is rather chosen randomly, via a known noise distribution. Nevertheless, Regev showed LWE to be as hard as the general case (of SVP), thereby providing strong evidence for it being hard, and allowing its (presumed) hardness to be extensively utilized for the purpose of establishing cryptographic security.

One may assume that the number of equations is rather large $m \gg n$ compared to the number of variables; thus, information-wise, there should be a unique solution for x that brings the linear sums close to the noisy evaluations of t. In other words, the algorithm must overcome the error ($e \sim \nu^m$) introduced when choosing t in order to find a vector x' that brings the product with the matrix M not too far from the public (slightly perturbed) target t, i.e., so that $\|Mx' - t\|$ is small.

This problem is shown to be as hard as solving worst-case version of lattice problems, hence, one can rather safely assume that it is hard and, consequently, that it is safe to construct cryptographic protocols whose security relies on the hardness of LWE.

3.6. Discrete subgroup—Lattices

So far, the natural cause for the set of selected vectors being sparse within the general domain is the difference in dimension $m \gg n$. Let us now generalize this notion to produce a natural subdomain regardless of that difference. In particular, one may even consider the cases where the matrix has full rank, namely a set of independent vectors where $m = n$. The computational problems as defined thus far become far easier since there is always a good solution to all equations.

Consequently, let us require the solution (the values assigned to the variables) be not arbitrary values in the field, but rather contained in a restricted sparse subset of the field. The resulting construct is a discrete subgroup, i.e., a lattice that is sparse within the general metric (over $\mathbb{F}^m$); if the subdomain of the assignment is discrete, the resulting subgroup is also discrete, even if the field is not.

In what follows, we assume the matrix is of full rank, for exposition purposes, so that it is clear that the sparseness of a lattice does not stem from the global dimension m being quite larger than the domain of the subgroup. Note, however, that lattices could be similarly defined more generally for non-full-rank matrices.

Definition 3.2. A *lattice* is defined by a (full-rank) matrix (called a basis) $M \in \mathbb{F}^{n \times n}$ over a field $\mathbb{F}$ and a subdomain $\mathcal{F} \subset \mathbb{F}$. The lattice comprises vectors of the form below:

$$\mathcal{L}[M] \stackrel{\text{def}}{=} \{Mx \mid x \in \mathcal{F}^n\}.$$

An extensive mathematical background on lattices appears in Appendix B. We may now consider computational problems over lattices, similar to that we have considered so far. These serve not only as a new perspective on the complexity of computational problems, but also provide further information about the complexity of problems in the average-case. This calls for fundamentally different mathematical analysis and has caused a tectonic shift in cryptography, which is always in need of hardness assumptions for average-case problems on which to base the security of protocols.

The computational problems call for a restricted type of assignments to satisfy the equations. Once the domain of x is sparse within the general domain, the computational and mathematical problems become more interesting. The most natural restriction in general (and in particular throughout our analysis of lattices below) is when the domain of M is the reals $\mathbb{R}$, while the domain of the assignment x comprises the integers $\mathbb{Z}$.

Computational problems over lattices, in particular CVP, consider a lattice plus a target vector $[M, t, \mathbb{F}, \mathcal{F}]$. By default, $\mathcal{F}$ is sparse in $\mathbb{F}$ and $\mathcal{F}$ conforms to $\mathbb{F}$'s structure, for instance, $\mathcal{F}$ could be a ring: the archetypal example is where $\mathbb{F} = \mathbb{R}$ and $\mathcal{F} = \mathbb{Z}$. Now, the distance of the closest vector to a target vector t in this context is the smallest radius of a ball around t which contains a vector of Mx; however, in this case, x is restricted to be a vector over $\mathcal{F}$:

$$\mathsf{val}[M, t, \mathbb{F}, \mathcal{F}] \stackrel{\text{def}}{=} \min_{x \in \mathcal{F}^n} \|Mx - t\|.$$

Restricting the domain and insisting on x taking this form allows the fantastic phenomena (as described extensively below) associated with lattices, in particular the intricate dependency of the computational complexity of problems on the ratio of approximation in quest.

Compute **12** (CVP-lattices). The *Closest-Vector problem*, over $\mathbb{F}, \mathcal{F}$, is as follows:

- **Input:** a lattice basis/matrix $M \in \mathbb{F}^{n \times n}$, a target vector $t \in \mathbb{F}^n$, and distance $\varepsilon > 0$.
- **Goal:** find a vector Mx for some $x \in \mathcal{F}^n$ within distance ε from t, that is, where $\|Mx - t\| \le \varepsilon$.

This small amendment to the parameter optimized by the algorithmic problem translates to a striking difference with regards to the complexity of these problems. It also diverts the mathematics involved in a completely different direction, opening new and exciting paths of research on interesting relationships between these problems and other variants, as well as on the complexity classes in which they reside.

3.7. Worst-to-average-case

Many classical problems, including some of the above mentioned, are known to be hard, typically NP-hard. Hence, an efficient solution for them implies an efficient solution for the entire class NP thus P = NP. Some classical problems, however, are not known to be NP-hard, nevertheless, no efficient algorithm is known for them. Their complexity is unclear and are thus a great source of fundamental open research questions regarding their complexity. One option not to be overlooked is that some of these problems may form their own class, which could lie between P (namely, efficiently solvable) and NP-hard. Some of these options will be discussed in the next section.

This, however, is all with respect to a problem's worst-case complexity. When it comes to hardness for the same problem's in the average-case, very little is known. It is unlikely that one could show NP-hardness for an average-case problem, as finding a probability distribution that assigns nonnegligible probability to the set of hard cases is a step towards solving the problem efficiently.

So how does one show hardness on average? For starters, as just observed, one should show hardness relative to a class or a problem different than NP, in particular, relative to a problem unlikely to be NP-hard. Still, what could be a good source for a hard problem that is not NP-hard?

What Ajtai suggested in his groundbreaking paper [4] is to assume that weak approximation of the classical computational problems regarding general lattice (such as CVP and SVP) are hard, and then come up with a *worst-to-average-case reduction*, which would establish that an efficient solution for an average-case problem (such as SIS) entails an efficient solution for a worst-case lattice problem.

Note that this a priori is a surprising concept: reductions from problem A to problem B utilize a procedure for B to solve A. It reads A's input and calls on a procedure for B with instances that encodes A's input so that altogether one can derive a solution for problem A on its input. Here, one is expected to reduce the problem, given an arbitrary input, to random instances that seemingly have no relation to the original input.

Ajtai, in his field-transforming mathematical breakthrough, showed that a version of SVP, the *unique-Shortest-Vector problem (uSVP)* (namely, when $\lambda_2 \gg \lambda_1$) can be reduced to a problem similar to the SIS-average-case problem:

Theorem 3.2 (SIS is hard [4, 66]). *SIS-average-case is as hard as uSVP* [4] *and SIVP* [66].

In terms of mathematical ideas and strategy, this emerging area relies heavily on a program laid out by the great Minkowski; this includes his original theorems, as well as their extensions. These are covered below: the mathematics involved is discussed in Appendix B while some conjectures that could greatly enhance our understanding and abilities in figuring out the complexity of these problems can be found in the next section, as well as in Section 4.3.

Regev's LWE problem. Ajtai's worst-to-average-case reduction to SIS is, nevertheless, not the most natural for applications to cryptography, which prefers computational problems more like CVP—so that the target vector can be kept secret while the matrix is public. One of the most useful theorems—whose applications are highly ubiquitous in cryptography—is the other fundamental worst-to-average-case reduction by Regev [77], who showed a reduction[3] of a similar lattice problem to the LWE problem. Regev came up with a reduction from GapSVP and SIVP (a lattice problem that computes a set of short linearly independent lattice vectors) to DGS problem—the problem of sampling from the Gaussian distribution over a lattice; he proceeded that with a quantum reduction from DGS to the LWE problem. One of the drawbacks, still a wonder, is that this reduction calls for the Quantum Fourier Transform—which is the basic advantage of quantum algorithms that Shor's algorithm **[81]** relies on.

By establishing hardness of LWE based on worst-case hardness, Regev constructed a simple and elegant public key cryptography system. Consequently, the security of many (if not almost all) important cryptographic primitives was proved building on this assumption: digital signatures **[63]**, fully homomorphic encryption **[20]**, and many more.

4. PLAN: IN A CLASS OF THEIR OWN

In the grand project of classifying computational problems into feasible or infeasible, one of the remaining fundamental tasks is to identify problems that are in-between, that is, problems that are neither efficiently solvable nor NP-hard. Let us hence define the class of computational problems:

Definition 4.1. The class *NP-intermediate* $\mathtt{NPI} \overset{\text{def}}{=} \{A$ unsolvable in polynomial-time and $A \notin$ NP-hard$\}$.

Our understanding of the principles that govern such problems and this class is limited at best. Here is the immediate caveat: if a problem is proven to be in that class, it would imply P $\neq$ NP; so our goal is to just give some evidence that a problem belongs in that class. The best evidence is if an apparently hard problem is in NP $\cap$ coNP, thus it being NP-hard implies NP $=$ coNP. Nevertheless, taking into consideration that some of the problems whose complexity is not yet settled could be in that class, one should develop methodologies to enable such evidences. Some of these problems may be extra related so as to make up a class of their own, in which case one hopes for a computational problem complete for that class. The plan is to identify such subclasses, if they exist, and find more connections between various such problems.

In particular, let us consider approximation problems. Some optimization problems are NP-hard, hence, one tries to approximate them. It is naturally the case that some weak ratios of approximation are known to be feasible while some stronger are known to be NP-hard. For most classical problems, improving even slightly on the ratio known to be feasible is NP-hard; while, for a selected few, there is no known precise ratio of approximation at which the problem immediately switches from feasible to NP-hard.

Computational problems over lattices could serve a crucial role in giving evidence to a problem in or out of the $\mathtt{NPI}$ class and in general exhibiting connections between problems:

3 Albeit, a quantum one—this is a bug not a feature: presuming LWE is found to be easy, this implies an efficient quantum algorithm, not a classical one.

- The first project is to establish up to which ratio these few still unresolved approximation problems are NP-hard, hopefully obtaining tight infeasibility results. This would most likely require novel PCP-reductions, as current technology has not shown any progress on this for nearly two decades.

- Once novel PCP-reductions are developed, these techniques may facilitate reducing apparently hard problems to weaker ratios of approximation for lattice problems, hopefully, to within ratios for which these lattice problems are known to be in coNP (beyond square-root of the dimension). If the reduction manages to cross that threshold, it would prove that the original problem is in coNP, which is the best evidence for not being NP-hard, therefore a great evidence to being in `NPI`.

- The next project is to extend the scope of the worst-to-average-case reductions. These would require stronger properties of lattices related to Minkowski's theory and, in particular, to reverse Minkowski theorems. With regards to computational problems, it may be quite productive to identify a restricted class of lattices, for which stronger Minkowski-type theorems are true, allowing strong connections with regards to the computational complexity of those problems.

- Lattice-based cryptography relies on the hardness of lattice problems and almost invariably on LWE presumed hard. Richer set of hardness assumptions could broaden and provide more general proofs of security for protocols.

- To complete the picture, note that it is quite possible that some of these lattice problems can be efficiently approximated to within polynomial ratios by a quantum machine (compared to the exponential ratio of the LLL algorithm). Shor's algorithm considers the multiplicative group modulo the number N and utilizes quantum Fourier transform to find cyclic subgroups of it, thereby factoring N. Minkowski developed lattice theory as the "Geometry of Numbers" as it generalizes phenomena related to numbers to high dimensions. It would be only fitting to apply quantum Fourier transform on a lattice $\mathcal{L}$ to find cyclic subgroups within it, thereby finding (or at least approximating) the best basis for the lattice $\mathcal{L}$. If true, this would drastically transform our perspective on compatibility, as the class of feasible computational problems would include much more than those solvable by efficient (random) algorithms. It would mean that the natural process of solving an algorithmic problem would be to translate it into approximating, say, SVP (possibly on average) then applying a quantum machine to solve that.

Let us therefore point out some conjectures and questions that could advance this grand plan. The most fundamental problem in this regards is the Unique-Games [47]. If proven NP-hard, it will immediately imply tight infeasibility results for a large class of approximation problems [51,56,71,73]. Furthermore, it will draw a very strict border of computational infeasibility of approximation and show that, for practically all classical problems, it is infeasible to improve even slightly on the best known approximation algorithm, assuming $P \neq NP$.

Moreover, just improving beyond the current $\frac{1}{2}$ barrier [28, 29, 52, 53] would be a tremendous achievement. Namely, showing that the following algorithmic problem is NP-hard: given a Unique-Games instance promised to be 0.51-satisfiable (namely, of value 0.49), satisfy a nonnegligible fraction of the constraints:

Question 4.1. Is there $c < \frac{1}{2}$, so that $\forall \varepsilon > 0$ there exists an alphabet size $q = q(\varepsilon)$ such that $\mathtt{PCP}_{1\leftrightarrow 1}[c, 1-\varepsilon]$ is NP-hard?

Already here, one of the options that should be given serious consideration is that the Unique-Games problem is not NP-hard, yet neither in P:

Question 4.2. Is there, $\forall \varepsilon > 0$, an alphabet size $q = q(\varepsilon)$ such that $\mathtt{PCP}^{q}_{1\leftrightarrow 1}[\varepsilon, 1-\varepsilon] \in \mathtt{NPI}$?

That is, one may be able to show that the problem is in coNP, which would be quite earth-shuttering:

Question 4.3. Is it true that $\mathtt{PCP}^{q}_{1\leftrightarrow 1}[\varepsilon, 1-\varepsilon] \in \text{coNP}$?

Another manner of giving an evidence of a problem residing in NPI—besides proving a problem is in coNP (as an evidence it is not NP-hard)—is to show connection between problems in that class by reducing one to another.

Thus, naturally, the question arises: For those problems in coNP and NP, how does one establish hardness?

A prime candidate to belong to the class NPI and show it is related to other problems is the canonical Factoring problem, where, given a (binary representation) of a natural number, the algorithmic goal is to find its representation via prime factorization. It is quite likely that this problem is in NPI, furthermore, one should hope that its complexity can be related to lattice problems [1].

Lattices. The complexity of computational problems over lattices is wide open, nevertheless, there are some striking results already established. These problems are interesting in their own right but, furthermore, have the capacity to facilitate results that give evidence to problems being in NPI. Beyond the Factoring problem above, other problems could benefit from being reduced to an approximation variant of a lattice problem.

There could be two purposes for such a reduction: either to establish hardness of the lattice problem, or to show the problem resides in coNP. To show that the lattice problem is hard, one reduces a problem known to be NP-hard to it. To show that the original problem is in coNP, reduce it to a lattice problem, for a weak enough approximation (square-root of the dimension)—these are known to be in NP ∩ coNP. The known complexity results for lattice problems are summarized in Figure 1.

The LLL algorithm [61], including Schnorr's refinement [79] and Babai's extension [15] to CVP, give the exponential upper bound on the right. The hardness for SVP is

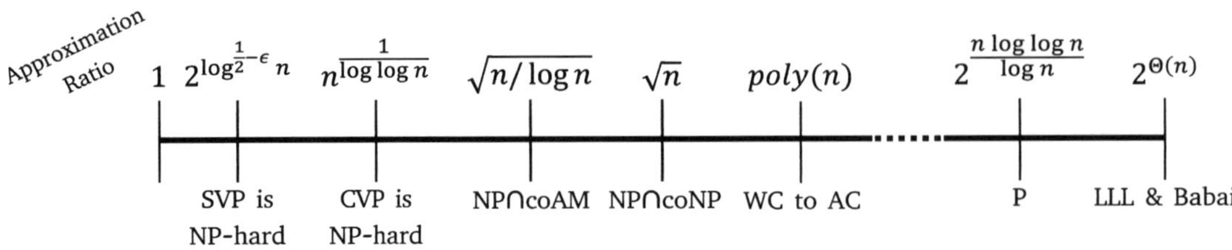

FIGURE 1
Complexity scale of lattice problems for approximation ratios

by Khot [49] following the original result by Micciancio [65]; while the hardness for CVP is from [30].

Let us go on the edge here and conjecture that both SVP and CVP are NP-hard to approximate to within a ratio of almost $\sqrt{n}$:

Conjecture 4.4. *Both CVP and SVP problems are NP-hard to approximate within a ratio of* $\tilde{o}(\sqrt{n})$.

Beyond the $\sqrt{n}$ ratio of approximation, these problems are in coNP[4] (and are clearly in NP) [2, 39]. Thus, Conjecture 4.4 asserts that CVP and SVP are NP-hard almost to the approximation ratio beyond which, if proven NP-hard, it would imply that NP = coNP. Proving these problems are hard almost to the ratios they are known to be in coNP is the holy grail of the field, nevertheless, this direction has seen no progress for almost two decades. A proof for any of these would be monumental and immensely extend our understanding of the complexity of lattice problems. It would also naturally require new types of PCP reductions:

Conjecture 4.5. *Approximating CVP/SVP to within $c\sqrt{n}$ (for some global constant c) is in* $\mathtt{NPI}$.

These conjectures are all important and with far-reaching consequences. Nevertheless, the connection between classical conjectures (such as UGC) and those related to lattices has not yet been established. Such potential connections are explored next—we discuss some conjectures that could illuminate the tight interplay between classical infeasibility and the complexity of lattice problems.

4.1. Break the soundness barrier

There are two major outstanding open research questions regarding classical gap problems whose resolution would greatly deepen our understanding of the complexity of approximation; these are the unique-games, $\mathtt{PCP}_{1\leftrightarrow 1}$, and 2-to-1-games, $\mathtt{PCP}_{2\to 1}$. There are two major versions for each with distinct characteristics, depending on the placement of the gap: either the assumption is that, in the "accept" case, all constraints are satisfied (also

4 Recall that coNP is the class of problems whose complements are in NP.

known as perfect completeness) or where the assumption is that all but a small fraction are satisfied. In $\text{PCP}_{1\leftrightarrow 1}$, when the assumption is that all are satisfied, the problem becomes trivially feasible (via Gaussian elimination). In the $\text{PCP}_{2\to 1}$ problem, when the assumption is that, in the "accept" case, only almost all constraints are satisfied, even satisfying a negligible fraction of the constraints (in the "reject" case) is NP-hard [53].

And thus, the only two versions to further investigate, regarding the "accept" case are:

(1) $\text{PCP}_{1\leftrightarrow 1}[\frac{1}{2}-\varepsilon, 1-\varepsilon]$,

(2) $\text{PCP}_{2\to 1}[0,\sigma]$.

Note that the $\text{PCP}_{1\leftrightarrow 1}$ problem has been discussed above and, despite the progress made in the last few years, it still seems wide open; as mentioned above, even improving the "accept" case beyond one-half is an open Conjecture 4.1.

Let us then direct our attention to the $\text{PCP}_{2\to 1}$ with perfect completeness. Its complexity is not only fundamentally interesting but—as will be discussed shortly—could prove fruitful in uncovering deep connections between classical PCP and lattice problems, which in turn could have a huge impact on cryptography.

Consider then the stronger promise regarding the "accept" instances, namely, that there is a solution satisfying *all* the constraints. Such a promise makes the problem easier, moreover, adapting the reduction of [53] (for the harder case where the promise is only that there is an assignment satisfying *almost* all constraints) fails fundamentally.

Khot conjectures that $\text{PCP}_{2\to 1}$ is hard even when trying to satisfy a small fraction of the constraints:

Conjecture 4.6 (Khot [47]). *For any field size q, there exists $\varepsilon(q) > 0$, where $\lim_{q\to\infty}\varepsilon(q) = 0$, for which $\text{PCP}_{2\to 1}[0, 1-\varepsilon(q)]$ is NP-hard.*

One should note here that, for σ close enough to 0, there is a folklore theorem indicating that the problem $\text{PCP}_{2\to 1}[0,\sigma]$ is NP-hard (in words, given an instance that can be fully satisfied, finding an assignment that satisfies almost all constraints is NP-hard). Nothing so far is known when the fraction of satisfiable constraints in the "reject" case becomes quite smaller than 1.

And here comes the central question in this context: What is the complexity of the $\text{PCP}_{2\to 1}$ problem when the algorithm needs to distinguish between the case where there is an all-satisfying assignment versus the case where not even a tiny fraction of the constraints can be simultaneously satisfied?

Let us boldly go against the "public opinion" here and suggest:

Conjecture 4.7. *For any alphabet size q, let $\varepsilon_P > 0$ be the largest number so that $\text{PCP}_{2\to 1}[0, 1-\varepsilon] \in P$; there is $\varepsilon_0(q) \gg \varepsilon_P$ so that for all $\varepsilon < \varepsilon_0(q)$,*

$$\text{PCP}_{2\to 1}[0, 1-\varepsilon] \in \text{NP} \cap \text{coNP}.$$

In words, it is clear that all these variants of $\texttt{PCP}_{2\to 1}$ are in NP; the conjecture claims that once the goal is to satisfy only a very small fraction, the problem could be in coNP as well.

If true, the $\texttt{PCP}_{2\to 1}$ problem with such parameters could be a perfect computational problem to assume being hard and build on it to prove other problems in $\texttt{NPI}$ are as hard. As discussed above, cryptography needs problems in NP $\cap$ coNP that can be safely assumed hard, hence, a plausible goal is a worst-to-average-case reduction relying on this hardness assumption.

4.2. Through lattices

Now, how does one go about proving such a statement? The most natural strategy—in light of progress in the last two decades—is to come up with a reduction from this problem to approximating CVP (or SVP) to within a ratio larger (weaker) than the square root of the dimension:

Conjecture 4.8. *Approximating* CVP *to within* $\Omega(\sqrt{n})$ *(where* n *is the lattice dimension) is as hard as solving the* $\texttt{PCP}_{2\to 1}[0, 1-\varepsilon]$ *for small enough* $\varepsilon < \varepsilon(q)$ *(*q *is the size of the field/alphabet).*

The reason Conjecture 4.8 suffices for proving Conjecture 4.6 is that approximating CVP to within such a ratio is known to be in coNP (and is trivially in NP) [2, 39].

Consequently, one may contemplate a more comfortable hardness assumption to base cryptography on, in particular to base the hardness of LWE on this type of assumption:

Question 4.9. Is LWE as hard as approximating 2-to-1-games? That is, assuming an efficient algorithm for LWE, can one solve $\texttt{PCP}_{2\to 1}[0, 1-\varepsilon]$ for small enough ε, in particular for any $\varepsilon < \varepsilon_{\mathrm{NP}}$ where $\varepsilon_{\mathrm{NP}}$ is the smallest so that $\texttt{PCP}_{2\to 1}[0, 1-\varepsilon_{\mathrm{NP}}]$ is NP-hard.

This would immediately allow cryptography to rely on the hardness of 2-to-1-games (albeit, with parameters as stated above) for security proofs. Furthermore, assuming such a reduction is exhibited, one could consider other cryptographic hardness assumptions (preferably those that are not yet established) and try prove them being as hard as 2-to-1-games.

The assumption that the 2-to-1-games problem is infeasible seems more natural, hence the security of a cryptographic protocol based on its worst-case hardness seems more reliable (than being based on the infeasibility of a problem like the unique-SVP). Still, even those who trust the latter assumption would welcome another option for proving security.

Here is an example of such a question: Could the existence of bilinear (or 3-linear) maps be based on the hardness of $\texttt{PCP}_{2\to 1}$?

4.3. What's hard with lattices?

The fundamental, mathematical principles involved in lattices are crucial in various distinct fields. The theorems proven by Minkowski [67] more than a century ago and the research program he laid out continue to be very relevant. Some aspects affect the complexity of computational problems over lattices, in particular their average-case complexity, with natural applications to cryptography.

Naturally, the same lattice may be represented by numerous bases. Some properties are abstract and true regardless of the particular basis while some others depend on the basis. From computation's perspective, the lattice is given via a particular basis and the complexity of the computational problem depends fundamentally on the specific basis given as input. If that basis is close to orthogonal, a problem such as SVP could be computed efficiently, while, given an arbitrary basis, the same problem may turn out to be infeasible.

In other words, the hardness of computational problems over lattices stems from the hardness of finding a "good" basis given an arbitrary one. How could one establish such a statement (without proving computational lower bounds)? One could show relative hardness (that is, a reduction) and prove that efficient algorithm for, say, SVP could be utilized to efficiently transform a bad basis for a lattice to the best one, or at least to one that approximates the best one.

One property of a lattice $\mathcal{L}$, which is clearly independent of the choice of basis, is its "orthonormality," that is, how close to orthonormal the "most orthonormal" basis of $\mathcal{L}$ is. For standard lattices—namely, lattices which have a basis that achieve the *successive minima* (see Definition B.8 below)—one could examine this property via the minima: the more similar the lengths of the basis vectors to the minima, the closer to orthonormal the lattice.

Best lattice basis. There are a few distinct (not necessarily equivalent) plausible definitions for the best basis for a lattice; these formulate different manners by which to assert that the basis is either *short* or close as possible to being *orthogonal*. Several options exist for how to define what it means for a basis to be short. It could be the length of the longest vector, the sum of lengths, or the product of lengths. The discussion below is quite abstract, hence could apply to any of these definitions.

Consider the following *greedy* algorithm (not necessarily efficient) to find such a basis: add to the basis the shortest vector in the lattice, and then the next shortest vector, albeit one that is linearly independent of previous basis vectors (that is, not in the span of current basis), etc. Those vectors achieve, in terms of norms, the successive minima. Nevertheless, they do not necessarily form a basis for the lattice; in fact, there are canonical examples of lattices that have no basis achieving the successive minima (any such set of vectors spans a sublattice). Naturally, one should look for a full characterization of nonstandard lattices.

Question 4.10. Is it possible to characterize *nonstandard* lattices, that is, lattices which have no basis achieving their successive minima?

Question 4.11. Let $\mathcal{L}$ be a nonstandard lattice and consider various sequences of linearly independent vectors achieving the successive minima. Is the sublattice $\mathcal{L}' \subset \mathcal{L}$ generated by each of those sequences the same?

Question 4.12. Can one bound from above the length of the vectors in the shortest basis, relative to the successive minima?

Such a characterization could greatly improve our understanding of nonstandard lattices. Moreover, they may facilitate improving our technique regarding compatibility issues.

Another attempt to construct the best basis for a lattice is the (not necessarily efficient) algorithm of Korkin–Zoltriev [60]. The resulting basis is indeed quite short, however, it is clearly not the shortest (whichever reasonable definition of "short" one is interested in): there exists a lattice $\mathcal{L}$ such that we can perform an elementary operation over the Korkin–Zoltriev basis which makes the basis shorter while spanning $\mathcal{L}$.

Therefore, one should first sort out the plausible definition of a global *shortest* basis. Once that has been clarified, the natural question is

Question 4.13. Is CVP/SVP as hard as finding the shortest basis (given an arbitrary one)? Or at least approximating the shortest one?

In reverse. Consider the *discrete-Gaussian* distribution over a lattice with parameter σ as follows. First, choose a random vector according to the Gaussian distribution, with mean 0 and standard deviation σ. Then, round each point in a parallelepiped to the *canonical* vertex of that parallelepiped.

If σ is large enough, the discrete nature of the lattice becomes negligible. That is, the probability of being outside a relatively large ball (large with respect to the standard deviation of the Gaussian) in the discrete-Gaussian is comparable to the Gaussian. A somewhat related phenomenon is that for a large enough σ, if one chooses a random point according to the Gaussian distribution, then, taking it modulo a fundamental parallelepiped, the resulting distribution is very close to uniform.

Worst-to-average-case reductions fundamentally rely on the latter property. This is how the magic of removing any reference to the original input takes place; one can call on the average-case algorithm only giving the vectors modulo the parallelepiped, and use the outcome, knowing the entire vectors.

Stepping back, one should investigate the mathematics of lattices to figure out how large σ needs to be for the Gaussian to satisfy that property (the complexity heavily depends on that stretch). This was addressed by Miccancio and Regev in their fundamental paper [66]. Reverse Minkowski theorems play a central role when trying to either approximate lattice problems, or, alternatively, when trying to translate solving them into another problem, to establish its relevant hardness (these two processes are similar).

Minkowski's convex body theorem [67] states that an upper bound on the determinant of a lattice implies a lower bound on short vectors, that is, that global density implies local density. A reverse Minkowski theorem, proven in [78], states that a lower bound on the determinant of a lattice and its sublattices (intersection of $\mathcal{L}$ with a lattice subspace) implies an upper bound on the amount of short points, i.e., that local density implies global density on a sublattice. Note that local density does not in general imply global density if the lattice is long and narrow. This leads us to define the class of stable lattices.

Definition 4.2. A *stable lattice* $\mathcal{L}$ is a lattice such that $\mathsf{det}[\mathcal{L}] = 1$ and for any $\mathcal{L}' \subset \mathcal{L}$ (intersection of $\mathcal{L}$ with a subspace) it holds that $\mathsf{det}[\mathcal{L}'] \geq 1$.

Theorem 4.1 (Reverse Minkowski [78]). *For any stable lattice $\mathcal{L} \subset \mathbb{R}^n$, $\rho_{1/t}(\mathcal{L}) \leq 3/2$, where $t := 10(\log n + 2)$. Here, for a lattice $\mathcal{L} \subset \mathbb{R}^n$ and $s > 0$, $\rho_s(\mathcal{L}) := \sum_{y \in \mathcal{L}} e^{-\pi \|y\|^2/s^2}$ is the Gaussian mass of the lattice with parameter s.*

This theorem allows one to derive a bound on the number of short vectors in the lattice—this is the reason this is referred to theorem as "Reverse Minkowski." Another corollary of that paper is with regards to the covering radius parameter of stable lattices.

Definition 4.3. For a lattice $\mathcal{L} \subset \mathbb{R}^n$, define the *covering radius* $\mu(\mathcal{L})$ as

$$\mu(\mathcal{L}) := \max_{t \in \mathbb{R}^n} \operatorname{dist}(t, \mathcal{L}) = \min\{r \mid \mathcal{L} + \mathcal{B}(r) = \mathbb{R}^n\}.$$

Regev and Stephens-Davidowitz [78] showed that for any stable lattice it holds that $\mu(\mathcal{L}) \leq 4\sqrt{n}(\log n + 10)$. One important open question regarding stable lattices is the Shapira–Weiss conjecture which aims to prove a stronger result.

Conjecture 4.14 ([80]). *Let $\mathcal{L} \subset \mathbb{R}^n$ be stable lattice. Then $\mu(\mathcal{L}) \leq \mu(\mathbb{Z}^n) = \frac{1}{2}\sqrt{n}$.*

4.4. Synergy

In summary, the above far-reaching plan includes several huge projects, each of which may prove highly difficult (if true). Putting together all the parts could, nevertheless, prove reformative.

The first part concerns PCP reductions that would establish NP-hardness of approximating CVP/SVP to within the square-root of the dimension—these problems have been open for nearly two decades [30, 49]. Extending that further, the hope is to manage reducing problems in `NPI` to CVP/SVP with larger (weaker) ratios of approximation, say, some polynomial in the dimension. If true (and proven), such a statement would establish that utilizing a computer that can approximate SVP/CVP to such ratios, one could efficiently solve problems beyond those currently known to have (even randomized) efficient algorithms.

One may interpreted such a result in two ways: first, as implying that trusting the security of a cryptographic protocol relative to such an approximation of SVP/CVP is safe. On the positive side, such reductions could show that a computer that solves the average-case of these problems can solve efficiently problems not necessarily in P (or BPP).

Moreover, one might be able to utilize a worst-to-average-case reduction to be able to trust that average-case SVP/CVP to within some polynomial ratios is hard. The last piece of the puzzle could be a quantum algorithm that actually efficiently solves SVP/CVP, to within such ratio of approximation, on average. This would be a radical transformation of our perspective of computation: If that entire plan is true, it would mean that computation is not an adaptation of mechanical decision making to electronics. It could mean that solving a computational problem would mean translating it to appropriate approximation of lattice problems, on average, then utilizing quantum effects to solve it efficiently.

5. ABOUT COMPUTATIONAL COMPLEXITY AND PCP

The most fundamental open research question of computer science is whether the class P of efficiently solvable problems differs from the class NP, a class that includes many natural problems with no known efficient solution. Almost half a century after research into computational complexity theory has began, we are still quite clueless regarding these general questions. The questions are referred to as *computation lower bounds*, that is, proving that some problem is not computable by some class of computation. Even some weakest models—which are presumed to include only very simple computational problems—are not known to not contain the entire NP class. In light of this fundamental deficiency, a substitute infeasibility proof establishes a problem to be NP-hard, implying that, assuming $P \neq NP$, it is infeasible.

5.1. The dawn of NP-hardness

Cook and Levin [23, 62] and Karp [45] proposed a framework by which to bypass this deficiency in proving computational lower bounds. Their suggestion was to prove that problems are NP-hard, so that any efficient algorithm for the problem would yield an efficient algorithm for the entire class NP. This serves as a conditional proof of infeasibility: if $P \neq NP$, then NP-hard problems are infeasible. Subsequent works have shown many algorithmic problems to be NP-hard. This research field, which has been going strong ever since, is by now the bread and butter of computer science, and is taught in undergraduate courses on the computational complexity theory.

As a result, the focus of research has shifted towards introducing ideas and methodologies by which to tackle NP-hardness. The most natural choice is to design approximation algorithms, that is, algorithms that are guaranteed to find close-to-optimal solutions. This line of research—approximation algorithms—has been carried with various degrees of success. Let us pick two classical approximation problems to demonstrate this with, namely Vertex-Cover and Max-Cut.

For Vertex-Cover, a simple algorithm due to Gavril and Yannakakis[5] guarantees a ratio-2 approximation, while an ingenious algorithm by Geomans and Williamson [38] achieves a surprisingly good approximation for Max-Cut. The algorithm of Geomans and Williamson introduces the roaring power of Semi-Definite-Programs *(SDP)*, a technique that has since become central in the design of approximation algorithms, especially for constraint satisfaction problems.

Despite this considerable progress with respect to the design of algorithms, there was still no way to tell whether a certain approximation ratio for a given problem is attainable efficiently or not, causing researchers to waste precious time in futile attempts to find better approximation algorithms.

5 Papadimitriou and Steiglitz [72] credit this algorithm to Gavril and Yannakakis.

5.2. The PCP era

While it was quite plausible at this point that some approximation solutions are unattainable efficiently, there was no known method to establish this fact, even for those problems for which no known algorithmic techniques could come close to solve.

Fortunately, the Probabilistic Checking of Proofs characterization of NP *(PCP)* was introduced [11, 12, 36]. The intuitive concept—and the reason behind the name—is to model the class NP via a protocol where a weak *Verifier* verifies the validity of a solution. In PCP, the Verifier is allowed to use randomness and sampling, albeit only to read a constant number of bits. It must accept a valid proof with probability 1, and reject an input with no valid solution with probability close to 1. The proof of the PCP characterization of NP is quite complex, involving sum-check, low-degree (Reed–Muller code) test, and the core engine that makes it all happen, namely composition/recursion [12]. Once established, however, one can assume a genesis gap-problem is NP-hard, and proceed to show other problems are NP-hard via gap-preserving reductions.

Subsequent research into PCP and infeasibility produced many other techniques, including the Long-Code [19], Raz's parallel-repetition theorem [76], and the Fourier-analytic framework by Håstad [43] that have all been utilized to prove numerous inapproximability results. In certain cases, the results were optimal, in the sense that they showed that there was no way to improve on the best known algorithm (for instance, [9, 35, 42, 43] among many other works).

An illuminating example. Let us consider a classical example of an NP-hard optimization problem and its approximation version, one that captures the essence of what is known and what is still unknown with respect to the infeasibility of approximation. Define, for a graph $G = \langle V, E\rangle$, the *Max-Independent-Set*,

$$\mathrm{MIS}[G] \stackrel{\mathrm{def}}{=} \max_{S \subseteq V : \forall u,v \in S, \langle u,v\rangle \notin E} \frac{|S|}{|V|}$$

and the natural gap problem

Compute **13** (Gap Max-Independent-Set). The gap-MIS$[\alpha, \beta]$ problem is

- **Input:** a graph $G = \langle V, E\rangle$.
- **Goal:** distinguish between
  - **Accept:** $\mathrm{MIS}[G] \geq \alpha$,
  - **Reject:** $\mathrm{MIS}[G] < \beta$.

Clearly, for a fixed α, as β decreases, the problem becomes computationally easier. The ultimate goal is to classify the algorithmic problems gap-MIS$[\alpha, \beta]$, for all pairs α, β's, into feasible vs. infeasible.

Using the Gavril–Yannakakis algorithm (and observing that a vertex-cover of a graph may be defined as a set of vertices whose complement is an independent-set) one can show that for every $\varepsilon > 0$, the problem gap-MIS$[\frac{1}{2} + \varepsilon, 2\varepsilon]$ can be solved efficiently.

The question of whether the problem gap-MIS$[\frac{1}{2}, \varepsilon]$ is NP-hard epitomizes the limits, goals, dreams, and obstacles of research in the field.

5.3. Fourier transform and analysis of Boolean functions

In his *transformative* paper in the 1990s [43] Håstad was the first to introduce Fourier transform into PCP, in order to achieve optimal analysis of local tests, which in turn allowed him to establish tight infeasibility for numerous fundamental approximation problems (such as MAX-3SAT and Linear-equations over $\mathbb{F}_2$). He also managed to prove the infeasibility, to within $7/6$, of Vertex-Cover. To that end, Håstad proved that gap-MIS$[1/4, 1/8]$ is NP-hard. Consequently, and quite insightfully, he proposed proving infeasibility for a ratio up to 2 for Vertex-Cover, as the most fundamental open question, whose solution will take the field to the next level.

The challenge proposed by Håstad eventually led to quite a complex proof [31, 32], which improves the infeasibility up to a ratio of $10\sqrt{5} - 21 \approx 1.3606$. This result again, follows from an even stronger infeasibility result for the MIS problem, namely, that gap-MIS$[1/3, 1/9]$ is NP-hard. The proof introduces the full scope of the analysis of Boolean functions via Friedgut's Junta theorem [37],[6] as well as several other fields of mathematics [3, 34]. One crucial hidden parameter in that complex proof was consequently brought up to the surface and under the spotlight by Khot.

5.4. The UGC revolution

Since the original PCP theorem was insufficient to establish numerous fundamental problems as infeasible, Khot postulated in 2002 the stronger *Unique-Games conjecture* for that purpose. This breaks the task into two: proving the conjecture, and using it to establish infeasibility for the remaining problems. The latter part has been immensely successful, leading to new algorithmic methodologies, and interesting connection between a priori distinct areas. In addition, many infeasibility results relying on the conjecture have been established. In sharp contrast, there has been little progress with the former part, that of actually proving the conjecture—up until recently, the only debate was whether there had been very little progress, or none whatsoever.

The *Unique-Games* conjecture [47] (abbreviated UGC henceforth) asserts that the UG problem, as defined above **Compute** 6 is (NP-)hard. Khot showed that assuming (a variant of) UGC, for every $\varepsilon > 0$ the problem gap-MIS$[1 - 1/\sqrt{2}, \varepsilon]$ is NP-hard. This result is interesting for at least two reasons:

(1) It is the first infeasibility result for the MIS problem with constant completeness and vanishing soundness (in contrast to Håstad's result and the Dinur–Safra result, where the soundness is constant).

6 This is the first time Friedgut's insightful and widely applied theorem was utilized. Friedgut told us once that till then, he assumed that this theorem will never find application.

(2) It implies that it is NP-hard to approximate the minimum Vertex-Cover of a graph within factor $\sqrt{2} - o(1) \approx 1.41$, improving the Dinur–Safra result.

This result, however, *relies on an unproven conjecture*, stronger than that $P \neq NP$.

Subsequently, proofs of infeasibility relied on the UGC, and little progress was achieved otherwise with respect to the infeasibility of fundamental approximation problems. The conjecture is very insightful in that it disconnects the basic task of proving the statement, from the corollaries that could be established as a consequence.

The reason UGC is so fundamental has to do with the core methodology that allows a PCP-Verifier to require only a ridiculously small number of reads, namely, the *composition of PCP proofs* [12]. The natural route by which to prove the UGC is to start with a $\mathtt{PCP}_{1\leftrightarrow 1}$ with not so good parameters—say, $\mathtt{PCP}_{1\leftrightarrow 1}[\varepsilon_a, \varepsilon_r]$ for $0 < \varepsilon_a \ll \varepsilon_r \ll 1$—then use the composition/recursion to improve those parameters while maintaining the restriction on the constraints. The problem is that no such NP-hardness has been established to start the process. Another route would be to try to establish the special structure of the constraints.

Postulating the conjecture proposed a two-part plan: *UGC-proof*—prove the conjecture; and *UGC-use*—show infeasibility assuming that the UGC is true. Until recently, little to no progress had been made on the UGC-proof part. On the UGC-use part of the plan, in contrast, research has flourished, and numerous tight infeasibility results were established based on the UGC (only a small selection of which can be mentioned here). Assuming the UGC, Khot and Regev [55] proved that it is NP-hard to approximate Vertex-Cover to within a ratio $2 - o(1)$, that is, the trivial algorithm mentioned above is optimal.

As to the other problem mentioned above, in 2005, Khot, Kindler, Mossel, and O'Donnell [51] proved that the ingenious algorithm of [38] for Max-Cut is optimal in that any improvement over it would refute the UGC. That paper [51] relied on another conjecture called "Majority is most stable," which, gaining extra motivation, was subsequently proven in [71]. These insights made the community realize that there may be a close connection between UGC and SDP algorithms.

Many consequences of UGC followed (cf. [13, 22, 41, 54, 57] to name a few). Remarkably, assuming UGC, it was shown in [73] that a generic Semi-Definite Program gives the best approximation algorithm for the general class of Constraint-Satisfaction problems *(CSP)*. In other words, Raghavendra proved that the connection between UGC and SDP algorithms is very tight and, assuming UGC, the best approximation algorithm for any CSP is SDP based. Studying UGC also extended to other fields, and led to interesting developments in analysis, geometry [70, 71], integrality-gaps, and metric-embedding [24, 56].

UGC has also provoked a large body of work on algorithms [8, 10, 17, 21, 40, 47, 59, 83] designed to refute the conjecture. Despite the failure of these attempts, the ideas therein motivated the study of new algorithmic techniques (such as the Lasserre Semi-Definite Programs hierarchy) and uncovered surprising connections, between problems and among concepts (e.g., between small-set expansion and UGC [8, 74]).

The conjecture has further led to a large body of work introducing new connections and many new ideas; for attempts at proving it, refuting it (many exciting new algorithmic ideas were introduced for that purpose), and showing NP-hardness assuming it to be true. This area of the field has seen tremendous activity in the last decade and a half.

Summary. While there has been impressive progress on the UGC-use part of the plan, with respect to the UGC-proof part, the UGC remains open despite immense efforts, leaving computer science's best researchers perplexed. We have recently managed considerable progress on the UGC-proof part, as discussed next.

5.5. Present and future

The 2-to-1-Games conjecture, proposed by Khot (in the same paper as the UGC) assumes a less restrictive structure regarding satisfying assignments of PCP: once the PCP-Verifier has read the first value, it can accept 2 values for the second item read (in contrast to the single value accepted in the Unique-Games). This makes the problem harder to solve, and therefore the conjecture that it is infeasible is weaker. Nevertheless, the two statements are quite related—best demonstrated by the fact that there is a trivial reduction from 2-to-1-Games (**Compute** 7) to 1-to-1-Games (**Compute** 8) with value $\frac{1}{2}$ in the accept case.

Only quite recently, Khot, Minzer, and Safra have achieved significant progress, establishing that the 2-to-1-Games problem, albeit with nonperfect completeness, is in fact NP-hard. This achievement has been recognized by the community as going halfway towards the Unique-Games conjecture. The proof is a culmination of a series of four publications [28, 29, 52, 53].

Some consequences of the newly established hardness of 2-to-1-Games are the *unconditional* NP-hardness of gap-MIS$[1 - 1/\sqrt{2}, \varepsilon]$ (establishing an unconditional infeasibility result for the gap-MIS with constant completeness and vanishing soundness), and NP-hardness of approximating minimum Vertex-Cover of a graph within factor $\sqrt{2} - o(1)$ (improving on the Dinur–Safra result).

The first paper [52] presented a new approach for proving infeasibility for the *non-standard* 2-to-1-Games as well as for Vertex-Cover. The proof relies on a careful combination of sophisticated PCP techniques (such as smooth-parallel-repetition, composition, and the biased-long-code) and a novel component in this context, namely, the Grassmann graph—the only construct we know that allows weak 2-to-1 tests to be shown NP-hard.

The second paper [29] extended the ideas and proposed a candidate reduction, which, unlike the first, could work for *standard* 2-to-1-Games. This approach relies on a combinatorial statement regarding the Grassmann graph of a type not explored before. The third paper [28] initiated that investigation. It was clearly necessary to first study *expansion* on the Grassmann graph (just in time, it was shown by [18] that it is also sufficient). We utilized spectral analysis, which is considerably more challenging to carry out here than on the binary hypercube. However, we still did not manage to prove the statement necessary so as to complete the project: the more ambitious the statements we were trying to prove, the messier the analysis got (to the point where any improvement would take weeks or even months of case analysis). The final piece in the puzzle came in [53], where we completed the plan as set out by [28] and proved the necessary combinatorial statement, thus completing the grand program as set by [52].

Border of feasibility. This theorem immediately implies tight infeasibility for several fundamental problems, in the sense that improving even slightly on the best known algorithm is impossible (unless P = NP). Furthermore, it implies that given a Unique-Games instance that is promised to be $\frac{1}{2}$-satisfiable, it is infeasible to efficiently come up with an assignment that satisfies even a negligible fraction of the constraints.[7] This in turn has dramatic implications, since all known algorithms, when applied to $\frac{1}{2}$-satisfiable instances, are guaranteed to perform as well as on the standard $(1-\varepsilon)$-satisfiable instances. Therefore, all known algorithmic techniques can be disqualified: if they were to solve the standard Unique-Games problem efficiently, they would also solve the nonstandard, promised to be $\frac{1}{2}$-satisfiable instance and thereby the 2-to-1-Games problem. Because the known algorithms have proved to be inadequate, researchers who still doubt UGC are now challenged to introduce new algorithmic methodologies. This is a win–win situation since it will either produce new algorithmic methods, or else the UGC will be positively resolved.

One of the central tools employed in these endeavors is the analysis of Boolean function, via Fourier transform and related techniques. This facilitates proofs for optimal local-to-global properties with regard to local-tests of binary error-correcting codes, such as Hadamard-code or the Long-code. While of independent interest, much of the extra motivation for that research relates to applications for infeasibility.

5.6. Related open problems

Small-set expansion. Assume a graph $G=(V,E)$ and a set of vertices $S \subseteq V$: the *edge expansion* of S, denoted by $\Phi(S)$, is the ratio of the number of edges inside S to the total number of edges adjacent to any vertex in S.

The *Small-Set Expansion* problem, denoted by $\mathsf{SSE}_\delta(\varepsilon, 1-\varepsilon)$, asks to distinguish between the case that a given graph has a set containing δ fraction of the vertices whose *edge expansion* is at most ε, and the case that any such set has edge expansion at least $1-\varepsilon$. The problem was first suggested as a possible step towards tackling UGC: Raghavendra and Steurer proved that if the Small-Set Expansion problem is NP-hard, then the UGC is true [74]. Subsequent work has shown that the hardness of SSE implies hardness of various problems not captured by UGC (cf. [64, 75])

Interestingly, as yet, the recent progress on UGC has not had any affect on our understanding of SSE.

Question 5.1. Is there $\beta < 1$ such that for any $\varepsilon > 0$ there is $\delta > 0$ such that $\mathsf{SSE}_\delta(\beta, 1-\varepsilon)$ is NP-hard?

7 This is why the community refers to it as going $\frac{1}{2}$ the distance towards the Unique-Games conjecture.

Extreme tests and how to prove the UGC. All PCP constructions have a tensor structure, and eventually rely on error-correcting codes with special properties such as local-testing and local-correcting. An important aspect by which the recent proof of the 2-to-1-Games conjecture deviates from previous PCP techniques is with respect to list-decoding properties. Prior to our recent work, the codes that were utilized were list-decodable, that is, any word that passes the local-test with noticeable probability *corresponds* to a short list of legal-codewords. By "corresponds" we mean that except for negligible probability, local values that pass the test must agree with one of the codewords from the short list. Our work, in contrast, utilizes the Grassmann code (along with the Grassmann test), which is not list-decodable in the traditional sense. Nevertheless, a much weaker list-decodability property does hold for the Grassmann code, one which, nonetheless, is sufficient to prove hardness for 2-to-1-Games. In the Grassmann test, any subgraph that is by itself a Grassmann graph could potentially be assigned arbitrary values. Despite this complication, sophisticated analysis does establish some weak notion of global consistency, fortunately, one that suffices for the proof of the conjecture.

Going beyond "list-decodable codes" appears to be a prerequisite to prove infeasibility for d-to-d-Games (for constant d). It is possible that an even weaker notion of decodability might be necessary in order to prove the Unique-Games conjecture. For the Unique-Games conjecture, properties similar to those of the Grassmann code must hold for the desired code and, furthermore, its local-tester should be a 1-to-1 test. It is quite plausible that the decodability notion of this code would have to be even weaker, while still allowing some form of global consistency to be established. It is quite possible that the key insight here is to extend that notion so as to make the claimed notion of global consistency even weaker. We believe that a more sophisticated statement, similar to that regarding the Grassmann graph, has the potential to break the barrier and finally succeed in going below the 2-to-1 barrier.

The conjecture has been tantalizing the best researchers for almost two decades, with little progress made until recently. Nevertheless, in light of recent progress [53] where we proved the closely related, yet weaker, *2-to-1-Games* conjecture—recognized by the community as going half the distance toward the Unique-Games[8] conjecture— this project seems to have a better potential to succeed than ever before.

6. DISCUSSION

The purpose of this survey is to examine basic questions regarding the mathematics of computing through the lens of computational problems over linear equations in general and lattices in particular. The goal is to shed light on some fundamental open research questions regarding the mathematics of computing and facilitate progress on solving them. The plan and related conjectures might turn out to be false, nevertheless, a solution whichever

---

8 The Unique-Games can also be referred to as 1-to-1-Games. The 2-to-1-Games problem is another variant of the general d-to-d-Games, as detailed below.

way could prove transformative. On the other hand, if all parts turn out to be true, this could change our vision with regards to computation, not only theoretically but possibly also with practical applications.

First, one may attempt at showing a lattice computational problem to be infeasible (NP-hard) to the extreme ratio, which is assumed to be $O(\sqrt{n})$.

This is likely to require PCP theorems beyond current methodology. The consequences of such progress may facilitate a solution to the 2-to-1-Games problem, $\mathrm{PCP}_{2\to1}[0, 1-\varepsilon]$,[9] proving it is reducible to weaker ratios of approximation for lattice problems and therefore placing it in coNP for nontrivial ratios (ratios not known to be NP-hard or computationally feasible). This would be a strong evidence for this problem not being NP-hard. Furthermore, it may lead to worst-to-average-case reductions that would in turn yield improved cryptography through richer infeasibility assumptions.

To that end, one should clearly pursue research on the intriguing and beautiful mathematics of lattices and error-correcting codes, including new, improved algorithms or, alternatively, establishing currently known algorithms (such as LLL) as optimal. One should also pursue improved theorems of the reverse Minkowski type, to establish the weight of points on the lattice according to Gaussian measures (of various standard deviation). Such theorems may require further research of the Fourier transform over lattices. Lastly, one should pursue tighter connections between hardness of computational problems over lattices and codes, in the worst as well as average case.

The last piece of the plan would be to try and come up with efficient quantum algorithms for approximating those lattice problems with the mentioned ratios, thereby broadening the scope of feasible computational problems to some subclass of NP ∩ coNP **[33]**.

ACKNOWLEDGMENTS

This survey could not have been written without the involvement and highly helpful remarks of the following, to whom I am very grateful: Dor Minzer, Subhash Khot, Zeev Rudnik, Wojtek Samotij, Itamar Rot, Yael Eisenberg, Oren Renard, Ori Petel, and Eli Putterman.

The author was supported by the European Research Council (ERC) under the European Union's Horizon 2020 research and innovation programme (grant agreement No 835152), and by an ISF grant and a BSF grant.

9 Let us mention here that, in contrast, $\mathrm{PCP}_{2\to1}[\varepsilon, 1-\varepsilon]$ has been proven to be NP-hard **[53]**, which, in words, says that given a 2-to-1-Games instance, promised to have an almost perfect assignment, satisfying even a small fraction of the constraints is NP-hard. The proof uses practically all the basic mathematical machinery invented for such purposes and some more. The main mathematical breakthrough was to prove an expansion structure theorem on functions similar to Boolean functions. The reduction there relies fundamentally on the imperfect completeness, as it starts with linear equations over $\mathbb{F}_2$ which are known to be NP-hard only for imperfect completeness (and are efficiently solvable with perfect completeness).

A. THE COMPLEXITY OF CVP

Computational problems regarding linear equations may be solvable, NP-hard, or in-between. The outcome is interesting regardless of the option, and could allow great insight into complexity of classes, in particular average-case complexity, with deep and wide implications.

A.1. CVP is NP-hard

To see that CVP is NP-hard, consider the Max-Cut problem reformulated as equation instead of graph.

Claim A.1 (Decision Max-Cut is NP-hard). *The decision problem Max-*Cut$[1-\varepsilon]$ *(which accepts if the best assignment satisfies $\geq 1-\varepsilon$) is NP-hard for small enough ε.*

Proof. Reduce from 3SAT, assuming p clauses over n variables, each occurring 5 times in distinct clauses (which one can easily translate into an instance satisfying this assumption).

Variables.

- 5 variables for each clause: $y_{i,j}$ for $i \in [p]$ and $j \in [5]$—one for each occurrence of a literal in each clause, corresponding to $j = 1, 2, 3$, plus two auxiliary variables for every clause, where $j = 4, 5$, altogether $5p$.
- Extra (ghost) variables, so that every original variable x_i has exactly 5 positive and 5 negative variables associated with it. Name them $x_{i,j,b}$ for $i \in [n]$ and $j \in [5]$ and $b \in \{0, 1\}$ with the caveat that some of the x's already appear as y's.
- 1 global variable y_F.

Equations.

- Between all opposing occurrences of same variable, namely, for any $x_{i,j,0}$ and $x_{i,j',1}$ for $j, j' \in [5]$ (that is, 5 by 5 combinations), have an equation

$$x_{i,j,0} + x_{i,j',1} = 1.$$

- For any clause C_i, we have an equation for every pair $(y_{i,j}, y_{i,j'})$ where $j \neq j'$,

$$y_{i,j} + y_{i,j'} = 1.$$

- Between y_F and all variables except the ghosts ($5m$ of the clauses variable—5 $y_{i,j}$ for each clause)—satisfied only if those variable have a different value than y_F,

$$y_{i,j} + y_F = 1,$$

  so the global variable y_F forms a clique of size 6 with any 5 variables $y_{i,j}$ for a clause $j \in [p]$.

The total number of equations is $25n + 15p$.

Completeness. A satisfying assignment to the original 3CNF can be translated to an assignment satisfying $25n + 9p$ of new equations (out of $25n + 15p$):

- Assign $y_F = 0$.
- Satisfy all 5 by 5 for each original variable, assigning those 0/1 in accordance to the value the satisfying assignment to the 3CNF gives it.
- Assign the auxiliary variables for each clause 0/1 values that ensure that exactly 3 of the 5 are assigned 1 (for each clause, at least one of the literals is 1, so the auxiliary two can always guarantee 3 that are 1).

Altogether $9p$ out of the $15p$ are satisfied, $3 + 3 \cdot 2$.

Soundness. First, observe that the maximum is achieved when all $25n$ variable equations are satisfied (the number of equations satisfied, if not split according to sign, is ≤ 20 of the 25). Then, notice that out of the 15 for every clause, at most 9 can be satisfied: 3 by 3 split of the clique of size 6 formed with y_F.

So to complete the proof, observe that in each clause, to ensure that 9 of the equations are satisfied, at least one of the original literals must be 1, to satisfy 3 of the equations against y_F. ■

Now, as to the approximation:

Claim A.2 (Gap Max-Cut is NP-hard). *The gap problem Max-*Cut$[1 - \varepsilon(1+?), 1 - \varepsilon]$ *(which accepts if the maximal assignment satisfies $\geq 1 - \varepsilon$ and rejects if it is $< 1 - (1+?)\varepsilon$) is NP-hard.*

Proof. Reduce from gap-E3SAT$[7/8 + \varepsilon, 1]$, which is NP-hard. ■

B. FOUNDATIONS OF LATTICES

Let us formally define a lattice and, for clarity, discuss for now only full-rank lattices, where the dimension and the column-rank are equal. Accordingly, we will assume from now on that the columns of a basis/matrix $M \in \mathbb{F}^{n \times n}$ are linearly independent.

The fact that the columns of the matrix are linearly independent implies that they form a basis for the vector space $\mathbb{F}^n$. Denote the ith column of a matrix M by M_i, so one can write $M = \{M_1, \ldots, M_n\}$ to specify the vectors of the basis, and Mx denotes the linear combination of the basis' column vectors.

In the early (code) versions of the computational problems above, the sets of vectors $\{Mx \mid x \in \mathbb{F}^n\}$ formed a sparse subspace of $\mathbb{F}^m$. This sparseness is due to the number of variables being smaller than the dimension m of the entire space ($n \ll m$). In our case, since the matrix is square and $n = m$, how can we expect the set to be sparse and not to span all $\mathbb{F}^n$? It is indeed a crucial aspect of lattices that the domain $\mathcal{F}$ of the variables $\vec{x}_i$'s is sparse within $\mathbb{F}$. The set of vectors of a lattice form a sparse vector space within the entire space $\mathbb{F}^n$. The target vector could be any vector in $\mathbb{F}^n$, and one looks for a vector close to it of the form Mx, where $x \in \mathcal{F}^n$ is a vector in the lattice. Throughout, we denote the domain of all $\vec{x}_i$ by $\mathcal{F} \subset \mathbb{F}$, and assume it to be a ring with respect to the same operations as $\mathbb{F}$.

Definition B.1 (Lattice). Given a field $\mathbb{F}$ and its restriction $\mathcal{F} \subset \mathbb{F}$, a *lattice* is defined by a matrix $M \in \mathbb{F}^{n \times n}$ as

$$\mathcal{L}_{\mathbb{F}|\mathcal{F}}[M] \stackrel{\text{def}}{=} \{Mx \mid x \in \mathcal{F}^n\}.$$

Note that we assume $\mathcal{F}$ forms a discrete ring within $\mathbb{F}$, assuming a natural metric over $\mathbb{F}$.

Our primary focus herein is the case where the domain of the matrix is $\mathbb{F} = \mathbb{R}$, and thus $M \in \mathbb{R}^{n \times n}$, while the domain of the ring comprises the integers, $\mathbb{Z}$, and thus $x \in \mathbb{Z}^n$,

$$\mathcal{L}[M] \stackrel{\text{def}}{=} \{Mx \mid x \in \mathbb{Z}^n\},$$

nevertheless, some of the statements below are true for more general settings.

A more intuitive perspective of such a formalism is a sparse vector space within $\mathbb{R}^n$, determined by a basis (the columns of the matrix M) while allowing only integral linear combinations of the basis' vectors

$$\mathcal{L}[M] \stackrel{\text{def}}{=} \left\{\sum a_i M_i \mid a_i \in \mathbb{Z}\right\}.$$

Observe that, because the domain $\mathcal{F}$ of the variables $\vec{x}$ being sparse, different bases for the same vector space $\mathbb{F}^n$ do not necessarily span the same lattice.

Here is an examples of a 2-dimensional lattice:

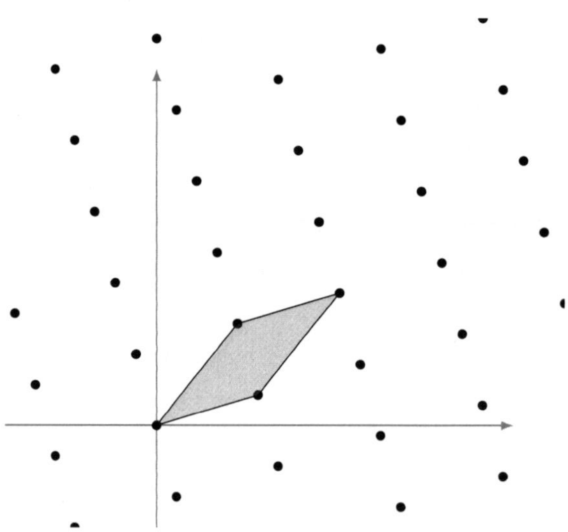

FIGURE 2
The lattice spanned by $(1, 0.3)$ and $(0.8, 1)$

B.1. Elementary, my dear

The same lattice can be generated by an infinite number of bases. A natural question is how one can determinate whether two bases generate the same lattice.

Claim B.1. *Let $M_1, M_2 \in \mathbb{R}^{n\times n}$ be full-rank matrices. Then $\mathcal{L}[M_1] = \mathcal{L}[M_2]$ iff there exists a unimodular matrix $U \in \mathrm{GL}_n(\mathbb{Z})$ such that $M_2 = \mathrm{UM}_1$.*

Invariance. In fact, any lattice basis can be transformed to a different basis of the same lattice by a sequence of elementary column operations: swapping columns, multiplying a column by -1, or adding an integer multiple of a column to another column.

Duality is quite an important concept in general, and here in particular: every lattice has a unique *dual lattice*.

Definition B.2 (Dual lattice). For a lattice $\mathcal{L}$, the lattice

$$\mathcal{L}^\vee \stackrel{\text{def}}{=} \{u \in \mathbb{R}^n \mid \forall v \in \mathcal{L} : \langle u, v\rangle \in \mathbb{Z}\}$$

is referred to as the *dual lattice* of $\mathcal{L}$.

Definition B.3 (Dual matrix). Given a full-rank matrix $M \in \mathbb{R}^{n\times n}$, its *dual matrix* is

$$M^\vee \stackrel{\text{def}}{=} M^{\top -1}.$$

Fact B.2. *The dual of the basis is a basis for the dual lattice, $\mathcal{L}^\vee[M] = \mathcal{L}[M^\vee]$.*

Trivially note that any lattice is a dual of its dual:

Fact B.3. $\mathcal{L}^{\vee\vee} = \mathcal{L}$; $M^{\vee\vee} = M$.

The following definition deals with the *stretch* of a lattice by some factor

Definition B.4. Given a lattice $\mathcal{L}$, for $q \in \mathbb{R}^+$, let $\mathcal{L}[M]$'s *q-scaled lattice* be

$$q \cdot \mathcal{L} \stackrel{\text{def}}{=} \{q \cdot v \mid v \in \mathcal{L}\}.$$

Fact B.4. $q \cdot \mathcal{L}[M] = \mathcal{L}[q \cdot M]$.

Another interesting fact regrading the relation between the stretched dual of the lattice and the dual of the dual is the following

Fact B.5. $(q \cdot \mathcal{L})^\vee = q^{-1} \cdot \mathcal{L}^\vee$.

The definitions so far relate to an abstract lattice, with no particular basis. Let us now examine some notions that are specific to a particular basis for a lattice.

Fundamentals. The *fundamental parallelepiped* of a lattice basis is the body defined by letting all x_i coefficients be between 0 and 1:

Definition B.5 (Fundamental parallelepiped). For a matrix M, define the *fundamental parallelepiped* as

$$\mathcal{P}[M] \stackrel{\text{def}}{=} \{Mx \mid \forall i,\ 0 \le x_i < 1\}.$$

Note that the fundamental parallelepiped tiles space when we shift it by a lattice vector, as it is half open.

Definition B.6. Given a matrix M and a vector $v \in \mathbb{R}^n$, the *rounding* of v according to the fundamental parallelepiped, denoted as $\lfloor v \rfloor_{\mathcal{P}[M]}$, is defined by

$$\lfloor v \rfloor_{\mathcal{P}[M]} \stackrel{\text{def}}{=} \sum_i \lfloor a_i \rfloor M_i \,,$$

where the coefficients a_i are the unique representation of v according to the basis given by the columns of M, namely $v = \sum_i a_i M_i$.

The mapping $v \to \lfloor v \rfloor_{\mathcal{P}[M]}$ can be thought of as a rounding function, mapping v to some lattice point in $\mathcal{L}[M]$ that is "closest to it" in terms of the coefficients.[10] Looking at the difference then, one gets a point from the fundamental parallelepiped $\mathcal{P}[M]$ (which could be thought of as the "fractional part" of v). We refer to this operation as modulo $\mathcal{P}[M]$, formally defined as

$$v \pmod{\mathcal{P}[M]} \stackrel{\text{def}}{=} v - \lfloor v \rfloor_{\mathcal{P}[M]} \in \mathcal{P}[M].$$

Clearly, different bases (for the same lattice) result in different fundamental parallelepipeds and consequently have different rounding values. This could affect the distance of the modulus vector.

Let us now use this notion of the fundamental parallelepiped to establish a method by which to figure out whether two matrices span the same lattice:

Lemma B.6. *Let $M, M' \in \mathbb{R}^{n\times n}$ be full-rank matrices so that $\mathcal{L}[M'] \subseteq \mathcal{L}[M]$ (equivalently, every column M'_i is a vector in $\mathcal{L}[M]$); then*

$$\mathcal{L}[M'] = \mathcal{L}[M] \iff \mathcal{P}[M'] \cap \mathcal{L}[M] = \{\vec{0}\}.$$

Proof. Observe that any vector $v \in \mathbb{R}^n$ can be written as $v = \sum_i a_i M'_i$ for $a_i \in \mathbb{R}$.

($\Longrightarrow$). By definition, $\mathcal{P}[M']$ includes only vectors v where all coefficients are $0 \le a_i < 1$ while any nonzero vector in $\mathcal{L}[M'] = \mathcal{L}[M]$ has at least one coefficients $|a_i| \ge 1$.

($\Longleftarrow$). To show that $\mathcal{L}[M] \subseteq \mathcal{L}[M']$, consider a vector $v \in \mathcal{L}[M]$; by definition, the vector $\lfloor v \rfloor_{\mathcal{P}[M']} \in \mathcal{L}[M'] \subseteq \mathcal{L}[M]$, and *since a lattice is closed under vector addition and negation*, the difference between those two, $v \pmod{\mathcal{P}[M']}$, is in $\mathcal{L}[M]$. On the other hand, $v \pmod{\mathcal{P}[M']} \in \mathcal{P}[M']$ and, by the premise, it must then be the case that $v \pmod{\mathcal{P}[M']} = \vec{0}$ and thus $v = \lfloor v \rfloor_{\mathcal{P}[M']} \in \mathcal{L}[M']$. ∎

Determinant. What is the volume of the fundamental parallelepiped?

Recall that the volume of the parallelepiped $\mathsf{vol}[\mathcal{P}[M]] = |\mathsf{det}[M]|$.

Fact B.7. *Given two matrices $M, M' \in \mathbb{R}^{n\times n}$, if $\mathcal{L}[M] = \mathcal{L}[M']$ then $|\mathsf{det}[M']| = |\mathsf{det}[M]|$.*

---

10 Observe that v could be very far from $\lfloor v \rfloor_{\mathcal{P}[M]}$.

This implies that the volume of the fundamental parallelepiped of a given lattice is independent of the choice of basis. This parameter is referred to as the determinant of a lattice.

Definition B.7 (Determinant). The *determinant* of a lattice $\mathcal{L}[M]$ is defined as

$$\mathsf{det}\big[\mathcal{L}[M]\big] \stackrel{\text{def}}{=} \mathsf{vol}\big[\mathcal{P}[M]\big] = \big|\mathsf{det}[M]\big|.$$

The determinant of a lattice is inversely proportional to its density: the larger the determinant, the sparser the lattice. Consider a ball of some (large) radius r around the origin and count how many points of the lattice are in that ball; it follows that the smaller the determinant, the more lattice points are in the ball,

$$\lim_{r\to\infty} \frac{|\mathcal{L} \cap \mathcal{B}[\vec{0}, r]|}{\mathsf{vol}[\mathcal{B}[r]]} = \frac{1}{\mathsf{det}[\mathcal{L}]}.$$

One of the most fundamental questions regarding lattices is when r becomes large enough for the value to become negligibly different from the limit. For a basis/matrix M that is close—in some reasonable sense—to an orthonormal basis, and the value converges quickly; in general, there is a nontrivial upper bound (established below) which is central to this theory.

The denser the $\mathcal{L}$, the less dense the $\mathcal{L}^{\vee}$, which can be summarized by the following:

Claim B.8. *For a lattice $\mathcal{L}$,* $\mathsf{det}[\mathcal{L}]\mathsf{det}[\mathcal{L}^{\vee}] = 1$.

Successive minima. Consider the ℓ^2-ball of radius r around the origin

$$\mathcal{B}[\vec{0}, r] \stackrel{\text{def}}{=} \{v \mid \|v\|_2 \leq r\}$$

and the smallest r such that $\mathcal{B}[\vec{0}, r]$ contains a set of vectors of $\mathcal{L}$ which spans an i-dimensional space:

Definition B.8. $\lambda_i[\mathcal{L}] \stackrel{\text{def}}{=} \inf\{r \mid \dim[\text{span}[\mathcal{L} \cap \mathcal{B}[\vec{0}, r]]] \geq i\}$.

Orthonormalization. Orthonormlization[11] is the process of taking a sequence of linearly-independent vectors and replacing it with an orthonormal basis, so that any prefix of the sequence spans the same space as the original prefix. The orthonormal basis is unique.

Definition B.9. Let $u_1, \ldots, u_n \in \mathbb{R}^n$ be a sequence of linearly-independent vectors. The *orthonormalization of* $u_1, \ldots, u_n$ is the sequence $u_1^{\perp}, \ldots, u_n^{\perp}$ so that

- $u_1^{\perp}, \ldots, u_n^{\perp}$ is an orthonormal basis.
- For $1 \leq i \leq n : u_i^{\perp} \in \text{span}[u_1, \ldots, u_i] \setminus \text{span}[u_1, \ldots, u_{i-1}]$.

Fact B.9. $u_1^{\perp}, \ldots, u_n^{\perp}$ *is unique (up to signs).*

Observe that the resulting basis, $u_1^{\perp}, \ldots, u_n^{\perp}$, is not necessarily a basis for the same lattice.

Denote by $M^{\perp}$ the matrix whose columns are $u_1^{\perp}, \ldots, u_n^{\perp}$.

---

11 This is the orthonormal version of the Gram–Schmidt orthogonalization: all vectors are of norm 1.

Regarding λ_1. One can immediately bound from below the first $\lambda_1[\mathcal{L}[M]]$ successive minimum—the length of the shortest vector—according to $M^\perp$:

Theorem B.10. *Let M be a basis for a lattice $\mathcal{L}$, and $M^\perp$ be its orthonormalization. Then*

$$\lambda_1\big[\mathcal{L}[M]\big] \geq \min_i \big|\langle M_i, M_i^\perp\rangle\big| > 0.$$

Proof. To bound the norm of any nonzero lattice vector from below, let us show a stronger claim: for any coefficient vector $z \in \mathbb{Z}^n$, it is the case that $\|Mz\| \geq |\langle M_k, M_k^\perp\rangle|$, where k is the maximal index so that $z_k \neq 0$. Note that the inner product $\langle Mz, M_k^\perp\rangle$ is independent of all indices other than k (larger indices are 0 and $M_k^\perp$ is orthogonal to $\operatorname{span}[M_1, \ldots, M_{k-1}]$):

$$\begin{aligned}\|Mz\| &= \|Mz\|\|M_k^\perp\| \geq |\langle Mz, M_k^\perp\rangle| = \left|\left\langle \sum_{i=1}^k z_i M_i, M_k^\perp \right\rangle\right| = |\langle z_k M_k, M_k^\perp\rangle| \\ &= |z_k||\langle M_k, M_k^\perp\rangle|,\end{aligned}$$

where the inequality is due to the Cauchy–Schwarz inequality; now, observe that z_k is a nonzero integer to complete the proof.

Hence, we have just shown that the inner product of the kth vector with its orthonormal basis vector is a lower bound on the shortest nonzero vector in $\mathcal{L}$. ■

Stretch and inclusion.

Claim B.11. *For an n-dimensional lattice $\mathcal{L}$, there exists a sequence of linearly-independent vectors $u_1, \ldots, u_n \in \mathcal{L}$ so that $\forall i$, $\|u_i\| = \lambda_i[\mathcal{L}]$.*

Furthermore, for any such sequence it holds that

$$\mathcal{B}(\vec{0}, \lambda_{i+1}[\mathcal{L}]) \cap \mathcal{L} \subseteq \operatorname{span}[u_1, \ldots, u_i],$$

that is, the intersection of the open ball of radius λ_{i+1} with $\mathcal{L}$ is spanned by the first i vectors.

Banaszczyk. Let us mention here the transference theorem by Banaszczyk, which is a far less obvious statement regarding the successive minima of a lattice and its dual:

Theorem B.1 (Banaszczyk). $1 \leq \lambda_1[\mathcal{L}]\lambda_n[\mathcal{L}^\vee] \leq n.$

Proof. For the first inequality, let s be the shortest in $\mathcal{L}$, and $u_1, \ldots, u_n \in \mathcal{L}^\vee$ be a linearly-independent set of vectors so that $\|u_i\| = \lambda_i[\mathcal{L}^\vee]$ (which exists due to Claim B.11). Then u_i is a basis of $\mathbb{R}^n$. Therefore, there must be an i for which $\langle s, u_i\rangle \neq 0$, and due to duality, $1 \leq |\langle s, u_i\rangle| \leq \|s\|\|u_i\| = \lambda_1[\mathcal{L}]\lambda_i[\mathcal{L}^\vee] \leq \lambda_1[\mathcal{L}]\lambda_n[\mathcal{L}^\vee]$. For the proof of the second inequality, see [84]. ■

B.2. Minkowski

Theorem B.12 (Blichfeldt). *For any lattice $\mathcal{L} \subset \mathbb{R}^n$ and a measurable set $S \subset \mathbb{R}^n$ of* $\operatorname{vol}[S] > \det[\mathcal{L}]$, *there must be $u \neq v \in S$ with $u - v \in \mathcal{L}$.*

Proof. Let M be a basis for $\mathcal{L}$. Note that we can break S according to the shift of the fundamental parallelepiped $S = \dot{\bigcup}_{w \in \mathcal{L}} S_w$ where

$$S_w \stackrel{\text{def}}{=} \{v \in S \mid \lfloor v \rfloor_{\mathcal{P}[M]} = w\} \quad \text{and} \quad \mathsf{vol}[S] = \sum_{w \in \mathcal{L}} \mathsf{vol}[S_w].$$

As shifting by a vector is measure-preserving and $\mathsf{vol}[S] > \mathsf{vol}[\mathcal{P}[M]]$, there must be two distinct points in S that are the same modulo the parallelepiped, that is, a pair $u \neq v \in S$ so that $u \in S_w, v \in S_{w'}$ for $w \neq w' \in \mathcal{L}[M]$ and

$$u \ (\text{mod } \mathcal{P}[M]) = v \ (\text{mod } \mathcal{P}[M]),$$

and thus

$$u - v = \lfloor u \rfloor_{\mathcal{P}[M]} - \lfloor v \rfloor_{\mathcal{P}[M]} = w - w' \in \mathcal{L}[M].$$

■

Theorem B.13 (Minkowski's convex-body theorem). *For any lattice $\mathcal{L} \subset \mathbb{R}^n$ and a centrally symmetric convex body $S \subset \mathbb{R}^n$ with $\mathsf{vol}[S] > 2^n \mathsf{det}[\mathcal{L}]$ there must be $u \neq \vec{0}$ so that $u \in \mathcal{L} \cap S$, in other words, $\mathcal{L} \cap S \neq \{\vec{0}\}$.*

Proof. Let $S_{\frac{1}{2}} \stackrel{\text{def}}{=} \{v \mid 2v \in S\}$ and observe that

$$\mathsf{vol}[S_{\frac{1}{2}}] = 2^{-n} \mathsf{vol}[S] > \mathsf{det}[\mathcal{L}],$$

and thus, by Theorem B.12, there must be $u \neq v \in S_{\frac{1}{2}}$ with $u - v \in \mathcal{L} - \{\vec{0}\}$. By definition, $2u, 2v \in S$ and since S is centrally-symmetric, so is $-2v \in S$ and by convexity

$$u - v = \frac{1}{2}(2u - 2v) \in S,$$

therefore, $u - v \in \mathcal{L} \cap S \setminus \{0\}$. ■

Theorem B.14 (Minkowski's second theorem). *For any $\mathcal{L} \subset \mathbb{R}^n$,*

$$\prod_{i=1}^{n} \lambda_i[\mathcal{L}] \leq 2^n \frac{\mathsf{det}[\mathcal{L}]}{\mathsf{vol}[\mathcal{B}[\vec{0}, 1]]}.$$

Proof. Consider a set of linearly-independent vectors $u_1, \ldots, u_n \in \mathcal{L}$ so that $\|u_i\| = \lambda_i[\mathcal{L}]$ (which exists due to claim B.11 above) and let $u_1^{\perp}, \ldots, u_n^{\perp}$ be their orthonormalization. Let us define the convex symmetric body (in fact, an ellipsoid)

$$\mathcal{E} \stackrel{\text{def}}{=} \left\{ v \in \mathbb{R}^n \mid \sum_{i=1}^{n} \frac{\langle v, u_i^{\perp} \rangle^2}{\lambda_i[\mathcal{L}]^2} < 1 \right\}.$$

Now, show

Claim B.15. $\mathcal{E} \cap \mathcal{L} = \{\vec{0}\}$.

Proof. The differences between the open balls of radius λ_{k+1} and λ_k for $k = 0, \ldots, n$ (where $\lambda_0 \stackrel{\text{def}}{=} 0$ and $\lambda_{n+1} \stackrel{\text{def}}{=} \infty$) partition $\mathbb{R}^n$, thus, for $u \in \mathcal{L} \setminus \{\vec{0}\}$ there must be a unique $k > 0$ so that

$$u \in \mathcal{L} \cap \left(\mathcal{B}(\vec{0}, \lambda_{k+1}) \setminus \mathcal{B}(\vec{0}, \lambda_k)\right).$$

By Claim B.11, $u \in \text{span}[u_1, \ldots, u_k]$ and thus

$$\lambda_k[\mathcal{L}]^2 \leq \|u\|^2 = \sum_{i=1}^{k} \langle u, u_i^{\perp} \rangle^2 \leq \lambda_k[\mathcal{L}]^2 \sum_{i=1}^{k} \frac{\langle u, u_i^{\perp} \rangle^2}{\lambda_i[\mathcal{L}]^2} < \lambda_k[\mathcal{L}]^2,$$

and it must be the case that the only lattice point in $\mathcal{E}$ is $\vec{0}$. ■

This, in turn, implies, by Minkowski's Theorem B.13 above, an upper bound on T's volume

$$2^n \det[\mathcal{L}] \geq \text{vol}[\mathcal{E}] = \text{vol}\big[\mathcal{B}(\vec{0}, 1)\big] \prod_{i=1}^{n} \lambda_i[\mathcal{L}],$$

which completes the proof. ■

The volume of a ball can be found here.

Corollary B.16 (Minkowski's first theorem). *For any $\mathcal{L} \subset \mathbb{R}^n$, $\lambda_1[\mathcal{L}] \leq (\det[\mathcal{L}])^{\frac{1}{n}} \sqrt{n}$.*

Dual lattice. The latter inequality implies some meaningful statements regarding the relation between a lattice and its dual:

Claim B.17. *For any (full-rank) lattice $\mathcal{L}$, $\lambda_1[\mathcal{L}] \cdot \lambda_1[\mathcal{L}^{\vee}] \leq n$.*

Proof. By Minkowski's bound, Corollary B.16,

$$\lambda_1[\mathcal{L}] \cdot \lambda_1\big[\mathcal{L}^{\vee}\big] \leq \sqrt{n} \cdot \det[\mathcal{L}]^{\frac{1}{n}} \cdot \sqrt{n} \cdot \det\big[\mathcal{L}^{\vee}\big]^{\frac{1}{n}} = n \cdot \big(\det[\mathcal{L}] \cdot \det\big[\mathcal{L}^{\vee}\big]\big)^{\frac{1}{n}}.$$ ■

B.3. Fourier transform over lattices, etc.

Fourier transform. It is defined (see [82]) as follows:

Definition B.10. The Fourier transform of a function $f \in L^1(\mathbb{R}^n)$ is a function $\hat{f} : \mathbb{R}^n \to \mathbb{R}$ defined as

$$\hat{f}(w) \stackrel{\text{def}}{=} \int_{\mathbb{R}^n} f(x) e^{-2\pi i \langle x, w \rangle} \, dx.$$

Claim B.18. *Some basic properties of Fourier transform:*

(1) $\widehat{f+g}(w) = \hat{f}(w) + \hat{g}(w), \widehat{\lambda \cdot f} = \lambda \cdot \hat{f}(w)$.

(2) *If* $h(x) = f(x+z)$ *then* $\hat{h}(w) = e^{2\pi i \langle w, z \rangle} \hat{f}(w)$.

(3) *If* $h(x) = f(\lambda x)$ *then* $\hat{h}(w) = \frac{1}{\lambda^n} \hat{f}(\frac{w}{\lambda})$.

(4) *If* $f(x) = f_1(x_1) \cdots f_n(x_n)$ *then* $\hat{f}(y) = \widehat{f_1}(y_1) \cdots \widehat{f_n}(y_n)$.

Definition B.11 (Gaussian). Denote by $\mathcal{N}_n[\sigma]$ the *Gaussian* (normal) n-dimensional (namely, over $\mathbb{R}^n$) *distribution* centered at $\vec{0}$ and of "width" (standard-deviation) σ.

$$\mathcal{N}_n[\sigma](x) = \frac{1}{(2\pi)^{\frac{n}{2}} \sigma^n} e^{-\frac{1}{2\sigma^2} \|x\|^2}.$$

Fact B.19. *The Fourier transform of the Gaussian distribution $\mathcal{N}_n[\sigma]$ is the Gaussian distribution $\big(\frac{\sqrt{2\pi}}{\sigma}\big)^n \mathcal{N}_n[1/\sigma]$.*

The Poisson summation formula. It asserts the following:

Fact B.20. *For any "nice" $f : \mathbb{R}^n \to \mathbb{R}$ and for any full-rank n-dimensional lattice $\mathcal{L}$,*

$$\sum_{v \in \mathcal{L}} f(v) = \frac{1}{\det(\mathcal{L})} \sum_{v \in \mathcal{L}^\vee} \hat{f}(v).$$

Altogether this gives

Theorem B.21. *For any full-rank n-dimensional lattice $\mathcal{L}$ and any $\varepsilon > 0$, there is $r(\varepsilon)$ so that*

$$\Pr_{v \sim \mathcal{N}_n[\lambda_n[\mathcal{L}] \cdot r(\varepsilon)]} \left[v \notin \mathcal{B}(\lambda_n) \mid v \in \mathcal{L}^\vee \right] \le \varepsilon.$$

B.4. The Korkin–Zolotarev basis

Let us consider the projection of a lattice $\mathcal{L}$ according to a given vector $u \in \mathcal{L}$.

Definition B.12. For a vector $u \in \mathcal{L}$ of a lattice $\mathcal{L}$, denote by

$$\pi_{\perp u}(v) \overset{\text{def}}{=} v - \frac{\langle u, v \rangle}{\|u\|^2} u,$$

which altogether gives the $(n-1)$-dimensional lattice, the result of projecting $\mathcal{L}$ on the hyperplane perpendicular to u, thus

$$\mathcal{L}_{\perp u} \overset{\text{def}}{=} \{\pi_{\perp u}(v) \mid v \in \mathcal{L}\}.$$

Definition B.13. A matrix M is a *Korkin–Zolotarev basis* **[60]** for $\mathcal{L}[M]$ if

- M_1 is shortest, $\|M_1\| = \lambda_1[\mathcal{L}[M]]$;
- $\pi_{\perp M_1}(M_2), \ldots, \pi_{\perp M_1}(M_n)$ is a Korkin–Zolotarev basis for $\mathcal{L}_{\perp M_1}$;
- For any $1 \le j < i \le n$, $\langle M_i^\perp, M_j \rangle \le \frac{1}{2} \langle M_j^\perp, M_j \rangle$.

Fact B.22. *The Korkin–Zolotarev basis always exists (constructively).*

B.5. The LLL algorithm

Recall the orthonormalization process described above (the orthonormal version of Gram–Schmidt) of Definition B.9.

Obviously, applying the orthonormalization process to a given lattice basis does not necessarily result in another basis for the lattice. This is the most fundamental problem with applying the orthonormalization process as is to a lattice basis.

Let us therefore try a similar process that preserves the lattice spanned by the basis, in exchange for a weaker notion of orthogonality. In other words, the inner product between any two vectors must not be too large and the angle between the vector and the span of previous vectors is within $[\pi/3, 2\pi/3]$, instead of being $\pi/2$.

There is also the issue of the order in which the process is carried out, that is, the order of the columns of M:[12] the algorithm and the outcome of the orthonormalization

12 Recall that order does not change at all the lattice spanned by the basis.

process depend strongly on the order of the columns in M. Ideally, one would be fantastically able to find an order according to which the projection $\langle M_i, M_i^{\perp}\rangle$ is monotonically nondecreasing, in which case it follows from Theorem B.10 that M_1 is the shortest vector. Unfortunately, the requirements of quasiorthogonality and nondecreasing projection are generally mutually exclusive.

The condition below is a relaxation of the nondecreasing order and represents a compromise, an order in which the next vector projection is at least a constant times the current. Unfortunately, this constant turns the algorithm into an approximation algorithm and, moreover, one with only an exponential ratio of approximation!

Let us therefore define an LLL-reduced basis **[61]**:

Definition B.14. A basis/matrix M is *δ-LLL-reduced* (think of $\delta \in (\frac{1}{4}, 1)$) if the following two properties hold:

- *(Close to orthogonal)*

$$\forall 1 \le j < i \le n, \quad |\langle M_j, M_j^{\perp}\rangle| \ge 2|\langle M_i, M_j^{\perp}\rangle|,$$

  that is, there is no way to add or subtract M_j from M_i to make them more perpendicular.

- *(Lovász condition)*

$$\forall 1 \le i < n, \quad \delta\langle M_i, M_i^{\perp}\rangle^2 \le \langle M_{i+1}, M_i^{\perp}\rangle^2 + \langle M_{i+1}, M_{i+1}^{\perp}\rangle^2.$$

To motivate the extra term in Lovász condition, take, for example, two vectors of length 1, so that the angle between them is $\pi/3$. The orthogonality condition holds, but $|\langle M_2, M_2^{\perp}\rangle| = \sqrt{\frac{3}{4}}$. Due to symmetry, however, changing the order of columns changes nothing. Therefore, the extra term guarantees that, if M_i is switched with M_{i+1} and M_i is adjusted to be as orthogonal as possible, there would be no reason to switch back (preferring the vector with smaller norm).

Fact B.23. *For a δ-LLL-reduced basis M, for any $1 \le i < n$,*

$$\langle M_{i+1}, M_{i+1}^{\perp}\rangle \ge \sqrt{\delta - \frac{1}{4}}\langle M_i, M_i^{\perp}\rangle.$$

Claim B.24. *Let M be a δ-LLL-reduced basis, then*

$$\lambda_1[\mathcal{L}[M]] \ge \|M_1\|\left(\frac{\sqrt{4\delta - 1}}{2}\right)^{n-1}.$$

Proof. First, note that, for any $1 \le i \le n$, applying Fact B.23 $(i-1)$-times (and observing that $\|M_1\| = \langle M_1, M_1^{\perp}\rangle$) implies

$$\langle M_i, M_i^{\perp}\rangle \ge \left(\delta - \frac{1}{4}\right)^{\frac{i-1}{2}} \|M_1\|,$$

which achieves its minimum when $i = n$, thus, by Theorem B.10 above,

$$\lambda_1[\mathcal{L}[M]] \ge \min_i |\langle M_i, M_i^{\perp}\rangle| \ge \left(\delta - \frac{1}{4}\right)^{\frac{n-1}{2}} \|M_1\|. \qquad \blacksquare$$

B.5.1. The algorithm

Algorithm 1: The LLL algorithm

```
Input: full-rank matrix M and δ ∈ (1/4, 1)
Output: δ-LLL-reduced basis for 𝓛[M]
repeat
    Compute orthonormalized M^⊥
    for i ← 2 to n;  j ← i − 1 to 1 do
        M_i ← M_i − ⌊⟨M_i, M_j^⊥⟩ / ⟨M_j, M_j^⊥⟩⌉ M_j                // Round
    end
    if ∃i : δ⟨M_i, M_i^⊥⟩^2 > ⟨M_{i+1}, M_i^⊥⟩^2 + ⟨M_{i+1}, M_{i+1}^⊥⟩^2 then
        M_i ↔ M_{i+1}                                                // Swap
    end
    else return M                                                    // Done
```

Observe that the "Round" and "Swap" steps do not change the lattice. Next we show that the algorithm terminates within polynomial time, and, furthermore, that once it does, the basis is reduced.

Running time. The correctness of the algorithm, as well as the fact that it runs in polynomial time, are proved simultaneously by introducing a parameter. This parameter is at most exponential in the dimension n at the start of the execution, and at each step is reduced by a constant factor < 1, while never going below the determinant multiplied by the minimal quanta (precision) of the vectors.

Fact B.25. *Given an integer basis M, consider*

$$\mathfrak{D}\mathfrak{D}[M] \stackrel{\text{def}}{=} \prod_{i=1}^{n} \left|\langle M_i, M_i^{\perp}\rangle\right|^{n-i+1}$$

which satisfies the following properties:

- $\mathfrak{D}\mathfrak{D}[M] \le (\max_{1\le i\le n} \|M_i\|)^{\frac{n(n+1)}{2}}$;
- $\mathfrak{D}\mathfrak{D}[M] \ge 1$;
- *$\mathfrak{D}\mathfrak{D}[M]$ is reduced by a factor $\sqrt{\delta}$ in each swap step of the LLL algorithm, and does not increase by an orthogonalization step.*

B.6. Extension

This algorithm can be improved to approximate CVP to within same ratio, and SVP to within $2^{n\frac{\log\log n}{\log n}}$.

B.7. CVP algorithm

Covering radius. Let us define yet another interesting parameter of a lattice. How far can a vector be from the lattice $\mathcal{L}$ (that is, the distance from its closest lattice vector)?

Definition B.15 (Covering tadius).

$$\mu(\mathcal{L}) \stackrel{\text{def}}{=} \max_{t \in \mathbb{R}^n} \operatorname{dist}(t, \mathcal{L}) = \inf\left\{ r > 0 \mid \mathbb{R}^n \subset \bigcup_{v \in \mathcal{L}} \mathcal{B}(v, r) \right\}.$$

Claim B.26.

$$\frac{1}{2}\lambda_n[\mathcal{L}] \leq \mu(\mathcal{L}) \leq \frac{1}{2}\sqrt{n}\lambda_n[\mathcal{L}].$$

B.7.1. Babai's nearest plane algorithm

Let us now present an algorithm for approximating CVP à la LLL **[16]**. Let $M_1, \ldots, M_n$ be the LLL-reduced basis for $\mathcal{L}$. Every vector $v \in \mathcal{L}$ can be represented as Mz for $z \in \mathbb{Z}^n$. Let $H_k \stackrel{\text{def}}{=} \{Mz \mid z_n = k\}$ for $k \in \mathbb{Z}$ which altogether partition all lattice vectors in $\mathcal{L}$, each belonging to some hyperplane H_k. Note that, to find an approximately closest lattice vector to t, one could find the closest hyperplane H_k to t; then, move t so it is at the same relation with H_0 as it was with H_k and continue recursively on the $(n-1)$-dimensional sublattice.

Algorithm 2: Babai's algorithm

Input: full-rank LLL-reduced matrix M, $t \in \mathbb{R}^n$
Output: A lattice vector v relatively close to t
$v = \vec{0}, t' = t$
for $n' = n$ *to 1* **do**
  $k = \lfloor \langle t', M_{n'}^{\perp} \rangle \rceil$ `// Find nearest hyperplane`
  $v = v + kM_{n'}$ `// Remember to add` $kM_{n'}$ `back`
  $t' = t' - kM_{n'}$ `// Continue w/ sublattice where coefficient of` $M_{n'}$ `is 0`
end
return v

Analysis. First, observe that the algorithm runs in a polynomial time and returns a lattice vector v. One observes that

(1) $t' + v = t$ always.

(2) After the stage in the loop when $n' = j$, it holds that

$$|\langle t', M_j^{\perp} \rangle| \leq \frac{1}{2} |\langle M_j, M_j^{\perp} \rangle|,$$

and that continues to be true throughout (as both sides never change).

Therefore, at the end of the run

$$\|t'\| \leq \frac{1}{2}\sqrt{\sum_{i=1}^{n}\langle M_i, M_i^{\perp}\rangle^2}.$$

Corollary B.27.

$$\mu(\mathcal{L}[M]) \leq \frac{1}{2}\sqrt{\sum_{i=1}^{n}\langle M_i, M_i^{\perp}\rangle^2}.$$

Claim B.28. *Running Babai's algorithm on $\frac{3}{4}$-LLL (where $\delta = \frac{3}{4}$) gives an error of*

$$\|t'\| \leq 2^{\frac{n}{2}-1}|\langle M_n, M_n^{\perp}\rangle|.$$

Corollary B.29. *Running Babai's algorithm on $\frac{3}{4}$-LLL is $2^{\frac{n}{2}}$-approximation algorithm for CVP.*

REFERENCES

[1] L. M. Adleman, Factoring and lattice reduction, draft March 16, 1995.

[2] D. Aharonov and O. Regev, Lattice problems in NP ∩ coNP. *J. ACM* **52** (2005), no. 5, 749–765.

[3] R. Ahlswede and L. H. Khachatrian, The complete intersection theorem for systems of finite sets. *Eur. J. Comb.* **18** (1997), no. 2, 125–136.

[4] M. Ajtai, Generating hard instances of lattice problems. In *Proc. 28th ACM Symposium on the Theory of Computing (Philadelphia, Pennsylvania, USA)*, pp. 99–108, ACM, New York, NY, USA, 1996.

[5] M. Ajtai, The shortest vector problem in L_2 is NP-hard for randomized reductions. In *Proc. 30th ACM Symposium on the Theory of Computing (Dallas, Texas, USA)*, pp. 10–19, ACM, New York, NY, USA, 1998.

[6] *49th Annual IEEE Symposium on Foundations of Computer Science, FOCS 2008, October 25–28, 2008, Philadelphia, PA, USA*, IEEE Computer Society, 2008.

[7] S. Arora, L. Babai, J. Stern, and Z. Sweedyk, The hardness of approximate optima in lattices, codes, and systems of linear equations. *J. Comput. Syst. Sci.* **54** (1997), no. 2, 317–331.

[8] S. Arora, B. Barak, and D. Steurer, Subexponential algorithms for unique games and related problems. *J. ACM* **62** (2015), no. 5, 42:1–42:25.

[9] S. Arora, E. Berger, E. Hazan, G. Kindler, and M. Safra, On non-approximability for quadratic programs. In *46th Annual IEEE Symposium on Foundations of Computer Science (FOCS 2005), 23–25 October 2005, Pittsburgh, PA, USA, Proceedings*, pp. 206–215, IEEE Computer Society, 2005.

[10] S. Arora, S. Khot, A. Kolla, D. Steurer, M. Tulsiani, and N. K. Vishnoi, Unique games on expanding constraint graphs are easy: extended abstract. In *Proceedings of the 40th Annual ACM Symposium on Theory of Computing, STOC 2008*, pp. 21–28, ACM, 2008.

[11] S. Arora, C. Lund, R. Motwani, M. Sudan, and M. Szegedy, Proof verification and the hardness of approximation problems. *J. ACM* **45** (1998), no. 3, 501–555.

[12] S. Arora and S. Safra, Probabilistic checking of proofs: A new characterization of NP. *J. ACM* **45** (1998), no. 1, 70–122.

[13] P. Austrin, Balanced max 2-SAT might not be the hardest. In *Proc. 39th ACM Symposium on the Theory of Computing (San Diego, California, USA)*, pp. 189–197, ACM, New York, NY, USA, 2007.

[14] P. Austrin, S. Khot, and M. Safra, Inapproximability of vertex cover and independent set in bounded degree graphs. *Theory of Computing* **7** (2011), no. 3, 27–43.

[15] L. Babai, Trading group theory for randomness. In *Proc. 17th ACM Symposium on Theory of Computing (Providence, Rhode Island, USA)*, pp. 421–429, ACM, New York, NY, USA, 1985.

[16] L. Babai, On Lovász' lattice reduction and the nearest lattice point problem. *Combinatorica* **6** (1986), no. 1, 1–13.

[17] B. Barak, F. G. S. L. Brandão, A. W. Harrow, J. A. Kelner, D. Steurer, and Y. Zhou, Hypercontractivity, sum-of-squares proofs, and their applications. In *Proceedings of the 44th Symposium on Theory of Computing Conference, STOC 2012, New York, NY, USA, May 19–22, 2012*, pp. 307–326 ACM, 2012.

[18] B. Barak, P. K. Kothari, and D. Steurer, Small-set expansion in shortcode graph and the 2-to-2 conjecture. *CoRR* (2018), arXiv:1804.08662.

[19] M. Bellare, O. Goldreich, and M. Sudan, Free bits, PCPs, and nonapproximability – towards tight results. *SIAM Journal on Computing* **27** (1998), no. 3, 804–915.

[20] Z. Brakerski and V. Vaikuntanathan, Efficient fully homomorphic encryption from (standard) LWE. *SIAM Journal on Computing* **43** (2014), no. 2, 831–871.

[21] M. Charikar, K. Makarychev, and Y. Makaryshev, Near-optimal algorithms for unique games. In *Proc. 38th ACM Symposium on Theory of Computing*, pp. 205–214, ACM, New York, NY, USA, 2006.

[22] S. Chawla, R. Krauthgamer, R. Kumar, Y. Rabani, and D. Sivakumar, On the hardness of approximating multicut and sparsest-cut. *Computational Complexity* **15** (2006), no. 2, 94–114.

[23] S. A. Cook, The complexity of theorem-proving procedures. In *Proceedings of the Third Annual ACM Symposium on Theory of Computing, STOC'71*, pp. 151–158, ACM, New York, NY, USA, 1971.

[24] N. R. Devanur, S. Khot, R. Saket, and N. K. Vishnoi, Integrality gaps for sparsest cut and minimum linear arrangement problems. In *Kleinberg* [58], pp. 537–546.

[25] I. Diakonikolas, D. Kempe, and M. Henzinger (eds.), In *Proceedings of the 50th Annual ACM SIGACT Symposium on Theory of Computing, STOC 2018, Los Angeles, CA, USA, June 25–29, 2018*, ACM, 2018.

[26] I. Dinur, The PCP theorem by gap amplification. In *Kleinberg* [58], pp. 241–250.

[27] I. Dinur, E. Fischer, G. Kindler, R. Raz, and S. Safra, PCP characterizations of NP: toward a polynomially-small error-probability. *Comput. Complex.* **20** (2011), no. 3, 413–504.

[28] I. Dinur, S. Khot, G. Kindler, D. Minzer, and M. Safra, On non-optimally expanding sets in Grassmann graphs. In *Diakonikolas et al.* [25], pp. 940–951.

[29] I. Dinur, S. Khot, G. Kindler, D. Minzer, and M. Safra, Towards a proof of the 2-to-1 games conjecture? In *Diakonikolas et al.* [25], pp. 376–389.

[30] I. Dinur, G. Kindler, and S. Safra, Approximating-CVP to within almost-polynomial factors is NP-hard. In *Proceedings 39th Annual Symposium on Foundations of Computer Science (Cat. No. 98CB36280)*, pp. 99–109, IEEE, 1998.

[31] I. Dinur and S. Safra, The importance of being biased. In *Proc. 34th Annual ACM Symposium on Theory of Computing*, pp. 33–42, ACM, New York, NY, USA, 2002.

[32] I. Dinur and S. Safra, On the hardness of approximating minimum vertex cover. *Ann. of Math. (2)* **162** (2005), no. 1, 439–485.

[33] L. Eldar and S. Hallgren, An efficient quantum algorithm for lattice problems achieving subexponential approximation factor, 2022.

[34] P. Erdős, C. Ko, and R. Rado, Intersection theorems for systems of finite sets. *The Quarterly Journal of Mathematics* **12** (1961), no. 1, 313–320.

[35] U. Feige, A threshold of $\ln n$ for approximating set cover. *J. ACM* **45** (1998), no. 4, 634–652.

[36] U. Feige, S. Goldwasser, L. Lovász, S. Safra, and M. Szegedy, Interactive proofs and the hardness of approximating cliques. *Journal of the ACM* **43** (1996), no. 2, 268–292.

[37] E. Friedgut, Boolean functions with low average sensitivity depend on few coordinates. *Combinatorica* **18** (1998), no. 1, 27–35.

[38] M. X. Goemans and D. P. Williamson, Improved approximation algorithms for maximum cut and satisfiability problems using semidefinite programming. *J. ACM* **42** (1995), no. 6, 1115–1145.

[39] O. Goldreich and S. Goldwasser, On the limits of nonapproximability of lattice problems. *Journal of Computer and System Sciences* **60** (2000), no. 3, 540–563.

[40] A. Gupta and K. Talwar, Approximating unique games. In *Proceedings of the Seventeenth Annual ACM–SIAM Symposium on Discrete Algorithms, SODA 2006, Miami, Florida, USA, January 22–26, 2006*, pp. 22–26, ACM Press, 2006.

[41] V. Guruswami, R. Manokaran, and P. Raghavendra, Beating the random ordering is hard: inapproximability of maximum acyclic subgraph. In *Proc. Annual IEEE Symposium on Foundations of Computer Science* [6], pp. 573–582.

[42] J. Håstad, Clique is hard to approximate within $n^{1-\varepsilon}$. *Acta Math.* **182** (1999), no. 1, 105–142.

[43] J. Håstad, Some optimal inapproximability results. *J. ACM* **48** (2001), no. 4, 798–859.

[44] D. Hilbert and W. Ackermann, *Grundzüge der theoretischen Logik*. Grundlehren der mathematischen Wissenschaften 27, Springer, Berlin Heidelberg, 1938.

[45] R. M. Karp, Reducibility among combinatorial problems. In *Complexity of Computer Computations: Proceedings of a symposium on the Complexity of Computer Computations*, pp. 85–103, Springer US, Boston, MA, 1972.

[46] S. Khot, Improved inaproximability results for MaxClique, chromatic number and approximate graph coloring. In *Proc. 42nd Annual Symposium on Foundations of Computer Science, FOCS 2001, 14–17 October 2001, Las Vegas, Nevada, USA*, pp. 600–609, IEEE Computer Society, NW Washington, DC, USA, 2001.

[47] S. Khot, On the power of unique 2-prover 1-round games. In *Proceedings of the 17th IEEE Annual Conference on Computational Complexity, Montréal, Québec, Canada, May 21–24, 2002*, IEEE Computer Society, NW Washington, DC, USA, p. 25, 2002.

[48] S. Khot, Hardness of approximating the shortest vector problem in high L_p norms. In *Proc. 44th IEEE Symposium on Foundations of Computer Science*, IEEE Computer Society, NW Washington, DC, USA, p. 290, 2003.

[49] S. Khot, Hardness of approximating the shortest vector problem in lattices. *J. ACM* **52** (2005), no. 5, 789–808.

[50] S. Khot, Inapproximability of NP-complete problems, discrete fourier analysis, and geometry. In *Proceedings of the International Congress of Mathematicians 2010*, pp. 2676–2697, Hindustan Book Agency, New Delhi, 2010.

[51] S. Khot, G. Kindler, E. Mossel, and R. O'Donnell, Optimal inapproximability results for MAX-CUT and other 2-variable CSPS? *SIAM J. Comput.* **37** (2007), no. 1, 319–357.

[52] S. Khot, D. Minzer, and M. Safra, On independent sets, 2-to-2 games, and Grassmann graphs. In *Proceedings of the 49th Annual ACM SIGACT Symposium on Theory of Computing, STOC 2017, Montreal, QC, Canada, June 19–23, 2017*, pp. 576–589, ACM, New York, NY, USA, 2017.

[53] S. Khot, D. Minzer, and M. Safra, Pseudorandom sets in Grassmann graph have near-perfect expansion. In *2018 IEEE 59th Annual Symposium on Foundations of Computer Science (FOCS)*, pp. 592–601, IEEE, 2018.

[54] S. Khot and R. O'Donnell, SDP gaps and UGC-hardness for Max-Cut-Gain. *Theory of Computing* **5** (2009), no. 1, 83–117.

[55] S. Khot and O. Regev, Vertex cover might be hard to approximate to within $2 - \varepsilon$. *J. Comput. Syst. Sci.* **74** (2008), no. 3, 335–349.

[56] S. Khot and N. K. Vishnoi, The unique games conjecture, integrality gap for cut problems and embeddability of negative-type metrics into l_1. *J. ACM* **62** (2015), no. 1, 8:1–8:39.

[57] G. Kindler, A. Naor, and G. Schechtman, The ugc hardness threshold of the l_p Grothendieck problem. In *SODA*, edited by S.-H. Teng, pp. 64–73, SIAM, 2008.

[58] J. M. Kleinberg (ed.), In *Proceedings of the 38th Annual ACM Symposium on Theory of Computing, Seattle, WA, USA, May 21–23, 2006*, ACM, 2006.

[59] A. Kolla, Spectral algorithms for unique games. *Computational Complexity* **20** (2011), no. 2, 177–206.

[60] A. Korkine and G. Zolotareff, Sur les formes quadratiques. *Math. Ann.* **6** (1873), no. 3, 366–389.

[61] A. K. Lenstra, H. W. Lenstra, and L. Lovász, Factoring polynomials with rational coefficients. *Math. Ann.* **261** (1982), 513–534.

[62] L. Levin, Universal search problems. *Probl. Peredachi Inf.* **9** (1973), no. 3, 115–116.

[63] V. Lyubashevsky, Fiat-Shamir with aborts: applications to lattice and factoring-based signatures. In *International Conference on the Theory and Application of Cryptology and Information Security*, pp. 598–616, Springer, 2009.

[64] P. Manurangsi, Inapproximability of maximum biclique problems, minimum k-cut and densest at-least-k-subgraph from the small set expansion hypothesis. *Algorithms* **11** (2018), no. 1, 10.

[65] D. Micciancio, The shortest vector problem is NP-hard to approximate to within some constant. *SIAM Journal on Computing* **30** (2001), no. 6, 2008–2035. Preliminary version in FOCS 1998.

[66] D. Micciancio and O. Regev, Worst-case to average-case reductions based on Gaussian measures. In *45th Symposium on Foundations of Computer Science (FOCS 2004), 17–19 October 2004, Rome, Italy, Proceedings*, pp. 372–381, IEEE Computer Society, 2004.

[67] H. Minkowski, *Geometrie der Zahlen*. B.G. Teubner, 1896.

[68] H. Minkowski, *Geometrie der Zahlen* 40, Teubner, 1910.

[69] D. Moshkovitz and R. Raz, Sub-constant error probabilistically checkable proof of almost-linear size. *Comput. Complex.* **19** (2010), no. 3, 367–422.

[70] E. Mossel, Gaussian bounds for noise correlation of functions and tight analysis of long codes. In *Proc. Annual IEEE Symposium on Foundations of Computer Science* [6], pp. 156–165.

[71] E. Mossel, R. O'Donnell, and K. Oleszkiewicz, Noise stability of functions with low influences: invariance and optimality. *Annals of Mathematics* (2010), 295–341.

[72] C. H. Papadimitriou and K. Steiglitz, *Combinatorial optimization: algorithms and complexity*. Prentice-Hall, 1982.

[73] P. Raghavendra, Optimal algorithms and inapproximability results for every CSP? In *Proceedings of the Fortieth Annual ACM Symposium on Theory of Computing, STOC'08*, pp. 245–254, ACM, New York, NY, USA, 2008.

[74] P. Raghavendra and D. Steurer, Graph expansion and the unique games conjecture. In *Proc. 42nd ACM Symposium on Theory of Computing*, pp. 755–764, ACM, New York, NY, USA, 2010.

[75] P. Raghavendra, D. Steurer, and M. Tulsiani, Reductions between expansion problems. In *Proceedings of the 27th Conference on Computational Complexity, CCC 2012, Porto, Portugal, June 26–29, 2012*, pp. 64–73, IEEE Computer Society, 2012.

[76] R. Raz, A parallel repetition theorem. *SIAM J. Comput.* **27** (1998), no. 3, 763–803.

[77] O. Regev, On lattices, learning with errors, random linear codes, and cryptography. *J. ACM* **56** (2009), no. 6.

[78] O. Regev and N. Stephens-Davidowitz, A reverse Minkowski theorem. In *Proceedings of the 49th Annual ACM SIGACT Symposium on Theory of Computing, STOC 2017, Montreal, QC, Canada, June 19–23, 2017*, edited by H. Hatami, P. McKenzie, and V. King, pp. 941–953, ACM, 2017.

[79] C. P. Schnorr, A hierarchy of polynomial-time basis reduction algorithms. *Theoretical Computer Science* **53** (1987), no. 2–3, 201–224.

[80] U. Shapira and B. Weiss, Stable lattices and the diagonal group. *Journal of the European Mathematical Society* **18** (2016), no. 8, 1753–1767.

[81] P. W. Shor, Algorithms for quantum computation: discrete logarithms and factoring. In *Proceedings 35th Annual Symposium on Foundations of Computer Science*, pp. 124–134, IEEE Computer Society, NW Washington, DC, USA, 1994.

[82] E. M. Stein and G. Weiss, *Introduction to Fourier analysis on Euclidean spaces.* Princeton Mathematical Series 32, 1971.

[83] L. Trevisan, Approximation algorithms for unique games. *Theory of Computing* **4** (2008), no. 1, 111–128.

[84] W. Banaszczyk, New bounds in some transference theorems in the geometry of numbers. *Mathematische Annalen* 296 (1993), no. 1. 625–635.

MULI (SHMUEL) SAFRA

Quantum Science and Technology, Tel Aviv University, P.O. Box 39040, Tel Aviv 6997801, Israel, safra@tauex.tau.ac.il

POLYHEDRAL TECHNIQUES IN COMBINATORIAL OPTIMIZATION: MATCHINGS AND TOURS

OLA SVENSSON

ABSTRACT

We overview recent progress on two of the most classical problems in combinatorial optimization, namely the matching problem and the traveling salesman problem. We focus on deterministic parallel algorithms for the perfect matching problem and the first constant-factor approximation algorithm for the asymmetric traveling salesman problem.

While these questions pose seemingly different challenges, recent progress has been achieved using similar polyhedral techniques. In particular, for both problems, we use linear programming formulations, even exponential-sized ones, to extract structure from problem instances to guide the design of algorithms.

MATHEMATICS SUBJECT CLASSIFICATION 2020

Primary 68W01; Secondary 68W20, 68W25, 68Q25, 68R10, 05C85

KEYWORDS

Approximation algorithms, combinatorial optimization, derandomization, linear programming, matchings, traveling salesman problem

INTERNATIONAL CONGRESS OF MATHEMATICIANS 2022 JULY 6–14

Proc. Int. Cong. Math. 2022, Vol. 6, pp. 4970–4994
DOI 10.4171/ICM2022/127

1. INTRODUCTION

[1] The matching problem and the traveling salesman problem are at the heart of combinatorial optimization. In the matching problem, the goal is to pair up the vertices using the edges of the given graph. Work on matchings has contributed to the development of many core concepts in modern computer science, including linear-algebraic, probabilistic, online, streaming, and parallel algorithms. Matchings has been used to show the limitations of several models of computation such as monotone circuits and linear programs (extension complexity). Edmonds was the first to give a polynomial-time algorithm for it. His landmark paper [16] from 1965 is often credited with the idea that polynomial-time is a good abstraction of time efficiency and thus, in turn, popularizing the central complexity class P. However, half a century later, we still do not have a full understanding of the matching problem in many relevant settings such as parallel and space-bounded algorithms.

The traveling salesman problem—of finding the shortest tour of n given cities—is perhaps the most famous benchmark problem for NP-hard optimization problems, similar to what the matching problem is for efficiently solvable problems (i.e., problems in P). Indeed, the study of the traveling salesman problem has played a key role in the development and evaluation of heuristics, integer programming solvers, and approximation algorithms. Already in 1954, Dantzig et al. [14] used a linear program to solve a 49-city instance. The strength of this linear program, often referred to as the subtour elimination relaxation or Held–Karp relaxation, is one of the most fascinating open problems in combinatorial optimization.

While making progress on these questions has been notoriously difficult, there have been recent exciting advancements that we overview in this article. In the first part (Section 2), we survey progress on finding the efficient deterministic algorithms that we *know should exist* ever since the beautiful randomized algorithms for matching problems by Lovász [33] and Mulmuley, Vazirani, and Vazirani [36] were discovered several decades ago. We focus on the breakthrough work by Fenner, Gurjar, and Thierauf [18, 19] that (almost) derandomized the Isolation Lemma of [36] in the special case of perfect matchings in bipartite graphs. We then explain the challenges that we needed to overcome to extend their result to general graphs [46]. Finally, while the work of [19] has inspired much progress, it remains elusive to completely derandomize the approach of [36]. In Section 2.4, we give some fascinating open questions in this line of work.

In the second part (Section 3), we consider the traveling salesman problem. Recent years have seen exciting progress on longstanding open problems: for the asymmetric version, we achieved the first constant-factor approximation algorithm in [48]; and for the symmetric version, Karlin, Klein, and Oveis Gharan [28] made the first improvement on the classic approximation algorithm by Christofides [10] and Serdyukov [43]. We note that our focus in Section 3 is almost exclusively on approximation algorithms for the asymmetric

1 Part of the writing in this overview article is taken from a grant proposal of the author, and some descriptions (as we point out) are taken from our works [46, 48].

traveling salesman problem. We do not discuss the recent results for the symmetric version in detail. We do, however, point out similarities between one approach for the asymmetric case and the breakthrough algorithm in [28]. Finally, in Section 3.5, we give some of our favorite open problems related to the traveling salesman problem.

While the matching problem and the traveling salesman problem pose seemingly different challenges, much of the recent progress has been achieved using similar algorithmic ideas. Specifically, *polyhedral techniques*—the study of the convex hull of integer solutions and finding/exploiting properties of it—have played a key role in both our works on the matching problem [46] and the asymmetric traveling salesman problem [48]. In this overview, we highlight one polyhedral property that is central in both results, the so-called laminar structure of extreme points. We have made an effort to keep notation light and to introduce concepts when they are used. For a comprehensive reading on polyhedral techniques for combinatorial optimization, we recommend the excellent book by Schrijver [41].

2. THE ALGORITHMS THAT MUST EXIST FOR PERFECT MATCHINGS

A central problem in combinatorial optimization is the (maximum weight) perfect matching problem. Recall that a perfect matching of a graph is a subset of the edges so that every vertex is incident to exactly one edge, i.e., all vertices are paired up using edges. The *maximum weight perfect matching* problem now simply asks us to find a perfect matching of maximum weight in a given edge-weighted graph. This problem has a long and beautiful history.

For bipartite graphs, the Hungarian method is often taught in algorithm classes as a prime example of the powerful primal–dual technique. It is referred to as the Hungarian method as it relies on ideas developed by the Hungarians König and Egerváry, although amazingly we now know that the algorithm was already discovered by Jabobi more than a century ago [26]! One advantage of bipartite graphs over general graphs is that the polyhedral structure, i.e., the exact linear programming formulation, of the perfect matching problem is significantly simpler for bipartite graphs. It was in 1965 that Edmonds developed the first polynomial-time algorithm for general graphs [16]. Alongside his algorithmic discovery, he also formulated his famous description of the perfect matching polytope [15].

We have thus known efficient, i.e., polynomial-time, deterministic algorithms for the maximum weight perfect matching problem in both bipartite and general graphs for over half a century. Surprisingly, the situation is dramatically different for the following slight changes to the objective:

- *k-Extendable matching:* find a perfect matching that maximizes the sum of weights of the k heaviest edges or, equivalently, find k edges of maximum weight that can be extended to a perfect matching.

- *Exact matching:* decide (or find if it exists) whether a given edge-weighted graph has a perfect matching of weight *equal* to a target W.[2]

The status of both these problems is very intriguing: *we have yet to discover efficient deterministic algorithms that we have known should exist for decades!* But how can we be so sure that the above mentioned problems should admit efficient deterministic algorithms? The answer to this question lies in the fundamental study of the power of randomization in algorithm design. There is strong evidence (see, e.g., [25]) that any problem that admits an efficient randomized algorithm also admits an efficient deterministic algorithm. This may at first be surprising as there are several problems for which only efficient randomized algorithms are known. However, complexity theory tells us that this discrepancy is very likely due to a lack of algorithmic techniques and not due to a fundamental difference in the power of randomized and deterministic polynomial-time computation.

If we allow randomization, there is an incredibly versatile algorithmic technique developed by Mulmuley, Vazirani, and Vazirani [36], building upon earlier work by Lovász [33]. Specifically, the algorithmic technique in [36] yields efficient randomized *parallel* algorithms for all the above-mentioned problems: maximum weight perfect matching, k-extendable matching, and exact matching. While the obtained randomized algorithms are relatively simple and even parallelizable, it is a notorious problem to remove the need for randomness, i.e., to derandomize the approach. This is true even for vanilla perfect matchings and devising an efficient *deterministic parallel* algorithm for the (unweighted) perfect matching problem is a long-standing open problem.

While it remains elusive to derandomize the algorithmic technique in [36] completely, there has been significant progress in the last few years starting with the groundbreaking work of Fenner, Gurjar, and Thierauf [18, 19]. We give an overview of this progress with a focus on parallel algorithms for the perfect matching problem. In Section 2.1 we briefly explain the main techniques in the mentioned randomized parallel algorithms. We then give a more detailed description of the elegant framework of [19] which was the starting point of much of the recent progress. In Section 2.3 we then explain the challenges and our ideas for generalizing the result of [19] from biparite to general graphs. Finally, in Section 2.4, we comment on open questions and why we think progress on the exact matching and the k-extendable matching problems requires major new ideas.

2.1. Randomized parallel algorithms for the perfect matching problem

The description in this section of randomized parallel algorithms for the perfect matching problem is mainly taken from [46]. For those algorithms, linear-algebraic techniques have been a very powerful approach. They rely on the Tutte matrix of a graph

2 We remark that for the exact matching problem to be tractable, we assume that the edge-weights are polynomially bounded; otherwise, NP-completeness follows immediately from subset sum. In fact, the problem can be reduced to the case when edge-weights only take values 0 and 1, which is referred to as the exact red–blue matching problem.

$G = (V, E)$, which is the $|V| \times |V|$ matrix defined as follows:

$$T(G)_{u,v} = \begin{cases} X_{(u,v)} & \text{if } (u,v) \in E \text{ and } u < v, \\ -X_{(v,u)} & \text{if } (u,v) \in E \text{ and } u > v, \\ 0 & \text{if } (u,v) \notin E, \end{cases}$$

where we ordered the vertices arbitrarily and $X_{(u,v)}$ for $(u,v) \in E$ are variables. Tutte's theorem [52] says that $\det T(G) \neq 0$ if and only if G has a perfect matching.

A natural algorithm based on the Tutte matrix, replaces each indeterminate by a random value from a large field and then computes the determinant. If the graph has a perfect matching, the Schwartz–Zippel lemma ensures that the value of the computed determinant is nonzero with high probability. Furthermore, as computing determinants can be done efficiently in parallel [7, 13], this yields an efficient *randomized* parallel algorithm for *deciding* whether a graph has a perfect matching [33].

A second approach that will be more amenable to derandomization, adopted by Mulmuley, Vazirani, and Vazirani [36] for the search version (see also [29] for another randomized parallel algorithm for the search version, i.e., for finding a perfect matching if it exists), is to replace the indeterminates by randomly chosen powers of two. Namely, for each edge (u,v), a random weight $w(u,v) \in \{1, 2, \dots, 2|E|\}$ is selected, and we substitute $X_{(u,v)} := 2^{w(u,v)}$. Now, let us make the crucial assumption that one perfect matching M is *isolated*, in the sense that it is the *unique minimum-weight perfect matching* (minimizing $w(M) = \sum_{e \in M} w(e)$). Then $\det T(G)$ remains nonzero after the substitution: one can show that M contributes a term $\pm 2^{2w(M)}$ to $\det T(G)$, whereas all other terms are multiples of $2^{2w(M)+1}$ and thus they cannot cancel $2^{2w(M)}$ out. The determinant can still be computed efficiently in parallel as all entries $2^{w(u,v)}$ of the matrix are of polynomial bit-length, and so we have a parallel algorithm for the decision version. An algorithm for the search version also follows: for every edge in parallel, test whether removing it causes this least-significant digit $2^{2w(M)}$ in the determinant to disappear; output those edges for which it does. The final ingredient of the randomized approach of [36] is that assigning random weights to edges does indeed isolate one matching with constant probability. This is known as the Isolation Lemma and is a powerful concept that turns out to be true in the much more general setting of arbitrary set families.

2.2. Fenner, Gurjar, and Thierauf's approach for bipartite graphs

The elegant framework introduced by Fenner, Gurjar, and Thierauf [19] has been the basis for many subsequent developments including our result on general graphs that we describe in the next section. Their starting point is the randomized algorithm of Mulmuley, Vazirani, and Vazirani [36]. It forms an attractive starting point for derandomization because the only randomized ingredient is the selection of the weight function w and there is a simple condition that guarantees its correctness: the algorithm succeeds if the weight function w is *isolating*, i.e., there is a unique minimum weight perfect matching with respect to w. Thus to find a deterministic parallel algorithm for the perfect matching it is sufficient to find (in parallel) an isolating weight function with polynomial values. In other words, we would like to derandomize the Isolation Lemma.

Fenner, Gurjar, and Thierauf's construction of isolating weight functions is actually oblivious to the considered graph. Specifically, for any $n \in \mathbb{N}$, they construct a family $\mathcal{F}_n$ of simple weight functions such that for *any* bipartite n-vertex graph there is an isolating weight function $w \in \mathcal{F}_n$. To completely derandomize [36], the weight functions in $\mathcal{F}_n$ should be simple (efficiently computable in parallel) and satisfy the following three conditions:

(1) For every n-vertex graph G, there is an isolating weight function $w \in \mathcal{F}_n$.

(2) The number of weight functions in $\mathcal{F}_n$ is at most a polynomial in n.

(3) Each weight function in $\mathcal{F}_n$ assigns integer weights that are bounded by a polynomial in n.

The last condition ensures that we can calculate the determinant of the Tutte matrix where we replaced X_e by $2^{w(e)}$ efficiently for every $w \in \mathcal{F}_n$. The first condition ensures that we are guaranteed to succeed if we try all weight functions in our family and the second condition says that we can afford to try all of them (polynomially many) in parallel.

We remark that the construction of $\mathcal{F}_n$ becomes trivial if we drop the second or last condition. Indeed, the family that contains all 0, 1-weight functions guarantees (1) and (3); and the family consisting of the single weight function w defined by $w(e_i) = 2^i$ (where pairs of vertices/edges are ordered in an arbitrary fixed order) guarantees (1) and (2). Prior to the work [19], no nontrivial bounds were known, and they almost managed to satisfy all criteria when restricted to bipartite graphs. Specifically, they construct such a family $\mathcal{F}_n$ for bipartite graphs where the polynomial bound on the size of $\mathcal{F}_n$ (the second condition) and the range of the weights (the third condition) were relaxed from polynomial to quasi-polynomial $2^{\log n^{O(1)}}$.

To construct $\mathcal{F}_n$, order the set $\{e_1, e_2, \ldots, e_{\binom{n}{2}}\}$ of potential edges in an n-vertex graph arbitrarily. It is not hard to see that the weight function w defined by $w(e_i) = 2^i$ is isolating. However, w does not have (quasi)polynomial values and it turns out that it is hard to find such a weight-function immediately. A key idea of [19] is to build the weight function in rounds, and at each step consider a much easier problem. In each round, a weight function is selected from a family $\mathcal{W}$ of $O(n^6)$ many weight functions defined by

$$\mathcal{W} = \{w^{(\ell)} \mid \ell = 1, \ldots, 2n^6\}, \quad \text{where } w^{(\ell)}(e_i) = 2^i \bmod \ell.$$

Theorem 2.1 ([19]). *For every n-vertex bipartite graph with at least one perfect matching, there is a selection of weight functions $w_1, w_2, \ldots, w_k \in \mathcal{W}$ with $k = O(\log n)$ such that there is a unique perfect matching M minimizing the lexicographic order of $(w_1(M), \ldots, w_k(M))$.*

Note that the above theorem implies the promised family of weight functions. Indeed, putting enough weight on the first weight function compared to the second and so on reduces lexicographic minimization to that of minimizing the total weight of a matching. So the family $\mathcal{F}_n = \{\sum_{i=1}^{k} n^{2(k-i)} w_i \mid w_1, \ldots, w_k \in \mathcal{W}\}$ satisfies the three criteria (1)–(3) where the polynomial bounds are replaced by the quasipolynomial bound $n^{O(\log n)}$.

Now, to prove Theorem 2.1, we first give a sufficient condition for a weight function to be isolating. We then explain how [19] ingeniously selects the weights w_i in rounds using girth (the length of the shortest cycle) as a progress measure.

Circulations: a sufficient condition for a weight function to be isolating. If a weight function w is *not* isolating, then there exist two minimum-weight perfect matchings, and their symmetric difference consists of alternating cycles. In each such cycle, the total weight of edges from the first matching must be equal to the total weight of edges from the second matching (as otherwise we could obtain another matching of lower weight). The difference between these two total weights is called the circulation of the cycle. Formally, the *circulation* of an even cycle C with respect to weight function w is defined by

$$\text{circulation}(C, w) = \left| \sum_{e \in M_1} w(e) - \sum_{e \in M_2} w(e) \right|,$$

where M_1 is the matching obtained by taking every second edge of C and M_2 is the matching $C \setminus M_1$. By the above, we can observe the following.

Observation 2.2. *If all cycles have nonzero circulation, then w is isolating.*

Weight function in $\mathcal{W}$ doubles the girth. As aforementioned, a key idea of [19] for proving Theorem 2.1 is to select the weight function in rounds, and at each step consider a much easier problem. Specifically, instead of trying to show that there is a weight function $w \in \mathcal{W}$ that assigns a nonzero circulation to *all* cycles (potentially exponentially many), let us start by finding a weight function $w \in \mathcal{W}$ that assigns a nonzero circulation to "short" cycles.

Suppose the considered bipartite graph $G = (V, E)$ has no cycle of length at most k. We will assign nonzero circulation to all cycles of length at most $2k$. To this end, we start by bounding the number of such cycles. For ease of notation, we bound the number of 8-cycles when G has no cycles of length 4. It will then be clear how to adapt the argument to the general case. The bound on the number of 8-cycles follows from a nice encoding argument. We associate a signature (a, b, c, d) with each 8-cycle, where a is the first vertex, b is the third vertex, c is the fifth vertex, and d is the seventh vertex when we traverse the cycle starting from one of the vertices. Now, if two 8-cycles have the same signature then it is easy to see that would yield a cycle of length 4. Hence, if G has no cycles of length 4, the number of 8-cycles is at most the number of signatures which is at most n^4. We can generalize this argument to get the following:

Lemma 2.3. *A graph with no cycles of length at most k has at most $2n^4$ cycles of length at most $2k$.*

We have thus bounded the number of cycles that we consider in this round by a polynomial in n. An easy argument shows that we can always find a weight function $w \in \mathcal{W}$ that assigns nonzero circulation to such a relatively small set of cycles.

Lemma 2.4. *Let $\mathcal{C}$ be a set of at most $2n^4$ even cycles, then there is a weight function in $\mathcal{W}$ that assigns nonzero circulation for every cycle in $\mathcal{C}$.*

Proof. Let w be the weight function defined by $w(e_i) = 2^i$. As already observed, w assigns nonzero circulation to all cycles. In particular, we have circulation$(C, w) \neq 0$ for every $C \in \mathcal{C}$. Furthermore, using that the w-weight of any edge is at most $2^{\#\text{edges}} \leq 2^{\binom{n}{2}}$, we also have circulation$(C, w) \leq 2^{n^2}$. We thus have

$$\prod_{C \in \mathcal{C}} \text{circulation}(C, w) \neq 0 \quad \text{and} \quad \prod_{C \in \mathcal{C}} \text{circulation}(C, w) \leq (2^{n^2})^{|\mathcal{C}|} \leq 2^{2n^6}.$$

That there is a weight function in $\mathcal{W}$ that assigns nonzero circulation to all cycles in $\mathcal{C}$, i.e., that there is an $\ell \in \{1, 2, \dots, 2n^4\}$ such that

$$\prod_{C \in \mathcal{C}} \text{circulation}(C, w) \neq 0 \bmod \ell$$

now follows from the fact that the least common multiple of $1, 2, \dots, 2n^6$ is greater than 2^{2n^6}, so not all these numbers can divide $\prod_{C \in \mathcal{C}} \text{circulation}(C, w)$. ■

Given a bipartite graph $G = (V, E)$ with no cycles of length at most k, we can thus find a weight function $w \in \mathcal{W}$ such that all cycles for length at most $2k$ has nonzero circulation. The following lemma ensures that the girth is a good progress measure for bipartite graphs. It shows that the subgraph $H = (V, E')$, where $E' \subseteq E$ is the union of perfect matchings minimizing w, has no cycle of length at most $2k$.

Lemma 2.5. *Consider the subgraph $H = (V, E')$ of G where $E' \subseteq E$ is the union of perfect matchings that minimize a weight function w. Then H does not contain any cycle C with* circulation$(C, w) \neq 0$.

Before giving the proof of this lemma, let us explain how it implies Theorem 2.1. Consider *any* n-vertex bipartite graph $G = (V, E)$. There are at most $n^4 \leq 2n^4$ cycles of length at most 4. So there is a weight function $w_1 \in \mathcal{W}$ that assigns nonzero circulation to these cycles by Lemma 2.4. Let $G_1 = (V, E_1)$ be the subgraph where $E_1 \subseteq E$ is the union of perfect matchings that minimize w_1. Then the above lemma says G_1 that has no cycles of length at most 4. Lemma 2.3 then says that G_1 has at most $2n^4$ cycles of length at most 8. This allows us to repeat the same argument to show that there is a weight function $w_2 \in \mathcal{W}$ such that the graph $G_2 = (V, E_2)$, where $E_2 \subseteq E_1$ is the union of perfect matchings of G_1 that minimize w_2, has no cycles of length 8. Note that G_2 contains those perfect matchings M that minimize the lexicographic order of $(w_1(M), w_2(M))$. By repeating this $k = \lceil \log_2 n \rceil$ steps, we select weight functions $w_1, \dots, w_k$ such that the graph G_k that contains those perfect matchings M that minimize the lexicographic order of $(w_1(M), \dots, w_k(M))$ has no cycles of length at most n. In other words, G_k has no cycles and there is therefore a unique perfect matching in G_k.

Proof of Lemma 2.5. The argument uses that the *bipartite* perfect matching polytope has the following simple characterization:

$$\begin{aligned} x(\delta(v)) &= 1 \quad \text{for } v \in V, \\ x_e &\geq 0 \quad \text{for } e \in E. \end{aligned}$$

Here, we used $\delta(v)$ to denote the set of edges incident to vertex v and $x(F) = \sum_{e \in F} x_e$ for a subset $F \subseteq E$ of edges. Let x^* be the convex combination of all perfect matchings that minimize w. By definition, x^* is in the perfect matching polytope, and its support equals E', i.e., $x^*_e > 0$ for every $e \in E'$.

Now suppose toward contradiction that E' contains a cycle C with circulation(C, w) $\neq 0$. In other words, if we let M_1 and M_2 be the unique partitioning of C's edges into two matchings, then

$$\sum_{e \in M_1} w(e) \neq \sum_{e \in M_2} w(e).$$

Suppose $\sum_{e \in M_1} w(e) < \sum_{e \in M_2} w(e)$ (the other case is symmetric). Then

$$y_e = \begin{cases} x^*_e + \varepsilon & \text{if } e \in M_1, \\ x^*_e - \varepsilon & \text{if } e \in M_2, \\ 0 & \text{otherwise} \end{cases}$$

is a feasible solution to the bipartite perfect matching polytope for a small enough $\varepsilon > 0$. Indeed, every vertex in C is incident to exactly one edge in M_1 and one edge in M_2 and so the degree constraints are maintained; we also have $y \geq 0$ by selecting $\varepsilon > 0$ small enough since $x^*_e > 0$ for every $e \in E'$. We further have that y has lower cost than x^* because $\sum_{e \in M_1} w(e) < \sum_{e \in M_2} w(e)$. A contradiction since x^* is a convex combination of perfect matchings minimizing w. ■

The proof of the above lemma completes the proof of Theorem 2.1 and the overview of the framework of Fenner, Gurjar, and Thierauf. A careful reader may have noted that the only place where we used that the graph was bipartite was in the proof of Lemma 2.5. In that proof, we used the simple structure of the bipartite perfect matching polytope. In the next subsection, we explain why replacing this lemma for general graphs is nontrivial and give a brief outline of the approach in [46].

2.3. Polyhedral techniques for general graphs

Parts of this section is adapted from [46]. To extend the argument to general graphs, it will be useful to look at the method explained in the previous section from a polyhedral perspective. We begin from the set of all perfect matchings, of which we take the convex hull: the perfect matching polytope. After applying the first weight function $w_1 \in \mathcal{W}$, we want to consider only those perfect matchings which minimize the weight; this is exactly the definition of a face of the polytope (e.g., face $F[1]$ in Figure 1). Recall that the goal is to show that for a small k, there exists $w_1, w_2, \ldots, w_k \in \mathcal{W}$ so that the lexicographic minimizer of $(w_1, w_2, \ldots, w_k)$ is unique. Now we need to have a smart progress measure to show that there is such a choice of $w_1, \ldots, w_k$ for a small k. It is in this part that a good polyhedral understanding plays a key role. Specifically, if we consider the set of solutions that minimize w_1 then that defines a face F_1 of the polytope (convex hull of solutions) and, then minimizing w_2 defines a subface F_2 of F_1, and so on. The goal is to show that there is a selection of

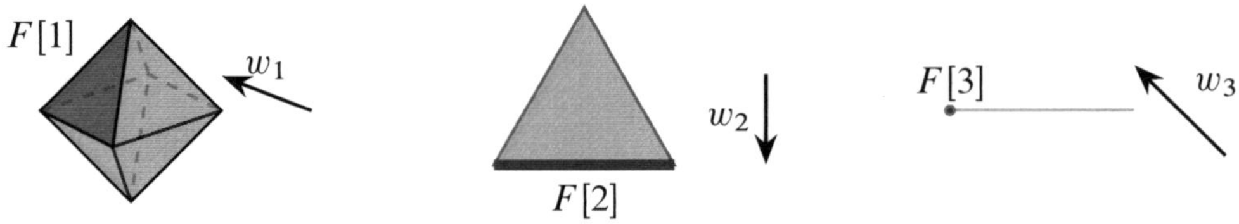

FIGURE 1
Polyhedral perspective on the construction of isolating weight function.

$(w_1, \ldots, w_k)$ so that the final face is of dimension 0, i.e., has a unique solution. In Figure 1, this happens after the choice of three weight functions.

In the bipartite case, any face is characterized by just taking a subset of edges (i.e., making certain constraints $x_e \geq 0$ tight). This simple structure of the bipartite perfect matching polytope was crucial in the proof of Lemma 2.5 and allowed [19] to have the girth as a simple progress measure as we described in the previous subsection.

In the nonbipartite case, the description of the perfect matching polytope is more involved. Namely, in addition to the degree constraints, Edmonds characterization [15] of the perfect matching polytope of a general graph $G = (V, E)$ also involves exponentially many odd-set constraints:

$$\begin{aligned} x(\delta(v)) &= 1 && \text{for } v \in V, \\ x(\delta(S)) &\geq 1 && \text{for } S \subseteq V \text{ with } |S| \text{ odd}, \\ x_e &\geq 0 && \text{for } e \in E. \end{aligned}$$

Recall that $\delta(v)$ denotes the set of edges incident to v and $x(F) = \sum_{e \in F} x_e$ for a subset $F \subseteq E$ of edges. Moreover, for a subset $S \subseteq V$ of vertices, we use $\delta(S)$ to denote the edges that cross the cut defined by S, i.e., the edges with exactly one endpoint in S.

Thus for general graphs, a face is not only defined by making certain constraints $x_e \geq 0$ tight but may also include tight *odd-set constraints* $x(\delta(S)) \geq 1$. This complicates our task, as depicted in Figure 2 (the same example was first given by [19] and then by [46] to demonstrate the difficulty of the general-graph case). Now a face is described by not only a subset of edges, but also a family of tight odd-set constraints. Thus we can no longer guarantee that any cycle whose circulation has been made nonzero will disappear from the support of the new face, i.e., the set of edges that appear in at least one perfect matching in this face (as, e.g., illustrated in Figure 2). Our idea of what it means to remove a cycle thus needs to be refined as well as the measure of progress we use to prove that a single matching is isolated after a few rounds.

Unfortunately, the current progress measure for general graphs is significantly more complex than for bipartite graphs and beyond the scope of this overview. Instead, we mention two crucial properties that allow us to deal with these odd-set constraints.

Decomposition into two sub-instances. The first property is easy to see: once we fix the single edge e in the matching which crosses a tight set S, the instance breaks up into two *independent* subinstances. That is, every perfect matching which contains e is the union of:

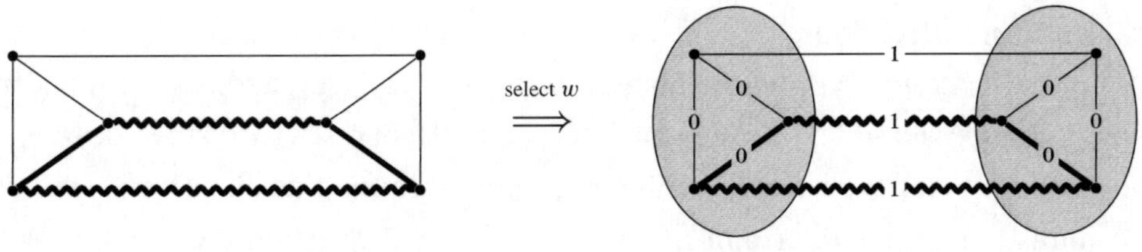

FIGURE 2
An illustration of the difficulty for general graphs. In trying to remove the bold cycle, we select a weight function w such that the circulation of the bold cycle is $|1-0+1-0| \neq 0$. By minimizing over w, we obtain a new, smaller subface—the convex hull of perfect matchings of weight 1—but every edge of the cycle is still present in one of these matchings. The vertex sets drawn in gray represent the new tight odd-set constraints that describe the new face (indeed, for a matching to have weight 1, it must take only one edge from the boundary of a gray set).

the edge e, a perfect matching on the vertex set S (ignoring the S-endpoint of e), and a perfect matching on the vertex set $V \setminus S$ (ignoring the other endpoint of e). Intuitively, this allows us to employ a divide-and-conquer strategy: to isolate a matching in the entire graph, we will take care of both subinstances and of the cut separating them.

One can see that the divide-and-conquer strategy would lead to a low depth recursion (and the selection of few weight functions), assuming that we could always find a balanced tight odd-set constraint. That is a tight odd-set constraint $x(\delta(S)) = 1$ with $|S| = \Omega(n)$ and $|V \setminus S| = \Omega(n)$. However, in general there is no reason to expect that we would always be able to find such a balanced cut and we combine the above strategy with a well-known structural property of the perfect matching polytope called laminarity.

Laminarity. The second crucial property that we utilize is that the family of odd-set constraints tight for a face exhibits good structural properties. Namely, it is known that at most $2n-1$ odd-set constraints are enough to describe any face and they have a very nice structural property called laminarity. A *family of sets is laminar* if any two sets in the family are either disjoint or one is a subset of the other. While this structural property is not very hard to prove, we omit it here as we will prove and exploit a very similar polyhedral fact in Section 3.4, where we discuss the asymmetric traveling salesman problem. The structure enables a scheme where we use the laminar family to define our progress measure and make progress in a bottom-up fashion. Combining this bottom-up approach with the techniques of [19] allows us to extend Theorem 2.1 to general graphs (albeit by increasing the number of weight functions by a logarithmic factor).

Theorem 2.6 ([46]). *For every n-vertex graph with at least one perfect matching, there is a selection of weight functions $w_1, w_2, \ldots, w_k \in \mathcal{W}$ with $k = O(\log^2 n)$ such that there is a unique perfect matching M minimizing the lexicographic order of $(w_1(M), \ldots, w_k(M))$.*

2.4. Future directions

The results described *almost* (instead of completely) derandomize [36] because of the quasipolynomial ($n^{(\log n)^{O(1)}}$) instead of polynomial bounds. To further develop these techniques to show that the perfect matching problem (even for bipartite graphs) has an efficient deterministic parallel algorithm remains a prominent question (see [3, 40] for recent progress in the special case of planar graphs where the decision and counting versions were previously known). This would most likely require an alternative approach as selecting logarithmically many weight functions in rounds naturally lead to quasipolynomial bounds.

The randomized algorithm in [36] is very versatile, and it can be used to solve several natural variants of the perfect matching problem, including the aforementioned k-extendable matching problem and the exact matching problem. Any progress on efficient (even sequential) deterministic algorithms for these problems would be very interesting. Indeed, no nontrivial results are known for general graphs, and there has only been progress on the exact matching problem in very special cases [21, 30, 54].

To make further progress on these questions, we believe that it is important to develop a good polyhedral understanding of the exact matching and the k-extendable matching problems. Indeed, all the work following the approach of [19] relied on our excellent polyhedral understanding of those problems [22, 23, 46]. A concrete step is to *determine the extension complexity*[3] of the exact matching problem and the k-extendable matching problem on bipartite graphs. We remark that it is important that we restrict ourselves to bipartite graphs as already the perfect matching problem for general graphs has exponential extension complexity [39]. This is in contrast to the perfect matching polytope for bipartite graphs, which has linear (in the number of edges) extension complexity. We therefore believe that the resolution of the above question would make significant progress towards understanding the additional difficulty posed by these variants.

3. THE (ASYMMETRIC) TRAVELING SALESMAN PROBLEM

The traveling salesman problem, of finding the shortest tour that visits n given cities, is one of the best-known optimization problems. It is a cornerstone NP-hard optimization problem that has played a central role in devising and evaluating techniques for overcoming NP-hardness (see, e.g., the books [4, 11, 32]). The difference compared to problems in P is that we do not expect NP-hard optimization problems to admit efficient algorithms that are guaranteed to find optimal solutions (unless P = NP). Therefore, when confronted with such an optimization problem, we need to relax our requirements on, e.g., optimality or reliability (that the algorithm is guaranteed to work on every input). If we relax reliability, we obtain heuristics where our goal is to devise algorithms with good performance on typical instances. If we relax optimality, we obtain approximation algorithms. *Approximation algo-*

---

3 An extension of a polyhedron P is a polyhedron Q such that P is the image of Q under a linear map. The extension complexity of P is the minimum number of facets (inequalities) of any extension of P.

rithms are efficient (i.e., polynomial-time) algorithms that are guaranteed to find a solution that is close—within a factor called the approximation guarantee—in value to an optimal solution. The study of approximation algorithms gives a mathematically rigorous way for (i) having a more fine-grained understanding of NP-hard optimization problems (some are easier to approximate than others) and (ii) in evaluating different algorithmic techniques. It is also a very intuitive notion as, in many situations, it is sufficient to find a solution that is close to optimal but not necessarily optimal.

As for the matching problem, polyhedral techniques have been central in the development of good algorithms for the traveling salesman problem. The most studied linear programming relaxation, which is often referred to as the Held–Karp relaxation (because of the seminal paper [24])[4] or the subtour elimination relaxation (because of the structure of the formulation), has puzzled researcher for decades. Indeed, a longstanding conjecture states that the Held–Karp relaxation approximates the value of any metric traveling salesman problem instance within a factor $4/3$ when distances are symmetric (the distance $\mathrm{dist}(i, j)$ of going from city i to city j equals the distance $\mathrm{dist}(j, i)$). Experimental evidence for the conjecture is given in [6]. A similar situation also holds in the asymmetric setting (when $\mathrm{dist}(i, j)$ does not necessarily equal $\mathrm{dist}(j, i)$) albeit with a much larger gap. Closing these gaps are considered major open problems in theoretical computer science.

Recently there have been significant advances on these questions. For the asymmetric traveling salesman problem, we obtained the first constant-factor approximation algorithm in [48]; see also the work by Traub and Vygen [50] who simplified the approach and obtained a better approximation guarantee. For the symmetric version, Karlin, Klein, and Oveis Gharan [28] obtained the first improvements on the classic $3/2$-approximation algorithm by Christofides [10] and Serdyukov [43]. We focus here on algorithmic approaches for the *asymmetric* traveling salesman problem. However, we briefly comment on the breakthrough work by [28] in Section 3.3, where we discuss a similar algorithm for the asymmetric problem.

3.1. Designing approximation algorithms for ATSP

An instance of the asymmetric traveling salesman problem (ATSP) is a tuple (V, dist) where V is the set of vertices/cities and dist gives the distances between vertices. In other words, an instance is a complete directed graph with edge-weights given by dist. A tour is a cycle that visits every vertex exactly once, i.e., a Hamiltonian cycle. The goal is to find a tour $F \subseteq E$ of minimum total distance $\mathrm{dist}(F) = \sum_{e \in F} \mathrm{dist}(e)$.

Without any assumptions on the distances, a simple reduction from the problem of deciding whether a graph is Hamiltonian shows that it is NP-hard to approximate the shortest tour to within any factor. Therefore it is common to assume that the distances satisfy the triangle inequality: the distance $\mathrm{dist}(i, k)$ from i to k is no longer than the distance $\mathrm{dist}(i, j)$ from i to j plus the distance $\mathrm{dist}(j, k)$ from j to k. All results that we mention refer to this

4 The relaxation in fact dates back to the earlier paper by Dantzig et al. [14] who solved a special 49-instance of the problem.

setting and we will assume that the distances satisfy the triangle inequality from now on (without explicitly stating it). One can see that this assumption is equivalent to allowing the tour to visit cities more than once; see the remark after Theorem 3.2. This viewpoint is very convenient when designing and analyzing algorithms for ATSP.

When designing an approximation algorithm, the task is to devise a polynomial-time algorithm that, for *any* instance (V, dist), outputs a tour whose length is guaranteed to be at most a factor c worse than the length of an optimal tour. The factor $c \geq 1$ is often referred to as the approximation ratio or the approximation guarantee. In the analysis, we thus face the problem of upper bounding our cost with the cost of a complex optimal solution that is even NP-hard to compute. A common technique to overcome this difficulty is to analyze the algorithm compared to a "simpler" lower bound that we can compute in polynomial time. Such a good lower bound then often also helps in the design of the approximation algorithms. In Sections 3.2 and 3.3, we see two complementary approaches that are based on two natural lower bounds: minimum cost cycle cover and minimum spanning tree, respectively. In Section 3.4, we then give an overview of the approach for achieving a constant-factor approximation algorithm and explain how the polyhedral structure of the Held–Karp relaxation allows us to reduce the general problem to very structured distances.

3.2. The repeated cycle cover approach

In this section, we explain the elegant "repeated cycle cover" approach by Frieze, Galbiati, and Maffioli [20]. A *cycle cover* of a directed graph is a subset C of the edges so that each vertex has in-degree and out-degree equal to one. In other words, C consists of a collection of cycles so that each vertex is in exactly one. A minimum cost cycle cover C in an edge-weighted graph is a cycle cover of minimum total distance $\text{dist}(C) = \sum_{e \in C} \text{dist}(e)$. It is not hard to see that we can compute a minimum cost cycle cover in polynomial time: it reduces to that of calculating a minimum cost perfect matching in a bipartite graph. Furthermore, we have the following observation, which follows by noting that an optimal tour is a cycle cover consisting of a single cycle, so the *minimum* cost cycle cover can only have a smaller cost.

Observation 3.1. *A minimum cost cycle cover costs at most the length of an optimal tour.*

The algorithm in [20] now ensures connectivity by the following procedure that repeatedly finds cycle covers:

(1) Find a minimum cost cycle cover.

(2) Select an arbitrarily proxy node for each cycle.

(3) Recursively solve the problem on proxies.

An illustration of the algorithm is given in the left part of Figure 3. First we find a cycle cover consisting of three cycles (depicted by solid edges). In each of these cycles, we select a proxy vertex (depicted in gray); and then in the next cycle cover instance, we find a minimum cost cycle cover on those proxy vertices a, c, and f (depicted by dashed edges). In this example,

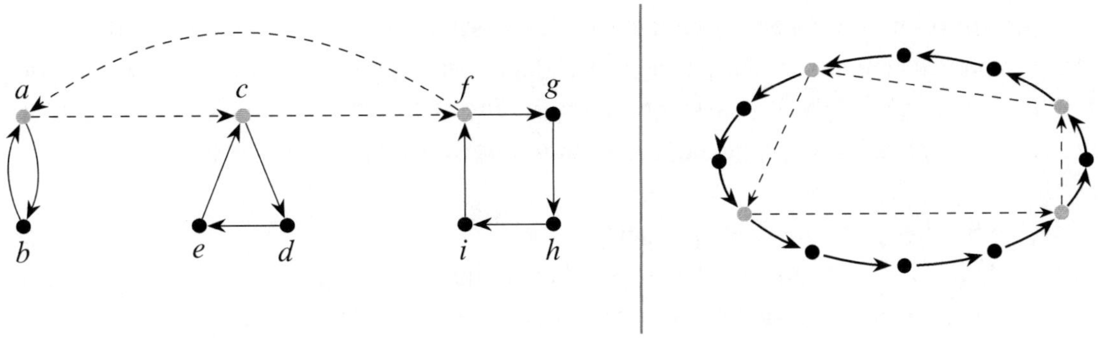

FIGURE 3
(Left) An illustration of the repeated cycle cover algorithm; (Right) Short-cutting does not increase the length of the tour by the triangle inequality.

we managed to connect the graph after only two iterations. In general, we have that each cycle cover at least halves the number of proxy vertices (they are exactly halved if each cycle in the cycle cover has length two). Thus, the algorithm selects at most $\log_2(n)$ cycle covers. We can furthermore upper bound the cost of each of these cycle covers by the length of an optimal tour. Indeed, let V_i be the set of proxy nodes when the ith cycle cover is found. Then, by Observation 3.1, the cost of the minimum cost cycle cover is at most the length of an optimal tour of V_i, which in turn is at most the length of an optimal tour that visits all vertices by the triangle inequality (see right part of Figure 3). The cost of the ith cycle cover is thus at most the length of an optimal tour. Combining the facts that we select at most $\log_2(n)$ cycle covers and each of them has cost at most the length of an optimal tour yields that the repeated cycle cover algorithm always finds a tour that is at most a factor $\log_2 n$ longer than an optimal tour.

Theorem 3.2 ([20]). *The repeated cycle cover algorithm is a* $\log_2(n)$*-approximation algorithm for ATSP.*

We remark that here we used the observation that finding a connected Eulerian (the in-degree of each vertex equals its out-degree) edge set is equivalent to finding a tour when distances satisfy the triangle inequality. Indeed, if the graph is connected and the in-degree of every vertex equals its out-degree, then we can efficiently find a so-called Eulerian tour that walks each edge exactly once. In the solution found in the left part of Figure 3, an Eulerian tour is $a - b - a - c - d - e - c - f - g - h - i - f - a$. Now any such tour can be short cut into a tour that visits each vertex exactly once by simply traversing the vertices in the same order as the Eulerian tour but not revisiting vertices (in the example this gives $a - b - c - d - e - f - g - h - i$). By the triangle inequality, this does not increase the length of the tour. Therefore, in the subsequent we will slightly abuse notation and refer to a connected Eulerian edge set as a tour.

The factor $\log_2(n)$ appears at first sight rather pessimistic. For the repeated cycle cover algorithm to return a tour with that approximation guarantee, basically all the found cycles must have length two and all cycle covers have a cost that equals the length of an optimal tour. However, it turns out that such worst-case instances do exist, and it is nontrivial to obtain improved guarantees. The papers [8,17,27] refine the approach to improve the constant in front of $\log_2(n)$ but the first asymptotic improvement on the logarithmic approximation guarantee was obtained by using another natural lower bound that we describe next.

3.3. The spanning tree approach

Another lower bound on the length of an optimal tour is the cost of a minimum cost spanning tree (where we forget the orientation of the edges). This is a lower bound since if we take a tour and remove a single edge, we get a tree whose cost is upper bounded by the length of the tour (minus the length of the dropped edge).

Using the spanning tree as a lower bound naturally leads to a complementary algorithm to the repeated cycle cover approach. Instead of ensuring that the graph is Eulerian and iteratively making it connected, we first connect the graph and then add edges to make it Eulerian:

(1) Find a minimum cost spanning tree T.

(2) Find a min-weight set of edges F so that $T \cup F$ is Eulerian.

For the symmetric case, this is the famous algorithm of Christofides [10] and Serdyukov [43]. As aforementioned, the cost of the tree T is at most the length of an optimal tour. Furthermore, for the symmetric case, one can show that the cost of the second step is at most half the optimum no matter the selected tree T. This yields the classic approximation guarantee of $3/2$ for the symmetric traveling salesman problem. In contrast, there is no hope to get a good upper bound on the cost of the second step in the asymmetric case if we start with an arbitrary spanning tree. Figure 4 depicts such an example. A minimum spanning tree is depicted by solid edges on the left. Extending that tree to an Eulerian graph must add four dotted edges of large cost as depicted in the middle. However, an optimal tour (depicted on the right) only uses one of these expensive (dotted) edges. By selecting the cost of the dotted edges to be large enough and increasing the number of vertices in the middle layer, we obtain an instance in which the tour obtained by extending a minimum spanning tree is a linear (in the number of vertices) factor more expensive than the optimal tour.

To overcome this difficulty, we use the Held–Karp relaxation which provides a stronger lower bound that can be seen to generalize both the minimum cost cycle cover and minimum weight spanning tree lower bounds.

Held–Karp relaxation. The Held–Karp relaxation has a variable $x_{(u,v)} \geq 0$ for every ordered pair of vertices (u, v). The intended solution is that $x_{(u,v)}$ should indicate whether the tour goes from u to v. We let E be the set of all such ordered pairs/edges. For a subset $S \subseteq V$ of vertices, we use $\delta^+(S) = \{(u, v) \in E \mid u \in S, v \notin S\}$ and $\delta^-(S) = \{(u, v) \in E \mid u \notin S, v \in S\}$ to denote the outgoing and incoming edges to S, respectively. We also let $\delta(S) = \delta^+(S) \cup \delta^-(S)$ be the "undirected" cut and for a subset

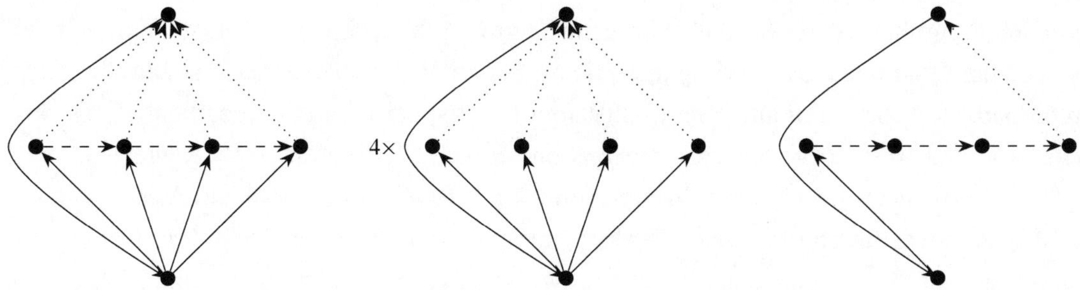

FIGURE 4
An example where correcting the degrees of a minimum spanning tree is expensive. Solid, dashed, and dotted edges have distances 1,2 and $M \gg 2$, respectively. The distances of the remaining pairs is the shortest path distance in this graph. The tour obtained by correcting a minimum spanning tree is depicted in center (using four expensive dotted edges) and an optimal tour is depicted on the right (using one expensive edge).

$F \subseteq E$ we let $x(F) = \sum_{e \in F} x_e$. With this notation, the relaxation is now defined as follows:

$$\begin{aligned} \text{minimize} \quad & \sum_{e \in E} x_e \cdot \mathrm{dist}(e) \\ \text{subject to} \quad & x(\delta^+(v)) = x(\delta^-(v)) = 1 \qquad v \in V, \\ & x(\delta(S)) \geq 2 \qquad \emptyset \neq S \subset V, \\ & x \geq 0. \end{aligned}$$

The first set of constraints says that each vertex should be visited once, so the in-degree and the out-degree should equal one for each vertex. The second set of constraints enforces that the solution is connected and they are sometimes referred to as subtour elimination constraints. We remark that although the Held–Karp relaxation has exponentially many constraints, it is well known that we can solve it in polynomial time either by using the ellipsoid method with a separation oracle or by formulating an equivalent compact (polynomial size) linear program.

Thin spanning trees. Consider a solution x to the Held–Karp relaxation and let $z_{\{u,v\}} = x_{(u,v)} + x_{(v,u)}$ be the solution where we dropped the orientation of edges. It is well known that $(n-1)/n \cdot z$ can be written as a convex combination of spanning trees (this can be seen by a simple calculation using Edmonds' characterization of the spanning tree polytope). In other words, there is a distribution μ over spanning trees satisfying

$$\Pr_{T \sim \mu}[e \in T] = \frac{n-1}{n} \cdot z_e \quad \text{for every } e \in E.$$

The selection of μ satisfying the above equality for every edge is not unique, and among all such distributions, [5] proposed to select the one of *maximum entropy*. In other words, instead of selecting a minimum spanning tree in the first step of the algorithm, we sample a tree T from μ. This randomized version of Christofides/Serdyukov algorithm has been very influential both for the symmetric and the asymmetric version. Although we focus on ATSP, let us mention that the recent major breakthrough [28] that presents the first improvement in more than four decades is a deep analysis of this algorithm. Their analysis heavily relies on properties of the maximum entropy distribution of spanning trees; specifically, that this distribution has very strong negative correlation properties (called strongly Rayleigh).

To analyze the randomized algorithm for ATSP, another influential contribution of [5] is the formulation of *thinness*: a clean sufficient condition for a tree T to have a low correction cost. A *spanning tree T is α-thin* with respect to a solution x to the Held–Karp relaxation if

$$\big|\{e \in T \mid e \text{ has exactly one endpoint in } S\}\big| \leq \alpha \cdot x\big(\delta(S)\big) \quad \text{for every } \emptyset \neq S \subsetneq V.$$

In words, the number of times the tree T crosses each cut is bounded by the (fractional) crossings of the linear program solution. Now using Hoffman's circulation theorem, they bound the correction cost of a thin tree leading to the following theorem:

Theorem 3.3 ([5]). *Given a spanning tree T that is α-thin with respect to an optimal solution to the Held–Karp relaxation, we can in polynomial time find a tour whose length is at most a factor $O(\alpha)$ longer than an optimal tour.*

Asadpour, Goemans, Madry, Oveis Gharan, and Saberi [5] then obtained their $O(\log n/\log\log n)$-approximation algorithm by showing that a tree sampled from the maximum entropy distribution is with high probability $O(\log n/\log\log n)$-thin (with respect to the Held–Karp solution). This analysis is tight in the following sense: it is known that there are instances so that, if we sample a tree from the maximum entropy distribution, the obtained tree is likely to be $\Omega(\log n/\log\log n)$-thin. However, it is conjectured that a $O(1)$-thin tree always exist. This has been proven for special graph classes, such as planar graphs and more generally bounded-genus graphs [37]. Major progress was achieved by Anari and Oveis Gharan [2], who showed that there always exists a $O(\log\log(n)^{O(1)})$-thin tree. The proof is highly nontrivial and it is not known to imply a polynomial-time algorithm. One key component in the proof is, e.g., a generalization of the celebrated proof of the Kadison–Singer problem. As we further discuss in Section 3.5, it remains an intriguing open question whether there always exists a $O(1)$-thin tree.

3.4. General distances are not that general: a constant factor approximation

In this section, we give a brief overview of the recent constant-factor approximation algorithm for ATSP that was given by [48] and then simplified and improved to an approximation guarantee of 22 by [50]. The algorithm is based on a series of reductions that harness strong structural properties from the Held–Karp relaxation. These reductions reduce the task to solving ATSP on highly specialized instances. For those instances, one can then adopt the techniques in [45] that solved similar special cases. Specifically, a key component is the constant-factor approximation algorithm for so-called node-weighted ATSP instances. We say that an ATSP is *node-weighted* if the distances dist is the shortest path metric of a directed graph $G = (V, E)$ with non-negative node-weights $\{y_v\}_{v \in V}$. That is, the weight of an edge

$(u, v) \in E$ in G is $y_u + y_v$, and the distance dist between a pair (a, b) of vertices is the shortest path from a to b in this graph. See the left part of Figure 5 for an example.

Theorem 3.4 ([45]). *Given a node-weighted ATSP instance, there is a polynomial-time algorithm that returns a tour whose length is at most a constant factor longer than the optimal value of the Held–Karp relaxation.*

The algorithm in [45] is an extension of the repeated cycle cover approach: it maintains an Eulerian subset of edges and iteratively adds new Eulerian sets of edges to connect the graph. However, the algorithm in [45] is more complex because the selection of new edges is done in a very careful and nontrivial manner as to not lose more than a constant-factor in the approximation guarantee.

The approach of [48] for general distances now performs a series of reductions to apply the techniques of [45]. While [50] made excellent progress in improving the simplicity and the guarantee of the algorithm, the complete algorithm remains rather complex and we refer the reader to conference version [47] for a longer overview of the approach. Here we focus on one key insight that allows us to focus on laminarly-weighted ATSP instances which generalizes node-weighted instances but still keep a similar structure. Interestingly, the techniques we use here are similar to what was used in Section 2.3 for the matching problem. Specifically, we will use that the optimal solution to the dual linear program can be selected to be a laminar family. Recall that a laminar family $\mathcal{L}$ of subsets is such that any two sets $A, B \in \mathcal{L}$ are either subsets of each other or disjoint (see right part of Figure 5). In order to simplify the dual, we use the fact that it is equivalent to finding a tour that visits every vertex at least once and to find a tour that visits every vertex exactly once (since we assume the triangle inequality). This allows us to drop the constraint that the in-degree and out-degree of a vertex are equal to 1. That is, we obtain an equivalent formulation of the Held–Karp relaxation by replacing $x(\delta^+(v)) = x(\delta^-(v)) = 1$ by $x(\delta^+(v)) = x(\delta^-(v))$ for every vertex $v \in V$. By associating variables $(\alpha_v)_{v \in V}$ and $(y_S)_{\emptyset \neq S \subset V}$ with the degree constraints and the subtour elimination constraints, respectively, we obtain the dual linear program:

$$\begin{aligned} \max \quad & \sum_{\emptyset \neq S \subset V} 2 \cdot y_S \\ \text{subject to} \quad & \sum_{S:(u,v) \in \delta(S)} y_S + \alpha_u - \alpha_v \leq \text{dist}(u, v) \quad \text{for } u \neq v \in V, \\ & y \geq 0. \end{aligned}$$

Now a key property of the dual is the laminarity of optimal solutions, which is similar to the structure that we used for the perfect matching problem in Section 2.3.

Lemma 3.5. *There exists an optimal solution (α, y) to the dual such that the support of y is a laminar family of vertex subsets.*

Proof. This proof is taken from [48]. We show the existence of a optimal laminar solution using a standard uncrossing argument (see, e.g., [12] for an early application of this technique to the Held–Karp relaxation of the symmetric traveling salesman problem). Select (α, y) to be an optimal solution to the dual minimizing $\sum_S |S| y_S$. That is, among all dual solutions that maximize the dual objective $2\sum_S y_S$, we select one that minimizes $\sum_S |S| y_S$. We claim that the support $\mathcal{L} = \{S : y_S > 0\}$ is a laminar family. Suppose not, i.e., that there are sets $A, B \in \mathcal{L}$ such that $A \cap B, A \setminus B, B \setminus A \neq \emptyset$. Then we can obtain a new dual solution $(\alpha, \hat{y})$, where $\hat{y}$ is defined, for $\varepsilon = \min(y_A, y_B) > 0$, as

$$\hat{y}_S = \begin{cases} y_S - \varepsilon & \text{if } S = A \text{ or } S = B, \\ y_S + \varepsilon & \text{if } S = A \setminus B \text{ or } S = B \setminus A, \\ y_S & \text{otherwise.} \end{cases}$$

That $(\alpha, \hat{y})$ remains a feasible solution follows since $\hat{y}$ remains nonnegative (by the selection of ε) and since for any edge e we have $\mathbb{1}_{e\in\delta(A)} + \mathbb{1}_{e\in\delta(B)} \geq \mathbb{1}_{e\in\delta(A\setminus B)} + \mathbb{1}_{e\in\delta(B\setminus A)}$. Therefore $\sum_{S:e\in\delta(S)} \hat{y}_S \leq \sum_{S:e\in\delta(S)} y_S$ and so the constraint corresponding to edge e remains satisfied. Further, we clearly have $2\sum_S \hat{y}_S = 2\sum_S y_S$. In other words, $(\alpha, \hat{y})$ is an optimal dual solution. However,

$$\sum_S |S|(y_S - \hat{y}_S) = \big(|A| + |B| - |A \setminus B| - |B \setminus A|\big)\varepsilon > 0,$$

which contradicts that (α, y) was selected to be an optimal dual solution minimizing $\sum_S |S| y_S$. Therefore, there can be no such sets A and B in $\mathcal{L}$, hence it is a laminar family. ■

We now use the laminar structure to argue that we can assume that our distances are very structured. Let x be an optimal primal solution and let (y, α) be an optimal dual solution with laminar support $\mathcal{L} = \{S \mid y_S > 0\}$. Consider the graph $G = (V, E)$ where the edge-set is the support of x: $E = \{e \mid x_e > 0\}$. Then by complementarity slackness, we have $\mathrm{dist}(e) = \sum_{S:(u,v)\in\delta(S)} y_S + \alpha_u - \alpha_v$ for every edge $e \in E$. Now note that for any cycle $v_0 \to v_1 \to \cdots \to v_{k-1} \to v_k = v_0$ (and thus Eulerian subset of edges), we have that its distance is given by y:

$$\sum_{i=0}^{k} \mathrm{dist}(v_i, v_{i+1}) = \sum_{i=0}^{k-1}\Big(\sum_{S:(v_i,v_{i+1})\in\delta(S)} y_S + \alpha_{v_i} - \alpha_{v_{i+1}}\Big) = \sum_{i=0}^{k-1}\Big(\sum_{S:(v_i,v_{i+1})\in\delta(S)} y_S\Big).$$

Hence, if we let the weight of an edge $e \in E$ in G be $\mathrm{dist}'(e) = \sum_{S:e\in\delta(S)} y_S$, then a tour has the same length in the shortest path metric of G obtained by using edge-weights $\mathrm{dist}(e)$ as in that obtained by using edge-weights $\mathrm{dist}'(e)$. This allows one to argue that we can replace the original distances with the shortest path metric of $G = (V, E)$ where each edge $e \in E$ has weight $\mathrm{dist}'(e)$ without loss of generality when designing approximation algorithms with respect to the Held–Karp relaxation. In other words, the distance from a vertex u to v is given by the shortest path in $G = (V, E)$ where the weight of each edge e is given by the total weight $\sum_{S\in\mathcal{L}:e\in\delta(S)} y_S$ of the sets it crosses in a laminar family. We refer to such instances as

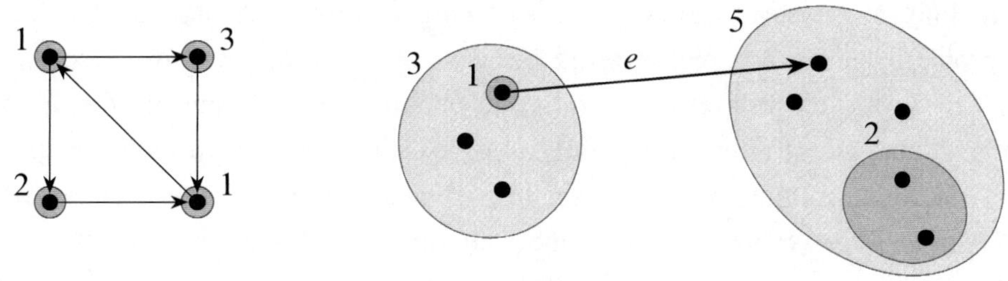

FIGURE 5
(Left) A node-weighted ATSP instance. The node-weights are depicted next to the vertices. The length of an edge is the sum of incident node weights and the length between two vertices is the shortest path distance in this graph. So the distance from the top-left vertex to the bottom-right vertex is $1 + 2 + 2 + 1 = 6$. (Right) An example of a laminarly-weighted ATSP instance. The sets of the laminar family are shown in gray, with their y-values written on their borders. We depict a single edge e that crosses three sets in the laminar family and has distance $1 + 3 + 5 = 9$.

laminarly-weighted (see right side of Figure 5). We remark that laminarly-weighted instances have some additional structure in [48] that we have simplified here.

Theorem 3.6. *Assume we have a polynomial-time algorithm that provides an α-approximation with respect to the Held–Karp relaxation for laminarly-weighted ATSP instances. Then there is a polynomial-time α-approximation algorithm with respect to the Held–Karp relaxation for the general ATSP problem.*

While laminarly-weighted instances have a similar structure to node-weighted instances, the approach in [48] performs several additional steps in order to use the techniques in [45]. In spite of the simplifications in [50], the overall algorithm remains complex and it is an interesting open problem to find a simple constant-factor approximation algorithm even for node-weighted instances. We discuss this and other open problems next.

3.5. Future directions

To give a tight analysis of the Held–Karp relaxation is a longstanding open problem (see, e.g., the two first open problems in the book [53] on approximation algorithms). We believe that the approximation guarantees given by this relaxation are $4/3$ and 2 for the symmetric and asymmetric traveling salesman problems, respectively. This would match the best known lower bounds [9].

The recent breakthrough in [28] opens up several promising directions for the symmetric version. Indeed, for the special case of unweighted shortest path metrics, the small improvement in [38] (using the same approach as in [28]) was quickly followed by more substantial improvements using different techniques [34,35,42]. It is interesting to know whether one can combine those techniques with the ones in [28]. Another exciting possibility is to exploit the laminar structure of the cost functions in the symmetric case as we did for ATSP.

The known constant-factor approximation algorithms for the asymmetric traveling salesman problem remain complex even in the case of node-weighted instances [45]. An important step for further progress is therefore to obtain a simpler constant-factor approximation algorithm. An intriguing possibility is to use the repeated-cycle cover approach. Instead of selecting a minimum cycle cover in each step, we would select one at random using the Held–Karp relaxation (similar to the randomized modification of the algorithm by Christofides/Serdyukov used in [28]). While the $\log_2(n)$ approximation guarantee is tight for the deterministic version, the approximation guarantee of the randomized version remains open.

In Section 3.3, we mentioned that the thin-tree conjecture implies a constant-factor approximation algorithm for ATSP. While we now know other methods for achieving a constant-factor approximation guarantee, the thin-tree conjecture is interesting in itself and it remains relevant for the asymmetric traveling salesman problem. In particular, it would imply a constant-factor approximation algorithm for the bottleneck version, where we wish to find a tour (Hamiltonian cycle) that minimizes the longest edge [1]. Finding a constant-factor approximation algorithm for bottleneck ATSP remains a challenging open problem and we believe that further progress is also likely to shed light on the thin-tree conjecture.

Finally, much of the work on the traveling salesman problem has been on analyzing the Held–Karp relaxation. But there are no real reasons to believe that we cannot achieve better algorithms using other lower bounds. In fact, the recent progress on the path traveling salesman problem is not with respect to a relaxation [49,51,55]. Moreover, we do not have any strong lower bounds on the relaxations obtained by using so called lift-and-project methods or hierarchies of relaxations [31,44].

ACKNOWLEDGMENTS

The author is very grateful to Jakub Tarnawski and László A. Végh. Jakub coauthored both [46] and [47] and László coauthored [47]. I also wish to thank my excellent mentors who have been central for my career, including my PhD advisor Monaldo Mastrolilli, postdoc advisor Johan Håstad, and my colleagues at EPFL.

FUNDING

This work was partially supported by the European Research Council (ERC) under the European Union's Horizon 2020 research and innovation programme (grant agreement 335288–OptApprox) and the Swiss National Science Foundation project 200021-184656 "Randomness in Problem Instances and Randomized Algorithms."

REFERENCES

[1] H. An, R. D. Kleinberg, and D. B. Shmoys, Improving Christofides' algorithm for the s-t path TSP. *J. ACM* **62** (2015), 34:1–34:28.

[2] N. Anari and S. Oveis Gharan, Effective-resistance-reducing flows, spectrally thin trees, and asymmetric TSP. In *FOCS*, pp. 20–39, IEEE Computer Society, 2015.

[3] N. Anari and V. V. Vazirani, Planar graph perfect matching is in NC. *J. ACM* **67** (2020), 21:1–21:34.

[4] D. L. Applegate, R. E. Bixby, V. Chvatál, and W. J. Cook, *The traveling salesman problem: a computational study*. Princeton University Press, 2006.

[5] A. Asadpour, M. X. Goemans, A. Madry, S. Oveis Gharan, and A. Saberi, An $O(\log n/\log\log n)$-approximation algorithm for the asymmetric traveling salesman problem. *Oper. Res.* **65** (2017), no. 4, 1043–1061.

[6] G. Benoit and S. Boyd, Finding the exact integrality gap for small traveling salesman problems. *Math. Oper. Res.* **33** (2008), no. 4, 921–931.

[7] S. J. Berkowitz, On computing the determinant in small parallel time using a small number of processors. *Inform. Process. Lett.* **18** (1984), no. 3, 147–150.

[8] M. Bläser, A new approximation algorithm for the asymmetric TSP with triangle inequality. *ACM Trans. Algorithms* **4** (2008), no. 4.

[9] M. Charikar, M. X. Goemans, and H. J. Karloff, On the integrality ratio for the asymmetric traveling salesman problem. *Math. Oper. Res.* **31** (2006), no. 2, 245–252.

[10] N. Christofides, Worst-case analysis of a new heuristic for the travelling salesman problem. Tech. Rep. 388, Graduate School of Industrial Administration, Carnegie-Mellon University, 1976.

[11] W. J. Cook, *In pursuit of the traveling salesman: mathematics at the limits of computation*, Princeton University Press, 2014.

[12] G. Cornuéjols, J. Fonlupt, and D. Naddef, The traveling salesman problem on a graph and some related integer polyhedra. *Math. Program.* **33** (1985), no. 1, 1–27.

[13] L. Csanky, Fast parallel inversion algorithm. *SIAM J. Comput.* **5** (1976), 618–623.

[14] G. Dantzig, R. Fulkerson, and S. Johnson, Solution of a large-scale traveling-salesman problem. *Oper. Res.* **2** (1954), 393–410.

[15] J. Edmonds, Maximum matching and a polyhedron with 0, 1 vertices. *J. Res. Natl. Bur. Stand.* **69** (1965), 125–130.

[16] J. Edmonds, Paths, trees, and flowers. *Canad. J. Math.* **17** (1965), 449–467.

[17] U. Feige and M. Singh, Improved approximation ratios for traveling salesperson tours and paths in directed graphs. In *APPROX*, pp. 104–118, Springer, 2007.

[18] S. A. Fenner, R. Gurjar, and T. Thierauf, Bipartite perfect matching is in quasi-NC. In *STOC*, pp. 754–763, ACM, 2016.

[19] S. A. Fenner, R. Gurjar, and T. Thierauf, Bipartite perfect matching is in quasi-NC. *SIAM J. Comput.* **50** (2021), no. 3.

[20] A. M. Frieze, G. Galbiati, and F. Maffioli, On the worst-case performance of some algorithms for the asymmetric traveling salesman problem. *Networks* **12** (1982), no. 1, 23–39.

[21] R. Gurjar, A. Korwar, J. Messner, and T. Thierauf, Exact perfect matching in complete graphs. *ACM Trans. Comput. Theory* **9** (2017), 8:1–8:20.

[22] R. Gurjar and T. Thierauf, Linear matroid intersection is in quasi-NC. *Comput. Complexity* **29** (2020), no. 2, 9.

[23] R. Gurjar, T. Thierauf, and N. K. Vishnoi, Isolating a vertex via lattices: Polytopes with totally unimodular faces. *SIAM J. Comput.* **50** (2021), no. 2, 636–661.

[24] M. Held and R. M. Karp, The traveling-salesman problem and minimum spanning trees. *Oper. Res.* **18** (1970), 1138–1162.

[25] R. Impagliazzo and A. Wigderson, $P = BPP$ if E requires exponential circuits: derandomizing the XOR lemma. In *STOC*, pp. 220–229, ACM, 1997.

[26] Jacobi's bound. https://www.lix.polytechnique.fr/~ollivier/JACOBI/jacobiEngl.htm, accessed: 2021-10-25.

[27] H. Kaplan, M. Lewenstein, N. Shafrir, and M. Sviridenko, Approximation algorithms for asymmetric TSP by decomposing directed regular multigraphs. *J. ACM* **52** (2005), no. 4, 602–626.

[28] A. R. Karlin, N. Klein, and S. Oveis Gharan, A (slightly) improved approximation algorithm for metric TSP. In *STOC*, pp. 32–45, ACM, 2021.

[29] R. M. Karp, E. Upfal, and A. Wigderson, Constructing a perfect matching is in random NC. *Combinatorica* **6** (1986), no. 1, 35–48.

[30] A. V. Karzanov, Maximum matching of given weight in complete and complete bipartite graphs. *Cybernetics* **23** (1987), no. 1, 8–13.

[31] J. B. Lasserre, An explicit equivalent positive semidefinite program for nonlinear 0-1 programs. *SIAM J. Control Optim.* **12** (2002), no. 3, 756–769.

[32] E. L. Lawler, J. K. Lenstra, A. H. G. R. Kan, and D. B. Shmoys, *The traveling salesman problem: a guided tour of combinatorial optimization*, Wiley, 1991.

[33] L. Lovász, On determinants, matchings, and random algorithms. In *FCT*, pp. 565–574, Akademie-Verlag, Berlin, 1979.

[34] T. Mömke and O. Svensson, Removing and adding edges for the traveling salesman problem. *J. ACM* **63** (2016), 2:1–2:28.

[35] M. Mucha, 13/9-approximation for graphic TSP. *Theory Comput. Syst.* **55** (2014), no. 4, 640–657.

[36] K. Mulmuley, U. V. Vazirani, and V. V. Vazirani, Matching is as easy as matrix inversion. *Combinatorica* **7** (1987), no. 1, 105–113.

[37] S. Oveis Gharan and A. Saberi, The asymmetric traveling salesman problem on graphs with bounded genus. In *SODA*, pp. 967–975, 2011.

[38] S. Oveis Gharan, A. Saberi, and M. Singh, A randomized rounding approach to the traveling salesman problem. In *FOCS*, pp. 550–559, IEEE Computer Society, 2011.

[39] T. Rothvoss, The matching polytope has exponential extension complexity. *J. ACM* **64** (2017), 41:1–41:19.

[40] P. Sankowski, NC algorithms for weighted planar perfect matching and related problems. In *ICALP*, pp. 97:1–97:16, LIPIcs. Leibniz Int. Proc. Inform. 107, Schloss Dagstuhl - Leibniz-Zentrum für Informatik, 2018.

[41] A. Schrijver, *Combinatorial optimization – polyhedra and efficiency*. Springer, 2003.

[42] A. Sebö and J. Vygen, Shorter tours by nicer ears: 7/5-approximation for the graph-TSP, 3/2 for the path version, and 4/3 for two-edge-connected subgraphs. *Combinatorica* **34** (2014), no. 5, 597–629.

[43] A. I. Serdyukov, O nekotorykh ekstremal'nykh obkhodakh v grafakh. *Upr. Sist.* **19** (1978), 76–79.

[44] H. D. Sherali and W. P. Adams, A hierarchy of relaxations between the continuous and convex hull representations for zero-one programming problems. *SIAM J. Discrete Math.* **3** (1990), no. 3, 411–430.

[45] O. Svensson, Approximating ATSP by relaxing connectivity. In *FOCS*, pp. 1–19, IEEE Computer Society, 2015.

[46] O. Svensson and J. Tarnawski, The matching problem in general graphs is in quasi-NC. In *FOCS*, pp. 696–707, IEEE Computer Society, 2017.

[47] O. Svensson, J. Tarnawski, and L. A. Végh, A constant-factor approximation algorithm for the asymmetric traveling salesman problem. In *STOC*, pp. 204–213, ACM, 2018.

[48] O. Svensson, J. Tarnawski, and L. A. Végh, A constant-factor approximation algorithm for the asymmetric traveling salesman problem. *J. ACM* **67** (2020), 37:1–37:53.

[49] V. Traub and J. Vygen, Approaching 3/2 for the s-t-path TSP. *J. ACM* **66** (2019), 14:1–14:17.

[50] V. Traub and J. Vygen, An improved approximation algorithm for ATSP. In *STOC*, pp. 1–13, ACM, 2020.

[51] V. Traub, J. Vygen, and R. Zenklusen, Reducing path TSP to TSP. In *STOC*, pp. 14–27, ACM, 2020.

[52] W. T. Tutte, The factorization of linear graphs. *J. Lond. Math. Soc.* **22** (1947), 107–111.

[53] D. P. Williamson and D. B. Shmoys, *The design of approximation algorithms*. Cambridge University Press, 2011.

[54] R. Yuster, Almost exact matchings. *Algorithmica* **63** (2012), no. 1–2, 39–50.

[55] R. Zenklusen, A 1.5-approximation for path TSP. In *SODA*, pp. 1539–1549, SIAM, 2019.

OLA SVENSSON

Ecole Polytechnique Fédérale de Lausanne (EPFL), School of Computer and Communication Sciences, CH-1015 Lausanne, Switzerland, ola.svensson@epfl.ch

This Page Intentionally Left Blank

MIP* = RE: A NEGATIVE RESOLUTION TO CONNES' EMBEDDING PROBLEM AND TSIRELSON'S PROBLEM

THOMAS VIDICK

ABSTRACT

Connes' Embedding Problem is a deep question on approximability of certain tracial von Neumann algebras by finite-dimensional matrix algebras. We survey the connections between operator algebras, quantum information and theoretical computer science that enabled the recent resolution of this problem. The resolution goes through an equivalent formulation, known as Tsirelson's problem, in terms of separating convex sets whose definition is motivated by the study of nonlocality in quantum mechanics. We construct an explicit separating hyperplane using the theory of two-player games from complexity theory.

MATHEMATICS SUBJECT CLASSIFICATION 2020

Primary 68A12; Secondary 46L10, 68Q15

KEYWORDS

Quantum interactive proofs, Tsirelson's problem, entanglement

ICM INTERNATIONAL CONGRESS OF MATHEMATICIANS 2022 JULY 6–14

Proc. Int. Cong. Math. 2022, Vol. 6, pp. 4996–5025
DOI 10.4171/ICM2022/71

1. INTRODUCTION

In the 1930s [58] von Neumann laid the foundations for the theory of (what are now known as) von Neumann algebras, with the explicit goal of establishing Heisenberg's matrix-based formulation of quantum mechanics on a rigorous footing. Following the initial explorations of Murray and von Neumann, the new theory progressively took on a life of its own, and von Neumann algebras now routinely make their apparition in areas as diverse as geometry, representation theory, free probability, statistical mechanics, and many others. In his 1976 paper completing the classification of injective von Neumann algebras [14], for which he received the 1982 Fields medal, Alain Connes made a casual remark that has become a central problem in the theory of operator algebras. Paraphrasing, Connes' remark was that any finite von Neumann algebra, i.e., one that has a finite trace, "ought to" be well approximated by finite-dimensional matrix algebras. Thanks to the work of other mathematicians, including Kirchberg and Voiculescu, the remark, now known as *Connes' Embedding Problem* (CEP), rose to prominence as one of the most important open questions in operator algebras. Quoting Vern Paulsen, "The reason that so many operator algebraists care about this conjecture is that it plays much the same role in operator algebras as is played by the Riemann hypothesis in number theory. There are many problems that we would know the answer to, if only Connes were true." For example, Kirchberg showed that CEP is equivalent to the *QWEP conjecture* about the equivalence of the minimal and maximal tensor products on the full group C^* algebra of a nonabelian free group [33]. Voiculescu gave a reformulation in terms of the existence of matrix microstates in free probability [56]. Rădulescu showed that a group is hyperlinear if and only if its group von Neumann algebra satisfies CEP [48]. Goldbring and Hart showed that CEP holds if and only if every type II_1 tracial von Neumann algebra has a computable universal theory [23]. Many more equivalent formulations are known (see, e.g., [11] for a survey).

In these notes we give an overview of an approach to CEP that arose from the study of the nonlocal effects of entanglement in quantum mechanics, and recently led a negative answer to the problem [27]. In the 1980s Boris Tsirelson was placing the study of quantum correlations, i.e., those families of distributions that can be generated from local measurements on a bipartite physical system, on a rigorous mathematical footing. In his work Tsirelson discovered that there was a freedom in deciding how locality should be reflected in the mathematical formalism, and asked if that freedom had observable consequences. Namely, Tsirelson realized that "locality" of measurements could be modeled either by requiring that the Hilbert space associated with the entire system factors as $\mathcal{H} = \mathcal{H}_A \otimes \mathcal{H}_B$, with observables on either system being localized to the corresponding subspace as $A \otimes \mathrm{Id}$ and $\mathrm{Id} \otimes B$, respectively, *or* by allowing the Hilbert space to remain arbitrary but requiring that observables associated with each system mutually commute, i.e., $[A, B] = 0$. While the two models are clearly different from an algebraic point of view, Tsirelson's Problem (TP) asks whether they lead to the same families of distributions, i.e., whether the algebraic distinction has any observable consequence.

Tsirelson's problem rose to prominence due to its relevance for a purely computational task: as we will see later, were TP to have a positive answer then the "largest quantum violation of a Bell inequality," a quantity of much interest to experimentalists, would be computable. (This "largest violation" determines how conclusive an experiment demonstrating the nonlocal effects of quantum mechanics may be.) This realization led to a further study of the problem and a proof of its equivalence with CEP, thus elevating it to the same status as the multitude of other equivalent formulations already mentioned. Moreover, it also led to a potential approach to a negative answer, by showing that the largest quantum violation of a Bell inequality is in fact *not* computable.

The goal of these notes is to explain the relation between CEP and TP, describe the approach to Tsirelson's problem through computability theory, and sketch how that approach eventually led to a resolution of the problem. Our main conceptual tool will be the theory of two-player games, a construct from classical complexity theory which rose to prominence in the 1990s through its connection with the PCP theorem, a sweeping reformulation of the complexity class NP, and applications to hardness of approximation for constraint satisfaction problems. Techniques developed in this study, entirely independent from quantum information, play an essential role in the resolution of CEP.

We start by giving a precise formulation of the two (equivalent) problems which we are concerned with, Connes' Embedding Problem (CEP) and Tsirelson's Problem (TP), in Section 2. In Section 3 we give a first hint of our approach to resolving these problems, which proceeds by constructing hyperplanes separating two convex sets introduced by Tsirelson. This will lead us to introduce nonlocal games as a rich class of hyperplanes to work with. We end the section by sketching a proof of the equivalence between CEP and TP that goes through nonlocal games and an algebra associated to them. In Section 4 we get to the heart of the matter, which is the construction of interesting two-player games and concrete requirements on them that suffice to answer our algebraic problems. It is in this section that complexity theory makes its apparition, as our requirements will push us into the design of very efficient "compression" procedures that find their inspiration in the efficient "proof checking" revolution that led to the PCP theorem in complexity theory. In Section 5 we explain how the complexity-theoretic techniques are combined with ideas from self-testing in quantum information and stability in group representation theory to complete the argument. We end with a brief outlook in Section 6.

2. PROBLEM STATEMENT(S)

We start by reviewing two equivalent, but rather distinct in flavor, formulations of the problem that is the focus of this article. The first formulation is due to Connes [14] and known as *Connes' Embedding Problem* (CEP). The second formulation is due to Tsirelson [53], and we will refer to it as *Tsirelson's Problem* (TP).

2.1. Connes' embedding problem

The standard formulation of CEP states that "every separable type II_1 von Neumann algebra has an approximate embedding into the hyperfinite factor $\mathcal{R}$." Shortly we reformulate this statement using more elementary language. Before doing so we clarify the terms used in Connes' formulation.

A (separable) von Neumann algebra $\mathcal{M}$ is a subalgebra of $B(\mathcal{H})$, the bounded linear operators on a (separable) Hilbert space $\mathcal{H}$, that contains the identity, is closed under taking adjoints (an operation which we denote *), and is closed in the strong operator topology.[1] A *state* τ on $\mathcal{M}$ is a positive linear functional such that $\tau(1) = 1$. A state τ is *tracial* if $\tau(xy) = \tau(yx)$ for all $x, y \in \mathcal{M}$. It is *normal* if the restriction of τ to the unit ball of $\mathcal{M}$ is continuous with respect to the strong operator topology. A *tracial* von Neumann algebra $(\mathcal{M}, \tau)$ is a von Neumann algebra $\mathcal{M}$ equipped with a faithful normal tracial state τ.

A *commutative* von Neumann algebra is isomorphic to $L^\infty(X, \mu)$ for some probability measure space (X, μ). For this reason tracial von Neumann algebras are often thought of as *noncommutative probability spaces*. A von Neumann algebra is a *factor* if it has a trivial center. von Neumann factors are classified in *types*. In their pioneering work on von Neumann algebras, Murray and von Neumann showed that every tracial von Neumann algebra decomposes as a product of type I_n factors, for $1 \le n < \infty$, and a type II_1 factor. While for any $1 \le n \le \infty$, a type I_n factor is always isomorphic to $B(\mathcal{H})$ for some separable Hilbert space $\mathcal{H}$ of dimension n, type II_1 factors are much harder to classify; in fact, there cannot be a classification up to isormorphism by countable structures [49], rendering the problem all but hopeless. (Connes received the Fields medal in 1982 for his work on the classification of type III factors, which are not tracial.)

Murray and von Neumann introduced a specific II_1 factor denoted $\mathcal{R}$ and referred to as *the hyperfinite factor*. Here the use of "the" is justified by the fact that $\mathcal{R}$ is characterized up to isomorphism as the unique separable II_1 factor that satisfies a strong form of approximability by matrix algebras. Namely, $(\mathcal{M}, \tau)$ is said to be *approximately finite-dimensional* (AFD) if for every finite subset F of $\mathcal{M}$ and every $\varepsilon > 0$ there is a $*$-subalgebra $Q \subset M$ such that $Q \simeq M_n(\mathbb{C})$ for some n and for every $x \in F$ there is $y \in Q$ such that $\|x - y\|_2 \le \varepsilon$.[2] It can be shown that there is a unique AFD II_1 factor, which is referred to as "the hyperfinite factor $\mathcal{R}$" when the specific isomorphism does not matter. Concretely, there are many possible definitions of $\mathcal{R}$. The most straightforward definition, which is also the original one, is as the completion of the algebra $\bigcup_{n\ge1} M_{2^n}(\mathbb{C})$, where each $M_{2^n}(\mathbb{C})$ isometrically embeds in $M_{2^{n+1}}(\mathbb{C})$ using diagonal blocks. The trace on M is the natural extension of the (dimension-normalized) matrix trace on each M_{2^n}, which we write as $\mathrm{tr}(\cdot)$. With this definition it is immediate that $\mathcal{R}$ is AFD.

There exist some nonhyperfinite tracial von Neumann algebras (we give an example below). CEP is the statement that every such algebra, nevertheless, has some form of weak

---

1 This is the topology generated by the seminorms $x \mapsto \|xv\|$ for $v \in \mathcal{H}$, with $\|\cdot\|$ the operator norm on $\mathcal{H}$.

2 The norm is given by $\|x\|_2 = \tau(x^* x)^{1/2}$.

approximation by finite-dimensional matrix algebras. The meaning of the second half of the statement of CEP, "has an approximate embedding into the hyperfinite factor $\mathcal{R}$," can be formalized by requiring a trace-preserving embedding into an ultrapower $\mathcal{R}^\omega$. Rather than defining ultrapowers, we give an equivalent formulation due to Voiculescu **[57]**. For $(\mathcal{M}, \tau)$ a tracial von Neumann algebra and $x_1, \ldots, x_n$ Hermitian elements of $\mathcal{M}$, we say that $(x_1, \ldots, x_n)$ *has matricial microstates* if for every $\varepsilon > 0$ and $N \geq 1$, there is an integer $d \geq 1$ and $A_1, \ldots, A_n \in M_d(\mathbb{C})$ self-adjoint such that for all $p \leq N$ and $i_1, \ldots, i_p \in \{1, \ldots, n\}$,

$$\left|\operatorname{tr}(A_{i_1} \cdots A_{i_p}) - \tau(x_{i_1} \cdots x_{i_p})\right| < \varepsilon.$$

Then CEP is the statement that for any tracial von Neumann algebra $(\mathcal{M}, \tau)$, every tuple (x_i) of self-adjoint elements in $\mathcal{M}$ has matricial microstates. With more work, Kirchberg **[33]** (see also **[16]**) showed using the theory of Jordan algebras that CEP is equivalent to the statement that for every tracial von Neumann algebra $(\mathcal{M}, \tau)$, every finite sequence of unitaries $u_1, \ldots, u_n$ in $\mathcal{M}$ and every $\varepsilon > 0$ there are an integer $d \geq 1$ and $U_1, \ldots, U_n$ unitaries in $M_d(\mathbb{C})$ such that for all $i, j \in \{1, \ldots, n\}$,

$$\left|\operatorname{tr}(U_i^* U_j) - \tau(u_i^* u_j)\right| < \varepsilon. \tag{2.1}$$

This last formulation may be appealing to the computer scientist as it states that every finite subset of the unitary group of M approximately embeds into a finite-dimensional matrix unitary group—a form of infinite-dimensional, nonquantitative Johnson–Lindenstrauss lemma **[29]** for operators.

The versatility of CEP arises from the many examples of tracial von Neumann algebras that are known. We give some examples coming from groups; for many more, see, e.g., **[1]**. We restrict our attention to discrete, countable groups. For G a countable discrete group, let λ be the left regular representation of G in $\ell^2(G)$. Then the strong operator closure of the linear span of $\lambda(G)$ in $B(\ell^2(G))$ is a von Neumann algebra called the *group von Neumann algebra* of G and denoted $L(G)$. Letting $(\delta_g)_{g\in G}$ be the natural orthonormal basis of $\ell^2(G)$ and $e \in G$ the unit, there is a natural trace φ on $L(G)$ given by $\varphi(x) = \langle \delta_e, x\delta_e\rangle$. One can check that this is a normal faithful tracial state, hence $(L(G), \varphi)$ is a tracial von Neumann algebra. Moreover, $L(G)$ is a factor if and only if G has the i.c.c. property, namely every nontrivial conjugacy class is infinite. Thus the group von Neumann algebra of an infinite i.c.c. group G is a II_1 factor. Some examples are $L(S_\infty)$, where S_∞ is the group of finitely supported permutations of the natural numbers, and $L(\mathbb{F}_n)$ for $n \geq 2$, with $\mathbb{F}_n$ the free group on n generators. It can be shown that $L(G)$ is isomorphic to $\mathcal{R}$ if and only if G is an i.c.c. amenable group. Thus $L(S_\infty)$ is isomorphic to $\mathbb{R}$, whereas $L(\mathbb{F}_n)$ for $n \geq 2$ is not. Connes **[14]** showed that $L(\mathbb{F}_n)$ satisfies CEP, i.e., it embeds in R^ω, and this discovery prompted his remark about all type II_1 factors.

A group G is *hyperlinear* if and only if for every finite $F \subseteq G$ and $\varepsilon > 0$ there are a $d \geq 1$ and a map $\theta : F \to U_d(\mathbb{C})$ that is an (F, ε)-almost homomorphism. Namely, if $g, h \in F$ are such that $gh \in F$ then $\|\theta(g)\theta(h) - \theta(gh)\|_2 < \varepsilon$, if $e \in F$ then $\|\theta(e) - \mathrm{Id}\,\|_2 < \varepsilon$, and if $x \neq y \in F$ then $\|\theta(x) - \theta(y)\|_2 \geq 1/4$. This formulation is due to Rădulescu **[47]** who introduced the terminology "hyperlinear." Later, Elek and Szabó **[18]** showed that the notion of soficity introduced by Gromov can be characterized in an equivalent manner, requiring θ to map to the symmetric group S_d. Radulescu showed that a countable group G is hyperlinear if and only if $L(G)$ embeds into R^ω, and he gave an example of G, different from $\mathbb{F}_n$, such that $L(G)$ is not hyperfinite but embeds into R^ω, thus giving another example of a nonhyperfinite II_1 factor that satisfies CEP. The conjecture whether every countable group is hyperlinear remains open (as does the stronger conjecture whether every countable group is sofic).

2.2. Tsirelson's problem

In the early 1980s Boris Tsirelson [53] wrote a series of papers laying out the mathematical formalism for the systematic study of the nonlocal properties of quantum mechanics. In quantum mechanics, the state of a physical system is represented by a unit vector $|\psi\rangle$ in a separable Hilbert space $\mathcal{H}$.[3] A measurement (or PVM, for projective-valued measure) is represented by a finite collection $\{P_1, \dots, P_k\}$ of projections on $\mathcal{H}$ such that $\sum_i P_i = \mathrm{Id}$. Here k is the number of outcomes that the measurement can have; according to the Born rule, the probability that the ith outcome is obtained when a system in state $|\psi\rangle$ is measured according to $\{P_i\}$ is given by $\langle\psi|P_i|\psi\rangle$.

Tsirelson was interested in modeling situations in which a physical system is composed of two isolated parts that can be measured independently, by observers present in separated locations.[4] Let us imagine that each observer can make one out of n possible measurements, each with k possible outcomes, on their share of the system. To model the statistical behavior that such an experiment might have, Tsirelson introduced the following subset of $[0,1]^{n^2k^2}$:

$$\begin{aligned} C_{qs}(n,k) = \big\{ & \big(\langle\psi|A_a^x \otimes B_b^y|\psi\rangle\big)_{x,y,a,b} : \mathcal{H}_A, \mathcal{H}_B \text{ Hilbert spaces}, |\psi\rangle \in \mathcal{H}_A \otimes \mathcal{H}_B, \\ & \big\||\psi\rangle\big\| = 1, \forall (x,y) \in \{1,\dots,n\}^2, \{A_a^x\}_{a\in\{1,\dots,k\}}, \\ & \{B_b^y\}_{b\in\{1,\dots,k\}} \text{ PVM on } \mathcal{H}_A, \mathcal{H}_B \text{ resp.}\big\}. \end{aligned} \tag{2.2}$$

Here the subscript qs stands for *quantum spatial* and refers to the presence of a tensor product in the expression $\langle\psi|A_a^x \otimes B_b^y|\psi\rangle$. This tensor product is natural if one accepts the rule for associating a Hilbert space to composite systems in nonrelativistic quantum mechanics, which proceeds by tensoring. Thus in the definition of C_{qs} it is understood that observer A's system is modeled using a Hilbert space $\mathcal{H}_A$, observer B's using $\mathcal{H}_B$, and the Hilbert space associated with them jointly is $\mathcal{H}_A \otimes \mathcal{H}_B$, the space in which the system state vector $|\psi\rangle$ lives. Continuing, Tsirelson observed that one could consider an a priori more general

---

3 We adopt Schödinger's bra-ket notation: a ket $|\psi\rangle$ is used to denote a vector $|\psi\rangle \in \mathcal{H}$, whereas a bra $\langle\psi|$ is used to denote a linear form $\langle\psi| : |\varphi\rangle \in \mathcal{H} \mapsto \langle\psi|\varphi\rangle = \langle\psi,\varphi\rangle \in \mathbb{C}$.

4 We do not make the notion of "separated locations" precise other than through the upcoming formalism; indeed, finding a formalization of it is the entire point of Tsirelson's work. For the moment, the reader can consider that we are only interested in the nonrelativistic scenario.

definition,

$$C_{qc}(n,k) = \{(\langle\psi|A_a^x B_b^y|\psi\rangle)_{x,y,a,b} : \mathcal{H} \text{ Hilbert space}, |\psi\rangle \in \mathcal{H}, \||\psi\rangle\| = 1,$$
$$\forall (x,y) \in \{1,\ldots,n\}^2, \{A_a^x\}_{a\in\{1,\ldots,k\}}, \{B_b^y\}_{b\in\{1,\ldots,k\}} \text{ PVM on } \mathcal{H}$$
$$\text{such that } [A_a^x, B_b^y] = 0 \forall (a,b) \in \{1,\ldots,k\}^2\}. \tag{2.3}$$

Here the subscript qc stands for "quantum commuting" and refers to the fact that in this definition spatial isolation is modeled by the constraint that measurement operators should commute, a condition which also allows for their joint measurability. This definition is more natural from a relativistic viewpoint, e.g., in algebraic quantum field theory, observables associated with space-time isolated regions are required to commute, but there is no a priori separation of the global Hilbert space into tensor products.

Each definition gives rise to a family of convex sets (convexity is easily verified by taking direct sums of PVMs and scaled vectors). Both provide reasonable models for the distributions, sometimes also referred to as *correlations*, that can be generated by an experiment of the form that Tsirelson envisioned. Moreover, in the case all Hilbert spaces are taken to be finite-dimensional, it is an exercise to show that the two sets coincide.[5] Possibly due to this observation, Tsirelson initially assumed that the sets coincide in general, and went on to prove results about the sets C_{qs}; in particular, he introduced techniques to bound certain facets of it. When asked for a proof of the equality, however, Tsirelson realized that it eluded him and posed the question as an open problem.[6]

Tsirelson's problem has two variants. The first, referred to as *Tsirelson's strong problem*, asks about strict equality between the two sets. This problem was answered in 2019 in a beautiful work by Slofstra **[51]**, who showed that the set $C_{qs}(n,k)$ is not closed for all large enough n,k. Since $C_{qc}(n,k)$ is easily verified to be closed, the sets cannot always be equal. Slofstra proved this result by introducing novel techniques relating approximation properties for groups to the suprema of linear functionals on these sets through the language of two-player games, which we will introduce in the next section. In his formulation of the problem, Tsirelson indicated that, if the sets were shown distinct, then an "even more important" problem would arise, which is referred to as the *weak Tsirelson's problem*: does $\overline{C_{qs}(n,k)} = C_{qc}(n,k)$ for all n,k? Here we will refer to this formulation directly as *Tsirelson's Problem* (TP).

While Tsirelson's problem may at first glance look like an arcane question in the foundations of quantum mechanics, there is a good reason why the authors of **[41]** asked Tsirelson for a proof of his claim regarding equality of the two sets. To explain their motivation, one should bear in mind that the problem of optimizing a linear functional over $C_{qs}(n,k)$ is of primary importance for experiments demonstrating the nonlocality of quantum mechanics, a key feature of the theory that has puzzled physicists and philosophers alike ever since the EPR thought experiment brought it to the fore. Unfortunately, even for small,

5 A slightly more difficult exercise is to show that they always coincide when $n = k = 2$.

6 See "Bell inequalities and operator algebras", available at https://www.tau.ac.il/~tsirel/download/bellopalg.pdf.

fixed n, k, direct optimization over $C_{qs}(n,k)$ seems intractable, as one has no a priori bound on the dimension of the space $\mathcal{H}$ that will lead to an (even approximately) optimal correlation. In their paper, Navascues et al. introduce a decreasing family of outer approximations of the set $C_{qs}(n,k)$ that are each represented as a positive semidefinite set, which implies that optimization over each set can be performed in time commensurate with its description size using semidefinite programming, an extension of linear programming. However, Navascues et al. were only able to show that their outer approximations converge to the set $C_{qc}(n,k)$, instead of $C_{qs}(n,k)$. If Tsirelson's (weak) problem had an affirmative answer, their work would lead to an algorithm for computing the supremum of a linear function over $C_{qs}(n,k)$, or equivalently, computing the largest quantum violation of a Bell inequality. Thus the original motivation for solving Tsirelson's problem is purely computational, and as we will see later, it is surprising also how the problem was eventually resolved.

Further motivation for resolving Tsirelson's problem arose when Fritz [20] and Junge et al. [30] independently showed that Tsirelson's problem follows from Kirchberg's QWEP conjecture, itself shown equivalent to CEP by Kirchberg. Later, Ozawa [43] established the equivalence between the three conjectures, thus tying TP to CEP and the many equivalent formulations of it. In Section 3.2 below we will sketch a different proof of the equivalence between TP and CEP that does not go through the QWEP conjecture.

3. SEPARATING HYPERPLANES AS NONLOCAL GAMES

The formulation of Tsirelson's problem as a question about equality of two convex sets provides a natural geometric approach to its resolution. For $n, k \geq 1$ and $\lambda \in (\mathbb{R}^{n^2k^2})^*$, a linear functional on $\mathbb{R}^{n^2k^2}$, we introduce the quantities (see also Figure 1)

$$\omega_{qa}(\lambda) = \sup_{p \in C_{qs}(n,k)} |\lambda \cdot p| \quad \text{and} \quad \omega_{qc}(\lambda) = \sup_{p \in C_{qc}(n,k)} |\lambda \cdot p|. \tag{3.1}$$

Here the subscript qa stands for "quantum approximate"; we write $C_{qa}(n,k)$ for the closure $C_{qa}(n,k) = \overline{C_{qs}(n,k)}$. We also define a quantity $\omega_{\text{loc}}(\lambda)$, where the supremum is taken over "local" correlations p (this is the case where all PVMs in (2.2) mutually commute, see (3.3) below for a precise definition and a justification of the term "local").

To give a negative answer to Tsirelson's problem, it suffices to find n, k and a λ such that $\omega_{qa}(\lambda) < \omega_{qc}(\lambda)$. In the foundations of quantum mechanics, an inequality of the form $\omega_{\text{loc}}(\lambda) \leq \alpha$ is called a *Bell inequality*, and an inequality of the form $\omega_{qa}(\lambda) \leq \beta$ is called a *Tsirelson inequality*. The best right-hand side in a Tsirelson's inequality is referred to as the "largest quantum violation" of the corresponding optimal Bell inequality. The design of functionals λ such that $\omega_{\text{loc}}(\lambda) < \omega_{qa}(\lambda)$ is relevant to the design of experiments witnessing the "nonlocality" of quantum correlations. Because of this, many functionals have been studied, such as the famous CHSH inequality $\omega_{\text{loc}}(\lambda_{\text{CHSH}}) < 2\sqrt{2}$ where $\lambda_{\text{CHSH}} \in (\mathbb{R}^{2^2 2^2})^*$ is a specific functional named after its inventors, who also showed that it satisfies $\omega_{qa}(\lambda_{\text{CHSH}}) \geq 4$ (Tsirelson later showed that this bound is tight [52]). How does one go about finding interesting λ? One can use guessing and physical intuition for how special quantum phenomena

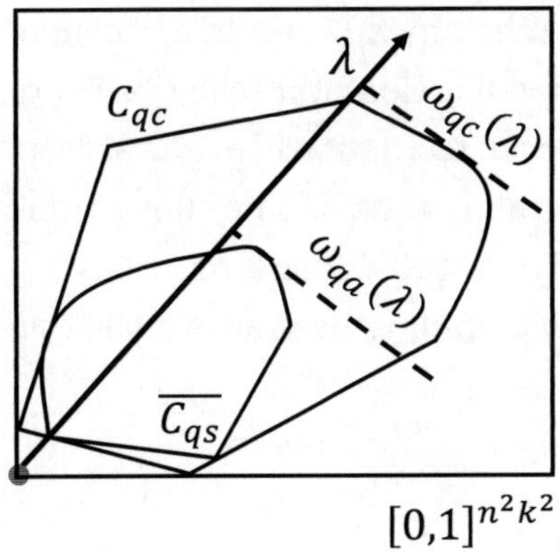

FIGURE 1
Separating convex sets

such as mutual incompatibility of observables might be "detected" by some λ. This, however, can be rather tedious due to the infinite search space: essentially no better algorithm for approaching ω_{qa} from below is known other than enumerating over progressively finer nets in increasing dimensions for the Hilbert space; for approaching it from above, slightly better candidate algorithms are known [41] that work well in practice but, as mentioned earlier, are not even known to converge to the right value—indeed, showing that they do led to formulating Tsirelson's problem, and it follows from the refutation of it that they do not.

In the 1990s, emerging collaborations between physicists and computer scientists stimulated by the nascent field of quantum computation led to the study of a subclass of functionals termed "nonlocal games" which we now introduce.

3.1. Nonlocal games

The idea for a nonlocal game is to interpret the supremum in (3.1) as the optimal winning probability in a certain cooperative two-player game. Let us start with an example of such a game. Fix an n-vertex graph H, as well as a target number of colors $k \geq 1$. The "coloring game" associated with H is played as follows. In the game, two cooperating, but noncommunicating, *players* (traditionally referred to as "Alice" and "Bob") interact with a *referee* as follows. The referee first selects a pair of questions by sampling two vertices of G, x and y, independently and uniformly at random. The referee sends the label x to Alice, and y to Bob. Each player is required to reply with a "color" represented by an integer $a, b \in \{1, \ldots, k\}$, respectively. The referee declares this run of the game as a win for the players if and only if whenever $x = y$ then $a = b$ and whenever (x, y) is an edge in H then $a \neq b$. (If $x \neq y$ is not an edge in H then all answers are accepted.) The players' goal is to maximize their winning probability, taken over the referee's choice of questions, in the game; they are allowed to coordinate their choice of strategy but not to communicate once the game starts.

This last sentence is rather informal; let us make it more precise. What is a valid strategy? For each pair of questions (x, y), the players provide answers according to some distribution $p(a, b|x, y)$. So a strategy specifies a correlation in the sense of Section 2.2. Physical restrictions on the players' actions translate into restrictions on the class of correlations that are allowed. The informal restriction here is that the players "cannot communicate" with each other. The most natural formalization of this requirement is that players are constrained to compute their answers "locally", using functions $f_A, f_B : \{1, \ldots, n\} \to \{1, \ldots, k\}$, respectively. For two players determining their answers in this way, the success probability is precisely

$$p_{\text{succ}} = \frac{1}{n^2} \sum_{x,y=1}^{n} \left(1_{x=y} 1_{f_A(x)=f_B(y)} + 1_{\{x,y\} \in E} 1_{f_A(x) \neq f_B(y)}\right),$$

where 1_S denotes the characteristic function of a set S and E is the edge set of the graph H. Clearly, this expression is 1 if and only if $f_A = f_B$ is a proper coloring of the graph, i.e., adjacent vertices never get assigned the same color. Thus the game has a local strategy which wins with probability 1 if and only if the chromatic number of H is at most k. This relation, between success probability in a game and a natural graph parameter, hints at rich connections between games and combinatorial optimization, with games providing a conceptual framework in which to study specific questions about combinatorial optimization such as *hardness of approximation*.[7]

Generalizing the preceding example, a (two-player, one-round) game is specified by integers n, k, the number of questions and answers per player in the game, respectively, a distribution π on $\{1, \ldots, n\}^2$ according to which questions are chosen, and a decision predicate $V : \{1, \ldots, n\}^2 \times \{1, \ldots, k\}^2 \to \{0, 1\}$ which identifies correct question–answer tuples. With this notation the maximum success probability of a local strategy, which we refer to as the "local value" of the game, is

$$\omega_{\text{loc}}(G) = \sup_{f_A, f_B} \sum_{x,y} \pi(x, y) \sum_{a,b} V(x, y, a, b) 1_{f_A(x)=a} 1_{f_B(y)=b}. \tag{3.2}$$

Defining $\lambda_G \in (\mathbb{R}^{n^2 k^2})^*$ by $(\lambda_G)_{x,y,a,b} = \pi(x, y) V(x, y, a, b)$ and introducing the polytope

$$C_{\text{loc}}(n, k) = \text{Conv}\left\{(1_{f_A(x)=a} 1_{f_B(y)=b})_{x,y,a,b} : f_A, f_B : \{1, \ldots, n\} \to \{1, \ldots, k\}\right\} \tag{3.3}$$

we have that

$$\omega_{\text{loc}}(G) = \sup_{p \in C_{\text{loc}}(n,k)} |\lambda_G \cdot p| = \omega_{\text{loc}}(\lambda_G),$$

justifying our abuse of the notation $\omega_{\text{loc}}(\cdot)$ in (3.2). To summarize, the maximum success probability of local strategies in a game G with n questions and k answers per player can be identified with the supremum of a certain linear functional derived from G over the convex set $C_{\text{loc}}(n, k)$. This connection having been made, a natural question arises: why not consider *quantum* strategies, in which the players would make local measurements on a shared quantum state in order to determine their answers? Instead of a pair of functions, a strategy

7 We emphasize that the games discussed here are entirely distinct from the games considered in the "game theory" of Nash equilibria, where there are two players playing against each other. There is little or no connection between the two areas.

is now modeled as a tuple $\mathscr{S} = (\{A_a^x\}, \{B_b^y\}, |\psi\rangle)$ of measurement operators (PVM) for each player and a shared state $|\psi\rangle$. The no-communication assumption can be implemented by requiring that the tuple satisfies the conditions introduced in the definition of $C_{qs}(n,k)$ in (2.2) (in which case we qualify the strategy as "quantum spatial") or of $C_{qc}(n,k)$ in (2.3) (in which case we qualify it as "quantum commuting").[8] This leads us to define

$$\omega_{qa}(G) = \sup_{p \in C_{qs}(n,k)} |\lambda_G \cdot p| \quad \text{and} \quad \omega_{qc}(G) = \sup_{p \in C_{qc}(n,k)} |\lambda_G \cdot p|. \tag{3.4}$$

Beyond a mere reformulation of the optimization problems (3.1), the framing of linear functionals as two-player (also called "nonlocal" to emphasize their use as witnesses of quantum "nonlocality") games suggests a particular mode of thinking about them, e.g., we can now use intuition about player strategies, questions and answers as opposed to arguably much dryer doubly-indexed families of PVMs.

Going back to the example of the coloring game, each of the quantities in (3.4) leads us to a variant of the chromatic number: for H a graph and G_H the coloring game associated to it, we define the *quantum spatial* (resp. *quantum commuting*) *chromatic number* of H as the smallest k such that $\omega_{qa}(G_H) = 1$ (resp. $\omega_{qc}(G_H) = 1$). Examples of graphs whose quantum spatial chromatic number is strictly smaller than their chromatic number have long been known **[10,21]**. The possible relevance of the study of the new chromatic numbers to TP and CEP is pointed out in **[45]**, who formulate some related quantities in terms of operator systems; multiple works have since explored further variants of the chromatic number **[44,50]** and introduced other classes of games that are connected to combinatorial parameters. For example, the coloring game was generalized in **[42]** to a *graph homomorphism game* whose study led the authors to associate a C^*-algebra with a game; we describe this algebra in the next section. In **[4]** the authors introduced a *quantum isomorphism game* and a related notion of "quantum isomorphism" of two graphs, and showed that there exist graphs that are quantum isomorphic, but not isomorphic. Further study of this notion led to connections with quantum groups **[38]** and a surprising characterization of quantum isomorphism in terms of homomorphism counts from *planar graphs* **[34]** (in contrast, Lovász characterized "classical" graph isomorphism in terms of homomorphism counts from any graph). To summarize, we find that the study of quantum strategies in two-player games has provided a rich framework in which to connect combinatorics and functional analysis, leading to valuable insights in both areas.

3.2. The game algebra

The connection between TP and CEP made in **[20,30,43]** goes through Kirchberg's QWEP conjecture. An arguably more direct route has more recently been found using nonlocal games. Rather informally, the idea is that a quantum strategy for the players in a game G, i.e., a collection of PVM operators, can be thought of as a certain kind of representation for

---

8 To show formally that both types of strategies do not imply communication, we compute the marginal distribution on one player's answers and observe that it is independent of the question to the other player.

an abstract algebra $\mathcal{A} = \mathcal{A}(G)$ associated with the game, whose generators are labeled by (question, answer) pairs and whose relations express the game constraints. The (non)existence of different types of successful strategies (quantum spatial, quantum commuting) in the game corresponds to the (non)existence of different kinds of representations for the algebra, thus tying a statement such as $\omega_{qa}(G) < 1 = \omega_{qc}(G)$ to representability properties of $\mathcal{A}$.

To introduce the game algebra more formally, we first describe the class of *synchronous* games to which the construction applies. A game is synchronous if $X = Y$, $A = B$, and for all x and $a \neq b$, $V(x, x, a, b) = 0$, i.e., identical questions always require identical answers. Informally, the synchronicity condition enables to "factor out" the bipartite structure of a game and focus on representing the strategy for a single player.

Definition 3.1. Let $G = (X, A, \pi, V)$ be a synchronous game. The game algebra $\mathcal{A}(G)$ is the abstract unital $*$-algebra generated by elements $\{e_{x,a}\}_{x,a \in X \times A}$ such that for all $x, y \in X$ and $a, b \in A$,

$$e_{x,a}^* = e_{x,a}, \quad e_{x,a}^2 = e_{x,a}, \quad \sum_a e_{x,a} = 1, \quad \text{and} \quad V(x, y, a, b) = 0 \implies e_{x,a} e_{y,b} = 0.^9$$

Note that the game algebra may be trivial; for example, if $V(x, y, a, b) = 0$ always then the constraints cannot be satisfied. To see the connection between representations of the game algebra and perfect strategies in G (we call a strategy *perfect* for a certain game if it leads to a winning probability of 1 in the game), as a first exercise one may verify that $\omega_{\text{loc}}(G) = 1$ (i.e., there exists a perfect local strategy for G) if and only if there is a unital $*$-homomorphism from $\mathcal{A}(G)$ into $\mathbb{C}$. (The "if" direction is easier; the synchronicity condition on the game is used for the "only if" direction.) This observation can be generalized as follows.

Theorem 3.2. *Let G be a synchronous game. Then*

(i) [32, COROLLARY 3.7] $\omega_{qa}(G) = 1$ *if and only if there is a unital $*$-representation of $\mathcal{A}(G)$ into $\mathcal{R}^\omega$;*

(ii) [44, COROLLARY 5.6] $\omega_{qc}(G) = 1$ *if and only if there is a $*$-representation of $\mathcal{A}(G)$ into a C^*-algebra with a tracial state.*

Similarly to Voiculescu's reformulation of CEP in terms of microstates or Radulescu's definition of hyperlinearity the condition (i) is equivalent to the existence of approximate representations of $\mathcal{A}(G)$ in finite-dimensional matrix algebras. The theorem implies that the existence of a synchronous game G such that $\omega_{qa}(G) < 1 = \omega_{qc}(G)$ is equivalent to the existence of a tracial C^*-algebra that does not embed into $\mathcal{R}^\omega$; the latter statement is easily seen to be equivalent to the negation of CEP.

We say a few words about the proof of Theorem 3.2. To show the "only if" direction for the second claim, given a commuting strategy $(\{A_a^x\}, \{B_b^y\}, |\psi\rangle)$ there is a natural state on $\mathcal{A}(G)$ given by $\tau(W) = \langle\psi|\varphi(W)|\psi\rangle$ where W is a polynomial in the $e_{x,a}$ and

9 The algebra does not depend on the question distribution π.

$\varphi(W)$ replaces $e_{x,a}$ by A_a^x in W. It is immediate that this is a state; that it is tracial follows (with some work) from the synchronicity condition. For the first claim, a priori the condition $\omega_{qa}(G) = 1$ only gives a sequence of finite-dimensional strategies whose success probability approaches 1. One can turn each such strategy in an approximate representation of $\mathcal{A}(G)$ into finite matrix algebras, eventually leading to a representation into some ultrapower of $\mathcal{R}$.

To show the "if" direction for the second claim, applying the GNS construction, we get PVMs for the first player from any tracial state on $\mathcal{A}(G)$. Constructing appropriate PVMs for the second player requires a little more work; essentially, one uses the trace to construct commuting left and right representations of the game algebra. For the first claim, our starting point is a sequence of approximate representations in finite dimensions. From this we immediately get a sequence of families of PVM for the first player. There is a natural definition for PVM elements for the second player which guarantees that PVM elements associated with different players commute. To conclude, the player's PVMs can be put into the required tensor-product form by appealing to the equivalence between spatial and commuting strategies in finite dimensions.

4. CONSTRUCTING NONLOCAL GAMES

To build intuition about nonlocal games and the associated game algebra, we first review a fundamental example, the "Mermin–Peres Magic Square game." In Section 4.2 we build on this example to construct a family of games whose game algebra has approximate representations into matrix algebras of increasing minimal dimension. In Section 4.3 we outline our approach for turning this family of games into a counterexample to TP. This forces us into complexity-theoretic considerations which we explore in Section 4.4.

4.1. The Magic Square game

We start with a classic example, the *Magic Square game* G_{MS} due to Mermin and Peres [36, 46]. This game is a synchronous game with $n = 6$ questions, which are best visualized as the three rows and three columns of a 3×3 square that can be pictured as follows:

$$\begin{array}{ccccc} y_1 & y_2 & y_3 & & +1 \\ y_4 & y_5 & y_6 & & +1 \\ y_7 & y_8 & y_9 & & +1 \\ & & & & \\ -1 & -1 & -1 & & \end{array}$$

In the game, each of the 6 questions has $k = 4$ possible answers, which are identified with the four possible $\{\pm 1\}$ assignments to the entries of the three squares in the row or column indicated by the question such that the product of the entries is as labeled on the picture, $+1$ for a row and -1 for a column. For example, possible answers to the question associated with the first row are $\{(1,1,1),(1,-1,-1),(-1,1,-1),(-1,-1,1)\}$, which are identified with the answer set $\{1,\dots,4\}$ in some arbitrary way. The game decision predicate V_{MS} enforces the constraint that, whenever the players are asked a row and column that intersect, the values that their respective answers assign to the intersection square(s) should be identical. For example, if x is associated with the first row and y with the first column then $V_{\mathrm{MS}}(x, y, (1,1,1), (1,1,-1)) = 1$ whereas $V_{\mathrm{MS}}(x, y, (1,1,1), (-1,1,1)) = 0$. Note that this constraint implies that whenever the players are asked the same question then their answers should be identical, hence G_{MS} is a synchronous game.

A local strategy for this game is a pair of functions $f_A, f_B : \{1, \dots, 6\} \to \{1, \dots, 4\}$; its success probability is the probability over $x, y \in \{1, \dots, 6\}$ chosen uniformly at random that $V_{\mathrm{MS}}(x, y, f_A(x), f_B(y)) = 1$. As an exercise, the reader may use the fact that not all constraints in the square can be simultaneously satisfied to show that $\omega_{\mathrm{loc}}(G_{\mathrm{MS}}) = 34/36$. This example illustrates the connection between games and constraint satisfaction problems that has proved so fruitful in complexity theory.

What is the game algebra $\mathcal{A}_{\mathrm{MS}} = \mathcal{A}(G_{\mathrm{MS}})$? Generators for $\mathcal{A}_{\mathrm{MS}}$ are six PVM with four elements each, $\{e_{x,a}\}_{a \in \{1,\dots,4\}}$ such that $\sum_a e_{x,a} = 1$ for all $x \in \{1, \dots, 6\}$. An equivalent presentation in terms of self-adjoint operators that square to identity can be found as follows. Let $\{e_{x,a}\}_a$ be the four orthogonal projections associated with the first row. Let $y_1 = e_{x,(1,1,1)} + e_{x,(1,-1,-1)} - e_{x,(-1,1,-1)} - e_{x,(-1,-1,1)}$ and similarly define y_2 and y_3. Then y_1, y_2, y_3 square to 1, pairwise commute, and satisfy $y_1 y_2 y_3 = 1$. Conversely, to any such triple, it is straightforward to associate a four-outcome PVM $\{e_{x,a}\}_a$. A similar construction can be employed for each row and column, a priori leading to 18 y_i operators. However, using the condition that $V(a, b, x, y) = 0 \implies e_{x,a} e_{y,b} = 0$ and the consistency condition enforced in G_{MS}, we get that y_i defined in this way from the PVM associated with the corresponding row must equal to y_i defined from the PVM associated with the column that y_i appears in.

To summarize, $\mathcal{A}_{\mathrm{MS}}$ is generated by elements $y_1, \dots, y_9$ such that $y_i^* = y_i$, $y_i^2 = 1$, any two y_i appearing in the same row or column of the magic square commute, and the y_i satisfy the magic square row and column constraints, e.g., $y_1 y_4 y_7 = -1$. Our observation that G_{MS} does not have a local strategy that succeeds with probability 1 implies that $\mathcal{A}_{\mathcal{S}}$ has no unital $*$-homomorphism into $\mathbb{C}$. What about homomorphisms in higher-dimensional algebras? With a little work, it is possible to show that there is no such (unital) homomorphism into $M_2(\mathbb{C})$ or $M_3(\mathbb{C})$, but there is one into $M_4(\mathbb{C})$ given by the following operators:

$$\begin{array}{ccc} I \otimes \sigma_Z, & \sigma_Z \otimes I, & \sigma_Z \otimes \sigma_Z, \\ \sigma_X \otimes I, & I \otimes \sigma_X, & \sigma_X \otimes \sigma_X, \\ -\sigma_X \otimes \sigma_Z, & -\sigma_Z \otimes \sigma_X, & \sigma_Y \otimes \sigma_Y, \end{array} \tag{4.1}$$

where

$$\sigma_X = \begin{pmatrix} 0 & 1 \\ 1 & 0 \end{pmatrix}, \quad \sigma_Z = \begin{pmatrix} 1 & 0 \\ 0 & -1 \end{pmatrix}, \quad \text{and} \quad \sigma_Y = i \sigma_X \sigma_Z$$

are the Pauli matrices. Moreover, homomorphisms from $\mathcal{A}_{\mathrm{MS}}$ into $M_d(\mathbb{C})$ for some d obey an interesting "rigidity" phenomenon. Let $Y_1, \dots, Y_9$ be the image of the generators under any such homomorphism. Then it is easy to verify that the row and column constraints imply that

$$\{Y_1, Y_5\} = \{Y_2, Y_4\} = 0, \quad \text{and} \quad [Y_1, Y_2] = [Y_1, Y_4] = [Y_5, Y_2] = [Y_5, Y_4] = 0, \tag{4.2}$$

where $\{A, B\} = AB + BA$ is the anticommutator. Conversely, any four self-adjoint matrices that square to identity and satisfy (4.2) can be extended to a $*$-homomorphism of $\mathcal{A}_{\mathrm{MS}}$. The rigidity phenomenon referred to above is that the algebra generated by any such (finite-dimensional) Y_1, Y_2, Y_4, Y_5 is isomorphic to $M_4(\mathbb{C}) \otimes M_{d'}(\mathbb{C})$ for some d', with $Y_1 \mapsto I \otimes \sigma_Z$, etc. (The other operators, Y_3, Y_6, Y_7, Y_8, Y_9 are uniquely defined from those four by the row & column constraints.) Thus *any* finite-dimensional unital $*$-representation of $\mathcal{A}_{\mathrm{MS}}$ is isomorphic to the representation given in (4.1), possibly tensored with the identity. This very special property allows us to use the fact that a correlation achieves a high success probability in a game, $\lambda_{\mathrm{MS}} \cdot p = 1$, to conclude that any realization of this correlation using PVMs acting on a Hilbert space must satisfy specific algebraic relations; this fact will be crucial to the eventual resolution of TP.

To summarize, the example of the Magic Square helps us demonstrate two important points. Firstly, it is possible to design a game such that $\omega_{qa}(G) = 1$ and, moreover, any strategy that witnesses this is of a certain minimal dimension—here, 4. Secondly, it is possible to force such strategies to have a certain rigid structure—here, the operators used as part of the strategy must contain two pairs of mutually anticommuting operators, such that operators from different pairs commute.

4.2. The Pauli braiding game

To bound $\omega_{qa}(G)$ for some game G it is useful to understand the structure of *approximately* optimal strategies in G. This is because, due to the nonclosure of $C_{qs}(n, k)$, we can have $\omega_{qa}(G) = 1$ without there being any perfect quantum spatial strategy for G, and so it will be convenient to develop techniques that are able to rule out the existence of not only perfect but also near-perfect strategies.

To get us started we state an important tool in the study of approximate group representations.

Theorem 4.1 ([24]). *Let G be a finite group and $f : G \to U_d(\mathbb{C})$ such that*

$$\mathop{E}_{x,y\in G} \operatorname{tr}\big(f(y)^* f(x) f(x^{-1}y)\big) \geq 1 - \varepsilon,$$

for some $\varepsilon \geq 0$ and where the expectation is taken over the choice of a uniformly random pair of elements from G. Then there is a representation $g : G \to U_{d'}(\mathbb{C})$ and an isometry $V : \mathbb{C}^d \to \mathbb{C}^{d'}$ such that

$$\mathop{E}_{x\in G} \big\| f(x) - V^* g(x) V \big\|_2^2 \leq 2\varepsilon.$$

A map f as in the theorem is called an *approximate representation* of the group G; indeed, the condition with $\varepsilon = 0$ is equivalent to that of being a representation. The theorem is an example of a *stability* result, stating that approximate representations are close to exact representations. Here, the measure of "approximate representation" is rather loose, since group relations are only required to hold on average and under the ℓ_2-norm $\|X\|_2 = \operatorname{tr}(X^*X)^{1/2}$ (as opposed to, say, for all relations and under the operator norm). The use of the ℓ_2-norm requires us to allow $d' > d$ in the conclusion of the theorem; that this is necessary is easy to see by "cutting off a corner" from a high-dimensional representation. We remark that Theorem 4.1 has been extended to the case of amenable groups, with appropriate modifications to allow for infinite-dimensional representations, in [15].

In some cases, such as the Magic Square game studied in the previous section, we can observe that the game algebra is "almost" a group algebra—in fact, it is isomorphic to a quotient of a group C^*-algebra. Namely, if we let $\mathcal{P}_2$ be the group generated by X_1, Z_1, X_2, Z_2, J satisfying $X_1^2 = Z_1^2 = X_2^2 = Z_2^2 = J^2 = 1, [J, X_1] = [J, Z_1] = [J, X_2] = [J, Z_2] = 1$, and $[X_1, X_2] = [X_1, Z_2] = [Z_1, X_2] = [Z_1, Z_2] = 1, [X_1, Z_1] = [X_2, Z_2] = J$ (where now $[a, b] = aba^{-1}b^{-1}$ denotes the group commutator) then it can be verified that $\mathcal{A}(G) \simeq \mathbb{C}(\mathcal{P}_2)/\langle J + 1\rangle$.[10] In particular, any (approximate) representation of $\mathcal{A}(G)$ "descends" to an (approximate) representation of $\mathcal{P}_2$ that (approximately) sends J to -1. Since $\mathcal{P}_2$ has a single (exact) representation that sends J to -1, Theorem 4.1 can be applied to deduce that near-perfect strategies in G_{MS}, i.e., strategies whose winning probability is close but not necessarily equal to 1, must be proportionately close to optimal strategies. In particular, it implies the existence of a constant $\varepsilon_0 > 0$ such that any quantum spatial strategy that succeeds with probability larger than $1 - \varepsilon_0$ in G_{MS} makes use of a Hilbert space for each player that has dimension at least 4; moreover, the algebra generated by the strategy's PVMs contains operators that are close, in the norm $\|\cdot\|_2$, to a representation of the group $\mathcal{P}_2$.

The connection between game algebra and quotient of a group C^*-algebra is quite general and extends to a large class of synchronous games introduced in [12,32] and referred to as *linear constraint system games*; this was shown in [22]. The tools introduced so far suggest the possibility of designing games whose game algebra is isomorphic to quotients of larger groups, such as, for example, the group $\mathcal{P}_N$ which is defined as $\mathcal{P}_2$ but with N pairs of anticommuting generators; this group has a unique irreducible representation sending J to -1, of dimension 2^N. Working out the rules for such a game leads to the following.

Theorem 4.2 ([39]). *There is an $\varepsilon_0 > 0$ and for every $N \geq 2$ a synchronous game $G_{\mathrm{PBT}}^{(N)}$ with $2^{O(N)}$ questions and $O(1)$ answers such that any quantum (spatial or commuting) strategy which succeeds in $G_{\mathrm{PBT}}^{(N)}$ with probability at least $1 - \varepsilon_0$ induces an approximate representation of $\mathcal{P}_N$ sending J to -1 and must have dimension at least 2^N.*[11]

The game from Theorem 4.2 is called the "Pauli braiding game," referring to how the defining (anti)commutation relations "braid" the group generators together. For succinctness, we do not describe this game in its entirety here. To design it, it suffices to find an appropriate decision predicate function that will enforce the group relations. The simpler case of $\mathbb{Z}_2^n$ is known in complexity theory as the "linearity test" of Blum, Luby, and Rubinfeld [8]. This test amounts to verifying that the players' answers a, b, and c to questions $x, y \in \mathbb{Z}_2^n$ and

10 The algebraic relations obtained from the stated relations by sending $J \mapsto -1$ are known as the Weyl–Heisenberg relations.

11 One might worry that Theorem 4.1 only guarantees closeness to a representation up to an isometry, which can change the dimension of the underlying space. This is true, and a little extra work which we skip here is needed to obtain the strict dimension bound mentioned in the theorem.

$x + y$, respectively, are related as $c = a + b$.[12] Blum et al. show that near-perfect *local* strategies are close to homomorphisms from $\mathbb{Z}_2^n$ to $\{-1, 1\}$, and this is extended to finite-dimensional matrix representations in [54]. For the case of $\mathcal{P}_N$, we combine the linearity test for testing the product rule between commuting elements in $\mathcal{P}_N$ and the Magic Square game for testing anticommuting elements. The stated number of questions, $2^{O(N)}$, follows from the number of group elements, which is $2 \cdot 4^N$, and is about quadratically larger due to the use of auxiliary questions that are associated with, e.g., a pair of group elements.

4.3. A fixed-point argument

At this point we have designed an infinite family of games $(G_{\mathrm{PBT}}^{(N)})_{N \geq 1}$ such that for all $N \geq 1$, $\omega_{qa}(G_{\mathrm{PBT}}^{(N)}) = \omega_{qc}(G_{\mathrm{PBT}}^{(N)}) = 1$. While this clearly does not provide a separation, there is more that we may hope to use. In particular, thanks to the rigidity (stability) arguments exposed in the previous section, we know that there is an $\varepsilon_0 > 0$ such that, for any $N \geq 2$ and any quantum spatial strategy for $G_{\mathrm{PBT}}^{(N)}$ that succeeds with probability at least $1 - \varepsilon_0$, the Hilbert space underlying the strategy must have dimension at least the dimension of the smallest representation of $\mathcal{P}_N$ that sends J to -1, i.e., 2^N. For a game G and $p \in [0, 1]$, we let $\mathcal{E}(G; p)$ be the smallest dimension of a strategy that succeeds in G with probability at least p; then, according to Theorem 4.2, we have that

$$\forall N \geq 1, \quad \mathcal{E}(G_{\mathrm{PBT}}^{(N)}; 1 - \varepsilon_0) \geq 2^N. \tag{4.3}$$

Equation (4.3) shows that any quantum strategy of dimension $< 2^N$ has success probability *bounded away from* 1 in $G_{\mathrm{PBT}}^{(N)}$. To complete our goal, it would suffice to create a *single* game G that satisfies this property *for every* $N \geq 1$. Indeed, if $\mathcal{E}(G; 1 - \varepsilon_0) \geq 2^N$ for all N then it follows that $\omega_{qa}(G) < 1$, because the optimal success probability of a quantum spatial strategy in G can be arbitrarily well approximated by finite-dimensional strategies. If, in addition, we are able to guarantee that $\omega_{qc}(G) = 1$ then we will have completed our negative resolution of TP, separating $C_{qa}(n, k)$ from $C_{qc}(n, k)$ for n and k being the number of questions and answers in G, respectively.

The key idea is to define the game G as the *fixed point* of a certain *compression procedure* that transforms families of games such as $(G_{\mathrm{PBT}}^{(N)})_{N \geq 1}$ into other families with comparable size but increased requirements in terms of the minimal dimension of near-optimal strategies. To make this precise, we first need a means of representing infinite families of games. Recall that a *computable function* is $f : \mathbb{N} \to \mathbb{N} \cup \{\bot\}$ such that, informally, there is an algorithm that on input $n \in \mathbb{N}$ returns $f(n)$ if $f(n) \in \mathbb{N}$; if $f(n) = \bot$ the algorithm does not terminate. A computable function is *total* if $f(n) \in \mathbb{N}$ for all n.[13] Computable functions are enumerable and can thus themselves be encoded as integers in a natural way (e.g., via some unambiguous encoding of a Turing machine that computes the function).

12 Here it seems like there are three players; a small variant of the test works with two players.

13 We wrote "roughly speaking" because we are not making the notion of algorithm precise. It is a major success of computability theory that essentially any reasonable notion of computability that has been formalized has been shown equivalent to the other notions. For concreteness, one can replace "algorithm" by "Turing machine."

Fix a canonical encoding of games as natural numbers; since the collection of all games (with, say, question distribution that has rational coefficients) is countable, this can be done in a straightforward manner. We say that a function $\mathcal{G} : \mathbb{N} \to \mathbb{N}$ *succinctly represents* the family $(G_N)_{N\geq 1}$ if $\mathcal{G}$ is computable and for every $N \geq 1$, $\mathcal{G}(N)$ is the representation of G_N. Now suppose that there exists a total computable function Compress that, given as input a succinct representation $\mathcal{G}$ for a family of games $(G_N)_{N\geq 1}$, returns a succinct representation $\mathcal{G}'$ for a family of games $(G'_N)_{N\geq 1}$ such that the following conditions hold for all $N \geq 1$:

(C.1) If $\omega_{qa}(G_{N+1}) = 1$ then $\omega_{qa}(G'_N) = 1$;

(C.2) $\mathcal{E}(G'_N; \frac{1}{2}) \geq \min\{\mathcal{E}(G_{N+1}; \frac{1}{2}), N\}$.[14]

In the next section we argue that the existence of such a "compression" procedure is fairly natural once one is familiar with the use of nonlocal games in complexity and cryptography, and in particular with the design of delegated computation protocols using the PCP theorem—buzzwords that will be explained later.[15] For the time being, let us assume that the map Compress exists. We will make use of an additional ingredient in the form of a *refutation procedure* NPA for the quantum commuting value. NPA is an algorithm that takes as input the integer representation of a (single) game G and halts if and only if $\omega_{qc}(G) < 1$. (If $\omega_{qc}(G) = 1$, then NPA(G) runs forever.) The existence of such a procedure follows from the results of Navascues et al. [41] that were already mentioned in Section 2.2, and we take it for granted.

Using these two procedures, Compress and NPA, let us define another function, call it F, that takes as input (the integer representation of) a succinct representation $\mathcal{G}$ for a family of games $(G_N)_{N\geq 1}$ and returns a succinct representation $\mathcal{G}'$ that is defined as follows. (We specify $\mathcal{G}'$ as an algorithm expressed in high-level language, which can ultimately be implemented by some computable function.) On input N, $\mathcal{G}'$ does the following:

(1) It computes the description of $G_1 = \mathcal{G}(1)$.

(2) It runs NPA on G_1 for N steps. If NPA halts, then it returns the description of a trivial game that always accepts.

(3) It computes $\mathcal{T} = \text{Compress}(\mathcal{G})$.

(4) It returns a description of the game $G'_N = \mathcal{T}(N)$.

We observe that, provided $\mathcal{G}$ and Compress are total computable functions, and Compress returns a total computable function when given one as input, F is also a total computable function. Applying a fundamental result in the theory of computable functions, Rogers' fixed point theorem, the map F has a fixed point, call it $\mathcal{G}_\infty$, that is a computable function. Let

---

14 Here the last "N" can be replaced by any unbounded function of N for the ensuing argument to work.

15 This is not to say that it is straightforward—indeed, the two conditions together already imply that executing Compress once on a trivial family of games that always accept yields an infinite family of games with increasing dimension requirement.

$G_\infty = \mathscr{G}_\infty(1)$ (more precisely, the game whose integer representation is $\mathscr{G}_\infty(1)$). We claim that $\omega_{qa}(G_\infty) < \omega_{qc}(G_\infty) = 1$, thus providing us with the desired separation. To show this, suppose first that $\omega_{qc}(G_\infty) < 1$. Then since $\mathsf{F}(\mathscr{G}_\infty) = \mathscr{G}_\infty$, and since NPA must halt on G_∞ after some number N_∞ of steps, for $N \geq N_\infty$ the game $G_\infty(N)$ is a trivial game that always accepts. By a straightforward induction using property (C.1) of Compress, it follows that $\omega_{qa}(G_\infty) = 1$, hence $\omega_{qc}(G_\infty) = 1$ as well, a contradiction. So $\omega_{qc}(G_\infty) = 1$ and at step 2 NPA never halts. We then get by induction using property (C.2) of Compress that $\mathscr{E}(G_\infty; \frac{1}{2}) \geq N$ for all N. This implies that $\omega_{qa}(G_\infty) \leq \frac{1}{2}$, because no sequence of finite-dimensional strategies can ever get a success probability larger than $\frac{1}{2}$.

The preceding argument shows that to refute TP it "only" remains to design the Compress procedure. This, of course, is the hard part. Before we tackle this task, we discuss a subtle point about the preceding argument.

4.4. Enter complexity

In the analysis of the fixed point $\mathscr{G}_\infty$ of the map F, we implicitly assumed that the game $G_\infty = \mathscr{G}_\infty(1)$ is well defined. What if $\mathscr{G}_\infty$ never halts on input 1? Rogers' fixed point theorem does not guarantee that the fixed point itself is a total function, and it need not be defined on all inputs. Even if it were a total function, there would not be an a priori guarantee that $\mathscr{G}_\infty$ returns well-formed outputs on every input—in general, it will return integers which, depending on our encoding procedure, may not all correspond to well-defined games. Indeed, we should detect that there is something suspicious in the entire setup. A function Compress satisfying all the requirements we have listed is easy to design; for example, $\mathscr{G}' = \text{Compress}(\mathscr{G})$ could on input N return a game that is a mixture of G_{N+1} and $G_{\text{PBT}}^{(N)}$,[16] and this would easily satisfy both (C.1) and (C.2) (indeed, with a stronger bound of 2^N instead of N in (C.2)).

Observe that by virtue of being a fixed point of F, $\mathscr{G}_\infty$, as a family of games, has a size (as a function of N) that is at least that of Compress($\mathscr{G}_\infty$), which has a size that is at least that of Compress(Compress($\mathscr{G}_\infty$)), etc. Therefore, for F to have a fixed point that is a well-defined family of games, it is *necessary* that the procedure Compress lives up to its name, i.e., satisfies the following additional requirement:

(C.3) The size of the game G'_N is smaller than the size of the game G_N.

Here, by "size" we mean the size of an explicit representation of the game, which we can approximate by the total number of questions and answers. In the next section we will see that a more refined notion of size, in terms of the running time of an algorithm computing the referee's questions and its decision, is needed.

While it may not be immediately clear at the level of the discussion, a proper formalization of (C.3), together with small modifications to the description of F (e.g., the intro-

16 What we mean is that the referee would flip a coin to decide which game is played, and inform the players of their decision; for both conditions to hold we'd place a higher probability on $G_{\text{PBT}}^{(N)}$ being played.

duction of a "time-out" condition that ensures that the output of F is always a well-defined family of games, whatever its input), leads to a procedure such that we are able to guarantee that any fixed point is a valid description of a family of games. Thus complexity-theoretic requirements on Compress arise naturally from our strategy based on identifying $\mathscr{G}_\infty$ as a fixed point, and this beyond the most elementary requirement that the map be computable. In the next section we give some of the main ideas that go in the design of Compress; as we do so, we will discover that more refined complexity-theoretic requirements are required for us to proceed with the construction.

5. COMPRESSION

So how do we implement a "compression" procedure such that (C.1), (C.2), *and* (C.3) hold? Although it has well-established parallels in classical complexity and cryptography, this is a relatively new question in the study of nonlocal games and comparatively few techniques are known for it [19,26,37]. Two main ideas have been used. The first is the idea of *efficient verification of computations*, which takes its origin in classical complexity theory in the 1980s (where it was studied under the name of "program checking" [7]) and received a huge boost when probabilistically checkable proofs (PCP) were discovered in the 1990s [2,3]. The second is the idea of *rigidity*, which we already encountered when analyzing the Magic Square game in Section 4.1 and whose relevance to quantum information and cryptography was first made explicit in work by Mayers and Yao who coined the term "self-testing" for it [35].

In this section we aim to give a flavor of both techniques and how they come together to implement compression. In the process we will see that more refined arguments about complexity make their apparition. As observed in Section 4.4, the design of a procedure which satisfies both (C.1) and (C.2) is relatively straightforward if one does not impose any requirement on how the size of family of games $\mathscr{G}' = \mathsf{Compress}(\mathscr{G})$ depends on that of $\mathscr{G}$. This leads us to reframe the question of implementing compression into one of reducing the size of a game given as input—given a game G (which we think of as $\mathscr{G}(N+1)$), how do we design G' (which we think of as $\mathscr{G}'(N)$) that has similar properties (same $\omega_{qa}(G)$, same dimension requirements) but a smaller number of questions and answers—since we are now talking about the Nth game in the family, and not the $(N+1)$th? We first discuss the problem of reducing the number of answers in a game, and then that of reducing the number of questions.

5.1. The PCP theorem and answer reduction

The colloquial formulation of the PCP theorem is that mathematical proofs can be written in a format such that the validity of the entire proof can be verified by looking only at a few randomly chosen locations of it. It will be useful to express this slightly more formally. First, we fix a *language*, which in general is a subset $L \subseteq \{0,1\}^*$ of strings of bits of any length, and for the example could be the set of all valid statements in, say, Peano arithmetic. Second, we fix a proof verification procedure D that takes as input a statement x and a proof

$\Pi \in \{0,1\}^*$ and returns a bit $D(x,\Pi) \in \{0,1\}$, with 1 indicating that the proof is valid. In the example, D would check that all the claimed steps in Π follow from an axiom and that the proof indeed establishes the statement x.[17] The PCP theorem states that from D it is possible to compute a D' such that D' takes inputs x and Π', is allowed to toss some random coins, but can only look at 10 bits of Π' and then returns a decision in $\{0,1\}$. It should be that (i) for any (x,Π) that D accepts there is a Π', which can be computed from Π, such that D' accepts (x,Π') (this is usually referred to as the "completeness" property) and (ii) for any x such that there is no Π such that D accepts (x,Π) there is also no Π' such that D' accepts (x,Π') with probability larger than $1/3$ (this is referred to as "soundness"—note the apparition of a small probability of error, which can be made arbitrarily small by allowing D' to make more queries to Π', but cannot in general be driven to zero).

There is a crucial requirement for the PCP theorem to apply that is worth emphasizing: the transformation described above is only possible in the case where D is *efficient*, in the sense that the time it takes to evaluate an input (x,Π) is a fixed polynomial in the length of x (in particular, it can only ever access polynomially many bits of Π, which can thus be truncated without loss of generality). This efficiency requirement is crucial to the proof of the PCP theorem, which first represents the entire computation done by D as a "tableau" with intermediate variables associated to each computation step, before finding an encoding of it that can be checked very efficiently; this last step uses techniques from the theory of error-correcting codes. The PCP theorem thus states that proofs that can be verified efficiently, in a number of computation steps polynomial in their length, can be encoded in such a way that verification can also be sparse—only a few bits need to be accessed in order to make a high-confidence decision.[18]

Why is this relevant to our task? Recall that, given a game G, our goal is to find a game G' that is smaller than G and such that (C.1) and (C.2) hold. Here we are concerned with reducing the size of answers in G'; we will address the size of questions in the next section. Fix a pair of questions (x,y) for G. We can think of the referee's task in the game as verifying the claim that "there exists a pair of answers (a,b), *that can be locally produced from* (x,y), such that $V(x,y,a,b) = 1$." Setting aside the italicized part, the referee's task amounts to verifying the existence of a proof, the pair (a,b), that passes some verification procedure, $V(x,y,\cdot,\cdot)$. The PCP theorem indicates that there is some "encoding" of (a,b), call it (Π^a,Π^b), that can be verified by only examining a few locations of it. We could then devise another verification procedure V', for the game G', that samples a pair of questions (x,y) as in G, as well as a few locations $(i_1,\ldots,i_5)$ that it needs to see in Π^a, $(j_1,\ldots,j_5)$ that it needs to see in Π^b, and would send $x' = (x,i_1,\ldots,i_5)$ and $y' = (y,j_1,\ldots,j_5)$ as its questions. The players would locally compute Π^a and Π^B, respectively, and respond with the requested locations. The PCP theorem would guarantee that this verification procedure

17 Note that here D should accept statements and proofs of any length. Formally, it could be modeled as a Turing machine with two input tapes.

18 "Sparse" is often called "local" in the literature. We use a different word to make the distinction with the notion of "locality" associated with the players in a two-player game, which in our context is distinct.

is essentially equivalent to the original one; however, now the length of answers has been reduced down to a constant.

While this is a plausible sketch for how answer reduction may be achieved, there are a number of major caveats that need to be addressed. Firstly, we implicitly assumed that the PCP encoding of the "proof" (a, b) would naturally take the form (Π^a, Π^b). However, in general a PCP encoding is calculated globally, and such a nice bipartitioning may not (in fact, cannot) hold. Secondly, it is essential for the soundness of the argument that we certify that (a, b) not only exist but can be produced locally from (x, y). Thirdly, again for soundness Π^a should not be allowed to depend on the queries $(i_1, \ldots, i_5)$ that are being made to it: we need to find a mechanism that forces the player to fix it independently of them, even though the referee will never see the entirety of it. Finally and crucially, as already mentioned the PCP theorem only applies to efficient verification procedures. To make use of it here, it is therefore essential that the verification predicate V used in G can be implemented by an algorithm that runs in time polynomial in the length of (x, y).

The last point forces us to rethink our approach. While we initially thought of games as some mildly restrictive formulation of linear forms, the desire to "compress" games puts us face to face with a new algorithmic requirement: we now have to keep track of the complexity of the verification predicate. As long as we do so, however, we have a plan for reducing the size of answers. While this plan raises specific challenges, all of them can be addressed using variations of techniques that have been developed in the decades-long history of using the PCP theorem to implement efficient proof verification in a variety of settings. In the next section we will see how reducing the size of questions prompts us to impose similar efficiency requirements on the procedure used to sample questions $(x, y) \sim \pi$ in G.

5.2. Rigidity and question reduction

In the previous section we saw how techniques developed for the study of PCPs could be leveraged to implement savings in the length of answers in a nonlocal game (at the cost of a small increase in the question length). The idea for reducing the length of questions appears in [40], where it is referred to as "introspection." While the PCP theorem takes its full meaning in a classical context, the idea of introspection makes essential use of quantum-mechanical features, and in particular the possibility to test that incompatible measurements have been made on a shared quantum state.

To explain the idea suppose first that the distribution π on questions in the game G is uniform over $\{(x, x) : x \in \{0, 1\}^N\}$. Suppose that G is modified into a game G' such that with probability $1/2$ the players are asked to play the game $G_{\mathrm{PBT}}^{(N)}$ introduced in Section 4.2 (and with probability $1/2$ they play the original G). Let us see how introducing the Pauli braiding game can be used to force the players to locally generate their own questions in exactly the same way as the referee would have.

For simplicity, let us assume that the players' strategy succeeds with probability 1 in $G_{\mathrm{PBT}}^{(N)}$, when it is played. Again for simplicity let us assume that the state $|\psi\rangle \in \mathcal{H} \otimes \mathcal{H}$ used as part of the players' strategy is a "maximally entangled" state, i.e., it satisfies

$\langle\psi|A \otimes B|\psi\rangle = \mathrm{tr}(AB^T)$ for any $A, B \in B(\mathcal{H})$.[19] In the game $G_{\mathrm{PBT}}^{(N)}$, there is a question associated with each element of $\mathcal{P}_N$, to which the answer is a single bit.[20] For each such question, in the strategy there is a two-outcome PVM $\{A_0^x, A_1^x\}$ that the player applies when receiving the question. We can write each such PVM as an observable $A^x = A_0^x - A_1^x$. Adapting our notation to the present situation, we conclude that as part of the strategy for every $a, b \in \{0,1\}^N$ there is an observable $A(a,b)$ that corresponds to the player's measurement on the question associated with the group element $X_1^{a_1} \cdots X_N^{a_N} Z_1^{a_1} \cdots Z_N^{a_N} \in \mathcal{P}_N$. Moreover, whenever the strategy has a success probability sufficiently close to 1 in the game, there is an isometry V such that $A(a,b) \simeq V^* \sigma_X(a) \sigma_Z(b) V$, where the approximation is meant in the sense of Theorem 4.1 and we introduced the shorthand $\sigma_X(a) = \sigma_X^{a_1} \otimes \cdots \otimes \sigma_X^{a_n}$ and similarly for $\sigma_Z(b)$.

Consider the following modification to $G_{\mathrm{PBT}}^{(N)}$. Introduce a pair of additional questions, labeled X resp. Z, on which the player is expected to perform the 2^N-outcome PVM that corresponds to a joint measurement of all observables $\{A(a, 0^N, 0)\}$ resp. $\{A(0^N, b, 0)\}$, which is possible since they commute (in a perfect strategy).[21] These two questions have much longer answers which can be used for "randomness generation," in the following sense. Using that $A(0^N, b, 0) \simeq V^* \sigma_Z(b) V$, it follows that the PVM applied on question Z is isometric to a rank-1 measurement in the joint eigenbasis of all $\sigma_Z(b)$ (possibly tensored with an irrelevant identity). Since all rank-1 projections have the same trace (recall that here we are working in a finite-dimensional matrix algebras, whose trace is unique) it follows that each answer is obtained with the same probability, $1/2^N$. Furthermore, using that the joint eigenbases of $\{\sigma_X(a)\}$ and of $\{\sigma_Z(b)\}$ are mutually unbiased, it follows that if both players are sent the same question (Z, Z) then they must provide the same uniformly distributed answer, while if the question is (X, Z) they must each provide uncorrelated uniformly random answers. Geometrically this observation corresponds to the statement that any p such that $\lambda_{G_{\mathrm{PBT}}^{(N)}} \cdot p = 1$ has a certain projection (e.g., to $x = Z$, $y = X$) that is proportional to the all-1 vector.

Thus using $G_{\mathrm{PBT}}^{(N)}$, suitably modified by the introduction of additional questions as described above, we can guarantee that any strategy which succeeds with sufficiently high probability in this part of the game must return uniformly random identical answers $(a, a) \in (\{0,1\}^N)^2$ to the question (Z, Z). In this way it is possible to enforce that the players locally generate a pair (a, a) that is distributed exactly as the questions that the referee would send them in the game G. Moreover, the "effort" in doing so is virtually trivial: each player was sent a single question that essentially reads "generate the same random value as your partner!"

---

19 This can be shown to hold without loss of generality whenever the game is a synchronous game, which is the case for all games considered here.

20 There are more questions, which are used to test the group relations; this explains the "$2^{O(N)}$" in Theorem 4.2.

21 The game should *enforce* that such a measurement is being made in any (near)-optimal strategy; this is not hard to achieve.

While this constitutes the main idea—using rigidity to force players to locally generate their own questions—there are many issues to address. Firstly, we need the player not to report the question that they generated, but use it in order to compute an answer that it then sends to the referee. For this there is a simple workaround applying ideas from the previous section: we can in a first step ask the player to report the generated question, as well as its answer, and in a second step to perform answer reduction. Secondly, the distribution of questions in the game need not be uniform over identical (a, a) or independent (a, b) answers. This requires extending the method described above. In particular, "complicated" distributions are likely to be harder to enforce, requiring more questions to a point where no question reduction is achieved—thus another complexity requirement creeps in, that the families of games we consider should have question distributions of bounded complexity, a requirement that should of course be formalized in an appropriate manner. Finally and most importantly, while the "useful" question (Z, Z) in the game is short, ensuring that the player performs the right action on it requires the referee to implement the entire game $G_{\mathrm{PBT}}^{(N)}$. As mentioned in Theorem 4.2, this game has $2^{O(N)}$ questions, which in general will be far larger than the number of questions in G! In the next section we address this issue by discussing a game that has similar guarantees in terms of testing but much smaller question size.

5.3. The quantum low-degree test

In the previous section we sketched how the task of question reduction can be completed, *provided* there is an analogue of the game $G_{\mathrm{PBT}}^{(N)}$ for testing the group $\mathcal{P}_N$ but with a reduced number of questions. Such a result is shown in [27], building on [39].

Theorem 5.1 ([28]). *There are an $\varepsilon_1 > 0$ and for every $N \geq 2$ a synchronous game $G_{\mathrm{LDT}}^{(N)}$ with $2^{\mathrm{poly}\log(N)}$ questions and answers such that any quantum (spatial or commuting) strategy which succeeds with probability at least $1 - \varepsilon_0$ in $G_{\mathrm{LDT}}(N)$ induces an approximate representation of $\mathcal{P}_N$ that sends J to -1 and must have dimension at least 2^N.*

The only difference with Theorem 4.2 is the number of questions and answers, which is now quasipolynomial instead of exponential. This difference hides a deeper difference in terms of how the game is structured. Recall that the Pauli braiding test is built on the linearity test of [8], interpreting the latter as a test for the group $\mathbb{Z}_2^N$ and extending it to a test for $\mathcal{P}_N$ whose analysis could be performed based on Theorem 4.1. With a much smaller number of questions it is no longer possible to have a question associated with each element of the group. Since $\mathcal{P}_N$ can be generated by $2N$ elements it is still possible to have a question per generator, and plausible to show that in optimal strategies the observables associated with each of these questions generate a group isomorphic to $\mathcal{P}_N$. This task, however, will clearly be rather arduous in the case of near-perfect strategies. This is because near-perfect strategies provide a presentation (observables associated to the generators) which satisfy, at best, a certain set of relations on the average to some small constant error (in the norm $\|\cdot\|_2$). Extending the generators to the entire group by taking products will quickly build up the error in a way that, if the only available tool is the triangle inequality, is likely to become unmanageable.

What is needed for Theorem 5.1 is an *efficient* stability result: a small (quasipolynomially many generators and relations) presentation of the group $\mathscr{P}_N$ such that any collection of operators that approximately satisfies the defining relations (in a similar sense as Theorem 4.1) is close to an exact representation—where crucially the closeness should depend on the initial approximation quality but not, or only very mildly, on the size of the group.

Theorem 5.1 is obtained as a quantum extension of the PCP theorem. The generating set it is based on is defined using the Reed–Muller error-correcting code, in a way that we do not have space to detail here. As far as we are aware, it is the only "efficient" (small number of generators) and "robust" (approximate representations are close to exact ones) stability result of its kind and may be a tool of independent interest in other areas.[22] Interestingly, a pared-down version of the result for the group $\mathbb{Z}_2^N$ is used in the analysis of the answer reduction procedure from Section 5.1. Indeed, while the use of the PCP theorem made in that section is a priori entirely classical the analysis needs to take into account quantum strategies for the players, and the classical soundness analysis is not sufficient.

5.4. MIP* = RE

The previous sections complete our sketch of the design of the compression procedure, and following the argument from Section 4.3 of the construction of a correlation separating C_{qc} from $\overline{C_{qs}}$. While we started off without making considerations of complexity, we were led to introduce such considerations due to (1) the requirements for applying the fixed-point argument, and, more crucially, (2) the necessity of using tools such as the PCP theorem to implement the game compression procedure.

A small modification of the definition of the fixed-point $\mathscr{G}_\infty$ leads to an interesting consequence in complexity theory itself. In the definition of F, replace the algorithm NPA used at step 2 by the execution of an arbitrary Turing machine M, i.e., replace the step by "Run M for N steps. If M halts then return the description of a trivial game that always halts." We claim that with this modification the game $G_\infty = \mathscr{G}_\infty(1)$ satisfies $\omega_{qa}(G_\infty) = 1$ if and only if M halts (in some finite number of steps), and $\omega_{qa}(G_\infty) < 1$ otherwise. This can be shown using very similar reasoning to that employed in Section 4.3. Suppose first that M halts. Then step 2 detects this for some large enough N, and, as in Section 4.3, we conclude that $\omega_{qa}(G_\infty) = 1$. If, however, M never halts then step 2 never completes, and, again as in Section 4.3, we quickly see that $\omega_{qa}(G_\infty) \leq \frac{1}{2}$.

Furthermore, it can be verified that the procedure which to M associates the corresponding game $G_\infty = G_\infty(M)$ can be implemented in time polynomial in the description of M. That is, to any Turing machine M we are able to associate a game $G_\infty(M)$ that has a perfect quantum spatial strategy if M halts, and no near-perfect quantum spatial strategy in case M does not halt. In complexity-theoretic terms this establishes a reduction from the halting problem to the problem of deciding between $\omega_{qa}(G) = 1$ and $\omega_{qa}(G) \leq \frac{1}{2}$ (here $\frac{1}{2}$ is an arbitrary positive quantity < 1). The halting problem is a complete problem for the class of recursively enumerable languages RE, while the latter problem is (once properly formu-

22 See, e.g., **[6]** for a discussion of some group stability results.

lated) complete for the class MIP* of languages that have "quantum multiprover interactive proof systems." Thus the argument establishes the equality MIP* = RE, which gives its title to [27]. From a purely complexity-theoretic standpoint, this equality is interesting because it relates two classes that are a priori defined in very different terms, and it is surprising because the class RE is very large and makes no reference to time complexity at all (the definition of the halting problem does not refer to how much time is allowed for the Turing machine to halt) while the class MIP* does impose efficiency requirements on the verification time, i.e., the time it takes for the referee in the game to generate questions and verify answers to them. It is notable that the equality parallels a celebrated result MIP = NEXP [5], a major stepping stone on the way to the proof of the PCP theorem which is now given a form of "quantum" or "noncommutative" extension. On a more philosophical note, the equality MIP* = RE vindicates the long-witnessed hardness of designing and analyzing interesting Bell inequalities, showing that the optimal quantum bound is in general an uncomputable function of the coefficients of the Bell functional.

6. OUTLOOK

We end with some brief remarks on future work. While in this document we have insisted on the role played by complexity theory in the design of a separating correlation, and hence indirectly in the design of an algebra that refutes Connes' Embedding Problem, we are not aware of a metaargument that would require this. In particular, while it can be formally shown that the complexity-theoretic equality MIP* = RE directly implies a refutation of Tsirelson's problem, the converse is not known to hold. It would be very interesting if a more direct argument, without making any reference to even computability theory, could be found. This has previously been the case, when Slofstra's proof that C_{qs} is not closed [51] (which was closely tied to a proof of undecidability) was later greatly simplified [13, 17], removing all references to computability.

The particular proof technique described here leads to some interesting follow-up questions. In the realm of complexity theory, it is interesting to study variants of complexity classes associated with quantum correlations and characterize their complexity; see [37] for recent work in this direction. In terms of group theory, we believe that the notion of efficient stability put forward in Section 5.3 deserves further study, as stability questions already have a rich history [9, 24, 25, 31]. Of course, an important open question is that of proving the existence of a nonhyperlinear or even nonsofic group; the work outlined in Section 3.2 provides a promising avenue towards this.

ACKNOWLEDGMENTS

I thank Henry Yuen and Vern Paulsen for comments. A small portion of the text of these notes is repurposed from an article by the same author published in the Notices of the AMS [55].

FUNDING

This work is supported by NSF CAREER Grant CCF-1553477, AFOSR YIP award number FA9550-16-1-0495, MURI Grant FA9550-18-1-0161 and the IQIM, an NSF Physics Frontiers Center (NSF Grant PHY-1125565) with support of the Gordon and Betty Moore Foundation (GBMF-12500028).

REFERENCES

[1] C. Anantharaman and S. Popa, An introduction to II_1 factors. Draft book available at https://www.math.ucla.edu/~popa/Books/IIun.pdf, 2017.

[2] S. Arora, C. Lund, R. Motwani, M. Sudan, and M. Szegedy, Proof verification and the hardness of approximation problems. *J. ACM* **45** (1998), no. 3, 501–555.

[3] S. Arora and S. Safra, Probabilistic checking of proofs: a new characterization of np. *J. ACM* **45** (1998), no. 1, 70–122.

[4] A. Atserias, L. Mančinska, D. E. Roberson, R. Šámal, S. Severini, and A. Varvitsiotis, Quantum and non-signalling graph isomorphisms. *J. Combin. Theory Ser. B* **136** (2019), 289–328.

[5] L. Babai, L. Fortnow, and C. Lund, Non-deterministic exponential time has two-prover interactive protocols. *Comput. Complexity* **1** (1991), no. 1, 3–40.

[6] O. Becker and M. Chapman, Stability of approximate group actions: uniform and probabilistic. 2020, arXiv:2005.06652.

[7] M. Blum and S. Kannan, Designing programs that check their work. *J. ACM* **42** (1995), no. 1, 269–291.

[8] M. Blum, M. Luby, and R. Rubinfeld, Self-testing/correcting with applications to numerical problems. *J. Comput. System Sci.* **47** (1993), no. 3, 549–595.

[9] M. Burger, N. Ozawa, and A. Thom, On Ulam stability. *Israel J. Math.* **1** (2013), no. 193, 109–129.

[10] P. J. Cameron, A. Montanaro, M. W. Newman, S. Severini, and A. Winter, On the quantum chromatic number of a graph. *Electron. J. Combin.* **14** (2007), no. R81, 1.

[11] V. Capraro, In *Connes' embedding conjecture*, pp. 73–107, Springer, Cham, 2015.

[12] R. Cleve and R. Mittal, Characterization of binary constraint system games. In *International colloquium on automata, languages, and programming*, pp. 320–331, Springer, 2014.

[13] A. Coladangelo, A two-player dimension witness based on embezzlement, and an elementary proof of the non-closure of the set of quantum correlations. *Quantum* **4** (2020), 282.

[14] A. Connes, Classification of injective factors. *Ann. of Math.* (1976), 73–115.

[15] M. De Chiffre, N. Ozawa, and A. Thom, Operator algebraic approach to inverse and stability theorems for amenable groups. *Mathematika* **65** (2019), no. 1, 98–118.

[16] K. Dykema and K. Juschenko, Matrices of unitary moments. *Math. Scand.* (2011), 225–239.

[17] K. Dykema, V. Paulsen, and J. Prakash, Non-closure of the set of quantum correlations via graphs. *Comm. Math. Phys.* **365** (2019), no. 3, 1125–1142.

[18] G. Elek and E. Szabó, Hyperlinearity, essentially free actions and l_2-invariants. the sofic property. *Math. Ann.* **332** (2005), no. 2, 421–441.

[19] J. Fitzsimons, Z. Ji, T. Vidick, and H. Yuen, Quantum proof systems for iterated exponential time, and beyond. In *Proceedings of the 51st annual ACM SIGACT symposium on theory of computing*, pp. 473–480, ACM, 2019.

[20] T. Fritz, Tsirelson's problem and Kirchberg's conjecture. *Rev. Math. Phys.* **24** (2012), no. 05, 1250012.

[21] V. Galliard, A. Tapp, and S. Wolf, The impossibility of pseudo-telepathy without quantum entanglement. 2002, arXiv:quant-ph/0211011.

[22] A. Goldberg, Synchronous linear constraint system games. *J. Math. Phys.* **62** (2021), no. 3, 032201.

[23] I. Goldbring and B. Hart, Computability and the Connes embedding problem. *Bull. Symbolic Logic* **22** (2016), no. 2, 238–248.

[24] W. T. Gowers and O. Hatami, Inverse and stability theorems for approximate representations of finite groups. *Sb. Math.* **208** (2017), no. 12, 1784.

[25] D. H. Hyers, On the stability of the linear functional equation. *Proc. Natl. Acad. Sci. USA* **27** (1941), no. 4, 222.

[26] Z. Ji, Compression of quantum multi-prover interactive proofs. In *Proceedings of the 49th annual ACM SIGACT symposium on theory of computing*, pp. 289–302, ACM, 2017.

[27] Z. Ji, A. Natarajan, T. Vidick, J. Wright, and H. Yuen, $\mathrm{MIP}^* = \mathrm{RE}$. 2020, arXiv:2001.04383.

[28] Z. Ji, A. Natarajan, T. Vidick, J. Wright, and H. Yuen, Quantum soundness of the classical low individual degree test. 2020, arXiv:2009.12982.

[29] W. B. Johnson and J. Lindenstrauss, Extensions of Lipschitz mappings into a Hilbert space. *Contemp. Math.* **26** (1984).

[30] M. Junge, M. Navascues, C. Palazuelos, D. Perez-Garcia, V. B. Scholz, and R. F. Werner, Connes' embedding problem and Tsirelson's problem. *J. Math. Phys.* **52** (2011), no. 1, 012102.

[31] D. Kazhdan, On ϵ-representations. *Israel J. Math.* **43** (1982), no. 4, 315–323.

[32] S.-J. Kim, V. Paulsen, and C. Schafhauser, A synchronous game for binary constraint systems. *J. Math. Phys.* **59** (2018), no. 3, 032201.

[33] E. Kirchberg, On non-semisplit extensions, tensor products and exactness of group C*-algebras. *Invent. Math.* **112** (1993), no. 1, 449–489.

[34] L. Mančinska and D. E. Roberson, Quantum isomorphism is equivalent to equality of homomorphism counts from planar graphs. In *2020 IEEE 61st annual symposium on foundations of computer science (FOCS)*, pp. 661–672, IEEE, 2020.

[35] D. Mayers and A. Yao, Self testing quantum apparatus. *Quantum Inf. Comput.* **4** (2004), no. 4, 273–286.

[36] D. Mermin, Simple unified form for the major no-hidden-variables theorems. *Phys. Rev. Lett.* **65** (1990), no. 27, 3373.

[37] H. Mousavi, S. S. Nezhadi, and H. Yuen, On the complexity of zero gap MIP. 2020, arXiv:2002.10490.

[38] B. Musto, D. Reutter, and D. Verdon, The Morita theory of quantum graph isomorphisms. *Comm. Math. Phys.* **365** (2019), no. 2, 797–845.

[39] A. Natarajan and T. Vidick, A quantum linearity test for robustly verifying entanglement. In *Proceedings of the 49th annual ACM SIGACT symposium on theory of computing*, pp. 1003–1015, ACM, 2017.

[40] A. Natarajan and J. Wright, NEEXP is contained in MIP*. In *2019 IEEE 60th annual symposium on foundations of computer science (FOCS)*, pp. 510–518, IEEE, 2019.

[41] M. Navascués, S. Pironio, and A. Acín, Bounding the set of quantum correlations. *Phys. Rev. Lett.* **98** (2007), no. 1, 010401.

[42] C. M. Ortiz and V. I. Paulsen, Quantum graph homomorphisms via operator systems. *Linear Algebra Appl.* **497** (2016), 23–43.

[43] N. Ozawa, About the Connes embedding conjecture. *Jpn. J. Math.* **8** (2013), no. 1, 147–183.

[44] V. I. Paulsen, S. Severini, D. Stahlke, I. G. Todorov, and A. Winter, Estimating quantum chromatic numbers. *J. Funct. Anal.* **270** (2016), no. 6, 2188–2222.

[45] V. I. Paulsen and I. G. Todorov, Quantum chromatic numbers via operator systems. *Q. J. Math.* **66** (2015), no. 2, 677–692.

[46] A. Peres, Incompatible results of quantum measurements. *Phys. Lett. A* **151** (1990), no. 3–4, 107–108.

[47] F. Rădulescu, The von Neumann algebra of the non-residually finite Baumslag group $\langle a, b | ab^3a^{-1} = b^2 \rangle$ embeds into $\mathbb{R}^\omega$. 2000, arXiv:math/0004172.

[48] F. Rădulescu, A non-commutative, analytic version of Hilbert's 17th problem in type II_1 von Neumann algebras. *Theta Ser. Adv. Math.* (2008), 93–101.

[49] R. Sasyk and A. Törnquist, The classification problem for von Neumann factors. *J. Funct. Anal.* **256** (2009), no. 8, 2710–2724.

[50] G. Scarpa and S. Severini, Kochen–Specker sets and the rank-1 quantum chromatic number. *IEEE Trans. Inf. Theory* **58** (2011), no. 4, 2524–2529.

[51] W. Slofstra, The set of quantum correlations is not closed. *Forum Math. Pi* **7** (2019).

[52] B. Tsirelson, Quantum generalizations of Bell's inequality. *Lett. Math. Phys.* **4** (1980), no. 2, 93–100.

[53] B. S. Tsirelson, Some results and problems on quantum Bell-type inequalities. *Hadronic J. Suppl.* **8** (1993), no. 4, 329–345.

[54] T. Vidick, *The complexity of entangled games*. Ph.D. thesis, UC Berkeley, 2011.

[55] T. Vidick, From operator algebras to complexity theory and back. *Not. Amer. Math. Soc.* **66** (2019), no. 10, 1618–1627.

[56] D. Voiculescu, The analogues of entropy and of Fisher's information measure in free probability theory, II. *Invent. Math.* **118** (1994), no. 1, 411–440.

[57] D. Voiculescu, Free entropy. *Bull. Lond. Math. Soc.* **34** (2002), no. 3, 257–278.

[58] J. von Neumann. *Mathematische Grundlagen der Quantenmechanik*. Springer, Berlin, 1932.

THOMAS VIDICK

California Institute of Technology, Pasadena, CA, USA, vidick@caltech.edu

This Page Intentionally Left Blank

LIST OF CONTRIBUTORS